THE TAPESTRY OF
MODERN ASTROPHYSICS

THE TAPESTRY OF MODERN ASTROPHYSICS

Steven N. Shore
Indiana University South Bend

A JOHN WILEY & SONS, INC., PUBLICATION

Copyright © 2003 by John Wiley & Sons, Inc. All rights reserved.

Published by John Wiley & Sons, Inc., Hoboken, New Jersey.
Published simultaneously in Canada.

No part of this publication may be reproduced, stored in a retrieval system or transmitted in any form or by any means, electronic, mechanical, photocopying, recording, scanning or otherwise, except as permitted under Section 107 or 108 of the 1976 United States Copyright Act, without either the prior written permission of the Publisher, or authorization through payment of the appropriate per-copy fee to the Copyright Clearance Center, Inc., 222 Rosewood Drive, Danvers, MA 01923, (978) 750-8400, fax (978) 750-4470, or on the web at www.copyright.com. Requests to the Publisher for permission should be addressed to the Permissions Department, John Wiley & Sons, Inc., 111 River Street, Hoboken, NJ 07030, (201) 748-6011, fax (201) 748-6008, e-mail: permreq@wiley.com

For general information on our other products and services please contact our Customer Care Department within the U.S. at 877-762-2974, outside the U.S. at 317-572-3993 or fax 317-572-4002.

Wiley also publishes its books in a variety of electronic formats. Some content that appears in print, however, may not be available in electronic format.

Library of Congress Cataloging-in-Publication Data:

Shore, Steven N. 1953-
    The tapestry of modern astrophysics / Steven N. Shore.
      p.cm.
    Includes index.
    ISBN 0-471-16816-5 (cloth : alk. paper)
    1. Astrophysics. I. Title.

QB461 .S4465 2002
523.01--dc21
                                                                                                         2002072158

Printed in the United States of America

10 9 8 7 6 5 4 3 2 1

CONTENTS

Preface   xxi

1 From Gases to Clusters: Concepts in Gravitation and Gas Laws   1
   1.1 Introductory Remarks 1
   1.2 Gravity 1
      1.2.1 The Two-Body Problem 2
      1.2.2 The Effects of Finite Size 3
      1.2.3 Tides 6
      1.2.4 Precession 7
         1.2.4.1 Rotational Precession 8
         1.2.4.2 Apsidal Motion and Orbital Precession 10
      1.2.5 The Three-Body Problem 11
      1.2.6 Resonances 13
         1.2.6.1 Trapping 14
      1.2.7 The N-Body Problem 15
   1.3 Thermodynamics and Statistical Mechanics 15
      1.3.1 Thermodynamic Quantities 16
      1.3.2 Microphysics and the Gas Laws 19
      1.3.3 Statistical Descriptions 19
      1.3.4 Entropy and Mixing 20
         1.3.4.1 The Maxwellian Distribution for Thermal Equilibrium 21
         1.3.4.2 Discretizing the States: Boltzmann Distribution 23
         1.3.4.3 The Partition Function 25
      1.3.5 The Planck Function and Blackbody Radiation 26
      1.3.6 Degenerate Equation of State 28
         1.3.6.1 The Fermi–Dirac Distribution Function 29
         1.3.6.2 Bose–Einstein Statistics 34
      1.3.7 The Saha Equation: Ionization Balance in Thermal Equilibrium 35
         1.3.7.1 The Concept of Local Thermodynamic Equilibrium (LTE) 36
   1.4 Collective Behavior: Fluids 37
      1.4.1 Merging Gravitation and the Gas Laws: Stellar Statistical Mechanics 37
         1.4.1.1 The Mean Path 38
         1.4.1.2 Dynamical Friction 38

- 1.4.2 When Is a Gas a Fluid? 39
- 1.4.3 Viscosity and Vorticity 43
- 1.4.4 Sound and Other Collective Fluid Instabilities 44
- 1.4.5 Plasmas and Magnetohydrodynamics (MHD) 46
  - 1.4.5.1 Magnetohydrodynamics and Alfven Waves 49
- 1.4.6 Diffusion 51
  - 1.4.6.1 Diffusive Separation in Gases 53
  - 1.4.6.2 Rate Balance, the Boltzmann Collision Integral, and Diffusion 54
- 1.5 The Virial Theorem 56
  - 1.5.1 Virialized Gas Bags 61
  - 1.5.2 The Gravithermal Catastrophe 62
  - 1.5.3 Evaporation of Star Clusters 62
  - 1.5.4 Tidal Limits for Star Clusters and Galaxies 64
- 1.6 The Fokker–Planck Equation 64
  - 1.6.1 The H Theorem 67
  - 1.6.2 Agglomeration Equations 68
- 1.A Gravitational Potential for Homogeneous Spheroids 70
- 1.B A Lightning-Fast Review of Hamiltonian and Lagrangian Mechanics 75
- 1.C General Relativity on the Cheap 77
  - 1.C.1 Special Relativity 77
  - 1.C.2 The Metric 81
  - 1.C.3 The Equivalence Principle and Mach's Principle 86
  - 1.C.4 Deriving the General Relativistic Equations of Motion 87
    - 1.C.4.1 The Riemann Curvature Tensor 87
    - 1.C.4.2 The Ricci Tensor and the Curvature Scalar 88
    - 1.C.4.3 Electromagnetism 89
    - 1.C.4.4 The Schwarzschild Solution 91
  - 1.C.5 Horizons and Black Holes 96
  - 1.C.6 Symmetries and Killing Observers 97
  - 1.C.7 Static Curvature: Gravitational Lenses 98
  - 1.C.8 Dynamic Curvature: Gravitational Waves 99
  - 1.C.9 Connections with Global Structures: A Comment 101

2 The Raw Material: Instruments and Observations 103

- 2.1 The Role of Instruments 103
- 2.2 Calibration 105
  - 2.2.1 Astronomy: Parallax and Proper Motion 106
    - 2.2.1.1 Statistical Parallax 108
    - 2.2.1.2 Biases in Parallax Data 108
  - 2.2.2 Luminosities 109
    - 2.2.2.1 Magnitudes 109
  - 2.2.3 Spectral Energy Distribution 112
    - 2.2.3.1 Satellite Observations 117

      2.2.4   Masses: Binary Stars   118
      2.2.5   Radii and Luminosity Ratios: Eclipsing Binary Stars   119
2.3   Photon Detectors   120
      2.3.1   Photographic Plates   120
      2.3.2   Photomultiplier Tubes (PMTs)   123
      2.3.3   Charge-Coupled Devices (CCDs)   124
      2.3.4   Multianode Microchannel Arrays (MAMAs)   125
      2.3.5   Bolometer Arrays and Infrared Detectors   126
      2.3.6   Higher-Energy Observations   128
      2.3.7   Radio Astronomy   128
2.4   Spectrographs   130
      2.4.1   Gratings and Resolution   131
      2.4.2   Correlation Spectrometers   132
      2.4.3   Multiobjects Spectrographs (MOSs)   132
2.5   Image Formation   134
      2.5.1   What Is an Image?   134
      2.5.2   The Nyquist Frequency and Resolution   136
      2.5.3   Interferometers   136
            2.5.3.1   Michelson Interferometer   137
            2.5.3.2   Fabry–Perot Interferometer   138
      2.5.4   Intensity Interferometers   139
      2.5.5   Aperture Synthesis   140
      2.5.6   Scintillitation: Atmospheric Seeing   142
      2.5.7   Speckle Inferometry   145
      2.5.8   Resolution   146
2.6   Image Reconstruction Methods   146
      2.6.1   The CLEAN Algorithm   149
      2.6.2   Bayesian Methods   150
      2.6.3   Maximum Entropy   151
      2.6.4   Wavelets   154
2.A   A Note on Cosmic Backgrounds across the Spectrum   156
2.B   Statistical Distributions   157
      2.B.1   Samples   157
      2.B.2   MEM and Maximum Likelihood   159
      2.B.3   Bias   160
2.C   Properties of the Fourier Transform and Convolutions   161
2.D   Implementation of Bayes' Theorem   163

3   Radiative Transfer and the Outer Layers of Stars   166

3.1   Introduction   166
3.2   The Phenomenon of Radiative Transfer   167
3.3   Transition Probabilities and Statistical Equilibrium   168
      3.3.1   Statistical Equilibrium   169
            3.3.1.1   Collision Rates   171
            3.3.1.2   Equilibrium Population   172

- 3.3.2 Strange Populations: Masers 175
- 3.4 Radiative Transfer 176
  - 3.4.1 Intensity, Flux, and Moments of the Radiation Field 176
  - 3.4.2 Setting up the Transfer Equation 179
    - 3.4.2.1 The Spherical Transfer Equation 180
  - 3.4.3 Some Solutions to the Transfer Equation 182
    - 3.4.3.1 Two-Stream Transfer 184
    - 3.4.3.2 Limb Darkening and Temperature Gradients 184
    - 3.4.3.3 Life in a Fog, or What Scattering Means 186
    - 3.4.3.4 Thermalization Length 189
  - 3.4.4 Diffusion Approximation 190
    - 3.4.4.1 A Note on Probability and Radiative Transfer 190
- 3.5 Opacity 192
  - 3.5.1 Bremsstrahlung, Ionization, and Recombination 192
    - 3.5.1.1 Thermal Bremsstrahlung or Free–Free Opacity 192
    - 3.5.1.2 Ionization and Recombination 194
  - 3.5.2 Line Processes: Bound–Bound Transitions 197
    - 3.5.2.1 Line Profiles 197
  - 3.5.3 Collisional Broadening and Lifetimes 199
    - 3.5.3.1 Doppler Broadening and the Voigt Profile 199
  - 3.5.4 Curve of Growth 201
  - 3.5.5 Types of Scattering 203
    - 3.5.5.1 Resonance Scattering 203
    - 3.5.5.2 Rayleigh Scattering 204
    - 3.5.5.3 Redistribution in Line Profiles 205
    - 3.5.5.4 Fluorescence and Raman Scattering 205
  - 3.5.6 Electron Scattering: Thomson and Compton Scattering 208
    - 3.5.6.1 The Compton Effect, the Kompaneets Equation, and Diffusive Transfer of High-Energy Photons 209
  - 3.5.7 Broadening Mechanisms for Line Profiles 214
    - 3.5.7.1 Hyperfine Structure: Effect of Atomic Structure and Nuclear Spin 214
    - 3.5.7.2 Zeeman Effect: Effect of Magnetic Fields 215
    - 3.5.7.3 Stark Broadening 216
    - 3.5.7.4 Van der Waals Broadening 221
    - 3.5.7.5 Rotational Broadening 221
    - 3.5.7.6 Doppler Imaging 223
- 3.6 Stellar Atmospheres 224
  - 3.6.1 The Justification for Classifying Spectra 225
  - 3.6.2 The Stellar Atmosphere Problem 227

      3.6.3 Radiative Equilibrium: What Goes in, Comes Out 230
      3.6.4 Determining the Thermal Structure 231
      3.6.5 Convection in Stellar Atmospheres 232
      3.6.6 LTE versus NLTE in Stellar Atmospheres 237
      3.6.7 Some Complications 238
              3.6.7.1 Gravity Darkening by Rotation 238
              3.6.7.2 External Illumination 239
3.7 Extended Envelopes and Outflows 241
      3.7.1 The Spherical Transfer Equation Yet Again 241
      3.7.2 Low-Frequency Observations of Extended Envelopes and Winds 243
      3.7.3 Stellar Winds 245
              3.7.3.1 Radiation Pressure as a Driving Mechanism 245
              3.7.3.2 Radiation-Driven Elemental Diffusion in Atmospheres 248
              3.7.3.3 Thermal Evaporation: The Parker Wind Solution 249
              3.7.3.4 Mechanical Driving Mechanisms 250
              3.7.3.5 Chromospheres and Coronae 251
      3.7.4 Escape Probabilities and Radiative Transfer in Flows 252
              3.7.4.1 The Effect of a Velocity Gradient on the Escape of Photons: Kinematics and Line Profiles 256
              3.7.4.2 Time-Dependent Flows: Novae and Supernovae and Spectrum Formation 259
3.A Quantum-Mechanical Interlude: Time-Dependent Perturbation Theory for Transition Strengths 260
3.B Radiative Transfer in the Lagrangian Frame 263
3.C A Brief Survey of Methods for Solving the Transfer Equation 265
      3.C.1 Diffuse Scattering as an Example of Probabilities: Invariant Embedding 265
      3.C.2 Integral Equation Methods for Radiative Equilibrium 269
      3.C.3 Discrete-Ordinates Methods 270
3.D A Walk through the Stellar Spectroscopic Zoo 272
      3.D.1 Stellar Classification and Types of Stars 272
              3.D.1.1 A Historical Digression 272
      3.D.2 The Classification Scheme 274
              3.D.2.1 O Stars 274
              3.D.2.2 B Stars 275
              3.D.2.3 A Stars 276
              3.D.2.4 F Stars 277
              3.D.2.5 G Stars 277
              3.D.2.6 K Stars 278
              3.D.2.7 M Stars 278
              3.D.2.8 L Stars: Methane Dwarfs 278
              3.D.2.9 Carbon Stars 278
              3.D.2.10 White Dwarf Stars 279

4 The Interiors of the Stars and Stellar Evolution   **280**

    4.1 Introductory Remarks   280
    4.2 Self-Gravitating Spheres   282
        4.2.1 The Virial Theorem in Stellar Structure   282
        4.2.2 The Kelvin–Helmholtz Timescale   284
    4.3 Thermodynamics and Equations of State   285
        4.3.1 Polytropic Equation of State   285
        4.3.2 The Radiation-Dominated Equation of State   286
        4.3.3 Effects of Partial Ionization   288
        4.3.4 Stability of Polytropes   289
        4.3.5 Degenerate Equations of State   291
    4.4 Equations of Structure   292
        4.4.1 Mass Continuity and Hydrostatic Equilibrium   293
        4.4.2 Polytropes and the Lane-Emden Equation   294
        4.4.3 Energy Transport in Stellar Interiors   299
            4.4.3.1 Radiation   299
            4.4.3.2 Conduction   301
            4.4.3.3 Convection   302
            4.4.3.4 Semiconvection and Double-Diffusive Processes   304
            4.4.3.5 Overshooting and the Mixing Length Concept   305
        4.4.4 Rotation and von Zeipel's Theorem   307
        4.4.5 Dimensional Analysis, Scaling Relations, and Homology   309
    4.5 Stellar Pulsation and Stability   311
        4.5.1 Observational Justification   311
        4.5.2 The Mechanism for Pulsation   314
        4.5.3 Stabilty Analysis: One-Zone Pulsation   315
        4.5.4 Probing the Stellar Envelope   319
            4.5.4.1 Helioseismology   321
            4.5.4.2 The Linear Pulsation Equation   323
    4.6 Energy Generation Mechanisms   328
        4.6.1 Gravitational Contraction Again   329
        4.6.2 Nuclear Reactions   330
            4.6.2.1 Nuclear Binding Energy   330
            4.6.2.2 The Reaction Rates   332
            4.6.2.3 Hydrogen Burning: The CNO Cycle   337
            4.6.2.4 Hydrogen Burning: The Proton–Proton (pp) Chain   338
        4.6.3 Helium Burning: The 3α Reaction   341
            4.6.3.1 Carbon Burning   342
            4.6.3.2 Higher-Order Nucleosynthesis and Equilibrium Processing   343
            4.6.3.3 Neutron Processing and Heavy-Element Nucleosynthesis   344

        4.6.4   Explosive Hydrogen Burning   348
        4.6.5   Abundance of the Elements in the Solar System   349
        4.6.6   Neutrino Process   351
                4.6.6.1   The Solar Neutrino Problem   352
                4.6.6.2   Plasma Screening Corrections   355
                4.6.6.3   The URCA Process   356
4.7   Stellar Evolution   357
        4.7.1   Observational Basis   357
                4.7.1.1   Historical Remarks: The Hertzsprung–Russell Diagram   357
                4.7.1.2   The Observer's HR Diagram   360
        4.7.2   Mass Determinations   360
                4.7.2.1   The Mass–Luminosity Relation   361
        4.7.3   Theory: The Principal Stages of Stellar Evolution   362
        4.7.4   The Earliest Stages of Evolution Protostars   364
                4.7.4.1   Main Sequence   366
                4.7.4.2   Ascent of the Red Giant Branch   369
                4.7.4.3   From the Giant Branch Tip to the Horizontal Branch   372
                4.7.4.4   The Asymptotic Giant Branch (AGB): Hot Onions   373
        4.7.5   Death   377
                4.7.5.1   How the Most Massive Stars End Their Lives   377
                4.7.5.2   Postasymptotic Giant Branch Evolution of Low-Mass Stars and Planetary Nebulas   378
                4.7.5.3   White Dwarf Stars: The Final State of Low-Mass Stars   380
        4.7.6   Core Collapse Supernovae Type II and Neutron Star Formation   385
                4.7.6.1   Classification of Supernovas   385
                4.7.6.2   Physics of Core Collapse and Shock Generation   388
                4.7.6.3   Direct Evidence for Nucleosynthesis in Supernovae   388
        4.7.7   Neutron Stars   389
                4.7.7.1   Neutron Star Interiors   391
                4.7.7.2   Neutron Star Cooling and Detectability   393
        4.7.8   Pulsars   394
                4.7.8.1   Emission Regions and Mechanisms   396
                4.7.8.2   Rotational Properties: Period Changes and Glitches   399
4.8   Isochrones   401
        4.8.1   The Initial Mass Function: A Result of Isochrones   404
4.A   Magnetic Dynamos and the Interplay between Turbulence and Rotation   405

4.A.1 The Dynamo Equations 406
    4.A.1.1 The Dynamo Number and Scaling Relations 409
4.B Calculation of Nuclear Reaction Networks 410

5 Structure and Evolution of Close Binary Stars   413

5.1 Introduction 413
5.2 Eclipses and Their Uses 414
5.3 Effect of Proximity: Tides and the Roche Surface 417
    5.3.1 The Roche Surface 418
    5.3.2 Tidal Interactions: Circularization and Synchronization 420
5.4 Evolution of Stars in Close Binaries 422
    5.4.1 Common Envelope Evolution 423
    5.4.2 Angular Momentum Consideration and the Roche Surface 426
    5.4.3 Period Changes 428
        5.4.3.1 Gravitational Radiation 428
        5.4.3.2 Magnetic Braking 430
5.5 Mass Transfer in Close Binaries 431
    5.5.1 Mass Loss by Winds 431
    5.5.2 Accretion from Stellar Winds 432
    5.5.3 Accretion by Streams and Roche Lobe Overflow 434
    5.5.4 Formation and Structure of Viscous Accretion Disks 435
        5.5.4.1 Viscous Accretion Disks 437
        5.5.4.2 Magnetorotational Instability and Viscosity in Disks 442
    5.5.5 Boundary Layers 445
5.6 Cataclysmic Variables and Compact Objects in Close Binaries 447
    5.6.1 Novae and X-ray Bursts: Surface Nuclear Explosions 448
    5.6.2 Accretion by Magnetized Stars 450
    5.6.3 Black Holes in Binary Systems 452
5.7 Formation of Binary Systems: A Comment or Two 453
    5.7.1 Blue Stragglers: Collisions and Captures in Clusters 456
5.A Some Hydrodynamic Details 457

6 The Interstellar Medium   459

6.1 Introductory Remarks 459
6.2 Gas 461
    6.2.1 Ionized Regions: Emission Nebulae 462
        6.2.1.1 Nebular Lines 463
        6.2.1.2 Recombination Lines 466
    6.2.2 Heating and Cooling 472
    6.2.3 Cosmic Ray Ionization and Charge Transfer 474

- 6.2.4 Absorption Lines 475
  - 6.2.4.1 Atomic Resonance Lines: H I Lyman α 1216 Å and Others 476
  - 6.2.4.2 H I 21 cm and Related Radio Lines 482
  - 6.2.4.3 Molecular Spectra 485
  - 6.2.4.4 Molecular Hydrogen (H_2) 490
- 6.2.5 Masers 491
- 6.2.6 Warm and Hot Diffuse Gas 492

6.3 Dust 495
- 6.3.1 Observations 497
  - 6.3.1.1 Optical and Ultraviolet Extinction 498
  - 6.3.1.2 Ultraviolet and Infrared Broad Features 500
- 6.3.2 Grain Optics 501
- 6.3.3 Infrared Observations 502
- 6.3.4 Big Grains 504
  - 6.3.4.1 Depletion of Metals 505
  - 6.3.4.2 Solar System Measurements: Meteorites and Heliospheric Dust 506
  - 6.3.4.3 Grain Charging 507
  - 6.3.4.4 Poynting–Robertson Drag and Accretion 508
- 6.3.5 Small Grains and Big Molecules 509
  - 6.3.5.1 Diffuse Interstellar Bands (DIBs) 512
- 6.3.6 Polarization and Grain Alignment in an External Magnetic Field 514
  - 6.3.6.1 Grain Size Distribution and Origin of the Dust 518
- 6.3.7 Reflection Nebulae and Light Echos 520
- 6.3.8 The Galactic Distribution of Dust 521

6.4 Molecular Clouds 522
- 6.4.1 Observational Mass Determination Using CO Lines 523
- 6.4.2 The Population of Clouds 526

6.5 Magnetic Fields 527
- 6.5.1 Cosmic Rays 527
- 6.5.2 Synchrotron Radiation 533
- 6.5.3 Propagation Effects 536
  - 6.5.3.1 Faraday Rotation and Large-Scale Field 536
  - 6.5.3.2 Electron Density by Scintillation and Dispersion Measure 540
  - 6.5.3.3 How to Search for Warm Gas 541
- 6.5.4 Magnetized Clouds 542
  - 6.5.4.1 Direct Observation of the Zeeman Effect 542
  - 6.5.4.2 Ambipolar Diffusion 542

6.6 Molecules and Astrochemistry 543
- 6.6.1 Observations 543
- 6.6.2 Interstellar Chemistry 544

- 6.6.2.1 Surface Chemistry: Formation of Molecular Hydrogen 544
- 6.6.2.2 Gas-Phase Chemistry 546
- 6.6.2.3 Ionization and Dissociation 547
- 6.6.2.4 Molecular Cooling Processes 548
- 6.6.2.5 H_3^+ as an Example of Ion Chemistry 549
- 6.6.2.6 Isotopic Fractionation 550
- 6.6.2.7 Polycyclic Aromatic Hydrocarbons (PAHs) and Fullerenes 552

6.7 Dynamical Gas Bags in the Interstellar Medium 554
- 6.7.1 An Introduction to Shocks 554
  - 6.7.1.1 The Rankine-Hugoniot Conditions for Shocks 555
  - 6.7.1.2 Magnetic Shocks 558
  - 6.7.1.3 Magnetic Shocks in a Partially Ionized Medium: J and C Shocks 561
  - 6.7.1.4 Radiative Shocks 562
  - 6.7.1.5 Shock Precursors 563
  - 6.7.1.6 Shock Chemistry 564
- 6.7.2 Ionization Fronts and Photodissociation Regions 565
- 6.7.3 Static H II Regions: Strömgren Spheres 567
  - 6.7.3.1 Time-Dependent Strömgren Spheres 568
  - 6.7.3.2 Compact H II Regions 571
- 6.7.4 Expanding H II Regions: Dynamics Driven by Radiative Heating 571
  - 6.7.4.1 Blisters, Champagne Flows, and Bubbles 574
- 6.7.5 Stellar Explosions and Their Remnants 574
  - 6.7.5.1 Free Expansion and Transition to the Sedov–Taylor Solution 577
  - 6.7.5.2 Stalled Shocks and Stagnation Pressure 577
- 6.7.6 Snowplow Phase 579
- 6.7.7 Stellar Wind Bubble 580
- 6.7.8 Producing the Hot Diffuse Gas 580
- 6.7.9 Pressure-Driven Expansion of a Planetary Nebula 581
  - 6.7.9.1 Breakup of the Shock Front 582

6.8 Instabilities and the Formation of Structure 582
- 6.8.1 Gravito-acoustic Waves: The Jeans Instability 583
  - 6.8.1.1 Magnetized Clouds, Ambipolar Diffusion, and the Jeans Criterion 586
  - 6.8.1.2 Equilibrium of Pressure-Bounded Spheres 589
  - 6.8.1.3 Relation to Star Formation 590
- 6.8.2 Thermal Instability and Multiple Phases 590
- 6.8.3 Pressure-Modified Gravitational Collapse 597
- 6.8.4 Angular Momentum 599

6.9 Large-Scale Distribution of the Gas   599
    6.9.1 Buoyancy and the Rayleigh–Taylor Instability   600
    6.9.2 The Parker Instability   600

6.10 Turbulence in the Interstellar Medium   602
    6.10.1 The Role of Dissipation: The Kolmogorov Spectrum   603
    6.10.2 The Role of Magnetic Fields: The Kraichnan Spectrum   606
    6.10.3 Driving the Turbulence   607
    6.10.4 Confronting Observations   608

6.A Synchrotron Spectra: Some Details   614

6.B The Parker Instability: Some Details   618

6.C Dimensional Analysis and Similarity Solutions of the Hydrodynamic Equations: Some Details   621
    6.C.1 Dimensionless Dynamical Equations   621

6.D The Velocity Correlation Tensor and Representation of Turbulent Flows: Some Details   624
    6.D.1 Time Dependence   627

7 Our Galaxy and Others as Stellar Systems   629

7.1 The Galaxy as a Stellar System: Introductory Remarks   629
    7.1.1 The Composite HR Diagram of Field Stars   631
    7.1.2 Aggregates: Open Clusters, OB Associations, and Globular Clusters   633
    7.1.3 Stellar Hydrodynamics   637
        7.1.3.1 The Stellar Velocity Distribution   640
        7.1.3.2 The Equations of Motion   642
        7.1.3.3 Modeling the Galactic Disk   644
    7.1.4 The Halo   647

7.2 Large-Scale Structure of the Galaxy   649
    7.2.1 Galactic Rotation   649
    7.2.2 Determination of Solar Galactocentric Distance   650
        7.2.2.1 Kinematics from Dynamics   651
        7.2.2.2 Observable Consequences: The Oort Laws for Galactic Rotation   655
    7.2.3 Spiral Structure   660
        7.2.3.1 Observations   660
        7.2.3.2 Departures from Symmetry: Density Waves and Bars   661

7.3 Chemical Evolution of the Galaxy   669
    7.3.1 Evidence for Metallicity Evolution   670
    7.3.2 Inputs   672
        7.3.2.1 The Initial Mass Function   672
        7.3.2.2 The Star Formation Rate   673

- 7.3.3 Stellar Populations and Population Synthesis 675
- 7.3.4 Models for Galactic Chemical Evolution 677
  - 7.3.4.1 Closed-Box Models 677
  - 7.4.3.2 Primary versus Secondary Elements 679
  - 7.3.4.3 Spallation: Cosmic Rays and Light-Element Synthesis 680
  - 7.3.4.4 The Lowest-Metallicity Fossils 682
  - 7.3.4.5 Feedback and Stimulated Star Formation 683
  - 7.3.4.6 Some Process Affecting Evolution of the Gaseous Component 685
- 7.3.5 Nucleocosmochronology 688
- 7.4 Galaxies and Clusters of Galaxies: Introductory Remarks 689
- 7.5 The Hubble Classification Scheme for Galaxies 690
  - 7.5.1 Spirals: Active Star Formation and Global Patterns 690
  - 7.5.2 Ellipticals: Frozen Populations 692
    - 7.5.2.1 The Fundamental Plane 694
  - 7.5.3 Irregulars and Dwarfs: Stochastic Experiments 695
- 7.6 Rotation Curves: Bright Flow Tracers and Dark Matter 697
- 7.7 The Complications: Peculiar Galaxies 700
  - 7.7.1 Active Galactic Nuclei (AGN) 700
    - 7.7.1.1 Taxonomy 700
    - 7.7.1.2 Supermassive Black Holes: The Central Engine 703
    - 7.7.1.3 Reverberation Mapping and Size of the Emitting Region 706
    - 7.7.1.4 Central Engine Luminosities 707
  - 7.7.2 Radio Galaxies 708
    - 7.7.2.1 Minimum Energy: Estimating Properties of the Emitting Regions 708
    - 7.7.2.2 Unsteady Synchrotron Sources: Time Evolution 711
    - 7.7.2.3 Jets and Lobes 713
    - 7.7.2.4 Superluminal Motions and Beaming 713
  - 7.7.3 Interacting Galaxies 715
- 7.8 Clusters of Galaxies 719
  - 7.8.1 The Local Group 720
  - 7.8.2 Applications of the Virial Theorem to Clusters 720
  - 7.8.3 X-Ray Emission from Clusters of Galaxies 723
    - 7.8.3.1 Evidence for Intracluster Gas from Radio Galaxies 728
  - 7.8.4 Gravitational Lensing by Clusters and Dark Matter 730
- 7.A Deprojection Methods: Abel's Equation and Inversion of Surface Measurements 731

8 The Biggest Picture: Cosmology   **734**

    8.1   Introductory Remarks   734
    8.2   The Distance Scale   740
         8.2.1   Overview   740
         8.2.2   The Cepheid Distance Calibration   742
         8.2.3   Outburst Calibrators   744
                8.2.3.1   Classical Novae and Supernovae   744
                8.2.3.2   Supernovae Type Ia   746
                8.2.3.3   Supernova Remnants   747
         8.2.4   Calibrators Based on Global Galactic Properties   747
                8.2.4.1   Tully–Fisher Linewidth–Luminosity Relation   747
                8.2.4.2   Faber–Jackson Velocity Dispersion–Luminosity Relation   748
         8.2.5   Maser Proper Motions   749
         8.2.6   Surface Brightness Fluctuations   749
         8.2.7   Luminosity Functions and Statistical Distance Calibrators: Internal Galactic Properties   750
                8.2.7.1   Luminosity Functions of Planetary Nebulae and Globular Clusters   750
                8.2.7.2   H II Region Sizes   751
                8.2.7.3   Schechter Function for Galaxy Clusters   751
    8.3   Fundamental Parameters: The Redshift and the Hubble Constant   752
         8.3.1   The Redshift (z)   752
         8.3.2   Current Results for the Hubble Constant   753
         8.3.3   Absolute Age Calibrations: Decay of Radioactive Elements   754
         8.3.4   Cosmological Corrections to Galaxy Properties and the Redshift   755
    8.4   Relativistic Cosmology   757
         8.4.1   Derivation of the Metric   758
                8.4.1.1   First Pass: Differential Geometry   758
                8.4.1.2   Second Pass: Symmetries and the Killing Vectors   759
                8.4.1.3   Third Pass: The Field Equation   760
         8.4.2   The Redshift from the Friedmann–Robertson–Walker Metric: Interpreting the Hubble Law   763
    8.5   Cosmological Models and Evolution of the Scale Factor   765
         8.5.1   The Friedmann–Robertson–Walker (FRW) Evolution Equations   766
         8.5.2   Choosing the Equation of State   768
    8.6   Cosmic Background Radiation (CBR)   769
         8.6.1   The Past   769

- 8.6.2 Current Status 770
  - 8.6.2.1 The Future 771
- 8.6.3 Coupling between Matter and Radiation: Formation of CBR 772
- 8.6.4 A New Complication: The Cosmological Constant (Λ) 774
  - 8.6.4.1 An Aside on Λ and the Physical Constants 775
- 8.6.5 The Density Parameter: Ω 776
- 8.6.6 World Models: Solutions to the FRW Equations 778
  - 8.6.6.1 Matter-Dominated Universe: The Friedmann Model 779
  - 8.6.6.2 Radiation-Dominated Universe: The Tolman Model 780
  - 8.6.6.3 Initial Conditions 781
- 8.6.7 Linking the Scale Factor to Observations and Tests of Cosmological Models 783
  - 8.6.7.1 Lookback Time, Proper Distances, and Horizons 783
- 8.6.8 Inflation 790
- 8.6.9 A Note on Radiative Transfer in an Expanding Universe 795
  - 8.6.9.1 Applications: The Gunn–Peterson Test for Lyman Continuum Absorption 796
  - 8.6.9.2 Applications: Cosmological Lyman α Absorption Lines 797
- 8.6.10 Primordial Nucleosynthesis 798
- 8.7 Large Structures in the Universe 805
  - 8.7.1 Observational Constraints on the Large-Scale Structure 806
    - 8.7.1.1 The Darkness of the Night Sky 806
    - 8.7.1.2 Tests of Distributed Properties: V/V_{max} and the Malmquist Bias 807
    - 8.7.1.3 The *Hubble Deep-Field* Survey and Related Studies 808
    - 8.7.1.4 Redshift Surveys 809
    - 8.7.1.5 Redshift Surveys and the History of Star Formation 810
    - 8.7.1.6 A Comment on Catalogs and Distributions 811
    - 8.7.1.7 Dark Matter and Catalogs 811
    - 8.7.1.8 Supernovae, Λ, and Dark Energy 814
  - 8.7.2 Finding Structures: Correlation Functions 815
  - 8.7.3 Forming Structures in an Expanding Universe 817

                8.7.3.1 Gravito-acoustic (Jeans) Instability in an
                        Expanding Universe   818
        8.7.4   Limits to Growth   822
                8.7.4.1 Radiative Scattering and Damping: The Silk
                        Mass   822
                8.7.4.2 Reionization of the Universe   823
                8.7.4.3 The Role(s) of Λ 826
        8.7.5   Nonlinear Evolution of Density Fluctuations   826
        8.7.6   Background Fluctuations and CBR Variations   830
                8.7.6.1 The Sachs–Wolfe Effect   832
                8.7.6.2 The Sunyaev–Zeldovich Effect   833
        8.7.7   Whence These Perturbations?   834
8.8 Cosmological Gravitational Lenses   836
8.9 Exeunt   841
8.A Gamma-Ray Bursts   842
8.B Newtonian Derivation of the Evolution Equations   844
8.C Galaxy Formation and ΛDM Scenarios   845

Index   **849**

PREFACE

> We have a habit in writing articles published in scientific journals to make the work as finished as possible, to cover up all the tracks, to not worry about the blind alleys or describe how you had the wrong idea first, and so on. So there really isn't any place to publish, in a dignified manner, what you actually did in order to get to do the work.
> —Richard Feynman

This is a large book about a very big subject. The scope of modern astrophysics is, quite literally, the whole cosmos and everything in it.

The order of the topics is dictated by necessity. Astrophysics requires remote sensing. You need basic physical concepts— gravitation, thermodynamics, statistical mechanics, and radiative processes—in order to understand how the bodies you're studying are structured. Nuclear physics, hydrodynamics, plasma processes and magnetohydrodynamics, and relativity play essential parts in the development of models. Increasingly, areas that have been thought to be separate in the curriculum come together in a beautiful cross-disciplinary synthesis.

Chapter 1 begins with the basic physical processes common to all cosmic bodies, those involving gravitation, thermodynamics, and the gas laws. My aim is to show the similarity of many of the methods that are used to tackle what may otherwise appear to be very different problems. For instance, rather than postpone the discussion of stellar statistical dynamics to the chapter on galaxies, it is introduced here as an application of statistical mechanics to gravitational interactions in a gas of stars. A special feature of this first chapter is an appendix that provides an overview of general relativity, since this is needed for subsequent discussions of black holes in binary systems and active galactic nuclei, not to mention cosmology.

Chapter 2 reviews some of the general aspects of observations. These include calibration, instrumentation, and image formation and reconstruction with the goal of acquainting you with the idea that no astronomical observation is entirely "theory-free," and that much of the information provided by observations must be recovered from the data by manipulation.

Chapter 3 deals with radiative transfer, mainly in the context of stellar atmospheres, but also includes material theory of line formation that will be important in Chapter 6 on the interstellar medium and Chapter 8 on cosmology. It also introduces concepts of spectral classification and discusses stellar atmospheres in a general way in order to describe the basis for analysis of stellar surface properties.

Chapter 4 deals with stellar structure, energy sources, evolution, and nucleosynthesis. A particular emphasis here is on pulsating stars as an example of a ubiquitous instability in stars. You will also find an appendix that briefly presents some basic dynamo theory, for which there are broad astrophysical applications.

Chapter 5 presents an *intermezzo* on binary stars emphasizing processes that take place when tides limit the radial growth of stars. The chapter also provides an introduction to accretion disks and related phenomena. My justification for treating these objects in a separate chapter is that they present many phenomena that combine the material of the preceding chapters and also set the stage for discussions of accretion and disk formation that are needed for understanding galactic nuclei.

Chapter 6 covers the interstellar medium (ISM). It returns to radiative processes, this time treating the ISM as the ultimate example of a radiating gas in NLTE. Since some of the most spectacular examples of gas dynamics, shocks of all varieties, occur in the ISM, this chapter also introduces ideas about shocks and similarity solutions and reviews some features of turbulence.

Chapter 7 treats the Milky Way as a stellar system and as the prototype for the study of extragalactic systems. I have chosen to combine this discussion with a review of the properties of galaxies of various types in order to emphasize that our understanding of our own stellar system is informed by the study of external galaxies and vice versa. This chapter also covers radio galaxies, clusters of galaxies, and active galaxies.

Finally, in Chapter 8, we come to cosmology, where we begin with a discussion of the observational constraints and, in Weinberg's phrase, *cosmography*, and then go on to discuss the formation of structure and galaxies within the context of solutions of the field equations for different equations of state and scenarios for the earliest epochs of the Big Bang.

The coverage is slanted heavily toward stellar and interstellar processes. For this, I make no apologies. You can't understand galaxies without a detailed examination of their constituents. While with each successive step in the distance scale the questions become, in a sense, bigger, our loss of resolution means our inferences become more tentative and this loss of information at each step compounds our ignorance of the governing processes. In teaching science, a top-down view may be magisterial but it is often misleading. In dealing with astrophysics it is ahistorical and, I believe, pedagogically unwise.

I have tried to produce a book that reflects how astrophysics is done rather than a stylized account of what is to be learned. There are occasional redundancies and recapitulations, but they are intentional, resulting, in part, from a deliberate effort to maintain the feel of lectures. For instance, some of the material that is included in Chapter 2 on instruments concerning calibration of absolute stellar properties is repeated in a different way in Chapter 5 on binary stars. By that point, you may have forgotten some of the ideas, and this way you will see it in a new context, with additional caveats on the application of the methods in pathological cases. Galactic dynamics, discussed in Chapter 7, draws heavily on material in Chapter 1, as does the derivation of the basic cosmological models in Chapter 8. Spectral types and Hubble galaxy classification are postponed until their respective chapters, but photometry is included in the chapter on instruments.

There is another feature of this book that I feel compelled to explain. You will frequently find extended qualitative discussion before getting to the analytical treatment. In fact, often I'll adumbrate a derivations in order to physically motivate the discussion. While some readers will find this an annoying habit, I hope you will bear with the presentation. Much of astrophysics *is* qualitative, and it is a tribute to the power of dimensional heuristic arguments that they serve as such effective guides when faced with diverse and complicated processes. This book isn't intended to be a formal axiomatic presentation of astrophysics. Progress often comes instead from the combination of observations and "horse sense," that elusive quality called *physical intuition* that the discussion aims at helping you develop. At the same time, the intuition is developed within a theoretical framework and this requires some lengthy justifications and derivations.

I sincerely hope that you find this book to be up-to-date and useful as it stands. I had in mind not only the student who is taking a course along with reading this book, or someone preparing for a PhD candidacy exam, but also the mythical interested reader who wants a point of entry into the subject as a whole. If this book proves useful for any of you, I will rest content!

During the writing of this book, I have struggled with the issue of completeness. It was very tempting to write a series of review papers instead of a monograph, but the urge to do so finally was supressed as the writing continued. As for the omissions, both of topics and references, all one can say is that there *are* stranger things on heaven and Earth than are dreamed of in your philosophies, and there was only so much space in the book. The bibliographic problem was equally difficult. I've chosen to use footnotes rather than a collected bibliography in order to place the reference immediately in context, even at the risk of redundancy. The literature is vast and an attempt to completely summarize any area would be folly. In any specialty, comprehensive lists quickly become impossibly long and cumbersome. On the other hand, I wanted to be sure that you are able to understand what is going on among the active areas in astrophysics and to have pointers to key papers in the field. In addition to journal articles, I have whenever possible referred to reviews in the most accessible literature, principally monographs[1] and *Annual Reviews* volumes, *Astronomy and Astrophysics Reviews*, *Fundamentals in Cosmic Physics*, *Physics Reports*, *Reviews of Modern Physics*, *Space Science Reviews*, and *Reports on the Progress of Physics*. For conferences, I've preferred those with extended reviews, especially the proceedings of schools. Many of the primary sources are cited in this secondary literature. Conference proceedings are too often ephemeral and redundant, containing material that later appears in the refereed literature[2] and I've tended to avoid these. And although a neo-Luddite, I've even slipped in the occasional uniform resource locator (URL) for an especially useful website if it is likely to remain stable for years to come. And for many of the original sources, electronic access is now available through the *ADS*, *JSTOR*, *PROLA* (American Physical Society), and other archives, an enormous change from the days when you had to rely only on interlibrary loan for a copy. Alas, the situation has not yet improved much for observatory publications to which I have made only infrequent reference, although these are becoming available through the ADS.

---

[1] In this regard, I want to note two superb general astrophysics books that have appeared since the mid-1990s: Carroll, B. W. and Oslie, D. A. 1996, *An Introduction to Modern Astrophysics* (Reading, MA: Addison Wesley) and Padmanabhan, T. 2001, *Theoretical Astrophysics*, 2 vols. (Cambridge, UK: Cambridge Univ. Press). I hope the reader will consult these in addition to reading the present work. In so rich a field, the more perspectives you can experience the better. However, I should add that the present work has been written independently of these monographs.
[2] An important exception is, however, the now rare conference that includes extended discussions, and these are referenced where appropriate.

The primary organizing principle of this work has been to discuss the physical basis of the astrophysical phenomena, not necessarily the details emerging from observations. There are two reasons for this. The first is that the observations have a history of developing within the time it takes the ink of any text to dry and, for this reason, you should be encouraged as soon as possible to consult the primary literature. The second is that the minute details that often reveal the operative mechanisms are understood through physical modeling. It is this latter process that I am hoping you will get to know as a result of this book. Finally, on the lack of problems, I'll close with an admonition. Consider the book itself to be one long problem set. I did.

I sincerely hope that you, the reader, will find herein moments of enjoyment as you traverse this broad terrain. This is just the beginning.

ACKNOWLEDGMENTS

You have to be a lunatic to start a project of this magnitude. Many dear friends and colleagues have said this repeatedly since 1996. They often went on to make helpful suggestions, and those from whom I benefitted from scientific discussions about aspects of this work deserve public acknowledgment. This is not just a formality. It is both a duty and a pleasure.

I begin with Federico Ferrini of the University of Pisa. Although his duties ultimately prevented his participation in finally writing the book, our discussions and collaboration were the impetus for this project. He contributed several draft sections of the interiors, extragalactic, and cosmology chapters. It's partly for this reason that the pronoun "we" appears throughout the text—he has been with me in spirit throughout the writing. Federico also provided invitations for extended visiting appointments for me at Pisa during the summers of 1997 and 1998, including graciously providing accommodations at Villa Ferrini.

Francesco Palla and Daniele Galli arranged for extended stays at Osservatorio Astrofisico di Arcetri during May–August 2000, December 2000, and June–August 2001. Words cannot convey the debt that I owe them for their companionship, collaboration, and friendship through the years. I hope they can read between these lines. I want also to thank Prof. Franco Pacini for his generous interest and support during these sojourns.

Greg Franklin who was acquisitions editor for physical science at John Wiley & Sons at the time this book was proposed, was instrumental in getting this whole project started. Without his encouragement, and the resolve and support of his successor, George Telecki, I doubt that this book would ever have been completed. I also thank my editor, Rosalyn Farkas, for her patient, professional guidance through the production process.

I want to express special thanks to Jason Aufdenberg, Ted LaRosa and Rosa Poggiani, who read and commented on virtually every chapter, and to those colleagues who provided extensive comments on individual chapters: Scilla Degl'Innocenti, Ivan Hubeny, Michael Kinyon, Loris Magnani, John Mariska, and Michael Wiescher. Their critiques, support, and wit made it possible to keep going through some of the bleakest parts of the writing. Lys Ann Shore not only read, with the keenest of eyes, but edited much of the manuscript as it was being born. *None* of them should be responsible for any errors that have slipped through despite their best efforts.

Colleagues have answered questions, listened as I tried out ideas, sent preprints, responded to email questions, sent figures and data for the text, shared offices, and provided counsel—too many to list individually, but some stand out for their generosity: Gene Avrett, Steve Balbus, Eric Blackman, Peter Bodenheimer, Butler Burton, Vittorio Canuto, Joe Cassinelli, Paola Castelli, Claudio Chiuderi, Don Cox, Alex Dalgarno, Kris Davidson, Mike DuVernois, Eli Dwek, Bruce Elmegreen, Edith Falgarone, Andrea Ferrara, Doris Follini, Adam Frank, Maryvonne Gerin, Paul Goldsmith, Eric Herbst, Jeff Hester, Paul Hodge, Roberta Humphreys, Namir Kassim, Scott Kenyon, Susan Kleinman, John Lattanzio, Alex Lazarian, Grant Mathews, Dimitri Mihalas, Jim Peebles, Saul Perlmutter, Mirek Plavec, Reuven Ramaty (deceased), Ravi Sankrit, John Scalo, Hans-Rudi Schild, Chris Sneden, Steve Stahler, Ruggero Stanga, Curt Struck, Ethan Vishniac, Rolf Walder, Jim Webb, Bob Williams, Ernst Zinner, and many others who I hope will know implicitly how deeply grateful I am.

To my research collaborators, Tom Ake, Bruce Altner, Jason Aufdenberg, Mike Corcoran, Rosa Diaz-Miller, Giorgio Einaudi, José Franco, Bob Gerhz, Peter Hauschildt, Ted LaRosa, Loris Magnani, Ron Polidan, George Sonneborn, Sumner Starrfield, Greg Schwarz, and Karen Vanlandingham, my sincere thanks. I have learned so much from you all and shared so many intense discussions and joyous moments of understanding. In an age where the act of teaching is derided, I want to publicly acknowledge my debt to the *maestri* of the art from whom I first learned the craft: Nandor Balazs, Tom Bolton, Bob Garrison, Don Goldsmith, Johannes Hardorp, Barry Lutz, John Percy, Deane Peterson, Ernie Seaquist, Frank Shu, Mike Simon, Steve Strom, and Sydney van den Bergh.

During the course of this writing, I found havens of discussion and companionship in the homes of Christophe Dupraz (Paris) and Urs Mürset (Zürich). Jan Palouš, Harry Nussbaumer, José Franco, have provided visiting positions at Center for Theoretical Studies and Charles University (Prague), ETH (Zürich), and UNAM (Mexico City), respectively, during which some parts of this book have been written and/or researched.

Special thanks are reserved for the Publications Board and the Council of the American Astronomical Society for my appointment as a scientific editor of the *Astrophysical Journal*. It has given me the privilege of working with the *capi di capi* Helmut Abt and Rob Kennicutt, their chief of staff, Janice Sexton, and the other scientific editors, and to get to know by correspondence researchers throughout the world. *Nothing* means more in my life than this position, and it has been a continuing inspiration while working on this book.

Students at Arizona State University, Case Western Reserve, New Mexico Institute of Mining and Technology, University of Pisa, Indiana University South Bend, and University of Notre Dame have been subjected to many parts of this book. Its organization parallels lectures and seminars, but the final structure reflects my thoughts on how an astrophysics course sequence *can* be organized. I especially want to thank those who by their comments and questions have helped shape my approach to teaching astrophysics: Kevin Crain, Jason Daly, Jon Darnel, Dean Hines, Shelly Lesher, Mark MacKinnon, Scott Michael, Amy Roberts,

Benoit Semelin, Scott Tipton, Grazia Umana, Giada Valle, Isabelle Waxin, and William Zech.

I thank my colleagues at IUSB, Dean Alvis, Mike Darnel, Jerry Hinnefeld, Michael Kinyon, Monika Lynker, and Rolf Schimmrigk, with whom various parts of this book have been discussed over lunches, in hallways and offices, and at our Friday morning math/physics seminars. It should indicate the scale of this effort that the writing of this book has spanned the tenure of three deans of Liberal Arts and Sciences at my university and I thank Elizabeth Scarborough, Lynn Williams, and Miriam Shillingsburg for their support and understanding during my absences when I should likely have been attending to departmental matters. George Walker, IU Vice President for Research, has always been an ardent supporter of science research at IUSB and I thank him as well for his personal kindness and encouragement.

Finally, I wish acknowledge that support for my research since 1990 and partial support for writing this book has come from NASA, Indiana University, the ETH-Zentrum, CNR, and the French and Italian Ministries of Education, the European Community, and the IUSB Distinguished Research Award.

Now, as I gaze at the mounds of pages of manuscript and notes, the dominant emotion is an overwhelming feeling of gratitude to my family. This book is dedicated to them:

To Lys Ann,
Who as clear-sighted editor, gentle critic, and beloved lifelong companion has contributed to every stage of the production of this book and has suffered so gracefully to see its birth:

If my slight Muse do please these curious days,
The pain be mine, but thine shall be the praise.
—William Shakespeare, *Sonnet 38*

And to the furry aliens who share our home and life,
Adso, Cody, Copernicus, Indiana, Juniper, and Violet,
through whose eyes the world becomes a mysterious and exciting place.

STEVEN N. SHORE

Department of Physics and Astronomy
Indiana University South Bend
May 2002

1 From Gases to Clusters: Concepts in Gravitation and Gas Laws

> Now would I have a book where I might see all characters and planets of the heavens, that I might know their motions and dispositions.
> —Christopher Marlowe, *Doctor Faustus*, Act II, Scene 5

1.1 INTRODUCTORY REMARKS

Faced for the first time with the enormous diversity of astronomical phenomena, you may be tempted to formulate explanations for each observation. After all, when you have only small numbers of examples of any type of object, and none of them look precisely alike, it is easiest to treat them individually. In the *Almagest*, the ancient mathematician Ptolemy, in the second century A.D., advised that the complications of the heavens were to be expected, and if the models had to be equally complex, well that is life. In contrast, nearly 1100 years later William of Occam admonished philosophically minded observers of nature to keep the model economical and to avoid introducing unnecessary complications in the explanations. He didn't specify what level would constitute *unnecessary*, but this prescription has been at the bottom of all scientific work since. It is one of those little miracles of the universe that this great variety can indeed be unified through a remarkably small set of physical laws, and chief among these are gravity and thermal physics. Gravity and the gas laws are inextricably linked in cosmic bodies. The masses of such bodies guarantee that gravitation structures them, and their internal pressure and temperature and density are set in response to this compression. Even more amazing, perhaps, is that only classical approaches are required for many of the most important questions. This chapter should serve mainly as a review of familiar concepts. Starting with gravitation, we cover the two body problem and then go into some of the consequences of tides, resonances, and the more complicated three- and N-body problems. Then we step back to cover thermodynamics for a first time, in order to provide the framework for statistical mechanics. Finally, we show how to combine the two in the fluid limit and how the statistical mechanics of stellar systems isn't really so different from that of a gas.

1.2 GRAVITY

It is impossible to work in astrophysics without, at some point, encountering the effects of gravitation. The huge masses of cosmic bodies means that self-gravity is important in structuring the masses and often dominates their stability. The great advantage we have is that many of the most interesting problems require only a classical picture of gravity.

1.2.1 The Two-Body Problem

The simplest astronomical problem was first discovered by Kepler in 1609, given as an exercise by Halley in 1684, and solved by Newton in 1686: given a central mass and an orbiting body, what is the form of the orbit? We begin with a mass $M \gg m$, where the second body is a test mass (we will generalize this later) and take as the Lagrangian[1] for the motion:

$$L = \frac{m}{2}\left(\dot{r}^2 + r^2\dot{\phi}^2\right) + \frac{GMm}{r}, \tag{1.1}$$

which yields the equation of motion:

$$\ddot{r} = -\frac{GM}{r^2} + r\dot{\phi}^2. \tag{1.2}$$

Now, for a two body orbit, the angular momentum is strictly constant because we have a central point mass (no extended spatial structure means that there is no torque):

$$r^2\dot{\phi} = j = \text{constant}. \tag{1.3}$$

This is the *specific angular momentum*, or the angular momentum *per unit mass* for the orbiting body. Substituting this for the angular velocity, we get

$$\ddot{r} = -\frac{GM}{r^2} + \frac{j^2}{r^3}. \tag{1.4}$$

Two steps remain. The first is to remove the singularity at the origin, which is accomplished by the coordinate transformation, $x = r^{-1}$. The second is to make use of the fact that the angular momentum is constant: you can use the angular coordinate, ϕ in place of the time for the orbital position, so that $d/dt = jr^{-2}d/d\phi$. With this change of variables, Eq. (1.2) becomes

$$\ddot{r} = -j^2 x^2 \frac{d^2 x}{d\phi^2} = -GMx^2 + j^2 x^3, \tag{1.5}$$

---

[1] We assume that you have seen this form of classical mechanics. A very brief outline of Lagrangian methods is given in Appendix 1.B since it will be very useful as a way of setting up dynamical problems later in this book.

which, using the prime to denote differentiation with respect to ϕ, reduces to

$$x'' = -x + \frac{GM}{j^2}. \tag{1.6}$$

This is a harmonic oscillator in ϕ, which is obtained as a function of time using Eq. (1.3). The solution for r is then given by

$$r = \frac{A}{1 + B\cos\phi}, \tag{1.7}$$

where A and B are constants of integration. Notice that $r(\phi)$ is the general equation for a conic section. The constants are found by setting the maximum distance to $a(1 + e)$, where e is the eccentricity, $e^2 = 1 - b^2/a^2$ for a semimajor axis a and semiminor axis b and the minimum separation to $a(1 - e)$, so

$$r = \frac{a(1 - e^2)}{1 + e\cos\phi}. \tag{1.8}$$

This is only an approximate solution, since both masses are really orbiting the center of mass. Actually, we need to use the reduced mass

$$\mu = \frac{M_1 M_2}{M_1 + M_2} \tag{1.9}$$

so the orbital angular momentum is[2] given by

$$j = \mu a^2 \omega.$$

The period of the orbit is found from Kepler's third law (also called Kepler's *harmonic* law)

$$\omega^2 a^3 = GM. \tag{1.10}$$

Here M is replaced by $M_1 + M_2$ when the large ratio approximation is not used. This solution permits the determination of the mass of a body for either resolved systems, those for which the semimajor axis can be directly measured, and unresolved systems for which the radial velocities of both components are known. Using only the velocity extrema for a circular orbit provides the mass ratio, and we will use this in more detail when we discuss the calibration of stellar masses in Chapter 2.

1.2.2 The Effects of Finite Size

The point mass approximation for the gravitational field has its limits—it applies only in the far field and only for the exterior solution. In order to solve for the interior structure of a massive body, or to find the gravitational force for a mass of

---

[2] Recall that using $\mathbf{r} = \Sigma_i M_i \mathbf{r}_i / \Sigma_i m_i$ for the definition of the center of mass coordinate, then $j = \Sigma_i m_i r_i^2 \omega_i$ and, since $\omega_i = \omega$ for all masses, the equation for j follows.

arbitrary shape for a closely orbiting body, we have to use the appropriate equations for the full potential. The field depends on the geometry of the mass distribution since each element produces an acceleration $d\mathbf{g} = -GdM(\mathbf{r})/r^2$. In proposition 70 and subsequent demonstrations in the *Principia*, Newton showed that for a homogeneous spheroid only the mass interior to some radius r produces a centripetal (gravitational) attraction. We can simplify the treatment by considering for the moment only a sphere. Then, since $dM/dr = 4\pi r^2 \rho$, where ρ is the density, we have by Gauss' theorem $\int \nabla \cdot \mathbf{g} \, d\mathbf{r} = \int \mathbf{g} \cdot \hat{\mathbf{n}} \, dS$, which gives the Laplace equation for the external potential, where M remains constant with increasing r:

$$\nabla^2 \Phi = \frac{\partial^2 \Phi}{\partial x_i \partial x_i} = 0. \tag{1.11}$$

The matching, or boundary, condition for this is that the field vanishes as r^{-1} as $r \to \infty$. In other words, no matter what the details of the mass distribution are, an infinitely distant body resembles a point mass. For a distributed mass the potential at an arbitrary point $\mathbf{r}$ is given by

$$\Phi(r) = -G \int_0^r \frac{\rho(\mathbf{r}')d\mathbf{r}'}{|\mathbf{r}-\mathbf{r}'|}. \tag{1.12}$$

For the near field, the most convenient coordinate system is often spherical, (r, θ, ϕ). In these, any function can be expanded in terms of spherical harmonics and radial functions of the form

$$\Phi(r, \theta, \phi) = \sum_{lmn} a_{lmn} R_{nl}(r) Y_{lm}(\theta, \phi) \tag{1.13}$$

where a_{lmn} are constant coefficients. The radial function, $R_{nl}(r)$, depends on the symmetry of the body, in particular whether it is a cylinder or a sphere. The subscripts are the eigenvalues (modes) of the solution of the Laplace equation. It is useful to note that any distributed quantity in spherical coordinates can be represented as a series in spherical harmonics. To find the internal gravitational force requires explicit knowledge of the internal mass distribution. For a sphere, take the radial component of the divergence for $\nabla \cdot \nabla \Phi$ to obtain

$$\frac{1}{r^2}\frac{\partial}{\partial r} r^2 \frac{\partial \Phi}{\partial r} = -\frac{1}{r^2}\frac{\partial}{\partial r} r^2 \left[\frac{GM}{r^2}\right] = -4\pi G\rho;$$

this generalizes to

$$\nabla^2 \Phi = -4\pi G \rho, \tag{1.14}$$

the Poisson equation. We will see in Chapter 4 that the difficult part of modeling the structure of a self-gravitating system is knowing how to self-consistently derive the density from mechanical and thermal equilibrium conditions. But for the moment, we can see that the solution for the potential is

$$\Phi(\mathbf{r}) = -G \int d\mathbf{r}' \frac{\rho(\mathbf{r}')}{R}, \qquad (1.15)$$

where $R^{-1} = |\mathbf{r} - \mathbf{r}'|^{-1}$ is the Green function for a spherically symmetric body.[3] The force follows from

$$\mathbf{F} = -\nabla \Phi = -G \int d\mathbf{r}' \frac{\rho(\mathbf{r}')(\mathbf{r} - \mathbf{r}')}{R^3}. \qquad (1.16)$$

In Appendix 1.A we extend this to ellipsoidal distributions. Cosmic masses, however, often require even more complicated distributions so that the field must be described in the full three dimensions. For the field outside a mass distribution, that is, for a vacuum, Eq. (1.14) reduces to the Laplace equation. We can turn this reasoning around and use measurements of the accelerations of an orbiting body to obtain the internal mass distribution through an inverse problem.[4] Equation (1.15) is actually a *convolution*, which can be seen by defining an operator, the Dirac δ, that selects the value of an arbitrary function f for at particular point r within a larger domain:

$$f(\mathbf{r}) = \int f(\mathbf{r}') \delta(\mathbf{r} - \mathbf{r}') d\mathbf{r}'.$$

Since $\rho(\mathbf{r}) = \int \rho(\mathbf{r}') \delta(\mathbf{r} - \mathbf{r}') d\mathbf{r}'$ we can define the *Green function*, $\mathscr{G}(|\mathbf{r} - \mathbf{r}'|)$, to be the solution of the equation $\nabla^2 \mathscr{G} = -\delta(\mathbf{r})$; for a sphere this is

$$\mathscr{G}(r) = \frac{1}{R} \quad \text{where} \quad R = |\mathbf{r} - \mathbf{r}'|,$$

which means that the integral of the Poisson equation can also be written as

$$\Phi(\mathbf{r}) = \int \mathscr{G}(|\mathbf{r} - \mathbf{r}'|) \rho(\mathbf{r}') d\mathbf{r}'.$$

This is very important for many applications of gravitation we will encounter, especially for the problem of distributed masses in stellar dynamics. We compute the potential through a convolution and then examine the motion of particles in such a field by differencing the value obtained at different points. This is the heart of several dynamical N-body algorithms. We will discuss applications of the Poisson equation to stellar structure in Chapter 4 and to galactic and cluster structure in Chapter 7. You can already see, however, that for extended bodies we will have to supplement Eq. (1.14) with some means for computing the mass distribution. This is where hydrostatic equilibrium and the equation of state enter, topics we will postpone until later chapters.

---

[3] Morse, P. and Feshbach, H. 1953, *Methods of Mathematical Physics*, vol. 1 (NY: McGraw-Hill).
[4] See Lambeck, K. 1988, *Geophysical Geodesy* (London: Oxford Univ. Press) for an especially clear introduction to the problem of computing the geopotential from satellite observations by inversion.

1.2.3 Tides

The point mass approximation is pretty good when the distance between objects is large compared with their radii. But in close encounters, or for close binaries or multiple systems, the finite extent of the mass adds a new wrinkle to the treatment of gravity, the tidal interaction. By assuming a point, you are saying that an external mass produces no differential force from one side of the body to the other. If, however, the distance r between the centers is small compared with R_i, the size of either body, then each mass element not only feels a slightly different force from the perturber but also accelerates in a slightly different direction depending on its location relative to the line of centers. To see this, imagine that the central body of mass M is spatially distended with a radius R. Assume that the body is momentarily at rest with respect to another mass, m, at a distance a from the center of M. We will assume that m is a very compact body compared with its neighbor. For the moment, assume that we are not dealing with an orbit, in the sense that the two masses have no angular momentum. Then M experiences a differential acceleration toward m such that the nearer side, at a distance $a - R$ accelerates more rapidly than both the far side, $a + R$, and the center, a. All equilibrium mechanical properties are referred to the center of M; consequently, there are oppositely directed accelerations as the first-order expansion of the motion of the surface with respect to the center. This is why there are apparently roughly oppositely directed gravitational forces that raise a (roughly) symmetric bulge. Now if the differential acceleration is comparable to the self-gravity, the body will not be able to remain coherent and stable. This condition is expressed by

$$F(a - R) - F(a + R) \geq \frac{GM^2}{R^2}, \qquad (1.17)$$

which means the body is unstable if its radius exceeds

$$R_{\text{critical}} = \left(\frac{GM^2}{F'(a)} \right)^{1/3} \qquad (1.18)$$

relative to the separation a between the components. Here $F'(a)$ is the differential gravitational force of the companion at a distance a from M. This is the *Roche limit* for a self-gravitating body, neglecting cohesive forces. The radius at which this occurs can be expressed in many ways, but since we have already written out the *harmonic law*, we can substitute Eq. (1.10) into Eq. (1.18) to obtain

$$R_{\text{critical}} = R_{\text{RL}} = \left(\frac{M}{m} \right)^{1/3} [G(M + m)]^{1/3} \omega^{-2/3} \qquad (1.19)$$

yielding finally a relation that depends on the mass ratio and the orbital period of the system. The separation drops out of consideration because of the conservation of angular momentum and energy.

The radius given by Eq. (1.19) is the maximum size that a *homogeneous* body can be before the tidal force disrupts it. For planetary moon and ring systems, this means that for some period, it is also impossible for a large body to form. Roche's original investigation concerned the origin of the Saturn ring system, which we know now is also dominated by the effects of resonances with the gravitational torquing of its moons, and the stability of the Earth–Moon system.[5] But for more centrally condensed bodies, the tidal limit is more easily reached. Another way of writing this—closer in fact to the one derived by Roche—is

$$R_{\rm RL} \approx \left(\frac{\rho_2}{\rho_1}\right)^{1/3} a. \qquad (1.20)$$

This means that the more centrally condensed the body is—in other words, the lower its mean density is with respect to the mean density of the system—the smaller its tidal radius.

The full solution for the tidal potential requires looking at the azimuthal and meridional as well as radial accelerations produced by the companion. For a fluid mass, these generate flows that redistribute angular momentum. In the Earth–Moon system, both the elasticity of the solid Earth and the viscosity of the oceans come into play. The lunar, and solar, tides should, in principle, remain stationary with respect to the line of centers between the Earth and the perturbing body. The planet is rotating, however, and this, coupled with the frictional ocean–crust coupling, causes the tidal bulge to slightly lead the Moon. Two different effects follow from this. The bulge torques the Moon and changes its angular momentum. As you now know, a positive torque on a Keplerian orbit increases the semimajor axis and increases the orbital period. At the same time, the spin–orbit coupling of the Earth's (and Moon's) rotation enforces the trend toward synchronism, the equality of the orbital and rotational periods. We will return to this in more detail when discussing close binary stars. Yes, these are elementary considerations, but the tide is a fundamental consequence of gravity and a ubiquitous astrophysical phenomenon. For instance, in the encounter between two galaxies, the fact that the bodies are not point masses and actually have rather low mean densities despite their enormous masses means that tidal interactions dominate the physics of any collision. The dramatic consequences are seen in the enormous stellar (and gaseous) tails that mark the occurrence of such events.

1.2.4 Precession

For dynamical purposes, as we've just done in the two body problem, the masses are usually assumed to behave like points. If, however, they are close enough, then this approximation breaks down. There are two consequences of this change. One

---

[5]There is another way to see what is happening here that uses an argument based on angular momentum. Recall that for the two body problem, the angular momentum of a circular orbit is $J_{\rm orbit} = \mu a^2 \omega$, where μ is the reduced mass. Imagine a body that is a bound swarm of particles, each of which is perturbed in its motion around the main mass, M and rigidly corotating at frequency ω with the mass. Particles closer to m, the companion, now have too low an angular momentum for the local circular orbit and therefore fall toward m. Those on the opposite side of the center of M have too large an angular momentum and therefore recede from it. Consequently, by the condition that any elemental mass would be attempting to achieve Keplerian motion, the body elongates relative to the center of m along the line of centers. This phenomenon is especially important for comets and other fragile bodies (i.e., Comet Shoemaker-Levy 9).

8   FROM GASES TO CLUSTERS: CONCEPTS IN GRAVITATION AND GAS LAWS

is the tide, and this relates to the rotational behavior as well. The other is that the gravitational field felt by an orbiting mass no longer looks pointlike.

1.2.4.1 Rotational Precession
A freely rotating body with an angular momentum **J** and a rotation frequency ω obeys the Euler equation

$$\frac{d\mathbf{J}}{dt} + \omega \times \mathbf{J} = \tau, \qquad (1.21)$$

where τ is the torque. In the absence of an external force, a body whose rotation axis is not coincident with its principal axes (body symmetry axis) will precess. Dissipationless motion produces a freely wobbling body, which is actually observed in cosmic bodies. For instance, the rotation axis of the Earth is not quite at its symmetry axis. Set a body spinning about the z axis with frequency ω_z. Then the freebody precession frequency is

$$\Omega = \frac{C-A}{A}\omega_z \qquad (1.22)$$

which for the Earth is observed to be about 430 days. Here A and C are the moments of inertia around the x (or y) axis around the z axis, respectively. This freebody precession, which is observed at a geodetic station as a variation of latitude, is called the *Chandler wobble*. It has a semiamplitude of only 0″.15 and is longer than solid-body rotation would predict since the moments of inertia differ by about 0.3% and should show a period of about 300 days. An interesting feature of the precession is the stochastic variation on top of the overall motion, from which the rigidity of the interior can be obtained.[6]

Consider a rotating body that feels the gravitational field of a distant point mass. Assume that the angle of inclination between the rotation axis and the orbital plane of the perturber is i and that the body has moments of inertia C around the rotational axis and A in the plane perpendicular to this. For the moment, because it is physically the most reasonable choice for a rotating mass, we will assume that M is an oblate spheroid, in which case $A < C$. The differential gravitational force from the perturber produces the tidal acceleration that precesses the body, since otherwise it would feel no torque. The invariance of angular momentum is analogous to the problem of momentum conservation with variable mass in the absence of force:

$$\frac{d}{dt}\mathbf{J} = \dot{J}_i \hat{n}_i + J_i \dot{\hat{n}}_i = 0, \qquad (1.23)$$

---

[6] For the history of S. C. Chandler's discovery (1891–1894), see Turner, H. H. 1904, *Astronomical Discovery* (London: Arnold). Although written in perhaps too flowery language, the account is an interesting glimpse into the work on dynamical astronomy at the end of the nineteenth century. For more physical discussion, see Munk, W. H. and Macdonald, G. J. F. 1960, *The Rotation of the Earth* (Cambridge, UK: Cambridge Univ. Press); Lambeck, K. 1980, *The Earth's Variable Rotation: Geophysical Causes and Consequences* (Cambridge, UK: Cambridge Univ. Press); Lambeck, K. 1988, *Geophysical Geodesy* (Oxford: Oxford Univ. Press).

where the angular momentum is written in terms of its components $\mathbf{J} = j_i \hat{\mathbf{n}}_i$. Now we need to examine the meaning of the second term, the change in the direction, a bit more closely. In the linear case, this is obvious and provides the definition of force from the second law of motion. For centripetal acceleration, where J_i is constant, this leads to the inverse square law, since

$$\frac{dn_i}{dt} = \omega \hat{z} \times \hat{\phi} \qquad (1.24)$$

points in the $-\hat{r}$ direction. Imagine now a rotating disk with a pivot, a gyroscope, in an external gravitational field. Take $\hat{J} \times \hat{g} \neq 0$, so that the two vectors are not initially parallel. Then $\hat{r} \times \hat{g} \neq 0$, so there is a torque pointing in the (x, y) plane (taking g parallel to z). Thus, to conserve angular momentum, there must be a net motion in $\mathbf{J}$:

$$\frac{d\mathbf{J}}{dt} = m \, \mathbf{r} \times \mathbf{g} \qquad (1.25)$$

We can imagine the dynamics of the rotating Earth and Moon as a gyroscopic problem so that, for a central field, the differential force across the Earth is $\delta F = F_{EM} \sim GMm \, \delta r/r^3$. The torque is then approximately $F_{EM} \, \delta r$ where we can now take $\delta r^2 \approx (C - A)$, the difference between the moments of inertia. Therefore, since $\omega_0^2 = GM/r^3$, we find that:

$$C\omega \dot{r} \approx \omega_0^2 (C - A) \qquad (1.26)$$

so that the precession frequency is given by

$$\Omega \approx \frac{\omega_0^2}{\omega} \left(\frac{C - A}{C} \right) \qquad (1.27)$$

You can also see this using dimensional analysis. Take

$$\text{dimension}[\omega \times J] \approx \text{dimension}[\tau]$$

and assume that the torque, τ is dimensionally the same as dimension$[\tau] \sim \delta I \omega_0^2$. Since the angular momentum of the perturbed body, in this case the Earth, is given by $\mathbf{J} = I\omega\hat{z}$, we obtain

$$\Omega \sim \frac{\delta I}{I} \frac{\omega_0^2}{\omega} \cos i \qquad (1.28)$$

where we have assumed that the rotation frequency for the perturbed body is ω and i is the inclination of the orbital plane to the equator of the perturbed body. We note that for the Earth–Moon system, this gives a very close approximation to the rate of driven precession, 5×10^4 yr since $\omega/\omega_0 \approx 30$ (remember the definition of one month!) and $\delta I/I \approx \frac{1}{300}$; the inclination of the orbit to the rotational symmetry plane is about 30°. Notice that this torque acts to align the rotational and orbital angular momenta. In addition, there are several accelerations in the system. First, the tidal bulge for the central body normally follows the perturber. It does not, however, respond instantaneously to the companion. Here we use the Earth as the body acted on and the Moon as the perturber, but in the more general case it could be a galaxy affected by a companion or a star in a close binary system.

The Earth–Moon system presents an interesting case of how the tide acts on the rotation of the perturbed body through a viscous fluid layer responding to a slow tidal forcing. You would naively expect that the tidal bulge lies along the line of centers of the Earth and the Moon. Due to the viscosity of the oceans and the boundary layer interaction with the solid planet (the frictional coupling), this bulge is actually rotationally advanced ahead of the Moon. The Earth therefore experiences a braking torque while the Moon gains angular momentum and increases in its distance from the planet. The rate of this lunar recession and tidally induced spindown depends on the viscosity and eventually the masses come to synchronism, when the rotational frequency of the Earth is the same as the orbital frequency of the Moon. The *Moon* is already in this state, having identical frequencies, but this was established long ago in its history. The relevance of this process for generalized cosmic bodies is that you would expect that the tidal perturbation would be radially dependent and that collisions (in galaxies) or viscosity (in stars) would redistribute the angular momentum with time. Stars in close binaries experience tides that lead, in the closer systems, to synchronism while the stars are on the main sequence. This will be discussed in more detail in Chapter 5. Clusters and galaxies, which are collisionless, nevertheless redistribute the angular momentum from the tidal bulge through perturbations in the gravitational potential that torque the stars, much like a density wave. We will return to this in Chapter 7. On a larger scale, precession also plays a role in clusters of galaxies. Galaxies are very large compared with their separations within typical clusters than are stars, and also much larger compared with the characteristic sizes of the clusters. Therefore, you would expect that they could be more easily deformed and would precess within the tidal field generated by the cluster. To estimate this effect, you can use the same formalism we just derived.

1.2.4.2 Apsidal Motion and Orbital Precession
An extended mass also complicates the two body orbital problem, introducing orbital precession or *apsidal motion*. The latter refers to the periodic or quasi periodic shift in the apsidal line that connects the two masses and passes through the center of mass of the system. To see this, take $M \gg m$. Then for a point mass, the gravitational acceleration on m would simply be $-GM/r^2$, where r is the distance from M. On the other hand, if M is physically large compared with the r, then there is a differential acceleration resulting from the mass distribution. The mass elements at a distance $r - R$ attract m more than those at $r + R$, just as we saw for the Roche limit. Further assume that the orbit of m is eccentric and, for the moment, take e to be large to emphasize the effect of variable distance from M. For the orbiting mass, this means that the acceleration looks more "pointlike" at apastron than at periastron. The result is that although the angular momentum is still constant and there is also a fixed orbital plane, since M is assumed to be spherical, there is now a precession of the orbit *within* the orbital plane, the

apsidal motion. Another way of stating this is through Bertrand's theorem[7] in a slightly altered wording—the only potentials admitting strictly simply periodic motion are harmonic oscillators. Since an extended mass produces higher-order terms in the expansion of the gravitational potential than a point mass (see Appendix 1.A), the orbits cannot close. Observationally, it is hard to distinguish orbital precession caused by a low mass and/or distant third body from true *apsidal motion* produced by an extended central mass unless the motion of the center of mass can be observed for the two-body system. Precession is even more important for orbits in galactic potentials. In these cases, although the mass distribution may be axisymmetric (hence torque-free), the orbits can be quite complicated and in general precess over time. This is, in effect, what happens to orbits in a binary system. A particle orbit around either mass must precess. If you put a swarm of particles in orbit, they may collide and settle into an accretion disk. We will encounter this picture in dealing with the transfer of mass in close binary systems. For the moment, however, you see the formation of disks is a dissipative consequence of the perturbed orbital dynamics of circulating bodies. Now on to the more general picture.

1.2.5 The Three-Body Problem

We have discussed how, for two bodies, a closed form is obtained for the equations of motion. It is significant that the system is completely invariant under rotation of the semimajor axis; since there is no torque in the system the angular momentum is not only strictly conserved, it's constant. Once we put a third body in the system, however, all bets are off: the angular momentum changes with time for a test mass because it sees two centers of attraction that are in mutual orbit. Let's look in more detail at what happens.

The most convenient way to write the equations of motion is through the Lagrangian, since this way you can see what happens when we add a second center. We treat the motion of the massive bodies in the so-called *restricted* approximation, assuming that they are equal mass and in a circular orbit. You will see that it is possible to get a feel for the generalization to an elliptical orbit, but in this case there are no simple solutions, and we will not treat this problem here. For two massive pointlike bodies, the corotating potential is

$$\Phi(x, y) = -\frac{GM_1}{r_1} - \frac{GM_2}{r_2} + \frac{1}{2}\Omega^2(x^2 + y^2) \qquad (1.29)$$

where Ω is the (two-body) orbital frequency of the binary and the distances r_1 and r_2 are taken from the locations of the respective masses. Transforming into the rotating frame introduces the Coriolis acceleration for slow rotation, which is the perturbation for slow rotation, and the centrifugal acceleration becomes important only for fast rotation for the equilibrium solution for the force. The equations of

---

[7]See Whittaker, E. 1924, *Analytical Mechanics* (Cambridge, UK: Cambridge Univ. Press); the usual statement is that for a central potential varying as $\Phi \sim r^{-n}$, the only values of n for which strictly period closed orbits are found are $n = -2$ and $n = 1$.

motion in rectangular coordinates become:[8]

$$\ddot{x} + 2\Omega\dot{y} = -\frac{\partial \Phi}{\partial x}, \tag{1.30}$$

$$\ddot{y} - 2\Omega\dot{y} = -\frac{\partial \Phi}{\partial y}. \tag{1.31}$$

The equilibrium points are those for which the right-hand sides vanish identically (imposing the condition that $\dot{x} = \dot{y} = 0$ is the reason for calling the surfaces that connect these points *zero velocity* surfaces), and the formal solution then gives five such locations in the (x, y) plane. For historical reasons, they are called *Lagrangian* points.[9] Three of the Lagrangian points are colinearly located along the line connecting the masses: L_1, lying between the masses at the point of force balance, and L_2 and L_3, that lie on either side. Two more, L_4 and L_5 lie perpendicular to the line connecting these. Their importance is their stability. It is possible to obtain stable orbits about these points, while none of the remaining three allows such motion. We mentioned a moment ago the importance of the Coriolis force. Here's why. The L_4 and L_5 points (hereafter simply $L_{4,5}$) are the points where the gravitational attraction of the central bodies balance the local centrifugal acceleration. This means that the only remaining acceleration is due to the Coriolis effect, and this produces helicity for any deflected body.

You can look at it this way. Suppose you set up a body at $L_{4,5}$ and give it a slight kick from rest. If it is deflected outward, away from the central masses, it will deflect to the right in the frame of rotation. Remember, this force depends not only on Ω but also on the velocity components, $(\dot{x}, \dot{y})$. So as the body accelerates (due to the excess centrifugal force moving outward and due to the excess gravitational force moving inward), it always deflects in the same sense. Hence, slow motion *around* $L_{4,5}$ is assured. The most important feature of the $L_{4,5}$ points is that they are local potential *maxima*; the other three loci are saddle points so there is no possible equilibrium. In particular, the L_1 point, lying between the two primary masses, is unconditionally unstable so any kick sends the test mass from one body to another, depending on the direction of the perturbation.

The frequency of this orbit is very pretty to calculate.[10] Since we are in the vicinity of a critical point, the derivatives $\Phi_{x_i} = \partial \Phi / \partial x_i$ vanish for both directions.

---

[8] You can derive the equations of motion in the noninertial system through a simple transformation. Take $R(\Omega t)$ to be the standard 2×2 rotation matrix such that $\mathbf{x}' = R\mathbf{x}$. Now taking the first derivative gives $\dot{\mathbf{x}}' = \Omega R'\mathbf{x} + R\dot{\mathbf{x}}$, and the second derivative becomes $\ddot{\mathbf{x}}' = -\Omega^2 R\mathbf{x} + 2\Omega R'\dot{\mathbf{x}} + R\ddot{\mathbf{x}}$. Here R' is the derivative of the rotation matrix with respect to Ωt. The first term is the centrifugal acceleration, the second is the Coriolis acceleration, and the last is the transformed secular acceleration. This may seem more complicated than just writing down the cross product as $(d/dt)_r = d/dt + \mathbf{\Omega} \times$, as usually done, but it illustrates the purely geometric origin of the Coriolis term and how to generalize to rotation in arbitrarily complex cases.

[9] Lagrange was the first to solve the restricted three-body problem, among his many other achievements. Specifically, he identified the nature of the fixed points of the solution and noted their applications in celestial mechanics. These results were more completely exploited by Laplace in the *Mecanique Celeste*.

[10] What follows is not just a formal academic exercise. Its generality will come in very handy when we discuss dynamics of stars within the Galaxy and also galactic interactions later in the book, and it serves as an introduction to perturbation methods for those of you who may not be too familiar with the methodology.

Here we will use the notation that a subscripted coordinate indicates partial differentiation with respect to that coordinate. Therefore, we can expand the potential as if it were locally a harmonic oscillator. It amounts to saying that any local perturbation is *always* equivalent to a period oscillation around the fixed point. Representing the perturbed coordinates by $\xi = x - x_0$ and $\eta = y - y_0$ we have

$$\Phi(\xi, \eta) \sim \Phi^0 + \tfrac{1}{2}\left(\Phi^0_{xx}\xi^2 + 2\Phi^0_{xy}\xi\eta + \Phi^0_{yy}\eta^2\right) = \Phi^0 + \tfrac{1}{2}\left(A\xi^2 + 2B\xi\eta + C\eta^2\right).$$

Then the perturbation equations are simple and linear:

$$\begin{aligned}\ddot{\xi} + 2\Omega\dot{\eta} &= A\xi + B\eta, \\ \ddot{\eta} - 2\Omega\dot{\xi} &= B\xi + C\eta.\end{aligned} \qquad (1.32)$$

We can search for stable solutions of this linearized system by assuming a single-mode expansion for each perturbed coordinate, $(\xi, \eta) \sim \exp i\omega t$. Substituting this into Eq. (1.32) and solving for the determinant of ω^2 gives

$$(\omega^2 - A)(\omega^2 - C) - (4\Omega^2\omega^2 + B^2) = 0, \qquad (1.33)$$

The stability requirement is $\omega^2 \geq 0$ since we're looking only for periodic solutions. Notice that there is a natural frequency in the system even if all of the derivatives vanished (straight-line motion). This is the rotation of the binary as a whole (even straight-line motion appears to rotate with a frequency 2Ω in the rotating frame). You see immediately that ω^2 is real only if

$$\left(\Phi_{xx} + \Phi_{yy} + 4\Omega^2\right)^2 \geq 4\left(\Phi_{xx}\Phi_{yy} - \Phi_{xy}^2\right). \qquad (1.34)$$

Let's pause for a moment. At the points for which the gradients vanish in the potentials, there are two possibilities. These points can be extrema, where $\partial^2\Phi/\partial x_i^2$ is either positive or negative, or they may be *saddle points*, where the second derivative also vanishes. In the latter case, the points are unconditionally unstable, as you can see from Eq. (1.34). Around the first, there can be stable orbits for small perturbations so it is possible to have bodies in orbit around the two Lagrangian points that lie off the line connecting the two bodies. But this isn't true at the collinear points. You cannot simply place a body at, say, L_1 or L_3 and expect that it will stay there without actively maintaining its position.

1.2.6 Resonances

Resonances occur in all dynamical where many periodicities are presented because of the orbital motions of the various masses. The solar system is the best case of this, where it has been long known as the *small divisors problem*. You can see why this name is applied from looking at a simple harmonic oscillator that has a periodic forcing

$$m\ddot{x} + \omega_0^2 x = F \sin \omega t \qquad (1.35)$$

The solution for the time-dependent amplitude now depends on the *difference* between the frequency of the force, ω, and the natural frequency of the oscillator, ω_0, through

$$x_\omega = -\frac{F_\omega}{\omega^2 - \omega_0^2}, \qquad (1.36)$$

where F_ω is the frequency-dependent forcing, so that as $\omega \to \omega_0$, the amplitude increases without bound. This is, of course, not observed to happen in nature, but it does illustrate the problem. Near resonances are very important and drive the amplitude up with attendant changes in the orbital characteristics. The period distribution in the observed orbits of asteroids[11] and the gaps in planetary rings are wonderful illustrations of this. Once a resonance is achieved, the body becomes very sensitive to the interactions with its perturbers and infinitesimal changes in the conditions produce large new deviations from the orbits. Orbital period locking can also occur. This happens, for instance, for several satellites in the Jovian system and also for the near resonance of the Jupiter–Saturn system.[12] An obvious application that we will see later occurs in disk galaxies, where the orbital motion of stars about the galactic center renders them sensitive to periodic forcing and promotes the growth of nonaxisymmetric collective motions.

1.2.6.1 Trapping

There is one more example of resonance that is found mainly in planetary rings, but such effects could also happen in galaxies when molecular cloud complexes gravitationally influence the bodies. Picture a small body, for instance, a rock, that moves around a planet in an orbit that is situated between two larger moons. Call its orbital frequency $\omega_0 \sim (\Phi_0')^{1/2}$. Now to first order, these three masses execute simple periodic motion that depends only on their distances from the central object. But the two moons exert a perturbation on the rock. The small body is therefore moved by some small δr either inward or outward. It consequently oscillates around its equilibrium position as if it were on a spring with frequency $\omega_g \sim (\Phi_0'')^{1/2}$. Now if the companion moons have a frequency separation $\Delta \omega$ that is of the same order as ω_g, then we have a resonance between the oscillation of the rock and the differential motions of the moons. Another way to understand this is that the orbits are torqued by the satellites that lie at the extrema of the orbits. The inner one increases the angular frequency, hence causing m to move outward, while the outer one slows it down, hence reducing the semimajor axis. The result is trapping between the two satellites. The freedom to place the companions anywhere in the plane of the orbits is why there are so many possibilities for trapping the oscillations of small bodies in ring systems. The best

---

[11] These are called the *Kirkwood gaps* and constitute the Indiana contribution to gravitational physics of the solar system. These are heliocentric radii that display a paucity of asteroidal bodies and, like the planetary ring systems, are likely dominated by collisional effects.

[12] This was first pointed out by Laplace; see Wilson, C. 1985, *Archiv. Hist. Exact Sci.*, **33**, 15 for a marvelously analytical history of this problem.

example of such trapping, which is called *shepherding*, is seen in planetary ring systems[13] but this also occurs at the equilibrium (Lagrangian) points in the three body problem. It may also be important in circumstellar disks in binary and protoplanetary systems (e.g., the dust and debris disks seen around stars like β Pic and perhaps even in younger systems where it could dominate the accretion process).

1.2.7 The *N*-Body Problem

A mass distribution is only virialized if it is gravitationally bound, so its total energy must be negative. For an ensemble of masses, this is equivalent to saying that there is a global distribution of the peculiar velocities that obey the Vlasov equation (if collisionless) or Boltzmann equation (if there are direct interactions between the particles). Another way to say this is that the bodies are moving in a stable way under their mutual gravitational interaction. We discuss the establishment of equilibrium more thoroughly in Section 1.5 in terms of the virial theorem, but for the moment we need to make a few remarks. The term *virialized* is much abused in astrophysics, but it generally means that there are no secular accelerations in the system. This means that $\ddot{I} = 0$. When solving for the mass of a cluster of stars or galaxies, based on the observed velocity *dispersion*, you must make an assumption about the applicability of the observed velocity distribution to the whole cluster.

One way to see what this means is the following. Suppose that you observe a cluster of stars and obtain the *line-of-sight* velocity for each individual member. Having only one dimension to work with, you must make some assumption about the unobservable motions. Of course, the simplest is that the distribution is isotropic, or in other words, if $f(v_i)$ is the distribution function in the observed dimension, then $f(\mathbf{v}) = f(v)$, or that the function is a scalar and the global f is the same as the one dimensional value. Consequently, on average, any line of sight through the cluster samples the same distribution as any other line of sight. This is a tall order for any astronomical system, since there are so many different processes that happen on such disparate timescales, and the sizes of the systems are usually quite large once you are above the size of a single star, but it is a standard assumption.

How can it break down? The system may be rotating. This can be diagnosed through the detection of systematic motion in spatially resolved systems. For instance, systematic velocity gradients across the face of a cluster are an indication, although not a unique one, that the mass may be rotating. Further, depending on whether the system has relaxed, it is possible that the mass is not spherically symmetric and there may be some systematic streaming motion within the ensemble. This *anisotropic* velocity dispersion, observed in the spheroidal components of some galaxies (see Chapter 7), complicates the application of the virial theorem when it is used to determine the mass of the cluster or galaxy.

1.3 THERMODYNAMICS AND STATISTICAL MECHANICS

We have been concerned only with systems whose energy remains constant, the so-called *Hamiltonian* cases. For most problems in orbital dynamics, this works

---

[13] See, for example Goldreich, P. and Tremaine, S. 1982, in ARAA, **20**, 249.

because the bodies do not collide or lose chunks or have internal dissipation (at least in idealized celestial mechanics without tides). Look at how many qualifiers we've needed! The Hamiltonian requirement simply this isn't true in many astrophysical applications, *even* for gravitational dynamics problems. You have already seen the first indication of this in the discussion of tides. Tides for a realistic star involve differential distortion with depth and shear and internal friction lead to dissipation of energy and drive orbital evolution. Further, mass loss changes the gravitational binding of the stars and radiation is a fundamental fact of life for hot self-gravitating systems. For these reasons, much of astrophysical calculation involves bookkeeping for the energetics of a system. The main tool for accounting for the transformation of energy between the different modes of a system is thermodynamics. The connection between the macroscopically measurable quantities, such as temperature and luminosity, and the description of the microphysics is provided by statistical mechanics. Here we review some of the quantities you will need for the rest of the discussion in this book.

1.3.1 Thermodynamic Quantities

Up to this point, we have concentrated on the dynamics of bodies. Now let's treat their energetics. We begin by defining U to be the internal energy of an ensemble of particles. How this quantity arises from a statistical treatment will be considered shortly. For now we take this to be a function of the temperature, which we will assume to measure the internal excitation of the system. The pressure is P and therefore, for any thermodynamic process, the work is defined mechanically as

$$dW = \mathbf{F} \cdot d\mathbf{s} = P\,dV = P d\frac{1}{\rho}, \tag{1.37}$$

where $\mathbf{F}$ is the force acting through a distance, s, and the pressure is defined as the force per unit area taken along the normal to the boundary[14] of the volume V and is a function of the internal properties of the system, the density, ρ, and, presumably, the temperature, T. A transformation is defined by the *first law of thermodynamics*, which states that the addition of a quantity of heat, Q, to the system produces a change in both the internal energy and work by the system against its surroundings:

$$DQ = dU + dW = dU - \frac{P}{\rho^2} d\rho \tag{1.38}$$

The use of DQ is symbolic because the "heat" is a loose concept and is defined only through a new quantity, the *entropy*, which is

$$DQ \equiv T\,dS. \tag{1.39}$$

---

[14] This means that the pressure is the diagonal component of a tensor, the shear stress is the off-diagonal term.

A key feature of classical thermodynamics is that all operations on a system are presumed to produce small changes. Thus, the equations are linear differential equations in the variables that describe the system. For instance, any change in the internal energy U is described by

$$dU = \left(\frac{\partial U}{\partial T}\right)_V dT + \left(\frac{\partial U}{\partial V}\right)_T dV$$

and because small changes are assumed to be continuous, we also have

$$\frac{\partial^2 U}{\partial V \partial T} = \frac{\partial^2 U}{\partial T \partial V}. \qquad (1.40)$$

For this particular case, the result seems rather trivial, but it has important implications for the thermal functions. We have several definitions for the thermodynamic variables in terms of the entropy because you see this is just an expansion of $S(T, V)$ in terms of the basic physical variables:

$$dS = \frac{1}{T}\left(\frac{dU}{dT}\right) dT + \frac{P}{T} dV \qquad (1.41)$$

gives, for instance

$$P = T\left(\frac{\partial S}{\partial V}\right)_T, \qquad \frac{dU}{dT} = T\left(\frac{\partial S}{\partial T}\right)_V. \qquad (1.42)$$

There are numerous other such transformations possible, depending on the variables we use to represent the quantities. Since the transformation we have written looks at the change in the internal energy at constant volume, we define the *specific heat*, c_V, to be $(dU/dT)_V$. Since we can also write $dW = d(PV) - V dP$, it follows that

$$T dS = \frac{\partial U}{\partial T} dT - V dP + d(PV) \qquad (1.43)$$

To preserve the definition of the change of the internal energy with respect to temperature, we can define the specific heat at constant pressure, c_P by

$$c_P = c_V + \frac{d(PV)}{dT}. \qquad (1.44)$$

We will see soon that for an ideal gas, this gives an especially simple relation between the two specific heats.

All other thermodynamic functions are defined in the context of specific processes that are performed on a system. We are quickly faced with a variety of ways of measuring the energy content depending on how we alter the state of the system. The simplest measure of the heat content of a system is the *enthalpy*,

$H = U + PV$. This quantity measures the work done by a thermodynamic transformation at constant pressure (notice from the first law that the work is entirely due to the change in the volume). The amount of energy the system has available from any isothermal transformation is the *free energy*, F (also called the *Helmholtz free energy*) $F = U - TS$. Finally, if we also allow a a change in the number of particles in the system (which does not necessarily mean a change in the density or pressure, but may be due to statistical properties of the ensemble as we will see in the text below), then the *Gibbs free energy* is $G = U - TS + PV$. The change of G or F with all state variables *except* the number (or mass) held fixed is given by μ, the *chemical potential*, a measure of the work required to change the number of particles in the system [i.e. $\mu = (\partial G/\partial N)_{P,V}$]. These thermodynamic functions are more often defined through differential transformations since they arise from changes to a system with specific state variables held constant.

Some important results follow from these simple definitions. It's been known since the seventeenth century that the pressure in a gas increases with increasing density, or decreasing volume, when kept at constant temperature. Further, for what we will hereafter call an *ideal* gas, the relation between P, V, and T is given by

$$PV = \frac{\mathcal{R}}{\mu} T, \qquad (1.45)$$

where we use, with some apologies for the ambiguity, the symbol μ to denote the mean molecular weight. A new constant has appeared, $\mathcal{R}$, the universal gas constant[15] Returning to the thermodynamic relations and using the specific heat at constant pressure $c_P = (dU/dT)_P$ we obtain

$$c_P - c_V = \frac{\mathcal{R}}{\mu}. \qquad (1.46)$$

Next, writing $\gamma \equiv c_P/c_V$, we can rewrite the enthalpy as

$$H = \frac{\gamma}{\gamma - 1} \frac{P}{\rho}. \qquad (1.47)$$

This is an especially convenient form since it leads to a simple expression for the entropy of an ideal gas:

$$T\,dS = \frac{1}{\gamma - 1} d\ln(P\rho^{-\gamma}), \qquad (1.48)$$

---

[15] This is directly related to k, the Boltzmann constant, through Avogadro's number or the mass of the proton, see the table of physical constants at the end of this chapter.

which is

$$P = K\rho^\gamma \tag{1.49}$$

for an *adiabatic transformation*, where K is a constant. This is the first example you will see of an *equation of state*, a relation between the properties describing the thermal and mechanical state of the medium.

Let's pause again for another short digression. Thermodynamics is often presented in a very formal way but we will avoid this approach, emphasizing instead its utility in many astrophysical arenas. The energy quantities we have defined, U, S, F, and G are all interrelated through the first law. It may be confusing that different quantities that measure the energy content of a system are used to describe phases of matter and the transformations between them. In the earliest stages of astrophysical research, the period between about 1860 and 1920, amazing progress was made in understanding relatively simple models that answered essentially mechanical questions of stability and structure. The problem was that the origin of the thermodynamic functions remained a mystery because there was no fundamental theory to connect the microphysical behavior of gases with their large scale manifestation. The clarification came from adopting an atomic rather than continuum approach, that connected the mean quantities, such as the internal energy, and the individual energies of an ensemble of interacting microscopic particles. This work was informed by studies of celestial mechanics and the N-body problem.

1.3.2 Microphysics and the Gas Laws

Atoms in gases and stars in clusters are really not different when you ignore the internal properties of the constituent "particles" and we will try here to put these together. For an ensemble of bodies moving under mutual forces, random changes in position result from myriad close encounters that occur in dense systems. These cause small changes in the motions of each of the bodies that, in steady state, produce a statistical distribution of the momenta. Although gases are dominated by shorter range fields than the one that structures clusters and galaxies of stars, namely, gravitation, they have much in common in the treatment of their motions. Keep this in mind because it will be the theme of the remainder of this chapter.

1.3.3 Statistical Descriptions

A container of gas is like a beehive, a locus of frenetic activity. It is really meaningless to describe the motion of every particle in the gas. Instead, we have the constraint that the collective behavior must produce the macroscopic quantities that we defined earlier. Statistical mechanics treats the medium as an ensemble. To determine the distributed properties of the collection, each sample becomes a separate experiment. You dip your probe into the system, pull out a random collection of bodies, and note their positions and momenta. Each time you do this, you obtain a completely uncorrelated and different sample that, in aggregate, gives you a fair representation of the state of the system. For the dynamical description, only two fundamental variables are needed for each mass in the sample: (1) its position, **x**, and (2) its momentum, **p**. The reason is connected with the equation of motion. With only these two quantities, we can completely describe the motion of classic ensemble. Ignoring the possible complications for non-pointlike masses, each body occupies an elementary volume in *phase space* $d\Gamma = d\mathbf{x}\,d\mathbf{p}$. If there is no dissipation, meaning that there is no energy loss from the system, the ensemble energy remains constant *in total*. The individual particles can, however, change their energy. If the bodies were truly individual and isolated, even this single particle energy would remain constant with time. This is described by a single function for each body, the Hamiltonian:

$$H(\mathbf{x},\mathbf{p}) = \frac{p^2}{2m} + \Phi(\mathbf{x}) \qquad (1.50)$$

The evolution of this function, since it is independent of time in the absence of dissipation, is given by

$$\frac{dH}{dt} = \frac{\partial H}{\partial t} + \frac{\partial H}{\partial x_i}\frac{dx_i}{dt} + \frac{\partial H}{\partial p_i}\frac{dp_i}{dt} = 0. \qquad (1.51)$$

The equations of motion result in simple form from this:

$$\begin{aligned}\frac{dx_i}{dt} &= \frac{\partial H}{\partial p_i} \\ \frac{dp_i}{dt} &= -\frac{\partial H}{\partial x_i}.\end{aligned} \qquad (1.52)$$

We will use these again soon, so it isn't just a formality to introduce them now. It is important to note that this function applies to *each* particle individually but only in the absence of interactions. This is what the Liouville theorem means. If the ensemble has constant energy, the Liouville theorem gives the change in the phase space density of the bodies as they vary in their individual energies under the constraint that the total energy remains constant.

1.3.4 Entropy and Mixing

When we defined the thermodynamic entropy, it appeared as function for describing the quantity of heat that cannot be recovered as work. There is a much deeper meaning to this quantity, but this only comes from a statistical description of matter. In examining this in the thermal context, we'll also be setting up the machinery for understanding the more general ideas of orbital chaos and phase space mixing. These will be very important for understanding ensembles of bodies in gravitational fields.

Two main properties are required for the entropy of an ensemble. It is additive, so we can take individual subsystems and linearly combine them, and it depends only on the volume of the phase space that is occupied by particles of a given energy.[16] Start with an ensemble of particles having a total energy E that is divided into two parts on the basis of their energy. Then the volume they occupy of phase space is $\Gamma(E)$ and that of each of the subsystems is $\Gamma(E_i)$ with $\Gamma(E) = \prod_i \Gamma(E_i)$. There are two constraints on the entropy. It is additive in the energy, that is, $S(E_1 + E_2) = S(E_1) + S(E_2)$, and $S(\Gamma(E)) = S(\Gamma[E_1]\Gamma[E_2]) = S(\Gamma[E_1]) + S(\Gamma[E_2])$, or more simply $S(\Gamma) = S_1(\Gamma_1) + S_2(\Gamma_2)$. Then taking the derivative with respect to the phase space volume, we have

$$\frac{\partial S(\Gamma)}{\partial \Gamma} = \frac{\partial S_1(\Gamma_1)}{\partial \Gamma} + \frac{\partial S_2(\Gamma_2)}{\partial \Gamma} = \frac{1}{\Gamma_2}\frac{\partial S_1}{\partial \Gamma_1} + \frac{1}{\Gamma_1}\frac{\partial S_2}{\partial \Gamma_2}. \qquad (1.53)$$

Since $S_j(\Gamma_j) = S(\Gamma_j)$ for any j, multiplying both sides of Eq. (1.53) by $\Gamma = \Gamma_1\Gamma_2$ gives

$$\frac{\partial S}{\partial \ln \Gamma} = k,$$

where k is a constant, in this case the Boltzmann constant. Using the additivity property gives

$$S = k \ln \Gamma = k \sum_j \ln \Gamma_j \qquad (1.54)$$

for the generalized form for the entropy in statistical mechanics. Since the phase space volume is the probability measure of finding a particle in the gas at some energy E, then if $p_j = \Gamma_j/\Gamma$ and $\sigma_j = \ln p_j$ we can use the linearity of the entropy to obtain

$$S = k \sum_j p_j \sigma_j = k \sum_j p_j \ln p_j, \qquad (1.55)$$

which is the Boltzmann form for the entropy. In other words, the entropy $S = \sum_i p_i S_i$ since $\sum_i p_i = 1$. In what follows, the probability p and the distribution function, f, will be interchangeable (to within a constant normalization factor).

1.3.4.1 The Maxwellian Distribution for Thermal Equilibrium

To treat the ensemble we have in a volume of a gas, we have to appeal to statistical methods. Assume that we have some probability density for finding a particle in the range $[x_i - dx_i, x_i + dx_i]$ in space and $[p_i - dp_i, p_i + dp_i]$ in momentum. Call

---

[16] The importance of the specific formalism that leads to the equation for entropy is that it is based on a procedure for subdividing a complex dynamical system into its parts based only on the energy. The derivation, which follows one originally due to Planck, uses the additivity of the energies and assumes that the interaction between subsystems is very weak. If there is strong mixing between the systems, then there are additional terms in the energy and the phase space for the particles is also altered. There is an especially nice discussion of the traditional derivation of entropy in Fermi, E. 1953, *Thermodynamics* (NY: Dover).

this function $F(\mathbf{x},\mathbf{p})d\mathbf{x}\,d\mathbf{p}$. The connection between F and the observable properties of the gas comes from its phase space integral with the total number of particles:

$$N = \int d\mathbf{x} \int d\mathbf{p}\, F(\mathbf{x},\mathbf{p}). \tag{1.56}$$

Now we appeal to the Hamiltonian. Since the energy is conserved for a dissipationless (closed) system, and the energy of each particle is redistributed throughout the gas through collisions between the bodies, we can assume that the chaotic molecular dynamics merely shuffles the same total energy of the system around among the constituents. This means that at any given moment, a body has only some probability of having a particular value of the energy, but over time it will, on average, have the mean energy of the system. In addition, if we take the gas to be isotropic and uniform on some microscopic scale, the momenta in the three spatial coordinates decouple. In other words, we assume that the motion in any spatial direction is uncorrelated with the motion in any other. We also assume that we can separate the probability density function into two parts, F_1 that describes space, and F_2 that covers the momenta:

$$F(\mathbf{x},\mathbf{p}) = F_1(\mathbf{x})F_2(\mathbf{p}). \tag{1.57}$$

Finally we *assert* that

$$F_2(p_1,p_2,p_3) = f(p_1)f(p_2)f(p_3). \tag{1.58}$$

In other words, we take each momentum component to have an identical but independent distribution. Now since we know that F_2 is also a function only of the energy, we can say that

$$f(p_1^2 + p_2^2 + p_3^2) = f(p_1^2)f(p_2^2)f(p_3^2). \tag{1.59}$$

There is only one simple function that satisfies this condition:

$$f(p_i) = c_1 \exp(c_2 p_i^2). \tag{1.60}$$

Here c_1 and c_2 are arbitrary constants. We're almost finished. The integral, Eq. (1.56), must be finite when taken over the whole range for each component of the momentum, $[-\infty, +\infty]$. It follows that c_2 must be negative. Then c_1 is simply the integral that normalizes the distribution function since f is a probability density and is bounded between 0 and 1. Thus

$$f(p)\,d\mathbf{p} = \frac{\exp(-c_2 p^2)\,d\mathbf{p}}{\int_{-\infty}^{\infty} d\mathbf{p}\exp(-c_2 p^2)} \tag{1.61}$$

which, for an isotropic gas becomes

$$f(p)\,d\mathbf{p} = \frac{\exp(-c_2 p^2) 4\pi p^2\,dp}{4\pi \int_{-\infty}^{\infty} dp\, p^2 \exp(-c_2 p^2)}. \tag{1.62}$$

The constant c_2 must have the dimensions of an inverse energy per unit mass, and since the most probable value for the energy is kT, we see that

$$c_2 = \frac{1}{2mkT}. \tag{1.63}$$

There you have it! This is the *Maxwell velocity distribution* function for an ideal gas.

Notice the assumptions we have made, because these are the principal differences from one astrophysical environment to another. We assumed independence and isotropy for the velocities. This works for a gas, but it may not hold for bodies moving in a strongly anisotropic external potential. We also assumed equilibrium. This may not hold if there are several populations of bodies with different velocity dispersions. However, for the moment, we can rest with the form for an ideal gas:

$$f(v)\,d\mathbf{v} = (2\pi mkT)^{-3/2} v^2 \exp\left(-\frac{mv^2}{2kT}\right) dv. \tag{1.64}$$

Alternatively, we can refrain from identifying the velocity dispersion in the gas with the temperature and think of it instead as a purely statistical quantity that measures the mean energy of a particle. In this sense, stars and atoms are the same when viewed as an ensemble, and a cluster is the same as a gas. The principal difference between them is the collective attractive potential formed by the interactions of the masses. We will return to this point shortly.

1.3.4.2 Discretizing the States: Boltzmann Distribution

An ideal gas, for instance one composed of electrons, consists of completely free particles. In atoms, however, the energies are constrained to take discrete values depending on the details of the interactions among the constituent electrons and the attractive electrostatic potential of the nucleus. This means that if the atom is immersed in a heat bath, its levels must also come into equilibrium. The resulting populations compose the *Boltzmann* distribution, which we now derive.

In strict thermal equilibrium, none of the populations depends on time and the total energy of the system, the total entropy, and the total number of particles are fixed. Each energy level in the atom is labeled by its statistical weight, g_i, which is the total number of particles that can be placed in it depending on their quantum numbers, and its energy ϵ_i. The number of particles is $N = \Sigma_i n_i$, which is given at the start. The entropy for the system is given by the statistical description, Eq. (1.55):

$$S = \sum_i n_i s_i = -k \sum_i n_i \ln n_i \tag{1.65}$$

to within a constant because the probability of occupying a state is the ratio n_i/N and N is constant. Finally, the total energy is $E = \Sigma_i n_i \epsilon_i$. Now in equilibrium, each of these is constant, so a small variation around thermal equilibrium for any of them yields no change. We can, however, combine the three using the technique of Lagrange multipliers:

$$\delta N + a_1 \delta S + a_2 \delta E = 0, \qquad (1.66)$$

where $a_{1,2}$ are unknown constants.[17] We have only one free variable, the level populations, so we take an arbitrarily small variation with respect to n_i to obtain

$$\sum_i \delta n_i [1 - k a_1 (\ln n_i + 1) + a_2 \epsilon_i] = 0. \qquad (1.67)$$

Since δn_i is arbitrary, the quantity in brackets must vanish to satisfy the equality. Therefore

$$n_i = \exp\left[-\frac{(1 - ka_1 + a_2 \epsilon_i)}{ka_1}\right] = A e^{-B \epsilon_i}, \qquad (1.68)$$

where we have simply conveniently lumped constants together to form A and B. Their precise physical significance is made clearer by dimensional analysis: B must be a *global* quantity with the same dimensions as n inverse energy, and A must be the normalization for the population. We then use the same reasoning for the discrete states that we used for the continuum. The normalization constant is $A = 1/N$, and the mean energy is kT. Since each state has a fixed value of g_i, we need to include the statistical weight in the summation over the levels:

$$N \equiv Z(T) = \sum_i g_i e^{-\epsilon_i / kT}. \qquad (1.69)$$

This distribution function strictly holds only for complete thermal equilibrium, in which case you have lost all information about how the particles got into the states. The population ratio for any two states is then

$$\frac{n_j}{n_i} = \frac{g_j}{g_i} e^{-(E_j - E_i)/kT} \qquad (1.70)$$

and only the temperature is needed to determine the relative populations. Here it is worth a moment's pause to discuss what it means to say that temperature alone determines the population ratios. This is the fundamental fact of thermal equilibrium, whether it is local or global. The Boltzmann distribution, like the Maxwellian, assumes that all physical processes, whether collisions with particles in the gas or absorption and/or emission of radiation, are all measuring the same temperature. Thus, if a balance between upward and downward collisions determines the populations, then the temperature measured from the velocity dispersion of the perturbers is the same as you find from taking the population ratios for an atom. This is to say that the *kinetic* temperature is the same as the *excitation* temperature. If radiation processes are involved, then the energy distribution produces a

---

[17] Since the sum of any vanishing quantities must vanish, the sum of any multiple of them must as well. In our case, we have only three such functions, but the method is easily generalized.

radiation temperature that is the same as the other two. This last point is easily seen because a hot gas radiates according to the Stefan–Boltzmann law, so the flux measured from this emission should be the same as you would find for the level population ratios. When the gas is dilute, so that collisions are rare, or when the radiation is not thermalized, then significant departures can occur from the population ratios given by Eq. (1.70). We will discuss this further in Chapters 3 and 6.

1.3.4.3 The Partition Function
In order to normalize the Boltzmann distribution, we must sum over the states. This is the *partition function* and is related to the total number of available states within the atom weighted by the probability of their occupation:

$$Z(T) = \sum_{j=0}^{\infty} g_j e^{-E_j/kT}. \tag{1.71}$$

You see that we allow each of the states to have a number of ways, g_j, of possibly packing particles into it, and the exponential gives the probability of occupying the state at some given temperature. Since the mean energy is given by the weighted average over the states, we have

$$\langle E \rangle = \frac{\sum_{j=0}^{\infty} g_j E_j e^{-\beta E_j}}{Z(T)}, \tag{1.72}$$

calling $\beta = 1/kT$ for convenience. Noting that

$$\langle E \rangle = -\frac{\partial \ln Z(T)}{\partial \beta}, \tag{1.73}$$

through the definition of the free energy, we arrive at

$$F = -kT \ln Z. \tag{1.74}$$

This relation will come in *very* handy in what follows. It is a basic connection between the microscopic description and the macroscopic thermodynamic observable.

Now let's return to the Boltzmann distribution and ask what happens if we go to the continuum limit. By this we mean that we drop the discrete approximation of individual states and imagine the gas to consist of free particles. Then we using the Maxwellian velocity law we just derived, the normalization of the distribution is the same as the partition function for a gas of free particles:

$$Z(T) = (2\pi mkT)^{3/2}. \tag{1.75}$$

The pressure is given by the ideal gas law:

$$P = nkT, \tag{1.76}$$

and the internal energy is simply proportional to the temperature, as it must be for a classical gas of uncorrelated particles (the key point here being their statistical independence).

1.3.5 The Planck Function and Blackbody Radiation

Photons, being massless, are a special case because they form the ultimate relativistic gas. Historically this was the first nonclassical distribution law:[18]

$$B_\nu \, d\nu = \frac{2h\nu^3}{c^2}(e^{h\nu/kT} - 1)^{-1} \, d\nu,$$

where ν is the frequency. This represents the monochromatic intensity of the radiation and is called the *Planck function*. We will use this extensively in later chapters. Remember, this is the distribution for radiation in strict thermal equilibrium with the matter, where the temperature, T, is an global, unique property of the gas. It may also be true locally, in the sense that in some volume the temperature may be uniform on the scale of a photon mean free path. Now we will see where it comes from.

The Planck function can be derived directly from the partition function, and it is rather simply motivated for photons. Thinking classically, the picture of thermal equilibrium for light is standing waves in an adiabatic box. There is no limit to the number of such waves that can be found with a given wavelength, since they are linear and therefore can be superposed. Planck's statistical guess was to say that the waves of energy $\epsilon = h\nu$ have an energy in a state j given by $E_j = n_j \epsilon_j$, where n_j is the number of photons in the state. Then the partition function is the sum over the ensemble of particles in each state j, given by

$$Z(T) = \sum_{n_1, n_2, \ldots, n_k} e^{-E_j \beta}. \tag{1.77}$$

The mean occupation number for the state as a function of temperature is

$$\langle n_j \rangle = \frac{\sum_{n_1, n_2, \ldots, n_k} n_j e^{-n_j \epsilon_j \beta}}{\sum_{n_1, n_2, \ldots, n_k} e^{-n_j \epsilon_j \beta}} \tag{1.78}$$

which is the same as

$$\langle n_j \rangle = -\frac{\partial}{\partial \epsilon_j} \ln Z(T). \tag{1.79}$$

This is where the critical difference comes in between massive and massless particles. If massless, there is no threshold for particle creation, so there is no constraint on the number of particles (e.g., you can have an essentially infinite number of particles at zero energy), and the numbers can range from none to infinite in a single state. On the other hand, particles with mass have a minimal

---

[18] See Kuhn, T. 1978, *Blackbody Radiation and the Quantum Discontinuity* (Chicago: Univ. Chicago Press); Kangro, H. 1972, *Planck's Original Papers in Quantum Physics* (London: Taylor and Francis Ltd.); Planck, M. 1914, *Theory of Heat Radiation* (NY: Dover Books).

energy and therefore there must be a fixed number of them, so each occupied state is some number less for the rest of the states. This means there is no constraint on $\Sigma_j n_j$ for the massless case and $N = \Sigma_j n_j$ for the massive ones. Thus, for massive particles there is some chemical potential that represents an "insertion energy" for a state. Continuing, for photons, the partition function is

$$Z(T) = \sum_{n=0}^{\infty} e^{-nh\nu\beta} = \left(1 - e^{-h\nu\beta}\right)^{-1}. \tag{1.80}$$

Then by Eq. (1.78) the mean occupation number is

$$n(\nu) = \left(e^{h\nu\beta} - 1\right)^{-1} \tag{1.81}$$

In other words, when you subdivide the system into bins of energy, for any given value you have many possible combinations of constituents. This is now the relative occupation number for an elementary volume so that the intensity of the radiation is $I_\nu = n(\nu)h\nu c/4\pi h^3$. This is given a special designation, one we have already used for the emissivity of a blackbody:

$$B_\nu d\nu = \frac{2h\nu^3}{c^2} \frac{d\nu}{e^{h\nu/kT} - 1}, \tag{1.82}$$

which is the Planck function. It is a unique functional description of radiation in thermal equilibrium. The central feature of this distribution, and its greatest initial success, is that it is directly connected to the thermodynamic description of the energy density for radiation. Specifically, the integral of the Planck function gives the Stefan–Boltzmann law:

$$\int_0^\infty B_\nu \, d\nu = \int_0^\infty B_\lambda \, d\lambda = aT^4. \tag{1.83}$$

The constant a is directly related to k and h, thus representing the first historical link between the gas laws and quantum mechanics. Of course, this is built into the distribution from the start since we assume a form for the relation between energy and frequency.[19] This is the same as the energy density, $u(T)$, and the pressure for an isotropic gas is $\frac{1}{3}u(T)$. Therefore, the density of a photon gas scales as T^3, since the energy is proportional to T, which again yields the extreme relativistic equation

---

[19] It is useful to see how the constants are related. Defining $x = h\nu/kT$, we have

$$\int_0^\infty B_\nu(T) d\nu = \frac{2k^4}{h^3 c^2} T^4 \int_0^\infty x^3 e^{-x} (1 - e^{-x})^{-1} dx.$$

The integral is actually

$$\sum_{n=0}^{\infty} \int_0^\infty x^3 e^{-(n+1)x} dx = \sum_{n=0}^{\infty} \frac{6}{(n+1)^4}.$$

The last series is the Riemann zeta function of 4. We'll compare this later to an alternate distribution for massless particles, neutrinos, so the details matter.

of state $P \sim \rho^{4/3}$. This means that a relativistic boson gas at constant entropy (adiabatic) varies as $\ln P\rho^{-4/3}$.

The low and high frequency limits of B_ν are often useful approximations to the spectrum. At low energies, this is called the *Rayleigh–Jeans* spectrum:

$$B_\nu^{\text{RJ}}(T) = \frac{2kT}{c^2}\nu^2 \qquad (1.84)$$

while the high energy limit is called the *Wien* spectrum:

$$B_\nu^{\text{Wien}}(T) = \frac{2h\nu^3}{c^2}e^{-h\nu/kT}. \qquad (1.85)$$

The peak of the Planck distribution follows the *Wien displacement law*:

$$\lambda_{\max} T = \text{constant} \qquad (1.86)$$

which is the simplest way to estimate the color temperature for a blackbody radiator (see table of physical constants). We will return to this in Chapter 3, where we'll discuss the connection between the Planck distribution and the creation term for radiation in a hot gas from the point of view of the radiative transfer. For now, it is enough to know that a gas in thermal equilibrium that is optically thick and a gray radiator will have a spectral peak that depends only on its temperature.

1.3.6 Degenerate Equation of State

As the temperature is reduced, a classical gas in an attractive potential (e.g., gravity) ultimately collapses. Since all motion ceases at 0 K, the velocity dispersion vanishes for a cold gas and so does the pressure. Recall that there are point particles in the classical world, but introducing quantum mechanics produces a dramatically altered picture: the gas still exerts a pressure even if there is no net motion of the gas simply from the intrinsic fuzziness of the bodies at the microscopic level. A simple application of the uncertainty principle leads directly to an equation of state because of this nonlocalization property, regardless of whether the exclusion principle prevents any two particles from occupying the same volume of phase space (this depends on the quantum statistics needed to describe the particles, as we'll see in a moment).

The *uncertainty principle* gives the minimum volume a free particle can occupy in phase space, $\Delta x_i \Delta p_i = h$. In other words, the momentum fluctuations never fall below some minimal value. This is the *Fermi momentum*, p_F. If the separation of the particles is, on average, given only by the number density, $\Delta x_i = n^{-1/3}$, then $p_F = hn^{1/3}$. Since the pressure for an isotropic gas is $P_{\text{deg}} = \frac{1}{3}u_F$, where u_F is the energy density, we can write $P \sim p_F^2 n/m$. Then we find for the equation of state:

$$P_{\text{NR}} = K_{\text{F,NR}}\, \rho^{5/3} \qquad (1.87)$$

where $K_{\text{F,NR}}$ is a scaling constant for *nonrelativistic* degeneracy. Notice that this is the same result as we obtained for the pressure of an ideal gas, but how the exponent is *not* γ, the ratio of specific heats. It is simply an index related to the compressibility of the medium. For increasing density, the momentum must also increase. It eventually becomes relativistic, at which point the increase in the pressure must be smaller since the apparent mass of the particle now increases as well. This has the effect of *appearing* to increase the mean molecular weight and therefore makes the medium more compressible.[20] The dimensional estimate is altered because of the change in the relation between the momentum and energy for a relativistic gas, $P_{\text{ER}} \sim p_F n$ and

$$P_{\text{ER}} = K_{\text{F,ER}} \, \rho^{4/3}, \quad (1.88)$$

where $K_{\text{F,ER}}$ is a different scaling constant for *extremely relativistic* degeneracy. The pressure now has the same density dependence as radiation, which means a very soft equation of state. The Fermi momentum is related to the chemical potential for the gas through ϵ_F, the Fermi energy. You will recall that the chemical potential, which gives the Gibbs free energy in thermodynamics, is the energy needed to insert a particle depending on the number of particles already in the system. With these qualitative guides to what we should expect from the formalism, we can now proceed to a more precise derivation of the distribution function for a degenerate gas.

1.3.6.1 The Fermi–Dirac Distribution Function

The foregoing considerations seem to apply equally well to any type of particle, and, you might think, lead to the analog of the Planck function. Yet there is a fundamental difference. Particles with half-integer spin obey the Pauli exclusion principle, which states that no more than a single particle can occupy an elementary volume in phase space. This is the same thing as saying that each particle in a quantized system that consists of antisymmetric constituents must have a unique set of labeling quantum numbers. For photons, and as we will see soon for bosons in general, there is no limit to the number of particles that can occupy a single state. In other words, electrons differ from photons not only by virtue of their mass but also because of their spin. No two electrons can have the identical set of all quantum numbers, hence no more than one can occupy a unique energy level (the values of dynamical variables for all degrees of freedom cannot be the same). This means that you can write the partition function as

$$Z(T) = 1 + e^{-(E-\epsilon_F)/kT} \quad (1.89)$$

for which the free energy is again found from $F = -kT \ln Z(T)$. Taking $\epsilon_F = \mu$

---

[20] Now why might the index be the same as that of an ideal gas? Think about how many degrees of freedom the particles have. For a nonrelativistic electron gas, although the accessible phase space volume is restricted there are no forbidden motions. The gas is ideal in the sense that all particles are free to move but they cannot move freely. Consequently, it isn't until you reach relativistic speeds that the inertia of the particle alters the phase space itself and changes how the momentum varies with energy.

to be the chemical potential for the gas, the number at a given energy becomes

$$n_{\rm FD}(E) = \frac{\partial F}{\partial \mu} = \left(e^{(E-\mu)\beta} + 1\right)^{-1}. \quad (1.90)$$

Note that the occupation number now depends on the density of the gas through the Fermi energy. The effect of degeneracy can be understood by comparing the volume in phase space occupied by particles in a thermal distribution with the quantum limit. The momentum of a classical particle varies as $T^{1/2}$, while in a degenerate gas it varies as $n^{1/3}$ with no dependence on the temperature. The ratio of the two volumes of phase space scales as $T n^{-2/3}$, so for any temperature, there exists a critical density above which the gas cannot remain Maxwellian.

How might such a state arise? At low temperatures, under normal laboratory conditions, electrons form degenerate systems within matter. An example is the electron distribution in metals. In astrophysical systems, such high densities as you find in everyday materials are encountered only in stellar interiors. These are usually quite hot (see Chapter 4), so at first glance you would expect that they should be ideal gases. However, a self-gravitating body obeys the virial theorem so as it cools, it contracts. The cooling can occur by photon or neutrino emission or by conduction. If this energy loss is efficient enough to maintain isothermal conditions, the medium becomes progressively denser and can cross the degeneracy limit. We will see in Chapter 4 that the consequence for the body is severe—there is an upper limit to the mass when it is supported only by degeneracy pressure. The limit was first described by Chandrasekhar (1933), who realized that it leads to an even more drastic consequence: the equation of state becomes progressively more compressible with increasing density because of relativistic degeneracy. This means that a massive enough star that is already contracting under its own weight will never be able to find a stable configuration and must ultimately collapse. We will postpone this discussion for a while until we meet it again in our discussion of stellar interiors.

There is yet another way to obtain this result by using a construction that is similar to Boltzmann's original procedure and explicitly invoking the Pauli exclusion principle for half-integer spin particles. If a state i is labeled by an ensemble of quantum numbers (such as the principal quantum number, spin, and orbital angular momentum), $\{q_i\}$, then $n_i = 0$ or 1 *only*. This means that for a continuum, one where the statistical weights of the individual states are just the phase space volume, the partition function is

$$Z(T) = \sum_{n_i=0}^{1} g_i e^{-n_i E \beta} = g_0(1 + \zeta e^{-E\beta}) \quad (1.91)$$

where ζ is called the *fugacity* or degeneracy of the state, which is again related to the chemical potential. Rather than using the free energy, we can use the variational principle to determine the occupation number for any state. If ϖ is the probability of having no particles in a state ϵ, then, according to the exclusion principle, the probability of having only one particle is $1 - \varpi$. The entropy is thus $S = \varpi \ln \varpi + (1 - \varpi)\ln(1 - \varpi)$ and the number of particles is $N = (1 - \varpi)$.

Following the same procedure as for the Boltzmann distribution, we get $\varpi(\epsilon) = (1 + \exp[-\beta(\epsilon - \alpha)])^{-1}$ where, α measures the *degeneracy* and represents a zero point for the energy of the system; this is again the Fermi–Dirac distribution. Notice that it has a few features that distinguish it from the Boltzmann distribution. First, at low temperature, it behaves differently. The states are filled to some level that is independent of temperature but depends on α. Only at high temperature does the gas behave as it would in classical thermal equilibrium. The particles are strongly correlated up to some critical energy level, which we now discuss, and this significantly changes the behavior of the equation of state.

The pressure is proportional to ρv_F^2, where now $v_F = p_F/m_e$ so that we obtain a limiting equation of state for this sort of "cold," nonrelativistic matter:

$$P_{NR} = K_{NR}\, \rho^{5/3} \qquad (1.92)$$

where K_{NR} is a constant. Notice that you found the same behavior for an ideal gas, but now there is no corresponding relation between temperature and density. This is very important. Temperature is irrelevant to the velocity distribution in a completely degenerate gas. The uncertainty of the particle momentum *alone* determines its velocity dispersion so that the distribution function does not resemble a Maxwellian. We will return to the consequences of this equation of state many times in subsequent chapters.

Let's extend this now to the highest-density limit. The Fermi momentum continues to increase if we compress the gas. Even for quantum-mechanical motion, however, the speed of the particle cannot exceed, c, the speed of light. We have assumed that the mass of the particle is constant in deriving the relation between pressure and momentum, but this is no longer correct as we approach the high-density, small-phase-space volume limit. The relativistic increase in the mass ultimately limits the volume, and, since the velocity dispersion cannot continue to increase, neither can the pressure. As the momentum becomes highly relativistic, the energy of the particle becomes $E \to p_F c$ when $p_F \gg m_e c$. Now taking $P = u/3$, where u is the energy density, we already know that $u = En$ so we obtain as a limiting equation of state:

$$P_{ER} = K_{ER}\, \rho^{4/3} \qquad (1.93)$$

where K_{ER} is another constant. What we now have is the most extreme form of equation of state a dense gas of massive fermions can achieve. Notice that we now have a softer equation of state than in the nonrelativistic limit, meaning this gas is more compressible. This results, in part, from the relativistic variation of the constituent particles' mass because as we compress the gas and the Fermi level rises, the effective inertia of the particles also increases, which offsets some of the phase space confinement (in the ultrarelativistic limit, the equation of state for fermions is qualitatively the same as for photons and magnetic fields) so compressed gasbag cannot maintain equilibrium forever. We shall see, when discussing stellar evolution, that this condition governs the final state of a star. But for the moment, let's go back to what it means for a gas to reach degeneracy.

The Fermi energy for nonrelativistic matter corresponds to a lower limit on the temperature, that value that the gas must exceed in order to restore the ideal-gas law. This is $p_F^2/2m_e = kT_F$, so $T_F \sim n^{2/3}$. Thus, for an adiabatic gas for which nT^3 is constant, there exists a unique density at which the gas becomes degenerate. Imagine that you take a star and begin to compress it, for instance by increasing the mean molecular weight in its core through nucleosynthesis. Then even without cooling, there is eventually a point where the gas supports itself purely by degenerate pressure, at which point it is stable.

The exact relations between the distribution function and the thermodynamic quantities follow from the partition function as they did for the Boltzmann distribution. Since the Fermi–Dirac distribution function is continuous, like the Maxwellian, we can use the definitions from the moments of the Vlasov equation to obtain the thermodynamic functions. To make the moments easier, we first change variables to $z = p^2/2mkT = \epsilon/kT$. We define the auxiliary function by

$$F_n(\alpha) = \int_0^\infty \frac{z^n \, dz}{e^{z-\alpha} + 1}, \qquad (1.94)$$

which is called the *Fermi–Dirac integral*. This is a function of the degeneracy parameter, α, and in compact form represents the departure of the gas from the ideal conditions of a Maxwellian. Then we obtain the partition function:

$$Z(T) = (2\pi mkT)^{3/2} \frac{2}{h^3} F_{1/2}(\alpha); \qquad (1.95)$$

we have normalized the phase space volume, to the uncertainty volume, and the factor of 2 comes from the spin. Notice that now the occupation of the states depends on both the density and temperature, in contrast to the ideal gas law. The density is

$$n = \frac{4\pi}{3h^3} (2mkT)^{3/2} F_{1/2}(\alpha). \qquad (1.96)$$

Then the pressure is the second moment of the distribution function:

$$P \sim \int p^2 f(p) p^2 \, dp$$

which becomes

$$P = \frac{8\pi}{3h^3} (2mkT)^{3/2} kT F_{3/2}(\alpha) = \frac{2}{3} nkT \frac{F_{3/2}(\alpha)}{F_{1/2}(\alpha)}. \qquad (1.97)$$

The difficulty this introduces into the calculation of thermodynamic properties for the gas comes from the density dependence of α, so the equation of state is very nonlinear in the number density.[21] In the relativistic limit, the single particle energy ϵ is

$$\epsilon^2 = p^2c^2 + m^2c^4 \tag{1.98}$$

so the limiting case for the internal energy of cold matter becomes

$$U = \frac{Vc^3}{h^3} \int_0^{p_F} p^2 (p^2 + m^2c^2)^{1/2} dp = \frac{Vc^3}{h^3} \left[x_F (2x_F + 1)(1 + x_F)^{1/2} - \sinh^{-1} x_F \right] \tag{1.99}$$

where $x = p/mc$. Remember that p_F depends on the density. For the intermediate case:

$$U \sim \int_0^{p_F} \frac{\epsilon^3 d\epsilon}{e^{(\epsilon - \alpha)\beta} + 1}. \tag{1.100}$$

The change in the exponent leads that the equation of state toward a much higher compressibility.

Neutrinos are a special case. Laboratory limits on the masses of the three known flavors, ν_e, ν_μ, and ν_τ are so low that these particles are assumed to be massless.[22] They represent the relativistic extreme for fermions. As with photons, their energy density varies as T^4, but their different spin means that the coefficient (the effective Stefan–Boltzmann constant) is different from that for photons because of the different forms of the partition functions. Choosing one neutrino flavor, ν_e, we can write the integral for a massless fermion as

$$\frac{u_{F,ER}}{u_{B,ER}} = \frac{\int_0^\infty \frac{x^3 dx}{e^x + 1}}{\int_0^\infty \frac{x^3 dx}{e^x - 1}} = \frac{\sum_{n=0}^\infty (-1)^n (n+1)^{-4}}{\sum_{n=0}^\infty (n+1)^{-4}}. \tag{1.101}$$

Since a is defined from the Planck function, then for massless neutrinos:

$$u_{\nu_e} = \tfrac{7}{8} aT^4, \tag{1.102}$$

neglecting the spin states (which changes the statistical weights). In a dense medium, they can decouple sooner than a photon gas because they interact only

---

[21] Further expansions of these integrals are given in Cox, J. and Guili, R. 1968, *Principles of Stellar Structure*, vol. 2 (NY: Gordon and Breach) provide extensive expansions for partially degenerate gases. See also Cloutman, L. D. 1989, *ApJS*, **71**, 677; Antia, H. M. 1993, *ApJS*, **84**, 101; Miralles, J. A. and Van Riper, K. 1996, *ApJS*, **105**, 407; Johns, S. M., Ellis, P. J., and Lattimer, J. M. 1996, *ApJ*, **463**, 1020.

[22] The most recent results from Superkamiokande and other solar neutrino observatories indicate that the neutrino is not massless but no strong limits can currently be placed directly on the rest mass. Instead what is measured is the mass splitting between species (in atmospheric and accelerator experiments) through neutrino oscillations and flavor mixing. The upper limits for m_{ν_e} remain small.

with matter through the weak force and don't thermalize the same way as photons. For instance, imagine a dense gas that is hot enough to form pairs via electron–positron annihilation. The neutrino pairs that are created can leave the medium while the photons will scatter and thermalize. This neutrino cooling is very important in dense environments, when photons are unable to freely escape. In general, neutrinos will not display a thermalized distribution unless their mean free path (see discussion below) is small enough that they scatter frequently before exiting the medium. Hence, we would expect that the general spectrum of neutrinos, which in astrophysical situations are usually the result of nuclear processes, will reflect the energy dependence of the weak interaction that produces them and not the temperature of the gas.

1.3.6.2 Bose–Einstein Statistics

We have dealt with photons as a separate case because they are massless. In the catalog of matter, however, there are symmetric massive particles that are not constrained by the Pauli exclusion principle. Because they are massive, these particles have a minimum energy the photon lacks. These *integer spin* particles are called *bosons*. Like fermions, these particles need not be elementary. For instance, ^{4}He has a total spin of zero and thus satisfies the criterion for Bose–Einstein statistics. There is no limit on the number that can occupy a given state and if the temperature is low enough or the density high enough, degeneracy effects and collective quantum behavior are possible. This is seen best in superconductivity and superfluidity, both of which are achievable in the laboratory at extremely low temperatures and that in stars can occur at enormous, near-nuclear, densities. Superposition is possible for any state provided the energy density is not so high that the particles begin to interact, so we can write the partition function as an infinite sum over the available states:

$$Z(T) = \sum_{n=0}^{\infty} e^{-n\epsilon_0/kT} = \left(1 - e^{-\epsilon_0/kT}\right)^{-1}. \tag{1.103}$$

Then using our previous notation, the occupation probability is

$$\varpi(E) = \left(e^{E/kT} - 1\right)^{-1}. \tag{1.104}$$

This is the Bose–Einstein (BE) distribution.[23] As usual, taking the logarithm and differentiating, we obtain F from which we derive the final form of the distribution function. The pressure for this type of gas is given by

$$P = (2\pi mkT)^{3/2} G_{3/2}(\eta) \tag{1.105}$$

---

[23] Huang, K. 1987, *Statistical Mechanics*, 2nd ed. (NY: Wiley); Landau, L. and Lifshitz, E. 1980, *Statistical Physics*, part 1 (Oxford: Pergamon).

where now we define the integral including the chemical potential as

$$G_n(\eta) = \int_0^\infty \frac{z^n\, dz}{e^{z-\eta} - 1} \quad (1.106)$$

in order to include the effects of the Bose–Einstein condensation, the analogy to degeneracy, when the gas distribution collapses to a state of zero momentum for extremely low temperature. This happens because even a boson gas has a zero-point energy due to quantum fluctuations. This is a purely quantum-mechanical effect that has now been obtained in the laboratory for both single particles and ensembles. Mesons, in particular pions, are integer spin particles that may, under sufficiently dense conditions, collapse as a Bose–Einstein condensate. This behavior is important for nuclear matter. An additional application of this effect was the search for a nonvanishing η for the cosmic background radiation in the formal fit of the Planck law to the COBE data (see Chapter 8), a cosmological constraint that requires the a zero mass for the photon.

1.3.7 The Saha Equation: Ionization Balance in Thermal Equilibrium

Statistical mechanics enters into astrophysics whenever the gas laws are needed, and especially when dealing with changes of state. Ionization and dissociation can also be treated using statistical mechanics. In thermal equilibrium, these processes are described by the *Saha equation*[24] which we now derive. To begin a physical picture, imagine an atom immersed in a radiation field. Absorption of a photon above the ionization energy leaves ion by ejecting a now free electron into the gas. The inverse process, *recombination*, refers to the capture of an electron by the ion that returns the atom to a neutral or lower ionization state. In equilibrium, these two processes balance so net rate for these competing reactions is independent of time. If the gas is in strict thermal equilibrium, photoionization exactly balances recombination as if it were a collision. In the simplest form, the fact that the kinetic equations depend only on temperature can be used for the analysis of chemical reactions in thermal equilibrium. Specifically, if you have a reaction of the form $A + B \leftrightarrow C + D$, then $(N_D N_C)/(N_A N_B) = K(T)$, as we will use for nuclear statistical equilibrium in Chapter 4.[25] Schematically, the rate for a particle to absorb a photon and release an electron, $\mathscr{I}$, is balanced by the inverse process of recombination, $\mathscr{R}$:

$$n_e N_{r+1} \mathscr{R} = N_r \mathscr{I}. \quad (1.107)$$

For thermal equilibrium, we can be more precise. The task reduces to counting the number of available states and the probability that a particle will go from one to

---

[24] The basic historical references are Saha, M. N. 1921, *Proc. Roy. Soc.* (London), **A99**, 135; Pannekoek, A. 1922, *Bull. Astr. Inst. Neth.*, **19**, 1; Fowler, R. H. and Milne, E. 1924, *MNRAS*, **83**, 403, **84**, 499; Payne, C. H. 1925, *Stellar Atmospheres* (Cambridge, UK: Harvard Observatory Monographs).

[25] In his 1920 Nobel address in chemistry, W. Nernst noted that "Eggert, Saha, and others have used my heat theorem (the theory of chemical combinations) successfully for the answers to astrophysical questions" [see, e.g., Farber, E. 1963, *Nobel Prize Winners in Chemistry: 1901–1961* (London: Abelard-Schuman)].

another. The only thing that matters in determining the number of particles in ionization state is its statistical weight, the number of available elementary phase space volume elements that the system occupies. This is provided by the partition function for each component of the system. For the reaction $A^{(r)} + \gamma \leftrightarrow A^{(r+1)} + e$, we need to count the number of states in the two atoms and the number in the continuum for the electrons. The fractional phase space volume filled by the electrons is due only to its translational degrees of freedom, which you already know from the Maxwellian distribution since the particle has no internal structure:

$$Z_{\text{trans},e}(T) = 2\frac{(2\pi mkT)^{3/2}}{n_e h^3}. \tag{1.108}$$

where the factor of 2 comes from the spin. For the atom $A^{(r)}$ and ion $A^{(r+1)}$, the partition functions are $Z_{\text{trans},r}Z_i(T)$ and $Z_{\text{trans},r+1}Z_{r+1}(T)$, respectively, where the translational parts are identical and formally the same as for the electron with $m_r = m_{r+1}$ substituted for m_e since the ions and electrons have the same temperature. The phase space volumes are formally the same as for the electrons with N_r and N_{r+1} substituted for n_e. Finally, the work function, the energy to go between the two atomic configurations, is the ionization energy χ_r so the relative population in equilibrium is given by

$$\frac{n_e N_{r+1}}{N_r} = \frac{Z_{\text{trans},e}(T)Z_{r+1}(T)}{Z_r(T)}e^{-\chi_r/kT}. \tag{1.109}$$

You will notice that the translational terms for the ions have canceled out. Substituting Eq. (1.108), we obtain

$$\frac{n_e N_{r+1}}{N_r} = \left(\frac{2\pi m_e kT}{h^2}\right)^{3/2}\frac{Z_{r+1}(T)}{Z_r(T)}e^{-\chi_r/kT}. \tag{1.110}$$

This is the Saha equation in its usual form. It is, strictly speaking, only an approximation. To illustrate its solution, consider a pure hydrogen plasma, for which it can be written more simply since such a gas has $n_e = n_p$, where n_p is the number density of protons so that $n_e^2 = n_H F(T)$ (we can work with total numbers or number densities). Since $2n_e + n_H = n$, we can solve for the ionization of hydrogen for each density as a function of temperature. It is easy to see that as the density increases, the equilibrium shifts to the neutrals, so that a low-density gas is inevitably more highly ionized than a higher-density medium. Our treatment must be modified for degenerate gases since the electron distribution is constrained by the density. For instance, the number of free states is reduced compared with the free-particle case. Rate equations that are in equilibrium are always of this form. For chemical reactions, the same basic formalism applies except that the rate balancing becomes more complex because of the form of the partition function (see Chapter 3).

1.3.7.1 *The Concept of Local Thermodynamic Equilibrium (LTE)*
The reason why the Saha equation works is made clearer by an appeal to the actual rates in LTE. We will defer the final treatment of statistical equilibrium

until Chapter 3, where we will relax the assumption of strict thermal equilibrium and include the nonlocal effects that arise from the radiation. If we increase the density of the gas, the recombination rate increases faster than the ionization rate. On the other hand, increasing the temperature at fixed density increases the collisional, hence photo-, ionization rate. The Saha equation therefore provides a direct connection between the electron density and temperature for a gas in LTE.

The most important feature is that we already know the level populations and the form of the radiation field. The assumption of thermal equilibrium means that the populations are given by the Boltzmann distribution, the radiation intensity is that of a blackbody with an emissivity that depends on the same temperature that provides the level populations, and all collisional processes occur at the same temperature. This means that the principle of *detailed balance* holds in the gas, all radiative rates balance *because* all collisional rates balance and the matter and radiation are in strict, local equilibrium.

1.4 COLLECTIVE BEHAVIOR: FLUIDS

1.4.1 Merging Gravitation and the Gas Laws: Stellar Statistical Mechanics

Most of the theory we have just developed for gases in thermal equilibrium can also be used for the treating self-gravitating ensembles of stars. First, we consider the Poisson equation. You know that we can write the density in terms of the distribution function, f, so that

$$\nabla^2 \Phi(\mathbf{x}) = -4\pi Gm \int f(\mathbf{x}, v) v^2 \, dv. \qquad (1.111)$$

This integro-differential equation shows how the velocity distribution governs the structure of the system. Now f evolves because of the potential, and this is the feedback that we had not yet included in the gas laws. The assumption for an ordinary gas is that collisions dominate. For gravity, "action at a distance" dominates the development of single-particle orbits, so that for each particle, the velocity depends on $\nabla \Phi$. Therefore, the orbit of any particle is given by Eq. (1.111) for $M(r)$, the mass contained within a sphere of radius r. Now consider a spherical distribution of stars. Individual stars *do not collide*, since separations are vast compared with their geometric radii,[26] so the usual idea of a gas doesn't seem to make sense. However, a "gas" of stars has a velocity dispersion, σ, that somehow comes from their initial conditions.

What makes gravitational "gases" simpler than plasmas is the comparatively small catalog of instabilities that appear. One reason for this is the force law itself. Gravity has only one sign. There is no self-shielding, and local gravitational analogs of electric fields cannot develop because of mass separation as they can in plasmas due to charges. Even as early as the second edition of the *Principia* in 1712, Newton realized that any ensemble of mutually gravitating masses will be unstable if its internal velocity dispersion is insufficient to prevent the collapse.

At any distance from the center of the mass distribution, there is a fictitious circular orbit that corresponds to a minimal angular momentum and energy. Then

---

[26]As we noted above in the discussion of tides, galaxies are far larger with respect to their mean free paths and therefore collisions are far more numerous.

particles with a range σ in their velocities also have a range in orbital angular momenta and, by Eq. (1.6), in the eccentricities of their orbits. Therefore, they move as if they are "supporting" themselves against their mutual gravitational interaction. Thus, in equilibrium, if there is no net expansion, contraction or rotation, we can write

$$\rho \nabla \Phi = \sigma^2 \nabla \rho. \tag{1.112}$$

Therefore

$$\nabla^2 \Phi = \sigma^2 \nabla^2 \ln \rho = -4\pi G \rho \tag{1.113}$$

so that we arrive at an equation for the equilibrium of this gas of stars:

$$\frac{1}{r^2} \frac{d}{dr} r^2 \frac{d \ln \rho}{dr} = -\frac{4\pi G}{\sigma^2} \rho \tag{1.114}$$

As we will soon show, this is the same as the structure equation for an isothermal gas sphere. The important thing to note at this stage is that we can associate a temperature with the velocity dispersion for the system.

1.4.1.1 The Mean Free Path

In any ensemble, there are two broad types of motion, the response to the long-range collective field of the bulk of the mass (or charge) and the local effects due to scattering. The first is the dynamical response to the potential. The second is related to the *mean free path*, the average distance that a particle moves before undergoing a scattering. To estimate this, assume that the cross section for the scatterer is $\Sigma(v)$. This may depend on energy, if the interaction is resonant or has a threshold, or it may depend on velocity because of the relative motion of the perturber and the particle. If the number density of the scatterers is n, then the mean time between collisions for a single particle of speed v is $\tau_c^{-1} = n \langle \Sigma v \rangle$ and the mean free path is $\lambda_{\rm mfp} = 1/(n\Sigma)$. For an ensemble, the the collision time is obtained by averaging over the velocity distribution through $\langle \Sigma v \rangle = \int f(v) v \Sigma(v) \, dv$. Depending on the details of the interaction, such as whether it is due to the gravitational or Coulomb force, the dependence on temperature or spectral index will differ.[27]

1.4.1.2 Dynamical Friction

Although stars in a cluster or galaxy form an essentially collisionless ensemble, they do deflect each other's trajectories because of the close encounters depending on their relative energies. This changes the number density at any point in space, which in turn alters the local gravitational acceleration felt by the stars. For a swarm of stars, this leads to a reaction force that is normally called *dynamical friction*. To see what happens, consider a point mass M and a test star of mass m.

---

[27]Although the details are too far removed from our discussion here, we should mention that the Langevin equation describes the random walk process rather, in the presence of a fluctuating force. See van Kampen, N. 1975, *Stochastic Processes in Physics* (Amsterdam: North-Holland) for an especially clear presentation.

The collision is defined by the deflection of the star:

$$\frac{1}{2}v^2 \approx \frac{GM}{b}, \qquad (1.115)$$

which also defines the *impact parameter*, b. The interaction cross section is πb^2 so that the collision frequency is

$$\tau_c^{-1} = N\pi \langle b^2 v \rangle, \qquad (1.116)$$

where N is the number density of stars of mass M and the result is averaged over the velocity distribution. We then substitute σ for v to obtain

$$\tau_c^{-1} \sim N\frac{(GM)^2}{\sigma^3}, \qquad (1.117)$$

and therefore each m feels a counteracting force as it orbits through the background:

$$F_{\text{drag}} \sim mv\tau_c^{-1}. \qquad (1.118)$$

and it slows it down. In effect, the background produces a drag—dynamical frictional— force on the mass. This means that it falls deeper into the gravitational potential of the other stars, thus increasing its velocity and increasing the drag. The star eventually spirals into the center of the mass distribution and redistributes its kinetic energy to the background stars.[28] The effect of this collective behavior, which is strictly reversible since the particles are collisionless and dissipationless akin to Landau damping, mimics a fluid because of the averaging over the velocity distribution of the background. The momentum of the body is redistributed throughout the background ensemble. This fluid is our next topic.

1.4.2 When Is a Gas a Fluid?

When is a gas a fluid? This question is especially important for astrophysics because of the enormous ranges of density and temperature that confront us in cosmic objects. In part, it is a matter of time. If collisions are unlikely on timescales much shorter than those of our observations, then single particle motion is an appropriate description. If, however, the particles move collectively and their distribution function changes in time, then a fluid description is most appropriate.

---

[28] Here are some basic references: Chandrasekhar, S. 1943, *Principles of Stellar Dynamics* (NY: Dover Books); Binney, J. and Tremaine, S. 1987, *Galactic Dynamics* (Princeton: Princeton Univ. Press); Kellogg, O. 1929, *Foundations of Potential Theory* NY: Dover; Huang, K. 1976, *Statistical Mechanics*, 2nd ed. (NY: Wiley); Kittel, C. 1976, *Thermal Physics* (San Francisco: Freeman); Shore, S. N., Livio, M., and van den Heuvel, E. P. J. 1994, *Interacting Binary Stars* (Berlin: Springer-Verlag); Chernoff, D. and Weinberg, M. D. 1990, *ApJ*, **351**, 121; Lee, H. and Ostriker, J. P. 1987, *ApJ*, **322**, 123; see also Gnedin, O. and Y. and Ostriker, J. P. 1997, *ApJ*, **474**, 223; Weinberg, M. D. 1986, *ApJ*, **300**, 93; see also Seguin, P. and Dupraz, C. 1994, *Astr. Ap.*, **290**, 709; Seguin, P. and Dupraz, C. 1994, *Astr.Ap*, **310**, 757; Spitzer, L. 1987, *Dynamics of Globular Clusters* (Princeton: Princeton Univ. Press).

First, begin with the equation for the evolution of the distribution function under ideal, collisionless conditions. Since this depends on the phase space, we work with those variables.

The meaning of Liouville's theorem for a gas is that the particles maintain a constant total volume in phase space, but collisions and chaotic mixing of trajectories cause fluctuations in the local density of particles within this volume. In general, it is akin to taking a packet of ketchup or mustard that you can obtain at any fast-food joint and squeezing it. It is incompressible, but its shape changes. You can expand one part while compressing another part, all without an overall change in the mass. We will do the same thing here. The redistribution of particles by collisions is an integral rate, depending on the probability of sampling the colliding particles at particular velocities and positions.

In the collisionless form, the evolution of the distribution function is provided by the Vlasov equation, the collisionless form of the Boltzmann equation:

$$\mathscr{L}f(x_i, v_i, t) \equiv \frac{\partial f}{\partial t} + v_j \frac{\partial f}{\partial x_j} + \dot{v}_j \frac{\partial f}{\partial v_j} = 0 \qquad (1.119)$$

The conditions on f are merely the result of ordinary experience. In a gas, the probability of encountering a particle with infinite speed is zero, and there is some ensemble mean that you expect at any instant, regardless of whether the distribution function is constant in time.

$$n(\mathbf{x}) = \int_{-\infty}^{\infty} f(\mathbf{x}, \mathbf{v}, t) d\mathbf{v} \qquad (1.120)$$

Now you see that the integral of the function over all possible velocity components is the same as a normalizing factor. Since the integral depends only on the position and time, and represents a probability of finding a particle within some volume V, the integral must be the number density, since $\int n(\mathbf{x}, t) dV = N(t)$, the total number of particles in the system as a function of time alone. The mean velocity is

$$n(\mathbf{x},t)\langle v_i(\mathbf{x})\rangle = \int_{-\infty}^{\infty} v_i(\mathbf{x},t) f(\mathbf{x},\mathbf{v},t) \, d\mathbf{v} \qquad (1.121)$$

$$n(\mathbf{x},t)\langle v_i(\mathbf{x}) v_j(\mathbf{x},t)\rangle = \int_{-\infty}^{\infty} v_i(\mathbf{x},t) v_j(\mathbf{x},t) f(\mathbf{x},\mathbf{v},t) \, d\mathbf{v} \qquad (1.122)$$

and higher mean values of multiplicative quantities can be taken as well. For instance

$$n(\mathbf{x},t)\langle v_i(\mathbf{x}) v_j(\mathbf{x},t) v_k(\mathbf{x},t)\rangle = \int_{-\infty}^{\infty} v_i(\mathbf{x},t) v_j(\mathbf{x},t) v_k(\mathbf{x},t) f(\mathbf{x},\mathbf{v},t) \, d\mathbf{v}, \qquad (1.123)$$

so that $n(\mathbf{x},t)\langle v_i(\mathbf{x}) v^2(\mathbf{x},t)\rangle$ represents the scalar product of the last two terms. Now we have added a set of correlations to the list of mean variables. These means of the physical variables are called the *moments of the distribution*, a term that has its origins in statistics and comes in very handy here. The second and third moments represent the momentum flux and the energy flux, respectively, while the first moment gives the mass flux. In other words, in order to remove the effects of the graininess of the distribution at the microscopic scale, you have to average over all possible velocities accessible to the system. As we will see shortly, this has a rather interesting consequence for self-gravitating ensembles.

Let us look at how you would evolve these quantities. For instance, start with the continuity equation. The first term is the explicit change of the density at some location with time:

$$\frac{\partial n}{\partial t} = \frac{\partial}{\partial t}\int f d\mathbf{v} - \int \frac{\partial f}{\partial t} d\mathbf{v} \qquad (1.124)$$

and the second is the divergence of the mass flux:

$$\frac{\partial n <v_j>}{\partial x_j} = \frac{\partial}{\partial x_j}\int f v_j \, d\mathbf{v} = \int \frac{\partial f v_j}{\partial x_j} d\mathbf{v} \qquad (1.125)$$

where we can now resolve the last term by

$$\frac{\partial}{\partial x_j} f v_j = v_j \frac{\partial f}{\partial x_j} + f \frac{\partial v_j}{\partial x_j} \qquad (1.126)$$

Integrating both sides of Eq. (1.125) over velocity eliminates the last term since there is no mean to the divergence (i.e., $<\nabla \cdot \mathbf{v}> = 0$). Collecting terms, you see that this is the same as $\int d\mathbf{v} \mathscr{L} f = 0$. Now, for the momentum equation, this is a little more complicated:

$$\frac{\partial}{\partial t} n\langle v_i \rangle = \frac{\partial}{\partial t}\int f v_i \, d\mathbf{v} = \int v_i \frac{\partial f}{\partial t} d\mathbf{v} \qquad (1.27)$$

since again $\langle \partial v_i/\partial t \rangle = 0$ and for the momentum flux:

$$\frac{\partial}{\partial x_j} n\langle v_i v_j \rangle = \frac{\partial}{\partial x_j}\int f v_i v_j \, d\mathbf{v} = \int v_i v_j \frac{\partial f}{\partial x_j} d\mathbf{v}. \qquad (1.128)$$

The last term is the hardest to reduce, the one dealing with the net acceleration. Here we have written na_i as $n\langle \dot{v}_i \rangle$, which comes from integration over the distribution function:

$$n\langle a_i \rangle = \int f \dot{v}_i \, d\mathbf{v} = \int f \dot{v}_j \delta_{ij} \, d\mathbf{v} = \int f \dot{v}_j \frac{\partial v_i}{\partial v_j} d\mathbf{v}. \qquad (1.129)$$

Collecting the terms, you see that everything we have done for the momentum equation is the same as $\int v_i \mathscr{L} f d\mathbf{v} = 0$. This integral is called the *first moment* of the Vlasov equation. The relation between the statistical and thermodynamic quantities is provided here by noting that the velocity v_i is the sum of ordered, V_i, and random, δv_i, components. For the ordered part, we have written $V_i = \langle v_i \rangle$.

For the random part, $\langle \delta v_i \rangle = 0$ but $\langle \delta v_i \delta v_j \rangle = \sigma^2 \delta_{ij}$. This is simply saying that the motions in any two directions are not correlated. Therefore we have

$$n\langle v_i v_j \rangle = nV_iV_j + n\sigma^2\delta_{ij}, \quad (1.130)$$

where the first term is called the *Reynolds stress* and the second is the classical pressure, as you may recall from Section 1.3.4.1. Therefore

$$n\left[\frac{\partial V_i}{\partial t} + V_j\frac{\partial V_i}{\partial x_j}\right] = -\frac{\partial P}{\partial x_i} + na_i. \quad (1.131)$$

The driving term for the dynamics can be combined into a single function, called the *stress tensor*. In keeping with our description of the gas

$$T_{ij} = \rho V_i V_j + P\delta_{ij}. \quad (1.132)$$

With this, the equations of motion can be put into an especially simple form:

$$\frac{\partial(nV_i)}{\partial t} + \frac{\partial T_{ij}}{\partial x_j} = na_i \quad (1.133)$$

This equation is called the *conservative form* of the equations of motion, a representation that is particularly useful for numerical computations since it is written in terms of currents and sources.

We still have to solve for the second moment. Now take take the product of $\mathscr{L}f$ with v^2 and integrate as before.[29] This gives

$$\frac{\partial}{\partial t}\left(\frac{1}{2}n\langle v^2\rangle\right) + \frac{\partial}{\partial x_j}\left(\frac{1}{2}n\langle v^2 v_j\rangle\right) - n\langle \dot{v}_k v_k\rangle. \quad (1.134)$$

You see that the averaging term is now more complicated because we have to include explicitly the third order terms. The second term is

$$\langle v_j v^2 \rangle = \left(V^2 + \langle(\delta v)^2\rangle\right)V_j + 2V_k\langle \delta v_k \delta v_j\rangle + \langle(\delta v)^2 \delta v_j\rangle.$$

This is where the Maxwellian distribution comes to the rescue. It truncates the order of the correlations required for the averages.

The reason for going through this discussion is that there are many problems that require only a continuum approximation. When the mean free path for particles is large enough, when collisions dominate the microscale of the system, then the momentum or energy gained by each body is rapidly distributed throughout the volume and the ensemble behaves like a fluid. The thermodynamic limit requires that the temperature, T, is a well-defined variable through the distribution function and that although it may change with space or time, the distribution from which it is drawn does not. This is what happens in a gas. Alternatively, it is

---

[29]Although a fully tensorial product, $v_i v_j$ could also be used this would not be as instructive.

plausible that in very dilute media, and a galaxy or star cluster are good examples of this, collisions are unlikely between individual members and each body obeys the dynamical equations separately, displaying global properties only through the gravitational field generated by the mass distribution as a whole. Thus, for stellar dynamical problems, a particle or kinetic approach is generally required. Galaxies in clusters are strongly tidally interacting, that is to say that the "particles" are collisional, but the sizes and masses of the individual galaxies are far greater than the constituent stars and the internal dynamics can be treated as a continuum. The choice is really one of scale. If the microphysics dominates the problem, then individual particle motions will have to be followed and collective (continuum) methods do not suffice. On the other hand, if the timescales are longer than the characteristic times for the internal processes and the behavior can be treated as a collective response to an external forcing, then the fluid approach is okay. We will return to this point when we discuss stellar hydrodynamics in Chapter 7.

You can immediately see several conserved quantities from Eqs. (1.134). For steady state flow, you can define a streamline on the basis of the mass flux, $J_i = \rho v_i$ along which

$$\mathcal{M} = \rho v^2 + P = \text{constant}. \tag{1.135}$$

This is the Bernoulli equation. In Chapter 6, this equation will reappear as a condition for the jump in the pressure, density, and velocity at the shock front. The fact that you have the mass flux means that if the flow is along the normal to an area A, the rate of mass flow through this area is $\dot{M} = \mathbf{J} \cdot \mathbf{A}$, again in steady state.

For incompressible flows you are quite used to the consequences of these equations since $vA = \text{constant}$ and therefore the pressure is a function only of the cross sectional area.[30]

1.4.3 Viscosity and Vorticity

The last step in obtaining the classical equations for a fluid is to recall the role of frictional coupling. Between two bodies, it has the effect of transferring momentum from one to another, accompanied by dissipation. For a fluid, this depends on the internal gradient of the velocity and dimensionally is represented by the *viscous* force:

$$\mathbf{F}_V = \eta \nabla^2 \mathbf{v} \tag{1.136}$$

Here η is the viscosity which, for most of our purposes, will be assumed to be

---

[30] Such flows also also satisfy $\nabla \cdot \mathbf{v} = 0$ as well as $\nabla \times \mathbf{v}$ if they are irrotational, and in this particular case you can add the property that the velocity is given by a potential, ϕ, such that $\nabla^2 \phi = 0$. For incompressible fluids, since $\nabla \times (\nabla P) = 0$ always, you can always remove the pressure from the equations of motion through a suitable transformation. In general, for compressible flows, this is not possible and actually is quite important in convection and rotating bodies. We will return to this in Chapter 4. In general, the velocity can be written for an incompressible medium as $\mathbf{v} = \nabla \times \psi + \nabla \phi$, where ψ is a vector potential and ϕ is a scalar.

constant. This added term leads to the Navier–Stokes equation:

$$\rho\left(\frac{\partial}{\partial t} + \mathbf{v} \cdot \boldsymbol{\nabla}\mathbf{v}\right) = -\nabla P + \rho \mathbf{a} + \eta \nabla^2 \mathbf{v} \qquad (1.137)$$

which is the basis of all viscous hydrodynamics. A dimensionless number, the *Reynolds number*, Re, measures the importance of the viscous forces relative to the dynamical (inertial) term in the equations of motion:

$$\text{Re} \equiv \frac{[\mathbf{v} \cdot \boldsymbol{\nabla}\mathbf{v}]}{[\eta \nabla^2 \mathbf{v}]} = \frac{UL}{\eta}. \qquad (1.138)$$

where U is a characteristic speed and L is a typical length scale for the flow. The viscosity also couples different parts of the fluid together in an irreversible way—that is, it is a dissipation term. To see how, take the product of Eq. (1.137) with the velocity:

$$v_i \frac{dv_i}{dt} = -v_i \frac{\partial P}{\partial x_i} + v_i \frac{\partial \Phi}{\partial x_i} + v_i \eta \frac{\partial^2 v_i}{\partial x_j \partial x_j}$$

and integrate over the volume, neglecting all surface integrals, to obtain

$$\frac{dE}{dt} + P\frac{dV}{dt} = -\int \eta \left(\frac{\partial v_i}{\partial x_j}\right)^2 dV \qquad (1.139)$$

which is always negative even if the surface is extended to infinity. The same is true if we define the *vorticity*, which measures circulation within the fluid on any length scale, as $\omega = \nabla \times \mathbf{v}$. For a fluid that is not irrotational, in other words, one for which we cannot define streamlines, the advection term in the equation of motion is expanded to provide

$$\mathbf{v} \cdot \boldsymbol{\nabla}\mathbf{v} = -(\nabla \times \mathbf{v}) \times \mathbf{v} - \nabla \tfrac{1}{2} v^2 = -\omega \times \mathbf{v} - \nabla \tfrac{1}{2} v^2 \qquad (1.140)$$

Now if the fluid is still treated as incompressible, so that $\nabla \cdot \mathbf{v} = 0$, then $\nabla^2 \mathbf{v} = -\nabla \times \nabla \times \mathbf{v}$. If we neglected the viscous term, we would arrive at the Kelvin circulation theorem, which states that the quantity $\Gamma = \int \omega \cdot d\Sigma = \int \mathbf{v} \cdot d\mathbf{l}$, called the *circulation*, is constant. Here Σ is a cross section to the circulation and we are taking the projection of the vorticity along the surface normal so $\mathbf{l}$ is around a vortical region, also called a *vortex line*. Thus, without viscosity, $d\Gamma/dt = 0$. With viscous coupling, however, $d\Gamma/dt < 0$ always.

1.4.4 Sound and Other Collective Fluid Instabilities

Now that you have the equations of motion in both the particle and fluid versions, let us do something with them. A continuous medium like a fluid can support collective modes. A periodic perturbation of one part propagates through the medium as a wave. It doesn't matter whether the medium is a gas of atoms or stars, the collective mode is qualitatively the same. To see this, imagine that you have a medium that is initially at rest with density ρ and that somewhere you increase the density by a small amount, $\delta\rho$. Although the details depend on the equation of state, this locally increases the pressure and produces a fuzzy piston that begins to expand into the surrounding medium. The medium responds with a velocity **v**. Because this is a collective response that depends on the thermodynamics and the continuum properties of the medium, you need to use the fluid form of the equations. First, perturb the continuity equation, since mass is conserved:

$$\frac{\partial \delta\rho}{\partial t} + \rho \frac{\partial v}{\partial x} = 0. \qquad (1.141)$$

For simplicity, we will stick with the one-dimensional form of the equations, but it can be generalized to three dimensions with just a bit of added mathematical complication, as we will see in a moment. The dynamics are described by the linearized form of the equation of motion:

$$\rho \frac{\partial v}{\partial t} + \frac{\partial \delta P}{\partial x} = 0. \qquad (1.142)$$

Now write $P = P(\rho)$ and rewrite Eq. (1.142) as

$$\frac{\partial v}{\partial t} + \frac{1}{\rho}\left(\frac{\partial P}{\partial \rho}\right)_T \frac{\partial \delta\rho}{\partial x} = 0, \qquad (1.143)$$

so combining Eqs. (1.141) and (1.143), we get

$$\frac{\partial^2 \delta\rho}{\partial t^2} + \left(\frac{\partial P}{\partial \rho}\right)_T \frac{\partial^2 \delta\rho}{\partial x^2} = 0. \qquad (1.144)$$

This is a wave equation for a pressure or density disturbance in a gas or fluid. It defines the *sound speed*:

$$c_s = \left(\frac{\partial P}{\partial \rho}\right)_T^{1/2}. \qquad (1.145)$$

You see that c_s depends on the equation of state and is only constant for an isothermal gas. This is also the linearized form of the equations, the response of the fluid to a very small perturbation. We will see in Chapter 7 what happens when the pressure and/or density variations become large, but here we can say that the propagation of a small disturbance will occur at a characteristic speed that is *independent* of the amplitude, an acoustic wave. This leads to the dispersion relation

$$\omega^2 - k^2 c_s^2 = 0 \qquad (1.146)$$

for the wave, if there is no dissipation, so that each point on the wave moves at a *phase* velocity, $\omega/k = c_s$, that is equal to the group velocity $v_g = d\omega/dk$. The latter is defined by expanding the wavenumber dependent phase of the wave through

$$\omega = \omega_0 + \frac{d\omega}{dk}k + \cdots$$

and then noting that

$$\delta\rho \sim \exp\left[i\left(-kx + \omega_0 t + \frac{d\omega}{dk}kt\right)\right]$$

so that the derivative must have the dimensions of a velocity. When the wave amplitude is large, these two differ and the phase speed is slower or faster than the collective motion, depending on whether $\delta\rho$ is positive or negative, respectively. The result is that the simple wave becomes distorted with the higher-density portion moving faster and the wavefront steepens. This generates shocks.

Finally, there are two simple fluid instabilities that you can immediately see from these equations and that are seen in virtually all astrophysical environments. The first is the Kelvin–Helmholtz instability, and we mention this now because it is directly connected with vorticity. The simplest way to see it is to take a fluid interface along which there is a velocity shear. Assume that the fluid is inviscid. Now, as you imagined for a sound wave, take a bump along the interface. Since circulation is conserved, then the fluid pressure drops above and below the bump and lift is generated. This increases the amplitude of the perturbation and the result is a local injection of vorticity into the flow that must, overall, remain the same as the original shear. The second is the Rayleigh–Taylor, or buoyancy, instability that occurs when two fluids of different densities are separated and vertically stacked under gravity.

1.4.5 Plasmas and Magnetohydrodynamics (MHD)

Having discussed the Vlasov equation and its use for stellar gases, we now turn briefly to plasmas. Whereas for a self – gravitating gas of stars you are dealing with particles of the same sign of the force, in a plasma screening effects are possible because of spatial variations in the charge. When you separate stars in a cluster, the gravitational field between them decreases and they feel more the environment than they do one another. For a charged gas, on the other hand, charge separation produces a local electric field whose magnitude depends on the effective charge of the medium. In one region, a charge excess of one sign can develop, and even a neutral gas can easily form very strong local electric fields if the charges fluctuate. This is especially true for a pure hydrogen plasma. Remember that protons are nearly 2000 times more massive than electrons and are sluggish in their adjustment to any electric fields when they first appear. It is the electrons that move to cancel out the charge excesses.

Take the case of a local neutral plasma and assume that the *equilibrium* number densities of electrons and ions are equal, so that $n_{e,0} = n_{p,0}$, with no neutral phase to complicate the picture. Then take a cloud of electrons and displace it from the protons by some distance $\delta\mathbf{x}$. Since the plasma is assumed to be perfectly conducting and there is initially no charge excess, only the perturbation enters the equation of motion for the electrons

$$\delta\mathbf{E} = -4\pi e\,\delta n_e\,\delta\mathbf{x} \tag{1.147}$$

which follows from Gauss' law

$$\nabla\cdot\delta\mathbf{E} = -4\pi e\,\delta n_e \tag{1.148}$$

assuming that the ions do not move. The electrons respond to this charge separation as if it were a harmonic oscillator potential with a spring constant that depends on the density of the charges:

$$m_e\ddot{\mathbf{x}} = -4\pi e^2 n_e\mathbf{x}. \tag{1.149}$$

The *plasma frequency*, ω_p, follows immediately from this equation. It is the characteristic response frequency of the charges (electrons) to any local electric fields. It is also the resonant frequency for a periodic forcing, for example the time varying electric field of a radio frequency photon:

$$\omega_{pe} = \left(\frac{4\pi e^2 n_e}{m_e}\right)^{1/2} \approx 50\,n_e^{1/2}\text{ kHz}. \tag{1.150}$$

This response frequency is lower for the ions because of their larger mass:

$$\omega_{pi} = \left(\frac{Z_i m_e}{m_i}\right)^{1/2}\omega_{pe} \tag{1.151}$$

For a magnetized plasma, there is a second characteristic frequency besides the plasma frequency, the cyclotron or *Larmor* frequency. This comes from the Lorentz force on an electron in a magnetic field:

$$m_e\ddot{\mathbf{x}} = \frac{e}{c}\mathbf{v}\times\mathbf{B} \tag{1.152}$$

so that, by dimension equivalence, you see that the electron orbits in the field with the following frequency:

$$\omega_B = \frac{eB}{m_e c}. \tag{1.153}$$

The same result holds for any ion of mass m_i and charge $Z_i e$, with the frequency replaced by the ion cyclotron frequency $(Z_i/m_i)\omega_B$. Another built-in timescale is the sound crossing time for any region of size L for a thermal speed $c_{s,e}$, which is $L/[(m_e/m_i)^{1/2}c_{s,e}]$. This is true for any ion and depends only on $T_e^{1/2}$, the electron temperature (this assumes that the ions and electrons are thermalized). This is what makes plasmas so much harder. They are able to respond collectively because the charges move and generate small-scale fields that produce a fluidlike response, but they are also dissipative because of the radiation of waves internally on the plasma frequency.

As we have said, screening is possible because the fluid is composed of oppositely charged particles. This affects the local electrostatic field of a point particle. In simple terms, there is a distance beyond which the ensemble of charges masks the effect of a single ion. This is called the *Debye length* and is derived by considering a thermal distribution of particles. The number density for the charges at any point in space depends on their energy, which in turn depends on the electrostatic potential ϕ. The potential comes from the Poisson equation, which depends on the charge excess:

$$\nabla^2 \phi = -4\pi e(n_e - n_i) \tag{1.154}$$

We can use the Boltzmann distribution to find the number of particles at a given energy since the work function is $\pm e\phi$, depending on the charge of the particle. Call n the total number density. Then the relative occupation of the states with potential ϕ becomes

$$n_e = ne^{-e\phi/kT}, \qquad n_i = ne^{e\phi/kT}$$

For weak potentials, that is, for a nearly neutral plasma, $\phi/kT \ll 1$, the expansion of the exponentials gives

$$\nabla^2 \phi = \frac{4\pi n e^2}{kT} \phi \tag{1.155}$$

and you see the emergence of a characteristic length scale for the field

$$\lambda_D = \left(\frac{kT}{4\pi n_e e^2} \right)^{1/2} \tag{1.156}$$

This equation defines the *Debye length*. The solution to Eq. (1.155) shows that the potential surrounding a point charge in a plasma falls off much more rapidly with distance than a simple Coulomb potential:

$$\phi(r) = \frac{1}{r} e^{-r/\lambda_D}. \tag{1.157}$$

Screening affects all of the kinetic properties of the plasma. For example, although the gravitational field of a mass has infinite range, this is not true for a point charge embedded within a plasma. Therefore, only particles within a distance λ_D have a perturbing effect on a charge and the number of particles within this sphere is given by $n_e \lambda_D^3$ for a fully ionized gas. Thus λ_D is the largest impact parameter that you would expect for collisions. This is important for computing a number of properties in a plasma, in particular the rate of bremsstrahlung emission (see Chapter 3).

The astrophysical environments in which we will be working cover an enormous range of densities and temperatures, and it is possible that the normal assumptions of the gas laws break down. By this we mean not just the effects of degeneracy, but simpler violations related to the distribution function. Remember that collisions maintain the distribution function. This is what we meant before by *thermalization.* As with stars in a cluster, the cross section for charges to interact is not that of hard spheres. It depends on the relative velocity of the particles and, therefore, the mean free path is a function of particle energy. This is where you run into trouble. If a local electric field is strong enough, the electrons will accelerate. Their interaction cross section with the ions drops and their mean free path increases. This means that the electric field grows with time, further accelerating the electrons that remain as the ones responsible for the field are lost from the region. Eventually, the field becomes strong enough that the particles accelerate faster than the plasma can heal the field and a runaway acceleration occurs. This limit is the *Dreiser field*. The electrons so produced cannot further thermalize and therefore can be regarded as a beam, streaming away from the acceleration site.

1.4.5.1 *Magnetohydrodynamics and Alfven Waves*

When a fluid approximation applies, the medium still responds to the presence of electric and magnetic fields, especially the latter. The electrons couple strongly to the ions, and the neutral fluid that results can nonetheless generate a current. This is the *magnetohydrodynamic* limit, where we forget about the microphysical kinetics and treat the plasma with the continuum. The equations that apply here are the same as for a current

$$\frac{\partial \rho}{\partial t} + \nabla \cdot \mathbf{J} = 0 \tag{1.158}$$

supplemented by the induction equation:

$$\frac{\partial \mathbf{B}}{\partial t} = \nabla \times (\mathbf{v} \times \mathbf{B}) \tag{1.159}$$

$$\nabla \cdot \mathbf{B} = 0 \tag{1.160}$$

$$\nabla \times \mathbf{B} = \frac{4\pi}{c} \mathbf{J}. \tag{1.161}$$

Since we have assumed that the medium is perfectly conducting and globally neutral, the internal comoving electric field vanishes and therefore $\mathbf{E} + \mathbf{v} \times \mathbf{B}/c = 0$. Finally, the equation of motion for the fluid is

$$\rho \frac{d\mathbf{v}}{dt} = -\nabla P + \frac{1}{c} \mathbf{J} \times \mathbf{B} + \rho \mathbf{g} \tag{1.162}$$

if we allow for an external acceleration **g**. The extra term, compared with our usual one, comes from the collective response of the fluid to the Lorentz force.

Within the plasma, a perturbation in the density produces a change in the current through the continuity equation, which in turn changes the magnetic field through the induction equation. It is thus possible to obtain both sound waves, which are motions of the gas independent of the magnetic field, and a collective mode that looks like a magnetic wave. To see this, let us perturb the magnetic field, $\mathbf{B}_0$, by a small amount, $\mathbf{b}$, so that $\mathbf{B} = \mathbf{B}_0 + \mathbf{b}$. This amounts to the same thing as changing the current. We will stay within the idealized limit of a perfect conductor and ignore viscosity. It is easiest to work with an incompressible medium for the moment, but remember that the current will be perturbed because it is the bulk average of the ions and electrons. If the fluid is isothermal and initially at rest (ignoring gravity), then

$$\rho \frac{\partial \mathbf{v}}{\partial t} = \frac{1}{c} \delta \mathbf{J} \times \mathbf{B}_0, \tag{1.163}$$

and since from the induction equation, $\delta \mathbf{J} = (c/4\pi) \nabla \times \mathbf{b}$, we have

$$\rho \frac{\partial \mathbf{v}}{\partial t} = \frac{1}{4\pi} (\nabla \times \mathbf{b}) \times \mathbf{B}_0 \tag{1.164}$$

and from the induction equation:

$$\frac{\partial \mathbf{b}}{\partial t} = \nabla \times \mathbf{v} \times \mathbf{B}_0, \tag{1.165}$$

you see that there is a characteristic speed for the response of the plasma:

$$v_A = \frac{B_0}{(4\pi\rho)^{1/2}}, \tag{1.166}$$

which renders the equations dimensionless. This is the *Alfven speed*, the rate at which a magnetic disturbance moves through the fluid. It is the magnetic analog of a sound wave, except it is a disturbance in the magnetic field (or current) that moves at this speed even if the medium is incompressible. This wave, which is transverse and incompressible, produces motion of the massive fluid because of the assumed complete coupling between the field and the plasma. Now we can put the density variations back into the picture. Instead of assuming that $\nabla \cdot \rho \mathbf{v} = 0$, we explicitly allow for density perturbations through

$$\frac{\partial \delta \rho}{\partial t} + \nabla \cdot \rho \mathbf{v} = 0, \tag{1.167}$$

although we assume that the field is initially constant so that $\nabla \times \mathbf{B} = 0$. The perturbed equations for a plane-wave expansion then give the following dispersion relation:

$$\omega^2 \mathbf{v} - \mathbf{k}(\mathbf{k} \cdot \mathbf{v}) c_s^2 + \frac{1}{4\pi\rho} (\mathbf{k} \times \mathbf{B})(\mathbf{k} \cdot \mathbf{v}) = 0. \tag{1.168}$$

There are now compressive as well as transverse modes for propagation of the waves. It is essentially the same thing as saying that the pressure perturbation is now of the order of $\delta P = (c_s^2 + v_A^2)\delta\rho$, resulting in a magnetosonic mode of propagation for a magnetic disturbance. Gas pressure produces a longitudinal restoring force while the cold gas, like a vacuum, responds only with the transverse $\mathbf{v} \times \mathbf{B}$ restoring force.

Now let us return again to the question as to when a gas behaves like a fluid. As we said, it depends on the collective response to a perturbation and on the mean free path. But there are other agents that can enforce collective motion, and we have just dealt with one of them: magnetic fields. Tangled fields scatter the electrons and ions, redirecting any streaming into random motion and coupling them at large distances, even when they would otherwise be expected to show beams. The condition for the field to move as if it were frozen into the plasma is seen from the following consideration. Take the continuity equation for the fluid and combine it with the evolution equation for an otherwise sourceless magnetic field:[31]

$$\frac{d\rho}{dt} + \rho \nabla \cdot \mathbf{v} = 0,$$
$$\frac{d\mathbf{B}}{dt} + \mathbf{B}\nabla \cdot \mathbf{v} = 0,$$
(1.169)

so that the field behaves as by $B \sim F\rho$, where F is a scalar function. This is the condition of a magnetic field that is *frozen* into the plasma and means that $\mathbf{B}$ increases when the fluid is compressed. This is a limiting case. We will return to it when dealing with interstellar molecular clouds since there the medium *has* a neutral phase, that is not supported by a magnetic field, and the field can have a weaker dependence on the density of the medium and still be frozen into the charged portion.[32]

1.4.6 Diffusion

Microscopic diffusion is the local response of a gas to an external field, in the astrophysical case due to local temperature variations or gravity. The flow is diffusive, in the sense of the Fokker–Planck equation (see Section 1.6), because of the small mean free path in many dense environments, such as the interior of the Sun or a stellar atmosphere. Turbulence, that is, random macroscopic motions, may keep the medium homogeneous. If such stirring is suppressed by, for instance, a magnetic field, diffusion will proceed. In interstellar clouds, this occurs mainly because of the separation of dust and gas in a very-low-density medium. The solar interior appears to require diffusion of helium toward the core, and separation of elements has been used to explain abundance anomalies in otherwise normal main

---

[31] We have used the expansion $\nabla \times \mathbf{v} \times \mathbf{B} = -\mathbf{v} \cdot \nabla \mathbf{B} + \mathbf{B}\nabla \cdot \mathbf{v}$ since $\nabla \cdot \mathbf{B}$ always vanishes.

[32] We note here that the separation of the neutral and ionized gas is not, strictly speaking, a separation of different fluids. It is the same gas in different states. When in the charged state, which is some fraction of the time depending on the ionization sources, the gas feels the Lorentz force. When it changes charge state, it can slip through the field without producing a counteracting electric field. In this way, the gas ultimately separates from the magnetic field. Under interstellar medium conditions, this has important consequences for star formation in the guise of ambipolar diffusion.

stars, such as the chemically peculiar main sequence stars. All of these will be discussed in their proper place. Here we will estimate how such effects can occur.

As we did for dynamical friction, let us look at the interaction between charges in a plasma. For a point source potential, the change in perpendicular component of the velocity, Δv, due to collision with an impact parameter b is

$$\Delta v = \frac{\pi Z_1 Z_2 e^2}{mvb} \qquad (1.170)$$

so the root mean square (rms) deflection is $(\langle(\Delta v)^2\rangle)^{1/2}$. Each collision statistically produces this small change in the particle energy but within an area A per unit time, $nv \, dA = \pi nvb \, db$, there will be N of them depending on the density. Thus the average rate of change of $(\Delta v)^2$ per unit time is

$$\frac{d}{dt}\langle \Delta v \rangle^2 = 2\pi \int_{b_{min}}^{b_{max}} (\Delta v)^2 nvb \, db = \frac{(\pi Z_1 Z_2 e^2)^2}{v} \ln\frac{b_{max}}{b_{min}} \qquad (1.171)$$

and the *collision frequency* becomes Eq. (1.171) divided by v^2:

$$\nu_c = \frac{2\pi^3 Z_1^2 Z_2^2 e^4 n}{m^2 v^3} \ln\left(\frac{b_{max}}{b_{min}}\right). \qquad (1.172)$$

The last factor, $\ln \Lambda$, is sometimes called the collision or Coulomb integral. You already know the maximum distance for Λ, the Debye length λ_D. The minimum distance is the minimum impact parameter for large deflection, which we found earlier for dynamical friction:

$$b_{min} \approx \frac{2Z_1 Z_2 e^2}{mv^2}.$$

Replacing v with the mean velocity gives $\nu_c \sim T^{-3/2}$, which is the temperature-dependent conductivity for the gas. An important feature of this rate is that it decreases with increasing temperature, allowing for the weak temperature and density dependence of the collision integral, and therefore high-energy particles have a long mean free path.

Diffusion is slow particle dynamics. It means replacing the acceleration by the collision timescale, and assuming that the collisions produce small uncorrelated velocity fluctuations, Δv. The motion of a particle is, therefore, approximately a random walk under a stochastic force

$$m\frac{\langle (\Delta v)^2 \rangle^{1/2}}{\tau_c} = \langle F \rangle, \qquad (1.173)$$

where now $\langle F \rangle$ is the mean force on the particle averaged over time and the particle distribution. We can use ν_c to approximate τ_c^{-1}. Now as an estimate of the characteristic speeds, assume a density of about 10^{17} cm^{-3}, which is about what you find within the upper layers of a stellar envelope, and a temperature of 10^4 K. the collision frequency is about 10^{12} s^{-1} so for a star like the Sun, for which the surface gravity is $g \approx 10^4$ cm s^{-2}, we find typical diffusion velocities of order 10^{-8} cm s^{-1}. This may be very slow, but highlights one of the advantages astrophysicists have over their laboratory-bound counterparts—the experiments last virtually forever.

1.4.6.1 Diffusive Separation in Gases
Binary diffusion occurs in the shadowy area between a gas and a fluid. The differential motion between two species coupled by collisions can be treated within a fluid approximation by looking at the continuity equation and partial pressure for each species and using a friction term for their interaction. As an example, let's deal with an ionized mixture of helium and hydrogen (and electrons, of course). First we write the continuity equation for each species:

$$\frac{\partial n_i}{\partial t} + \nabla \cdot n_i \mathbf{v}_i = 0. \qquad (1.174)$$

We then add the equations of motion under a background acceleration, **a**:

$$n_i \frac{\partial \mathbf{v}_i}{\partial t} = -\nabla P_i + n_i(Z_i e \mathbf{E} + m_i \mathbf{a}) + \mu_{ij} n_i \nu_{c,ij} \mathbf{v}_{ij}, \qquad (1.175)$$

where m_i and Z_i are the mass and charge of species i, $\nu_{c,ij}$ is the collision frequency between species i and j with the mass replaced by the reduced mass, μ_{ij}, when the interacting species are both massive, and $\mathbf{v}_{ij}$ is the differential velocity of the two components. We need to include the electric field **E** because of possible charge separation and therefore also require the Gauss equation for the electric field:

$$\nabla \cdot \mathbf{E} = 4\pi e \sum_j Z_j n_j. \qquad (1.176)$$

We will work in spherical coordinates. This makes the computation easier because it renders the problem one-dimensional. For diffusive separation of two gases, the total density remains approximately constant so that neglecting the electrons:

$$n_i v_i + n_j v_j \approx 0$$

The electric field is mainly supported by the electron pressure gradient:

$$E = -\frac{1}{n_e} \frac{\partial P_e}{\partial r} \qquad (1.177)$$

In steady state, the diffusion speeds are equal and opposite for the two species, $v_{ij} = -v_{ji}$. An important application for stellar physics is the relative separation of hydrogen and helium and we will use this as an illustrative case for binary diffusion.[33] In what follows, we will assume that the acceleration may be due to radiation pressure and gravity, since both can be important in stellar atmospheres and interiors. For H and He, the relative mass fractions are defined by

$$X = \frac{m_p n_H}{\rho}, \quad Y = \frac{4 m_p n_{He}}{\rho}$$

so the partial pressures are given by

$$P_H = \frac{4X}{5X+3} P, \quad P_{He} = \frac{2(X+1)}{5X+3} P$$

and the diffusion velocities are

$$\frac{v_{He}}{v_H} = -\frac{X}{Y}. \tag{1.178}$$

Since the electrons balance the electric field, while the bulk pressure balances the acceleration, a, we obtain for the diffusion velocity:

$$v_{14} = -\frac{4 m_p T}{\mu \rho Y v_{c,14}} \left[\frac{1}{P_e} \frac{\partial P_e}{\partial r} + \frac{1}{P_H} \frac{\partial P_H}{\partial r} - m_p a \right] \tag{1.179}$$

This velocity is then put back into the continuity equation:

$$\frac{\partial X}{\partial t} + \frac{1}{r^2} \frac{\partial}{\partial r} (r^2 X v_{14}) = 0 \tag{1.180}$$

to obtain the depth-dependent hydrogen profile, which immediately provides the helium distribution by hypothesis.[34]

1.4.6.2 Rate Balance, the Boltzmann Collision Integral, and Diffusion

We will use the diffusion to illustrate the application of the Boltzmann collision integral and how rate equations are solved for a multicomponent gas. First, let us set up the equations for a single component medium. The rate for a transition from a state i to an arbitrary state j is given by the probability $P(j;i)$, while the inverse rate is $P(i;j)$. Assume that the distribution functions cover the phase space volumes Γ_i and Γ_j. Then the change in f_i with time is given by the balance of the

---

[33] For further discussions of diffusion in stellar interiors, see, for instance, Watson, W. D. 1971, *Astr. Ap.*, **13**, 203; Montmerle, T. and Michaud, G. 1976, *ApJS*, **31**, 489; Cowley, C. R. and Day, C. A. 1976, *ApJ*, **205**, 440; Noerdlinger, P. 1978, *ApJS*, **36**, 259; Iben, I. Jr. and Macdonald, J. 1985, *ApJ*, **296**, 540; Bahcall, J. N. and Loeb, A. 1990, *ApJ*, **360**, 267; Michaud, G. 1991, *Ann. Physique*, **16**, 481.

[34] For general discussions of the diffusion problem, see also Chapman, S. and Cowling, T. G. 1970, *Mathematical Theory of Non-uniform Gases*, 3rd ed. (Cambridge, UK: Cambridge Univ. Press); Burgers, J. M. 1969, *Flow Equations for Composite Gases* (NY: Academic Press).

rates of transitions into Γ_i compared to the rates out of it:

$$\frac{df_i}{dt} = [\text{rate out}] - [\text{rate in}] = -\int P(j;i)f_i\,d\Gamma_j + \int P(i;j)f_j\,d\Gamma_j \quad (1.181)$$

Now let us look at a two component gas in which the species are collisionally coupled to each other. Assume that the distribution functions are statistically independent although in LTE. The population distributions will be taken here to be continuous functions, as we assumed for the Saha equation[35] so the initial state has $F(\mathbf{p}_1, \mathbf{p}_2) = f(\mathbf{p}_1)f_2(\mathbf{p}_2) = f_1 f_2$ and the final state is $F'(\mathbf{p}'_1, \mathbf{p}'_2) = f(\mathbf{p}'_1)f(\mathbf{p}'_2) = f'_1 f'_2$. The collision rate depends on the differential velocity of these species and we assume that momentum and energy conservation hold so that $E_1 + E_2 = E'_1 + E'_2$ and $\mathbf{p}_1 + \mathbf{p}_2 = \mathbf{p}'_1 + \mathbf{p}'_2$. Assume that the probability of a transition from the initial state i to the final state j is given by $P(j;i)$, which have the dimension of a rate (number per unit time) so that the rate for transitioning from a state Γ_{12} to $\Gamma_{1'2'}$ is

$$R_{12,1'2'} = \int P(1'2';12)f_1 f_2\,d\mathbf{p}_2\,d\mathbf{p}'_1\,d\mathbf{p}'_2 \quad (1.182)$$

and the inverse rate is

$$R_{1'2',12} = \int P(12;1'2')f'_1 f'_2\,d\mathbf{p}_2\,d\mathbf{p}'_1\,d\mathbf{p}'_2, \quad (1.183)$$

We still need to specify the probabilities P for the collisions. The collision time, as you now know, depends only on the *relative* speed and not on the order in which the collisions occur. This means that we can take $P(ij;i'j') = P(i'j';ij)$ so that the generalized form of Eq. (1.181) becomes:

$$\frac{\partial f_1}{\partial t} = \int P(12;1'2')(f_1 f_2 - f'_1 f'_2)\,d\mathbf{p}_2\,d\mathbf{p}'_1\,d\mathbf{p}'_2 \quad (1.184)$$

This is the *Boltzmann collision integral*, which forms the basis of the statistical picture for collisional redistribution of particles within the phase space available to a gas. Substituting the definition for the mean free path into Eq. (1.184) then gives

$$\left(\frac{\partial f_1}{\partial t}\right)_{\text{coll}} = \int d\Omega\,d\mathbf{p}_2\,d\mathbf{p}'_1\,d\mathbf{p}'_2 \left[\frac{d\sigma}{d\Omega}|\mathbf{v}_{12} - \mathbf{v}'_{12}|(f_1 f_2 - f'_1 f'_2)\right] \quad (1.185)$$

Diffusive separation can also be driven by temperature gradients even in an otherwise uniform gas. One way to see this is to take the thermal flux, which is $\kappa \nabla T$ and compare this with the rate of diffusive heat transport by particle motions, $c_P \lambda_{ij} \nu_{c,ij}$. In a statistical mechanical sense, this is the origin of the diffusive cooling law.

---

[35] In Chapter 3 we will quantize the distributions and deal with the case for levels in atoms, but here we are looking at the thermal distributions of free particles.

56 FROM GASES TO CLUSTERS: CONCEPTS IN GRAVITATION AND GAS LAWS

1.5 THE VIRIAL THEOREM

The Red Queen in *Alice Through the Looking Glass* explains that "you have to run twice this fast just to stay in the same place," and in effect, this is an appropriate dictum for motion in a gravitational field. For bound two-body systems, we have seen that the angular momentum and the gravitational potential combine to produce a unique orbit for which the velocity is known at each moment. For a more complicated mass distribution, or for a cluster of masses, the mutual attraction of all of the bodies produces a global velocity field, one that governs both the motion of each constituent mass *and* the configuration in space of the ensemble. The resulting distribution is unique, provided we have included all forms of interaction between the bodies (we will see in a little while that the global condition also holds if we allow for collisional interactions between the masses).

We thus arrive at the moment you've been waiting for, when we combine these two apparently diverse subjects of gravitation and thermodynamics into a single subject. The virial theorem provides a powerful global constraint on the stability of a gravitationally bound system and it is central to many astrophysical problems. We will first derive the dynamical representation, starting with the phase space approach that we have been using for statistics, and then pass to the continuum limit that is characteristic of fluid media. In this way, you will see how very disparate phenomena are unified within a single picture.

The virial theorem originated in mechanics. The prime example is the motion of ensembles of point masses moving under the influence of their collective gravitational field. You treat the force as the result of the entire collection of masses but each particle moving independently and without collisions. We'll go into the details in a moment, but for now, let's assume that we have a set of particles that are bound within a background potential. The particles are assumed to be collisionless, and the motion is solved for each individually. Otherwise, we place no constraints on the energy of the bodies. The equations of motion are written in component form as

$$m \frac{d^2 x_i}{dt^2} = F_i \tag{1.186}$$

where, for simplicity, we're assuming that the particles have the same mass. Now if we take the scalar product with the displacement, $x_i - x_{i,0}$ and look for a scalar quantity that is analogous to the work, we find that

$$m(x_i - x_{i,0}) \ddot{x}_i = \frac{1}{2} m \frac{d^2 x^2}{dt^2} - m \dot{x}_i^2 = W. \tag{1.187}$$

Notice that the first term is the second time derivative of the moment of inertia, which measures the rate of distortion of the body, and the second term, mv^2, is twice the kinetic energy, $2K$. The last term is the net potential energy. If for the moment we restrict our attention to the steady state in which the symmetry of the body and the global mass distribution is unaltered, then

$$2K + W = 0 \tag{1.188}$$

This is the *virial theorem*, which was first derived by Clausius in the mid-nineteenth century. It says, in effect, that there is a limit to the amount of work that can be obtained from a bound system. To be more specific, let's now restrict our attention to a gravitational field. In this case, the total mass is simply the volume integral of the density and, for simplicity, we will assume that the spatial density is constant. Then the gravitational potential energy is

$$\Omega = -\frac{3\,GM^2}{5R} = -2K \tag{1.189}$$

where M is the total mass, R is the mean radius of the distribution of masses, and the kinetic energy is $K = \frac{1}{2}M\langle v \rangle^2$, where $\langle v \rangle$ is the mean velocity. The problem we will face in the application of the theorem to astronomical observations of real systems is immediately clear. Observers, in general, have access to only one component of the velocity, the radial velocity, although the application of the virial theorem requires complete information about the three-dimensional motion. This is a problem. For a gas, as we will see later, this is not too serious a deficiency. You can usually assume that the motions are random and that the distribution is isotropic. On the other hand, for stars in galaxies, and galaxies in clusters, this is more problematic.

We now revert to the statistical description of ensembles, except now we do it for more massive bodies. First, let's ignore the statistical aspects of the velocity distribution and simply look at the gross aspects of the equilibrium. Consider that we have an aggregation of masses dm so that at a point within the ensemble we have

$$V(\mathbf{x}) = -G\int \frac{dm(x')}{|\mathbf{x}-\mathbf{x}'|} \tag{1.190}$$

From the equation of motion, taking the same scalar multiplication only this time with $(\mathbf{x}-\mathbf{x}')$, we obtain on integration over the entire body:

$$\frac{1}{2}\frac{d^2 I}{dt^2} = 2T + \Phi \tag{1.191}$$

where

$$\Phi = -G\int d\mathbf{x}\,d\mathbf{x}' \frac{\rho(\mathbf{x})\rho(\mathbf{x}')}{|\mathbf{x}-\mathbf{x}'|} \tag{1.192}$$

which we can generalize to a tensor formalism by using

$$W_{ij} = -G\int d\mathbf{x}\,d\mathbf{x}' \frac{\rho(\mathbf{x})\rho(\mathbf{x}')(x_i - x'_i)(x_j - x'_j)}{|\mathbf{x}-\mathbf{x}'|^3}, \tag{1.193}$$

in the notation introduced by Chandrasekhar (1967). Then using

$$I_{ij} = \int d\mathbf{x}\,\rho(\mathbf{x})(x_i - x'_i)(x_j - x'_j)$$

and

$$T_{ij} = \tfrac{1}{2}\int d\mathbf{x}\, \rho(\mathbf{x})\dot{x}_i \dot{x}_j$$

we arrive at the tensor form of the virial theorem:

$$\frac{1}{2}\frac{d^2}{dt^2} I_{ij} = 2T_{ij} + W_{ij}, \qquad (1.194)$$

a form that can be extended to higher order moments as needed. Remember, you always start with the equations of motion so the system is always satisfying momentum conservation. The higher moments are conveniences, and the system of equations that result are additional global constraints on the equilibrium. This use of the tensor virial equations is especially important if the body or ensemble is anisotropic—for instance, it may be rotating or possess a velocity distribution that is very far from a relaxed state—and the equations that result provide both the stability criteria and body oscillation frequencies (through the time derivative of the moment of inertia tensor).

Return for a moment to the picture of stars in a cluster, each of which moves in its own orbit about the center of mass. Each body has some momentum p and energy ϵ. Now, as we've seen, we can write the entropy as the probability of finding any one of the particles in some range Δp. This means that if all of the particles have the same mass, they occupy a region of phase space that is $d\Gamma = 4\pi p^2\, dp\, dV$ for a uniform velocity distribution. Then the number density of particles is

$$n(\mathbf{x}) = \int f(v)\, d\mathbf{v} = 4\pi \int f(\mathbf{x}, v) v^2\, dv. \qquad (1.195)$$

Although for a cluster we can actually measure the motions of the individual "particles" (i.e., the stars or galaxies), we are usually interested in ensemble averages just as we were in the thermodynamic limit. This means that we need to define further quantities that can be tracked for the whole gaggle and evolve only in time and space. Since the distribution function is a statistical representation of the particle velocities, we can use the usual method of taking moments. Thus, the mean velocity component in the ith direction is given by

$$\langle v_i \rangle = \frac{\int f(v) v_i\, d\mathbf{v}}{\int f(v)\, d\mathbf{v}}, \qquad (1.196)$$

and the next moment gives the mean stress tensor, the Reynolds stress, for the ensemble:

$$\langle v_i v_j \rangle = \frac{\int f(v) v_i v_j\, d\mathbf{v}}{\int f(v)\, d\mathbf{v}}. \qquad (1.197)$$

Notice that the velocity dispersion is related to these moments through $\sigma^2 = \langle v^2 \rangle - \langle v \rangle^2$ and is simply the diagonalized Reynolds stress for a spherically symmetric system viewed moving with the mean velocity. Recall that the averaging is taken over n, the number density of the particles.

The evolution of the distribution is assumed to occur in space, time, and momentum through simple motions. Let's go back to the Vlasov equation and take the moments with respect to the spatial distribution, This means averaging over the spatial volume to obtain n, the total number of particles. Then

$$\int x_i \left[\frac{\partial}{\partial t} n \langle v_i \rangle + \frac{\partial}{\partial x_j} n \langle v_i v_j \rangle - n g_i \right] dV = 0, \quad (1.198)$$

which, as you see, now depends *only* on the time and otherwise yields global quantities averaged over the bulk of the mass distribution. Thus

$$\frac{\partial}{\partial t} \int dV n x_i \langle v_i \rangle - \int dV n \langle v_i \rangle \langle v_i \rangle - \int dV n x_i g_i + \int dS_j n x_i \langle v_i v_j \rangle$$

$$- \int dV n \langle v_i v_j \rangle \delta_{ij} = 0. \quad (1.199)$$

You will notice that we now have a surface term, which arises from the integral of the divergence over the volume,[36] a characteristic feature of this Eulerian form of the equations. Then, for the collected terms, we get

$$\frac{1}{2} \frac{d^2 I}{dt^2} = 2 \langle K \rangle + \langle \Omega \rangle, \quad (1.200)$$

which, you see, is once again the virial theorem for the ensemble, only this time we use mean quantities to describe the energies.

Let us return to that cluster of stars. If the individual stars have almost no angular momentum, they will follow nearly radial orbits as they move within the cluster potential. The timescale is the same as for a harmonic oscillator with the mean density of the cluster, the freefall time. Since we assumed at the start that the total energy of the system was negative, in other words the stars are bound to the mass of the cluster, the random assortment in orbital phase is sufficient to insure that the shape of the cluster remains constant with time. In other words, we can neglect the $d^2 I/dt^2$ term in the virial equation. When we observe the motions, because the system is assumed to be transparent—meaning we can see all of bodies in the cluster without interference—then the spherical mass distribution still permits the use of the radial velocity as a good measure of the total kinetic energy. On the other hand, if the distribution is not spherical, such as you would find in a tidally distorted cluster or one that was unstable to the formation of a bar (see Chapter 7 for further discussion), or if there is some rotation in the system as a whole, then the use of radial velocities as estimators of the total kinetic energy is

---

[36]Although this surface term can usually be ignored unless there is a bounding medium, its importance in the virial estimates has been stressed by Zweibel, E. and McKee, C. 1995, *ApJ*, **439**, 779.

biased. There is another problem, however, in the actual use of the virial theorem for mass distributions. The radius we use depends not only on the *projected* appearance of the cluster but also on the assumed geometry. Just because something looks spherical doesn't mean that it is, since we could be observing the ensemble at some angle that obscures a flattening or elongation of the whole cluster. The value of R depends on the assumed spatial structure, which in turn affects our considerations of the mass distribution. Ultimately, we should really write the radius as $\langle R \rangle$, weighted by the mass distribution with projection factors removed.

This provides a constraint on the kinetic energy of the system given a potential energy, but also indicates that the motion is restricted to bound states for which the total energy, $E = K + \Omega$, is given by

$$E = \tfrac{1}{2}\Omega, \qquad (1.201)$$

which is the gravitational binding energy of the system. Notice that we have made no use of the random motions. In this case, the assumption is that individual particles move within the gravitational potential of the system and that they are dispersionless—each orbit is individually subject to the same global constraint.

Let us now look at what happens if we take a continuum limit, using the conservative form of the equations of motion:

$$\frac{\partial}{\partial t}\rho v_i + \frac{\partial T_{ij}}{\partial x_j} = \rho \frac{\partial \Phi}{\partial x_i}. \qquad (1.202)$$

We've included gravity here and also assumed that the form for a fluid flow. Taking the scalar product of Eq. (1.202) with **x** and integrating over the spatial volume gives

$$\int dV \left[\frac{\partial}{\partial t}(\rho x_i v_i) - \rho v_i v_i \right] + \int dS_j x_i T_{ij} - \int dV T_{ij} \delta_{ij} = \int dV \rho \Phi \qquad (1.203)$$

Now you notice that the first term is the change in the moment of inertia we spoke about in the particle case, and the second term is twice the kinetic energy in the flow. The most important difference here is that the pressure comes into the equation through the stress tensor, which also includes the additional advection terms, so that we get

$$2K + \Omega + 3\int P\, dV = 0, \qquad (1.204)$$

which is a *global* constraint for the flow. This is the full fluid form of the virial theorem and the one that we will be using most often. It expresses the connection between the work done by a flow and the action of external forces on the fluid. This comes to the same result we obtained for the general distribution function, as it must, but now the various terms have been identified with their fluid equivalents.

1.5.1 Virialized Gas Bags

Now we come to the simple and surprising result of our merger of mechanics and thermodynamics. You will recall that we took the equation of state to be polytropic when deriving the equilibrium structure of a simple gas acting only under its own gravitational field. We can write the kinetic energy of this gas in terms of the temperature as $dK = \frac{3}{2}kT\,dN = \frac{3}{2}(\mathcal{R}/\mu)T\,dM$, where N is the total number of particles in the mass. The internal energy is $dU = c_v T\,dM$. The connection between the two comes from the relation between the specific heats, $c_v(\gamma - 1) = \mathcal{R}/\mu$. On substitution, we get

$$K = \tfrac{3}{2}(\gamma - 1)U. \tag{1.205}$$

Now the total energy of the body is $E = U + \Omega$, but $\Omega = -2K$ so that the total energy is

$$E = (4 - 3\gamma)U = -\frac{4 - 3\gamma}{3(\gamma - 1)}\Omega \tag{1.206}$$

and you see that $E \to 0$ as $\gamma \to \frac{4}{3}$. In other words, the system becomes unbound if the equation of state is too soft (and you'll notice that there is another difficulty waiting in the wings when $\gamma = 1$, the isothermal case). Notice that we made *no* assumptions about the geometry of the body.

A simple consequence of the virial theorem is that a self-gravitating gasbag that is losing energy must contract if its density remains constant through its interior. The rate of energy loss, $\dot{E} = L$, where L is the luminosity, so that

$$\frac{R}{\dot{R}} = \frac{3}{10}\frac{GM^2}{RL} \tag{1.207}$$

This is called the *Kelvin–Helmholtz contraction time*. In more convenient units, this is

$$t_{\text{KH}} = 2 \times 10^7 \text{ yr}\left(\frac{M}{M_\odot}\right)^2\left(\frac{R}{R_\odot}\right)^{-1}\left(\frac{L}{L_\odot}\right)^{-1} \tag{1.208}$$

The timescale is very wrong for the age of the Sun, but it is a fair estimate of the contraction of a star before the onset of nuclear reactions when only the heat from the initial collapse is being radiated away. We can also use this very simple result to estimate how long it takes for an optically thick layer to cool. If the layer has a mass ΔM, its *thermal timescale* is

$$t_{\text{th}} = \frac{GM\,\Delta M}{RL} \tag{1.209}$$

This will be very important later when we discuss the problem of mass accretion in binary systems and novae (see Chapter 5).

1.5.2 The Gravitothermal Catastrophe

Since the virial theorem relates the internal velocity dispersion and total energy, it is easy to see that

$$K = -E. \qquad (1.210)$$

This seemingly innocuous relation has a profound consequence for the evolution of self-gravitating systems, regardless of the number of bodies involved. It means that the *specific heat is negative for gravitationally bound systems*. In consequence, if the system cools, it becomes *more tightly bound*—its total energy becomes more negative. This means that σ, the velocity dispersion, increases. The final state of the system is a collapsed core in which the density increases without bound. One way of saying this is that a self-gravitating system can never lose energy rapidly enough to cool.[37] We will encounter this idea many times in the discussions ahead. It i fundamentally important for understanding stellar structure, and also the structure of globular clusters, and best examined in the context of the Lane–Emden equation. It does not even help to reduce the mass of the ensemble. If $\dot{M}$ is the mass loss rate, the body's internal energy varies as

$$\dot{E} = -\frac{1}{2}\frac{GM}{R}\dot{M} \qquad (1.211)$$

and the internal velocity dispersion consequently increases. This depends on how the mass is lost.

We have yet to include an important feature of gravitational collapse, the time dependence of the gravitational potential produced by the change in the mass distribution. Although we may start out with a velocity distribution, $f(v)$, at the start of the collapse, the density determines Φ. Imagine that the particles have some set of initial angular momenta at the start of the collapse so that they orbit some mass $M(r)$. If the potential changes on a timescale $\Phi/\dot{\Phi}$ that is short compared with the orbital or collisional timescales for the constituent masses, a particle can experience a nearly impulsive change in its kinetic energy. For the ensemble, if the collapse is fast enough, $f(v)$ becomes isotropic as a result of the random orbital phases of the particles. This process is called *violent relaxation*.

1.5.3 Evaporation of Star Clusters

As an application of the virial theorem, let us examine the effect of finite gravitational binding energy on a cluster of stars. Any mass has a finite escape velocity. This means that the velocity distribution for the ensemble has an upper cutoff because any particle with $v \geq v_e$ leaves the system. The first application of this goes back quite far, 1912 to be precise, when Jeans argued that this is the

---

[37] Lynden-Bell, D. and Wood, R. 1968, *MNRAS*, **138**, 495; Lynden-Bell, D. 1967, *MNRAS*, **136**, 101; Lynden-Bell, D. and Lynden-Bell, R. M. 1977, *MNRAS*, **181**, 405; Antonov, V. A. 1962, *Vest. Leningrad Univ.*, **7**, 135, translated in Goodman, J. and Hut, P. *Dynamics of Globular Star Clusters: IAU Symp. 113* (Dordrecht: Reidel), p. 525; Binny, J. and Tremaine, S. 1987, *Galactic Dynamics* (Princeton: Princeton Univ. Press).

mechanism for the loss of an atmosphere by a planet.[38] The problem is connected with the maintenance of an equilibrium velocity distribution in the presence of loss of particles. Consider the adiabatic problem, where you have a lid on the atmosphere. Assume that the gas is collisionally dominated, so that the velocity distribution can be maintained throughout the loss process. Open the box. Those atmospheric particles that are moving at the escape velocity, or faster, leave. The flux of particles will be $J = n\langle v \rangle$ which is given by

$$J = \frac{\int_{v_{\text{esc}}}^{\infty} f(v) v^3 \, dv}{\int f(v) v^2 \, dv} \qquad (1.212)$$

Notice that this is a very small number compared with the mean velocity of the distribution, at least it seems to be. But if the temperature of the atmosphere is sufficiently high, in other words, if $\langle v \rangle \to v_{\text{esc}}$, then a substantial fraction of the atmosphere will be lost. Now if this process takes place adiabatically, so collisional repopulation of the high velocity tail of the distribution occurs and the temperature is maintained, then the flux will be steady and the atmosphere will eventually disperse.

This same process happens, in effect, for stars in self-gravitating systems. If the distribution function is maintained, then collisions from the stars in the core repopulate the tail of the distribution and the cluster shrinks. This increases the internal velocity dispersion and increases the rate of loss of stars. Consider the case of adiabatic evolution. The total energy remains constant, $E(t) = E(0)$, so that $M(t)^2/R(t) = M(0)^2/R(0)$, as we have just discussed. Now the collision timescales as $t_c^{-1} \sim n\Sigma\sigma$, 1 where $n \sim MR^{-3}$ is the number density, $\sigma \sim (M/R)^{1/2}$ is the velocity dispersion from the virial theorem, and $\Sigma \sim \sigma^{-4}$ is the cross section we have discussed for collisions between gravitationally interacting masses. Therefore, assuming the adiabatic scaling, we obtain $t_c^{-1} \sim MR^{-3}R^{3/2}M^{-3/2} \sim M^{-7/2}$. The evolution equation for mass loss is consequently

$$\dot{M} = K_c M^{-5/2}, \qquad (1.213)$$

where K_c is a scaling constant depending on the initial conditions of the cluster and the central concentration at $t = 0$. This means that the mass of the cluster has the following time dependence:

$$M(t) = M(0) \left(1 - \frac{t}{t_r} \right)^{2/7}. \qquad (1.214)$$

This is an essentially exponential decrease with a characteristic timescale of t_r. Mass can be expelled adiabatically from the cluster through stellar winds. Another removal mechanism is collisional stripping of stars from the periphery of the

---

[38] The most complete discussion of this problem is in Jeans, J. H. 1927, *Dynamical Theory of Gases* (Cambridge, UK: Cambridge Univ. Press).

cluster (those that are the most loosely bound are the ones that are the most easily lost and carry away the least momentum from the ensemble):

$$\frac{\delta E}{E} = \frac{\delta M}{M} - \frac{\delta R}{R}. \qquad (1.215)$$

The formation of binaries has a different effect on the cluster than the mere loss of members. For instance, in a molecular gas, the recombination of the constituents releases energy that heats the remaining ensemble. The same thing happens with a cluster. Close binary formation means that kinetic energy is released as the cluster forms tight pairs. These heat the cluster and cause it to expand. The effect is just opposite to what you get from loss of stars and reduction of the gravitational potential.

1.5.4 Tidal Limits for Star Clusters and Galaxies

As we discussed earlier, the presence of a large perturbing mass reduces the effective gravitational binding energy of an ensemble and promotes escape of particles. This is another way of looking at the Roche problem, except now we are dealing with its effects on a collection of masses with finite temperature[39] so that we now include the internal velocity dispersion. Where we have only the mass of the primary to contend with, evaporation leads to a slow readjustment. When a companion is present, its attraction can substantially reduce the binding energy and more quickly shift the internal velocity distribution closer to the critical escape velocity. We will see this *again* for stellar winds, where radiation pressure, an internal process, does the same thing.

A final point concerns our assumption of dynamical isotropy. For a collisionally dominated gas, you normally don't have to worry about anisotropic distributions. This is equivalent to saying that the gas has different temperatures depending on direction, which simply doesn't happen under normal conditions. For self-gravitating and/or collisionless systems, however, this is far from impossible. You will see this again when we discuss the Galaxy (i.e., the Milky Way Galaxy) because the velocity distribution function is very different for stars depending on whether you look in the plane or in the vertical direction. Another way to say this is that in a gravitational field of a galaxy, the stars are never truly thermalized and therefore there is no "temperature" for the distribution. Many models for clusters of stars and galaxies are computed using isothermal equations of state, but this cannot be the general rule.

1.6 THE FOKKER–PLANCK EQUATION

We will discuss two different approaches to diffusive motion of a kinetic distribution. The first involves motion in space and time of an ensemble, the change in volume density as a result of random processes. The other focuses on the change

---

[39] Our original discussion dealt only with the relative gravitational forces due to self-attraction and the companion. Now we can extend this treatment to include the effect of the internal pressure. This isn't only important for clusters—it is the process that promotes mass transfer in binary star systems.

in the velocity distribution resulting from collisions. As an illustration of the methods, consider a collection of stars that have a number density in space and time $n(x,t)$ and a distribution function $f(\mathbf{v})$. We are first interested in how to shuffle these from one place to another by random processes. Take $P(\mathbf{x}, \mathbf{x} + \delta\mathbf{x})$ to be the probability of a collision moving a particle from position $\mathbf{x}$ to $\mathbf{x} + \delta\mathbf{x}$. In what follows, we will use only scalar quantities, but the equations can be generalized. Particles starting out anywhere within a volume V wind up at a point $\mathbf{x}$ through random walks that are described by

$$n(x,t) = \int n(x - \delta x, t - \delta t) P(x - \delta x, x) d\delta x d\delta t. \tag{1.216}$$

Expand these terms to second order in space, to represent diffusion, and to first order in time to obtain

$$n(x - \delta x, t - \delta t) = n(x,t) - \frac{\partial n}{\partial x}\delta x - \frac{\partial n}{\partial t}\delta t + \frac{1}{2}\frac{\partial^2 n}{\partial x^2}\delta x^2, \tag{1.217}$$

for the number density and

$$P(x - \delta x, x) = P - \frac{\partial P}{\partial x}\delta x + \frac{1}{2}\frac{\partial^2 P}{\partial x^2}\delta x^2 \tag{1.218}$$

for the probability. Integration of Eq. (1.216) then gives

$$\frac{\partial n}{\partial t} = -\frac{\partial}{\partial x}\left[\frac{\langle \delta x \rangle}{\delta t}n\right] + \frac{\partial}{\partial x}\left[D\frac{\partial n}{\partial x}\right] \tag{1.219}$$

We have defined the mean step size by

$$\langle \delta x \rangle = \int P \delta x d\delta x, \tag{2.220}$$

and the dispersion in the step size as $\langle (\delta x)^2 \rangle$ similarly. We also have obtained an estimate of the diffusion coefficient

$$D = \frac{\langle (\delta x)^2 \rangle}{\delta t} \tag{1.221}$$

in terms of the expectation of a stochastic walk. This is why a gaussian process provides the best model. If there is no net drift, that is if $\langle \delta x \rangle = 0$, then the equation for the density has the form $\partial_t n = D\nabla^2 n$, which is a classical diffusion equation, and is also called *Fick's law*. The Fokker–Planck equation in spatial variables provides a model for the diffusion of a tracer in a gas and you will see this again in Chapter 4 when we discuss the diffusive mixing of elements in stellar interiors.

Now you know that the same will be true for a particle in velocity, since collisions that change the motion produce a random walk among the accessible velocities by the assertion of microscopic chaos in the distribution function. This form of the equations, which we will state by analogy with Eq. (1.184), is the *Fokker–Planck equation* and it is the basis of much modern work on simulations of gravitational dynamics within ensembles of particles, such as star clusters.[40] Let us examine what happens when we shuffle particles in velocity rather than in space. The Boltzmann equation is the fully collisional version of the evolution equation for the distribution function. In equilibrium, when f does not change in time, we achieve detailed balance between the rate that carry particles from one to another. The general form of the evolution equation is, however

$$\mathscr{L}f(\mathbf{x},\mathbf{v},t) = \mathscr{C}(f), \quad (1.222)$$

where the right-hand side is the collision integral. In analogy with Eq. (1.185), we can write this as

$$\mathscr{C}(f) = \int [f(\mathbf{v} - \Delta\mathbf{v})P(\mathbf{v} - \Delta\mathbf{v}, \Delta\mathbf{v}) - f(\mathbf{v})P(\mathbf{v}, \Delta\mathbf{v})]d\,\Delta\mathbf{v}, \quad (1.223)$$

so we are taking particles into and out of the velocity bin $\mathbf{v}$ by some amount $\Delta\mathbf{v}$. The probability of this happening depends on the collision cross sections and the relative velocities and masses of the particles. In the same spirit as Eqs. (1.185), however, we will be able to ignore many of the details by averaging over the probability function.

We need collect together only terms of the same order in $\Delta\mathbf{v}$, so on expansion we get (dropping the boldface for ease of notation):

$$\text{Integrand} \sim -\left[f\frac{\partial P}{\partial v} + P\frac{\partial f}{\partial v}\right]\Delta + \frac{1}{2}\left[\frac{\partial^2 f}{\partial v^2}P + 2\frac{\partial f}{\partial v}\frac{\partial P}{\partial v} + f\frac{\partial^2 P}{\partial v^2}\right]\Delta^2. \quad (1.224)$$

We expand here only to second order. In addition, we need a couple of definitions. The average change in the velocity averaged over the probability function of a collision is

$$\langle(\Delta v)^n\rangle = \int_{-\infty}^{\infty}(\Delta v)^n P(v, \Delta v)d\,\Delta v \quad (1.225)$$

For a Gaussian process, $\langle \Delta v \rangle = 0$ and $\langle (\Delta v)^2 \rangle^{1/2}$ is the root mean square (rms) step size in velocity resulting from the spectrum of collisions. Since we can write

---

[40] The most complete reference is Risken, H. 1989, *The Fokker–Planck Equation: Methods of Solution and Applications,* 2nd ed. (Berlin: Springer-Verlag). The classic astrophysical introduction is Chandrasekhar, S. 1943, *Rev. Mod. Phys.,* **15**, 1 on stochastic processes in astronomy—a reference you are urged to consult. For applications specifically geared to stellar dynamics, see Cohn, H. N. 1979, *ApJ,* **234**, 1036; Spitzer, L. 1987, *The Dynamical Evolution of Globular Star Clusters* (Princeton: Princeton Univ. Press); Spurzem, R. and Takahashi, K. 1995, *MNRAS,* **272**, 772.

the operator in the Vlasov equation as the total time derivative, we get

$$\frac{df}{dt} = -\frac{\partial}{\partial v_j}\left[\frac{\langle \Delta v \rangle}{\delta t}f - \frac{1}{2}\frac{\partial}{\partial v_j}\left(\frac{\langle (\Delta v)^2 \rangle}{\delta t}f\right)\right]. \qquad (1.226)$$

The first term is a "current" or streaming velocity, a steady response of the flow to a background force. The second term represents the diffusion in velocity, the random walk resulting from the collisions with the fluctuating background.[41]

If you think of this as the description of stars moving in a cluster it may help you visualize what is happening. Every star attracts every other star. That is what we were asserting at the very start of this chapter and here it becomes an even more important way of distinguishes between a normal (terrestrial) gas and a gas of stars. As each moves, the gravitational field it feels changes because of its altered position relative to all the other stars in the ensemble and vice versa. Thus the dynamical feedback between the gravitational field and the particle distribution function occurs because at the same time as the bodies are moving with respect to each other, they are producing a time variable global gravitational field. A body at very large distance from the cluster doesn't know that any of this is happening, though. It sees the cluster as a homogeneous entity whose shape changes are more like a fluid than a swarm of individual stars. Hence we have the difference between the hydrodynamic and kinetic approaches to stellar dynamics, and ultimately to the physics of all self-gravitating bodies. Even for a star, we have the same situation.[42]

1.6.1 The *H* Theorem

As an example of how in a dynamical system collisions act to redistribute particles within the phase space, consider for a moment the entropy function as we derived it for a distribution function

$$S = \int d\mathbf{v}\, f(\mathbf{v})\ln f(\mathbf{v})$$

and now assume the redistribution of particles in velocity is due only to elastic collisions with the total energy of the system remaining constant. Then, from

$$\frac{dS}{dt} = \int d\mathbf{v}\, \frac{\partial f}{\partial t}[\ln f + 1], \qquad (1.227)$$

---

[41] The individual rates are computed, for instance, from a Langevin equation for a fluctuating force. Take $\mathscr{g}$ to be a random variable representing the acceleration. Then the equation of motion is $dv/dt = \mathscr{g}$, which yields Δv for some interval of time. For more details, see Gardiner, C. W. 1985, *Handbook of Stochastic Methods*, 2nd ed. (Berlin: Springer-Verlag) and Risken (1989) and the brief discussion in Shore, S. N. 1992, *An Introduction to Astrophysical Hyrdodynamics* (San Diego: Academic Press).

[42] In a wonderful exchange in the 1950s at a meeting of the Royal Astronomical Society, Hoyle advanced a variation of a comment, originally by Eddington (1927, *The Internal Constitution of the Stars* (Cambridge, UK: Cambridge Univ. Press), p. 393), that "someday we should be able to understand something as simple as a star": "...Fundamentally, a star is a pretty simple structure." To this came the retort from the audience: "You'd look simple too at a distance of ten parsecs"; the story is confirmed by Fellgett, P. 1995, *Obs.*, **115**, 93 and citations therein.

we can substitute Eq. (1.185) for the collision term to obtain

$$\frac{dS}{dt} \sim \int d\mathbf{v}_1\, d\mathbf{v}_2 [f_1(\mathbf{v}_1')f_2(\mathbf{v}_2') - f_1(\mathbf{v}_1)f_1(\mathbf{v}_1)][\ln f_1(\mathbf{v}_1) + 1]\alpha_{12} \quad (1.228)$$

where now α_{12} is the rate $\sigma|\mathbf{v}_1 - \mathbf{v}_1'|$. Since we are integrating over all possible incoming velocities for the two colliding distributions of particles, the subscripts can be interchanged ($1 \leftrightarrow 2$) and the two added to give

$$\frac{dS}{dt} \sim \frac{1}{2}\int d\mathbf{v}_1\, d\mathbf{v}_2 [f_1(\mathbf{v}_1')f_2(\mathbf{v}_2') - f(\mathbf{v}_1)f(\mathbf{v}_2)][\ln f_1(\mathbf{v}_1)f_1(\mathbf{v}_2) + 2]\alpha_{12}. \quad (1.229)$$

Subtracting the same integral except now taken over the *final* states (the primed velocities), we get

$$\frac{dS}{dt} \sim \frac{1}{4}\int d\mathbf{v}_1\, d\mathbf{v}_2\, d\mathbf{v}_1'\, d\mathbf{v}_2' [f_1(\mathbf{v}_1')f_2(\mathbf{v}_2') - f_1(\mathbf{v}_1')f_2(\mathbf{v}_2')]$$
$$\times \left[\ln\left(\frac{f_1(\mathbf{v}_1)f_2(\mathbf{v}_2)}{f_1(\mathbf{v}_1)f_2(\mathbf{v}_2)}\right)\right]\alpha_{12}. \quad (1.230)$$

Now notice that we have an integrand of the form

$$(A - B)\ln\left(\frac{A}{B}\right),$$

which has the property that if $A \geq B$, then the integral is positive or zero, and if $A < B$, it still is. In other words, $dS/dt \geq 0$. This is the fundamental, unanticipated result of the Boltzmann H theorem[43] and this is the connection between the purely mechanical result for the collisions and the macroscopic thermodynamical result that entropy never decreases. Collisions redistribute particles in such a way as to smooth out the fluctuations in phase space and completely fill the available slots open to the particles.

This result is of fundamental importance in considering both gas kinetics and stellar or galactic interactions when treated in N-body systems. Even without perturbations, the system will relax in its internal distribution to redistribute the random motions among all the particles according to the overall evolution of the distribution function. In the absence of external perturbations, the gas will completely homogenize regardless of the graininess of the distribution.

1.6.2 Agglomeration Equations

Up to this point, we have dealt exclusively with problems in which the masses of the constituent particles remain constant. In fact, we have generally been even more restrictive in asserting that, except for reactions between particles, the total number of particles remains constant. Agglomeration is a growth process by the

---
[43] The term H was applied by Boltzmann as a capital eta, for entropy, not as the roman letter H.

addition of particles having lower mass to form larger particles. It is a process that finds wide applicability in astrophysics ranging from the kinetic growth and destruction of dust grains to the change in the spectrum of galactic molecular clouds. We will examine the process in the context of growth and destruction of a mass m, but we could equally use the kinetic problem of taking particles at a velocity v and changing it by some amount δv. Call the initial mass distribution $n(m)$. Then $\alpha(m, m - \delta m)$ is the collisional probability of going from $m - \delta m$ to m and $\beta(m - \delta m, m)$ is the destruction probability of going from m to $m - \delta m$. A source term, $S(m)$, denotes particle creation with some arbitrary mass dependence.[44] With these definitions, the time development of the mass spectrum is given by

$$\frac{\partial n}{\partial t} = \frac{1}{2} \int \alpha(m, m - m')n(m - m')n(m')dm'$$

$$- \int \beta(m - m', m')n(m')dm' + S(m). \qquad (1.231)$$

The factor of $\frac{1}{2}$ prevents double counting and we assume only two-body encounters for collisions. This is a variant of the Smoluchowski equation. Notice that we are maintaining a constant volume for the particles so there are no spatial dependences in the equations. As written, we are considering only a local process but the time derivative can be changed from partial to advective if there are migration and spatial diffusion processes in addition to temporal changes. Now for an example. Suppose we have a process for which $\alpha = \beta = $ constant and that we inject particles at a constant rate. Now call $S_0 = \int_0^\infty S(m)dm$ and $N(t) = \int_0^\infty S(m, t)dm$. Then Eq. (1.231) becomes

$$\frac{dN}{dt} = -\frac{1}{2}\alpha N^2 + S_0$$

which has the analytic solution

$$N(t) = \left(\frac{2S_0}{\alpha}\right)^{1/2} \tanh\left(\frac{\alpha S_0 t^2}{2}\right)^{1/2}, \qquad (1.232)$$

so you see that the total number of particles eventually saturates with time. When the rates depend on the mass, the creation rate is a convolution over the mass spectrum, so the equation requires a transform for its solution.[45] Notice that in order to examine how the spectrum develops with time we must work with the full integral equation or use transforms to convert the convolutions to products.

---

[44] This is also called the *injection spectrum*. The same equation holds for sputtering, for instance, or for diffusion-limited aggregation, both processes that are connected with the growth and destruction of dust grains.
[45] See Norman, C. A. and Silk, J. 1980, *ApJ*, **238**, 158, Appendix B for a very nice discussion of the various approximate solutions of this equation for constant rates.

TABLE 1 Crib Sheet: Useful Physical and Astrophysical Constants and Conversion Factors

Symbol	Meaning	Value	Units	Comments
G	Newtonian gravitational constant	6.6726×10^{-8}	cm^3 g^{-1} s^{-2}	
c	Speed of light	2.9979×10^{10}	cm s^{-1}	
k	Boltzmann constant	1.3087×10^{-16}	erg K^{-1}	
h	Planck's constant	6.6261×10^{-27}	erg s	
b	Wien law constant	0.28978	cm K	
σ	Stefan–Boltzmann constant	5.6705×10^{-5}	erg cm^{-2} K^{-4}	$\pi^2 k^4/(60\hbar^3 c^2)$
e	Electron charge	4.8032×10^{-10}	esu	
		1.6022×10^{-19}	C	
m_e	Electron mass	9.1094×10^{-28}	g	
		0.511	MeV c^{-2}	
m_p	Proton mass	1.6726×10^{-24}	g	
		938.272	MeV c^{-2}	
amu	$\frac{1}{12}$ ^{12}C mass	1.6605×10^{-24}	g	
$\hbar/m_e c$	Electron Compton radius	3.8616×10^{-11}	cm	
a_0	Bohr radius	0.529×10^{-8}	cm	
$M_\odot$	Solar mass	1.9889×10^{33}	g	
$L_\odot$	Solar luminosity	3.846×10^{33}	erg s^{-1}	
$R_\odot$	Solar radius	6.96×10^{10}	cm	
M_E	Earth mass	5.973×10^{27}	g	
R_E	Earth radius	6.378×10^{8}	cm	Equatorial
arcsec	Arcsecond	1/206265	rad	
Jy	Jansky	10^{-23}	erg s^{-1} cm^{-2} Hz^{-1}	
AU	Astronomical unit	1.495×10^{13}	cm	
pc	Parsec	3.085×10^{18}	cm	3.026 ly
day	Day	86400	s	
yr	Year	3.156×10^{7}	s	
Å	Ångstrom	10^{-8}	cm	
fm	Fermi	10^{-13}	cm	
b	Barn	10^{-24}	cm^2	Cross section
T	Tesla	10^{4}	gauss	
N	Newton	10^{5}	dyne	kg ms^{-2}
J	Joule	10^{7}	erg	kg m^{2}s^{-2}
eV	Electron volt	1.6022×10^{-12}	erg	
K	Kayser	1	cm^{-1}	
R	Rayleigh	$\frac{1}{4\pi} \times 10^{6}$	photons s^{-1} cm^{-2} sr^{-1}	Surface brightness
S_{10}		1.23×10^{-5}	cm$^{-2}\mu$m^{-1} at 5500 Å	Surface brightness
Dy	Debye	1.1125×10^{-37}	esu cm	Dipole moment
		3.3356×10^{-30}	C m	
Ry	Rydberg unit	13.595	eV	1 eV = 8069 cm^{-1}
atm	Atmosphere	101.325	Pa	Pressure
torr	Torricelli unit	133.322	Pa	Pressure

1.A GRAVITATIONAL POTENTIAL FOR HOMOGENEOUS SPHEROIDS

Hurl that spheroid down the field!

—Tom Lehrer, *Fight Fiercely Harvard*

Self-gravitating spheroidal figures, which were first studied in the eighteenth century by Maclaurin, Clairault, and Euler, have been enduring objects of interest in astrophysics for over 200 years.[46] The catalog of investigators reads like a who's who of applied and fundamental mathematics: Jacobi, Riemann, Dedekind, Poincaré, Darwin, Cartan, and Chandrasekhar. In part because of their application to planetary and stellar structure, they have been used since the mid-nineteenth century as test models for theories of the response of ideal equations of state to rotational and tidal deformation, especially by Kelvin and Tait, Darwin, and Jeans. There are several reasons for this continuing attention. First, models of liquid, that is, incompressible, stars mimic stiff equations of state that are often encountered in degenerate matter and are useful for the study of neutron star structure. In addition, because they can be addressed with a minimum of assumptions, hydrostatic figures are actually equipotential structures and therefore provide insights into the stability of more complicated self − gravitating objects. There are also similarities to the properties of collisionless systems, such as elliptical galaxies and galactic spheroids, and these have been widely exploited as models. Finally, the bifurcation properties of these figures are reminiscent of fission scenarios for binary and multiple star formation and therefore serve as test cases for codes. In light of these applications, let us examine some of the basic properties of such ellipsoids in detail as a prologommena for further applications.

Consider a point (x, y, z) within a homogeneous body whose shape can be described as a triaxial ellipsoidal figure. Take a point situated relative to a spherical coordinate system:

$$\xi = x + r \cos \theta \cos \phi$$
$$\eta = y + r \cos \theta \sin \phi \qquad (1.233)$$
$$\zeta = z + r \sin \theta$$

such that the ellipsoid that passes through (ξ, η, ζ) is given by

$$\frac{\xi^2}{a^2} + \frac{\eta^2}{b^2} + \frac{\zeta^2}{c^2} = 1 \qquad (1.234)$$

where the denominators are the relevant semimajor axes. We want to find the radius of the surface that passes through the chosen point, given the parameters (a, b, c) for the ellipsoid, and with that in tow we want to compute the gravitational potential for an arbitrary point. The radius, r_1, is the solution of a quadratic equation that is found by substituting Eq. (1.233) into Eq. (1.234) and solving for the radius:

---

[46] The early history, from Newton's *Principa* through Laplace's *Mecanique Celeste* is thoroughly covered by Todhunter, I. 1873, *History of Mathematical Theories of Attraction and the Figure of the Earth* (NY: Dover Books). The Bowditch translation of Laplace's *Mecanique Celeste* is available as a reprint (or the 19th century original), and vol. 3 contains a remarkable set of notes on the early development of the problem of computing the potential and of figures of equilibrium. As a brief list, the following works illustrate the development of the subject: Thomson, W. (Lord Kelvin) and Tait, P. G. 1903, *Treatise on Natural Philosophy*, vol. 2 (Cambridge, UK: Cambridge Univ. Press); Jeans, J. H. 1927, *Astronomy and Cosmogony* (Cambridge, UK: Cambridge Univ. Press); Webster, A. G. 1942, *The Dynamics of Particles and of Rigid, Elastic, and Fluid Bodies* (NY: G. E. Stechert and Co.) (a work that essentially recapitulates Thomson and Tait with some of the steps filled in); Ramsey, A. S. 1940, *An Introduction to the Theory of Newtonian Attraction* (Cambridge, UK: Cambridge Univ. Press); Chandrasekhar, S. 1969, *Ellipsoidal Figures in Equilibrium* (NY: Dover Books).

72   FROM GASES TO CLUSTERS: CONCEPTS IN GRAVITATION AND GAS LAWS

$$\left(\frac{\cos^2\theta\cos^2\phi}{a^2} + \frac{\cos^2\theta\sin^2\phi}{b^2} + \frac{\sin^2\phi}{c^2}\right)r^2$$

$$+ 2\left(\frac{\cos\theta\cos\phi}{a^2}x + \frac{\cos\theta\sin\phi}{b^2}y + \frac{\sin\theta}{c^2}z\right)r$$

$$+ \left(\frac{x^2}{a^2} + \frac{y^2}{b^2} + \frac{z^2}{c^2} - 1\right) = Ar_1^2 + 2Br_1 + C = 0 \quad (1.235)$$

The *positive* root for r is then

$$r_1 = \frac{(B^2 - AC)^{1/2} - B}{A}, \quad (1.236)$$

in abbreviated notation. In what follows, keep in mind that $C(x, y, z)$ is independent of the angular coordinates (θ, ϕ) but r_1 depends on these angles. The gravitational potential inside the mass is given by

$$\Phi = G\int_0^{r_1} \frac{dM}{r} = \frac{1}{2}G\rho\int r_1^2 \sin\theta\, d\theta\, d\phi = \frac{1}{2}G\rho\int_\Omega r_1^2\, d\Omega \quad (1.237)$$

where the density, ρ, is assumed to be constant. For an interior point, the limits on the integral are $(0, \infty)$, while for an exterior point, the limits on the integration are given by (r_1, ∞). Note that by Newton's theorem for homogeneous symmetric bodies, only points interior to r_1 attract mass farther out.[47] We can now write this potential using the functions (A, B, C) as follows. First, squaring r_1

$$r_1^2 = 2\frac{B^2}{A^2} - \frac{AC - 2B(B^2 - AC)^{1/2}}{A^2},$$

we substitute this into Eq. (1.237). You can show that

$$\int_\Omega \frac{B(B^2 - AC)^{1/2}}{A^2} d\Omega = 0$$

so we then find

$$\Phi = \int \left(\frac{B}{A}\right)^2 d\Omega - \frac{1}{2}C\int \frac{d\Omega}{A} \quad (1.238)$$

[47] We recommend the discussion in Chandrasekhar, S. 1995, *Newton's Principia for the Common Reader* (London: Oxford Univ. Press), although it should be supplemented by the new, and definitive, translation of the *Principia* by Cohen, I. B. and Whitman, A. 1999, *Isaac Newton: The Principia* (Berkeley: Univ. California Press). See also Westfall, R. S. 1971, *Force in Newton's Physics* (NY: American Elsevier).

for the potential. To simplify the treatment, we can define an auxiliary function

$$\mathscr{W} \equiv \frac{1}{2}\int \frac{d\Omega}{A} \tag{1.239}$$

from which it follows that

$$\Phi = G\rho\left[\left(\frac{\partial \mathscr{W}}{\partial a^2}\right)x^2 + \left(\frac{\partial \mathscr{W}}{\partial b^2}\right)y^2 + \left(\frac{\partial \mathscr{W}}{\partial c^2}\right)z^2\right] + C\mathscr{W}. \tag{1.240}$$

If we call $M = \cos^2\theta/a^2 + \sin^2\theta/c^2$ and $N = \cos^2\theta/b^2 + \sin^2\theta/c^2$, then

$$\mathscr{W} = \frac{1}{2}\int \sin\theta\, d\theta \int \frac{d\phi}{M\cos^2\phi + N\sin^2\phi} = 2\pi\int_0^{\pi/2} \frac{\sin\theta\, d\theta}{(MN)^{1/2}}$$

or, writing the denominator out completely:

$$\mathscr{W} = 2\pi abc^2 \int_0^{\pi/2} \frac{\sin^2\theta\, d\theta}{\left[(a^2\sin^2\theta + c^2\cos^2\theta)(b^2\sin^2\theta + c^2\cos^2\theta)\right]^{1/2}}. \tag{1.241}$$

On substituting

$$\cos\theta = \frac{c}{(s+c^2)^{1/2}}$$

we obtain

$$\mathscr{W} = \pi abc \int_0^\infty \frac{ds}{\left[(s+a^2)(s+b^2)(s+c^2)\right]^{1/2}} = \pi abc \int_0^\infty \frac{ds}{\Delta(a,b,c)}. \tag{1.242}$$

The gravitational potential is therefore given by

$$\Phi = \pi G\rho abc \int_0^\infty \left(1 - \frac{x^2}{a^2} - \frac{y^2}{b^2} - \frac{z^2}{c^2}\right)\frac{ds}{\Delta}. \tag{1.243}$$

Now we return to the polar form for r_1:

$$r_1^{-2} = \sin^2\theta\left(\frac{\cos^2\phi}{a^2} + \frac{\sin^2\phi}{b^2}\right) + \frac{1}{c^2}\cos^2\theta. \tag{1.244}$$

Just to be confusing, we'll reiterate that (a,b,c) are the semimajor axes of the mass distribution. For a uniform homogeneous mass density, also called a *homeoid*, the potential is given by

$$\Phi = G\int \rho r_1^2\, d\Omega = G\rho \int_0^{\pi/2} \sin\theta\, d\theta \int_0^{2\pi} d\phi\, r_1^2. \tag{1.245}$$

Make the transformation that $y = \tan \phi$ so that Eq. (1.245) becomes

$$\int r^2 \, d\Omega = 2 \int_0^{\pi/2} \int_0^\infty \left[\frac{\sin^2 \theta}{a^2} + \frac{\cos^2 \theta}{c^2} + \left(\frac{\sin^2 \theta}{b^2} + \frac{\cos^2 \theta}{c^2} \right) y^2 \right]^{-1} \sin \theta \, d\theta \, dy. \tag{1.246}$$

To simplify this integral, we can use the auxiliary expression

$$\int_0^\infty \frac{dy}{\alpha + \beta y^2} = \frac{\pi}{2} (\alpha \beta)^{-1/2}$$

and the substitution of $x = c^2 \tan^2 \theta$ to obtain the gravitational potential in a more compact form:

$$\Phi = \pi G a b c \rho \int_0^\infty \left[1 - \frac{x^2}{u + a^2} - \frac{y^2}{u + b^2} - \frac{z^2}{u + c^2} \right]$$

$$\times \frac{dx}{\left[(x + a^2)(x + b^2)(x + c^2) \right]^{1/2}}. \tag{1.247}$$

We now have the gravitational potential for a homogeneous density distribution. If we define the integral

$$A_i \equiv abc \int \frac{dx}{(x + a_i^2)\left[(x + a^2)(x + b^2)(x + c^2) \right]^{1/2}} = abc \int \frac{dx}{(x + a_i^2) \Delta}, \tag{1.248}$$

a notation due originally to Poincaré, then we find that

$$\Phi = \pi G \rho \left(A_1 x^2 + A_2 y^2 + A_3 z^2 \right) \tag{1.249}$$

For a rotating body, in particular for one with constant angular frequency Ω, we find two possible equilibrium figures, one oblate and one prolate. Let us concentrate on the more familiar one, the oblate Maclaurin ellipsoid, for which $a = b > c$, an oblate spheroid. In this case, we are faced with the integral

$$I = a^3 (1 - e^2)^{1/2} \int_0^\infty \frac{du}{(u + a^2)^2 (u + c^2)^{1/2}}$$

On substitution of $z = (u + a^2)^{-1}$ and taking $x = z^{1/2}$, after some rearrangement the integral in Eq. (1.248) can be put in a standard form:

$$\int \frac{x^2 \, dx}{(a^2 - x^2)^{1/2}} = -\frac{x}{2} (a^2 - x^2)^{1/2} + \frac{a^2}{2} \sin^{-1} \frac{x}{|a|} \tag{1.250}$$

which, on substitution using $c = a(1 - e^2)^{1/2}$ gives

$$A_1 = A_2 = \frac{(1 - e^2)^{1/2}}{e^3} \sin^{-1} e - \frac{1 - e^2}{e^2}$$
$$A_3 = \frac{2(1 - e^2)^{1/2}}{e^3} \sin^{-1} e - \frac{2}{e^2}$$
(1.251)

We consequently find from the equation for an equipotential surface that

$$\frac{\Omega^2}{2\pi G\rho} = \frac{2(1 - e^2)^{1/2}}{e^3}(3 - 2e^2)\sin^{-1} e - \frac{6}{e^2}(1 - e^2). \qquad (1.252)$$

Notice that $\Omega^2/(2\pi G\rho)$ is a continuous function of the axial ratio but it has a maximum value of about 0.23 for an eccentricity of 0.82. The angular momentum continues to increase up to $e = 1$, but the angular frequency cannot increase beyond this value as long as the body remains incompressible. What happens beyond this point we will delay considering until we get to our discussion of the origin of binary stars in Chapter 5.

Why can so much insight be gained with such comparatively simple tools? First, a rotating hydrostatic body for which we can treat the density as constant is isobaric on equipotentials. As we will see in Chapter 4, this is not true once radiative transfer is included because of the simultaneous requirements of thermal and mechanical balance (von Zeipel's theorem). But as long as we can treat the body as isothermal and either incompressible or constant density (the "liquid star" so popular as a toy model in the first half of the last century), we can proceed almost geometrically. You see, if you maintain constant density then the gravitational potential varies only because of changes in the shape of the body, not its internal mass distribution. The model is not as naive as it seems. A distribution of stars in a galaxy, a collisionless fluid, can be approximated as a classical ellipsoid. Even without rotation, the gravitational potential of the mass distribution can be solved and its stability evaluated. We will see more of this in Chapter 7.

1.B A LIGHTNING-FAST REVIEW OF HAMILTONIAN AND LAGRANGIAN MECHANICS

Begin with the equations of motion:

$$m\frac{d^2 x_i}{dt^2} = \frac{\partial \Phi(x_i)}{\partial x_i}, \qquad (1.253)$$

where Φ is a scalar potential function of the coordinates only. We can now write this slightly differently:

$$m\, d\dot{x}_i + \frac{\partial \Phi}{\partial x_i} dt = 0.$$

Now replace dt with $dx/\dot{x}$. Then you see that we can combine these into a single function that depends on time, coordinates, and velocity but that is conserved in this system:

$$H = T + \Phi = \text{constant}. \tag{1.254}$$

This is the Hamiltonian function. Here T is the kinetic energy, $T = \frac{1}{2}m\dot{x}^2$ and Φ is the potential function. Now we can also write a function that gives the equations of motion directly:

$$dL = \frac{\partial L}{\partial x_i} dx_i + \frac{\partial L}{\partial \dot{x}_i} d\dot{x}_i. \tag{1.255}$$

This is the *Lagrangian* function, where $L = T - V$. The key point here is that the potential does not depend on the velocity. This is *not* the way the function is usually obtained. Normally you define a quantity called the *action*:

$$S = \int_{t_1}^{t_2} L(x, \dot{x}, t)\, dt, \tag{1.256}$$

which you see has the dimensions of an impulse, and then assert that a mechanical system will follow a trajectory, in the presence of a field, that minimizes this quantity—the *principle of least action*, such that

$$\delta S = \delta \int_{t_1}^{t_2} L\, dt = \int_{t_1}^{t_2} \delta L\, dt = 0; \tag{1.257}$$

at least this guarantees that the path is extremal although it does not insure that it is minimal. Now expand δL with respect to its dependent coordinates, keeping the time as an external variable:

$$\delta L = \frac{\partial L}{\partial x_i} \delta x_i + \frac{\partial L}{\partial \dot{x}_i} \delta \dot{x}_i \tag{1.258}$$

and use $\delta \dot{x} = d\delta x/dt$ to take the variation of the velocity components. The integration then yields

$$\delta S = \int_{t_0}^{t_1} \left\{ \left(\frac{\partial L}{\partial x_i} \right) \delta x_i - \frac{d}{dt} \left(\frac{\partial L}{\partial \dot{x}_i} \right) \right\} \delta x_i\, dt + \left(\frac{\partial L}{\partial \dot{x}_i} \right) \delta x_i \Big|_1^2, \tag{1.259}$$

where the last term is the integrated function evaluated at the endpoints of the motion. Here, however, $\delta x_i = 0$ for all components, so the last term vanishes. The equations of motion now take the *Euler–Lagrange* form:

$$\frac{\partial L}{\partial x_i} = \frac{d}{dt} \frac{\partial L}{\partial \dot{x}_i}. \tag{1.260}$$

You may think that all of this formalism gains us nothing, but it has several substantial advantages in the construction of equations for arbitrary potentials and coordinates. First, it is a function that is not required to be constant, unlike the Hamiltonian. The second is that it is very general; the function is a scalar so it is invariant under changes in the coordinate representation. Finally, it is linear in the velocity components so we can write down the kinetic terms in any coordinate system. There is another justification, although a more subtle one. This is the standard approach now used in field theory and the sooner you get used to seeing this formalism, the better.[48]

1.C GENERAL RELATIVITY ON THE CHEAP

Much of modern astrophysics deals with extreme environments and objects where gravitational fields are large and velocities approach that of light. Under such circumstances, the insights provided by classical Newtonian physics may be limited or even misleading. As part of your theoretical toolbox, you will therefore need to have the equipment to treat such situations and to develop an intuition for the physics involved. Therefore, we will here provide an introduction to the standard results of non-Newtonian gravitation theory, general relativity.[49] Any time the velocities are comparable to c, or the gravitational potential energy of a body is comparable to its rest mass, you will be entering this non-classical realm. This happens remarkably often in the cosmos, from close binary systems (cataclysmic variables) to neutron stars and pulsars, to cosmology.

1.C.1 Special Relativity

From the Lorentz transformations, you know that an accelerated observer sees a progressively more distorted view of the world compared with an inertial companion. This means that there must be a way of taking any observations made in one

---

[48] There are countless books on classical mechanics, but we have our favorites. Chief among these is Landau, L. and Lifshitz, E. 1985, *Mechanics*, 3rd ed. (Oxford: Pergamon). This terse volume is notable for its insistence on using Lagrangian mechanics from the outset and for its discussion of perturbation problems, which are especially important in astrophysics. A more comprehensive treatment is provided by Goldstein, H. 1980, *Classical Mechanics*, 2nd ed. (Reading: Addison-Wesley). Sommerfeld, A. 1946, *Mechanics* (NY: Academic Press) has a wonderful treatment of the relativistic Kepler problem, mimicking the one used in the old quantum theory to explain fine structure.

[49] Review collections include Hawking, S. W. and Israel, W. eds. 1979, *General Relativity: An Einstein Centennial Survey* (Cambridge, UK: Cambridge Univ. Press); Hawking, S. W. and Israel, W. eds. 1987, *Three Hundred Years of Gravitation* (Cambridge, UK: Cambridge Univ. Press). Some superb textbooks and monographs are available, including Landau, L. D. and Lifshitz, E. M. 1971, *Classical Theory of Fields* (Oxford: Pergamon); Adler, R., Bazin, M., and Schiffer, M. 1975, *Introduction to General Relativity*, 2nd Ed. (NY: McGraw-Hill); Weinberg, S. 1972, *Gravitation and Cosmology* (NY: Wiley); Misner, C., Thorne, K., and Wheeler, J. A. 1973, *Gravitation* (San Francisco: Freeman); Hawking, S. W. and Ellis, G. F. R. 1973, *The Large Scale Structure of Spacetime* (Cambridge, UK: Cambridge Univ. Press); Schutz, B. 1983, *A First Course in General Relativity* (Cambridge, UK: Cambridge Univ. Press); Wald, R. 1983, *General Relativity* (Chicago: Univ. of Chicago Press); Stewart, J. 1995, *Advanced General Relativity* (Cambridge, UK: Cambridge Univ. Press); Ludvigsen, M. 1999, *General Relativity: A Geometric Approach* (Cambridge, UK: Cambridge Univ. Press); Chandrasekhar, S. 1982, *The Mathematical Theory of Black Holes* (London: Oxford Univ. Press).

frame and transforming it into the other, if causal connections and covariance of the physical laws is preserved, This is the basis of *general* relativity, the theory finalized in 1916 by Einstein. In dropping the restrictive assumption of constant relative velocity, general relativity allows for accelerated observers to find a correspondence rule for any mutually observed events. There are, however, several taxes that are paid for this lovely jewel. The first is comparative complexity of the mathematics. The second is the realization that a theory must include geometry as its basis.

You should already be familiar with the essential idea of special relativity, that two observers can bring their measurements of an event into correspondence, through the principle of simultaneity, by applying the Lorentz transformations to the space and time coordinates:

$$
\begin{aligned}
x' &= \gamma(x - vt) \\
y' &= y \to x'^2 = x^2 \\
z' &= z \to x'^3 = x^3 \\
t'^0 &= \gamma\left(t^0 - \frac{v}{c^2}x\right) \to x'^0 = \gamma\left(x^0 - \frac{v}{c}x^1\right),
\end{aligned}
\qquad(1.261)
$$

where here we write $x^0 = ct$ and

$$\gamma = \left[1 - \left(\frac{v}{c}\right)^2\right]^{-1/2} = (1-\beta^2)^{-1/2}.$$

The velocity addition equation for relative motion $u = dx/dt$ is

$$v' = \frac{u-v}{1 - uv/c^2}, \qquad(1.262)$$

which follows from the postulate that the speed of light is the same for all observers regardless of their relative motion. The transformation of coordinates is linear and given by the components of the matrix Λ^i_j. The form of the Lorentz transformation matrix follows from rewriting Eq. (1.261) as parallel and perpendicular coordinates with respect to the velocity such that $\mathbf{x} = \mathbf{x}_\parallel + \mathbf{x}_\perp$

$$
\begin{aligned}
\mathbf{x}'_\perp &= \mathbf{x}_\perp \\
\mathbf{x}'_\parallel &= \gamma(\mathbf{x}_\parallel - \mathbf{v}t) \\
x'^0 &= \gamma\left(x^0 - \frac{1}{c}\mathbf{v}\cdot\mathbf{x}\right),
\end{aligned}
$$

or, in component form

$$\Lambda^i_j = \delta^i_j + \frac{v^i v_j}{v^2}(\gamma - 1) \qquad(1.263)$$

so that the coordinates transform as

$$x'^i = \Lambda^i_j x^j. \tag{1.264}$$

The most important feature, for our purposes, is that we obtain a relation between the distances measured by the two observers, and we assert that they get the same result (this is required if the speed of light is taken to be the same in the two frames):

$$(dx'^0)^2 - (dx')^2 = (dx^0)^2 - (dx')^2 = ds^2 \tag{1.265}$$

where ds is now an invariant of the motion. This is another way of expressing the velocity addition formula. This is the *proper distance*, the separation in the spacetime between two events. Since in one frame, the primed one, we can take the velocity to vanish and the separation between events to be $d\mathbf{x}' \cdot d\mathbf{x}' = 0$, then we can define the *proper time* to be $d\tau^2 = dt^2 - (d\mathbf{x})^2$ so that

$$d\tau = \gamma^{-1} dt. \tag{1.266}$$

The proper separation is therefore a *Lorentz invariant*, a characteristic that defines a 4-vector. The 4-momentum, p, results from the 4-velocity $(c, \mathbf{v})$ through $p = \gamma(c, \mathbf{v})$, and obeys the invariance relation:

$$p \cdot p = (p^0)^2 - \mathbf{p} \cdot \mathbf{p} = \frac{1}{c^2} E^2 - \mathbf{p} \cdot \mathbf{p} = m^2, \tag{1.267}$$

where m is now the rest mass of the particle. Suppose you begin accelerating relative to a stationary observer, or at least someone who you both *agree* is stationary. At constant relative speed, the proper distance as measured between the two of you is ds. Therefore, the light cone is defined simply by the usual relation for a particle that has a consistently null proper time. Now allow one of the observers to accelerate. The transformation is Lorentzian *at each instant* for each observer, but as the speed is increasing with time, the transformation also changes—in other words, the trajectory followed by the accelerated observer relative to the stationary (or inertial one, it does not matter) becomes increasingly distorted compared with what you would have expected at constant velocity.

For relativistic dynamics, one statement of the principle of covariance is that the form of the equations of motion remain the same for all observers. This is tantamount to saying that if you are in a moving frame, the equation you would use to describe your physical experience will have the same form, using the same physical quantities, as in any other reference frame. You see the equations of motion in their Newtonian form:

$$\frac{dp^i}{dt} = F^i, \tag{1.268}$$

where F^i is the force ($F^0 = 0$) and the components of the momentum are defined by $p^i = m\dot{x}^i$. Since the mass is not constant between frames, the more familiar form $m\ddot{x}_i$ does not suffice to provide the equations of motion. In an arbitrary frame, replacing dt by $d\tau = \gamma^{-1} dt$ and assuming that the force transforms as

$$F'^i = \Lambda^i_j F^j = F^i + \frac{\mathbf{v} \cdot \mathbf{F}}{v^2} v^i$$

leads directly to the expressions for the relativistic momentum components:

$$p^i = \gamma m v^i \quad (i = 1, 2, 3), \qquad p^0 = \frac{E}{c}. \tag{1.269}$$

The energy of a particle, which is invariant under Lorentz transformations, is given by

$$E^2 = p^2 c^2 + m^2 c^4 \rightarrow E = \gamma m c^2 \tag{1.270}$$

The mass is now essentially defined through the 4-momentum. This E is *not*, however, the kinetic energy, K, which is given by $K = (\gamma - 1)mc^2$. We can now define the 4-momentum components from Eq. (1.269) to be E/c and $\mathbf{p}$, where $\mathbf{p}$ are the 3-space components of the momentum (an ordinary vector) such that $p'^i = \Lambda^i_j p^j$ is the transformation law for the momenta.[50] From Eq. (1.270) it follows that an extremely relativistic particle has $E = pc$. We will return to this again in Chapter 3 when we discuss the Compton effect.

The historical origin of the special (or restricted) theory of relativity was the invariance property of the Maxwell equations under the Lorentz transformations and these results frequently apply to astrophysical problems.[51] Take the motion parallel to the x axis, for simplicity, so that

$$\begin{aligned} E'_x &= E_x \\ E'_y &= \gamma \left(E_y - \frac{v}{c} B_z \right) \\ E'_z &= \gamma \left(E_z + \frac{v}{c} B_y \right) \\ B'_x &= B_x \\ B'_y &= \gamma \left(B_y + \frac{v}{c^2} E_z \right) \\ B'_z &= \gamma \left(B_z - \frac{v}{c^2} E_y \right) \end{aligned} \tag{1.271}$$

For the electric field, for instance, we now have the relativistic generalization of the familiar Lorentz force $\mathbf{E}' = \mathbf{E} + (1/c)\mathbf{v} \times \mathbf{B}$. From this, since we have the

---

[50] Notice that since we write the scalar product in four dimensions as $a \cdot b = a_0 b_0 - \mathbf{a} \cdot \mathbf{b}$, we do not include the negative sign explicitly for the definition of the components.

[51] You might want to look at Lorentz, H. 1952, *The Theory of Electrons*, 2nd ed. (NY: Dover Books) and Miller, A. 1982, *The Emergence of Special Relativity: 1902–1911* (Reading: Addison-Wesley) for the background on this topic. Of course, any textbook on electromagnetic theory includes a discussion of the Lorentz transformations, but it is also instructive to see the historical context.

nonrelativistic form $\mathbf{B} = \nabla \times \mathbf{A}$ and

$$\mathbf{E} = -\frac{1}{c}\frac{\partial \mathbf{A}}{\partial t} - \nabla\phi,$$

we can write potential as a 4-vector, $A = (\phi, \mathbf{A})^T$, from which it follows that

$$A'^i = \Lambda^i_j A^j \tag{1.272}$$

is the generalized Lorentz transformation for the potentials.

1.C.2 The Metric

Let's suppose you work in curvilinear coordinates, let's say cylindrical. Then the kinetic energy of a free particle is

$$T = \tfrac{1}{2}m(\dot{r}^2 + r^2\dot{\phi}^2 + \dot{z}^2) \tag{1.273}$$

which, for our purposes, is the same as the Lagrangian (in the absence of a field). From now on, we will set the mass of the test particle m to unity and ignore it. Now the equations of motion, through the Euler–Lagrange equation, take the following form:

$$\ddot{r} - r\dot{\phi}^2 = 0, \quad \ddot{\phi} + r^{-2}\dot{r}\dot{\phi} = 0, \quad \ddot{z} = 0. \tag{1.274}$$

In other words, the equations have the form

$$\frac{d^2 x_i}{dt^2} + \sum_{jk} A_{ijk}\frac{dx_j}{dt}\frac{dx_j}{dt} = 0, \tag{1.275}$$

where the coefficient A_{ijk} is a function of the set of coordinates. This does not seem to be too useful, except that it recasts the equations in a very general form. But this is for the simple Newtonian case, where we don't have to worry about causality and the connections within a *spacetime*. In relativity, the scalar product explicitly includes the time because the passage of time is different for different observers depending on their velocities relative to one another. Thus, we generalize the kinetic energy to

$$T = \tfrac{1}{2}g_{ij}\dot{x}^i\dot{x}^j. \tag{1.276}$$

The quantity g_{ij} is called the *metric*. It is an invariant of the transformation, and is the means whereby we form the scalar product. The coordinates, x^i, are written with superscripted indices to denote the coordinates in the "world at large," that is, the ones relative to which the body is moving. The change in perspective comes with the statement that the dot in Eq. (1.276) means the derivative with respect to the proper time, the time in the observer's frame, not the time relative to the external observer. The coordinates are called *contravariant*, while if the subscript is used it is *covariant*. We will return to this shortly, the distinction is not just notational.

The metric now depends on the coordinates. You saw this in Eq. (1.275). The same form appeared, only in this case it was in three dimensions. The line element we used was:

$$ds^2 = dr^2 + r^2 d\phi^2 + dz^2.$$

This is the Pythagorean theorem in differential (local) form, more familiar from

$$ds^2 = dx^2 + dy^2 + dz^2,$$

while for a spherical coordinate system we would use

$$ds^2 = dr^2 + r^2 d\theta^2 + r^2 \sin^2\theta \, d\phi^2,$$

where θ and ϕ are the meridional and azimuthal angles, respectively. For *special relativity*, we need to add the time:

$$ds^2 = c^2 dt^2 - dl^2, \tag{1.277}$$

where dl is the spatial line element and, as usual, c is the speed of light. The generalization of this is formally

$$ds^2 = \eta_{ij} \, dx^i \, dx^j. \tag{1.278}$$

This equation also introduces a notation, called the *Einstein summation convention*; we will sum over repeated indices *unless we explicitly say that we are not doing so* and repeated indices are treated as arbitrary—you can substitute symbols for them as you want (also called "dummy indices," as in computer programming). The quantity η_{ij} is called the *Minkowski metric*, a diagonal matrix whose coefficients you can read off from Eq. (1.277) using $x^0 = ct$. The causal connection between events is insured by the sign of the coordinates, called the "signature" of the metric $(+, -, -, -)$. This is *not* a simple Euclidean metric where we don't have a specific connection between the space and time coordinates. The Lorentz invariance forces us to generalize the equation of motion.

The condition that the metric exists, and that we can start with a local coordinate system and extend it gradually to the whole of the spacetime, is the same as saying that the spacetime is a manifold. This means that we can create a little chart, a local coordinate patch (think of the area covered by a typical city map compared with the size of the Earth) and then, by a set of overlapping charts, build up an atlas that can describe the whole surface. We are sure you are familiar with this, because you have used geographical atlases having started with just a small piece. Globes are not constructed this way, but they could be—by placing tiny dots on the surface each of which is locally flat, and piecing them together with sufficient overlap, you could construct the whole surface. You need rules to tell you how to rotate the specific directions relative to the rest in order to properly reproduce the angular relations between cardinal directions (the direction toward which a compass points, for instance), but those rules can be specified.

It is the same with the spacetime. At least we assert that it is. There are no discontinuities, no walls, no holes, nothing that prevents our making general statements about continuity over all of the spacetime based on the local behavior. The physical way of saying this is that the *physical laws are covariant*, that is, the same everywhere, and that *spacetime is a manifold*. Therefore, we can use the tools of differential geometry to extend the dynamical concepts we are developing in this section. Eventually, in Chapter 8, we will need to return to this question about whether the assumption that spacetime has a manifold structure really makes sense, when dealing with everything there is, but for the moment, we can certainly use the tools for relatively local problems.

Return now to the Lagrangian for a free particle. In these generalized coordinates, we can write the same equations for the motion:

$$\frac{d}{d\tau} \frac{\partial T}{\partial \dot{x}^k} = \frac{\partial T}{\partial x^k} \tag{1.279}$$

using τ to represent the proper time for an observer. The metric does not depend on the time explicitly, but it does depend on the coordinates. This means that

$$\frac{d}{d\tau} g_{ij} \dot{x}^i \dot{x}^j = g_{ij} \left(\ddot{x}^i \delta_m^j + \ddot{x}^j \delta_m^i \right) + \frac{\partial g_{jk}}{\partial x^l} \dot{x}^l \left(\dot{x}^i \delta_m^j + \dot{x}^j \delta_m^i \right) \tag{1.280}$$

and, since

$$\frac{\partial \dot{x}^k}{\partial \dot{x}^l} = \delta_l^k$$

where δ_l^k is the Kronecker delta symbol, a unit matrix that vanishes unless $k = l$, we obtain

$$\frac{d}{d\tau} g_{ij} \left(\dot{x}^i \delta_k^j + \dot{x}^j \delta_k^i \right) - \frac{\partial g_{ij}}{\partial x^k} \dot{x}^i \dot{x}^j = 0 \tag{1.281}$$

and now take

$$\frac{dg_{ij}}{d\tau} = \frac{\partial g_{ij}}{\partial x^l} \dot{x}^l \tag{1.282}$$

so with this, we find that

$$\frac{d^2 x^i}{d\tau^2} + \Gamma^i_{jk} \dot{x}^j \dot{x}^k = 0, \tag{1.283}$$

a form that is virtually identical to Eq. (1.275) but much more general because it explicitly provides a way for calculating what are now called the *connection coefficients* or *Christoffel symbols*:

$$\Gamma^i_{jk} = -\frac{1}{2} g^{im} \left(\frac{\partial g_{jk}}{\partial x^m} - \frac{\partial g_{jm}}{\partial x^k} - \frac{\partial g_{mk}}{\partial x^j} \right). \tag{1.284}$$

The quantity g^{ij} is the inverse of the metric, in this case the contravariant form of g_{ij}. To find this requires computing g, the determinant of g_{ij}, but since we are working in components this is the same as finding the determinant of a 4×4 matrix.

A tensor has the following transformation properties between coordinate bases:

$$g'_{ij} = g_{lm} \frac{\partial x^l}{\partial x'^i} \frac{\partial x^j}{\partial x'^m} \qquad (1.285)$$

so that[52]

$$g' = g \left| \frac{\partial x^l}{\partial x'^i} \right| \left| \frac{\partial x^j}{\partial x'^m} \right| = g \left| \frac{\partial x^l}{\partial x'^i} \right|^2. \qquad (1.286)$$

Therefore, for volume integrals and transformations, we have

$$dV' = dx'^0 \cdots dx'^3 = \left| \frac{\partial(x'^0, \ldots, x'^3)}{\partial(x^0, \ldots, x^3)} \right| dV = J \, dV, \qquad (1.287)$$

where J is the Jacobian transformation,[53] and therefore from Eq. (1.286) we have

$$(-g')^{1/2} \, dV' = (-g)^{1/2} \, dV. \qquad (1.288)$$

This is the *proper volume*, and it transforms as a scalar quantity. The negative sign on g ensures that the proper volume is always positive. This result is an important simplification and also leads to several definitions that will prove to be very useful. First

$$g^{ij} = \frac{1}{g} \frac{\partial g}{\partial g_{ij}}. \qquad (1.289)$$

The contraction of the Christoffel symbol is

$$\Gamma^i_{ik} = \frac{1}{2} g^{im} \frac{\partial g_{im}}{\partial x^k}, \qquad (1.290)$$

so that

$$\Gamma^i_{ik} = \frac{1}{2} \frac{1}{g} \frac{\partial g}{\partial g_{im}} \frac{\partial g_{im}}{\partial x^k} = \frac{1}{2} \frac{\partial}{\partial x^k} \ln |g| = \frac{\partial}{\partial x^k} \ln(-g)^{1/2}. \qquad (1.291)$$

This will be very handy when reducing the order of tensors yet to come. The second result is that we have a way of writing the divergence, which for a vector is

---

[52] We use here the usual law for determinants of products of matrices that $|AB| = |A||B|$.
[53] See, for example, Jeffreys, H. and Jeffreys, B. 1953, *Methods of Mathematical Physics* (Cambridge, UK: Cambridge Univ. Press); Morse, P. and Feshbach, H. 1953, *Methods of Mathematical Physics* (NY: McGraw-Hill).

a scalar quantity, in a more general form

$$\nabla \cdot \mathbf{F} = \frac{1}{(-g)^{1/2}} \frac{\partial}{\partial x^k} \left[(-g)^{1/2} F^k \right] \quad (1.292)$$

where now $F^k(-g)^{1/2}$ is called the *density* (or if **F** is a tensor, this is called the *tensor density*).[54] What we have just found is the same as

$$F^k_{;k} = \frac{1}{(-g)^{1/2}} \frac{\partial}{\partial x^k} \left[(-g)^{1/2} F^k \right] \quad (1.293)$$

the covariant form of the divergence (note the semicolon for the covariant derivative; see Section 1.C.4).

Yes, all this looks truly horrible in general form, but wait a moment before you panic. You've already explicitly computed the coefficients for several coordinate systems just a moment ago! You see that, for cylindrical coordinates, the Christoffel symbols can be read off directly from the equations of motion:

$$\Gamma^1_{11} = 0, \quad \Gamma^1_{22} = -r, \quad \Gamma^2_{12} = \Gamma^2_{21} = \tfrac{1}{2} r^{-2} \quad (1.294)$$

and all other coefficients vanish (look, for example, at Γ^3_{mn} for all m, n). This equation is called a *geodesic* and is the motion for a free particle. These are the same, for instance, as the coefficients resulting from transformations of unit vectors[55] from one coordinate system to another.

Now what this means is that we have a geometric way to describe a particle trajectory. If the right hand side of Eq. (1.283) were *not* zero, this is where the force, $\partial \Phi / \partial x^i$ would go. Now you see that the coordinate forces, the ones that appear if you are not in an inertial frame (in other words, the centrifugal and Coriolis forces) come from the deviation of the trajectory from straight line motion. It was this step that led Einstein—with substantial mathematical direction from M. Grossmann—to put in the mathematical details, to the realization that gravitation can be described in essentially geometric terms. The proper choice of frame removes gravity completely, and in that frame special relativity (inertial motion) applies; this is what is meant by a freely falling observer. In this frame, the force vanishes and only the trajectory, a purely geometric feature, remains.[56] But since we can *always* find such a frame, it is possible to generalize the equations of

---

[54] If you need to compute a divergence in non-Cartesian coordinates, this is a very useful equation to remember.

[55] There are far more elegant ways of writing all this using differential forms [e.g. see Flanders, H. 1962, *Differential Forms, with Applications to the Physical Sciences* (NY: Academic Press) and references in n.48], but the index notation is closely connected with the actual computations. Another point is that the "classical" literature, especially Eddington (1936), makes extensive use of this formalism. Ultimately, the point is that no matter how you say it, it is still a theory of of gravity that we're dealing with, not the mathematical apparatus.

[56] There is a simple demonstration of this, one you should know very well. Take a paper cup filed with water and punch a hole in the bottom. Hold onto the cup. A stream of liquid immediately issues from the bottom. Now, keeping your finger over the hole, refill the cup and then, being careful not to tip it, drop the cup by removing your finger. What happens to the water?

motion to an invariant form. This is the physical origin of *general relativity*. The field does not explicitly depend on velocity, and with the proper choice of coordinate system it simply disappears. The problem becomes finding a rule for transforming *back* to the world of the stationary observer. After all, we know that gravity exists, there is a force there even if we can seem to get rid of it (you may think you are in a frame with no gravity, but eventually the ground informs you that you're mistaken).[57]

1.C.3 The Equivalence Principle and Mach's Principle

The basis of the general theory is the *equivalence principle*, which states that all accelerated frames are the same. Put more precisely, there is no difference between gravitational and inertial mass. This has been verified by the Eötös experiment, which measures the gravitational constant as a function of position of the laboratory relative to the Sun.[58] In other words, it does not matter whether a body is in freefall in a gravitational field or undergoing acceleration for some other reason. It is always possible to transform into a frame of reference that is, locally, inertial and to find the evolution of the system in that frame. A nice demonstration of this is to hang a mass on a spring and, suddenly, move the spring upward. The mass remains initially unaccelerated.[59]

The equivalence principle is closely allied with another that was first discussed by E. Mach in the nineteenth century. In the *Principia*, Newton presented the analysis of the corotating motion of a fluid within a swinging bucket, a familiar first-year physics example. In this problem, the fluid is forced to the bottom of the bucket by the centrifugal force felt in the noninertial frame. One way of stating this is "how does the fluid 'know' that it is supposed to remain inertial?" The explanation, according to *Mach's principle*, comes from the distribution of matter in the universe at large. The overall inertial frame of all gravitating matter provides the frame of reference for the circulating fluid, relative to which its inertia is defined. This form of the principle played a role in Einstein's and Grossman's thinking in the early years of general relativity, and it has since led to the formulation of several attempts to connect the local inertial behavior of bodies with the large-scale distribution of mass through a scalar field (the so-called Brans–Dicke theory of gravitation), but none of these attempts has yielded a final picture of what inertia actually means. It is significant, however, that even at this opening stage in the discussion of general relativity theory (GRT), you are forced to take the behavior of bodies, formulated in a local region of spacetime, with respect to the cosmos at large. The connection between general relativity (hence, gravitation) and cosmology is therefore inevitable and we will exploit this in Chapter 8.

---

[57]Adams, D. 1987, *The Hitch-hiker's Guide to the Galaxy* (NY: Horizon).

[58]See Dicke, R. H. 1965, *Theoretical Significance of Experimental Relativity* (NY: Gordon and Breach); Will, C. M. 1993, *Theory and Experiment in Gravitational Physics* (Cambridge, UK: Cambridge Univ. Press) for further discussion of this and related tests of the principle.

[59]A more comic version is well known from cartoons. Imagine that you are standing on a scale with someone on the floor below you sawing through the floor around it. At an instant, you see that you have suddenly lost all weight! As with the dropped cup, the mass and the measuring device fall at the same rate. Your mass was indicated by the compression of the springs in response to a local gravitational acceleration which has been removed by your conversion into a freely falling observer.

1.C.4 Deriving the General Relativistic Equations of Motion

Now let's derive the generalized force law in light of the equivalence principle. As with ordinary dynamics, you have a choice of observing frame. The comoving one, the same as the Lagrangian coordinate system, moves with the mass and in that frame everything behaves inertially. But in the external (contravariant) frame, that's not true. So we need a way of getting from one coordinate system to another, and that is again the one that uses the connection coefficients:

$$DA^i = \partial A^i + \Gamma^i_{jk} A^j \delta x^k. \qquad (1.295)$$

This is usually written as

$$A^i_{;k} = A^i_{,k} + \Gamma^i_{jk} A^j, \qquad (1.296)$$

which defines the notation where the semicolon denotes the covariant derivative and the comma denotes the ordinary or contravariant derivative. This is the same as the convective derivative in ordinary dynamics, and means that we're on the right track to getting a formalism for general trajectories. Finally, look at what happens if we compare two different paths around a surface. You can see it this way. Imagine a boat floating on a spherical "water world" and displace it always parallel to itself around a closed path. As the boat sees it, the path is on a flat surface, but when you compare its orientation at the end, you see that a rotation results—the ship is not pointing in the same direction as when it started. This idea of parallel transport, extensively described by Fermi (1927), was one of the starting steps in deriving a completely physical picture of relativistic dynamics. The point is that the curvature or the surface has introduced a *torsion* in the system such that

$$A^i_{;j;k} - A^i_{;k;j} = R^i_{jlm} A^l \delta x^m \qquad (1.297)$$

where now R^i_{jlm} is the *Riemann curvature tensor*.[60] It is invariant under transformation of coordinates.[61]

1.C.4.1 The Riemann Curvature Tensor

The derivation of the component representation of the Riemann tensor proceeds as follows. First, it helps to have the identity

$$A^i_{j;k} = A^i_{j,k} + \Gamma^i_{jm} A^m_k + \Gamma^m_{jk} A^i_m \qquad (1.298)$$

so that you can now take

$$A^i_{;j;k} = \left(A^i_{;j}\right)_{,k} + \Gamma^i_{jl} A^l_{;k} + \Gamma^l_{jk} A^i_{;l}. \qquad (1.299)$$

---

[60] You should prove for yourself that this is indeed a tensor, even though the Christoffel symbols are not.

[61] Another point is that the metric is a divergenceless quantity. This is an alternative way of saying that there are no boundaries to the space, and that there are transformations relative to whom the metric is conserved. These are the so-called *Killing vectors*. See Robertson, H. P. and Noonan, T. 1969, *Relativity and Cosmology* (Philadelphia: Saunders) and Peebles, P. J. E. 1970, *Physical Cosmology* (Princeton: Princeton Univ. Press) for particularly clear explanations of how these are derived. We discuss them in Section 1.C.6 below.

The rest of the derivation involves more index manipulation, but you will eventually arrive at the definition of R^i_{jkl} in terms of the Christoffel symbols:

$$R^i_{jmk} = \Gamma^i_{jm,k} - \Gamma^i_{km,j} + \Gamma^i_{jl}\Gamma^l_{km} - \Gamma^i_{kl}\Gamma^l_{jm}. \tag{1.300}$$

Since the Christoffel symbols are symmetric with respect to interchange of the lower indices, there are several elementary properties for the Riemann tensor in terms of permutations of the indices. An important feature of the curvature is that is nonlinear in the connection coefficients and, consequently, the equations that describe even the simplest motion are often quite complicated in appearance. This tensor has an important symmetry, called the Bianchi identities:

$$R^i_{jkl;m} + R^i_{jmk;l} + R^i_{jlm;k} = R^i_{j(kl;m)} = 0 \tag{1.301}$$

where $(\cdots)$ is a frequently used convention that indicates symmetric permutation of the indices. This follows from the definition of the Riemann tensor in terms of the Christoffel symbols.

1.C.4.2 The Ricci Tensor and the Curvature Scalar

The final quantities we have to introduce are the Ricci tensor and the curvature scalar. The raising and lowering of indices is accomplished by projections through the metric tensor, in other words by scalar products. The *Ricci* tensor, R_{ij}, is the contraction of the Riemann tensor:

$$R_{ij} = R^n_{inj}, \tag{1.302}$$

where we sum over the repeated indices as usual. In terms of the Christoffel symbols, this is given by

$$R_{ij} = (-g)^{-1/2}\left[(-g)^{1/2}\Gamma^k_{ij}\right]_{,k} - \left[\ln(-g)^{1/2}\right]_{,ij} - \Gamma^k_{mi}\Gamma^m_{jk} \tag{1.303}$$

recalling that $\Gamma^k_{ik} = [\ln(-g)^{1/2}]_{,i}$. Finally, taking the scalar product with the metric gives us the trace of the Ricci tensor:

$$R = g^{jk}R_{jk} = R^m_m. \tag{1.304}$$

This is a scalar quantity, the *Riemann scalar*, that is more simply called the *curvature*. Because it is a scalar, it is invariant under change of coordinate systems. Now return to the Bianchi identities in Eq. (1.301). Since $g^{ij}_{;j} = 0$, it follows that

$$R_{;m} - R^i_{j;m} - R^i_{m,j} = 0 \tag{1.305}$$

and also that

$$R_{;m} = \tfrac{1}{2}g^l_m R_{;l} = R^l_{m;l} \tag{1.306}$$

With just a bit little more index manipulation we arrive at a fundamental identity:

$$\left(R_{ij} + \tfrac{1}{2}g_{ij}R\right)_{;j} \equiv G_{ij;j} = 0. \tag{1.307}$$

This is actually the equation for the gravitational field of a vacuum, when there are no source terms: $G_{ij} = 0$. It is purely geometric and actually tells you nothing about the possible effects of matter. Thus, it is similar to the Laplace equation, for which the source of the field is assumed to be pointlike compared with the space in which we make the field measurement. It is historically important to note that this is not how Einstein derived the field equations. In fact, it is just backward to what he did. Einstein recognized the importance of the relation between the classic condition for the stress tensor and a purely geometric quantity. He also knew that for a perfect fluid the conservation equation can be written in covariant form:

$$T^{ij}_{;j} = 0 \tag{1.308}$$

since it is a tensor relation and therefore formally invariant to coordinate transformation. This gives an expression for the *curvature in terms of the stress tensor*:

$$R_{ij} = \frac{8\pi G}{c^4}\left(T_{ij} - \frac{1}{2}g_{ij}T\right), \tag{1.309}$$

where T is the trace of the stress tensor. You have now come full circle, seeing how the curvature measures the stress and how the mass–energy stress tensor produces curvature. It has been a long slog, but this is the final result—a field equation for the gravitational field. For the simplest case of a vacuum solution, where the density of matter vanishes, we simply have $R_{ij} = 0$ for all i and j.

1.C.4.3 *Electromagnetism*

The stress tensor for an electromagnetic field, which renders the Maxwell equations explicitly covariant and as compact as possible notationally, can be derived rather simply. To begin, take the electromagnetic field equations to be

$$\begin{aligned}
\frac{1}{c}\frac{\partial \mathbf{B}}{\partial t} &= \nabla \times \mathbf{E} + \frac{4\pi}{c}\mathbf{J} \\
\frac{1}{c}\frac{\partial \mathbf{E}}{\partial t} &= -\nabla \times \mathbf{B} \\
\nabla \cdot \mathbf{E} &= 4\pi\rho \\
\nabla \cdot \mathbf{B} &= 0
\end{aligned} \tag{1.310}$$

and write the equations explicitly in Cartesian coordinates. Then, noting that the metric in this case is η_{ij}, the Minkowski metric, we can write the first equation, for instance, as

$$\frac{\partial B_1}{\partial x^0} - \frac{\partial E_3}{\partial x^2} + \frac{\partial E_2}{\partial x^3} = J^2.$$

By extension, we arrive at the formal representation

$$\frac{\partial F^{ij}}{\partial x^j} = F^{ij}_{,j} = s^i \tag{1.311}$$

where s^i is the current density:

$$s = (\rho, \mathbf{J})^T$$

The advantage of this formalism is that F^{ij}, the field, is a tensor and therefore Eq. (1.311) holds for any generalized metric coordinate system; that is, $F_{lm} = g_{li}g_{mj}F^{ij}$ and, in addition, we can employ the covariant derivative for the currents:

$$F^{ij}_{;j} = s^i. \tag{1.312}$$

Since the Maxwell equations also contain constraint conditions, the divergence terms, and the individual field components are given by derivatives of potentials, A_i, you can show that

$$F_{ij} = \frac{\partial A_i}{\partial x^j} - \frac{\partial A_j}{\partial x^i} = A_{i,j} - A_{j,i} \tag{1.313}$$

such that:

$$F_{(ij;k)} \equiv F_{ij;k} + F_{ki;j} + F_{jk;i} = 0, \tag{1.314}$$

introducing the new notation that $F_{(ij;k)}$ is the even permutation of the indices. The equations of motion for a charged particle become

$$mc\frac{dv^i}{d\tau} = \frac{e}{c}F^{ij}v_j \tag{1.315}$$

where v_i is the component of the 4-velocity and that from the dynamical equations for the field (the wave equation is, for instance, a four-dimensional operator on the stress). To find the stress tensor, write the Lagrangian for the field as

$$L = -\frac{1}{16\pi}F_{ij}F^{ij}$$

and take the variation with respect to the potentials A_m, using Eq. (1.313) for the definition of the field tensor. From the variational equation you will arrive at the final form for the stress tensor:

$$T^{ij} = \frac{1}{4\pi}\left[-F^{ik}F^j_k + \frac{1}{4}g^{ij}F_{mn}F^{mn}\right]. \tag{1.316}$$

Recall that raising and lowering of indices is accomplished through the metric. There isn't really anything new here, but the formalism is now generalized to an arbitrary metric (and can be extended further using nonmetric formalism) and, besides a more compact notation, highlights the physical role of the stress tensor.

1.C.4.4 The Schwarzschild Solution

For a centrally symmetric gravitational field, the external solution is a vacuum state. You already know this from the classical Laplace equation that describes the gravitational field around a mass. The field gets progressively weaker, so the departures from inertial trajectories becomes vanishingly small with increasing distance. For the relativistic treatment, you say the same thing geometrically. The curvature vanishes as one goes far enough away from the central mass, so that $g_{ij} \to \eta_{ij}$ as $r \to \infty$. It greatly simplifies the calculation of the metric, since the coefficients depend only on the radial coordinate, r, for an isotropic spacetime. They must be independent of time. The angular terms do not matter—the spherical symmetry of a point mass is guaranteed by meridians taken around the mass (trajectories at minimum angular momentum are circular orbits) and angular momentum is conserved (as intuitively it must be in a two body problem). Therefore

$$ds^2 = A(r)dt^2 - B(r)dr^2 - r^2 \, d\Omega^2, \tag{1.317}$$

where $d\Omega^2 = d\theta^2 + \sin^2\theta \, d\phi^2$, and A and B are simple functions of the radial coordinate. Such metrics crop up repeatedly in astrophysical problems. The most important is the Schwarzschild metric, the solution for an isolated, nonrotating point mass in a vacuum. The more general description of a uniform distribution of mass, one satisfying global homogeneity and isotropy and also constant curvature (the same as uniform density) is the Robertson–Walker metric, which is the basis of the line element in Big Bang cosmology (see Chapter 8).

It is, of course, possible to derive the Schwarzschild metric in the usual way by substituting Eq. (1.317) into the Euler–Lagrange equations and solving the resulting set of equations. A more intuitive approach uses a lovely demonstration by Sommerfeld[62] that makes use of the equivalence principle in a neat way. As originally applied to the hydrogen atom, but later generalized to the motion of planets, you can think of the metric connection with gravitation as coming from the binding energy of the orbiting body and its change in effective mass as a function of velocity. As a result, the angular velocity variations within an elliptical orbit change the mass as a function of orbital phase and looks just like a variation in the specific angular momentum with phase. It is strictly periodic, but the effect is to produce an apparent torque on the orbiting mass, which causes the orbit to precess even for the Kepler problem. This connects with the gravitational potential through the virial theorem, since the mass is related to the rest mass by $m = \gamma m_0$ and the Lorentz factor is given by

$$\gamma \to \left(1 - \frac{2GM}{rc^2}\right)^{-1/2}. \tag{1.318}$$

Then through the Lorentz transformations, we have the simple approximation for

---

[62] Sommerfeld, A. 1946, *Electrodynamics* (NY: Academic Press).

the proper interval:

$$ds^2 = \left(1 - \frac{2m}{r}\right)dt^2 - \frac{dr^2}{1 - 2m/r} - r^2\,d\Omega^2, \qquad (1.319)$$

where now $m = GM/c^2$ is called the *Schwarzschild radius*, since the quantity m is either a geometric mass or length unit. The problem with the metric comes from the radial dependence of the metric coefficients for the time and radial coordinates. There is a singularity here, where the proper length becomes infinite and, even for a finite time interval in the covariant frame, the external observer sees an infinite time interval. This is the *Schwarzschild singularity*, the origin of the concept of a *black hole*.

The details are as follows. Following Schwarzschild's original notation. we write

$$ds^2 = e^{2\nu}\,dx^{02} - e^{2\lambda}\,dr^2 - r^2\,d\theta^2 - r^2\sin^2 d\phi^2 \qquad (1.320)$$

with the requirement that ν and λ depend only on r, not on time. This formal choice is merely a convenient form. We could have used any two functions of radius alone. For this metric, the only nonvanishing components of the Christoffel symbols are

$$\begin{aligned}
\Gamma^0_{01} &= \Gamma^0_{10} = \nu' \\
\Gamma^1_{00} &= \nu' e^{2(\nu-\lambda)} & \Gamma^1_{11} &= \lambda' \\
\Gamma^1_{22} &= -re^{-2\lambda} & \Gamma^1_{33} &= -r\sin\theta\, e^{-2\lambda} \\
\Gamma^2_{12} &= \Gamma^2_{21} = \frac{1}{r} & \Gamma^2_{33} &= -\tfrac{1}{2}\sin 2\theta \\
\Gamma^3_{13} &= \Gamma^3_{31} = \frac{1}{r} & \Gamma^3_{23} &= \Gamma^3_{32} = \cot\theta.
\end{aligned} \qquad (1.321)$$

Since we're looking for the vacuum solution, we can use $R_{ij} = 0$ and therefore

$$\begin{aligned}
R_{00} &= \left[-\nu'' + \lambda'\nu' - \nu'^2 - \frac{2\nu'}{r}\right]e^{2(\nu-\lambda)} = 0 \\
R_{11} &= \left[\nu'' - \lambda'\nu' + \nu'^2 - \frac{2\lambda'}{r}\right] = 0 \\
R_{22} &= [1 + (\nu' - \lambda')]e^{-2\lambda} - 1 = 0 \\
R_{33} &= R_{22}\sin^2\theta = 0;
\end{aligned} \qquad (1.322)$$

the last Ricci component is really redundant and can be ignored. Now we use the equations of motion to derive the limit for this last set of equations ($R_{22} = 0$). Take the weak-field limit of the metric. In this case, $g_{ij} \to \eta_{ij} + h_{ij}$. Then the

Christoffel symbols become

$$\Gamma^k_{ij} \to -\tfrac{1}{2}\eta^{km}(h_{ij,m} - h_{jm,i} - h_{im,j}).$$

Since the radial equation far from the mass must be given by the Newtonian formalism

$$\frac{d^2 r}{dt^2} = -\frac{\partial \Phi}{\partial r}, \qquad (1.323)$$

we can translate this into the geodesic form for an affine parameter s through

$$\ddot{x}^1 + \Gamma^1_{00}\dot{t}^2 = 0,$$

where:

$$\Gamma^1_{00} = -\tfrac{1}{2}\eta^{11}h_{00,1}.$$

By comparison with the equation of motion, we find that the metric perturbation is $h_{00} = 2\Phi/c^2$ so that

$$g_{00} \approx 1 + 2\Phi = 1 - \frac{2GM}{rc^2}. \qquad (1.324)$$

Now returning to $R_{00} + R_{11}$, we have

$$\nu' + \lambda' = 0, \qquad (1.325)$$

which provides the relation between the metric coefficients, yielding finally

$$ds^2 = \left(1 - \frac{2GM}{c^2 r}\right) dx^{0\,2} - \left(1 - \frac{2GM}{c^2 r}\right)^{-1} dr^2 - r^2 d\theta^2 - r^2 \sin^2\theta\, d\phi^2, \qquad (1.326)$$

which fills out our more heuristic derivation. Notice that because we are looking at the static vacuum field outside a point mass, M is constant. For the interior solution, the problem is complicated by the need for an auxiliary condition on the mass since it is now a function of radius. The detailed derivation follow the same lines we've just used. We will postpone them until Chapter 4, where the interior solution will be needed to describe neutron star structure.

Notice that for a Schwarzschild metric, the equations of motion are altered from those in a classical two-body problem. First, there is a new term, one that does not appear in the dynamics as we described them in section 2, above which looks like a tidal force that is of order $g(r)m/r$. Here $g(r)$ is the point gravitational acceleration. The effect of this perturbation is to produce a precession of the isolated two-body orbit, which is again a manifestation of Bertrand's theorem (the restrictive conditions necessary for a closed orbit in a two-body problem) since the potential now deviates from r^{-1}. It is a small effect and can be added as a perturbation for the solar system because the mass of the Sun is so small and the orbital semimajor axes are so large compared with m. Also, the solar radius is so large, of order 10^5 m. For collapsed bodies, however, where $R \sim m$ such as neutron stars, this is no longer true and general relativistic effects dominate the dynamics of such bodies even when their masses are relatively small. You can write the Lagrangian for the motion in this coordinate system as

$$L = \frac{1}{2}\left(\frac{ds}{d\tau}\right)^2 - \Phi(r) \tag{1.327}$$

relative to a proper time τ, with Φ representing the central gravitational potential. Then the equations for planar motion are

$$\frac{d}{d\tau}\left(1 - \frac{2m}{r}\right)\dot{t} = 0, \tag{1.328}$$

since the time measured in the inertial and orbiting frame must differ and be brought into correspondence. For the radial acceleration we obtain

$$\frac{d}{d\tau}\left(1 - \frac{2m}{r}\right)^{-1}\dot{r} = \frac{1}{2}\left[\dot{t}^2\frac{2m}{r^2} - \left(1 - \frac{2m}{r}\right)^{-2}\frac{2m}{r^2}\dot{r}^2\right] - r\dot{\phi}^2 - \frac{\partial\Phi}{\partial r}, \tag{1.329}$$

taking $\theta = \pi/2$ and assuming that $\dot{\theta} = 0$. Finally, the azimuthal equation is simply the conservation of orbital angular momentum, as it must be for a central field:

$$\frac{d}{d\tau}r^2\dot{\phi} = 0. \tag{1.330}$$

Although these look formidable, there is a beautiful feature in the radial equation. The time dilation depends on the gravitational potential through $2m/r$, although this is a very small number for bodies in the solar system (even for a body orbiting at $1R_\odot$, this is only 2×10^{-6}, which qualifies as small), and it adds a small perturbation to the classical equation for the radial acceleration of that depends on r^{-3}. This looks precisely like a torque, although the angular momentum is still conserved, and means that the orbit does not exactly close after one cycle. The deviation is just the observed value for Mercury, about 41 arcsec per century, for the excess precession compared with the rate expected for an isolated two-body system.

This derivation also yields the *gravitational redshift*, one of the first classical tests of general relativity. The change in the frequency of a photon depends on the Doppler equation through $\omega' = \omega(1 - \boldsymbol{\beta} \cdot \mathbf{n})$, and therefore, for $v \to c$, $\Delta\omega = \Phi/c^2$. In other words

$$\frac{\Delta\lambda}{\lambda} = \frac{GM}{rc^2} \tag{1.331}$$

for a point mass. For the Sun, this is about 10^{-5}, a measurable quantity. For more compact stars, such as white dwarfs, the measurement becomes more straightforward, amounting to about 42 km s^{-1} as a systematic shift for a star of 1 $M_\odot$ and 0.01 $R_\odot$. Let's see how two observers at different points relative to the central mass should see a redshift for a light signal propagating outward from the depths of the gravitational field. This happens along trajectories having $ds = 0$, called *null geodesics*, that are the paths traveled by massless particles in the spacetime. Photons qualify, of course, but so may neutrinos and other massless fields (although more recent experimental evidence points strongly to a small but not zero rest mass for neutrinos). Then time dilation results from propagation

$$dt' = \left(1 - \frac{2m}{r}\right)^{1/2} dt, \qquad (1.332)$$

in the same way that the kinematic theory produces a Doppler shift. Another consequence for such particles is that they should be deflected by the gravitational field, due to the curvature, of order

$$\Delta \theta = \frac{4m}{r} \qquad (1.333)$$

This is the physical basis of a *gravitational lens*, although in this form we can only apply it directly to a point mass. We will return to the more general case of extended mass distributions when we come to cosmology, in Chapter 8, but we can already make some use of the concept. Newtonian theory also predicts a deflection for a body of arbitrarily small (but not zero) rest mass, but the value is only half of the relativistic result, of order 0.87 arcsec at the solar limb. The detection of gravitational deflection of starlight during the 1919 solar eclipse by Eddington and Dyson provided the first astronomical test of general relativity.[63]

The reduction of the order of the equations gets us from the general curvature tensor to the one immediately connected with gravitation. To see why this happens, recall that the Poisson equation relates the inverse of the curvature to the density of matter:

$$\frac{\partial^2 \Phi}{\partial x^i \, \partial x^i} = -4\pi G \rho. \qquad (1.334)$$

A simple way to see where the connection come in for mass is to consider the equations of motion with a gravitational field:

$$\frac{d^2 x^i}{dt^2} = -\frac{\partial \Phi}{\partial x^i}. \qquad (1.335)$$

---

[63] See Pais, A. 1982, '*Subtle is the Lord*' (Oxford: Oxford Univ. Press) for the most complete description of this observation and Will (n.69) for additional description of more recent gravitational deflection experiments. We should add, however, that the correct derivation of the known result for Mercury's orbital precession excess was *really* the first test of the theory. Here we would like to make a philosophical aside. It sometimes happens in astrophysics that theoretical predictions are possible. In general, however, the real successes come from general theories that contain the explanations of previously paradoxical results. Relativity is one of these, the Saha equation is actually another in its explanation of the systematics of stellar spectra and the determination of elemental abundances (Payne 1927). Yet another is the explanation of radio emission from the Crab nebula by synchrotron emission (Shklovsky 1952). Predictions are much less frequent, for instance, the 21-cm line (van der Hulst 1943), gravitational collapse (Chandrasekhar 1933, Oppenheimer and Volkoff 1939), Big Bang nucleosynthesis and the cosmic background radiation (Gamow 1946; Alpher, Follin, and Herman 1953), and gravitational lenses (Zwicky 1933). A third type of model, the phenomenological sort, is constructed for specific purposes—to explain observed effects in terms of scaling properties to known objects—but also has predictive importance in providing the seed material for new observations. Most astrophysical theory falls into this last category.

1.C.5 Horizons and Black Holes

For a null geodesic, the Schwarzschild metric contains a surface at a radius $2m = 2GM/c^2$ at which the redshift seen by an inertial observer becomes infinite. We will denote this radius as R_s, called the Schwarzschild radius. Although the proper time for a freely falling observer is finite, and this poor soul will die rather quickly, any time interval seen by the person watching safely from the inertial frame will require an infinite time to pass before this event is observed. This is called the *event horizon*. A signal will not be able to reach an observer no matter how close she is to the horizon. The object enclosed within this event horizon is a *black hole*. Notice that the horizon is not a "material" surface. It represents a deceleration only to an inertial observer, but this means that the velocities in the vicinity of this radius become comparable to the speed of light. An approximate way of stating the condition for the body is that its gravitational binding energy becomes comparable to its rest mass. If we define a dimensionless number, which we will call the Einstein number, to be this ratio:

$$\mathscr{E} = \frac{1}{2} \frac{|\Omega|}{Mc^2}, \qquad (1.336)$$

then for a classic value of the binding energy for a nonrotating point mass, $\Omega = -GM^2/2R_s$, then $\mathscr{E} \approx 4$. It can be shown that the last stable circular orbit possible in this geometry is at $r = 6m = 3R_s$, so for a tighter orbit that results from viscous dissipation accretion onto the central body is inevitable. In effect, there must be a last stable orbit because if the periasteron is too close to the horizon the change in the particle's mass and angular momentum cause the orbit to spiral.

For a rotating point mass, that is one that is completely contained within R_s, the spacetime near the event horizon is affected by the rotation. In particular, it is now impossible for a body to remain at rest with respect to a distant observer. This is the Kerr solution,[65] discovered nearly 50 years after Schwarzschild's solution. The rotation also changes the structure of the horizon. At the poles, there is only one event horizon, which is identical with the particle crossing R_s. At the equator, there are now two separate regimes, $r_{\pm} = m \pm (m^2 - a^2)^{1/2}$, where a is the ratio of the rotational kinetic energy of the central mass to its gravitational potential.

---

[64] Price, R. H. 1982, *Amer. J. Phys.*, **50**, 300; Frankel, T. 1979, *Gravitational Curvature: An Introduction to Einstein's Theory* (San Francisco: Freeman); Misner, C., Thorne, K., and Wheeler, J. A. 1973, *Gravitation* (San Francisco: Freeman). For the most recent measurement of the solar redshift, see Lopresto, J. C., Schrader, C., and Pierce, A. K. 1991, *ApJ*, **376**, 757.

[65] See Chandrasekhar, S. 1983, *The Mathematical Theory of Black Holes* (London: Oxford Univ. Press) for a very complete discussion of the dynamical solutions in this metric. The discovery papers are Kerr, R. P. 1963, *Phys. Rev. Lett.*, **11**, 237 and Kerr, R. P. in Robinson, I. and Schücking, E. 1965, in *Quasistellar Sources and Gravitational Collapse* (Chicago: Univ. Chicago Press). This last reference is one of the most interesting historical documents of modern astrophysics, recording the proceedings of the first of what would become known as the Texas Symposia on Relativistic Astrophysics.

This intermediate region is called the *ergosphere*. It is a region from which a body can escape, having extracted rotational energy from the hole.[66]

We will return to the more detailed astrophysical implications of the existence of black holes in Chapters 5 and 8, but here we should point out one important feature of this inner radius. Imagine a Keplerian disk surrounding a black hole. In the comoving frame of the disk matter, collisions will occur relativistically. Therefore the energy released through such collisions, given the depth of the potential well, will have a temperature of order

$$T \approx \frac{GM}{kR_s} \approx \frac{c^2}{2k} = \text{constant}, \qquad (1.337)$$

independent of the mass of the collapsed body. All properties that depend on the surface area now depend only on the mass of the hole.[67] For our purposes, and for many applications in astrophysics, the only important feature of black holes is that they are collapsed objects for which the gravitational potential at the event horizon, or Schwarzschild radius, is independent of the mass and is always $\frac{1}{2}c^2$.

1.C.6 Symmetries and Killing Observers

This title of this section seems bloodthirsty, but it's not.[68] The metric is a function of the coordinate system, but there are some special trajectories along which it remains invariant. These have special significance for general relativity because they correspond to inertial observers. Recall that the metric, which is a tensor, transforms as

$$g'_{kl} = g_{ij} \frac{\partial x^i}{\partial x'^k} \frac{\partial x^j}{\partial x'^l}. \qquad (1.338)$$

Imagine that we choose our coordinate system such that $\mathbf{x} = \mathbf{x}' + \boldsymbol{\zeta}$ is a displacement (note that here the boldfacing does not restrict the coordinate system to three dimensions). Now we can find the class of trajectories that will leave the metric invariant to any observer. This is called an *isometry* and represents the path along with the distances between points remains constant:

$$g_{ij}(\mathbf{x}) = g_{ij}(\mathbf{x}' + \boldsymbol{\zeta}) = g'_{kl}(\mathbf{x}). \qquad (1.339)$$

Expanding g_{ij} we obtain $g_{ij}(\mathbf{x}' + \boldsymbol{\zeta}) = g_{ij}(\mathbf{x}') + g_{ij,m} \zeta^m$ and using Eq. (1.338), we

---

[66] This is called the Penrose process. See Penrose, R. and Floyd, R. M. 1971, *Nature Phys. Sci.*, **229**, 177 and Misner, Thorne, and Wheeler (n.65). for additional discussion.

[67] This is beautifully summarized in a remark by Chandrasekhar (n.65) that "Black holes are the most perfect macroscopic bodies there are in the universe: the only elements in their construction are our concepts of space and time." To this one should add that their properties are also governed by an amazingly small set of physical parameters, the mass, charge, and angular momentum of the body, and nothing else.

[68] Burke, W. L. 1985, *Applied Differential Geometry* (Cambridge, UK: Cambridge Univ. Press); Crampin, M. and Pirani, F. A. E. 1986, *Applicable Differential Geometry* (Cambridge, UK: Cambridge Univ. Press). The name refers to Wilhelm Killing (1847–1923). See Hawkins, T. 1982, *Arch. Hist. Exact Sci.*, **26**, 126 for a historical discussion of Killing's work on Lie algebras.

obtain (using $x^i_{,j} = \delta^i_j$):

$$g_{kl,m}\zeta^m + g_{ml}\zeta^m_{,k} + g_{km}\zeta^m_{,l} = 0. \qquad (1.340)$$

These are Killing's equations. We can greatly simplify Eq. (1.340) by noting that $\zeta_i = g_{ij}\zeta^j$. Then

$$g_{ij,k}\zeta^k + g_{lj}\zeta^l_{,i} + g_{il}\zeta^l_{,j} = 0. \qquad (1.341)$$

Applying the chain rule of differentiation and the transformation from contravariant to covariant coordinates, you will arrive at

$$\left(g_{lj}\zeta^l\right)_{,i} + \left(g_{il}\zeta^l\right)_{,j} + (g_{ij,l} - g_{lj,i} - g_{il,j})\zeta^l$$
$$= \zeta_{i,j} + \zeta_{j,i} + 2(g_{ij,k} - g_{il,j} - g_{lj,i})g^{lk}\zeta_k, \qquad (1.342)$$

which, using definition of the Christoffel symbol, becomes

$$\zeta_{i;j} + \zeta_{j;i}. \qquad (1.343)$$

This form connects the isometry represented by the Killing vectors with the Riemann tensor and curvature:

$$\zeta_{i;k;j} - \zeta_{i;j;k} = R^l_{ijk}\zeta_l. \qquad (1.344)$$

In many astrophysically interesting cases the metric is diagonal such as, for instance, the Schwarzschild solution for a point mass. In addition, as in that case, the metric may be independent of time and depend only on r and θ. In such a case, the Killing equations become

$$g_{rr,r}\zeta^r + 2g_{rr}\zeta^r_{,r} = 0, \qquad (1.345)$$

$$g_{\theta\theta,r}\zeta^r + 2g_{\theta\theta}\zeta^\theta_{,\theta} = 0, \qquad (1.346)$$

and

$$g_{rr}\zeta^r_{,\theta} + g_{\theta\theta}\zeta^\theta_{,r} = 0. \qquad (1.347)$$

For a spherically symmetric distribution of mass, the ϕ coordinate always provides an isometry. You can see this by thinking of the displacement of a mass around a parallel of latitude, or any great circle, on a sphere. The curvature is invariant along such a trajectory and, therefore, g_{ij} must also be invariant for displacement in the ϕ direction. We will return to this point in Chapter 8 when we derive the cosmological solution for an isotropic homogeneous spacetime, the Friedmann–Robertson–Walker metric.

1.C.7 Static Curvature: Gravitational Lenses

Classical gravitation allows light to follow curved trajectories provided the mass of a "particle" of light doesn't vanish. The calculation is essentially the same as any hyperbolic trajectory for a body deflected by a point mass. A straightforward estimate of the deviation is provided by the ratio $\Delta p_\perp/p$ at the solar limb, which is $2GM/c^2$ for a body of arbitrarily small mass traveling at velocity c. For general relativity, the metric connects acceleration to variations in both the space and time coordinates, and even a massless particle undergoes deflection. The geodesic equation is

$$\ddot{x}^i + \Gamma^i_{00} c^2 = 0, \tag{1.348}$$

where we use the Schwarzschild metric to represent the field of a point mass, leading to a deflection of

$$\Delta \theta = \frac{4GM}{c^2}, \tag{1.349}$$

At 1 $R_\odot$, this is 1.75 arcsec, precisely twice the Newtonian value. Note that this is a unique value because of the point source approximation. For viewing of a source precisely on-axis, every trajectory grazing the stellar limb will be deflected by the same angle and a ring appears, sometimes in the literature called an *Einstein ring*. For larger impact parameters, the ring becomes an arc and for large enough distances, the source appears to be displaced but essentially undistorted. An extended mass can be treated similarly if it is spherically symmetric. We use again the property of uniform spheroids, that only the interior mass affects a trajectory at impact parameter r. This means that at any specified distance from the deflecting mass, several trajectories may converge. In other words, an extended mass can form multiple images. We will return to this in Chapters 7 and 8. We note here that for a mass with density ρ, the larger the impact parameter of a trajectory and the larger its deflection. Therefore, the radius of the ring decreases with increasing distance. An off-axis trajectory may produce multiple images depending on the mass distribution.

1.C.8 Dynamic Curvature: Gravitational Waves

So far, we have been dealing with static mass distributions. To a distant observer, however, once a mass moves, it creates a time-variable field that should be a way of transferring momentum and energy. The process is similar to a moving charged particle. Because of the finite speed of light, there is a delayed signal that looks like a wave for a periodic motion.

The gravitational perturbation looks like a local change in the inertial frame of the observer.[69] In effect, it produces the sort of change you would have in the structure of a local patch of spacetime if you produced a new small scale length gradient in the gravitational field. It is a tidal acceleration, a differential force across any mass, that is carried by the curvature tensor. That is why we spent so much time on the derivation of this term. Even a vacuum must feel the wave, although it is only through the response of a nonspherical mass distribution that this wave can be detected. You can begin by assuming that the metric now consists of a static and time-dependent component, $g_{ij} \rightarrow g_{ij} + \epsilon\, h_{ij}$, where the static field

---

[69] Weber, J. 1961, *General Relativity and Gravitational Waves* (NY: Interscience); Bonazzola, S. and Marck, J.-L. 1994, *Ann. Rev. Nucl. Part. Sci.*, **44**, 655; Will, C. M. 1994, *Theory and Experiment in Gravitational Physics*, 2nd ed. (Cambridge, UK: Cambridge Univ. Press); Thorne, K. S. 1987, in *Three Hundred Years of Gravitation*, eds. S. Hawking and W. Israel, (Cambridge, UK: Cambridge Univ. Press); Thorne, K. S. 1980, *Rev. Mod. Phys.*, **52**, 285, 299.

at large distance from a mass is, by assertion, asymptotically flat, η_{ij}. This means that the field equations, normally terribly nonlinear in form, are linearized. You can already guess what will happen. The Riemann tensor depends on the second derivatives of the metric perturbation, so you expect that a wave equation results. This time dependent solution of the gravitational field therefore resembles an outwardly propagating wave supported by the gravitational field of the Universe at large by Mach's principle.

We will need a few quantities to simplify the calculation. The inverse of the metric depends, as we have said, on the metric scalar, defined by $\mathrm{Det}(g_{ij})$ which, for the perturbed metric we are adopting, becomes:

$$\mathrm{Det}(g_{ij}) = |g_{ij}| = 1 - \epsilon Tr(h_{ij}) \tag{1.350}$$

to order ϵ. Thus $\ln h \approx -\epsilon Tr(h_{ij})$ to first order. For the evaluation of the Christoffel symbols you can use $g^{ij} = \eta^{ij}$. To first order in ϵ, the Riemann tensor is

$$R^i_{jkl} = \Gamma^i_{kj,l} - \Gamma^i_{jl,k}$$
$$= -\tfrac{1}{2}\eta^{im}\left[h_{mj,k,l} + h_{mk,j,l} - h_{jk,m,l} - h_{mk,l,k} - h_{ml,k,k} + h_{jl,m,k}\right] \tag{1.351}$$

since the zeroth-order term is the constant Lorentz metric η_{ij}. The Ricci tensor is

$$R_{jl} = \Gamma^s_{sj,l} - \Gamma^s_{jl,s}. \tag{1.352}$$

Remember that the metric is being evaluated far from the source of the gravitational radiation and in the linearized approximation. At the origin, where the field is strong, the detailed waveform is more difficult to compute and we will not show that result here (see the references at the beginning of this section). Since a gravitational field cannot radiate as a dipole, the next order is the quadrupole. Omitting the details,[70] the source luminosity is given by

$$\frac{dE}{dt} = -\frac{G}{45c^5}\left(\frac{d^3}{dt^3}Q\right)^2, \tag{1.353}$$

where Q is the quadrupole moment:

$$Q_{ij} = \int \rho\left(3x_i x_j - \delta_{ij}x^2\right) dV. \tag{1.354}$$

The displacement produced in a test mass from the passage of a gravitational wave is transverse to the wavevector and two-dimensional. There are two states of polarization, called $+$ and $\times$ because of their orientation. Gravitational collapse, as well as motion of a very short period binary system, will look to an external observer like a variation in the local gravitational field. Because the amplitude at a distance is expected to be small, even though extremely massive bodies are responsible for generating the motion in the first place, sensitive detectors are required.

---

[70] An especially clear derivation is provided in Landau, L. and Lifshitz, E. M. 1971, *The Classical Theory of Fields* (Oxford: Pergamon Press), p. 323ff.

The first attempts to detect gravitational waves used monolithic cylindrical detectors with piezoelectric sensors that were designed to directly observer the changes in the shape of the cylinder due to the polarization of the incoming perturbation. These detectors, following a design originally due to Weber (see n.69), were hampered by low sensitivity and the difficulty in isolating them from extremely small-scale kicks and knocks from terrestrial sources. The fundamental limitation in these detectors is the unknown nature of the noise at the level of deformation that must be detected. The noise must be very different than gaussian and therefore is very hard to remove from the data. To get around the problems of the single detector and to extend the range of frequencies to which the detector is resonant, several newer designs are now being implemented. Most are based on long baseline laser interferometry, in which the detector is essentially a Michelson interferometer where the length of the arms physically varies and the signal of this change is a fringe shift in two beams directed along the orthogonal arms. Several projects are currently under construction to detect such signals, with the Laser Interferometric Gravity Wave Observatory (LIGO) and VIRGO (for the Virgo cluster of galaxies) being the largest and closest in design. LIGO has two 4-km-long arms and has detectors sited at Hanford, Washington and Livingston, Louisiana. VIRGO has two 3-km-long arms and is being constructed outside Pisa. GEO600, a two-arm 0.6-km array near Hamburg, Germany, is already operational. Eventually these will all be operated in joint observing mode for confirmations. Two additional test facilities are also under construction, one 0.4 km detector in Australia near Perth and a 0.3 km detector, TAMA 300, near Tokyo. Studies are also under way for LISA, a space based six satellite mission with an effective baseline of 5×10^6 km in heliocentric orbit but this is in the future (after around 2015, if ever). Ultimately, as in the case for neutrino detectors, the cooperative functioning of these detectors is essential. Perhaps more than any other field of astrophysics, gravitational wave astronomy beautifully emphasizes the international nature of contemporary astrophysics.[71] But what of the timescales? Since the signal from ordinary accretion onto black holes is unlikely to be detected, something extraordinary and catastrophic must occur. For a coalescing neutron star binary (see Chapter 5 for how this might happen), the emitted signal will be a chirp in the 0.1–1-kHz range with increasing frequency as the stars spiral inward and tidal forces become dominant in their structure. The very-low-frequency emission that accompanies the orbital motio of cataclysmic variables would not be detected (this is below the seismic frequency and dominated by local effects below ~ 10 Hz). Similarly one would expect to see the signature of merger of black holes and anything else at about the same frequency. And when first formed, neutron stars and pulsars should emit in this range.

1.C.9 Connections with Global Structures: A Comment

Although GRT deals with local changes and provides a description of gravity by building outward from a local field, Einstein realized the cosmological implications of the work very early (Einstein 1917). The application of isotropy and homogeneity to the catalog of assumptions about the mass distribution produces a way of

---

[71] Recent reviews include Robertson, N. A. 2000, *Class. Quant. Grav.*, **17**, R19; Ju, L., Blair, D. G., and Zhao, C. 2000, *Rep. Prog. Phys.*, **63**, 1317.

extending the construction of a metric to the entire spacetime. As we will discuss in Chapter 8, you pay a price for this freedom with a very large number of possible solutions. Even worse, you have an additional degree of freedom that was absent in the theory up to this point. Since you know that the metric is divergenceless, you can add a term Λg_{ij} to the field equations without any penalty, provided Λ is constant. This is the point at which the *cosmological constant* appeared in the theory and the difficulties wrought by this step we will describe when we go through the derivation of the Friedmann–Robertson–Walker solution for a homogeneous universe in Chapter 8.[72]

---

[72] Additional references: Beem, J. K. and Parker, P. E. 1990, *J. Math. Phys.*, **31**, 819; Mashhoon, B. 1977, *ApJ*, **216**, 591; Mashhoon, B. 1975, *ApJ*, **197**, 705.

2 The Raw Material: Instruments and Observations

It is a capital mistake to theorize before one has data.
—Arthur Conan Doyle, *Scandal in Bohemia*.

It is also a good idea not to put overmuch confidence in the observational results that are put forward until they are confirmed by theory.
—A. S. Eddington

2.1 THE ROLE OF INSTRUMENTS

Setting up standards for astrophysical bodies is rendered harder because they are outside the atmosphere and can be observed only after the effects of the atmosphere have been removed. Astrophysics is the art of extracting sensible conclusions from noisy data. It's also a science for which nearly all information is gathered by remote sensing rather than by laboratory experiments, other than those nature runs in unscheduled ways. Consequently, in order to set the rest of our discussion in context, this chapter deals with some aspects of instruments and observations. We don't pretend to comprehensively cover the subject in depth, that is not the point of the discussion. We want instead to expose the similarity between different areas of observational astrophysics. In this chapter, you will get some introduction to the methods and their limitations to help clarify various rules for theory testing and methods used for devising phenomenological or fundamental explanations for cosmic phenomena.[1]

Instruments have been employed by astronomers practically from the beginning of the discipline. We hesitate to use the word *science* to describe the earliest activity, since beyond its origins in maintaining the calendar, the function of astronomical observation was largely spot checking of positions to provide appropriate parameters for geometric models. Quantitative astronomical measurements

---

[1] For the general theory of individual instruments, see Born, M. and Wolf, E. 1999, *Principles of Optics*, 7th ed. (Cambridge, UK: Cambridge Univ. Press); Wolf, E. and Mandel, L. 1994, *Coherence in Optics* (NY: Cambridge Univ. Press); Kuiper, G. P. and Middlehurst, B., eds. 1960, *Telescopes: Stars and Stellar Systems*, vol. 1 (Chicago: Univ. Chicago Press); Hiltner, W., ed. 1962, *Astronomical Techniques: Stars and Stellar Systems* vol. 2 (Chicago: Univ. Chicago Press); Lena, P. 1988, *Observational Astrophysics* (Berlin: Springer-Verlag); Schroeder, D. J. 1987, *Astronomical Optics* (San Diego: Academic Press); Walker, G. A. 1987, *Astronomical Observation: An Optical Perspective* (Cambridge, UK: Cambridge Univ. Press); Smith, R. C. 1995, *Observational Astrophysics* (Cambridge, UK: Cambridge Univ. Press); Rieke, G. H. 1996, *Detection of Light: From the Ultraviolet to the Submillimeter* (Cambridge, UK: Cambridge Univ. Press); Kitchin, C. R. 1998, *Astrophysical Techniques*, 3rd ed. (Philadelphia: Institute of Physics Publ.).

began with the simplest geometric devices used by the Egyptians, Babylonians, and Chinese observers mainly for positional and chronometric purposes. Greek astronomy concentrated on building self-consistent models within the kinematic framework, using an untestable dynamical principle to account for the motions—a point to which we will return in a moment—but systematic observation was not one of its concerns.[2] Such scrutiny of cosmic bodies did not begin until well into the second millennium with the establishment of observatories in Beijing (ca. 1200), Samarkand (by Ulugh Begh ca. 1400), India (by Jai Singh ca. 1700), and Hven in Denmark (by Tycho Brahe ca. 1580).[3] Positional astronomy, and its modern incarnation—astrometry—provides a foundation on which rests much of astrophysics. It has, however, become progressively more dependent on theory and as such merges with the second type of observation, *testing predictions* of specific models.

Only kinematic properties of cosmic bodies could be studied before the introduction of the telescope by Galileo in 1609. The rate of discovery represented by the use of this new instrument was staggering, allowing for the first time the exploration of physical properties of celestial objects. Within three years, he described the terrestrial appearance of the lunar surface, including almost all of the now accepted selenological features, the existence and periodicity of the Jovian moons, the rings of Saturn, sunspots, and most important of all, the resolution of the Milky Way and clusters into component stars.[4] Improvements in this instrument were rapid. It is interesting to note that Galileo also realized that geometrical imperfections in the telescope objective could limit the resolution of the instrument and developed the first aperture stop to mask the edges of his lenses. Although it reduced the brightness limit, the quality of the image was significantly improved. This procedure was subsequently adopted in all early telescopes, permitting the very large focal lengths ultimately achieved by the end of the seventeenth

---

[2] For a superb, interpreted translation of the masterpiece of ancient astronomy, see Toomer, G. 1979, *Ptolemy's Almagest* (Princeton: Princeton Univ. Press). The best introductions to pre-mediaeval Western mathematical astronomy are Evans, J. 1998, *The History and Practice of Ancient Astronomy* (London: Oxford Univ. Press). and Neugebauer, O. 1959, *The Exact Sciences in Antiquity* (NY: Dover Books). For instruments, see Laird, E. and Fischer, R. 1995, *Pèlerin de Prusse on the Astrolabe* (Binghampton, NY: Mediaeval and Rennaisance Texts and Studies) and the excellent introductory article by North, J. D. 1974, *Sci. Am.*, **230**(1), 96. A more complete, though more technical, treatment is Zinner, E. 1956, *Deutsche und Niederländische Astronomiche Instrumente des 11.-18. Jahrhunbderts* (München: C. H. Beck'sche Verlagsbuchhandlung). For an introduction to early Chinese astronomy, start with Needham, J. 1957, *Science and Civilization in China*, vol. 3: *Science of Heaven and Earth* (Cambridge, UK: Cambridge Univ. Press).

[3] See Thoren, V. 1991, *Lord of Uraniborg* Cambridge, U.K.: Cambridge Univ. Press; Raeder, H. and Strömgren, B., eds. 1946, *Tycho Brahe's Description of His Instruments and Scientific Work: Astronomiae Instauratae Mechanica* (Copenhagen: Enjar Munksgaard). See also Christensen, J. R. 1999, *On Tycho's Island: Tycho Brahe and His Assistants, 1570–1601* (Cambridge, UK: Cambridge Univ. Press).

[4] Galileo's observational writings, collected in Drake, S. 1957, *Discoveries and Opinions of Galileo* (NY: Anchor Books), provide polemical justifications for experimental procedure that are still well worth reading, and his description of the observations is quite exciting. For further discussion of his instruments and methods, see Rosen, E. 1947, *The Naming of the Telescope* (NY: Henry Schuman); McMullan, E. ed. 1967, *Galileo: Man of Science* (NY: Basic Books); and especially the biography by Drake, S. 1978, *Galileo at Work: A Scientific Biography* (Chicago: Univ. Chicago Press). Thomas Harriot, at the same time, constructed and used astronomical telescopes but his observations are only now being appreciated and were not published in the seventeenth century.

century. The extension of focal lengths and increased lens quality dominated the first half of the seventeenth century. The improved images permitted, for example, the resolution of surface features on Mars and Jupiter, the solution to the geometry of the Saturn ring system, and discovery of Titan. This century of discovery culminated with the invention of the reflecting telescope by Newton in 1675. With the improved image quality that resulted from the better optical figuring of metallic surfaces, elimination of chromatic aberration, and the ease of achieving large apertures, the limiting magnitude and resolution of astronomical observations quickly improved.[5]

We want to stress that Galileo introduced a much more important procedure into astronomy than just putting a piece of glass in front of your eye at the end of a leather tube and looking up. His proof of the veracity of the images produced by the telescope under terrestrial conditions, the famous demonstration to the senators of Venice, showed that the instrument could be trusted to work under *known* conditions, and therefore could be believed under *unknown* conditions. In other words, it was a tool for *discovery*, not just demonstration. This procedure has remained at the core of all scientific instrumentation, but especially in astronomical work since we generally do not know the appearance and properties of the objects we are studying in advance of their detection. In other words, the foundations of all astrophysical observation are *instrument calibration* and the *establishment of laboratory and cosmic standards* against which to compare instruments and observations in order to achieve physically useful data. There are some general techniques that span the various special techniques of observation and these will be the center of our attention here. Increasingly, observational work is done using archival data that have been obtained and reduced before the astrophysicist attacks them. Sometimes this is a danger, since without a substantial understanding of the techniques connected with data acquisition and reduction, artifacts can be mistaken for data. We hope that a broad overview of the methods required for data analysis suffices to make you, the potential future user of such packages and methods, aware of the pitfalls in blind acceptance of the results of specific algorithms.

2.2 CALIBRATION

How do you establish standards when you can't perform direct controlled experiments? This is another way of asking: how do you calibrate astronomical sources when they're inaccessible to laboratory measurement? Nothing in astrophysics is so precious as an *absolute* number. Here we summarize some of the problems with determination of fundamental physical parameters for cosmic objects. The quest for such numbers lies at the core of the history of the field, and we hope you will not mind some historical digressions along the way.[6]

---

[5] See especially King, H. C. 1955, *The History of the Telescope* (NY: Dover Books).

[6] A fine place to start your hunt through the literature is with the papers *and* extensive recorded discussions in Hayes, D. S., Pasinetti, L. E., and Philip, A. G. D., eds. 1985, *Calibration of Fundamental Stellar Quantities: IAU Symp. 111* (Dordrecht: D. Reidel). This symposium is the most complete available collection regarding calibration of fluxes, radii, temperatures, masses, and radial and rotational velocities for stars that are, after all, the basic objects of our business.

2.2.1 Astrometry: Parallax and Proper Motion

To begin with, the scale of the sidereal realm is known geometrically only in the relatively nearby universe. First, we have the mean distance from the Earth to the Moon. This was obtained very early in the history of astronomical observations and was exploited by Ptolemy in the *Almagest* in the second century to set limits on the scale of the distance to the planets[7] and the scale of the universe. His value, approximately 60 Earth radii, augmented the determination by Eratosthenes (ca. 200 B.C.E.) of the radius of the Earth.[8] Aristarchus (ca. 150 B.C.E.) attempted to determine the solar distance using a measurement of the lunar quadrature, but this is a terribly hard measurement involving minutely small angles that cannot be adequately resolved by naked eye, and his value was far short of correct. The innovation presented by Copernicus in 1543 was not just the displacement of the Earth from the center of the world, but the introduction of a way of scaling the sizes of the orbits of the planets around the Sun, a technique that we will see again in Chapter 7.[9] Although its absolute value was unknown at the time, it was subsequently called the *Astronomical Unit* (AU). Its size was successfully set, during the 1761 transit of Venus across the solar disk, by Mason and Dixon.[10] It has since been refined many times, but is of order 1.5×10^{13} cm. With this comes another extremely important number, the *solar radius* $R_\odot$, which is 6.95×10^{10} cm. Considerable use was made of both these numbers in tests of gravitation theory during the eighteenth century.

In the interim between Copernicus and Newton, another determination of distance set a new length standard. In an attempt to accurately time eclipses of the Jovian satellites, a method that seemed a good prospect for the determination of longitude, Roemer found a difference of approximately 1000 sec in the timing between conjunction and opposition. In other words, although now cited as the first reasonably accurate determination of the speed of light, you can say that Roemer set the size of the Astronomical Unit at about 500 light seconds. Considering the modern value, about 8.23 minutes, this isn't bad at all. This determination of c merges with Bradley's discovery of aberration due to the orbital motion of the

---

[7] In Aristotle's physical picture, the world was divided into sub- and supralunary parts. Those regions interior to the lunar sphere were terrestrial and, therefore, transient. Those in the celestial part of the world were different, as best illustrated by the periodicity of the planetary luminaries. While this grossly oversimplifies a beautiful and complex picture—the first attempt at a complete physical construction of a worldview—it suffices for our purposes.

[8] The method is simple, in principle, and was extensively used in the fifteenth and sixteenth centuries to place minimum distance limits for comets and *new stars* by Georg von Peurbach (1472) and Tycho Brahe (1572, 1577). You observe the rising point of the Moon relative to a fixed star. Then, knowing its angular speed, you observe the position again when it is on the meridian. The difference in the angular positions, (observed) − (calculated), is the parallax. Given the resolution that a keen-eyed observer can achieve, of order 10 arc min or so, this is achievable with fair accuracy. It does, however, require a good determination of the mean motion, which has deviations that are nearly the same order of magnitude and that limit the accuracy of the measurement.

[9] The law for galactic rotation, discovered by Lindblad and Oort in the 1920s, makes use of essentially the same construction except it uses the orbital *velocities* of the stars about the galactic center rather than the maximum angular separations of the interior planets from the Sun, but the effect is similar.

[10] Yes, these are the same ones for whom the famous line in northern Maryland is named; they were the team who led the survey of the northern boundary of the Fairfax claim in 1767. See Woolf, H. 1959, *The Transits of Venus: A Study of Eighteenth-Century Science* (Princeton: Princeton Univ. Press).

Earth, which at first was thought to be due to parallactic shifts and complicated the search for true stellar parallax.

The distances to stars then became a scaling problem. Newton and Halley argued that, if the Sun is taken as a standard candle, the nearest stars must be extremely distant but not infinite. The calibration of the magnitude scale, accomplished in the nineteenth century by Herschel and finalized by Pogson, sets the ratio of brightness between stars of different apparent magnitude.

Even the unit used to designate astronomical distances records the method, the *parsec*, a neologism from *par*allax *sec*ond. It is the distance at which the Earth's mean orbit subtends 1 arcsec, which is 206,265 AU (the number of seconds of arc in one radian). It follows that this distance is of order 3.08×10^{18} cm. Such small angular measurements, while routine now, required heroic efforts to achieve in the prephotographic era and were not accomplished until Bessel succeeded in measuring the parallax of 61 Cyg, about 0.314 arcsec.[11] The *Hipparcos* satellite, over a 3-year span (1989–1992) achieved the highest accuracy yet, of order 5 *milli*arcseconds, and extended the geometric measurement of stellar distances to more than 100 pc.[12] This is far enough to include most types of stars and clusters that are used for the next step of the distance scale.

The distance to an object can also be estimated from its motion relative to the observer. Not its radial velocity, the projection of the velocity vector along the line of sight, but its *transverse*, or *proper* motion is used. Assume that the observed angular motion is μ, usually measured in arcsec yr^{-1}. This angular displacement, observed over very long timescales, is often hard to obtain for anything but the fastest moving stars with any reasonable accuracy from ground-based observations. The situation has dramatically improved with the *Hipparcos* and *Tycho* catalogs. Assuming that you know the parallax for the star, the transverse velocity component is:

$$v_T = \frac{4.74\mu}{\pi} \text{ km s}^{-1}, \qquad (2.1)$$

where the constant comes from the units we have used for the respective quantities for μ (arcsec yr^{-1}), the parallax π (pc^{-1}), and the velocity (km s^{-1}). This is combined with the radial velocity of the star, v_R, which is obtained from spectroscopic observations, to obtain the space velocity, $v^2 = v_T^2 + v_R^2$. Now suppose you observe a group of stars that are moving toward, or away from, some point in the sky, like watching snowflakes against your headlights when you drive in a storm. The snow is coming down at some speed v, but it appears, because of the motion of the car, that the flakes are coming from some point ahead of you and diverging

---

[11] See the historical survey by Fernie, J. D. 1975, *JRASC*, **69**, 153. The original announcement can be found in 1838, *MNRAS*, **4**, 152. The star was chosen because of its high proper motion, 4.1 arcsec yr^{-1}, made it a likely candidate among the bright stars for a detectable parallax. Its parallax is now known to be 0.292 arcsec.

[12] Perryman, M. A. C. et al. 1992, *Astr.Ap.*, **258**, 1 (a special issue devoted to the first results of the mission) presents the first in-orbit performance assessment of this satellite and provides superb background information on the project and its history. A general review of modern astrometry is provided by Kovalevsky, J. 1998, *Rep. Prog. Phys.*, **61**, 77 and we recommend this as a supplement to standard (older) treatments.

as you move through them. For a cluster of stars, or a group that is comoving through the local region of space occupied by the Sun, this method is called the *convergence point*. Call the angular distance of the star from this point on the sky λ. Then the radial velocity component is the true space velocity projected along the line of sight, $v_R = v \cos \lambda$ and the transverse velocity component is $v_T = v \sin \lambda$. We can substitute this into Eq. (2.1) to obtain

$$\pi_i = \frac{4.74 \mu_i}{v_{R,i} \tan \lambda_i} \qquad (2.2)$$

for each star i in the cluster. This provides a statistical measure of the distance when we take the ensemble average. It isn't required that each star has the same space motion, just that each is not significantly deviant from the mean motion. The method has been exploited for one cluster in particular, the Hyades, to obtain a distance of 46.3 ± 0.3 pc using *Hipparcos* observations. This cluster is the principal calibration for the distance scale, in part because of its well populated main sequence and the presence of several red giants as cluster members. It provides a representative distribution of types of stars, and therefore provides *relative* calibration of stellar luminosities.

2.2.1.1 Statistical Parallax

For more distant objects, other techniques can be applied. One of these uses the observed space motions of stars of similar type (how you know that they are similar will be discussed in several parts of the book, especially the appendix in Chapter 3 on stellar classification). The moving cluster method is only useful provided you know that the stars are, in fact, moving through space together. It is possible, however, to obtain a statistical measure of stellar properties by exploiting the same idea if you have some other indication of intrinsic stellar properties. This is the *statistical parallax* method.

2.2.1.2 Biases in Parallax Data

Statistical parallaxes are of prime importance for the determination of absolute stellar properties. Thus, anticipating their use in measuring luminosity, we should mention at the outset one of the problems with interpretation of catalogs. There is a systematic bias in deriving the luminosity of a sample of stars due to the effect of parallax errors.[13] You take a group of stars with observed brightness and measured parallax, where each has an associated uncertainty $\delta \pi$. Assume you can select for the type of star, although how you do this will be deferred to Chapter 3, where we discuss spectroscopic classification criteria. For a given measured parallax, π, stars with $\pi' + \delta \pi$, with $\pi' < \pi$, and stars with $\pi'' - \delta \pi$, with $\pi'' > \pi$ both contribute to the sample. The number of stars of a given parallax increases as sampled volume increases, so $N \sim \pi^{-3}$, while the apparent brightness of the stars decreases as π^2. The result is to derive a lower intrinsic luminosity for the sample than is actually the case, regardless of other sources of systematic error such as

---

[13] Trumpler, R. J. and Weaver, H. F. 1953, *Statistical Astronomy* (NY: Dover Books); Lutz, T. E. and Kelker, D. H. 1973, *PASP*, **85**, 875; Koen, C. 1992, *MNRAS*, **256**, 65; Oudmaijer, R. D., Groenwegen, M. A. T., and Schrijver, H. 1998, *MNRAS*, **291**, L41.

intervening absorbing material (interstellar extinction; see Chapter 6) or unresolved companions. There are simply more apparently faint stars that are shifted into the band $\pi - \delta\pi$ than are apparently brighter ones. As a result, distances subsequently derived from comparisons of observed stars or clusters or even galaxies will be systematically underestimated.

This humble and obvious effect has a profound consequence, to which we will return in Chapter 8. Many of the fundamental distance calibrators employed in cosmological studies are based on galactic stars whose intrinsic luminosities are derived from parallax measurements but for which there are no other means for obtaining such properties other than statistical samples.

2.2.2 Luminosities

Once you know the distance, it is necessary to have some standard for the measurement of radiant flux. The first is a relative measure, set up as a ranked ordering of stars: the magnitude system.

2.2.2.1 Magnitudes

The magnitude system, a relic of the original brightness ordering from the Greeks, is now more than a qualitative judgment of the relative brightness of celestial bodies. Herschel's ratio, a factor of 100 per five magnitudes, is a relative measure determined by estimating the brightness of unresolved illuminated apertures of different areas compared to stars of the standard sequence. This is, however, a very unsatisfactory procedure because it depends on the subjective judgment of an observer. The more complete and quantitative calibration was achieved by Pogson[14] who determined the interval $\Delta m = 1$ to be a factor of 2.512 in intensity. We will see shortly how photometers can be used to obtain physical measures of brightness, but let's finish with the magnitude scale first.[15] Given the logarithmic dependence of the *subjective* brightness of objects, the *definition* of the magnitude scale is

$$\Delta m_\lambda = -2.5 \log \frac{I_\lambda}{I_{\lambda,\,std}} \qquad (2.3)$$

where the reference intensity, $I_{\lambda,\,std}$, is an agreed-upon standard object relative to which all others should be measured. This object—historically for optical measurements the relatively bright northern hemisphere star α Lyrae (also known as Vega, or HD 172167)—is *defined* as a star with *uniform* zero apparent magnitude for a specified wavelength, in this case the Johnson V filter. In principle it could be defined as a magnitude of zero at *all* wavelengths but this is impractical. It is required, then, to know the physical flux associated with any wavelength region of this star in order to calibrate both Vega and all of the secondary standards that have been set up around the celestial sphere for relative measurements. This step is achieved by using lamps that have been calibrated relative to a *blackbody* or, at

---

[14] Pogson, N. 1856, *MNRAS*, **17**, 12.
[15] This definition is a historical artifact but it is universally employed in the astronomical, and this is not the place to propagandize for its abolition.

ultraviolet and X-ray wavelengths, with a synchrotron light source or monochromatic spectral lamp. Recently, and especially in the UV, model atmospheres have been used for normalization rather than direct measurements (the standard G191B2B is calibrated in this way for STIS), lending support to Eddington's remark at the head of this chapter. We recommend looking through some of the references to get an understanding of how the bootstrapping works.[16] As with most astronomical measurements, a heavy dose of theory is required *before* the numbers can be obtained, and the steps will be filled in with the subsequent chapters. First, the effects of the atmosphere must be removed by observing the changes in stellar brightness with varying zenith distance. This is done spectrophotometrically in order to obtain the best fit for the extinction as a function of wavelength. Then the observed brightness is compared at each wavelength relative to the calibrated lamp or furnace. For instance, the standard calibrations are performed against blackbodies operating at the melting point of platinum (2042 K), copper (1358 K), and gold (1338 K). Alternatively, since Vega has both a known parallax and measured radius, its flux can be modeled using stellar atmospheres codes (see Chapter 3) that can be compared with the observations, but this is a very theory dependent method and not yet thoroughly reliable (although it has been rapidly improving) Finally, measurements from space using on-board standard lamps permit the determination of the brightness independent of a model for the atmospheric extinction.[17]

Since we have just discussed standards, an aside is on order on the reduction of flux to standard units. This can be tortuous, so here are some pointers. A *lumen* is defined as the flux of a blackbody at the melting point of platinum (2042 K) passing through a projected area of $(60\pi^2)^{-1}$ cm^2. At 5550 Å, the peak of the radiation, 1 W = 680 lumens. As a unit of intensity (surface brightness), 1 *stilb* = 1 lumen cm^{-2} sr^{-1}. Finally, 1 *phot* = 1 lumen cm^{-2} and a star with an apparent visual magnitude of zero outside the Earth's atmosphere is 2.54×10^{-10} phot, and a surface brightness of one $m_V = 0$ star deg^{-2} equals 6.90×10^{-7} stilb. Using the definition of $m_V = 0$ also provides for conversion of intensity to *candela* (cd), which corresponds to 2.45×10^{29} cd. Surface brightness for extended

---

[16] The Vega spectrophotometric calibrations are discussed in detail Oke, J. B. and Schild, R. E. 1970, *ApJ*, **161**, 1015 and Hayes, D. and Latham D. 1975, *ApJ*, **197**, 587. The most complete discussion of absolute calibration is provided by Hayes, D. S. 1985, in *IAU Symp. 111* see n.6 for Vega from 3200 Å to 4μ and is *essential* reading. See also Oke, J. B. 1965, *ARAA*, **3**, 230. For the solar calibration, see Labs, D. and Neckel, H. 1968, *Z.Ap.*, **69**, 1; Tüg, H., White, N. M., and Lockwood, G. W. 1977, *Astr. Ap.*, **61**, 679; Neckel, H. and Labs, D. 1984, *Solar Phys.*, **90**, 205; Neckel, H. 1984, *Space Sci. Rev.*, **38**, 187. Blackwell, D. E. et al. 1983, *MNRAS*, **205**, 897 present the calibrations for Vega using a standard lamp (furnace) and this is a very useful reference for the technique. Discussion of infrared standards is provided by Rieke, G. H., Lebovsky, M. J., and Low, F. J. 1985, *AJ*, **90**, 900; Hammersley, P. L. et al. 1998, *Astr. Ap. Suppl.*, **128**, 207; Beichman, C. A. 1987, *ARAA*, **25**, 521; Cohen, M. Walker, R. G., Barlow, M. J., and Deacon, J. R. 1992, *AJ*}, **104**, 1650.

[17] The same procedure is, in effect, used for all wavelength regimes. In the radio, absolute flux determinations are achieved with standard radiometers. In the infrared, cold sources are used with heterodyne detectors to measure differential power, such as those on the COBE spacecraft. Catalogs are maintained of wavelength specific astronomical standards, and you should check specific references for observatories using their World Wide Websites and Cox, A. N., ed. 2000, *Astrophysical Quantities*, 4th ed. (Berlin: Springer-Verlag).

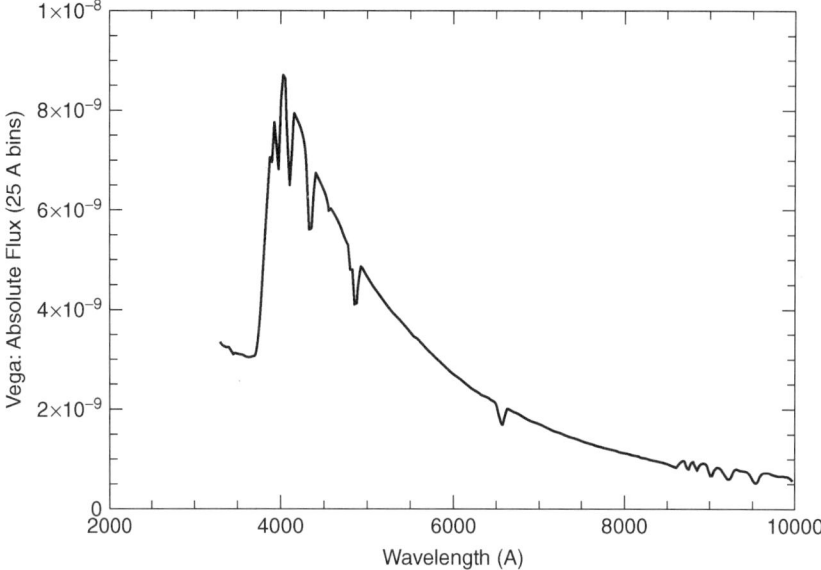

FIGURE 2.1. Absolute spectrophotometry for Vega (α Lyr); units of erg s^{-1} cm^{-2} Å^{-1} (Hayes 1985 calibration).

astronomical sources is often quoted in mag arcsec^{-2} and, for solar system sources, is often quoted in rayleighs (or the bizarre unit, related to conventional sky brightness measurements: $S(10)$, the number of $V = 10$ G2 V stars per arcsec^2; see Table 1.1).

There are two more types of astronomical magnitude. One is the called the *absolute magnitude*, which can be directly related to the monochromatic luminosity, which is the magnitude at a standard distance, *chosen* to be 10 pc for convenience. Not coincidentally, this was for some time the practical distance limit for high accuracy radius and parallax measurements and also the approximate distance to Vega (the absolute energy distribution is shown in Fig. 2.1). In the V filter, for instance, this is denoted M_V. The *bolometric absolute magnitude*, M_{bol}, depends on the integrated flux over the *entire* spectrum. For most cosmic bodies, this is only an approximately determined quantity, mainly because of the difficulty in obtaining comprehensive multiwavelength observations of objects over their entire range of emission. The absolute magnitude provides an additional quantity, the *distance modulus* or $m_\lambda - M_\lambda$, that is often used to express the relative distances d between objects:

$$m_\lambda - M_\lambda = 5 - 5\log\left(\frac{d}{\text{pc}}\right)$$

in the absence of extinction and corrected to outside the atmosphere. The only advantage to this number is that it provides a direct link with the measured flux of an object in a standard filter. It is, however, no more fundamental than the luminosity.[18]

The observation of a stellar flux takes place in only a portion of the spectrum and must be corrected in order to obtain the bolometric value, hence the stellar luminosity, a quantity called the *bolometric correction*. There are several ways to do this. One is to assume that the observed brightness is a part of a blackbody. This is, however, only a formal solution to the problem and is generally a miserable assumption that leads to gross errors. Another is to use *standard candles* or *taxonomic equivalents* of the object in question, but this relies on the comparative similarities between the previously observed object and the one you are working on. Diversity is more common in the cosmos than similarity and superficial concordance in one wavelength region does not imply identity throughout the spectrum. It is, nevertheless, possible to find cases where this holds—there really are standard objects out there—hence, it is plausible to use this technique to obtain the corrections. Finally, there is the model dependent route, but this also still has a relatively low reliability for many bodies.

2.2.3 Spectral Energy Distributions

The most straightforward way to determine the emitted energy as a function of wavelength or frequency for a wide variety of objects is to use standard filters. A catalog of the various color indices is beyond the scope of this book, but many have been created for specialized applications. The first spectroscopically calibrated system and still the most widely used is the *UBV* color system. The bandpasses are centered at 3600, 4400, and 5500 Å, with widths of order 400 Å. It has several advantages; the principal one is the high throughput of the filters and the very large number of objects that have been measured using this system. It was originally intended as a fast way to select among stars of different temperatures. For reference, the zero point of this system, $V = 0$, corresponds to a flux (for a point source) of 3.47×10^{-6} erg s^{-1} cm^{-2} integrated over the filter bandpass. In general, the observed flux is given by

$$F_{\text{obs}} = \int_0^\infty F_\lambda S_\lambda \, d\lambda \qquad (2.4)$$

where S_λ is the filter throughput (see Fig. 2.2) (not a normalized function since its peak depends on the opacity of the filter). The two *color indices* in this system are $U - B$ and $B - V$. Remember that the more negative the magnitude, the brighter the star. This means that $B - V < 0$ corresponds to a *brighter B* flux. The spectral sensitivity, S_λ, is really a product of S_λ^{det}, the sensitivity of the detector and optical

---

[18] One of the reasons for the survival of the magnitude system is the enormous range of fluxes and luminosities of astrophysically interesting objects. Taking the Sun as a standard luminosity, a red dwarf has a luminosity of order 10^{-2} $L_\odot$ while the most luminous stars in the Galaxy are at about 10^6 $L_\odot$. A typical galaxy has a luminosity of about 10^{11} $L_\odot$, so you can see the comparative advantage in a logarithmic system, however cumbersome.

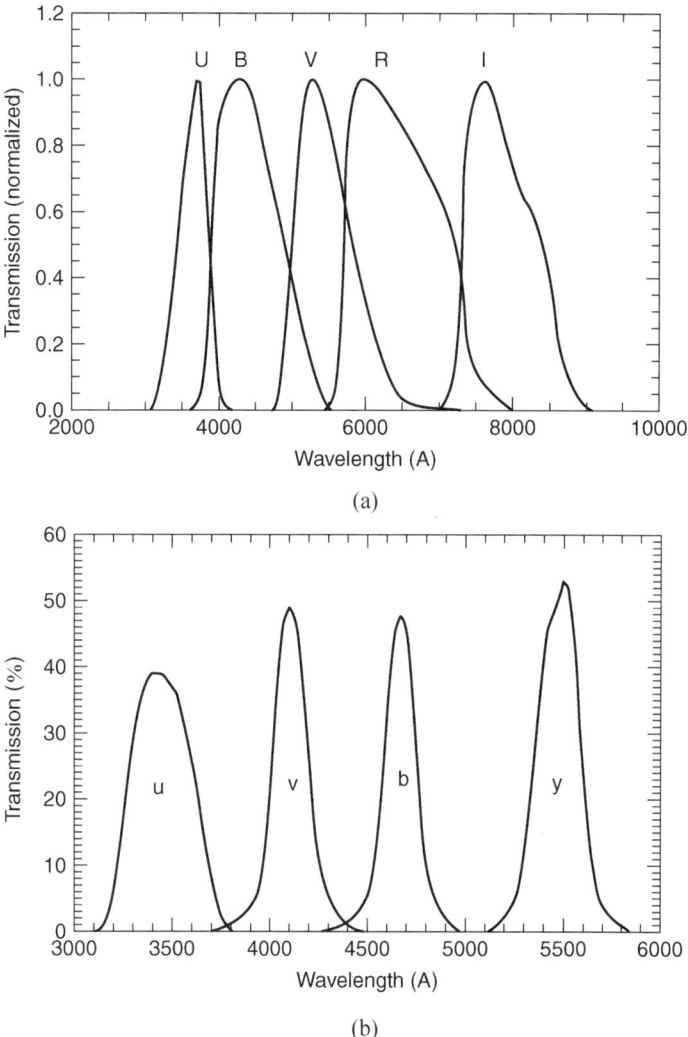

FIGURE 2.2. (a) UBVRI and (b) *uvby* filter profiles.

train of the telescope, and $S_\lambda^{\text{filter}}$, the throughput of the specific filter combination.[19] The Hayes[20] calibration for the absolute flux for α Lys is $F_\lambda = 4.65 \times 10^{-9}$ erg s^{-1}cm^{-2} Å^{-1} at $\lambda 5000$ Å and 3.44×10^{-9} erg s^{-1} cm^{-2} Å^{-1} at $\lambda 5556$ Å. In this scale, the apparent magnitude and color for the Sun are $V = -26.75 \pm 0.06$ and $(B - V)_0 = +0.661 \pm 0.03$. A comparison between the Sun and Vega shows

---

[19] See Bessel, M. S. 1990, *PASP*, **102**, 1181 for examples of how the sensitivity curves are derived for the *UBVRI* system. A detailed description of the calibration process is provided in Stone, R. P. S. 1996, *ApJS*, **107**, 423 for CCD observations with *UBVRI* filters. A superb database is maintained by the Geneva Observatory, accessible through *http://obswww.unige.ch/gcpd/system.html*, which also lists inter-system filter transformations and provides transmission data and references.

[20] Optical spectrophotometry is normalized to 5000 Å, while the measurements have been made at 5556 Å.

114 THE RAW MATERIAL: INSTRUMENTS AND OBSERVATIONS

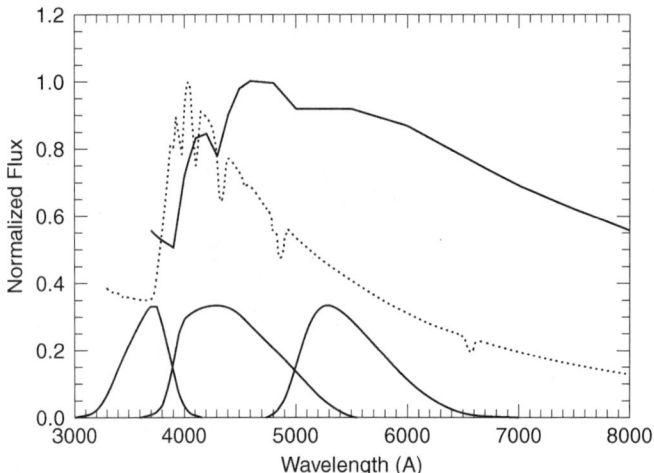

FIGURE 2.3. Normalized solar —— and Vega ----- continua with superimposed *UBV* bandpasses. Note the *U* excess of Vega compard with the Sun.

clearly the origin of the color index (Figure 2.3). See Tables 2.1[21] and 2.2[22] for selected broadband photometric and narrowband line filters, sample calibrations are listed in Table 2.3.

With the wide proliferation of photometric systems, due mainly to space instruments such as WFPC and WFPC2, the standard is to use the *AB* magnitude system[23] to quote fluxes. This is defined as

$$AB = -2.5 \log F_\nu - 48.60 \qquad (2.5)$$

where F_ν is the monochromatic flux in erg s^{-1} cm^{-2} Hz^{-1} and the zero point is from the definition of zero magnitudes within the filter.

The problem with photometric measurements is that the profile of the filters may be very complicated. Given your ignorance of the continuum distribution of the object you're observing, the resulting integrated flux may be contaminated by intrinsic emission from far outside the bandpass. This is one reason for

---

[21] The list in Table 2.1 includes only the standard filters that have been used to produce large catalogs. (*uvgr*): Thuan, T and Gunn, J. 1976, *PASP*, **88**, 543, Joergensen I. 1994, *PASP*, **106**, 967; (*UBV*): Johnson, H. L. and Morgan, W. W. 1953, *ApJ*, **117**, 313, Sandage, A. and Smith, L. L. 1963, *ApJ*, **137**, 1057, Bessel, M. S. 1990, *PASP*, **102**, 1181; Geneva: Rufener, F. G. and Nicolet, B. 1988, *Astr. Ap.*, **206**, 357; *uvby*: Strömgren, B. 1966, *ARAA*, **4**, 433; Washington, DC: Geisler, D., Claria, J. J., and Minniti, D. 1991, *AJ*, **102**, 1836. The table does not include the filter sets for, for instance, the WFPC2 or FOC imagers on HST, which are cross calibrated to the *UBV* system using photometric standards, nor are the 13-color filters included. See especially Golay, M. 1974, *Introduction to Astronomical Photometry* (Dordrecht: Reidel); Low, F. J. in *Methods of Experimental Physics*, vol. 12A, *Astrophysics*, N. Carleton, ed. (NY: Academic Press).
[22] Here we again concentrate only on the most widely used filters. Specific applications often require special filters, but these are the ones for which the most extensive data sets exist.
[23] Oke, J. B. and Gunn, J. 1983, *ApJ*, **266**, 713; see also Hamuy, M. et al. 1992, *PASP*, **104**, 533. Oke, J. B. 1990, *AJ*, **99**, 1621 provides a list of spectrophotometric standards for *V* and *AB*(λ5460 Å) (all of which are north of $-30°$).

TABLE 2.1 Some Broadband Standard Photometric Filters

System	Filter	λ_0 (Å)	Bandpass (Å)	System	Filter	λ_0 (Å)	Bandpass (Å)
Johnson	U	3620	650	Strömgren	u	3500	380
	B	4420	1000		v	4100	200
	V	5400	800		b	4700	100
Cousins	R	6800	950		y	5500	200
	I	9000	200				
Infrared	J	1.25 μ	0.2 μ	Geneva	U	3458	170
	K	2.20	0.4		B	4248	283
	L	3.40	1.2		V	5508	298
Thuan–Gunn	u	3530	400		B_1	4022	171
	v	3980	400		B_2	4480	164
	g	4930	700		V_1	5408	202
	r	6550	900		G	5814	206
Washington	C	3910	1100				
	M	5085	1050				
	T_1	6330	800				
	T_2	8050	1500				

TABLE 2.2 Some Narrowband Line Filters

System	Filter	λ	$\Delta\lambda$	Feature	System	Filter	λ	$\Delta\lambda$	Feature
DDO	[35]	3458	370	BJ	Wing	C1	7117	53	TiO(0, 0)
	[38]	3796	172	BJ		C2	7545	50	Cont
	[41]	4166	83	CN, Fe		C3	7806	42	Cont
	[42]	4257	73	CN		C4	8122	43	CN $\Delta v = 2$
	[45]	4517	76	CH, Fe		C5	10392	55	Cont
	[48]	4863	186	CH, Fe		C6	10544	58	VO
Crawford	β(W)	4861	150	H β		C7	10800	74	Cont
	β(N)	4681	30	H β		C8	10968	73	CN(0, 0)

TABLE 2.3 Sample Calibrations for Standard Fluxes for $m_\nu = 0$

Filter	λ (μ)	F_ν^{*a}	Filter	λ (μ)	F_ν^{*a}
U	0.36	1810	H	1.65	980
B	0.44	4260	K	2.20	620
V	0.55	3540	L	3.40	280
R	0.70	2870	L'	3.74	252
I	0.90	2250	M	4.80	150
J	1.25	1670	Q	20.0	10

[a] Units of Jy (10^{-23} erg s^{-1} cm^{-2} Hz^{-1}).

choosing white dwarfs and dO stars as standards in the optical and UV, their continua are much less complicated. As such, filter construction and use requires very careful calibration. This undesirable feature of filters is caused by the way they are constructed. Layers are deposited on a substrate or combinations of different glasses are used. This means that there are sometimes wavelengths that are far from the center of the filter that can contribute to the measured flux (called *leaking*). The detector sensitivity adds to this problem if it happens to be more responsive at the contaminating wavelength than the central one, and it is necessary to make the filters as sharp as possible within the range of interest.[24]

Several filter systems have been developed to account for special features of spectral energy distributions. The DDO system is adapted to measure the strengths of the CH, CN, and metallic lines in stars ranging from F through K. The Wing filter system is designed to measure the strength of taxonomically important features in the near infrared, out to 1.5μ. For galaxies, emphasizing the contribution of late type stars, the Mg_2 index has been spectrophotometrically defined to cover Mg H 5174 Å, and Mg b 5177 Å, similar to one of the DDO bandpasses. For hotter stars, narrow line He I 4026 Å photometry has been used for the B stars. Chemically peculiar stars show an enhanced absorption in the middle part of their optical spectra, centered at around 5200 Å and a special filter has been developed for this purpose. Finally, for measurement of global properties of galaxies, a set of broadband filters has been developed by Thuan and Gunn. In other words, provided the calibration is careful, the filter well specified, and the application appropriate, there are *many* ways to dissect a spectrum without actually performing high resolution spectrophotometry.

It is important to reemphasize that the hardest part in establishing the use of any of these color systems is the initial calibration. All of these have specific requirements and peculiar sensitivities to variations in the properties of the objects to which they are applied, and none is universally applicable. Their main utility comes from the high throughput attained with relatively broadband measurements. In general, for any arbitrary spectral distribution, a filter system defines an index using

$$C_1 - C_2 = -2.5 \log \frac{\int S_{1,\lambda} F_\lambda \, d\lambda}{\int S_{2,\lambda} F_\lambda \, d\lambda} \qquad (2.6)$$

so that the detailed match of the stellar continuum to the filter profile is critical. The increasing availability of stellar atmospheres codes and published synthetic spectra makes it possible now to try proposed filter systems on theoretical models before drawing conclusions from the observations. Large catalogs are available for several photometric systems. For optical measurements, among them are Geneva, Strömgren ($uvby\beta$), Johnson, and the so-called 13-color system (Tonanzintla), HST

---

[24] Several of the filters used on the Wide Field Camera 1 and 2 on HST have this, for example. The problem is usually worst at short wavelength, where the transparency of the filters to longer wavelengths, coupled with the sensitivity of the detectors to longer wavelengths, makes absolute flux determinations difficult.

WFPC filter system. For IR colors, these include *JHKLM* and IRAS (12, 25, 60, 100μ).[25]

2.2.3.1 Satellite Observations

Satellite observations are strictly required for all wavelengths below ~ 3000 Å because of atmospheric extinction from ozone and the overall high energy opacity of the gas. At long wavelengths, beyond the visible, the near infrared is compromised, except in some windows, by CO, CO_2, and H_2O absorption shortward of 10μ. In the longer wavelength IR, satellites are again required, and it is generally the case that no spectral region other than the optical and centimeter radio is entirely free of atmospheric effects. For the near IR, shortward of 25μ, high-altitude sites such as Mauna Kea (Hawaii) and Perenal (Chile) provide some solution. By placing the site above a sufficient column density of the atmosphere, the window opacities in the infrared are substantially reduced *but not eliminated*, and there is at any rate no significant reduction in the UV opacity. A similar behavior is found for the millimeter and submillimeter wavelengths. The obvious solution, although by no means the cheapest, is to place the instruments in orbit. Some space projects since 1980 are listed in Table 2.4.[26] The list is not complete, but the satellites listed have produced the largest archival databases and will form

TABLE 2.4 Some Space Astronomy Projects: 1980–2000

Name and Acronym	Wavelength	Instruments	Mode
Infrared Astronomy Satellite (IRAS)	12–100μ	P, SL	Survey
International Ultraviolet Observatory (IUE)	0.1–0.3μ	SL, SH	Obs
Cosmic Background Explorer (COBE)	1–300μ	P, SL	Survey
Hubble Space Telescope (HST)	0.1–2μ	SL, SH, I, Astr	Obs
Infrared Space Observatory (ISO)	1–200μ	P, SL, SH, I	Survey, Obs
Extreme Ultraviolet Explorer (EUVE)	0.01–0.1μ	P, SL	Survey, Obs
Far Ultraviolet Spectrographic Explorer (FUSE)	0.07–0.12μ	SL, SH	Obs
ROSAT	XR	SL, I	Survey, Obs
Chandra Advanced X-ray Astronomy Facility	XR	SL, SH, I	Obs
Compton Gamma Ray Observatory (CGRO)	γ	Various	Survey, Obs
Hipparcos	UBV	Astr, P	Survey

---

[25] There are two large databases now available for general use that access all of the catalogs. One, SIMBAD, is operated by the *Centre des Donnes Stellaires* (CDS) at University of Stassborg with an American mirror site at *Center for Astrophysics* at Harvard at the URL: *http://simbad.u-strasbg.fr/sim-fid.pl*. The other, NASA Extragalactic Database (NED), is operated by NASA through *IPAC* at Caltech at URL: *http://nedwww.ipac.caltech.edu*. Both provide access to the relevant literature as well as data for individual objects.

[26] In Table 2.4, for the Hubble Space Telescope, although not *specifically* designed for any particular wavelength region, and in fact dominated by the optical and near IR, the practical wavelength limits for this instrument range from 0.1 to about 2μ. There are two multifilter imaging instruments (WFPC2 and FOC), two imaging spectrographs (STIS and NICMOS, although the latter is used mainly as an imager), the corrective optical mirrors (COSTAR), and the astrometric/guidance set of FGSs. Abbreviations in the table: (P) photometry; (SL) low-resolution spectrophotometry; (SH) high-resolution spectrophotometry; (I) imaging; (Astr) astrometry.

2.2.4 Masses: Binary Stars

One advantage presented by binary stars, even without knowing anything about the physical origin and evolution of these systems, is that you can use them to determine some fundamental physical parameters from comparatively simple observations. The first is the monochromatic luminosity *ratio* for the stars. The second is the mass *ratio*. In systems that are spatially resolved, this is relatively simple. Noting that the stars are at the same distance from you, regardless of the systemic parallax, you can measure the relative flux of the two stars immediately. The broader the spectral coverage, the closer you come to the ratio of the bolometric luminosities. For unresolved systems, the stars can be separated by their relative motions through spectral line shifts and, again, the ratio of their flux contributions can be obtained at each wavelength.

For the masses, it is possible to apply simple dynamics to the problem. To simplify matters, let's assume that the orbit is circular. For short period binaries with periods less than a couple of weeks, the velocity amplitudes will be large enough and the orbits will usually be pretty close to circular, so this method can be used almost without modification.[27] Two stars, M_1 and M_2, in orbit about a common center of mass, exhibit a ratio of maximum radial velocity, K_1 and K_2, that is inversely proportional to their mass ratio:

$$\frac{K_1}{K_2} = \frac{M_2}{M_1} \qquad (2.7)$$

if both stars are visible in the spectrum. Noting the spectra contributions for each star through comparison of their lines, and looking at template single stars (to obtain the standard colors for stars matching the distribution of spectral lines, or, in other words, having the same spectral type), you can obtain the mass ratio and the luminosity ratio directly.

For the case where only one star is visible, you can still place limits on the mass of the companion because the binary is gravitationally bound and therefore has a unique relation between its mass, separation, a, and orbital period, P

$$\frac{a^3}{P^2} = \frac{G}{4\pi^2} M, \qquad (2.8)$$

where M is the sum of the masses. Since $K_1 = M_2 a/M$, assuming this to be the

---

[27] There may be other problems, connected with how normal the stars are in such systems, or how representative of the properties of single stars, because of evolutionary effects and dynamical interactions between the components. We will return to this problem in Chapter 5. For now, keep it in the back of your mind as a potential problem.

visible star, you find for circular orbits with inclinations i

$$f(M) \equiv \frac{M_2^3 \sin^3 i}{(M_1 + M_2)^2} = 1.04 \times 10^{-7} \times K_1^3 P^{-2} \qquad (2.9)$$

for circular orbits with K in km s^{-1} and P in days. The left hand side is called the *mass function* (the constant depends on the units used for the observables and also on the orbital eccentricity; see Chapter 5). The problem is that we have a new uncertainty introduced by uncertainties in i, the removal of which requires either an astrometric orbit, both velocities, or eclipses. The *observed* maximum radial velocity is used to determine this quantity, and this is related to the intrinsic properties of the binary through the relative orientation of the orbital plane with respect to the observer. Although this provides plausible limits for the masses of the unseen component, there is nothing like the direct detection of the spectra of both stars to be sure of the mass. The alternative is to know the inclination. This is where eclipsing binaries are so important, to which we now turn.

2.2.5 Radii and Luminosity Ratios: Eclipsing Binary Stars

If you know that the two stars eclipse, you can be reasonably sure that they lie in the same line of sight and therefore $i \approx \pi/2$. Even without knowing most of the details of the spectral formation and other processes in the system, geometrical aspects of the eclipse provide important information about the stars. First, consider a light curve of the sort shown in Figure 2.4. This is a total eclipse; you know this from the bottoming out of the light curve. The luminosity ratio comes from the ratio of the minima, since the same area is occulted in both cases. The relative radii come from the timings of the first contact, t_1 and second contact, t_2, or more

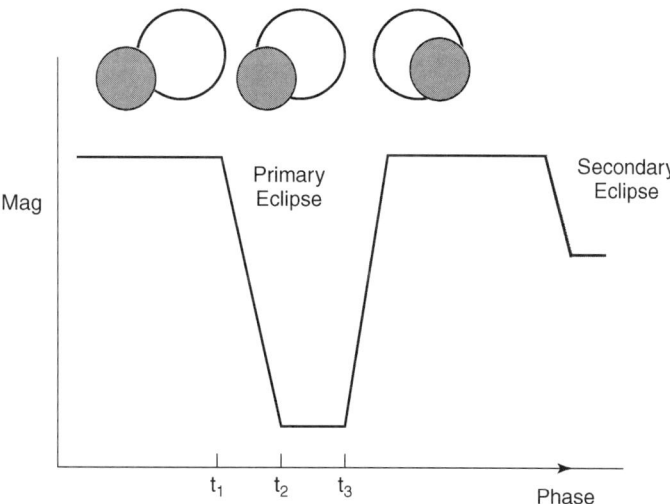

FIGURE 2.4. Schematic of the phases and timing of an eclipsing binary light curve. The cartoon shows a detached system at these phases.

specifically $\Delta t_1/\Delta t_2 = (r_1 + r_2)/(r_1 - r_2)$, where r is the size of the component relative to the semimajor axis. This trivial result is profoundly important in that the ratio of stellar surface brightnesses depends on temperature. Therefore, from the luminosity ratio and relative radii, the ratio of temperatures can be approximately specified.

For real stars, this is not quit so easy. As we will discuss in Chapter 3, the surface brightness of a star does not approximate that of a bowling ball. Because the atmosphere has a rather steep temperature gradient, the stellar limb is generally darker than the center. This smooths out the artificially discontinuous onset of a purely geometric eclipse. In addition, mass transfer can easily introduce absorbing matter between the stars and throughout (or even surrounding) the system. Finally, tidal distortion of the stars introduces further geometric complications that are only approximately understood (although this effect can be modeled comparatively easily). Models for incompressible self-gravitating bodies, of the sort we discussed in Chapter 1, were tools for treating ellipsoidal variations and tides in light curve synthesis work even as early as the 1920s. These topics will be explored further in Chapter 5.

2.3 PHOTON DETECTORS

With the exception of cosmic rays and neutrinos, which require quite specialized treatment, all astronomical observations are quintessentially about detecting photons from radio through gamma ray energies. It is appropriate that as the 20th century opened with the *Cartes du Ciel* photographic project, the *National Geographic Palomar Sky Survey* greeted the mid-century, and this century opens with the *Sloan Digital Sky Survey*. A more remarkable contrast is, however, provided by the wavelength coverage available to astronomers at the start of the two eras. While the dawn of the last century saw the extension of photographic capability to cover the entire optical window, complete sky coverage is now available from γ-ray through centimeter wave radio through the combined application of ground-based and space-based observing platforms.

2.3.1 Photographic Plates

Photographic plates were, before the mid-1980s, the only archival medium available to astronomers. They are *still* the longest-lived, in the sense that once developed they require no special reduction programs to extract some useful data. In addition, new techniques have been devised to push the dynamic range of the plates to obtain more information in the partially saturated portion of the plate.

The first astronomical photography began in the mid-1860s with the pioneering efforts of W. and M. Huggins, H. Draper, and D. Rutherfurd.[28] These were "wet" emulsions, so named because the imaging took place with the emulsion in a moist state. There is little of permanent quantitative value that can be obtained from

---

[28] See Clerke, A. 1903, *A Popular History of Astronomy in the Nineteenth Century* (London: Blackie); Hearenshaw, J. 1986, *The Analysis of Starlight—One Hundred and Fifty Years of Astronomical Spectroscopy* (Cambridge, UK: Cambridge Univ. Press).

these early efforts, although they do provide some spectral information for stars on a timescale of about 120 years. The observations were hampered by the combination of an insensitive medium and small apertures, but for solar work they provided important information that could not be obtained through visual observations. Serious arguments about the veracity of the images and the data that they could provide continued for nearly 30 years until the introduction of the dry emulsion. Widespread use of large field photography was also spurred by Barnard's Atlas of the Milky Way and Ritchie's photographs of nebulae, both of which made cataloging and morphological studies possible with relatively large apertures (using the Yerkes refractor and the Lick reflector, respectively). The construction of the 100 inch Hooker telescope at Mount Wilson was especially designed take advantage of the rapid technological improvements in photography that were happening during the first decade of the twentieth century.

Perhaps the key development for photographic work was the invention of the Schmidt telescope. This wide-field instrument can be produced with a large aperture at comparatively low cost because of its spherical mirror and thin corrector plate and is capable of providing wide fields, of order 10°, that are virtually aberration free in the inner $\frac{2}{3}$ of the field of view.[29] Although efforts were made in the nineteenth century to begin a massive photographic all-sky atlas, the *Carte du Ciel*, it never reached completion because of problems with systematics between different observatories and political differences among the participating institutions. Kapteyn was able to carry out a large-scale survey of the *Selected Areas* that are still the basic set of observations for photographic proper motions, and there were several other photographic projects of similar magnitude mounted over the years. The most complete archival records of the sky began in 1948 with the installation of the 48 in Schmidt at Mt. Palomar. This was the *Palomar Sky Survey*, which was completed with the support of the Carnegie Foundation and the National Geographic Society. Along with the Whiteoak extension, it reaches a southern declination limit of nearly $-60°$ in two effective colors, the blue (Kodak 103a-O emulsion) and the red (103a-E)[30] (see Table 2.5[31]).

We speak of this detection method in the past tense because it is virtually unused in modern work for the acquisition of *new* data. Large archives of such material exist, however, and it is therefore important that you have an understanding of the uses and limitations of these data. For almost a century of astrophysical research, these are the only impersonal, quantitative records that exist of photometry and spectra. Any time you need a *long* baseline, you are likely to encounter some need to use the actual plates from an earlier period or interpret the published record. Hence our extended discussion. The detector quantum efficien-

---

[29] See Hiltner, W., ed. 1962, *Astronomical Techniques* Chicago: Univ. of Chicago Press; Eccles, M. J., Sim, M. E., and Tritton, K. P. 1983, *Low Light Level Detectors in Astronomy* (Cambridge, UK: Cambridge Univ. Press).

[30] The roman numeral indicates grain size; higher numbers are finer-grained—*a* means that it is a specifically designed astronomical emulsion, and the letter denotes the spectral range. In this case, O is approximately 4000–5000 Å, E is 4500–700 Å, and intermediate D and red F emulsions were also produced although not used in the Palomar sky survey. The ESO southern survey used the finer grained IIIa-J and IIIa-F, as did the second Palomar survey, both of which were in advance of the *Guide Star Selection Catalog* for the Hubble Space Telescope project.

[31] In Table 2.5, R is the resolving power of the plate measured in line pairs mm^{-1}.

TABLE 2.5 Properties of Principal Astronomical Photographic Emulsions

Emulsion	λ_{min}	λ_{max}	R	Designation
103a-O	3000	5000	80	Blue
103a-E	6600	—	80	Panchromatic
IIa-O	3000	5000	87	Blue
IIa-F	4500	6800	100	Panchromatic
IIIa-J	4500	5500	200	Orthochromatic
IIIa-F	4500	6800	200	Panchromatic
I-N	6800	8900	100	Infrared
I-Z	8900	11800	125	Infrared

cies (DQEs) of photographic plates were usually in the range of 1% to a few percent. Effective S/N ratios with special sensitizing methods were able to reach 25–50, but these often required extraordinary efforts. For instance, the plates had to be prepared in either nitrogen or hydrogen ovens and baked for extended periods.

The various attempts to use film as a direct recording medium in place of the more rigid plates ultimately were abandoned. Contact copies are, however, the basis of many of the collections that are now available, and nearly all astrometric work for large field identification is still performed on the Palomar and ESO sky survey plates. Although these have been digitized, and are available online (through the Space Telescope Science Institute Website), there is still more on the original plates than could be recorded in the reductions. The transformation of the qualitative photographic record to a quantitative digital format is achieved through the use of microdensitometers. These are scanners that illuminate the plate with a laser whose intensity is read by a photomultiplier and transformed through an analog-to-digital converter. Two dimensional images are reassembled by cross correlation of individual scans. The method was the basis of the *Guide Star Selection System* catalog of HST.[32]

Photographic emulsions have several severe disadvantages for quantitative work. The response curve for the emulsion is very nonlinear, called the *H & D curve*, which is the relation between photographic density and transmission. The light must pass through the emulsion before it can be recorded by the detector, and the emulsion is on a glass plate of much greater thickness. The result is a diffuse transmission that must be corrected in order to obtain a linear scale for opacity of the emulsion. This intrinsic graininess is also a problem for the image formation, since light is scattered during the exposure process and the irregularity of the grains makes it impossible to predict precisely where on the grain the image will form. Finally, unlike the digital imaging process, there is an intermediate processing step, the development in a chemical bath, that produces variations in the clear plate (the photographic equivalent of a flat field) that change from one exposure to another. These deficiencies are made up for, in part, by the large formats that are available for plates that are only now being approximated with mosaics of charge-coupled devices (CCDs). We again emphasize that for much large scale structure

[32] Lasker, B. M. et al. 1988, *ApJS*, **68**, 1; Lasker, B. M. et al. 1990, *AJ*, **99**, 2019; Morrison, J. et al. 2001, *AJ*, **121**, 1752.

and morphology survey work, such as cosmological galaxy counts, and for long time baseline studies such as are required for variable stars, photographic archives represent the principal resource data.[33]

There is an interesting survival of the photographic era still present in astrophysics. In Chapters 4 and 6, we will introduce the concept of the equivalent width of a spectral line. This is the effective width that a line appears to have on a photographic plate, and is normalized to the level of the continuum. This is an artifact of the photographic process, as is the habit of normalizing spectrophotometric data to an arbitrarily chosen continuum when comparing model atmospheres to observed line profiles.

Finally, we should note that once the photographic images have been digitized, all the deconvolution methods we will describe later in this chapter can be used to derive improved data from the images. In addition, there are purely photographic methods that can be used to enhance the image contrast, in particular *unsharp masking*,[34] although these methods generally fail to preserve the photometric integrity of the data. They are intended only to enhance structural information and can also be duplicated with digitized images.[35]

2.3.2 Photomultiplier Tubes (PMTs)

The photoelectric effect was exploited very early in the twentieth century for the measurement of brightnesses of astronomical bodies.[36] When a photon with sufficient energy is absorbed by a metal, and this level is set by the *work function* of the material, an electron is emitted. The process is essentially one of counting the photons. A window is coated with a film of the metal to form the photocathode. The emitted electron is accelerated through a large potential difference between the photocathode and the collector and strikes a series of secondary anodes that amplify the signal by creating a burst of secondary electrons. In this way, a single photon can be counted as a bunch of charge that is seen as a pulse at the collecting anode. The limit on the count rate is set by the counting electronics, but typical

---

[33] Some of the most important sources for modern cosmological and extragalactic studies are photographic in origin. The Zwicky catalog of compact galaxies, the Shane–Wirtanen survey, the Jagellonian catalog, the *Morphological Catalog of Galaxies* and the Arp, and Arp and Madore, atlases of interacting galaxies are all based on plates taken with a variety of telescopes and emulsions.

[34] This technique uses a reverse print of the image as a blocking filter for an illuminated negative that is photographed through the reverse image. The method is a sort of edge enhancement and is often able to recover very weakly exposed portions of the plate.

[35] Originals and copies of the sky surveys are located at Carnegie Institution of Washington, Space Telescope Science Institute, European Southern Observatory. Large Schmidt collections exist at Palomar, Kitt Peak, Cerro Tololo, ESO, Case Western Reserve University; extensive spectroscopic collections exist at Mt. Wilson, Kitt Peak National Observatory, Dominion Astrophysical Observatory, Lick Observatory, David Dunlap Observatory; large photographic archives for photometry exist at Sternberg (Moscow), Harvard College Observatory, Yerkes Observatory, Allegheny Observatory, and Lowell Observatory.

[36] Some of the older references regarding photometric data acquisition and analysis are still helpful for understanding both the older published data and the limitations or uses of the techniques. See, *e.g.* Henden, A. E. and Kaitchuck, R. H. 1982, *Astronomical Photometry* (Princeton: van Nostrand Reinhold); reviews by R. H. Hardie and H. L. Johnson in Hiltner, W., ed. 1962, *Astronomical Techniques; Stars and Stellar Systems*, vol. 2 (Chicago: Univ. Chicago Press) are succinct, although dated, and still helpful.

rates are of order several hundred kilohertz. The disadvantage of these detectors is that they emit a high noise unless cooled because of the thermal emission of electronics from the photocathode. Although superseded in most applications by charge-coupled devices, photomultiplier tubes (PMTs) are still extensively used in laboratory applications, where signal strength is not a limiting factor, and for purposes of single photon counting. Before about 1990, almost all photometric studies were carried out using these detectors; indeed, the Johnson-Morgan photometric system was originally defined with respect to the response curve for an RCA 1P21 photomultiplier tube. The High Speed Photometer (HSP) (one of the first-generation instruments on HST that was removed when the corrective optics were installed in Dec. 1993) was a battery of such devices, each equipped with its own filter. This instrument was able to obtain count rates in the megahertz range with special application to the optical and UV study of very fast phenomena such as pulsar light curves.

The principle behind the operation of a PMT is the cascade generated by the photoelectric effect. A coating with a spectrally dependent (effective) quantum efficiency, Q_{eff}, is chosen for the specific application, depending on the work function. This layer is placed on a photocathode that serves as the electron emitting surface. High voltage, usually several kilovolts, accelerates the liberated electrons, which are unfocused but maintain a relatively small solid angle. On striking another electroemissive plate, and these are usually arranged as vanes within a high-vacuum tube, a cloud of electrons is generated. If $\delta > 1$ is the emissivity of each surface (its amplification factor), then $Q_{eff} \delta^n$ is a measure of the amplification for an n-vane system. The arrangement of the plates varies with tube design, but the usual is a "venetian blind" configuration (the one used in the RCA 1P21, the original astronomical PMT workhorse). The PMT is the basis of a number of other devices, in particular the multianode microchannel array (MAMA) detectors (see below).[37]

2.3.3 Charge-Coupled Devices (CCDs)

The primary modern detector is the *charge coupled device* (CCD).[38] It is a multistage electron transfer device, hence the name. A photon is absorbed in a silicon diode that is sandwiched between electrodes, creating a well in which charge is stored. The electrons accumulate in this well for some prespecified

---

[37] The is another instrument that mimics many of the characteristics of the PMT, although using only a single amplification stage and solid state throughout. This is the *reticon* or *digicon* detector. Although used historically during the period of crossover between vidicons and the more modern CCD and MAMA detectors, these devices provide true photon counting capabilities. An emitted electron is accelerated through about 10–20 kV, but with a focusing magnetic field (as in a vidicon). It then strikes a silicon diode array that is readout into a register that is queried by the system after a preset integration interval. Image tube front ends can be used to intensify the target and increase the overall quantum efficiency of the system, but the advent of fully two dimensional photon counting devices, such as multianode microchannel array (MAMA) detectors, renders these one-dimensional linear arrays less attractive for most applications.

[38] It is instructive to read the earliest review, Mackay, C. D. 1986, *ARAA*, **24**, 255, to get an idea of what a revolution was represented by this new device. The book by Rieke (n.1) is a superb summary of the basic properties. Individual observatory handbooks and satellite descriptions should be consulted for current instrumental details.

integration period, after which they are linearly transferred out of the detector by a clock-stepped sequence of pulses. The transfer efficiency is very high, and the result is that the detector is linear above some minimum threshold. The problems that continually plagued astronomical imaging before the invention of solid state devices are absent for CCDs—in particular, the inherent nonlinearity of photographic plates and the inability to use photoelectric detectors in array format. There are two sources of error that are intrinsic to the detector. The first is fixed pattern or *flat-field* error, due to imperfections in the material from which the chip has been fabricated. Fortunately, this is very stable and the gain variations produced by these large-scale glitches can be removed by a simple normalization once a standard image of the unfocused chip has been accumulated. The other is the problem of local imperfections in the device, due to impurities in the silicon. These produce delays in the transfer of charge, resulting in image distortions that are different from one exposure to another. They are local—that is, they dominate only one transfer line at a time—and therefore can be eliminated through multiple exposures. This procedure known as *flat fielding*. It involves forming a "superflat," which is the a set of exposures obtained of a uniform defocused field, such as the inside of the dome or a plain white surface held near the aperture. An observed image is then normalized by division to the superflat which removes the stable blemishes and irregularities. This contrasts with our previous discussion of photographic plates. Flat fielding is not possible for photographic images since the background is largely produced by chemical fogging that varies from one plate to another.

These minor problems are a small price to pay for the most important feature of the CCD as a detector: it is a linear device that has an enormous dynamic range and very high S/N capability. Peak efficiencies reach nearly 100%, and the large formats (greater than 2048×2048 pixel^2) that are routinely produced with readout noise of only a few electrons makes these the detectors of choice. More recent advances have included an enormous extension of the sensitivity of these devices. They are now routine in infrared observations (around 0.2–1 eV) and have been used on *Chandra* and *XMM-Newton* in the X-ray range for imaging and spectroscopy, above 1 keV). Their sole limitation at present is that they are not photon counting devices.

The *reticon* was an earlier solid-state detector that had many of the advantages of the CCD and was a real time photon counting device. It was used on two of the first-generation instruments on HST, the *Goddard High Resolution Spectrograph* (GHRS) and the *Faint Object Spectrograph* (FOS).[39] Its disadvantage was the relatively large size of the pixels and problems in generalizing it to a two-dimensional imager. These have been largely overcome with the introduction of the MAMA detectors and the improvement in readout noise from CCDs that permits faster scan rates. The GHRS replacement instrument, the *Space Telescope Imaging Spectroscope* (STIS), uses both CCDs and MAMA detectors.

2.3.4 Multianode Microchannel Arrays (MAMAs)

These devices are a class of detectors that combine the photon counting capacity of photoelectric detectors with the array properties of CCDs. The principle is

---

[39] See the calibration summary for the GHRS by Heap, S. R. et al. 1996, *PASP*, **107**, 871.

really quite simple. A photocathode liberates electrons on exposure to light and the electrons strike a microchannel plate consisting of pores that are coated with a photo-emissive surface, like the vanes of a photomultiplier. These channels amplify the signal, ultimately producing a burst of 10^5–10^6 electrons that strike an anode array underneath the plate. There they are detected as a pulse by an array proportionality counter that registers the location and amplitude of the signal in two dimensions as a function of time. In this way, an image is built up over time with each individual photon time-tagged for its arrival and location. The result is that the noise in the resulting image is truly dominated by photon counting statistics. As with a CCD, the flat field must be accumulated over time using unfocused images but again, it is stable and can be removed by normalization. Very high count rates can be achieved, of order 1–10 MHz.[40] Very high ultraviolet quantum efficiencies have been achieved with MAMA detectors, and they are now used on the STIS and the *Far Ultraviolet Explorer* (FUSE) for observations below 3000 Å. Their principal limitation is that they are very sensitive to high exposure levels (not simply count rates).[41]

2.3.5 Bolometer Arrays and Infrared Detectors

Bolometers are one of the most antique, and at the same time most versatile, photodetectors. In a sense, they are classical devices, since they do not make use of the photoelectric effect. Bolometers are circuit elements whose resistance changes depending on temperature. The standard choice is a Ga:Ge crystal whose resistance varies as $R_{bol} \sim T^4$, where T is the equilibrium temperature of the device when exposed to infrared radiation. Cooling significantly improves the performance, and it also reduces the background within the detector assembly. Arrays have been constructed, for instance, the *Infrared Astronomical Satellite* (IRAS), which had a complex focal plane of scanning detectors.[42]

More recently, CCD and solid-state detectors have been added to the complement of devices available for imaging. The *Infrared Space Observatory* (ISO) and *Near Infrared Camera and Multiobject Spectrograph* (NICMOS) on HST have made use of these arrays. They are still comparatively small relative to their optical counterparts, typically only a few hundred pixels on a side, but the technological improvements have been quite rapid. The key problem with the IR is that photon energies are small compared with the bandgap in most semiconductors, for instance, Si, and doping with metals, such as germanium, is required to enhance the infrared performance. Finally, photovoltaic detectors, usually InSn diodes, are also extensively used. These can also be incorporated into arrays and imaging detectors.

---

[40] Timothy, J. G., Morgan, J. S., Slater, D. C., Kasle, D. B., Bybee, R. L., and Culver, H. E. 1989, in *SPIE Proc. Vol. 1158, Ultraviolet Technology III*, vol. III R. E. Huffmann, ed. (Bellingham, WA: SPIE) p. 104; Horch, E., Morgan, J. S., Giaretta, G., and Kasle, D. B. 1992, *PASP*, **104**, 939.
[41] For more complete, current discussions consult the various instrument handbooks for HST and FUSE. The development of MAMA and related area detectors is occurring at a rapid pace and it is best to keep up in real time with the literature by following individual new instruments.
[42] See Beichmann 1987, *ARAA*, **25**, 521. For a comparison with the *Infrared Space Observatory*, ISO, see Genzel, R. and Cesarsky, C. J. 2000, *ARAA*, **38**, 761.

Self-scanned arrays, also called multiplexed arrays, use InSn detectors that are continuously read out to separate accumulators, unlike CCD detectors in which the individual pixels integrate for some period and are then read out sequentially. Instead, they are similar to MAMA detectors or Digicons (self-scanned silicon arrays). Each pixel is stored in a separate register which is then dumped after a specified interval. As such, these detectors produce photon statistics.[43]

A key feature of infrared observations is that everything in the environment radiates, from the telescope superstructure to the mirrors to the atmosphere. Depending on the wavelength, the combined effects of atmospheric opacity due to molecular absorption (especially from H_2O, CO_2, and CO) and background significantly affect the detectability of cosmic sources. Two solutions have been found, neither of which is especially easy. One is to go as far up in the atmosphere as possible. For groundbased observations, this requires altitudes greater than about 3 km in order to remove the atmospheric water vapor as much as possible. The most widely exploited sites are Mauna Kea in Hawaii and Perenal in northern Chile. Long duration balloon flights of single-purpose instruments have long been used at far infrared wavelengths, usually at altitudes of order 30 km. Finally, high altitude aircraft, in particular the *Kuiper Airborne Observatory* (KAO) and the soon to be available *Stratospheric Observatory for Infrared Astronomy* (SOFIA), permit limited duration flights with multi-instrumented facilities for the middle portion of the infrared that is completely unavailable from even mountaintop sites.

The second solution is to bypass the terrestrial sites entirely and go to space. For the infrared, this is a more drastic step than for any other wavelength region. The primary reason is that the detectors must be cooled in order to increase their sensitivity and also to reduce detector background, and the satellites must be carefully shielded from both solar and terrestrial irradiation[44] This limits the working lifetime of satellites due to loss of cryogen. To date, three large scale missions have been accomplished: IRAS, ISO, and the *Cosmic Background Explorer* (COBE). By the time this book appears, both balloon and satellite follow-on missions to COBE will have been launched. We will discuss some of these in Chapter 8. The *Space Infrared Telescope Facility* (SIRTF) is prominent among these.

Calibration of infrared detectors requires the ability to measure the brightness of standard objects. In general, this means the use of objects that are known to emit blackbody spectra. Planets and asteroids are most frequently employed, but standard stars are available for point source sensitivity calibration. Chopping against sky, since the emission from all backgrounds is approximately constant for small angular displacement, provides a variable signal that includes $(S + N)$, source and noise, in one integration and N in the other with subtraction of the signals yielding the source flux. For extended sources this is not always feasible and requires a standard "cold load" in the focal plane. One of the major advances in

---

[43] We thank Susan Kleinmann for discussions of these detectors. She has also pointed out an important feature of their operation, related to the high backgrounds encountered in the infrared. Although InSn detectors have comparatively high readout noise, it is generally quite small compared to the contribution from sky and therefore the statistical properties of the signals are very close to the Poisson distribution.

[44] Recall that the high reflectivity of the Earth's atmosphere, its *albedo*, means that a significant part of the incident solar radiation is scattered back, making the Earth a bright source subtending a large solid angle for any orbiting observatory.

recent space observatory design has been the inclusion of standard blackbody radiators as a part of the instruments to yield absolutely calibrated fluxes. Asteroids are especially useful calibrators, although they are moving particularly useful standard, although they are moving targets. Because they are more or less bricks glowing in the infrared, they are as close as you can reasonably get to blackbody radiators and, since their distances from the Sun and satellite are precisely known, so are their temperatures and intensities.

2.3.6 Higher-Energy Observations

The atmosphere is opaque below about 3000 Å because of ozone absorption, although from high altitude sites it is possible to reach nearly 2700 Å. Therefore, for any high energy photons, there are only airborne (balloon) and satellite alternatives available for observations. In the X ray, several satellites have provided both all-sky surveys and pointed high resolution observations.

The basis of X-ray detection is the pulse height analyzer, the working end of all other instruments. Its physical basis is different from other schemes for photon or particle detection, and this gives us an excuse to discuss it in light of the growing databases from the ASCA, ROSAT, *Chandra*, and XMM missions.

We'll use the example of an incoming charged particle in a gas. The mean free path is a function of energy and therefore the number of liberated electrons due to ionization, the number of particles being proportional to the energy of the incoming particle. Therefore, *within a counting time interval*, the amplitude of the pulse is proportional to the number of liberated electrons which, in turn, is proportional to the energy of the ionizing particle. Should the brightness of the source be high enough, however, a phenomenon called *pileup* occurs in which the arrival rate of photons exceeds the saturation level of the detector. In this case, low-energy photons are shifted to higher energy bins and a spectral distortion results. A problem with high-energy detectors, especially at γ-ray energies, is their relatively poor positional sensitivity. This makes them susceptible to spurious background events, such as charged particles and atmospheric γ events.

Imaging is difficult because of the transparency of the optical materials. One method to improve the location of sources at high energies is coded aperture masks such as those employed on GRANAT/Sigma. Here the pulse location at the two-dimensional proportional counter is cross-correlated with the aperture, which is partially filled with randomly spaced opaque patches.

2.3.7 Radio Astronomy

As ground-based instruments, radio telescopes have historically followed a very different development from optical instruments. This comes, in part, from the early participation of electrical engineers who, in the 1950s, viewed the opening of this wavelength regime as a great opportunity for instrument development. Radar and radio detection and signal processing techniques had explosively expanded in the years immediately following World War II and the centimeter spectral range was where much of this work had centered.

Single radio antenna apertures are around 100 m or smaller and therefore have beams that subtend very large solid angles on the sky compared to optical telescopes. The two exceptions are the Arecibo telescope in Puerto Rico and the Nancay Krauss-type parabola in France. These are not steerable dishes. They operate instead in a modified transit mode. For the Nancay telescope, the secondary reflector, which is mounted vertically, can be tilted through a limited range. The Arecibo telescope employs a steerable secondary that scans an upwardly pointing fixed reflector. This wide acceptance cone leads to source confusion in crowded regions and reduces the effective intensity of a point source. Neither limitation is specific to the millimeter or centimeter regime, but they are more severe because of the long wavelength. Therefore what is actually measured is an *antenna temperature*, T_A, which is the equivalent blackbody luminosity of the source weighted by the solid angle of the telescope. There is an additional problem that affects infrared observations as well, which we include in Appendix 2.A on backgrounds. The receiver is sensitive to emissions from artificial sources and the Rayleigh–Jeans tail of the thermal emission from the Earth. It is also sensitive to the emission from the ionosphere. Some of these backgrounds are reduced with interferometers, especially any incoherent source of radiation that cancels out in the correlation of the antennas of an array, but the coherent signals persist.[45]

Radio observations are usually quoted with reference to a source in thermal equilibrium at the observed wavelength, called the *brightness* temperature, T_B. Defined by

$$F_{\text{source}} = B_\lambda(T_B) = \frac{2kT_B}{\lambda^2}, \qquad (2.10)$$

it is the temperature that a thermal radiator would have *if it emitted the observed monochromatic flux*. The Rayleigh–Jeans approximation is used because, in general, centimeter wavelengths permit this limit. It is an artificial concept, except where you actually expect that the emission process depends on temperature (as, for instance, detection of line radiation from neutral hydrogen or molecules in the interstellar medium). It is not the same as T_A but requires a proper calibration of the point spread function for the telescope. For resolved sources, however, the distinction between antenna and source temperature is not needed and the intensity of the object is quoted in *janskys*, Jy, equal to 10^{-23} erg s^{-1} cm^{-2} Hz^{-1} (or 10^{-26} W m^{-2} Hz^{-1}). The final temperature used to characterize radio observations is the *system* temperature, T_{sys}. This includes all sources of noise from the detector and all of the "downstream" electronics. This is given by

$$T_{\text{sys}} = T_{\text{back}} + T_{\text{sky}} + T_{\text{ground}} + T_{\text{loss}} + T_{\text{cal}} + T_{\text{rec}}. \qquad (2.11)$$

---

[45] Two fine overviews of the field are Kraus, J. D. 1986, *Radio Astronomy*, 2nd ed. (Powell, OH: Cygnus-Quasar); Rolfs, K. and Wilson, T. L. 1996, *Tools of Radio Astronomy*, 2nd ed. (Berlin: Springer-Verlag). More specifics regarding telescopes and detection methods are found in Kraus, J. D. 1950, *Antennas* (NY: McGraw-Hill); Thompson, A. R., Moran, J. M., and Swenson, G. W. 2000, *Interferometry and Synthesis in Radio Astronomy*, 2nd ed. (NY: Wiley); Christansen, W. N. and Högbom, J. A. 1985, *Radiotelescopes*, 2nd ed. (Cambridge, UK: Cambridge Univ. Press). The compilation by NRAO astronomers and engineers, Verschuur, G. and Kellermann, K. I., eds. 1988, *Galactic and Extragalactic Radio Astronomy*, 2nd ed. (Berlin: Springer-Verlag) is particularly comprehensive in scope.

130   THE RAW MATERIAL: INSTRUMENTS AND OBSERVATIONS

Here T_{back} is the contribution from all cosmic backgrounds above the atmosphere, including the cosmic background radiation (CBR; see Chapter 8), and the Galactic thermal and nonthermal background (see Chapter 6). The terrestrial sources include the ground irradiation of the telescope, since the Earth is roughly a 300 K blackbody, T_{ground}, and thermal emission from the sky and ionosphere and nonthermal from the magnetosphere, T_{sky}. The detector electronics have resistive losses in the lines feeding the receiver, T_{loss}, and the calibration sources inject a quantity T_{cal}. Finally, the receiver has a finite resistance and so contributes T_{rec}. The total noise power is therefore

$$P_{\text{sys}} = GkT_{\text{sys}}\Delta\nu \qquad (2.12)$$

where $\Delta\nu$ is the bandpass of the receiver and G is the gain. The telescope sensitivity for a single antenna depends on the signal to noise ratio for the source and on the integration time (see Appendix 2.B on statistics) and is assumed to be due to thermal fluctuations so that the noise depends on the integration time through $(\Delta t)^{1/2}$. Thus, the sensitivity for an antenna with an area A and beam efficiency η (which depends on the solid angle subtended by the antenna and detector combination) is given by

$$F_{\min} = \frac{2kT_{\text{sys}}}{AG\lambda^2(\Delta t\,\Delta\nu)^{1/2}} \qquad (2.13)$$

Typical system temperatures, depending on wavelength, run from tens of degrees to about 1000 K[46] depending mainly on the sky contribution (especially from water vapor at millimeter wavelengths). Calibrators must be observed frequently to obtain the sky emissivity and chopping between source and sky is also used to reduce the effects of the background. Infrared observations suffer from the same contributions since, in general, the sky and other terrestrial emissions are important radiation sources. In addition, the radiation from the components of the telescope itself, such as the mirrors, contribute to the backgrounds.

2.4   SPECTROGRAPHS

There are two main types of spectrographs in use in modern observations. One is designed to give stable, high spectral resolution in order to provide line profiles and wavelengths with the highest possible throughput, and the other is designed to measure velocities. The first, usually using holographic or echelle gratings, is now capable of reaching spectral resolutions of order $R = \lambda/\Delta\lambda$ of order 10^6, while the latter can measure velocities of order 10 m s^{-1}.

---

[46] The usual way that these numbers are quoted is as the *noise-equivalent power* (NEP) in units of W Hz$^{-1/2}$.

2.4.1 Gratings and Resolution

The echelle grating was invented by Michelson[47] as a high throughput method for achieving very high spectral resolution at low cost when it is impossible to rule a very high resolution grating. Otherwise, the grating consists of a surface of grooves that have been mechanically ruled or holographically etched. The quality of the groove profile determines the scattered light contribution to the background. The surfaces need not be planar. For example, Rowland circle gratings, routinely employed in laboratory spectrographs, are used for the *Far Ultraviolet Spectrographic Explorer* (FUSE). From simple interference principles you can obtain the *grating equation*:

$$m\lambda = d(\cos r - \cos i), \qquad (2.14)$$

where m is the *order* of the dispersed image, d is the groove spacing, and i and r are the angles of incidence and reflection, respectively. For optical gratings, $m \geq 0$. In X-ray spectroscopy, where grazing incidence is required because of the transparency of the grating, $m \leq 0$. The undispersed, or *zeroth-order* image can be used along with filters for imaging or any of the orders can be redispersed to achieve a higher resolution, as you would for an echelle. Specific orders are selected by the shape of the grooves, called *blazing*; most operate in first or second order. This reduces order overlap, as do blocking filters, but sometimes does not eliminate it and care should be exercised by observing standard sources (e.g., in second order you may see the first order image of a shorter wavelength part of the spectrum).

In general, the spectral resolution is set by the combination of the grating spacing and the projected slit width. For precise spectrophotometry, the slit should be matched as closely as possible to the point spread function of the optical system and sky, in order to ensure that you have not lost flux. Slitless systems are used in some cases for obtaining a set of monochromatic images of the source, such as planetary nebulae, or for low resolution work on wide fields. One method is to use a prism mounted in front of the telescope aperture, the *objective prism* method. This is like walking down the street with a prism in front of your eyes. You see everything with a low resolution dispersed spectrum. As you can imagine, the confusion problem can be extreme in crowded fields, but it has the advantage of very large throughout for survey work. A more compact implementation of this is a *grism*, the combination just before the focal plane of the detector of a transmission grating mounted on a thin prism for separation of the spectra (usually with $\lambda/\Delta\lambda \sim 100 - 10^3$. Holographic gratings and echelles can reach $\sim 10^6$ or higher. Satellite UV spectroscopy is now routine between 10^4 and 10^5. X-ray spectral resolutions are still improving but routine observations with 10^3 are now available.

---

[47] For a useful introduction, see Jenkins, F. A. and White, H. E. 1976, *Fundamentals of Optics*, 4th ed. (NY: McGraw-Hill) and Born, M. and Wolf, E. 1999 (n.1).

2.4.2 Correlation Spectrometers

The correlation technique, first described by Griffin[48] uses a template spectrum and the entire wavelength region of the spectrum without making use of the spectrum itself. The dispersed spectrum of a star is matched to the focal plane resolution of the template and is then imaged onto a photomultiplier. The template is shifted by the resolution of the grating and the variation of the throughput is recorded. In this way, a cross correlation function, $C(\Delta\lambda)$, is obtained, the peak of which gives the shift between the stellar and laboratory spectrum. In its simplest form, this can be done without a template at all. Differential velocities can be directly measured using a single high signal to noise spectrum accumulated for either a standard of known radial velocity or the target itself by digitally performing the correlation. In this way, if the spectrum has especially narrow features, statistical resolutions far exceeding that of the spectrum can be obtained. Figure 2.5 shows how this is performed. Several automated instruments now exist for such measurements, and for faint sources the advantages of such techniques have proved to be enormous. Applied to halo and globular cluster stars, it has provided the first measurement of long period binaries while it is the backbone of the large scale structure surveys for galaxies carried out by several groups during the 1990s and beyond. Profound results have emerged from the long term application of this method. Planetary companions to solar type stars, thought initially unmeasurable only a decade ago, are now being detected in increasing numbers through the application of correlation techniques. The addition of modern detectors and the high throughput available with CORAVEL and similar instruments following decades of data collection can now detect radial velocity amplitudes of only a few hundred meters per second with periods of order years, typical of Jupiter-sized companions.[49] The newer methods employ a fundamental standard gas cell through which the stellar spectrum is projected. Although the first implementation by the DAO group used hydrogen fluoride, modern implementations use iodine as the wavelength standard, achieving accuracies of order 0.05 km s^{-1} over long timescales.

2.4.3 Multiobject Spectrographs (MOSs)

Beginning in the late 1980s, through the combined improvements in computer control and optical fiber technologies, a new class of instruments began to appear. These use the high transmission of modern fibers to virtually eliminate the slit from the spectrograph. Instead, individual fibers serve as the apertures and these are aligned at an exit, essentially a synthesized slit, to produce a stacked set of input images for dispersion by a grating. The advantage of this arrangement is that dozens of objects in the focal plane can be observed *simultaneously*. The multiplexing advantage of this arrangement, facilitated by precise astrometry before setting

---

[48] Griffin, R. F. 1967, *ApJ*, **148**, 465. See also specific descriptions of CORAVEL by Baranne, A., Mayor, M., and Poncet, J. L. 1979, *Vistas in Astr.*, **23**, 279, and other implementations of the basic idea by Griffin, R. F. and Gunn, J. E. 1974, *ApJ*, **191**, 545 and Fletcher, J. M. et al., 1982, *PASP*, **94**, 1017, among others.

[49] See reviews by Marcy, G. W. and Butler, R. P. 1998, *ARAA*, **36**, 57; Marcy, G. W. and Butler, R. P. 2000, *PASP*, **112**, 137.

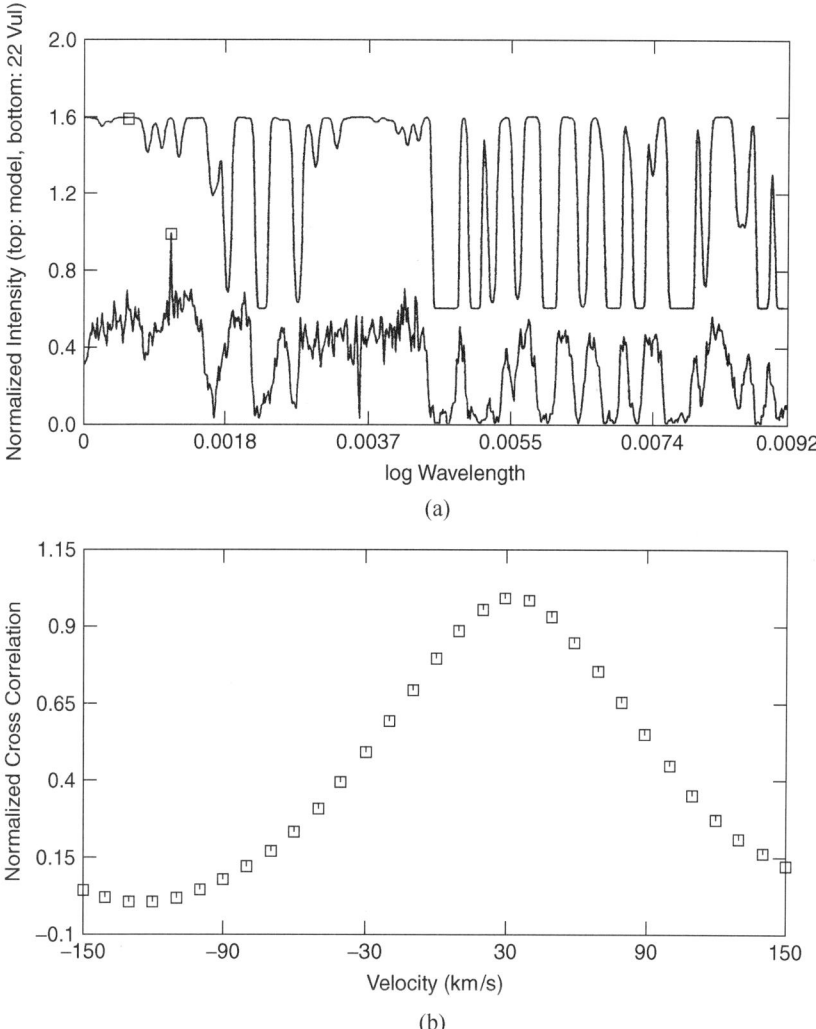

FIGURE 2.5. (a) Sample spectrum (in this case the ζ Aur binary 22 Vul, lower spectrum) and a theoretical template spectrum (laboratory rest wavelengths, upper spectrum). (b) Correlation function showing the peak detection. This is the basis of CORAVEL and other correlation spectrometers.

the fibers on top of the images, means a small increase in the work at the outset, but results in tremendous reductions in the time required to observe many faint objects. The prototype instruments required much more labor in advance of the observations, in that the observer was required to construct a template, based on the positions of the objects, to which the fibers were glued. Now, multiarm or *hydra*, instruments allow the observer to steer the fiber to the object and eliminate the disadvantages of the cumbersome plates. The technique is routinely used now for groundbased observations of galaxies in large scale surveys, where as many as 100 spectra can be obtained in a single exposure. This dramatically contrasts with the productivity of conventional, nonmultiplexed spectrographs that could produce only one spectrum in the same exposure time. It is, however, still true that a clear night at the telescope is one of the most precious commodities known and not even a second of that time should *ever* be wasted.

2.5 IMAGE FORMATION

All telescopes are ultimately Fourier transform devices, meaning that the formation of an image is essentially a result of phase combination in the focal plane. There are two different types of telescope apertures. Filled apertures are monolithic or segmented reflectors. Synthesized apertures are produced by interferometers, which are arrays of filled apertures. Whether filled or synthesized, the aperture is dominated by a point spread function that results from interference at the focus of different parts of the wavefront that undergo differential phase delays through the optical components of the instrument. We will discuss this at some length because it is the basis of much of the modern work on telescope design and because the restoration of images has become a major ingredient of astrophysical investigations.

2.5.1 What Is an Image?

An image is the two dimensional focal plane distribution of photons produced by an optical element (i.e., telescope) combined with some sort of detector. The formation of an image of an *astronomical* source can be separated into two stages: (1) the transfer of the photon through the atmosphere and (2) its transfer through the telescope and detector combination to the focal plane. In both cases, the resulting image is the accumulated history of arrival locations for the photon. The transfer function is the probability distribution intrinsic to the medium or telescope that determines the angular distribution throughout the system. For a telescope, the integration over the entrance is achieved at the focus. Light from different parts of the mirror arrives at the focal plane with slightly different pathlengths, hence with slightly different phases. Summing the contributions of each area element of the mirror is really a sum over spatial frequencies and, consequently, is a Fourier transform of the aperture. Light of wavenumber $k = 2\pi/\lambda$ and direction $\hat{\mathbf{k}}$ is incident at a directional cosine $\hat{\mathbf{n}} \cdot \hat{\mathbf{k}}$, where $\hat{\mathbf{n}}$ is the surface normal to the mirror of the telescope. The focal plane image is formed by integration of the amplitude A over the aperture with radial distance r from the center of the mirror of aperture D, and the intensity as a function of angle θ on the sky is therefore given by

$$I(\theta) = \left| \int A_0(r) e^{ikr\hat{\mathbf{n}} \cdot \hat{\mathbf{k}}} 2\pi r \, dr \right|^2. \tag{2.15}$$

For a circular aperture, this integral reduces to

$$I(\theta) = \left| \frac{J_1(kD\theta)}{kD\theta} \right|^2 \tag{2.16}$$

where θ is now the angular size of the source and $J_1(x)$ is the first ordinary Bessel function. The resolution is set by the first zero of this function. This zero gives the diffraction limit of the aperture, which is approximately $\theta \approx \lambda/D$. For optical wavelengths, you see that the angular spread, also called the *point spread function* (psf), is extremely small, of order 10^{-6} rad or of order a few arcseconds for even very modest apertures. In principle, even a 1 m telescope can achieve 0.1-arcsec resolution. For centimeter and millimeter wavelengths, this is not at all good. You see enormous blocks of sky for apertures less than a few tens of meters, and such large telescopes are prohibitively expensive to construct—hence the requirement of aperture synthesis. On the other hand, optical telescopes cannot achieve this resolution in practice. The turbidity of the atmosphere is a fundamental limit to image formation. That is why the *Hubble Space Telescope* was put in space and why such heroic efforts are now being made to produce adaptive optical apertures. There is always a tradeoff in observational work. What you gain in flux with filled apertures you pay for with the technical complexity of the solution. With synthesized apertures from arrays, you pay with complicated point spread functions, the need to perform systematic image reconstruction, and loss of flux, but you gain enormously in angular resolution.

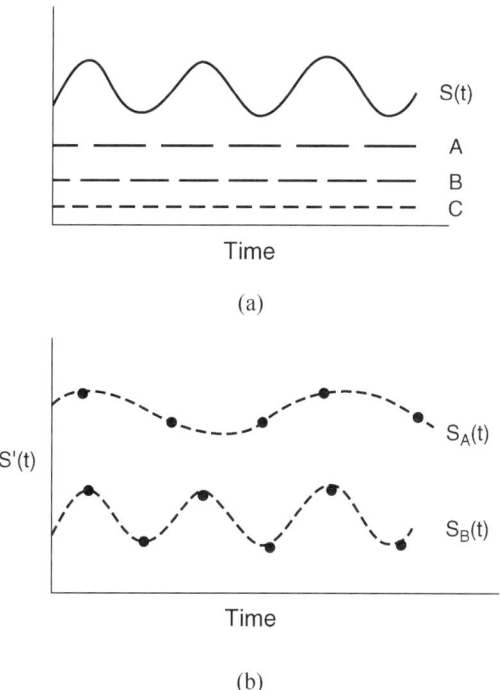

FIGURE 2.6. Illustration of the effects of sampling on the reconstruction of a signal. (a) Three windows are shown: undersampled (A), approximately critical (Nyquist) sampling (B), and oversampled (C). (b) The approximate inferred signal is shown for curves (A) and (B). Notice how the Nyquist frequency sample is sensitive to phase (what happens if you shift the window?) and the slight jitter in the actual window function actually facilitates detection of the signal.

2.5.2 The Nyquist Frequency and Resolution

The Nyquist frequency represents the limiting sampling rate for a signal and forms the basis of all sampling theory. Simply put, you cannot resolve a frequency in the data, whether it is the spatial or temporal frequency, unless you have at least two samples per period. This is the fundamental limit to the ability of any device to resolve a signal. One way to see this is to imagine a periodic signal consisting of a series of pulses of duration τ each with amplitudes $+1$ and -1 in succession. If you sample less frequently than twice per cycle, the mean value you will get is lower than the amplitude *and* you will get the wrong period (see Fig. 2.6). This is the effect of *aliasing*, the bane of all signal processing. If, on the other hand, you have sampled the signal more often than once per τ, you will get the correct frequency although it may not have precisely the right amplitude. The more times you *oversample* the signal, the better off you are. Thus, the Nyquist frequency is

$$\nu_{\text{Nyquist}} = \frac{2}{\Delta t} \qquad (2.17)$$

for a temporal signal, and similar limits result for spatial sampling. You must exceed ν_{Nyquist} to resolve the signal. For instance, when imaging a source by combining individual pointings from a single aperture, such as constructing an image from scans, there must be some degree of oversampling of the projected size of the aperture (the *primary beam*). This is usually accomplished by at least half-beam separations between pointing centers. In aperture synthesis, the Nyquist spatial frequency is the limit beyond which a source cannot be resolved by the reconstructed point spread function.

2.5.3 Interferometers

There is no fundamental difference between the formation of an image by a telescope, the separation of a light signal into different wavelengths by a diffraction grating, and the operation of an interferometer. In all cases, the key feature is that there is a phase difference between signals taking different paths that, when combined, produce a variation in the intensity at the detector.[50]

The first practical stellar interferometer was first constructed around 1900 by Michelson, and this is still the workhorse of all such devices. A variant was used to detect the radius of the supergiant α Orionis by placing two mirrors on arms at the entrance of the 100 in Mt. Wilson telescope in the 1920s and observing the separation at which fringes disappeared. These instruments have two separate uses in astronomical observations. They provide extremely high spectral resolution

---

[50] The most widely used reference for matters optical is Born and Wolf (1999, n.1), although it is worthwhile taking a look at the venerable classic by Michelson, A. A. 1927, *Light Waves and Their Uses* (Chicago: Univ. of Chicago Press) and Michelson, A. A. 1890, *Phil Mag.*, **30**, 1 for a general picture of how it all started. See also Jenkins and White (1974, n.47), which contains an excellent description of the echelle grating and Fabry-Perot interferometers. See also Steel, W. H. 1983, *Interferometers*, 2nd ed. (Cambridge, UK: Cambridge Univ. Press); Cook, A. H. 1971, *The Interference of Electromagnetic Radiation* (Oxford: Oxford Univ. Press).

and can be used in a nonimaging mode, with the exception of speckle techniques. They have also been adapted to imaging in order to achieve the highest spatial resolution, especially Michelson and intensity interferometers. When used this way, they can be one-dimensional devices, used simply to obtain radii, or can be used in combination as synthetic apertures.

2.5.3.1 Michelson Interferometer

In an attempt to gain the highest spectral resolution possible, the Michelson interferometer uses two interfering paths that produce fringing when the path difference is introduced through the motion of one mirror. This signal is therefore an intensity as a function of the path difference between the arms of the interferometer. This is the Fourier transform of the intensity as a function of wavenumber, or spatial frequency (which is the conjugate variable). Thus

$$I(k) = \frac{1}{2\pi}\int_{-\infty}^{\infty} I(\Delta x) e^{ik\Delta x} d\Delta x \qquad (2.18)$$

for a linear detector. In constructing a spectrograph, the traverse Δx determines the effective resolution in wavenumber Δk. The same is true for the fringes observed for separated mirrors where the beam combination is achieved by a telescope at its focus.[51] The fringe visibility is essentially the cross correlation function of the amplitudes:

$$\Gamma(u,v) = \int I(\xi, \eta) \exp(i[u\xi + v\eta]) \, d\xi \, d\eta, \qquad (2.19)$$

where (u, v) is the dual coordinate system to that on the sky. We will return to this point in a moment. Suppose you are observing a single star. Then the combination of the two beams gives a variation of the intensity:

$$I_{12} = I(1 + 2V \cos kL\hat{\mathbf{b}} \cdot \theta). \qquad (2.20)$$

The visibility V depends on the model adopted for the source. For a point source, $V = 1$. For a uniform disk of angular diameter $\Delta \theta$, $V = 2J_1(kL\,\Delta\theta/2)/kL\,\Delta\theta$. The principle is the same as the one we have used for the basic Michelson interferometer. If you are observing a binary star, the visibilities are given by

$$\Gamma_{12}^2 = \frac{1}{(1+R)^2}\left[\Gamma_1^2 + (R\Gamma_2)^2 + 2R\Gamma_1\Gamma_2 \cos kL\,\Delta\theta\hat{\mathbf{b}} \cdot \theta\right],$$

where the Γ_i factors are the individual visibilities, R is the intensity ratio between the components, and $\Delta \theta$ is the angular separation between the components. The

---

[51] This first achieved in the 1920s by Michelson and Pease at Mt. Wilson, see Michaelson, A. A. and Pease, F. G. 1921, *ApJ*, **53**, 249; reprinted with commentary in Townes, C. H. 2000, *ApJ*, **525**, 148, the AAS Centennial Edition of the *Astrophysical Journal*.

angular separation is reached at the first zero of Γ_{12} as the source rotates across the baseline.[52]

2.5.3.2 Fabry–Perot Interferometer

This interferometer is used mainly for imaging in very narrow bandpasses. Unlike the Michelson-type device, it has the advantage of two-dimensional throughput but it cannot achieve the highest resolution except for relatively narrow emission lines. For high spectral resolution, it has been exploited on stars for interstellar and stellar line profiles with $\lambda/\Delta\lambda > 10^6$. With its high throughput and good spectral coverage, it has been used for galaxies with increasing success now that CCD arrays have grown sufficiently large and with high enough spatial resolution to be useful.[53] This device uses multiple interference between separated plates in a slitless mode. For an entrance angle θ, two adjacent rays have a path difference of $2ct\cos\theta$. The PEPSIOS system, used since the mid-1960s, changes the index of refraction of gas placed between *fixed* etalon plates by varying the pressure, so there are no constraints from the mechanical stability of the system. Two of the spectrographs on ISO have used this spectrograph, which has the key advantage of throughput for long wavelengths.

The operational principle is quite straightforward. Take ψ_n to be the nth mode of interference and a partial wave has a phase $\phi_n = n\phi$. Take r to be the reflectance of the etalon, and assume the transmission is $1 - r$. The individual amplitudes add as

$$A_\omega(t) = \sum_{k=0}^{\infty} (1-r) r^k e^{i\omega t} e^{-2n\phi k} = \frac{1-r}{1 - r\exp{-2in\phi}} e^{i\omega t}. \qquad (2.21)$$

The intensity, $I = A^*A$, is given by

$$I = \frac{(1-r)^2}{1 + r^2 - 2r\cos 2\phi n}, \qquad (2.22)$$

so for half-integer values of n, I is minimum while for integer values it is maximum. Then, if the amplitudes are independent of mode n, the intensity

---

[52] See the review by Shao, M. and Colavita, M. M. 1992, *ARAA*, **30**, 457; Shao, M. et al. 1988, *Astr. Ap.*, **193**, 357; Tango, W. J. and Twiss, R. Q. 1980, *Prog. in Optics*, **17**, 239; Hanbury-Brown, R. and Twiss, R. Q. 1956, *Nature*, **178**, 1046 (see also Mandel, L. and Wolf, E., eds. 1970, *Selected Papers Coherence and Fluctuations of Light*, vol. 1, 1850–1960 (NY: Dover Books); Hanbury-Brown, R. 1974, *The Intensity Interferometer* (London: Taylor and Francis); for the infrared, see Dyck, Benson, and Ridgway, S. 1993, *PASP*, **105**, 610. For binary star observations, see Pan, X. et al. 1992, *ApJ*, **384**, 624.

[53] Strong, J. 1958, *Concepts of Classical Optics* (San Francisco: Freeman Press); Hernandez, G. 1986, *Fabry-Perot Interferometers* Cambridge, UK: Cambridge Univ. Press; Cook, A. H. 1971, *Interference of Electromagnetic Radiation* (Oxford: Oxford Univ. Press); Vaughan, J. M. 1989, *The Fabry-Perot Interferometer: History, Theory, Practice, and Applications* (Bristol: A. Hilger); Hobbs, L. 1967, *ApJ*, **157**, 135 describes the PEPSIOS system for high-resolution spectroscopy, Meaburn, J. 1968, *Ap. Space Sci.*, **2**, 117 provides a good introduction to nebular spectral imaging.

produced by the sum of the N reflections is

$$I \sim \left| \sum_{n=0}^{N} \psi_n e^{i\phi_n} \right|^2 \sim \frac{|1 - e^{i(N+1)\phi}|^2}{|1 - e^{i\phi}|^2}. \tag{2.23}$$

For real ϕ we obtain

$$I \sim \frac{\sin^2(N+1)\phi/2}{\sin^2 \phi/2}, \tag{2.24}$$

so the first zero is at $\phi = 2\pi/(N+1)$, where $\phi = k\,\Delta x$ for an etalon spacing Δx. For the case where each reflection has a coefficient r, then $\psi_n \sim r^n$ and

$$I \sim \left| \frac{1 - re^{-i(N+1)\phi}}{1 - re^{-i\phi}} \right|^2, \tag{2.25}$$

which, in the limit of $N \to \infty$, again becomes

$$I \sim (1 + r^2 - 2r\cos\phi)^{-1}. \tag{2.26}$$

The Fabry–Perot is therefore capable of producing very-high-resolution line profiles,

$$\frac{\lambda}{\Delta\lambda} \approx 3n\frac{r^{1/2}}{1-r},$$

especially if a filter is used to isolate the spectral region of interest (a monochrometer can also be an upstream interferometer to serve as an initial dispersing element that then is isolated using a slit assembly). The most advanced systems are based on this design, made possible with the improvement in CCD imaging detectors (e.g., the TAURUS spectrometer). Gravitational wave interferometers also employ the Fabry–Perot method to lock laser signals and to extend pathlengths.

2.5.4 Intensity Interferometers

When you combine two beams of light, the signals interfere in a rather simple way. Hanbury-Brown and Twiss (1956) first realized that the amplitudes were not the only way to do this, and that a correlation of the intensity from each beam would also interfere. This was the basis for the intensity interferometer, which is really an optical version of the modern aperture synthesis radio telescope and the new generation of optical interferometers, such as the European Southern Observatory VLT (four 8 m telescopes that can be operated independently or interferometrically), and the Keck I/II pair of 10 meter telescopes on Mauna Kea. The idea is to take the full aperture intensity from two widely separated inputs and use the telescope, in effect, as a light bucket for collecting as many photons as possible from the wavefront. There is no fundamental difference between the operation of this sort of array and any other that is based on interference.

2.5.5 Aperture Synthesis

Conventional telescopes are filled apertures; that is, at every point in the solid angle of the instrument there is a reflecting surface from which an image can be formed. The essential idea behind *aperture synthesis* is the simulation of a filled aperture through the use of a finite number of elements. The technique was pioneered in radio astronomy by necessity. Radio wavelengths require enormous apertures to achieve any reasonable resolution since λ is typically of order tens of centimeters to many meters. The largest single steerable dishes, like the Green Bank Telescope (GBT) and Effelsberg, are prohibitively expensive to construct on scales larger than 100 m, and it is very hard to maintain an adequate surface figure in the presence of gravitationally induced distortions for anything larger. The largest single dish, Arecibo, only partly circumvents this. With a nearly 300-m aperture, it is a fixed telescope mounted in a very large karsed hole in uplifted terrain. Operated in a modified transit mode, the telescope is restricted to viewing the local zenith and a limited angle away from it through a moving secondary that views different parts of the focal plane. Yet these telescopes still provide only limited angular resolution, and none can compete with even comparatively small optical telescopes.

Instead, one can make use of the interference between signals as a function of separation of the telescopes viewed from the sky. This is the (u, v) plane, the projection of the baseline of a pair of antennas in an interferometer onto the celestial sphere. Filled aperture telescopes can be thought of as a continuum of interferometers, since they are basically Fourier transform devices, that take sample a finite solid angle and combine them in the focal plans. An interferometer does much the same thing, only the signal is really a correlation between different detectors that view the source with their individual solid angles, Ω. Take a wavefront incident on a pair of telescopes. They are pointed in a direction $\hat{\zeta}$ on the sphere. Then the projected baseline is $L\hat{n} \cdot \hat{\zeta}$, where $\hat{n}$ is the baseline normal vector. Now this means that there is a time delay between the signals detected at the two telescopes that depends on the directional cosine, with the result that the combination of the two signals, $V_1(t)$ and $V_2(t)$, produces a beat with a phase shift $cL/\lambda \sin \Theta$, where $\Theta = \cos^{-1} \hat{n} \cdot \hat{\zeta}$. Thus, as the source moves across the interferometer, there is a point where the wavefronts coincide. Proper delays must be built into the system to ensure coherent propagation of the detected signal from each antenna to the central combining unit, the *correlator*. The detected signal has an intensity $\langle V_1(t)V_2^\star(t)\rangle$, where the brackets indicate some time average over an integration period of the detector for the combined amplitudes.

Let's be a bit more precise. Modern aperture synthesis is based on the ability to track the center of the field with the ensemble of telescope apertures. This is the *phase center*, the location in the sky to which all antennas continue to point. Relative to this point, there will be slight phase differences seen on each baseline because of the angular separation of the emission regions from the center of the field. If we call these deviant positions (x, y), then the intensity seen by the interferometer will be

$$I(u,v) = \int I(x,y) \mathsf{P}(x,y) e^{-i(ux+vy)} \, dx \, dy \qquad (2.27)$$

for an antenna pattern (the beam of the interferometer) P. You see only those parts of the wavefront that interfere coherently, and this means that the resulting map is particularly sensitive to structured rather than diffuse emission, a point to which we will return in a moment. This is a Fourier transform of the intensity distribution on the sky, again emphasizing that the image formed by a telescope, whether filled aperture or an interferometer, is a phase-summed intensity distribution.

You now have several options. One is to shift the antennas, analogous to a Michelson interferometer, to change their separation in order to sample a range of baselines. This is cumbersome, to say the least, when the dishes are 25 m or more across. The other is to use *aperture synthesis*. The trick in aperture synthesis[54] is to let the Earth do the work of moving the baselines around the sky. As the baseline rotates, or in the frame of the interferometer as the sky rotates, the projected baselines sweep out trajectories that cover the sky. There are $\frac{1}{2}N(N-1)$ such paths, where N is the number of elements in the array. An important feature of the method is that *the number of baselines exceeds the number of elements provided $N > 3$*. A technique called *phase closure* also applies here, in which noise from the individual elements in the array cancels on correlation if there are ≥ 3 elements. The spatial frequency is sampled by the elements of the interferometer, so that the effective resolution is given by the longest baseline:

$$\Delta\theta \sim \frac{\lambda}{L}. \qquad (2.28)$$

Fundamentally there is no difference between a synthesized aperture and its filled cousin, except that it is as if pieces are missing from the phase map of the sky. Some parts of the sky will not have been observed by any instrument and the synthesized beam is always much smaller than that of the individual telescope. Therefore, we have to use restoration methods to recover the sky distribution from the synthetic image.

The principal instruments currently available for millimeter and centimeter wavelength aperture synthesis observations are the very large array (VLA) (27 identical 25-m telescopes in a variable configuration, maximum resolution approximately 0.1 arcsec), the very long baseline array (an additional 10 similar antennas distributed on a continental baseline with a resolution of order 10 μarcsec), the Australia Telescope (AT, a southern hemisphere array of design similar to that of the VLA), and MERLIN. The first user-oriented arrays to be constructed were both linear arrays, Westerbork (still in use) and the Green Bank Interferometer. For shorter wavelengths, the main instruments are IRAM (millimeter multielement array of 30-m telescopes), Owens Valley Radio Observatory (OVRO), and the Berkeley-Illinois-Maryland Array (BIMA). Still to be con-

---

[54] Thompson, A. R., Moran, J. M., and Swenson, G. W. 2000, *Interferometry and Synthesis in Radio Astronomy*, 2nd ed. (New York: Wiley); Burke, B. F. and Graham-Smith, F. 1997, *Introduction to Radio Astronomy* (NY: Cambridge Univ. Press); Perley, R. A., Schwab, F. R., and Bridle, A. H. eds. 1989, *Synthesis Imaging in Radio Astronomy* (San Francisco: ASP Conf. Vol. 6).

structed is the large Chilean Millimeter Array (ALMA). Extending this method to continental and intercontinental scale baselines for VBLI observations requires considerable refinement of time standards and recording devices. Instead of using a realtime on-site correlator, the signals must be recorded on magnetic tape with time stamping and then combined after the fact. A complication introduced by the separation is that the array cannot be treated as a plane relative to the sky because of the radius of the Earth. The sky sampling is also much sparser, and the telescopes are far from identical, so the synthesized bean is much more complex.

2.5.6 Scintillation: Atmospheric Seeing

Every photon that enters the atmosphere runs an obstacle course. Turbulent eddies, resulting from small-scale shear flows and local temperature differences in the medium, and convective cells formed in response to the vertical temperature gradient, produce random variations of the refractive index along a given path. In turn, this means that all photon paths suffer both phase delay and geometric dispersion. To further complicate matters, remember that n, the index of refraction, depends on wavelength, so refraction introduces a chromatic dependence to the size and appearance of the image of a source that is outside the atmosphere. Perhaps the best example of this is the appearance of the *Space Shuttle* during launch. At a distance of about 10 km from the launch site it is a resolved body, although with a very small angular diameter. By 150 or 200 km, it is completely unresolved and two effects appear. The first is that the images of the shuttle separate depending on wavelength—that is, there are multiple images of the spacecraft.[55] The second is the appearance of *scintillation* or twinkling, resulting from the line of sight turbulence.

The buoyant thermal fluctuations in the Earth's atmosphere produce fluctuations in the paths of individual photons over both space and time. The individual paths are different because the phase speed, $v_p = c/n$, is inversely proportional to the refractive index. Thus, there is a phase shift that is proportional to the rms of the density and/or temperature fluctuations. This follows from linearizing expansion of n in terms of T and ρ, assuming that the turbulence is locally in pressure balance this is similar to the mixing length theory in the Schwarzschild approximation that we discuss in Chapter 3. The image varies because the focus combines waves that are alternately in and out of phase, hence constructively and destructively interfering. The result is a temporal variation of the amplitude, and there-

---

[55] See Greenler, R. 1983, *Rainbows, Halos, and Glories* (Cambridge: Cambridge Univ. Press) for further examples of this and other striking atmospheric effects. For more on the atmospheric structure for astronomical observation, see Goody, R. 1995, *Principles of Atmospheric Physics and Chemistry* (London: Oxford Univ. Press); Goody, R. and Yuan, Y. L. 1990, *Atmospheric Radiation: Theoretical Basis*, 2nd ed. (NY: Oxford Univ. Press). For some earlier techniques and historical development, see Middleton, W. E. K. 1952, *Vision through the Atmosphere* (Toronto: Univ. of Toronto Press). The classic treatments of turbulent transfer are Tatarski, V. I. 1961, *Wave Propagation in a Turbulent Medium* (NY: McGraw-Hill) and Tatarski, V. I. 1971, *The Effects of the Turbulent Atmosphere on Wave Propagation* (Springfield VA: National Tech. Info. Center).

fore of the intensity.[56] Atmospheric fluctuations produce the seeing disk even before the light reaches the aperture through distortion of the incoming wavefront. Consider the amplitude, A, to be a function of angle θ and write

$$A(\theta) = \int d\Omega_0 \, A(\theta_0) G(\theta - \theta_0) \qquad (2.29)$$

for the amplitude after it has passed through the distortion of the atmosphere via some transfer function, G. In other words, the image you see is a convolution, resulting from adding up all of the parts of the image at a point θ that come from the different paths. Thus, $A = G \star A_0$ or, in the Fourier domain, $\hat{A} = \hat{G}\hat{A}_0$, using the convolution theorem. This is the same effect as for the image formation by any focusing optical system, for instance, for a telescope, but in the atmosphere we have a stochastic medium and the *transfer function*, or psf, is not strictly stable. The formation of the structure of an image in the focal plane of an optical instrument results from the interference between the component wavefronts of the image due to the finite solid angle of the instrument, and is the resolution of the telescope and its detectors. For the atmosphere, this changes as a function of time and produces a disk for the star that is stable only on long timescale when many of the individual images are added together.

Now let us see how a stochastic transfer function works. The phase of a wavefront is given by $\phi = (\omega/c)\int n \, ds$, where n is the index of refraction for a pathlength s through the atmosphere. For a plane wave, if $\mathbf{x}$ is the direction of the source, then

$$A(\mathbf{x}) = \exp k \int_z^{z+\delta z} n(\mathbf{x}, s) \, ds,$$

where k is the wavenumber. Now we must make a choice for the fluctuation spectrum. The simplest one to assume a Gaussian distribution of statistically independent fluctuations, although when we discuss turbulence in Chapter 6, you will see that this is not the best picture for the behavior of the density and velocity field of a turbulent medium. For a Gaussian, there are two parameters, the mean and dispersion. If we assume that the mean fluctuation of A vanishes, then we can define two moments of the amplitude:

$$\langle \delta A(\mathbf{x}) \rangle = 0, \qquad G(\mathbf{x}) = \langle \delta A(\mathbf{x}) \delta A^*(\mathbf{x} + \boldsymbol{\zeta}) \rangle \qquad (2.30)$$

where the latter is the correlation function for the amplitude. This depends on the

---

[56] The dry atmosphere index of refraction is a function of P, the atmospheric pressure, and T, the temperature as well as the wavelength:

$$n - 1 = 7.76 \times 10^{-5} \frac{P_{\text{mbar}}}{T} \left(1 + 7.52 \times 10^{-3} \lambda_\mu^{-2}\right)$$

where λ_μ is in microns.

fluctuation in the phases, so that it can be replaced by

$$G(\mathbf{x}) = \exp\langle i[\phi(\mathbf{x} + \boldsymbol{\zeta}) - \phi(\boldsymbol{\zeta})]\rangle \qquad (2.31)$$

We now use the expansion for some random variable u

$$\langle e^{iu}\rangle \approx 1 + i\langle u\rangle - \tfrac{1}{2}\langle u^2\rangle + \cdots$$

and note that $\langle u\rangle = 0$ to obtain on substitution

$$G(\mathbf{x}) \approx \exp\left[-\tfrac{1}{2}D_\phi(\mathbf{x})\right], \qquad (2.32)$$

where we have used

$$D_\phi = \langle |\phi(\mathbf{x} + \boldsymbol{\zeta}) - \phi(\boldsymbol{\zeta})|^2\rangle \qquad (2.33)$$

for the the *phase dispersion*. In turbulence theory, this is called the *structure function* for the phases. In general, in a turbulent medium, the more distant the fluctuations, the less correlated they become. You see that the phase dispersion arises from the fluctuations in the refractive index along a path, which, in turn, depend on the turbulence spectrum. Anticipating the discussion of turbulence in Chapter 6, we assert that the energy in the turbulent velocity fluctuations has the *Kolmogorov spectrum* $E(k) \sim k^{-5/3}$. This is assumed to be the same as for the density and so also for the index of refraction fluctuations. Now call $\hat{D}_\phi(k)$ the Fourier transform of the structure function. For the Kolmogorov spectrum the largest power is at the largest spatial scale, called the source length or the *outer turbulence scale*. Then we find

$$\hat{D}_\phi(k) \sim C_n^2 k^{-11/3}, \qquad (2.34)$$

where C_n is a constant for the refractive index evaluated for homogeneous isotropic turbulence. A point source seen through the atmosphere encounters very few large turbulent cells and two points close enough together on the sky are likely to remain coherent if they are separated by a distance $\theta \approx r_0/z_0$, where z_0 is the "thickness of the atmosphere" (about 20–30 km) and r_0 is called the *Fried coherence length*. This latter is the typical cell size, for our atmosphere it's around 1 m, and is the scale of the aperture over which the wavefront is coherent. Because of the relatively weak dependence of the index of refraction on wavelength r_0 is not steeply dependent on λ but it depends on the altitude of the observatory as well as wind and humidity conditions aloft.[57] The reason we have spent so much time on this is that scintillation is not just a phenomenon limited to the formation of an image in the Earth's atmosphere. In the interstellar medium, we get the same

---

[57] Beran, M. J. and Oz-Vogt, J. 1994, *Prog. Optics*, **33**, 319 provides a superb modern overview of turbulent modifications of images by the atmosphere. See also Lena, P. 1999, *Observational Astrophysics*, 2nd ed. (revised) (Berlin: Springer-Verlag).

effect because of refractive variations in the medium due to electron density fluctuations. This is distinct from diffractive variations. The latter can be seen very easily if you look at the shadow cast by branches of a tree against a bright point source—the fluctuations in the intensity of the shadow are like those due to changes in point sources observed at radio wavelengths.

2.5.7 Speckle Interferometry

"When life hands you lemons, you make lemonade," a cliché that we can take as a description of the concept of *speckle interferometry*. It makes a virtue of atmospheric scintillation by using coherent imaging through the stochastic transfer function. If you sample the atmosphere rapidly enough, at about 100 Hz or faster, each sample will produce an image that is essentially diffraction limited. You see the intrinsic image of each turbulent element, and by Fourier transform and subsequent combination of the images, you can reconstruct the incident image. The technique was pioneered by Labeyrie in the 1970s. Its great advantages are high throughput and broadband sensitivity provided images can be obtained in very short intervals, within the coherence timescale of the atmnospheric fluctuations. In its original photographic implementation, the technique was inherently limited to bright stars and required very large aperture telescopes for light collection. With the use of low-noise digital imagers, television-based systems, CCDs, and MAMAs, the method has finally become practical.[58] A variety of image adding methods are used, we'll just mention one. It is called *shift-and-add*, which, in effect, has been used in all speckle processing at some level since the technique was first introduced. You take a very short exposure of an object and assume that the brightest point in the image is the statistical center of the kth image:

$$I_k(\boldsymbol{\theta}) = \mathscr{T} \star I_0 + N_\mathbf{k}(\boldsymbol{\theta})$$

for some turbulent atmospheric transfer function $\mathscr{T}$ and intrinsic intensity I_0. Then you shift the images to register on the maximum intensity point without rotating the field, coadd, and average to reduce the noise (which is assumed to be uncorrelated):

$$I(\boldsymbol{\theta}) = \frac{1}{N} \sum_{k=1}^{N} I_k(\boldsymbol{\theta} + \boldsymbol{\theta_k}),$$

and this becomes the image you treat with Fourier techniques or some other form of image processing. It isn't quite speckle in the sense that each image is not analyzed separately but only in ensemble, but the technique does remove image motion and reduce the noise in the final reconstructed image.

---

[58] See, for example, Labeyrie, A. 1970, *Astr. Ap.*, **6**, 85; McAlister, H. A. 1985, *ARAA*, **23**, 59; Bates, R. H. T. and Fright, W. R. 1982, *MNRAS*, **198**, 1017; Dainty, J. C. 1973, *Opt. Comm.*, **7**, 129; Dainty, J. C. 1976, *Prog. in Optics*, **14**, 1; Roddier, F. 1988, *Phys. Rep.*, **170**, 97.

2.5.8 Resolution

Any interferometric observation has finite resolution, determined by the sampling rate of the data and the longest interval of time or space that is sampled. To explain this point, we first begin with a standard description of the resolution. Consider that you have a gaussian signal in, say, wavelength λ; that is, assume that

$$P(\lambda) \sim \exp - \left(\frac{\lambda - \lambda_0}{\Delta \lambda}\right)^2 \qquad (2.35)$$

such that $\Delta \lambda$ is the full width at half maximum (FWHM) of this line. Now take the Fourier transform of this (see Appendix 2.C for details) to obtain

$$\hat{P}(k) \sim \exp - \left(\frac{k - k_0}{\Delta k}\right)^2. \qquad (2.36)$$

The connection between these two dispersions is

$$\Delta \lambda \, \Delta k \geq 1. \qquad (2.37)$$

This is a fundamental result. It shows that the larger the displacement you obtain in your interferometer, or the broader the wavelength coverage, the higher the resolution in the conjugate variable. Without this feature of Fourier transform systems, related directly to the phase, you could not obtain the resolution of a spectrograph. This is a form of uncertainty principle, that the resolutions are reciprocal. To obtain very high spatial (wavelength) resolution requires the largest possible aperture or the largest separation of the analyzing elements in an interferometer. Prism systems, or variable filter devices that work because of the wavelength dependence of the index of refraction, do not have constant effective resolution. For gratings, on the other hand, $\Delta \lambda / \lambda$ is constant.

2.6 IMAGE RECONSTRUCTION METHODS

Reconstructing astrophysical data requires that you introduce the minimum spurious variation possible in the flux and the spatial distribution. By this, we mean that you seldom know, in advance, what the thing should look like or what its flux may be. It's *not* like looking for a polar bear in a snowstorm—you know what it looks like and can use other, non-flux-conserving tricks, to find it.[59] The complication comes from the intrinsic properties of the medium through which the light is propagated, usually the atmosphere, and the particular problems with the com-

---

[59] See Bates, G. A. 1994, *Digital Image Processing* (NY: Wiley), which provides an excellent example of the sorts of things that astronomers do not do to data, especially showing what you *can* do when you don't have to worry about spectrophotometric fidelity of the image. For the observer's toolbox, see, for example, Kuhn, J. R. 1982, *AJ*, **87**, 196; Bracewell, R. N. 1979, *ARAA*, **17**, 113.

bined telescope and detector. Several methods have been developed to perform this task, all of which are usually used as black boxes.[60] Our aim here is to introduce them.

If you are only interested in finding the period of some variation, or some characteristic length, there are some simple techniques. Perhaps the most obvious is to fold the data repeatedly on finite intervals which phases the time series on a large set of closely spaced periods. You look for the period that minimizes the dispersion in some phase interval. But this only gives you the period, not the amount of power in that mode or at that length scale, which may be necessary to understand the underlying cause of the variability. For these questions, more sophisticated methods have to be used and this brings us closer to the reconstruction of an image.

Keep in mind that anything that works for two dimensions is also likely to work in one; that is, the same methods that are useful for the restoration of an image can be used for a time series or a spectrum.[61] Assume you've made some observations over a period of time T at the times $\{t_k\}$. The lowest frequency in the data is therefore $2\pi/T$, while the highest is $\min(t_k - t_{k\pm 1})$ for some pair. The discrete Fourier representation of the signal is

$$S(\omega) = \sum_{k=-\infty}^{\infty} X_k e^{-i\omega t_k}.$$

The power spectrum is the square modulus of this quantity, so

$$P(\omega) = |S(\omega)|^2 = \frac{1}{N}\left[\left(\sum_k X_k \cos \omega(t_k - \tau)\right)^2 + \left(\sum_k X_k \sin \omega(t_k - \tau)\right)^2\right],$$

where the phase shift for a specified frequency ω is

$$\tan 2\omega\tau = \frac{(\Sigma_k \sin 2\omega t_k)}{(\Sigma_k \cos 2\omega t_k)}.$$

---

[60] We should mention here one of the best cures for the black-box approach. There is a single journal, *Inverse Problems* that is devoted to algorithms and problems involving reconstruction of distributions from all sorts of data (seismic, tomography, radar imaging, and of course astrophysics). It is an outstanding resource for the most recent work and excellent reviews. Some additional material is published in several of the *SIAM* journals, and we also recommend *Waves in Random Media*. If we may indulge in a pedagogical harangue, remember that it is important in astrophysical research to be open to techniques that have been developed in completely unrelated fields where similar *types* of problems have been investigated.

[61] An amazingly early paper on periodogram analysis is Schuster, A. 1906, *Proc. Roy. Soc.* (London), **77**, 136 in which he anticipates many of the issues of image as well as time series reconstruction as its title highlights: "The Periodogram and its Optical Analogy." Van der Klis, M. 1989, in *Timing Neutron Stars* eds. Ögelman, H. and van den Heuvel, E. P. J. (Dordrecht: Kluwer, NATO ASI series), p. 27 provides one of the most complete introductions to Fourier methods for period analysis.

This is the periodogram. It isn't necessary that the times be regularly spaced, in fact it is often to your advantage to randomize them, so there is no restriction on the sum.[62] In fact, the principal problem is that astronomical time series are usually obtained at irregular intervals, a problem that has plagued astronomical observation since time immemorial.[63] Essentially, the mathematical problem with time series or with images, is that you are sampling a *continuous* function at at a discrete set of times or positions. To be more specific, let us use the times $\{t_k\}$ so that the observed signal, $g(t)$, is

$$g(t) = \sum_k^N f(t_k) W(t - t_k). \qquad (2.38)$$

Here W is the *window* function, the quantitative representation of the sampling. For point sampling, $W(t - t_k) = \delta(t - t_k)$ but this is generally not true for astronomical data. What we actually obtain is the signal accumulated for some interval of time, the resolution or response time, around t_k. This is a convolution so that a Fourier transform of this signal can, which in principle, pick out the period if one exists. However, this method suffers from the structure of the window function. Now you can see the advantage of a random series, and its perils. The windowing function transform is also a random variable in reality. In effect, since there is noise accompanying any real observation this is a modulation in time of the noise. So if you look at its power spectrum, you have a random variable against which any peak in the real data should be compared. This is how the false alarm probability used in periodogram-based analyses. The modification introduced by Scargle uses this explicitly. The cosine and sine components of the periodogram are weighted by the transform of the window which consists of all observations k in the time interval T:

$$P(\omega) \equiv \frac{1}{2}\left[\frac{[\sum_{k \in T} X_k \cos \omega(t_k - \tau)]^2}{\sum_{k \in T} \cos^2 \omega(t_k - \tau)} + \frac{[\sum_{k \in T} \sin \omega(t_k - \tau)]^2}{\sum_{k \in T} \sin^2 \omega(t_k - \tau)}\right] \qquad (2.39)$$

Now assume that noise in the observations, pure noise, would generate peaks in $P(\omega)$. The probability distribution of this noise is assumed to be exponential, so that

$$\text{Pr}(p < P(\omega) < p + dp) = e^{-p} dp, \qquad (2.40)$$

[62] Deeming, T. J. 1975, *Ap. Space Sci.*, **36**, 137, **42**, 257; Lomb, N. R. 1976, *Ap. Space Sci.*, **39**, 447; Scargle, J. D. 1982, *ApJ*, **263**, 835.

[63] It is instructive to consider, for instance, the determination of the Venus cycle by the Mayan astronomers before the seventh century or the Sothic cycle of the Egyptians, but any of the ancient calendrical cycles will do. The weather is far better in Egypt than the Yucatan, and the rising of Sirius just before the Sun is a simpler observation than the position of Venus, the Sun, and the Moon. The window for an observation is determined by the weather, and although the phenomenon is periodic, it is not long in duration and a small error in the date translates into large potential error in the period. This is even more easily seen in the determination of the periodic variability of β Per by Goodricke and Piggott in 1782. Although the variability had likely been noted by other observers, as had the variation of Mira (o Cet), the period is about 2.8 days and requires regular sampling in order to obtain a correct period. If the sample is taken less frequently than the Nyquist frequency, then an alias results. Random sampling actually helps to reduce the aliasing problem, but you see that the window function fundamentally limits what periodicities can be derived from the data.

so the cumulative probability of the power spectrum having a measured value less than some peak, p, is

$$\Pr(P(\omega) < p) \equiv F_P(p) = \int_0^p \Pr(p' < P(\omega) < p' + dp')\, dp' = 1 - e^{-p}. \quad (2.41)$$

This can be generalized from probability theory. If there is one frequency present among a possible N peaks, then

$$F_N(p) = (1 - e^{-p})^{N-1} F(p). \quad (2.42)$$

The probability p is chosen by the observer as a measure of the accuracy required for the identification of a particular peak.[64] There are still limitations with this method, due principally to the effects of large gaps (very low frequency null intervals) that both broaden the peaks and mix with the signal in the periodogram. Aid is, however, available with alternative analysis techniques that have been developed for other applications, which we now discuss.

2.6.1 The CLEAN Algorithm

Used mainly in radio astronomy,[65] where it was first introduced for the reconstruction of images from incomplete sampling by multielement interferometers, the technique called CLEAN has been generalized to several other types of data sets. In particular, it has been used for time series analysis,[66] as a way of eliminating spurious periods from the periodogram, and has been implemented in several other areas of astronomy. Its basis is the response of a psf to a point source, with which it is possible to construct a so-called *dirty beam*. This term is used by radio astronomers to describe the angular response of an incompletely sampled celestial sphere by the elements of the interferometer but it can just as easily refer to the window function in a stochastic or sparse time series.

The image obtained by an interferometer is described in the (u, v) plane that is incompletely sampled because of the distribution of the antennas of the array. Thus, what should be a continuous function, $V(u, v)$, is actually a discrete function weighted by the intensity at each point on the sky at which a sample has been taken:

$$V_{\text{disc}}(u, v) = \frac{1}{2N} \sum_{i=0}^{N} [V(u, v)\delta(u - u_i)\delta(v - v_i)$$
$$+ V^*(u, v)\delta(u + u_i)\delta(v + v_i)], \quad (2.43)$$

---

[64] See Scargle 1982, n.62; Horne, J. H. and Baliunas, S. L. 1986, *ApJ*, **302**, 757 for further discussions. See also Shore, S. N., Brown, D. N., and Sonneborn, G. 1987, *AJ*, **94**, 373, who show examples in an appendix of the effects of large data gaps.

[65] Högben, J. A. 1974, *Astr.Ap.Suppl.*, **15**, 417; Schwarz, U. 1978, *Astr.Ap.*, **65**, 345. See also Perley, R. A., Schwab, F. R., and Bridle, A. H. eds. 1982, n.54.

[66] Roberts, D. H., Lehár, J., and Dreher, J. W. 1987, *AJ*, **93**, 968.

where V_{disc} is the discrete form of the visibilities as detected by the interferometer. The aim is to reconstruct the image as it is on the sky

$$M(x,y) = \sum_{i=0}^{N} \left[V_i e^{i(u_i x + v_i y)} + V_i^* e^{i(u_i x + v_i y)} \right] \qquad (2.44)$$

with the window function

$$W(x,y) = \frac{1}{2N} \sum_{i=0}^{N} \left[e^{i(u_i x + v_i y)} + e^{-i(u_i x + v_i y)} \right] = \frac{1}{2N} \sum_{i=0}^{N} \cos(u_i x + v_i y). \qquad (2.45)$$

This function is the response of the system to a point source, and it is assumed that the individual point sources distribute over the map because of the wings of this function. In general, the *map* is the reconstructed image, $M(x,y)$, which is effectively a model for the appearance of the source that is the beam convolved image on the sky, which is the Fourier transform of the observed image in the (u,v) plane:

$$M(x,y) = W(x,y) \star I(x,y). \qquad (2.46)$$

This is the convolution of the window function with the image. The inverse transform then has peaks corresponding to the spatial frequencies of the point sources.

2.6.2 Bayesian Methods

The basic problem of deconvolution for a spectrum is to determine, given the optical transform that the telescope plus spectrograph represents, the form of the input spectrum given an observed spectrum:

$$D(x) = \int \Pi(x - x') S(x') \, dx', \qquad (2.47)$$

where Π is the total psf, including all contributions. One way to do this was first suggested by Richardson (1973) and Lucy (1974) and has become a standard for reconstructing images using techniques drawn from basic probability theory, Bayes' theorem.[67]

Simply stated, Bayes's theorem prescribes how data modify a hypothesis H in light of available evidence or data D and, perhaps, previous knowledge K:

$$P(H|E \cdot K) = \frac{P(H|K) P(D|H \cdot K)}{P(D|K)}. \qquad (2.48)$$

---

[67] The history of this theorem is discussed in detail in Todhunter, I. 1949 (rpt), *A History of the Mathematical Theory of Probability* (NY: Chelsea). Bayes' original paper is available in facimilie in Bru, B., ed. 1988, *Cahiers d'Histoire et facsimile des Sciences*, **18**, 1 or through JSTOR. It is still a method that has a hotly debated theoretical basis. Nevertheless, the applications of the theorem in astronomical data analysis have been multiplying.

The convention is to call $P(D|K)$ the *expectedness* or expectation of the data, $P(D|H \cdot K)$ is the *likelihood* of the data given the hypothesis, $P(H|K)$ is the *prior probability* of the hypothesis, and, most important, $P(H|D \cdot K)$ is the *a posteriori probability*. In general, K can be assumed and dropped from the formalism rendering the *prior* $P(H)$. You should keep in mind, however, that biases come into the formulation of this hypothesis (a term synonymous with *model*), the adjustment of which is the intent of theorem. Stated more generally, the a priori probability of a value x is $P(x)$ and that we know in advance the conditional probability of obtaining some value y given x (i.e., we know the *conditional* probability $P(y|x)$). Bayes' theorem is an intuitive way of calculating the inverse conditional probability, $P(x|y)$, the probability that a result x will be obtained given the input data y (also called the a posteriori probability). Then the joint probability of obtaining both x and y as a result of some measurement is

$$P(x, y) = P(x)P(y|x) = P(y)P(x|y). \qquad (2.49)$$

Then, since $\Sigma_y P(y|x) = 1$, we have

$$P(y|x) = \frac{P(y)P(x|y)}{\left(\Sigma_y P(y|x)P(x)\right)} \rightarrow \frac{P(y)P(x|y)}{\int P(y')P(x|y')\,dy'}, \qquad (2.50)$$

where the transition from the summation to the integral assumes a continuous probability distribution function. This equation, the statement of the theorem, is the method for determining the a posteriori probability of obtaining a value y given x. The ratio of the conditional probability to the integral is also called the likelihood.[68] We discuss in Appendix 2.D how this can be implemented for the reconstruction of images in the presence of noise and/or restructuring by the detector plus telescope where, for instance, $P(x)$ is the psf of the entire system.

2.6.3 Maximum Entropy

In a statistical sense, the "information content" of a signal is defined by its entropy, a quantity that provides a relative measure of how probable some configuration (or state) is. You know this from thermodynamics. A uniform medium has a high thermal entropy as the states approach equal probability of occupancy. With a signal, it is just the opposite. Pure noise, uniformly filling an image or whitening a signal, contains no information. In 1949, Shannon drew this analogy by defining a quantity that is like thermodynamic entropy but of opposite sign; if p_j is the probability of a signal j, then the individual measure of the entropy is essentially the number of bits in the signal; $\ln p_j$ and the weighted

---

[68] Edwards, A. W. F. 1984, *Likelihood* (Cambridge, UK: Cambridge Univ. Press); Lucy, L. B. 1974, *A. J.*, **79**, 745. See also, for instance, Tsumuraya, F., Miura, N., and Baba, N. 1994, *Astr. Ap.*, **282**, 699; White, R. 1994, *Restoring HST Data* (Baltimore: STScI); Jeffreys, H. 1983, *Theory of Probability*, 3rd ed. (NY: Oxford Univ. Press); Jeffreys, H. 1957, *Scientific Inference*, 2nd ed. (London: Oxford Univ. Press); Jeffreys, W. H. and Berger, J. O. 1992, *Am. Sci.*, **80**(1), 64; Berger, J. O. and Berry, D. 1988, *Am. Sci.*, **76**, 159.

average of the signal is then the total Shannon entropy over N samples:

$$S = - \sum_{j=1}^{N} p_j \ln p_j. \qquad (2.51)$$

So far, this is just an extension of the concept of information that originated with Boltzmann's treatment of statistical mechanics. But what is $\{p_j\}$? Again we appeal to our idea that a picture is generated by passing light through an optical system by a stochastic process, that the psf of the (universe plus telescope + detector) system measures the chance of a photon hitting some part of the detector having started somewhere else on the sky. A good guess is that the probability of seeing a photon at some point j in the image is measured by the relative intensity of the image at that pixel, $p_j = I_j/I$ where I is the mean intensity of the image, for N pixels (we don't specify the format of the image which could be linear or multidimensional). Since a probability must satisfy $0 \le p_j \le 1$ for all j, it is quite reasonable to use the relative intensity. The image becomes a map of the frequency with which photons arrive at a specific point in the image (we will use an image here, but it is any photon-created data set of any dimension, for instance a spectrum), hence it measures the probability of finding a photon at some point $\mathbf{x}$ in the field. In addition to the psf, you have to consider what the probability is that the photon will have arrived at that spot through a random process, connected with σ, the noise in the data.

The thing in the sky (object) is somehow distorted (screwed up) in every detection (image) is somehow compromised. We don't see the *sky*, we see an *image of the sky*. This is more than an epistemological concern. What we believe we are trying to achieve in image reconstruction is to get as close as we can to what the object would look like were it not for the limitations of the observational procedure. But the entropy alone is not enough to tell us what the most likely model should be for the source. For if we merely maximize S, we would obtain $\delta S = -\delta(\sum_{i=0}^{N} p_i \ln p_i) = 0 \to \ln p_i = -1$, which is a uniformly gray image. Noise makes this worse. Somehow we need additional criteria to tell us when we're getting close to the "thing in itself." In order to reconstruct the image, or signal, in the presence of noise, the entropy requires a constraint. The usual measure of the quality of a fit of a model $\mathcal{M}$ to the data $\mathcal{D}$ is provided by minimizing the χ^2 statistic:

$$\chi^2 = \sum_{j=1}^{N} \frac{(M_j - I_j)^2}{\sigma_j^2}, \qquad (2.52)$$

where M_j is the prediction for the point j of the model and I_j is the value of the data with the associated noise being σ_j. The maximum entropy method (MEM) uses the entropy as a constraint on the minimization, which is also called a "penalty" in some of the literature. Here we make use of Lagrange multipliers. If the variance of any quantity vanishes, it can be added with a constant multiplier to any other quantity whose variance is also zero. Therefore, the maximum entropy method maximizes the function

$$Q = S - \lambda \chi^2 \qquad (2.53)$$

for a constant Lagrange multiplier λ. Since you presumably know the psf, it's possible to write a representation of the image—the model, $\mathcal{M}$—in terms of the source you are trying to reconstruct, $\mathcal{S}$, with added noise, $\mathcal{N}$, whose intrinsic distribution you know:

$$M_k = T_{kj}S_j + N_k, \tag{2.54}$$

where T_{kj} is the response matrix, which is the transfer function for the combination of the atmosphere and the optical system. If we are treating a convolution, then this can be modified to include the psf or beam profile $M_k = P_{k-j}S_j$. You might wonder at our separating the noise from the signal. The success of this procedure depends on the nature of the source. If it is very bright, if we are not limited to Poisson statistics, then the sky and detection system are the main contributors to the noise and we may remove the sky contribution by chopping or some other differential measurement. On the other hand, the noise *may* depend on the intensity of the source or not be completely removed by sky subtraction if the observation is background limited. In the limit where the source itself is contributing to the noise because of low count rate, then the fluctuations will have passed through the same optics as the source and reconstructing the image is more complex. For now, let's assume a bright object and linear noise.

One of the principal uses of the method is the reconstruction of an image for an interferometer that incompletely samples the Fourier (u, v) plane. Since you usually sample the beam only over a limited range of the Fourier plane, T is the *dirty beam* we discussed for CLEAN. With MEM, χ^2 maximization is performed with a set of trial model sets, each of which is assumed to be close to the actual image, using the constraint equations that determine the goodness of fit. Assume that the sampled region is $\mathcal{S}$ which does not necessarily include the zero spacing —but may if you have a large-aperture single-antenna measurement without serious sidelobe problems—and cannot extend to infinite baselines. This image is discretized, and therefore so is its Fourier transform. If we have some quantity A for which the measurements are made at specific points x_j, then

$$A(\hat{k}_l) \equiv \hat{A}_l = \sum_j A_j e^{-ik_l x_j},$$

so taking j as the index and $l/N\Delta x$ as the wavenumber, we would translate the quantities (e.g., the model) to

$$M_k = \sum_j m_j e^{-2\pi ijk/N}.$$

Take I_i to be the intensity at some point on the sky, and take M_i to be the model Fourier transform. Then the quantity that must be minimized includes the χ^2 statistic in the *Fourier* domain:

$$Q(\lambda) = -\sum_{i=0}^{N} m_i \ln m_i + \lambda \sum_{k \in \mathcal{S}} \frac{|M_k - F_k|^2}{\sigma_k^2}, \tag{2.55}$$

where F_k is, in this case, the Fourier transform of the sky intensity as viewed by the interferometer. Taking $\partial Q/\partial m_i = 0$ and keeping in mind that the quantities for the χ^2 are the Fourier transforms of the data set, we get

$$-(\ln m_i + 1) - 2\lambda \sum_k \frac{M_k - F_k}{\sigma_k^2} \frac{\partial M_k}{\partial m_i} = 0,$$

which becomes

$$-(\ln m_i + 1) + 2\lambda \sum_k \frac{M_k - D_k}{\sigma_k^2} e^{2\pi i jk/N} = 0. \quad (2.56)$$

The last term comes from taking the derivative of the transform:

$$\frac{\partial M_k}{\partial m_j} = e^{2\pi i jk/N}.$$

Now the model is iterated between the image and transform domains by inverse transforms. This can be very time consuming for large images, and it would be impossible without the availability of the Fast Fourier transform (FFT). The technique has been used to reconstruct many different types of data, from images to spatial distributions of galaxies in clusters and gravitational lenses to the structure of accretion disks.[69]

2.6.4 Wavelets

The newest technique to be applied to image reconstruction, the wavelet method is like a scale dependent Fourier Transform that makes use of a local analyzing function to determine the structure of an image (see references in footnotes 70–72). One way to look at this is to consider the properties of the Fourier transform. It requires an infinite interval over which to sample the data. Sharp edges lead to serious problems (these are known as the *Gibbs effect*, the failure of the transform to converge at edges because it doesn't have compact support). There are several solutions to this. One is that you can smooth the data, for instance by multiplying and summing over a filter that removes the edges. This

---

[69] Two basic papers providing introductions to the method are Frieden, B. R. 1972, *JOSA*, **62**, 511 and Gull, S. F. and Daniell, G. J. 1978, *Nature*, **272**, 686. For treating the effects of Poisson noise see Frieden, B. R. and Wells, D. C. 1978, *JOSA*, **68**, 93. One of the first papers to apply maximum entropy analysis to radio data is Abels, J. G. 1974, *Astr. Ap. Suppl.*, **15**, 383. Bryan, R. K. and Skilling, J. 1980, *MNRAS*, **191**, 69; Narayan, R. and Nityananda, R. 1986, *ARAA*, **24**, 127; Horne, K. 1985, *MNRAS*, **213**, 129; Marsh, T. R. and Horne, K. 1988, *MNRAS*, **235**, 269; Roddier, F. 1988, *Phys. Rep.* (n.58); Skilling, J. and Bryan, R. K. 1994, *MNRAS*, **211**, 111; Fougère, P. F. ed. 1990, *Maximum Entropy and Bayesian Methods* (Dordrecht: Kluwer); Skilling, J., ed. 1989, *Maximum Entropy and Bayesian Methods* (Dordrecht: Kluwer). Many of the founding papers are in Jaynes, E. T. 1989, *Papers on Probability, Statistics, and Statistical Physics*, 2nd ed. (Dordrecht: Kluwer) and also Jaynes, E. T. 1957, *Phys. Rev*, **106**, 620, **108**, 171. A very lovely paper, far removed from the usual astrophysical literature but quite relevant, is Uffink, J. 1996, *Stud. Hist. Phil. Mod. Phys.*, **27**, 47 on the constraint rule in MEM, and we urge you to take a look at this paper for the general background it provides.

changes the data, and if the filtering is performed by a convolution, you have not really removed the problem. An alternative is to use a representation of the data that does not span the entire range of frequencies from $[-\infty, \infty]$. This is the idea behind the *wavelet transform*. The mathematical way of saying this is that the analyzing kernel for the transform has compact support, which means it spans only a finite range and vanishes outside the interval. Wavelets have two features. One is that they are functions of the scale. That is, like changing the ocular on a microscope, you focus in on progressively finer details. The second is that they allow for possible changes in the intrinsic signal by including a shift. Therefore, keeping the notation from Appendix 2.C, the transform is *two*-dimensional:

$$(W_g f)(a, b) = \int_{-\infty}^{\infty} f(t) g^*\left(\frac{t-b}{a}\right) dt, \qquad (2.57)$$

where a is the magnification and b is the shift; g is called the wavelet kernel. In the simplest applications, b can be ignored and the transform is one only on scale. The power of this technique comes from the fact that it can be inverted in the same way as we operated with the Fourier transform. First, you will notice that this is a convolution with a variable scale. Therefore, if we take the Fourier transform of both sides, we get

$$(W_g f)(a, b) = \frac{a}{i\pi |a|^{1/2}} \int_{-\infty}^{\infty} \hat{f}(\omega) e^{ib\omega} g^*\left[a\left(\omega - \frac{\omega^*}{a}\right)\right] d\omega, \qquad (2.58)$$

which can be inverted using

$$f(t) = C_g^{-1} \int_{-\infty}^{\infty} db \int_{-\infty}^{\infty} \frac{da}{a^2} (W_g f)(a, b) g^*\left(\frac{t-b}{a}\right), \qquad (2.59)$$

where the normalization is defined as the integral of the weighted power of the analyzing kernel:

$$\int_{-\infty}^{\infty} \frac{|\hat{g}(\omega)|^2}{\omega} d\omega. \qquad (2.60)$$

The choice of kernel is dictated by the problem, a disadvantage from the perspective of not knowing a priori what the kernel means (unlike the partial wave decomposition represented by the standard Fourier technique), but it has the great advantage in allowing you to detect structure in an image or signal by an impersonal method. Notice that we *still* make use of the Fourier transform, but since you have chosen g to vanish outside of a particular interval, this doesn't matter. The transform will extend only over a finite domain. A typical sampling kernel used for two-dimensional images is

$$g(r, a) = \left[2 - \left(\frac{r}{a}\right)^2\right] e^{-r^2/2a^2} \qquad (2.61)$$

which is a symmetric "Mexican hat," for which $b = 0$. Here r is the radial distance from the center of the kernel. This has the advantage, as all such kernels should, of having zero mean and returning a flat image for an image with a vanishing gradient. Since r is strictly positive, there is no worry about asymmetries and *chirp* introduced by the transform. It is not, however, the best one to use for time series analysis. This is because one of the features you may be looking for in a time series (e.g., the photometry of a variable star or the flux of a gamma burster) is a change in the shape of the variation. Then it is useful to have the full wavelet transform, including the shift, so that you can study the effects of changes in the intrinsic properties of the time series, not just clean up the variations.[70]

2.A A NOTE ON COSMIC BACKGROUNDS ACROSS THE SPECTRUM

Astronomical noise consists mainly of emission from terrestrial and cosmic diffuse sources. All require physical modeling to understand their origins, and yesterday's discovery is often today's tool and tomorrow's noise.

The infrared is dominated by almost everything outside the detector. Every element of the telescope, the primary mirror, the superstructure of the telescope, the secondary mirror and its supports, the dome, and the sky all emit thermal radiation. In addition, specific molecular transitions, especially from water vapor, also radiate at these wavelengths. For sources between 10 and about 30μ, dust in the solar system, confined mainly to bands within about 20° of the ecliptic plane, emit thermally. This is the zodiacal light, known originally as an optical phenomenon seen within about 45° of the Sun and due to scattering; the same grains form an emitting population at their thermal wavelengths. The galaxy emits in this near-IR wavelength mainly from thermal fluctuations of small grains (a phenomenon we will come back to in Chapter 6, where it will be of more physical importance than a nuisance) while high latitude dust illuminated by the diffuse interstellar radiation field is mainly responsible for the far IR background. The galactic nonthermal emission, due to cosmic ray synchrotron, contributes to the background at centimeter wavelengths as does emission from the magnetosphere. For terrestrial observations, the thermal background also dominates the submillimeter and millimeter wavelengths. In general, for detectors intended for IR observations longward of about 5μ, the instrument must be cooled with liquid N_2 or liquid He. Optical detectors are also sky-limited; the emission is mainly in bands due to auroral and ionospheric emission lines. It is therefore comparatively isolated in wavelength, although artificial light sources contaminate sites that are not sufficiently isolated. In space environments, the optical and UV are rather

---

[70] For some more recent examples of how wavelets are used in practice, see Starck, J. L. et al. 1997, *ApJ*, **482**, 1011: spectral analysis; Foster, G. 1996, *AJ*, **112**, 1709: time series analysis; Escalera, E. and Mazure, A. 1992, *ApJ*, **388**, 23, and Jones, C. et al. 1995, *ApJ*, **445**, 607 for cosmological applications; Starck, J.-L., Bijaoui, A., Lopez, B., and Perrier, C. 1994, *Astr.Ap.*, **283**, 349: aperture synthesis. General theoretical references are the monographs by Daubechies, I. 1992, *Ten Lectures on Wavelets* (Providence: SIAM); Meyer, Y. 1992, *Wavelets* (Providence: SIAM); and Chui, C. K. 1992, *An Introduction to Wavelets* (San Diego: Academic Press). A marvelously useful review is presented for turbulence by Farge, M. 1992, *Ann. Rev. Fluid Mech.*, **24**, 395, and we urge you to consult this.

dark and the background is dominated, as it is at shorter wavelengths, by unresolved objects.[71]

2.B STATISTICAL DISTRIBUTIONS

Astronomical observations rely almost exclusively on statistical stability of samples. We intend this as a brief review just to serve as a quick refresher and to keep the discussion self-contained, of a few of the principal statistical methods that are usually employed in astronomical observations.

2.B.1 Samples

For a sample of a variable X with a normalized distribution $p(x)$[72] for the measured values, the *mean* value of the sample is given by

$$\langle X \rangle = \bar{x} = \int p(x) x \, dx. \tag{2.62}$$

The *variance* of the sample is given by

$$\text{Var}[X] = (\sigma[X])^2 = \langle (X - \langle X \rangle)^2 \rangle, \tag{2.63}$$

and σ is called the *standard deviation*. If you are looking at several variables at once, the *covariance* is a matrix given by

$$\langle X_i X_j \rangle = \langle (X_i - \langle X_i \rangle)(X_j - \langle X_j \rangle) \rangle = \langle X_i X_j \rangle - \langle X_i \rangle \langle X_j \rangle. \tag{2.64}$$

The mean is the same as the average for small numbers and the *central-limit theorem* states that as the sample size grows, the average approaches the mean arbitrarily closely.

An important question is, when you have sampled for long enough, or have a large enough pool with which to work, that you can distinguish any variance in your sample from pure chance? This is a basic problem in astronomy because of the comparatively small numbers used in most statistical arguments. We therefore work within the Poisson limit and derive the signal to noise ratio.

Consider a binomial problem of a coin flip. You want to find out the probability of having a number of trials N of which precisely n come up heads. If the intrinsic probability of a single coin flip is $\frac{1}{2}$, assuming that the coin is alright, then the

---

[71] Zissis, G. J. 1993, *The Infrared and Electro-Optical Systems Handbook*. vol. 1, *Sources of Radiation* (Bellingham, WA: SPIE Press). See especially Chapter 3 on natural sources by D. Kryskowski and G. H. Suits.
[72] This has the same properties as a probability distribution in being always positive and bounded between 0 and 1.

probability if having n out of N trials with heads is

$$P_n = \frac{N!}{(N-n)!\,n!}\left(\frac{1}{2}\right)^n\left(\frac{1}{2}\right)^{N-n}, \qquad (2.65)$$

which we can generalize to

$$P_n(p) = \frac{N!}{(N-n)!\,n!}p^n(1-p)^{N-n}. \qquad (2.66)$$

This is the binomial distribution. The Poisson probability of an event is the small number limit of Gaussian statistics. First, consider a very weak source, one for which the mean count rate for photons is λ so that, on average, in an interval Δt you would expect to count $\langle n \rangle = N = \lambda\,\Delta t$. Then, given this mean arrival rate, the question you can ask is this: suppose you have counted some number m of photons during Δt and $m \neq N$. Has the source varied or is this consistent with the fluctuations you would expect in the count rate? The probability of counting some number m is

$$P_m(\Delta t) = \frac{N^m}{m!}e^{-N} = \frac{(\lambda\Delta t)^m}{m!}e^{-\lambda\Delta t}. \qquad (2.67)$$

It is easy to verify that the mean rate is $\langle m \rangle = \Sigma_m P_m m$ and that the expected fluctuation in this is

$$\Delta m = \langle m \rangle^{1/2}. \qquad (2.68)$$

In other words, in the small number limit, a situation that is all too familiar to any astronomical observer, the signal to noise ratio grows as

$$\frac{S}{N} = S^{1/2} \rightarrow \frac{S}{N} \sim (\Delta t)^{1/2}, \qquad (2.69)$$

where now Δt may be thought of an integration time. As we discussed in Chapter 1, this is the same law for fluctuations that we obtain from Bose–Einstein statistics so that *photon statistics* represent the Poisson limit. In other words, if you want to improve the quality of a signal by, say, a factor of 2, you must quadruple the integration time. Often this is impossible, so other methods must be used to obtain more information out of the data than is possible by simply increasing the integration time.[73]

As a multivariant generalization, the Poisson distribution can be written as

$$P(x_1, x_2, \ldots) = \prod_{i=0}^{N} \frac{a_i^{x_i}}{x_i!} e^{-a_i}. \tag{2.70}$$

For a Gaussian process, the large number limit of a Poisson distribution, the probability of a process X having some measured value x is

$$P(x) = (2\pi\sigma^2)^{-1/2} \exp\left[-\tfrac{1}{2}(x-\bar{x})^2 \sigma^{-2}\right], \tag{2.71}$$

where $\bar{x} = <X>$ as defined in Eq. (2.62).

2.B.2 MEM and Maximum Likelihood

An illustration of MEM method is to realize that you are really looking for evidence of deviation of the principle first enunciated by Bayes and expanded on by Laplace, the principle of maximum ignorance. Imagine that you have a coin, or a die, and you have a set of possible outcomes. In the absence of any indication of cheating on the part of the person who is flipping the coin or rolling the die, you can assume that the entropy is $-\sum_{i=1}^{n} p_i \ln p_i$, as usual, and that the condition for independent trials is that $\sum_{i=0}^{N} p_i = 1$. Then you obtain

$$Q(\lambda) = S + \lambda \sum_{i=0}^{N} p_i,$$

so that the variation gives

$$\ln p_i + 1 - \lambda = 0,$$

---

[73] Let us look at an example of how you would use this distribution in practice. Suppose you have a number of counts, N, at some wavelength λ in a bandpass $\Delta\lambda$ in a time T. We will further suppose that the count rate is quite low even though N may be large. For Poisson counting, you would expect a fluctuation in the number to be $N^{1/2}$ and the mean arrival rate is $r = N/T$. Then if you are looking for a transient event, one lasting for an interval Δt, you would expect that the probability of seeing n counts in this interval is given by $P_n(\Delta t)$, so it is

$$P_n(\Delta t) = \frac{(r\Delta t)^n}{n!} e^{-r\Delta t},$$

which measures the likelihood of a detected excess. This is compared with the expected fluctuations to provide a confidence level that the excess is real. There are, however, some important conditions that must be checked before applying this method. For instance, the events must be uncorrelated. This means that you know that the noise sources are either Poisson or Gaussian. This requires careful characterization of the detector before making measurements. In addition, you must be sure that the detector is linear, that there are no lower thresholds for counting.

so that all p_i are the same and constant. This is exactly what you would have expected from the start if you had applied Laplacian reasoning—for a die you would have estimated that the probability was $\frac{1}{6}$ of any side appearing in any trial, or $\frac{1}{2}$ in any coin flip. In this case, we have no model for the data, just the constraint that the trials are independent. For the image case, we seek a goodness-of-fit statistic to quantify how close we are coming to fitting the data set so there is a separate statistic χ^2. The likelihood function, $\mathscr{L}$ provides an estimation procedure for the parameters of a model given the data. Call $\{x_j\}$ is the set of data points and $\{\beta_j\}$ are the parameters of the model. Assume the parameters are drawn from some probability distribution. Then

$$\mathscr{L}_x(\beta_1, \ldots, \beta_m) = \prod_{j=1}^{m} p_{\beta_1, \ldots, \beta_m}(x_j)$$

with the maximum likelihood being given by

$$\frac{\partial}{\partial \beta_k} \ln \mathscr{L}_x(\beta_1, \ldots, \beta_m) = 0 \qquad k \in (1, \ldots, m).$$

Should the parameters behave according to Poisson statistics. Now assume that n_i is the observed counts in a bin so you would expect N_i in the same bin, given by the likelihood function

$$e^{-\mathscr{L}} = \Pi_k \frac{N_k^{n_k}}{n_k!} e^{-N_k}, \qquad (2.72)$$

and therefore, the probability of the data given the model (or hypothesis) is

$$P(D|H) \sim e^{-\mathscr{L}}. \qquad (2.73)$$

The probability of the model is given by the entropy, which is

$$S = -\sum_{k=1}^{\max} p_k \ln p_k. \qquad (2.74)$$

2.B.3 Bias

We will discuss bias further in Chapters 7 and 8, when we discuss the Malmquist bias in detail.[74] It is sufficient to note at the moment that all catalog selection criteria are a form of bias. For instance, you may choose to determine all of the stars in a cluster that have emission lines, but the sample may be biased on the basis of the underlying stellar spectrum (the same is true for a sample of emission line galaxies in a cluster of galaxies). One example of observer bias that comes to mind, the one that led to the recognition of rotation as a controlling factor for Be stars, comes to mind. Struve, in the 1930s and later, separated early-type stars into

---

[74] See, for example, Jaschek, C. and Murtagh, F., eds. 1990, *Errors, Bias, and Uncertainties in Astronomy* (Cambridge, UK: Cambridge Univ. Press).

those whose narrow lines were easily measured and those whose broad lines were very hard to deal with. The latter often showed emission lines. The discovery of binarity was harder for these, for instance because the broad lines made it difficult to detect low-amplitude radial velocity variations. The same is true for the detection of stellar magnetic fields. Techniques of measurement often produce selection effects that limit their effectiveness in testing theories. Here we mention just a few things that you should consider when evaluating data sets. First we have to mention catalogs. In a survey, you typically set a limit to which the data will be taken. This may be coordinate boundaries, but it is more likely a brightness level. We have already discussed bias in parallax surveys (Section 2.2.1.2 above). Proper motion surveys are defined with respect to some error in the displacement of the objects on the sky, and redshift surveys are often limited by the radial velocity, but in general the most frequent limit is one of signal strength. This is the kind of survey that often suffers from the Malmquist bias. Although we will discuss this at greater length later, this is what happens when you have noise in the system or photon fluctuations and background. If you set a cutoff at some minimum brightness and the signal has an intrinsic fluctuation ΔF for some flux F, then the random source at $F_{min} - \Delta F$ will not be included in the catalog while $F_{min} + \Delta F$ will be. The data are therefore slanted toward thinking that the faintest sources are actually brighter than they really are. To correct for this requires some knowledge of the intrinsic distribution of the errors or fluctuations in F, which is not always known, but it *biases* the survey at the faint end. This means that the sample is *incomplete* at the fainter end. If you already know the distribution function for F, then this can sometimes be determined by the departure of the source counts from the expected distribution (e.g., a systematic deficit of the number of faint sources.). In galactic structure surveys, or the determination of intrinsic emissivity of galaxies in a sample, this may be very important. Low-surface-brightness objects are always undercounted, and bias is introduced by using small apertures in determining the integrated brightness of extended objects in the presence of sky emission.

2.C PROPERTIES OF THE FOURIER TRANSFORM AND CONVOLUTIONS

We have referred many times to the Fourier transform and its applications in this chapter, and will make heavy use of it in many of the sections to follow, so let's digress for a moment to derive some of its properties. This is not just a mathematical exercise. Taking the Fourier transform of a physical quantity that has a spatial or temporal dependence is to make a statement about what you think it is composed of; you think what you are observing is a superposition of individual fluctuations in space and/or time at many different frequencies.[75]

---

[75] Standard references are Bracewell, R. N. 1986, *The Fourier Transform and its Applications*, 2nd ed. (NY: McGraw-Hill); Papoulis, A. 1965, *Probability, Random Variables, and Stochastic Processes* (NY: McGraw-Hill); Jennison, R. C. 1961, *Fourier Transforms and Convolutions for Experimentalists* (NY: Pergamon); Titchmarsh, E. C. 1967, *Introduction to the Theory of Fourier Integrals*, 2nd ed. (Oxford: Oxford Univ. Press).

Imagine that you are driving over a bad road or flying through a storm. To characterize the jostling you're experiencing, you can use two different descriptions. One is to look at the interval in space or time between the bumps. The other is to ask how frequently they occur. These are complementary descriptions and, either way, you're getting knocked around by the same cause, and therefore the power in the two descriptions must be the same. The same is true for a sound. You hear a signal that is actually a pressure wave producing a complex oscillation, in time, in your ear (the detector), but you interpret this as a pitch. On top of this, there is a modulation that allows you to hear different components of the spectrum contributing more or less at different times. These examples are the basis of the use of the Fourier transform in signal processing.

Take a signal to be $f(t)$.[76] This is created by a superposition of signals at different frequencies with individual amplitudes. To simplify matters, we will use the complex representation:

$$f(t) = \sum_{i=0}^{\infty} \hat{f}(\omega_i) e^{i\omega_i t}. \tag{2.75}$$

Now take the product of both sides with $\exp -i\omega t$ and perform the integral over all possible times:

$$\int_{-\infty}^{\infty} f(t) e^{-i\omega t} dt = \sum_{i=0}^{\infty} \hat{f}(\omega_i) \int_{-\infty}^{\infty} e^{i(\omega_i - \omega)t} dt. \tag{2.76}$$

The integral on the right-hand side is a very sharply peaked operator, one that selects out only the point at which $\omega_i = \omega$. In fact, this is the Dirac δ function[77] and obtain

$$\hat{f}(\omega) = \frac{1}{2\pi} \int_{-\infty}^{\infty} f(t) e^{-i\omega t} dt. \tag{2.77}$$

This is the Fourier transform of the time signal into its component frequencies, and it gives the amplitude of a wave of frequency ω to the overall time variable signal $f(t)$.

The transform has several very interesting properties. The first is called *Parseval's theorem* (although it was also discovered by Rayleigh in his study of the resolution of an optical instrument):

$$\int_{-\infty}^{\infty} |f(t)|^2 dt = \int_{-\infty}^{\infty} |\hat{f}(\omega)|^2 d\omega, \tag{2.78}$$

where $|\cdots|^2$ denotes the squared modulus of the function. This is, essentially, a statement of conservation of energy. The total power is the same regardless of the

---

[76] We deal here with time, but the spatial case is similar.

[77] Properly speaking, this is not a true function at all since it is defined only at a point and has no continuous derivative; it is actually a distribution. See Lighthill, M. J. 1957, *Introduction to Fourier Analysis and Generalized Functions* (Cambridge, UK: Cambridge Univ. Press) for more properties of generalized functions, including the δ function.

representation you choose for the signal. The second comes from the comparison of two signals. When you hear a sound, your ears receive the signal at slightly different phase because they are a finite distance apart. The central accumulator, your brain, matches the signals by correlating them. You have already seen this in action with the description of an interferometer, and it forms the basis of aperture synthesis. The correlation is performed by taking one signal, $f(t)$, and shifting it in time with respect to another signal, $g(t)$, and adding the product together as a function of the shift:

$$C(\tau) = \int_0^\infty f^*(t) g(t+\tau) \, dt. \tag{2.79}$$

This is called the *cross-correlation* for reasons that will become clear in a moment. When we discussed correlation spectrometers, this was the operation performed by taking the template and shifting it with respect to the source spectrum and detecting the changes in the throughput with a photomultiplier. If we think of this as another signal in time, we can take the Fourier transform of the function $C(\tau)$ to obtain:

$$\hat{C} = \hat{f} \star \hat{g}, \tag{2.80}$$

so that *the transform of the correlation function is the simple product of the Fourier transforms of the individual signals*. You can then see that for the *autocorrelation*:

$$A(\tau) = \int_0^\infty f(t) f^\star(t+\tau) \, dt, \tag{2.81}$$

and Parseval's theorem leads to the result that

$$\int A(\tau) \, d\tau = \int |\hat{f}|^2 \, d\omega = \text{power}, \tag{2.82}$$

so that the Fourier transform of the autocorrelation function is called the *power spectrum*. This is also known as the *Wiener–Khinchine* theorem.

2.D IMPLEMENTATION OF BAYES' THEOREM

The relation of Bayes' theorem to the convolution problem follows from the analogy between the psf and a probability distribution. The original spectrum is the a priori probability of observing a photon at some wavelength, if you are dealing with a spectrum, or of recording an event at some part of the field of view. The psf is a continuous redistributing operator. Said differently, given $D(x)$ as the observed data, and given $\Pi(x)$ as the psf, we wish to find the input spectrum $S(x)$, which is consistent with the data given that we know the conditional probability of any observed value having been contributed from some other portion of S. Especially since the psf has a finite range (or in mathematical language, it has *compact support*), the integration has to extend only over a finite portion of the spectrum. Therefore, you can write

$$D(x) = \int \Pi(y|x) S(y) \, dy, \tag{2.83}$$

since $\Pi(y|x)$ is the probability that y comes from the interval $x < y < x + dx$. Now assume that Π is normalized and nowhere negative. This is a good model for a true psf in an optical system and also the condition for a probability distribution function. It then follows from Bayes' theorem that we can define a weighting function

$$W(y|x) = S(x) \Pi(x|y) \Big/ \int S(y') \Pi(x|y') \, dy', \tag{2.84}$$

which is the fraction of the spectrum contributing to an observed point x. By substitution, we have

$$S(y) = \int D(x) W(y|x) \, dx, \tag{2.85}$$

so that finally we obtain an integral equation for the reconstructed source:

$$S(y) = \int \frac{D(x) S(x) \Pi(x|y) \, dx}{\int S(y') \Pi(x|y') \, dy'}. \tag{2.86}$$

For those of you with a yearning for new jargon, this is a nonlinear Fredholm equation of the first kind with eigenfunction S. Since we know the data, assume it as a first guess (the a priori) so that $S^0 = D$. The algorithm then substitutes this into Bayes' theorem to produce a new guess for model for S. If we call this $S^{(k)}$, after k trials, and the previous result for the reconstructed spectrum as $S^{(k-1)}$, then we have the general form

$$W^{(k-1)}(y|x) = \frac{S^{(k-1)} \Pi(x|y)}{D^{(k-1)}}, \tag{2.87}$$

where

$$D^{(k-1)}(x) = \int S^{(k-1)}(y) \Pi(x|y) \, dy. \tag{2.88}$$

Thus we have the final form of the algorithm:

$$S^{(k)}(y) = S^{(k-1)}(y) \int \frac{D(x)}{D^{(k-1)}(x)} \Pi(x|y) \, dx. \tag{2.89}$$

If the solution converges, then $S^{(k)} \to S$ as $k \to \infty$, typically 10 or 20 iterations. There is no guarantee of this, but the method (like maximum entropy) has the advantage that the iterative process has natural criteria for stopping the algorithm.[78] The primary criterion for convergence is that the convolved solution comes as close as possible to the original data:

$$\|D - \Pi \star S^{(k)}\| \to 0 \qquad (k \to \infty) \qquad (2.90)$$

by some metric. The convergence criterion used by Lucy is a minimized least-squares solution.[79]

---

[78] To write the algorithm for numerical computation, the a posteriori and a priori probabilities have to be taken as single variable functions, while the conditionals are taken as convolution kernels. This is where the connection lies between the deconvolution methods and the RL procedure:

$$f(x) = \int g(y) P(x|y) dy \to \int g(y) K(x-y) \, dy = K \star g$$

for any f, g, and K.
[79] Lucy, L. B. 1974, *A. J.*, **79**, 745; Lucy, L. B. 1994, *Astr. Ap.*, **289**, 994; Craig, I. J. D. and Brown, J. C. 1986, *Inverse Problems in Astronomy* (Bristol: A. Hilgar).

3 Radiative Transfer and the Outer Layers of Stars

If the Sun shine by Day, and the Stars by Night,
why, we shall know one another's Faces without the help of a Candle,
and that's all the Stars are good for.
—William Congreve, *Love for Love*, Act 2, Scene 5

3.1 INTRODUCTION

Nearly everything we know quantitatively about the Universe beyond the solar system comes from the study of spectra. Modern astrophysics began with spectroscopy.[1] Yet no photon comes to us without a history, and it is imperative to understand the effects of radiative transfer in order to disentangle the contributions from intervening media and the source of the radiation. Finally, since every absorption, scattering, or emission leaves some signature, the spectrum presents tremendous diagnostic possibilities for understanding the physical state of the gas through which the photons have passed and from which they have emerged. In this chapter, we will be surveying two related problems. The first is the thermal equilibrium of the outer layers of bodies that are illuminated from below by some amount of energy and that thermally and hydrostatically adjust to that energy input. To do this, we will first examine how radiation is transferred through a medium and then how the structure adjusts. Then we will go on to treat the formation of the spectrum and how the inverse problem is solved. In the next chapter, we will use these ideas again, in the deep and opaque medium of stellar interiors, to show how the internal structure of a star is determined in counterbalance to its self-gravitation.

Radiative processes of astrophysical importance can loosely be divided into two groups, those that involve transitions between bound states and those that involve a free (unbound) electron in either the initial or final state (or both). The former are spectral line processes, the latter are continuum processes. We will use the

---

[1] The *Astrophysical Journal* was originally founded as a journal of astronomical physics and spectroscopy. All propagandists for astrophysics at the end of the nineteenth century emphasized the primacy of spectroscopy as the great method for discovery and analysis. Its main attraction was its link directly to the laboratory through atomic and molecular structure and chemistry, and its ability to show dynamics through Doppler measurements that would be otherwise inaccessible to astrometric measurements.

quantum processes[2] to show how, through a very simple picture of the interaction of radiation with matter, you arrive at a formalism for the general problem of atomic level populations in the presence of radiation. First, we will set the stage for how you get photons through a gas. In order to do this, w first have to introduce a few basic concepts: emission and absorption coefficients and the basic equation of transfer.[3]

3.2 THE PHENOMENON OF RADIATIVE TRANSFER

Imagine a beam of photons illuminating a slab of gas from below at some angle to the surface. For the moment, assume that the layer, which has a thickness, dz, has a constant density, ρ. You can further assume that the gas is at some temperature, T, which is also a constant. We have to define a few basic terms. The monochromatic intensity of the radiation, I_ν, is the amount of energy passing through an element of surface, dA, per unit solid angle, $d\Omega$, per unit time, dt, in the frequency internal $d\nu$ in a direction $\hat{\mathbf{n}}$. This factor, $\hat{\mathbf{n}} \cdot d\Omega$, in the definition means the intensity is the same as a surface brightness and is constant for a spatially resolved source. For an unresolved (point source) the intensity decreases with distance since the solid angle remains constant (the point spread function of the detector sets the apparent size).

We can assume that there are processes that remove radiation from the beam, and those that add to it. Photons are removed from the line of sight by being scattered and through absorption. Photons are added due to scattering, essentially by redirection from some other path into the line of sight, and by emission (creation of photons) from the medium along the line of sight. The amount of

---

[2] For orientation, we briefly review some aspects of time-dependent perturbation theory in Appendix 3.A, but it is beyond the scope of this book to delve into the details. Instead, we will assume that you have some ready reference at hand and plow ahead with applications.

[3] Some very useful monographs for radiative transfer and stellar atmospheres are listed here. Some are also referenced along with the appropriate discussion in the text that follows, but are included here for completeness: Athay, R. G. 1972, *Radiation Transport in Spectral Lines* (Dordrecht: Reidel); Böhm-Vitense, E. 1989, *Introduction to Stellar Astrophysics*, Vol. 2, *Stellar Atmospheres* (Cambridge, UK: Cambridge Univ. Press) Cannon, R. 1985, *Radiation Transfer in Spectral Lines* (Cambridge, UK: Cambridge Univ. Press); Chandrasekhar, S. 1950, *Radiative Transfer* (NY: Dover); Cowley, C. R. 1968, *Theory of Stellar Spectra* (NY: Gordon and Breach); Gray, D. 1993, *The Observation and Analysis of Stellar Photospheres* (Cambridge, UK: Cambridge Univ. Press); Jeffries, J. T. 1968, *Spectral Line Formation* (Boston: Blaisdell); Kalkofen, W., ed. 1984, *Methods in Radiative Transfer* (Cambridge, UK: Cambridge Univ. Press); Mihalas, D. M. 1978, *Stellar Atmospheres*, 2nd ed (San Francisco: Freeman); Mihalas, D. M. and Milahas, B. W. 1984, *Foundations of Radiation Hydrodynamics* (NY: Dover Books); Shu, F. H. 1992, *The Physics of Astrophysics*, vol. 1 (Sacramento: University Science Books); Sobolev, V. V. 1962, *Treatise on Radiative Transfer* (Princeton: van Nostrand); Stenflo, J. O. 1994, *Solar Magnetic Fields: Polarized Radiation Diagnostics* (Dordrecht: Kluwer). The history of spectroscopic analysis is comprehensively reviewed by Clerke, A. 1902, *Problems in Astrophysics* (London: Blackie) through the beginning of the twentieth century. This is a remarkable survey, the first to treat the whole vista of spectroscopy as a tool for the analysis of cosmical bodies, and many of the problems outlined were not solved for decades. A complementary survey at the end of the same century is provided by Hearnshaw, J. B. 1986, *The Analysis of Starlight: One Hundred and Fifty Years of Astronomical Spectroscopy* (Cambridge, UK: Cambridge Univ. Press).

energy removed from the beam depends on the distance traversed through the layer and the density and temperature of the gas. The same is true for emission. We can write the loss of radiation as $-\kappa_\nu \rho I_\nu \, ds$, where now ds is the actual pathlength along the beam through the layer in a direction $\hat{n}$. For a planar layer, this is at an angle θ relative to the surface normal. Radiation is added to the beam at a rate $j_\nu \rho \, ds$. Here κ_ν and j_ν are called the *absorption* and *emission* coefficients, respectively. Quantitatively, we can write this process as

(Change in intensity in direction $\hat{n}$)

= +(all creation processes per unit distance)

+(scattering into line of sight per unit distance)

−(destruction processes per unit distance)

−(scattering out of line of sight per unit distance)

This "formula" translates directly into the *equation of radiative transfer*. To do this, we lump all the creation terms into j_ν, whether they are due to scattering or intrinsic emission. The details will occupy us for much of this chapter, but for now it is enough to write down the formal representation of the schematic process we have just outlined:

$$\frac{1}{c}\frac{\partial I_\nu}{\partial t} + \hat{n}\nabla I_\nu = -\kappa_\nu \rho I_\nu + j_\nu \rho. \tag{3.1}$$

Since the beam may vary explicitly with time, we have included it. In practice, however, the time-independent form is generally used. Much of the discussion of this chapter centers on this deceptively compact equation, whose solution depends critically on the specific properties of the environment such as the geometry, velocity field, and thermodynamic state of the gas. First, we need to find some way of making sense of the rates of photon creation and destruction, the emission and absorption coefficients. Then, having a physical idea of what is involved, we will proceed to the formal solution of the transfer equation and discuss some of its consequences.

3.3 TRANSITION PROBABILITIES AND STATISTICAL EQUILIBRIUM

The picture presented by quantum mechanics for the interaction of light with matter is that of a random process, the rates for which are set by the properties of the matter. This was the original point of view taken by Einstein at the onset of the modern work on atomic structure. The approach is similar to the one we used in Chapter 1 for thermodynamics, where we just look at the bottom line of the energy budget and leave the details to be sorted out at a lower level. In other words, as a start, we can assume that we can either measure the rates required for the various processes of absorption and/or emission, or we can calculate them, but we don't need them to describe the basic picture of the interaction. This will set up the transfer equation, and we will then proceed to study how it enters into the problem of determining the structure of a stellar atmosphere.

3.3.1 Statistical Equilibrium

Consider first what happens if an incident photon with a frequency ν hits an atom. For simplicity, we will treat only a two state system, but this will be generalized at the end. Imagine that there are two atomic states, E_i and E_j with $j > i$. To produce a transition between them, the atom must radiate or absorb a photon with an energy $h\nu_{ij} = \Delta E_{ij}$. If the energy of the incident photon is not sufficiently high, or if it is too high, it can be scattered by the atom without any change in the internal excitation, a process first described by Rayleigh (to which we will return shortly).[4] Although it changes direction, the process is elastic and the frequencies of the incident and final photons are identical. However, if there is a close correspondence between the energy of the incident photon, $h\nu$, and the separation of two energy levels, ΔE_{ij}, then the system can resonate with the electromagnetic field of the photon and the atom is excited. Because this is analogous to the typical conditions for resonance, in which there is a characteristic response frequency for driving a periodic system, the process is known by as resonant absorption or emission.

Now suppose that some fraction of the atoms, n_i/N, are in the lower energy level i, where N is the total number of atoms of the species, and suppose that the incident photons have an intensity I_ν at a frequency ν. On the analogy of a *response function* of a classical harmonic oscillator, you can assume that the state has some broadband frequency response to the radiation. Although we don't yet know how to compute it, let's label this as $\phi_i(\nu)$ and call it the *line profile function*. Since we are going to treat the simplest case, we will take a two-level system, although any level may be coupled to many other states. The upper energy level, j, has an initial fractional population n_j/N and a response function $\phi_j(\nu)$. The levels need not have different frequency responses, but they might.

The state of the system is uniquely determined for this system by the interaction of the *pair* of states i and j with the radiation. Only some fraction of the incident photons are even seen by the atom, so we can regard the absorption and emission processes as a stochastic problem where the rate of absorption depends on some probability, B_{ij} that the photon produces a transition $i \to j$. The *rate* of this process depends on the intensity of the incoming light, which is the number of incident photons per second n_ν. This is called a *stimulated* process because it requires an external radiation field in order to happen. The probability of absorbing a photon at the central frequency ν_{ij} is B_{ij}. Then the *rate* of upward transitions is the number of incident photons times the probability that one of them will be absorbed by the atom:

$$R_{ij}^{\text{stim}}(\nu) = \frac{n_i B_{ij} \int_{\text{line}} n_\nu \phi_i(\nu) h\nu \, d\nu}{4\pi}. \quad (3.2)$$

It is also possible to *induce* the inverse process by the incident radiation field. This is called *stimulated emission* because it too depends on the intensity of the

---

[4] There is an alternative, however, involving a change in the photon energy. This is Raman scattering, which we will also discuss soon.

radiation field. The rate is similar to Eq. (3.1):

$$R_{ji}^{\text{stim}}(\nu) = \frac{n_j B_{ji} \int_{\text{line}} n_\nu \phi_j(\nu) h\nu \, d\nu}{4\pi}, \qquad (3.3)$$

where you may notice that the response or profile function for this process may not be the same as for the upward transition. This is because the state that is responding is the upper one. Finally, there is also a chance that the atomic system, once excited, re-radiates *spontaneously* with some probability A_{ji} since any excited state has a finite lifetime. The rate of spontaneous transitions depends only on the properties of the state, so it is written as:

$$R_{ji}^{\text{spont}}(\nu) = \frac{n_j A_{ji} \int_{\text{line}} \phi_j(\nu) h\nu \, d\nu}{4\pi}, \qquad (3.4)$$

Notice that R_{ji}^{spont} is independent of I_ν. It leads to the creation of a photon. As such, it requires that the system must already be excited by some process. We will see shortly that this process is collisional excitation.

Now, where does the radiation come from? If we immerse the system in a heat bath with an arbitrary energy density, we expect that the level populations then evolve with time because of the competing rates of emission and absorption:

$$\begin{aligned}\frac{dn_i}{dt} &= -R_{ij}^{\text{stim}} + \left(R_{ji}^{\text{stim}} + R_{ji}^{\text{spont}}\right) \\ &= -\sum_{j>i} n_i \int_{\text{line}} I_\nu B_{ij} \phi_i(\nu) \, d\nu + \sum_{j>i} n_j \left(\int_{\text{line}} I_\nu B_{ji} + A_{ji}\right) \phi_j(\nu) \, d\nu, \quad (3.5)\end{aligned}$$

where we have allowed for the possibility of coupling other levels to i through their time dependences and integrated over the line profile to remove the frequency dependence of the rates. We have not yet specified the spectrum of the radiation field. If it does not originate within the gas (e.g., if it is a flash incident from the outside), its spectrum may be very different from what the local conditions would generate in *strict thermodynamic equilibrium*. In this case, bizarre populations may result, as we will see shortly. However, in order to define what these transition probabilities are, we can simplify the problem to the one that can be solved exactly, the case of *thermal equilibrium*. There is no compelling reason to think that the frequency responses of the two states are different, so we can take $\phi_i(\nu) = \phi_j(\nu)$, an assumption called *complete redistribution*. If the atomic system is in equilibrium with the radiation, its level populations won't depend on time, $dn_i/dt = 0$, and look only at the coupling between two levels:

$$\frac{n_j}{n_i} = \frac{B_{ij} I_\nu}{B_{ji} I_\nu + A_{ji}}. \qquad (3.6)$$

Thus, when the spontaneous probability is small, or when the intensity is high, the population ratio is independent of the intensity of the radiation field and depends only on the ratio of the stimulated probabilities. We cannot, however, make much use of the level populations per se. It is much more useful to be able to compute the radiation field because that is what we ultimately see.

Let us now get more realistic. The atom is not really isolated. It experiences collisions with neighboring electrons, atoms, ions, and molecules, and these dominate the populations. The radiation is just a small effect, where the energy density is low compared with that in the gas itself. So we assume that there are two other rates that act, the collisional excitation rate $n_i C_{ij} N_p$ and deexcitation rate $n_j C_{ji} N_p$. Here N_p is the number of background perturbers, whatever they are. If the collisions dominate the populations, then

$$\frac{n_j}{n_i} = \frac{C_{ij}}{C_{ji}}. \tag{3.7}$$

This is really very nice because it is independent of the number of perturbers and gives a complete solution to the *intensity* of the radiation that will be in equilibrium with the collisions; in other words, the radiation that is formed under *local thermodynamic equilibrium* (LTE) conditions. You have here the whole quantum-mechanical picture of blackbody radiation; it is the rate of emission and/or absorption that leaves the level populations precisely as they would be if they were in thermal equilibrium.

3.3.1.1 Collision Rates

The collision rates are defined with respect to the distribution function of the perturbing particles. The excitation rate has a threshold, ΔE_{ij}, which is the same as the photon energy for the transition, so the excitation is given by

$$q_{ij} = \int_{\Delta E_{ij}}^{\infty} \sigma(v) v f(v) v^2 \, dv. \tag{3.8}$$

Here $\sigma(v)$ is the velocity (energy) dependent collision cross section and v is the random velocity of the perturbers, whose distribution is $f(v)$. We have already covered in Chapter 1 why f is likely to be a Maxwellian distribution. Because there is a minimal energy required to excite the transition, only a fraction of the total potential perturbers will actually produce a radiationless upward transition. There is, however, no minimum energy for the de-excitation. The cross sections are scaled to the geometric value:

$$\sigma_{ij} = \pi \left(\frac{\hbar}{mv}\right)^2 \frac{\Omega(i,j)}{g_i}, \tag{3.9}$$

where g_i is the statistical weight of the lower state. The rate for electron-atom collisions is given by

$$n_i C_{ji} = n_i n_e q_{ij} = \left(\frac{2\pi}{kT}\right)^{1/2} \frac{\hbar^2}{m^{3/2}} \frac{\Omega(i,j)}{g_j}, \tag{3.10}$$

where n_e is the electron density, and the {collisional efficiency} factor is defined assuming a thermal velocity distribution:

$$\Omega(i,j) = \int_0^\infty \Omega(i,j;E) e^{-E/kT} \frac{dE}{kT}. \qquad (3.11)$$

We should add that collisional *ionization* also occurs if the gas is hot enough. In this case, the threshold energy is replaced by χ, the ionization energy.[5]

3.3.1.2 Equilibrium Populations

We can now appeal to gas theory. In thermal equilibrium, which means that the atomic states are strongly coupled to the colliding electrons and that the radiation is in equilibrium with both, the level populations are determined by the Boltzmann distribution, as we derived in Chapter 1. This means that

$$\frac{n_j}{n_i} = \frac{g_j}{g_i} e^{-\Delta E_{ij}/kT}. \qquad (3.12)$$

We also obtain from Eq. (3.12) the relation between the upward and downward collision rates using Eq. (3.5). Since we made the assumption that the radiation is in equilibrium with the atom, the rates of emission and absorption are strictly those that result from precise balance of the radiative rates alone. This permits the calculation of the emitted intensity at any frequency:

$$I_\nu = \frac{A_{ji}/B_{ji}}{(n_i B_{ij}/n_j B_{ji}) - 1} = \frac{A_{ji}}{B_{ji}} \left(e^{\Delta E_{ij}/kT} \frac{g_i B_{ij}}{g_j B_{ji}} - 1 \right)^{-1}. \qquad (3.13)$$

The resulting energy distribution for the radiation must the Planck function because the conditions we have used are those of thermal equilibrium:

$$I_\nu \equiv B_\nu(T) = \frac{2h\nu^3}{c^2} (e^{h\nu/kT} - 1)^{-1}. \qquad (3.14)$$

Knowing this, we have a means for phenomenologically defining the stimulated transition probabilities without any further physical assumptions:

$$g_i B_{ij} = g_j B_{ji}, \qquad (3.15)$$

---

[5]The detailed computation of the collision strengths depends on detailed solution of the Schrödinger equation. Resonance states often mix with the continuum and there can be complex energy dependence for the interaction. Useful compilations are found in Osterbrock, D. E. 1989, *Astrophysics of Gaseous Nebulae and Active Galactic Nuclei* (Sacramento: Univ. Science Books) along with references to the computations. See also Mott, N. and Massey, H. 1965, *Theory of Atomic Collisions*, 3rd ed. (Oxford: Oxford Univ. Press); Moisewitch, B. L. and Smith, S. J. 1968, *Rev. Mod. Phys.*, **40**, 238; Massey, H. and Burhop, H. S., and Gilbody, H. B. 1969, *Electronic and Ionic Impact Phenomena*, vol. 1 (Oxford: Oxford Univ. Press); Seaton, M. J. 1968, in *Nebulae and Interstellar Matter: Stars and Stellar Systems*, vol. 8 Middlehurst, B. M. and Aller, L., eds. (Chicago: Univ. Chicago Press).

and we also have a means for connecting the spontaneous and stimulated probabilities:

$$\frac{A_{ji}}{B_{ji}} = \frac{2h\nu^3}{c^2}. \tag{3.16}$$

This is the final step in the derivation of theory for line transitions. At the end, you see that we have not only derived a set of rate equations but also found relations between the rate of stimulated absorption and the spontaneous emission probabilities.

As we said, spontaneous emission creates photons. This means that the rate of emission for a gas whose populations are determined only by collisions is given by

$$j_\nu = 4\pi h \nu_{ij} n_j A_{ji}. \tag{3.17}$$

This is the emissivity of the atom. The emission and absorption coefficients are defined, for equilibrium, by Kirchhoff's third law:[6]

$$\frac{j_\nu}{\kappa_\nu} = B_\nu(T). \tag{3.18}$$

For any arbitrary pair of energy levels the absorption coefficient is given by

$$\kappa_\nu = h\nu_{ij}(n_i B_{ij} - n_j B_{ji}) = h\nu_{ij} n_i B_{ij}(1 - e^{-h\nu_{ij}/kT}) \tag{3.19}$$

This last form for κ_ν includes the correction for the stimulated emission. The assumption in Eq. (3.6) is very important to keep in mind—the condition for thermal equilibrium requires that the ratio of emission to absorption rates by a body depends only on the temperature, and for this reason I_ν is the same as the blackbody emissivity when collisions dominate and the medium is in thermodynamic equilibrium.

The derivation of the equations for the transition probabilities from time dependent quantum mechanical perturbation theory is left to Appendix 3.A.[7] The transition matrix, $\langle m|V(0)|n\rangle = V_{mn}$ is the source of the dramatic difference in intrinsic line strength of transitions between different states. Its value depends on the coupling between the levels, in particular on the symmetries and other attributes of the constituent states. You know that in atomic hydrogen, transitions are impossible for dipole radiation between states that have the same angular momentum. This is easy to see because there is no dipole radiation possible without a change in the angular symmetry of a state. So we can say that a transition with $\Delta l = 0$ is strictly forbidden by the rules of dipole radiation. This is equivalent to saying that no radiative transition can be excited of the form 1s → 2s, but it *is* possible to induce transitions like 1s → 2p.

---

[6]The laws, more properly called the *Kirchhoff–Bunsen laws*, relate to the emissivity of gases: (1) a tenuous hot gas overlying a cold source produces emission lines; (2) a dense gas or a colder gas overlying a hot source produces absorption lines at the same wavelength, and (3) the ratio of the emissivity to absorption in any gas in thermal equilibrium is a function of only temperature.

[7]In order to convince you to refer to it, we will use the notation defined there.

The strength of a transition depends on the selection rules and the perturbation responsible for inducing it. For instance, the strongest (lowest-order) perturbation is the electric dipole. However, some states have vanishing coupling to this order, so the next ones are electric quadrupole and magnetic dipole. These all arise from the expansions of the electromagnetic potential in the interaction term. For instance, the perturbation amplitude depends on $V = -e\mathbf{x} \cdot \mathbf{E}$ for the dipole, so that the transition probability depends on the intensity of the electric field and the square of the dipole matrix. From our previous definition, you can see that the Einstein coefficients are the same as the Fermi Golden Rule probabilities and can actually be computed, not just observed in the laboratory. It is often the case, however, that a particular transition cannot be easily computed and must instead be treated empirically. Most complex atoms—and there are times when this means more than two electrons—require enormously complex programs to determine the atomic structure and then to compute the overlap integrals required for the transition probabilities.

Atomic states having similar quantum numbers may arise from the same electron configuration and have populations that are nearly identical. Within a coupling scheme these sets of states are called *multiplets*. Within the atom, interactions between electron states depend on spin and angular momentum, as well as principal quantum numbers. The single-electron wavefunctions form the basis for calculating the total energy of a state, depending on how the electrons are assumed to couple to each other. States arising from the splitting of a single-electron configuration by $L - S$, or spin−orbit (or Russell−Saunders) coupling, have a total orbital angular momentum $\mathbf{L}$ and a total spin $\mathbf{S}$, and these couple to form a *total* angular momentum $\mathbf{J} = \mathbf{L} + \mathbf{S}$. The coupling follows the normal vector rules.[8] The separation of the substates in J depends on the strength of the spin−orbit coupling, which is much smaller than the strength of the nuclear potential. Consequently, these states represent fine structure compared to the separation of the energies based exclusively on the principal quantum number. Collisions rapidly equilibrate the populations in these states under the normal conditions in stellar atmospheres, but in tenuous environments the populations of substates can depart radically from equilibrium (as happens, for instance, with masers and lasers).

The immediate connection with atomic structure is through the spontaneous transition probabilities A_{ji}, which depend only on intrinsic properties of the states. Any level has a finite lifetime if there is some perturbation, V, that permits it to make a transition to another state. There are a number of available modes, electric dipole and quadrupole and magnetic dipole and quadrupole, for instance, depending on the intrinsic moments of the states, but in general the electric dipole transition will be the strongest. The probability of such a transition always depends on the strength of the interaction averaged over the participating states. Thus, if

---

[8] See, for example, Herzberg, G. 1944, *Atomic Spectra and Atomic Structure* (NY: Dover Books); Condon, E. U. and Shortly, G. E. 1953, *Theory of Atomic Spectra* (Cambridge, UK: Cambridge Univ. Press); Condon, E. U. and Odabasi, H. 1980, *Atomic Structure* (Cambridge, UK: Cambridge Univ. Press); Cowan, R. D. 1981, *The Theory of Atomic Structure and Spectra* (Berkeley: Univ. California Press).

V_{ji} is the strength of the perturbation, then

$$B_{ij} = \frac{8\pi^2}{3\hbar^2 c}|D_{ij}|^2 \qquad (3.20)$$

for a dipole transition. This means that the spontaneous rate is given by

$$A_{ji} = \frac{64\pi^4 \nu^3}{3\hbar c^3} S(i,j), \qquad (3.21)$$

which defines the *strength of a multiplet*. If the particular perturbation, V, happens to vanish on average, and this happens in many instances, then the transition is said to be *forbidden* by the selection rules that couple the states.

3.3.2 Strange Populations: Masers

Energy-level populations come into equilibrium when the only photons available for excitation and/or deexcitation have been locally generated and when collisions dominate the excitation process. Often, though, we find cases where the exciting photons are not coming from the medium but are incident from some outside source whose temperature may be markedly different. In this case, stimulated processes can become very important. Start with the absorption coefficient $\kappa \sim n_i B_{ij}(1 - g_i n_j/g_j n_i)$. To obtain this from Eq. (3.19), we have used the relation between the stimulated absorption coefficients. The population ratio, n_j/n_i, may become larger than the ratio of the statistical weights *if* the radiative process should overpopulate the upper state relative to the lower state. The minimal requirement for this is a three level system, where the intermediate state can be bypassed by some absorption process that favors the coupling of the lowest and highest states. In addition, we require that collisions do not strongly depopulate any of these levels. The mechanism is essentially fluorescent in character, as we will see below, but the basic physical process is clear: if the population ratio inverts, then κ may become *negative*. This means that the signal is *amplified* as it propagates, a condition that is known as *light amplification by the stimulated emission of radiation* or *laser*.

Astrophysically, molecules present sufficiently complex systems, and even at low energy they can be very efficient amplifiers of this sort. The principal reason is that there are many transitions available for excitation and many that can get "stuck." They operate at radio (microwave) frequencies, and are thus known as *masers* (in fact, the first suggestion for this nonequilibrium process used NH_3 as a model system, see vol. 3 of the *Feynman Lectures* for a very pretty discussion of the physics). The requirement of low density is easily met in a circumstellar envelope or in interstellar molecular clouds, and the requisite radiation field can also easily result from high-intensity IR emission from dusty circumstellar envelopes. It is only necessary that the molecules have the right tuning of their coupled levels. This is met by a surprising number of species, for instance OH, CH, SiO, NH_3, H_2O, and

HCN.[9] The details of how this operates will be discussed in Chapter 6 for the interstellar medium. It requires a solution to the transfer equation. But you can already see how it comes about through this consideration of the level populations; a population inversion, characteristic of a maser, results in amplification of the signal within a narrow bandpass because the absorption coefficient turns negative.

3.4 RADIATIVE TRANSFER

To start with the simplest problem, let us use the formalism we have just developed to look at the transfer of radiation through a medium in equilibrium, or as it is usually called *thermal equilibrium*. The outer atmosphere of stars, the portion from which we see the light emerging, varies in radial extent depending on the type of object we treat. Stars on the main sequence, for example, have comparatively high surface gravities and small pressure scale heights, so the radial extent of the optically thin layer above their photospheres is also small. In other words, the fact that the thickness is small compared with the stellar radius means that they don't know about the curvature of the star—they can be treated as plane parallel layers. In contrast, low-surface-gravity objects like giants and supergiants have extensive regions over which they are optically thin because they have quite large density and pressure scale heights. The neglect of curvature—which, as we will see, means that the atmosphere never sees an infinite optical depth—severely limits the interpretation of these bodies. This is an important feature to remember for a large variety of astrophysically important objects, not just for supergiants. The temperature of an atmosphere is like the temperature of your skin; it is regulated by the rate of energy loss compared with the rate of supply. If an atmosphere can more easily radiate its energy to space, it will be cooler than if it has some direction where it is effectively infinitely opaque.

3.4.1 Intensity, Flux, and Moments of the Radiation Field

We need to define some quantities that we will use to describe the radiation field. The first is the *intensity*, I_ν, we have already defined. The monochromatic intensity is a scalar quantity, related directly to the number of photons in a given energy range measured in some arbitrary direction. If we are interested not in the direction, but only in the energy density at some depth z in a medium, then we can write the average or *mean intensity* as an integral over the solid angle:

$$J(z) = \frac{\int I \, d\Omega}{\int d\Omega} = \frac{1}{4\pi} \int_0^{2\pi} d\phi \int_0^{\pi} d\theta \sin\theta I(z, \theta, \phi). \qquad (3.22)$$

Since we will treat mainly thin plane parallel layers, we now define the directional

---

[9] Elitzur, M. 1982, *Rev. Mod. Phys.*, **54**, 1225; Elitzer, M. 1992, *Cosmic Masers* (Cambridge, UK: Cambridge Univ. Press); Goldreich, P., Keeley, D., and Kwan, J. 1973, *ApJ*, **179**, 111.

cosine, for convenience, as $\mu = \cos\theta$. Then Eq. (3.22) becomes

$$J(z) = \tfrac{1}{2}\int_{-1}^{1} d\mu\, I(z,\mu). \tag{3.23}$$

For instance, for normally incident oppositely directed beams of radiation through a plane parallel layer, mean intensity is approximated as $J = \tfrac{1}{2}(I_+ + I_-)$.

A word on procedure is in order here. Stars are point sources, so the information that is contained in the full transfer equation is not usually available to us, with very few exceptions such as the Sun and planets. This means that we must find a way of treating the transfer equation so as to produce observable quantities. This involves averaging over the visible hemisphere of the body. This is the origin of the moments of the transfer equation and the reason why the problem must be solved as a coupled set of equations rather than solved for the quantities individually. In effect, each measures a different angular aspect of the radiation. The mean intensity gives the energy density, the flux gives the rate of divergence, and the K moment gives the relative anisotropy of the radiation. For instance, the mean intensity is what you actually see if you are standing in a very dense fog. You cannot distinguish direction for the radiation source, since it is scattered many times, and seems to be completely filling any field of view. The denser the environment, for scattering, the more uniform the radiation background appears to be. This uniform appearance of the source is typical of an extremely opaque environment, whether through scattering or emission and absorption processes. Thus J must be related to the number of photons in a unit volume, or the radiative energy density, by J/c.

The monochromatic *flux*, F_ν, is the net energy flow rate through a surface in the normal direction at some depth:

$$F_\nu(z) = \int I(z,\mu)\hat{\zeta}\cdot\hat{n}\, d\Omega, \tag{3.24}$$

where $\hat{\zeta}$ is the direction of the beam, and $\hat{n}$ is the normal to the surface. This is actually the difference between the outgoing and incoming photon intensities through a surface element, so that

$$F_\nu(z) = 2\pi\int_{-1}^{1} I_\nu(z,\mu)\mu\, d\mu. \tag{3.25}$$

Again using the two stream analogy, the flux is approximated by the difference in the two streams, $F = (I_+ - I_-)$. You've already encountered this in the conservative form for the energy balance of a gas in Chapter 1. There, the rate of energy loss is given by the divergence of the flux of energy, so that when the equation of energy conservation is written in conservative form you have

$$\frac{\partial E}{\partial t} + \nabla\cdot\mathbf{F} = 0. \tag{3.26}$$

For a gas in thermal equilibrium, the divergence of the integrated or *bolometric* flux, F, must vanish because the energy is constant. We will return to this point in

Section 3.6.3 when we discuss energy balance in atmospheres. The radiation from such a system is described by the Planck function and since it is isotropic, the mean intensity is related to the energy flux through the Stefan–Boltzmann law:

$$F = \int_0^\infty F_\nu \, d\nu = \frac{4\pi}{c} \int_0^\infty B_\nu(T) \, d\nu = aT^4, \qquad (3.27)$$

where a is the Stefan–Boltzmann constant. This leads directly to an important astrophysical quantity. The *bolometric* or total luminosity, found by integrating the entire spectral distribution of flux over the surface of an opaque body, defines the temperature that a blackbody radiator would need to radiate at the observed luminosity. This is called the *effective temperature* and it is defined from the luminosity of the source:

$$L = 4\pi\sigma T_{\text{eff}}^4 R^2. \qquad (3.28)$$

It is important to remember that we have made several critical assumptions to derive this quantity. For a brick, the surface is well defined and the object closely approximates a blackbody. A star, however, is fuzzy and has no real "surface" in a bolometric sense. In other words, since the opacity is strongly wavelength-dependent, there can be a large difference between the physical conditions sampled by the emerging radiation even in very short wavelength intervals. This means that T_{eff} is only a reference value to characterize the total flux, but it may have little to do with what you actually see from the star. Always keep in mind that, except in a *very* few instances (such as the Sun), we do not see resolved surfaces for stars and we do not observe them over their entire spectral range. We attempt to infer their properties from a (usually narrow) spectral window. The *bolometric correction* is consequently one of the most important products of stellar atmospheres calculations. It is defined as the factor by which the monochromatic flux must be multiplied to obtain the integrated value as a function of T_{eff}.

We have defined the depth-dependent quantifies, the mean intensity and the flux, by integrating over the solid angle (or, for the plane parallel case, integrating over the directional cosine). You may have noticed that this is analogous to taking the moments of the dynamical equations. The zeroth-order moment is the mean intensity, J_ν. We now define the next moments by analogy. The first moment is

$$H_\nu = \tfrac{1}{2} \int_{-1}^{1} I_\nu \, \mu \, d\mu, \qquad (3.29)$$

which is called the *Eddington flux*. Notice that this is *not* the same as the quantify we defined earlier as the flux. This one is a formal convenience for solving the transfer equation. Finally we define

$$K_\nu = \tfrac{1}{2} \int_{-1}^{1} I_\nu(z, \mu) \mu^2 \, d\mu \qquad (3.30)$$

which is related to the radiation pressure. You notice that for an *isotropic* radiation field, $K = \tfrac{1}{3} J$. This factor of $\tfrac{1}{3}$, called the *Eddington factor* which measures the anisotropy of the radiation. The deepest layers should always have this ratio between the moments. As you move toward the surface, progressively more of the radiation "sees" the boundary and escapes, so this ratio changes. These moments are introduced here for the same reason we encounter them in the Boltzmann equation, namely, to address a closure problem. In order to solve the equation, we must somehow connect the moments through a formal mechanism because the gradient of the intensity depends explicitly on the directional cosine.

3.4.2 Setting Up the Transfer Equation

Now we return to the radiative transfer equation. Unlike our first pass, in Section 3.1, we now explicitly include scattering as well as absorption and specialize the equation to a thin layer.[10] For a thin layer (i.e., for a plane parallel layer), what you are really interested in is how far you can see into the medium. This is the *optical depth*, τ, which is a dimensionless quantity that measures the relative opacity of the medium in terms of the average distance a photon travels between successive interactions:

$$d\tau = -\kappa \rho \, dz. \tag{3.31}$$

Here we make two assumptions: (1) we are looking vertically into the layer and (2) we are looking from the outside. Thus an optical depth of order unity is the distance over which a photon has a virtual certainty of being scattered or absorbed. Using this definition for the optical depth, we can now recast the equation of transfer:

$$\mu \frac{dI_\nu}{d\tau_\nu} = I_\nu - \frac{j_\nu}{\kappa_\nu} = I_\nu - S_\nu. \tag{3.32}$$

Here we have defined the *source function* for the radiation field by analogy with Kirchhoff's law:

$$S_\nu(\tau) = \frac{j_\nu(\tau)}{\kappa_\nu(\tau)}. \tag{3.33}$$

In strict thermal equilibrium, whether local or global, the source function is $B_\nu(T)$, which depends only on temperature. If, however, scattering is an important source for local photons in the medium, then S_ν will not uniquely reflect the local

---

[10] There is a historical precedent for this choice. The earliest spectroscopic observations of the solar atmosphere, by Frauenhofer around 1810, showed that most of the radiation appears to be a continuum with a number of absorption lines at what appeared to be irregular intervals. Once the radiation laws had been formulated by Kirchhoff and Bunsen about 40 years later, it was realized that the layer forming the absorption was thin and overlying the hot, opaque solar surface. Eclipse observations of the chromosphere seemed to agree with this picture, although this region was in emission. It was, however, seen in projection against dark sky and could still be consistent with thermodynamic laws. The first solutions of the transfer equation, by Schuster in 1905 for the continuum and by K. Schwarzschild in 1905 for lines, kept this picture of a *reversing layer* as a convenience. It is an appropriate treatment for geometrically thin atmospheres, or as an approximation in an imbedded layer problem, but for many astrophysical applications it represents only a local and not a global picture of the radiating medium.

temperature. This is the difference between the LTE assumption and the more general approach to the photon creation term which is, oddly enough, called *non-local thermodynamic equilibrium* (NLTE). We will soon deal with other differences, connected mainly with the calculation of ionization equilibria and state populations, but this approximation of the source function is important. It states that any radiation you observe is *assumed* to be due to the temperature of the gas. It implies that the observed intensity is due to a temperature gradient that couples directly to the pressure, and hence uniquely to the atmospheric structure with depth, through the hydrostatic equation. On the other hand, if NLTE is assumed, the emergent intensity may be from nonlocal processes that significantly change the *inferred* properties of the gas compared with the actual conditions.

To illustrate the meaning of the optical depth, let's look at the case of a purely absorbing medium that has no internal sources. This corresponds to a light source seen through a slab of cold gas, such as a star seen through an interstellar cloud. In this case, there is no connection between the state of the absorber and the radiative source, and the energy density of the radiation is so low that it has no effect on the physical state of the cloud gas. This is also called *extinction*. On integrating Eq. (3.32) with no source term, we find that

$$I(z, \mu) \approx I_0 e^{-\kappa \rho z / \mu}, \qquad (3.34)$$

where I_0 is the unabsorbed intensity. This was the first known solution of the transfer equation for an absorbing medium and is still called *Beer's law*. If you think of the beam of radiation as a spatial distribution of the photons, then the probability of a photon reaching you decreases with increasing optical depth. Alternatively, the number of photons created along the line of sight increases if the source function is equal to a constant, S_0. Then the intensity varies as

$$I(z, \mu) = S_0 (1 - e^{-\kappa \rho z / \mu}). \qquad (3.35)$$

At small optical depth the intensity is simply $I_\nu \sim \tau S_0$. We will use this approximation soon for thermal radio emission from extended envelopes.

3.4.2.1 The Spherical Transfer Equation
Before simplifying the transfer equation, we should write it in spherical coordinates to show how the different reduction assumptions change the physical picture. We take the coordinate system (r, θ, ϕ) with ϕ being the azumithal angle. The gradient is given by

$$\nabla \to \frac{\partial}{\partial r} \hat{\mathbf{r}} + \frac{1}{r} \frac{\partial}{\partial \theta} \hat{\theta} + \frac{1}{r \sin \theta} \frac{\partial}{\partial \phi} \hat{\phi},$$

so the full equation of radiative transfer becomes

$$\mathbf{n} \cdot \nabla I_\nu = \mu \frac{\partial I_\nu}{\partial r} - \frac{1 - \mu^2}{r} \frac{\partial I_\nu}{\partial \mu} = -\kappa_\nu \rho I_\nu + j_\nu. \qquad (3.36)$$

The solution of the transfer equation can proceed using the moments of the radiation field and, just as we found for the dynamical variables in the Vlasov equation, there are always more moments in the transfer equation than we know what to do with. This is true regardless of the geometry, and ultimately we require some *closure condition* to complete the set of equations to be solved. Once again, we will deal only with frequency-independent quantities. Define the n*th* moment of the intensity:

$$M_n = \int_{-1}^{1} I\mu^n \, d\mu, \tag{3.37}$$

where $n \geq 0$. This expression does not precisely match any of the physical quantities we have just defined for the radiation field, but this form makes the moment equations simpler to evaluate. The source function is $S = j/\kappa$ as usual. We will take this to be independent of angle for the moment, which entails assuming that the two coefficients have the same angular dependence so $\int_{-1}^{1} S\mu^n \, d\mu = 0$ for any odd value of n. Multiplying Eq. (3.36) by μ^n and integrating over angle,[11] we obtain:

$$\frac{\partial M_{n+1}}{\partial r} + \frac{n+2}{r} M_{n+1} - \frac{n}{r} M_{n-1} = -\kappa\rho\left(M_n - \frac{2}{n+1}\delta_{n,e}S\right) \tag{3.38}$$

Here the symbol $\delta_{n,e}$ vanishes if n is odd and embodies the isotropy assumption we made for S. Equation (3.38) provides the general transfer equation and illustrates the closure problem. You see that the nth moment depends on the $(n+1)$st, so some connection between different moments must be assumed. The contact with our earlier physical definitions of the moments comes from $M_0 = 2J$, $M_1 = \frac{1}{2}F$, and $M_2 = 2K$. The equations for the moments now read

$$\frac{\partial F}{\partial r} + \frac{2}{r}F = 0 \tag{3.39}$$

$$\frac{\partial K}{\partial r} + \frac{3K - J}{r} = -\frac{1}{4}(F - 2S). \tag{3.40}$$

As the curvature of the atmosphere is reduced, these equations come progressively closer to the plane parallel case:

$$\mu \frac{dI(\tau, \mu)}{d\tau} = I(\tau, \mu) - S(\tau, \mu). \tag{3.41}$$

---

[11] To perform this integral, you need to take

$$\int_{-1}^{1} d\mu \left\{ \frac{\partial}{\partial \mu}\left[(1 - \mu^2)\mu^n I\right]\right\},$$

which vanishes at the limits of the integral.

Notice that now the optical depth becomes a unique measure of the distance to which we can see in the medium because of the disappearance of the curvature-dependent terms in the formal transfer equation. Equation (3.41) can be formally integrated along any path at constant angle since, for the plane parallel case, τ is independent of μ. Applying the *no-influx* upper boundary condition (no radiation impinging from outside, something that astrophysicists often assume, although sunbathers know otherwise), we find for the emergent surface intensity:

$$I_\nu(0, \mu) = \int_0^\infty S_\nu(\tau_\nu) e^{-\tau_\nu/\mu} \frac{d\tau_\nu}{\mu}. \tag{3.42}$$

Since this is at the surface, Eq. (3.42) depends only on angle. We are looking at a planar surface and only at the last step in the emission process. But there is another interesting feature of the equation: it is essentially the Laplace transform of the radiative source function. This is no minor technical point, it means that all of the information about the depth dependence has been compressed into the directional cosine μ. Put differently, it means that the angle and depth are conjugate variables for the planar problem (and *only* for this geometry!). Another way of saying this is that the emergent intensity as a function of viewing angle is a map of the source function with depth. So in principle it is possible to invert the angular distribution of the radiation to obtain the depth dependence of the source function.

3.4.3 Some Solutions to the Transfer Equation

From Eq. (3.42) we have the formal solution for the emergent intensity at the surface. But the most important feature of the transfer equation is that it provides the means for computing the depth dependent intensity of the radiation. If you are flying through a cloud, this is something you may want to know—for example, how bright will the environment be at some arbitrary depth? There is, however, a more basic use in astrophysics. Since radiation is the primary means of energy transfer in many cosmic gases, the intensity of the radiation determines the local thermal properties of the medium. This is where the assumption of LTE comes in: Because we have made this assumption, we find that the temperature and radiation density gradients are the same thing. To begin with, we need to examine the formal solution to the transfer problem.

There are several important connections between the physical moments of the radiation field. For example, the integrated intensity is directly related to the source function through Eq. (3.37):

$$J(\tau) = \frac{1}{2} \int_\tau^\infty \int_{-1}^1 S(t) e^{-(t-\tau)/\mu} \frac{d\mu}{\mu} dt. \tag{3.43}$$

We can integrate over the angles to obtain

$$J(\tau) = \int_\tau^\infty S(t) E_1(|t-\tau|) \, dt \equiv \Lambda_\tau[S(t)]. \tag{3.44}$$

This formally defines an operator, Λ_τ, that maps the source function onto the mean intensity. The function $E_1(x)$, called the *exponential integral*, is defined by

$$E_n(x) = \int_1^\infty e^{-xy} y^{-n}\, dy, \qquad (3.45)$$

where n is an integer and $E_n(0) = 1/(n-1)$ for $n > 1$. We won't go into the formal details,[12] but we should explain what Eq. (3.44) physically means. You can see this terms of the integral equation for the flux:

$$F(\tau) = 2\left[\int_\tau^\infty S(t) E_2(t-\tau)\, dt - \int_0^\tau S(t) E_2(\tau-t)\, dt\right], \qquad (3.46)$$

which follows from the definition that the flux is the difference between outward- and inward-directed radiation. These are more than formalities; they indicate that the radiation emerging at some optical depth τ depends on the transfer from greater depths, the efficiency of which depends on the exponential integrals. In this sense, the integral representation is a convolution of the transfer function with the local source function. It is also important to note that these representations are strictly defined only for plane parallel—that is, geometrically thin—layers.

If the radiation is due only to scattering, Eq. (3.44) reduces to a unique integral equation:

$$J(\tau) = \Lambda_\tau[J(t)]. \qquad (3.47)$$

This is the Milne equation. It was first derived for an isotropic radiative medium and is independent of its thermal properties. In fact, this is precisely the cloud problem, and in this context its importance goes beyond its role in stellar atmospheres. Any scattering medium with an isotropic source function obeys this equation, whether it is the outer layer of a star or a cloud in the interstellar medium. From Eq. (3.46) we have

$$F(0) = 2\int_0^\infty S(t) E_2(t)\, dt. \qquad (3.48)$$

Substituting in Eq. (3.44) for the source function, we find that

$$F(0) = 2E_3(0)\cdot(\text{constant}) + \frac{3F}{4}\left[\frac{4}{3} - 2E_4(0)\right], \qquad (3.49)$$

and since $F(0) = F$, this yields

$$J(\tau) = \tfrac{3}{4} F(0)\bigl(\tau + \tfrac{2}{3}\bigr).$$

---

[12] See discussion and tables in Erdyli, A., Magnus, W., Oberhettinger, F., and Tricomi, F. G., eds. 1953, *Higher Transcendental Functions*, vol. 1 (NY: McGraw-Hill); Abramowitz, M. and Stegun, I. A. 1965, *Handbook of Mathematical Functions* (NY: Dover Books). Consult Kourganoff, V. 1952, *Basic Methods in Transfer Problems* (NY: Dover Books); Busbridge, I. W. 1960, *The Mathematics of Radiative Transfer* (Cambridge, UK: Cambridge Univ. Press) for examples connected with radiative transfer.

This is an exact solution for a thin layer. Since the flux is constant (absent curvature effects, the condition that the flux is divergenceless is the same as saying it is constant) the source function reaches a limiting value at vanishingly small optical depth. For a thermal function, the temperature (the fourth root of the source function) reaches approximately approximately 84% of the effective temperature. It is also the origin of the idea that the photosphere is defined at a geometric height slightly above $\tau = 1$, in this case at $\tau = \frac{2}{3}$.

3.4.3.1 Two-Stream Transfer

Now let's use the example of two streams of radiation in order to show how the moment equations can be interpreted. Consider a layer with a uniform, constant internal source function, one that does not depend in any way on the radiation losses. We can separate the problem into two opposing beams of photons by writing $J = \frac{1}{2}(I_+ + I_-)$. Assume that each comes in at normal incidence. Then the transfer equation can be written as two coupled equations:

$$\frac{dI_+}{d\tau} = I_+ - B, \quad -\frac{dI_-}{d\tau} = I_- - B. \tag{3.50}$$

Suppose we take the layer to be thin enough for B to be constant. In this case, we can combine the equations to form a single second-order equation just as we did for the flux in the plane parallel atmosphere:

$$\frac{d^2 J}{d\tau^2} = J - B \rightarrow J = B + ae^{-\tau} + be^{\tau}, \tag{3.51}$$

where a and b are constants. If there is no external radiation incident at the top of the atmosphere, so that $I_-(0) = 0$, and if we assume that the radiation becomes isotropic at depth, $J(\tau) \rightarrow B$ as $\tau \rightarrow \infty$, then the solution to Eq. (3.50) is

$$I_+(\tau) = B, \quad I_-(\tau) = B\left(1 - \frac{1}{2}e^{-\tau}\right). \tag{3.52}$$

This solution suffices to illustrate the problem of thin layers in a planetary atmosphere, although in that case we would have assumed that $I_-(0)$ is finite, and it is especially well suited to treating problems where there is flux incident at the top of the layer. It can also be generalized to the case of a finite layer. For instance, suppose again you are flying through a cloud that is illuminated from above by an external source. The local source function is due solely to scattering. Since the layer has a finite thickness, the losses are maximal at the two bounding surfaces. There the source function is smallest and the cloud is dark. However, in the middle, at about $\tau = 1$, the scattering isotropizes and the medium is extremely bright. This is a way of seeing what K/J means, since the more this departs from $\frac{1}{3}$, the more anisotropic—meaning streaming—the radiation becomes.

3.4.3.2 Limb Darkening and Temperature Gradients

Since it is so close and its surface is resolved, the Sun provides a useful justification for several approximations that lead to closed form solutions of the transfer equations. One results from the assumption of a linear source function. You know that such approximations are generally valid only as a perturbation on some solution, and this case is not an exception. You are, in effect, saying that a star has a hard surface but with a little bit of fuzz around it, and this superficial layer is relatively optically thin. It is then plausible to say that the surface, which occurs at a fixed optical depth $\tau_{\nu,0}$, corresponds to the monochromatic photosphere, and that is observed through a thin *reversing layer*, $\Delta\tau_\nu$. This was the original approximation for the part of the solar atmosphere that forms its absorption spectrum when seen against the photosphere. If the atmosphere is in LTE, then the source function depends only on temperature. We perform a linear expansion of $B_\nu(T)$ in optical depth:

$$B_\nu(T) \approx B_\nu(\tau_{\nu,0}) + \left(\frac{dB_\nu(T)}{dT}\right)_0 \frac{dT}{d\tau_\nu} \Delta\tau_\nu \equiv a_\nu + b_\nu \Delta\tau_\nu, \qquad (3.53)$$

where a_ν and b_ν are constants and the temperature gradient is expressed in terms of the optical depth rather than the metric pathlength through the layer. Substituting this into Eq. (3.42) gives the angular dependence for the emerging radiation:

$$I_\nu(0,\mu) = a_\nu + b_\nu \mu = I_\nu(0,0)(1 - u_\nu + u_\nu \mu). \qquad (3.54)$$

Here we have used the conventionally adopted form for the surface intensity variation and introduced u_ν, the *monochromatic limb-darkening coefficient* (see Fig. 3.1a, b). You can see from Eq. (3.53) that this parameter is directly related to the temperature gradient in an atmosphere that is in thermal equilibrium. It would, perhaps, be better to say that it is related to the gradient in the *source function* with optical depth. There are many reasons why the limb might be darker than the center of a disk and not all of them relate to temperature. In fact, it is not impossible for the limb to appear brighter than the center of the disk. Think, for instance, of an isolated cloud that appears in front of the Sun on an otherwise clear day. Because scattering and absorption reduce the intensity through the thickest part of the cloud, the limb, from which the photons are scattered into the line of sight, is brighter than the center. It isn't that the cloud is hotter as you move away from the center, it is simply that the photons escape more easily. This renders interpretation of the limb darkening a difficult problem if the medium is both absorbing and scattering. In LTE, the source function is uniquely dependent on temperature so that the limb darkening (or limb brightening) for a plane parallel medium measures the decrease (or increase) in emitted energy due to the temperature gradient. For a body like the Sun, viewed as a resolved star from close up, we see a change in surface brightness as we look toward the limb. In general, we would expect the surface to be cooler than the interior because it is easier to cool in the low density regions over the photosphere than in the denser lower layers. Thus we normally anticipate $dT/d\tau > 0$, $u \geq 0$, and $I(0,\mu) \leq I(0,0)$. Actually, this works amazingly well.[13]

The formal solution to the equation of radiative transfer yields the emergent spectrum as a function of depth and angle. The key, however, is that you must

---

[13] Neckel, H. and Labs, D. 1994, *Solar Phys.*, **153**, 91 present an analysis of the $\lambda\lambda$ 3030–10990 Å limb darkening obtained over a five year period. This is an excellent source for earlier literature.

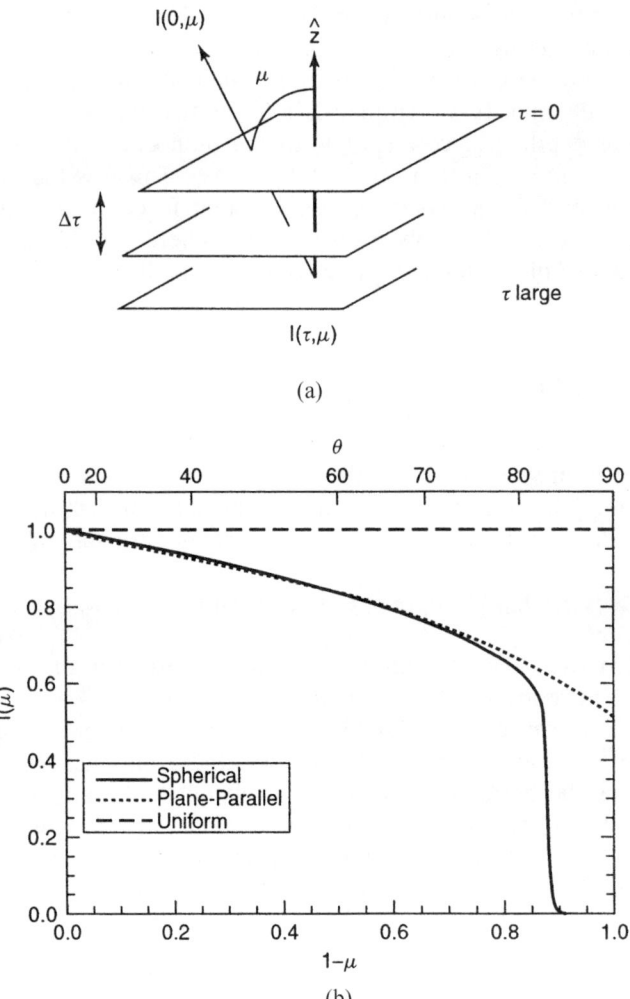

FIGURE 3.1. (a) The integration path for a plane parallel atmosphere. (b) Monochromatic limb darkening. Comparison of intensity for a uniform disk, a plane parallel atmosphere, and a spherical atmosphere (credit: J. P. Aufdenberg, CfA).

already know the structure of the atmosphere. Without knowing the intrinsic emissivity and absorption as a function of depth—or in other words, the density and temperature at each point in the atmosphere—there is no way to integrate the differential equation and determine the emergent spectrum.

3.4.3.3 Life in a Fog, or What Scattering Means

The removal or destruction of photons is a *local* process and happens only in the direction of the source. Thus, in order for true absorption to occur, you have to place the obscuring medium between you and the radiative source. This removes a major ambiguity in reconstructing the nature of the source and the intervening medium. Understanding the distribution of the emission is the conceptually more

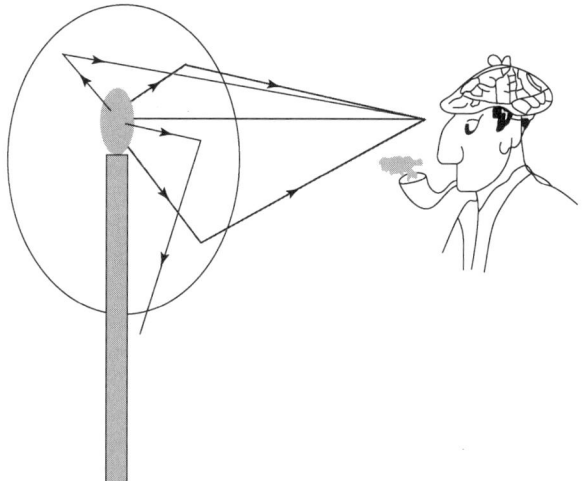

FIGURE 3.2. Cartoon illustration scattering. Notice that Sherlock Holmes sees the street lamp dimmed by scattering through the intervening fog, while the scattering produces an extended halo. Some scattering paths are indicated (and you might consider how each is weighted by the phase function).

challenging problem because the radiation may come from diverse sources and the local conditions may be far removed from those that create the incident radiation in the case of scattering. For instance, let's return to the case of a fog. Imagine that you are standing under a lamp. You see the bulb, but as you look away from it you also see a bit of a halo. This isn't emission in the air—it is some of the photons that were initially not directed straight at you being redistributed in direction. The bulb appears a bit fainter than it would have without the intervening material because of the scattering of light out of the direct line of sight, but the halo makes up for at least some of the brightness. In effect, the surface brightness of the main photon source goes down while that of the surroundings goes up, as you see in Figure 3.2. As you move farther away, the brightness and dominance of the central light source decreases compared with this halo, until there is just a soft glow where the bulb had been. Finally, you observe even the light from the halo merge into the background completely and you can no longer distinguish where the bulb is. You see, this isn't local at all. Photons from many different directions, some initially directed in the opposite way, are reaching your eye. So scattering acts as a source for photons along the line of sight. It does so, though, at the expense of all other directions. The spectrum also has nothing (well, almost nothing) to do with the thermal properties of the scattering medium. It reflects the source. Alternatively, absorption and redistribution through collisions and reradiation is a local effect that directly reflects the conditions prevailing in the absorbing gas. This is the process that we talked about earlier, *LTE*, and in this case the source of the radiation is everywhere dependent only on the gas temperature, with which the radiation is in equilibrium.

In this fog problem, the source function is not due to intrinsic emission from the medium. The only reason why you see a photon from any direction other than toward the central source is by redirection so the source function is the same as the mean intensity, $S = J$. For the moment, let's forget about the photon's energy and just take a monochromatic beam. This is the same thing as assuming that all the scattering processes are gray and coherent. It also allows us to drop the frequency subscript, or to treat the moments and absorption coefficient as frequency integrated quantities. Then the equation of transfer becomes

$$\mu \frac{dI(\tau, \mu)}{d\tau} = I(\tau, \mu) - J(\tau). \qquad (3.55)$$

The mean intensity already has its angular dependences removed by definition, and its use as the source function requires that S is strictly isotropic. Thus, we can integrate Eq. (3.55) over μ:

$$\int_{-1}^{1} \mu \frac{dI(\tau, \mu)}{d\tau} d\mu = \frac{d}{d\tau} \int_{-1}^{1} I\mu \, d\mu = \frac{1}{2} \frac{dF}{d\tau} = \int_{-1}^{1} I \, d\mu - J \int_{-1}^{1} d\mu = 0. \quad (3.56)$$

In other words, the flux through this planar layer is constant. This is what radiative equilibrium means. In three dimensions, the condition would be expressed by the continuity equation, $\nabla \cdot \mathbf{F} = 0$. To continue, for the next moment we obtain

$$\int_{-1}^{1} \mu^2 \frac{dI(\tau, \mu)}{d\tau} d\mu = \int_{-1}^{1} I\mu \, d\mu - J \int_{-1}^{1} \mu \, d\mu = \frac{d}{d\tau} \int_{-1}^{1} I\mu^2 \, d\mu \qquad (3.57)$$

so that

$$\frac{dK}{d\tau} = \frac{1}{4} F. \qquad (3.58)$$

This is the first step. It says that if the flux is constant then the source function is a linear function of the optical depth. You should not be surprised by this. It just means that in a plane parallel isotropic medium, the only way that the radiation can emerge is through the surface. As long as there is no density gradient other than a surface, the solution follows immediately.

The flux depends only on temperature, through the optical depth, if the medium is in LTE. In this case, the radiation source is the source function $S \to J = B(T)$, the frequency integrated Planck function. Since temperature governs the emissivity for strict LTE, the temperature gradient follows the flux and the deepest layers are the hottest. Where the radiation can escape to the outside world, the source function drops. It is important, however, that the emission of a blackbody is very efficient; only a comparatively small change in the local temperature is required to cause the source function to decrease by large amounts.

Two simple alterations serve as examples of how the source function behaves when we drop the plane parallel assumption. One is to consider an isotropic medium with *two* boundaries, one above and one below, our earlier cloud example. In this case, the source function will peak in the center and decrease because of losses at either boundary. The other is a spherical layer. Here the flux is varying because of the divergence of the area of the surface, and the source function drops faster than it would in the linear case. Just keep in mind, however, that in none of these cases have we said *anything* about the structure, composition, or even the stability of the medium! We are simply following the photons. The stellar atmosphere problem is when we couple the solution of the transfer equation, which is essentially the solution of the energy equation in hydrostatics, to the hydrostatic equation, which solves the momentum conservation condition for a static medium.

3.4.3.4 Thermalization Length

In our fog problem, as you move gradually away from the source you notice that it not only fades but also becomes progressively less distinct. Eventually it fades into the background. At this point, you have passed the *thermalization length*, which is an important concept for understanding how thermal equilibrium enters into radiative transfer. Consider a medium where the source function is internal and isotropic, and so is independent of μ, but not uniquely due to scattering. The optical depth includes all extinction processes operating in the medium so the total extinction coefficient is $\kappa_T = \kappa + \sigma$, where κ is the pure absorption coefficient and σ is the scattering coefficient. Either or both may be frequency dependent. Since $S \neq J$, the zeroth moment of the transfer equation is

$$\frac{dH}{d\tau} = J - S$$

and the first moment is

$$\frac{dK}{d\tau} = H.$$

Applying the Eddington approximation, $K = \frac{1}{3}J$, we get

$$\frac{1}{3}\frac{d^2 J}{d\tau^2} = J - S. \tag{3.59}$$

For a stellar atmosphere, since it is hot even though perhaps a bit tenuous, it is reasonable to assume that there are two components to the source function. Assume that one is the scattering and the other is from actual thermal emission so that

$$S = \frac{\kappa}{\kappa_T}B(T) + \frac{\sigma}{\kappa_T}J. \tag{3.60}$$

Let us make one further approximation by way of introducing what radiative diffusion means. Suppose you find yourself sitting deep in the atmosphere, so that the medium is very optically thick. Then you can use a *linear* approximation for B because $d^2B/d\tau^2 = 0$; in this case, allows Eq. (3.59) to be written as

$$\frac{d^2(J - B)}{d\tau^2} = 3\lambda(J - B), \tag{3.61}$$

where $\lambda = \kappa/\kappa_T$. This equation has a characteristic scale length of $\tau_0 = (3\lambda)^{-1/2}$. This is the diffusion or *thermalization* length for the atmosphere. It is the characteristic optical depth into which you can see the source distinctly, and in deeper layers it merges into the background thermal emission.

3.4.4 Diffusion Approximation

If the photon mean free path is small, it is said to diffuse outward. Therefore, no large gradients of the temperature should occur because the rate of diffusion is slow. Put differently, unlike at the surface, there will only be small local changes in the source function. We can approximate the source function, or in our case the Planck function, by Eq. (3.53), the one that gave the linear limb darkening law from the formal solution for the equation of transfer. Then putting this into the definition of the flux gives $F = \frac{1}{3}b$, where $b = \int b_\nu \, d\nu$ is the integrated limb darkening coefficient from Eq. (3.54), so that

$$F = \frac{1}{3}\frac{dB}{d\tau} = \frac{-4ac}{3\kappa_R \rho} T^3 \frac{dT}{dz}. \qquad (3.62)$$

In order to arrive at this form of the integrated flux, we have introduced an averaging procedure for the opacity

$$\frac{1}{\kappa_R} \equiv \int_0^\infty \frac{1}{\kappa_\nu} \frac{dB_\nu(T)}{dT} d\nu \bigg/ \int \frac{dB_\nu(T)}{dT} d\nu, \qquad (3.63)$$

which is called the *Rosseland mean opacity*. It is a harmonic mean that was first defined for radiation pressure in a stellar interior, as we will show in Chapter 4, but it also enters for the flux in an optically thick environment. This equation has the same form as the normal conductive heat flux that you know from Newton or Fourier, that the flux depends on the gradient of the temperature in a medium:

$$F = -K\frac{dT}{dz} = -K\hat{z} \cdot \nabla T. \qquad (3.64)$$

This defines the *radiative diffusion coefficient*, K a quantity that we will use extensively in our discussion of stellar interiors in the next chapter. Notice that the opacity is a *harmonic* mean weighted by the frequency-dependent source function. The temperature dependence of K, which is significantly different than for particle conductivity, depends on the frequency distribution of the opacity.

3.4.4.1 A Note on Probability and Radiative Transfer
The appearance of a diffusion equation suggests another approach to the radiative transfer problem. Notice that the rate of emergence of a photon from a nonemitting layer varies exponentially with traversed distance. Call this p so that $I = pI_0$ with $0 \le p \le 1$. The rate of re-emission varies with $q = 1 - p$. In this picture, the transfer equation describes the probability of a photon arriving at some point in the medium through a variety of processes. The way we have written the equation

does not include the change in the direction of the photon, but that is what happens in scattering. It is the probability that after going some distance dz through the medium in some initial direction μ, the photon is redirected to some direction μ', possibly with some change $\Delta\nu$ in the frequency because of the process. The Compton effect is a familiar example of the inelastic scattering process, where the change in the frequency of the photon depends on the difference between the initial and final directions, while Thomson scattering by electrons is an example of an elastic process. There is some phase function, $p(\mu, \mu'; \nu, \nu')$, that describes the process, but otherwise the basic probabilistic formalism remains unchanged. The radiative transfer problem thus reduces to a probability calculation, the chance that a photon will random walk its way out of a medium through some series of collisions. Because of this, the optical depth can be seen to be related to the number of scatterings and/or absorptions, and the number of steps is related to τ by $\tau \sim N^{1/2}$. In effect, it defines the rms distance that a photon will go compared with the geometric thickness of the medium. If the number of scatterings is small, the medium can cool.

Let us try a simple approach as an illustration. Since the phase function is the probability that a photon is in a direction μ' having been incident in a direction μ, rotate the coordinate system so that $\mu = 0$. To make the example concrete, we will take a dipole function, $p(\mu'; 0) = a(1 + b\mu'^2)$ where a and b are constants and the function is normalized and assume that the scattering is coherent. Now generate a pseudo-random number $0 \leq \varpi \leq 1$, which you set equal to $p(\mu'; 0)$. Thus, by inversion, the directional cosine for the emergent photon after scattering is $\mu' = ([\varpi/a] - b)^{1/2}$. If the probability of absorption is $p_{abs} = e^{-\Delta\tau/\mu'}$, then another random number ϖ' gives the distance traveled by the photon until the next interaction, $\Delta\tau = -\mu' \ln \varpi'$, and so on. In this way, you can follow the distribution of the radiation as a function of both depth and angle through its walk to the surface, which is defined as a boundary in the calculation. The problem with this approach, which is called a *Monte Carlo* simulation, is that it requires following a very large number of "photons" in order to ensure high statistical accuracy for the result. Its great advantage is in allowing you enormous flexibility in choices of physical processes, such as inelastic scattering, geometry, and other physical properties of the medium, such as temperature and/or density gradients. As such, it is often exploited in astrophysics, especially in complex radiative environments, when analytic or finite difference techniques are too cumbersome or even impossible.[14]

We dwell on this point because it is so useful in thinking about how to derive the properties of the radiating gas from the emergent spectrum. We have already talked about the *thermalization* length and the concept of optical thickness, and here it is even more obvious. We are really asking, at each moment, what the

---

[14] This technique, named after the famous Riviera casino, was developed by Metropolis and Teller at Los Alamos during World War II for computing neutron transfer in complex geometries. The problem is essentially the same as for photons, as emphasized by Davison, B. 1957, *Neutron Transport Theory* (Oxford: Oxford Univ. Press). Good introductions are provided by Hammersley, J. M. and Handscomb, D. C. 1964, *Monte Carlo Methods* (London: Methuen); and Spanier, J. and Gelbard, E. M. 1967, *Monte Carlo Principles and Neutron Transport Problems* (Reading: Addison-Wesley). Monte Carlo methods are also frequently used to simulate instrumental properties, such as detectors and optics, and to create sample data sets with which to check analysis software.

probability of interaction is between the photon and the atom, and what the relative probabilities are for the different processes. As a final example, consider how we can treat line radiative transfer in a probabilistic formalism.[15] Recall the definition of the profile function, ϕ_ν. It acts as a probability distribution in frequency, just as the phase function does in the direction cosine. If we are very near a line center, then resonance absorption and/or scattering is most likely. As we go progressively farther from line center, Rayleigh and Raman scattering become progressively more important. Finally, if the energy is high enough, the electron may be removed entirely from the atom. All these processes compete with collisions for producing the final distribution of photons with energy. Also, there is a tendency for a photon to scatter *through* a line profile by redistribution, since it also diffuses outward spatially by going to the core of the line profile.

3.5 OPACITY

Opacity governs the rate of flow of energy from within a radiating medium to the universe at large. The term covers any process that redistributes the photon in frequency or direction, such as scattering where the photon is not destroyed, or absorption, where it is through collisional de-excitation of the absorber. In this section we will first discuss the important continuum processes that dominate the emission from astrophysical plasmas. While it is arbitrary to separate them out from line processes, since atoms are after all quantized systems regardless of the final destination of the electrons, they have to be treated somewhat differently. Continuum processes can even happen in a fully ionized hydrogen plasma, in particular *bremsstrahlung* emission and absorption, while line radiation requires that at least a few bound states be accessible. Thus, at the highest temperatures or at the highest densities, continuum processes can be expected to dominate the spectrum. We will then address the interaction of photons with bound states through line transitions.[16]

3.5.1 Bremsstrahlung, Ionization, and Recombination

3.5.1.1 Thermal Bremsstrahlung or Free–Free Opacity

Thermal continuum radiation from an optically thin gas occurs because of collisions between ions and electrons. Being more mobile, the electrons whiz around in a more nearly static background of protons and other ions. When they are deflected, they radiate a photon whose frequency depends on the interaction timescale, which means on the relative speed of the interacting particles and on the distance between them at the time. The closer the approach, the greater the energy and the higher the frequency. The interaction potential is especially easy to treat, since it is just the coulomb (electrical) interaction. Also, the electron makes a transition between two continuum states, so it is relatively simple to cast the

---

[15] Gu, Y., Lindsey, C., and Jeffries, J. T. 1995, *ApJ*, **450**, 318; Boisse, P. 1990, *Astr. Ap.*, **228**, 483.

[16] For the best available introduction to general theory of radiation–matter interactions, you cannot do better than consult Rybicki, G. B. and Lightman, A. P. 1979, *Radiative Processes in Astrophysics* (NY: Wiley); see also Tucker, W. H. 1975, *Radiation Processes in Astrophysics* (Cambridge, MA: MIT Press).

process in quantum mechanical terms. To be thermal, the interaction must take place at speeds near that of sound, and the spectral distribution of the radiation must approach that of a blackbody as the density is increased to where the gas is opaque.

Let us first try for a simple order of magnitude estimate. The physical picture is very similar to our discussion of dynamical friction. Take the interaction potential to vary with distance as $V = -Ze^2/r$, where Z is the ionization of the massive perturber. The electron's trajectory, $r(t) = (b^2 + v^2t^2)^{1/2}$ is given by a small kick due to a close encounter with the ion, where b is the distance of closest approach or *impact parameter*. Then the normal impulse given to the electron is

$$\Delta p_\perp = \int_{\text{path}} F(t)\, dt = Ze^2 \int_{-\infty}^{\infty} \frac{dt}{(vt)^2 + b^2} = \frac{\pi Ze^2}{bv}. \tag{3.65}$$

The change in the dipole moment as a function of frequency is $\dot{d}_\omega = (e/2\pi) \int e^{i\omega t} v(t)\, dt$. Thus

$$\dot{d}_\omega \approx \frac{e}{m_e} \Delta p_\perp = \frac{\pi Ze^2}{bv}. \tag{3.66}$$

Then the rate of emission is given by

$$\frac{dE}{dt} \sim \frac{Z^4 e^6}{m_e^2 c^4 b^2 v^2}. \tag{3.67}$$

Now the rate of energy loss depends on the effective dipole of the interaction, so the radiated power varies as $\Delta p_\perp^2$ —in other words, the radiation varies as v^{-2} for each collision. The particles, however, have a Maxwellian velocity distribution that depends on the temperature, and the total rate is the weighted sum of all of the contributors. This means that we take the emission rate for the particle and average it over the encounters within a cylinder of radius b. The closer the encounter, the higher the frequency of the emitted radiation. This has an important consequence. Since there are few particles in the high-velocity part of the thermal distribution, there will also be lower emission with a cutoff that should depend on kT, the mean energy of the electrons. But most of the particles are in the lower energy part of the curve, and there is also a slower encounter and a larger cross section. The two factors enhance the lower energy part of the emission compared with the higher energy part, and compared with a blackbody distribution. Only as the opacity increases does the low-energy part diminish. Of course, this almost complete independence of frequency cannot go on to zero frequency. Somewhere there must be some opacity, and at that point the distribution follows the Rayleigh–Jeans portion of the Planck function.

The final emissivity for an optically thin plasma is given by

$$j_{\text{ff}} = n(Z)n_e \left(\frac{2m_e}{3\pi kT}\right)^{1/2} \left(\frac{32\pi^2 Z^2 e^6}{3m_e^2 c^3}\right) g_{\text{ff}}(\nu) e^{-h\nu/kT}. \tag{3.68}$$

Here g_{ff} is called the *Gaunt factor* and represents a quantum mechanical correction to the otherwise classical process. It is only weakly dependent on frequency and for an optically thin plasma at $\nu \ll kT/h$ contributes a frequency dependence of $\nu^{-0.1}$.[17] The relatively weak dependence of the emissivity on frequency means that the absorption coefficient, which is given by $\kappa_\nu = j_\nu/B_\nu(T)$, is

$$\kappa_{ff} \approx \nu^{-3}(1 - e^{-h\nu/kT}). \tag{3.69}$$

This form has two consequences. The first is that an optically thick bremsstrahlung source ultimately relaxes to a Planck distribution, so that at long enough wavelength any plasma emits as $F_\nu \sim \nu^2$. The second is that for an extended source, as we will see (Section 3.7.2), the angular size of the source depends on the frequency. This is of special importance in the study of radio emission from extended atmospheres. A magnetic field complicates this emission process because the acceleration is changed compared with the Larmor equation for a coulomb interaction. In particular, you have to include the Lorentz force in the emission mechanism. We will deal with this in more detail in Chapter 6 when we discuss synchrotron (nonthermal) emission.

3.5.1.2 Ionization and Recombination

Ionization is the transition from any bound state to a continuum state; the reverse is recombination. It thus has a threshold energy, χ, called the ionization energy (see Table 3.1). For hydrogen, this is given by $\chi_{H,n} = 13.595$ eV$/n^2$ or $\lambda_I =$ (91.2 Å) n^2 for quantum number n. For reference, the $n = 1 \rightarrow \infty$ transition is called the *Lyman limit*, that at 3648 Å is called the *Balmer jump*, and the others are known by the hydrogen line series for which they are the limiting frequency; the continuum that is produced at higher frequencies is labeled by the series as well. We have already used the idea when computing the ionization fraction of a gas in thermodynamic equilibrium. However, in the context of radiative processes, this takes on a new significance. It is the minimum energy a photon must have to produce such a transition.

The strongest interaction occurs near resonance, when $h\nu \approx \chi$. In quantum mechanical terms, this is equivalent to a $B_{n\kappa}$ for a stimulated absorption coefficient between the bound atomic state i and the continuum state characterized by the wavenumber κ for the electron of energy $\hbar\kappa/c$. The absorption coefficient is computed analogously to any bound transition. For a hydrogenic atom (an ion with

---

[17] To be more precise, this factor for bremsstrahlung is related to the Coulomb integral and is approximately

$$g_{ff} = \left(\frac{3}{\pi^2}\right)^{1/2} \ln \frac{(2kT)^{3/2}}{13.5 e^2 m_e^{1/2} \nu} \approx 1.38 T^{-0.34} \nu_{GHz}^{-0.11}$$

as discussed in Kaplan, S. A. and Pikel'ner, S. B. 1970, *The Interstellar Medium* (Cambridge, MA: Harvard Univ. Press). See also Hjellming, R. M. 2000, in *Astrophysical Quantities*, 4th ed., Cox, A. N. ed. (Berlin: Springer-Verlag), Chap. 6.

TABLE 3.1 Atomic Ground-State Configurations and Ionization Potentials

Element	Z	Ground State	χ_0	χ_1	χ_2	χ_3
H	1	$1s\,^2S_{1/2}$	13.595	—	—	—
He	2	$1s^2\,^1S_0$	24.59	54.40	—	—
Li	3	(He)$2s\,^2S_{1/2}$	5.39	75.62	122.42	—
Be	4	(He)$2s^2\,^1S_0$	9.32	18.21	155.85	217.66
B	5	(He)$2s^22p\,^2P^\circ_{1/2}$	8.30	25.15	37.92	259.30
C	6	(He)$2s^22p^2\,^3P_0$	11.26	24.38	47.87	64.48
N	7	(He)$2s^22p^3\,^4S^\circ_{3/2}$	14.53	29.59	47.43	77.45
O	8	(He)$2s^22p^4\,^3P_2$	13.62	35.11	54.89	77.39
F	9	(He)$2s^22p^5\,^2P^\circ_{3/2}$	17.42	34.98	62.65	87.14
Ne	10	(He)$2s^22p^6\,^1S_0$	21.56	41.07	63.5	97.02
Na	11	(Ne)$3s\,^2S_{1/2}$	5.14	47.29	71.72	98.88
Mg	12	(Ne)$3s^2\,^1S_0$	7.65	15.03	80.14	109.29
Al	13	(Ne)$3s^23p\,^2P_{1/2}$	5.99	18.82	28.44	119.96
Si	14	(Ne)$3s^23p^2\,^3P_0$	8.15	16.34	33.49	45.13
Ca	20	(Ne)$3s^23p^64s^2\,^1S_0$	6.11	11.87	51.21	67
Fe	26	(Ne)$3s^23p^64s^23d^64s^2\,^5D_4$	7.90	16.18	30.64	56.8

charge $Z - 1$), this absorption coefficient is

$$\kappa_\nu = \frac{64\pi^4 Z^4 m_e e^{10}}{3^{3/2} c h^6} n^{-5} \nu^{-3}, \qquad (3.70)$$

where n is the quantum number of the state, and $\nu \geq \chi_n/h$.[18] Collisional ionization may occur if the state population is sufficiently high, but it is difficult to have this occur normally from the ground state. This is especially true of hydrogen in equilibrium. Photoexcitation followed by collisional ionization is more the rule. If the gas is locally in thermal equilibrium, however, the balance between radiation and collisions is guaranteed and the Saha equation again describes the ionization state of the gas.

A good first estimate for the cross section for any photon process is to πr_e^2, where r_e is the Thomson radius for the electron. This always approximates a cross section for an electromagnetic interaction, and since the overall size of the atom depends on the coupling strength, the charge Z always enters into the final result. For any bound-bound process, a good first guess is that the cross section is πa_n^2, where a_n is the approximate Bohr model radius of the equivalent hydrogenic orbit. The threshold hydrogenic photoionization cross section from of the nth state of neutral hydrogen is $\kappa_n \approx 8 \times 10^{-18} n$ cm^2.

---

[18] Loudon, R. 1983, *The Quantum Theory of Light*, 2nd ed. (NY: Oxford Univ. Press); Mihalas, D. 1978, *Stellar Atmospheres*, 2nd ed. (San Francisco: Freeman); Griem, H. 1997, *Principles of Plasma Spectroscopy* (Cambridge, UK: Cambridge Univ. Press).

Collisions are usually frequent enough in stellar atmospheres, and always in stellar interiors, that we can use the Saha equation to determine the ionization balance. In dilute media, such as outermost layers of stellar atmospheres or interstellar nebulae, this breaks down and radiative processes are important. Then the ionization rate depends on the match between the cross section for bound-free transitions and the spectral distribution of the radiation. Recombination, on the other hand, depends on the coincidence between the continuous energy distribution of the electrons and the capture cross section. The former depends only on frequency while the latter depends only on temperature. Because the rates are integrated over the velocity distribution of the particle and the frequency dependence of the radiation field, the equilibrium in LTE depends only on temperature and the number of ions and electrons, which are the only remaining free parameters. The radiative ionization rate depends on the match between the absorption coefficient and the spectral energy distribution:

$$R^{\text{ion}} = 4\pi n_i \int_{\nu_0}^{\infty} \kappa_\nu J_\nu \frac{d\nu}{h\nu} = n_i \beta \qquad (3.71)$$

while the recombination rate is given by

$$R^{\text{rec}} = n_{i+1} n_e \alpha(T). \qquad (3.72)$$

Here κ_ν is the continuous absorption coefficient and $\alpha(T)$ is the recombination coefficient averaged over the thermal velocity distribution and including all radiative processes. This is really a formal definition that depends on the details of the recombination process. In thermal equilibrium, this is just the Saha equation because the electron distribution is thermalized. When the densities are low enough, this no longer holds because the radiation need not be in equilibrium with the temperature of the gas. The best example is the one we will encounter in Chapter 6 under interstellar conditions where the radiation for an H II region comes from an embedded star.

Level populations are altered by ionization in the same way they are by line transitions and can be treated similarly in statistical equilibrium. Whenever collisions are infrequent compared with the timescale for radiative processes, the conditions of strict LTE break down and the full rate equations must be solved. To include ionization and recombination is not conceptually different, but it is more complicated because of the coupling between levels introduced by the recombination process. The ionizing rates are of the form

$$n_i R_{ik} = 4\pi n_i \int_{\nu_0}^{\infty} \kappa_\nu J_\nu \frac{d\nu}{h\nu}, \qquad (3.73)$$

where, for LTE, the integrated intensity would be replaced by the Planck function. The inverse radiative processes, radiative recombinations, are represented by

$$n_c R_{ci} = 4\pi n_c \int_{\nu_0}^{\infty} \kappa_\nu J_\nu e^{-h\nu/kT} \frac{d\nu}{h\nu} \quad n_k R_{ci} = 4\pi n_c \int_{\nu_0}^{\infty} \kappa_\nu B_\nu (1 - e^{-h\nu/kT}) \frac{d\nu}{h\nu} \qquad (3.74)$$

$$n_i B_{ic} = n_k \int_{\nu_0}^{\infty} B_{ci} \left(\frac{A_{ci}}{B_{ci}} + J_\nu \right) \frac{d\nu}{h\nu} = n_c \int_{\nu_0}^{\infty} \kappa_\nu \left(\frac{2h\nu^3}{c^2} + J_\nu \right) e^{-h\nu/kT} \frac{d\nu}{h\nu}, \qquad (3.75)$$

using the definition of the transition probabilities and correcting the absorption coefficient, through B_{ci}, for NLTE. In final form the statistical equilibrium leads to the condition that

$$\int_{\nu_0}^{\infty} \kappa_\nu J_\nu \frac{d\nu}{h\nu} = \int_{\nu_0}^{\infty} \kappa_\nu \left[\frac{2h\nu^3}{c^2} + J_\nu \right] \frac{d\nu}{h\nu}. \qquad (3.76)$$

The recombination coefficients are derived from the condition of detailed balance, since radiative recombination can be thought of as spontaneous emission of continuum radiation from a hot plasma. The recombination coefficient is normally summed over all lower levels, including the cascade.[19] There is one process that straddles the line between bound–bound and bound–free transitions. In some strongly atoms, the mixing between states is so large that there is actually a repulsion of one of the resultant states into the continuum while the other remains bound. Because of the overlap between this state and the continuum, a quasibound system is formed and one of the electrons may actually leak out of the atom *without* any radiative loss. In effect, this is a two-electron process where the photon is not emitted to the outside world. The process is called *autoionization* (Allen 1929) and was first described quantum mechanically by Fano.[20] Its inverse, *dielectronic recombination*, involves recombination that produces internal excitation of bound electrons. It is especially important in low density, highly ionized plasmas, such as the solar corona, because the resonances change significantly the cross sections for higher-energy transitions. We will see more of this in the discussion of ionized interstellar gas in Chapter 6.[21]

3.5.2 Line Processes: Bound–Bound Transitions

When the initial and final states are bound, we are talking about line transitions. First we will cover line profiles and then scattering processes that depend on the presence of some resonance energy in the atomic/molecular system. It is important to note that bound processes are coherent in the comoving frame of the atom, and fundamentally differ from those that form the continuum.

3.5.2.1 Line Profiles
The classic equation of motion for a damped harmonic oscillator interacting with an oscillatory electric field **E** is

$$m\ddot{\mathbf{x}} = -\omega_0^2 \mathbf{x} - \gamma \dot{\mathbf{x}} + e\mathbf{E}, \qquad (3.77)$$

---

[19] A lovely summary of processes is given in Appendix 1 of Kaplan, S. A. and Pikel'ner, S. B. 1970, n.17.
[20] Fano, U. 1961, *Phys. Rev.*, **124**, 1866; Fano, U. and Cooper, J. W. 1968, *Rev.Mod.Phys.*, **40**, 441. Good overviews can be found in Temkin, A., ed. 1966, *Autoionization: Astrophysical, Theoretical, and Laboratory Experimental Aspects* (Baltimore: Mono Books); Temkin, A., ed. 1985, *Autoionization: Recent Developments and Applications* (NY: Plenum Press); Fano, U. and Fano, L. 1972, *Physics of Atoms and Molecules: An Introduction to the Structure of Matter* (Chicago: Univ. Chicago Press); Cowan, R. D. 1981, *Theory of Atomic Structure and Spectra* (Berkeley: Univ. California Press).
[21] On a lighter note, one way you can think of this is that if recombination is like standing in the outfield during a baseball game and having the ball drop into your glove, dielectronic recombination is having to jump for it.

where $\mathbf{E} = \mathbf{E}_\omega \exp(i\omega t)$. Taking a single mode expansion gives the frequency dependent dipole moment for a frequency near resonance, where $\omega \approx \omega_0$:

$$|\mathbf{d}_\omega|^2 = e^2|\mathbf{x}_\omega|^2 = \frac{4e|\mathbf{E}_\omega|}{\left(\omega^2 - \omega_0^2\right)^2 + \gamma^2\omega_0^2}. \qquad (3.78)$$

The emission line profile function is

$$\omega^4|\mathbf{x}|_\omega^2 = \omega^4|\mathbf{E}_\omega|^2\left(4\omega_0^2(\omega - \omega_0)^2 + \gamma^2\omega_0^2\right)^{-1}. \qquad (3.79)$$

The Poynting vector is $\mathbf{S} = (c/4\pi)\mathbf{E} \times \mathbf{B}$, and this has an average value

$$P(\omega) = \frac{2}{3}\frac{e^2}{c^3}|\dot{v}|^2 = \frac{2e^2}{3c^3}|\ddot{x}_\omega|^2. \qquad (3.80)$$

If $\mathbf{F}$ is the force damping the oscillator, then the energy radiated during a single oscillation is $\int \mathbf{F} \cdot \mathbf{v}\, dt$. Since this must be the same as $\int P(\omega)\, d\omega$ over the same interval, and $\mathbf{F} = -m\gamma\mathbf{v}$ by definition, we find the damping constant *for radiation alone* to be

$$\gamma_{\text{rad}} = \frac{2e^2\omega_0^2}{3mc^3}. \qquad (3.81)$$

This is the minimal broadening for an energy level. Perturbations, like collisions with environmental particles, broaden the profile so this is the minimum value for a classical oscillator. For quantum mechanical estimates, the damping is due to the finite lifetime of the level. Thus $\gamma \sim A_{ji}$, and it also depends on the branching distribution for transitions to lower energy levels. Recall that the lower *and* upper states contribute to the line profile.

The cross section is therefore a function of frequency:

$$\sigma_\omega = \frac{\pi e^2}{mc} \frac{\gamma}{(\omega - \omega_0)^2 + (\gamma/2)^2} \qquad (3.82)$$

and the total cross section, $\sigma = \int \sigma_\omega\, d\omega$, is given by

$$\sigma = \frac{\pi e^2}{mc}. \qquad (3.83)$$

This fundamental result is the classical strength of an absorption or emission line. It is a constraint on the quantum-mechanical result, such that the *oscillator strength*, f_{ij}, is defined as the ratio of the quantum to the typical cross section; in other words

$$\sigma_{ij} = \frac{\pi e^2}{mc} f_{ij}. \qquad (3.84)$$

Although the quantum mechanical result is derived differently than the classical oscillator model we have used here, it must produce formally the same result, as we discuss in Appendix 3.A. The Lorentzian line profile is expected as the profile for an atomic transition *in the absence of additional broadening by environmental perturbations*. We will return to this last point in discussing Stark and Van der Waals broadening.[22]

3.5.3 Collisional Broadening and Lifetimes

Collisions lower the number of photons when they de-excite a state. The lower the frequency of collisions, the less coupled the electrons are to the radiation, and the photons may cease being in exact thermal equilibrium with the matter. Then the conditions for Eq. (3.13) break down. In your daily life, this is not as unusual as you think. Much of the radiation in daily life is due to nonequilibrium processes, like fluorescent bulbs, lasers, and even lightning. In a sense, even the color of the daytime sky betrays this condition, because the color of the Sun is that of a nearly 6000 K blackbody while the equilibrium temperature of the environment is only about 300 K. In other words, the source of the energy may have a spectral distribution very different from that emitted by the body. This is the primary task of radiative transfer theory, to find the connection between the distribution of the light we see and the conditions in the medium from which it originates.

We have spent so much time on this because it is essential for understanding the interaction of matter with radiation. The appearance of the spectrum of a cosmic body is the only way we ever really know anything about its physical structure, its thermal state, and its abundances. Unfortunately, this comes only after some assumptions have been made about how to model the source of the radiation, the danger of any inverse problem. We always have to assume something about the medium, be it the geometry, symmetry, opacity, or whatever.

3.5.3.1 Doppler Broadening and the Voigt Profile
The absorbing and emitting atoms are in motion because the gas is hot. In this moving frame, they see the incoming photons shifted from rest wavelengths. However, this is essentially one-dimensional motion because the individual atom sees only one photon at a time. Since the electron distribution in LTE is isotropic and Maxwellian, there is no net shift in the line center due to the scattering or absorption. Therefore, in equilibrium, the velocity probability distribution immediately transforms into a probability distribution in the observed wavelength of the photon:

$$G(\Delta \lambda) = \exp - \left(\frac{\Delta \lambda}{\Delta \lambda_D} \right)^2 \qquad (3.85)$$

In other words, a δ-function line profile will be seen through the gas as having a width corresponding to the one dimensional thermal speed. Whatever the specific frequency response of the atomic state, it will no longer be absorbing monochromatic light, and in complete redistribution the same is true for the emission

---

[22] A very useful review on laboratory measurements is presented by Huber, M. C. E. and Sandeman, R. J. 1986, *Rep. Prog. Phys.*, **49**, 397.

profile. Each frequency is equivalently broadened, so that the overall line profile is a *convolution* of the Doppler Gaussian and the intrinsic profile function, $\phi(\Delta\lambda)$:

$$\kappa(\Delta\lambda) \sim \int_{-\infty}^{\infty} G(\Delta\lambda')\phi(\Delta\lambda - \Delta\lambda')\, d\Delta\lambda' = G \star \phi. \tag{3.86}$$

An additional effect comes from the presence of disordered motions on a scale that is similar to the atomic collision mean free path. Usually called *microturbulence*, this is simply a way of including superthermal motions into the line profile that are disordered and whose source is spatially unresolved. If the "turbulence" is isotropic and has a Gaussian distribution, then the line width is given by

$$\Delta\lambda_D = \frac{\lambda_0}{c}\left(v_{th}^2 + \xi_t^2\right)^{1/2}. \tag{3.87}$$

This microturbulent broadening has the effect of desaturating the line profile, reducing the optical depth by spreading the absorption out over a broader wavelength range. The microturbulence is required to produce superthermal broadening and its value relative to the sound speed is an important consideration. If the gas is thermalized (i.e., if the energy levels have LTE populations), then the linewidth *should* yield the same (kinetic) temperature as the excitation and ionization. In many environments it does not. For instance, molecular clouds, which have very low temperatures of order a few tens to hundreds of degrees, are often observed to have much broader lines than 0.1 km s^{-1}. This is usually referred to as *turbulent broadening*; the departures of the line profile from a Gaussian shape are sometimes quite large. However, as we shall see (e.g., for interstellar gas), the interpretation of these linewidths is much more complicated. At present, although disparaged as a fudge factor,[23] the required velocities are often small and may indicate the presence of true dissipative hydrodynamic processes in many astrophysical environments.

When the intrinsic profile is Lorentzian, the resulting convolution is called a *Voigt profile* (Fig. 3.3). There are now two characteristic profile widths, one from the atomic velocity distribution and the other from the atomic transition itself. We can form a dimensionless ratio, $a \equiv (\Delta\lambda)_{\text{damp}}/\Delta\lambda_D$ and normalize all wavelengths steps to the Doppler width such that $v \equiv \Delta\lambda/\Delta\lambda_D$, then

$$H(a,v) = \frac{\pi^{1/2}}{a}\int_{-\infty}^{\infty} \frac{\exp -y^2}{(v-y)^2 + a^2}\, dy. \tag{3.88}$$

In more compact form, the absorption coefficient becomes

$$\kappa(\Delta\lambda) = \kappa_0 H(a,v), \tag{3.89}$$

---

[23] This term has been used far too frequently to describe physical parameters in astrophysical models. Microturbulence may not be properly interpreted at present, it may not have anything to do with hydrodynamic turbulence, but it does appear to be required in order to explain the observed line profiles. In the absence of any alternative broadening mechanism for the line cores, the inclusion of some kinetic field in the line forming region that exceeds the thermal velocity points to a deep physical problem that should be addressed, not dismissed.

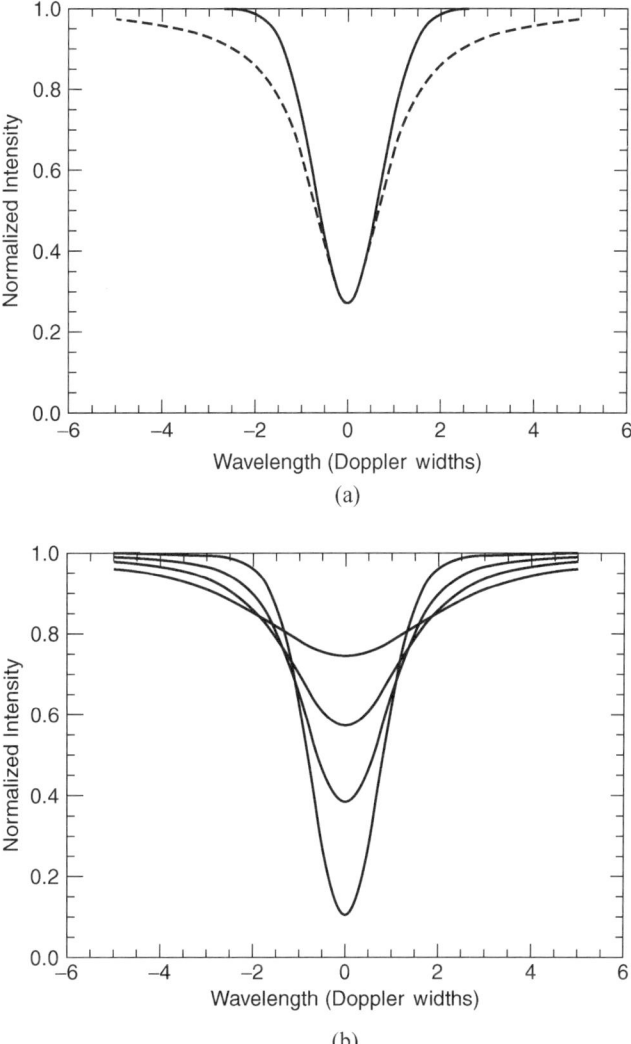

FIGURE 3.3. (a) Comparison of a Gaussian and Lorentzian profile ($a = 1.0$) for identical central intensities. (b) Voigt profile variation for $a = 0.1, 0.5, 1.0$, and 2.0. Notice the growth of the wings relative to the Gaussian core as the damping parameter is increased.

where κ_0 is the average absorption coefficient. Again, it does not matter what the intrinsic width of the state is. The broader the thermal distribution, the broader the line and the lower the opacity at line center.

3.5.4 Curve of Growth

It is only by knowing how the opacity of a line increases as a function of column density that we can determine how much of an absorber we have in the line of sight. In general, when gradients in the temperature or density are present, the problem requires a model of the medium. This is the point of computing model atmospheres, as we shall soon discuss. However, it is instructive to look at the case of a uniform homogeneous absorbing medium, because it actually seems to occur in nature in the form of interstellar, or even intergalactic neutral hydrogen clouds. Keep this in mind for when we come to the discussion of the Lyman α forest of extragalactic absorbers.

The inspiration for developing the concept of the equivalent width, and the curve of growth, was the use of photographic plates. Since these could not be absolutely calibrated, and the sensitivities were usually poor, it was necessary to come up with some way of quantifying the "blackness" of a line profile. On photographic plates, the line was formed by the projection of the slit, weighted by the line profile. The equivalent width is defined as the width, in wavelength units, that a perfectly black top-hat profile would have to have in order to produce the equivalent flux deficit in the line. To calculate this, we normalize the flux at any wavelength to that of the continuum and define the residual intensity as

$$r_{\Delta\lambda} = \frac{F_{\Delta\lambda}}{F_{c,\Delta\lambda}}. \tag{3.90}$$

The absorbance is $a_{\Delta\lambda} = 1 - r_{\Delta\lambda}$ and the effective width of the line is the absorbance-weighted displacement. This means that, since $F_{\Delta\lambda} = F_{c,\Delta\lambda}\exp - \tau_{\Delta\lambda}$, we are actually weighting the probability of a photon emerging from the continuum at some displacement from line center and then integrating over the line to find the mean width or *equivalent width* at the central wavelength λ:

$$W_\lambda = \int_{\infty}^{\infty} a_{\Delta\lambda}(\tau_0)\, d\Delta\lambda. \tag{3.91}$$

The spectral line is described by $H(a,v)$. In general, because the damping factors are comparatively small for many lines relative to the Doppler widths, the wings of the damped part of the intrinsic profile are so weak as to be essentially invisible for low column densities and the Doppler core dominates the profile. The condition for a weak line is that $a_{\Delta\lambda}$ is replaced by $\tau_0 G(\Delta\lambda/\Delta\lambda_D)$. Integration of Eq. (3.41) shows that the equivalent width for such a line is linearly proportional to both the Doppler width and the optical depth at line center:

$$W_\lambda \sim \tau_0 \Delta\lambda_D, \tag{3.92}$$

and is also linearly proportional to the column density, N, of the absorbers. For increasing optical depth the wings of the profile contribute progressively more of the absorption because the opacity at line center *saturates*. Depending on both thermal and possible nonthermal motions, the Gaussian Doppler profile eventually ceases to increase in its effective absorption at large distances from line center. This is why the intrinsic broadening of the atomic levels is so important. For this intermediate region, the equivalent width depends on the optical depth as

$$W_\lambda = \Delta\lambda_D(\ln \tau_0)^{1/2}, \tag{3.93}$$

which increases only very slowly with increasing column density and is called the *saturated* portion of the curve of growth. Finally, with this saturation of the Doppler core of the profile, the damping wings take over as the primary contributors to the opacity for increasing column density. Since the gaussian part of the profile is now irrelevant, the profile depends on $(\Delta\lambda)^{-2}$ and the equivalent width varies as

$$W_\lambda \sim \tau_0^{1/2}. \tag{3.94}$$

In this regime, we reach the so-called *damped* portion of the curve of growth. The details of the intrinsic line profile are now of prime importance. The greater the damping width, the faster the onset of this portion of the curve. Since, as we shall see, the width of a line reflects the interaction between the atom and its surroundings, the onset of the damping portion of the curve depends not only on the number of absorbers but also on the thermal and physical properties of the absorber's surroundings.

In a stellar atmosphere, this is a crude but useful initial measurement of the total line absorption. Since the medium is not isothermal, a full atmospheric model is required to compute the opacity at each wavelength so that the equivalent width is a much less useful parameter for determining the column density than matching the full line profile. It is now feasible to use this since CCD and other electronic detectors are linear and can produce absolute flux measurements. For interstellar lines, the individual absorbing regions are physically unassociated with one another and are dominated by turbulent broadening. With the individual densities being so low, the absorption at one part of the profile is essentially decoupled from another and the comparison of individual line profiles as a function of wavelength permits the determination of column densities and ionization fractions, as well as abundances. We will have more to say about this in Chapter 6.

3.5.5 Types of Scattering

Scattering involves redirection of the photon and, possibly, redistribution in energy, but not its destruction. Following an absorption, the atom radiatively de-excites. There are several outcomes of this process, which depend on the energy of the incoming photon and the details of the energy-level structure for the atom. The nomenclature distinguishes between them based on the energy of the incident photon relative to the level separation. Assume that the target atom is in its ground state. If the level separation is $\Delta E_{0,j}$, then *resonant scattering* occurs if $h\nu = \Delta E_{0,j}$. If $h\nu < \Delta E_{0,j}$, this is *Rayleigh scattering*. If, however, $h\nu > \Delta E_{0,j}$, and if $\Delta E_{0,j} > \Delta E_{ij}$ for some $0 < i < j$, then the process is called *Raman scattering*. Line scattering is mediated by the same profile function as an absorption process. Continuum processes such as the scattering by free electrons and dust grains, are not resonant and have broad frequency dependence.

3.5.5.1 Resonance Scattering
Resonance scattering implies that the incoming photon is within the line core, or within the inner damping wings, of the profile: $\omega \approx \omega_0$. This is the case we have

treated in detail for line formation. The photon is either coherently scattered or partially redistributed into the wings of the profile. At any rate, this is line scattering. The importance of this process is that it provides a momentum transfer precisely at the wavelength of the transition because only the direction of the photon is changed. Thus, unlike true absorption and re-emission, scattering is coherent and preserves polarization. Thus, the signature of a resonant scattering medium is the polarization of the emission lines. This may also result from the medium having some other scattering agent, like electrons, that have a nearly frequency-independent response across the line, but there will usually be other signatures of this. For instance, the continuum and the line should be acted on equally by a gray scatterer so that the *relative* line polarization should remain constant. Line scattering is especially important at low density since collisional depopulation of the upper state is mitigated when $n_e C_{ji}/A_{ji} < 1$. Then there must be sufficient population in the lower state to scatter, a condition most easily met with ground-state transitions. This is why the other processes we discuss here are most efficient from resonance transitions.

3.5.5.2 Rayleigh Scattering

When the frequency is significantly lower than the resonance, the scattering depends only on the width of the atomic state relative to the photon energy. In effect, this scattering is related to resonance because it occurs in the far wing of the line. Rayleigh scattering is a coherent process that depends on the dipole moment and leaves the polarization of the photon unchanged. The observed polarization can, however, change because of the deviation of the photon's direction. In the limit where $\nu \ll \nu_0$, the scattering coefficient from Eq. (3.82), reduces to the following. In the limit we have just stated

$$\sigma_R = \sigma_e f_{ij} \frac{\nu^4}{\nu_0^4}. \tag{3.95}$$

The scattering agent can be any transition, provided it has a strong enough coupling (electric dipole transitions are favored), but the intensity of the scattered radiation depends entirely on the number of scatterers along a line of sight. In general, atomic and molecular hydrogen are the prime agents in astrophysical environments by virtue of their abundance. Neutral and ionized helium may in some cases play the same role. The scattering with which you are most familiar is Rayleigh scattering by N_2 in the Earth's atmosphere. This is the answer to the ultimate *blue sky* question. From Eq. (3.85) you see that the scattering cross section varies as ω^4, which is why a yellow star produces a blue sky. It is also instructive to take a pair of polarizing sunglasses and look at various angles from the sun, rotating the glasses axially to change the plane of polarization. You will notice that the degree of polarization depend on the scattering angle, α, between the sun and the observer. This is an example of the phase function, $p(\alpha)$, which is the angular distribution of the emission for a scattering angle α. For Rayleigh scattering, this is a dipole:

$$p(\alpha) = 1 + \cos^2 \alpha. \qquad (3.96)$$

Thomson scattering (nonrelativistic electron scattering, see text below) has the same phase function. This also provides the probability distribution for the scattering process, which is very peaked in the forward and reverse direction.[24]

3.5.5.3 Redistribution in Line Profiles

In the theory of line formation, you know that absorption of a photon at a frequency $\Delta \nu$ from line center does not necessarily lead to a re-emission of the photon at the same frequency. If the process is independent of the initial energy, that is, if the photon is completely redistributed in frequency stochastically, the profile becomes a probability distribution for the radiation. It is important, however, to recall that the line wings and core are actually formed differently. The core in a Voigt profile is dominated by the Gaussian that results from coherent scattering in the atomic frame of the incoming photon and the only randomness is due to the velocity distribution of the gas. The wings, however, are formed by the combined quantum-mechanical processes of spontaneous reemission and collisional and Stark redistribution of the photon. The probability of the emergence of a photon in the core is not the same as in the line wings. The differential process of re-emission is called *partial* redistribution. This has two effects. The first is that the level populations are changed depending on the wing opacity. This is quite as expected. The second effect is, however, that the line formation is more complicated and sensitive to the environment depending on the strength of the wings.[25]

3.5.5.4 Fluorescence and Raman Scattering

In *fluorescence*, a line is excited by a photon at shorter wavelength and the atom is left in a different final state from the initial state. For instance, C IV 1550 Å happens to coincide with a number of Fe II and Fe III multiplets that can be excited by absorption and that reradiate the UV photons at visible wavelengths. The O I 8446-Å and 1.287μ lines are pumped by Lyβ. In fact, coupling of lines of different species can ultimately produce emission at other wavelengths through absorption by and subsequent radiative decays from the upper state if it is not collisionally de-excited. It is an essentially resonant process in which the exact, or extremely close, coincidence of two lines of different species produces an enhanced population of excited levels relative to absorption from the continuum.

---

[24] Bohren, C. F. and Huffman, D. 1983, *Absorption and Scattering of Light by Small Particles* (NY: Wiley); some interesting examples of atmospheric scattering are covered in a delightful collection of essays by Bohren, C. F. 1991, *What Light through Yonder Window Breaks* (NY: Wiley). See also the discussion in Chandrasekhar, S. 1950, *Radiative Transfer* (NY: Dover); Young, A. 1982, *Physics Today*, **35**(1), 42.

[25] Woosley, R. and Stibbs, D. 1953, *The Outer Layers of a Star* (London: Oxford Univ. Press); Mihalas, D. 1978, *Stellar Atmospheres*, 2nd ed. (San Francisco: Freeman); Oxenius, J. 1986, *Kinetic Theory of Particles and Photons* (Berlin: Springer-Verlag).

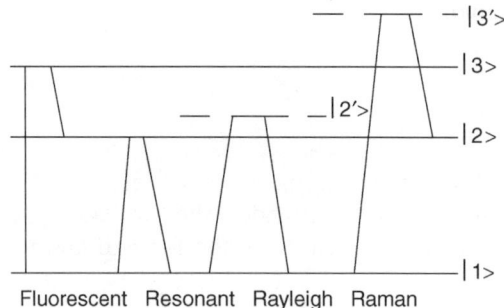

FIGURE 3.4. Energy-level diagram for line scattering mechanisms. Fluorescence leaves the atom in an excited state, unlike resonance scattering. Rayleigh and Raman scattering differ in the final state of the atom. See text for further discussion.

Another way of saying this is that the brightness temperature is elevated relative to the equivalent (thermal continuum) brightness temperature.[26]

In *Raman scattering*, this process is extended to the farther wings of these fluorescent levels. This often neglected process is related to Rayleigh scattering.[27] Unlike Rayleigh scattering, for which the initial and final quantum numbers are the same, Raman scattering is an inelastic process that occurs when $\omega \approx \omega_0$, but is slightly above or below the resonance energy, depending on the level structure. Like fluorescence, it leaves the affected atom or molecule in an excited state. Unlike Rayleigh scattering, which takes place in the very far wing of a line profile, here we have a close coupling between an excited upper state and a lower, still excited, energy. The comparison of these excitations is shown in Figure 3.4. Quantitatively, the process occurs when the two states have energies E_i for the initial and E_f for the final and the photons have frequencies ν_i and ν_f, respectively:

$$h\nu_f = h\nu_i - \Delta E_{ij}. \tag{3.97}$$

The frequency width of the line is substantially increased since it is essentially the

---

[26] A striking example of this process is seen in symbiotic binaries where a very hot compact companion orbits within the wind of a red giant. The [O I] 1641 Å and [O I] 6300 Å lines are particularly strong, excited in the red giant atmosphere by fluorescent ultraviolet (FUV) emission from the companion. See, for example, Shore, S. N. and Aufdenberg, J. P. 1991, *BAAS*, **23**, 942; Shore, S. N. and Aufdenberg, J. P. 1993, *ApJ*, **416**, 355. In any low-density medium of sufficient UV opacity, you will see optical and infrared fluorescent emission. Prominent examples are the [Fe II] lines seen in active galaxies (Seyferts and quasars) and novae. See, for example, Hines, D. C. and Wills, B. J. 1995, *ApJL*, **448**, L69. Continuum fluorescence is used as a conversion process for some photon detectors when high-energy photons excite optical emission, a technique widely exploited in mineralogical analysis, room lighting, and "black light" displays and T-shirts.

[27] The discovery papers are reprinted in Ramaseshan, S., ed. 1988 *Scientific Papers of C. V. Raman*, vol. 1, *Scattering of Light* (Bangalore: Indian Academy of Sciences). See Loudon *op. cit.*; Stenflo, J. 1994, *Solar Magnetic Fields: Polarized Radiation Diagnostics* (Dordrecht: Kluwer) and the general review article and conference proceedings in Stenflo, J. 1996, *Solar Phys.*, **164**, 1.

same as the excess energy of the incident photon above the energy of the lower state divided by h.[28] Although used for decades in laboratory studies of molecules, atomic Raman scattering has only relatively recently been exploited in astrophysics mainly in cool stellar atmospheres and symbiotic binary systems. In particular, a breakthrough came with the identification of two strong emission lines at 6830 and 7088 Å with Raman scattering of the O VI 1032 Å doublet. The line width depends on the mismatch between the exciting transition and the bound state of the absorbing ion so, in general, these lines are quite broad (as are those in symbiotic and related systems). The cross section for hydrogenic atomic scattering is given by

$$\sigma_{\text{Raman}} = \frac{1}{16}\sigma_e \nu_i \nu_f^3 \left\{ \sum_m \left[\frac{(gf)_{im}(gf)_{fm}}{\nu_{im}\nu_{fm}} \right]^{1/2} \frac{\nu_{im} + \nu_{fm}}{(\nu_{im} - \nu_i)(\nu_{fm} - \nu_f)} \right\}^2. \quad (3.98)$$

Here the sum is taken over all intermediate states m, with ν_{im} and ν_{fm} representing the frequency for the intermediate transitions. For many transitions it is sufficient to use only ν_i and ν_j. An important physical attribute of the process is that Raman scattering is coherent and consequently produces polarized light. It does not involve the redistribution of the photon's energy over the width of the excited state. The photon comes out at precisely the difference between its local energy and that of the next available state.

As Raman noted in the first experimental reports of this emission, this scattering is an intrinsically weak process, like Rayleigh scattering, and requires a very strong source and/or a large column density to be observable. Unlike Rayleigh scattering, however, this process involves line transfer because it is a "near resonance" process, so it also requires a near coincidence between some strong line and an intrinsic transition in the scattering atom or molecule. For this reason, the process was thought for a long time to be irrelevant in astrophysics. The discovery of strongly polarized lines in the red spectra of symbiotic stars due to the scattering of O VI 1032 Å by Lyβ suggests that it is a new probe of thermally complex cosmic environments and should be more exploited. It now appears that in many strong X-ray sources, where high ionization lines are formed in the vicinity of the accreting compact body in a close binary system, the atmosphere of the companion is illuminated by these EUV transitions and produces Raman scattering lines. These have now been detected in many systems, regardless of the effective temperature of the irradiated star, and are an essential diagnostic of the hottest parts of the systems. Remember that you will not see Raman scattering in most systems if there is no strong line at significantly higher energy that can pump the transition. It has recently been recognized to provide an important spectral diagnostic for the outer solar atmosphere.

---

[28] Nussbaumer, H., Schmid, H. M., and Vogel, M. 1989, *Astr. Ap.*, **211**, L27; Schmid, H. M. 1992, *MNRAS*, **275**, 227.

3.5.6 Electron Scattering: Thomson and Compton Scattering

Classical electron scattering is independent of frequency and has a constant cross section that you can derive immediately from Eqs. (3.78) when $\omega_0 = 0$ and $\gamma = 0$:

$$\sigma_e = \frac{8\pi e^4}{3m_e^2 c^4} = 6.65 \times 10^{-25} \text{ cm}^2. \tag{3.99}$$

This is the *Thomson or electron scattering cross section*. It is the simplest scattering mechanism for a couple of reasons. First, an unbound electron has no resonance frequency. It moves immediately once it has been hit by any photon and oscillates with precisely the same frequency. Second, its coupling to a photon depends only on its charge to mass ratio, and its mass is quite small. The oscillation is a dipole induced at any frequency ω. The light scattered by free electrons is polarized. However, the electrons in a gas have a far higher thermal speed than the ions, by a factor of $(m_p/m_e)^{1/2}$, and so their scattering of line radiation becomes a broadening mechanism! Again, it is important to note that electrons, if they are cold (nonrelativistic) do not care about the frequency of the photon they are scattering (unless the photon energy is about $m_e c^2$ or higher). This is the ultimate nonresonant process.

There is a limit to this freedom. If the electron is accelerated by a photon whose energy is comparable to the rest energy of the free scatterer, the dipole moment lags behind the excitation. In effect, the reason is that the electron is being forced to oscillate relativistically and now knows that its inertia is velocity dependent. This introduces a critical frequency, $\omega_0 \approx m_e c^2/\hbar$, above which classical (Thomson) electron scattering must break down and is especially important for X and γ radiation. This is the familiar Compton effect, originally a demonstration of the existence of the photon but also a type of scattering of very-high-energy photons (or in the inverse process, the scattering of lower-energy photons by very relativistic electrons). Compton scattering has the effect of redistributing the photon energy as a function of the incident angle, θ. We will need this later,[29] so let us derive the general formalism for treating relativistic scattering. For a relativistic electron, the relation between energy, ϵ, and momentum, $\mathbf{p}$ is given by $\epsilon^2 = \mathbf{p} \cdot \mathbf{p} + m_e^2$, where for convenience we set c and $\hbar$ equal to unity (in the final expression we will have to put this back in). Let ω and $\hat{\mathbf{n}}$ be the frequency and direction of the incoming photon and ω' and $\hat{\mathbf{n}}'$ be the same for the outgoing (scattered) photon, and let $\mathbf{p}'$ be the momentum of the electron after scattering. Then the energy and momentum conservation conditions are given by

$$\omega + \epsilon = \omega' + \epsilon'$$
$$\omega \hat{\mathbf{n}} + \mathbf{p} = \omega' \hat{\mathbf{n}}' + \mathbf{p}'. \tag{3.100}$$

---

[29] You will see this again in Chapter 8 when we discuss the Sunyaev–Zel'dovich effect in X-ray-emitting clusters of galaxies, the scattering of cosmic background photons by hot electrons. The scattering of high-energy photons in accretion disks around neutron stars and black holes also requires this treatment. Any time the radiation had to random walk out of the environment, a diffusion approximation is called for.

Then $(\epsilon' - \epsilon)(\epsilon' + \epsilon) = p^2 - p'^2 = (\mathbf{p} - \mathbf{p}') \cdot (\mathbf{p} + \mathbf{p}')$. Then for $\mathbf{p}$, substitute $\mathbf{p} + \omega \hat{\mathbf{n}} - \omega' \hat{\mathbf{n}}'$ and for ϵ' substitute $\epsilon + \omega - \omega'$. Calling $\Delta = \omega' - \omega$ we find that

$$\Delta = \frac{\omega \mathbf{p} \cdot (\hat{\mathbf{n}} - \hat{\mathbf{n}}') + \omega^2 (1 - \hat{\mathbf{n}} \cdot \hat{\mathbf{n}}')}{\mathbf{p} \cdot \hat{\mathbf{n}}' - \epsilon - \omega(1 - \hat{\mathbf{n}} \cdot \hat{\mathbf{n}}')}. \tag{3.101}$$

For the interaction in which an electron at rest is accelerated by the collision with a photon, the relation between the photon and electron energy is

$$\omega' = \frac{\omega}{1 + (\hbar \omega / m_e c^2)(1 - \cos \theta)}$$

so that $\hspace{10em}$ (3.102)

$$\Delta \lambda = \frac{\lambda}{\lambda_c} (1 - \cos \theta),$$

where $\lambda_c = \hbar / m_e c$ is the *Compton wavelength*. Calling $x = \hbar \omega / m_e c^2$, the angle-averaged Compton cross section is related to that for Thomson scattering by

$$\sigma_{\text{Compton}} = \sigma_e \frac{1}{x} \left[\left(1 - \frac{2(x+1)}{x^2}\right) \ln(2x+1) + \frac{1}{2} + \frac{4}{x} - \frac{1}{2(2x+1)^2} \right], \tag{3.103}$$

which varies as $\ln x / x$ in the strongly relativistic limit. This equation is the *Klein–Nishima cross section*.[30] The inverse process, which happens for relativistic electrons colliding with low-energy photons, redistributes the photon energy in the same fashion. In particular, in an application we will meet again when discussing the cosmic background radiation, hot electrons scatter the microwave photons up to EUV and X-rays with the consequent decrease in the intensity of the background source. We emphasize, though, that this is *not* an absorption process. It conserves photon *number* but not the number in any frequency interval.

3.5.6.1 The Compton Effect, the Kompaneets Equation, and Diffusive Transfer of High-Energy Photons

The evolution equation for the effect shows that it is essentially a random walk of the electrons in energy, in other words it is described by a Fokker–Planck equation. The derivation of the inverse scattering process is due to Kompaneets (1957). The Compton effect is the quantum-mechanical scattering of a photon by a free electron. For relatively low energy, we will treat this as non-relativistic (i.e., $\epsilon < m_e c^2$). If the energy of the electron is less than the incident photon, $\hbar \omega < \epsilon$, the photon loses energy to the scatterer, the normal Compton effect. If the particle kinetic energy is greater than that of the incident photon, the radiation is shifted

---

[30] Its derivation is discussed in detail in Bjorken, J. D. and Drell, S. D. 1964, *Relativistic Quantum Mechanics* (NY: McGraw-Hill). See also Blumenthal, G. R. and Gould, R. J. 1970, *Rev. Mod. Phys.*, **42**, 237.

up in energy at the expense of the kinetic energy of the particle. This is the *inverse Compton effect*. This latter process is very important for a wide variety of cosmical problems. It affects the X-ray emission observed from cataclysmic variables, especially those with central black holes, through scattering by the surrounding accretion disk. For active galactic nuclei, scattering occurs from the circumnuclear matter. In a cosmological context it is required when treating scattering of cosmic background infrared and microwave photons by hot electrons within galaxy clusters. But any time you have to deal with multiple Compton scatterings some sort of photon diffusion treatment is required. In the rest frame of the electron, the photon energy is $\gamma \hbar \omega (1 + \beta \cdot \hat{n})$, and the scattering produces an additional boost of $\gamma(1 + \beta \cdot \hat{n})$ in the observer's frame.

As we just found, restating Eq. (3.102):

$$\frac{\omega'}{\omega} = \frac{\mathbf{p} \cdot \hat{\mathbf{n}} - \epsilon}{\mathbf{p} \cdot \hat{\mathbf{n}}' - \epsilon + \omega(1 - \hat{\mathbf{n}} \cdot \hat{\mathbf{n}}')}. \qquad (3.104)$$

The change in the energy of the photon is given by Eq. (3.101). In the electron's rest frame we can further simplify this expression to

$$\Delta \approx \frac{\omega \mathbf{p} \cdot (\hat{\mathbf{n}} - \hat{\mathbf{n}}')}{m_e c}. \qquad (3.105)$$

The maximum frequency of the emergent photon is then given by

$$\omega' = \gamma^2 (1 + \beta \cdot \mathbf{n})^2 \omega \approx 4\omega \left(\frac{kT}{m_e c^2} \right)^2 = 4x, \qquad (3.106)$$

which sets the scale for the change in photon energy per collision and defines a variable we'll need momentarily.

For the evolution of this spectrum, we turn to the Boltzmann equation. The electron distribution remains thermalized and is described by the usual Maxwellian function $f(\epsilon)$ before collision and $f(\epsilon')$ afterward. Then, since the number of photons is strictly conserved, the scattering that produces a photon with energy $\hbar \omega'$ *removes* one with $\hbar \omega$. The distribution then evolves according to

$$\left(\frac{\partial n}{\partial t} \right)_c = \int d\mathbf{p} \int d\mathbf{p}' \, d\omega' [\sigma(\mathbf{p}, \omega | \mathbf{p}', \omega') f(\epsilon') n_\omega (1 + n_{\omega'})$$
$$- \sigma(\mathbf{p}, \omega | \mathbf{p}', \omega') f(\epsilon) n_{\omega'} (1 + n_\omega)]. \qquad (3.107)$$

Here stimulated processes have been included. To see how this happens, recall that the rate for emission from a discrete state is $n_u(A_{ul} + B_{ul} I_\nu)$ and that $B_{ul} = A_{ul}(c^2/2h\nu^3)$. We then rewrite I_ν in terms of the photon number and take the rate of emission to be proportional to the scattering cross section to obtain the terms $n(\omega)(1 + n(\omega'))$ and $n(\omega')(1 + n(\omega))$ in the Boltzmann equation.

We proceed along a similar line to the one we used for the Fokker–Planck derivation in Chapter 1. Start with the detailed balance condition for the thermalized photon distribution function:

$$\frac{1}{\hbar\omega}B_\omega \sigma(p,\omega,\Omega|p',\omega',\Omega')\left[1+\frac{c^2}{8\pi^2\omega'^3}B_{\omega'}\right]$$
$$=\frac{1}{\hbar\omega'}B_{\omega'}\sigma(p',\omega',\Omega'|p,\omega,\Omega)\left[1+\frac{c^2}{8\pi^2\omega^3}B_\omega\right], \quad (3.108)$$

where B_ω is the Planck function. Now, for convenience, call the *forward* reaction on the *lhs* σ_+, and call the *reverse* reaction on the right-hand side (RHS) σ_-. Substituting the explicit equation for B_ω and using Eq. (3.108), we obtain

$$\omega^2 \sigma_+ = \omega'^2 \sigma_-. \quad (3.109)$$

We need this for the Boltzmann integral. Now substituting this result back into Eq. (3.107), we find that

$$\frac{I_\omega}{\omega^3} = n_\omega = \text{constant}, \quad (3.110)$$

which is the same statement as conservation of photon number and a basic result we will need for much of the discussion in moving media. Now on to the Boltzmann equation for the evolution of n_ω, the photon number. First, note that the shift for a single scattering, if small, leads to a Taylor series expansion:

$$n(\omega') = n(\omega) + \frac{\partial n}{\partial \omega}\delta\omega + \frac{1}{2}\frac{\partial^2 n}{\partial \omega^2}(\delta\omega)^2 + \cdots. \quad (3.111)$$

A similar one follows for $f(\epsilon')$ in terms of $f(\epsilon)$ but here, because we assume the distribution remains Maxwellian, we find that

$$e^{-\epsilon'/kT} = e^{-\epsilon/kT}\left(1+\delta+\tfrac{1}{2}\delta^2+\cdots\right), \quad (3.112)$$

where $\delta \equiv (\epsilon' - \epsilon)/kT$. We can now collect terms of similar order in δ:

$$\text{Integral} \approx -\left[\frac{\partial n}{\partial \omega}+n(1+n)\right]\langle\delta\rangle - \frac{1}{2}\left[\frac{\partial^2 n}{\partial \omega^2}+2(1+n)+n(1+n)\right]\langle\delta^2\rangle. \quad (3.113)$$

The brackets around δ^m signify averages over $f(\epsilon)$. The second-order term is obtained using the assumption of nonrelativistic scattering, otherwise we would have to include full energy dependence of the Klein–Nishima equation. The $\langle \delta^2 \rangle$ term is really

$$\langle \delta^2 \rangle = \int dp f(\epsilon) \int d\sigma (1 - \boldsymbol{\beta} \cdot \hat{\mathbf{n}}) \delta^2. \tag{3.114}$$

Substituting the approximate form for δ from Eq. (3.105) and using $d\sigma = \tfrac{1}{2}\sigma_T(1 + \mu^2) d\Omega$ for Thomson scattering, we obtain

$$\langle \delta^2 \rangle = n_e x^2 \sigma_e \frac{kT}{m_e c^2}. \tag{3.115}$$

The key to performing the averaging integral is to realize that in the integrand in Eq. (3.114) all odd terms in the directional cosine, μ, cancel on integration. Now, rather than perform the first order integration by brute force, we follow Danse and de Zotti and use a very clever trick.[31] Notice that the evolution equation for the photon number is second order in energy and first order in time, making it a parabolic form—in other words, we will obtain a diffusion equation in energy rather than in space. We can rewrite it as

$$\frac{\partial n}{\partial t} = \frac{1}{x^2} \frac{\partial}{\partial x} x^2 j \tag{3.116}$$

in terms of some current, j, given by

$$j = g(x) \left[\frac{\partial n}{\partial x} + n(1 + n) \right], \tag{3.117}$$

where g is some trial function. Since we know the explicit form of $\langle \delta^2 \rangle$, we substitute this into Eq. (3.113) and equate the coefficients of the n, $\partial n / \partial x$, and $\partial^2 n / \partial x^2$ terms. This step yields

$$\frac{1}{x^2}[g(1 + 2n) + g'] = 2I_1 + 2x^2(1 + n) \tag{3.118}$$

for the coefficient of the first derivative, and $x^{-2}g = x^2$ for the second derivative. We've rewritten the integral as I_1. Then, trivially, $g = x^4$, and from Eq. (3.118) we find that

$$I_1 = \frac{1}{2} x(4 - x) n_e \sigma_e \frac{kT}{mc^2}. \tag{3.119}$$

This gives the mean change in the photon energy for a single scattering, preserving the characteristic that the change in the energy depends on whether it is to the

---

[31] Zel'dovich and Sunyaev 1969, *Ap. Space Sci.*, **4**, 301; Wright, E. L. 1979, *ApJ*, **232**, 348; Rephaeli, Y. 1995, *ARAA*, **33**, 541; Sunyaev, R. and Zel'dovich 1972, *Comments Astron. Ap.*, **4**, 173; Danese, L. and de Zotti, G. 1977, *Riv. N. Cimento*, **7**, 277; Birkinshaw, M. 1999, *Phys. Rep.*, **310**, 97.

high- or low-frequency side of the peak of the flux distribution. Now define the dimensionless time variable, y, by

$$y = \frac{kT_e}{mc^3} \int n_e \sigma_e \, dl$$

for a pathlength $dl = dt/c$. We can do this for photons because they travel at a fixed speed, c, so it is possible to interchange the time and distance. This new variable, y, is really the same as a scattering optical depth. On collecting terms, the final form of the evolution equation is

$$\frac{\partial n}{\partial y} = \frac{1}{x^2} \frac{\partial}{\partial x} \left[x^4 \left(\frac{\partial n}{\partial x} + n(1+n) \right) \right], \tag{3.120}$$

a result called the *Kompaneets equation*. This is a nonlinear diffusion equation, where the nonlinearity comes from the stimulated emission that we included explicitly at the start.

Having gone to all the trouble of including it, however, we now drop the stimulated term in order to progress toward a solution. This is appropriate when the energy density of the radiation is low compared with that of the thermalized electrons, and in many astrophysical applications it isn't such a bad assumption. On taking this step, we have for Eq. (3.120) the reduced form:

$$\frac{\partial n}{\partial y} = \frac{1}{x^2} \frac{\partial}{\partial x} x^4 \frac{\partial n}{\partial x}. \tag{3.121}$$

Calling $u = \ln x$, then

$$\frac{\partial n}{\partial y} - 3 \frac{\partial n}{\partial u} \equiv \frac{\partial n}{\partial \xi}$$

for $\xi = u + 3y = \ln x + 3y$. The diffusion equation, Eq. (3.121), now simplifies to

$$\frac{\partial n}{\partial y} = \frac{\partial^2 n}{\partial \xi^2}, \tag{3.122}$$

which has the standard solution:

$$n(y, \xi) = (4\pi y)^{-1/2} \exp\left(-\frac{\xi^2}{4y}\right). \tag{3.123}$$

Using the definitions of y and ξ, we can then rewrite Eq. (3.123) as an explicit function of x and t. More importantly, we find the change in the shape of the spectrum, such that the photons at lower energy are boosted up to higher energy by the inverse Compton process. Notice that we have said nothing about whether we are dealing with high- or low-energy electrons. In a cold gas, the electron distribution is heated and high energy photons are downshifted. The detailed incorporation of the Klein–Nishima formula makes the description of the process more precise, but the process is qualitatively the same.

3.5.7 Broadening Mechanisms for Line Profiles

Spontaneous emission alone produces a Lorentzian profile, but there are several additional mechanisms that make a line broader than this. We will broadly divide these into *intrinsic* and *extrinsic* categories. Among the intrinsic mechanisms, two are caused by atomic structure alone: hyperfine structure and the Zeeman effect. The former is inescapable since it depends only on the quantum structure of a particular atomic state. The latter requires an external magnetic field, but both are time-independent, not stochastic, and the "broadening" is due to the instrumental resolution being insufficient to resolve the individual components. Two other intrinsic mechanisms depend, however, on thermal fluctuations within the gas, and these are "true" broadening agents. The Stark and van der Waals effects act statistically because the fluctuation timescales for local electric fields occur on timescales that are comparable to the lifetime of the atomic level. No matter what the wavelength resolution, they result in a broad profile. Turbulence and pulsation are harder to separate this way. These are due to velocity gradients over the line forming region and affect the coupling within the line between depths. Turbulence is a random process and if it has a Gaussian spectrum it acts just like the Doppler broadening from thermal motions (in effect, it is a spectrum of nonthermal motions). We used microturbulence as a fudge factor, but large-scale motions are also possible, and each parcel of gas is jostled with respect to its neighbors by an amount that depends on the turbulence spectrum. Pulsation is ordered motion, but in addition to shifting the line with respect to the observer it also produces a gradient in the velocity and desaturates the line at any depth, thereby changing its intrinsic strength (we will see the same effect from a stellar wind). Finally, when observing a star, we sit in an inertial frame. The stellar surface is, generally, unresolved. If the star is rotating, unresolved Doppler shifts from different parts of the surface are summed to produce a purely extrinsic broadening. Unless there are structural changes in the atmosphere as a result of the rotation (to which we will return in the next chapter), this broadening merely redistributes the photons and leaves the integrated line profile invariant. Put differently, the equivalent width for a rotating atmosphere is unaffected by the broadening (to within the signal to noise ratio of the observations).

3.5.7.1 Hyperfine Structure: Effect of Atomic Structure and Nuclear Spin

The interaction of an atomic state with the magnetic moment of the nucleus perturbs energy levels by a small amount. Because this comes about through a magnetic dipole interaction, it involves spin and orbital angular momenta and leads to a splitting of multiplet states. The induced energy separation is generally small, but it is intrinsic and not something that depends on the medium in which the radiating atom is situated. Hyperfine broadening is most important for complex atoms, especially among odd-Z rare-earth ions. It has an effect similar to increasing the microturbulence for the line, delaying the onset of saturation compared to a transition that is unaffected by the broadening. Another way of putting this is that each line is really a complex of absorbers with the total transition oscillator strength distributed over the components.[32]

3.5.7.2 Zeeman Effect: Effect of Magnetic Fields

The Zeeman effect acts as a coherent broadening mechanism unless the magnetic field depends on time or is chaotic along the line of sight. It is due to the removal of degeneracy from atomic levels through the interaction between the external field and the atomic magnetic moment. To estimate its magnitude, take the magnetic dipole moment to be μ and the external field strength as $\mathbf{B}$. Then the splitting due to the field is $\Delta E = \langle \mu \cdot \mathbf{B} \rangle$ where the bracket indicates the expectation value for the level. The splitting is linear in the field strength and depends on the state spin and orbital angular momenta. The detailed solution depends on the coupling scheme, but for LS coupling we find

$$E_{JM} = E_J + \mu_0 g M_J B. \tag{3.124}$$

Here $\mu_0 = e\hbar/2mc$ is the Bohr magneton and g is called the *Landè factor*:

$$g = 1 + \frac{J(J+1) + S(S+1) - L(L+1)}{2J(J+1)} \tag{3.125}$$

for *LS* coupling. Transitions with $\Delta M_J = \pm 1$ are called $\sigma_\pm$ components, and those with $\Delta M_J = 0$ are called π components. When more than one component of each substate is present—and this is generally the case—then the individual intensities depend on the dipole moments of the subtransitions. However, the magnetic field does introduce one new complication that affects the interpretation of the emergent spectrum. Once levels have been separated by a magnetic perturbation, they "know" about the azimuthal quantum number because of the spatial anisotropy introduced by $\mathbf{B}$. Thus the emitted photons carry angular momentum—they are polarized. Not all lines are broadened. *Null lnes* have fortuitously vanishing Landè factors and can be used as templates for determining disk-integrated magnetic fields.[33] Although this normally is not too important, especially if there is a small optical depth in the region of line formation, it does mean that a proper treatment of the transfer must include the circular polariza-

---

[32] See Kurucz, R. L. 1992, *Rev. Mex. Astr. Ap.*, **23**, 45, 187.

[33] See, for instance, lists of such lines are found in Landstreet, J. D. 1969, *PASP*, **81**, 896; Adelman, S. J. 1973, *PASP*, **85**, 227 and 1974, *PASP*, **86**, 486. The technique is not exploited extensively but remains one of the best ways to determine magnetic broadening.

tion. A detailed discussion is beyond the scope of this book, you should consult Stenflo (1994) for a comprehensive guide to the literature.[34]

The *Hanle effect* is related to spectral line formation in a magnetic field and has become an important diagnostic tool in astrophysics. It is actually a by-product of the effects of weak magnetic fields on the splitting of the levels. If the collision rates are low enough, specifically if the populations of the sublevels are not mixed, the individual m states maintain their coherence. The lines are polarized and don't mix states. However, there is a characteristic resonance among the magnetic sublevels when $\omega_B/A_{ul} \approx 1$, such that any polarized light will be resonantly transmitted through the lines without mixing. Spectropolarimetry will then detect the subcomponents even when Doppler broadening smears the Zeeman components together because $\Delta\lambda_Z < \Delta\lambda_D$. The effect is most easily observed in resonance lines where the transition probabilities are high and resonant scattering occurs in NLTE. The scattering coefficients are essentially the same as those for Rayleigh scattering.[35]

3.5.7.3 Stark Broadening

In any interaction of the sort that gives rise to thermal bremsstrahlung, bound states are also perturbed. In the ion that is producing the change in an electron's path there may be bound electrons in excited states. These interact with the time-dependent electric field of the perturber. The two processes are intimately

---

[34] The direct measurement of weak magnetic fields has been steadily improving, and we should mention one instrumental method that serves as a prototype for many other techniques—*Zeeman polarimetry*, see Landstreet, J. D. 1992, *Astr. Ap. Rev.*, **4**, 35. It uses the normal Zeeman effect, the one displayed by neutral hydrogen. If you are looking along the direction of the field, the emitted radiation is right and left circularly polarized. This means that with a quarter-waveplate polarizer, you can chop back and forth across the line and look for small changes in the intensity of the features. The method works like this. The wavelength splitting is $\Delta\lambda_Z = g\mathbf{B}\cdot\hat{\mathbf{n}}\lambda^2$. Now find a position on a line wing where the slope of the line is large. Between the right- and left-handed polarizations there is a centroid shift for the line. If you sit at one wavelength, this changes the intensity. You can measure this photometrically by isolating one part of the line profile and chopping at very high rate between opposite senses of circular polarization using a Pockles cell so that

$$\delta I_{\lambda_0} = \frac{1}{I_0}\frac{dI}{d\lambda}\bigg|_{\lambda_0} \Delta\lambda_Z.$$

This means that you can measure the field strength by looking at the modulation of the signal once the line profile is known. The technique is very powerful and is a modification of others that have been used for solar work. The key is that it is photometric and can tolerate a large bandpass. It can also measure the variations on lines where the Stark effect completely dominates the Zeeman separation of the subcomponents. The only drawback to the method is that, for stars, it can only provide the mean longitudinal field. If the star has a large magnetic dipole moment, the method works rather well. Conversely, it permits only the measurement of the magnetic field projected in the line of sight. Hence it smooths out any field that has fine structure over the stellar photosphere. Using lines formed at different optical depth sometimes helps with this, but most stellar atmospheres have such small geometric depths that the technique does not really resolve any of the field structure this way.

[35] House, L. L. 1970, *JQSRT*, **10**, 1171; Stenflo, J. O. 1996, *Solar Phys.*, **164**, 1; Landi-Degl'Innocenti, E. 1996, *Solar Phys.*, **164**, 21; Ignace, R., Cassinelli, J. P. and Nordseik, K, H. 1999, *ApJ*, **520**, 335. The most complete treatment is Stenflo, J. 1994 (n.27).

linked when an ion other than a proton is involved. The perturbation follows the same pattern as we used for the transition probability computation, or for the Zeeman effect. The difference is that the perturbation is impulsive in time. The interval spent in the vicinity of the atom is roughly $\Delta t \approx b/v$, so the energy perturbation is of order $\Delta E \sim \hbar v/b$. There are few high-energy close collisions, while there are many that are lower energy. So the statistical effect of the interactions should, like *free–free* transitions, be heavily weighted toward the statistically most likely encounters. These produce profiles that have very extended Lorentzian wings. The damping factor now depends on the plasma density and temperature, $\Gamma \sim n_e^{2/3} T^{1/2}$, since the broadening depends on $\langle \mathbf{d} \cdot \mathbf{E} \rangle$. The dipole moment is $\mathbf{d}$, $\mathbf{E}$ is the electric field strength, and the bracket indicates average over the time variations of the ambient ion spatial distribution. This interaction produces a huge broadening compared those due to the simple radiative damping from the transition probability alone.

One way to see what is happening is to concentrate on what the local charge distribution looks like around a radiating atom. The electric field due to the neighboring ions fluctuates only at low frequency. The ions move slowly relative to the cloud of electrons because of their mass. This means that fast close encounters are extremely rare. This statistical effect should dominate the cores of the lines and the inner wing because the strength of the interaction is not too large and the impacts occur slowly enough for the adiabatic approximation to hold. If collisions by electrons are included, however, this is no longer true.[36] The states change on a short timescale, and the energy of the states cannot change adiabatically—transitions may be induced between states, and this affects both the core and the wings of the line. The line wings have lower opacity than the core and permit observations of deeper layers. The strength of the hydrogen line wings, and those of abundant atoms, depend strongly on the local density and temperature conditions through the Stark broadening. Thus, they actually provide a way of tracing the depth dependence of the physical state of the atmospheric plasma. By transforming the intensity at any wavelength, which is a function of *optical* depth, you can extract from the intensity profile the run of density and temperature with *geometric* depth.

Picture a single hydrogen atom imbedded within a fluctuating neighborhood of uncorrelated ions and electrons. The strongest perturbation is from the nearest ambient particle so that, assuming Poisson statistics, the probability of a single perturber being at a distance r is:

$$dP(r) = 4\pi \left(\frac{r^2}{\langle r \rangle^3} \right) e^{-(r/\langle r \rangle)^3} dr, \qquad (3.126)$$

---

[36] Griem, H. 1964, *Plasma Spectroscopy* (NY: McGraw-Hill); Griem, H. R. 1974, *Spectral Line Broadening in Plasmas* (NY: Academic Press); Smith, E. W., Cooper, J., and Vidal, C. R. 1969, *Phys. Rev.*, **185**, 140; Vidal, C. R., Cooper, J., and Smith, E. W. 1970, *JQSRT*, **10**, 1011 and 1971, ibid, **11**, 263. See also the tables for Stark broadening in Vidal, C. R., Cooper, J., and Smith, E. W. 1973, *ApJS*, **25**, 37. An updated survey of the literature is provided by Griem, H. 1997, *Principles of Plasma Spectroscopy* (Cambridge, UK: Cambridge Univ. Press).

where $\langle r \rangle = n_i^{-1/3}$ is the mean distance between ions for a number density n_i. The strength of the resultant electric field, $\mathscr{E} \sim r^{-2}$, is then also a function of density so $P(r)$ is actually a probability distribution for electric field strengths:

$$dP(r) \to dW(\mathscr{E}) \sim \left(\frac{\mathscr{E}}{\langle \mathscr{E} \rangle}\right)^{-5/2} e^{-(\mathscr{E}/\langle \mathscr{E} \rangle)^{-2/3}} d\mathscr{E}. \qquad (3.127)$$

Since the energy perturbation for the the linear Stark effect is proportional to $\mathscr{E}$, $\Delta\lambda$ will be as well. Thus, $W(\mathscr{E})$ actually becomes a distribution for the line opacity as a function of distance from line center leading to a profile of the form

$$S(\Delta\lambda) \sim (\Delta\lambda)^{-5/2}. \qquad (3.128)$$

This is called the *Holtzmark profile* for the quasistatic ionic field. Because it is a mean field approximation for a thermal distribution, it is an essentially classical result.[37]

This formulation is still an incomplete treatment of what must be occurring in a realistic plasma. For one thing, it doesn't agree with your intuition since it ignores the *frequency* of interactions. The greater the rate of collisions relative to the radiative transition probability for a state, the broader it should be since its lifetime is progressively closer to the collision interval. This broadens the spectral line. The reason we get this temperature-independent profile is that we have chosen to use only the ions, which are sluggish and produce a quasistatic charge perturbation, to provide the broadening. The electrons are far more mobile and prove to be the temperature-sensitive broadening agents. They produce the close, strong, fast interactions. The line wings are affected as well as the core since the timescale for the interaction is related to the frequency shift through the uncertainty principle; although infrequent, strong collisions produce large shifts. Since the probability of such events is small, the wings will reflect this as they do for the Holtzmark treatment.

The collisional broadening is treated using time-dependent perturbation theory. As we mentioned earlier, these fast interactions cannot be treated adiabatically. They may induce nonradiative transitions among atomic substates whose degeneracies are temporarily lifted as a result of the impulsively varying electric field, and they can reshuffle the level populations. Without going into the details of the process, it is possible to understand qualitatively how the *impact approximation* alters the treatment of the broadening. We cannot simply average over a slowly fluctuating background of ions that produce a mean perturbing field that depends only on density. This process depends on temperature as well through the electron velocity distribution. The theory requires the correlation function for the time-dependent wavefunctions, the Fourier transform of which is the frequency-dependent line profile. We can briefly describe how the calculation proceeds.

---

[37] Since the ions are partially screened by the electrons, this final broadening depends on temperature through the Debye length, $\lambda_D \sim (T/n_e)^{1/2}$. The ratio of the mean separation to λ_D scales as $T^{-1/2} n_i^{1/6}$, which is generally small.

The time development of a state is described by an operator, T_a such that if we know the state of a system at time $t = 0$, we can use time-dependent perturbation theory to find its condition at any future t, $|a;t\rangle = T_a|a;0\rangle$. Its evolution is given by

$$\Delta T_a^\dagger(t,0)T_b(t,0) = T_a^\dagger(t+\Delta t, 0)T_b(t+\Delta t, 0) - T_a^\dagger(t,0)T_b(t,0)$$
$$= T_a^\dagger(t+\Delta t, t)T_b(t+\Delta t, t)T_a^\dagger(t,0)T_b(t,0) - T_a^\dagger(t,0)T_b(t,0)$$
$$= [T_a^\dagger(t+\Delta t, t)T_b(t+\Delta t, t) - 1]T_a^\dagger(t,0)T_b(t,0) \quad (3.129)$$

There is nothing special—yet—about this formalism; it is basically the description of any collision. The averaging is over the thermal distribution of electrons in the plasma. Now, however, comes the approximation. We have a product of two pairs of operators, one that acts on the interval $(t, t + \Delta t)$ and the other on the previous interval. We *assert* that the two intervals are not only independent but that the impact takes place over a vanishingly small Δt, hence the name *impact approximation*. On averaging over collisions, this is reasonable if the plasma density is small enough since the close interaction will take place for only a negligible time compared with the motions in the atomic system or the gas. Then we convert $\Delta T_a^\dagger(t,0)T_b(t,0)$ into a differential equation

$$\frac{d}{dt}\langle T_a^\dagger(t,0)T_b(t,0)\rangle = \Phi_{ab}\langle T_a^\dagger(t,0)T_b(t,0)\rangle, \quad (3.130)$$

replacing the first operator with a symbolic transition operation, the impact operator Φ_{ab}. Then

$$T_a^\dagger(t,0)T_b(t,0) = e^{\Phi_{ab}t}. \quad (3.131)$$

Because the timescale for the interaction is short, the frequency shift produced will be large. This separation permits us to write the operator as

$$e^{\Phi_{ab}t} = e^{i(H_{0,a}-H_{0,b})t/\hbar}\Phi_{ab}\,e^{-i(H_{0,a}-H_{0,b})t/\hbar},$$

where the unperturbed states are given by H_0. This is found from the solution of the time-dependent perturbation problem, allowing intermediate states to be excited during the interaction and integration over the paths of the incident electrons and averaging over the thermal distribution function. The correlation function is found by summing over all intermediate states for an interaction time τ:

$$C(\tau) = \left[\sum_{iji'j'}\langle i|\mathbf{d}^\dagger|j\rangle\langle|T_b^\dagger|j'\rangle\langle j'|\mathbf{d}|i'\rangle\langle i'|T_a|i\rangle\right] \to \left[\mathbf{d}_{i'j'}^*\langle i'j'|T_a^*T_b|ij\rangle\mathbf{d}_{ij}\right].$$

This is just a compact way of writing a very nasty set of integrals over the various bound states. The operator T is formally an exponential in the perturbing Hamiltonian but is actually a sum over successive approximations to the path of the perturber. The frequency dependent line profile is given by the Fourier transform of the correlation function over time:

$$\phi(\omega) = \frac{1}{\pi} \text{Re} \int_0^\infty d\tau \, e^{-i\omega\tau} C(\tau). \tag{3.132}$$

This gives the formal solution:

$$\phi(\omega) = -\text{Re} \sum_{ii'jj'k} \langle i|d_k|j\rangle \langle j'|d_k|i'\rangle \left\langle ij \left| \left(\frac{i}{\hbar}(\omega - [H_{0,a} - H_{0,b}]) + \Phi_{ab} \right) \right| i'j' \right\rangle, \tag{3.133}$$

which, you may see, is a modified Lorentzian type profile with a shift in the line center and a damping width that depends on the interaction potential. This is where all the effort goes in such calculations, determining how the interaction changes the widths of the levels when averaged over the thermal distribution of the impacting electrons and the fluctuations in the heavier-ion background. The temperature dependence for this broadening arises from the integral of the interaction potential over the velocity distribution of the electrons, and this is where different approximations produce different predictions for the final line profile.

Now consider a very strong absorption line. The line core is very opaque, hence the emergent light comes only from shallow layers in the atmosphere. Radiative broadening of a line, especially hydrogen, doesn't produce very extended wings, so for a normal spectral line the depth of formation varies rapidly across the profile, and, in effect, each line samples one depth in the atmosphere. For the strongest lines of the most abundant species, especially for hydrogen, the Stark wings extend very far from line center and have a relatively weak dependence on the wavelength. Therefore, the range in depth of formation of the profile spans a large variation of temperature and pressure. Through pressure, in hydrostatic equilibrium the wings are sensitive to the surface gravity. The profile wings also provide a probe of the ionization of the atmosphere. This is *almost* independent of abundance for the principal diagnostic species, hydrogen. This is because it is the predominant species.[38] The intensity as a function of position from line center is a measurement of the electron pressure scale height and therefore depends on the surface gravity.

[38] There is an interesting dependence, however, on the abundance of H and the H/He ratio that seems opposite to intuition. Imagine comparing two stars of identical temperature and surface gravity except one has a larger H/He ratio than the other. Because the pressure depends on the mean molecular weight, a hydrogen-depleted atmosphere has a larger pressure at the same optical depth as the normal atmosphere and, therefore, a *greater* Stark broadening. The line is broader and stronger (larger equivalent width) for a lower abundance of hydrogen. It is important to bear in mind that the Stark effect, which is a local effect, measures only the pressure and environmental broadening, and is only *related* to surface gravity, not a direct measurement to it.

You can understand the diagnostic role played by this broadening by thinking of the action of an environmental ion on the energy of an atomic level. The perturbation of the level's energy is linearly proportional to the field strength. For nearby ions, assuming homogeneous distributions, the field is strong and the perturbation is large. The same holds true for close impacts by electrons. Yet such collisions are comparatively rare, so their effect on the level is weighted by the probability of such an event. This is a random distribution, depending on the velocity distribution of the ions and electrons and also on their density. Since $\Delta\lambda \sim \langle\delta\mathscr{E}\rangle$ the distribution of field strengths maps into a line profile. This means that once we know the electron density, n_e and the temperature, T, we also know the broadening (sample Balmer lines profile for Vega and Lyman lines for the white dwarf HZ 43 are shown in Figure 3.5). For non-hydrogenic species, for instance, He I, the electron collisions also produce line broadening through the *quadratic* Stark effect, which depends on E^2. Again, we won't go into the details, but again deflect your attention to the references.

3.5.7.4 Van der Waals Broadening

Neutrals also contribute to the line wings, a process called *van der Waals broadening*. It operates, in principle, just like the Stark effect. The difference is that it is a very local broadening, with far smaller perturbations, because of the intrinsic weakness of the interaction that is due to an induced dipole moment which has a potential varying as r^{-6}. This is the main way to determine the depth-dependent density of a planetary atmosphere, by looking at the broadening of the scattered line radiation. In cool stars and planetary atmospheres, where the chief broadening agents are molecular and neutral hydrogen and neutral helium, this is the dominant environmental broadening agent.

3.5.7.5 Rotational Broadening

Rotational broadening is caused by the projection of the rotational velocity of each element of a stellar photosphere into the unresolved line of sight to the star from a stationary external observer. As such, it changes the observed profile, but maintains the equivalent width because it has nothing to do with processes that are occurring in the atmosphere. For a spherical star in rotation, the local surface velocity depends only on the distance of that point from the axis and therefore only on the latitude. On the other hand, the projection of that surface element toward a distant observer depends on the longitude—that is, whether the particular surface element is approaching or receding from the observer. Otherwise, in the rest frame of the surface, nothing happens. Like the Doppler broadening of an intrinsic profile, the profile resulting from rotation is a convolution, but of a different sort. It depends on the rotational width

$$\Delta\lambda_{\rm rot} = \frac{v_{\rm eq}\sin i}{c}\lambda_0,$$

where $v_{\rm eq}$ is the equatorial rotational velocity and i is the inclination of the axis to the line of sight. Each element of the surface has a unique radial velocity that is weighted by the projected area at that velocity, $v_{\rm rad} = v_{\rm eq}\sin i\cos\theta\sin\phi$ taken

222 RADIATIVE TRANSFER AND THE OUTER LAYERS OF STARS

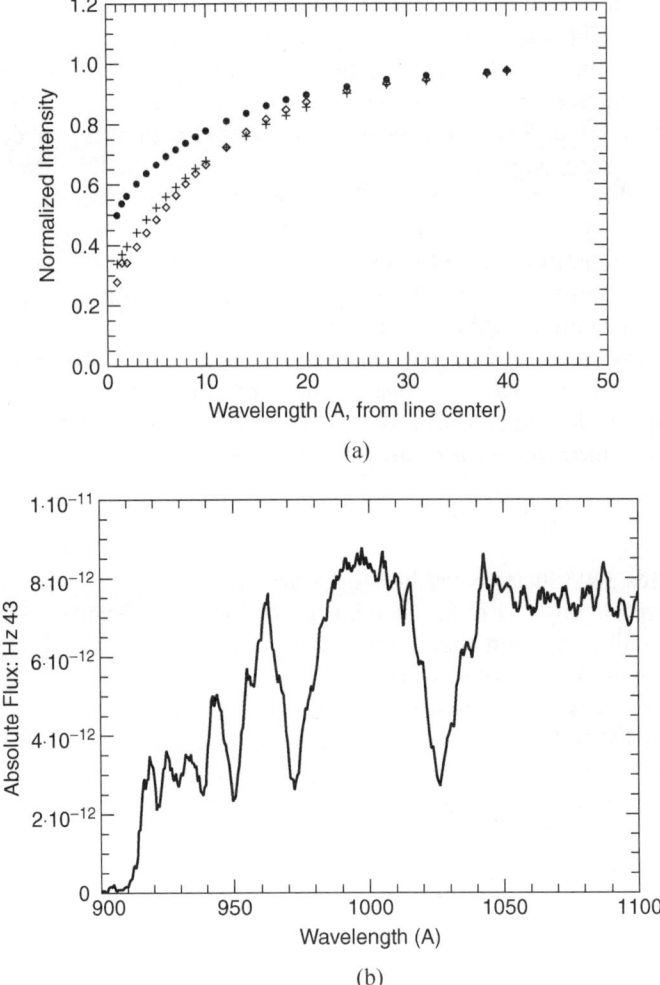

FIGURE 3.5. (a) Hα (large dots), Hβ (crosses), and Hγ (open diamonds) for Vega (data from Peterson, D. M. 1969, *SAO Special Rep.* 293). (b) The HUT spectrum of the white dwarf HZ 43 showing the Lyman series from Lyβ, illustrating the strength of the Stark wings on the hydrogen lines. Notice the weak, sharper photospheric lines from other sources.

relative to the center of the disk, and the isovelocity image of the stellar surface is a strip of width $\delta\phi$. For a uniform surface brightness, the rotational profile is given by

$$R_{\rm rot}(\Delta\lambda) = 1 - \left(\frac{\Delta\lambda}{\Delta\lambda_{\rm rot}}\right)^2. \qquad (3.134)$$

while including limb darkening, so that there is a cosine dependence to the limb, we obtain

$$R_{rot}(\Delta\lambda) = \frac{c}{\Delta\lambda v \sin i} \frac{\frac{2}{\pi}\left[1-\left(\frac{\Delta\lambda}{\Delta\lambda_{rot}}\right)^2\right]^{1/2} + \frac{u}{2}\left[1-\left(\frac{\Delta\lambda}{\Delta\lambda_{rot}}\right)^2\right]}{1+\frac{2u}{3}}. \quad (3.135)$$

This profile is actually a map of the projected *shape* of the star, weighted by the limb darkening, into a velocity. You can see how this happens because we have a unique identification of every point on the visible disk with a radial velocity relative to the observer as long as the star is rigidly rotating. This is no longer true, of course, if the star has a more complicated rotation law, or if the surface has a complex shape.[39] The observed profile is given by a convolution of the rotational profile with the intrinsic line profile, $\phi_{int}(\Delta\lambda)$, at each point on the surface:

$$I(\Delta\lambda) = R_{rot} \star \phi_{int}(\Delta\lambda). \quad (3.136)$$

The more limb-darkened a star is, the narrower its profile because the extremal limb velocities will be weighted lower as a result of the low surface brightness. The line peaks at the center because the meridian has the largest area[40] while the maximum velocity that you will ever see is the $v_{eq} \sin i$ (Figure 3.6). Hence, unlike any of the broadening mechanisms we have discussed here, the kernel of the rotational convolution is strictly bounded and vanishes beyond a finite value, a very distinctive line profile.

3.5.7.6 Doppler Imaging

For a rotating star whose surface is not uniform, such as a spotted object, the periodicity of the rotation means that a map can be produced of the surface. This is the technique of *Doppler imaging*. The whole technique for mapping the surface structure of stars is predicated on the same assumption as rotational line broadening, that there is a one-to-one relation between the colatitude and colongitude of each point on the stellar photosphere and the radial velocity of the point relative to the observer. In principle, this is easily understood for rigidly rotating stars,

---

[39] See, for instance, Collins, G. W. II and Sonneborn, G. 1977, *ApJS*, **34**, 41; Collins, G. W. II and Smith, R. C. 1985, *MNRAS*, **213**, 519; Smith, R. C. and Collins, G. W. II 1992, *MNRAS*, **257**, 340; Collins, G. W. II and Truax, R. J. 1995, *ApJ*, **439**, 860; Pérez, H. F. et al., 1999, *Astr.Ap.*, **346**, 586.
[40] To see this, consider the strip $\sin\theta \, d\theta \, d\phi = d\mu \, d\phi$ on the surface, where the radius is set equal to unity. This strip of width $d\phi$ is weighted meridionally by the limb darkening, $I(\theta) = 1 - u + u\mu$ so that the integration is taken over the strip:

$$R_{rot}(\phi) = \frac{\int_0^{\mu_\star} I(x,y)d\mu}{\pi \int_0^{\mu_\star} I(\theta)d\mu}.$$

To recover the line profile, note that $\sin^2\theta_\star + \cos^2\theta_\star \sin^2\phi = 1$ provides the direction cosine at the limb $\mu_\star$. Then use $\Delta\lambda = \lambda(v \sin i \cos\theta \sin\phi)/c$ for the conversion to wavelength in the final integral.

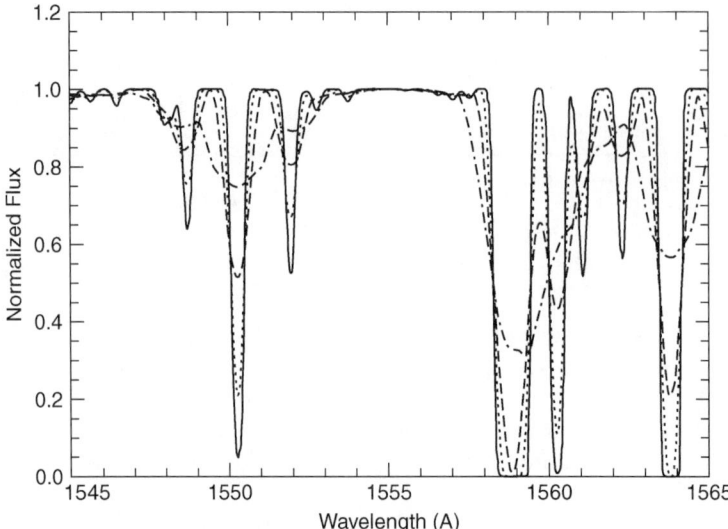

FIGURE 3.6. Rotational broadening (50, 100, and 200 km s^{-1}) for a sample theoretical slab spectrum of iron peak elements with $n_e = 10^{10}$ cm^{-3} and $T_{\text{eff}} = 5000$ K for $u = 0.5$. This illustrates the range encountered in v_{rot} for rapidly rotating cool stars.

since then Ω, the rotation frequency, is identical for every point and $|\mathbf{v}|$ varies only depending on latitude. One of the early applications in starspot modeling was the oblique rotator,[41] which assumes a formal expansion for the surface brightness in terms of spherical harmonics, Y_{lm}: $I(\theta, \phi) = \Sigma_{lm} a_{lm} Y_{lm}(\theta, \phi)$, that differentially weights parts of the surface in a rotating frame that is inclined to the line of sight and also to the rotational axis. A more general scheme employs maximum entropy reconstruction[42] in a nonparametric way while exploiting the same basic principle of velocity mapping to the surface. There is an ambiguity introduced by limb darkening and inclination, but the redundant information provided by line profile variations helps resolve this. Specifically, while the reconstruction of the surface distribution is not well constrained from photometric variations alone, line profile changes provide precise velocity information.

3.6 STELLAR ATMOSPHERES

What we have been setting up with all this formalism is a way to derive a reliable model for the thermal conditions in the outer layers of stars with which to determine other important physical properties, such as abundances, masses, and radii. One of the most important astrophysical discoveries of the nineteenth century was that stellar spectra can be distinguished by types, based on the

---

[41] This was originally applied to periodic Ap magnetic spectrum variables for which profile and Zeeman data are available by Deutsch, A. 1970, *ApJ*, **159**, 985. Modified approaches are found in Mihalas, D. 1973, *ApJ*, **184**, 851; Rice, J. B., Wehlau, W. H., and Khokhlova, V. L. 1989, *Astr. Ap.*, **208**, 179.

[42] See, for instance, Piskinov, N. E. and Rice, J. B. 1994, *PASP*, **105**, 1415 and references therein.

complexity of the line distributions and the specific species that are present. One of the greatest contributions of the 20th century was the explanation of why this is not a simple linear temperature and/or abundance sequence.

3.6.1 The Justification for Classifying Spectra

Spectra are the *lingua franca* of astrophysics and if radiative transfer is the grammar, spectral classification is the vocabulary (well, at least the nouns). But without some physical basis, how far can you get? In the middle of the nineteenth century, in the grip of a general taxonomic mania, elaborate systems appeared for separating stars by their spectral morphology. The first broad schemes by Secchi (1864) and Huggins (1865) distinguished four basic types of stars. These were based on prototypes, like Rigel (β Ori), Sirius (α Canis Majoris), and Arcturus (α Boötis), and on colors (especially striking for the Secchi class IV stars). There were similarities noted, for instance between the spectra of especially red stars and laboratory benzene and carbon flames, and the hydrogen lines were easily observed, but otherwise these were purely descriptive exercises. Further attempts to find a continuous sequence were made by Lockyer (who used the sequence to support a scenario for stellar evolution) and Vogel, with some additional refinement as resolution improved and photographic spectra were employed. The final ordering, based on all purely morphological considerations, was achieved by Maury, who corrected the initial ordering of the Harvard sequences. Maury's main contribution, largely ignored at the time, was to introduce a linewidth criterion into the classification the nature of which became observationally apparent when Hertzsprung found that her narrow (hydrogen) line stars were systematically farther and, therefore, both more luminous and larger in radius than stars of the same basic (temperature) class but with broader lines. The now universal classification scheme was adopted in 1910 by the International Solar Physics conference and became the undisputed basis of all further schemes after the completion of the Henry Draper (HD) catalog of stellar spectra and positions, a task involving the classification of over 250,000 stars accomplished virtually singlehanded by A. J. Cannon.[43] The types were separated on the basis of specific absorption features, those that were prominent in the Frauenhofer spectrum of the Sun. The spectral types were named on the basis of the strongest solar features, like K (Ca II) and B (Hα). The ordering was refined quickly with the discovery of the optical series of He II by Pickering and the O stars made their first appearance in the classification scheme (see Appendix 3.D for a brief description of the modern criteria for spectral classification). It is significant that the basis of the Harvard system predated both Planck's derivation of the blackbody law and Bohr's theory of atomic structure, although it was based, in part, on Wien's law for the temperature dependence of the color of a star.

---

[43] The repeat performance of this virtuoso feat has been N. Houk's reclassification of the HD stars using the more elaborate MK system, the first volume of which appeared in 1975. See the latest volume of the series, Houk, N. and Swift, C. 1999, *Michigan Catalog of Two-Dimensional Spectral Types for the HD Stars*, vol. 5 (Ann Arbor: Univ. Michigan Press) and references therein. In effect, at last, the HD list will have the elaboration Maury sought.

As usual, the application of discrete classifications to continuously variable characteristics is artificial and often even misleading.[44] The problem of setting up meaningful standards depends on many purely instrumental considerations like spectral resolution and on many other observational constraints. For this reason, long after the introduction of the refinements of the MK system, which included luminosity criteria, impersonal photometric methods were introduced to succinctly quantify spectral properties without the artificiality of taxonomic discreteness. The first successful system, the *UBV* system, was introduced by Johnson and Morgan in 1953. This broadband method, which we briefly discussed in Chapter 2, is most sensitive to global variations in metallicity (through the complex absorption line spectrum) and the hydrogen lines. Several refinements of the basic idea have employed narrower bandpasses, in particular the Strömgren system, and more filters to isolate specific features of the energy distribution, like the *DDO* system that concentrates on the CN feature or the Barbier and Chalonge system that isolates the Balmer jump. These have all been driven by the need to obtain as much information as possible on progressively fainter stars, thereby extending the spatial domain in the Milky Way and even external galaxies that is accessible to quantitative study.

The principal reason for the success of the photometric program was its superiority over photographic plates as a way of studying spectra. Photographic recording was inherently limited as a means for obtaining absolute fluxes because of the nonlinearity of the intensity response of the emulsions. The plates and films were also comparative insensitive relative to photoelectric devices, although by mid-twentieth century this was partly overcome with image intensifiers. These factors combined to limit the quantitative information that could be extracted from a spectrum to relative, not absolute, properties. Thus, precise fluxes and colors were really out of the question, but velocities and line profiles, and their variations, could be obtained. These dominated observations before the advent of two-dimensional imaging detectors, like vidicons and CCDs, and also drove theoretical developments. There was no need to compute absolute stellar atmosphere models, and even computed line profiles were normalized to a mean continuum. Even without absolute fluxes, much can be done. One line at a time still contains a lot of information. But it is not the same as a full radiative transfer solution and the quantitative information that is available from currently available spectroscopy.

Spectral classification provides a *starting guess* for the properties of the object you are studying. However, the accuracy of this guess depends on whether you have played by the rules of the system. For instance, differential extinction by the interstellar medium, due to the wavelength dependence of dust absorption and scattering, changes the colors of a star and compromises the use of photometric classification methods. Spectral types have the advantage that, although qualitative, they are independent of extinction because they depend on the visibility of lines relative to the *local* continuum. In this context, stellar population syntheses have made use of analogous procedures. Specific mass ranges can be isolated based on the strengths of particular spectral features, as we will presently discuss.

---

[44]A superb review of the taxonomic issues can be found in Morgan, W. W. 1950, *Publ. Univ. Mich. Obs.*, 10, 33. See also Garrison, R. F. Jr., ed. 1984, *The MK Process and Stellar Classification* (Toronto: Univ. Toronto Press), especially the review by Mihalas on the interface between atmospheres and spectral classification.

However, because any feature's visibility is weighted by the contribution of its population to the total luminosity, what one obtains as a final is biased. A single high luminosity star whose spectral characteristics are loosely the same as a star of lower luminosity can overwhelm the fainter population, or produce a peak in the population distribution of the fainter population, depending on the goals of the synthesis.

These morphological classification systems are gradually being superseded as new detectors become available. In fact, where spectral types are most often used is where it is, perhaps, most dangerous to apply them. The faintest objects you can study photometrically are galaxies at high redshift. Alas, these are composite systems consisting of stellar populations that are inhomogeneous in type and widely distributed in the interior of the galaxy. Individual stars, and groups of stars, may have very different environments. If these are spatially unresolved, the photometric methods cannot distinguish among these possibilities. So while a number can be assigned to the body, something like a color or a total brightness, it is impossible to unambiguously invert this to provide a picture of the stellar population.[45]

3.6.2 The Stellar Atmosphere Problem

The *atmosphere problem* is how to simultaneously satisfy two substantially different equilibrium conditions: hydrostatic and radiative. In an atmosphere, the problem is rendered especially difficult because of the variable rate of energy loss to the world at large and the very strong wavelength dependence of the opacity. This latter effect is due mainly to the relatively low density of the atmosphere. The radiation does not diffuse, but instead sees very different opacities from superficial layers depending on the wavelength. This in turn feeds back into the source function and thus changes the distribution of radiation with frequency. To see this, consider the hydrostatic law for a thin layer:

$$\frac{dP}{dz} = -\rho g \rightarrow \frac{dP}{d\tau} = \frac{g}{\kappa}. \tag{3.137}$$

Now, which wavelength do we use in order to connect this with the transfer equation? Clearly we can scale between spectral regions:

$$d\tau_\lambda = \frac{\kappa_\lambda}{\kappa_0} d\tau_0, \tag{3.138}$$

so that at different wavelengths we see into different depths of the atmosphere. This is shown most dramatically in the accompanying picture of Jupiter taken some years ago with ground-based infrared detectors near 2μ (Fig. 3.7). There you observe the deeper layers as brighter, while in the optical it is the other way around. Of course, being a planet, the illumination is from the outside but the point is still that different wavelength regions produce different opacities and

---

[45] This is one of the reasons for spending so much time in this book on stars—they are the key to understanding galaxies and the evolution of the galaxian population because they are the principal constituents of these bodies.

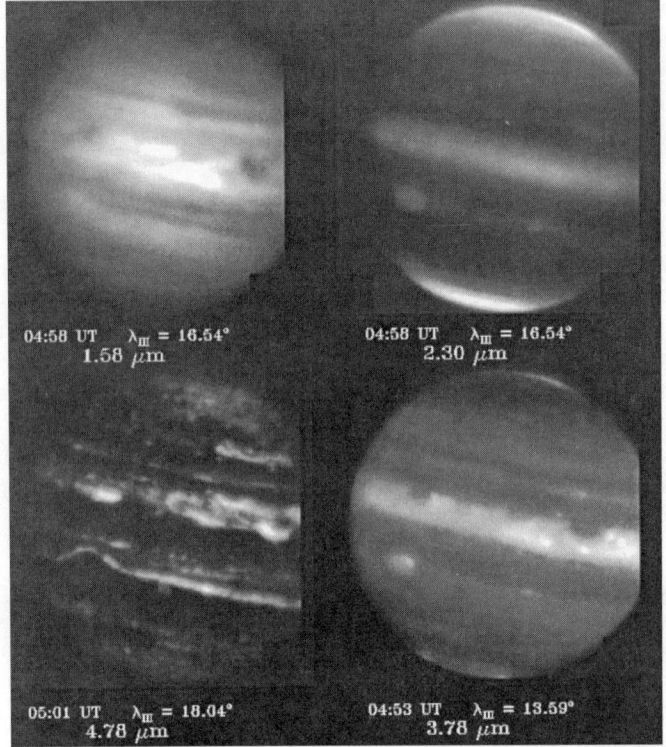

FIGURE 3.7. Images of Jupiter taken with near-infrared filters showing the effects of differential absorption (Dec. 24, 2000). This phase ($\lambda_{III} = 16.5°$) shows the Great Red Spot in the lower left (credit: Galileo/International Jupiter Watch, R. Beebe, NMSU). The darkness at long wavelengths (4.78μ) of the equatorial cloud deck (bright at 1.58μ) shows the contrast between thermal emission and scattering.

visibility into the atmosphere. For a star, it is even easier to see this. Wherever you look into the atmosphere at an opaque wavelength, you see only the outermost layers. These are necessarily cooler, hence fainter. This could happen, for instance, when you look at an absorption line. The core of the line is darker if the atmosphere is cooler in its outer layers in LTE.[46] The optical depth is also not the same as the scale height. For an isothermal atmosphere, for instance, the pressure scale and density scale heights are identical for an ideal gas. Both come from the condition that a pressure (sound) wave can propagate across a layer within one freefall time so pressure balance can be maintained against gravity in a layer of this thickness. The optical depth is proportional to the column density, which is approximately $\kappa \langle n \rangle H_P$, where $\langle n \rangle$ is the mean number density. Thus the depth to which you see in the atmosphere depends on the wavelength as much as the scale height.

---

[46] For a scattering line, this simply means that the source function is reduced. Always keep in your mind that the intensity is not directly related to temperature in the case of NLTE.

Another important point to keep in mind when examining spectra is that the outer layers of an atmosphere need not produce only *dark* lines. If there is a source for scattering, or if the temperature doesn't change monotonically with height, it is possible for emission lines to appear. The line profile reflects the *source function*, not uniquely the thermodynamic state of the medium. In general, the temperature decreases with height. This does not mean, however, that a temperature inversion cannot happen. Of course, anyone living in Los Angeles would be quite familiar with this, since temperature inversions are a fact of life in the Earth's atmosphere. The electron temperature of the ionosphere is so high because of ionization by incoming solar ultraviolet. So if the electron temperature, and through collisions the excitation temperature of the atoms, increases with increasing height, the source function will invert and an emission line will occur. A major difficulty in interpreting combined emission and absorption spectra, as, for instance, for stellar winds, is to distinguish between these two alternatives.

To solve the atmosphere problem, you are trying to determine the run of pressure and density that are consistent with the gravity and the temperature gradient produced by the stellar luminosity. This must be consistent with the opacity distribution that results from the composition. Combining the gravity and luminosity with the stellar mass determines the radius and the effective temperature. This in turn sets the flux, and for a thin atmosphere that is all you need to get started. The condition of radiative equilibrium guarantees that everywhere through the atmosphere you have just the right run of thermodynamic variables to ensure stability. Or rather, your model does. This also requires that you are able to solve self-consistently the populations of the atomic levels and the ionization structure in order to obtain the opacity, which then feed back into the model through the transfer equation to determine the temperature structure. From the equation of state, a variation in local temperature readjusts the density and pressure and changes the opacity from the ionization and population equations. In principle, the whole set of calculations converges with some tolerable accuracy to an equilibrium model.

The competition between energy generation and energy transfer dominates stellar matter. The radiation emerges from the deep interior at some rate and, if the star is to remain thermally stable, must pass completely through the outer layers and into space. So the temperature gradient adjusts to maintain this flow. This is already clear from the diffusion approximation. But it also means that any increase in the opacity must steepen the temperature gradient. A simple way of seeing this is to imagine that the flux is diffusive and take the LTE approximation for the source function. This last step is crucial, since there is otherwise no direct relation between the source function gradient and that of the temperature. If there are two regions that have opacities κ_1 and κ_2, in this case the Rosseland means, then

$$\left(\frac{dT}{dz}\right)_2 \approx \frac{\kappa_2 T_1^3}{\kappa_1 T_2^3}\left(\frac{dT}{dz}\right)_1. \qquad (3.139)$$

If the temperatures are similar, then the higher opacity will have the larger gradient and the temperature will rise in the interior layer and fall in the overlying, more opaque layer. You know this effect from pulling a blanket over yourself in the winter. Your body serves as a constant heat source, the room is the constant heat sink, and the covers control the rate of loss of heat through conduction. Eventually, the system equilibrates. This *backwarming* is important for interpreting the emergent spectrum. Those regions that have a high opacity will appear cooler than what you would have expected from the local emissivity for a blackbody corresponding to the effective temperature (or for a continuum model not employing spectral lines). In contrast, wherever the opacity is low, the emission will be much higher than you might have expected. This is because the flux leaks out from the deeper layers at whatever wavelengths have the lower opacity, and the effect of windows is to produce pseudo-emission features whose brightness is far larger than the equilibrium continuum or blackbody emissivity.

Normally this is not a insurmountable obstacle. For normal young main-sequence stars, all of which have approximately the same abundances, the pattern is complicated but can be calibrated. Where it causes problems is when you are trying to compare a stellar model with an unknown to determine the temperature and gravity, perhaps for studying the abundances. The reason is the following. If the temperature is too high compared with the expectations of the model, then the abundance of trace species may have to be increased. This is because of the temperature dependence of the ionization equilibrium. A signal of this is that the color and ion temperatures will differ. The color temperatures are determined from continuum fitting, while the ion temperatures come from solving for the abundances of the same element from different ions. This is even true for the line profiles. The line core has a higher opacity than the wings; hence it is formed in shallower layers of the atmosphere. For the hydrogen and helium lines, by virtue of their elemental abundances, this is especially important. In an atmosphere that suffers substantial line blanketing, the profiles will differ from those expected for a lower metallicity. Conversely, a metal poor atmosphere is more transparent and will *appear* to be hotter (bluer). This is the basis, for example, for separation of populations in a galaxy. The lower metallicity stars have higher ultraviolet emission than their otherwise identical metal-richer counterparts because of the change in the UV opacity.

3.6.3 Radiative Equilibrium: What Goes In, Comes Out

For the energy flow to completely balance so that the temperature of the medium remains constant with time, we impose the condition of radiative equilibrium. This means that the frequency-integrated, or *bolometric*, flux remains constant, although the spectral distribution of the monochromatic flux may change drastically throughout the medium. Thus

$$\frac{d}{dz}\int_0^\infty F_\nu \, d\nu = \frac{dF}{dz} = \int_0^\infty \kappa_\nu (J_\nu - S_\nu) \, d\nu = 0. \qquad (3.140)$$

The requirement in LTE is that the re-emission be the same as that of blackbody, so that the total energy in the mean radiation field can never exceed the emissivity from a blackbody when the density becomes high enough.

This is more obvious for a nonlocal source. Take a planetary surface, like an asteroid, at some distance from the Sun, and assume that the radiation arrives with a color temperature of about 5800 K. Most of the flux is in the visible range. But it is very dilute because of the distance of the body from the Sun, so that $J_\nu = (1 - A_\nu)W(r)J_{\nu,\odot}$. Here $W \sim (R_\odot/d)^2$ is the dilution factor and A_ν is the albedo (the fraction that is reflected by the surface). The absorption leads to heating that comes into equilibrium *at an inevitably lower temperature* because of W. Even if the re-radiation doesn't look like it, it will still locally be the absorption coefficient-weighted rate, $J_\nu = \kappa_\nu B_\nu(T_{eq})$, where now T_{eq} is the equilibrium temperature from solving Eq. (3.140).

In view of what's coming in the next chapter, you should always keep in mind that radiative equilibrium of an atmosphere is the same as thermal equilibrium *unless* mass motions are required to transport the flux, as in convection. Photons are the only energy loss mechanism even though stellar winds sometimes carry some of the energy. However, a stellar interior has other means for losing energy, in particular neutrinos. This is a fundamental difference.

3.6.4 Determining the Thermal Structure

The problem of computing an atmosphere is that the assumed temperature gradient is not the one that a particular composition produces. Only a few analytical solutions are known for the transfer equation, and these often fail to adequately predict the emergent spectrum. The problem has been treated by a variety of temperature correction methods, all of which require iteration of the populations and source function within the radiative transfer, coupled with the hydrostatic requirement, in order to achieve a solution. We will not go into these in detail here[47] but instead just show how the simplest one works. If you have an initial guess for the source function—and it may be a Planck function—at a given temperature for a prespecified temperature structure,[48] then you can write a perturbation expansion in the form

$$B_\nu(T) = B_\nu(T_0) + \left(\frac{\partial B_\nu(T)}{\partial T}\right)_{T_0} \delta T. \qquad (3.141)$$

The requirement of radiative equilibrium determines the change in temperature because

$$\int_0^\infty \kappa_\nu J_\nu \, d\nu = \int_0^\infty \kappa_\nu B_\nu(T_0) \, d\nu + \delta T \int_0^\infty \kappa_\nu \frac{\partial B_\nu}{\partial T} \, d\nu. \qquad (3.142)$$

---

[47] See Mihalas (1978); Kalkofen, W., ed. 1987, *Numerical Radiative Transfer* (Cambridge, UK: Cambridge Univ. Press); Kurucz, R. L. 1970, *SAO Special Report* 305 (this is the description of the ATLAS model atmosphere code and is a particularly useful guide to the methods used in actual computations); Crivellari, L., Hubeny, I., and Hummer, D., eds. 1991, *Stellar Atmospheres: Beyond Classical Models* (Dordrecht: Reidel).
[48] For instance, you might take the gray atmosphere solution for which $T(\tau) \sim (\tau + \frac{2}{3})^{1/4}$.

In general, your initial guess will not satisfy the radiative equilibrium condition after you have solved the formal equation of transfer. Because you can compute mean intensity using $J_\nu = \Lambda_\tau[B_\nu]$ from the source function alone, the iterative scheme, called *Lambda iteration*, is written as

$$\delta T = \frac{\int_0^\infty \kappa_\nu \left(\Lambda_{\tau_\nu}[B_\nu(T_0)] - B_\nu(T_0) \right) d\nu}{\int_0^\infty \kappa_\nu (\partial B_\nu / \partial T) \, d\nu}. \tag{3.143}$$

As written, this procedure converges very slowly and is never used in this form in practice. But the basic idea serves as the starting point for many iterative methods for correcting the temperature structure while requiring flux constancy. You can also see why this works at all for a plane parallel atmosphere: both the flux and gravitational acceleration are the same throughout the atmosphere.

As a starting solution, let's return for a moment to the formal solution of the transfer equation. If we adopt a thermal frequency integrated source function, $B = aT^4$, then

$$T^4(\tau) = \tfrac{3}{4} T_{\text{eff}}^4 \left(\tau + \tfrac{2}{3} \right) \tag{3.144}$$

where we have used the definition of the effective temperature to substitute for $F(0)$ in Eq. (3.49). In general, if both scattering and absorption are included, this need not be true, but it holds generally for a thermalized medium. This is an exact solution for a thin layer in LTE and emphasizes the point that the temperature varies slowly with optical depth for the grey plane parallel case.

3.6.5 Convection in Stellar Atmospheres

The atmosphere is a layer of gas, illuminated from below, and supported by pressure gradient against gravity. If it were free to move, the pressure gradient could take on any value and the medium would be able to dynamically adjust to any rate of energy input. This is not possible, however, if we constrain the medium to remain mechanically stable, and this is where the radiative transfer enters. Compare two optically thick layers, so that we can use the diffusion approximation to simplify the comparison, with opacities $\kappa_1 < \kappa_2$. If we require that they pass the same flux, then the ratio of their temperature gradients must correspond to the ratio of the opacities:

$$\frac{\kappa_2}{\kappa_1} = \frac{\left(\dfrac{dT}{dz}\right)_2}{\left(\dfrac{dT}{dz}\right)_1}.$$

In other words, as we have discussed earlier, the temperature gradient steepens to "push" the photons through a more opaque medium. There is, however, a limit to how large dT/dz can become and this is due to gravity.

Consider a parcel of gas that has a slightly greater temperature by an amount δT compared to the temperature of the surrounding gas, T. For an ideal gas, this means that at pressure equilibrium, the density must be lower by an amount

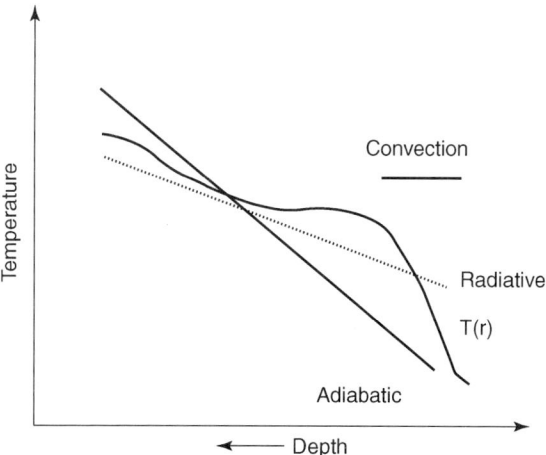

FIGURE 3.8. The principal temperature gradients for the basic mixing length argument. This figure compares the adiabatic and radiative unstable layer (consider how this relates to convection zones in the outer atmospheres of late-type stars).

$\delta \rho = -(\rho/T)\, \delta T$. This density decrease makes the parcel buoyant with respect to the rest of the layer because of the force $F_b = \delta \rho g$, and it begins to accelerate upward. It therefore rises into a region of the atmosphere at some distance L where the temperature is lower by an amount $(dT/dz)L$. In addition, because the pressure drops with height, the parcel expands and cools adiabatically. If it can exchange heat with its surroundings, it will mix or cool and lose buoyancy; it then sinks, and there is no net heat transfer. The parcel is stable. You can see that the critical temperature gradient is given by the condition that the parcel moves *adiabatically*, otherwise there is no net force for the motion. If, however, the gradient required to radiatively transport energy is steeper than the adiabatic one, then the parcel continues to move and there is a net *convective* transfer of energy. This is the fundamental process underlying energy transport by fluid motion in stars, whether in their atmospheres or interiors. The requirement that the layer remain mechanically stable under gravity and yet accommodate to the amount of flux coming from the interior sets the limit to the steepness of the thermal gradient.

Another way to see this is diagrammatically as in Figure 3.8. Notice that if we start with a superadiabatic gradient and move the blob upward any distance *adiabatically* and in pressure balance—assuming no radiative or conductive losses and only density accommodation to the change in external pressure—then it winds up hotter than the adiabatic atmosphere and continues to rise at *every* location. Similarly, for adiabatic motion in a subadiabatic atmosphere, the blob is everywhere negatively buoyant for upward displacement and returns to its original location. Only for the adiabatic gradient is the motion neutrally buoyant, but then there is a need to invoke some magical agent that forces mixing and deceleration without explicitly stipulating the cause of viscosity.

The first solution of the stability problem for a fluid layer was found by Rayleigh for the incompressible case. Without going into the detailed solution of the stability problem[49], we can gain insight through dimensional analysis of the problem as follows. Imagine that the layer transports energy by diffusion, or conduction, with a conductivity σ. Assume as well that the fluid is viscous with coefficient η. A parcel that begins moving feels a retarding force from viscosity, and so the timescale for slowing down is $t_v = L^2/\eta$. The time to conductively transport energy is $t_c = L^2/\eta$. The buoyant rise time for the blob is $t_b = L/v_c$ so there will be net motion of the fluid provided the buoyant timescale is shorter than *both* the viscous and conductive timescales, or in other words when

$$t_b < (t_v t_c)^{1/2},$$

the geometric mean of the two. The velocity of a parcel is given dimensionally by the equation of motion $\rho v_c^2 / L \approx \delta \rho g$ so that

$$\frac{\delta \rho g L^3}{\rho \eta \sigma} > 1. \qquad (3.145)$$

Since you already know that the density perturbation is related to the temperature gradient, you can write

$$\delta \ln \rho = \left(\frac{d \ln T}{dz}\right) L,$$

which gives the dimensionless criterion for net energy transfer. This is, essentially, the Rayleigh number. It suffices to say, without knowing its precise value, that when the number is large enough, the fluid must convect. But how does this relate to the atmosphere problem? After all, the medium has a large range in opacity and here we have ignored this by saying that the energy transfer has some conductivity, σ. You can see that this σ is really κ^{-1}, so that for a layer that has a high opacity, the medium is more unstable to convective transport. We will meet this again in the context of stellar interiors where the gas is so dense and opaque that the diffusion approximation is certainly valid. For the atmosphere, however, this is less certain and requires a more detailed treatment.

The picture of convection we have just developed is of a continuous flow, the so-called Rayleigh–Benard problem, in which we have assumed that pure circulation occurs between fixed boundaries in an incompressible medium. This means ignoring the pressure and density variation within the layer, so it can only hold strictly for a very thin region. Thermal expansion is responsible for the buoyancy, which is also called the *Boussinesq approximation*. Although we used the dimensional argument for a parcel of gas, it is actually not quite right, at least within the picture used for solving the fluid equations. Astrophysical convective flows occur in nearly inviscid fluids, so the Rayleigh number is usually extremely large. This results in turbulence with many different length scales contributing to the transport of energy in the fluid. The picture of slow, steady circulation must be replaced

---

[49] See also Chandrasekhar, S. 1961, *Hydromagnetic and Hydrodynamic Stability* (London: Oxford Univ. Press; rpt. Dover Books); Shore, S. N. 1992, *An Introduction to Astrophysical Hydrodynamics* (San Diego: Academic Press) and references therein.

with something like the random motion of buoyant eddies. This is achieved, at least in part, by introducing a concept taken from turbulence theory, the *mixing length*.

As originally developed in aerodynamics and for turbulent fluids, the mixing length is the mean distance an eddy travels before it dissolves into the background. We used this in setting up the instability criterion when we said that the parcel moves adiabatically. In fact, there is always some heat exchange with the background, and we need some other means for computing how the parcel changes its internal temperature. Assume that the temperature difference between the blob, ∇, and its environment is $\delta T = (\nabla - \nabla_{\text{env}})\delta z$ over some small displacement δz. For compactness of the notation, even though it is one dimensional only, we will use the convention $\nabla_{\text{Ad}} \equiv (d \ln T/d \ln P)_{\text{Ad}}$ and similar notations to represent the other temperature gradients. This distance is essentially the same as L, which we will identify with the mixing length. There are two important temperature differences in this picture, $\nabla_{\text{rad}} - \nabla_{\text{Ad}}$ and $\nabla - \nabla_{\text{Ad}}$, the first of which determines the inability of radiation to carry the energy and the second, the *superadiabatic gradient*, governs the rate of transport by convection. Radiation is always present, so it must always be included in the total flux. The radiative gradient is defined as the temperature gradient required to carry the incident flux:

$$\nabla_{\text{rad}} = \frac{3\kappa F P}{4 a c T^4 g}, \tag{3.146}$$

and the adiabatic gradient is given by the equation of state. The actual temperature gradient is different depending on the stability of the atmosphere.

The energy flux in the convective elements is

$$F_c = \rho v_c c_P \, \delta T = \rho v_c c_P \, (\nabla - \nabla_{\text{Ad}}) \frac{d \ln P}{dz} L, \tag{3.147}$$

so the principal unknown is v_c, the velocity of the blob. Since buoyancy determines the blob velocity, we find that

$$\tfrac{1}{8} \rho v_c^2 = \delta \rho g L, \tag{3.148}$$

where the factor of eight is from the assumption that the convective velocity is the time average of the slow motion of the blob. Then, since

$$v_c = 2\sqrt{2} \, (\nabla - \nabla_{\text{Ad}})^{1/2} \frac{g}{H_P} L,$$

we obtain the convective flux:

$$F_c = 2\sqrt{2} \, c_p (\nabla - \nabla_{\text{Ad}})^{3/2} g^{1/2} H_P^{-3/2} \frac{\rho}{T} L^2. \tag{3.149}$$

Again, keep in mind that ∇ measures the steepness required for the atmospheric temperature gradient for the flux, F, that is incident on the medium:

$$F = F_{\text{rad}} + F_c. \tag{3.150}$$

The convective flux can be written as

$$F_c = F\left(1 - \frac{\nabla}{\nabla_{\text{rad}}}\right) = \mathscr{A}(\nabla - \nabla_{\text{Ad}})^{3/2}.$$

$\mathscr{A}$ is now a function that combines the terms in Eq. (3.149) that depend on the local temperature and pressure through the scale height. The temperature gradient then comes from solving a cubic equation:

$$\mathscr{A}\nabla_{\text{rad}}(\nabla - \nabla_{\text{Ad}})^{3/2} = \nabla_{\text{rad}} - \nabla. \tag{3.151}$$

This is the end result of the mixing length picture with one addition. The choice of L has been left open. Yet it is the whole point of the calculation that the transported flux depends critically on how far a blob goes before it mixes with the background. For the Rayleigh–Benard problem, this is a completely free parameter. All of the motions are assumed to take place within a layer whose properties are specified at the outset, so the Rayleigh criterion for convection is a dimensionless number that allows for a scaling of the size of the layer with the imposed temperature gradient. That is not the case for a stellar interior, and even less so in a stellar atmosphere. Not only does the medium have a temperature gradient, but the pressure and density vary as well. Intuitively, you might think to take $L = H_P$, the pressure scale height. Since the blob remains in pressure equilibrium with the background gas, it must expand and cool during its rise. Thus, since the temperature gradient would not be strongly superadiabatic in an optically thick medium, you might anticipate that one scale height would be about as far as the blob would move before it reached the level of *neutral buoyancy*. This concept works for planetary atmospheric clouds, which are just convective plumes, but for a radiating medium such as a stellar atmosphere it is only an approximation. Overshooting of the neutrally stable layer usually results. The blob doesn't know to come to a crashing halt when it has traversed one scale height. Thus, it is usual to take $\alpha = L/H_P$, to be ≥ 1 so that the flux becomes

$$F_c = 2\sqrt{2}\,(gH_P)^{1/2}\frac{\rho}{T}c_P(\nabla - \nabla_{\text{Ad}})^{3/2}\alpha^2. \tag{3.152}$$

If the medium is optically thin at the end of the blob's motion—and you can check on this from the atmospheric structure—then it may cool even more rapidly than it would by an adiabatic expansion and neutral stability will occur earlier. An even more exact picture would be to assume that α is a function of temperature and density in the atmosphere, but this makes the computation of the convective transport far more difficult.

A slight modification to this formalism is to include the effect of a change in molecular weight due to the state of the blob. For instance, in an atmosphere this could result from changes in the ionization of the gas. For a stellar interior, as we will see, this can result from chemical composition gradients. The effect is easily included in the formalism because of the requirement that the blob remains in pressure equilibrium throughout its motion. Then

$$\frac{\delta \rho}{\rho} = -\frac{\delta T}{T} + \frac{\delta \mu}{\mu},$$

where μ is the mean molecular weight. If $\delta \mu$ is due only to temperature, then

$$\frac{\delta \rho}{\rho} = -\frac{\delta T}{T}\left(1 - \frac{\partial \ln \mu}{\partial \ln T}\right), \qquad (3.153)$$

and you can see, by comparison with Eq. (3.148), that the principal effect of this change is to alter the buoyant force and the velocity of the blob. For stellar atmospheres, this is enough detail to understand the role of convection. For interiors, we will need to return to the issue because of the role of overshooting in the evolution of stars.

3.6.6 LTE versus NLTE in Stellar Atmospheres

One issue that often arises in quantitative analysis of stellar spectra is when and how to apply the LTE assumption. It is not merely an academic argument, since the choice governs every derived quantity from a stellar atmosphere. It is obvious that one condition that must be met is that the collisions be sufficiently frequent, compared with the radiative timescales, that the populations obey the Boltzmann distribution and the ionization ratios should be given by the Saha equation. In other words, for all dominant transitions, $A_{ji} < n_e C_{ji}$ and otherwise the principle of detailed balance applies. This latter situation directly implies that the source function should be $B_\nu(T)$ so $J_\nu = B_\nu$ at all depths in the atmosphere. We also assume, in effect, that $K/J = \frac{1}{3}$ in LTE. Finally, the LTE assumption means that there is a unique parameter, the temperature, that characterizes the electron, atomic level, and ion energy distributions and that is the temperature. Since T is directly linked, through the equation of state and the Saha equation, to the pressure, we can solve the structure of the atmosphere and with it determine the emergent intensity at every wavelength through the formal solution to the transfer equation.

But where and when does this break down? One place is in the outermost layers of the atmosphere. The density drops sufficiently that detailed balance may not be satisfied. This situation does not affect the weak lines, which require large column densities to be detected and are formed rather deep in the atmosphere; it will mainly produce deviations in the strengths of the strong lines, mainly in the line cores. Specifically, the lower-level populations of resonance transitions are generally greater than would be expected for LTE. This also means that higher excitation lines that are above the ground state but well below the thermal energy of the continuum may be weaker than expected. This, in turn, changes the abundances derived from these lines (one signature is that the abundance is a function of the excitation energy). In addition, when the density is low, recombinations are less frequent than predicted by the Saha equation. This alters the ionization ratios. For a trace species, if you cannot measure more than one ionization state, the derived abundance may not be correct.

The presence of emission does not necessarily mean that the medium is in NLTE. A rise in the temperature gradient with increasing height above the photosphere, which can result from a variety of nonradiative heating mechanisms (turbulent acoustic input, magnetic reconnection, magnetic shocks, etc.) will produce emission simply because the source function can follow the temperature rise until the density falls to where collisions fail to couple the radiative transitions to the radiation. When this happens, the line profile can reverse and form deep absorption cores even though the temperature continues to rise outward. Scattering marks the most important departure from LTE, but here polarization is an important clue to the physical processes forming the lines.

In general, always keep in mind that LTE is an *approximation*, just as plane parallel atmospheres are. High-gravity stars (main sequence, subdwarfs, and white dwarfs) are well treated by LTE. These are also stars for which the plane parallel approximation holds rather well. Yet even in these cases, the cores of the strongest lines (e.g., hydrogen Balmer and Lyman transitions and the He I triplets) show strengths that depart from LTE. The chromosphere and coronae of cool stars are, however, poorly treated by LTE. The atmospheres of giants and supergiants require the spherical treatment and, in general, NLTE treatment of both continuum and line formation. The dividing line is not well defined, but a rule of thumb is that the resonance lines of the dominant ions are likely formed under NLTE for $\log g < 3.5$ or so for most T_{eff}. Nebulae in particular, and the interstellar medium *in general* require full solution of the equations of statistical equilibrium, as do most stellar winds (especially Wolf–Rayet stars), nova and supernova ejecta, and other moving media.

3.6.7 Some Complications

3.6.7.1 Gravity Darkening by Rotation

We have already discussed the broadening that an external observer sees in spectral lines coming from a rotating surface. The rotation also affects the structure of the stellar interior and atmosphere. The reason is connected with the requirement of hydrostatic equilibrium. Rotation forces cylindrical symmetry in what was an otherwise spherical star. But rapid rotation can actually distort the star by altering the potential

$$\Phi = \Phi_{\text{grav}} - \tfrac{1}{2}\Omega^2 r^2 \sin^2\theta, \qquad (3.154)$$

where Ω is the rotation frequency, so the surface gravity is now a function of latitude. This means the pressure scale height at the equator is increased, and the temperature gradient reduced, compared with a nonrotating star. In addition, the angle between the radial and normal directions on the surface is altered so the effective limb darkening is increased. The combined effects are called *gravity darkening*. We will discuss this further in Chapter 4, but for now let's examine the consequences of this variation of the effective temperature for the appearance of the integrated stellar spectrum.

Since the total surface flux remains unaltered compared with a nonrotating star of the same physical properties, the increased effective area at the rotational equator means that the flux can be reduced, while to conserve total energy the flux at the poles increases. The star consequently gets brighter and hotter at the poles. One way of seeing this is that the flux varies inversely as the surface area, so for fixed mass $F \sim g$. Thus the effective temperature goes like $T_{\rm eff} \sim g^{1/4}$, which is called the *von Zeipel gravity darkening* law. The critical parameter, as you can see from the potential, is the ratio of the rotational to gravitational energies:

$$\lambda_\Omega = \frac{R_\star^3 \Omega^2}{GM}, \qquad (3.155)$$

where $R_\star$ is the stellar radius. When this number is of order unity the star is significantly distorted.

This change in the local surface gravity has a dramatic consequence for observations. A rapidly rotating star looks different depending on the viewer's perspective. Lines that become stronger with temperature are enhanced at the poles and weakened at the equator relative to a nonrotating star. Keep this in mind when you are too readily interpreting a spectrum—what you see is emerging from an unresolved surface whose characteristics may be quite far from uniform. The key difference is that the rotational broadening also depends on aspect, so the lines that are enhanced at the equator will also be broader when seen at high inclination of the rotation axis. However, when viewed nearly pole on, such lines are not only narrow (since then there is essentially no rotational broadening) but also more severely limb-darkened through the combined effects of viewing angle and reduced effective temperature. The integrated spectrum can vary greatly in its properties. This can lead to serious taxonomic inconsistencies when applying the usual classification criteria to rapidly rotating stars. Spectrophotometric continuum observations will be at odds with the individual line strengths, and the placement of the star in the HR diagram can be seriously altered.

Tides produce much the same effect as rotation. By altering the local acceleration, they distort the shape of the equipotential surface and produce local changes in the atmospheric structure. Stars that are constrained to roughly full their Roche surfaces, that is when they are in close binary star systems, are likely to have large variations in their surface properties. There is some debate about whether a convective atmosphere shows such effects, because the temperature gradient is relatively insensitive to changes in the local physical conditions. It is likely, however, that any radiative star will show some form of this effect.

3.6.7.2 External Illumination

An external source causes changes in the source function at small optical depth, in particular a local heating. If it is due only to scattering, then the local temperature will not be affected. But if there is any absorption, then a temperature inversion may result. This has the effect of producing absorption at some part of the line profile, depending on the distribution of opacity. Since the outer atmosphere will generally be the most strongly heated, the largest changes are expected in the line core. It is possible, however, that self-absorption reduces the strength of this part of the profile relative to the wings. This depends on the abundance of the absorber. In the Earth's atmosphere, solar irradiance in the ultraviolet causes a substantial temperature increase in the ionosophere and exosphere due to ionization compared with the lower layers. There is also a temperature rise in the middle atmosphere due to absorption of UV by ozone. In a stellar atmosphere, a close hot companion is likely to drive an inversion in the temperature gradient and may even substantially heat the photosphere. An example of this is Her X − 1 = HZ Her, where X-ray illumination of an F star by a companion accreting neutron star increases the temperature of the side facing the hot source by almost 20%. The illuminated face displays an A-type spectrum. There are many less dramatic examples, usually called *reflection effect* for binary stars (see Chapter 5). The fact that some radiation has been thermalized from the companion and is emitted from the surface, in other words, that the color temperature of the emitted radiation is cooler than the illuminating source, is an indication that the radiation is, at least, partially thermalized. Thus the atmospheric temperature gradient has been changed.

A final example of external illumination is α Orionis, or Betelguese, the dominant red supergiant in Orion. This star's envelope, which is very rich in molecules such as CO, has a finite boundary despite its extensive outflow. The explanation lies again in external illumination. The interstellar radiation field, the cumulant radiative intensity of all stars in the Galaxy, has a spectral distribution similar to an A star. Therefore, there is a portion of the incoming radiation that has energies above the photodissociation energy for CO.[50] This truncates the emission region by destroying the emitters. There is no real difference between this and the problem of a molecular cloud near a star forming region. In all cases, the medium solves the problem of thermal balance due to a nonlocal heat source whose effective temperature may be very different from the local equilibrium radiation temperature.

The problem becomes more acute for planets since their atmospheres are completely structured by incoming radiation. Before the 1990s this was an area limited exclusively to the solar system. Now, with the discovery of extrasolar planetary systems, it has become increasingly important that astrophysicists of all stripes be at least familiar with the basic physics of planetary atmospheres. To treat the problem simply, we show in Appendix 3.C how to solve the scattering problem for an atmosphere. You have already begun this with the first pass through the equation of transfer. A complication is added, however, by the fact that the illuminating radiation is always at a very high color temperature relative to the reradiated light from the planet. This is, of course, the source of the *greenhouse effect*. Heating occurs in the upper atmosphere because of the absorption of ionizing UV. For the Earth, the principal opacity source shortward of 0.3μ is O_3, so the ionization of this species produces a local heating at high altitude. At longer, visible, wavelengths, the atmosphere is nearly transparent. But absorption does occur, and the surface has a relatively low albedo. The match between opacity and incoming light for planetary atmospheres is dominated by molecular opacity which is greatest for rovibrational transitions in the red and infrared and in electronic transitions in the UV. For a solar type star illuminating a planet with a

---

[50] See, for example, Glassgold, A. E. and Huggins, P. J. 1986, *ApJ*, **306**, 605; Rodgers, B. and Glassgold, A. E. 1991, *ApJ*, **382**, 606.

Jovian type mixture (e.g., H_2, CH_4, NH_3, CO, CO_2, H_2O), the opacity minimum leads to efficient heating even to very large depths, until the scattering optical depth is unity. On the other hand, local cooling occurs in the same way we found for a dilute medium. The radiation energy density is reduced with distance, but not its spectral distribution, and therefore the equilibrium temperature at the top of the atmosphere would be low. This means, however, that any radiation that is absorbed at depth will require re-radiation at an inefficient wavelength, likely within strong molecular absorption bands, and the temperature must rise to accommodate the reduced cooling efficiency. Ultimately, of course, thermal balance is re-established but at the expense of an increase in the temperature over that occurring in a purely scattering medium.

3.7 EXTENDED ENVELOPES AND OUTFLOWS

In the classical theory of stellar atmospheres, the absorbing or emitting layer is assumed to be very thin. In many interesting problems, however, this assumption is quite misleading. Any low surface gravity object, especially if it is hot, has a large scale height. Although it may not be as much as a stellar radius in size, the region may be extended enough that it cannot be treated as either planar or geometrically thin. In this case there are two effects. The first is that the transfer equation must include the full geometry, especially the curvature. The second is that the flux is no longer constant, a consequence of the geometry, and the radiation field becomes progressively more dilute with increasing distance. This means that the energy density drops faster than would be observed in a "normal" atmosphere and the temperature will differ markedly from the classical treatment.

3.7.1 The Spherical Transfer Equation Yet Again

For an extended atmosphere, the plane parallel version of the transfer equation does not provide enough information and we must go back to the full equation in spherical coordinates:

$$\mathbf{n} \cdot \nabla I_\nu = \mu \frac{\partial I_\nu}{\partial r} - \frac{1 - \mu^2}{r} \frac{\partial I_\nu}{\partial \mu} = -\kappa_\nu \rho I_\nu + j_\nu. \quad (3.156)$$

Our one simplification is to take an isotropic source function. In this case, the first moment of Eq. (3.156) gives (see Eq. 3.38)

$$\nabla \cdot \mathbf{F} = \frac{1}{r^2} \frac{\partial}{\partial r} r^2 F = \frac{dF}{dr} + \frac{2}{r} F = 0, \quad (3.157)$$

so that $F = F_0(R_\star/r)^2$. It means that the condition for radiative equilibrium now provides the *dilution factor* for the radiation field in a natural way because of the radial dependence of the incident flux. The absorber sees a lower local energy density than in the plane parallel case, and therefore the local reemission can occur at lower temperature. Because of this radial dependence of the flux, the extended case differs in a very important way from the planar problem: *The temperature of an extended atmosphere approaches zero as the distance increases toward infinity.* For a plane parallel problem, however, you will recall that the temperature goes to a constant value. This is because F remains constant in the thin atmosphere case, and this is a fundamental limitation of the approach.

For an extended envelope, radiative equilibrium means that whatever the effective temperature may be for the illuminating body, the local temperature will always be only smaller. For instance, take the case of a dusty shell illuminated by a hot star but located some distance, r, from its surface. The incident radiation, $F_\nu = W(r) F_{\nu,\star}$, is absorbed by a dust grain of radius a with an absorption coefficient $\kappa_\nu = \pi a^2 Q_\nu^{\mathrm{abs}}$. We chose the example of a dust grain because there are a sufficient number of modes to ensure that the emission is in thermal equilibrium and that the emissivity is given by $B_\nu(T_g)$, which is the Planck function at the grain temperature T_g. Invoking radiative equilibrium gives

$$\int_0^\infty \pi a^2 Q_\lambda^{\mathrm{abs}} W(r) \pi F_{\star \lambda} d\lambda = \int_0^\infty \pi a^2 Q_\lambda^{\mathrm{em}} W(r) \pi B_\lambda(T_g) d\lambda. \quad (3.158)$$

For gray grains, this means that $T_g \sim L_\star^{1/4} r^{-1/2}$ from the integrated Planck function. For Mie-type grains,[51] where $Q_\lambda^{\mathrm{em}} \sim (a/\lambda)^2$, the emission has a stronger dependence on temperature, $B(T_g) \sim T_g^6$, so the grain temperature varies more slowly with distance from the source, $T_g \sim L_\star^{1/6} r^{-1/3}$. Thus, since $L_\star \sim T_{\mathrm{eff}}^4$, the grain temperature will always be a small fraction of the effective temperature even though it is in thermal balance.

The thermal balance of the gas is especially sensitive to the match between the emission and absorption coefficients. For instance, the predominant heating of dust grains near hot stars is by absorption of the ultraviolet Balmer continuum. On the other hand, the grains reradiate in the infrared. Astrophysical graphite, so named because it doesn't really resemble laboratory carbon-rich compounds but has somewhat the same composition, has a high absorption efficiency in the UV, especially in a broad absorption band at about 0.2μ, and a structureless continuum in the IR. In contrast, the infrared emissivity for silicates is dominated by several strong bands, the most prominent of which is at about 10μ. The UV absorption for silicates is nearly structureless and less efficient than graphite, so radiative equilibrium means graphite grains will be hotter for the same irradiation than silicates, all other things being equal. The integrated IR emission thus depends critically on the composition of the grains. You would have thought that the traditional definition of blackbody emission would preclude this, but the problem is that the bolometric reemission is seldom observed and is rarely at a single temperature or distance from the radiative source.

Scattering by lines produces emission lines in extended atmospheres. This merely reflects the solid angle of the absorbing source *vs.* the extended portion of the atmosphere.[52] The ratio of absorption to emission depends on the ratio of the solid angle available for the scattering compared with that subtended by the stellar photosphere. Remember, you see absorption *only in front of the photosphere* if the

---

[51] See van der Hulst, H. 1957, *Scattering of Light by Small Particles* (NY: Dover Books), for a very clear discussion of basic scattering theory for dielectric and metallic bodies. See also our discussion of grain optics in Chapter 6.
[52] This was the original suggestion by Schuster (1908) for the emission lines in the Wolf–Rayet and related stars!

medium is otherwise optically thin. As the column density of the absorbers is increased, it is possible for self-absorption to occur, in which case the minimum source function can never fall below the local continuum.

3.7.2 Low-Frequency Observations of Extended Envelopes and Winds

One of the best examples of the effect of extension on an atmosphere is provided by the analysis of the radio emission from ionized regions. One can think of this as a problem of either a static or dynamical environment, although the latter is usually invoked. In the general case, you begin with a spherical system having some density gradient. If the medium is an emitting stellar wind, this is rather easy because the continuity of mass enforces a relation for steady outflow between density, velocity, and area. If the velocity is zero, the density must be known from some other specification.

The key point is that the photospheric radius is a function of frequency so that the optical depth at which a given frequency becomes unity samples a specific depth in the atmosphere. To be more precise, the frequency at which the spectral slope changes from being optically thin to optically thick is a measure of the total absorbing column toward the "photosphere." The solid angle subtended by the optically thick surface increases with increasing wavelength and partially offsets the drop in the photon escape rate. For bremsstrahlung emission, the continuum at radio frequencies varies as $j_\nu \sim \kappa_\nu T_e \nu^2$. As we have discussed, the absorption coefficient is given by

$$\kappa_\nu = 3.69 \times 10^8 n_e^2 g_{ff} T_e^{-1/2} \nu^3 (1 - e^{-h\nu/kT_e}).$$

This becomes

$$\kappa_\nu \approx 8.44 \times 10^{-28} \left(\frac{\nu}{10\,\text{GHz}}\right)^{-2.1} \left(\frac{T_e}{10^4\,\text{K}}\right)^{-1.35} n_e^2 \;\text{cm}^2. \quad (3.159)$$

at radio frequencies, where n_e is the electron density, T_e is the electron temperature (which is assumed to be the same as T), and g_{ff} is the Gaunt factor for free–free (thermal bremsstrahlung) emission. Now assume that the density varies as $n_e \sim r^{-n}$. For instance, note that the result for a wind that has a constant mass loss rate and is expanding at terminal velocity has $n_e \approx \dot{M}/(v_\infty r^2)$). The optical depth varies as $\tau \sim \tau_0 r^{-(2n+1)} \nu^{-2}$.

The frequency at which the emission departs from optically thin bremsstrahlung provides a measure of the optical depth of the source. In effect, there is a direct correspondence between this frequency and the radius within the envelope or wind. The optical depth is the line integral of the opacity *and* the density so that

$$\tau_\nu = \int \kappa_\nu n \, dr. \quad (3.160)$$

Then the radius at which the source becomes optically thick at each frequency is obtained when the following condition is satisfied:

$$\nu^{-2} r_\star^{-2n+1} = \text{constant}. \quad (3.161)$$

Here the constant depends on the density law. We will only present here the scaling argument, the details are left as an exercise for you. Since the envelope is just marginally opaque at this critical radius, we can solve the equation of transfer in a straightforward way:

$$F_\nu = j_\nu(T_e) D^{-2} \int_0^\infty (1 - e^{-\tau}) 2\pi r \, dr \approx C_{ff} \nu^{-0.1} \int \tau_\nu(r) r^2 \, dr, \quad (3.162)$$

where C_{ff} is a numerical coefficient and we assume that the optically thick core is very small in comparison with the envelope. Thus, using the approximate form for the bremsstrahlung opacity, we get the following approximate relation:

$$F_\nu \sim \nu^{2(2n-3)/(2n-1) - 0.1}. \quad (3.163)$$

For a blackbody at long wavelength, this flux should vary as $F_\nu \sim \nu^2$. But the flux from this tenuous extended atmosphere actually varies as $F_\nu \sim \nu^{0.6}$, for a constant velocity outflow or an inverse square density law, much less steeply than we would have expected for blackbody emission. It is just the effect that the emissivity is not very sensitive to frequency for a hot plasma while the opacity depends strongly on both density and temperature. The increase in the relative surface area at low frequency as one sees progressively larger parts of the envelope offsets the decrease in the emissivity as a function of frequency.

Why isn't this seen in an ordinary stellar atmosphere? Actually, it is for supergiants in both the infrared and radio spectral regimes. The scale height for a more compact star is so small, and the density changes so quickly, that the effect plays no role. The radius of the atmosphere at low frequency is always extremely close to that of the shorter wavelengths. Only when the atmosphere becomes really distended does this effect predominate. Another point is that we have assumed here that the medium is isothermal. While it may be a good approximation for envelopes, and for H II regions as we will discuss in Chapter 6, it is not at all appropriate for normal stellar atmospheres. Small changes in the opacity permit you to view deep regions where the temperature, hence the emissivity, can become quite high. This increases the flux and again steepens the spectrum. The isothermal assumption is normal for interpretations of optically thin plasmas, but it is one that you should apply with care. The temperature structure of a radiating medium can only be known through a solution of the radiative equilibrium condition coupled to some hydrodynamic assumptions. Giants and supergiants have both extended atmospheres and often possess strong outflows, yet they may be cool enough that the ionization is depth-dependent and the temperature profile far from isothermal. Again, care must be applied when attempting to extract mass loss rates from the observed fluxes since, although the dynamical requirements of our formalism may be satisfied, the radiating component may not vary as r^{-2}. Instead, a variable ionization introduces a different radial dependence, one that may be either steeper or even not monotonic with distance, and this dependence can be obtained only through model atmosphere analysis. Merely asserting that the dependence is some alternative power law begs the question of origin, although it is an interesting exercise to compute the change in the spectrum for such a choice and we leave that for the reader.

3.7.3 Stellar Winds

A wind is a nonstatic atmosphere in the same way that an extended atmosphere is a static wind. If it is thermally stable, it still respects radiative equilibrium, but it no longer has to satisfy the hydrostatic equation. Current models for driving mass outflows separate into roughly three types: those that depend on the stellar luminosity through radiation pressure, those that heat the outer atmosphere and drive a sort of evaporation, and those requiring some kind of mechanical input from pressure, turbulence, or magnetic fields. We will briefly discuss each of these here.[53]

3.7.3.1 Radiation Pressure as a Driving Mechanism

The photon transports momentum as well as energy. You know this because of the Compton effect, where a scattering process changes the energy of the photon depending on its angle. So how does this kick the scattering or absorbing atom? Let's be heuristic for a moment. The momentum per photon is $p = h\nu/c$. They arrive per unit area at a rate $\pi F_\nu/h$. The acceleration produced by integrating over all photon frequencies and over the absorption (scattering) cross section is

$$g_{\rm rad} = \frac{\pi}{c} \int_0^\infty \kappa_\nu F_\nu \, d\nu$$

The radiative flux is therefore actually also a momentum flux. Thus, because of its intrinsic anisotropy, the flow of energy from the interior to space creates a radiation pressure gradient. The higher the flux, the steeper the gradient. However, absorption or scattering of the radiation is required to transfer some portion of its momentum to the surrounding medium. Thus, the efficiency of coupling between the gas and photons depends on the match between the monochromatic flux and the opacity. The radiation pressure gradient is found by simply redefining $g_{\rm rad}$ through

$$\frac{1}{\rho} \frac{dP_{\rm rad}}{dr} = \frac{\pi}{c} \int_0^\infty \kappa_\nu F_\nu \, d\nu. \qquad (3.164)$$

We expect that a strong outward acceleration is produced in the atmospheres of high-luminosity stars when the stellar opacity is high, either through lines or continuum scattering. The limit for an optically thin medium is when the acceleration produced by a succession of single scatterings is sufficient to levitate the atmosphere against gravity. This is known as the *Eddington limit*[54] when $\pi F \kappa_e/c = g$; since $F = L/4\pi r^2$ and $g = GM/r^2$, the critical luminosity is independent of

---

[53] A superb guide to the general theory of outflows from stars is provided by Lamers, H. J. G. L. M. and Cassinelli, J. P. 1999, *Theory of Stellar Winds* (Cambridge, UK: Cambridge Univ. Press). See also Kahn, F. 1974, *Astr.Ap.*, **37**, 149.

[54] Eddington was actually looking at the problem of the stability of a hot luminous star with an optically thick interior in which the ratio of radiation to gas pressure approached unity. In the *Internal Constitution of the Stars* (1927), he derived an upper stable mass for a polytrope of index $n = 3$. We will return to this in the next chapter.

distance because both of the accelerations vary as r^{-2}:

$$L_{\text{Edd}} = \frac{4\pi GMc}{\kappa_e}. \qquad (3.165)$$

The effect of radiation pressure is to destabilize the atmosphere. If both the luminosity and opacity are high enough, nothing prevents the material from leaving the star and being accelerated to velocities well in excess of escape. A substantial portion of the momentum efflux of the radiation is thus transformed into bulk fluid motion.

This particular critical luminosity has been rediscovered many times. The use of electron scattering is to provide an *ultimate* limit, since this is the smallest opacity you would expect for an atmosphere. There are much larger possible values depending on the scatterer. For instance, Kahn (1974) found that the limiting luminosity for dust precludes the direct accretion of more than about 60 $M_\odot$ of gas to form the central star. This is based on the assumption that the agent large and has a high scattering cross section. This is also a point worth emphasizing repeatedly throughout the rest of this discussion. Accretion and outflow are essentially the same problem with the boundary conditions reversed. Accretion, if it is to effectively settle on the star as we will discuss in Chapter 5, must traverse the sonic point on the way in while, as we will see, a wind does it on the way out as it accelerates to the escape velocity. It is thus expected that the same parameters should dominate the two cases.

The hydrostatic atmosphere is altered as follows. Assume that the initial state of the atmosphere is a plane parallel, thin layer with surface gravity g. Then the steady state momentum equation is then

$$\rho v \frac{dv}{dz} = -\frac{dP}{dz} + \rho a, \qquad (3.166)$$

where v is the velocity and a is any (generalized) acceleration (negative in the same sense as the surface gravity). For the moment, we will treat the isothermal case, taking $P = \rho c_s^2$, where c_s is the sound speed, but this will be generalized. For a steady-state planar flow, the continuity equation yields a constant mass flux, $J = \rho v$. Writing $P = Jc_s^2/v$ and substituting into Eq. (3.166), we obtain:

$$\frac{1}{v}(v^2 - c_s^2)\frac{dv}{dz} = a, \qquad (3.167)$$

with the condition that when $v \to 0$, we are at the stellar surface. Let's concentrate on how to evaluate the acceleration, a. If it is the surface gravity, g, then for a planar medium it is constant everywhere and $v \leq c_s$. We cannot maintain a positive velocity gradient everywhere because it would eventually have to cross $v = c_s$. But since g is constant, this is not possible. As you might expect, the layer establishes hydrostatic equilibrium and diffuse flows only are possible. With a temperature gradient, this can be changed, and if the sound speed is large enough, it is possible to establish an evaporative wind, but we postpone that discussion for a moment.

Since we have included radiation pressure, we really have $P = P_g + P_{rad}$ and the radiation pressure gradient is actually the same as an outward acceleration. This is the driving mechanism that makes a stellar wind possible. If a differs only slightly from g, the radiative acceleration can still have a small effect on the atmosphere but won't necessarily produce a net outflow. This is the basis for selective elemental diffusion theory. Imagine that the radiative acceleration is small and that the atom moves only a small distance before it collides with a background particle. These motions produce a random walk of the particle that mimics diffusion with a drift velocity $v_{diff} = g_{rad}\tau_c$, where τ_c is the collision timescale. Radiation pressure acts against both gravity and viscous coupling with the (hydrostatic) background to produce an upward-directed Brownian drift. The general theory of radiative diffusion, first discussed by Milne (1927)[55] has been greatly extended. When radiation pressure is sufficiently great, however, g must be replaced by $a = g_{eff} \equiv -g + g_{rad}$. Then the equilibrium condition now changes completely. Even for a plane parallel atmosphere, the radiative acceleration is depth dependent because of the opacity and the change in the spectral energy distribution. From Eq. (3.164) you see that the greatest driving force is transferred to the gas when the flux and opacity maxima closely correspond and there may be some depth in the atmosphere where $g = g_{rad}$. At such a point, by Eq. (3.166), $g_{eff} = 0$. By Eq. (3.167), at that depth the flow becomes sonic, $v = c_s$. Beyond that, since the radiative acceleration can continue to grow outward, the velocity gradient continues to be positive and the flow accelerates. In other words, once you reach the Eddington luminosity, nothing can stop the flow from reaching *at least* the escape velocity. Treating the atmosphere in a spherical geometry does not fundamentally alter this picture, although the flux now varies as r^{-2} as goes the surface gravity. The solution for the flow is fixed at the sonic point, r_s, since for constant mass loss we know that $\dot{M} = 4\pi\rho_s c_s r_s^2$. In addition, the sonic point is found from the condition that g_{eff} vanishes, and the flow can be solved by inward and outward integration of the radiative transfer and dynamical equations from that point. The details depend, however, on how the radiative force is specified, and a number of prescriptions have been developed for calculating the radiation pressure from an ensemble of lines. The most frequently employed prescription[56] (also called the *CAK* solution) assumes that the lines are distributed throughout the spectrum with some statistical dependence on the velocity law because of their mutual shadowing. To be more precise, the optical depth for a *single* line in the Sobolev approximation scales as $\tau \sim \Delta\nu_D/(dv/dz)$, where dv/dz is the gradient of the radial velocity. Since for a weak line the radiative acceleration depends on $\tau_\nu F_\nu$, it also scales with the velocity gradient. This renders the dynamical equation highly nonlinear since now $g = g_{rad}$ depends on dv/dz. Computing the line formation in a comoving frame treatment, as in Appendix 3.B, allows for a more realistic prescription, since the driving is due to millions of individual transitions, not single lines, and especially in the vicinity of the sonic point there is strong radiative coupling between the different depths and lines.

---

[55] Milne, E. A. 1927, *MNRAS*, **87**, 697; Michaud, G. 1970, *ApJ*, **160**, 641; Richer, J. et al. 1998 *ApJ*, **492**, 833.
[56] Castor, J., Abbott, D., and Klein R. 1975, *ApJ*, **195**, 157.

A further complication is the inherent instability of such flows. Although we enforced a steady state condition, does this really apply? Imagine a flow that is slightly perturbed in either density or velocity. For the moment, for simplicity, assume a single driving line that is optically thin. Then the radiative acceleration increases for a positive density perturbation and, in the supersonic portion of the wind, the differential motion produces internal shocks. Any fluctuation from the photosphere that loads the flow with waves will see them steepen on propagation. The growth of the instability is mitigated somewhat by scattering, so the more optically thick the wind, the less likely it is to catastrophically disrupt. Nonetheless, one would expect that all radiatively driven winds should show signatures of this instability in time-variable line profiles, and this is indeed observed among the high luminosity stars for which time series have been obtained.

3.7.3.2 Radiation-Driven Elemental Diffusion in Atmospheres
We discussed the process of diffusive elemental separation in a star due to gravitational settling in Chapter 1. Now we include the effects of radiation pressure. Two scenarios are usually invoked for altering the abundance pattern in a stellar atmosphere. One is due to an external source thats dump some matter with altered abundances onto the stellar surface. This is observed in many mass transferring binary stars. The other depends on the properties of the star itself. As opposed to gravitational settling, which generally produces atmospheric depletion, overabundances can result from a "failed" stellar wind. The mechanism requires a stable medium, but turbulence only lengthens the timescale through homogenization. There is no way to stop some form of diffusive separation of elements within a star since it is a kinetic effect[57] just as the Fokker–Planck equation predicts mass segregation in clusters. Producing *overabundances* of high-mass species is more difficult, and this appears to require some form of radiative mechanism to drive *upward* diffusive motion.

If the star is slowly rotating and/or possessing a strong magnetic field, differential radiative acceleration can act on species depending on their spectral line distribution. This produces a driven diffusive elemental separation in which the atmosphere remains in hydrostatic equilibrium but each atom experiences a momentum kick from photons, Upward motion continues as long as $g_{rad} \geq 0$. As an atom moves into progressively higher in the atmosphere, its ionization increases and the radiative acceleration drops as the driving lines disappear. By the continuity equation, the turnover in the velocity gradient produces an abundance enhanced layer at relatively small optical depth, hence there is a change in the line spectrum with little change in the spectrophotometric distribution of the photosphere. The mechanism requires that the separation occur relatively upward, and overabundances of factors of 100 to 10^4 can be produced by scouring the atmosphere to density increases of about the same magnitude relative to the depth at which the layer is formed. Diffusive separation of particular isotopes is assumed to

---

[57] Models for the solar interior now require settling of helium toward the core and even planetary interiors are affected by diffusion (the best example is Saturn, where the slow sinking of helium has depleted the envelope of the planet and produced net heating).

3.7.3.3 Thermal Evaporation: The Parker Wind Solution

Thermal evaporation results whenever the sound speed becomes comparable to the escape velocity. We have already discussed a variant of this in the section on Jeans escape of an atmosphere when discussing the escape of stars from a cluster (see Chapter 1). The idea is that the high-energy tail of the Maxwell distribution function causes a small fraction of the particles to have velocities exceeding that required for escape. The estimate of the speed is rather simple. If the mean thermal speed, v_T is approximately v_{esc}, then a wind will result. An evaporative flow occurs with a mass flux proportional to

$$J \sim \int_{v_{esc}}^{\infty} f(v) v \, dv,$$

where f is the particle distribution function, so that the closer v_T comes to v_{esc}, the larger J will be. For low temperatures such as those normally encountered in an atmosphere, there is no dynamical outflow caused by thermal pressure excesses. The analysis is nearly the same for a spherical isothermal evaporative wind as for the radiative case:

$$v \frac{dv}{dr} = -\frac{1}{\rho} \frac{dP}{dr} - \frac{GM}{r^2}, \quad (3.168)$$

but now the mass loss rate and not the mass flux is constant, since there are no restrictions on the thickness of the accelerating region:

$$\dot{M} = 4\pi r^2 \rho v = \text{constant}. \quad (3.169)$$

so that

$$c_s^2 \frac{d \ln \rho}{dr} = -\frac{2}{r} - \frac{d \ln v}{dr}. \quad (3.170)$$

---

[58] The alternative for producing chemical effects is an actual change in the composition of the star through accretion of *extrinsic* material. For the λ Boo stars (see Appendix 3.D), this may result from accumulation of gas from the interstellar medium or circumstellar disks in which the heavy elements have been depleted onto dust grains (see Chapter 6). The grains are driven away through radiation pressure, in the same way that a stellar wind forms, but the metal-poor gas accretes onto the star. For the Ba stars, the abundance pattern is similar to that seen in s-process nucleosynthesis (see Chapter 4) and it is likely that these stars have been polluted by material from an evolved companion through Roche lobe overflow accretion (see Chapter 5). The strongest support for this picture comes from the Algol systems, for which CNO abundances are altered for the main sequence star by accretion of processed material from a companion (see Chapter 5).

On substitution into Eq. (3.168), we find an analog to Eq. (3.167)

$$\frac{1}{v}(v^2 - c_s^2)\frac{dv}{dr} = -\frac{2c_s^2}{r} - \frac{GM}{r^2} \qquad (3.171)$$

for the equation of radial motion. The critical condition that the two sides vanish simultaneously yields a characteristic radius for the sonic point:

$$r_s = \frac{GM}{2c_s^2}, \qquad (3.172)$$

which, because $r_s \geq R_\odot$, sets a limit of order $T > 10^6$ K for the temperature of the gas at the sonic point. The essential feature of an evaporative wind is that the high temperature of the gas requires an internal kinetic energy that is progressively closer to the gravitational potential. The flow is constant along streamlines:

$$\frac{1}{2}v^2 - c_s^2 \ln v - \frac{2c_s^2}{r} - \frac{GM}{r} = \text{constant}. \qquad (3.173)$$

This solution is called the *solar wind*.[59] It inevitably has a low mass loss rate because it originates at small optical depth and low density, unlike radiation driven flows that may arise quite deep in a stellar atmosphere.

A magnetized wind has a second critical point. This is because the Alfven speed, $v_A = B/(4\pi\rho)^{1/2}$, is for most astrophysically interesting cases quite large compared with the sound speed. As the density drops, the magnetic field more effectively controls the dynamics because its energy density drops more slowly, so in order for the wind to truly escape the star, its speed must exceed not only c_s but also v_A. The field has an additional effect on the wind if it is attached to a rotating underlying star. By the strong coupling when $v < v_A$, the field constrains the wind to (more or less) rigidly corotate with the surface. This means that the star loses angular momentum to the wind and ultimately spins down at the expense of the mass loss. This happens because the effective moment arm is r_A, the Alfven radius at which $v = v_A$, is generally larger than r_s. Even for comparatively weak fields like those present in the solar wind, the field plays the dominant role in torquing down the solar rotational velocity.

3.7.3.4 Mechanical Driving Mechanisms

Alternative mechanisms have been suggested for driving the winds observed in stars, mainly cooler giants and supergiants. These generally involve some kind of wave—either pressure waves like shocks generated by photospheric pulsation or waves generated by a coupling of fluid motions with a magnetic field. In those cases that use a continuum of *Alfven waves*, a pressure results from the turbulence that produces an outward acceleration through a pressure gradient. Depending on

---

[59] See, for instance, the classic discussions by Parker, E. N. 1963, *Interplanetary Dynamical Processes* (NY: Interscience) and Brandt, J. 1970, *Introduction to the Solar Wind* (San Francisco: Freeman). See also Cassinelli, J. and Lamers, H. (n.53).

where the strongest coupling between the atmospheric gas and the waves is achieved, the effective critical point can be quite far out in the wind, and the velocity can be quite slow. On the other hand, prompt deposition of the waves, which in some models results in heating the plasma as well, can produce very large mass loss rates and fast terminal velocities. Otherwise, the basic equations for an isothermal case are the same as for an evaporative wind:

$$\frac{1}{v}(v^2 - c_s^2)\frac{dv}{dr} = -\frac{2c_s^2}{r} - \frac{4\pi r^2 v}{\dot{M}}\frac{dP_t}{dr} - \frac{GM}{r^2}, \qquad (3.174)$$

so the critical point is moved outward by an amount that depends on the turbulent/wave pressure gradient. Notice that the driving enters like radiation pressure; the gradient may be large and extend far from the photosphere. As a result, gas is continually accelerated within a gravitational field whose strength decreases so eventually it may be possible to push the material over the edge and drive its escape. The principal effect of turbulence, or any other distributed acceleration, is that it can act quite far from the star where radiation would be ineffective because of the low densities.

3.7.3.5 Chromospheres and Coronae

There is only one problem with this picture and it is best illustrated by the Sun. The solar wind is an observed outflow at high velocity from the Sun. It reaches speeds of order v_{esc} and has a very low associated mass loss rate, of order $10^{-14} M_\odot$ yr^{-1}. However, it cannot arise in the photosphere because the temperature there is simply insufficient to drive evaporation. Obviously, from our estimates above, radiation pressure also fails to provide sufficient driving for a wind. In order to get any mass loss, the gas must somehow reach rather high temperatures, certainly well in excess of 10^4 K. Fortunately, this happens. The outer solar atmosphere, the chromosphere, is an emission line forming region above the temperature minimum (where the local gas temperature approximately equals the boundary temperature). Seen against the photosphere, this region shows itself by the production of emission line cores in many of the strongest lines, for instance, Ca II H and K. The core, in this case, reflects an increase in the source function due to collisional populations of the upper states of the transition relative to the photosphere. In other words, the line core isn't formed only by scattering (this is clear when you look at a pencil beam coming from the solar surface, so scattering isn't plausible). It must reflect a *real* rise in the temperature, at least in this case.

This clears up two points. The first is that the solar wind requires a high temperature, and this is achieved only in the lower-density outer stellar atmosphere. The second is that the mass loss rate in the solar wind is very low. This must mean that the density at the critical radius is low since the mass loss is fixed by the sound speed at that point. On the other hand, we are confronted with a new question: Why is the outer atmosphere hotter than the photosphere? The answer *appears* to lie in the direction of stellar activity and surface convection. The original presumption was that acoustic (sound) waves generated by the turbulence of the upper convection zone steepen into shocks as they propagate into the upper atmosphere.⁶⁰ These are strongly damped by viscosity and radiation and ultimately don't get very far, although that dissipation may be important in the chromospheric temperature rise. Another candidate that produces many of the same effects with longer propagation lengths is Alfven wave heating. These have both compressible and incompressible modes, some of which damp quickly and others of which can reach considerable heights in the chromospheric layer. These may be important in driving winds in late-type high-luminosity stars, where the connection with convection comes from the stirring required for the generation of the waves in the first place. The most recent studies of the Sun with the *Solar Heliospheric Observatory* (SOHO) and the *Transition Region and Coronal Explorer* (TRACE) show that the most probable mechanism for heating the outer layers of the Sun and cool stars is that in situ magnetic field annihilation in the complex fields of the coronal regions. This reconnection process releases the energy needed to create the extreme temperatures observed in the outermost part of late-type stellar atmospheres, the corona. This region in the Sun reaches about 2×10^6 K. In some more evolved stars, the observed X-ray emission requires a factor of 10 higher temperature. This is an electron temperature, *not* a brightness temperature, and therefore is not simply indicating some nonthermal or nonequilibrium radiative process. The X-ray emission comes from very hot gas. In the lowest-density environments, it is possible that the electron distribution function develops a non-Maxwellian high-energy tail and the usual concept of temperature does not apply, but emission lines indicate that at least some thermalization is taking place. Magnetic fields are a characteristic feature of convective envelopes, as we will outline in the next chapter. If this is ubiquitous, turbulence driven magnetic reconnection is an effective way to accelerate ions and electrons that, through collisions, heat the gas.⁶¹

3.7.4 Escape Probabilities and Radiative Transfer in Flows

A velocity gradient alters the shape of a spectral line seen by a distant observer, but it also affects the line formation by reducing its central optical depth. This doesn't happen in the continuum because of the slight frequency dependence of the opacity over the available velocity interval. But for a line, whose width is usually due just to thermal Doppler or radiative and thermal broadening, a shift of several linewidths is more than enough to substantially reduce the opacity. The simplest treatment for this effect is known under several names depending on which astrophysicist is describing it. The one most frequently used in stellar problems is the *Sobolev method*.⁶² The picture assumes that along a line of sight, a velocity shift results because of the differential motion of the absorbing layers. The absorbers each have a characteristic width, say, $\Delta \lambda_D$, and the velocity gradient is given by $v' = dv_{\rm rad}/ds$ along some line of sight s. Along this direction, a distance

---

⁶⁰See also references in Narin, U. and Ulmschneider, P. 1991, *Space Sci. Rev.*, **54**, 377; 1996, *Space Sci. Rev.*, **75**, 453.

⁶¹Parker, E. N. 1995, *Current Sheets* (NY: Oxford Univ. Press); Parker, E. N. 1979, *Cosmical Magnetic Fields* (NY: Oxford Univ. Press). An interesting discussion of the possible role of superthermal electrons and ions for the energy balance of tenuous stellar plasmas is Scudder, J. D. 1994, *ApJ*, **427**, 446.

⁶²See Sobolev, V. V. 1960, *Moving Envelopes of Stars* (Cambridge, MA: Harvard Univ. Press) provides the best summary of the otherwise difficult to find early literature on this problem.

ds corresponds to a shift in the line's central wavelength of $\Delta\lambda/\lambda_0 = v'ds/c$. Thus the optical depth is reduced by a factor of

$$\tau(\Delta\lambda) \approx \tau_0 \frac{c\Delta\lambda_D}{\lambda_0 v'} \exp - \left(\frac{\Delta\lambda + \lambda_0 v(s)/c}{\Delta\lambda_D}\right)^2, \qquad (3.175)$$

assuming that the intrinsic profile is just due to Doppler broadening. The simplest version of the transfer equation then takes the escape probability $\exp -\tau$ for each depth and shifts the profiles for the overlying depth along the line of sight to compute the emergent profile. Where the velocity gradient is large, the line in the superior layer is shifted out of the core of the lower level and also the mean free path for photon absorption is changed by the velocity gradient. This has the effect of increasing the opacity in the wings of the line asymmetrically toward the blue and decreasing it within the core.

Desaturation has two effects. The first is that one can see deeper into the outflow near the rest wavelength so in general the rest wavelength is nearly transparent to line radiation. This, of course, depends on the acceleration of the flow. The second is that at high velocity, if the wind gradually reaches terminal

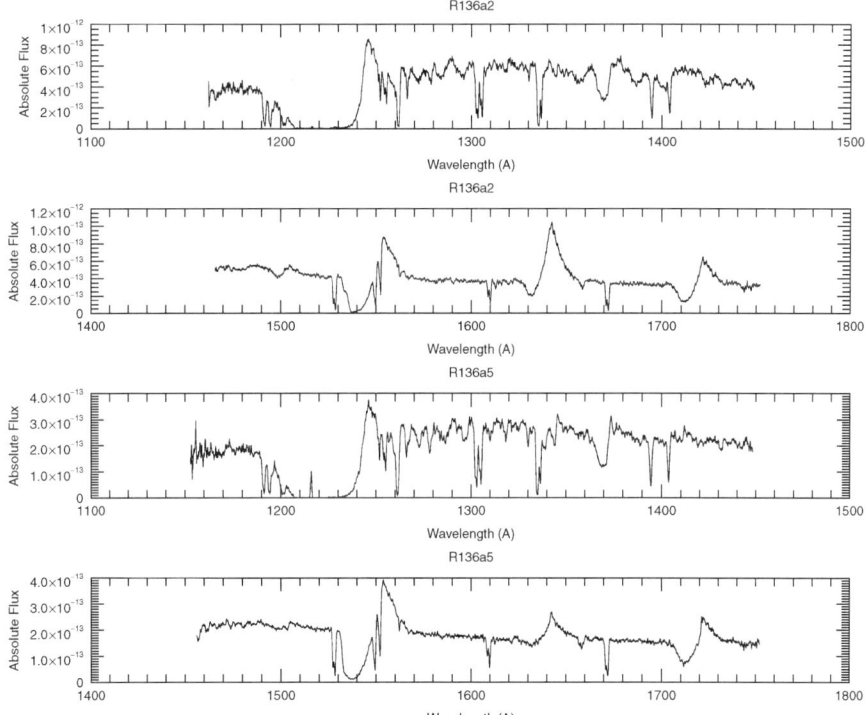

FIGURE 3.9. GHRS spectra of four stars in the core of R136, a young massive cluster in the LMC (see Chapter 7). Notice the strong P Cygni absorption troughs for the resonance lines (Si IV 1400, C IV 1550) and the weaker P Cygni profile on He II 1640 and N IV] 1718.

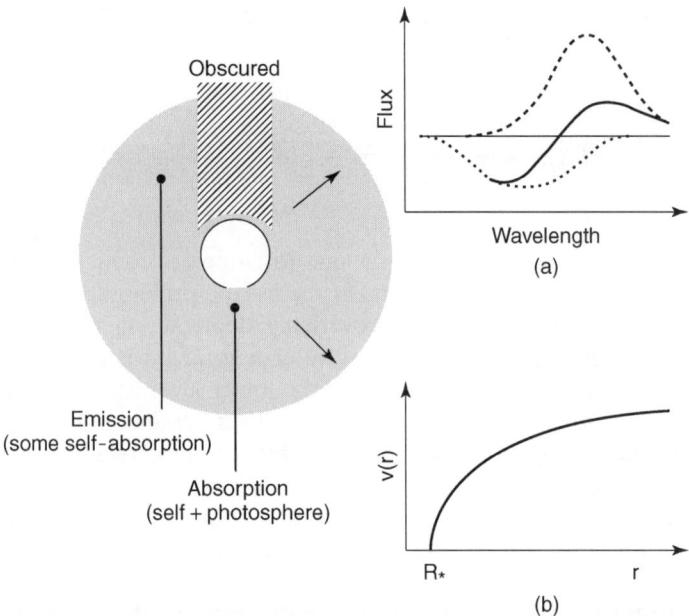

FIGURE 3.10. Schematic of how to form a P Cygni profile in an expanding atmosphere. The inserts show (a) the differential to the line profile and (b) the schematic velocity as a function of radius.

velocity, the decrease of the velocity gradient substantially increases the atmospheric opacity and one sees only the most peripheral parts of the wind. The absorption optical depth is so large that the underlying photosphere at the terminal wavelength may be completely concealed. The result is called a *P Cygni profile*, after the luminous star in whose spectra these lines were first recognized in the nineteenth century. It has an absorption trough that extends to the terminal velocity at negative (approaching) radial velocities, and an emission component that extends up to about the same recessional velocity (Fig. 3.9). The absorption extremum is formed in the outermost part of an accelerating wind, so you might wonder why, in this very low density regime, the profile should be blackest. The answer comes from the velocity gradient. The periphery of the flow has the smallest gradient. If the line is intrinsically strong enough, this is precisely the portion of the wind that will have the largest *effective* optical depth since the lines will not be desaturated by kinematic displacement with respect to the underlying layers.

The absorption profile arises only in that part of the wind seen in projection against the stellar photosphere and for those portions of the extended enveloping gas that are opaque (Fig. 3.10). This is a source function effect and is not related to the local temperature, which may actually be quite high. The photosphere is simply brighter. However, as you saw for an extended envelope, the emission does not mean that the volume is filled with hot gas. It means only that at the velocity of the line wing, there is no effective photosphere—the solid angle subtended by the emitting gas is much larger than the stellar surface. Thus scattering from a moving extended environment has the same effect in producing emission lines, except that the lines will be shifted because of the motion of the scatterers with respect to the observer.

The transfer depends on knowing the velocity as a function of radius, and this comes from some solution of the hydrodynamic equations. The problem is that for radiatively driven outflows, the velocity depends on knowing the solution of the transfer equation because the main source of radiation pressure is the scattering and absorption by spectral lines. The line opacity of the atmosphere at the photosphere thus sets the initial acceleration, which feeds back through the velocity gradient into the optical depth. The resulting flow can, in principle, be arrived at self-consistently, but to date this has not been achieved. Notice that the problem is actually quite similar to the static atmosphere case. The difference is that the hydrostatic equation has been replaced by a hydrodynamic one, so it is more complicated. You have to calculate the effective acceleration of the atmosphere in order to know what the lines themselves will be doing to drive the flow. The usual approximation is called a β *law*:

$$v(r) = v_0 + v_\infty \left(1 - \frac{R_\star}{r}\right)^\beta, \qquad (3.176)$$

where v_0 is usually the sound speed or a turbulent velocity, but may be larger depending on whether the photosphere is truly a static surface, and v_∞ is some terminal velocity. This velocity, and β, are free parameters of the model and are determined by fitting model profiles to the observed ones. There really is no firm dynamical justification for this choice, but it is a convenient approximation (yet should be used with caution!). Observations of luminous stellar winds show that they are also variable. These variations take many forms. There are the overall changes in the gross structure of the line profile. In the Be stars, and also many more luminous stars like the *luminous blue variables* (LBVs), the emission or absorption components can disappear entirely, indicating that the mass loss rate may be time-dependent. This is hard to understand with radiative mechanisms since the stars do not always show sufficiently large photometric changes to explain the change in the wind structure. Smaller scale changes also occur. These are variously labeled *discrete absorption components* (DACs) or *narrow absorption components* (NACs) or changes in the blue edge of the wind features. Their explanation is qualitatively connected with the inherent instability of line driving, that as the velocity gradient increases, the desaturation of otherwise strong lines, which perform the bulk of the work in driving the flow, also increases. This increases the radiation pressure throughout the flow, simply because more light gets out, and the result is that the lower-density regions actually move faster than the denser, more opaque background. The flows can quickly become supersonic in the wind frame, resulting in shocks. There is also some evidence that surface (photospheric) oscillations can produce pressure waves that travel outward in the wind, resulting in steep gradients in the velocity and ultimately shocks as they propagate. The details are, however, still to be decided among these schemes.[63]

---

[63] See Owocki, S., and Puls, J. 1999, *ApJ*, **510**, 355; Kudritzki, R. P. and Puls, J. 2000, *ARAA*, **38**, 613, and references therein for discussion of the theory. Several massive ground-based and UV monitoring campaigns were mounted to study this variability in hot star winds; see, for instance, Massa, D. 1995, *ApJL*, **452**, L53; Kaper, L. et al. 1997, *Astr. Ap.*, **327**, 218.

A linear velocity gradient produces a different profile from one that reaches a terminal velocity. The reason is connected with the velocity gradient—a linear law does not have a characteristic deceleration length on which scale the optical depth increases, while a driven wind does. Explosions, like those observed in novae, produce linear laws by the initial distribution of particle velocities. Those that initially move the fastest get out farthest in a fixed time. It is not a steady-state flow and has no characteristic timescale. It is, therefore, self-similar. On the other hand, the accelerated flow does have a length on which it decelerates, so the opacity reaches a maximum at some characteristic position in the flow. We will return to this in Section 3.7.4.2, when we discuss spectrum formation in a time-dependent flow.

3.7.4.1 The Effect of a Velocity Gradient on the Escape of Photons: Kinematics and Line Profiles

If the layer has a velocity gradient, then at each wavelength the line core is encountered at a different frequency. Recall that the optical depth is an integral over column density so the velocity gradient, in effect, reduces the column density. It also makes the pathlength a function of frequency. If you assume that the velocity gradient is a constant, just for a moment, then you can replace dz with

$$dz \to \frac{dv}{dv/dz} = \frac{v_0}{c} \frac{d\nu}{dv/dz}.$$

Then because of the velocity shift, the line profile $\phi(\nu)$ becomes $\phi(\nu_0 - \Delta \nu)$. Define a new variable:

$$dy = \phi(\nu) \, d\nu.$$

Then, if the opacity at line center is τ_0, the escape probability for the profile is the mean intensity:

$$J = \frac{1}{4\pi} \int \phi(\nu) P(\nu) \, d\Omega \, d\nu. \tag{3.177}$$

With our change of variable this now reads

$$J = \tfrac{1}{2} \int_{-1}^{1} d\mu \int_{-\infty}^{\infty} dy \, e^{-\tau_0 y}. \tag{3.178}$$

so that the probability of observing a photon is

$$\beta = \int_{-1}^{1} \left(\frac{1 - e^{-\tau_0}}{\tau_0} \right) d\mu. \tag{3.179}$$

Since we have taken a plane parallel medium, the velocity has been assumed to be only along the normal direction. This means that the radiation escaping along large angles sees a reduced gradient and a very large optical depth compared with the normal, and the result is that the photons become more peaked in the forward direction. This changes the Eddington factor for the medium, but it's a result you expected.

The emergent line profile is quite sensitive to bulk motion or flows. This is also true for the continuum, although not to the same degree. The line optical depth is reduced at any wavelength by a velocity gradient. This can be understood as follows. Consider an atmosphere with an underlying continuum. If the optical depth at line center is τ, then the probability for a photon to escape from the layer in question along a direction $\hat{s}$, with a directional cosine μ is

$$P_{esc}(\tau, \mu) = e^{-\tau/\mu} \qquad (3.180)$$

and from the two layers is the product $P_{esc}(\tau, \mu)P_{esc}(\tau', \mu')$. Now the average over all possible lines of sight for a given optical depth provides the net escape probability, which we denoted as β. This probability is given by

$$\beta(\tau) = \int_{-1}^{1} e^{-\tau/\mu} d\mu = \frac{1}{\tau}(1 - e^{-\tau}). \qquad (3.181)$$

This is the estimate for the rate of escape of photons from the medium, as we just found. You can see that if the optical depth vanishes, the escape probability approaches unity. On the other hand, it falls to zero only as $1/\tau$, not too rapidly. Bulk motion in the medium, however, shifts the absorption lines from the upper layer away from the rest wavelength. This decreases the line of sight opacity at line center, but increases it in the line wing. The photon would normally have a higher chance of escaping from the line wings and, for an optically thick medium at rest, would be trapped in the line core. Now the radiation has a higher chance of escaping from the Doppler core of the line. The consequence of this is that the upper layer now absorbs against the blueward (or redward) continuum relative to line center and the emergent line, which may have the same equivalent width, has a reduced central depth. If the shift is actually larger than the intrinsic width of the line profile, then the photons that are absorbed by the upper layer will never have passed through the lower layer line core; thus the emergent line is determined by the conditions in each layer independent of what the underlying (or overlying) layers are doing.

The feature of any atmosphere that produces both absorption and emission in the line is that its source function in a given layer may be lower or higher than that in the deeper layer. If the source function is lower, photons are absorbed at the local frequency ν and either scattered or lost through collisional de-excitation of the upper level. Scattering diminishes the number of photons along the line of sight, although the rate of scattering depends mainly on the column density of the target atoms and not on their excitation conditions. Absorption process ultimately thermalizes the photons, redistributing their energy via collisions throughout the continuum, and depends on the local thermal state of the gas. Thus the absorption line depends on the column density to the continuum source, but the emission portion of the line depends on both the optical depth through the medium and its geometry. It is also important to remark that the equivalent width of a scattering line remains constant since the photons are never thermalized.

A P Cygni profile is a unique signature of an outflow, although the dynamics of the flow may not uniquely reflect in that profile. The ability of a photon to escape the core of the profile is determined by both the local density and temperature conditions (which determine the opacity at line center in the rest frame) and also the velocity gradient. While for a steady-state flow the density is related directly to the radius by the velocity field (and that is specified independently in most problems) and the mass loss rate, the absorption coefficient depends on the ionization state of the gas as well as the local mass density. It is therefore necessary when computing the line profiles from a radiative flow to determine both the solution to the equations of ionization *and* statistical equilibrium at each point in the concurrently moving frame. This can seldom be done exactly. Treatments have been attempted in recent years, called *unified stellar atmospheres*, that couple the comoving frame treatment of the radiative transfer with the calculation of line blanketed model atmospheres. But the treatment has limited successes (albeit impressive ones) and is tremendously computer-intensive (at least in this first decade of the twenty-first century). Instead, let's look at a simplified version of the problem.

We return to the case we have already outlined above of a radially expanding atmosphere whose outflow velocity depends only on the radius and whose velocity law is known in advance. We assume that the opacity at line center is fixed by specifying some ionization parameter as a function of radius for the ith species, $X_i(r)$, and the dilution factor for the radiation field $W(r)$. The optical depth at line center is

$$\Delta \tau_{0,i} = \kappa_{0,i} X_i(r) \rho(r) \Delta l, \qquad (3.182)$$

where Δl is the pathlength. The pathlength is determined by the diffusion of photons out of the line core. For a static atmosphere, the photon random walks its way out toward the wings of the line whence it escapes from the medium. In a moving medium with an internal velocity gradient, the scale for this diffusion is set by velocity shear. If the medium has a large enough gradient, ∇v determines l. The natural width of the line, the thermal speed v_{th}, sets the lower bound for the velocity gradient along the photon's path:

$$\Delta \tau_{0,i} = \kappa_{0,i} X_i(r) \rho(r) \frac{\Delta v}{dv/dl}, \qquad (3.183)$$

and $\Delta v = \Delta \nu_D c / \nu_0$, where $\Delta \nu_D$ is the Doppler width of a line with rest frequency ν_0. For a large velocity gradient along a line of sight, the lower layers of the atmosphere contribute strongly to the emergent line profile and may even form emission at the local blueshift (or redshift) corresponding to that layer. In a decelerating flow with a monotonic velocity gradient the reduction in the velocity gradient causes a rapid increase in the line opacity *even though the density may have dropped so far as to render an equivalent static line profile completely optically thin.* The velocity gradient is given along each ray. The velocity law, however, is given in terms of the distance from the center of the star, so we need to transform between these in order to make any sense out of the problem. Call $z = r\mu$. Then the radial gradient transforms into one along the line of sight by

$$\frac{dv_z}{dz} = (1 - \mu^2) + \mu^2 \frac{d \ln v(r)}{d \ln r}, \qquad (3.184)$$

which can then be inserted into the escape velocity formalism to compute the line profiles. Besides being called the *Sobolev method*, this treatment of line transfer has many discipline-related names. In molecular line work, either for the interstellar medium or stellar envelopes, this is called the *large velocity gradient* (*LVG*) approximation. It is also known as the *on-the-spot* approximation, and even as the *escape probability* method. Always the basic assumption is the same, that the velocity gradient is so large that the mean free path of a photon is very large compared with the a static atmosphere. Thus, two distant regions of the atmosphere are completely decoupled from each (in Appendix 3.B we derive the concurrently moving frame equations without the restrictive assumptions we've been imposing here).

In this approximation, you compute the line formation along each line of sight through the atmosphere. Since the radial velocity of a layer relative to the observer is a function of both angle and distance from the central star, it is possible to associate each point in the line profile with one line of sight through the stellar atmosphere. To see this, consider a spherically symmetric, outwardly accelerating atmosphere. The distance along a ray at a distance R to the central star, is $z = (r^2 - R^2)^{1/2}$ for rad $= \mathbf{v} \cdot \mathbf{n}$, where $\mathbf{n}$ is the unit vector that lies along z. For a spherical outflow, v_{rad} is bilaterally symmetric, identical across the meridional plane passing through the star and changing sign in the plane of the sky. Therefore, within the line, the entire back half of the expanding atmosphere is radiatively decoupled from what happens in the front half along a line-of-sight. This will not hold in a complex turbulent flow because of shear and line-of-sight projections of the random motions. Differential motion of a dense line spectrum may cause lines rather distant from each in frequency to overlap, even if formed in very distant portions of the atmosphere that are otherwise mechanically distinct. For this reason, we normally resort to treating one line at a time, although newer Monte Carlo procedures can include very large numbers of transitions.[64]

3.7.4.2 *Time-Dependent Flows: Novae and Supernovae and Spectrum Formation*
Novae and supernovae provide wonderful illustrations of what happens in a time-dependent outflow. These are explosive events, the details of which we will defer until Chapters 4 (supernovae) and 5 (novae). For now, the most important thing to consider is that the dynamical timescale is short compared with the radiative and sound crossing times for the ejecta. A shock propagates through the stellar envelope, whether it is due to core collapse (SN II), the explosion of a white dwarf (SN Ia), or the explosive ejection of an accreted layer by a white dwarf in a binary system. All that matters is that total ejecta mass is finite and is expanding.

---

[64] See, for instance, Abbott, D. A. and Lucy, L. B. 1985, *ApJ*, **188**, 679; Lucy, L. B. 1999, *Astr. Ap.*, **344**, 282.

Now this means that the optical depth in the shell goes through several changes. First, the shell expansion occurs rapidly compared to the thermal timescale for the material, so the adiabatic expansion quickly lowers the temperature. How fast this happens depends on the expansion velocity and the total mass. As the ejecta thin out, the optically thick surface begins to retreat in the comoving frame. Since the losses are reduced for the deeper layers compared with the outer, fastest zones, the reduced opacity causes the effective temperature of the material to increase, aiding the fall of opacity in the outer layers. Compressional heating results from the passage of a shock through the ejecta, and the temperature can become quite high since the shock speed is much greater than the pre-explosion sound speed (e.g., in an accreting nova the shock speed is several hundred to several thousand km s^{-1}, which produces temperatures of order 10^6 K or higher). Thus, the initial spectrum will appear to be much cooler than the pre-explosion object, but this will change on the expansion timescale as the optical depth decreases from these two effects.

There are two fundamental consequences of the change in the spectrum. The first is flux redistribution. This is produced by the large line opacities, produced by relatively low ionization stages of heavy elements (Fe II, Cr II, Mn II, etc.), that are important contributors when the temperature falls below 10^4 K. For a star that is initially at a temperature of order 10^5 K, for example, this low value of $T_{\rm eff}$ is reached with only a modest increase in the radius, of order $\Delta R(t)/R_0 \approx 10$. For a white dwarf that is the seat of a nova explosion, for instance, this takes only a few days at most for expansion velocities $\geq 10^3$ km s^{-1}. For supernovae, this may be even faster depending on the temperature of the envelope at the start of the expansion. The peak of the flux distribution shifts to long wavelengths where the opacity is lower, precisely as we discussed for the effect of line blanketing in a static atmosphere on the temperature gradient, and the object appears to increase its brightness at these wavelengths at the expense of the shorter wavelengths. In SN 1987A, for instance, the ultraviolet flux declined by a factor of about 10^3 while the optical flux increased by a similar amount in only a few days. Once this occurs, the long wavelengths remain bright until the expansion begins to reduce the opacity, although not the temperature. Then the deeper layers come into view, ionizing the gas when the temperature of the pseudo-photosphere (the optically thick surface, which is not a static site in the ejecta) climbs to $T_{\rm eff} > 15,000$ K or so. At this point, the shorter wavelengths begin to increase at the expense of the longer and the visual and IR fluxes decline. Since the outer layers also occupy a progressively larger solid angle relative to the opaque surface, emission lines appear and, eventually, the spectrum turns to that of a fast-moving nebula as the opaque surface drops out of the bottom of the mass shell.

3.A QUANTUM-MECHANICAL INTERLUDE: TIME-DEPENDENT PERTURBATION THEORY FOR TRANSITION STRENGTHS

There are innumerable treatments at virtually all levels of quantum mechanics, even for astrophysics. Our intent is to keep the approach of this chapter as self-contained as possible. This pedestrian derivation of time-dependent theory is to jog your memory, or warn you of the bumps in the road ahead, since we assume you have seen this before. The Schrödinger equation for any time-dependent quantum mechanical system is given by

$$i\hbar \frac{\partial}{\partial t}|\psi\rangle = H|\psi\rangle, \tag{3.185}$$

where $|\psi\rangle$ is the wavefunction of the Hamiltonian H. Now suppose that the energy of the system comes from two contributors, one of which is the static atomic or molecular system, represented by H_0, and the other, which depends on time and comes from some external perturber. We will assume that this enters purely as an external, time dependent potential $V(t)$ so that $H = H_0 + V(t)$ is the total Hamiltonian for the system. If H_0 represents a bound system, like an atom, then the states are quantized with some set of discrete quantum numbers that label them and some fixed energy. These are the eigenstates of the time-independent problem. Let us further assume that any state of the full system, H, can be separated into time-dependent and time-independent parts and that it is possible to expand the state representation in terms of the eigenstates states of H_0 via $H_0|n\rangle = E_n|n\rangle$. Although it is concealing an enormous amount of work to say it, the wavefunctions, $|n\rangle$ are known from structure calculations. The expansion treats the potential V as a small change, so that the gross structure of the system H_0 is left largely unchanged, but the populations of the states, or rather the probability of being in one of them, varies during the interaction. In other words, the presence of a perturber changes the relative energies slightly and so also mixes populations, but it does not fundamentally alter the structure of the atom. The states are modified by amplitudes that depend on time. Thus each state has an amplitude and phase as well as a spatial distribution:

$$|\psi\rangle = \sum_n a_n(t) E^{-iE_n t/\hbar}|n\rangle \tag{3.186}$$

so that we assume that, due to some external variable perturbation, the atom may find itself "in between" states, or in other words that some mixing of the states takes place during the interaction. After it is all over, the system will be in one or another of the stationary states, so it makes a difference how quickly the interaction occurs. We know also that $H_0|m\rangle = E_m|m\rangle$ for some arbitrary eigenstate m because we set the system up that way. Thus, multiplying by the conjugate state $\langle m|$ and integrating over space, we obtain

$$i\hbar \frac{da_m}{dt} = \sum_n \langle m|V(t)|n\rangle a_n e^{-i(E_n - E_m)t/\hbar}. \tag{3.187}$$

Now assume that the initial state is $|n\rangle$ and the final state is $|m\rangle$. We define $\hbar\omega_{mn} = E_m - E_n$ (the sign isn't important because it enters into the exponential as a phase) so that for a two-state system

$$a_m(t) = \frac{1}{i\hbar}\int_0^t \langle m|V(t')|n\rangle e^{i\omega_{mn}t'}\, dt'. \tag{3.188}$$

In the limit $t \to \infty$, you see that the amplitude is the Fourier transform of the perturbation matrix and so it depends on the frequency. You can also now see how resonance happens. If the perturbation is an incoming photon, which is an electromagnetic field that oscillates at a frequency ω, then

$$a_m(t) = -\frac{1}{\hbar}\langle m|V(0)|n\rangle \frac{\exp[i(\omega_{mn} - \omega)t] - 1}{\omega_{mn} - \omega}, \qquad (3.189)$$

which peaks in the limit $\omega \to \omega_{mn}$. Because this is an undamped system, there is a strong resonance peak at $E = E_{mn}$. The probability of a state is, as always in quantum mechanics, given by

$$P_{mn} = \int |a_m(t)|^2 \rho_n(E)\, dE_m, \qquad (3.190)$$

where $\rho_m(E)$ is the density of final states (since the integral is taken over all final state energies). This integral is strongly peaked to the resonance. Now this also means that there is a strong peak within the time interval $\Delta t = 2\pi/\Delta\omega$, so that we obtain

$$W_{mn} = P_{mn}/\Delta t = \frac{2\pi}{\hbar^2}|\langle n|V|m\rangle|^2 \rho(E_n). \qquad (3.191)$$

This is the equation for the transition probability for a specific potential $V(t)$. This equation is the basis for all time-dependent theory and was first derived by Fermi (it is called the *Golden Rule* after a comment he made around 1950). Another approach that also helps in seeing what happens during a collision, especially for broadening is the *interaction* picture, in which we look at the phases and ignore the time independent part of the problem. Consider that a system starts out in a state $|0\rangle$ and evolves such that $|t\rangle = T(t,0)|0\rangle$. The integral of the time dependent Schrödinger equation is given now by

$$|t\rangle = |0\rangle + \frac{1}{i\hbar}\int_0^t V(t')|t'\rangle\, dt', \qquad (3.192)$$

which you see is now an integral equation. The time development of the state depends on the history of its interactions. This equation is iteratively (it is actually a Fredholm equation), and what we get is an integration over all possible interactions at all times prior to the present. This is also the basis of the Feynman diagram for nonrelativistic quantum mechanics.[65] For calculating collisional excitation, you can use the Born approximation. It assumes that the change in the momentum of the incident, free, electron is not substantially affected by the interaction and that the collision induces a phase shift for the incident plane wave.

---

[65] One of the clearest introductions is Feynman's own: Feynman, R. P. and Hibbs, A. 1965, *Path Integrals and Quantum Mechanics* (NY: McGraw-Hill).

This can be written as

$$\langle j|V|i\rangle \sim \int e^{i(\mathbf{k}_i-\mathbf{k}_j)\cdot\mathbf{r}} V(r) r^2 \, dr \, d\Omega, \quad (3.193)$$

where **k** is the wavenumber. The integral over angle is performed first:

$$\int V(r) e^{ikr\mu} r^2 \, dr \, d\mu = -\frac{2}{k} \int V(r) \sin kr \, r^2 \, dr, \quad (3.194)$$

which is strongly peaked around $kr \approx \pi/2$. Our key result is that the cross section becomes a function of energy, or momentum, by a Fourier transform of the potential for the interaction. This is why finite range interactions, mediated by the exchange of a massive particle, have a threshold in the cross section. One must create the particle in order for the interaction to occur.[66]

3.B RADIATIVE TRANSFER IN THE LAGRANGIAN FRAME

When treating the transfer of radiation in a wind or expanding medium, it is often more convenient to work in the *Lagrangian* framework, which is called the *comoving frame*. This is distinct from the *observer's frame*, in which the atmospheric structure is usually described. The derivation of the transfer equation in a moving medium is may seem like just an exercise in Lorentz transformations but there is a lot of physical insight along the way and it is useful to see how this works.[67] There are several important applications for this formalism. The first, and most immediately obvious, is the problem of radiative transfer in a stellar wind where the radial velocity gradient introduces changes in the apparent direction (aberration) and frequency of the photons. Another is the transfer of radiation in a cosmological framework where the change in the scale parameter introduces a redshift (see Chapter 8). In the latter case, however, the equations do not require keeping the curvature terms. Call the flow velocity $v = \beta c$ where, as usual, $\gamma = (1-\beta^2)^{-1/2}$. This is the local velocity of the comoving frame and it depends on the radial distance r. Start with the Doppler transformation for the frequency:

$$\nu' = \gamma\nu(1-\beta\mu). \quad (3.195)$$

The superscripted variables are comoving and the unmarked variables are observer's frame values. We want to write the entire transfer equation in comoving or

---

[66] The general theory for Green functions in collisions makes use of this. We recommend Messiah, A. 1962, *Quantum Mechanics vol II* (NY: Wiley; Dover Books); Massey, H. and Burhop, 1972, *Atomic and Ionic Impact Phenomena* (London: Oxford Univ. Press) (this is a multivolume encyclopedia that is always worth consulting before diving into the literature); Moisiewitsch, B. A. and Smith, S. J. 1968, *Rev. Mod. Phys.*, **40**, 1; Mott, N. and Massey, H. 1968, *The Theory of Atomic Collisions*, 3rd ed. (London: Oxford Univ. Press). Although none is very recent, they present a lot of the technical details that are useful to know in reading the more recent work on collision cross section calculations.

[67] Castor, J. I. 1972, *ApJ*, **178**, 779; Mihalas, D. 1980, *ApJ*, **237**, 574; Hauschildt, P. H. and Wehrse, R. 1991, *JQSRT*, **46**, 81; Mihalas, D. and Mihalas, B. 1984, *Foundations of Radiation Hydrodynamics* (NY: Oxford Univ. Press; Dover Books).

Lagrangian coordinates (we will treat only steady state flows). Recall that in the Lorentz transformation, because of the contraction in the longitudinal direction, an aberration angle is introduced for any source of radiation not along the direction of motion. The direction cosine for this angle is

$$(1 - \mu'^2)^{1/2} = \frac{(1 - \mu^2)^{1/2}}{\gamma(1 - \beta\mu)}. \tag{3.196}$$

From Eq. (3.156) you can already see why the spherical equation of transfer couples the frequency and angular components of the intensity. From the invariance of the photon number density [recall that we needed this to derive the Kompaneets equation; see Eq. (3.110)] the intensity scales as

$$I(\nu, \mu) = \left(\frac{\nu'}{\nu}\right)^3 I(\nu', \mu') \tag{3.197}$$

so we have

$$j = \left(\frac{\nu'}{\nu}\right)^2 j' \tag{3.198}$$

for the emissivity. From Kirchhoff's third law and dimensional agreement of I and the source function, it follows that the opacity scales as

$$\kappa = \left(\frac{\nu}{\nu'}\right)\kappa'. \tag{3.199}$$

Thus if we call the standard spherical operator $\mathscr{S} = \hat{\mathbf{n}} \cdot \nabla$, then

$$\left(\frac{\nu}{\nu'}\right)\mathscr{S}I'(\mu', \nu') - 3\left(\frac{\nu}{\nu'^2}\right)I'(\mu', \nu')\mathscr{S}\nu' = j'(\nu') - \kappa'(\nu')I'(\mu', \nu'), \tag{3.200}$$

where now $\mathscr{S}$ is evaluated in the comoving (primed) coordinate. To do this, we need the inverse transforms from the observer's frame to the comoving system

$$\mu = \frac{\mu' - \beta}{1 + \beta\mu} \tag{3.201}$$

$$(1 - \mu^2)^{1/2} = \frac{(1 - \mu'^2)^{1/2}}{\gamma(1 + \beta\mu')} \tag{3.202}$$

$$\nu = \gamma\nu'(1 + \beta\mu'), \tag{3.203}$$

which will be substituted into $\mathscr{S}$ after the transformations. Notice that the term $\mathscr{S}\nu'$ comes into the equations in the same way that an absorption coefficient enters, as a multiplier of the intensity. This is essentially the same as the Sobolev term, except that here we are explicitly solving for the full geometry and making no assumptions about the line profile.

In setting up the formal solution, notice that you require the partial derivatives of ν' and μ' with respect to r and μ. This is where the coupling of the moments becomes more complicated than we normally encountered in the stationary spherical problem. The radial velocity gradient couples the frequency, radial displacement, and direction cosine. For instance

$$\frac{\partial}{\partial r} \to \left(\frac{\partial}{\partial r}\right)' + \frac{\partial \nu'}{\partial r}\frac{\partial}{\partial \nu'} + \frac{\partial \mu'}{\partial r}\frac{\partial}{\partial \mu'},$$

where the prime on the radial derivative means that the terms are evaluated in the comoving frame. The angular term becomes:

$$\frac{\partial}{\partial \mu} \to + \frac{\partial \nu'}{\partial \mu}\frac{\partial}{\partial \nu'} + \frac{\partial \mu'}{\partial \mu}\frac{\partial}{\partial \mu'}.$$

The full spherical equation reduces to

$$\gamma(\mu' + \beta)\frac{\partial I'}{\partial r} + \gamma(1 - \mu'^2)\left[\frac{1 + \beta\mu'}{r} - \gamma^2(\mu' + \beta)\frac{d\beta}{dr}\right]\frac{\partial I'}{\partial \mu'}$$

$$- \gamma\left[\frac{\beta(1-\mu'^2)}{r} + \gamma^2(\mu' + \beta)\right]\nu'\frac{\partial I'}{\partial \nu'}$$

$$= -\left(3\gamma\left[\frac{\beta(1-\mu'^2)}{r} + \gamma^2\mu'(\mu' + \beta)\right] + \kappa'\right)I' + j', \quad (3.204)$$

which for $\beta \to 0$ reduces to the usual spherical equation. Notice, however, that even in the plane parallel case, $r \to \infty$, we still have terms that depend on the angular derivatives, unlike the static case, because of the aberration terms in the Lorentz transformations.

3.C A BRIEF SURVEY OF METHODS FOR SOLVING THE TRANSFER EQUATION

There are many roads to the Nirvana of a solution of the transfer equation. Our intent in this appendix is to at least expose you to some of them.

3.C.1 Diffuse Scattering as an Example of Probabilities: Invariant Embedding

In the main text, we've concentrated on stellar atmospheres and interiors, where the energy is coming up from below and there are distributed sources within the medium. For planets, however, the problem is the other way around—the radiative equilibrium of such an atmosphere is governed by energy incident at the topmost layer. Although it is out of the mainstream of the text, here we outline how such a problem is treated. Its relevance for many astrophysical problems should be clear. In the interstellar medium, for instance, scattered light dominates the dusty locales of recent star formation. For planets, such as the Earth, temperature inversions often occur because of this diffuse radiation field from scattering and absorption. And now we also face analyzing extrasolar planetary atmospheres.

The problem is posed as follows. Consider an incident flux, πF, at the top of the atmosphere. Take the layer to have a finite optical depth, $\Delta\tau$, and the radiation is a pencil beam incident with a direction cosine μ_0.[68] First, let's see how to treat this in one dimension. Then we can generalize it to the case of an anisotropic source and an angle-dependent scatterer. The *principle of invariance* is simplicity itself. It states that in a medium through which you want to transfer radiation, you can concentrate on a thin layer that is added to either the top or bottom of a thicker slab. It doesn't matter where the extra layer has been placed. First described by Helmholtz, the technique was generalized by Ambartsumian and Chandrasekhar and now forms one of the foundations on which the treatment of scattering radiative transfer rests.

Imagine a slab of optical depth τ, on top of which we add a small layer, $d\tau$. Assume an intensity I_0 falls on the top. Now call the backscattering, or reflection, coefficient for the layer R and the forward, or transmission coefficient T. The medium can send the light in two directions, forward or backward. If backward, the total reflection coefficient for the medium is $R(\tau + d\tau)$ and the reflected intensity is $I_1 = RI_0$ by definition. Similarly, the total transmitted intensity is $I_2 = TI_0$. Now comes the application of the principle of invariance. Take the thin layer and add it to the top of the slab. Then within this layer, we imagine two beams. One crosses the interface going inward, I_3 and the other, I_4, is directed outward. The geometry is shown in Figure 3.11. Label the coefficients in the layer $d\tau$ with subscript 1 and those in τ with subscript 2. Then:

$$I_3 = R_1 I_0$$
$$I_4 = R_2 I_3$$
$$I_2 = TI_0 = R_1 I_4 + T_2 I_3.$$

Then we arrive at the total transmission coefficient:

$$T = \frac{T_1 T_2}{1 - R_1 R_2}. \qquad (3.205)$$

Now for the total reflection coefficient, solve for I_3 and substitute into I_1 to obtain

$$R = \frac{T_1^2 R_2}{1 - R_1 R_2} + R_2. \qquad (3.206)$$

---

[68] See Sobolev, V. V. 1963, *A Treatise on Radiative Transfer* (Princeton: Van Nostrand), one of the clearest treatments of the formalism. You should also consult vol. 2 of Chandrasekhar's *Collected Papers* and Chandrasekhar, S. 1950, *Radiative Transfer* (NY: Dover Books) and Ambartsumian, V. 1957, *Theoretical Astrophysics* (Oxford: Pergamon Press). See also Bellman, R. and Wing, G. M. 1975, *Introduction to Invariant Imbedding* (NY: Wiley); van der Hulst, H. C. 1980, *Multiple Light Scattering: Tables, Formulas, and Applications* (NY: Academic Press); Periaiah, A. 1999, *Space Sci. Rev.*, **87**, 465.

A BRIEF SURVEY OF METHODS FOR SOLVING THE TRANSFER EQUATION 267

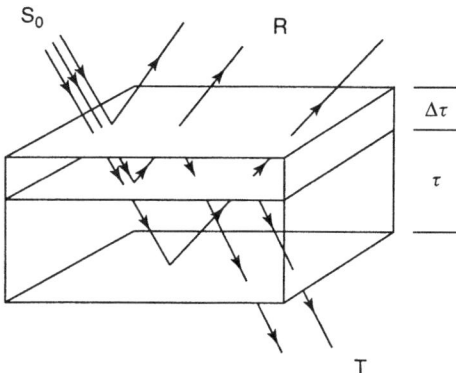

FIGURE 3.11. Sample photon trajectories for invariant embedding calculation (one-dimensional and two-dimensional).

You will notice that the lower boundary is assumed to have no incident flux from the outside. The principle of invariance can now be stated simply—identification of the substructure with the labels is arbitrary. Let's write the reflection coefficient explicitly in terms of functions rather than subscripts:

$$R(\tau + d\tau) = \frac{T^2(d\tau)R(\tau)}{1 - R(\tau)R(d\tau)} + R(d\tau). \tag{3.207}$$

Now apply the principle of invariance, interchanging τ and $d\tau$, and expand each coefficient in a Taylor series, calling $t = T(d\tau)$ and $r = R(d\tau)$. Both t and r can be assumed to be constants, for simplicity. Linearizing, we obtain

$$\frac{dT}{d\tau} = (rR - t)T \tag{3.208}$$

$$\frac{dR}{d\tau} = rT^2. \tag{3.209}$$

This is a *Ricatti* equation whose solution is found in terms of

$$TdT - \left(R - \frac{t}{r}\right)dR = 0, \tag{3.210}$$

whose solution is

$$T^2 = R^2 - 2\frac{t}{r}R + 1 \tag{3.211}$$

since $T(0) = 1$ and $R(0) = 0$. This solution can then be used to obtain R and T separately as functions of the optical depth, depending on the assumed values for r and t.

Now let's assume that the input flux is uniquely from one direction but that the scattering may be more complicated than unidimensional. Then the same set of invariant embedding equations become

$$I(\mu, \mu_0)\left(\frac{1}{\mu} + \frac{1}{\mu_0}\right)\Delta\tau = \frac{\varpi\Delta\tau}{4\mu}F + \frac{\varpi\Delta\tau}{2}\int_0^1 I(\mu, \mu')\frac{d\mu'}{\mu'}$$

$$+ \frac{\varpi\Delta\tau}{2}\int_0^1 I(\mu', \mu_0)\frac{d\mu'}{\mu_0} \quad (3.212)$$

$$+ \varpi\Delta\tau \int_0^1 \frac{I(\mu, \mu'')}{F}\frac{d\mu''}{\mu''}\int_0^1 I(\mu', \mu_0)\frac{d\mu'}{\mu_0}.$$

These look formidable, but they are just the angular generalizations of the equations we've used in the one-dimensional approximation. The left-hand side is the emergent intensity while the right hand side is the sum of the successive orders of scattering to second order. We now assert that

$$I(\mu, \mu_0) = F\rho(\mu, \mu_0)\mu_0, \quad (3.213)$$

so defining the auxiliary function

$$\Phi(\mu) \equiv 1 + 2\mu\int_0^1 \rho(\mu, \mu')d\mu', \quad (3.214)$$

we obtain an integral equation:

$$\Phi(\mu) = 1 + \frac{\varpi}{2}\mu\Phi(\mu)\int_0^1 \frac{\Phi(\mu)}{\mu + \mu'}d\mu'. \quad (3.215)$$

Now taking a first approximation, $\Phi_0 = 1$, we find

$$\Phi_1(\mu) = 1 = \frac{\varpi}{2}\mu\ln\frac{1+\mu}{\mu}, \quad (3.216)$$

from which we can compute the scattered intensity by differentiating Eq. (3.214) with respect to the direction cosine. This is essentially a limb darkening/brightening law for a purely scattering atmosphere. In the forward sense of figuring out the appearance of an illuminated layer, we're done. Inverting an observation of such a layer to obtain its internal properties is much more difficult, and we pass over that task in silence.

3.C.2 Integral Equation Methods for Radiative Equilibrium

Integral equation methods are best illustrated by the solution of the Schwarzschild–Milne equation for a purely scattering medium:

$$J(\tau) = \int_0^\infty J(t) E_1(|t - \tau|)\, dt \equiv \Lambda_\tau[J(t)]. \tag{3.217}$$

Wiener and Hopf[69] noted that the formal solution to the transfer equation is really a Laplace transform of the source function. Therefore, Eq. (3.217) is a convolution that can be solved using Laplace transforms. Calling $j_+(s)$ the Laplace transform of J in the interval $(0, \infty)$, and $j_-(s)$ the transform in $(-\infty, 0)$, we obtain

$$j_+(s) + j_-(s) = s \int_0^\infty J(t) e^{-st}\, dt \int_{-\infty}^\infty E_1(|t - \tau|) e^{-s(\tau - t)}\, d\tau,$$

which becomes

$$j_+(s) + j_-(s) = j_+(s) \int_{-\infty}^\infty E_1(|x|) e^{-sx}\, dx. \tag{3.218}$$

We want to digress for a moment here because the solution was an early application of operator methods, especially the Laplace transform, and provided a closed-form solution for the scattering problem. Define the Laplace operator, $\mathscr{L}_s$ as

$$\mathscr{L}_s\{f(t)\} = s \int_0^\infty e^{-st} f(t)\, dt.$$

This form of the transform is peculiar to radiative transfer because of the $1/\mu$ dependence of the formal solution [Eq. (3.42)]. Using Eq. (3.217), it also follows that

$$\int_0^\infty f(\tau) \Lambda_\tau\{g(t)\}\, d\tau = \int_0^\infty g(\tau) \Lambda_\tau\{f(t)\}\, d\tau. \tag{3.219}$$

Now comes the radiative transfer-specific step. From Eq. (3.217), calling $f(t) = e^{-st}$ and $g(t) = B(t)$ and substituting into Eq. (3.219) gives

$$\mathscr{L}_s \Lambda_\tau[J] = \tfrac{1}{2} s \int_1^\infty J(\tau) [\mathscr{I}_1 + \mathscr{I}_2]\, d\tau,$$

---

[69] Wiener, N. and Hopf, E. 1931, *Sitzber. Preuss. Akad. Wiss. Berlin, Kl. Math. Phys. Tech.*, 696 [see also 1964, *The Selected Papers of Norbert Wiener* (Cambridge, MA: MIT Press), p. 361].

defining the following integrals

$$\mathscr{I}_1 = \int_1^\infty \frac{e^{-s\tau} du}{u(s+u)} \qquad (3.220)$$

$$\mathscr{I}_2 = \int_1^\infty \frac{e^{-u\tau} du}{u(s-u)} \int_1^\infty \frac{e^{-s\tau} du}{u(s-u)}.$$

To obtain these requires a substitution of the explicit form for the $E_1(x)$ function. Then using Eq. (3.219), we see that $\mathscr{L}_s\{\Lambda_\tau[J]\} = j(s)$, where $j(s) = \mathscr{L}_s\{B\}$. This gives a nonlinear integral equation for the Laplace transform of the source function:

$$j(s) = sj(s)\int_1^\infty \frac{du}{u(u^2-s^2)} + \frac{1}{2}s\int_1^\infty \frac{j(u)\,du}{u^2(s-u)}, \qquad (3.221)$$

from which a closed-form solution for $j(s)$ can be obtained. The source function, $J(\tau)$ is then obtained from the one sided inverse Laplace transform, using the inversion described above. Having a closed-form solution for the transfer equation, however lovely, was not immediately useful for stellar atmosphere computations. Not until after interest grew in planetary atmosphere and nuclear reactor problems did this method see extensive applications.[70]

3.C.3 Discrete-Ordinates Methods

The subsequent development of the moment equations was largely formal, spurred by the introduction of the discrete-ordinates method by Chandrasekhar. Here you expand the intensity in terms of its angular components. In principle, the intensity distribution as a function of optical depth can be separated into τ-dependent and angular components

$$I(\tau,\mu) = \sum_{l=0}^\infty A_l(\tau)P_l(\mu),$$

where the coefficients A_l depend on optical depth and $P_l(\mu)$ are the Legendre polynomials. The discrete-ordinates method, however, assumes instead that the scattering source function J can be written as

$$J(\tau) = \tfrac{1}{2}\int_{-1}^1 I\,d\mu = \tfrac{1}{2}\sum_{j=-n}^n a_j I(\tau,\mu_j), \qquad (3.222)$$

where the weights a_j are determined from the orthogonality conditions for the $P_l(\mu)$ for critical angles μ_j, which are the zeros of $P_j(\mu)$ for a finite expansion of

---

[70] Busbridge, I. W. 1960, *The Mathematics of Radiative Transfer* (Cambridge, UK: Cambridge Univ. Press); Noble, B. 1965, *Methods Based on the Wiener-Hopf Technique for the Solution of Partial Differential Equations* (NY: Pergamon); Kourganoff, V. 1952. *Basic Methods in Transfer Problems* (NY: Dover Books).

order n. The choice of Gaussian weights is dictated by the form of the integral, $\int_{-1}^{1} f(x)dx \approx \Sigma_i a_i f(x_i)$. Noting that $\int_{-1}^{1} P_m P_l d\mu$ vanishes for $m \neq l$, the scattering equation can be expanded in terms of the individual coefficients

$$\mu_i \frac{dI_i}{d\tau} = I_i - \frac{1}{2}\sum_j a_j I_j, \qquad (3.223)$$

and, assuming $I_j = g_j \exp - k\tau$, Eq. (3.223) is reduced to an eigenvalue problem. First, on substitution, you will find that

$$g_i(1 + \mu_i k) = \frac{1}{2} \sum_{j=-n}^{n} a_j g_j.$$

Noting that the sum over j is now independent of j, we find that

$$\frac{1}{2}\sum_j a_j g_j = \text{constant}.$$

It follows that

$$\frac{1}{2}\sum_j \frac{a_j}{1 + \mu_i k} = 1.$$

Because of two symmetry conditions—namely, that $\mu_j = -\mu_{-j}$ and $a_j = a_{-j}$—the problem is reduced to finding the roots of

$$\frac{1}{2}\sum_{j=1}^{n} \frac{a_j}{1 - \mu_j^2 k^2} = 1 \qquad (3.224)$$

subject to the constraint that $\sum_{j=1}^{n} a_j = 1$. On substitution, the intensity takes the form

$$I_i = b(\tau + q_i(\tau)),$$

where q_i is called the *Hopf function* in the literature. The importance of this solution for the stellar problem was that it provided a simple means for computing the temperature in a radiatively dominated atmosphere. It still provides the starting point for numerical models. Continuing, the roots of Eq. (3.224) are $\pm k_\alpha$ ($\alpha = 1, \ldots, n-1$). These yield a closed-form solution for the limb darkening law:

$$I(0, \mu) = \frac{\sqrt{3}}{4} FH(\mu),$$

where the function $H(\mu)$ is defined by

$$H(\mu) = \frac{1}{\mu_1 \cdots \mu_n} \frac{\prod_{i=1}^{n}(\mu + \mu_i)}{\prod_{i=1}^{n-1}(1 + k_i \mu)}, \qquad (3.225)$$

which was defined by Chandrasekhar. The function $\Phi(\mu)$ that we found from Eq. (3.215) is equivalent to the $H(\mu)$ function.

3.D A WALK THROUGH THE STELLAR SPECTROSCOPIC ZOO

> At the border of knowledge, similitude is that barely sketched form, that rudimentary relation which knowledge must overlay to its full extent, but which continues, indefinitely, to reside below knowledge in the manner of a mute but ineffaceable necessity.
> —Michel Foucault, *The Order of Things*, Chapter 3, Section V

3.D.1 Stellar Classification and Types of Stars

3.D.1.1 A Historical Digression

The still-common nomenclature that distinguishes hot and cool stars preserves an historical artifact. It is consequently worth a moment's reflection on the origin of the classification scheme used now for stellar spectra as a case study. In the first rush of ideas about stellar evolution, Lockyer and later Russell suggested that stars evolve along the main sequence from the hotter to cooler end. The meteoritic hypothesis (Lockyer) posited that the increased spectral complexity toward cooler temperatures was caused by continued accretion of metal-rich material over time, while the contraction hypothesis (Russell) attributed the distribution of colors and luminosities in the HR diagram to a cooling sequence. Although different in detail, these ideas have the same basic trend of evolution—toward cooler, more compact stars with progressively more complex spectra. As a result, hot stars were initially called "early types," and those cooler than the Sun were called "late type." The names stuck, so now stars hotter than roughly class F are referred to as early-type and the cooler ones are still called late-type stars.[71]

The basis of the Morgan-Keenan or *MK* types, is the relative visibility of specific absorption lines in spectra and their ordering.[72] It is independent of

---

[71] The basis of the distinction is now known to be wrong, the main sequence is a *mass* ordered sequence of hydrostatic stars undergoing core hydrogen processing, and the complexity of the spectra is an ionization effect; see Chapter 4, below. See also Jaschek, C. and Jaschek, M. 1987, *The Classification of Stars* (Cambridge, UK: Cambridge Univ. Press); Kahler, J. 1989, *Stars and Their Spectra* (Cambridge, UK: Cambridge Univ. Press) for general overviews of the various spectral types.

[72] Morgan, W. W., Keenan, P. C., and Kellman 1943, *An Atlas of Stellar Spectra* (Chicago: Univ. Chicago Press); Morgan, W. W., Abt, H. A., and Tapscott, J. W. 1978, *Revised MK Spectral Atlas* (Tucson: Kitt Peak National Observatory); Keenan, P. C. and McNeil, R. C. 1976, *An Atlas of the Spectra of Cooler Stars* (Columbus: Ohio State Univ. Press); Garrison, R., ed. 1985, *The MK Process and Stellar Classification* (Toronto: Univ. Toronto Press).

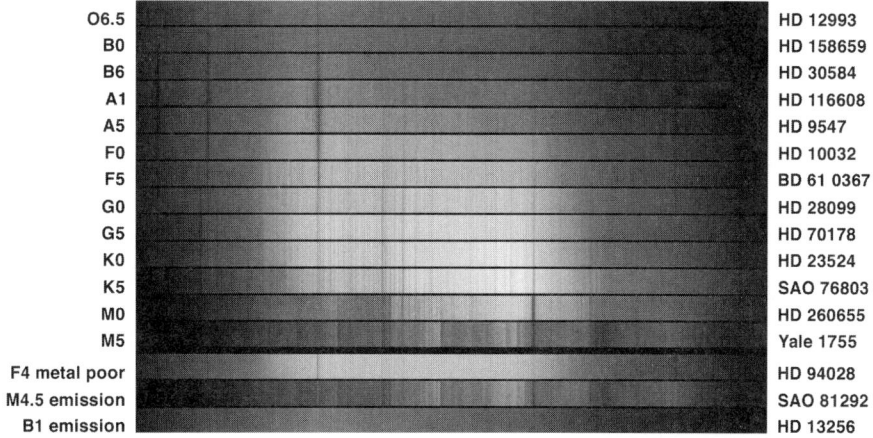

FIGURE 3.12. Sample optical spectra $\lambda\lambda 3800$–5000 Å for main sequence stars labeled by spectral type. (Credit: NOAO/AURA/NSF).

theory, as emphasized many times by its chief practitioners.[73] The system was, however, originally motivated by some early concepts of spectrum formation. The Harvard system, which was initially based on the complexity of the spectra and drew much of its inspiration from the ideas of Lockyer (those, in particular, that connected age and metallicity with temperature) and the relative strengths of lines of specific ionization. This one-dimensional system was an ordering in temperature only, although there were indications that some stars had different linewidths (Maury used *c* to indicate such narrow lines—contrasted with broader *a* and *b* types—which was later adopted in the Mount Wilson notation for giant, where *d* was used for "dwarf"). The physical basis of the MK system, which added the dimension of "luminosity" to the original one dimensional Harvard temperature classification system, was based on two fundamental ideas. One was ionization equilibrium as was the original HD system. The other was, however, based on pressure broadening of hydrogen lines due to the Stark effect.

Line ratios reflect several things about the star—its temperature and gravity through the ionization balance and particular line strengths and broadening, and composition. The majority of bright stars you see in the night sky with the naked eye have been formed comparatively recently, so composition may not play too much role. But when one looks at fainter stars, sampling a wider range of the Galaxy in space and time because they are statistically farther away and also possibly intrinsically fainter, then the requirement of a homogeneous population breaks down. Actually, it is not clear what spectral types thus really mean, although they certainly provide some handle on the properties of the stars, because the internal dispersion of physical characteristics defining spectral morphology depends on the class. Spectra for the main sequence temperature classes are

---

[73] For instance, a B2 V star will always have that classification, even though its effective temperature and surface gravity may be assigned differently by different analyses, a comment due to Nolan Walborn, whose work on the extension of the MK system to the ultraviolet is referenced later in this section.

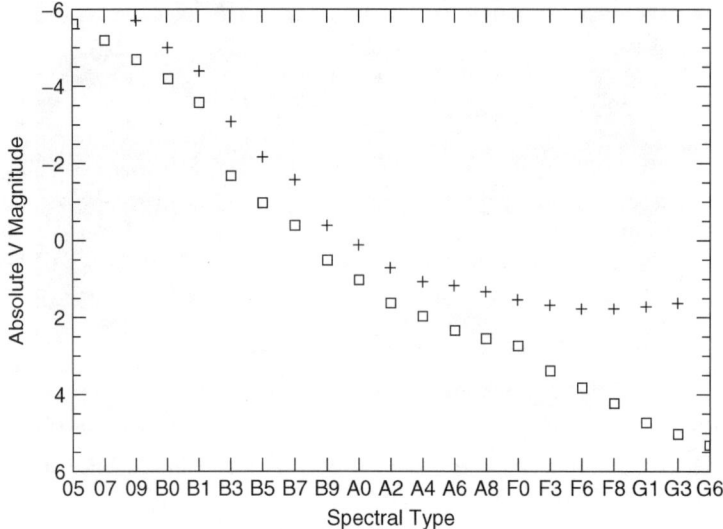

FIGURE 3.13. Calibration of M_V versus MK spectral type for luminosity class V (open squares) and III (crosses) (data from N. Houk, Univ. Michigan).

illustrated in Figure 3.12. The calibration of absolute magnitude *vs.* MK type is shown in Figure 3.13 for dwarf (V) and giant (III) stars.

3.D.2 The Classification Scheme

3.D.2.1 O Stars

Among the normal stars, the O stars are the most massive and hottest. Their spectrum is dominated by helium absorption or emission, and the primary discriminants of spectral subtypes depend on either He II/He I ratios, or of highly ionized species like Si IV or N III. Specifically, the He II 4686/He I 4471 line ratio discriminates temperature along with Si IV λ4089 and C III λλ4068, 4647, 4651. The changes can easily be understood in terms of ionization equilibria. The luminosity effects are, however, harder to understand. They depend on the fact that the stars have winds, and that their atmospheres show progressively stronger evidence of their low density environments with decreasing surface gravity. In particular, He II λ4686 (and λ1640) turn into emission features at low enough gravity, as does the N III multiplet λλ4634, 4640. The highest luminosity objects that show this are called Of stars, to distinguish them from Oe stars, which show emission only for the Balmer series. An excellent place to start is Conti, P. S. and Underhill, A. B., eds. 1988, *The O Star and Wolf-Rayet Stars* (Washington, DC: NASA SP-497).

Perhaps the strangest residents of the HR diagram, the *Wolf–Rayet* (WR) stars, appear to be related to the Of and O stars, but the relation is still phenomenological. In fact, there is even disagreement over what constitutes a WR spectrum. These stars are intrinsically hot, hence most of their flux emerges below the atmospheric cutoff in the far ultraviolet. As a result, it has been only relatively recently that their properties have been understood. The UV is dominated by resonance absorption lines from ionized species, in particular N V $\lambda1240$, Si IV $\lambda1400$, and C IV $\lambda1550$. In addition, lines formed from recombination, O V] $\lambda1371$, He II $\lambda1640$, and N IV] $\lambda1718$ are important. The He II line is especially significant, since it is the ionized helium analog to Hα; it is in emission in WR and Of stars. The resonance lines develop dramatic P Cygni absorption components with increasing luminosity (decreasing surface gravity), indicating that the highest luminosity members of the class have strong mass loss. The terminal velocities of these lines are often quite high, often greater than 2000 km s^{-1}. The WR stars are similar in the UV except that they show differentiation based on whether they are in the nitrogen or carbon sequence, in particular whether N IV] $\lambda1486$ and N III] $\lambda1750$ are present vs. C III $\lambda1175$ and C III] $\lambda1910$.[74]

3.D.2.2 B Stars

The key feature identifying these stars is the absence of lines of He II. The key spectral identifiers are C II $\lambda4267$ (which peaks in the middle of the spectral sequence), Si II $\lambda\lambda4128,4130$/He I $\lambda4144$, and Mg II $\lambda4481$ (which peaks toward the cooler end of the sequence). The He I spectrum becomes stronger relative to the Balmer lines toward the hotter end of the spectral sequence. There are several luminosity discriminants, [Si IV and He I]$\lambda4119$/He I $\lambda4144$ (increases with decreasing surface gravity), N II $\lambda3995$/He I $\lambda4009$, He I $\lambda4121$/He I $\lambda44144$, and Mg II $\lambda4481$/He I $\lambda4471$, all of which increase with decreasing surface gravity (and increasing luminosity). The ultraviolet luminosity effects are particularly interesting. The B stars do not show strong intrinsic mass loss. Thus the presence of emission among these stars is unusual and suggests an astrophysically interesting process may be operating. As the luminosity increases, there is a systematic trend for stellar spectra to show P Cygni profiles. This indicates that mass loss is strong for some stars. The dominant spectral lines are again due to strong resonance transitions, only this time of somewhat lower ionization stages: C II $\lambda1335$, Si IV IV $\lambda1400$, Al II $\lambda1672$, and Al III $\lambda1860$. Si III $\lambda1300$, a complex of several lines, is especially useful as a temperature indicator and also as a probe of mass loss, as is N IV] $\lambda1718$.

A class of emission line objects, the Be stars, occurs in this region. These are main-sequence stars, often binaries, with strong hydrogen emission lines and rapid rotation. A special subclass of chemically anomalous main sequence objects, the Bp and helium peculiar stars, are associated with magnetic fields and stellar winds. Until recently, these were thought to be confined to later than B0 V. The discovery of θ^2 Ori C, an O star with variable Balmer and UV resonance lines, has changed this picture. The star has a detectable magnetic field and appears in most behaviors to resemble the helium strong stars.[75] A good place to start is Underhill, A. B. and Doazan, V., eds. 1982, *The B Stars with and without Emission Lines* (Washington, DC: NASA SP-456).

---

[74] See Kudrizski, R. P. and Hummer, D. G. 1990, *ARAA*, **28**, 303; Walborn, N., Nichols-Bohlin, J., and Panek, R. J. 1985, *International Ultraviolet Explorer Atlas of O-type Spectra from 1200 to 1900 Å* (NASA Ref. Publ. 1155); Germany, C. D., ed. 1990, *Properties of Hot Luminous Stars: ASP Conf.* 7 (San Francisco: Astr. Soc. Pacific). This was the first of the *Boulder–Munich* conferences, an ongoing series.

[75] Roundtree, J. and Sonneborn, G. 1993, *Spectral Classification with the International Ultraviolet Explorer: An Atlas of B-type Spectra* (NASA Ref. Publ. 1312); see also: Roundtree, J. and Sonneborn, G. 1991, *ApJ*, **369**, 515 for further detail and. Walborn, N., Nichols-Bohlin, J., and Panek, R. J. 1995, *International Ultraviolet Explorer Atlas of B-type Spectra from 1200 to 1900 Å* (NASA Ref. Publ. 1363).

3.D.2.3 A Stars

Temperature classification is rather hard for these stars. They have temperatures at which metals are mostly ionized and there are no strong temperature dependences. The hydrogen spectrum is strongest for these stars and Ca II $\lambda 3933/$[He and Ca II]$\lambda 3968$ increases with decreasing temperature. The strength of the Mg II $\lambda 4481$ line is also used. The strongest metallic lines are otherwise due to Fe II, but these are quite weak, except in a very small fraction of stars. A key discriminant is the complete absence of He I lines.

The A stars best demonstrate the importance of the Stark wings on the hydrogen Balmer lines. With relatively few transitions available in the optical, the visual appearance of the wings and core of the Balmer series are the primary luminosity classifiers. These depend on a proper interpretation of the broadening of the lines since electron scattering is also capable of producing broad wings or shallow cores. In addition, because of the strength of the hydrogen series, it is especially easy to see dilution effects among the A stars due to low density environments far from LTE.

The most luminous A stars, such as α Cygni, show emission at Hα. Most others show some indication of mass outflows, but otherwise only the supergiants show any strong evidence for extension effects or curvature. An overview is provided by Wolff, S. 1983, *The A-Type Stars: Problems and Perspectives* (Washington, DC: NASA SP-463).[76]

---

[76] Among stars in the middle part of the HR diagram, the A stars show an unusually large variety of spectroscopically anomalous objects. It is not entirely clear whether this is a discovery effect, that the peculiarities are most easily seen in the temperature range 8000–10,000 K, or whether processes conspire to produce these uniquely in this region. It's worth a moment's discussion of these freaks. *Magnetic peculiar stars* (Ap and Bp stars): These are main sequence objects in the mass range 1.5 to around 10 $M_\odot$ that show strong integrated surface magnetic fields (> 100 G) that are periodically variable. The coolest stars, at around 8000 K, show enhanced lines of rare earths (mainly the lighter elements, such as Ce, La, Sm, Eu, and Dy) and enhancements of Sr, Y, Zr and iron peak species (such as β CrB and α^2CVn). The hotter stars, above $\sim 10,000$ K, show strongly enhanced Si (hence they are called *silicon stars*, such as HD 34452 and 56 Ari). The surface distribution of these elements appears to be patchy, indicated by the observed spectral, photometric, and radial velocity variations that occur on a rotational timescale. In the hottest stars, He is deficient for temperatures below about 15000 K (He weak stars, such as 3 Cen A and CU Vir) and *enhanced* for hotter stars (He strong stars, such as σ Ori E and HD 184897). In general, these stars are anomalously slow rotators compared with spectroscopically normal main-sequence stars of the same mass and effective temperature. *Metallic line stars* (Am, HgMn, and Mn stars): This group, which was first identified by the enhanced abundances of the Fe-peak elements, shows several differences compared with the Ap stars. Among the HgMn stars, mercury is isotopically peculiar: in the best studied example, χ Lup, only ^{204}Hg is present in the spectrum, although the solar mixture is weighted toward isotopes of lower mass (see e.g., Leckrone, D. S. et al. 1999, *AJ*, **117**, 1454; Brandt, J. C. et al. 1999, *AJ*, **117**, 1505). Several, especially HR 7775 and κ Cnc, show enhanced Pt that also shows an isotopic anomaly (Wahlgren, G. M. et al. 2000, *ApJ*, **539**, 908 and references therein). Calcium appears to be depleted compared with the solar abundance, and as with the Ap stars, there is a systematic trend that the heavier elements are more enhanced compared with the solar abundance than the light species. Sensitive searches for mean magnetic fields have produced only upper limits of less than 50 G. These stars are slow rotators, and most are embers of close binary systems with periods less than a few weeks (Sirius, α CMa, is an exception to this latter rule). One, AR Aur, is a member of an eclipsing binary. λ *Bootis stars*: These are the exceptions to the overabundances usually seen among the chemically peculiar stars. They show depletions of heavy elements that seem to follow the trend for depletion from the interstellar medium (Venn, K. A. and Lambert D. L. 1990, *ApJ*, **363**, 234; Paunzen, E. and Gray, R. O. 1997, *Astr. Ap. Suppl.*, **126**, 407). The indications are that the atmospheres of these stars are externally altered by accretion but they remain a mystery of the mid-main sequence.

3.D.2.4 F Stars

These stars are classified based on the change in the relative line strengths of low ionization metals compared with hydrogen. The following line ratios are the most commonly employed: Fe I $\lambda 4065/H\delta$; Mn I $\lambda\lambda 4030 - 34/$Si II $\lambda\lambda 4128,4130$; Ca I $\lambda 4226/H\delta$; and CH $\lambda 4300/$Fe I $\lambda 4385$. In all cases, the ratio increases with decreasing temperature. Notice that the first molecule appears at this spectral type, the CH feature. The Ca II/Ca I ratio decreases with increasing temperature, but the absorption line strength ratios change rather more slowly because of the enormous optical depth of the Ca II resonance lines. For luminosity effects, Mg II, Fe II, and Ti II lines are normally used. Ti II, Mg II, and especially Sr II $\lambda 4077/H\delta$, Sr II $\lambda 4077/$Fe I $\lambda\lambda 4045$, and Ca II/Hδ all increase with increasing luminosity.

A most important feature of this spectral class is the first appearance of chromospheric indicators. At the approximate temperature of these stars, surface convection is at least partly responsible for energy transport. The correlation observed between ultraviolet emission line fluxes and X-ray emission and stellar effective temperature is a good, although not unique, indication of a connection between this convection and heating of the outer stellar atmosphere. The emission is observed even for main-sequence stars. It is also important to note that the identification of chromospheres and coronae in the late-type stars is not a result from Balmer line emission. There are, as you have seen, many ways for that to occur without any change in the temperature gradient of the outer atmosphere. A gas will emit X-rays, on the other hand, only if it is either highly relativistic (inverse Compton scattering), which is very unlikely for stars, or if the gas is *actually* hot. Thus, chromospheric activity appears to be typical of stars on the cooler part of the spectral sequence.[77] A comprehensive review of these low-temperature stars is found in Cram, L. and Kuhi, L. (eds.) 1989, *The FGK Stars and T Tauri Stars* (Washington, DC: NASA SP-502).

3.D.2.5 G Stars

The main sequence stars of this spectral type are much like the Sun, which is defined as the G2 V standard, although it has proven rather hard to find a precise solar analog. The G stars are classified according to the strength of two molecular bands, CN $\lambda 4216$ and CH $\lambda 4300$. In addition, several line ratios are important, in particular Fe I $\lambda 045/H\delta$, Ca I $\lambda 4226/H\delta$, Cr I$\lambda 4254/$Fe I $\lambda 4250$, and Cr I $\lambda 4274/$Fe I $\lambda 4271$. Luminosity effects are harder to quantify. A very fine introduction, even if dated, to the basic physics of the Sun as a quintessential "normal" star is Jordan, S., ed. 1981, *The Sun as a Star* (Washington, DC: NASA SP-450). See also Stix, M. 1991, *The Sun: An Introduction* (Berlin: Springer-Verlag) for a comprehensive overview of solar physics. The Sun is simply too close! It would require a complete book just to describe the phenomenology, but the Sun *is*, after all, a star. A detailed discussion of efforts to treat it as such, that is to determine

---

[77] For the theory of modeling cool star atmospheres, which means later than A in this case, see the reviews by Gustofsson B. 1989, *Ann. Rev. Astr. Ap.*, **29**, 701; Gustofsson B. and Jorgensen, U. G. 1994, *Astr. Ap. Rev.*, **6**, 19.

its spectrophotometric twins, provides a good example of how hard it is to connect the properties of otherwise presumably similar objects across vast differences in resolution and sensitivity. See, for instance, Hardorp, J. and Tomkin, J. 1983, *Astr. Ap.*, **127**, 277; Cayrel de Strobel, G. 1996, *Astr. Ap. Rev.*, **7**, 243.

3.D.2.6 K Stars

These are stars that are cooler than the Sun and are defined by the dominance of the neutral metal lines. The hydrogen lines are weak and the ionized metallic spectrum weakens with decreasing temperature. The key temperature discriminants are the Mg I $\lambda 5174$ line and similar ratios of neutral species to those seen in the G stars. At the lowest temperature end, some oxides appear, in particular TiO. Luminosity classification is based on the relative strength of Ca I, which decreases with increasing luminosity, and MgH $\lambda\lambda 4780, 5211$, which also decreases. One important result is that the infrared triplet of Ca II near 8200 Å appears in emission and increases toward higher luminosity. This is important evidence for the occurance of chromospheres in cool stars and the onset of mass loss. An important feature of this class is the Wilson–Bappu effect, a classification criterion related to the equivalent width of the infrared Ca II emission lines.

3.D.2.7 M Stars

The spectra of these stars are dominated by molecular bands of TiO, ZrO, and VO, and by very strong absorption from neutral metals. The molecular lines are the surest indication of such stars. The best available overview of the M type and cooler (non-brown-dwarf) stars is by Johnson, H. and Querci, F., eds. 1986, *The M Stars* (Washington, DC: NASA SP-492) and the more recent review by Allard, F., Hauschildt, P. H., Alexander, D. R., and Starrfield, S. 1997, *ARAA*, **35**, 137.

3.D.2.8 L Stars: Methane Dwarfs

The prototype of this class is GD 165B, but the relatively high-resolution near-infrared spectrum of Gl 229B makes this star an even better candidate for benchmarking model atmosphere computations. The spectrum looks astonishingly like a Jovian planet. The TiO bands are absent, presumably because depletion of Ti by particulate formation in the high-pressure, low temperature environment. Neutral resonance lines are visible, in particular K I at 7660 and 7700 Å and also the Na I doublet. VO, CaH, and CrH are also observed, and H_2O becomes stronger with decreasing color temperature. The effective temperature lies in the range 1500–2500 K although at the coolest end the classification merges into the brown dwarfs.

3.D.2.9 Carbon Stars

As a general class, these are giants that show enhanced molecular absorption of key species, notably C_2 and *CN*, and also evidence for dust and *s*-process nuclear products. The spectral types have been defined according to the absorption strength of the C_2 Swann bands.

3.D.2.10 White Dwarf Stars

The rough types are discriminated by the relative strengths of the helium and hydrogen lines.[78] The DA stars show only Balmer lines, DB only He I, DC only continuum, DO only He II, DZ only weak metals, and DQ show absorption by carbon. The problem is that there is no good calibration for what this spectral classification means. It is enough to say that it provides a qualitative description of whether the degenerate has an extensive hydrogen envelope.

---

[78] Sion, E. M. *et al.* 1983, *ApJ*, **269**, 253.

4 The Interiors of the Stars and Stellar Evolution

See the ball I hold.
Let chemists toil like asses,
Our fire their fire surpasses,
And turns all our lead to gold.[1]
—John Gay, *The Beggar's Opera*

4.1 INTRODUCTORY REMARKS

The most important things to know about stars are that they are massive and they have no walls. The first means that they must reach very high internal temperatures in order to generate sufficient pressures to guarantee their stability against collapse. The second means that they have outer boundaries that communicate to space all of the excess energy that they are able to generate. Since the Universe is so large and empty and cold, the absence of insulation requires all internal heat to flow outward. Thus, any star ultimately cools and no star can last in a hot stage forever. In addition, though enormous, only a finite amount of energy is available. The gravitational potential energy can be tapped only at the expense of disequilibrating the body, either slowly as in contraction or catastrophically fast as in collapse. Alternatively, you could imagine that the rest energy could somehow be used, but this requires nuclear processes to occur and still only a fixed amount of energy can be obtained through nucleosynthesis. No matter how much energy you try to release, it leaves the star, never to be recovered. So stars must die. This, in a nutshell, is the driving condition of stellar evolution.[2]

---

[1] You will notice that the nucleosynthesis envisioned in the eighteenth century is very different from the neutron processes now known to be required in the production of these elements. But the line is sung by highwaymen and not astrophysicists.

[2] The founding, classic monographs of stellar evolution are Chandrasekhar, S. 1939, {*Introduction to the Study of Stellar Structure* (Chicago: Univ. Chicago Press; Dover Books); Eddington, A. S. 1927, *The Internal Constitution of the Stars* (Cambridge, UK: Cambridge Univ. Press, rpt. Dover Books); Schwarzschild, M. 1957, *Structure and Evolution of the Stars* (Princeton: Princeton Univ. Press; rpt. Dover Books); Struve, O. 1950, *Stellar Evolution* (Princeton: Princeton Univ. Press). A unique historical survey is found in Struve, O. and Zebergs, V. 1962, *Astronomy in the 20th Century* (NY: Macmillan), and a snapshot of the field in mid-twentieth century is found in the volume from the 50th anniversary of the Yerkes Observatory, Hynek, J. A., ed. 1951, *Astrophysics: A Topical Symposium* (NY: McGraw-Hill). This last reference is especially interesting, and sobering, reading since there are still many areas addressed there that remain basically open. Some monographs with a primarily nuclear bent are Audouze, J., Chiosi, C., and Woosley, S. E. 1986, *Nucleosynthesis and Chemical Evolution: 16th Advanced Course, Saas-Fee* (Geneva: Geneva Observatory); Clayton, D. D. 1968, *Principles of Stellar*

Work on the structure and evolution of the Sun goes back to mid-nineteenth century thermodynamicists, specifically Kelvin in 1857, Helmholtz in 1857, and Lane in 1867.[3] Using only arguments based on thermodynamics and the newly developed theory of ideal-gas equations of state, they examined the implications of hydrostatic equilibrium for a hot, cooling self-gravitating spherical gaseous body. The Sun, they argued, is stable on short timescales such as years. This is much longer than the freefall time, as we will see, and therefore the Sun appears to be in hydrostatic equilibrium. It doesn't appear, they said, to have internal energy sources such as coal fire plants in the core, but is likely hot because it started as a very extended body and contracted to its present size some time ago. Then how much must it contract to maintain its observed luminosity while maintaining hydrostatic equilibrium? In other words, the Sun has an energy source in its enormous gravitational potential resulting from its mass. As the star contracts, it releases potential energy, some of which does work in maintaining equilibrium and some of which is radiated away as heat. If it radiates at a constant rate, the contraction must be able to maintain this condition. How does it do this, and how long does it have? The problem was not to solve the dynamical structure of the Sun; that is taken care of by the assumption of quasi-static contraction. The energy generated throughout the star does work throughout the star and flows away only through the surface. Therefore, the timescale is given by the energy content, which is the self-gravitating potential energy, and the rate of energy loss or the bolometric luminosity.

The consensus was that the characteristic timescale for the solar contraction was about 10^7 to 10^8 years. Yes, this is a large number, compared with the publication time between the first papers and now, for instance, but it is unacceptably short compared with any of the known constraints on the age of the Earth and life. Darwin was already well aware of these contradictory timescales when he proposed far greater ages for some geologic properties found on Earth, like the salinity of the oceans.[4] In addition, there is another important constraint from the history of life on Earth. To derive energy from gravity requires that the star

---

Evolution and Nucleosynthesis (NY: McGraw-Hill; rpt. Univ. Chicago Press); Rolfs, C. E. and Rodney, W. S. 1988, *Cauldrons of the Cosmos* (Chicago: Univ. Chicago Press). An encyclopedic yet leisurely survey is provided by Cox, J. and Giuli, R. 1968, *Principles of Stellar Structure* (NY: Gordon and Breach). Among the notable recent monographs, we recommend Kippenhahn, R. and Weigert, A. 1990, *Stellar Structure and Evolution* (Berlin: Springer-Verlag); de Loore, C. and Doom, C. 1992, *Structure and Evolution of Single and Binary Stars* (Dordrecht: Kluwer); Schatzman, E. L. and Praderie, F. 1993, *The Stars* (Berlin: Springer-Verlag); Hansen, C. J. and Kawaler, S. D. 1994, *Stellar Interiors: Physical Principles, Structure, and Evolution* (Berlin: Springer-Verlag). Finally, we enthusiastically recommend reading Iben I. Jr. 1991, *ApJS*, **76**, 551, which is the published version of his Russell Prize lecture.

[3] The interested reader would do well to consult the footnotes in Chandrasekhar's 1939 monograph *Introduction to the Study of Stellar Structure*, still available in reprint, which is the best historical bibliographic source available. See also the brief historical survey by Arny, T. 1990, *Vistas Astr.*, **33**, 211 and Gingerich, O. 1999, *Astrophys. Space Sci.*, **267**, 3.

[4] A very useful reference for the history of work on stellar evolution is Darwin, *Origin of Species*, Chap. 10: "Geological Arguments." In addition, see Butterfield, J. D. 1975, *Lord Kelvin and the Age of the Earth* (Chicago: Univ. Chicago Press); Kahl, R., ed. 1971, *Selected Writings of Hermann von Helmholtz* (Middletown, CT: Weslyan Univ. Press); Kelvin 1891, *Popular Lectures and Addresses* (London: Macmillan).

actually contract. If the same luminosity is maintained, then as R decreases, T must increase. This has two important consequences: the star must have started out very large, and it must have been very cool. Both of these are problems because over the known history of life on the planet, there seems to to have been no major change in the effective temperature of the Sun. Either the rate of energy generation has been decreasing to match the change in radial contraction, or something is fundamentally wrong with this explanation.

The solution comes from the very thing that the early work neglected. If the star is in hydrostatic equilibrium, because it is compressible, it has a denser and much hotter core than does the surface. This never entered the early work because of the model of "liquid stars" that dominated the early studies of stellar structure. Because the matter was assumed to be incompressible[5] one could solve the Poisson equation for the potential while avoiding the pressure equation. Regardless of what the equation of state is, stars are ultimately compressible, and this dominates their internal structure and evolution.

4.2 SELF-GRAVITATING SPHERES

It is a beautiful fact of nature that gravity is always attractive. The overall energy of a self-gravitating body must, at worst, equal zero and at best is less than zero. This means that if a pressure-supported body begins to cool, it contracts and heats up. Lynden-Bell has expressed this as the *negative specific heat* paradox: a hot self-gravitating body can never heat up rapidly enough to stop itself from losing energy. Because stars are hot and radiate, they must eventually change their state. Either, as Kelvin and Helmholtz realized, they compensate for energy loss by contraction, or they generate energy by fusion reactions and alter their internal composition.

4.2.1 The Virial Theorem in Stellar Structure

In the next few sections we will derive several conditions that govern the global equilibrium and then discuss the thermodynamics you will need to understand the equations of structure and energy generation.[6] We begin with the equation for the structure of a pressure-supported star:

$$\ddot{\mathbf{r}} = -\frac{1}{\rho}\nabla P + \nabla \Phi. \qquad (4.1)$$

Here Φ is the internal gravitational potential, a point to which we will return in a

---

[5]See, for instance, Jeans, J. H. 1929, *Astronomy and Cosmogony* (Cambridge, UK: Cambridge Univ. Press).
[6]Some you have already seen in Chapter 1 in the context of general thermodynamic principles. Their importance to stellar interiors is, however, so basic that we feel it is appropriate to repeat them in context, to reinforce the physical ideas.

moment. The time derivative may be advective

$$\frac{d}{dt} = \frac{\partial}{\partial t} + \mathbf{v} \cdot \nabla \qquad (4.2)$$

if, for instance, there is bulk motion.[7] The procedure is familiar by now. You take the spatial moment of this equation and integrate over mass to obtain a global scalar equation. First, the left-hand side of Eq. (4.1) becomes

$$\int dM\, \mathbf{r} \cdot \ddot{\mathbf{r}} = \frac{d^2}{dt^2} \int dM\, \mathbf{r} \cdot \mathbf{r} - \int dM\, \dot{r}^2 = \frac{1}{2}\ddot{I} - 2K. \qquad (4.3)$$

Here K is the kinetic energy and I is the moment of inertia of the body. This K refers again to bulk motions, including rotation and other flows within the body. It does not involve motions of the center of mass within an external potential, only those bounded by the stellar surface. The term involving the moment of inertia refers to bulk changes in the mass distribution, for instance, to radial or nonradial pulsations. The right-hand side becomes

$$-\int dM \frac{1}{\rho} \mathbf{r} \cdot \nabla P + \int dM\, \mathbf{r} \cdot \nabla \Phi = 3 \int P\, dV + \Omega. \qquad (4.4)$$

Here Ω is the gravitational potential energy:

$$\Omega = -\int \frac{GM}{r}\, dM. \qquad (4.5)$$

This may include tidal terms due to the influence of a close companion or time-independent changes in the shape of the star caused by rotation (the centrifugal potential). We actually need the "no walls" assertion for the next step. The pressure integral normally involves a surface term, but we can drop this in light of our assumption since by Gauss' theorem:

$$\int dV\, \nabla \cdot (\mathbf{r}P) = \int dS\, \hat{\mathbf{n}} \cdot \mathbf{r}P, \qquad (4.6)$$

where $\hat{\mathbf{n}}$ is the surface normal. Since the pressure vanishes at the surface ($S \to \infty$), this integral is zero. It may, however, still be required if any external perturbations affect the star's geometry. For example, interacting binary stars can show radial truncation of either (or both) components by the gravitational attraction of its companion, the Roche radius. But let's continue:

$$\frac{1}{2}\frac{d^2 I}{dt^2} = 2K + \Omega + 3\int P\, dV. \qquad (4.7)$$

---

[7] For example, we could have a potential for a rotating star, in which case we would also need to include the Coriolis and centrifugal terms arising in the corotating frame.

We have now arrived, yet again, at the virial theorem. The time derivative of I vanishes for a stable body, or is at worst oscillatory. As long as the changes in the stellar mass distribution are slow or average to zero, the left-hand side vanishes. This is the chief equilibrium condition for a self-gravitating body. The total energy is $E = K + \Omega$ so that, in the absence of pressure support, $E = \frac{1}{2}\Omega < 0$. Pressure increases the internal energy, or decreases the gravitational potential, with the result that the body is more prone to instability.[8]

4.2.2 The Kelvin–Helmholtz Timescale

Amazingly, we can already say something about the evolution of a star knowing nothing except Eq. (4.7). Since the total energy is always negative, a star that loses internal energy it must contract. When this happens, only a portion of the energy is converted into heat. The luminosity of the body in the case of spherical symmetry and constant (mean) density is

$$L = -\frac{dE}{dt} = \frac{3}{10}\frac{GM^2}{R}\frac{\dot{R}}{R}. \tag{4.8}$$

We may require additional geometrical factors to describe the mass distribution, but this is close enough for our purposes. Now assume that all the luminosity of the star derives from the contraction. Thus, a constant luminosity requires a constant contraction rate, and we obtain a characteristic time for the process:

$$t_{KH} = \frac{3}{10}\frac{GM^2}{RL}. \tag{4.9}$$

This is the *Kelvin–Helmholtz contraction time*, the rate at which a hot self-gravitating body must contract to maintain a steady luminosity in the absence of any alternative internal energy mechanism. It was derived independently in the late 1850s by Kelvin and Helmholtz from identical considerations of the age of the Sun and the Earth. Now we can turn this around, in light of the discovery of nuclear reactions as the chief energy source within stars, and say that it is the characteristic thermal timescale for a radiating star in the absence of alternative energy sources. Because it is the thermal timescale, it is essentially the rate at which energy losses, from diffusion, produce structural changes in the stellar interior. This timescale will reappear when when we examine pre-main-sequence stellar evolution. It dominates the early history of stars before they initiate core nuclear burning. It is interesting, however, that such an important constraint follows immediately from the virial theorem.

---

[8]Since we can assume, for the moment, that there are no *bulk* motions, we take $2K = 0$. The internal energy is $U = \int P\,dV$. If rotation, expansion, or any other velocity field is present other than molecular chaos or turbulence, we have to include it in the kinetic energy. We will return to this point in Chapter 6 when discussing the stability of molecular clouds.

4.3 THERMODYNAMICS AND EQUATIONS OF STATE

In stellar interiors, it is often more convenient to think in traditional thermodynamical terms about processes. This is because the opacity is so high, with the exception of neutrino processes, that thermal equilibrium is rapidly established and we easily achieve the continuum limit of statistical processes. Thus, we now examine the thermodynamic variables, in order to complete our picture of the star and its physical description.

4.3.1 Polytropic Equation of State

In Chapter 1, we derived the basic thermodynamic relations. Now we recall them in order to establish an equation of state for a stellar interior. We first require the specific heat at constant pressure in terms of c_v, the specific heat at constant volume:

$$c_p = c_v + \frac{\mathcal{R}}{\mu}. \tag{4.10}$$

It follows that for an ideal gas $c_p/c_v > 1$. Clearly from the definitions of the specific heats, we can derive a more general relation using $U = U(V, T)$:

$$c_p - c_v = \left[\left(\frac{\partial U}{\partial V}\right)_T + P\right]\left(\frac{\partial V}{\partial T}\right)_P. \tag{4.11}$$

The importance of these relations is that in nonadiabatic portions of the star, there may be large changes in the equation of state and in the relation between U and P. This is expressed as changes in the difference between the two specific heats, or alternatively, in a change in their ratio. Now we define a new variable, the ratio of specific heats $\gamma \equiv c_p/c_v$. Notice that by Eq. (4.10), γ is always ≥ 1.

This formalism works well as long as the gas behaves "perfectly," but this doesn't happen at the temperatures and densities encountered just below the surface of any star. For one thing, the mean molecular weight changes in regions undergoing nuclear processing. Radiation pressure also modifies the equation of state. Finally, partial ionization results from variation in the temperature of the interior, which often exceeds the ionization potential of the most abundant species. All of these alter the relations among the thermodynamic variables. For a perfect gas, the kinetic energy depends only on the temperature and number of particles, N:

$$dK = \frac{3}{2}kT\,dN = \frac{3}{2}\frac{\mathcal{R}}{\mu}T\,dM, \tag{4.12}$$

and the internal energy is $dU = c_v T\,dM$. Then we find

$$dK = \tfrac{3}{2}(\gamma - 1)\,dU. \tag{4.13}$$

It is important to connect what we have just done with the dynamical version of the virial theorem from Chapter 1. The equipartition law states that $K = \tfrac{3}{2}NkT$ and requires an isotropic velocity distribution of the presumably independent constituent particles in phase space. This is likely a good assumption in the depths of a star, except where the equation of state breaks down (as in degenerate gases), but in our more general introductory discussion of the virial theorem, we did not require this assumption.

We now show that it is possible to write the equation of state in terms of γ that provides a simple way to connect the thermodynamics with the hydrodynamic balance. Take the case of an adiabatic medium for which $dS = 0$:

$$c_v \, dT + \frac{\mathscr{R} T}{\mu V} dV = 0, \qquad (4.14)$$

which leads directly to $TV^{\gamma-1}$ = constant. We thus arrive at a unique closed form for the equation of state, which we will exploit as far as possible in what is to come:

$$P = K\rho^{\gamma}, \qquad (4.15)$$

where K is usually called the *entropy constant* since it value depends only on the entropy. This is called a *polytropic* gas. It is the most generally used form for the equation of state, being grounded firmly in the thermodynamic state of the medium. For a perfect gas, one composed of identical particles with only translational degrees of freedom and no additional correlations in the distribution function, the ratio of specific heats is $\frac{5}{3}$. This comes from the fact that the kinetic energy for each free motion is $\frac{1}{2}kT$. For a system with f degrees of freedom, $\gamma = 1 + 2/f$.

4.3.2 The Radiation-Dominated Equation of State

How do the thermodynamic variables change when we include radiation pressure? For any isotropic photon gas, the energy density is related to the pressure through $P = \frac{1}{3}u$, as for any totally relativistic perfect gas. You already know this from the Eddington moments of the transfer equation. (Remember the factor K_ν/J_ν from our discussion of the radiative transfer equation?) For a closed system in thermal equilibrium, we know that the internal energy density depends only on T so that $U = u(T)V$. We also know that $dS = 0$, since LTE is exactly adiabatic. Thus the first law of thermodynamics yields an exact differential[9] having the integration condition that

$$\frac{1}{T}\frac{\partial}{\partial V}\left(\frac{\partial U}{\partial T}\bigg|_V\right) = \frac{\partial}{\partial T}\left(\frac{1}{T}\left[P + \frac{\partial U}{\partial V}\bigg|_T\right]\right). \qquad (4.16)$$

Substituting for the pressure and energy density, we find that

$$\frac{du}{u} = 4\frac{dT}{T} \qquad (4.17)$$

---

[9]This means that when $df(x, y) = a(x, y)dx + b(x, y)dy$ vanishes, then $\partial a/\partial y = \partial b/\partial x$.

and hence $u = aT^4$; the resulting pressure law is

$$P_{\text{rad}}(T) = \tfrac{1}{3}aT^4, \tag{4.18}$$

where the integration constant a is the Stefan–Boltzmann constant and the factor of $\tfrac{1}{3}$ comes from the isotropy of the radiation in optically thick conditions. As you know, this equation is also a constraint on the LTE integrated source function and the exact integral of the Planck function. It completely specifies the radiation pressure in an optically thick medium. In those regions of low opacity, such as the outer atmosphere, it is this requirement of anisotropy that fails, and consequently the radiation pressure departs from this formalism. Comparing Eq. (4.18) with Eq. (4.15), we see that a radiation dominated gas is one that has $\gamma = \tfrac{4}{3}$. This is a very soft equation of state, in fact as we will see the softest that is permitted for stability of a self-gravitating fluid, and is due to the relativistic nature of the "particles" in the gas. For astrophysical problems, this is most useful. We often must deal with radiation-dominated fluids, or at least those in which the radiation cannot be completely neglected. The total energy is now

$$U = \frac{3}{2}\frac{\mathscr{R}}{\mu}T + aT^4 V = c_v T + aT^4 V, \tag{4.19}$$

and the total pressure becomes

$$P = \frac{\mathscr{R}}{\mu}TV^{-1} + \frac{1}{3}aT^4. \tag{4.20}$$

We will define the ratio of gas to total pressure as

$$\beta \equiv \frac{P_g}{P} \tag{4.21}$$

to simplify the notation in what follows. You should note, however, that β is generally *not* a constant in the stellar interior, although this was the main assumption of Eddington's model for stellar structure. Since $P_{\text{rad}} = P - P_g$, Eq. (4.20) becomes

$$U = \left(\tfrac{3}{2}\beta + 3(1-\beta)\right)PV = \tfrac{3}{2}(2 - 3\beta)PV, \tag{4.22}$$

so taking the derivatives we obtain

$$dU = \left[\frac{3}{2}\frac{\mathscr{R}}{\mu} + 4aT^3 V\right]dT + aT^4\, dV. \tag{4.23}$$

From Eq. (4.23), we can read off the specific heat at constant volume:

$$c_v = \frac{3\mathscr{R}}{2\mu}\frac{8 - 7\beta}{\beta} \tag{4.24}$$

and from Eq. (4.10), we immediately obtain c_p. Note that a relation between T and ρ follows from P_{rad}/P_g in terms of only β:

$$\frac{a}{3}\frac{\mu}{\mathscr{R}}\frac{T^3}{\rho} = \frac{1-\beta}{\beta}. \tag{4.25}$$

Thus, since $P = \mathscr{R}\rho T/\mu\beta$, we obtain

$$P = \left(\frac{3}{a}\frac{1-\beta}{\beta^4}\left(\frac{\mathscr{R}}{\mu}\right)^4\right)^{1/3} \rho^{4/3}. \tag{4.26}$$

You will notice a new complication has appeared in our description of a radiative gas. In the general form of the equation of state, a single γ does not suffice to specify the relations between the constituent thermodynamic variables, and it is convenient to use a set of three adiabatic exponents. These are *defined* analogously to the polytropic equation of state:[10]

$$\begin{aligned} \Gamma_1 &\equiv \left.\frac{\partial \ln P}{\partial \ln \rho}\right|_s \\ \frac{\Gamma_2}{\Gamma_2 - 1} &\equiv \left.\frac{\partial \ln P}{\partial \ln T}\right|_s \\ \Gamma_3 - 1 &\equiv \left.\frac{\partial \ln T}{\partial \ln \rho}\right|_s. \end{aligned} \tag{4.27}$$

and are simply convenient and compact representations of the equation of state. You can now derive all the adiabatic indices for the equation of state where radiation pressure is included. For an ideal gas all indices are identically equal to γ. Another advantage of a genuine polytropic equation of state is that, by definition, a perfect gas that undergoes no state changes (e.g., for which μ remains constant) has all of its specific heats equal to the adiabatic values.

4.3.3 Effects of Partial Ionization

Ionization changes the internal energy of the gas. The number of degrees of freedom increases because there are now additional free particles (electrons), thus lowering γ. Also, ionization has a threshold, χ_H. To make life simpler, we will use a purely atomic hydrogen gas so that there is a single ionization potential and a simple relation between the ionization and the number density of the species, namely, $n_p = n_e$. The internal energy is

$$U = \tfrac{3}{2}(1 + x)NkT + Nx\chi_H, \tag{4.28}$$

---

[10] They should not be taken as having more significance than the γ that represents the actual ratio of the specific heats and the one that enters into the stability criterion.

where x is the ionization fraction. The specific heat at constant volume changes because the ionization fraction depends on the temperature through the Saha equation:

$$\frac{x^2}{1-x} = \frac{2(2\pi mkT)^{3/2}}{h^3 n Z_H(T)} e^{-\chi_H/kT}, \qquad (4.29)$$

where $Z_H(T)$ is the neutral hydrogen partition function. Since all we need for the computation of c_v is the partial derivative with respect to *temperature*, the calculation is tractable:

$$c_v = \left(\frac{\partial U}{\partial T}\right)_V = \frac{3}{2}\mathcal{R}\left[(1+x) + \left(1 + \frac{2\chi_H}{3kT}\right)\frac{\partial x}{\partial \ln T}\right]. \qquad (4.30)$$

You can see here how the ionization affects c_v from the second term. A similar change can be calculated for c_P. Here we are not merely displaying formalism. The change in the heat capacity of a gas affects its mechanical equilibrium through the hydrostatic and energy transport equations, and thus determines the stability of the mass.

4.3.4 Stability of Polytropes

We now return to the virial theorem and derive a profound result from truly minimal physical assumptions. In the absence of bulk motions, the kinetic energy is the thermal energy, and by Eq. (4.13) this can be substituted into the virial theorem to relate the internal energy to the gravitational potential energy:

$$3(\gamma - 1)U + \Omega = 0. \qquad (4.31)$$

Now since the total energy E is given by $U + \Omega$, we have

$$E = -(3\gamma - 4)U = \frac{3\gamma - 4}{3(\gamma - 1)}\Omega. \qquad (4.32)$$

Stability requires that the mass must remain bound so the total energy must be negative. Unfortunately, you have found that it vanishes if $\gamma = \frac{4}{3}$. In other words, it is possible to have a polytrope for which the matter *cannot* be stable. Keep this in mind because you will need it shortly. The virial theorem leads to a most interesting astrophysical consequence.[11] If a self-gravitating body contracts, the

---

[11]Actually, this does not require that the equation of state have any particular form, as long as we can take global averages over the mass to obtain the various contributions to the total energy. The most useful feature of Eq. (4.32) is that it provides a *general* constraint on the stability of *any* gaseous self-gravitating body, regardless of its equation of state. We should emphasize also that the γ used for this result need not be the ratio of specific heats. It can be a general index, as you will see from the structure equation.

total energy increases. This is the *negative specific heat* problem that was mentioned above, which leads to the *gravothermal catastrophe*. A radiating star must evolve as long as its equation of state remains ideal. If, however, the gas reaches degeneracy, the contraction may stop. You can see how this leads to gravitational collapse. As a self-gravitating body cools it is forced to contract and heat up, increasing its energy losses (because of the increase in T the surface loss terms go up). The contraction rate must increase to compensate for the losses. If, however, the equation of state is independent of temperature, as it is for a degenerate gas, the energy losses do not produce a contraction. For this reason, white dwarf stars are thermodynamically stable and normal stars are not.

State changes in the interior also affect the local stability of the gas. For instance, as a consequence of partial ionization, γ can be lower than $\frac{5}{3}$, and the adiabatic exponents, Γ_i. are no longer the same. In particular, Γ_1, the critical exponent for global stability, decreases from partial ionization because of the increase in the number of degrees of freedom in the gas and becomes smaller than for complete ionization. Such partial ionization zones, which are also regions of high opacity, can *locally* have $\Gamma_1 < \frac{4}{3}$. Since for the whole star to maintain this condition is impossible, the partial ionization zone must be confined to a comparatively small mass range in the interior. The integral condition for global stability $\int (\Gamma_1 - \frac{4}{3}) P \, dV > 0$ can still be satisfied over the whole structure, notwithstanding this *local* failure. This lack of mechanical stability in the ionization zone will be the subject of our discussion of a fundamental property of some stars: this region serves as the thermal source for the work required to drive pulsation as an ordered motion.

Since stability depends so critically on the equation of state, let us take a moment to examine what this means. What does γ really measure? The heat capacity of a gas depends on two things, the kinetic distribution of the free particles and the number of internal degrees of freedom. In pressure equilibrium, the greater the number of energetic modes available to the constituents, the larger the heat capacity will be for the gas. The smaller γ means the medium is more compressible. The cooling leads to a greater contraction of the mass for lower γ, so the system radiates more gravitational potential, thus cooling more, and leading eventually to a catastrophic collapse. As we just mentioned, state changes affect the stability of the medium. In an ionization zone, the specific heat changes rapidly because of a change in the number of free particles. The free energy changes rapidly and the opacity rises. The medium has become more opaque, and thus has a higher specific heat. Since the ionization depends on pressure as well as temperature, both c_p and c_v change. The ratio of the two, because of the increase in the entropy of the zone, decreases and approaches the dangerous region near $\frac{4}{3}$. The overall result is that convection has a destabilizing effect, and since this occurs whenever the temperature gradient is steep, it also occurs when the γ value is small.

We should add one final point before continuing our discussion of the thermodynamics in stellar interiors. Although we know that stars are not really polytropes and that the Lane–Emden equation (Section 4.4.2) has more illustrative than predictive significance, such equations of state are nevertheless very important in developing your feeling for how self-gravitating objects behave. Their great utility is that such models provide exact solutions for the structure that will be seen to provide good starting guesses for more precise calculations. The nonlinearity of the structure equations when radiative transfer and energy generation are included requires that we start with some initial model that is, at least approximately, representative of the interior. Yet for many applications, especially for studying the stability of stars, such a simple equation of state proves useful. How marvelous that even a very antiquated physical picture still provides insights; it leads us to hope that in the end stars may truly be simple. Well, we'll see.

4.3.5 Degenerate Equations of State

In an ideal gas, the phase space available for the particles is determined by their density and their speed. They have the same velocity distribution and have sufficiently high energies, or low enough densities, that a collisionless gas assumption essentially works. Elastic collisions are not constrained. As the density rises, however, we reach some point at which this is not true. One way of looking at this is as follows.

We treat the world of daily life in classical mechanical terms because its relevant parts are large compared with the quantum scale. Billiard balls are not fuzzy bodies. They have hard edges, move as solids, and cannot occupy the same volume as their neighbors. Their center of mass motion is unrestricted by anything but the collisional mean free path. If the density increases, there is a maximum pressure that comes from this mechanical exclusion rule that is very hard. Now consider what happens in a very dense medium for a *quantum* billiard ball. A particle has a de Broglie wavelength given by $\lambda = h/p$. With a number density n, the mean distance between particles is $n^{-1/3}$. As long as the density is low enough, the only way a particle of momentum p knows about its neighbors is through collisions. But as the density is increased, a critical point is reached at which the mean particle separation becomes approximately the same as the de Broglie wavelength:

$$n_c p^{-3} h^3 \approx 1. \tag{4.33}$$

The volume of phase space that a particle occupies is given by $(\Delta x \, \Delta p)^3$. Now the point is that there is a minimum value that any body has, even at zero temperature. It is given by the uncertainty principle. The particle momentum is determined by its temperature, in the nonrelativistic case $p \sim (kT)^{1/2}$. Thus, to find out if a gas truly obeys the ideal equation of state, we ask whether its phase space volume is larger than this minimum, which is h^3; in other words, we must have

$$\frac{\Delta x \, \Delta p}{h} > 1 \tag{4.34}$$

in order to apply the ideal-gas law. Otherwise, it does not matter what the temperature is, the pressure will not respond to changes in the velocity dispersion of the particles. The governing parameter formed from Eq. (4.33) can be expressed in terms of the electron density and temperature:

$$\alpha \equiv \frac{T^{3/2}}{n_e h^3} \qquad (4.35)$$

and describes the behavior of any particle in a gas as the temperature is lowered. Now we come to the application of quantum statistics: electrons are fermions and obey the *exclusion principle*. At zero temperature, there can be no more than one particle in each elementary volume of phase space. The parameter α measures the pressure of a fermion gas in the limit of very high density *or* very low temperature. For this reason, α is called the *degeneracy parameter*[12] because it measures the departure of a gas from an ideal equation of state.[13]

At every density there is some temperature below which a fermion gas does not "know about" the thermal energy or excitation of the gas. The equation of state is independent of temperature. Interestingly enough, nonrelativistic electron degeneracy has the same polytropic approximation as an adiabatic gas, $P \sim \rho^{5/3}$. The principal difference is that now we have a unique value for the proportionality constant in terms of Planck's constant, an indication that we are dealing with a quantum-mechanical phenomenon. As we discussed in Chapter 1, a relativistic degenerate electron gas behaves asymptotically as $P \sim \rho^{4/3}$. You will recall that this form of the equation is both soft and dangerous. In the virial theorem, we see that this is the same γ that we found for a photon gas. In the extreme relativistic limit, the particles behave as if they are "radiation-like."

4.4 EQUATIONS OF STRUCTURE

The stellar interior problem differs in several important ways from that of computing an atmosphere. Although it is necessary to properly treat hydrostatic equilibrium and thermal balance, the energy generation mechanisms that operate in the interior make this a complicated task: unlike an atmosphere, an interior has both gravitational and nuclear internal energy sources, so the condition of constant flux through a thin layer does not apply; the gravitational acceleration is not constant, nor is the mass, and these have to be solved together; the mechanisms that act to release energy in stars involve a transformation of the abundances with time—it is therefore necessary to include changes in the equation of state for changing composition; and there are avenues for energy loss that don't affect normal atmospheres, in particular neutrinos. Thus it is possible for the thermal timescale of a star to change very quickly, depending on the dominant energy loss mechanism, and this does not happen in a stellar atmosphere.

The timescales involved in a stellar interior span a very large range. The shortest is the neutrino loss time, since stellar matter is normally transparent to these particles, and they leave the site of energy generation virtually unaltered by

---

[12] Usually, the degeneracy parameter is written as $\alpha^{1/3}$, but we will use the convention of volume ratios.
[13] For bosons this limit at zero temperature does not apply. A Bose–Einstein condensate has zero volume in phase space and can be thought of as a completely stable, infinitely cold body with no zero-point energy.

interaction. Thus, for the usual densities at which such processes first operate, the star can cool instantaneously if the neutrino luminosity exceeds that for the photons. Then there is the gravitational collapse time, how long it takes for the body to undergo freefall. This is usually a matter of minutes to hours, but is still very short compared with the thermal timescale for photon diffusion, the Kelvin–Helmholtz timescale. Finally, there is the longest timescale, that of energy generation by nuclear burning. The range spanned by these intervals, from the order of seconds to aeons, accounts for the difficulties encountered in creating models for stellar evolution.

4.4.1 Mass Continuity and Hydrostatic Equilibrium

A star is a compressible body whose density, ρ, depends on radius. For a spherical mass in steady state, this reduces to the condition that $\int \rho \, dV = \Delta M = \int \rho dS \, dr$, where S is the surface area. Thus follows the most elementary structure equation, the one that gives the mass distribution:

$$\frac{dM(r)}{dr} = 4\pi r^2 \rho(r), \quad (4.36)$$

which is sometimes called the continuity equation (although it is merely the definition of the differential mass in a thin shell). This equation has another role in stellar structure. Inverted, it permits the use of the mass as the independent variable, providing $r(M)$. For many applications to evolutionary models, where the radius changes rapidly over time, this is the more convenient form.

The next condition that must be satisfied for most stars is hydrostatic equilibrium. We have already used this in constructing an atmosphere, but with the proviso that the gravitational acceleration is constant.[14] Hydrostatic equilibrium is the state reached when balance is achieved between the pressure within a layer and the weight of the overlying mass, $\Delta F = S \Delta P = -g(r)\Delta M$, where g is the local acceleration of gravity. Unlike the atmosphere, however, we cannot assume that g is constant because of the change in the internal mass fraction:

$$\frac{dP(r)}{dr} = -\frac{GM(r)}{r^2}\rho(r). \quad (4.37)$$

Using the mass as the independent variable, we can express this as

$$\frac{dP(r)}{dM(r)} = -\frac{GM(r)}{4\pi r^4}. \quad (4.38)$$

Hereafter we will drop the explicit dependence of the quantities on r where appropriate. This approach has the advantage of allowing us to work in units of the mass fraction without worrying about radial distance, for a spherical, static star. The structure equations written in terms of the mass variable are called the

---

[14] The distinction between normal and extended atmospheres is, in fact, whether we need to include the radial dependence of the acceleration or can treat it as a constant.

Lagrangian form. A layer is labeled by its mass fraction and followed as its radial position changes with time. This is especially useful in those stages of evolution that are dynamical or where the radius undergoes significant inflation (e.g., on the asymptotic giant branch; see text below). In fluid mechanics, the term is reserved for comoving derivatives, and the analogy holds well for dynamical problems in interiors.

4.4.2 Polytropes and the Lane–Emden Equation

Models of the thermal structure of a hot self-gravitating gas bag are almost as old as thermodynamics. In 1867, Homer Lane published in the relatively obscure *American Journal of Science* a comprehensive adiabatic model for the core temperature and envelope structure of a convective self-gravitating spherical mass. This was substantially extended by Robert Emden, who wrote the first monograph on stellar and solar interiors (in 1907) and established the basic equations for an arbitrary equation of state.[15]

First, we combine the hydrostatic and mass continuity equations and take the radial derivative of both sides to obtain

$$\frac{d}{dr}\left[\frac{r^2}{\rho}\frac{dP}{dr}\right] = -G\frac{dM}{dr} = -4\pi G\rho r^2. \qquad (4.39)$$

Now you see that the equation of hydrostatic equilibrium is really the just the Poisson equation for an extended mass distribution. This is not a surprise because the pressure gradient is interchangeable with the gradient of the gravitational potential, $\rho\nabla\Phi = \nabla P$, whenever the system is in mechanical equilibrium as long as we don't need to worry about the energy transfer. Thus

$$\nabla \cdot \frac{1}{\rho}\nabla P = -4\pi G\rho, \qquad (4.40)$$

independent of the coordinate system. This step is actually very delicate and physically extremely important. We assume that the body is spherical with constant density surfaces corresponding to those of constant pressure. This is certainly true for a polytropic equation of state, but it is not *necessarily* true for any arbitrary gaseous body. This is the *barotropic* condition, the statement that hydrostatic and thermal equilibrium hold simultaneously. It implies that isobaric surfaces are also isotherms. This is true, for instance, in nonrotating stars but clearly begins failing

---

[15] Lane's paper dealing with the internal temperature of the Sun is available in reprint with commentary in Meadows, A. J. 1970, *Early Solar Physics* (Oxford: Pergamon). Emden made important contributions to meteorology and balloon aviation as well as astrophysics. His monograph, Emden, R. 1907, *Gaskugeln: Anwendungen der Mechanischen Wärmetheorie auf Kosmologische und Meteorlogische Probleme* (Berlin: B. G. Teubner-Verlag), concerned the application of a generalized pressure law to gaseous bodies in equilibrium, including atmospheres. Lane more specifically addressed the problem of the solar interior. The great advantage provided by a polytropic equation of state was the ability to solve exactly the mechanical and thermal structure of a medium without requiring an equation for the energy.

as soon as the rotation rate, measured by λ_Ω (see Chapter 3 and Section 4.4.4, below) becomes large.

By choosing the polytropic equation of state, which we express as $P = P(\rho)$, we have reduced the complexity of the problem since the model will deal only with the density as a function of radius. It follows that the temperature can be computed from the thermal condition that the gas ultimately obeys the ideal-gas law. The most obvious choice for the equation of state is the one we have just derived, a polytrope with constant γ and with the scaling constant depending on the central pressure and density $K = P_c/\rho_c^\gamma$. Thus

$$\frac{1}{r^2}\frac{d}{dr}r^2\left(\frac{K\gamma}{\gamma-1}\frac{d\rho^{\gamma-1}}{dr}\right) = -4\pi G\rho. \quad (4.41)$$

We now formally simplify the equation by assuming that the derivative is written linearly:

$$\theta_n(r) \equiv \rho^{\gamma-1}, \quad (4.42)$$

with the *polytropic index*, n, given by $\gamma = 1 + 1/n$ as the defining index of a model for a given pair of central pressure and density. The structure equation also contains a characteristic scale length for the radius:

$$\lambda_n = \left[\frac{(n+1)K}{4\pi G}\right]^{1/2}. \quad (4.43)$$

Thus, the equilibrium radius of a polytrope of index n depends on both the central properties and n. Taking $x = r/\lambda_n$, we arrive at the dimensionless form of the *Lane–Emden equation*:

$$\frac{1}{x^2}\frac{d}{dx}x^2\frac{d\theta_n(x)}{dx} = -\theta_n(x)^n. \quad (4.44)$$

We now require some boundary conditions. Fortunately, these come quite naturally. First, the density should obviously be greatest at the center of the star. Since θ_n and ρ are really the same, $\theta_n = 1$ at $x = 0$ by definition and $d\theta_n/dx = 0$ at $x = 0$. In addition, since $x \geq 0$, any series for $\theta_n(x)$ should be positive for all x and the first derivative should be negative for $x \neq 0$ and vanish for $x = 0$; hence

$$\theta_n(x) = 1 - \sum_{j=1}^{\infty} a_j x^{2j} \equiv 1 - \Sigma(x). \quad (4.45)$$

Also, since $0 \leq \theta_n \leq 1$ everywhere, then $\Sigma \leq 1$. These obvious conditions, however, leave the surface unspecified. Yet the surface should be the radius at which the density or pressure vanishes so that $d\theta_n/dx \leq 0$ throughout the interior of the body and that $\theta_n = 0$ at some $x = x_{n\star}$. Each polytropic index has a different value of $x_{n\star}$, so the stellar radius depends uniquely on the equation of state. Knowing R, you can compute the total mass of the body. This comes from scaling the integral:

$$M(R) = 4\pi \int_0^R \rho(r)r^2 dr = 4\pi \lambda_n^3 \rho_0 \int_0^{x_{n\star}} x^2 \theta_n^n \, dx. \qquad (4.46)$$

However, from Eq. (4.44) you can substitute for the integrand and use the boundary conditions to obtain

$$M(R) = -4\pi \lambda_n^3 \rho_c \left(x^2 \frac{d\theta_n}{dx} \right)\bigg|_{x_{n\star}}. \qquad (4.47)$$

The scaled mass is unique for each polytropic model depending only on the index n. There are analytic solutions for a few cases ($n = 0, 1, 5$), but in general you have to solve the equations numerically with the boundary conditions we have specified. Thus, $x_\star$ is the eigenvalue of the Lane–Emden equation. Since you know ρ throughout the interior, you also find the run of temperature, which combined determine the properties of the interior matter: its ionization and most important, its opacity. As a starting point to a more realistic stellar model this is not bad, but to really understand a star requires that we bring in more information about energy generation and transfer. But let's wait on that for just a moment.

The twentieth century opened with an extensive set of investigations on the structure and stability of incompressible and polytropic bodies by Poincaré, Darwin, Dedekind, Jeans, Schwarzschild, and Cartan for the stability of liquid stars. Much of this was driven by an interest in the stability of rotating bodies as model problems for the formation of binary and planetary systems. For these questions, incompressible toy models were quite useful. But for the structure of realistic stars, they failed badly. Eddington realized that it was possible to obtain a simple solution for a luminous star. His picture was based on radiative equilibrium and required that massive stars be dominated by radiation pressure because of their high internal temperatures. In detail, the model is wrong. But it's name, the *standard model*, indicates how important it was in the early days of stellar structure calculations.

Consider the case $n = 3$, which corresponds to the star in radiative equilibrium. To simplify the computation, take the ratio of gas to radiation pressure to be constant *throughout* the interior and use Eq. (4.26) for the pressure. From this equation of state, the virial theorem shows that the star becomes progressively less stable as $\beta \to 0$ since in this limit, $\gamma \to \frac{4}{3}$. This yields a maximum luminosity of such a configuration. Rewriting the flux equation now in terms of the radiation pressure gives

$$F = -\frac{c}{\kappa_R \rho} \frac{dP_{\text{rad}}}{dP} \frac{dP}{dM} 4\pi r^2 \rho, \qquad (4.48)$$

Then, since $dP_{\text{rad}}/dP = 1 - \beta$ and $L = 4\pi r^2 F$, Eq. (4.48) shows that the luminosity of a radiation dominated star is given by

$$L = \frac{4\pi GMc}{\kappa_R}(1 - \beta). \qquad (4.49)$$

You should recognize this result. For the limit $\beta \to 0$, this is the Eddington luminosity we derived in Chapter 3. Notice that the temperature gradient has disappeared from the problem. This is because the radiation pressure depends only on T and its gradient establishes hydrostatic equilibrium. Of course, the limitation of this model is that it works only if the *entire* star is dominated by radiation with constant β, but the most massive stars come rather close to this. The result actually *predicts* a mass-luminosity relation $L \sim M$, whose consequences we will explore soon. The standard model is thus a very instructive test case. It approximates nicely the structure of hot nearly isothermal bodies, perhaps the early self-gravitating large-scale structures in the early Universe. Now let's get more specific. From Eq. (4.47) we have

$$K = \left[\frac{3}{a} \frac{1-\beta}{\beta^4} \left(\frac{\mathscr{R}}{\mu} \right)^4 \right]^{1/3}. \qquad (4.50)$$

If β and μ are constant throughout the star, a numerical integration of Eq. (4.44) gives

$$-x_{3\star}^2 \frac{d\theta_3}{dx}\bigg|_{x_{3\star}} = 2.018,$$

so from Eq. (4.47) we find a relationship between β and the stellar mass:

$$M \approx 18 \frac{(1-\beta)^{1/2}}{\mu^2 \beta^2} M_\odot. \qquad (4.51)$$

If stars were really polytropes, we could invert this equation to find the fractional contribution of radiation pressure to the support of the interior for any stellar mass. From Eq. (4.47) we can also find the concentration of the interior from the ratio of the average to central density:

$$\frac{\langle \rho \rangle}{\rho_c} = -\frac{3}{x_{3\star}} \frac{d\theta_3}{dx}\bigg|_{x_{3\star}} = \frac{1}{54.2}. \qquad (4.52)$$

In the early work on polytropes, this connection between the central density and structure was very important as a motivation for studying the long-term dynamics of binary star systems. Eccentric orbits in close binaries precess because of the multipole gravitational field produced by the finite size of the component stars. As a result, because of tidal distortion, the higher-order terms for the gravitational potential depend on $\rho_c / \langle \rho \rangle$ and can be determined from the precessional period.[16]

---

[16] For instance, see Kopal, Z. 1959, *Close Binary Stars* (NY: Wiley); Batten, A. 1974, *Binary and Multiple Star Systems* (Oxford: Pergamon). The most complete discussion of polytropic models is Chandrasekhar (n.2).

Another fundamental application concerns the case of a completely degenerate and extreme relativistic electron gas. The equation of state has the structure of a polytropic of index $n = 3$, but now the scaling constant is $K = 1.25 \times 10^{15}/\mu_e^{4/3}$ dyn cm^{-3} *independent* of β and temperature. It is one of those remarkable non-coincidences of astrophysics that the two equations of state produce the same thermodynamic model, since they both deal with relativistic matter and the mass dependence is subsumed in p. If μ_e is constant throughout the star, the stellar mass is

$$M = \frac{5.80}{\mu_e^2} M_\odot \qquad (4.53)$$

The fact that there is a constant mass for such an equation of state that depends only on the composition is a profoundly important result. Equation (4.53), called the *Chandrasekhar mass*, was the first indication that stars at the ends of their lives face the possibility of catastrophic gravitational collapse; if relativistic electron degeneracy is a limiting equation of state, this is the largest mass that can be supported. You can also see that *any* polytropic fermion gas will show the same sort of maximum value except it will be lower if the constituents are more massive than electrons. Even without including general relativity, if support for a star comes from degeneracy there must be a "cold" mass above which it is impossible to find a stable final state.

The *isothermal* problem is traditionally treated separately from the normal polytrope, although there is no reason why this must be so. For instance, since the isothermal case is the limit of $n \to \infty$, we have $(\theta_n(x))^n = (1 - \Sigma(x))^n \to \exp - \Sigma$ and $\Sigma(0) = 0$ and $d\Sigma/dx = 0$ at $x = 0$. The alternate way to get the structure equation is to take the temperature constant in the ideal-gas law and write the hydrostatic equation in terms of the gravitational potential, $\nabla \Phi = (\mathscr{R}T/\mu)\nabla \ln \rho$; now $\ln \rho$ and Φ are interchangeable. Interpretation of the Lane–Emden variable shifts from representing the density to the gravitational potential, but otherwise the derivation is the same:

$$\frac{1}{x^2}\frac{d}{dx}x^2\frac{d\psi}{dx} = -e^{-\psi}. \qquad (4.54)$$

In changing the labels, ψ is the scaled potential, while the radius is now normalized to the isothermal pressure scale height. The mass interior to some radius is found analogously to the usual method, but the total mass is actually unbounded since there is no distance at which ψ vanishes (think of how $\theta_n = 0$ defines the surface).

The physical weakness with the polytropic treatment and the Lane–Emden model of the interior is that we have left out all the information about energy generation and transfer. The assumption that the body rapidly establishes a uniform interior state might hold for a planetary interior, provided it does not solidify or undergo phase changes, but it does not work for stars. The reason is, in part, due to the variable thermal timescale in the different layers and also the steep temperature gradient established in the interior. There must be a gradient in the opacity and, as we discussed in Chapter 3, so the radiative gradient must vary in order to follow the changes in local opacity. Thus the polytrope is an excellent model whenever you need something to start with, and we shall see that it also works in an unexpected way for degenerate configurations, but it fails in the essential job of describing the thermal state of a stellar interior.

4.4.3 Energy Transport in Stellar Interiors

4.4.3.1 Radiation

We have already described in Chapter 3 most of the conditions of radiative and convective equilibrium required for the thermal balance of a stellar interior. In this regard, interiors are simpler than atmospheres for photon transfer because they are so dense and opaque. For instance, the source function is always given by $B_\nu(T)$, the Planck function. Then the opacity can be calculated without having to solve the detailed rate equations and can, in fact, be integrated over frequency. The Rosseland mean opacity is a harmonic mean over the Planck function and depends only of temperature, composition, and pressure:

$$\kappa_R = \kappa_0 \rho^n T^{-s}, \tag{4.55}$$

where κ_0 is a constant that depends on the elemental composition. The opacities used for realistic interiors calculations are tables within which the appropriate values are interpolated rather than such simple fitting formulas. Nevertheless, Eq. (4.55) is valuable for deriving scaling relations. For instance, a hot optically thick plasma radiates like a blackbody whose integrated source function is $B = \int_0^\infty B_\nu(T) d\nu = aT^4$. The frequency integrated emissivity is given by $j = \int_0^\infty \kappa_\nu B_\nu(T) d\nu$ and for thermal bremsstrahlung varies with temperature as by $j \sim T^{1/2}$. This yields an approximate form for the opacity that was first derived by Kramers:

$$\kappa_K = \kappa_{0,K} \rho T^{-7/2}. \tag{4.56}$$

At the limit of very high temperature, when ion–electron collisions are unimportant and all the matter is completely ionized, the opacity approaches the Thomson limit of electron scattering and becomes independent of temperature:

$$\kappa_T = \kappa_{0,T} \rho. \tag{4.57}$$

As we discussed in Chapter 3, radiative transfer is complicated by the variety of contributors to the real opacity of a gas. The schematic formalism used for Kramers opacities ignores the extreme temperature and density sensitivities of the many line and continuum absorbers. Remember that every species that has a maximum opacity at any depth near the local peak of the thermal distribution will be a very efficient absorber and heat the medium. Lines and resonances near continuum edges are narrow and therefore very temperature-sensitive. The physical point we have stressed, and will continue to use in discussions yet to come, is that the most efficient coupling between radiation and matter occurs when the peak of the spectrum happens to coincide with an absorption peak and anywhere the opacity changes rapidly with wavelength, and this also means that the coupling depends sensitively on chemical composition and temperature. For the stellar interior especially, because the temperatures are so high and so many ionization edges are accessible to the radiation, resonance absorbers produce a steep temperature dependence in the resulting opacity. As we will see when discussing pulsation, this feature of the opacity is the mechanically destabilizing feature of radiative transfer.[17] Photon transfer in stellar interiors is essentially diffusive (a circular statement that really defines the distinction between what we will call the interior or "envelope," and the atmosphere). We can write the flux as $\mathbf{F} \cdot \hat{\mathbf{r}} = -D_{\text{rad}} \hat{\mathbf{r}} \cdot \nabla T$. Since, however, you already know that the transfer equation in an optically thick medium is given by $F = -(4acT^3/3\kappa\rho)dT/dr$, we obtain the radiative diffusion coefficient:

$$D_{\text{rad}} = \frac{4ac}{3\kappa_R \rho} T^3, \tag{4.58}$$

so the explicit final form the flux equation is

$$L(r) = -\frac{16\pi acr^2}{3\kappa_R \rho} T^3 \frac{dT}{dr} = -\frac{64\pi^2 acr^4}{3\kappa_R} T^3 \frac{dT}{dM(r)}. \tag{4.59}$$

The Rosseland mean depends on the gradient of the source function as a function of frequency as much as its temperature. Consequently, at some depths in the interior, where the temperature is about the same as χ/k, where χ is an ionization energy for an important species, the opacity changes rapidly with temperature. On the other hand, in those parts of the star dominated by electron scattering, the Rosseland mean depends only on density. Thus regions where ionization is partial should have high opacities and require large temperature gradients, as we have already discussed for atmospheres. This implies that ionization zones are also likely to be convectively unstable. This effect is added to the one discussed earlier concerning the specific heat of a partially ionized zone. The two effects are reinforcing.

The opacity is the principal cause of the metallicity sensitivity of stellar models. The small effect it has on the pressure through the mean molecular weight is comparatively insignificant. A low metallicity star is more transparent, so it cools more easily and for the same flux requires a lower temperature gradient in its outer envelope. As a result, it must have a smaller radius than its higher metallicity counterpart. This shows up in evolutionary calculations. It is important to note that the elemental abundance differences between stars in the Magellanic Clouds and the Galaxy alter their structures at the same mass and also change their evolutionary histories, especially mass loss. Individual evolutionary and structure sequences must be computed for different chemical mixtures. Not only the atmospheres are

---

[17] A massive project has been undertaken by several groups, notably at Lawrence–Livermore (the OPAL project) and University of London and collaborators [called the Opacity Project or (OP)], to compute new opacities for stellar interiors. The opacities have changed largely because of resonances and autoionization effects near ionization edges. You can see from the Rosseland mean how this happens, because these are where the opacity is strongly frequency dependent and heavily weighted to specific temperatures when the Planck peak coincides with the opacity maxima. See Rogers, R. J. and Iglesias, C. A. 1992, *ApJ. Suppl.*, **79**, 507 for the OPAL compilation of opacities for a variety of stellar mixtures and pointers to the more recent literature.

4.4.3.2 Conduction

changed but also the stellar lifetimes and even the zero points of the mass–luminosity calibrations.

Conduction, although unimportant in the more transparent outer layers, is a very important mode of heat transfer in the opaque interior plasma. It requires that the density be sufficiently high for collisions to completely dominate over radiative or buoyant processes. Conventional convection theory treats the energy transfer competition between the conductive and convective fluxes for an incompressible fluid by assuming that the density is only a function of temperature. The electrons are mainly responsible for the heat transfer since they have the highest collision frequency. However, and this cannot be neglected when treating the interiors of white dwarfs, degeneracy effects are important in the collisions. As the degeneracy increases, the lowest momentum states become progressively less accessible and the conduction ceases to be efficient.

If the density is high enough, particle kinetics rather than radiation can be the principal mode of energy transfer. A temperature gradient is the same as a gradient in the mean particle kinetic energy so collisions transfer the energy diffusively outward:

$$\frac{\partial E}{\partial t} = \nabla \cdot D_{\text{cond}} \nabla T, \qquad (4.60)$$

where D_{cond} is the diffusion coefficient, given by $\frac{1}{3} c_v v_e \lambda$, taking v_e as the thermal velocity and λ as the mean free path for collisions. To put this on the same footing as diffusive radiative transfer, since the processes are in thermal equilibrium, from Eq. (4.60) we write

$$\kappa_{\text{cond}} = \frac{K_{\text{rad}}}{D_{\text{cond}}}. \qquad (4.61)$$

The total opacity is the harmonic mean of the conductive and Rosseland opacities:

$$\frac{1}{\kappa} = \frac{1}{\kappa_{\text{cond}}} + \frac{1}{\kappa_R}, \qquad (4.62)$$

since both processes operate in dense environments. Since the electron speed varies as $T^{1/2}$ and the mean free path varies as $v_e^{-3} \sim T^{-3/2}$, the scaling for the diffusion coefficient is T^{-2}. For nondegenerate matter the conduction coefficient has the same temperature dependence as the collision frequency, ν_c. The conduction coefficient is $\lambda_D^2 \nu_c$, where λ_D is the Debye length that was described in Chapter 1. The result is

$$\nu_c \approx 2 \times 10^{-4} Z^{-4} T^{5/2} \ln \Lambda \qquad (4.63)$$

and Λ is the Coulomb integral [Eq. (1.172)], given by the ratio of the Debye length to the minimum distance that an electron gets to an ion of charge Z:

$$\Lambda = \frac{\lambda_D}{b} \approx 1.3 \times 10^4 T^{3/2} n_e^{-1/2}. \tag{4.64}$$

As ν_c increases, the conductive *opacity* decreases. Conduction, like thermal bremsstrahlung opacity, is modified when the electrons become degenerate. The final scattering states are restricted only to those above the Fermi momentum, so the density of final states is reduced. This increases the mean free path for scattering, and the mean momentum is then p_F, so the medium has a very low opacity and electron conductivity dominates.

4.4.3.3 Convection

You know that any location where the gas locally has a high opacity and a low γ is likely to pose trouble: it is where convection probably occurs. We have already discussed this for the atmosphere where you saw that the steepness of the temperature gradient in the partially ionized portions of the envelope produce mass motions. In the interior, the problem appears somewhat simpler. Here we find that the star rapidly establishes a fully adiabatic temperature gradient:

$$\frac{dT}{dr} = \frac{dT}{dr}\bigg|_{Ad} = T\left(\frac{d \ln T}{d \ln P}\right)\frac{d \ln P}{dr} = \frac{T}{P}\left(1 - \frac{1}{\Gamma_2}\right)\frac{dP}{dr}, \tag{4.65}$$

and the equation of state uniquely connects the pressure and temperature uniquely within the convective region.[18]

The interior gas is so opaque that there is little chance for a rising element to radiate, and the viscosity is high enough that it always moves under essentially adiabatic conditions. The temperature gradient should then be ∇_{Ad} inside the star whenever $\nabla_{rad} > \nabla_{Ad}$. The conventional approach is to set $\nabla = \nabla_{Ad}$ within the convection zone. Another approach can also be used for stellar interiors, where the situation approaches the typical laboratory environment far closer than the outer layers of the star. This uses the Rayleigh number, Ra. You will recall from Chapter 3 that there is a critical value, Ra_c at which the gas begins to convect. Let's concentrate for a moment on the effects of the temperature gradient, radius, and gravity. If we keep the compressibility, viscosity, and conductivity constant, we can write the critical Rayleigh number schematically as

$$Ra_c \sim MR^2 \langle \nabla T \rangle, \tag{4.66}$$

---

[18] L. Biermann (1936, *Astr. Nachr.*, **257**, 269), in the founding paper on convection in stellar interiors, noted that the variation in the opacity of stellar material would cause the radiative gradient in the *standard model* of Eddington (1926) to eventually exceed this limit. He proceeded to show that regions of partial ionization were the most likely places for convection to occur and that the gas would transfer energy on distances of the order of the pressure scale height. He used some of the ideas of the Göttingen aerodynamicists, especially Prandtl and Schmidt, to justify this assumption and thus introduced the idea of a mixing length into astrophysics. The work was greatly refined much later by E. Vitense (1953, *Z. Ap.*, **32**, 135) who developed the first practical method for computing the buoyancy of the gas in an atmosphere.

where M and R are the stellar mass and radius, respectively, and $\langle \nabla T \rangle$ is the *mean* temperature gradient for the star. Substituting the radiative gradient

$$(\nabla T)_{\rm rad} \sim \frac{\kappa_R \rho}{T^3} \frac{L}{R^2}, \qquad (4.67)$$

we find that a *critical luminosity* must exist at which convection will succeed radiation as the primary mode of energy transport for a star of a given mass and radius. To see this, equate the two representations of ∇T and get

$$L_c \sim \frac{T^3}{\kappa_R} \cdot \frac{R^3}{M^2}. \qquad (4.68)$$

Thus, at each mass, the star becomes more *unstable* to the onset of global convection if the opacity is increased. Take, for instance, main sequence stars. Combining the mass-radius scaling, $R \sim M^{1/2}$ and the mass–luminosity relation, $L \sim M^3$ shows that the critical variation of the temperature with mass becomes $T \sim M_c^{7/6}$. Thus, if M dwarf stars are completely convective, B stars should not be. We can also look only at the envelope, setting the characteristic length equal to the *pressure scale height*, $H_p = \mathscr{R}T/\mu g$. The Rayleigh number now becomes

$$\mathrm{Ra}_c \sim \frac{T^4}{g^3} \frac{dT}{dr} \qquad (4.69)$$

Now assume that the mass and temperature of the star are known. The density then scales as $g^{3/2}$ so that we can substitute in the luminosity for $T^3 \nabla T$:

$$L_c \sim \frac{g^{1/2}}{\kappa_R}. \qquad (4.70)$$

Equations of this sort can be generalized to give a critical luminosity as a function of location in the Hertzsprung-Russell diagram or as a function of spectral type. The higher the opacity or the lower the surface gravity, the lower the critical luminosity, all other things being the same, so giants and supergiants are likely to be convective than their main-sequence counterparts or have deeper convective envelopes when convection develops. We encounter this convective stage during the ascent of the giant branch and also in the pre-main-sequence contraction of a protostar. The track on the HR diagram can be understood using this global approach to the Rayleigh number; these are the paths in the HR diagram followed by homologous fully convective stars, stars of high enough luminosity at a given temperature and gravity that they must develop deep convective envelopes. Notice that even without the detailed computation of the interior models, we can predict that such stars must exist simply on the basis of the Rayleigh criterion for convection. Like any similarity technique, this one must be applied with caution to be certain that you are not comparing the properties of physically different objects, but when properly constrained, this is a powerful guide.

We can gain more insight from this dimensional analysis. Call T_{BCZ} the temperature at the base of the envelope convective zone. Since a convecting envelope must have a critical or supercritical Rayleigh number, let $\Delta T \approx T_{\text{BCZ}} \gg T_{\text{eff}}$ and substitute this into Ra_c. Then $gT_{\text{BCZ}}R^3 \approx$ constant. Using mean values for g this yields a simple mass-radius relation for the envelope $MR \approx$ constant. Thus, for stars that maintain constant basal temperature, as determined by the nuclear source, the radius of the outer layers increases in response to a decrease in the stellar mass. It is the interplay between buoyancy and the temperature gradient that sets this up. Whenever convection dominates the envelope it expands to accommodate the adiabatic temperature gradient. This is independent of any detailed dynamical theory of convective energy transport and results only from the condition that heat is efficiently carried by fluid motions throughout the convecting region. Although the exact relation is different, its qualitative behavior is the same.[19]

Convection is mass motion, and so is pulsation. They are linked through the instability of partial ionization zones of the stellar envelope. We will discuss how the envelope selects the organized pulsational modes, but here we should make some comments about how convective transfer occurs during pulsation. If the surface gravity is high enough, the buoyant motions will be fast and virtually instantaneous adjustment of the convective zone will occur throughout the cycle. There is, however, a lag between the maximum radius and minimum temperature introduced in the pulsation of stars due to the variation in the ionization, and hence the opacity, through the cycle. If the gravity is low, as in giant and supergiant stars, the convective motions may not be fast enough over even one mixing length to keep up with the change in the environment. This time-dependent form of convection is still poorly understood and requires a full solution to the fluid equations for a turbulent medium. As you've probably guessed, at the moment that is out of the question. Instead, plausible workarounds have been developed for nonlocal mixing of gas that simulate the conditions that are expected during a pulsational cycle.

4.4.3.4 Semiconvection and Double-Diffusive Processes

Semiconvection, the transport of mass and entropy due to compositional gradients, is actually more of a misnomer than mixing length. You already know this for an ideal gas, since even in the Schwarzschild scenario of a blob moving buoyantly through a background medium with which it maintains pressure equilibrium:

$$\frac{\delta P}{P} = \frac{\delta \rho}{\rho} + \frac{\delta T}{T} - \frac{\delta \mu}{\mu} = 0, \qquad (4.71)$$

so the buoyancy is determined not only by ∇T, the temperature gradient, but also by $\nabla \mu$, the composition gradient. This also means that a medium can become

---

[19]An important application comes from binary star evolution. In a binary system that has a loser with a deep convective envelope, the offloading of mass onto the gainer causes the star to relax to a greater size, maintaining contact for a longer time with the Roche surface. The mass loss occurs at the thermal rate (that is, on the thermal timescale). We discuss this in Chapter 5.

"semiconvective" if the compositional gradient is large enough. This mechanism for heat (entropy) transport was first discussed by Ledoux in 1949, and the criterion is so named. Semiconvection is also connected with the concept of "overshooting," and we will discuss these together.

There is an analogy to this process in everyday experience, known as "salt fingers." The process works like this. Take a medium that is otherwise thermally stable but that has a composition gradient. Assume that the mixing is due to both diffusion and advection so the evolution equation for this scalar quantity is

$$\frac{\partial X}{\partial t} + v_i \frac{\partial X}{\partial x_i} = \frac{\partial}{\partial x_j}\left[D_X \frac{\partial X}{\partial x_j}\right]; \qquad (4.72)$$

D_X is a diffusion coefficient for the tracer X. This representation comes from the Fokker–Planck equation for a stochastic mass transport (see Chapter 1) and has been used as a macroscopic diffusive term in nuclear reaction networks.[20] The problem is then to couple this with the normal equations of fluid motion. Let's look at a simple version of how this works. In the classical Rayleigh–Benard problem, the competition between buoyancy and viscosity is handled by the thermal diffusion equation. Now for simplicity, look at the incompressible approximation again, only now assume the density to vary with both composition *and* temperature:

$$\frac{\delta \rho}{\rho} = \alpha_T \frac{\delta T}{T} + \alpha_X \frac{\delta X}{X}. \qquad (4.73)$$

Here α_T and α_X are assumed to be constants. The fluid equations of motion are supplemented now by Eq. (4.72). This is the case of *double diffusive convection*. As in the thermal problem, we can form from the linearized equations a "double-diffusion Rayleigh number" using both κ and D. Notice that the signs of α_T and α_X may be different. From Ra you see that the effective conductivity is the harmonic sum of the two contributions while the buoyant coupling term is the direct sum. Even for thermally subcritical regimes, it is possible to induce mass mixing by large enough composition gradients.

4.4.3.5 Overshooting and the Mixing Length Concept

The usual Schwarzschild criterion for convection assumes the existence of a single characteristic scale for the mass motions. If the pressure scale height is a reasonable limit for the mixing length, any process that produces more mixing than you would expect from such a length scale is treated in the somewhat artificial way we have just discussed. There are, however, good reasons to think that convection in a stellar interior is turbulent and thus has no single length. The overshooting is actually more likely the result of the spectrum of this turbulence, and detailed calculations are yet to be carried out for the transport of mass across large

---

[20] See also Langer, N. 1986, *Astr. Ap.*, **164**, 45; Brüggen, M. and Hillebrandt, W. 2001, *MNRAS*, **320**, 73.

abundance gradients.[21] It seems we are begging the question here: Does convection actually take place according to the mixing length prescription?

We skirted this issue when discussing stellar atmospheres, but the interior is a different problem. For one thing, turbulent mixing not only transports mass and nuclear products but also produces a dynamical viscosity that can transport angular momentum. The effect of mixing in a region of variable chemical composition is also not negligible for stellar evolution since the abundances in any part of the interior affect both the local energy generation and opacity. In stars, the simplest answer to the question of whether there is a mixing length is "yes, because we say there is" but this is far from satisfactory. In fact, the answer must ultimately be "no."[22] The Schwarzschild criterion, or its extended form according to the Ledoux criterion, does not accurately represent the physical process that transports mass and heat in a turbulent medium. The mixing length scenario is, essentially, particle transport. A parcel of gas is moved some distance, determined by the lapse in pressure in the surrounding medium, and then mixes with that background and dissolves. It deposits its energy locally, having been in a sufficiently opaque medium before that, and one that is superadiabatic, that the transfer of its internal heat cannot take place by either conduction or radiation. However, turbulence is a field theory for a continuous medium. The variables ρ and $\mathbf{v}$ as well as T must satisfy the fluid equations that describe the entire medium, not just individual blobs. The idea that turbulent eddies remain intact and move with one characteristic length scale until they stop moving is a historical conceptual artifact; hence the reason for this polemic. The calibration of α, the parameter that describes the number of pressure scale heights traversed by a fluid element until it undergoes mixing, is a formal convenience. It describes a global average of properties of the turbulence, averaged over a layer. The adiabatic temperature gradient is only an average quantity. Models of turbulent energy transport[23] show that the nonlinear terms due to the gradients of the convected quantities are coupled to the velocity and tend to concentrate flows into very thin sheets and plumes that can seriously violate the equilibrium conditions required by the Schwarzschild criterion. The result is that the mixing length prescription *forces* the fluid to behave as if it were an ensemble of essentially identical blobs, which is only a gross approximation of reality. Is it truly useless to maintain this fiction? Not at all. The theory is able to provide an efficient means for calculating the energy transport and is not entirely wrong. It is true, however, that some of the apparent paradoxes of mixing required to explain details of late stages of stellar evolution and nucleosynthesis may eventually vanish when a proper theory of turbulent convection is available.

---

[21] Castellani, V., Giannone, P., and Renzini, A. 1971, *Ap. Space. Sci*, **10**, 340, 1971, *Ap. Space Sci.*, **10**, 355; Castellani, V. 1985, *Fund. Cosm. Phys.*, **9**, 317; Iben I. Jr. 1974, *ARAA*, **12**, 213; Iben, I. Jr. and Renzini, A. 1983, *ARAA*, **21**, 271; Ledoux, P. 1947, *ApJ*, **105**, 305; Roxburgh, I. 1965, *MNRAS*, **130**, 223.

[22] This is hardly a new viewpoint. You will find the discussions about this very early in the development of the field, recorded in Thomas, R. N., ed. 1960, *Aerodynamic Phenomena in Stellar Atmospheres: IAU Symp.* 12 (Bologna: Zanichelli) [also published as 1961, *Suppl. Nuovo Cimento*, **22(1)**, 1]. You should consult the review in this symposium by E. Böhm-Vitense and the discussion by W. Malkus, H. Liepmann, and F. H. Clauser that follow.

[23] See Canuto, V. and Christensen-Dalsgaard, J. 1998, *Ann. Rev. Fluid Mech.*, **30**, 167.

4.4.4 Rotation and von Zeipel's Theorem

We have repeatedly stressed that stars must satisfy two requirements for stability: one mechanical and the other thermal. Rotation distorts the figure of a star and alters its interior hydrostatic structure, but it has another effect because of the thermal constraint. Early in the work on stellar structure, von Zeipel showed that a star has a very hard time remaining in steady state even for comparatively slow rotation, a result we will now describe in more detail.

In a barotropic star, pressure and density surfaces are coincident and the structure can be solved uniquely by a polytropic equation of state. This is one reason why we could ignore the temperature when solving polytropic stars. In fact, the Lane–Emden equation was set up assuming barotropicity. But now suppose that the star is rotating *and* maintaining both hydrostatic and thermal equilibrium. Its surface is no longer spherical, even if for slow enough rotation this distortion is negligible, so we must include this to properly treat the structure. Main sequence stars often have rather high observed rotation rates, with $\lambda_\Omega \approx 0.1$ for the fastest rotators. Accreters in binary systems may be even more severely affected. Assume that the gravitational potential in the *corotating* frame, Φ, now includes the centrifugal term. The equation of hydrostatic equilibrium, $\nabla P = \rho \nabla \Phi$, is formally unchanged. For a barotropic star, we would be assured that

$$\nabla \frac{1}{\rho} \times \nabla P = 0$$

to all orders, which was the requirement for a polytropic equation of state to apply. Yet now we have a star whose surface gravity varies with latitude. We can use the equipotentials as our reference coordinate system instead of the radius. The flux equation now reads

$$\mathbf{F} = -\frac{4ac}{3\kappa_R \rho} \frac{dT}{d\Phi} \nabla \Phi = -K \nabla \Phi. \tag{4.74}$$

We take the star to be thermally stable as well, so that we would expect that $\nabla \cdot \mathbf{F} = 0$. Unfortunately, if the star is rotating, its potential becomes

$$\Phi = \Phi_G - \frac{1}{2} \Omega^2 \varpi^2, \tag{4.75}$$

where Φ_G is the spherical gravitational potential, ϖ is the axial distance and Ω is the rotation frequency. For simplicity, and in keeping with the original derivation by von Zeipel, we will assume rigid rotation so that Ω is constant throughout the star. From the Poisson equation

$$\nabla^2 \Phi = -4\pi G \rho \left(1 - \frac{\Omega^2}{2\pi G \rho}\right), \tag{4.76}$$

and Eq. (4.74) becomes

$$\nabla \cdot \mathbf{F} = \nabla K \cdot \nabla \Phi + K \nabla^2 \Phi = \frac{dK}{d\Phi} \nabla \Phi \cdot \nabla \Phi + K \nabla^2 \Phi, \quad (4.77)$$

which cannot vanish everywhere. Remember that K is a function of density and temperature. Locally, Eq. (4.77) means that the star adjusts its pressure gradient to accommodate the variation of the effective potential due to rotation. A simple way of seeing this is that the surface area per unit solid angle is greater at the rotational equator than at the poles because of the centrifugal bulge. In radiative equilibrium the same flux would have to be transported through this larger surface area so the temperature drops locally in order to balance out the equatorial and polar flux; the polar flux increases for similar reasons. A decreased surface gravity also reduces the pressure gradient required to maintain equilibrium. This locally reduces the temperature gradient and since the diffusion limit for radiative transfer, this also reduces the flux. Thus, the temperature is not constant along equipotential surfaces and the flux varies with latitude:

$$F \sim g \rightarrow T_{\text{eff}} \sim g^{1/4}, \quad (4.78)$$

the effect we introduced in Chapter 3 as *gravity darkening*. The drop in effective temperature does not, however, occur over the whole surface. At the poles, the gravitational acceleration increases relative to the spherical star by compression so the effective temperature increases.

This is not what we would want to see! It requires that a rotating star must be *baroclinic*, even in its deep interior.[24] There must be currents within the star, regardless of the energy source, that redistribute energy between the poles and the equator. Let's stop for a moment to examine this, since we will need the ideas again in our discussion of close binary stars. Written in vector form, the equations of motion are

$$\frac{d\mathbf{v}}{dt} = -\frac{1}{\rho} \nabla P + \nabla \Phi. \quad (4.79)$$

Since we are now including rotation, we have

$$\frac{d\mathbf{v}}{dt} = \frac{\partial \mathbf{v}}{\partial t} + \mathbf{v} \times (\nabla \times \mathbf{v}). \quad (4.80)$$

Taking the curl of Eq. (4.80) then gives

$$\frac{\partial \omega}{\partial t} = -\nabla(1/\rho) \times \nabla P, \quad (4.81)$$

---

[24] Tassoul, J. L. 1978, *Theory of Rotating Stars* (Princeton: Princeton Univ. Press); Zahn, J.-P. 1992, *Astr.Ap.*, **265**, 115; Maeder, A. and Zahn, J.-P. 1998, *Astr.Ap.*, **334**, 1000. There is an enormous literature here, but these are especially useful guides.

where we have defined the vorticity as $\omega = \nabla \times \mathbf{v}$. The conditions that we required for polytropic models have now broken down. Heat must flow from the pole to the equator on the surface and, to maintain steady state, must be compensated by a flow along the axis toward the pole. Remember that this flow is along equipotentials. The direction of the flow is, of course, the opposite of what you see on Earth, where, because the atmosphere is illuminated from outside, the basic meridional temperature gradient makes the poles colder than the equator. But the principle is the same. The circulation timescale is very long, and the characteristic timescale is given by

$$t_{\text{circ}} \approx \frac{t_{KH}}{\lambda_\Omega} = \left(\frac{GM}{\Omega^2 R^3}\right) t_{KH} \tag{4.82}$$

where t_{KH} is the thermal timescale. Intermediate mass main sequence stars are often rapid rotators, with the most rapid having periods of order one or a few days. Then $t_{\text{circ}} > 10^2 t_{KH}$, which is longer than the main-sequence time for even these stars. Nonetheless, some global mixing can occur because of these mass motions in the interior. In addition, it does not penetrate the convective regions of the star where the temperature gradient is adiabatic. But the important result we obtain from this analysis is that *any* energy generation mechanism must ultimately satisfy thermal balance everywhere. Notice that this effect occurs even for single stars, and it does not require tidal mechanisms that drive internal motions in close binaries. It is also interesting to note that this will occur even in the deep interior, depending on the rotation frequency.

4.4.5 Dimensional Analysis, Scaling Relations, and Homology

All physical quantities describing equilibrium states of a star depend on only two variables: mass and radius. This is because the structure is completely specified once the conditions of hydrostatic and thermal equilibrium have been imposed. We can find algebraic relations between the various thermodynamic variables that govern the interior structure. For instance, let's take the total mass, M to be a fixed number. If the radius is R, then the central density, ρ_c scales like MR^{-3}. Clearly, the gravitational acceleration scales like $g \sim MR^{-2}$. By Eq. (4.38), the central pressure, P_c, and central temperature, T_c, scale as

$$P_c \sim \rho \frac{GM}{r} \sim M^2 R^{-4}, \qquad T_c \sim MR^{-1}. \tag{4.83}$$

To find a scaling equation for the radiative luminosity, we require the power law approximation for the opacity:

$$\kappa \sim \rho^n T^{-s} \sim M^{n-s} R^{-3n+s}, \tag{4.84}$$

and from Eq. (4.59), the radiative luminosity scales as

$$L \sim R T^{4+s} \rho^{-(1+n)} \sim M^{3+s-n} R^{-s+3n}. \tag{4.85}$$

Even before we come to the explanation of the various energy generation mechanisms, we can make one more step in the scaling laws. We guess the form for the energy generation, writing it analogously to the opacity as $\epsilon = \epsilon_0 \rho^\alpha T^\beta$ per unit mass and insert this into the luminosity equation:

$$L \sim M^{\alpha+\beta+1} R^{-3\alpha-\beta}. \tag{4.86}$$

This provides a mass–radius relation:

$$R \sim M^{(\alpha+\beta-2-s+n)/(3\alpha+\beta+3n-s)}. \tag{4.87}$$

This discussion is not merely an exercise in exponential gymnastics. There is some important physics here. The temperature and pressure scalings are derived from force balance but the mass–radius relation is requires this and the combined effects of energy generation and transfer. Changing the opacity or the energy generation changes Eq. (4.87). Since, for example, electron scattering depends only on density, in this case $n = 1$ and $s = 0$. For energy generation by the proton–proton (pp) process, as we will see, $\alpha = 2$ and $\beta = 4$ so that $R \sim M^{0.38}$. This isn't bad as a representation for the main sequence, although the solar interior is not really well represented by this opacity. For hotter stars, for which electron scattering opacity is closer to reality, the temperature sensitivity of the nuclear reactions is much steeper. Consequently, the hottest stars have a different mass–radius relation than the lower main sequence. Perhaps the most important result of this kind of argument is that it explains one of the most basic observations for stars on the main sequence: the mass–luminosity relation. Since, by Eq. (4.86), L depends on mass and radius, once you have the $M - R$ relation, you obtain a $M - L$ relation. Remember, it is the interplay between energy generation and energy loss along with the requirement that the star remains hydrostatically stable that gives these relations. Therefore, any stable stage in the star's life, when gravity is not a contributor to the luminosity, should have some sort of $M - L$ relation.[25]

This form of scaling is called *homology*. As in biology, where we compare samples from the same species but perhaps different ages or diets or environments, we are looking at the general scaling of properties between stars of the same *genus*. This has meaning astrophysically because the interior nuclear processes. Stars that have the same opacity sources and the same nuclear burning have essentially the same internal structure. For simple computations of evolutionary models, this is an ideal place to start.[26]

---

[25] As we will discuss for horizontal branch evolution, there is a mass–luminosity relation here, and this persists throughout the stages of core nuclear burning when the timescales are long compared with the dynamical times for the star.

[26] Observationally, an example of how this can be used is the comparison of a star with its Roche radius in a close binary system. Although we will discuss this further in Chapter 5, you can imagine that from an eclipsing binary you might want to know something about the internal structure of a star when it is of a given size and how its properties might be affected by a close companion. The Roche potential is another example of a scaling law because, for specified binary system parameters, it provides a mass–radius relation.

4.5 STELLAR PULSATION AND STABILITY

A marvelous feature of single stars is that at certain times in their lives, the interior conditions are right for the production of a global pulsation. The entire star, or a large part of its envelope, can be set into resonant oscillation through the interplay of opacity, energy generation, and the mass fraction lying above that part of the star where mass motions naturally occur: the convection zone. This is a generic behavior of stars. Where it occurs depends on details of the energy generation and opacity but is otherwise possible at many epochs during the evolution of the star.[27]

4.5.1 Observational Justification

There are two causes for periodic stellar variability, both of which are due to gravitation. One is extrinsic. The star may be a member of a binary system where the companion affects its atmosphere through illumination and its potential through tides. In addition, depending on the orbital inclination, it may eclipse along your line of sight. The other is an intrinsic cause. The star may pulsate. It is interesting to note that the earliest work on periodic stellar variability in the late eighteenth century by Pigott, Goodricke, and Herschel actually confused the two.[28] Pigott and Goodricke hypothesized that β Persei (Algol) is an eclipsing close binary with an orbital period of about 3 days that consists of a dark star of roughly the same radius as the visual primary. We know that this is correct. However, they and Herschel also identified η Aql as such a system,[29] and we know now that it is not. As work on stellar variations proceeded through the nineteenth century, it became clear that a very large number of stars are variables. The foremost among them is δ Cephei.

The breakthrough came with Levitt's[30] discovery of the *period-luminosity relation* for the Cepheid variables in the Magellanic Clouds: the brighter the variable, the longer its period. It was already clear that these stars were different than those found in globular clusters: (1) their periods were systematically longer, often by more than a factor of 10; (2) they were a different spectral type. Where the RR Lyr stars are F stars, the Cepheids range from G to K; and (3) their light curves were like only one of the two types of RR Lyr curves, showing steep rises and slower declines. The interpretation of a period-luminosity relation is difficult, but not

---

[27] ee Rosseland, S. 1949, *The Pulsation Theory of Variable Stars* (Oxford: Oxford Univ. Press; rpt. Dover Books); Cox, J. 1983, *Pulsation Theory of Variable Stars* (Princeton: Princeton Univ. Press); Unno, W., Osaki, Y., Ando, H., Saio, H., and Shibahashi, H. 1989, *Nonradial Oscillations of Stars:*, 2nd ed. (Tokyo: Univ. Tokyo Press).

[28] There is a superb historical discussion on the origins of variable star research by Hoskins, M. 1982, *Stellar Astronomy: Historical Studies* (London: Science History Publ.). The authorship question for the Algol model, and many other aspects of the early work on variable stars, is far more complex than the usual textbook descriptions. Several of the most important variables were discovered before 1810 by Pigott, R Scuti (an RV Tau type), R CrB, and δ Cep. In addition, he was actually responsible for much of the work on β Per, η Aql, β Lyr, and the interpretations of o Cet and P Cyg. This marvelous paper on pre-nineteenth-century astronomy is a fine example of how to present scientific history.

[29] Although binarity was invoked in several cases, there was some dispute before the end of the nineteenth century about whether spots and rotational modulation might not be a better model for the variability—another prescient suggestion.

[30] This was part of a large scale effort to study variability that shared the attention of the Harvard astronomers along with the Henry Draper catalog since the early 1870s.

impossible, within the context of a binary star model because of Kepler's third law. There clearly is a relation between the period and separation of the stars, and between the maximum brightness that a star of a given temperature can have and the orbital period. However, there is a far simpler interpretation, that of a gravitationally bounded oscillating atmosphere.

Pulsation is found in many stages of stellar evolution but only in limited regimes in the HR diagram. The stars are classified by their prototypes. On the main-sequence stars at spectral type F, and similarly for low luminosity subgiants, the δ Sct stars have periods ranging from 0.04 to 0.2 days. Among G and K subgiants and giants, the δ Cep variables range from 1.5 to around 60 days for population I and the population II W Vir and BL Her stars, a parallel sequence at lower luminosity, range from around 1 to 50 days. The RR Lyr stars sit among the G stars on the horizontal branch with periods of 0.4–1 day. Among red giants, the coolest variables are the o Cet or Mira type with periods between 100 and 500 days and M spectral types. Irregular variations are known for higher luminosity stars, such as α Ori (an M supergiant) and α Cyg (an A supergiant). The RV Tau stars are population II G and K giants with periods of 40–150 days, overlapping the longest period W Vir stars, and the UU Her stars are population II F supergiants with periods of 40 to around 70 days that span the midrange of the W Vir stars. Pulsation is not confined to normal stars, however. Hypergiants, called luminous blue variables (LBVs), with $M_V < -8$ undergo both pulsation and large-scale mass ejection and display variable massive winds that change their spectral energy distributions from O to A or later, are among the most extreme examples of nonlinear pulsators. Less extreme examples of the same pulsation-related mass ejection are the R CrB stars, which are carbon stars that are in the F–G giant range with intervals between eruptions of years and pulsation periods that are about the same as the UU Her stars. At the lowest luminosity end, there are also sequences of pulsating white dwarf stars whose periods are measured in minutes rather than months. The PG 1159 stars have periods of order 500–2000 s and lie in the temperature range 7.5×10^4 to 1.4×10^5 K. The ZZ Cet stars are DA white dwarfs with $10^4 < T_{\text{eff}} < 1.3 \times 10^4$ K and periods between about 100 and 1000 s. In short, there is clearly something to explain here—given the right conditions, stars are able to develop organized large-scale periodic motion that seems to be connected to their surface gravity (luminosity) and effective temperature. Sample light curves are shown in Figure 4.1.

You know that a self-gravitating mass has a characteristic dynamical timescale that is a multiple of the freefall time:

$$t_{ff} = \frac{1}{(G\rho)^{1/2}} \sim R^{3/2}.$$

This is the fastest mechanical timescale available to the body. We can also define a new parameter, $Q = \sqrt{P\bar{\rho}}$, using the period, P, and *mean* density that accounts in t_{ff} for centrally condensed configurations. We can also assume that it is typical of the period of an oscillator driven against gravity only by pressure variations. The luminosity is $L = 4\pi R^2 \sigma T_{\text{eff}}^4$. Now take two stars with similar masses and temperatures but different radii. Then the luminosity is related to the timescale by

$$L \sim P^{4/3}, \tag{4.88}$$

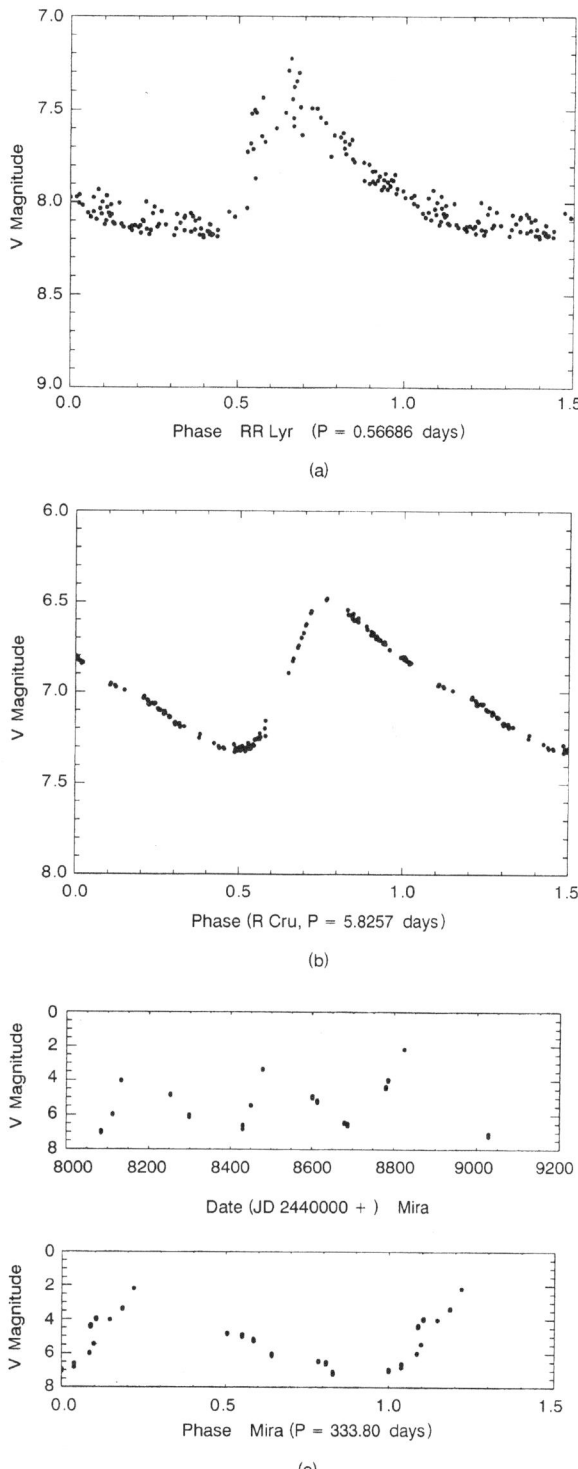

FIGURE 4.1. Sample light curves for pulsating variable stars from the Hipparcos database. (a) RR Lyr; (b) R Cru (δ Cep variable); (c) Mira (o Cet) (long period variable).

where P is now the period. There may be some variation in surface temperature during the cycle. In fact, there should be if we have a sphere adiabatically oscillating, so Eq. (4.88) may have a dependence on color as well (and through this on the star's effective temperature). This has little effect on our overall argument, so we will ignore it. We have arrived at a law that allows us to find the intrinsic luminosity of a pulsator knowing only its period *provided we are looking at the same driving mechanism*. In principle, once we have the distance to a single calibrator, we should be able to unambiguously determine the distance to the pulsator. If it is in a cluster, or more interestingly if it is in a galaxy, the period and apparent magnitude yield the distance to the host. This was one of the fundamental discoveries at the beginning of the modern era of astrophysics, since it provides a way of determining the absolute magnitude of a star knowing *only* what type of variable it is and its period. The zero point of the relation is *still* controversial, but it is possible to determine observationally. We will discuss this further below. For the moment, however, it should serve to justify our interest in the problem.

This nonetheless leaves open some fundamental physical questions. Why would a star pulsate? Is it the entire star, or is it only a superficial layer that moves? We already know that the coupling of radiation to matter through opacity can produce small-scale, turbulent, motion in convection zones, and we have also seen that the interplay between the requirements of radiative equilibrium and hydrostatic equilibrium can produce large-scale, albeit slow, circulation currents. Yet these are secular effects. One is turbulent and so statistically in steady state. The other is just bounded flow that transports angular momentum and matter through a star. Here we are faced with a more difficult question: What maintains a periodic oscillation in a self-gravitating *radiating* body?

4.5.2 The Mechanism for Pulsation

The pulsation of a star is a mechanical response to some departure from simple thermo-mechanical equilibrium. You might already suspect that it must be closely related to the presence of convection, in that it is a mechanical instability produced by the inability of the medium to remain in mechanical equilibrium and transfer energy by radiation or conduction. This section will support and extend that prejudice. Let's start with a schematic star, one that has a constant energy source in its deep interior and a thin layer on which we lavish our attention. Assume that this layer is especially opaque to radiative transfer, yet tenuous enough that conduction plays no role. The layer heats up and begins to expand. So far, this sounds a lot like the introductory scenario for convection. The main difference is that now we are looking at the whole layer, not just a little parcel of gas or a local analysis. We're attempting to determine whether a global dynamical response is possible. Everything hangs on how the opacity changes during a perturbation. If the layer expands and adiabatically cools when its opacity *decreases*, then the layer radiatively loses energy, the pressure drops, and it recontracts. Should the opacity now rise on recompression and heating, then the cycle can start again. This is what is meant by a pulsational instability. As in convection as expressed by the Rayleigh number, we see that the instability results from a competition between the thermal and mechanical timescales for a gaseous layer.

The restoring force is gravity, not cooling and conduction, so comparing the freefall time to the sound travel time across a layer of thickness l

$$t_{\text{grav}}/t_s \sim c_s l^{-1/2} g^{-1/2} \tag{4.89}$$

shows that the layer becomes unstable when its size becomes comparable to one scale height.

We can get some idea about what might cause this instability by concentrating on just one mass zone somewhere in the star and assuming that it is so geometrically thin—roughly one pressure scale—that no thermodynamic variables have any internal gradients. This also allows us to linearize the structure equations. Take the zone's mass to be ΔM and keep it fixed so any fluctuations in the state variables of the layer depend only on time. The analysis of pulsation by a local treatment has a fundamental limitation—it cannot provide dynamical models or give the radial dependence of the amplitude. It does, however, indicate what physical conditions likely lead to pulsation and what the frequencies will be.

4.5.3 Stability Analysis: One-Zone Pulsation

We can now write an approximate solution for the pulsation problem in terms of a thin layer. The reason we do this is not just as an exercise to flex your algebraic muscles. It requires that you think about the physical processes that we have discussed throughout this chapter. Schematic constructions like one-zone models are testing grounds of physical theories, somewhere between toys and reality. With them, you can isolate the essential physics in simple ways. So now we ask the question: If you kick a slab of gas that is required to remain in global hydrostatic and radiative balance during a perturbation, how will it respond?

First, let's look at the thermal response of a gas that is periodically compressed by a piston, the steam engine analogy of Eddington. Imagine a slab whose internal pressure, P, and volume, V, vary with time. For convenience, imagine the variation of the volume to be $\delta V = \alpha \sin \omega t$, where ω is the frequency of the driver. Next, assume that the flux through the slab varies as $\delta F = F_0 \sin(\omega t + \phi)$, taking the input flux, F_0, to be constant but allowing allowing for a phase lag, ϕ, between the thermal and mechanical response of the gas. We will denote averages over a single cycle with a bar. Since

$$\delta P = -\gamma \frac{\bar{P}}{\bar{V}} \delta V,$$

the total work done during a pulsational cycle is

$$\Delta W = \oint \delta P \frac{d\delta V}{dt} dt = -\gamma \frac{\bar{P}}{\bar{V}} \delta V \alpha^2 \omega \oint \sin \omega t \cos \omega t \, dt,$$

which vanishes for an adiabatic pulsation. Now if $\bar{F}$ is constant during a cycle such that the average perturbation of the flux vanishes, then the energy equation gives

$$\delta F = \frac{\gamma}{\gamma - 1} \bar{P} \frac{d\delta V}{dt} + \frac{1}{\gamma - 1} \bar{V} \frac{d\delta P}{dt},$$

so solving for the work, you obtain

$$\Delta W = -\frac{\gamma - 1}{\bar{V}} F_0 \alpha \oint \cos(\omega t + \phi) \cos \omega t \, dt = -\pi F_0 \alpha \frac{\gamma - 1}{\bar{V}\omega} \cos \phi, \quad (4.90)$$

which shows that for ΔW to be positive, the phase lag between the flux and the volume variations must be between $\pi/2$ and $3\pi/2$. The argument is of only limited use because we assumed that the radiation escapes the thin layer instantaneously. This doesn't work well in the interior but it is actually useful for atmospheric pulsations, such as those seen in δ Sct stars. Historically, this argument was important for descriptions of the observed phase shift between the velocity and light curves for pulsating variables and formed the basis of many of the early arguments about the mechanism responsible for driving the pulsations in Cepheids.[31] If powered only by opacity, the pulsations are called κ-excited. If nuclear energy generation fluctuates with positive feedback, the pulsation is ϵ-excited.

With this prelude in mind, let's begin the first act. The physical variables will be expanded to first order, $x(M, t) = x_0(M) + \delta x(t)$. We further assume that all perturbed variables can be analyzed using a single-mode expansion in time of the form $e^{\omega t}$ with ω independent of mass. This really simplifies the algebra so that you can concentrate on the physics. If the system has simply periodic solutions, ω will be a purely imaginary number. *Overstability* means that it is complex with a positive real part, and *dynamical instability* means that it is real and positive. The terminology is due to Eddington. In what follows we will examine only the linear perturbation problem, so we are really making two approximations: we are assuming that things vary relatively slowly, at least slower than the dynamical timescale (and in effect we are assuming periodicity is possible and looking for it); and we are assuming that all the perturbations are small. Thus we cannot solve the full dynamical problem—we are not able to determine the motion of the layer, only its frequency responses. This limitation is important. The one-zone treatment cannot be used to analyze the motions in the interior of a star; that requires knowing how the amplitude depends on depth and can be obtained only if we know how the gradients behave. This is act 2. Instead the one-zone model permits us to ask questions about stability based on some local interior properties provided the amplitude of the stellar radial variations is small.

We will go through more details here than usual in order to develop techniques that you will need later in this book for some related stability problems. We start

---

[31] One-zone thermodynamic treatments have been extended to the fully nonlinear regime; see, for example, Buchler, J. B. 1993, *Astrophys. Space Sci.*, **210**, 9. The simplified thermodynamic argument we're using here due to Whitney in the discussions reported at *IAU Symp.* 12, to which we referred earlier (n.22).

with the mass equation:
$$\frac{\partial r}{\partial M} = \frac{1}{4\pi r^2 \rho}.$$

Keeping both zeroth- and first-order terms for the moment, we have

$$\left(1 + \frac{\delta r}{r_0}\right)\frac{\partial r_0}{\partial M} + r_0 \frac{\partial}{\partial M}\frac{\delta r}{r_0} = \frac{1}{4\pi r_0^2 \rho_0}\left(1 - 2\frac{\delta r}{r_0} - \frac{\delta \rho}{\rho_0}\right). \quad (4.91)$$

Now if we assume that the zone moves up and down uniformly for a small ΔM, then

$$\frac{\partial}{\partial M}\frac{\delta r}{r_0} = 0,$$

so changes in the density and the width of the zone are directly coupled:

$$\frac{\delta \rho}{\rho_0} = -3\frac{\delta r}{r_0}. \quad (4.92)$$

The equation of motion

$$\ddot{r} = -\frac{1}{\rho}\frac{\partial P}{\partial r} - \frac{GM}{r^2} \quad (4.93)$$

has an equilibrium solution

$$\frac{\partial P_0}{\partial M} = -\frac{GM}{4\pi r_0^4},$$

which we can now use with along the first-order perturbation to obtain

$$r_0 \omega^2 \frac{\delta r}{r_0} = -4\pi r_0^2 \left[\left(2\frac{\delta r}{r_0} + \frac{\delta P}{P_0}\right)\frac{\partial P_0}{\partial M} + P_0 \frac{\partial}{\partial M}\frac{\delta P}{P_0}\right] - 2\frac{GM}{r_0^2}\frac{\delta r}{r_0}.$$

By ignoring the gradient in the pressure perturbations within the zone, Eq. (4.93) becomes

$$\left[\omega^2 - 4\omega_0^2\right]\frac{\delta r}{r_0} = -\omega_0^2 \frac{\delta P}{P_0}. \quad (4.94)$$

Here the inverse gravitational freefall timescale has been written as a frequency, $\omega_0^2 \equiv GM/r_0^3$. We now require an equation of state. Were we to choose a polytrope with $P \sim \rho^{\Gamma_1}$, we would have

$$\omega^2 = (4 - 3\Gamma_1)\omega_0^2, \quad (4.95)$$

which is negative when $\Gamma_1 > \frac{4}{3}$. This is the stability condition you already know. Recall, however, that when we obtained this in Section 4.3.4, it was for static models only. But by choosing this equation of state, we have neglected a vital piece of reality; this pressure law holds exactly only if we fail to examine how the layer satisfies radiative equilibrium during the pulsation. We can remedy this by taking a more general prescription:

$$\frac{\delta P}{P_0} = \alpha \frac{\delta \rho}{\rho_0} + \beta \frac{\delta T}{T_0},$$

where α and β will be assumed to be constants within a thin shell. We then use the diffusion equation for the radiative transfer:

$$L = -\frac{64\pi^2 ac}{3\kappa} T^3 \frac{\partial T}{\partial M}$$

and assume the variations in the opacity depend only on temperature and density through

$$\frac{\delta \kappa}{\kappa} = \left(\frac{\partial \ln \kappa}{\partial \ln T}\right) \frac{\delta T}{T_0} + \left(\frac{\partial \ln \kappa}{\partial \ln \rho}\right) \frac{\delta \rho}{\rho_0} = \kappa_T \frac{\delta T}{T_0} - 3\kappa_\rho \frac{\delta r}{r_0}$$

to arrive at the perturbation for the radiation through the layer:

$$\frac{\delta L}{L_0} = (4 + 3\kappa_\rho)\frac{\delta r}{r_0} + (4 - 3\kappa_T)\frac{\delta T}{T_0}. \tag{4.96}$$

We still lack the thermodynamic constraint of the first law, and this turns a second-order equation into third order, thereby allowing for complex roots. Its time-dependent form

$$T\frac{dS}{dt} = \epsilon - \frac{\partial L}{\partial M} = c_v \frac{dT}{dt} - \frac{P}{\rho^2}\frac{d\rho}{dt}$$

becomes Eq. (4.97), on taking the first-order perturbations

$$-\frac{L_0}{\omega}\frac{\partial}{\partial M}\left(\frac{\delta L}{L_0}\right) = c_v T_0 \frac{\delta T}{T_0} + 3\frac{P_0}{\rho_0}\frac{\delta r}{r_0} \tag{4.97}$$

and using Eq. (4.92) for the density perturbation. Why have we left the mass gradient of the luminosity when all other gradients have otherwise been neglected? This is a way to simulate the "leakage" of energy through the zone. Approximate this by taking ΔM to be the mass thickness of the zone:

$$\frac{\partial \delta L}{\partial M} \approx \frac{\delta L}{\Delta M}.$$

Substituting this into Eq. (4.97) and rearranging terms, we find that the final polynomial for ω is a cubic:

$$\omega^3 - \omega^2(\kappa_T - 4)\frac{L_0}{c_v T_0 \Delta M} - \omega \frac{\omega_0^2}{c_v T_0}\left[(4 - 3\alpha) - 3\beta\frac{P_0}{\rho_0}\right]$$

$$+ \omega_0^2 \frac{L_0}{c_v T_0 \Delta M}\left[(3\kappa_\rho + 4)\beta + (4 - 3\alpha)(\kappa_T - 4)\right] = 0 \quad (4.98)$$

For the pulsations to grow requires $\text{Re}(\omega) > 0$. You see that that this layer has three characteristic frequencies for the radial perturbation. One is the obvious one, the inverse of the freefall time. The other two depend on the opacity gradients. These dominate the mechanical coupling of the temperature to the pressure. It means that in places where there are large values of κ_T and κ_P, we are very prone to instability on the local radiative diffusion timescale. So ω_0 is indeed involved in the pulsation, but is reduced by the heat capacity and opacity of the medium. Because this is a cubic system, you are not guaranteed that these other roots are strictly real. Overstability, the appearance of complex roots, may occur if the loss terms have phase relation to the compressional heating.

When we ignore energy generation and assume purely adiabatic oscillations, we recover the original stability criterion. This is because the system is now only quadratic in frequency, the driving term coming from the hydrostatic equation alone. The essential feature of the linear adiabatic equation is that we don't use *any* information about the energy generation or transfer. We assumed that the amplitude is small and that there is no change in the thermal properties of the medium during the pulsation. Historically, the equation we have arrived at was derived by Rosseland and Eddington in the early part of the twentieth century. In fact, the assumptions do not require us to consider anything about the energy generation and accounts for the early derivation of the equation; it is a purely classical result which could just as easily have been written down by Darwin, Emden, or Rayleigh had they chosen to consider the problem of radial pulsation of polytropes. Interestingly enough, none of these investigators seem to have been seriously interested in the problem (although Emden does briefly mention the possibility for atmospheric oscillations). There was more concentration on stability of liquid stars because of the dominant interest in the origin of the solar system and binary stars. Pulsation only became central to astrophysics after the discovery of the period–luminosity relation.

4.5.4 Probing the Stellar Envelope

The current effort to use variability to probe the interior stellar structure requires a number of physical inputs. You now have the tools for understanding these. The opacity is crucial. Without a detailed understanding of the energy transfer, it is impossible to use the oscillation spectrum to determine the internal temperature gradient, which is essential for coupling surface properties with the inferences about interior nuclear processing. For the Sun this is especially critical because from Earth we can measure the core conditions through neutrino observations. No measurement of electron neutrino emission from the Sun has yet yielded the *expected* flux from any portion of the pp chain (see Section 4.6.6.1). One of the many solutions that made use of purely astrophysical interpretations was to argue that the temperature gradient in the convection zone and in the core were different than the models assumed.

To check this one can resort to the pulsation characteristics of the envelope, a process known as *helioseismology*. For stars in general it is called *asteroseismology*. This technique uses the spectrum of high-order nonradial oscillations excited by the convection to probe the gradient of the sound speed through the stellar envelope. If you have ever stood at a lunch counter and seen the oscillation in the surface of a coffee cup resting on a vibrating counter, you've seen something like it. The counter has a very complex set of frequencies. One or a few modes is (are) amplified by the boundary conditions in the cup; the rest interfere and disappear. If you change the surface with cream, or alter the shape of the cup, you get different modes for the same driving spectrum. The Sun behaves in the same way. Depending on the input spectrum, only a limited number of modes finally appear at the surface. The surface is bounded in depth because of the finite thickness of the convection zone.

The driving is internal. The convection zone creates acoustic waves that are trapped within it by the temperature gradient. The problem is essentially one of geometric ray tracing. The larger the temperature gradient, the larger the effective refractive bending of the wavefront. A lovely feature of helioseismology is that as in the eigenvalue problem for standing waves in the acoustic cavity of the convection zone, the amplitude of the waves is *always* small. The oscillations are generated from convection by subsonic motions that in the dense interior cannot turn into shocks (although they may turn nonlinear when they approach the outer stellar atmosphere). Ultimately, a linear wave equation can be used to study the modes, and the problem of finding resonance is one of finding the eigenfrequencies. Complications arise when computing the gradient in the sound speed, which is the feature that changes the index of refraction.

The acoustic index of refraction, n, is defined as it is for light, as a ratio of the phase to group velocity of a wave. In an optical medium, the path is defined by the time so that the phase shift is given by $\phi = \int n\, dl$ for a pathlength l. The problem of computing the resulting spectrum is much like the emission of sound by a brass instrument, say, a trumpet. A broad turbulence spectrum is injected through a mouthpiece (for a trumpet this is the lip buzz that is imposed in the jet of air coming from the player's lips), and the gas in the chamber oscillates in response to the pressure fluctuations. The acoustic impedance of the system is determined by the rate of pressure drop at the bell of the instrument and the flare of the horn, and different resonances are achieved through the feedback of the reflected wave and the production of standing waves within the instrument. The trapped modes drive the radiation by the medium, which ultimately damps the oscillations if they are not continually driven.[32] The reason for using the horn analogy rather than,

---

[32] A superb introduction to theoretical acoustics of musical instruments is Rössing, T. and Fletcher, N. 1993, *Physics of Musical Instruments* (Berlin: Springer-Verlag). See also Ben-Menahem, A. and Singh, S. J. 2000, *Seismic Waves and Sources*, 2nd Ed. (NY: Dover Books).

say, an organ pipe or flute is that the refractive index for the waves depends on the temperature gradient. The horn of a brass instrument flares, analogous to the temperature gradient. Each mode is, in principle, trapped at its resonance, which is of order kH_p for a wavenumber k, but the sound speed is a function of depth (hence the analogy with seismology), and the trapped modes are not regularly spaced. The spectrum is quite dense.

The one zone treatment cannot suffice for a detailed analysis of the acoustic spectrum. The wavelengths are large, probing deep into the stellar envelope, so the waves sample a large range of thermal conditions. It is still true that the observed frequencies are eigenmodes, but the gradients of temperature and pressure must be included. For the Sun, this is easier than for global stellar pulsation. The waves are all linear, having extremely small velocity amplitudes of order 0.01 c_s or less, and in the first approximation are also adiabatic. It is possible then to derive an equation for the depth-dependent amplitude in terms of the thermodynamic quantities alone and thus to probe not only the structure but also the equation of state, of the outer solar envelope, the part at or above the convection zone. To do this, we must modify the approach that we have taken to the pulsation problem and look at the Lagrangian variations in the physical variables. This is especially useful in setting up the eigenvalue equations for standing waves, which is, after all, what we have in pulsation, and also in deriving the spectrum for the helioseismology problem.

4.5.4.1 Helioseismology

The Sun does not pulsate. This obvious observation has a deep significance. The physical conditions preclude resonance throughout the solar convection zone and large scale ordered motion is impossible. By the mid-1960s, however, it had become clear that there was some sort of ordered motion, albeit on a relatively small scale, in the solar atmosphere. Doppler imaging of the surface displayed a characteristic timescale of several minutes, dubbed the 300-*s oscillation*, that could be analyzed as a very-high-order (large horizontal wavenumber) mode of oscillation. The radial extent of this motion was not known at first, but it was clear that the convection zone must be playing a role, just as it does in the global oscillations of more evolved stars.

Imagine that a local noise source is set up in a room. The source has a broad acoustic spectrum, but the room has a characteristic length and therefore possesses a set of resonant frequencies. Certain wavelengths are selectively amplified because of the reduced impedance of the chamber. This is the same phenomenon that produces the characteristic sound of a wind instrument. A turbulent input has a very broad spectral power distribution, $S(\omega)$. The musical instrument, like the room, has a specific set of resonant frequencies that serve as filters, $F(\omega)$, such that the amplification occurs even for very small power, as long as the oscillating system performs work on the medium. Closely linked to this acoustic phenomenon is the now current idea of *stochastic resonance*, in which a noisy system can achieve resonance through its broadband excitation of a responsive medium. It is a remarkably efficient mechanism precisely because a noisy source is bound to excite something and the amplitude will grow as long as the noise is input. The solar convection zone is just such a noise source. It generates a broad spectrum of pressure waves due to the turbulent motions of gas. Although local in origin, the individual modes can couple to the stellar envelope in a comparatively limited range of frequencies because there is a characteristic response for the envelope. This is set by the temperature gradient, which governs the buoyancy frequency for the gas, and by the condition that the disturbances are of such small amplitude that they do not alter the hydrostatic structure of the envelope.

Now let's introduce a new analogy. The propagation of these pressure waves in the solar envelope resembles seismic waves that move through the Earth's interior after an earthquake. With depth in a planet, the increased density causes an increase in the effective sound speed for a seismic wave. Thus, the index of refraction, defined by $v_p = c_s/n$, *decreases* and the wave continually refracts *away* from the normal. In a stellar interior, since the temperature increases with depth, so does the sound speed. Thus, you encounter the same propagation geometry—the index of refraction for the waves steadily decreases as you go deeper, with the result that the wavefront follows a curved trajectory that is concave upward.

Qualitatively, a resonance is achieved when there are an integer number of intersection points with the surface around the solar circumference. Waves of different frequency will have different numbers of such nodes and this selects out a countable set from the continuous input spectrum, unlike an earthquake, where the damping is fast and the source is not continually active. The effective depth of penetration of the waves is the mean depth of the convection zone, H_p and therefore, we would expect that a resonant condition is set by $\lambda/H_p = m/n$, where m,n are integers. Note that these are standing waves within the broad spectrum of traveling waves. This is the basis of *helioseismology*, that the solar atmosphere and envelope act in concert to form a resonant system that couples to specific frequencies.[33] Because we can resolve the solar surface, we can actually spatially extract this spectrum.

Quantitatively, a plane wave propagates with a wavenumber $\mathbf{k}$ such that $\mathbf{k} = \nabla \phi$, where ϕ is the phase of the wave. The basis of ray optics is that these directions depend on the sound speed in the medium. To see how, assume that $\phi = \mathbf{k} \cdot \mathbf{x}$, where $\mathbf{x}$ is some distance along the ray. Then since $k = \omega/c_s$, Snel's law of refraction is recovered. Defining $\hat{\mathbf{k}} = \mathbf{k}/|\mathbf{k}|$, then

$$\frac{\hat{\mathbf{k}} \cdot \hat{\mathbf{n}}}{c_s} = \text{constant}. \tag{4.99}$$

As the sound speed changes with depth, the rays are progressively deviated away from the normal direction to the interface (for a two-layered medium we have $\cos\theta_1/c_{s,1} = \cos\theta_2/c_{s,2}$, where $\cos\theta = \hat{\mathbf{k}} \cdot \hat{\mathbf{n}}$). Resonance is achieved when the wave, which is so deviated, reaches the surface again at the same phase as another

---

[33] This field has generated an enormous literature, and we don't intend to review it here. A good place to start is with some of the earlier reviews, among which are Deubner, F.-L. and Gough, D. O. 1984, *ARAA*, **22**, 593; Bahcall, J. N. and Ulrich, R. K. 1988, *Rev. Mod. Phys.*, **60**, 297; Christensen-Dalsgaard, J. et al. 1996, *Science*, **272**, 1286. Brown, T. M. and Gilliland, R. L. 1994, *ARAA*, **32**, 37 provide a superb overview of asteroseismology, the application of these ideas to stars other than the Sun for which only the surface averaged modes are observable.

horizontally displaced oscillation. In this way a random ensemble of pressure waves can achieve a discrete spectrum of resonant modes.

Notice that the trajectory depends on the wavelength because of the depth dependence of the temperature gradient. Specific frequencies (wavelengths) are resonantly selected by this process, and the spectrum is consequently a sensitive, albeit complicated, probe of the depth-dependent temperature and density gradient. One important point is that these waves are *not* radial propagation modes, so they resemble very-small-amplitude nonradial pulsation modes. The same phenomenon is encountered in pulsating stars—the higher azimuthal modes are normally present with very small amplitude only if the thermal structure of the envelope is right. In general, the lower the frequency the greater the depth to which the mode is sensitive so only for the longest time series can the regions be probed near the stellar core. These are achieved with considerable coordination skill by the *Whole Earth Telescope* and GONG, the two large-scale global multisite monitoring campaigns mounted to date.

4.5.4.2 The Linear Pulsation Equation

Why don't the Sun or solar-type stars show organized low-mode-number radial pulsations? Put another way, why isn't the Sun a Cepheid variable? The answer depends on the depth of the convective envelope and mass fraction of the overlying layer. The opacity of the outer envelope determines the rate of escape of energy released during a pulsation and governs whether the pulsation will damp. For the solar question, the restoring force due to the mass of the layer above the driving zone is too small and the opacity of the envelope is too low to achieve a stable radial mode—the solar convection zone is too near the surface. For stars of lower effective gravity, and larger radius, the convection is deeper and pulsation is maintained through the combined effects of opacity and overlying mass.

We now extend our one-zone treatment to take the explicit depth dependences into account. As before, we take the perturbations of the pressure and radius and linearize the results:

$$\frac{\partial}{\partial M}\left[P\frac{\delta P}{P}\right] = 4\frac{GM}{4\pi r^4}\left(\frac{\delta r}{r}\right) + \omega^2 r \frac{\delta r}{r} \qquad (4.100)$$

$$\frac{\partial}{\partial M}\left[r\left(\frac{\delta r}{r}\right)\right] = -\frac{1}{4\pi r^2 \rho}\left(2\frac{\delta r}{r} + \frac{\delta \rho}{\rho}\right),$$

which, using $\xi = \delta r/r$, reduces to

$$P\frac{\partial}{\partial M}\frac{\delta P}{P} = \left(\frac{4\omega_0^2}{4\pi r^2} + r\omega^2\right)\xi \qquad (4.101)$$

$$r\frac{\partial \xi}{\partial M} = -\frac{1}{4\pi r^2}\left(3\xi + \frac{\delta \rho}{\rho}\right).$$

To progress, we again must assume something about the energy balance during a pulsation cycle. For adiabatic pulsation, the problem is straightforward because we know the equation of state at the start:

$$\frac{\delta P}{P} = \Gamma_1 \frac{\delta \rho}{\rho}. \qquad (4.102)$$

Including radiative losses and/or energy generation for the fully nonadiabatic treatment requires introducing the temperature and a more complete rendition of the equation of state, as we had for the one-zone treatment:

$$\frac{\delta T}{T} = (\Gamma_3 - 1)\frac{\delta \rho}{\rho}, \qquad (4.103)$$

and we also need to include the equation for radiative balance. We will neglect variations in the nuclear energy generation rate although such effects should be kept in mind when the star may go through a stage of off-center nuclear burning in multiple shells late in its life.

The Lagrangian perturbation of the pressure is

$$\frac{\Delta P}{P} = \Gamma_1 \frac{\Delta \rho}{\rho}. \qquad (4.104)$$

The variation at any radius is, however, the combined effect of local variations and transport (advection), so we replace the Lagrangian Δ with

$$\Delta = \delta + \xi \frac{\partial}{\partial r}.$$

Note that this comes from taking the comoving time derivative [see Eq. (1.31)] and integrating with respect to time; δ represents the Eulerian variation. Then

$$\delta P = P\Gamma_1 \left(\frac{\delta \rho}{\rho}\right) + \xi P \left[\Gamma_1 \frac{\partial \ln \rho}{\partial r} - \frac{\partial \ln P}{\partial r}\right]. \qquad (4.105)$$

The first-order equation of motion becomes

$$\rho \ddot{\xi} = -\frac{\partial \delta P}{\partial r} + \rho \frac{\partial \delta \Phi}{\partial r} + \delta \rho \frac{\partial \Phi}{\partial r}. \qquad (4.106)$$

The perturbed form of the strictly radial spherical[34] Poisson equation is

$$\frac{1}{r^2}\frac{\partial}{\partial r}\left[r^2\frac{\partial \delta\Phi}{\partial r}\right] = -4\pi G\delta\rho, \qquad (4.107)$$

showing that the density variations drive changes in the gravitational acceleration. Finally, we add in the linearized continuity equation:

$$\frac{\partial \delta\rho}{\partial t} + \frac{1}{r^2}\frac{\partial}{\partial r}\rho r^2\frac{\partial \xi}{\partial t} = 0. \qquad (4.108)$$

Here the velocity that produces the volumetric (divergent) changes in the density are due only to the displacements. The medium is presumed to be otherwise at rest. Thus, Eq. (4.108) can conveniently be integrated with respect to time to give

$$\delta\rho = -\frac{1}{r^2}\frac{\partial}{\partial r}r^2\rho\xi. \qquad (4.109)$$

Now you've made progress since this expression, substituted into the Poisson equation, allows you to integrate this equation to obtain the perturbed gravitational *acceleration*:

$$\frac{\partial \delta\Phi}{\partial r} = 4\pi G\rho\xi. \qquad (4.110)$$

The energy equation is

$$T\frac{dS}{dt} = \frac{dQ}{dt} = \frac{dU}{dt} - \frac{P}{\rho^2}\frac{d\rho}{dt}, \qquad (4.111)$$

where the time derivatives represent Langrangian variations, and

$$\frac{dQ}{dt} = \frac{1}{\rho(\Gamma_3 - 1)}\left[\frac{dP}{dt} - \frac{\Gamma_1 P}{\rho}\frac{d\rho}{dt}\right]. \qquad (4.112)$$

From the continuity equation, we can reduce the perturbation in the heat function to

$$\delta Q = \frac{1}{\rho(\Gamma_3 - 1)}\left[\Delta P + \frac{\Gamma_1 P}{r^2}\frac{\partial r^2\xi}{\partial r}\right]. \qquad (4.113)$$

---

[34] If we were to include nonradial terms, this equation would become

$$\frac{1}{r^2}\frac{d}{dr}\left(r^2\frac{d\Phi}{dr}\right) - \frac{l(l-1)}{r^2}\Phi = -4\pi G\rho,$$

where l is the angular order of the spherical harmonics with which the horizontal terms are expressed. We would also have to treat the horizontal displacements from the divergence in the continuity equation. We will confine our discussion only to radial modes to contrast it with the one zone treatment. For more details, see references listed at the end of this section.

326　THE INTERIORS OF STARS AND STELLAR EVOLUTION

Setting $\delta Q = 0$ and combining terms, we obtain

$$\frac{1}{r^3}\frac{d}{dr}\left[P\Gamma_1 r^4 \frac{d}{dr}\frac{\xi}{r}\right] + \left(\omega^2 \rho + \frac{1}{r}\frac{d}{dr}[P(3\Gamma_1 - 4)]\right)\xi = 0, \qquad (4.114)$$

which is the *linear adiabatic wave equation* for pulsation.[35] Any energy losses incurred during a cycle will damp the outer layers and drive the interior. There is a close match between loss and driving for stably pulsating stars, as Eddington first realized. Stars pulsate in only comparatively isolated locations in the HR diagram, a display of the thermodynamical fine tuning required to achieve ordered motion in low-resonance modes. To calculate when a star of some mass may undergo pulsation requires an unperturbed model that provides $(\rho(r), P(r), \Gamma_1)$ as functions of the radius. Even if large-scale motion is not achieved, the wave equation is not restricted to only one wavelength—the spectrum of oscillation frequencies *and* amplitudes depend on r and probe the thermal state of the interior. Supplemented by appropriate boundary conditions, which we will discuss in a moment, the distribution of frequencies depends sensitively on the state variables. Temperature variations affect the equation of state, and these are driven entirely by nonadiabatic effects that are closely tied to the opacity and convection theory. The next step, as we did for the one zone treatment, is to include the variation of the flux through a pulsation cycle, only this time to consider it as a function of radius within the envelope. Again, although we have been discussing large-scale motions of the stellar envelope, we wish to emphasize that virtually all stars pulsate at some level. The large scale periodic oscillations, however, require special resonance conditions that can only be met at specific locations in the HR diagram.

[35] Details of the derivation are as follows. As we did for the one-zone model, write the mass conservation equation as

$$\frac{\partial r}{\partial r_0} = \frac{r_0^2 \rho_0}{r^2 \rho}$$

from the Lagrangian condition that you sit on a mass zone throughout its motion. Then, defining $r(M,t) = r_0(1 + \xi)$, and all other Lagrangian perturbations accordingly, you will find that

$$r_0 \frac{\partial \xi}{\partial r_0} = -3\xi - \frac{\delta \rho}{\rho_0}.$$

Now for the equation of state, assume an adiabatic process with

$$\frac{\delta P}{P_0} = \Gamma_1 \frac{\delta \rho}{\rho_0}.$$

From the equation of motion:

$$\ddot{r} = -\frac{1}{\rho}\frac{\partial P}{\partial r} - \frac{GM}{r^2},$$

and perturb it using $\partial P_0/\partial r_0 = -\rho_0 GM/r_0^2$ so that

$$\rho_0 r_0 \ddot{\xi} = -\left(4\xi + \frac{\delta P}{P_0}\right) - P_0 \frac{\partial}{\partial r_0}.$$

In general, the models predict rather well where stars of different mass and evolutionary state may become pulsationally unstable.[36] Yet, to extract detailed predictions of how the cycles appear and precise period–luminosity–mass correlations requires detailed knowledge of the opacities and equation of state. This is how one of the major astrophysical projects of the twentieth century began. Evolutionary calculations show that for a star to enter the instability strip as a classic Cepheid variable requires a mass between around 5 and 10 $M_\odot$. Pulsation models computed with more sophisticated treatments than ours and employing full interior models had for many years predicted *pulsation masses* falling outside this range. This became known as the "Cepheid mass problem" and dominated much of the discussion of later stages of stellar evolution. A remark by Simon[37] finally catalyzed a large effort to recompute stellar opacities, the *Opacity Project* to which we referred in the previous chapter. The result was an increase in the opacity in precisely the right way, the inclusion of resonances at critical temperatures. Series limits had previously been treated only approximately in most opacity tables. Another instance came from the attempt to understand the existence of the β

Substituting the equation of state for $\delta\rho/\rho_0$ (and dropping the zero subscript, which is no longer needed once you have converted all of the derivatives $\partial/\partial r$ into $\partial/\partial r_0$) gives:

$$\xi\frac{\partial}{\partial r}[(3\Gamma_1 - 4)P] - 3\xi P\frac{\partial\xi}{\partial r} + \Gamma_1 r\frac{\partial P}{\partial r} + 3P\frac{\partial}{\partial r}(\Gamma_1\xi) + P\frac{\partial}{\partial r}\left(r\Gamma_1\frac{\partial\xi}{\partial r}\right).$$

Adding and simplifying the second and fourth terms gives

$$2 + 4 = 3\Gamma_1\frac{\partial}{\partial r}(P\xi)$$

and for the third and fifth terms, you will obtain

$$3 + 5 = \frac{\partial}{\partial r}\left(r\Gamma_1 P\frac{\partial\xi}{\partial r}\right).$$

Now adding the terms together (the second through fifth) gives

$$2 + 3 + 4 + 5 = 4P\Gamma_1\frac{\partial\xi}{\partial r} + r\frac{\partial}{\partial r}\left[\Gamma_1 P\frac{\partial\xi}{\partial r}\right] = \frac{1}{r^4}\frac{\partial}{\partial r}\left[r^4\Gamma_1 P\frac{\partial\xi}{\partial r}\right],$$

so that dividing by $r\rho$ gives and assuming $\xi \sim \exp i\omega t$ gives

$$\frac{1}{\rho r^4}\frac{d}{dr}\left[r^4\Gamma_1 P\frac{d\xi}{dr}\right] + \frac{1}{r\rho}\left(\frac{d}{dr}[(3\Gamma_1 - 4)P]\right)\xi = -\omega^2\xi,$$

and *voilà*—thus appears the *linear adiabatic wave equation*.

[36] See Bono, G., Marconi, M., and Stellingwerf, R. F. 1999, *ApJS*, **122**, 167 for sample calculations.

[37] This was in a meeting on stellar variability, Simon, N. 1987, in *Stellar Pulsation: Lecture Notes in Physics No.* **274**, Cox, A. N., Sparks, W., and Starrfield, S., eds. (Berlin: Springer-Verlag). See also the discussion by Christensen-Dalsgaard, J. 1994, *Frontiers of Astrophysics: Proceedings of the Rosseland Centenary Symposium*, Lilje, P. B. and Maltby, P.), eds. (Oslo: Norwegian Acad. Sciences) on how the opacity affects the pulsation periods.

Cephei stars. These stars are periodic, although not necessary radial, pulsators that lie completely outside the normal instability region of the HR diagram. They were a persistent challenge to explain because they cannot be driven by the hydrogen convection zone that is responsible for most pulsations of massive stars. However, Stellingwerf[38] suggested that an increase in the Rosseland opacity at a temperature roughly corresponding to the ionization potential of helium could produce pulsational instability in the B-type β Cep stars. Again, the results from the Opacity Project and OPAL[39] calculations support this conclusion. An example of how differences in the internal structure for different stars at the same position in the HR diagram affects their pulsational properties comes from the Herbig Ae/Be stars, pre-main-sequence stars that cross the lower instability strip[40] The period depends on the central concentration and the stars show the same period at very different luminosities for pre- and post-main sequence evolution.

4.6 ENERGY GENERATION MECHANISMS

Since thermal timescale for the Sun is much shorter than the age of life on Earth, as must be true for other stars, we can no longer avoid examining those processes responsible for supplying energy to the interior. The luminous output of a star affects its environment in many complex ways. The effective temperature, the spectral energy distribution of the surface emission, depends on the luminosity and radius of the star. These depend on its evolutionary state, its metallicity, and many other factors that we have covered or will discuss. This luminous energy, and possibly kinetic energy lost in the form of a wind, restructures the medium around the star out to considerable distances. We will go into this more in Chapters 6 and 7 when covering star formation and galaxy evolution. You should remember that all large scale structures, galaxies and clusters of galaxies, are internally composed of stars. How they generate their energy is crucial to understanding how they evolve, hence how the galaxy that contains them evolves.

Now with that motivation, we can start. The thermal energy content of a star is

$$E_{\text{thermal}} = \int \frac{3\mathscr{R}}{2\mu} T(r) dM(r), \qquad (4.115)$$

assuming that it behaves as an ideal gas, and the energy evolution equation becomes

$$\frac{d}{dt}\left(\frac{3\mathscr{R}}{2\mu}T\right) - \frac{P}{\rho^2}\frac{d\rho}{dt} + \epsilon - \frac{1}{4\pi r^2 \rho}\frac{dL(r)}{dr} = 0, \qquad (4.116)$$

---

[38] Stellingwerf, R. 1978, *AJ*, **83**, 1184.
[39] Seaton, M. J., Yan, Y., Mihalas, D., and Pradham, A. K. 1994, *MNRAS*, **266**, 805; Rogers, F. J. and Iglesias, C. A. 1992, *ApJS*, **79**, 507.
[40] Marconi, M. and Palla, F. 1998, *ApJL*, **507**, L141; Marconi, M. et al. 2000, *Astr.Ap.*, **355**, L35.

where ϵ is the rate of energy generation and $L(r)$ is the radiative loss. This equation describes the overall internal thermal balance of the star without specifying the energy source.

4.6.1 Gravitational Contraction Again

If a self-gravitating gas bag contracts, it radiates away a part of its released gravitational potential. Assuming that there are no other internal energy sources, the resulting luminosity is

$$\frac{\partial L}{\partial M} = \epsilon_{\text{grav}} = -\frac{\partial U}{\partial t} + \frac{P}{\rho}\frac{\partial \ln \rho}{\partial t}. \qquad (4.117)$$

Notice that this works because of the star's compressibility[41] through the equation of state. Substituting

$$\frac{1}{\Gamma_3 - 1} = \rho \frac{\partial U}{\partial P} \qquad (4.118)$$

into Eq. (4.117) and using the definition of Γ_1, we find

$$\epsilon_{\text{grav}} = -\frac{1}{\Gamma_3 - 1} \frac{P}{\rho} \frac{\partial}{\partial t} P\rho^{-\Gamma_1}. \qquad (4.119)$$

In other words, the increase in the entropy of the body that produces the radiation comes from the work done by compression. You can also see here that the softness of the equation of state enters through the adiabatic exponents. The luminosity is

$$L(t) = -\int_0^M \epsilon_{\text{grav}} \, dM(r) \qquad (4.120)$$

for constant stellar mass. The change in the radius is responsible for releasing gravitational potential. Here we can use the homology relations to determine that

$$P\rho^{-\Gamma_1} = \left(\frac{R}{R(0)}\right)^{3\Gamma_1 - 4}, \qquad (4.121)$$

so we can write the luminosity as

$$L = \frac{3\Gamma_1 - 4}{\Gamma_3 - 1} \frac{\dot{R}}{R} \int P \, dV = \frac{3\Gamma_1 - 4}{2(\Gamma_3 - 1)} \Omega \frac{\dot{R}}{R}. \qquad (4.122)$$

The reason why the time derivative of the radius lies outside the mass integral is explained by the assumption of homology—that is, that the internal structure of

---

[41] This also works for the heating of a galaxy undergoing contraction or collapse, in the sense that the velocity dispersion of the stars is increased by the change in the gravitational potential, as we will see in Chapter 7.

the star is not changed by the quasi-static contraction. In effect, you assume that the star maintains hydrostatic equilibrium throughout the contraction, just as we did when deriving the Kelvin–Helmholtz timescale. Notice that the efficiency depends on the compressibility of the material, in other words, on the equation of state, and that the critical condition is $\Gamma_1 = \frac{4}{3}$, as we found for a uniform polytrope.

Gravitational energy generation is the defining feature of nearly all stages of pre-main-sequence evolution. Any massive body that evolves on the thermal timescale, provided $t_{\text{evol}} \approx t_{\text{KH}} > t_{ff}$, will slowly contract on the Kelvin–Helmholtz timescale until stability is reached, either by the initiation of nuclear burning or by degeneracy or a change in the equation of state. We now examine how nuclear reactions occur in a stellar interior.

4.6.2 Nuclear Reactions

Contraction does not continue indefinitely. In a nondegenerate mass, it increases the internal temperature and density and a new energy source can eventually be tapped *if* the threshold for thermonuclear reactions is crossed. This is the critical dividing line separating stars from nonstars such as brown dwarfs and planets. From Eq. (4.116) it follows that if $\epsilon_{\text{nuclear}} > \epsilon_{\text{grav}}$ the energy release can stabilize the star for a timescale that is long compared with the thermal timescale.[42] Inevitably, once nuclear reactions begin in a star, its life is forever changed and so is the future of the galaxy that hosts it. All reactions produce a systematic change in the stellar composition, building toward the most stable nuclei, so the process of energy release must also be one of chemical transformation.

4.6.2.1 Nuclear Binding Energy

The *binding energy* of a nucleus is determined by the strong interaction, mediated by pion exchange, between the constituent nucleons. It appears as a mass deficit for the nucleus compared with the individual particle masses:

$$B(Z, A) = (M(Z, A) - A)m_p c^2. \qquad (4.123)$$

This binding energy depends—in a complex way, which is still not thoroughly understood—on the number of nucleons. In particular, for nuclei with $A < 56$, it increases with the number of added nucleons. This is true for elements lighter than iron and nickel. However, the nuclear force saturates in this mass range and thereafter B/A decreases. For the temperatures and densities encountered in normal stellar matter, the addition of a particle to a nucleus releases energy (called *exoergic*) for $A < 56$ and requires additional energy if $A > 56$. This division is the key to understanding how nuclear energy generation feeds back into the stability of stars.

We also need to introduce another quantity, called the Q value, that defines how much energy can be released in a reaction. Consider a nuclear reaction with

---

[42] Excellent review articles to start with are Käeppler, F., Thielemann, F.-K., and Wiescher, M. 1998, *Ann. Rev. Nucl. Part. Sci.*, **48**, 175; and Arnould, M. and Takahashi, K. 1999, *Rep. Prog. Phys.*, **62**, 393.

incoming particle i, target nucleus T, product nucleus P, and outgoing particle(s) o. Then by mass-energy conservation we can know that

$$(KE)_{i,T} + (M_i + M_T)c^2 = (KE)_o + (M_o + M_P)c^2. \quad (4.124)$$

Here the kinetic energies are measured in the center of mass frame for the reaction. Using the binding energy, we have a mass deficit ΔM for each participant in this reaction; hence

$$Q \equiv [\Delta M_i + \Delta M_T - \Delta M_o - \Delta M_P]c^2. \quad (4.125)$$

This is the amount of energy that any exoergic nuclear reaction produces. Thus, once we know the reaction *rate*, we can compute the luminosity of the burning site. The usual values for Q are a few mega electron volts (MeV) per nucleon, or a few times 10^{18} erg g^{-1}. The binding energy is therefore of order $Q/m_p c^2 \sim 10^{-3}$ so the solar luminosity can be maintained easily for more than 10^{10} yr.[43] The timescale for nuclear burning is far longer than for cooling simply because of the enormous amount of energy released by any reaction, *not* because it is so efficient.

In the deep interior of a star, the temperature is sufficiently high that all principal species have been reduced to bare nuclei, with the exception of the screening provided by the environmental electrons they have shed. The primary interacting species will be particles of like charge. For simplicity, we will consider what happens if two nuclei, of mass and charge (M_A, Z_A) and (M_B, Z_B), interact. Since the charges of the incoming particle and target have the same sign, they repel at large distance. At much smaller distance, they attract because of the strong interaction and can form bound states. However, in order to get to such a configuration, the two have to overcome their electrostatic repulsion or *Coulomb barrier*. For scaling, the size of the atomic nucleus is of order

$$R(A) \approx 1.3 \, A^{1/3} \text{ fm},$$

typical of the range of the strong force. As a consequence of the small energies which nuclear particles have under stellar conditions only s waves (zero angular momentum) are important. Recall that in the atomic case, if R is the nuclear radius and k the relative momentum, then $kR \ll 1$. A second effect of the low energies is the great importance of the Coulomb barrier in the case of reactions between charged particles. The height of the Coulomb barrier scales as

$$E_C = 1.440 \, \frac{Z_A Z_B}{R_{\text{fm}}} \text{ MeV} \quad (4.126)$$

---

[43]A historical point of interest is that Harkins, W. D. 1917, *J. Am. Chem. Soc.*, **39**, 856 had already noticed a correlation between the elemental abundances and the binding energy, which was vastly extended by Jensen, J. H. D. and Suess, H. E. 1947, *Naturwissenschaften*, **34**, 131 and became a fundamental point in subsequent discussions of equilibrium versus nonequilibrium nucleosynthesis.

and the typical thermal energy, kT is

$$E_{\text{th}} = 0.086\, T_9 \text{ MeV}, \tag{4.127}$$

where $T_9 = T/10^9$ K. With these scalings we now proceed to examine the various stages of nuclear processing available to stars of different mass.

4.6.2.2 The Reaction Rates

The picture commonly used to describe stellar nuclear reactions is a two-step process, consisting of the formation of a compound nucleus by the two initial particles that subsequently decays into the final states. The probability that a proton penetrates the nuclear electrostatic repulsion depends on its kinetic energy. If $E = 1/2 m_B v_B^2$ is greater than $E_C = Z_A Z_B e^2 / r$ at some distance, the incoming particle simply passes through the repulsion and continues with some phase shift to the wavefunction. On the other hand, if $E \ll E_C$, the particle will be repelled by the target. There is some optimum distance, hence energy, at which the two nuclei essentially overlap. This distance depends on the momentum of the incoming particle through its de Broglie wavelength, $\lambda = h/(E/2m)^{1/2}$. One way to think of the nucleus is that it is fuzzy on this scale, so that the incoming particle has a cross section of order $\pi \lambda^2$, which varies as E^{-1}. Thus, the higher the incident energy, the smaller the cross section. However, at the Coulomb barrier, if the overlap is about the same size as λ we have a probability that the particle makes it through the repulsion by tunneling. The ratio $\eta \equiv E_C/E$ of the two energies at a distance λ is

$$\eta = \frac{2\pi Z_1 Z_2 e^2}{\lambda E} = \frac{2\pi Z_1 Z_2 e^2}{\hbar v}. \tag{4.128}$$

Thus, the penetration probability[44] can be understood simply from the uncertainty principle. A particle with momentum p occupies a spatial extent λ so that if the distance of closest approach is $a = Z_1 Z_2 / E$ then $\exp - \lambda/a$ is the probability of finding a particle isolated within this region. This energy dependent factor is written as $\exp - \eta$, and the nuclear cross section can be written as

$$\sigma(E) = \frac{S(E)}{E} e^{-\eta}; \tag{4.129}$$

but note that this is really as much a definition of the parameters as a formal law. The factor $S(E)$ has been added because of the complication of resonances, as we have already discussed for atomic systems. The nuclear reaction rate depends on the distribution of the bound states and whether there are resonances, screening, or other complications affecting the particular process. In ignorance, we characterize these with an energy-dependent factor that we will examine momentarily. This

---

[44] This is also called the *Gamow* or *Sommerfeld factor*; see Gamow, G. 1937, *Structure of Atomic Nuclei and Nuclear Transformations* (Oxford: Oxford Univ. Press). An especially complete and detailed discussion of penetration factors is found in Breit, G. 1959, *Handb. der Physik*, **41 / 1**, 1; although this is not the most modern treatment, it is a truly majesterial survey of the subject of reaction rate calculations.

cross section, $\sigma(E)$, has an interesting energy dependence, even if S is constant. At high energy it decreases rather fast, and at low energy it exponentially vanishes because $\eta \sim E^{-1/2}$. But there is some energy at which the reactions peak. This is the key to understanding nuclear reactions in stars, and the reason why *thermonuclear* processes depend so sensitively on the stellar mass.[45]

We are interested in the rate of energy generation in a thermal plasma, so we must include the velocity distribution of the reacting particles. Do do this, we write the reaction rate in terms of the Maxwellian functions for particles i and j. First, we need to transform to the center of mass frame and use relative velocities for the reaction rates. If $\mathbf{v}_i$ and $\mathbf{v}_j$ are the respective particle velocities, then with the following definitions we can make the coordinate change. The relative velocity is $\mathbf{v} = \mathbf{v}_i - \mathbf{v}_j$ and the center of mass velocity is

$$\mathbf{V} = \frac{m_i \mathbf{v}_i + m_j \mathbf{v}_j}{m_i + m_j}.$$

Therefore, $\mathbf{v}_i = \mathbf{V} + m_j \mathbf{v}_i / M$ and $\mathbf{v}_j = \mathbf{V} - m_i \mathbf{v}_j / M$. Using the Jacobian for the transformation, we find

$$d\mathbf{v}_i d\mathbf{v}_j = \left| \frac{\partial (\mathbf{v_i}, \mathbf{v_j})}{\partial (\mathbf{V}, \mathbf{v})} \right| d\mathbf{V} d\mathbf{v} = d\mathbf{V} d\mathbf{v}. \qquad (4.130)$$

For the integrand of the joint velocity distribution, we need

$$E = \tfrac{1}{2}\left(m_i v_i^2 + m_j v_j^2 \right) = \tfrac{1}{2} M V^2 + \tfrac{1}{2} \mu v^2. \qquad (4.131)$$

The reactions occur in the center-of-mass frame and the cross section must depend only on the *relative* velocity. The center-of-mass contribution averages out so that

$$\lambda_{ij} = 4\pi n_i n_j \left(\frac{\mu}{2\pi kT} \right)^{3/2} \int_0^\infty \sigma(v) v \exp\left(-\frac{\mu v^2}{kT} \right) v^2 dv. \qquad (4.132)$$

There is a tuning of the energy of the peak in the reaction rate to the thermal energy distribution of the reacting particles. Since the peak energy of the Maxwellian distribution depends on the core temperature, which in turn depends on M/R, the reaction rate is a function of temperature. Rewritten in terms of the energy in the center of mass frame, the number of reactions occurring per second is

$$\lambda_{AB} = N_A N_B \int_0^\infty \sigma_{AB}(E) v_A(E) f(E) \, dE, \qquad (4.133)$$

---

[45] A large-scale project has begun to maintain an online database of nuclear reaction rates of astrophysical interest, *NACRE* (Nuclear Astrophysics Compilation of Reaction Rates). This can be accessed through *http://www-astro.ulb.ac.be/nacre.htm*. Nuclear structure and decay data can also be found through this Website.

which becomes, on substituting a Maxwellian for the incoming nucleus

$$\lambda_{AB} = N_A N_B (2\pi\mu_A kT)^{-3/2} \int_0^\infty S(E) e^{-\eta(E) - E/kT} dE, \qquad (4.134)$$

where μ_A is the reduced mass $A_1 A_2/(A_1 + A_2)$ for the reacting particles. From these you may estimate two quantities that give the importance of the reaction in the stellar evolution: the lifetime of the nuclei B: $\tau_B = N_B/\lambda_{AB}$, and the power developed per gram, if Q_{AB} is the energy released in the reaction $\epsilon_{AB} = \lambda_{AB} Q_{AB}/\rho = N_B/\tau_B \times Q_{AB}/\rho$. The peak of the exponent occurs at $kT = -1/\eta'$, which is

$$\frac{2\pi Z_A Z_B m_A^{1/2} e^2}{\sqrt{2}\hbar} \frac{1}{E^{3/2}} = \frac{E}{kT}. \qquad (4.135)$$

This provides the match between the "optimum" energy and the temperature and thus provides the scaling for the reaction rate. Using $\eta = \eta_0 E^{-1/2}$, this equation becomes

$$E_0 = \left(\tfrac{1}{2}\eta_0 kT\right)^{2/3}. \qquad (4.136)$$

The width of the peak is also determined from the integral expression for the thermonuclear reaction rate, given by the second derivative of the exponential:

$$\Delta = \left(\frac{3\eta_0}{4} E_0^{-5/2}\right)^{-1/2}. \qquad (4.137)$$

Now define:

$$\frac{1}{\overline{A}} = \frac{1}{A_2} + \frac{1}{A_{\text{prod}}}, \qquad (4.138)$$

where A_{prod} is the mass of the product nucleus from the compound state. Scaling the energy to MeV and the temperature to 10^9 K, which are particularly useful normalizations for reaction rates encountered during advanced stages of evolution, we obtain

$$E_0 = 0.12 \text{ MeV}(Z_1 Z_2)^{2/3} \overline{A}^{1/3} T_9^{2/3}, \qquad (4.139)$$

$$\Delta = 0.749 \text{ keV}(Z_1 Z_2)^{1/3} \overline{A}^{1/6} T_6^{5/6}. \qquad (4.140)$$

The reaction rate can thus be approximated by its value at the peak energy:

$$\lambda_{AB} \approx \lambda_{AB}^0 \left(\frac{T}{T_0}\right)^n e^{-a_{AB} T^{-1/3}}, \qquad (4.141)$$

and n and a_{AB} depend on the particular reaction channel by fitting. We must remind you here that this critical temperature is due to the combined effects of the nuclear structure and the thermal velocity distribution but comes primarily from the height of the repulsive barrier. Everything else, resonances and other structure dependent specifics of the reactions, gets lumped into the reaction strength. This is really analogous to the atomic case *with the exception that atomic systems have a weaker electrostatic repulsion term*.

Nuclear structure enters through the level distribution of the compound nucleus. The nuclear potential is attractive within the distance $R(A)$ due to pion exchanges. Thus the bound states depend on the depth of the potential and the coupling, though the angular momenta, of the individual nucleons. This can be described in terms of the nuclear *shell model*, which is the analog of the atomic case where two flavors of particles are involved.[46] The potential has a finite positive energy portion that comes from the Coulomb potential. As in the atomic case, the width of a nuclear level is given by the lifetime of the states involved in the reactions. If the capture is through resonance and not a direct process, the resulting cross section depends on the level profile, which is called the *Breit–Wigner resonance* profile:

$$\sigma = \pi \lambda^2(E) \omega \frac{\Gamma_{\text{in}} \Gamma_{\text{out}}}{\left[(E - E_r)^2 + \Gamma_{\text{tot}}/2\right]^2}. \tag{4.142}$$

Here the statistical weight for the state is

$$\omega = \frac{(2J_{\text{in}} + 1)}{(2J_{\text{target}} + 1)(2J_{\text{out}} + 1)} \tag{4.143}$$

and Γ_{tot} is the total width of the level, which is just the direct sum of all contributing widths. This is virtually identical with the Lorentz profile that you already know from the spectroscopic problem. The integrated profile is analogous to the case of complete redistribution. The reaction rate, if the energy of the reacting particles is very close to the resonance energy, depends now on the width of *both* the entrance and exit channels:

$$\lambda_{ij} \sim \frac{\Gamma_{\text{in}} \Gamma_{\text{out}}}{\Gamma_{\text{tot}}}. \tag{4.144}$$

---

[46] We won't go into details here. There are many wonderful nuclear physics monographs, for example; Blatt, J. M. and Weisskopf, V. 1952, *Theoretical Nuclear Physics* (NY: Dover Books); Bethe, H. and Morrison, P. 1956, *Elementary Nuclear Physics* (NY: Wiley); Preston, M. A. 1962, *Physics of the Nucleus* (Reading, MA: Addison-Wesley); de Shalit, A. and Feshbach, H. 1974, *Theoretical Nuclear Physics* (NY: Wiley); Williams, W. S. C. 1991, *Nuclear and Particle Physics* (London: Oxford Univ. Press); Feshbach, H. 1992, *Theoretical Nuclear Physics: Nuclear Reactions* (NY: Wiley); and the lovely little book by Landau L. and Smorodinsky, Ya. A. 1959, *Lectures on Nuclear Theory* (NY: Dover Books). A series of reviews of nuclear reactions in astrophysics began with Fowler, W. A., Caughlan, G. R., and Zimmermann, B. A. 1967, *ARAA*, **5**, 525 and extended through two additional updates. Of course, there are *many* good books on nuclear physics in general and even more recent works, but these are among our favorites. You should always consult *Annual Reviews of Nuclear and Particle Science*, *Reviews of Modern Physics*, *Reports on the Progress of Physics*, and *Physics Reports* since many comprehensive supplements updating the standard textbook treatments appear there. In addition, tables of reactions can be found in the journals *Atomic and Nuclear Data Tables* and *ApJ Supplements*, among others. Several online databases are now available, particularly the NACRE compilation.

This is primarily because of the difference in the decay paths than the capture paths. This happens less frequently in atomic systems, but it is certainly not impossible.

There are three main types of reactions. *Nonresonant capture* actually occurs on the wing of some resonance, far enough from E_r that the cross section is a very slowly varying function of energy. For this capture process

$$S(E) = S(E_0) + \left(\frac{dS}{dE}\right)_{E_0} (E - E_0). \qquad (4.145)$$

For a *compound nucleus*, there are several steps to computing the reaction. First, you need to know the binding energies of the incident and target nuclei, which determine whether the reaction is endo- or exothermic. Consider, for instance, the last step in the *ppI* chain (see text below), $^3\text{He}(^3\text{He}, 2p)\alpha$. The binding energy of the two helium isotopes are 14.93 and 2.42 MeV, respectively, and for the hydrogen it is 7.29 MeV (on a scale where ^{12}C has $B(12, 6) = 0$). This releases 12.86 MeV, in addition to any excess kinetic energy. This also means that the formation of the compound state is substantially above the ground state for the final ^{4}He nucleus,[47] as you would expect because it is an exothermic reaction. The states that are populated by charged particle reactions in stars are *not* the ground states, unlike the atomic problem. They lie above the free kinetic energy of the incident particle, at least at the thermal energy.

The *triple-α* (or 3α) reaction is an excellent example of the role played by a near-resonant reaction in powering a star. For this reaction, the density of states in the compound nucleus, ^{8}Be, is very low. There is, however, a resonant state that is about 0.1 MeV above the thermal energy of the incident ^{4}He nuclei at the temperatures that are typical of the stellar core at the tip of the red giant branch, of order 10^9 K. It is possible to get some beryllium resonantly formed, although unstable, that immediately captures another α to produce carbon (Salpeter 1952). Another example is $^{12}\text{C}(p, \gamma)^{13}\text{N}$, the initial reaction in the CNO cycle. There is a broad resonance in ^{13}N that is at 0.46 MeV with $\Gamma/E_0 \approx 0.1$ (an excitation energy of 2.34 MeV relative to the ground state of the nucleus) for proton capture. At the temperatures of main-sequence stars, this dominates the reaction rate and becomes even more important as the mass (core temperature) increases. The existence of even a "distant" resonance makes the capture possible. There are many other examples, several of which we will discuss in the next sections.

Reactions are slow for stars at the comparatively low core temperatures found on the zero age main sequence. These involve only light nuclei that have a very low *density of states*, so they have rather small cross sections. This feature is vital for your very existence: if the cross section for the CNO cycle, and for the *pp* chain,

---

[47]Clayton (n.2) gives the example of a $^7\text{Li}(p, \gamma)^8$Be reaction that has an observed resonance at around 0.4 MeV. This corresponds to an excitation of the state at 17.6 MeV, coming from the difference in the binding energies of the input and final channel. The laboratory detection of any emitted γ will not be at the excitation energy but at the relative energy compared with the differences in binding energies. This is the reason why mass determinations are so critically important in the computation of reaction cross sections. Often a small shift in the location of an excited state can alter the cross section, hence the reaction rate, by many orders of magnitude.

were increased, you would likely not be here, a sobering example of the interdependence of nuclear and biological processes.[48] The main sequence exists for two reasons—the fuel source is plentiful *and* the reaction rate is very low. The abundance alone would not produce so great a contrast between the numbers of stars in different portions of the HR diagram.

For heavy nuclei, the density of states is very large even at low energy. This results from the enormous number of combinations of angular momenta possible once there are many nucleons. It is more similar to the molecular problem than the atomic because of the possibility that high angular momentum actually deforms the nuclear potential. This is because the constituent nucleons are all very much more massive than the electrons of an atom, and their moment of inertia contributes significantly to the total energy of the system. The problem encountered in a polyatomic molecule, with a triaxial symmetry, is quite similar to what happens in very heavy nuclear states: the energy levels are nearly a continuum.[49]

4.6.2.3 Hydrogen Burning: The CNO Cycle

Historically, the CNO cycle was the first stellar thermonuclear reaction system studied by von Weisäcker (in 1938), Gamow (in 1938), and Bethe (in 1939) in the founding papers on the subject.[50] Its initiating steps display the competition between proton capture and β decay:

$$^{12}C(p, \gamma)^{13}N(\bar{e}\nu)^{13}C(p, \gamma)^{14}N.$$

Here protons are consumed to convert ^{12}C into ^{14}N so in time, the N/C ratio rises and ^{12}C exponentially decreases with a timescale of $\lambda_{12,13}$. The proton capture continues, building progressively heavier nuclei along the β stable line:

$$^{14}N(p, \gamma)^{15}O(\bar{e}\nu)^{15}N(p, \alpha)^{12}C.$$

Thus protons are *catalyzed* by ^{12}C to form ^{4}He. There are two bottlenecks in this flow, one at nitrogen and the other at oxygen, because these are nonresonant proton captures that proceed slowly under main-sequence conditions. But the timescales are still very short compared with the usual reaction rates available for

---

[48] We admit to verging on an explanation like the *Anthropic Principle*, although with no intention of endorsing this circular view of the Universe. It is nonetheless fascinating that several features of life on this planet around this particular star are rather fortuitously fine-tuned. It is worth here quoting a remark by E. Öpik from 1938 in one of the founding overviews of stellar structure: "We are here confronted with a dilemma: if the reaction (note: meaning the initiating step of the *pp* sequence) were detectable experimentally, the sun would blow up from the immense energy generation, or rather, the sun could exist only as a diffuse star with spectrum M of about nine times its present radius.... If, however, the reaction takes place in the sun, there would be an upper limit to the probability of the reaction that makes it undetectable experimentally." This almost unknown paper, Öpik, E. 1938, *Publ. Obs. Astr. Univ. Tartu*, **30**(3), 1, contains many of the fundamental ideas of the nuclear reactions that ultimately drive the star away from the main sequence.

[49] Although we won't enter into the details, the *Hauser–Feshbach* method takes advantage of this by averaging the Gamow factor over these levels. The approximation is described in more detail in de Shalit and Feshbach (n.46). See also Rauscher, T., Thielemann, F-K., and Kratz, K-L. 1997, *Phys. Rev.*, **C56**, 1613 and references therein.

[50] von Weisäcker, C. F. 1938, *Physik. Z.*, **39**, 633; Gamow, G. 1938, *Phys. Rev.*, **55**, 718; Bethe, H. 1939, *Phys. Rev.*, **55**, 434.

stars, which would be the pp cycle, and the only thing this does is set the maximum abundance that any one of the intermediate nuclei can achieve. Notice that the nitrogen is passed through by this process, its abundance does not increase substantially because of the processing. Instead, the main product of this sequence is ^{4}He. Energy is also lost in the form of photons and neutrinos. The latter are not important for the overall energy balance since their emission rate is so low compared with the photon energy loss, even though their energies are comparable. There is also an alternate ending for this cycle, depending on the temperature, that proceeds from ^{15}N:

$$^{15}N(p,\gamma)^{16}O(p,\gamma)^{17}F(\bar{e}\nu)^{17}O(p,\alpha)^{14}N$$

Again, the key point is that these reactions all involve relatively heavy target nuclei processing light fuel. Thus, they are linear equations and can be studied rather simply (see Appendix 4.B).

4.6.2.4 Hydrogen Burning: The Proton–Proton (pp) Chain

All the mechanisms that convert four protons into a helium nucleus have a common feature—they involve two reactions that occur via weak interactions. These are necessary to transfer two units of charge from hadrons to leptons; two positrons are created and the conservations of the lepton number requires that two neutrinos appear in the final state. The presence of two weak interactions ensures that the process has a sufficiently long life to account for stellar ages. The cross sections for neutrino capture are typically of the order $\sim 10^{-42}$ cm^2; the mean free path of neutrino in stellar matter is $l_\nu = 1/n_N \sigma_\nu$, where n_N is the density of nuclei in the matter. For solar-type stars, with $\rho_c \sim 10^2$ g cm^{-3}, $n_c \sim 10^{26}$ cm^{-3}, this gives $l_\nu \sim 10^{16}$ cm, and neutrinos escape from the star. The positrons emitted with the neutrinos promptly annihilate against electrons and provide a way of transferring part of the energy released in the nuclear reactions into electromagnetic energy. Keep this step in mind, we will return to it in Section 4.6.6.1 when discussing observations of the solar neutrinos.

The initiating step of the pp reaction is proton capture followed by a β decay. The target and incident nucleus are identical, thus the rate is nonlinear in the hydrogen abundance. The network involves the sequence (PPI)

$$p(p,\bar{e}\nu_e\gamma)^2D(p,\gamma)^3He(^3He,2p\gamma)^4He.$$

The hydrogen abundance varies as

$$\frac{d[H]}{dt} = -\frac{1}{2}\lambda_{1,1}([H])^2 - \lambda_{1,2}[H][D] + \frac{1}{2}\lambda_{3,3}([^3He])^2. \qquad (4.146)$$

Notice the factor of $\frac{1}{2}$ in front of the quadratic (nonlinear) terms; this precludes double-counting of the reacting nuclei. For deuterium, we write

$$\frac{d[D]}{dt} = \frac{1}{2}\lambda_{1,1}[H]^2 - \lambda_{1,2}[H][D] \qquad (4.147)$$

and for ^{3}He we have

$$\frac{d[^3\text{He}]}{dt} = \lambda_{1,2}[\text{H}][\text{D}] - \frac{1}{2}\lambda_{3,3}[^3\text{He}]^2. \qquad (4.148)$$

This step must occur twice for the network to proceed to ^{4}He, which cancels the factor of $\frac{1}{2}$ on the quadratic term. Finally, in this simplified version of the reaction network, ^{4}He evolves according to

$$\frac{d[^4\text{He}]}{dt} = \frac{1}{2}\lambda_{3,3}[^3\text{He}]^2. \qquad (4.149)$$

This is not a catalytic reaction. There is a flow *through* deuterium and ^{3}He resulting in a net production of core ^{4}He and depletion of core H. The reactions cannot proceed rapidly because of the low rate of capture of protons by the initiating step, and the weak interaction required for deuterium formation, but thereafter the reaction must be very fast indeed. In equilibrium, the hydrogen abundance may not change substantially, but the equilibrium abundance of deuterium is minuscule because of the near balance of creation and destruction:

$$\frac{[\text{D}]}{[\text{H}]} \approx \frac{1}{2}\frac{\lambda_{1,1}}{\lambda_{1,2}}. \qquad (4.150)$$

This is a function of temperature and also proton exposure time. It is important not only for stellar nucleosynthesis but also as a thermometer of nuclear processing during the Big Bang (see Chapter 8). There is one more very important feature to notice for the pp chain. It involves only hydrogen and all other elements are formed as primary yields. Thus, a star formed in the earliest history of the Galaxy, with only Big Bang–processed matter, will behave differently than for the present-day abundances. Even for high-mass stars, you would expect that CNO burning would not occur. This complicates the interpretation of stellar properties in galaxies observed at high redshift. Because the metallicity changes the nuclear energy generation as well as the opacity, the mass–luminosity relation changes for main sequence stars. You must include these effects when analyzing such things as the rate of star formation and the mass distribution of the stellar population. We will see more of this in subsequent chapters. Now let this serve as a *caveat*.

The consumption of protons by light nuclei does not stop with the production of helium. The core temperature may be high enough for further processing to occur in spite of being too low to initiate or sustain a substantial CNO contribution to the luminosity. While we will return to this when discussing observations of solar core neutrinos (Section 4.6.6.1) for completeness we give here (see also list in Table 4.1) the PPII chain as

$$^3\text{He}(^4\text{He},\gamma)^7\text{Be}(e,\nu_e)^7\text{Li}(p,\alpha)^4\text{He}$$

TABLE 4.1 Nuclear Reaction Yields

Name	Reaction	Q (MeV)	ν_e Loss (MeV)
PPI	$p(p, \bar{e}\nu_e)^2D$	1.442	0.263
	$p(pe, \gamma)^2D$	2.486	1.44
	$^2D(p, \gamma)^3He$	5.493	—
	$^3He(^3He, 2p)^4He$	12.859	—
PPII	$^3He(\alpha, \gamma)^7Be$	1.586	—
	$^7Be(e, \nu_e)^7Li$	0.861	0.80
	$^7Li(p, \alpha)^4He$	17.347	—
PPIII	$^7Be(p, \gamma)^8B$	0.135	—
	$^8B \to {}^8Be + \bar{e} + \nu_e$	17.98	7.2
	$^8Be \to 2{}^4He$	0.095	—
CNO	$^{12}C(p, \gamma)^{13}N$	1.943	—
	$^{13}N \to {}^{13}C + \bar{e} + \nu_e$	2.221	0.710
	$^{13}C(p, \gamma)^{14}N$	7.551	—
	$^{14}N(p, \gamma)^{15}O$	7.297	—
	$^{15}O \to {}^{15}N + \bar{e} + \nu_e$	2.761	1.000
	$^{15}N(p, \alpha)^{12}C$	4.965	—
3α	$^4He + {}^4He \to {}^8Be$	−0.09	—
	$^8Be(\alpha, \gamma)^{12}C$	7.367	—
α reactions	$^{12}C(\alpha, \gamma)^{16}O$	7.162	—
	$^{16}O(\alpha, \gamma)^{20}Ne$	4.930	—
Carbon	$^{12}C(^{12}C,\gamma)^{24}Mg$	13.933	—
Oxygen	$^{16}O(^{16}O,\gamma)^{32}S$	16.542	—

and the third branch, the PPIII chain:

$$^7Be(p, \gamma)^8B(\bar{e}\nu_e)^8Be(\alpha)^4He.$$

Both these reaction sequences end by producing two ^{4}He nuclei, by either proton capture or proton breakup. The reaction rates for the weak decays are much slower than the rest of the network, so these rapidly reach equilibrium, which means that ^{7}Be and ^{7}Li quickly reach equilibrium abundances.

The pp chain is considerably lower in both efficiency and energy yield than the CNO cycle, 26.73 MeV for the entire cycle of $4p + 2e^- \to {}^4He + 2\nu_e + \gamma$, and rapidly loses out to the CNO burning with an increase in the stellar mass. There are two important differences between the reactions. First, the temperature sensitivity of the pp chain is far lower than the CNO cycle. This means the nuclear reactions are more distributed over the stellar core, so hydrogen is consumed over a larger mass fraction of the star. A relatively smaller molecular weight gradient develops in such a lower mass star, and the core can never truly become isothermal because it does not have a uniform molecular weight. The second difference concerns the consequences of the energy release for the thermal structure of the core. In the lower-mass stars, the core remains radiative. On the other hand, CNO processing is more centrally condensed and, as we have discussed in Section 4.4.3.3, this leads to core convection. The resulting mixing is fast and ensures that the core maintains a uniform chemical composition out to the boundary of the core. Therefore, it is subject to the limitations of an isothermal homogeneous core and the contraction is faster than for the lower main-sequence stars. This produces a barrier at the core boundary to any circulatory mixing resulting from rotation because the mean molecular weight of the core is higher than the envelope, a zone called the *μ-barrier*. Thus, there is a fundamental structural difference for the whole of a star's life that depends exclusively on its mass. Those stars that are higher mass than the limit for CNO processing, roughly 1.5 $M_\odot$, evolve as if composed to essentially two stars, one of core composition and the other the envelope, compared with lower-mass stars. It's worth noting that the white dwarf remnants left behind by the higher-mass stars will have a different in composition (C- and O-rich for reasons you'll soon see) than lower mass corpses (which will be He white dwarfs).

4.6.3 Helium Burning: The 3α Reaction

Helium processing is more complex than the hydrogen reactions because of a peculiarity of nuclear structure. There are no stable nuclei at $A = 5$ (check the nuclide table if you don't believe it), and it is not possible to form stable nuclei by the continued capture of protons by helium and the reactions cycle back to hydrogen, as you have seen. This was the cataract in the light elements through which the flow of nucleons could not pass in the first calculations of cosmological nucleosynthesis of the heavy elements. It finally forced the conclusion that stars are the only viable site for their formation (see Chapter 8). In a pure helium environment, it *is* possible to get past this gap by the formation of a short-lived compound nuclear state, ^{8}Be with subsequent α capture to produce the stable nucleus, ^{12}C. But it is not that simple. If the density is high enough that an α capture can occur within the ^{8}Be lifetime, it either breaks the nucleus apart or leads to stability depending on the existence of a nearly resonant capture with a state in ^{12}C. Called the *triple-α process*, the reaction proceeds as follows:

$$\alpha(\alpha,\gamma)^8\text{Be}(\alpha,\gamma)^{12}\text{C}^*,$$

where the second α capture leaves the carbon in an excited state, which then decays via an electromagnetic channel. The inverse capture reaction is actually the preferred mode for the decay of this excited state, but the radiative transition is rapid enough to ensure that some stable nuclei are formed.[51] The capture occurs on the wing of a resonance at 7.644 MeV whose width is 8.3 eV, while the (^{8}Be, α) state is at 7.366 MeV. The width of the ground state of Be is 2.5 eV, which corresponds to a life time of 2.6×10^{-16} s, a very short time indeed, but larger than the 2α reaction time. The reaction rate is, consequently, very sensitive to temperature and for the $\alpha + \alpha$ reaction increases with temperature. Thus, the concentration of ^{8}Be increases since the effect of Coulomb repulsion in ^{8}Be + α is reduced. At $T \sim 10^8$ K, this three-body process is quite efficient. The Coulomb

---

[51] Historically this was a serious problem for nuclear processing theory. Many of the basic problems were first aired at the 1953 Liege Astrophysics Symposium, *Les Processus Nucléaires dans les Astres*, the proceedings of which are a goldmine of historical insights, especially the discussions.

barrier can be breached by the now abundant α particles and other reactions become important:

$$^{12}C(\alpha,\gamma)^{16}O(\alpha,\gamma)^{20}Ne(\alpha,\gamma)^{24}Mg(\alpha,\gamma)^{28}Si.$$

The first three reactions are rather more efficient than the last two and, in general, helium burning produces a mixture of ^{12}C and ^{16}O, plus some ^{20}Ne for the highest temperatures.

You will recall that ^{14}N is a major byproduct for the CNO cycle. This nucleus can now capture an α particle, leading to a new series of reactions:

$$^{14}N(\alpha,\gamma)^{18}F(e^+,\nu)^{18}O(\alpha,\gamma)^{22}Ne(\alpha,n)^{25}Mg$$

The contributions of these reactions, in terms of number of new nuclei, is much lower than for 3α burning, but these heavier nuclei have important consequences later in the star's life: a neutron is available for processing ^{14}N coming from the CNO cycle. Since neutrons do not experience the electrostatic barrier, they can react with nuclei in the stellar core to synthesize new elements. Such reactions dominate nucleosynthesis in late stages of the evolution of stars around 5–10 $M_\odot$ on the asymptotic giant branch and, as we will see, produce observable consequences.

4.6.3.1 Carbon Burning

The 3α fusion enriches the core in ^{12}C and heavier nuclei. Since nuclei with higher Z have more repulsive coulomb barriers, higher temperatures are required to ignite the fusion of the heavier species. The star always tends to increase its central temperature because of continued core contraction due to self-gravity and energy losses. It is only required to have an abundant species and a small compression for the initiation of burning for the heavier elements. Thus, at $T \sim 7 - 8 \times 10^8$ K, carbon ignition occurs with a variety of possible outcomes:

$$^{12}C + ^{12}C \rightarrow ^{24}Mg^*$$
$$\rightarrow ^{20}Ne + \alpha$$
$$\rightarrow ^{23}Na + p$$
$$\rightarrow ^{23}Mg + n$$
$$\rightarrow ^{24}Mg + \gamma$$
$$\rightarrow ^{16}O + 2\alpha$$

The $^{24}Mg^*$ is an excited resonant capture state with a rather large width. For its decay, as for many complex nuclei, the most probable channels involve fragmentation with emission of an α particle. Further increase of the temperature leads first to photodisintegration of ^{20}Ne via $^{20}Ne(\gamma,\alpha)^{16}O$, with the accompanying α

capture, ^{20}Ne$(\alpha, \gamma)^{24}$Mg. At $T \sim 1.4 \times 10^9$ K, oxygen burning is reached:

$$^{16}O + {}^{16}O \rightarrow {}^{32}S^* \rightarrow {}^{28}Si + \alpha$$
$$\rightarrow {}^{31}P + p$$
$$\rightarrow {}^{31}S + n$$
$$\rightarrow {}^{32}S + \gamma$$

At $T \sim 5 \times 10^9$ K, α particles released by photodisintegration can be subsequently capture by, for example, ^{20}Ne, producing a near equilibrium reaction network. Starting from Mg and Si, ^{32}S, ^{36}A, ^{40}Ca, ^{44}Sc, ^{48}Ti, ^{52}Cr, and ultimately ^{56}Ni, are produced, finally leading to the production of iron, the final step in the exoergic reactions in the star.

4.6.3.2 Higher Order Nucleosynthesis and Equilibrium Processing

In addition to neutron processing, there are other entrance and decay channels by which reactions can proceed. For instance, α decay is the transformation $(A, Z) \rightarrow (A-4, Z-2)$, and, of course, the inverse can also occur. Another example concerns the modification of CNO chain at high temperature. The scheme of reactions can be modified drastically, essentially for two reasons. The elements that present β instability are assumed to decay before they could capture a proton, a condition that may not hold at the high temperatures of very massive stars or during explosions (novae and supernovae), when rapid proton capture is very favored. The α capture also becomes rather likely in competition with the proton capture. Other reactions, such as heavy nuclear interactions that occur in later stages of evolution in the approach to nuclear statistical equilibrium, involve nuclei that may be part of the network as separate species. The result is that the reaction network is much like the atomic rate equations:

$$\frac{dN(A,Z)}{dt} = \sum_{A',Z',A'',Z''} \lambda(A',Z',A'',Z'';A,Z)N(A',Z')N(A'',Z'') \quad (4.151)$$
$$- \sum_{A',Z'} \lambda(A,Z,A',Z')N(A,Z)N(A',Z').$$

The notation is awkward, but it captures the basic features. There are *many* avenues for destruction of a nucleus just as the rate equations in the atomic and molecular case permit a large number of entrance and exit channels.

The equations are rather complicated to integrate because of their inherent nonlinearity. However, we have the advantage that they are rather stiff, they have rates that are essentially constants, there are few large traverses from the β-stable nuclei, and there is little cycling once we arrive at the heavier elements. The result is that very large networks can be integrated over a wide range of densities and temperatures in reasonable times. One of the most widely used schemes is described in Appendix 4.B. The explicit version of the initial value problem is limited by the fastest reaction in the network,[52] but the implicit scheme referenced here is not so constrained (like the explicit/implicit integrators for hydrodynamic problems).

When photodisintegration processes balance reactions—which is when the reacting nuclei reach thermal equilibrium with the radiation bath—the nuclei reach the condition called *nuclear statistical equilibrium*. In this case, the forward and reverse processes balance, just as we saw for ionization and dissociation equilibria, so that $A + B \leftrightarrow C + D$. Then the relative abundances of the reactants are determined only by the ratio of their partition functions, the same as for the Saha equation, and

$$\frac{N_A N_B}{N_C N_D} = \frac{U_A U_B}{U_C U_D} e^{-Q/kT}, \qquad (4.152)$$

where U_k is the (temperature-dependent) partition function for the kth species. The main difference for nuclear reactions is that we have to explicitly treat all the levels within the reacting masses, while in the atomic case we assumed a Maxwellian for the electrons. Here lies the main difficulty with statistical equilibrium. If the particles mediating the reactions are photons and protons, then this is exactly like the Saha equation for ionization equilibrium. The Q factor in this case is the energy required to liberate a proton from the system by photodisintegration. There is an enormous number of energy levels in complex nuclei, and their population in equilibrium depends on temperature. You already know the problem from our discussion of the hydrogen atom, that the partition function does not strictly converge near the ionization limit because of the divergence of the statistical weights for the states. The same problem occurs here. Some model distribution function is required for the energy levels of the constituent nuclei in order to complete the abundance calculation.[53]

4.6.3.3 Neutron Processing and Heavy-Element Nucleosynthesis

Unlike protons, neutrons do not know about the electrostatic barrier. Hence they can slip through any "crack," inserting into any nuclear state into which they can be fit by statistics. The main limitation is whether the state is allowed. In other words, since pairing is important for neutrons, which are fermions, the reaction rate density of final states is determined by the exclusion principle and the shell structure. It is generally safe to assume that neutron reactions can occur at any

---

[52] There is an alternative procedure that has yet to be widely exploited in astrophysics, that of Monte Carlo simulations. For large networks under complex physical conditions, it offers the most flexible alternative to the large storage requirements for complex matrices. We describe the method in Appendix 4.B. See also Gillespie, D. 1976, *J. Comp. Phys.*, **22**, 403 for a very interesting and underexploited method for time-dependent reaction networks with wide ranges of reaction rates.

[53] An early, but very illuminating discussion, is found in Truran, J. W., Cameron, A. G. W., and Gilbert, A. A. 1966, *Can. J. Phys.*, **44**, 576. Of course, there have been many improvements, mainly in the detection and classification of levels. A good reference for neutron reactions is Lynn, J. E. 1968, *The Theory of Neutron Resonance Reactions* (Oxford: Oxford Univ. Press) (particularly chaps. 4 and 5). For more recent results for level calculations for both charged and neutral reactions, see Rauscher, T., Thielemann, F.-K., and Kratz, K.-L. n.46.

temperature, although excited states may be required for some special cases. Although neutron reactions are exothermic, they are never a significant energy source for the star. There is never a sufficient supply of free neutrons available compared with protons and other nuclei. Instead, they are agents of nucleosynthesis in stellar interiors and their reactions provide the principal means for producing the bulk of the heavy elements above the iron peak.

Observationally, the first clear indication of neutron processing in stellar interiors was the discovery of the lines of the technetium isotope ^{99}Tc, a very short-lived species ($t_{1/2} \approx 2 \times 10^5$ yr) in spectra of cool stars.[54] In particular, those stars that show large abundances of heavy elements that can be formed only through neutron capture are the same ones that show systematic enhancements of ^{13}C, Ba, Zr, and rare-earth elements.

Neutron reactions can occur at essentially zero temperature. Since they have no charge, they do not freeze out like proton reactions. There is no optimum temperature to which the reaction rates are tuned. Instead, the real issue is whether the rate of neutron capture is fast or slow with respect to the lifetime of the target nucleus. If the nucleus captures a neutron, ignoring internal excitation, it becomes progressively more unstable. The neutron is an unstable particle and the more neutron rich a nucleus is, the less stable the nuclear neutrons are to β-decay (in other words, there are fewer filled proton shells at the same energy to prevent the decay). The rate of capture compared with the rate of either β or α emission determines the maximum neutron-enrichment achieved by a reaction sequence. This is not so much a function of temperature as of composition, at least at this general level.

Neutron capture is separated somewhat arbitrarily into *slow*, or *s*-process, and *rapid*, or *r*-process, captures. This nomenclature traces back to the fundamental paper in 1957 by Burbidge, Burbidge, Fowler, and Hoyle.[55] This depends on which part of the periodic table you are considering. The temperature dependence enters through the thermal sources for the neutrons and also for the density of the environment. The neutron capture rate is fast compared with the shortest lifetime of any of the neutron-rich isotopes of a species (A, Z) at high temperature and high density. This means an element can add neutrons until it reaches an isotope that is unstable on a timescale short compared with λ_n^{-1}. Then the system undergoes a decay to $(Z + 1, N - 1)$ and the process continues. In this way, the

---

[54] The laboratory identifications were performed by Meggers, W. and Scribner 1950, *J. Res. Nat. Bur. Standards*, **45**, 476. Its identification by Merrill, P. W. 1952, *ApJ*, **116**, 21 in the S stars showed its close association with other neutron-rich nuclei.

[55] This paper, Burbidge, E. M., Burbidge, G. R., Fowler, W. A., and Hoyle, F. 1957, *Rev. Mod. Phys.*, **29**, 547, is universally known as B^{2}FH. It is a beautiful, comprehensive paper that set the paradigm for all subsequent work on the synthesis of the elements. The one glaring question left in the study was the site of the *r*-process and this was identified by Hoyle and Fowler (in 1960) as supernova explosions. See Fowler, W. A. 1974, *Quart. J. RAS*, **15**, 82 for a superb review of high temperature reactions, especially nuclear statistical equilibrium. The fact that different parts of the abundance distribution originate in different processing sites was among the greatest astrophysical insights of the twentieth century. It pointed to the need to understand the temperatures in evolved stellar cores and ultimately to the importance of mass loss and recycling. As an indication of the continuing influence of this paper, see Wallerstein, G. et al. 1997, *Rev. Mod. Phys.*, **69**, 995. You should not consider yourself prepared for comprehensive exams without having read at least parts of these papers.

s-process stays near the so-called β-stable line (for light elements this is approximately $Z = N$, while for heavier elements it departs slightly toward the neutron-rich side of the line) while the *r process* always stays on the more neutron-rich side of the line. This also means that at any density and temperature, since we assume that the velocity of the neutrons is due to their thermal distribution.

Neutron processing typically proceeds through rather simple networks.[56] The two processes of neutron capture and β-decay are the main competing mechanisms for generating nuclei:

$$\frac{dN(A,Z)}{dt} = \lambda_{(A-1,Z)} N_n N(A-1,Z) - \lambda_{(A,Z)} N(A,Z), \qquad (4.153)$$

assuming that fission and more complicated reactions can be ignored. Actually, neutron processing is a wonderful example of a stochastic queueing process. The critical parameter is the *exposure*, which is the time average of the neutron flux, and the process is a *single-server* queue with multiple branch points. The essential picture is that a nucleus (A, Z) can go to $(A + 1, Z)$ through the addition of a neutron provided the timescale for the β decay is slow enough for the target to exist from the previous step in the chain.

The capture of thermal neutrons is almost the same as the random process that gives rise to Poisson or noise statistics, so we expect the distribution of neutron-rich nuclei to evolve with time much like a single server Poisson process.[57] You see that the abundance of any isotope depends on the creation and destruction rates. The production of any isotopic species depends on the assumed exposure since the solution to Eq. (4.153) demands a functional form for the initial conditions. Multiple exposures to neutron fluxes produces a different history for the abundance with time than a single exposure, since the seed for a second exposure may be produced by the first processing. The simplest form for the solution of Eq. (4.153) is to take an initial neutron history of the form $\rho(\tau) = fN(^{56}\text{Fe})\tau_0^{-1}\exp(-\tau/\tau_0)$, where τ is the total neutron exposure up to a time t, f is a scaling parameter that is adjusted from the observed abundance curve, τ_0 is the characteristic timescale for the neutron source, and $N(^{56}\text{Fe})$ is the number of

---

[56] Mathews, G. J., and Ward, R. S. 1985, *Rep. Prog. Phys.*, **48**, 1371; Käeppeler, F., Beer, H. and Wisshak. K. 1989, *Rep. Prog. Phys*, **52**, 945; Meyer, B. S. 1994, *ARAA*, **32**, 153; Cowan, J. J., Thielemann, F.-K., and Truran, J. W. 1991, *Phys. Rep.*, **208**, 267.

[57] If the creation and destruction rates are identical for any nucleus and its neighbor, that is, if

$$\frac{dN_i}{dt} = \lambda(N_{i-1} - N_i),$$

then its abundance exactly follows a Poisson distribution. Saaty, T. L. 1961, *Elements of Queuing Theory* (NY: McGraw-Hill; rpt. Dover Books) provides an outstanding survey of the mathematics and we also recommend Gardiner, C. W. 1985, *Handbook of Stochastic Methods*, 2nd ed. (Berlin: Springer-Verlag) for general stochastic processes. This area is often neglected in astrophysical training.

seeds, which are presumed to be ^{56}Fe, so that[58]

$$\sigma(A,Z)N(A,Z) = \rho(\tau) = \frac{fN(^{56}\text{Fe})}{\tau_0} \prod_{k=56}^{A} \frac{1}{1+\lambda_k \tau_0}. \quad (4.154)$$

Explicitly, the complete network including the decays is written as

$$\frac{d}{dt}N(A,Z) = \phi_n \sigma_{n,\gamma}(A-1,Z)N(A-1,Z)$$
$$+ \lambda_\beta(A,Z-1)N(A,Z-1)$$
$$- [\phi_n \sigma_{n,\gamma}(A,Z) + \lambda_\beta(A,Z)]N(A,Z), \quad (4.155)$$

where ϕ_n is the neutron flux and $\sigma_{n,\gamma}(A,Z)$ is the neutron capture cross section [so that $\lambda_{(A,Z)} = \phi_n \sigma_{n,\gamma}(A,Z)$]. Notice that the β decays limit the neutron enrichment of a nucleus, since the nuclear force saturates for sufficiently neutron-rich nuclei and those at the neutron drip line ultimately β decay. We have not included fission, but this may also occur in sufficiently massive target nuclei. It adds an interesting complication to the network that mimics the behavior we discussed for proton capture and so is worth a moment's discussion. While neutron processing walks up the mass sequence, wandering more or less far from the β-stable nuclei depending on the exposure, fission recycles nuclei and presents a barrier to further buildup. For example, neutron capture on ^{235}U produces a recycling by fission back to the rare earths at about half the mass. This strong coupling between very distant regions of the periodic table on top of the usual neutron capture queue, builds up the abundances of the rare earths at the expense of continuing to the very highest masses.

Input to the queue is provided by the assumed exponential distribution for the injected neutrons. There are, however, pathways that are not this simple because the s-process does not take place at low temperature, when no neutron source is available. The high-temperature conditions that are expected for stellar interiors mean that highly excited states are accessed through the reactions, even for the relatively rare high energy channels that are in the tail of the Maxwellian distribution, and the excitations are often in the $\geq$ 10-keV range. The magic number nuclei are closed shells in the sequences and have high final abundances. The s nuclei follow a regular abundance pattern with nearly constant abundance for a given exposure level except at the magic nuclei.

Neutrons for the s-process are available only at special times in a star's life, especially in the outer core and shell burning regions of asymptotic giant branch (AGB) stars (see Section 4.7.4.4) through the reaction:

$$^{12}\text{C}(p,\gamma)^{13}\text{N}(\beta^+ \nu)^{13}\text{C}(\alpha,n)^{16}\text{O}.$$

---

[58] See Clayton, D. D. 1968, *Principles of Stellar Evolution and Nucleosynthesis* (NY: McGraw-Hill). The process is treated as a single server queue with an exponential arrival time.

Additional neutrons can be provided by proton capture on ^{22}Ne, which is formed by the reaction:

$$^{14}\text{N}(\alpha, \gamma)^{18}\text{F}(\beta^+\nu)^{18}\text{O}(\alpha, \gamma)^{22}\text{Ne}.$$

Neutrons are liberated by α capture on ^{22}Ne. These reactions require rather high temperatures and helium-rich environments because of the coulomb barrier for the incoming channel, but once started they provide sufficient fluxes at high enough energies to process Fe-peak seed nuclei up to very high masses. The r-process is altogether different and supernova explosions are the most likely event, although there have been suggestions of neutron star winds and possibly the inner accretion disks of the central engines in active galactic nuclei. The neutron density required can only be provided once some sort of neutrino process has leptonized the matter and produced a sufficient flux of neutrons to bypass the magic nuclei.

4.6.4 Explosive Hydrogen Burning

This class of reactions is a charged particle analog of rapid neutron capture and is called the *rp process*.[59] Formally, these reaction networks are identical to the r process, a competition between (p, γ) reactions and β decays, but because of the electrostatic barrier to the reactions, they require significantly higher temperatures than the equilibrium pp or CNO chains (proton capture analogs of s-processing). In our discussion of core hydrogen burning, we emphasized the relatively low temperature required for such processes and that the star's thermal timescale is long compared with any decay time in the network. There are instances, however, where the burning region is not in mechanical or thermal equilibrium, particularly when the nuclear burning ignites explosively (see our discussion of novae and X-ray bursts in Chapter 5).

The normal CNO cycle leaves ^{12}C as a catalyst, producing ^{4}He from four protons. At temperatures above 10^8 K and densities above 10^3 g cm^{-3}, however, the proton capture time becomes shorter than the β decay lifetime of the product nuclei (which is 598 s for ^{13}N and 122 s for ^{15}O). The reaction sequence begins with

$$^{12}\text{C}(p, \gamma)^{13}\text{N}(p, \gamma)^{14}\text{O}(\beta^+\nu)^{14}\text{N}(p, \gamma)^{15}\text{O}.$$

If the temperature is of order $T_9 \approx 0.2$, the sequence is completed by $^{15}\text{O}(\beta^+\nu)^{15}\text{N}(p, \alpha)^{12}\text{C}$ as usual. If, however, the temperature exceeds $T_9 \approx 0.4$, several alternative branches are accessible including $^{14}\text{O}(\alpha, p)^{17}\text{F}(p, \gamma)^{18}\text{Ne}$. Notice that the availability of helium in the burning region allows the nuclear processing to build significantly heavier elements. Ultimately, the reactions can break out of the CNO region, producing ^{21}Na from neon if $T_9 \geq 0.8$. You will notice that the proton processing leads to isotopes on the proton-rich side of the β-stable species and, in analogy to r-processing for neutrons, is limited by the drip line. At the moment, the greatest uncertainties for these networks are the binding energies and the cross sections.

---

[59]See Champagne, A. E. and Wiescher, M. 1992, *Ann. Rev. Nucl. Part. Sci*, **42**, 39; Görres, J., Wiescher, M., and Thielemann, F.-K. 1995, *Phys. Rev.*, **C51**, 392; Schatz, H. et al. 1998 *Phys. Rep.*, **294**, 167 for more complete reviews.

4.6.5 Abundances of the Elements in the Solar System

Armed with a model atmosphere and accurate laboratory wavelengths and oscillator strengths, the abundances of the elements can be obtained with high precision for one star, the Sun. These are listed in Table 4.2.[60] The pattern shows that the CNO elements, the iron peak, and the magic number nuclei are peaks in the distribution, and the general trend is for the lighter elements to have the higher abundances. Those produced in stars over the longest timescales, proton and helium burning, are also the most abundant, especially compared to the rare earths and heavier species. Several elements have not been observed directly in the photosphere; the most important are the noble gases which are taken from chromospheric and coronal determinations and are somewhat less certain.

The table also contains the abundances determined from meteorites, which provide a frozen record from the solar nebula of the conditions at the time of formation of the Sun. Not all types of meteoritic samples are equally useful for this purpose since some have undergone high temperature reprocessing, but the carbonaceous chondrites appear to have formed in a low temperature stage and not to have lost all of their volatile elements.

The discovery of SiC and AlO grains in meteorites and the determination of their isotopic composition has to be counted as one of the most exciting recent developments in experimental astrophysics. It is now possible to sample *single* grains and to place limits on their origins. The method requires the use of ion microprobes and obtains multiple samples from individual sites on the grains[61] to determine individual isotopic compositions.

Several subgroupings have been identified among the SiC grains (see Table 4.3[62]), which indicate different origins for the material. For instance, the so-called X-type grains have very high $^{12}C/^{13}C$ ratios, as large as 800 (the solar value is about 90) and show very neutron rich isotopes of intermediate mass species such as Si (e.g., $^{30}Si/^{28}Si$ can be quite high compared with normal solar values). The A and B-type grains show much lower $^{12}C/^{13}C$ ratios, less than 10, and large $^{14}N/^{15}N$ values compared with solar. Clearly, these come from not only different stellar sources but even different nuclear processes. The extreme silicon ratios argue for an environment that is not hot enough to have a large population of free neutrons, the main sources for which are p and α capture in red giant nuclear burning zones. The release of these anomalous composition grains, incorporating the signature of the nuclear processing, comes from the combined effects of deep mixing of the stellar envelope and transportation of the matter away from the stars in a grain-condensing wind. Some of these grains, those with anomalous oxygen abundances, bespeak supernova progenitors, while those with high ^{26}Mg, formed as the decay product of the short-lived isotope ^{26}Al, may come from Wolf–Rayet

---

[60] The values in Table 4.2 are logarithmic (base 10) values normalized to hydrogen, which is 12.00. From Gervesse, N. and Sauval, A. J. 1998, *Space Sci. Rev.*, **85**, 161. Some of these data are also available in Grevesse, N., Noels, A., and Sauval, A. J. 1996, *Cosmic Abundances*, Holt, S. S. and Sonneborn, G., eds. (San Francisco: ASP Conf. Ser. 99), p. 117.

[61] See Zolensky, M. E., Pieters, C., Clark, B., and Papike, J. J. 2000, *Meteor. Planet. Sci.*, **35**, 9 for a very complete description of all available techniques for microanalysis from meteoritic samples and eventually of samples collected in situ from cometary nuclei and asteroids.

[62] Data in Table 4.3 are from Zinner, E. 1998, *Ann. Rev. Earth. Planet. Sci.*, **26**, 147.

TABLE 4.2 Solar System Abundances

Element	Photosphere	Meteorites	Element	Photosphere	Meteorites
1 H	12.00	—	42 Mo	1.92 ± 0.05	1.97 ± 0.02
2 He	10.93 ± 0.004	—	44 Ru	1.84 ± 0.07	1.83 ± 0.04
3 Li	1.10 ± 0.10	3.31 ± 0.04	45 Rh	1.12 ± 0.12	1.10 ± 0.04
4 Be	1.40 ± 0.09	1.42 ± 0.04	46 Pd	1.69 ± 0.04	1.70 ± 0.04
5 B	2.55 ± 0.30	2.79 ± 0.05	47 Ag	0.94 ± 0.25	1.24 ± 0.04
6 C	8.52 ± 0.06	—	48 Cd	1.77 ± 0.11	1.76 ± 0.04
7 N	7.92 ± 0.06	—	49 In	1.66 ± 0.15	0.82 ± 0.04
8 O	8.83 ± 0.06	—	50 Sn	2.0 ± 0.3	2.14 ± 0.04
9 F	4.56 ± 0.3	4.48 ± 0.06	51 Sb	1.0 ± 0.3	1.03 ± 0.07
10 Ne	8.08 ± 0.06	—	52 Te	—	2.24 ± 0.04
11 Na	6.33 ± 0.03	6.32 ± 0.02	53 I	—	1.51 ± 0.08
12 Mg	7.58 ± 0.05	7.58 ± 0.01	54 Xe	—	2.17 ± 0.08
13 Al	6.47 ± 0.07	6.49 ± 0.01	55 Cs	—	1.13 ± 0.02
14 Si	7.55 ± 0.05	7.56 ± 0.01	56 Ba	2.13 ± 0.05	2.22 ± 0.02
15 P	5.45 ± 0.04	5.56 ± 0.06	57 La	1.17 ± 0.07	1.22 ± 0.02
16 S	7.33 ± 0.11	7.20 ± 0.06	58 Ce	1.58 ± 0.09	1.63 ± 0.02
17 Cl	5.5 ± 0.3	5.28 ± 0.06	59 Pr	0.71 ± 0.08	0.80 ± 0.22
18 Ar	6.40 ± 0.06	—	60 Nd	1.50 ± 0.06	1.49 ± 0.02
19 K	5.12 ± 0.13	5.13 ± 0.02	62 Sm	1.01 ± 0.06	0.98 ± 0.02
20 Ca	6.36 ± 0.02	6.35 ± 0.01	63 Eu	0.51 ± 0.08	0.55 ± 0.06
21 Sc	3.17 ± 0.10	3.10 ± 0.01	64 Gd	1.12 ± 0.04	1.09 ± 0.02
22 Ti	5.02 ± 0.06	4.94 ± 0.02	65 Tb	—	0.35 ± 0.02
23 V	4.00 ± 0.02	4.02 ± 0.02	66 Dy	1.14 ± 0.08	1.17 ± 0.02
24 Cr	5.67 ± 0.03	5.69 ± 0.01	67 Ho	0.26 ± 0.16	0.51 ± 0.02
25 Mn	5.39 ± 0.03	5.53 ± 0.01	68 Er	0.93 ± 0.06	0.97 ± 0.02
26 Fe	7.50 ± 0.05	7.50 ± 0.01	69 Tm	—	0.15 ± 0.02
27 Co	4.92 ± 0.04	4.92 ± 0.01	70 Yb	1.08 ± 0.15	0.96 ± 0.02
28 Ni	6.25 ± 0.04	6.25 ± 0.01	71 Lu	0.06 ± 0.10	0.13 ± 0.02
29 Cu	4.21 ± 0.04	4.29 ± 0.04	72 Hf	0.88 ± 0.08	0.75 ± 0.02
30 Zn	4.21 ± 0.04	4.29 ± 0.04	74 W	1.11 ± 0.15	0.69 ± 0.03
31 Ga	2.88 ± 0.10	3.13 ± 0.02	75 Re	—	0.28 ± 0.03
32 Ge	3.41 ± 0.14	3.63 ± 0.04	76 Os	1.45 ± 0.10	1.23 ± 0.02
33 As	—	2.37 ± 0.02	77 Ir	1.35 ± 0.10	1.37 ± 0.02
34 Se	—	3.41 ± 0.03	78 Pt	1.8 ± 0.3	1.69 ± 0.04
35 Br	—	2.63 ± 0.04	79 Au	1.01 ± 0.15	0.85 ± 0.04
36 Kr	—	3.31 ± 0.08	80 Hg	—	1.13 ± 0.08
37 Rb	2.60 ± 0.15	2.41 ± 0.02	81 Tl	0.9 ± 0.2	0.3 ± 0.04
38 Sr	2.97 ± 0.07	2.92 ± 0.02	82 Pb	1.95 ± 0.08	2.06 ± 0.04
39 Y	2.24 ± 0.03	2.23 ± 0.02	83 Bi	—	0.71 ± 0.04
40 Zr	2.60 ± 0.02	2.61 ± 0.02	90 Th	—	0.09 ± 0.02
41 Nb	1.42 ± 0.06	1.40 ± 0.02	92 U	< -0.47	-0.50 ± 0.04

TABLE 4.3 Properties of SiC Grains in Meteorites

Subgroup	$^{12}C/^{13}C$	$^{14}N/^{15}N$	$\delta(^{29}Si/^{28}Si)$	$\delta(^{30}Si/^{28}Si)$	Site(?)
Mainstream	15–100	0.3–20	15–20	272	AGB
A	< 3.5	272	≥ 0	< 220	AGB, J-type C stars
B	3.5–15	272	≥ 0	< 220	AGB, J-type C stars
X	20–8000	10–200	≥ 0	< 220	SN II
Y	≫ 100	> 300	≥ 0	< 220	High-mass AGB
Z	2–10	< 100	< 0	≥ 220	Low-mass AGB?

stars or supernovae. The grains with high ^{44}Ca almost surely come from supernova ejecta, since more recent observations by CGRO have found emission lines from ^{44}Ti, the precursor species for this calcium isotope, in several supernova remnants (most notably, Cas A, the remnant of a supernova in the late seventeenth century). The prospects are bright in the near future for vast improvements in the available sample sizes of analyzed material and for the application of even more sensitive techniques to isotopic abundance determinations.

4.6.6 Neutrino Processes

The densest state reached by a normal stellar core also has the highest temperatures, and photons begin to lose out to neutrinos.[63] This happens in the post-AGB stage when core temperatures and densities are at their highest before cooling. First, neutrinos are produced by pair annihilation:

$$e^+ + e^- \rightarrow \nu + \bar{\nu}.$$

Photoneutrino emission occurs via:

$$\gamma + e^- \rightarrow e^- + \nu + \bar{\nu}.$$

There is one other emission mechanism that is unique to dense matter. Actually, pair production was investigated very early in the modern era.[64] Collective plasma oscillations, called *plasmons*, can behave as if they have an effective mass given by

$$m_{\text{eff}} = \frac{\hbar \omega_p}{c^2}.$$

This quasiparticle has a decay channel that leads to neutrino emission:

$$\gamma \rightarrow \nu + \bar{\nu}.$$

---

[63] Bahcall, J. N. 1989, *Neutrino Astrophysics* (Cambridge, UK: Cambridge Univ. Press).
[64] Chandrasekhar, S. and Rosenfeld, L. 1935, *Nature*, **135**, 999; Temesváry, S. 1953, *Liege Astrophysical Symposium* (n.51, p. 122). Neither was, however, concerned with the neutrino emission. Instead the work focused on the generation of what were called "pairgas" under degenerate conditions at temperatures above 6×10^9 K that could add to radiation pressure, but it is instructive to see how often a process originally envisioned in one context can become important with improved physics.

This is the primary coolant in the last stages of evolution since the neutrinos are able to stream out of the stellar interior while the photons can only diffuse. It requires a density high enough for the plasma frequency to exceed the rest mass of the electron, creating a virtual pair that then annihilates to produce the neutrinos. You can use Eq. (1.150) to estimate this critical density, $n_{e,\star}$, from ω_p:

$$n_{e,\star} = \frac{m_e^3 c^4}{\pi \hbar^2 e^2}$$

corresponding to a mass density of $\approx 10^7$ g cm^{-3}. Since ω_p is *independent of temperature*, this emission is allowed as long as the Fermi energy is $\epsilon_F < m_e c^2$. In addition, although relativistic degeneracy reduces the emissivity because the density of intermediate electron states is reduced, it cannot entirely cancel the process and even white dwarf interiors cool by this emission.

A dramatic effect of this cooling is seen from our discussion of the helium flash. Remember that the reason for the flash is the onset of core electron degeneracy as the star approaches the red giant tip. Since the core density exceeds 10^6 g cm^{-3}, the energy loss by plasmon neutrinos is important relative to purely radiative processes since the interior is essentially transparent to the neutrinos. The reduction in the temperature has the effect of increasing the density, and therefore the mass of the core increases due to increased CNO processing in the surrounding shell and the degeneracy of the core increases relative to what you would find neglecting the neutrino losses. This feature of the cooling actually plays an important role in testing properties of the neutrino,[65] since anything that increases the efficiency of ν losses increases the final He core mass and also increases the luminosity of the He core burning stage, the horizontal branch, and the luminosity of the RR Lyr stars.

4.6.6.1 The Solar Neutrino Problem

There is certainly no question that the Sun and stars derive their luminosity from nuclear processes, but the unique signature of these is the emission of neutrinos.[66] Photons must diffuse through the opaque interior and are seen only once they can freely escape from the surface with the color temperature of the last layers with which they interact, while it was originally thought that ν_e, emitted from ordinary nucleosynthesis, would manage to leave the core of the star virtually unaffected, with an expected total flux of about 6×10^{10} cm^{-2} s^{-1}, a number that is instructive to compare with the photon flux that is around 10^7 times higher. Though small, it was clear that nuclear capture cross sections of readily available

---

[65] See, for example, Castellani, V. and Degl'Innocenti, S. 1993, *ApJ*, **402**, 574; Raffelt, G. G. 1996, *Stars as Laboratories for Fundamental Physics: The Astrophysics of Neutrinos, Axions, and Other Weakly Interacting Particles* (Chicago: Univ. of Chicago Press); Raffelt, G. 1999, *Ann. Rev. Nucl. Part. Sci.*, **49**, 163.

[66] The best available introduction to the basic physics is Bahcall, J. N. 1989 (n.63). A collection of the founding papers is found in Bahcall, J. N., Davis, R., Parker, P., Smirnov, A., and Ulrich, R., eds. 1994, *Solar Neutrinos: The First Thirty Years* (Reading, MA: Addison-Wesley).

materials such as ^{37}Cl are large enough that reasonable sized radiochemical detectors could be constructed. Beginning in the early 1960s with R. Davis' chlorine capture experiment deep underground in the Homestake mine, a number of observatories have been constructed to detect these particles and, from their flux, determine the *instantaneous* core properties of the Sun. Direct measurements of solar neutrinos have been proceeding now for more than 30 years and a variety of experiments have yielded firm results. These include not only the detection of ν_e from different reactions but also the direct confirmation that the emission is indeed coming from the Sun through position sensitive scattering measurements. The results are simultaneously encouraging and distressing. The neutrinos are there. This is a *good thing*. But *none* of the experiments observe the number predicted by any of the solar models that contain the agreed-on *standard* physics. This is a *bad thing*, and we now briefly review the situation in order to place the problem in overall perspective for stellar evolution.

In the pp chain, three key points determine the flux and spectrum of emitted neutrinos. The first is that the rate of the $p(p, e^+\nu)^2$D initiating reaction, the slowest in the PPI sequence, is determined uniquely by the central temperature. The second is that for the branching ratio between ^{3}He(^{3}He, 2p)^{4}He and ^{3}He(^{4}He, γ)^{7}Be, the relative rates of these two reactions determine what fraction of resulting neutrinos arises from the high-energy neutrinos (from ^{7}Be or ^{8}B). Finally, for the branching ratio between ^{7}Be(e, ν)^{7}Li and ^{7}Be(p, γ)^{8}B, the relative rates determine which fraction of the high energy neutrinos are due to ^{8}B decay. As we have discussed, interiors models set strict limits on the contributions from processes other than the pp chain to solar energy generation, placing an upper limit of about 1.5% for the CNO cycle component. Neutrinos have extremely small cross sections for interaction with matter, at least in the standard theory of weak interactions, and thus have a mean free path, much larger than the solar radius. When detected at the Earth, they reveal the conditions in the solar hearth and provide the strongest available constraint on the *contemporaneous* conditions in the solar interior.

To date, four experiments[67] have produced hard results for the solar neutrino spectrum. The GALLEX and SAGE[68] experiments used ^{71}Ga as a target to detect mainly the PPI neutrinos, and the Homestake Mine ^{37}Cl experiment, which detected neutrinos from ^{8}B, ^{7}Be, and pep steps in the reaction chain, had no directional sensitivity and measured purely total exposure for some period of time. The Kamiokande and Superkamiokande detectors are energetically blind to the ^{8}B neutrinos. The overall result for all of the detectors is that between a half and a

---

[67] For a tutorial overview of the various solar neutrino experiments, see Kirsten, T. A. 1999, *Rev. Mod. Phys.*, **71**, 1213. The history of the field is embedded in its units of measurement. The sensitivity of the experiment is indicated by the standard unit used for measuring the flux, the *solar neutrino unit* (SNU), which is 1 capture per 10^{36} nuclei of ^{37}Cl per day (recall that the typical cross section for these weak processes is of order 10^{-44} cm^2 and depends on the nuclear parameters for the equivalent oscillator strength). The SNU was originally defined for the Homestake Mine experiment, which uses chlorine captures, but is a standard unit now employed for comparisons with interiors models.

[68] See final reports in 1999, *Phys.Lett.*, **B447**, 127 (GALLEX collaboration) and 1999, *Phys. Rev. Lett.* **83**, 4686 (SAGE collaboration).

TABLE 4.4 Solar Neutrino-Producing Reactions and Sensitivities (Percent)

Reaction	GALLEX/SAGE	Kamiokande	Homestake
$p(p, e^+ \nu_e)^2 D$	57	100	0
$^3He(^4He, \gamma)^7Be(e, \nu_e)^7Li$	26	0	15
$^7Be(p, \gamma)^8B \rightarrow {}^8Be + e^+ + \nu_e$	4	0	76

third of the rates predicted by the so-called Standard Solar Model[69] predicted rates are observed. There are three neutrino-contributing steps, listed in Table 4.4.[70]

In summary, the GALLEX result is 78 ± 8 SNU, and the SAGE experiment finds close agreement, 67 ± 8 SNU. Determining the contributions of the different parts of the PPI–III reaction is not entirely independent of models since the relative contributions depend on the spectrum of the emitted neutrinos. Taking the Standard Model as a baseline predicts for both SAGE and GALLEX that about 60% of the detected signal comes directly from the pp reaction, 26% from ^7Be decays, and about equal amounts of the remainder from ^8B and CNO. Both measure about half the predicted rate (about 130 ± 7 SNU). In contrast, the Homestake experiment, for which the ^8B decay contributes over three-fourths of the flux, is nearly blind to the pp neutrinos but about a sixth of the detections come from ^7Be capture neutrinos. The observed value, 2.56 ± 0.23 SNU, is about a third of the predicted rate. Kamokande and Superkamiokande detections are essentially all from the first step of the reactions. Their data yield 2.8 ± 0.4 and 2.4 ± 0.1 SNU, respectively, both of which are again about half the predicted value. There are two fundamental problems here. The first is that the overall rate seen by all of these detectors is far lower than any reasonable model of the solar interior has been able to achieve. This is the more famous of the two issues, but the latter may be more serious in the long run and has been dubbed "the second solar neutrino problem". Even though the rates observed for the ^8B decay are low, they cannot be fit into the results from the other detectors without a serious discrepancy in the spectrum.

Regardless of the conundrum presented by this discrepancy between theory and observation, the direct detection of neutrinos that are unambiguously coming from the Sun is an outstanding confirmation that solar energy production proceeds by hydrogen burning. The beautiful feature of the water detectors is the direct locating of the neutrino source through Cerenkov emission. This isolates the source toward the Sun to within about 10 degrees. Also in its various incarnations, the Kamiokande detectors operated over a 9-year period that spanned a solar activity minimum. No modulation was seen.

We won't spend more time on the neutrino problem here. It is an evolving area as new experiments begin (for instance, the Sudbury Neutrino Observatory (SNO)

---

[69] Bahcall, J. and Pinsonneault, M. H. 1995, *Rev. Mod. Phys.*, **67**, 781; Castellani, V., Degli'Innocenti, S., Fiorentini, M. L., and Ricci, B. 1997, *Phys. Rep.*, **281**, 301; Bahcall, J. N., Basu, S., and Pinonnseault, M. H. 1998, *Phys. Lett. B*, **433**, 1.

[70] The totals do not add to 100% because a small residual contribution may come from the $p(pe, \nu_e)^2 D$ and CNO reactions, less than a few percent in each case. For the PPI chain, 99.8% of the neutrinos are produced by the first step, only 0.2% are generated by pep.

is now operating and several more will come online within the near future). Instead, we will here underline a few aspects that are clearly connected with the theory of nuclear cross sections useful to astronomy, to the correlated plasma effects, which naturally arise in the typical astronomical "matter," and to the general difficulty in extrapolating laboratory physics into an astrophysical context. This is a nice example on how astrophysical problems provoke fundamental advances in other areas of physics.

One source of uncertainty in predicting the neutrino fluxes are the cross sections for nuclear reactions in the pp chain. You already know that these take place under extreme conditions (with respect to the laboratory) of temperature, density and hence energy. Generally, all you can do is just to extrapolate the measured values down to the energy regime interest for the Sun and main-sequence interiors. This means that possible subbarrier resonances, bound states that can affect the tunneling cross section, may exist yet be missed by the extrapolation procedure. Most laboratory accelerator measurements are confined to energies above 2–5 MeV, well above those that are important for the pp-network. One exception is the $^3\text{He}(^3\text{He}, 2p)^4\text{He}$ reaction, the closing step in the PPI chain, which has been directly determined at and slightly below the energy of the solar Gamow peak[71] without any indication of resonances at low energy that would alter the capture rates.

But there is another possible answer. The neutrino may not behave as we've assumed. In 1998, the Superkamiokande group also announced the observation of a difference in the flux of atmospheric neutrinos seen through the Earth or directly above the detector. This has since been confirmed, and extended, by the SNO group. This is the signature of a mass difference between neutrino species. While the actual mass cannot be determined from these or accelerator measurements, the mass splitting, $\Delta m \approx 5 \times 10^{-6}$ eV, argues for some small rest mass for all three species. This points to the need to modify the Standard Model of electroweak theory. The effect on predicted solar neutrino emission is profound. If it is massive, a neutrino can flip between states as it traverses the dense matter of the interior.[72] This reduces the rate of detection since the experiments are sensitive only to ν_e. Although it is beyond the scope of this book to go into more detail, it seems now that the solution to the solar neutrino problem does not require any modifications in the nuclear or atomic physics needed to compute stellar structure models. For particle physics the Sun may have provided the sort of insight that the discovery of helium gave atomic physics—the indication of a new world in the stars.

4.6.6.2 Plasma Screening Corrections
Nonresonant charged particle reaction rates are determined by the rate of tunneling through the Coulomb barrier and the resulting cross section. Any ambient

---

[71] This is a very difficult energy range in which to work because it is so low. See Arposella, et al. 1996, *Phys.Lett.*, **B389**, 452; a comprehensive survey of the available cross sections is provided by Adelberger, E. C. et al. 1998, *Rev. Mod. Phys.*, **70**, 1265.

[72] A fine review of the astrophysical setting is found in Bahcall's monograph. Subsequent developments are reviewed by Haxton, W. C. 1995, *ARAA*, **33**, 459, and Fisher, P., Kayser, B., and McFarl, K. S. 1999, *Ann. Rev. Nucl. Part. Sci.*, **49**, 481.

electrons will affect the electrostatic interaction, since they reduce the effective charge of either the target nucleus, the incoming particle, or both. The traditional Debye treatment for a plasma was discussed in Chapter 1, where we found the characteristic screening length for the plasma, the *Debye length*:

$$\lambda_D = \left(\frac{kT}{4\pi n Z_1 Z_2 e^2} \right)^{1/2}. \tag{4.156}$$

This can be neglected as a correction for most stellar nuclear reactions, but for the solar interior, and those of relatively low-mass stars, Debye screening may actually be important.[73] This is an important limiting factor in the low-energy extrapolation for laboratory measurements.[74] At higher than pp or CNO energies, the screening becomes negligible for normal stellar interiors. In the solar interior, nuclei are either totally or almost completely ionized, and electrons cluster around nuclei. The average Coulomb potential between ions is much smaller than kT, and the screening can be approximated by the Debye–Hückel model, by replacing the bare nuclei Coulomb potential by the screened potential:

$$V(r) = \frac{Z_1 Z_2 e^2}{r} \exp\left(-\frac{r}{\lambda_D}\right). \tag{4.157}$$

For a dense plasma, we need to include the ions:

$$\lambda_D = \left(\frac{kT}{4\pi e^2 \rho N_A \xi} \right)^{1/2} \tag{4.158}$$

and

$$\xi = \sum_i (Z_i^2 + Z_i) \frac{X_i}{A_i}. \tag{4.159}$$

Since we are using the weak screening limit, where the separation between the two colliding nuclei is much smaller than the Debye radius, the interaction potential can be approximated by $V_D = Z_1 Z_2 e^2 / \lambda_D$ and the nuclear reaction rates in the solar plasma are enhanced by a factor $\exp(V_D/kT)$, which is equal to approximately 1.05 for the pp reaction and 1.22 for ^{3}He + ^{3}He.

4.6.6.3 The URCA Process

Nucleosynthesis during late stages of stellar evolution is dominated by the weak interaction. This is, in part, because of the steady buildup in the abundance of

---

[73] The first discussions of this effect are Schatzman, E. 1954, *ApJ*, **119**, 464; Salpeter, E. E. 1954, *Austr. J. Phys.*, **7**, 373; Salpeter, E. E. 1999, *Rev. Mod. Phys.*, **71**, S220. There is still considerable controversy about its precise treatment and the role of screening in the production of neutrinos for the Sun, but the classical (weak screening) treatment is the one that is most widely used for stellar interiors. See also Brown, L. S. and Sawyer, R. F. 1997, *Rev. Mod. Phys.*, **69**, 411; Brüggen, M., and Gough, D. O. 1997, *ApJ*, **488**, 867.

[74] See, for instance, the analysis of the reaction ^{3}He(D, p)^{4}He by Engsler, S. et al. 1988, *Phys. Lett.*, **B202**, 179, where the measured S(E) factor is compared with theoretical calculations.

heavy elements. As the nuclear fuel ignites in progressively denser and hotter environments, free electrons play a progressively larger role in the energy loss. The process of electron capture accompanied by beta decay was first discussed as an energy loss mechanism for evolved stars by Gamow and Schönberg (in 1941), who dubbed it the URCA process.[75] It is a deceptively simple prescription:

$$(A, Z) + e \to (A, Z - 1) + \nu$$
$$(A, Z) \to (A, Z - 1) + \nu_e + \bar{e}$$
$$(A, Z) \to (A, Z + 1) + \bar{\nu}_e + e \qquad (4.160)$$
$$(A, Z - 1) \to (A, Z) + \bar{\nu}_e + e, \qquad (4.161)$$

followed by

$$(A, Z + 1)e \to (A, Z) + \nu_e. \qquad (4.162)$$

There is no net change in the composition of the gas in this idealized cycle, but energy is continually lost because of the neutrino transport. The primary result of the process is that the nuclear burning region cools through the emission of electron neutrinos. This depends on the density of the medium, because at sufficiently high densities inverse beta decay is suppressed through degeneracy. The most important evidence for the role played by URCA cooling is the detection of neutrino emission from the core collapse event that preceded SN 1987A in the Large Magellanic Cloud. We will return to this when discussing supernova explosions.

4.7 STELLAR EVOLUTION

4.7.1 Observational Basis

4.7.1.1 Historical Remarks: The Hertzsprung–Russell Diagram

Up to now, we have discussed the interior processes in stars without direct reference to their observable properties. You already know that spectral classification provides some qualitative information about the luminosity and effective temperature. In the early Harvard system, the spectral morphology was taken as a temperature sequence alone, but there were already indications by the first decade of the twentieth century that another factor governed detailed spectral morphology. Maury, as an extension of the HD methodology, had proposed a second dimension for the classification of stellar morphology, the *class c* stars. These were stars that she noted with especially narrow hydrogen absorption lines. The Mount

---

[75] Gamow explains this odd acronym in his autobiography, *My World Line*. It was invented to trick the Physical Review editor, E. U. Condon, into accepting the term, reasoning it would be more acceptable as an acronym. It really derives from the Casino de Urca in Rio, where Schönberg watched patrons lose vast quantities of money—they went in with plenty and came out with nothing. This prompted Gamow to name the process after the place, which Chiu, H.-Y. 1968, *Stellar Physics. I* (Boston: Blaisdell) records as having been closed by the Brazilian government in the 1960s. It is instructive, in this time time of catchy acronyms, to know about one that isn't.

Wilson system, developed by Adams, extended the application of this morphological distinction to more stars. The decisive step occurred when Hertzsprung, using a color index for the Pleiades and Hyades clusters and contemporaneously Russell, using field stars with proper motions and parallaxes, announced that the luminosities of the *class c* stars were higher than other stars of the same temperature. Their data was limited, there were few proper motions or parallaxes available, but the result was to stand the test of time. They introduced a diagram that has since become the standard interperative plane for all stellar models, now known as the *Hertzsprung–Russell* diagram.

The original diagram is shown in Figure 4.2. In their papers, Hertzsprung and Russell distinguished essentially two groups of stars. The bulk of the stellar population formed the *main sequence*, while stars at the same effective temperature but higher luminosity clearly possessed larger radii and so were called *giants*.[76] Within a decade, the sequence of *white dwarf* stars had been added through the work of Adams and Fowler on the companion to Sirius. It took nearly 20 years for the interpretation of this diagram to emerge as an evolutionary snapshot of the galactic stellar population, largely due to Russell, Payne, and Strömgren, and involved a delicate interplay between stellar atmospheres and interiors. One of the most important steps was the realization that hydrogen is the main constituent in stars.[77] Without a handle on the basic physical properties of stars, their masses, luminosities, and radii, it is impossible to make quantitative statements about likely energy generation mechanisms. For instance, Einstein remarked that nuclear processes could be a source of energy (in his 1905 paper on inertia and energy) and Russell, Dugan, and Stewart[78] noted that nuclear processes might be responsible for powering stars, but the *rate* of the reactions and which nuclei are involved lie at the heart of the evolutionary calculations. The basics of thermonuclear reaction calculations were worked out by Atkinson and Houtermans, Wilson, and Gamow soon after the development of quantum mechanics.[79] This was left to von Weisäcker, Bethe, and Gamow in the *annus mirabilis* of 1938 to show that hydrogen nuclear processing is responsible for the main sequence and for Strömgren to explain the observed sequence as a mass-sequence of equilibrium configurations. The explanation of post-main sequence evolution and the understanding of the internal structure of stars after hydrogen core exhaustion that followed from the work of Chandrasekhar and Schönberg (1943), on the maximum mass fraction allowed for an isothermal core, required the later calculations of evolution by Schwarzschild and Hoyle (1953) and Salpeter (1953).

[76] Some useful historical references are deVorkin (1977, 1978), Herenshaw (1985), Struve and Zerbegs (1960). This diagram is of such central importance and so completely dominates the field of interior and stellar evolution that it is very useful to see what went into it.

[77] There is an excellent account of the controversy on the composition of stars in C. Payne-Gaposchkin's autobiography. Her work, and that of H. Shapley on the interpretation of eclipsing binary geometric properties and C. Moore on stellar mass determinations were essential empirical input to a general theory of stellar structure.

[78] Russell, H. N., Dugan, R. S., and Stewart, J. Q. 1927, *Astronomy. II* (Boston: Ginn and Co.). See also Russell, H. N. 1919, *PASP* **31**, 205.

[79] d'Atkinson, R. and Houtermans, F. G. 1929, *Z. Phys.*, **54**, 656; Wilson, A. H. 1931, *MNRAS*, **91**, 283. These references made extensive use of the *standard model* and Eddington's 1927 monograph. The comprehensive early survey by Gamow, G. 1937, *Structure of Atomic Nuclei and Nuclear Transformations* (Oxford: Oxford Univ. Press) is a very good place to start for the history of reaction calculations.

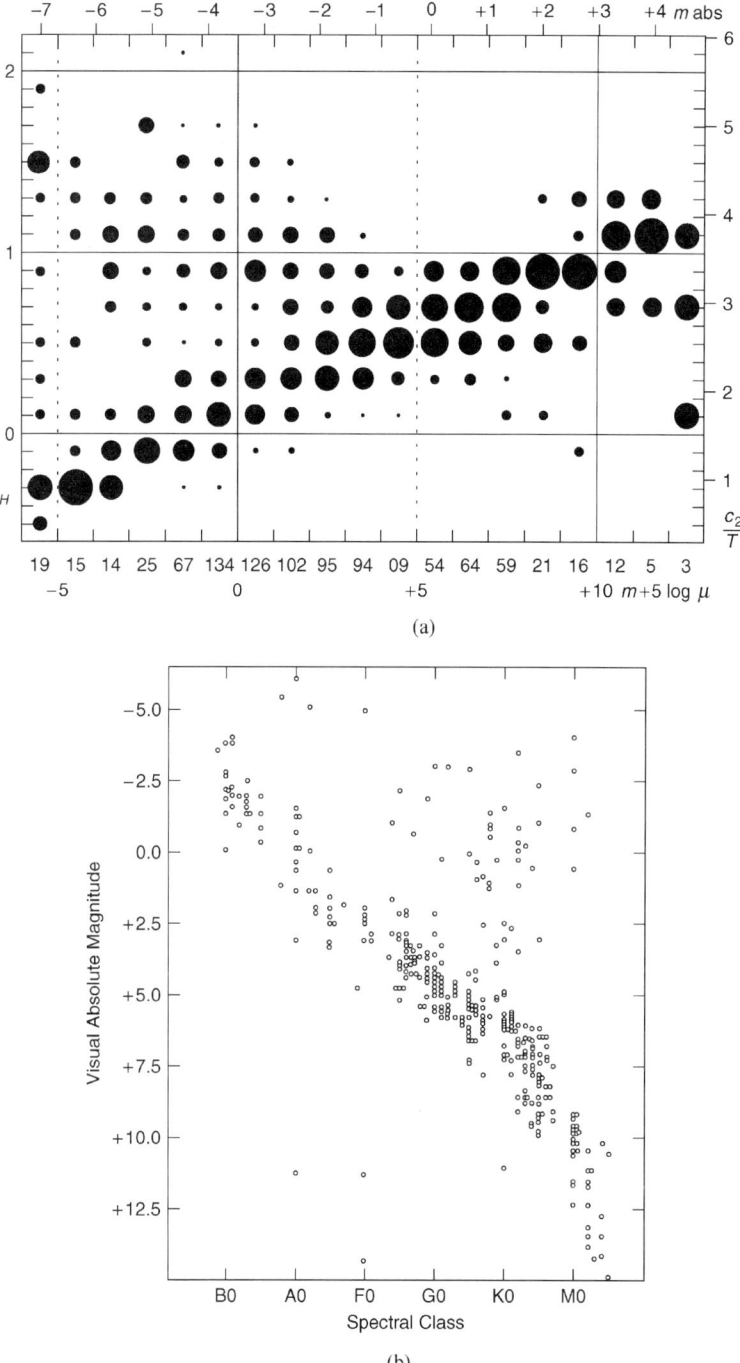

FIGURE 4.2. (a) The original diagram of color versus absolute magnitude from Hertzsprung, E. (1913, *BAN*, **19**, 91). Notice that the color index (I_H) is plotted in on the abscissa and the absolute magnitude, on the ordinate. (b) The HR diagram according to Russell, Dugan, and Stewart (1926) in its more familiar incarnation.

TABLE 4.5 Sample Detached Eclipsing Main-Sequence Binary Stars

System	Period (days)	Spectral Types	$M_1/M_\odot$	$M_2/M_\odot$	$R_1/R_\odot$	$R_2/R_\odot$	$M_{V,1}$	$M_{V,2}$
EM Car	3.41	O8 V, O8 V	22.9	21.43	9.3	8.3	-4.6	-4.3
CW Cep	2.73	B0.5 V, B0.5 V	13.5	12.1	5.7	5.2	-3.2	-2.9
QX Car	4.48	B2 V, B2 V	9.3	8.5	4.3	4.05	-2.3	-2.1
U Oph	1.68	B5 V, B6 V	5.2	4.7	3.4	3.0	-1.1	-0.4
DI Her	10.55	B5 V, B5 V	5.2	4.5	2.7	2.5	-0.5	-0.1
ζ Phe	1.67	B6 V, B8 V	3.9	2.6	2.9	1.9	-0.6	0.9
β Aur	3.96	A1 V, A1 V	2.4	2.3	2.6	2.4	0.8	1.0
V1647 Sgr	3.28	A1 V, A1 V	2.2	2.0	1.8	1.7	1.4	1.7
SZ Cen	4.11	A7 V, A7 V	2.3	2.3	4.6	3.6	0.3	0.6
CW Eri	2.73	F2 V, F2 V	1.5	1.3	2.1	1.5	2.4	3.3
DM Vir	4.67	F7 V, F7 V	1.45	1.46	1.76	1.76	3.1	3.1
UX Men	4.18	F8 V, F8 V	1.24	1.20	1.35	1.27	3.8	4.0
EW Ori	9.82	G0 V, G5 V	1.19	1.16	1.41	1.45	4.4	4.5
YY Gem	0.81	M1 V, M1 V	0.59	0.59	0.62	0.62	9.0	9.0

4.7.1.2 The Observer's HR Diagram

The original version of the HR diagram relied on spectral classification to provide a qualitative temperature. The resulting diagram is artificially quantized into vertical strips because of the discreteness of the scheme. This has since been replaced by quantitative, calibrated photometric measurements of luminosity and effective temperature. This *color–magnitude* diagram is most effectively studied for clusters, and the interpretation of these populations forms the fundamental test of stellar evolution theory. A key role is played now by stellar atmospheres models. In order to compare the theoretical diagram with an observed color-magnitude diagram, whatever the bandpass of the observations, it is necessary to transform from the relatively small bandpass of the color system to the bolometric spectral distribution. This requires a full treatment of the stellar radiative transfer to provide the bolometric correction. For example, an O star has a visual luminosity that is only a few percent of its total luminosity, and the B–V is almost completely insensitive to the effective temperature. In addition, although the visual colors are not substantially affected by metallicity, the bolometric correction is dominated by effects in the ultraviolet of the metallic lines and continua and a significant error can be made in the comparison if you do not take proper account of this. Modern stellar atmospheres calculations now provide the necessary tools for such corrections.

4.7.2 Mass Determinations

Although the existence of giants and dwarfs was clearly demonstrated by the cluster data, the key detail was provided by observations of eclipsing binary stars (some of which are listed in Table 4.5[80])—to be precise, of *double-line* binaries.

---

[80]Sources for Table 4.5 are Popper, D. M. 1980, *ARAA*, **18**, 115; Andersen, J. 1991, *Astr. Ap. Rev.*, **3**, 91; Harmanec, P. and Schoenberner, D. 1995, *Astr. Ap.*, **294**, 509.

These are systems where the two components have nearly equal luminosities. It does not mean that we obtain the best view possible of the population because these stars may not be entirely representative. It is enough that we can observe their orbits and apply gravitational theory to determine their masses, as we did in Chapter 1. Let's again state the definition of the mass function, which is the really the observable quantity that results from any radial velocity study:

$$f(M) = \frac{M_2^3 \sin^3 i}{M_1 + M_2} = (G\omega)^{-1} K_1^3 = 1.04 \times 10^{-7} \left(\frac{K}{\text{km s}^{-1}}\right)^3 \frac{P}{\text{days}} (1-e^2)^{3/2} M_\odot \tag{4.163}$$

Even for systems where you see both components and can obtain a mass ratio from the velocities, you are still constrained by the inclination, i, of the system in obtaining each mass separately, an ambiguity that is removed for eclipsing systems and for those binaries that have been resolved by interferometry. On the other hand, if the spectrum of one of the stars is normal, the mass ratio can be used to obtain the secondary star's mass. Further, for single-lined systems, even if they don't eclipse, this can be used to constrain the mass of the unseen companion, as one frequently must for systems with high-M/L companions such as degenerates, X-ray binaries, or large mass ratios.

Any characterization of the stellar population based on the qualitative information provided by spectral classification, as we have discussed, is inherently limited. It is not possible to extract precise information about masses and energetics from this sort of data. Refinements of stellar atmospheres calculations permit the determination of effective temperatures, and distances are provided by parallax and more indirect bootstrap calibrations, mainly using clusters. However, in light of the mass function, all of our knowledge about stellar properties ultimately rests on the close eclipsing binary stars. These are the only systems for which it is possible to simultaneously and unambiguously obtain all of the required global physical parameters. Regardless of distance, a spectroscopic binary permits two numbers to be unambiguously determined, the mass *ratio* and the luminosity *ratio*. The luminosity ratio requires broad spectroscopic coverage, perhaps extending into the UV and infrared to be precise, but the ratio of bolometric luminosities can be obtained. It is the *zero point* of the calibration that remains unknown. This is the value of spectral classification. Once there is a standard for any class whose distance and other absolute properties are known, it is possible to obtain absolute values for all class members. This is the hardest part of observational astronomy because the ambiguity of the distance scale plagues us at every level. The original intention of the classification system was just the proper ordering of a population, an almost mediaeval goal. It is delightful to find, however, that it provides so much more.

4.7.2.1 The Mass–Luminosity Relation

Armed with the mass ratios and spectral classifications, we return to the HR diagram. Already in the 1930s, mass measurements from eclipsing binaries showed that the main sequence is neither a cooling sequence nor an evolutionary track; *it is a mass sequence*. The lower the mass of the star *on the main sequence*, the lower its luminosity *and* the smaller its radius. We cannot emphasize this point too strongly; the *observed* locus in the color–magnitude diagram is formed by equilibrium states of different stellar masses. Once you realize that each location on the main sequence is determined by the independent variable of mass, then the duration of a star in the evolutionary stage must determine the probability of catching the star in a particular portion of the HR diagram. Hence, the number of stars in a given stage of the diagram indicates the relative evolutionary timescales for stages of evolution other than the main sequence, a frequently used argument from opinion polling and population statistics.

This reasoning is most dramatically supported by observations of clusters of stars, where one assumes simultaneous formation with some initial distribution of masses. Clusters provide much of the same information as binaries, although with some more theoretical massaging required to extract it. All the stars are at the same distance, and they have all been formed at essentially the same time and out of the same material. Cluster studies do not provide absolute numbers, but they provide relative information that requires only a single accurate calibrator to put the ratios on an absolute scale. The difference among HR diagrams for different clusters has to be ascribed to differences in the age, rotation, binary fraction, and the chemical composition of stars, assuming that the mass functions of the clusters are essentially similar. The HR diagram for field stars is of comparably limited use because of the relatively small data set of absolute calibrators. Because you have no information on the distances to most of the stars, you must use their spectral types to obtain absolute properties, and these are merely statistical, a method also called *spectroscopic parallax*.

No single power law fits the entire main sequence. Instead, the exponent depends on the stellar mass:

$$\begin{aligned} L &\sim M^{1.9}, & M &\leq 0.4\, M_\odot \\ L &\sim M^{4.75}, & 0.5 &< M < 7 M_\odot \\ L &\sim M^3, & M &\geq 7 M_\odot \\ L &\to L_{\text{Edd}} \sim M, & M &\to \text{lots of } M_\odot. \end{aligned} \qquad (4.164)$$

For the low-mass stars, those with deep convective envelopes, the slope of the mass–luminosity relation is determined by the opacity. The slope is, therefore, very sensitive to the stellar metallicity. For the intermediate mass stars, taking 5 $M_\odot$ as a fiducial value, the temperature dependence of the CNO cycle is the main agent responsible for the slope of the relation. There is only a weak dependence on metallicity through the abundance of the catalytic nucleus, ^{12}C. The opacity is a less important contributor because of the core convection. Finally, for the highest-mass stars, the Eddington limit becomes the asymptotic state of the star—the luminosity adjusts in equilibrium to maintain the stability of the star—so the relation becomes $L \sim M$ in the highest luminosity limit.

4.7.3 Theory: The Principal Stages of Stellar Evolution

The goal of stellar structure theory is the interpretation of the HR diagram. So far, we have considered only the ingredients—the opacity, energy generation mechanism, and thermal and mechanical balance that permit the computation of a single

TABLE 4.6 Stages of Energy Generation during Evolution

T_c (10^{7}K)	Source	Yield (MeV/nucleon)	E_n/N (MeV)	L_γ/L_{tot}	L_ν/L_{tot}
0–1	Gravitation	1	—	1.00	—
1–3	4H → ^{4}He	—	6.7	0.95	0.05
3–10	Gravitation	9	—	1.00	—
10–30	3 ^{4}He → ^{12}C	—	~ 0.7	1.00	—
30–80	Gravitation	90	—	0.50	0.50
80–110	2 ^{12}C → ^{24}Mg	—	0.3	—	~ 1.00
110–140	Gravitation	50	—	—	~ 1.00
140–200	2 ^{16}O → ^{32}S	—	~ 0.3	—	~ 1.00
200–500	→ Fe	250	0.4	—	~ 1.00

model. Stellar evolution is the history of the changes wrought in the interior structure of a self-gravitating mass by the competition between energy loss and energy generation while attempting to maintain mechanical equilibrium in the face of its self-gravitation. Internal nuclear energy release by compensating for radiative losses, slows the contraction induced by initial cooling and eventually stops and even reverses it. But the nuclear processing occurs at the expense of the particle number, all the while increasing the mean molecular weight of the matter in the processed regions.

It was the hope of the twentieth century that a star's history could be predetermined by just two parameters: its mass and initial chemical composition. Over time, this was enshrined as the Vogt–Russell "theorem." You can see broadly how this *might* work by thinking about all the processes we have discussed, although with tongue almost planted in cheek. A star's mass and radius determine its central temperature and density. The radius is determined by the opacity and flux, which in turn depends on the rate of internal nuclear reactions. These depend on composition and temperature. So you see that we keep coming back to two basic inputs. But there are additional considerations, attributes that we cannot ignore. The star may rotate. It may have a magnetic field. It may be inhomogeneous. Most of all, if it loses mass, when it does so affects its whole subsequent life. The point of the "theorem" had been that a star's location in an observational HR diagram could be transferred directly and uniquely onto a theoretical evolutionary track and this turns out not to be the case. Initial masses can arrive at similar locations in the HR diagram after very different histories and with very different internal structure.[81]

Table 4.6 summarizes the successive stages of energy release by gravitational and nuclear processes through the course of a typical star's life. Phases of

---

[81] The clearest statement of the premise is in the introductory text by Russell, H. N., Dugan, R. S., and Stewart, J. Q. 1927, *op. cit.* See also Russell, H. N. 1931, *MNRAS*, **91**, 951; Kahler, H. 1972, *Astr.Ap.*, **20**, 105; Kahler, H. and Weigert, A. 1974, *Astr.Ap.*, **30**, 431. There was considerable interest in the Vogt–Russell theorem during the 1970s when work was starting on models for late stages of evolution. It was hoped that linear interpolation sequences could be used for some of the more complex stages, such as core helium burning and double-shell phases, without worries about the stability of the models. Computations soon showed the folly of this procedure, and advancing technology and computational improvements made the schemes unnecessary. Nonetheless, some wonderful insights were gained by these studies and you should not neglect this little corner of the literature!

gravitational contraction lead to successive ignitions of the various species, eventually producing iron in the most massive stars. The nuclear-generated steps last for a time that depends on the condition that nuclear energy balances surface losses. Hydrogen burning is the longest phase in stellar evolution, as it is the least temperature-sensitive and involves the most intrinsically abundant fuel. Each successive stage is faster, in part because progressively less of the stellar mass is involved and in part because the rate of fuel consumption increases. You already know that the number of stars actually observed in any evolutionary phase is proportional to the duration of that phase so you can anticipate that an unbiased survey that is complete for a large spatial volume of the Galaxy will find a large fraction of stars burning hydrogen, a smaller fraction that are burning He, and scant evidence for stars in the later combustion phases. We will use this again in the discussion of the HR diagram for the Galactic field stars, where continuing star formation maintains an almost steady state population of different regions of the diagram.

4.7.4 The Earliest Stages of Evolution: Protostars

We begin at the star's beginning: its formative stage and earliest evolution. We will separate the pre-main-sequence development into two stages based on the timescale and the disequilibrium of the nascent star. The first is the *protostellar* stage. During this dynamical phase of evolution, we barely have a star, since the core that will eventually become a stable object gains most of its energy through infall. This is followed by a more quiescent phase, the *pre-main-sequence* phase of evolution, during which the nearly hydrostatic star evolves toward the conditions that end with the initiation of core hydrogen burning on the main sequence.

Star formation starts with the collapse of an overdense self-gravitating mass within a molecular cloud. We will assume you already have a collapsed object in the center of a more diffuse medium postponing until Chapter 6 our discussion of where the central mass comes from. Once the protostellar core becomes opaque and ceases to collapse, it is still accreting matter from the surrounding cloud. The core is locally much denser, and there is really no way to turn off the infall short of building a barrier, something the star actually does in an unusual way. First, you can get a simple estimate of the accretion rate by knowing the velocity distribution within the surrounding gas. In effect, the cloud matter acts like an inverse wind, which was actually the problem with which accretion theory started. The rate at which material falls onto the core depends on where it starts. If the cloud is pressure supported, the infall produces a rarefaction wave that moves outward from the growing protostar at the sound speed (or the turbulent velocity, depending on how the mass is being transported across the accretion interface). The radius at which the freefall time equal the sound travel time is $r = GM/c_s^2$. Now the rate of mass accretion is $\dot{M}$ so if the mass increases as $M = \dot{M}r/c_s$ and

$$\dot{M} = \frac{c_s^3}{G} = 2.4 \times 10^{-4} \left(\frac{T}{100 \text{ K}}\right)^{3/2} M_\odot \text{ yr}^{-1} \qquad (4.165)$$

is the *constant* mass accretion rate. By the virial theorem, half of the potential

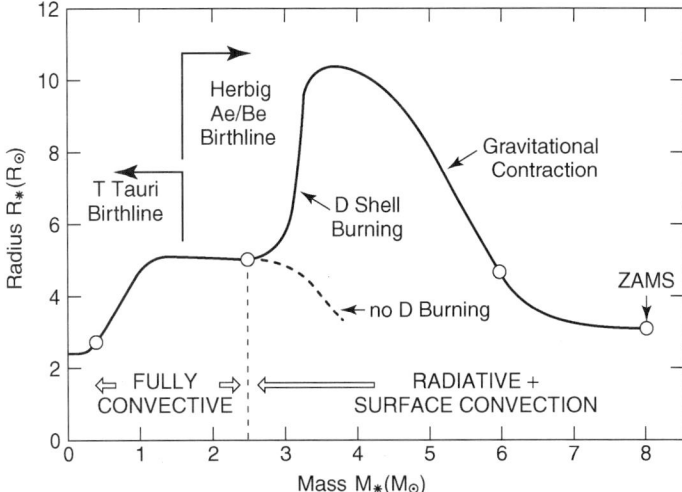

FIGURE 4.3. Growth of a protostellar core by accretion for a final 8 $M_\odot$ main-sequence star (credit: F. Palla, Osservatorio Astrofisico di Arcetri).

energy released by the accretion flow is radiated. Since the core is stationary with radius R_c, the accretion forms a standing shock at the surface that produces a luminosity:

$$L_s = \frac{GM\dot{M}}{2R_c}. \qquad (4.166)$$

This is the basic luminosity observed for the protostar at the critical mass for gravitational collapse of an isothermal body (the Jeans mass; see Chapter 6). The growing core eventually crosses the threshold for deuterium burning. You will recall the reaction ^{2}D(p, γ)^{3}He occurs at very low temperature, only about 10^6 K. With an energy release of about 2 MeV per reaction (about 2.6×10^{18} erg g^{-1}) and with an initial (interstellar) [D/H] of order 2.5×10^{-5}, the accretion-fed luminosity amounts to approximately $L \approx 10(\dot{M}/10^{-5}\ M_\odot\ \text{yr}^{-1})L_\odot$. Convection acts within the star as long as the flux exceeds the radiative flux which, for stars with Kramers-type opacities, means $L_{\text{rad}} \sim M^{11/2}R^{-1/2}$. The Kelvin–Helmholtz timescale is then equal to $t_{\text{KH}} = GM^2/(RL_{\text{nucl}})$ which is numerically about the same as $M/\dot{M}$ because the yield per unit mass for deuterium burning is coincidentally about the same as the gravitational binding energy.[82] The growth of the protostellar core is shown in Figure 4.3.

---

[82] The first numerical hydrodynamic simulations of protostellar collapse were made by Larson, R. B. 1969, *MNRAS*, **145**, 271; Penston, M. V. 1969, *MNRAS*, **144**. 425 (and references therein). These showed the distinctly non-homologous nature of the core formation and showed that the accretion forms a self-similar envelope. Although in principle it demolished the prevailing Hayashi treatment, it took considerable time for this picture to finally become the standard model. See Stahler, S. W. 1994, *PASP*, **106**, 337; Stahler, S. W. 1988, *ApJ*, **332**, 804; Palla, F. and Stahler, S. W. 1993, *ApJ*, **418**, 414; Palla, F. 1994, in *Infrared Astronomy* (eds. Mampaso, A., Prieto, M., and Sánchez, F.) (Cambridge, UK: Cambridge Univ. Press); Palla, F. and Zinnicker, H. 2002, *Physics of Star Formation in Galaxies: Saas-Fee Advanced Course 29* (Berlin: Springer-Verlag).

Since the core is increasing in mass by accretion, and also growing in radius, the gravitationally generated luminosity gradually decreases relative to the ^{2}D ignition and subsequent burning that occurs in the core. Independent of the environment, which is likely an accretion disk, the core evolution is dominated by the nuclear luminosity, but *not* hydrogen core fusion. The core reaches rough hydrostatic equilibrium and is completely convective; hence has a larger radius than predicted by the radiative track (since it expands to reach an adiabatic temperature gradient), and the final stage of contraction does not occur until the nuclear burning ends. This is relatively fast since the deuterium is rapidly exhausted.

When convection ends, turbulent mixing ceases to supply the core and the its energy loss can only be compensated by contraction. The binding energy, however, now increases so the accreting layer at the surface eventually reaches the temperature of nuclear ignition and initiation of ^{2}D processing swells the outer layers and for a short time produces another deep convection zone. This layer is, however, unable to support itself against contraction and eventually the star settles onto the hydrostatic contraction that leads to core hydrogen ignition. This is the *birthline* for protostars. Thereafter the star contracts radiatively to the main sequence and a star is born.[83] The difficulty with this picture comes from the observation that mass accretion during protostellar evolution is regulated by the formation of an optically thick disk.[84] Detailed models for the interior evolution of these protostars have thus far only treated spherical accretion except for the inflow of matter. The matching between the protostellar core and the boundary layer of the disk, and the reaction of the disk to the variations of the growing central mass and its various stages of nuclear processing, remain to be worked out. This is the barrier we mentioned above. In the accreting layer, there is an enormous dissipation of kinetic energy and considerable torque. Precisely what occurs here is as uncertain as it is for close binary stars (see discussion on boundary layers in Chapter 5).

4.7.4.1 Main Sequence

We first encountered this stage when we discussed the pp chain and the CNO cycle. Here we note that the time to central hydrogen exhaustion, the measure of the main-sequence lifetime, is easily estimated from the mass–luminosity relation. For the bulk of stars on the main sequence, $L \sim M^3$ so that $t_{ZAMS} \approx 10^{10}$ yr $(M/M_\odot)^{-2}$. It is possible to understand a few properties of the evolution of stars in their main sequence phase without the need to integrate models for stellar structure. For example, the correlation of central temperature with the stellar mass: you wait for a larger T_c as M increases, since the energetic need of the structure is more important. As M increases, the energy generation passes to the CNO cycle, which requires higher temperatures, even higher if the metal content Z is lower, since the efficiency of the cycle would be reduced. The behavior of the central density is more complex, since it depends on the details of the equation of state, but on the average it decreases as the mass increases.

---

[83] D'Antona, F. and Mazzitelli, P. 1994, *ApJS*, **90**, 46.

[84] The distribution of the planets within our solar system had already given a hint of this. The Jovian planets, in particular, and the asteroids form a nearly planar distribution. Comets are, however, more nearly spherically distributed and suggest that the collapse became more flattened with time. See also Bertout, C. 1989, *ARAA*, **27**, 351; Hartmann, L. 1998, *Accretion Processes in Star Formation* (Cambridge, UK: Cambridge Univ. Press).

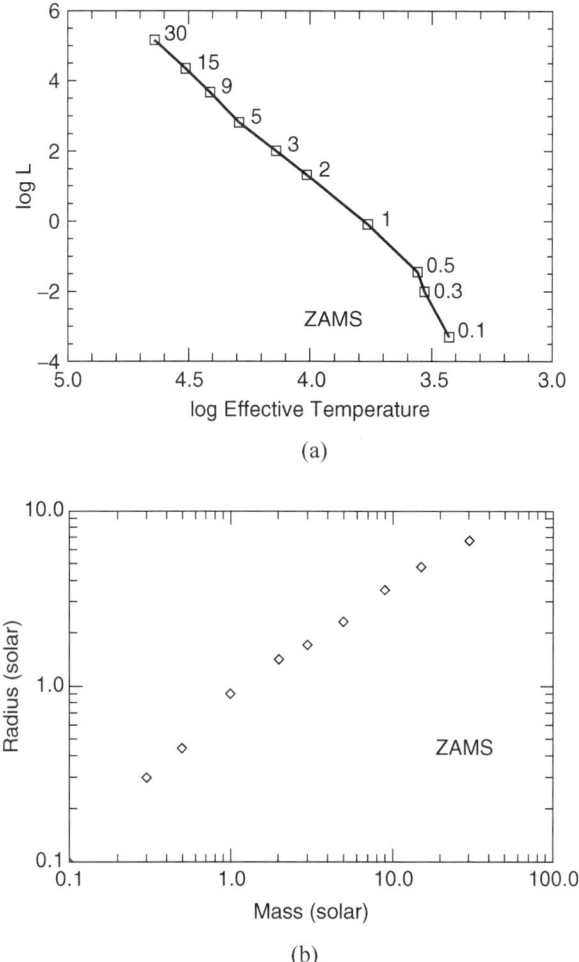

FIGURE 4.4. (a) The theoretical HR diagram for ZAMS models; the masses are indicated. (b) Mass–radius relation for ZAMS models.

Low mass stars occupy the faint, cool end of the main sequence, stars with radiative cores and thick convective envelopes. The convection deepens as the stellar mass decreases toward 0.3 $M_\odot$, by which point the interior becomes completely convective. By contrast, the upper main sequence is inhabited by massive stars possessing convective cores and radiative envelopes. Although the overall properties of the main sequence result from hydrogen core burning, you know that the details depend on mass. Figure 4.4a shows the locus of masses along the ZAMS (from which you can see the mass–luminosity relation) and Figure 4.4b shows the mass–radius relation. The variation of the slope of the main sequence in the HR diagram between 1 and 2 $M_\odot$ results from a change-over in the dominant nuclear energy generation process from the pp chain at the low mass end, to CNO burning for the more massive, hotter stars. Variations of the two compositional parameters, Y and Z, shift the location of the sequence. On lowering Z, the

TABLE 4.7 Representative Theoretical ZAMS Model Parameters

$M/M_\odot$	$\log L/L_\odot$	$\log T_{\rm eff}$	$R/R_\odot$	$T_c/(10^7\,{\rm K})$	ρ_c
0.3	-1.90	3.55	0.3	0.8	125
0.5	-1.42	3.59	0.44	0.9	74
1.0	-0.15	3.75	0.9	1.4	84
2.0	1.22	3.97	1.6	2.2	68
3.0	1.91	4.09	2.0	2.4	41
5.0	2.74	4.24	2.6	2.8	21
9.0	3.61	4.38	3.6	3.1	10.5
15.0	4.32	4.51	4.7	3.4	6.2

ZAMS becomes less luminous because of the reduction in the efficiency of CNO burning for the higher mass end and from changes in the opacity at the low-mass end. The same effect is obtained by increasing Y.[85] Because of the strong temperature dependence of the CNO cycle, $\epsilon_{\rm CNO} \sim T^{15}$, the core is convective for $M \geq 1.5\,M_\odot$ and is surrounded by a radiative envelope. For lower mass, the central core is radiative and the envelope is convective. This structure is maintained throughout the main sequence phase and plays the dominant role in the subsequent evolution of the different mass stars. Those of lower mass, like the Sun, undergo slower differential core contraction as hydrogen burning progresses and their core composition displays a slowly steepening gradient with time. More massive stars are homogenized by their core convection and show a jump of increasing magnitude in molecular weight as they consume the available fuel. Their cores contract uniformly and quickly once the nuclear fuel turns off and their evolutionary tracks reflect this, a feature we will examine in the next section. (See zero-age main-sequence (ZAMS) model parameters in Table 4.7.[86])

Another feature that distinguishes the most massive stars from the rest of the stellar population is that they might not display a main-sequence stage in the normal sense we have just discussed. Stars above about 40 $M_\odot$ may never be completely free of their accreting placental protostellar environment until some core hydrogen burning has already begun. But is there ever really a genuine observable zero-age main sequence? After all, there are different timescales for arrival on the ZAMS so in any cluster you don't see all the stars at the same stage, depending on their mass. In addition, for the most massive stars—whose strong winds turn on even before they reach the main sequence—there is even a question of whether a ZAMS exists at all for a single mass.[87] This suggests that the main sequence observed in OB associations, such as R136 in the LMC, are not zero-age and that models constructed assuming hydrostatic equilibrium may fail. Indeed, this is the fundamental limitation of all pictures of star formation. While low-mass stars can form even by spherical accretion, the same phenomenon that produces

---

[85] Iben, I. Jr. 1967, *ARAA*, **5**, 571; Schaller, D., Schaerer, D., Meynet, G., and Maeder, A. 1992, *Astr. Ap. Suppl.*, **96**, 269.

[86] Sources of Table 4.7 are Bodenheimer, P. 1989, *Encyclopedia of Astronomy and Astrophysics*, Meyers, R. and Shore, S. N., eds. (San Diego: Academic Press); Drilling, J. S. and Landolt, A. U. 2000, in *Astrophysical Quantities*, 4th ed. Cox, A. N., ed. (Berlin: Springer-Verlag), p. 395.

[87] For further discussion, see Bernasconi, P. A. and Maeder 1996, *Astr. Ap.*, **307**, 829.

radiatively driven winds in massive stars will act on the environmental gas to initiate an *outflow*. The limitation of the accretion rate to stars has been partially offset by realizing that the accretion can occur in a disk, in which case the Eddington-type limits do not apply, but this severely limits the rate of growth of the protostellar core. An alternative view, in which the massive stars form by agglomeration of lower-mass fragments, has been advocated as the way to circumvent this problem, but at the moment it remains an open issue.

There is a bottom to the main sequence. Below about 0.08 $M_\odot$, the core is not hot enough to ignite proton burning. These objects, expected from collapse calculations down to about 0.02 $M_\odot$, are called *brown dwarfs* in analogy with red dwarfs and because they closely resemble the Jovian planets. Although below this mass, agglomerative collisions between small mass parcels of matter, *planetesimals*, are required to build the bodies, above this limit the formation is quite similar to ordinary stars. These bodies are further distinguished between those that never get hot enough to ignite ^{2}D burning, even for a brief time, and those that never undergo any nuclear processing. The main sources of interior and atmospheric opacity are molecular, but in the coolest parts of the atmosphere it is necessary to use complete chemical networks that include particulate formation. In other words, dust and droplets are both opacity sources, and another strong contributor is H_2O, hot steam. These are no longer theoretical entities. All-sky infrared surveys, in particular the 2MASS project which is working at 2μ, is discovering these in increasing numbers. The spectra of several have been obtained—the first and still most famous is Gliese 229B—and as expected they resemble Jupiter very closely. This is another area that will certainly develop rapidly in the coming years.

4.7.4.2 Ascent of the Red Giant Branch

The helium core becomes nearly isothermal by the end of the main-sequence phase.[88] For massive stars, there is a molecular weight difference between the well mixed core and the envelope because of the convective mixing, although the core boundary is potentially blurred by nonlocal effects in the convective motions or *overshooting* of individual turbulent cells.[89] For lower-mass stars, a smoother composition gradient marks the core boundary where some of the hydrogen has been only partially consumed. The core has been contracting during this whole time, but now this accelerates. There is no longer an active nuclear source and, because its radius is so small, the core's thermal timescale is shorter than for the envelope. Its dynamical timescale is also shorter because of the higher density. If the star is to maintain equilibrium, the outer layers must expand. This drops the

---

[88] Surveys of main sequence and post-main-sequence evolution are found in Iben, I. 1974, *ARAA*, **12**, 213; Iben I. and Renzini, A. 1983, *ARAA*, **21**, 271; Iben, I. 1991, *ApJS*, **76**, 55.

[89] Here we encounter one of the consequences of assuming a mixing length theory. Strictly speaking, "overshooting" is a stellar interior analog of the generation of a convective plume or cloud in the Earth's atmosphere. When released, a single convective element rises to its level of neutral buoyancy but, if there is a strong enough temperature gradient and low enough viscosity, overshoots this point and oscillates with a decreasing amplitude around the top of the convective layer. This analogy, however, ignores the fully turbulent nature of the stellar case and, moreover, the broad spectrum of kinetic energy distributed over the individual length scales, as we discussed in Section 4.4.3.5. Thus, although models are routinely computed with various mixing length parameters, α, the phenomenological interpretation of the terminal main sequence evolution is better described as possibly requiring a more massive helium core than can be produced without some degree of envelope-core mixing.

effective temperature and also the envelope temperature, and new opacity sources appear, specifically molecules and H^-. The combined effect of the core contraction and the increased opacity of the outer layers forces convection to occur progressively deeper, eventually extending throughout the envelope. This follows from the large radiative temperature gradient that would be required to transport energy through the outer layers. The star reaches the path that it followed in the last stages of initial contraction as a completely convective body, the *Hayashi track*.[90] The result is that the star rapidly increases in radius to become a red giant.[91]

We now return to the different evolutionary histories of stars of different initial mass. The exhaustion of the central hydrogen in main-sequence stars is different at the low-mass end or high-mass end, that is if the core is radiative or convective, as a consequence of how the burning of hydrogen proceeds in the stellar interiors. In a radiative interior, the hydrogen is continuously consumed from the interior outward. The combustion moves from the core to a shell surrounding the core. In the interval before and after the exhaustion of H in the center, the stars attain the maximum efficiency of the CNO cycle and also a relative maximum of T_{eff}, a location in the HR diagram called the "main-sequence turn-off." When the convection is active, the mixing imposed by it is such that the hydrogen going to be fused is extracted over the whole region interested to convection, not only in the limited zone where burning is effective. The central source of nuclear energy is exhausting, and the star renews the gravitational contraction that will end only when the temperature induces efficient H combustion in a shell. On the HR diagram the evolution will be qualitatively as in Figure 4.5. The dynamic phase occurs on a very short timescale, and its consequence is to leave very few or no stars observable in this part of the HR diagram (Fig. 4.6. shows the color–magnitude diagram of an moderately old open cluster, NGC 752).

Hydrogen shell ignition is accompanied by an expansion of the external layers, while the luminosity remains approximately constant. The effective temperature decreases and the envelope overlying the shell quickly becomes convective due to the increased opacity. The star does not precisely follow the Hayashi track, however, since the helium core is centrally isothermal. As a general rule, *central combustions place the stellar models toward high effective temperatures, while combustions in shell places the star on the Hayashi track*. For a star leaving the upper main sequence, the helium core continually increases in mass, remaining nearly isothermal since there are no internal energy sources. This cannot last indefinitely since hydrostatic equilibrium requires a higher pressure and hence a contraction. Depending on the stellar mass, the core can ultimately reach temperatures that are sufficient to ignite the 3α reaction.

---

[90] Hayashi, C., Hoshi, R., and Sugimoto, D. 1962, *Prog. Theor. Phys. Suppl.* (*Japan*), **22**, 1. It should be noted that this track was originally found for red giant branch models and only later was it argued that the reverse situation would characterize the pre-main sequence stage of a star.

[91] There has been a lot of discussion in the literature about why this happens. See Iben I. Jr. 1974, *ARAA*, **12**, 215; Bhaskar, R. and Nigam, A. 1991, *ApJ*, **372**, 592; Frost, C. and Lattanzio, J. 1992, *Proc. Astr. Soc. Austr.*, **10**, 125; Renzini, A., Greggio, L., Ritossa, C., and Ferrario, L. 1992, *ApJ*, **400**, 280; Sugimoto, D. and Fujimoto, M. 2000, *ApJ*, **538**, 837.

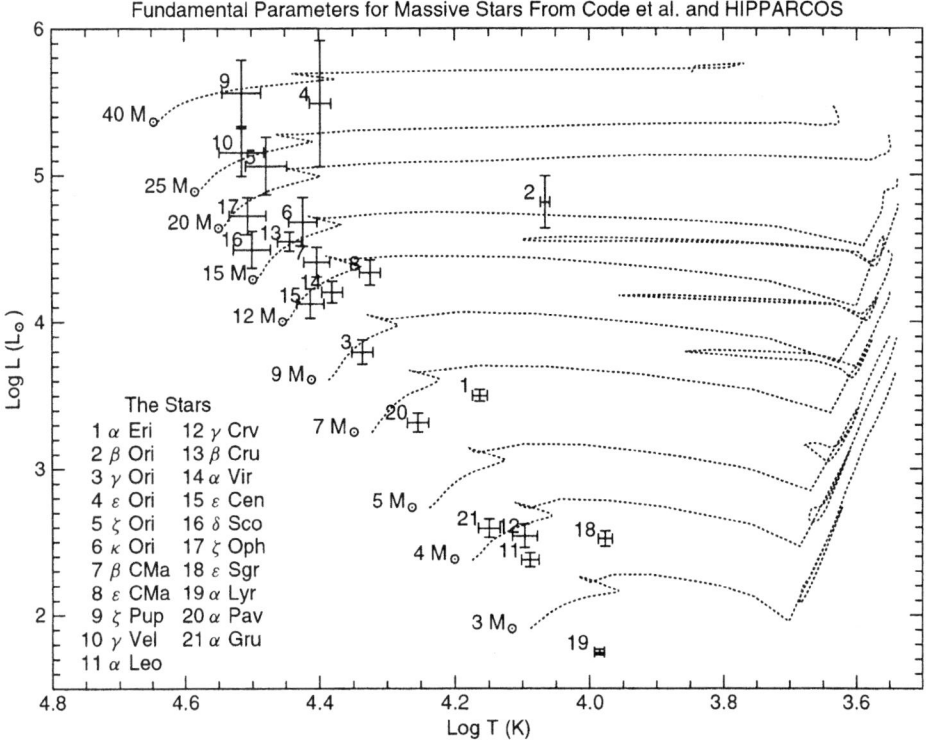

FIGURE 4.5. Evolutionary tracks for stars with solar composition from the main sequence through core helium exhaustion with comparison to absolute temperatures and luminosities for select hot stars based on the OAO-2 calibration (credit: J. P. Aufdenberg, CfA; evolutionary tracks from A. Maeder and G. Meyer, Geneva Observatory).

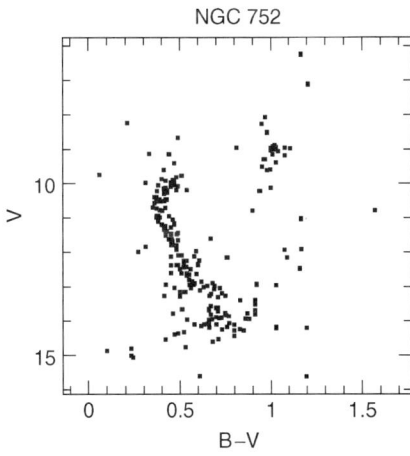

FIGURE 4.6. Color–magnitude diagram (CMD) of an old galactic open cluster NGC 752. Notice the gap between the main-sequence turnoff and the red giant clump (credit: J.-C. Mermilloid, Univ. Lausanne).

4.7.4.3 From the Giant Branch Tip to the Horizontal Branch

Core helium ignition is one of the most dramatic and wonderful events in stellar evolution. Most importantly, it separates the high- from low-mass stars in a way that depends entirely on the equation of state. A high-mass star, say, one of order 10 $M_\odot$, is fairly hot at the time its hydrogen core turns off. Contraction does not have to increase the temperature or density very far before the core reaches the conditions required for triggering the 3α reaction. Since the core is nearly isothermal and chemically homogeneous, the reaction that starts in the center rapidly engulfs the whole core and the increase in energy release stops the contraction throughout. The star readjusts its structure on the dynamical timescale because the envelope is convective and quickly reaches hydrostatic equilibrium. The luminosity produced by 3α is high the reactions begin in a more or less ideal-gas environment.

Low mass stars, such as the Sun, do not experience such a quiet ignition of helium core burning. They are cooler when their core hydrogen is exhausted and somewhat more centrally condensed because of the lower role of radiation pressure in their internal structure. Although they have initiated CNO burning in a thin shell surrounding the core during their ascent to the red giant branch, like the high mass stars, this is also exhausted rapidly. Their cores continue to contract. The core density crosses the threshold for partial degeneracy, despite the high temperature, before the critical temperature is reached for core helium ignition and becomes progressively more degenerate as the contraction continues. Ultimately, this slows the contraction, but it means that the thermodynamic environment is not that of an ideal gas at the moment of core helium ignition. The temperature begins to rise when the 3α reaction starts, in this case largely because of the density and *not* only the temperature, but the pressure changes more slowly than for a more massive, nondegenerate core. The result is a *thermonuclear runaway* (TNR). The increase in temperature accelerates the reaction rate which in turn increases the temperature, all without a dynamical readjustment of the core due to changes in pressure. Finally, when the degeneracy lifts, the pressure suddenly increases and the core rapidly expands. During the initial phase, though, the energy generation rate has increased enormously compared with the same stage in the higher mass stars. This is the *helium flash*, a nuclear process that is specific to low mass stars. It is important to note that the core degeneracy is not unique to these stars, nor is it unique to this stage of evolution. Each nuclear fuel can ignite under degenerate conditions for progressively more massive stars because of the Coulomb barrier. In addition, the temperature dependence and Q value of successive stages of nucleosynthesis causes the runaway reaction to become progressively more violent. The helium flash is disruptive but not catastrophic. In contrast, carbon burning can begin under similarly unstable conditions, but it can lead to the explosion of the star. The most spectacular example of the TNR is when the reactions begin on the surface of a star, where the overlying mass is not available to contain the explosion nor to diffuse the radiation. This is the case with classical and X-ray novae, and perhaps γ bursts, as we will discuss in Chapter 5.

The strong temperature dependence of 3α burning, $\epsilon_{3\alpha} \sim T^{40}$, induces core convection and readjustment of the star's structure occurs upon ignition of this new central energy source. The core expands and the star's track shifts from the

TABLE 4.8 Representative Evolutionary Timescales (yrs) through He Core Burning

Mass ($M_\odot$)	t_{PMS}	t_{MS}	t_{shell}	t_{He}
0.3	1.8×10^8	7.1×10^{11}	–	–
0.5	1.0×10^8	1.5×10^{11}	–	–
1.0	3.2×10^7	9.2×10^9	2.5×10^9	1.1×10^8
2.0	8.5×10^6	8.4×10^8	3.1×10^8	1.8×10^8
3.0	2.0×10^6	2.2×10^8	2.9×10^7	6.6×10^7
5.0	2.3×10^5	6.5×10^7	4.8×10^6	1.6×10^7
9.0	(1.5×10^5)	2.1×10^7	9.1×10^5	3.8×10^6
15.0	(6.2×10^4)	1.0×10^7	3.0×10^5	1.6×10^6

Hayashi track toward higher T_{eff}. The lower the stellar mass, the greater the reduction in its luminosity from the beginning of the core ignition to the standard double source structure. The star now has two nuclear burning zones, the core and a shell surrounding the core in which CNO processing takes place. The most important contribution to the energy production comes from the highly exoergic He fusion. The stellar luminosity is directly proportional to the He core mass, M_c, and hence stars with the same M_c even with different total mass M will be located in the HR diagram on an approximately horizontal strip, the zero-age horizontal branch (ZAHB). The envelope mass will, however, influence the external temperature of the star: the lower the envelope mass (and so at fixed M_c the total mass M), the higher T_{eff} is. (See also Table 4.8.[92])

4.7.4.4 The Asymptotic Giant Branch (AGB): Hot Onions
Following the central exhaustion of helium, the star ignites the He-rich shell surrounding the CO core and moves again toward the Hayashi track. The envelope convection deepens and can bring to the surface the products of previous combustions from the deep interior. This is the second time that low mass stars suffer a phenomenon now called "dredge-up"; the first occurred during H shell burning. The subsequent history of the star is, as always, governed by the opposition of energy release by fusion against gravity. The central helium is consumed and the star is burning He and H in two contiguous shells. With the exhaustion of central He, the contribution of 3α fusion lacks sufficient luminosity to maintain the hydrostatic structure of the star and energy is supplied by a consequent contraction and an increase of the efficiency of H shell burning. In the HR diagram this means that the star is going back on its previous path, retracing its steps up the giant branch toward decreasing effective temperature and ascending parallel to the Hayashi track. The He shell burning dominates over the H (CNO) shell burning and the star continues the evolution along the asymptotic giant branch (AGB), increasing progressively in luminosity. The core is now composed of ^{12}C, the direct product of 3α burning, and of ^{16}O, which is formed copiously by the reaction $^{12}C(^4He, \gamma)^{16}O$, which is the dominant reaction over 3α in the final stages of core

---

[92] Data in Table 4.8 are courtesy of Bodenheimer, P. 2002, *Encyclopedia of Physical Science and Technology*, 3rd ed. Meyers, R., ed. (San Diego: Academic Press). Parentheses indicate estimates based on older calculations.

fusion. The core becomes increasingly degenerate, due both to contraction and cooling by the increasing efficiency of neutrino emission.[93]

Along the AGB the situation is similar the one you met on the red giant branch. The main difference now is that the stars maintain He and CNO shell burning at the same time as the core begins processing carbon. This is where the *onion* structure comes in. The interior is now divided into several chemically distinct zones, and the stratification becomes progressively more extreme with each new stage of core processing. Low-mass stars ($M \leq 1\ M_\odot$) have degenerate cores and only massive stars are able to reach the central temperatures necessary to ignite carbon. The He shell lies immediately outside the core at very small mass fraction, typically around 20% of the stellar mass. Low-mass stars only undergo He shell burning, producing an extensive convection zone that moves toward the surface. Presumably, at this stage, the high luminosity drives mass loss and the erosion of hydrogen envelope reduces the mass and opacity overlying the He burning shell. When the mass of the shell becomes smaller than about 0.1 $M_\odot$, it cannot attain the pressure necessary to induce the combustion, and the low-mass stars leave the AGB.

The AGB is complicated by the fact that two very temperature-sensitive nuclear fuels burn simultaneously in thin shells. In fact, this noncentral processing is a unique and unstable stage of the evolution. The breakthrough in understanding the consequences of this energy generation first came with the discovery of thermal pulses.[94] The analysis of the instability is instructive because it is related to the thermal instability that you will see soon in the interstellar medium. When the timescale for energy release from a thin nuclear burning layer is fast compared with the sound crossing time for the region, and faster than the radiative timescale, the layer must expand. To see this, take the energy generation to vary as $\epsilon \sim T^n$ so that for a temperature perturbation δT, we find the variation in ϵ to be

$$\frac{\delta \epsilon}{\epsilon} = n \frac{\delta T}{T}$$

For the radiative equilibrium condition in a thin mass shell of width ΔM having a temperature gradient $\Delta T / \Delta M$, the one-zone equation for flux conservation becomes

$$\delta \frac{\partial L}{\partial M} = 4 \left(\frac{L}{\Delta M} \right) \frac{\delta T / T}{\Delta T / T}.$$

---

[93] Busso, M., Gallino, R., and Wasserburg, G. J. 1999, *ARAA*, **37**, 239.

[94] Ledoux, P. 1958, *Encycl. der Physik*, **51**, 605; Schwarzschild, M. and Härm, R. 1965, *ApJ*, **142**, 855. The Schwarzschild–Härm paper is one of the most important instances of prediction by theoretical models of stellar interiors. The story is illustrative of how theoretical discoveries are, or should be, made. The models they were computing for the double shell phase developed convergence problems, in the sense that they were unable to achieve stable models that did not vary in structure both through iterations and from one timestep to the next. In tracing down the possible sources of the problem, they assumed first that it was a numerical instability. Only on realizing the intrinsic thermal instability of the double-shell phase, due to the extreme temperature sensitivity of the two principal reactions, did they come to understand that the phenomenon was real.

Finally, the energy conservation equation for an ideal gas is

$$T\frac{dS}{dt} = \frac{3}{2}\frac{P}{\rho}\frac{d}{dt}\left(\ln\frac{P}{\rho^{5/3}}\right) = \epsilon - \frac{\partial L}{\partial M},$$

where S is the entropy, as usual. The perturbed equation becomes

$$\frac{3}{2}\left(\frac{P/\rho}{L/\Delta M}\right)\frac{d\delta S}{dt} = \left(n - 4\frac{T}{\Delta T}\right)\frac{\delta T}{T},$$

so that for $\Delta T/T > 4/n$, the change in the entropy will be positive and the system will be thermally unstable; this is a constraint on the magnitude of the stable temperature gradient for a specific reaction. Since CNO burning varies as T^{15}, it is clear that shell burning, if perturbed, will produce a positive change in the entropy. This is a thermal pulse. The timescale for nuclear energy release is $\tau_{\text{nucl}} \sim \Delta M/\epsilon$, while the timescale for energy transport is $c_V T/L$, so that when the nuclear energy generation increases too rapidly, the layer cannot transport the energy fast enough and must expand. This is essentially the ϵ-excitation mechanism if the envelope is not overstable. The analysis by Schwarzschild and Härm follows this line of reasoning. There are two separate equilibrium conditions that must now be satisfied. One is that the layer remains hydrostatic, that the star does not catastrophically disrupt from the energy release. In the helium flash, this was possible because the nuclear burning starts is in the deep interior and the mass of the overlying layers damps the expansion when the TNR begins. For a shell that ignites under nondegenerate conditions, the expansion that results from the change in temperature can be damped simply by the expansion because the temperature drops as the pressure falls. The other condition is radiative equilibrium, which depends on the photon diffusion time.

During a thermal pulse, material is processed at a mass farther out than the equilibrium position of the He shell. Once the pulse ends, the envelope relaxes and the burning region recedes toward the core, leaving behind carbon-enriched material. This pocket is below the level of the normal, equilibrium convective region of the envelope. However, convection that develops on the thermal timescale in response to the increased luminosity of the shell moves inward and *may* mix this processed material into the envelope. This is called the *third dredge-up* stage and is the most important in production of the observed s-process species.[95] The thermal pulses produce pockets of ^{13}C as a result of helium processing. Each relaxation of the envelope leaves this layer at the base of the convection zone unmixed, but subsequent increased pulses, attended by deepening convection, produces local mixing of this fundamental neutron source into the lower envelope from which neutrons can be produced by α capture. An alternative source is ^{22}Ne, but this requires higher temperatures. A beautiful set of observational constraints can act as a thermometer of the process, and these are practically unique among the nuclei. Isotopes of Zr are readily observable in these cool stellar spectra and the ^{95}Zr/^{96}Zr ratio is a very sensitive probe of the temperature of the neutron

---

[95] See especially Straniero, O. et al. 1997, *ApJ*, **478**, 332; Gallino, R. et al. 1998, *ApJ*, **497**, 388.

source. If ^{22}Ne is mainly responsible, this ratio dramatically decreases because, for the s process, the ^{96}Zr serves as a bottleneck—it has a very low neutron capture cross section—and there is pileup here if the neutron exposure and temperature are high. In addition, carbon excesses result from the nucleosynthesis and mixing brings the products to the surface, thereby serving as a formative mechanism for carbon stars. So far, however, the burning on the AGB is so intertwined with the treatment of envelope convection and fine tuning of the envelope structure that the picture must be called a scenario rather than a full theory. The details are not at all finalized.

One possible source for additional nucleosynthesis is *hot-bottom hydrogen burning*. Once the hydrogen envelope reaches the layer that has been processed by the previously active CNO shell, the products are mixed into the convective region and subsequently transported to the stellar surface on a hydrodynamic timescale. The details are here quite vague, even after several decades of knowing that the event occurs. In fact, there are likely several episodes of deep mixing by the convection zone. The first ascent of the AGB is usually accompanied by *first dredge-up*, and there may be a second when the star is in the process of reclimbing the giant branch. This *second dredge-up* stage does not occur for all mass ranges, however, and appears to be confined to the higher mass AGB stars. The *third dredge-up* is the most important from the standpoint of nucleosynthesis. It occurs during the thermal pulse stage, when the highest abundances of ^{13}C are produced by the hydrogen burning shell and the highest temperatures are reached at which neutron processing can occur. One way to see how mixing alters the results is to go back to the Fokker–Planck equation in its spatial form (Chapter 1) and take mass as the independent variable:

$$\frac{\partial X_i}{\partial t} = \left(\dot{X}_i\right)_{\text{nucl}} + \frac{\partial}{\partial M_r}\left(D_m \frac{\partial X_i}{\partial M_r}\right),$$

where D_m is a diffusion-like coefficient that depends on the physical parameters of the medium and represents the turbulence that is expected from convection. This form explicitly includes the nuclear processing in the first term on the right-hand side. The usual way of expressing the diffusion coefficient is to dimensionally combine the convective velocity, v_c and the mixing length, l:

$$D_m = \tfrac{1}{3}\rho v_c l.$$

Frankly, this is only a guess, but it has important consequences for the nuclear reactions. In the picture in which convection proceeds by the motion of gas parcels, this diffusion represents the overshooting of a blob past the usual boundary provided by the Schwarzschild criterion. This is where the nonlinearity of this approach shows up. The convective velocity depends on the temperature gradient (assumed to be adiabatic), which in turn depends on the equation of state, and the diffusion coefficient depends on the mixing length, which in turn depends on the local thermodynamic conditions. There have also been suggestions of shear mixing[96] driven by gravity wave instabilities. The diffusion coefficient depends on the rotational gradient and is due to the competition between shear and buoyancy. Scaled to the thermal diffusivity, K, the diffusion coefficient is

$$D_{\text{shear}} = K \frac{\Omega}{N} \left(\frac{d \ln \Omega}{d \ln r} \right)^2. \qquad (4.167)$$

Here N is the Brunt-Väisälä frequency, which depends on the radial density gradient and measures the buoyancy of the medium. It is a general feature of all late stages of evolution that some form of deep mixing is required by the observations. In the most massive evolved stars, the *luminous blue variables*, nitrogen is systematically enhanced.[97] These stars do not undergo dredge-up because they never reach the giant or AGB stage of evolution. Instead, they develop nearly catastrophically high-mass outflows that strip the outer layers and bring processed matter to the surface. It is now presumed that this eventually leads to the Wolf–Rayet stage for a single star. The stellar mass is considerably reduced by this process of mixing and mass loss, finally removing as much as half of the total. Two constraints are available for this phase of evolution, and these are almost unique for any portion of a star's lifetime. Since the timescale for the dredge-up is short, especially for the third stage, and potentially within the tenured lifetime of a typical faculty member, the nuclear products can be mixed to the surface and produce observable changes in the envelope composition. Such an event was first seen in the early 1960s for FG Sge.[98]

4.7.5 Death

4.7.5.1 How the Most Massive Stars End Their Lives

Gravity is a major energy source in massive stars, especially at the ends of their lives, and is essentially the only balance to neutrino losses once the nuclear fuels are exhausted.[99] Before that happens, however, there are several mechanisms that can actually destroy the star. The star is able to attain the temperature for the ignition of ^{12}C in a shell at about $M(r)/M \simeq 0.1$. The parameter governing the C ignition is the dimension of the degenerate carbon–oxygen core. A large core is needed to furnish the energy required for carbon burning in the face of neutrino losses. The minimum mass is $M_{\text{C+O}} \approx 1.1\ M_\odot$, and the He core mass depends on the mass of the convective core from previous stages of evolution and on the total stellar mass. As we found for the helium core flash, sufficiently massive stars ignite

---

[96] See Zahn, J.-P. 1974, in *Stellar Instability and Evolution: IAU Symp.* 59 Ledoux, P., Noels, A., and Rogers, A. W., eds. (Dordrecht: Reidel); Schatzman, E. 1993, *Astr. Ap.*, **279**, 431; Zahn J.-P. 1998, *Space Sci. Rev*, **85**, 79 and references therein. Although we won't derive the instability here, you should look at the discussion of the magnetorotational instability in the next chapter for more details. See also Shore, S. N. 1992, *An Introduction to Astrophysical Hydrodynamics* (San Diego: Academic Press).

[97] Walborn, N. 1982, *ApJ*, **256**, 452; Maeder, A. and Meynet, G. 1987, *Astr. Ap.*, **182**, 243; Humphreys, R. M. and Davidson, K. 1994, *PASP*, **106**, 1025.

[98] This was the first such post-AGB star discovered. See Jorcsik J. and Montesinos, B. 1999, *New Astr. Rev.*, **43** 415. It has since been joined by V4334 Sgr (also known as Sakurai's Object) and V605 Aql. See Duerbeck, H. W. et al. 2000, *AJ*, **119**, 2360.

[99] Maeder, A. and Conti, P. S. 1994, *ARAA*, **32**, 227.

^{12}C quietly since the core does not become degenerate. The history is different for less massive stars but more massive than those for which the He shell exhausts near the surface.

You recall that for stars around 1 $M_\odot$, He core ignition occurs under degenerate conditions that produce a thermonuclear runaway, the He flash. Despite higher temperatures, stars in the intermediate mass range of about 4–10 $M_\odot$ encounter similar conditions following He core exhaustion. Because the stars are unable to continue through ^{12}C$(\alpha, \gamma)^{16}$O, the core cools and contracts, powering some additional He and H (CNO) shell processing, but with the carbon core approaching ignition. This can take place under degenerate conditions, and since the binding energy is substantially higher results in a much greater energy per nucleon being released in the process and can unbind the matter of the envelope. Instead of simple core expansion on a thermal timescale, the violence of the ignition produces an explosion of the core that is powered by nuclear reactions instigated by the passage of the shock through the core. The requirement is that the more must be degenerate, which, for a mean molecular weight of 12, means a mass of around 0.5 $M_\odot$ or about 5 $M_\odot$ for the star. The lower limit of mass to end as a supernova, as above, is more difficult to define, since it may differ from the initial stellar mass, whether mass loss is active during the evolution as it is certainly true during the AGB phase. The Chandrasekhar mass (see below) is the absolute upper limit for global stability of a completely degenerate core, if all the mass of the envelope is removed. For stellar masses a little larger than at the end of C burning, the degenerate nuclei of O, Ne, and Mg can capture electrons, lowering the central pressure (by reduction of particle number and of degeneracy) so increasing the probability of further electron capture. This is a process that leads again the star toward a catastrophe, an implosion, destructive, perhaps leaving a dense, compact and dead remnant, a neutron star.

Higher mass stars, $M \geq 8\ M_\odot$, continue the combustion chain, with evolution timescales for each combustion becoming progressively shorter. Take as example a star of 25 $M_\odot$, a typical main sequence O star. The nuclear timescales vary from a few thousand years for C core burning to a few months for O burning, to 1 *day* (!) for Si burning. This accelerating rate, which for example forces these massive stars to ignite He before becoming red giants, finally leads them to their doom. Following production of ^{56}Ni, which decays to ^{56}Fe, the enormous core temperatures lead to photodestruction of iron, ^{56}Fe$(\gamma, 4n)13\ ^4$He. This step produces core neutrons and initiates collapse, the trigger of a core collapse supernova.

4.7.5.2 Postasymptotic Giant Branch Evolution of Low-Mass Star and Planetary Nebulae

Substantial mass loss must occur during AGB evolution. These stars have large circumstellar envelopes, often very rich in complex molecules and producing maser emission from the intense illumination by the photospheric infrared. They also show dust emission from low temperature matter at large distance from the central star.

Once the mass loss becomes sufficiently high that the star is actually stripped of its hydrogen envelope, life starts to get very interesting. The core is assumed to reach the Eddington luminosity, at which point it drives extensive mass loss. Models show that the star evolves are nearly constant bolometric luminosity,

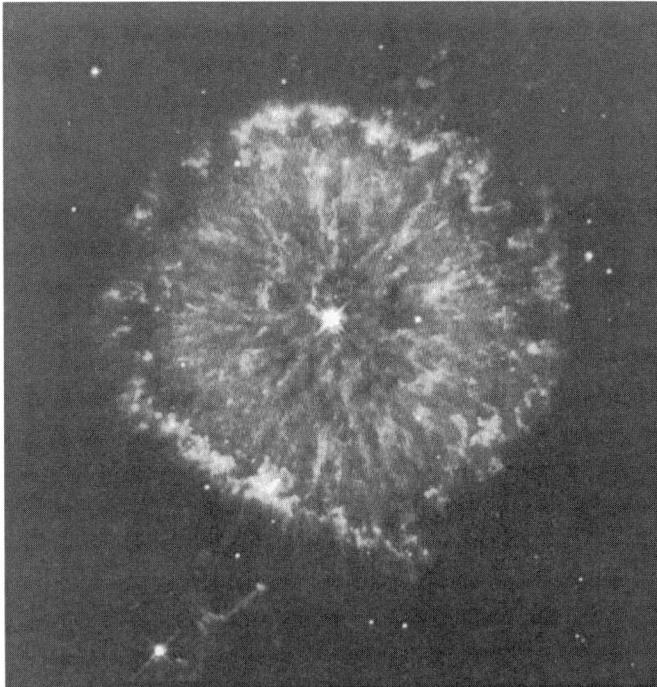

FIGURE 4.7. HST WFPC2 image of the young planetary nebula NGC 6571. Notice the knotted structure to the outer region, where the interaction occurs between the old wind of the AGB star and the fast wind from the newly exposed core (the star visible at the center of the image) (credit: NASA/STScI).

essentially at the Eddington limit, until the He shell burning is exhausted, after which cooling sets in. This leads to contraction and, because of the virial constraint, increases its surface effective temperature until degeneracy is reached. Thereafter, the star evolves at roughly constant radius to become an equilibrium white dwarf.

In the meantime, the most dramatic evidence for mass loss leading to the white dwarf stage is provided by *planetary nebulae*. These objects display a star merging back into the interstellar gas from whence it came. They can only be described as breathtakingly beautiful objects (see Fig. 4.7) formed by the ionization of the inner parts of the circumstellar environment produced by the AGB wind. The enormous variety of shapes comes from the line formation. The nebulae are optically thin, rendered visible by forbidden and recombination line emission from the inner part of the older wind. This slower neutral (or molecular) wind is roughly axisymmetric, and the planetary nebula is the ionized interior and photodissociated portion of the denser circumstellar environment. We will detail the line formation in Chapter 6. But it is important to note here that the slow AGB wind, whose innermost part is radiating, represents a significant contribution of processed material back to the interstellar medium. From the emission lines (see Chapter 6), it is possible to find abundance ratios that are consistent with the carbon enhancement expected for the AGB phase. Most models use the observationally determined relation between the mass loss rate and the stellar luminosity:

$$\dot{M} = -4 \times 10^{-13} \eta \frac{LR}{M} M_\odot \text{ yr}^{-1} \qquad (4.168)$$

where η is a scaling factor that is usually introduced in the modeling as a free parameter but generally of order unity.[100] The wind is an ideal site for grain formation. It reaches a low terminal velocity, but more important, it self-shields the outer part of the flow. Thus, the region where the temperature is low also has a relatively low drift velocity and comparatively high density, ideal for accretion of material onto nucleated cores. The discovery of SiC grains in a few carbonaceous chondritic meteorites, in particular the Murchison, support this contention.

Finally, there is a last gasp for some stars before they become white dwarfs. If sufficient helium remains outside the core, as the star undergoes a final post-AGB contraction this fuel can ignite. The difference is its effect on the overlying envelope, of which little may remain after the formation of a planetary nebula. A wind resulting from this flash will be completely free of hydrogen, and the ejecta will show almost pure helium and heavier knots of material. The mass range is restrictive, as with the helium flash, because of the specific ignition requirements so it should be a relatively rare occurrance among planetary nebulae.[101] However, two such candidate planetaries, Abell 30 and Abell 78, both among the hottest and luminous central stars, may have experienced this process. Clearly there are wonderful mysteries lurking in the last gasp of low-mass stars on their march toward the grave.

4.7.5.3 White Dwarf Stars: The Final State of Low-Mass Stars

There is a limit to the mass that can be supported by degeneracy pressure alone. White dwarfs, as such objects are called, are actually simpler than ordinary stars. They are in a virtually zero temperature configuration, so we can completely ignore energy transfer and generation. This is the same thing we found for polytropes! A mass–radius relation results from substituting the nonrelativistic equation of state into the hydrostatic equation,

$$MR^3 = \text{constant}, \qquad (4.169)$$

which shows that a degenerate star shrinks in response to an increase in mass. You would expect this sort of behavior since the only way a degenerate star can increase its internal pressure under mass loading is by increasing its mean density.

---

[100] The relation is due to Reimers, D. 1977, *Astr. Ap.*, **61**, 217. In follow-up work, Kudritzki, R. and Reimers, D. 1978 *Astr.Ap.*, **70**, 227 study the absolute rate and attempt to fix the coefficient, $\approx (5.5 \pm 1) \times 10^{-13}$ using model atmosphere analyses. This should make you very nervous. The use of empirical relations for stellar interior models may make the process consistent with the input algorithms and physics, but the assumptions are not always derived from a priori physical calculations. In this case, in particular, the AGB stage of evolution requires many significant assumptions that come from nature, not from first principles, and it is likely that many of the details will improve in the near future. There is a wide territory open in the area of late stages of evolution of intermediate mass stars.
[101] Iben, I. 1983, *ApJ*, **284**, 605.

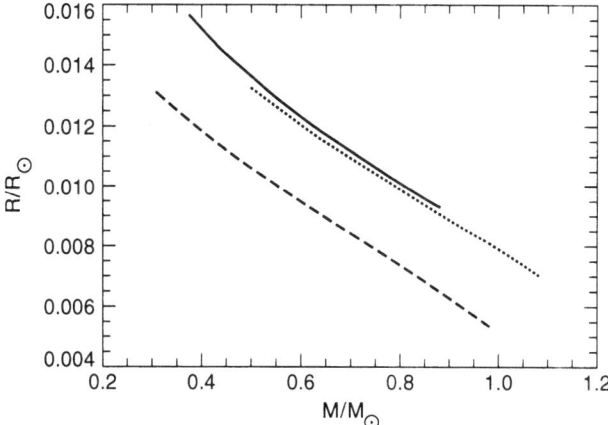

FIGURE 4.8. Mass–radius relation for helium (solid line), carbon (dashed line), and iron (dotted line) white dwarfs.

It has to contract. The release of gravitational potential is substantial, but its pressure does not increase with increasing heat input because of the degeneracy. Ultimately, there is a limit to how far the star can contract. This is the Schwarzschild radius, which, you may recall, depends only on the stellar mass. There must be a limiting mass above which no stable degenerate configuration is possible. This sort of qualitative argument was used by Landau[102] to great effect when first solving the equations for a degenerate neutron gas, and also by Oppenheimer and Volkoff (in 1938), who examined the general relativistic form of the hydrostatic equation.

The equation of state for a degenerate electron gas depends only on the density due to the quantum effects. It is essentially a zero-temperature state. Such an object, if T_{eff} were actually zero, would emit no light in spite of remaining in hydrostatic equilibrium. But you see white dwarfs, so it is reasonable to investigate the dependence of the temperature on the mass and the radius for partially degenerate configurations. For instance, among the closest stars, α CMa has a visible companion with an effective temperature that is about 10^4 K with a radius of only about 0.02 $R_\odot$. The mass–radius relation is shown in Figure 4.8 for white dwarfs of various compositions.

A mass that shrinks under its own gravity becomes progressively denser and warms up. We are interested now in the last phase of this process, when the electrons are almost degenerate. The structure of the star depends on the equation of state alone, since there are no nuclear energy sources to supply the radiative and neutrino losses. The nuclei remain ideal while the electrons are partially degenerate. We can place some limits on the temperature of the baryons that will prove to be useful. From virial equilibrium $3(\Gamma_1 - 1)U + \Omega = 0$, but now we have two species with different values of Γ_1. The ions are an ideal gas so $U_N = 3N_N kT/2$. The electrons are either nonrelativistically ($\Gamma_1 = \frac{5}{3}$) or relativistically ($\Gamma_1 = \frac{4}{3}$) degenerate and we will assume $\epsilon_F \gg kT_e$. Since the polytropic index is related to

---

[102] Landau, L. 1932, *Phys. Z. Sowiet.*, **1**, 285 (see also *Collected Papers* p. 60).

the exponent in the gas law through $\Gamma_1 = 1 + 1/n$, we will write

$$U_N + U_e + \Omega = \frac{3}{2}N_N kT + N_e \epsilon_F - \lambda_n \frac{GM^2}{R} = 0, \qquad (4.170)$$

since the mass distribution, required for the gravitational potential, depends on n. Recall that $\mu_e \approx 2/(1 + X)$ for complete ionization and since we expect hydrogen to be completely consumed, $\mu_e = 2$. For a fully ionized medium $N_e = \mu_N/\mu_e N_N$ so the virial equation becomes

$$\frac{3kT}{2\mu} = \lambda_n \frac{GM}{R} - \frac{3}{(n+1)\mu_e}\epsilon_F \qquad (4.171)$$

using $\mu^{-1} = \mu_N^{-1} + \mu_e^{-1}$. Now for some scaling arguments. The Fermi energy scales as $\epsilon_F \sim n_e^{1/n} \sim M^{1/n}R^{-3/n}$. We then see for $n = 3$ from Eq. (4.171) that

$$T \sim \frac{GM}{R}\left(1 - c_3 M^{-2/3}\right)$$

with a constant c_3. This yields a maximum mass, regardless of the mass-radius relation, for the ions to maintain positive temperatures.[103] Finally for $c_3^{1/3} M > 1$ the temperature tends to infinity for $R \to 0$. In this case the decreasing gravitational energy cannot be balanced by the increase in the electron energy, so the nuclear gas becomes steadily hotter. Turning again to the virial theorem, the total energy is

$$E = \tfrac{1}{2}\Omega$$

If the radius shrinks by an amount ΔR, this changes by

$$\Delta E = -\frac{1}{2}\Omega\frac{\Delta R}{R} = \frac{1}{2}\Delta\Omega. \qquad (4.172)$$

Since $\Omega < 0$, a contraction decreases the total energy. The internal energy is $U = U_e + U_N = -E$, hence a decrease of E corresponds to an increase of U:

$$\Delta E = -\Delta U. \qquad (4.173)$$

These relations are common to all stars and you know them from our discussion of the gravitothermal catastrophe in Chapter 1. There is a new phenomenon for a white dwarf due to the different radial dependence of U_e and U_N. You recall that

---

[103] This was basically the argument used by Chandrasekhar to demonstrate the existence of an upper mass to the white dwarf sequence, and the one that led to the infamous exchanges between him and Eddington on several occasions. See Wali, K. C. 1991, *Chandra: A Biography of S. Chandrasekhar* (Chicago: Univ. Chicago Press).

in the nonrelativistic case, $U_e \sim 1/R^2$, so

$$\Delta U_e = -U_e \frac{\Delta R}{R} = 2U_e \frac{\Delta \Omega}{\Omega}. \qquad (4.174)$$

When the stellar radius is very small, the internal energy of nuclei is much lower than for the electrons, and therefore

$$\Delta U_e \simeq -\Delta \Omega = -2\Delta E. \qquad (4.175)$$

Combining this with Eqs. (4.172) and (4.174) it follows that $\Delta U_e \sim -2\Delta U_N$. Hence, the temperature of the nucleon gas *decreases* and the white dwarf gets *cooler* when E decreases. In other words, in this regime the star has a *positive specific heat*! The gravitational energy released during the contraction is absorbed by the electron gas because its Fermi energy increases at the same time as the thermal energy of the nucleons decreases. The total energy, E, decreases and is lost in form of radiation. When all the thermal energy of nuclei has been used up, the temperature is zero, the degeneracy is complete, and any further contraction is impossible.

The cooling is very significant for the nucleons because they, unlike the electrons, can freeze. That is, because of the electron screening, the cold nucleons can form a crystal as their lowest energy configuration with the attendant emission of latent heat. The rapid energy loss once the condensation begins reduces the core temperature and changes the overall heat capacity of the star. One of the interesting observations for nearby white dwarfs, for which proper motions and parallaxes are available, is that there is a rather sharp lower luminosity of the population of field white dwarfs. This is also true for cluster members. The cutoff is likely explained by this change in the equation of state for the nucleons. The electron gas cannot radiate, since a completely degenerate gas fills all its lowest energy states and only those at the top of the Fermi sea are able to scatter. However, the nucleons retain thermal energy until they begin to solidify. Then they rapidly lose the latent heat of fusion and, as a consequence, cool very quickly. The energy is easily conducted to the surface by the electrons, and the star quickly turns off. You can therefore see how this produces a lower limit for the white dwarf (WD) luminosity; when the core temperature is lower than the threshold required for crystallization, it solidifies and the star cools rapidly.

The degenerate electron gas has a high thermal conductivity and therefore the central part of a white dwarf is essentially isothermal.[104] The star behaves like a metal sphere surrounded by a nondegenerate atmosphere. It is a very simple structure, for which analytic calculations can be performed by making a few simplifying assumptions: the atmosphere behaves like a perfect gas, its mass is negligible compared to the total stellar mass, and it is in radiative and hydrostatic equilibrium. If L is the stellar luminosity, which is now derived solely from cooling, taking a power law absorption coefficient, $\kappa = \kappa_0 \rho T^{-\nu}$, and the equation of state

---

[104] Mestel, L. 1950, *Proc. Cambridge Phil. Soc.*, **46**, 331 provides an early and very instructive example of the conductivity calculation.

for perfect gas to eliminate the density, the interior thermal equation is

$$\frac{dP}{dT} = \frac{16\pi^3}{45} \frac{k^5}{\mu \kappa_0} \frac{GM}{L} \frac{T^{4+\nu}}{P}, \quad (4.176)$$

where we neglected the variation of mass and luminosity with r inside the atmosphere. You can integrate this equation inward from the stellar radius to the radius of the surface dividing the degenerate interior region from the nondegenerate atmosphere. On this surface the pressure equilibrium corresponds to two different regimes on the two faces, and considering that T is the temperature on the surface:

$$P \sim T^{5/2} \sim \left(\frac{M}{L}\right)^{1/2} T^{5+\nu/2}, \quad (4.177)$$

which leads to

$$L = \frac{L_0}{M_0} M T^\nu. \quad (4.178)$$

Here M_0 and L_0 are constants. The luminosity is independent of the radius. The high thermal conductivity of the interior connects the temperature at the base of the atmosphere to the central temperature, $T = T_c$, so the luminosity of a white dwarf is directly related to the central temperature. Since its heat is purely relic rather than actively generated, the star's luminosity becomes

$$L = \frac{d}{dt}\left(\frac{3}{2} \frac{M}{\mu m_p} kT\right). \quad (4.179)$$

This gives the cooling time for the star:

$$t_{\text{cool}} = \frac{3}{5} \frac{k}{\mu m_p} \left(\frac{M_0}{L_0}\right)^{2/7} \left(\frac{M}{L}\right)^{5/7}. \quad (4.180)$$

For instance, for a $M = 1\,M_\odot$ star, t_{cool} ranges from 0.4×10^9 yr for $L = 10^{-2} L_\odot$ to 10^{10} yr for $L = 10^{-4}\,L_\odot$. Notice how this differs from the Kelvin–Helmholtz timescale since now the energy is not supplied by contraction. The process more closely resembles the freezing of a puddle of water on a winter night than the formation of a star. We have already discussed plasmon neutrino cooling and, in the early stages, this is the dominant mechanism for white dwarf interiors since the initial core temperature exceeds 10^8 K. Note, however, that such processes have very steep temperature dependences, of order T^6. As the temperature falls, so does the rate of neutrino loss. On the other hand, the neutrinos exacerbate the initial drop in temperature and are efficiently emitted from the interior, so the system cools quickly, on timescales of order 10^4 yr. With the loss of kinetic energy, since the nucleons are collisionally coupled to the electrons, the temperature may eventually fall below the crystallization point, T_c, and the charged nucleons arrange themselves in a lattice. Accompanying this phase transition is a sharp decrease in the specific heat and further loss of internal energy that propagates diffusively to the surface. After a final increase in the luminosity, the white dwarf simply turns off.

4.7.6 Core Collapse Supernovae Type II and Neutron Star Formation

For stars more massive than those able to quietly reduce their masses below the Chandrasekhar limit, a more spectacular end awaits: core collapse. This does not, however, necessarily mean a collapse of the entire star. The gravitational potential of the central mass is so high that even just the release of the binding energy of the iron core, $\leq 2\, M_\odot$, can stop the infall of the envelope if it is efficiently transferred outward and can even expel the matter. The shock produced during this explosion not only heats the envelope but may even initiate nuclear reactions. This is, very schematically, the heart of the spectacular events called *core collapse*, or type II, supernovae.

There are two fundamental stages in such core-collapse-induced explosions. The first is when the core, on reaching spectacularly high densities, suddenly stops collapsing. The second is when matter falling onto the core and radiation coming from the core drive a shock wave into the envelope, compressing and heating the matter and accelerating it beyond the escape velocity. First, we look at what happens in the core. The combination of URCA-generated neutrino cooling and photodisintegration of the core ^{56}Fe leads to a large net mass fraction of free neutrons relative to electrons and nuclei. Normally, a free neutron decays in about 10 min, through n $\rightarrow$ p + e + $\bar{\nu}_e$, since $m_n > m_p + m_e$. The inverse reaction does not occur under laboratory conditions since the energy gap of about 1 MeV would require very high temperatures ($\simeq 10^{12}$ K). However, at the density reached during core collapse, the electron energies must be at least ϵ_F, so neutron decay is possible if $m_n c^2 > m_p c^2 + (m_e c^2 + \epsilon_F^2)^{1/2}$. Since $m_n - m_p \simeq 2.5 m_e$, the decay of neutron will be possible only when $(m_e c^2 + \epsilon_F^2)^{1/2} \leq 2.5 m_e c^2$, which yields a critical density, $\rho_N \approx 2 \times 10^7$ g cm^{-3}. When $\rho > \rho_N$, neutron decay is suppressed while the inverse reaction produces free neutrons.

The current status of our picture of core collapse supernovae, that is SN II, is a lot better off than in 1991.[105] The fundamental process is now connected with the formation of a neutron star, and is driven by ν emission from the core and the effects of a wind. In effect, the problem is that once the core stops collapsing, there is still a large envelope that has yet to know it. The infall creates an accretion shock at high density and photon processes are completely dominated by the neutrinos. The entire process takes only between 0.01 and a few seconds!

4.7.6.1 Classification of Supernovae

Observationally, two broad types of supernova exist. The distinctions are originally due to Baade and Zwicky, before 1960, on the basis of their optical lightcurves and

---

[105]A stark contrast is provided by comparing Shklovsky, J. 1966, *Supernovae* (NY: Interscience) with Arnett, W. D. 1997, *Supernovae and Nucleosynthesis* (Princeton: Princeton Univ. Press). See also, for example, Burrows, A., Hayes, J., and Fryxell, B. A. 1995, *ApJ*, **450**, 830.

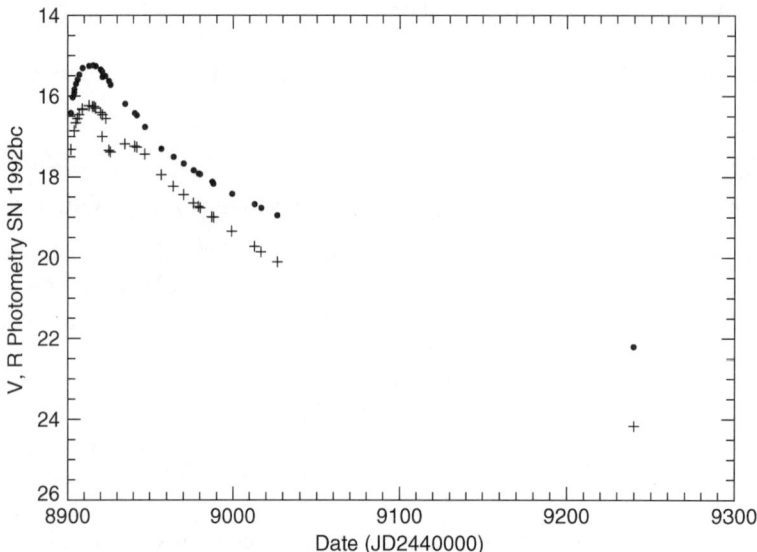

FIGURE 4.9. Sample SN Ia light curves (SN 1992bc, 1992al). Filled circles show V magnitudes; crosses show R magnitudes displaced by +1 mag for clarity (data from Hamuy, M. et al. 1996, *AJ*, **112**, 2408).

spectral characteristics, but have since been significantly refined. Before proceeding, we should note that all of the classifications are based on optical photometry and spectroscopy. At this point, there is no comparable classification scheme based on either the infrared or the ultraviolet. This is important because of the effects of the interstellar medium on the properties of extragalactic supernovae. Observed through deep extinction layers, or within molecular clouds, the properties of a supernova will appear significantly different. Also, we should say that until the late 1980s, most of the work on extragalactic events was based on sparse data that were often obtained quite long after visual maximum. Automated and dedicated supernova searches began at that time and are now fully mature, producing dozens of events per year. These can be followed almost immediately by spectrophotometry, so the classification is based on both the lightcurve behavior and the spectroscopic morphology.

In the broadest sense, type I events are associated with an older stellar population and are generally seen in galactic halos and elliptical galaxies. They are lacking hydrogen, and their spectra are dominated by extremely broad bands of Si II, Fe II, and iron peak elements. What we see in the lightcurve is a very regular variation with a characteristic decay time of order 60 days following a relatively short-lived peak (Fig 4.9). Type Ia supernovae are considered to be cosmological *standard candles*, much like Cepheid variables.[106] They have remarkably similar optical lightcurves, nearly the same maximum absolute visual magnitude, and a characteristic rate of decline that makes their photometric identification rather

---

[106] See Ruiz-Lapuente, P., Canal, R., and Isern, J., eds. 1997, *Thermonuclear Supernovae* (Dordrecht: Kluwer); Leibundgut, B. 2000, *Astr. Ap. Rev.*, **10**, 179.

simple. This uniformity of the class is profoundly important. It means that the whatever the progenitor, it must possess a very limited range of physical parameters. For instance, the maximum energy release, 10^{51} erg, over the outburst has less than a 10% dispersion. The energy could easily come from core collapse, but it requires a very narrow range of masses and radii in order to achieve this. There is a natural choice for the critical condition: a white dwarf at or near the Chandrasekhar mass. Since this mass has a unique radius, the gravitational potential, GM/R, has a characteristic value. Hence, the energy released by collapse will be the same within a small dispersion.

The spatial distribution of SN Ia events is more nearly isotropic in a galaxy than other types, so they appear to have a parent population that is older and more uniformly distributed. Hence, it is likely that they come from lower-mass, lower-metallicity stellar progenitors. In addition, they have very little or no hydrogen left in their spectra and clearly form ejecta of very low mass, less than 1 $M_\odot$, a substantial fraction of which is processed material. Their ejection velocities are exceedingly high, and the matter becomes optically thin almost immediately. The lightcurves observed in SN Ia events are so uniform that it can be assumed that there is a simple relation between the amount of ejected mass and the decay time. Furthermore, the light curve quickly enters an exponential decay stage with a characteristic timescale of about 2 months. This coincides with the half-life of ^{56}Co, a product of Fe nucleosynthesis. As we discussed, nucleosynthetic signatures, in particular the highest atomic mass r-process nuclei, provide important clues to the nature of the presupernova star. We will return to this after the discussion of type II events, but note here that the largest fraction of iron and iron peak elements are ejected through the SN Ia events. All of these, coupled with a characteristic photometric decay time of order 2 months, suggest that the parent object is likely a white dwarf that has been pushed over the Chandrasekhar limit and collapses to form a neutron star.

A white dwarf with a mass just below M_{ch} is stable but it need not remain so. Mass accretion from the interstellar medium or a companion in a close binary system may suffice to push it over the edge. This *accretion-induced collapse* scenario is still speculative but it is not untestable. The lack of hydrogen in the ejecta means that any mass accreted from the companion must either be hydrogen-poor or cannot be a large amount. One plausible configuration is that the companion from which mass is accreted is itself degenerate, the so-called *double-degenerate* systems, and that it is extremely deficient in hydrogen. There is also still a question about the nature of the thermonuclear event. If accretion triggers collapse, *and we must emphasize that this process is very different from massive self-induced core collapse*, then two variants of the event have been proposed. In one, the accretion produces an explosion in much the same way that it does in a nova (see Chapter 5), where a critical temperature is reached and a very temperature-sensitive reaction, such as carbon burning, ignites explosively in the white dwarf's envelope. This ignition creates a detonation wave that propagates outward in the star and triggers the outer layers, resulting in the explosion. Alternatively, a deflagration or flamelike wave, one that is dominated by conduction, can be produced by the ignition. Both types of waves propagate by causing continued consumption of fuel, the released energy serving to further the progress of the burning front. Either way, the liberation of nuclear binding energy must be sufficient to lift precisely the right amount of matter off of the white dwarf.

Type II supernovae are now better understood since the explosion of SN 1987A in the Large Magellanic Cloud (LMC). This was the first supernova event for which the progenitor was clearly identified, and actually studied, before the explosion. That star, Sk $-69^o 202$, was an early B supergiant in a small group of massive stars near the bar of the LMC. Its spectrum was obtained in the mid-1960s by Sanduleak and about a decade later by Fehrenbach, both in surveys of OB stars in the LMC. Neither spectrum indicates anything strange about the progenitor that would have pointed to its imminent demise, although subsequent observations of the illuminated environment show that the star was surrounded by a substantial circumstellar nebula that was unobservable before being ionized by the explosion. The B star probably had about 25 $M_\odot$ on the main sequence, having lost a significant fraction of that by the time of the explosion. The fact that the star was a B rather than M supergiant is very important, not so much for the nucleosynthesis, but for the optical light variations, as we will now discuss.

4.7.6.2 Physics of Core Collapse and Shock Generation

It was expected, early in the development of theoretical model for supernovae, that core collapse and the subsequent formation of a static neutron star core would provide enough energy to power an expanding shock that would lift the rest of the envelope. Detailed calculations during the decade 1970–1980, however, systematically failed to achieve this. Instead, the initially fast shock came to a halt within the overlying envelope. The way around this dilemma was provided by neutrinos. Under the density and temperature conditions encountered immediately after collapse, photon processes play almost no role in either energy or momentum transport. Neutrino emission is, however, copious and able to play the same role of radiatively accelerating matter. Electrons can scatter neutrinos since the cross section depends on the neutrino energy through E_ν^{-2}. Although no such processes occur in Fermi theory, electroweak interactions of this sort are possible and mimic electron–photon scattering. No neutronization of the medium need occur, and the result is that neutrinos exert a radiation pressure that is proportional to $\int n_e \sigma_{\nu_e}(E) F_{\nu_e}(E)\, dE$. The neutrino flux comes from the newly formed neutron star whose "neutrinosphere" forms the optically thick surface in the same way that the photosphere would in a normal star.

As the shock progresses outward from the core, it encounters the infalling mass of the envelope and heats it. Regardless of the dynamical details, the radial mass distribution of the envelope determines the rate of decline of the optical light by the conversion of the high-temperature postshock emission to optical. The more optically thick the envelope is, the longer the expanding gas will radiate and the slower the rate of decline. The identification of the events with the passage of a shock through a red supergiant envelope was based, initially, on this expectation. Much of the dispersion among the properties of SN II light curves is likely due to the way the shock traverses the circumstellar medium and whether the atmosphere is compact or extended (see Fig 4.10).

4.7.6.3 Direct Evidence for Nucleosynthesis in Supernovae

Two observations serve as key constraints on the amount of nuclear processing that occurs in a supernova explosion. One is the observation of the decay of the Co II

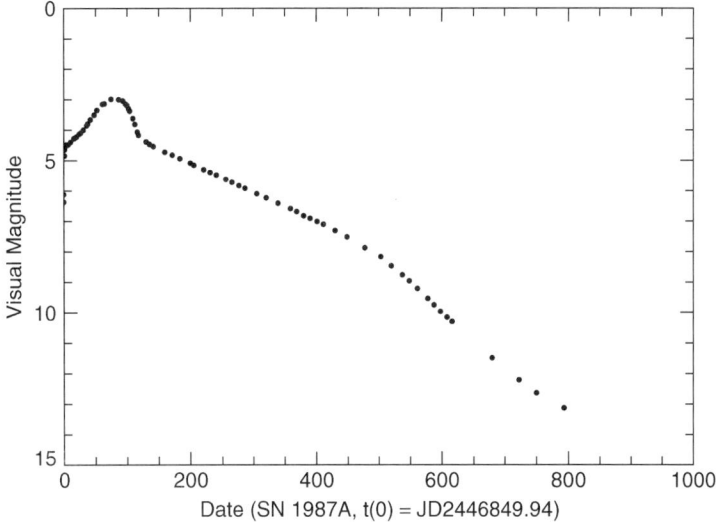

FIGURE 4.10. The V light curve for SN 1987A, a type II supernova.

infrared line and the detection of the γ-ray line due to the decay of ^{56}Co to its ground state in SN 1987A. Unlike the isotopic ambiguity that dominates optical and infrared observations, the nuclear lines are unambiguous indicators of the isotope. The other is more recent and provides a far more sensitive check on the nuclear production during the explosion. The direct detection of the decay line of ^{44}Ti at 1.157 MeV by COMPTEL for Cas A, a supernova that occurred around 1670, shows clearly that heavy elements are formed during a supernova event.[107]

4.7.7 Neutron Stars

The cinder that remains after core collapse is the most extreme physical environment known, a neutron star. In such an object we encounter the final unification of classical and nuclear physics, a body so dense that the nuclear equation of state provides support. These stars, in their manifestation as pulsars, provide one of the best handles on the equation of state at extreme densities.

The radius of this body can be estimated from the hydrostatic equation using dimensional analysis and a nonrelativistic degenerate equation of state for a neutron gas. Although neutrons are chargeless, they are fermions. They obey the same equation of state as a degenerate electron gas and have the same mass–radius

---

[107] This observation is, however, not as straightforward to interpret as would seem from this brief description. The ^{44}Ti decay actually proceeds through two intermediate nuclei, ^{44}Sc (which is formed by electron capture) and then ^{44}Ca, which is formed from electron capture and β^+ decay of ^{44}Sc whose half-life is about 4 h. It is the ^{44}Ca deexcitation that provides the observed line. For the supernova remnant to have an age of about 320 years, a precise lifetime for ^{44}Ti must be known in order to determine the initial mass formed of this isotope during the explosion. Laboratory measurements give $t_{1/2} = 60$ years, which is essential for obtaining the production yields from the line intensities.

relation. There is one important difference, however, resulting from their larger mass. The radius scales as

$$\frac{R_{NS}}{R_{WD}} \sim \left(\frac{m_e}{m_n}\right)\left(\frac{\mu_{e,NS}}{\mu_{e,WD}}\right)^{5/3}.$$

Since $m_n/m_e \sim 1.9 \times 10^3$, $\mu_{e,WD} = 2$, and $\mu_{e,NS} = 1$, the radius should be about $10^{-3} R_{WD}$. Taking as a fiducial number a WD radius of about 10^4 km, a neutron star radius will be about 10 km for a 1 $M_\odot$ star. This stable polytropic radius is so small that the gravitational potential is a fair fraction of the star's binding energy. The classical equation of hydrostatic equilibrium is insufficient for determining the structure—a fully relativistic treatment is required although the estimate is quite successful given its limitations regarding both the equation of state and treatment of structure. A detailed derivation of the modified structure equations is, however, not necessary. You can use the techniques we employed in Chapter 1 to extend the stress tensor and metric from the Newtonian limit to the relativistic framework. You can actually understand what happens by realizing that length scales are affected by the Schwarzschild metric and that you need to modify the mass density to include the energy density.

The structure is essentially that given by the interior Schwarzschild solution (see Chapter 1) supplemented by the equation of state, called the Tolman–Oppenheimer–Volkoff (TOV) equation[108]

$$\frac{dP}{dr} = -\frac{G(M + 4\pi r^3 P/c^2)}{r^2} \frac{\rho + P/c^2}{(1 - 2GM/rc^2)}, \qquad (4.181)$$

where the mass–energy density, $\epsilon = \rho c^2$. The $(1 - 2GM/rc^2)$ term comes from the Schwarzschild metric. The mass conservation equation then requires the mass–energy density but otherwise looks the same as before:

$$\frac{dM}{dr} = 4\pi r^2 \rho. \qquad (4.182)$$

As usual, we require this pair to be supplemented by an equation of state, and this is the chief source of difference among the various models for a neutron star interior. The first structure calculations for such stars took a very conservative view of the equation of state. Oppenheimer and Volkoff (in 1938) assumed that the neutrons support themselves by their degeneracy alone and chose a rather soft equation of state, essentially $P \sim \rho^{5/3}$. Their basic result is more or less the same as for a white dwarf except the instability sets in sooner because of the decreased

---

[108] Shapiro, S. A. and Teukolsky, S. L. 1983, *Black Holes, White Dwarfs, and Neutron Stars: The Physics of Compact Objects* (NY: Wiley); Glendenning, N. K. 1997, *Compact Stars: Nuclear Physics, Particle Physics, and General Relativity* (Berlin: Springer-Verlag).

radius of the final configuration and the relativistic alteration to the equation of hydrostatic equilibrium. With increasing mass, the body compresses and its radius comes progressively closer to the critical limit, the Schwarzschild radius. Their mass limit, a *maximum* mass rather than asymptotic value, was only 0.7 $M_\odot$. This is less than the Chandrasekhar mass for a white dwarf. Were this correct, it would be quite astonishing, indicating that all massive stars, at least those more massive than the Sun, would ultimately be unstable to gravitational collapse to a black hole. This contradicts the mass determinations for neutron stars in X-ray emitting binary systems, for which the mean mass is 1.4 $M_\odot$, just at the Chandrasekhar limit for a white dwarf but above the limit for a simply degenerate neutron equation of state.

The stability limit comes from the way the metric in Eq. (4.181) changes the gravitational binding energy (the effective mass) as the star becomes more compressed. In particular, as $E_{\text{grav}} \to M$, the combined effects of the redshift and the binding energy reduce the effective radius and the star is unable to achieve a stable point. The calculations are made more physically plausible by the incorporation of more realistic interactions for the constituent particles. Meson exchange between the neutrons hardens their equation of state. In particular, some models for the strong force within the nucleus have an infinitely repulsive core so that at high compression, the pressure rises extremely rapidly and the star is able to maintain a larger radius than for softer cores. The easiest way to understand this is that the pressure is proportional to the gradient of the interaction potential, V, so that the larger $\partial V / \partial \rho$ is, the greater the pressure. Several conclusions follow from this: the stiffer the equation of state, the larger the radius for a given mass; and the stiffer the equation of state, the larger the moment of inertia, since the star need not be as centrally condensed as for soft laws. The binding energy is also lower and the minimum stable rotation period is longer. This latter point has an important observable consequence: it sets the maximum allowable rotation frequency. The maximum rotation rate for a compact object is far lower than c/R, since the star becomes unstable when its rotational parameter $\lambda_\Omega \sim 1$, so that in terms of the Keplerian frequency, $\omega < \Omega_K$. This limits both the radius and moment of inertia of the star. Since the mass-radius relation for a neutron star depends on the equation of state, through both the central concentration and the radius, there is a relation between I and M.

4.7.7.1 Neutron Star Interiors

Neutron stars are genuinely weird objects and one of the great challenges in astrophysics is to elucidate their structure. A description of their physical structure is truly incredible: they are essentially *self-gravitating nuclei*.[109] The equation of

---

[109] The first survey of the consequences for relativistic stars of cold dense equations of state is Harrison, B. K., Thorne, K. S., Wakano, M., and Wheeler, J. A. 1965, *Gravitation Theory and Gravitational Collapse* (Chicago: Univ. Chicago Press); see also Bethe, H. A. 1971, *Ann. Rev. Nucl. Sci.*, **21**, 93; Pethick, C. J. and Ravenhall, D. G. 1995, *Ann. Rev. Nucl. Part. Sci.*, **45**, 429; Heiselberg, H. and Pandharipande, V. 2000, *Ann. Rev. Nucl. Part. Sci.*, **50**, 481. For sample calculations of nuclear equations of state for neutron stars, see Friedman, B. and Pandharipande, V. R. 1981, *Nucl. Phys.*, **A361**, 502; Glendenning, N. K. 1998, in *The Many Faces of Neutron Stars*: *NATO Advanced Summer Institute* Alpar, M. A., Buccheri, R., Ögelman, H., and van Paradijs, J., eds. (Dordrecht: Kluwer). The most comprehensive textbooks on the astrophysics of compact stars are Shapiro, S. L. and Teukolsky, S. A. 1983 (n.108) and Glendenning, N. K. 1997 (n.108).

state depends on interactions among the excited states of the nucleons and on the appearance of more exotic states such as hyperons and possibly free quarks. These particles of mass m_j become progressively more important with rising Fermi energy, since they are favored by the chemical potential when $\epsilon_F \geq m_j c^2$.

To see how the interaction potential affects the stellar structure, consider the relation between the potential and the pressure in a classical system. The gradient of the potential with separation is similar to the change in the energy per nucleon as a function of density:

$$P(\rho) = \rho^2 \frac{\partial E(\rho)}{\partial \rho}, \tag{4.183}$$

and E depends on the internucleon potential. For instance, the degenerate fermion case considered first by Oppenheimer and Volkoff gives a very "soft" equation of state. The star responds with significant compression for a rather small increase in the mass, and the stability limit is reached when $R \sim 2GM/c^2$. More realistic nuclear matter models are now available with a stronger repulsive core of the internucleon potential that yield stiffer equations of state. These permit a star to remain larger than its limiting radius for greater central densities, so the mass limit is also higher. The moment of inertia of a rotating neutron star is thus sensitively dependent on the equation of state, and so is the minimum rotational period that such a star can have. Remember, the more compact the star, the higher its maximum allowed rotation frequency. Wait until we get to pulsars, and you'll see that there are observable consequences of this apparently academic discussion!

Because of their extreme degeneracy, the outer layer of neutron stars are expected to form a solid for the same reasons we saw in white dwarf interiors. Even with strong magnetic fields, some lattice arrangement of the nuclei is energetically favored once the temperature is below $\sim 10^8$ K. Also in the crust, even with these high temperatures, the electrons pair and form a superconductor. Although initially formed with temperatures of order 10^9 K, this is still below the Fermi energy for densities of 10^{10} g cm^{-3} or higher. The neutrons also form pairs at higher densities than the crust. They feel an attractive potential at low energy. In this state, they display correlated motion seen in laboratory superfluids.[110] This change in the state of the interior from free fermions has a profound, observable effect. In rotating systems, we know that the angular momentum is quantized and by the Taylor–Proudman theorem for an incompressible inviscid medium, they conserve angular momentum on cylinders or vortex lines. The vortices pin to the crust and only the Magnus force, the attractive interaction between adjacent vortices, disturbs this equilibrium.[111]

---

[110] For excellent surveys of vortex dynamics in superfluids, see Putterman, S. 1971, *Superfluid Hydrodynamics* (Amsterdam: North-Holland); Donnelly, R. J. 1993, *Ann. Rev. Fluid Mech.*, **25**, 325.

[111] Pines, D. and Alpar, A. 1985, *Nature*, **316**, 27. There has been a lot of additional work, but the basic model is very well explained here and little of the big picture has changed.

4.7.7.2 Neutron Star Cooling and Detectability

Neutron star interiors cool mainly by neutrino emission, at least in their first few hundred years.[112] Surface cooling by photons is initially irrelevant to the overall structure of the star. The change in the structure of the interior does, however, significantly affect the visibility of the star during its early lifetime since the conductivity of nuclear matter is very large and the ultimate photon emission depends on it.

For a long time, it was thought that a neutron star would radiate essentially like a blackbody, albeit with gravitational redshifting of the photons. It is now realized that the star actually has an atmosphere, if you can call a region no more than one meter thick an atmosphere. Continuum absorption coefficients are not as easy to compute for such a gas, since the pressures preionize the material easily and the expected opacity is quite low for anything except electrons, but it does modify the observed spectrum sufficiently in the X-ray to explain some of the lack of detections of these stars in all-sky surveys (such as ROSAT and EINSTEIN). Two processes you have seen before are involved. The first is the single-neutron variant of the URCA process:

$$n \to p + e + \bar{\nu}_e, \qquad p + e \to n + \nu_e,$$

and the other is the so-called modified URCA process:

$$N + p + e \to N + n + \nu_e, \qquad N + n \to N + p + e + \bar{\nu}_e,$$

where N is either a proton or neutron. Any process of this type involving the weak decay of a baryon and subsequent interaction with the leptons so produced will yield similar results—energy is lost by the neutrinos. Both of these are extremely temperature sensitive above the threshold for reactions. For the direct URCA, the loss rate is

$$\epsilon_{DU} = 4 \times 10^{27} \left(\frac{Y_e n}{0.16 \text{ fm}^{-3}} \right)^{1/3} T_9^6 \text{ erg cm}^{-3} \text{ s}^{-1} \qquad (4.184)$$

and for the modified process, it is:

$$\epsilon_{MU} \approx 10^{23} \left(\frac{Y_e n}{0.16 \text{ fm}^{-3}} \right)^{1/3} T_9^8 \text{ erg cm}^{-3} \text{ s}^{-1}. \qquad (4.185)$$

In both cases, you will notice the very steep dependence of the emissivity on temperature. This is the key point to understanding the thermal evolution of the interior. Once T drops below about 5×10^7 K, photons provide the main energy

---

[112] Pine, D., Tamagaki, R., and Tsuruta, S., eds. 1992, *The Structure and Evolution of Neutron Stars* Reading: Addison-Wesley; Pethick, C. 1992, *Rev. Mod. Phys.*, **64**, 1133; Prakash, M. 1994, *Phys. Rep.*, **242**, 297; Page, D. 1998, in *The Many Faces of Neutron Stars: NATO Advanced Summer Institute* Alpar, M. A., Buccheri, R., Ögelman, H., and van Paradijs, J., eds. (Dordrecht: Kluwer), p. 539; Bignami, G. F. and Caraveo, P. A. 1996, *ARAA*, **34**, 331; Prakash, M., Bombaci, I., Prakash, M., Ellis, P. J., Lattimer, J. M., and Knorren, R. 1997, *Phys. Rep.*, **280**, 1; Tsuruta, S. 1998, *Phys. Rep.*, **292**, 1.

loss. Depending on the details of the equation of state and opacity, this happens sometime between 10^5 and 10^6 years after neutron star formation. The surface is then expected to show the effect of a magnetic field. Neutron stars can have fields in excess of 10^{12} G. Remember that a magnetic field is not isotropic so the electron conductivity will differ for longitudinal and transverse propagation. Since conductivity along the field will be more efficient, the polar region would be expected to be hotter and brighter than the equator, so isolated rotating neutron stars should pulse in thermal X rays.

4.7.8 Pulsars

Neutron stars entered the realm of the empirical with the discovery of CP 1919 + 21, the first radio pulsar. In 1967, while preparing observations of radio source scintillation from the solar wind as a way of studying the angular diameter of sources below the resolution limits of the existing radio telescopes, J. Bell and A. Hewish identified a strong variable radio source the first of the *Cambridge pulsars* (hence CP). Most bizarre was the combined periodicity, of order one second, and very short duty cycle. Rather than being a stochastic variable, CP 1919 + 21 was a *periodic*, pulsing source. It was quickly dubbed a *pulsar*, a neologism that is due to Hoyle. The task of localizing the source was a hard one, but it was eventually determined not to be terrestrial environmental interference (a real problem in the days before interferometers and fast Fourier analysis with aperture synthesis instruments). After eliminating other explanations, T. Gold proposed that a rotating neutron star would be able to produce the periodic emission detected from this source if the frequency was produced by an object rotating within its light cylinder. In other words, if $R < c/\Omega$, then the implied radius must be of order several kilometers. No object other than a neutron star is known that can satisfy this criterion and be stable. Black holes, of course, can be considerably smaller, but then the periodic emission requires very special emission mechanisms and fine tuning. A neutron star requires neither.[113]

There are essentially three different categories of neutron stars. Two types have now been detected in isolation. The first type is pulsars associated with a supernova remnant, of which the Crab and Vela pulsars are the premier examples (see Table 4.9[114]). The second type, of which Geminga is the best example, are isolated objects known mainly by their γ- and X-ray emissions.[115] In general, these are not associated with supernova remnants. The only type for which physical properties can be unambiguously determined, though, are pulsars or neutron stars in binary systems. The most frequently discovered are accreting neutron stars, which are detected by X-ray emission produced from accretion of mass from a close companion. These stars often have strong, pulsarlike magnetic fields that control the accretion flow and funnel matter toward the poles of the oblique field, producing X-ray binary pulsars. Nonaccreting neutron stars have also been discovered by

---

[113] For current information about pulsar properties, see the Princeton Pulsar Group Website, http://pulsar.princeton.edu. Also available through this site you will find the current version of the frequently updated Pulsar catalog.

[114] A comprehensive critical catalog of galactic supernova remnants is maintained by D. A. Green. See http://www.mrao.cam.ac.uk/surveys/snrs/ for periodic updates.

[115] Bignami, G. F. and Caraveo, P. A. 1996, *ARAA*, **34**, 331; Treves, A. et al. 2000, *PASP*, **112**, 297.

TABLE 4.9 Supernova Remnants with Associated Pulsars / Point X-ray Sources

SNR	Position	Size (arcmin)	Type	α	Comments
G184.6-5.8	0531 + 21	7 × 5	F	0.30	Crab, SN 1054, 2 kpc
G263.9-3.3	0833 − 45	255	C	—	Vela, 0.5 kpc
G320.4-1.2	1509 − 58	35	C	0.4	RCW89, 4.2 kpc, SN 185?
G332.4-0.4	1610 − 50	10	S	0.5	3.3 kpc
G5.4-1.2	1757 − 24	35	C?	0.2	> 4.3 kpc
G8.7-0.1	1800 − 21	45	S?	0.5	ROSAT
G343.1-2.3	1706 − 44	32	C	0.5	ROSAT
G34.7-0.4	1853 + 01	35 × 27	S	0.3	3 kpc

radio searches in much longer period systems. These evolve in isolation from their companions except for the gravitational interaction, such as PSR 1916 + 21. The final group, the millisecond pulsars, have periods of order 1.5–10 msec and are presumed to be members of close binary systems. The extreme periods are thought to be due to spinup during a now-ended stage of mass accretion.

This third category provides a special datum—pulsar masses are among the best determined for any cosmic body because they can be observed as members of nonaccreting close binary systems. Pulse frequencies substitute for direct radial velocity measurements and allow the mass function to be obtained.[116] Five systems have two neutron stars and these systems are quite eccentric, $0.25 \leq e \leq 0.68$. Others have white dwarf companions (many are millisecond pulsars) and nearly all have nearly circular orbits (PSR B1820-11 is a notable exception wih $e \approx 0.8$). Two with main sequence companions have enormous eccentricities, around 0.8–0.9. These highly eccentric systems are especially important because their orbits precess due to general relativistic effects and provide additional constraints on the orbital properties and masses (see Chapter 5 for discussion of gravitational radiation in binary systems). The derived mass range is very small, and for about 50 systems the neutron star mass is $M_{pulsar} = 1.35 \pm 0.04\ M_\odot$, a number that is consistent with the Chandrasekhar mass for a possible neutron star progenitor.

These observations beg the fundamental question: Where does the field come from? Amplification of about the right amount for a pulsar can result from collapse of a reasonably large object, order 1 $R_\odot$) to a radius of about 10 km if the seed field is about 1 G. This produces a neutron star magnetic field of about 10^{12} G if the flux is frozen during collapse. The process must take place so rapidly that the field does not have time to dissipate through reconnection or decay, and in this simple picture dynamo activity is ignored during or after the collapse. The fact that we see pulsars shows that the field doesn't have a strictly axisymmetric aligned dipole geometry. Therefore, you might suspect that some other mechanism is needed to produce the observed amplification, one that may act after the collapse in the protoneutron star and that there is no reason to believe that the field

---

[116] When first observed, PSR 1913 + 16 displayed unusually large variations in its 59-μs pulse period, changing by as much as 8 μs over as short as a 5-min interval as much as 80 μs from day to day. This led to the discovery of the first binary pulsar: Hulse, R. and Taylor, J. 1975, *ApJL*, **195**, 51. See also Manchester, R. N. and Taylor, J. H. 1977, *Pulsars* (San Francisco: Freeman). For a comprehensive mass compilation for radio pulsars, see Thorsett, S. E. and Chakrabarty, D. 1999, *ApJ*, **512**, 288.

survives intact from the main sequence to the collapse stage. Direct observation of the field strength and orientation is possible for pulsars in binary systems, where the cyclotron lines formed in the polar accretion column have been directly observed in X rays, with fields up to about 10^{12} G.

4.7.8.1 Emission Regions and Mechanisms

The stumbling blocks, then as now, are not what pulsars are (neutron stars), or even how they pulse (they are oblique rotators), but how they radiate and why their emission is so intense and collimated.[117] The enormous observed brightness temperatures, of order 10^{12} K, signal a coherent process as the source of the emission. High time resolution observations, with resolutions of order microseconds, have shown that the emission region is structured on very small scale lengths, of order centimeters or meters, and that the individual subpulses are not stable from one rotation cycle to the next. The striking feature is that the *mean* pulse profile, formed by the combination of hundreds or thousands of individual rotations, is rock steady.

The basic model for aligned pulsar magnetospheres was first developed by Goldreich and Julian (in 1969) and Mestel (in 1968) and elaborated for oblique fields by Arons, Fawley, and Scharlemann (in 1979); detailed discussions are found in the references. These studies showed that a magnetized rotating neutron star cannot remain surrounded by a vacuum. As a result, an emissive plasma can appear and the surface can become charged due to an electric field generated by the star's rotation. We will discuss the aligned case as the prototype, however be aware that this *cannot* apply to *pulsars* because axisymmetric magnetospheres cannot pulse as seen by a distant observer. Nevertheless, it is worth examining this beautiful model because it so perfectly highlights the physical requirements to be satisfied by any final model for pulsars.

Actually, this picture of a rotating magnetic star existed long before pulsars were even a dream in the theorists' imagination. Its root is the nineteenth-century concept of the *unipolar inductor*.[118] Let us follow the same basic picture for a moment. Assume that the neutron star's intense magnetic field is dipolar and axisymmetric. This is a great way to explore the basic physics, but it is a lousy explanation for a pulsar. Nevertheless, we have for **B**:

$$B_r = \frac{2M}{r^3} \cos\theta, \qquad B_\theta = \frac{M}{r^3} \sin\theta, \qquad (4.186)$$

where M is the magnetic dipole moment at $r = R_0$, the stellar radius. Our second assumption is that the star's interior is a perfect conductor. This follows from the

---

[117] See Michel, F. C. 1991, *Theory of Pulsar Magnetospheres* (Chicago: Univ. Chicago Press); Mészáros, P. 1992, *High-Energy Radiation from Magnetized Neutron Stars* (Chicago: Univ. Chicago Press); Manchester, R. H. and Taylor, J. 1977, *Pulsars* (San Francisco: Freeman); Lyne, A. and Graham-Smith, F. 1998, *Pulsar Astronomy*, 2nd ed. (Cambridge, UK: Cambridge University Press); Blandford, R. D., Hewish, A., Lyne, A. G., and Mestel, L., eds. 1993, *Pulsars as Physics Laboratories* (London: Oxford Univ. Press) [originally published as 1992, *Phil. Trans. Roy. Soc.* (London), **A341**, 1]; Weber, F. 1999, *Pulsars as Astrophysical Laboratories for Nuclear and Particle Physics* (Philadelphia: IOP Press).

[118] A wonderful introduction to this fascinating phenomenon of classical electrodynamics is Miller, A. I. 1986, *Frontiers of Physics*: 1900–1911 (Boston: Birkhäuser).

density of nuclear matter and the low temperature of the crust. Imagine such a body to be spinning with a frequency ω in vacuo. Then the external induced electric field is produced only by the dipole rotation so that

$$\mathbf{E} + \frac{1}{c}\mathbf{v} \times \mathbf{B} = 0 \tag{4.187}$$

and since for a rigid rotator $\mathbf{v} = \omega \times \mathbf{r}$, we obtain:

$$E_\theta = -\frac{\omega B R^5}{2cr^4}\sin 2\theta$$

$$E_r = -\frac{\omega B R^5}{2cr^4}(3\cos^2\theta - 1). \tag{4.188}$$

In addition, $\mathbf{E} = \nabla\Phi$, where now Φ is the electrostatic potential (notice that a net charge is not required for this field to exist, it is the same as a unipolar inductor) so that:

$$E_r = \frac{\partial \Phi}{\partial r}, \quad E_\theta = \frac{1}{r}\frac{\partial \Phi}{\partial \theta}, \quad E_\phi = 0. \tag{4.189}$$

The induced electric potential is quadrupolar:

$$\Delta\Phi \sim 3\cos^2\theta - 1, \tag{4.190}$$

which changes sign at the latitude $\theta_\star = \cos^{-1}(\frac{1}{3})^{1/2}$. The pole-to-equator potential drop is enormous, of order $\Delta\Phi \approx 10^{14}$ V for typical pulsar rotation periods, the consequences of which we will examine momentarily. Finally, the magnetosphere is bounded by the light cylinder, $R_L = c/\omega$, the distance at which the corotational speed is the speed of light. For a dipole field, there is a critical field line that just touches R_L, while at at lower latitudes, the field lines close within the light cylinder:

$$\sin\theta_c = \left(\frac{R_0\omega}{c}\right)^{1/2}. \tag{4.191}$$

Thus, trapped particles are confined to lower latitudes and the polar regions are open—particles experiencing a potential drop due to the induced electric field can freely stream away from the neutron star. From Eq. (4.191), particles below $\theta_\star$ trapped have a sign different from those that are lost from the poles and the star begins to develop a local surface charge. What happens near R_L requires a fully relativistic treatment, but we will concentrate on the near-surface properties of the magnetosphere. The corotating magnetospheric charge density is obtained from Gauss' law:

$$4\pi\rho = \nabla \cdot \mathbf{E} = -\frac{1}{c}\nabla \cdot [(\omega \times \mathbf{r}) \times \mathbf{B}]. \tag{4.192}$$

Substituting Eq. (4.188) into Eq. (4.192), we obtain the space charge density:

$$\rho = -\frac{\omega \cdot \mathbf{B}}{2\pi c}. \qquad (4.193)$$

Now the rate at which charges flow out of the star can be computed. Knowing that at the light cylinder the streaming speed is of order c, it follows that:

$$e\dot{N}_e = \rho c \pi r_p^2, \qquad (4.194)$$

where $r_p = R\theta_c$. Therefore, we have the rate of transport of energy across R_L:

$$\dot{E} = e\dot{N}_e \Delta\Phi. \qquad (4.195)$$

Finally, the angular momentum loss resulting from the radial drift of this charge across field lines is

$$\frac{dJ}{dt} = \frac{2}{\omega}\frac{dE}{dt} \sim \omega^4 B^2, \qquad (4.196)$$

just as we found from the simple radiation of energy by a rotating dipole. *This is not a coincidence.* The Poynting vector that is responsible for the transport of momentum from the star also drives the loss of charge. For characteristic pulsar parameters, the induced potential difference greatly exceeds the gravitational binding energy for electrons in the polar region, and these flow away from the surface and transit the light cylinder, leaving the ions behind and producing a net charge. When the local electric field becomes strong enough, the polar gap produces pair creation. The electrons flow back toward the surface and generate γ-ray emission from annihilation and from curvature radiation; the latter have frequencies that scale as $\nu \sim \gamma^3 \nu_B \sim E^3 B$ for relativistic particles.

While the reason they pulse is clear, the pulsar's emission mechanism(s) is much less so and has remained one of the most persistent problems since their discovery. Yet there are important clues. Pulsars show enormous radio brightness temperatures. They are also completely polarized, in both circular and linear modes, and the polarization angle changes through the pulse envelope. Pulses usually occupy only a small part of the total period (small duty cycle), which indicates that the emission region occupies only a small fraction of the surface area. High-time-resolution observations show that the pulse structure is only a mean envelope for much finer-structure emission. The pulse is extremely stable when integrated over a few hundred or a few thousand rotations (this may be only a matter of seconds for the shortest rotation periods!), although individual pulses vary greatly in intensity. Subpulses as short as a few microseconds have been resolved in radio observations. The elementary emission region is extraordinarily small, only a few tens of meters across, but this variation takes place within a very stable structure that may occupy several square kilometers of the surface. Each pulsar has a unique pulse shape and pulse variation, which further complicates our analysis of the emission mechanism. The overall pulse stability provides the basis for timing the spindown and meandering of the rotation. Within the envelope, however, individual subpulses sometimes regularly coherently drift in phase over many rotations. They can also turn on or off, or the entire pulse can *null* as they collectively disappear. Yet the emission region remains intact and recovers *within at most a few rotation periods*. Coupled with the enormously high brightness temperatures observed within a single pulse or subpulse, the emission region must maintain coherence over a significant part of the light cylinder.

4.7.8.2 *Rotational Properties: Period Changes and Glitches*

Although pulsars keep spectacular time, they are not perfect clocks. Within about one year of their discovery, observations showed that the periods of most pulsars systematically increase with time. This had been theoretically predicted in 1967, *before* these objects were discovered, by F. Pacini. In attempting to explain the power supply for relativistic synchrotron-emitting electrons in the Crab nebula, he applied a solution for a rotating dipole in vacuo developed by Deutsch[119] to the rotational spindown of a central magnetized neutron star. Pulsars were discovered a year later, but it took more than 2 years before the direct optical identification of the Crab pulsar. The spindown is computed by equating kinetic energy losses to the radiation rate for a magnetic dipole:

$$\frac{dE_{\text{rot}}}{dt} = \frac{d}{dt}\left(\frac{1}{2}I\omega^2\right) \sim M^2\omega^4, \qquad (4.197)$$

where I is the moment of inertia, so the *predicted* frequency change is a *spindown* with $\dot{\omega}/\omega \sim B^2\omega^3$. This can be tested. We *observe* secular increases in the rotation period for pulsars. These define an observed quantity called the *braking index, n,* for a generalized spindown mechanism:

$$\frac{1}{2}\frac{d}{dt}I\omega^2 = -A\omega^{n+1}, \qquad (4.198)$$

where A is a mechanism-specific constant. If I is not a function of ω, this gives the observable quantity:

$$n \equiv \frac{\omega\ddot{\omega}}{\dot{\omega}^2}, \qquad (4.199)$$

which must be modified if the star varies in shape as it changes its rotation frequency. Irrespective of the emission mechanism, the regularity of the statistical profile for pulsars means that the mean index can be determined from sufficiently long observing intervals. Observationally determining n requires both the first and second time derivatives of the rotation frequency. For magnetic dipole radiation $n = 3$, while for gravitational wave emission, $n = 5$. For dipole radiation, the spindown rate yields the magnetic dipole moment and thus the strength of the

---

[119] Deutsch, A. 1955, *Ann. d'Astrophys.*, **18**, 1. This is one of the best examples of a crossover solution in astrophysics. Deutsch was attempting to explain the very slow rotation of magnetic chemically peculiar main sequence A stars. His solution involved emission of magnetic dipole radiation depending on the rotation period of the star. Although the explanation probably fails for its intended purpose, the application to pulsar spindown has been spectacularly successful. The fundamental application to pulsars is Pacini, F. 1967, *Nature*, **216**, 567.

surface magnetic field. For gravitational radiation, it depends on the mass quadrupole moment. The pulsar age is estimated by $\omega/\dot\omega$, and its surface field is estimated by $(\dot\omega/\omega^3)^{1/2}$. Typical period derivatives are around $\dot P/P \approx 10^{-15}$, and the mean period of Galactic pulsars is around 1 s, corresponding to surface field of order 10^{12} G (10^8 T) and ages of order 10^6 years.[120] One cautionary note here is that period derivative does not give an unambiguous estimate of the pulsar age *if* the obliquity of the magnetic field changes secularly with time. At the moment, this is a controversial point. Should the magnetic field align with time, the rotating neutron star ceases to be an observable pulsar and its period derivative decreases with no change in the field *strength*. If, on the other hand, the magnetic geometry remains fixed but the field decays, the pulses will weaken and the period derivative will decrease. Which picture is correct remains an open question.

On top of the steady nature of the period decline, eight relatively young pulsars have displayed virtually discontinuous period spinups, called *glitches*, at irregular intervals. These have amplitudes $10^{-8} < \Delta\omega/\omega < 10^{-5}$, and $10^{-4} < \Delta\dot\omega/\dot\omega < 10^2$; the largest changes are observed for PSR 0833-45, the Vela pulsar; this pulsar has undergone eight such events since 1969. The Crab pulsar, PSR 0531 + 21, has undergone four such events. PSR 1737-30 has had five small events over a 3-year timescale, but the general separation between successive glitches is around 3–10 years. Recovery times for the rotation frequency range from about one week for the Crab to around one year for Vela. As with the pulse properties, each glitch event is unique and no two pulsars show the same behavior. Several models propose to explain this behavior. The most successful invokes discontinuous readjustment of the superfluid during a glitch event, with vortex lines depinning from crustal lattice sites and then repinning, thereby producing a redistribution of internal angular momentum. It is assumed that the crust, which is strongly coupled to the magnetic field, is spinning down by the braking torque. When the differential internal rotation is large enough, the superfluid readjusts to accommodate the new distribution of pinning sites.[121]

Neutron stars that are accreting matter in close binary systems can be spun up to nearly the Keplerian frequency since the boundary of the magnetosphere forms the inner part of the disk. An application of this is to the millisecond pulsars. It is easy to make a low magnetic field effective in the stellar spinup because in this case, the magnetopause is moved inward toward the surface. The weak field seen in these objects provides one of the most important inputs to the scenario, since these fast for isolated pulsars. The fastest rotators have periods that are perilously close to the Keplerian limit (the shortest periods known are of order 1.5–3 ms, a factor of only ~ 2 or ~ 3 above the limit, depending on the equation of state). This must result from accretion of material from a companion via a disk that spins

---

[120] Millisecond pulsars are exceptional. Their period derivatives are extraordinarily low, $< 10^{-17}$. This appears to be due to the fact that they possess systematically lower surface magnetic fields. From the same analysis, the field for the known sources is as small as 10^2 MG. Such comparatively low magnetic field strengths are not likely to be produced by free resistive decay of the stellar dipole, and for this reason, among others, this class of pulsar is thought be recycled neutron stars that have been spun up through accretion.

[121] Alpar, M. A., Anderson, P. W., and Pines, D. 1984, *ApJ*, **276**, 325; Pines, D. and Alpar, M. A. 1985, *Nature*, **316**, 27; Link. B., Epstein, R. I., and Baym, G. 1993, *ApJ*, **403**, 285; Alpar, M. A., ed. 1995, *Lives of Neutron Stars* (Dordrecht: Kluwer).

the star up to near corotation with the innermost part of the accretion disk. So-called *antiglitches* have been observed in Her X − 1 = HZ Her, which is a magnetized neutron star with a 1.7 s rotation period in a 1.4 day period binary with an A-type companion. The pulsar period is also decreasing with time. This supports the idea that spinup, rather than spindown, results from close binary evolution.[122]

4.8 ISOCHRONES

We now reach the *Holy Grail* of this and the previous chapter. With the description in hand of stellar interior physics, you should be able to evaluate and use computed models for the production of isochrones. This activity requires not only the interior model but also stellar atmospheres with which to transform between the "theorist's" and "observer's" HR diagram. Recall that the problem is the treatment of the outer layer of a star. The interiors modeler knows that there is no real influence of the optically thin region on what happens in the nuclear burning or even pulsational portion of the star—well, at least not for most of the stellar lifetime. On the other hand, the observer knows that the surface is all you will ever see. This final step in the comparison actually connects our last two chapters. Everything you know about the properties of the star comes from its surface, from its spectrum and colors. Here is where the most critical step occurs. Unless you have the mass, radius, and absolute luminosity, and in general you don't for cluster stars, you must transform the theoretical values of L and T_{eff} to some observational standard filter set such as *UBVR* or *uvbyβ*. This introduces a completely new set of problems that we must stop for a moment to examine.

In principle, there is a unique transformation between the (L, R) plane and $(T_{\text{eff}}, \log g)$ but this is very dependent on the model atmosphere analyses. There are two different pieces of information that the atmospheres must provide. One is the color, which places the star in temperature. The other is the bolometric correction. This transforms L into, say, M_V, and is a much more delicate matter. In particular, it depends critically on the abundance mixture used for the computation of the stellar spectral energy distribution. Filter photometry in the standard systems is generally quite broadband and covers a large number of absorption and emission lines as we discussed in Chapter 3. Atmosphere models can produce synthetic spectra that must then be folded through the transmission profiles for the individual filters to produce colors. For hot main sequence stars, the color transformation is not a particularly serious problem. In the optical, the colors are not severely affected by line blanketing and the photometry samples only only the long-wavelength tail of the energy distribution. For this reason, it is generally difficult to determine the properties of individual stars on the basis of colors only. Extinction plays an important role in altering, say, $B − V$ and differential effects in young clusters can hamper the placement of the main sequence turnoff point. The bolometric correction is, on the other hand, far less straightforward since the

---

[122] The extraordinary stability of millisecond pulsar periods appears to result from their weak magnetic fields. After the accretion event ends, no further emission of magnetic dipole radiation affects the pulsar period.

major part of the flux emerges in the ultraviolet. Here NLTE effects and details of radiative corrections change the temperature gradient and alter the emissivity in the parts of the spectrum that are inaccessible from the ground. A few stars hotter than B0 have been calibrated in this way,[123] and this does not include the very hottest stars for which emission lines and winds are important.

For cooler stars, although the bolometric corrections are smaller, the effects of metallicity are more severe. For temperatures below 15000 K, the spectrum is dominated by lines of neutral and singly ionized heavy elements, in particular those in the range $20 \leq Z \leq 30$ for which there are millions of lines. Although most of the optical transitions are weak, their sheer number compensates to make the spectrum very sensitive to the precise abundances. Older clusters are most affected by this comparison. For globular clusters, this is made worse by the need to include molecules in the catalog of absorbers when computing synthetic spectra. The placement of the cool horizontal branch, the giant branch, and the asymptotic giant branch all depend on the accuracy of the color transformations.

At the simplest level, the evolutionary structure model produces a unique way of obtaining the aggregate age for a cluster. This has been well understood since the founding work of Hoyle, Schwarzschild, and Sandage in the 1950s. The shortest-lived star on the main sequence is a function of mass, and the last star at any instant to leave the main sequence is the most massive star still in the hydrogen coreburning stage of evolution. At the next level, there are numerous and important complications.[124] Sample isochrones using the Geneva code are shown in Figure 4.11.

One of the most interesting uses for isochrones comes from the comparison between the predicted properties of stars in close, but completely detached, binary systems.[125] The basic idea is quite simple. You know that the two stars were formed at precisely the same time, although how they formed and what set their masses during their formation is another matter. They were formed out of the same material; hence they *must* have the same metallicity.[126] Thus the binary system evolves with the two stars on a single isochronal line in the HR diagram. If both spectra are visible and the stars can be separately analyzed, their separate radii and luminosities and effective temperatures can be determined from eclipse

---

[123] Aufdenberg, J. 2001, *PASP*, **113**, 119.

[124] See, for instance, Vandenberg, D. A. and Bell, R. A. 1985, *ApJS*, **58**, 561; Vandenberg, D. A. 1985, *ApJS*, **58**, 711; Green, E. M., Demarque, P., and King, C. R. 1987, *The Revised Yale Isochrones and Luminosity Functions* (New Haven: Yale Univ. Press); Maeder, A. 1990, *Astr. Ap. Suppl.*, **84**, 139; Maeder, A. and Meynet, G. 1994, *Astr. Ap.*, **287**, 803 (there are many papers from the Geneva group covering a wide range of masses); D'Antona, F. and Mazzitelli, I. 1994, *ApJS*, **40**, 467. The data tables for many of the sequences listed here are available through the SIMBAD database, http://simbad.harvard.edu. and references are available through the ADS, http://adsabs.harvard.edu,.

[125] See Claret, A. and Gimeñez, A. 1993, *Astr. Ap.*, **277**, 487 for an especially enlightening discussion of the problem of how to use the apsidal motion test to study the interior of stars. Recall our discussion of the Lane–Emden equation. One of the results you obtain for free is $\langle \rho \rangle / \rho_c$, the degree of central concentration. Tidal distortion, which is proportional to this parameter, produces orbital precession for binary stars just as it does for the Earth–Moon system. Since for detached systems mass transfer does not alter the angular momentum, the apsidal motion provides the tidally induced departure from spheres in the moment of inertia of the stars.

[126] This, too, is an open matter, since there are some close binaries that differ dramatically in their observed metallicities. But these cases are fortunately few and need to be further understood.

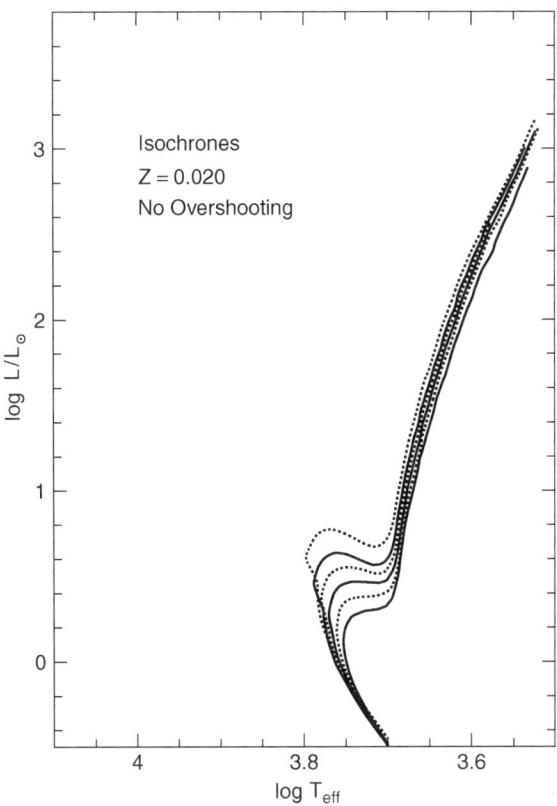

FIGURE 4.11. Sample isochrones for solar composition ($Z = 0.02$). (Credit: A. Maeder and G. Meynet, Univ. Geneva Observatory). The ages cover $9.6 \leq \log t \leq 10.1$ for masses from 0.8 to 1.5 $M_\odot$.

solutions, and their masses can also be determined individually. Therefore, only their compositions and ages are unknown, and it is straightforward to use interiors models to determine these. When the stars are on the same isochrone, the age has been determined. It can then be compared with the other properties; for instance, the metallicity of the stars can be predicted from this test and compared with the results from stellar atmosphere analyses.

The interior structure of the stars is predicted by the models. The degree of central concentration can then be used to predict the apsidal motion of the system. It is instructive to think about why. The idea for this test, and for many others of evolution and stellar properties, goes back to the 1920s and H. N. Russell. This was a period when absolute photometric measurement was impossible and quantitative data were very hard to come by. Instead, one used photographic plates and often only relative (sometimes even visual) measurements of brightness. Eclipsing binaries provided the only tool with which to study stars. They give relative parameters that are independent of distance, and even interstellar extinction, and provide information about luminosity and radius *ratios* immediately from photometry alone. Even the relative shapes of the component stars can be determined from the details of the eclipse ingresses and egresses. Radial velocities yield mass ratios for noneclipsing systems, and absolute masses when we know the inclinations, so we have all we need: mass and luminosity ratios, relative radii, and relative temperatures as functions of stellar mass. The idea is again simple, in principle. The more extended a mass distribution, the larger its quadrupole moment because of tides and the shorter the apsidal period of the binary. This is because the mass distribution now differs from a point mass and therefore the orbit precesses from the higher gravitational perturbations. The more centrally condensed the star, the longer the period and the smaller the amplitude of this perturbation. The problem is that the precession will always be quite slow, and one must be very careful to ensure that it is due only to the steady-state internal structure of the stars and not to mass loss or transfer effects. It also depends on the eccentricity of the binary system, and, as we will see, close binaries are expected to circularize and synchronize rather rapidly because of tides.

4.8.1 The Initial Mass Function: A Result of Isochrones

Once you have both isochrones and evolutionary timescales in hand, it is possible to address another fundamental property of stars, their mass distribution.[127] It is one thing to say we see the HR diagram of a cluster populated by individual stars from which we determine such properties as the age and metallicity and distance. It is another step to recognize that we are dealing with a sample population! From the spread of the main sequence and the tightness of their spatial location and velocities, clusters represent single, short-lived episodes of star formation. It does not matter for our purposes here whether we are discussing OB associations, open clusters, or globular clusters. The fact that the stars are formed at the same time out of the same material is enough. From the turnoff mass for the main sequence and knowing the abundances by fitting evolutionary tracks to the stars and possibly also from model atmosphere analysis of select cluster members, masses can be associated with the main-sequence luminosities. One complication is interstellar extinction and reddening, the discussion of which we postpone to Chapter 6, but it is possible to correct for distortions in the photometric properties of the stars by any intervening dust. Another complication, not so easily addressed, is the need to correct for unresolved binaries and anomalous abundances that also affect the photometry. With these limitations in mind, it is then possible to take a population census of the cluster, determining the number of stars per unit mass. The stars located fainter than the main sequence turn-off point are in their first stage of nuclear processing and essentially unaltered from their initial hydrostatic state. We know that mass loss rates for such stars are relatively low and understand pretty well their internal structure. While not necessarily actual ZAMS objects, these stars should nonetheless preserve the mass distribution that characterizes their formation. This is the *initial mass function*. No one cluster suffices to construct this, but it can be pieced together from a large enough sample of individual clusters. The paradigm is that each is a snapshot, a realization, of a statistically (stochastically) stable process. With enough clusters having collectively enough individual stars, it should be possible to see whether the assumptions are correct that the process is indeed stable and universal.

---

[127] See Miller, G. E. and Scalo, J. M. 1979, *ApJS*, **41**, 513; Scalo, J. 1986, *Fund. Cosmic Phys.*, **11**, 1.

We will discuss the mass function at some length when we talk about galactic chemical evolution and star formation in Chapter 7. However, it is important to outline here how calculations of stellar structure and evolution influence its determination. The main sequence timescale is fixed by the nuclear timescale for core hydrogen burning, as we discussed in Section 4.7.4.1. The star's luminosity slowly increases as it consumes its core supply, but its locus in the HR diagram barely changes, especially given usual observational uncertainties. As you know, stars of different mass process hydrogen at different rates, by the mass–luminosity relation, so we would expect that the most massive stars on the main sequence would be those with core hydrogen burning lifetimes just equal to the age of the cluster. However, we would like to know the relative proportion of mass tied up in high- versus low-mass *unevolved* stars. This is $\phi(M)$, the *initial mass function* (IMF). The argument for its determination goes back to Salpeter's 1955 study of clusters. Look at the relative number of stars as a function of mass for progressively younger clusters, and this gives you the number of stars on the main sequence as a function of mass, independent of age. His initial results suggested a simple power law:

$$\phi(M)dM \sim M^{-2.35} dM, \qquad (4.200)$$

but it is clear that this only approximates a more complicated distribution. It appears that the IMF is actually a universal property of star formation. The cluster mass functions differ depending on the population size, but the IMF seems to be a real stationary distribution. The important point for you to remember is that the IMF is not a directly observable quantity except for the lower-mass stars in the youngest clusters. It otherwise requires the use of stellar models, and assumptions about mass loss, to trace the stars back to their ZAMS locations. There is another problem that we have not yet fully faced. Many, if not most, stars are binary or multiple systems or have low-mass companions,[128] and the history of such stars can be very different from the single-star models we have been discussing here.

4.A MAGNETIC DYNAMOS AND THE INTERPLAY BETWEEN TURBULENCE AND ROTATION

The discovery of the activity cycle of the Sun, and of stars, supports the contention that magnetic fields in cosmic objects must be continually regenerated. The decay of the field, due to the finite electrical conductivity of matter, is sufficiently rapid to require that if a field is observed at all, it must not be primordial. In what follows, we will sketch the processes whereby stars generate magnetic fields by dynamo action. The problem is a general one, and we will discuss it generally. Yet it is the Sun, the only star for which we have detailed time-dependent information about the surface magnetic field structure by direct measurement, that best serves to illustrate the operation of the mechanisms we'll discuss here.

---

[128]A. Boesgaard put this very well in saying "Three out of every two stars is a binary." We do not know the frequency of binary and multiple systems among massive stars, nor is the mass ratio distribution well constrained. Because of the importance of binary stars in altering the picture we have been developing of stellar evolution, we will return to this point, among others, in Chapter 5.

4.A.1 The Dynamo Equations

To put this in perspective, consider that the plasma of the solar interior is in constant motion due to the convective instabilities we have already discussed. Since it is a charged medium (although electrically neutral), it must generate local currents. These, in turn, should create a dynamo by induction.

The electric field in a moving medium is due to two contributing physical processes: one from the induced field due to the motion if there is already a magnetic field present, and the other due to the finite rate of dissipation of the field from the potential difference and finite electrical conductivity. Any current can be assumed to be due to a locally generated electric field, so

$$\mathbf{E}' = \mathbf{E} - \frac{1}{c}\mathbf{v} \times \mathbf{B}, \tag{4.201}$$

which by Ohm's law for a medium with conductivity σ is related to the current:

$$\mathbf{E} = \frac{1}{\sigma}\mathbf{J}. \tag{4.202}$$

Now we return to the magnetohydrodynamic (MHD) equations from Chapter 1. For $\mathbf{E}$, the rate of change of the field with time is

$$\frac{\partial \mathbf{E}}{\partial t} = c\nabla \times \mathbf{B} - 4\pi \mathbf{J}, \tag{4.203}$$

while for the magnetic field we have

$$\frac{\partial \mathbf{B}}{\partial t} = -c\nabla \times \mathbf{E}. \tag{4.204}$$

In other words, the two effects that induce the formation of a magnetic field are the displacement current and the free charges that form $\mathbf{J}$, while the decay of the magnetic field is due to the generation of a spatially variable electric field. If we assume that the material is in motion, then the form of the equations is invariant, but the motion of a charged medium through a magnetic field produces the appearance of an electric field, thus adding to the decay term for the field because the generation of the electric field is at the expense of the magnetic field already present in the medium.

Assuming that the electric field in the moving medium is vanishingly small—that the conductivity is very high (although not infinite)—we obtain

$$\frac{\partial \mathbf{B}}{\partial t} = \nabla \times \mathbf{v} \times \mathbf{B} - \frac{c}{\sigma}\nabla \times \mathbf{J} = \nabla \times (\mathbf{v} \times \mathbf{B}) + \eta \nabla^2 \mathbf{B} + \nabla \times \mathscr{E}. \tag{4.205}$$

Here the magnetic *diffusion coefficient* is $\eta = 4\pi c^2/\sigma$, and $\mathscr{E}$ is the electromotive force. This last equation, the so-called *dynamo equation*, is of the most interest to us. The first term is that of a generator; the fluid motions are used to produce a magnetic field from the pre-existing field. The latter is a diffusive term, resulting from the transport of magnetic field through the fluid by random motions.

Any attempt to maintain an internal electric field in a perfectly conducting medium will be thwarted by the mobility of the charges, which immediately move to cancel any potential difference. The timescale for this cancellation is very short in comparison with the timescales for the field to begin building in the medium; that is, they take place on times short in comparison with the actual fluid motions, so there will be no net electric field. The effect of the diffusive term for the evolution of a magnetic field can thus be explained more clearly by thinking about the effect of mass motions in a magnetic field. For a net current to result from the fluid motion in some medium, there must be uncompensated potential differences that manage to survive within the fluid. In a highly conducting medium, there is a simpler amplification mechanism for the field that acts even without recourse to a dynamo. The magnetic field appear to move as if "frozen" into the medium, the spatial energy density of the magnetic field precisely following the fluid density. So if the density increases locally, so does the magnetic field.

To see this, assume that σ is the value usually quoted for the conductivity of stellar plasma, $> 10^{15}$ s^{-1}. This conductivity is large enough to ensure that the magnetic diffusion term effectively vanishes. Then, recalling Section 1.4.5.1

$$\frac{d\mathbf{B}}{dt} = -\mathbf{B}\nabla \cdot \mathbf{v}. \qquad (4.206)$$

From the continuity equation, we have

$$\frac{d\rho}{dt} = -\rho \nabla \cdot \mathbf{v}. \qquad (4.207)$$

Here, we have used the convective, or comoving, derivative: $d/dt \equiv \partial/\partial t + \mathbf{v} \cdot \nabla$. The magnetic flux is thus simply a scalar multiple, f, of the mass density, or $B \approx f\rho$. Thus, if the density is locally increased, so is the magnetic field strength. The magnetic field seems to move with the fluid; hence the appellation "frozen."

Now imagine that there is a small deviation from perfect conductivity. As the fluid moves, there will be some slippage of the mass through the field. This appears to change the magnetic field strength in the comoving fluid. At the same time, the fact that the field is changing induces the formation of a potential difference that, in a finite-conductivity environment, induces the generation of a current. All of this is at the expense of the magnetic field. The field decreases in local strength and will do so everywhere throughout the fluid in time. In fact, the form of the magnetic field decay is the same as that of the heat equation, so the field can be said to be diffusively lost. The energy simply goes over into heat, since the field generates dissipative currents that lose their energy through collisions throughout the fluid, and the result is the gradual fading away of the field with time.

The characteristic timescale for the decay of the field depends on the scale length for the field generation and the dissipation scale for the currents. To see this, look at the dimensionless form of the dynamo equation, but now ignore the effects of the fluid motions. The equation is linear in the field strength, which means that we can scale it by any arbitrary value of the field at any time. The change with time is related to the second derivative in space, so that we can relate the timescale for the change in the field strength, τ to the length scale L by

$$\tau = \frac{4\pi L^2}{\eta} \to 4\pi L^2 \frac{\sigma}{c^2}. \qquad (4.208)$$

Again, notice that in the case of an infinitely conducting medium the time for the decay is infinitely long. We have used only one possible representation for the magnetic diffusion coefficient, however, and we shall shortly see that this is one of the longer estimates.

The field throughout a magnetized body will therefore decay with time, unless it is regenerated or the medium undergoes steady collapse at a rate that is rapid compared with the decay rate. Eventually, regardless of the artifice, the body will be unable to prevent the dissipation of its internal field without constant mechanical, that is dynamo, input. Rotation is a natural means to drive the dynamo. Since most plasmas have high conductivities, the amplification of an external field by collapse or compression (such as shocks), coupled with rotation should generate a strong Lorentz force. This field is not, however, supported for indefinite lengths of time. Instead, Cowling's theorem—also known as the *antidynamo* theorem—states that no stationary (time-independent) axisymmetric dynamo is possible. But shear is essential to breaking the spherical symmetry.

The symmetry is further broken, and the effect of the rotation translated into a poloidal field, through the combined action of circulation and turbulence. An initially axisymmetric field is sheared by differential rotation, and if it is initially cylindrical (B_z) or poloidal (B_r, B_θ), then an azimuthal field (B_ϕ) results. Here r and θ are the radius and latitude, respectively. A poloidal field results from a toroidal potential field, $\mathbf{B}_p = \nabla \times \mathbf{A}_\phi$, so that the toroidal magnetic field results from a distortion of the poloidal field. Finally, in order to convert the toroidal field back into a toroidal potential, some additional symmetry breaking is required. Turbulence in a rotating medium has vorticity, or handedness, which is parallel to the local angular velocity vector and neither radial nor even hemispherically symmetric. In an electrically conducting fluid, buoyantly rising turbulent cells produce a helical twist to the toroidal field, and induce a poloidal component to the field. This is the basis of the $\alpha - \Omega$ dynamo model. The electromotive force is $\mathscr{E} = \alpha \mathbf{B}$, where α is related to the velocity correlation function and essentially measures the amplitude of velocity fluctuations in the fluid. In a fluctuating medium, the velocity breaks into a mean component, $\mathbf{V}$, plus a fluctuating part, $\mathbf{u}$, which has a vanishing mean value, but for which $\langle u^2 \rangle$ does not vanish. Here the brackets $\langle \cdots \rangle$ represent ensemble averages over the turbulent spectrum of the eddies in the fluid. The magnetic field evolution now depends on both the mean field $\mathbf{B}$ and the fluctuating part $\mathbf{b}$, and the dynamo equation becomes

$$\frac{\partial \mathbf{B}}{\partial t} = \nabla \times (\mathbf{V} \times \mathbf{B}) + \nabla \times \mathscr{E} + \eta \nabla^2 \mathbf{B}, \qquad (4.209)$$

where $\mathscr{E} = \langle \mathbf{u} \times \mathbf{b} \rangle = \alpha \mathbf{B}$. Thus α represents the fluctuations in the fluid, and describes schematically the way that this feeds back into the magnetic field strength. The evolution of the fluctuating part of the field is given by

$$\frac{\partial \mathbf{b}}{\partial t} = \nabla \times (\mathbf{u} \times \mathbf{B} + \mathbf{V} \times \mathbf{b}) + \nabla \times (\mathbf{u} \times \mathbf{b} - \mathscr{E}) + \eta \nabla^2 \mathbf{b}. \quad (4.210)$$

Then $\alpha \sim l_0 \langle u^2 \rangle / \eta$, which depends on the velocity fluctuation spectrum. The turbulence therefore controls the small-scale structure, and the differential rotation (shear) controls the large-scale ordered field and provides the symmetry breaking necessary to generating the dipole.[129] Both buoyancy (gravity) and shear act to drive the turbulence while rotation provides the requisite helicity.

4.A.1.1 The Dynamo Number and Scaling Relations

A schematic estimate of the strength of the dynamo components, and an approximate scaling law, results from the quantitative side of this picture. Differential field stretching causes poloidal to toroidal conversion, which takes place at a rate $\delta v_\phi / L$. Vortical motion of rising convective eddies transforms toroidal to locally poloidal field at a rate Γ, which is a pseudoscalar quantity whose sign depends on the hemisphere. The dynamo equations simplify by dimensional analysis. For the poloidal field, which is given by a vector potential field:

$$\mathbf{B}_p = \nabla \times \mathbf{A}_\phi \rightarrow \frac{A_\phi}{L}. \quad (4.211)$$

The rate at which poloidal field dissipates by turbulence and diffusion is, in the steady state, balanced by the conversion of toroidal to poloidal flux:

$$-\eta \nabla^2 \mathbf{A}_\phi \approx \Gamma \mathbf{B}_\phi \rightarrow -\eta \frac{A_\phi}{L^2} \sim -\Gamma B_\phi. \quad (4.212)$$

Finally, the poloidal field is wrapped to form the azimuthal field at a rate that depends on the shear $\Delta\Omega$:

$$-\eta \nabla^2 \mathbf{B}_\phi \approx \nabla \times (v_\phi \times \mathbf{B}_\mathbf{p}) \rightarrow -\eta \frac{B_p}{L^2} \sim \frac{\delta v_\phi}{L} B_\phi \sim \Delta\Omega B_p. \quad (4.213)$$

The dimensionless number

$$\mathscr{D} = \frac{\Gamma \Delta \Omega L^2}{\eta^2}, \quad (4.214)$$

called the *dynamo number*, serves as the scaling parameter for the generation by the $\alpha - \Omega$ process. Notice that $\mathscr{D}$ is independent of the magnetic field, and that in the steady-state case it is of order unity. For large values, the field will not be steady (what is usually meant by an "active" dynamo).

---

[129] Moffatt, H. 1978, *Magnetic Field Generation in Electrically Conducting Fluids* (Cambridge, UK: Cambridge Univ. Press); Parker, E. N. 1979, *Cosmical Magnetic Fields* (Oxford: Oxford Univ. Press); Biskamp, D. 2000, *Magnetic Reconnection in Plasmas* (Cambridge, UK: Cambridge Univ. Press); Schrijver, C. J. and Zwaan, C. 2000, *Solar and Stellar Magnetic Activity* (Cambridge, UK: Cambridge Univ. Press); Lazarian, A. and Vishniac, E. T. 1999, *ApJ*, **517**, 700.

Observationally, it appears that X-ray emission from late-type stars correlates with a slightly different measure of the dynamo activity. This is the Rossby number, which actually measures the convective (buoyant) timescale compared with the rotation rate: $Ro = \Omega \tau_c$, where τ_c is the convective turnover time for some large depth in the stellar envelope. When Ro is large, either the rotation is very rapid or the turnover line is large (the gravity is low and the stellar envelope is very distended). Empirically, UV emission line strengths and L_{XR} correlate with Ro, although the precise law is still debated. It seems to reasonable that the Rossby number measures at least something of the α-Ω dynamo activity. The Γ factor in $\mathscr{D}$ is a measure of the vorticity in the buoyant cells, so depends on both Ω and τ_c. The turbulent diffusion coefficient, η, scales as $L^2 \tau_c^{-1}$ for some characteristic length L. However, the determination of an empirical scaling relation, drawn as it is from very indirect measures of the magnetic field strength and its rate of generation and structure, does not provide a serious constraint on the dynamo theories. In general, the most that can be said from the measurement of activity in late-type stars is that a dynamo must be acting, that the field must be constantly emerging anew at the photosphere, and that the mechanism must depend on surface gravity, effective temperature, and rotational frequency.

There have been several attempts to modify the dynamo equations, incorporating the idea that the induction term saturates at some critical value of the magnetic field. In effect, this is the same as saying that as the field builds up, it feeds back into the turbulence and suppresses the growth of the α term in the dynamo equation:

$$\frac{\partial \mathbf{B}}{\partial t} = \nabla \times \alpha \mathbf{B} + \nabla \times \mathbf{V} \times \mathbf{B} + \eta \nabla^2 \mathbf{B}, \qquad (4.215)$$

where α is the term arising from the mean field part of the turbulent interactions, the mean self-induction or electromotive force. It is important to notice that the processes we have just discussed occur independently of the energy *generation* mechanism for a star or any other magnetiized medium. Dynamo activity can just as easily occur in differentially rotating circumstellar accretion disks, pre-main-sequence stars, neutrino-dominated protoneutron stars, or the differentially rotating turbulent gas in the interstellar medium. Since $\mathscr{D}$ governs the scaling without regard to the details of the medium, we would expect that the theory as developed for stars can be translated to explain the origin of magnetic fields in a wide range of astrophysical environments.

4.B CALCULATION OF NUCLEAR REACTION NETWORKS

Once you know the composition and the thermal conditions, the next step is to figure out what reactions can take place. In principle, you would put in an initial mixture and just turn on a network. In practice, this is much to large a computation to be carried out feasibly without approximations. The basic method we will outline here, for details you should look at some of the more recent discussions of the individual reactions.

The simplest network is a linear system. Assume that there are a set of reactions of the form[130]

$$A + a \rightarrow B + b$$
$$B + c \rightarrow D + d. \qquad (4.216)$$

The rate equations are

$$\frac{dA}{dt} = -\lambda_{ab} A$$

$$\frac{dB}{dt} = \lambda_{ab} A - \lambda_{cd} B \qquad (4.217)$$

$$\frac{dD}{dt} = \lambda_{cd} B.$$

In general, such a network can be written as

$$\frac{d}{dt}\mathbf{X} = \Lambda \mathbf{X}, \qquad (4.218)$$

where $\mathbf{X}$ are the abundances of the reacting species and Λ is the reaction rate matrix. There is a special simplification in this case because we have a system of linear equations. We can assume that any species abundance varies as $\exp \lambda t$ and the system can be solved as an eigenvalue problem for the timescales and thus for the abundances. If, however, $A(A,b)B$ is the reaction of interest, we would have to divide the rate for λ_{AA} by 2! in order to ensure that we do not double count the reaction (see discussion below) and the equations become nonlinear in the abundance of A and the system given by Eq. (4.218) would have to be solved iteratively.

When the system is in equilibrium, the abundances are static, and one says that they *freeze out*. Since the system of rate equations is linear, it can be solved in closed form. First, however, look at what you expect. Species A is only destroyed so its abundance should exponentially decrease with time while species B increases. However, since B is also being consumed to form D, the competition between the creation and destruction rates sets the maximum abundance one would expect to see and eventually the nuclear abundance is reprocessed into D. The rates as we have them here are linear in the constituent nuclei, assuming that a, b, c, and d are background particles.

For the more complicated or more extensive networks, another method that works well and fast is available although it requires some matrix inversion (not to

---

[130] In what follows, we will usually write this sequence as $A(a,b)B(c,d)D$, although for now we use the more explicit form to show how the equations are set up.

worry, there are routines available for this).[131] Call the one-body rates λ_j^n (for $j \to n$), the two-body rates $[n; j, k]$ (for $j + k \to n$), and the three-body rates $[n; j, k, l]$ (for $k + j + l \to n$). Then the abundance of the nth species X_n at a time t is

$$\frac{dX_n}{dt} = \sum_j \lambda_j^n X_j + \sum_{j \geq k} [n; j, k] X_j X_k + \sum_{j \geq k \geq l} [n; j, k, l] X_j X_k X_l, \quad (4.219)$$

which is the network for the n constituents. Now discretizing this system using $X_n = X_{n,0} + \Delta X_n$, we obtain

$$\frac{\Delta X_n}{\Delta t} = \sum_j \lambda_j^n (X_{j,0} + \Delta X_j) + \sum_{j \geq k} [n; j, k](X_{j,0} + \Delta X_j)(X_{k,0} + \Delta X_k)$$
$$+ \sum_{j \geq k \geq l} [n; j, k, l](X_{j,0} + \Delta X_j)(X_{j,0} + \Delta X_k)(X_{l,0} + \Delta X_l), \quad (4.220)$$

which on linearization (for small Δt) reduces to the matrix equation:

$$\frac{dX_n}{dt} = M_{np} \Delta X_p, \quad (4.221)$$

where we define the **M** rates by

$$M_{ij} = -\frac{1}{\Delta t} \delta_{ij} + \lambda_j^i + \sum_k [n; j, k] X_{0,k} + \sum_{k \geq i} [n; j, k, l] X_{0,k} X_{0,l} \quad (4.222)$$

or the first timestep; the procedure has an obvious generalization to subsequent times t^m using backward differencing so that $\Delta X_i = X_i(t^m + \Delta t) - X_i(t^m)$. You use the full network at the mth timestep to update the rates $\dot{X}_n$ and then find ΔX_n by matrix inversion. When three-body reactions are important, an iterative procedure is required, and this is described further in the reference.

---

[131] Delano, M. D. and Cameron, A. G. W. 1971, *Ap. Space Sci.*, **10**, 203. The procedure is fast and accurate and can be adapted to a variety of uses, including explosion calculations. The standard reference for numerical procedures is Press, W. H., Teukolsky, S. A., Vetterling, W. T., and Flannery, B. P. 2001, *Numerical Recipes in Fortran* 90 (Cambridge, UK: Cambridge Univ. Press). A version is also available in C++.

5 Structure and Evolution of Close Binary Stars

We are the twin Stars, and cannot shine in one Sphere; When he rises, I must set.
—William Congreve, *Love for Love*, Act. 3, Scene 3

5.1 INTRODUCTION

The discovery of binary stars was a triumph for Newtonian gravitation. As early as 1803, William Herschel argued on the basis of statistics that the frequency of close pairs seemed to require the existence of physical double stars, and similar arguments were presented by G. Boole.[1] The detection of orbital proper motion in ξ Gem finally clinched the argument and provided simultaneously a test of the universality of gravitation (the first non-solar system test) and a means for direct determination of stellar masses. This was an extremely hard observational task without parallaxes for the components, but the procedure worked well enough that it provided a means of determining parallax by an inverse method (knowing the spectral types of the stars, you can use the angular separation to provide the distance). At the very least, this discovery meant that it was finally possible to use comparative methods to determine ratios of stellar properties and supported the results obtained from the analysis of eclipsing binary lightcurves. Its additional importance is that the stars can be observed directly, so it doesn't require inferring their properties from photometry alone since their *ratios* are independent of distance and/or extinction.

Spectroscopic work on binary stars began with the discovery of composite spectra by Maury and Pickering during the HD project and with Hartmann's observation of orbital motion through Doppler shifts for ζ UMa. The breakthrough came, however, with the discovery of periodic shifts in the line spectrum of the photometric variable star β Per by Vogel and Pickering and for the long-period variable ζ Aur by Maury, who also produced a binary model to explain the spectral peculiarities of β Lyr. These observations of Algol and β Lyr confirmed the eighteenth-century eclipse explanation for the lightcurves by Goodricke and Pigott and opened the way for the direct determination of stellar masses that we

---

[1] Herschel, W. 1803, *Phil Trans. Roy. Soc.*, **93**, 339. See Boole, G. 1851, *Phil. Mag.*, Ser 4, **1**, which is reprinted in Boole, G. 1952, *Studies in Logic and Probability* (London: Watts). The paper, entitled "On the Theory of Probabilities, and in Particular on Mitchell's Problem of the Distribution of the Fixed Stars," bears a striking resemblance to later, twentieth-century, arguments adduced by Neyman and Scott to support evidence for galaxy clustering and by Page for binary galaxies and heralds the era of statistical astronomy that was so fruitful for galactic and large scale structure work.

discussed in Chapter 2 by combining lightcurves and radial velocities. It also demonstrated that the light variations were produced by transits of a low-luminosity companion since only one spectrum was detected. It is, therefore, somewhat odd that the work on evolution of binary systems should be so comparatively late in the study of these systems. Although in the 1940s Kuiper emphasized the critical role played by the Roche limit in the structure of close binary stars, and tidal interactions had even been suggested early in the century as a model for the origin of the solar system, the effect of mass transfer and mass loss were not explored until nearly 20 years later. One reason may be the comparatively slow development of stellar modeling codes and the difficulty of calculating models for single stars until the mid-1960s. The standard methods require simple boundary conditions and, as we will see, these do not apply for stars in close binary systems. Expansion of numerical methods and improvements in computing facilities also made major advances possible, and by 1967 work had begun in earnest on the problem.

Our decision to treat binary stars as a separate chapter is due to the enormous variety of processes that they exhibit, which span the different topics covered in this book. Many general processes are better studied in close binaries than anywhere else in the Universe. We have been repeatedly emphasizing that the problems encountered in astrophysics are often qualitatively the same from one arena to another, differing only in scales of length, mass, and time. The great advantage you have in studying stars is that you have much more information about the basic environment in which the phenomena occur. In the end, for instance, there isn't much difference between the accretion of mass onto a compact body in a close binary system and the infall of matter into a galactic scale black hole in an active galactic nucleus.

5.2 ECLIPSES AND THEIR USES

Eclipses and orbits for close binaries provide the main data for calibrating stellar masses, radii, and luminosities. In light of the discussion in Chapter 3, however, there is another feature of these systems that provides fundamental information about stellar structure. If the two stars are well separated so there is no dynamical interaction beyond orbital motion, and if there is little or no radiative interaction, then observing an eclipse provides information about the structure of the stellar atmospheres. Since the atmospheric temperature gradient produces limb darkening for a single star, it must have observable consequences for an eclipse. This means that stars depart from the billiard ball interpretation that we used to derive the basics of geometric eclipses. On ingress or egress, the lightcurve softens at the contact points because of the variation if surface brightness of the components. We had previously assumed that $I(\theta, \phi)$ was constant. If instead we take the Eddington–Barbier approximation that $I(\theta, \phi) \rightarrow I(\mu) = I(0)(1 - u + u\mu)$, where μ and u are the directional cosine and limb darkening coefficient, respectively, then the first contact point produces a smaller relative change in the light of the combined system at the start of the eclipse. Therefore, the effective radius obtained for the star will be slightly smaller, and the estimate of the surface gravity slightly higher, than the geometric model.

In addition, if the larger object (which is often neither the more massive star in the system nor the brighter component) has an extended atmosphere, then the eclipse can be used to determine the structure of the optically thin parts of the envelope. The method is the same as used in stellar occultations by planets, for instance, Jupiter, or for transits of Venus and Mercury across the solar disk. For a star seen through the atmosphere of a Jovian planet, the relative distances from the observer ensure that the eclipse samples the atmosphere with a pencil beam. Therefore, you can use $I(t) = I_0 \exp - \tau(t)/\mu$ (t), where both optical depth and directional cosine are functions of time and the envelope structure is determined by the inversion of the lightcurve through a model atmosphere. For a model with constant opacity, the simplest approximation, the lightcurve yields the optical depth variation into a density scale height, H_ρ, which converts to the pressure scale height H_P through the equation of state. With this transformation, it is possible to obtain the mean molecular weight provided you know the temperature gradient by spectroscopy. Measurements of Jupiter performed in situ by the Galileo probe confirmed the method. In particular, these data obtained the He/H ratio determined from stellar occultations. Of course, there are limitations of this remote determination, due mainly to uncertainties in the equation of state, but for stars this is less of a problem because of the comparative simplicity of the atmospheres relative to planets, where multiple phases are present (there are probably no clouds in stellar atmospheres, for instance). A recent development links the solar system to extrasolar planets with the discovery of observable transits of the Jovian mass companion in HD 209458. This single observation establishes both the mass of the companion, 0.69 ± 0.05 M_J, and also its radius, about $1.5 R_J$, both in Jovian units. It also opens the possibility for seeing absorption lines from the outer planetary atmosphere against the visible star's G0 V photosphere. And because the planet subtends such a small portion of the stellar disk, a bright point at mid-eclipse may occur and probe the transparency of the atmosphere (this is seen in transits in the solar system, for example, caused by focusing in an atmosphere with a depth-dependent index of refraction).

One class of binaries, the ζ Aurigae stars, has been extensively studied in this way. These are long-period (several years) binaries consisting of a red giant[2] and an A- or B-type companion, usually a main-sequence star. Here the eclipse is often not total, but the extended atmosphere of the giant still produces observable effects in the B-star spectrum when the impact parameter is large. The absorption is mainly due to lines, predominantly of the iron peak (see Fig. 5.1).[3] If the early-type component is cool enough, and its radiation diluted sufficiently by the orbital separation, the passage of the radiation through the red giant is strictly passive—there is no change in the ionization along the line of sight as a result of the illumination. Therefore, you obtain a column density of the absorber as a function of impact parameter, as shown in Figure 5.1. In this way, relatively simple transfer theory can be applied because the atmospheric structure is determined by the radiative equilibrium from the photosphere of the giant and not by the companion, and abundances can be obtained by an analog of the curve-of-growth method, precisely in the same way that you would observe an interstellar cloud

---

[2] The notable freak among these stars is ϵ Aur, for which there are no spectroscopic detections of the cool component that is now thought to be surrounded by a disk.

[3] We used the ζ Aur system 22 Vul for the CORAVEL model in Chapter 2, now you see why there might be problems with the interpretation of line shifts. In complex systems with extended atmospheres or complicated circumstellar environments, nonorbital motions complicate the interpretation of radial velocity curves.

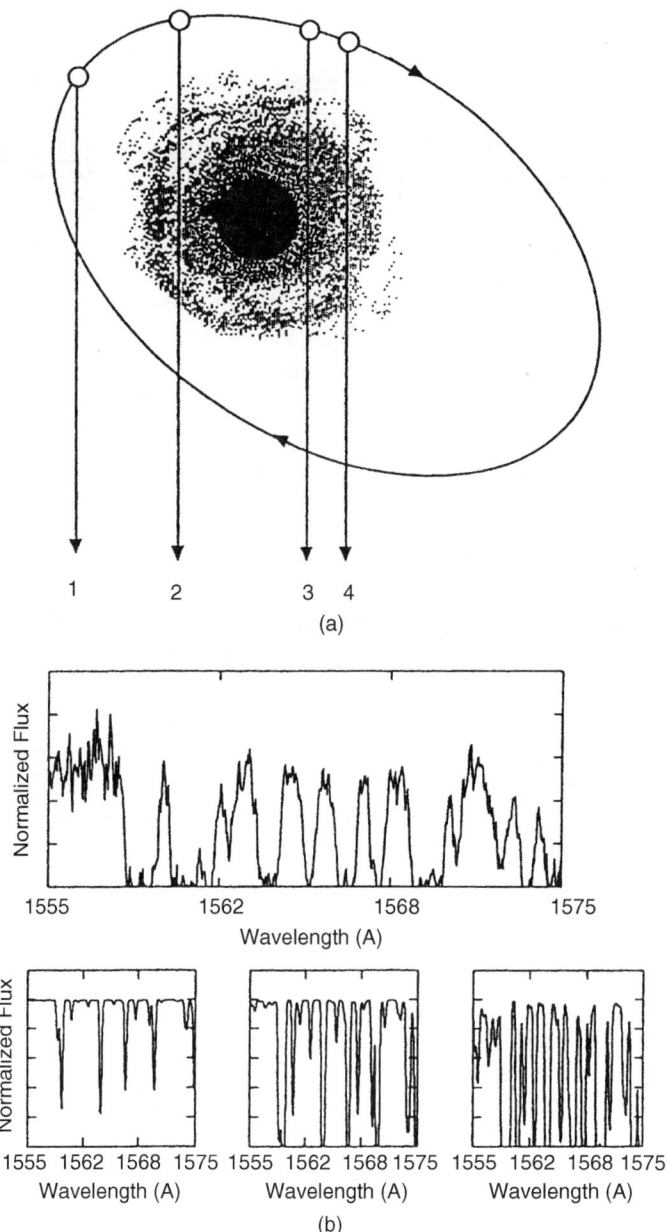

FIGURE 5.1. Development of the absorption spectrum during an eclipse for a ζ Aur system. The model spectra correspond roughly to the lines of sight marked 1, 2, and 3 (respectively) and the observed (top) spectrum is roughly at 3.

(Chapter 6) or the Ly α forest (Chapter 8).[4] We should also mention that the continuum version of this process has been observed at radio frequencies. The binary system PSR 1957 + 20 contains a millisecond pulsar and a low-mass companion and shows extended atmosphere effects for a different reason. Heating of the secondary atmosphere by the pulsar evaporates the atmosphere and produces an extended environment that attenuates the pulsar signal. Similar methods can be used in polarization for the study of circumstellar environments such as accretion disks.

A purely geometric property of eclipses can be used to study stellar rotation using the *Rossiter–McLaughlin effect*. This is a shift in the mean wavelength and change in the line profile of the visible star due to the differential occultation of its surface by a companion. You can easily see how this works. Assume the two stars have their spin angular momenta aligned with the orbital angular momentum. As the eclipse proceeds, during which time there is essentially no change in the radial velocity of either component, the approaching and then the receding portions of the visible hemisphere are obscured, producing a centroid wavelength shift in line in the opposite sense. Normally, we are able to see only the line formed over the whole stellar surface and must be content with measuring only the projection of the equatorial velocity, $v_{eq} \sin i$. The eclipse provides strict limits on i, the axis inclination to the line of sight. By using lines formed at different optical depths, and recalling the discussion of limb darkening from Chapter 3, it is even possible in principle to obtain information about the differential rotation of the surface. Again, this technique can be applied to analysis of *any* orderly motion around either star, and accretion disk mapping makes use of this as a straightforward extension of Doppler imaging.

Finally, besides providing information about fundamental structural parameters, the fact that the stars are formed at the same time means that well-separated systems provide a mini-isochrone for testing stellar evolution models. Two problems encountered in studying star clusters are avoided, that the individual stars may themselves be unresolved binaries and that any two may not have formed at precisely the same moment. We will return to this point shortly.

5.3 EFFECT OF PROXIMITY: TIDES AND THE ROCHE SURFACE

We began the discussion of stellar interiors by asserting that stars don't have walls, but now this remark needs to be qualified—although *isolated* stars don't have walls, when in a binary system they can behave as if they do if they have a close enough companion. We need to be precise now about what *close enough* means.

Let us start with an observational puzzle. You know now that the more massive a star is, the faster it evolves. You would then expect that in a binary, where the stars were presumably formed simultaneously, the more massive star in the system would be the more evolved. Ah, but this is not so for β Persei, perhaps the best studied eclipsing binary known. The system consists of a G giant and a B8 main

---

[4]The column densities are comparable to, or larger than, the *damped* Ly α systems, so this provides a useful introduction to the technique and has the advantage of time dependence, providing the exact geometry of the eclipse and the sampling line of sight.

sequence star, yet the mass of the optically more luminous B star, is normal while the giant is only 0.8 $M_\odot$. Although this was known observationally for some time, it was called the *Algol paradox* only after mature results began to emerge from stellar interior calculations. If you don't have a model for stellar evolution, or if you think that the HR diagram is a cooling sequence, there is no problem; the more extended star could easily be lower-mass and actually *less* evolved. But once you know the role of nuclear energy generation in stellar structure and evolution, Algol and systems like it *become* strange. The explanation for this evolutionary state of affairs demonstrates the effect of tidal limits to how large any star can become in a closed system.

5.3.1 The Roche Surface

The key element in binary star evolution is the role of the Roche limit, which was first introduced when we studied the three-body problem. It is known in the celestial mechanics literature as a *zero-velocity surface*, but we will treat it as the bounding equipotential for a self-gravitating star:

$$\Phi(r) = -\frac{GM_1}{r_1} - \frac{GM_2}{r_2} - \frac{1}{2}\omega^2\left[(x - \mu a)^2 + y^2\right], \tag{5.1}$$

where $r_1^2 = x^2 + y^2$ and $r_2^2 = (x - a)^2 + y^2$ for a circular orbit with the stellar separation a and reduced mass μ. This is the potential in the corotating frame with frequency ω. There are five equilibrium points ($\nabla\Phi = 0$), three of which are along the line of centers. Two are peripheral to the masses and lie as the critical points along equipotentials that envelope both stars. These are saddle points for which particle trajectories are unconditionally unstable. Two other points, Lagrangian points L_4 and L_5, lie diametrically opposite each other perpendicular to the line of centers. These are quasiequilibrium points for which local orbits are possible because of the Coriolis acceleration. The Roche lobe is the equipotential that passes through L_1, also called the *inner* Lagrangian point, that lies between the masses along the line of centers (see Fig. 5.2). Mass loss is inevitable for the star that is contacting its Roche surface. As a star's radius approaches the Roche surface, the body becomes more distorted and eventually, when it contacts L_1, the inner Lagrangian point, it loses mass to the companion and possibly from the system. This actually occurs before the photosphere reaches R_{RL} in the absence of magnetic fields or other constraints on the flow.

Several analytic approximations have been derived for the radius of the equivalent sphere whose volume equals that of the appropriate lobe. In general, the Roche radius depends on the mass of the components and q through

$$R_{RL} = f(q)a, \tag{5.2}$$

where $f(q)$ is provided by functional approximate fits to the exact calculations.[5]

---

[5]See, for instance, Plavec, M. and Kratochvil, P. 1964, *BAC*, **15**, 165; Paczynski, B. 1971, 1979, *ARAA*, **9**, 183; Eggleton, R. P. 1983, *ApJ*, **268**, 368; Mochnaski, S. W. 1984, *ApJS*, **55**, 551 (see also 1985, ibid., **59**, 445).

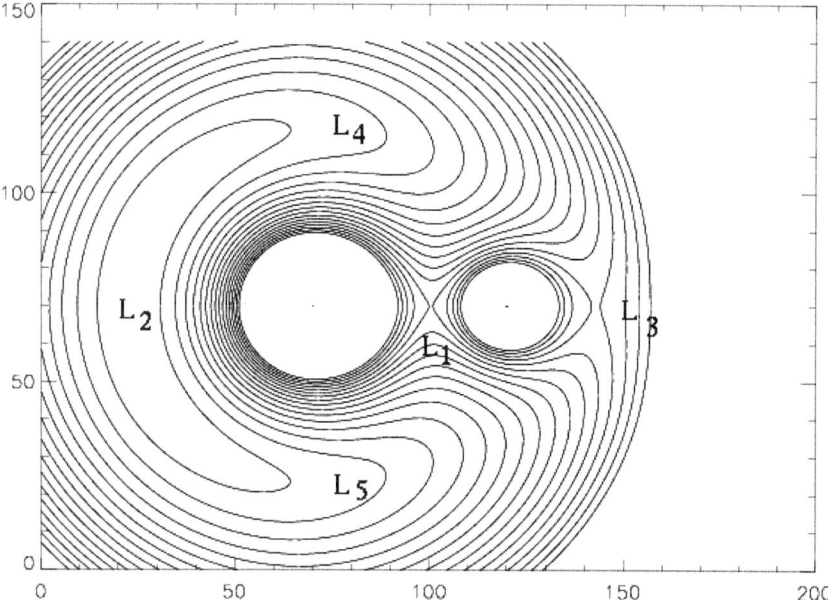

FIGURE 5.2. The Roche surface for $q = 0.3$. The Lagrangian points are indicated.

Two convenient, though somewhat restricted, approximations are

$$f(q) = 0.2 + 0.38 \log q, \qquad (5.3)$$

which holds for $0.5 \leq q \leq 20$, and

$$f(q) = 0.462 \left(\frac{q}{1+q} \right)^{1/3} \qquad (5.4)$$

for $0 \leq q \leq 0.5$. The most general result is

$$f(q) = \frac{0.49 q^{2/3}}{0.6 q^{2/3} + \ln(1 + q^{1/3})}, \qquad (5.5)$$

which is due to Eggleton.

Binary systems are theoretically distinguished relative to the Roche surface,[6] although, as the Bard says, one star in its lifetime plays many parts. Stars that are in mutual contact with, or overflowing, their critical equipotential surfaces are called *common envelope* or *contact* systems. When only one star's radius equals its Roche radius, the system is called *semidetached*. Finally, if both stars are separate, however much they may be distorted, the binary is *detached*. Geometrical methods

---

[6] Kuiper, G. 1941, *ApJ*, **93**, 133; Kopal, Z. 1959, *Close Binary Stars* (NY: Wiley).

for treating lightcurves have been developed based on this idea[7] each treating the shape of the star and the distribution of temperature over the surface more or less phenomenologically (by scaling model atmospheres to the local conditions on tidal ellipsoids or Roche surfaces and piecing together the surface). Such models provide fundamental parameters for the stars, but they cannot deal with the evolution or the interior structure. The complication introduced by tidal distortion is similar to, but more extreme than, the rotating case, where the *local* effective temperature varies over a nonspherical star because of the simultaneous requirements of radiative and hydrostatic equilibrium—von Zeipel's theorem—and that the flux, F, is given by the local plane parallel approximation, $F \sim g^{1/4}$, where $g = \mathbf{g} \cdot \hat{\mathbf{n}}$ is the local surface gravity for the normal $\hat{\mathbf{n}}$. Since $\mathbf{g} \sim \nabla \Phi$, where Φ is the total potential, including the effects of rotation and/or tidal distortion, the gravity will be lower or higher depending on the colatitude. Although rotational distortion is axisymmetric, tides are not, and therefore g varies with azimuth as well. Even for aligned angular momenta, there will be a phase dependence in the observable properties of the stars. This distortion affects the spectral characteristics of the star, which now cannot be fit into a single taxonomic category. It also alters the surface brightness as a function of latitude and, through such modifications, changes the limb darkening and the eclipse characteristics. The tidal distortion is also important for the internal structure of the stars and must be handled in a more detailed way than the axisymmetric case, but similar problems arise nonetheless.

5.3.2 Tidal Interactions: Circularization and Synchronization

In the Earth–Moon system, the Moon rotates with the same period as its orbit—that is, synchronously—while the Earth rotates more rapidly. Any point on Earth must then experience a periodic tidal acceleration since, in the rotating frame, any locale on the planet is carried through the alternating extrema of the perturbing force. This slowly alters the lunar orbit. The solid Earth is not distorted sufficiently to dissipate its rotational energy; hence the rotation only slowly approaches the lunar orbital period through tidal friction. The physical mechanism must be something like this. When we examined tidal interactions between extended masses in Chapter 1, we ignored the effects of the distortion on the interior structure of the bodies. Stars are (essentially) fluid and fluids yield to shear. Hence, the induced tidal distortion produces flows because the gravitational potential develops gradients along equipotential surfaces: $\nabla \Phi$ has components in the θ and ϕ directions, and these lead to flows. One way to say this is that surfaces of constant density and pressure are no longer parallel, $\nabla \rho \times \nabla P \neq 0$, so flows will be induced on the thermal timescale. In this regard, the setup is the same as we used for the von Zeipel problem in rotating stars, but the source of the acceleration is different because it can be isothermal. These flows transport momentum in the rotating frame, leading toward solid-body rotation and synchronism. Thus the

---

[7]See Wilson, R. E. 1994, *PASP*, **106**, 921 and Kallrath, J. and Milone, E. F. 1999, *Eclipsing Binary Stars* (NY: Springer-Verlag) for general reviews of lightcurve models and Binnendijk, L. 1977, *Vistas in Astr.*, **21**, 359. King, I. 2000, *ApJ*, **525** (IC), 61, the AAS Centennial Issue, provides an introduction to the Russell method (ellipsoids) for approximating eclipsing binary stars and reprints of the founding papers.

tidal perturbation consists of both the static response of the body, the deformation that leads to the Roche surface, and a dynamical wave driven by the relative motion of the perturber and the primary.

If the stars are not synchronously rotating, the tidal perturbation moves relative to the angular momentum. For the Earth–Moon system, the rotation causes the distortion to lead the position of the Moon by viscous drag of the fluid on the Earth. For stars, we can expect the same effect. There are two approximations to this condition. One is an equilibrium tidal surface, which is instantaneously in hydrostatic equilibrium everywhere in the star. The baroclinicity of the surfaces, as in the rotationally distorted case, induces slow circulation that ultimately redistributes angular momentum as well as energy. This produces circularization by loss of orbital angular momentum through viscosity, and is most efficient for stars with deep convective envelopes because the turbulence acts like an effective viscosity.[8] But the problem is, unfortunately, that the angular momentum transport depends on the internal viscosity, the precise nature of which is currently not known. Even for our planet, whose interior is known remarkably well from *geo*seismology and direct sampling, the coupling between the oceans and continents is not precisely understood. Nonetheless, we can identify the important quantities. First is the ratio of the tidal to self-gravitational force, the parameter that governs the Roche limit,

$$\lambda_T = \frac{M_2}{M_1}\left(\frac{R}{a}\right)^3,$$

where M_2 is the perturber's mass. The total energy of the binary is $E = \frac{1}{2}I\Omega^2 + \Phi + \frac{1}{2}\mu a^2\omega^2 = \frac{1}{2}I\Omega^2 - \frac{1}{2}\mu a^2\omega^2$ since the system is bound and we are neglecting the rotational kinetic energy of the secondary. The total angular momentum is $J = I\Omega + \mu a^2\omega$, and we can relate ω and a through the third law. Then for a dissipative system we have (see Appendix 5.A) $\dot{E} \sim \nu v_c^2/R^2$, where v_c is the velocity induced from the tidal distortion, analogous to the circulation current set up in a rotationally distorted star. From the energy variation, if J = constant, we find that

$$\frac{dE}{dt} = -(\omega - \Omega)I\dot{\Omega}, \qquad (5.6)$$

substituting for $\dot{\omega}$ from the angular momentum constraint. This leads to the scaling relation for the synchronism time:

$$t_s^{-1} \equiv \frac{1}{(\Omega - \omega)}\frac{d\Omega}{dt} \sim \nu q^2 \lambda_T^2 \sim \nu k_2 q^2\left(\frac{R}{a}\right)^6. \qquad (5.7)$$

For circularization, the orbital moment of inertia is needed instead of I, which

---

[8]Zahn, J.-P. 1977, *Astr. Ap*, **57**, 383; Tassoul, M. 2000, *Stellar Rotation* (Cambridge, UK: Cambridge Univ. Press); Zahn, J.-P. 1989, *Astr. Ap.*, **220**, 112; Rieutord, M. and Zahn, J.-P. 1997, *ApJ*, **474**, 760; Rocca, A. 1989, *Astr. Ap.*, **213**, 114.

contributes another factor of $(R/a)^2$:

$$t_c^{-1} \equiv -\frac{de/dt}{e} \sim k_2 q(1+q)\left(\frac{R}{a}\right)^8. \tag{5.8}$$

In both cases, the constant k_2 is the *apsidal motion constant*, which refers to the relative central concentration of the star's mass; you saw how to compute this for polytropes in the last chapter. The exact timescales depend on the details of the stellar models and change as the star evolves. Also, the viscous coupling varies depending on the fraction of the interior that is radiative since turbulent convection acts more efficiently than circulation currents to transport angular momentum.[9] Binaries are now known spanning the entire age range from 1 Myr to about 4 Gyr, especially in clusters. In principle, by using isochrones for the two components, having absolute dimensions from eclipses and astrometry, you can even sidestep the requirement of cluster membership for such tests. In general, synchronism is the rule for relatively short period systems, of order a few weeks or less, and circular orbits are also typical of short-period systems.

In cool star binaries, those with deep convective envelopes, synchronization appears to dramatically enhance the signatures of dynamo related activity. In several classes, particularly the RS CVn stars, all the chromospheric and coronal signature—X-ray emission, flaring, enhanced Ca II H and K emission, and the like—are increased compared with single stars of the same mass and evolutionary status. The more massive component is usually a subgiant G or K star and would be expected to have a rotation period of months, while in close systems, less than a few weeks, the star is actually spun up compared with its normal rate. As you saw in the last chapter, a dynamo depends on differential rotation, the stellar rotation frequency, and its surface gravity. So the evidence that the RS CVn binaries also show spots and long-term cycles (decades) in a wave of photometric distortion inspires confidence in the dynamo explanation and tidal interaction as the efficient cause. Similar effects are seen in the dMe stars and W UMa systems, an interesting point since the latter are common envelope systems.

5.4 EVOLUTION OF STARS IN CLOSE BINARIES

Our earlier treatment of the stages of evolution presumed that there are no limits on the star's radius. The Roche surface changes this, so you would expect that it also changes the star's history. The precise effect depends on what moment this limit is reached. We first need a bit of nomenclature. On reaching the Roche surface, the stellar envelope is presumed to become unbound and that mass loss or

---

[9] The most complete exploration of the evolutionary effects, see Claret, A. et al. 1995, *Astr. Ap.*, **299**, 724; Claret, A. and Cunha, N. C. S. 1997, *Astr. Ap.*, **318**, 187; Claret, A. 1999, *Astr. Ap.*, **350**, 56. For reference, the circularization timescales are about 10^8 yr for a ZAMS 5 $M_\odot$ + 5 $M_\odot$ system with periods greater than about 2 days, shorter for 10 $M_\odot$ and as short as 3×10^8 yr up to about 10^{10} yr for 1 $M_\odot$ + 1 $M_\odot$ systems for the same range. For the details, you should consult the references.

transfer is initiated on a hydrodynamic timescale. The star is generically referred to as the *loser* or *donor*, terms usually applied in the case of mass transfer between the binary components. The companion is called the *gainer*, which implies some amount of accretion. The alternative, to call one star the *primary* and the other the *secondary*, is based on the relative contributions to the combined spectroscopic and/or photometric properties and does not capture the physical nature of the interaction. For detached systems on the main sequence these terms also describe the respective masses but that correspondence breaks down once the stars begin their ascent of the HR diagram.

The taxonomic distinctions for the different evolutionary cases are based on the stage at which the nomenclature applies. *Case A* occurs before the terminal main sequence, when the loser is still undergoing core hydrogen burning. This is a slow nuclear stage, as you know, and not very sensitive to the stellar mass. For mass transfer to occur requires very small orbital separation because of the small stellar radii, and the timescale for stellar expansion is very slow. *Case B* occurs after hydrogen core exhaustion and during a relatively rapid stage of radial expansion, either in the traverse across the Hertzsprung gap or on first ascent of the giant branch but before helium core ignition. *Case C* is a late stage, when the star has developed a helium core and is on or near the giant branch or AGB. The designations are originally due to Kippenhahn and Weigert (1967).[10]

In *conservative* mass transfer, the process can be studied in a straightforward way because we neglect any net mass or angular momentum losses. The Roche surface would recede into the loser were it not for the increase in the separation between the components that results from the change in the mass ratio. The loser maintains contact with R_{RL} until contact is broken by the expansion of the system and thereafter the more evolved star continues as if it were a single star. In the event of mass loss from the gainer, the process of mass transfer may be reinitiated at some later stage, but this is unlikely. Dropping these assumptions of constant binary mass and total angular momentum makes the scenario more realistic but leads to a dizzying range of phenomenological models, each marked by the adoption of specific mechanisms for breaking constancy of one or both of these quantities.

5.4.1 Common Envelope Evolution

Since the Roche surface represents the limit of a set of bounding equipotentials, it is a surface along which flows can occur but that a star can maintain as an

---

[10] With the expansion of evolutionary codes in the mid-1960s, a number of groups investigated the process of mass transfer in close binary systems and formed the basic literature within a very brief time. These groups were in Warsaw (Paczynski), Munich (Kippenhahn, Weigert, and Gionanne), and Prague (Plavec, Harmanec, and Križ). The first conference on the subject was held in Copenhagen in 1968 and the scenarios have since formed the basis of all discussions of mass transfer and its effects on stellar evolution in close binaries. See Kippenhahn, R. and Weigert, A. 1967, *Z.Ap.*, **65**, 251; Paczynski 1971, *ARAA*, **9**, 183; Batten, A. H. 1973, *Binary and Multiple Systems of Stars* (Oxford: Pergamon); Thomas, H.-C. 1977, *ARAA*, **15**, 127; Plavec, M. 1970, *PASP*, **82**, 957; de Loore, C. W. H. and Doom, C. 1992, *Structure and Evolution of Single and Binary Stars* (Dordrecht: Kluwer); Shore, S. N., Livio, M., and van den Heuvel, E. P. 1994, *Interacting Binaries: 22nd Saas-Fee Advanced Astrophysics Course* Nussbaumer, H. and Orr, A., eds.. (Berlin: Springer-Verlag); Batten, A. H. 1995, *Rep. Prog. Phys.*, **58**, 885.

equilibrium shape. If the radius of the outer layers exceeds R_{RL}, matter does not have to catastrophically flow toward the companion. Instead, if both stars are in contact with L_1, a low-speed circulation can, in principle, be established because of the pressure reaction from the companion's outer layers. The resulting optically thick common envelope should, for contact systems, behave as if the layer sits on top of a very strange equipotential surface, one where the surface gravity depends on both colatitude and colongitude.[11] There is an outer bounding surface through which some mass is certainly lost, through the L_2 and L_3 points, but this is small compared with the flow that must pass through L_1 to maintain thermal balance. This is the observed situation in the W UMa stars. Marked by continuous light variations, these stars appear to be surrounded by a common atmosphere, but they have otherwise stable cores that place them on the main sequence. This is where the paradox arises. The mass–radius relation for main sequence stars is approximately $R \sim M^{1/2}$, while a star in contact with its Roche surface will satisfy the approximate relation $R \sim M^{1/3}$. In a common envelope system, the two stars will need to simultaneously both conditions. Their equilibrium radii give by the nuclear burning conditions must be the same as their respective Roche surfaces, $(M_1/M_2)^{1/2} = (M_1/M_2)^{1/3}$ (within a correction factor that depends on q) so the stars should have equal masses and, on the main sequence, they should also have equal luminosities.[12]

This is precisely what is *not* seen. The envelopes have uniform temperature, the $B-V$ variation through an orbit is vanishingly small, but their luminosities and masses differ by up to a factor of 10. Such a large difference in the luminosity implies that heat transport must be occurring in the common envelope but that the overall interior structure of the star is otherwise unaffected. To make matters worse, the coolest W UMa stars must have common *convective* envelopes, where the temperature gradient adjusts to the large variation in the surface gravity due to the angular gradients in the equipotential. How this happens has been a controversial question for decades[13] and remains an important unsolved problem in stellar hydrodynamics. The consensus presently is that the envelopes are never precisely in thermal equilibrium and that the mass transfer fluctuates between the components. Much of the mass transfers through the L_1 point, which also funnels the bulk of the energy from one star to another. Why this is not observed to be a hot spot in the systems remains a vexing question. One way to picture this is that the mass transfer rate exceeds the thermal timescale for the entire envelope that drives thermal oscillations that periodically overfill the Roche lobe of the gainer, or rather the star that is accreting at that moment.

Main sequence contact systems are only one example of evolutionary stages where common envelopes occur. Any circumstellar matter that completely engulfs the companion and is optically thick is, in effect, a common envelope. If the accreting material completely envelopes the gainer such that it is optically thick in

---

[11] For a general overview of common envelope systems, see Paczynski, B. 1976, in *IAU Symposium 73: Structure and Evolution of Close Binary Systems*, Eggleton, P. P., Mitton, S., and Whelan, J. eds. (Dordrecht: Reidel), p. 75; de Kool, M. 1990, *ApJ*, **358**, 189; Iben, I. Jr. and Livio, M. 1993, *PASP*, **105**, 1373; Taam, R. and Sandquist, E. L. 2000, *ARAA*, **38**, 113.

[12] Kuiper, G. P. 1941, *ApJ*, **93**, 133; Lucy, L. B. 1968, *ApJ*, **151**, 1123.

[13] See, for instance, Lucy, L. B. 1976, *ApJ*, **205**, 208; Kähler, H. 1989, *Astr. Ap.*, **209**, 67; Hazelhurst, J. 1996, *Astr. Ap.*, **313**, 487.

all directions when the gas is confined within the Roche lobe of the gainer, it becomes almost a matter of semantics whether to refer to the environment as a common envelope or a very thick accretion disk. For instance, consider for a moment the evolution of a short period pre-main sequence binary system. The separation is comparable to the stellar radius at the birthline for periods of order one week for stars more massive than about 1 $M_\odot$. This assumes that the evolution can be treated in the same way we described in Chapter 4 for isolated protostars. How this affects the final mass of the system and the mass ratio of the components is an open question. There are an increasing number of young binaries being found from speckle interferometry of star forming regions, especially in the Orion association, but these are wide systems that are unaffected by mass transfer during contraction. On the other hand, some pre-main-sequence binaries, in particular TY CrA and BM Ori, are known among the Herbig Ae/Be stars with these short periods and their structure *may* differ from those of isolated stars.

Cataclysmic binaries, to which we will return later in this chapter, are extremely close systems with periods of less than 1 day and at least one degenerate component. Their origin is linked in current hypotheses to a relatively late stage in the evolution of one of the more massive components, since there are no main sequence progenitors with these characteristics. There are two obvious ways to form a white dwarf. One is to invoke magic and drive the envelope off during planetary nebula formation in the post-AGB phase, as we discussed in the previous chapter. The other occurs in a common envelope. If one star engulfs the other as it evolves—and this can happen for virtually any system with orbital periods of less than a few weeks on the main sequence—the lower-mass companion will find itself orbiting within a dense circumstellar environment produced by the more massive star. Differential motion leads to heating, which scales as $v_{\rm orb}^3 a^{-1} \sim a^{-5/2}$, *and* transfer of angular momentum between the lower-mass component and the engulfing envelope. Consequently, if the heating is sufficient, the more massive star is stripped of its envelope with the resulting loss of binding energy, and the companion spirals inward. The result is a white dwarf with a much less evolved companion in a very-short-period system.

A final example of a short-lived but important common envelope stage is found during nova explosions. We discuss this in more detail in Section 5.6, but here we should mention that the white dwarf on which the outburst has occurred expands to the point where it may envelope the companion, forming a thick environment in which the companion star orbits. Another, more benign, version of common envelope evolution is observed in the symbiotic stars. The companion, usually a white dwarf or post-AGB object, orbits within the wind of the red giant, but there is no strong dynamical interaction that ultimately affects the evolution of the binary. For novae, and close binaries in late stages of mass transfer, the situation is more extreme. This common envelope stage is required in almost all evolutionary scenarios for the formation of cataclysmic binaries.[14] Viscous drag and the subsequent shear heating are then thought to be responsible for the ejection of the envelope.

---

[14] See review by Livio, M. 1994, in Shore, S. N. et al. 1994, *op. cit.*.

5.4.2 Angular Momentum Considerations and the Roche Surface

Now let's look at what happens when the masses of the stars are changed. Using Kepler's law, $GM = \omega^2 a^3$, where $M = M_1 + M_2$, we find that the orbital frequency varies as

$$2\frac{\delta\omega}{\omega} + 3\frac{\delta a}{a} = \frac{\delta M}{M}. \tag{5.9}$$

Therefore, if the mass transfer is conservative, we find that $\delta\omega = -\frac{3}{2}\delta a/a$. For a circular orbit, you know that the angular momentum of the system is

$$J = \mu a^2 \omega = G^{2/3} M^{-1/3} M_1 M_2 \omega^{-1/3}, \tag{5.10}$$

where $\mu = M_1 M_2/M$ is the reduced mass of the binary. From Eq. (5.9) you see that even if M is constant, the mass transfer changes μ, thereby changing the angular momentum of the components. What happens to the orbital period, an easily observed property of the binary, if the angular momentum of the system varies? It simplifies the calculation to take the variation of the logarithm of Eq. (5.10):

$$\frac{\delta J}{J} = \frac{\delta M_1}{M_1} + \frac{\delta M_2}{M_2} - \frac{1}{3}\frac{\delta M}{M} - \frac{1}{3}\frac{\delta\omega}{\omega}. \tag{5.11}$$

In conservative mass transfer, $\dot{M} = 0$, so $\dot{M}_2 = -\dot{M}_1$. Using $\delta P/P = -\delta\omega/\omega$, we obtain

$$\frac{\dot{P}}{P} = 3\left(\frac{1-q}{q}\right)\frac{\dot{M}_1}{M_1} \tag{5.12}$$

for the period change, written in terms of the mass ratio:

$$q \equiv \frac{M_2}{M_1}.$$

You see that the period decreases if M_1 is initially more massive, while it increases if the opposite is true. This change in the period can be directly measured for many close binaries. We here assume, however, that $\dot{J} = 0$, so it is clear that this is not a unique explanation for any observed change. But there is another, more fundamental, consequence of the mass transfer, even in the restrictive conservative case. A variation in the separation of the components, as implied by the period change, moves the "wall." The Roche surface is related to *each* star, so the competition between dynamical processes in the binary and the evolutionary timescales for the components alters the outcomes of stellar evolution compared with the isolated case.

For simplicity we will in what follows use Eq. (5.4) for the approximate form of the Roche radius. Then, if the angular momentum and mass ratio can change but M is kept constant, R_{RL} is given by:

$$\left(\frac{\delta R}{R}\right)_{RL} = -\left[\frac{5-2q}{3(1+q)}\right]\frac{\delta M_2}{M_2} - 2\frac{\delta J}{J} \tag{5.13}$$

An example of this scenario is the torquing of the star by formation of an accretion disk during conservative mass transfer. Since we assume that $\delta M_1 < 0$, the Roche lobe recedes from the surface of the mass losing component over time, unless variations in the angular momentum intervene. In other words, the mass transfer eventually stops for the system and the components evolve as essentially single stars thereafter. The evolutionary tracks for such stars are extremely interesting. The initially more massive star, called hereafter the *loser*, expands to reach its Roche surface and then evolves at constant Roche radius for some time, sliding roughly parallel to the main sequence while the gainer moves up in mass. The regulation is provided by the evolutionary expansion of the loser, resulting in the timescale for mass transfer at this stage equating approximately the thermal timescale for the stellar envelope. This is the basis for the quasistatic calculations of the three cases of mass transfer. If the expansion is on a nuclear timescale, and the loser does not have a deep convective envelope, the envelope is gradually stripped and the loser shrinks inside its Roche lobe faster than the lobe contracts. On the other hand, for case C, a convective envelope expands when the stellar mass is decreased; hence the star maintains contact with its Roche surface. In addition, in the more evolved stages of evolution, the core is hotter and has a molecular weight discontinuity at the core boundary. Therefore, once the envelope is removed, the core evolves toward the Wolf–Rayet stage and the mass loss takes place on the thermal timescale of the envelope.

Successively deeper layers of the mass loser should be exposed as mass transfer proceeds. Nucleosynthesis products should mix into the matter that flows onto the companion, whose mass is steadily increasing in this conservative picture, and the chemical composition of the gainer is altered. Signatures of this process are seen, for instance, in the changes in the CNO abundances in Algol systems, where the transfer occurs from a giant onto a main sequence star. For later stages of mass transfer, once the loser has reached the giant or supergiant stage, neutron-processed material may be transferred. For instance, the barium stars show evidence of matter that has been exposed to *s*-processing having been transferred to the companion. In fact, the dating of mass transfer events is even possible in some systems, those that are *s*-process-rich, since the absence of ^{99}Tc means the mass exchange likely took place long enough ago that the isotope has decayed.

Mass accretion by the gainer cannot take place arbitrarily rapidly. While the loser may dump mass onto its companion on a thermal timescale, the rate at which the gainer can assimilate the material quasi-statically depends on its thermal timescale. If the rate of mass transfer exceeds this, the envelope is forced to expand rapidly to accommodate to the additional heating, possibly leading to a common envelope configuration. Recall that the thermal timescale for the envelope is $t_{\rm th} \sim GM\Delta M/R$ for a parcel ΔM added to the surface. The accretion luminosity is $L_{\rm acc} = GM\dot{M}/2R$, and the accretion timescale is $t_{\rm acc} = \Delta M/\dot{M}$. If $t_{\rm acc} < t_{\rm th}$, the superficial layer cannot cool rapidly enough to establish hydrostatic equilibrium and it expands on the dynamical timescale. This thermally unstable condition can lead to a common envelope.

Roche lobe overflow through the L_3 point or mass loss by winds from one or both stars can drastic affect the evolution of the binary. Not only is the mass lost from the system, but the outflow must be transporting away orbital angular momentum. As with rotating stars, the efficiency of the net torque depends on the degree to which the lost mass remains coupled to the motion of the binary, which in turn points to the need to properly include magnetic fields and dynamical torques through accretion disks and jet outflows. Consider as well what happens when we move a companion through this common envelope. The gas is gravitationally stirred by the binary motion, especially if the envelope is more slowly rotating than the binary, and therefore differential motion leads to heating and drag. The models are still speculative, but the picture is at least consistent in leading to a short-period binary with loss of the common envelope.

5.4.3 Period Changes

Our discussion so far shows that a period change can be produced rather generally by any process that changes the angular momentum of either component, the angular momentum of the system, or binding energy. It can be due to mass loss or transfer, but it can also result from a variety of more indirect mechanisms. As we just saw, the controlling parameter for system evolution is the angular momentum loss.

5.4.3.1 Gravitational Radiation

For extremely short-period systems, typical of cataclysmic variables, period changes result from the emission of gravitational radiation.[15] This is not really surprising, if you think about what it means to take two stars of about 1 $M_\odot$ each and have them in an orbit with a period of order one hour! The system acts like an enormous barbell whose binding energy is a fair fraction of the total mass energy of the system. In general relativity, there is no dipole emission. Conservation of momentum ensures this because the dipole emission varies as $|\ddot{D}|^2$, where D is the dipole moment, and this vanishes for isolated masses. The wave emission is due to the next order, the quadrupole moment Q, and, as we discussed in Chapter 1, is given by

$$\left(\frac{dE}{dt}\right)_{GW} = -\frac{G}{45c^5}\left|\frac{d^3Q}{dt^3}\right|^2. \quad (5.14)$$

Thus, if we take the energy to be the orbital kinetic energy (related to the binding energy, of course, through the virial theorem) and expand Q as a single mode, we see that $\dot{\omega} \sim \omega^5$. This drives the components toward each other because of the negative binding energy of the system. As the semimajor axis decreases, the frequency of the orbit causes the radiation rate to increase steeply, ultimately driving the binary toward coalescence. From the virial theorem, you know that

---

[15] There is an interesting history behind this idea. The equation for binary evolution with emission of gravitational waves appears in the 1951 edition of Landau and Lifshitz's *The Classical Theory of Fields* as a worked problem for the student. It came to the attention of astrophysicists when the dwarf nova WZ Sge was shown to have orbital period about 81 min, based on radial velocity and eclipse measurements. Kraft, R. F., Mathews, J., and Greenstein, J. 1962, *ApJ*, **136**, 313 then proposed that the evolution of such systems is dominated by gravitational wave emission, years before the discovery of binary pulsars.

$E = \frac{1}{2}E_G = -GM_1M_2/2a$ so that

$$\frac{\dot{a}}{a} = \frac{\dot{E}_{GW}}{E_G}, \qquad (5.15)$$

and the period must decrease even without mass transfer or mass loss. To be more precise, for a binary system, the quadrupole moment substituted into Eq. (5.14) gives

$$\frac{dE}{dt} = -\frac{32G}{5c^5}\left(\frac{M_1M_2}{M}\right)^2 a^4\omega^6, \qquad (5.16)$$

so the gravitational "radiative" loss from the system depends steeply on the separation, and the period. Since $\dot{a}/a = -\frac{2}{3}\dot{\omega}/\omega$, we obtain the timescale for any period change:

$$t_{GW} = 9.72 \times 10^{11} \text{yr} \left(\frac{a_0}{1\,R_\odot}\right)^4 (M_1M_2M)^{-1}, \qquad (5.17)$$

writing a_0 for the initial separation and scaling all masses to $M_\odot$. The result, regardless of how long the timescale, is an inevitable drift of the system toward merger that leads to an increasing rate of convergence of the two components. For double-degenerate systems (both stars being white dwarfs) this can be very short, of order $\leq 10^8$ yr. As such, while this emission should play relatively little evolutionary role for normal binaries, it can dominate the orbital evolution for compact systems and provide significant luminosity for black hole and neutron star systems just before merger. For example, double-neutron star binaries are expected to be important gravitational wave sources for the interferometric detectors described in Chapter 1. For the last stages of coalescence of compact bodies, whether they are components in a cataclysmic system or, for instance, a binary neutron star, gravitational radiation is the dominant driver once the separation is of order $a \sim c/\omega$ and the stars inevitably collide. This is one of the scenarios for producing γ-ray bursts.

The confirmation of secular orbital evolution due to emission of gravitational waves is seen in the period change in PSR 1913 + 16, the Hulse–Taylor binary pulsar. This is a neutron star in a short period (7.75 h), eccentric ($e = 0.617$) binary system and was the first binary pulsar to be discovered, in 1974.[16] The pulsar has a 59-ms pulse period and a mass of 1.44 $M_\odot$ (the system mass is 2.83 $M_\odot$). The observed orbital period change is $\dot{P}/P = -2.43 \pm 0.03 \times 10^{-12}$, within about 1% of the predicted rate from gravitational wave emission. The same system also displays the predicted relativistic orbital precession to very high precision, 4°.22 yr^{-1}. Both components are compact objects, and there are no problems with tidal

---

[16] For the discovery, and for demonstrating the effects of general relativity on its orbital parameters, Hulse and Taylor were awarded the Nobel prize for physics in 1993; see Hulse, R. A. 1994, *Rev. Mod. Phys.*, **66**, 699, and Taylor, J. H. ibid, 711.

interactions that could obscure the GRT effects, so this and related binaries are laboratories for testing theories of gravitation other than GRT.[17]

5.4.3.2 Magnetic Braking

The effect of a magnetic field on one of the companions can be important in the evolution of the system through the action of magnetic breaking. This phenomenon depends on the tidal coupling between the rotational and orbital angular momenta in the system, and results because a rotating magnetized body loses angular momentum through dipole radiation at the rotation frequency, Ω. For a vacuum boundary condition, a rotating magnetized star with a dipole moment $\mathbf{M}$, as we saw with neutron stars, the energy loss is given by the Larmor formula with the time derivative $\dot{\mathbf{M}} \sim \Omega \mathbf{M}$ so that $\dot{E}_{rot} \sim -|\mathbf{M}|^2 \Omega^4$. In a close binary, where the orbital and spin angular momenta are strongly coupled, this stellar spindown forces a decrease in the orbital period of the system even without mass transfer. Its dependence on period is weaker than we found for gravitational radiation, but the coupling constant depends on the magnetic field strength and can be important if the field is strong enough, on the order of a few megagauss.

Another effect of the field is to alter the spin through torquing of the star by the mass *outflow*, even in completely detached systems. Matter is constrained by a magnetic field to corotate with the stellar surface out to the Alfven radius, $R_A \Omega \approx v_A$, and as you saw for the solar wind, is accelerated at the expense of the star's rotational angular momentum. The Alfven speed, v_A, depends on the *local* magnetic field, and hence on the surface field configuration, and because it also depends on the density it depends on the mass loss rate. For a steady-state mass loss rate, $\dot{M}$, at a terminal velocity v_∞, we obtain

$$R_A = \left(\frac{B_0^2 v_\infty}{\dot{M} \Omega^2} \right)^{1/2n} R_\star \quad (5.18)$$

for a field varying as $B \sim r^{-n}$. The angular momentum then varies as

$$\frac{\dot{J}_s}{J_s} = \frac{\dot{M}}{M} \left(\frac{R_A}{R_\star} \right)^2. \quad (5.19)$$

If tidal coupling between the stars somehow manages to enforce synchronism, this sink of angular momentum causes the binary to contract and results in a secular decrease in the orbital period. Magnetic braking is thought be important for the period variations of cataclysmic variables, especially AM Her–type stars, where the surface fields reach about 10–100 MG. Its effect has also been observed in the RS CVn stars, for which enhanced magnetic dynamo activity is generally found. The presumption is that the activity is enhanced by the higher rotation frequency compared with an isolated star of the same properties.

---

[17] See, for instance, Will, C. M. 1993, *Theory and Experiment in Gravitational Physics*, 2nd ed. (Cambridge, UK: Cambridge Univ. Press); Will, C. M. 1999, *Phys. Today*, **52**(10), 38; Peebles, P. J. E. and Wilkinson, D. T. 2000, *PASP*, **112**, 1141.

5.5 MASS TRANSFER IN CLOSE BINARIES

Now we come to the physics that underlies the evolutionary scenario, but first, a caveat. The Roche surface has acquired enormous significance since the 1960s or so as the limiting factor in the evolution of close binary stars. It plays a fundamental role as a limiting surface, but it does not behave like a cliff in the system. One has the impression that mass is bound up to the point where it contacts the critical radius and then falls toward the companion. While this is a valid approximation, stars do not have the sharp edges that this picture implies. Instead, as a star gets closer to R_{RL}, its mass becomes less stable and begins to flow. For real stars, this happens long before the *photosphere* comes into contact. In addition, because of the limited mass zone resolution in stellar interior codes, a comparatively arbitrary approximation is used for the removal of mass. Stars may show mass transfer effects even when their photospheres are well within the Roche radius provided they have sufficiently distended atmospheres, and this transfer can power accretion effects even before the major mass loss occurs.

5.5.1 Mass Loss by Winds

Wolf–Rayet stars are often, but not always, found in close binary systems and it is likely that the Roche surface plays a significant role in their formation.[18] The role of radiation pressure in driving the winds of massive stars is similar to the tidal interactions between the stars. Either way, the star is stripped down to a bare core. There is growing evidence, for instance, that the final structure is virtually the same for a wide range of initial binary separations and mass ratios. When we add radiation pressure to the Roche problem, it has the same effect as a reduction in the mass of the luminous star. Whether or not there is a wind, the size of the Roche lobe decreases proportional to $M_{\text{eff}} = (1 - \Gamma)M$, where Γ is the now familiar ratio of radiative to gravitational accelerations, g_{rad}/g. Several new effects enter when treating mass loss by a stellar wind in a close binary. One is that the total mass of the system changes because the material that is capable of leaving the surface of one of the stars is also able to leave the system as a whole, thus reducing the gravitational binding energy. On the other hand, depending on *how* the mass is lost, it may transport angular momentum, and this also alters the separation of the components. Isotropic mass loss at high speed will have little effect on J, while altering M. This is also true for jetlike mass outflows, especially if they are aligned with the orbital and rotational angular momentum axes. If, however, the mass is lost mainly from the orbital plane or if there is significant disk formation, then the torque may be important. Either effect ultimately enters into the calculation of the Roche radius, resulting in feedback.

---

[18] Langer, N. 1995, in *Wolf–Rayet Stars: Binaries, Colliding Winds, Evolution: IAU Symp. 163* van der Hucht, K. and Williams, P. M., eds. (Dordrecht: Kluwer), p. 15; Vanbeveren, D. 1993, *Space Sci. Rev.*, **66**, 327; de Loore, C. and Vanbeveren, D. 1994, *Astr. Ap.*, **292**, 463. An especially important point about the change caused in end states of stellar evolution is discussed by Wellstein, S. and Langer, N. 1999, *Astr. Ap.*, **350**, 148 who point out that binaries substantially alter the progenitor mass limits for the formation of black holes and neutron stars.

Proximity ensures interaction if both stars happen to possess strong stellar winds.[19] Since you know that terminal wind speeds for massive stars exceed 1000 km s^{-1}, such dissipative collisions must generate X-rays in excess of the emission from the individual stars. The wind luminosity is $L_w \approx \frac{1}{2}\dot{M}v_\infty^2$, which is converted to radiation by the hard collision that brings the flows to rest at the stagnation point where the dynamical pressures balance. For constant mass loss rates from each star, this scales as $\dot{M}_1 v_{\infty,1} r_1^{-2} = \dot{M}_2 v_{\infty,2} (1-r_1)^{-2}$, where the distances are scaled to the binary separation. The shock thickness increases with distance from the orbital plane and, because each wind emerges spherically from the stellar surface, there is a dynamical pressure *along* the front that accelerates the flow outward. A variety of instabilities have been seen in numerical simulations, including those due to cooling and shear, and the flow cools from advection. Is this just an exercise, or can such collisions be observed? The most likely place to find this interaction is Wolf–Rayet binaries. In many of these systems, one component is a highly evolved massive star with $\dot{M} \approx 10^{-5} M_\odot$ yr^{-1}, and the other is an O star that also has a significant outflow. The best studied examples are V444 Cyg, γ Vel, and CQ Cep. Young binaries may also have such interactions, and this has been suggested for massive pre-main-sequence Herbig Ae/Be binaries. Some short period O star binaries that have nearly identical components, such as AO Cas and ι Ori, display the signatures of disturbed wind lines and high X-ray fluxes. You can see that an extended version of this effect will also occur in compact clusters where individual windblown bubbles collide, but this problem is more properly treated in the next chapter.

5.5.2 Accretion from Stellar Winds

An accreting radially infalling flow in steady state, for which $\dot{M} = 4\pi r^2 \rho v$, satisfies the same equation as a stellar wind, but with different boundary conditions. Now we place the origin of the flow at infinity and look at what happens near the star. The solution for an isothermal wind is just an isothermal accretion flow in reverse (e.g., Parker's solution for the solar wind). This assumption is plausible for a wind because the transition through the sonic point takes place without a shock. It is plausible to assume that the matter, once it leaves the star, will continue to be pressure-accelerated toward the (supersonic) terminal velocity. Here, however, the inner boundary condition is reversed. We are going from *supersonic* to *subsonic* flow, and it is clear that we cannot do this smoothly—a shock is inevitable. In addition, we have the possibility of compressional heating of the matter as it converges onto the central star. Therefore, we will have to be a little more careful. The dynamical equation is, as usual

$$v\frac{dv}{dr} = -\frac{1}{\rho}\frac{dP}{dr} - \frac{GM}{r^2}. \tag{5.20}$$

Now we need to include the equation of state. Take a polytropic law. For an

---

[19] Shore, S. N. and Brown, D. N. 1988, *ApJ*, **334**, 1021; Stevens, I., Blondin, J., and Pollack, A. 1992, *ApJ*, **386**, 265; Usov, V. V. 1992, *ApJ*, **389**, 635.

adiabatic flow, the (constant) energy flux along a streamline is

$$E = \frac{c_s^2}{\gamma - 1} + \frac{1}{2}v^2 - \frac{GM}{r} = E_\infty = \text{constant}, \qquad (5.21)$$

so we obtain the critical radius at which $v = c_{s,c}$:

$$r_c = \frac{GM}{2c_{s,c}^2} \qquad (5.22)$$

just as before. But wait, there is a difference. The sound speed is now not constant and equals its value at the critical point, $c_{s,c}$. This gives a critical condition as well for the sound speed:

$$\left[\frac{1}{2} + \frac{1}{\gamma - 1}\right]c_{s,c}^2 - \frac{GM}{r_c} = \frac{c_{s,\infty}^2}{\gamma - 1}, \qquad (5.23)$$

so that

$$c_{s,c} = \left(\frac{2}{5 - 3\gamma}\right)^{1/2} c_{s,\infty}. \qquad (5.24)$$

You see that we have a problem for an ideal gas, one you will see this many times again since the same condition applies for accretion in a cooling flow in a cluster of galaxies and for spherical accretion onto a compact object in an active galactic nucleus. The matter cannot accrete too rapidly because, since it is hot from compression, it may radiate at or above the Eddington limit. This implies that steady flow can occur only if

$$\eta \frac{GM\dot{M}}{r_c} \leq L_{\text{Edd}}, \qquad (5.25)$$

where η is some efficiency for conversion of the released gravitational potential energy into radiation. The details of how that is computed are not important here. For a black hole, this is especially interesting since the accretion shock is near the Schwarzschild radius, which depends only on the mass of the gainer, and the Eddington luminosity also depends only on this mass. Therefore, we find a scaling relation between the rate of mass accretion and the stellar mass $\dot{M} \sim M$.[20] For accretion from a wind, the situation is much like the one we saw for dynamical friction. The gravitational deflection by the gainer determines the radius for the

---

[20] Bondi, H. 1952, *MNRAS*, **112**, 195; Chakrabarti, S. K. 1990, *Theory of Transonic Astrophysical Flows* (Singapore: World Scientific); Chakrabati, S. K. and Sabu, S. A. 1997, *Astr. Ap.*, **323**, 382; see also Shore, S. N., Livio, M., and van den Heuvel, E. P. 1994 (n.10).

accretion flow, called *Bondi–Hoyle accretion*, is

$$r_a = \frac{2GM}{(V_{\mathrm{orb}}^2 + v_W^2)}, \quad (5.26)$$

where V_{orb} is the orbital velocity. The rate of accretion is $\dot{M} \approx 4\pi\rho r_c^2 v_W$, if the wind speed is large compared with the orbital velocity, or put differently

$$\dot{M} = 16\pi G^2 M^2 \rho_W v_W^{-3}. \quad (5.27)$$

For a spherical wind from the companion, the density scales with the binary separation, a, through $\rho_W = \dot{M}_\star / 4\pi a^2 v_W$, where $\dot{M}_\star$ is the stellar wind mass loss rate. Finally, we arrive at the scaled rate with respect to measurable parameters of the binary system:

$$\dot{M} = \frac{(4\pi^2 GM)^{4/3}}{v_W^4} P_{\mathrm{orb}}^{-4/3} \dot{M}_\star, \quad (5.28)$$

where P_{orb} is the orbital period. The luminosity produced by the accretion shock is therefore of order:

$$L_{acc} \approx \frac{GM}{R}\dot{M} \sim M^{7/3} v_W^{-4} P_{\mathrm{orb}}^{-4/3} \dot{M}_\star. \quad (5.29)$$

The wind carries angular momentum, so an accretion disk may form in shorter-period systems. A bow shock is formed that has an aberration angle depending on the ratio of V_{orb}/v_W. Finally, because of the relative slip velocity of the wind and the post-shock flow, the accretion produces a trailing train of vortices in two and even three dimensions that change the local angular momentum at the gainer. This is actually the same effect as the von Kármán vortex sheet that forms in a symmetric flow around a body. The Kelvin circulation theorem states that if there is no initial vorticity in the flow, none can be formed through symmetric flow past a body. Therefore, drag produces a set of symmetrically circulating vortices on the training side of the gainer. This is the basis of the "flip-flop" instability that has appeared in two- and three-dimensional simulations of wind accretion flows.[21] In addition, if the star is rotating, the wind may either spin the gainer up or down depending on the relative torque.

5.5.3 Accretion by Streams and Roche Lobe Overflow

Mass loss must occur whenever a star comes into contact with its Roche surface as a stream directed toward the companion. One way to see why is to look at a one-dimensional flow problem using a local expansion in the gravitational potential around L_1, as we did to find the stability of the Lagrangian points in the first

---

[21] Bondi, H. and Hoyle, F. 1944, *MNRAS*, **104**, 273; Davidson, K. and Ostriker, J. P. 1971, *ApJ*, **179**, 855; Theuns, T. and David, M. 1992, *ApJ*, **384**, 587; Livio, M. 1994, in Shore et al. (n.10).

place. Along the line of centers, the gravitational acceleration, $\partial\Phi/\partial x = 0$ at L_1. Keeping only the second-order term $(\partial^2\Phi/\partial x^2)_{L_1}\xi$, we obtain for the fluid equation of motion an analog to the wind solution with nonconstant surface gravity from Chapter 3:

$$\frac{1}{v}(v^2 - c_s^2)\frac{dv}{d\xi} = -\Phi''_{L_1}\xi. \tag{5.30}$$

The sound speed on the left-hand side comes from the pressure gradient. The tidal acceleration changes sign at L_1 so the mass accelerates there through the sonic point and toward the companion. This isn't really surprising, since the flow is infalling once it crosses the point at which gravity vanishes in the binary frame and therefore the pressure-driven acceleration at the sonic point pushes the mass through the equilibrium point. The result is a highly supersonic flow that launches toward the companion with the sound speed from the L_1 point and is deflected by the Coriolis acceleration. The stream orbits the companion and collides with itself, again supersonically, forming an oblique shock that eventually deflects it into an orbit.[22] Ultimately, a disk is formed, the structure of which we now treat.

5.5.4 Formation and Structure of Viscous Accretion Disks

As we have just discussed, mass lost through the L_1 point generally carries angular momentum and therefore cannot fall directly onto the companion. Even if it were coming from the precise center of mass, the Coriolis acceleration in the corotating frame causes the stream to deviate from radial infall. Consequently, some sort of disk must form around the gainer in a close binary. Here we will describe some of the basic processes that are included in models for such disks. In Figure 5.3 we show an example of the sort of evidence eclipsing Algol systems provide for the existence of an accretion disk.

The condition for disk formation is that the angular momentum must be sufficient to send material into orbit around the gainer. The equivalent circular orbit is given by equating the angular momentum at the L_1 point with the equivalent Keplerian orbit around the gainer. The specific angular momenta are $j = \omega(m_2 a/M - R_{RL})^2$ at L_1 and $j_K = (Gm_2 r)^{1/2}$ around the gainer, where r is the radius of the circular orbit around m_2. If the angular momentum is too small, the stream may directly impact the gainer. This produces local shock heating and a deflection of the stream by the gainer's atmosphere. If a disk does form, this interaction point is moved out, but it is still present. The reason is that the stream, which is flowing supersonically, cannot adjust its structure on the slower sonic timescale. As a result, it slams into the circulating material and forms a standing shock. Since it is an oblique impact, this region refracts the shock and produces a contact, slip, discontinuity at the boundary.

---

[22] See, for example, Lubow, S. H. and Shu, F. H. 1975, *ApJ*, **198**, 383; Shu, F. H. and Lubow, S. H. 1983, *ARAA*, **19**, 277.

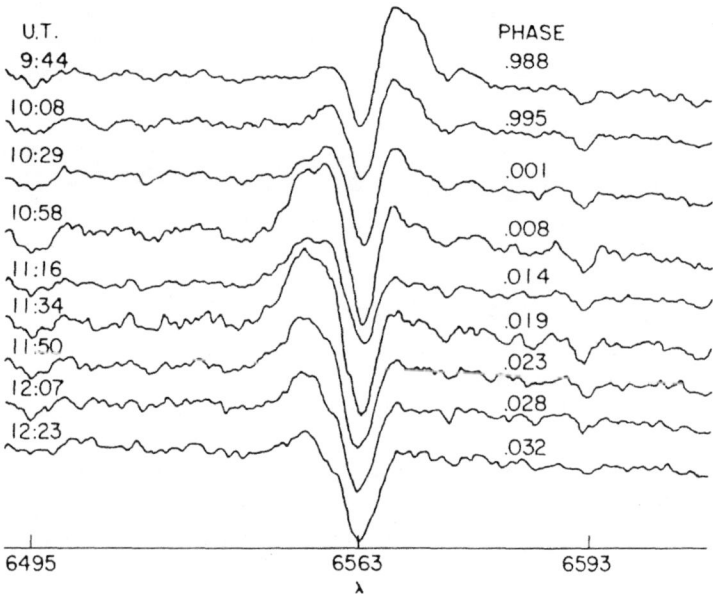

FIGURE 5.3. Variation of Hα through eclipse for the Algol system U Cep. Notice the obscuration of the blue line wing on ingress and the red wing on egress. This is an example of both the Rossiter–McLaughlin effect (the same thing happens for the absorption lines of the rotating photosphere) and the evidence for an extended circumstellar disk (credit: R. Polidan, NASA/GSFC, and M. Plavec, UCLA).

Assuming hydrostatic equilibrium, the disk's vertical structure is governed by the tidal component in the acceleration:

$$\frac{1}{\rho}\frac{dP}{dz} = -\frac{GM}{r^2}\frac{z}{r}. \qquad (5.31)$$

The simplest estimate of the disk thickness comes from assuming that it is vertically isothermal, so $P = \rho c_s^2$. Noting that the vertical gravitational acceleration can be written as

$$g_z = \frac{v_\phi^2}{r} \cdot \frac{z}{r},$$

the density shows a Gaussian vertical profile with a scale height as follows:

$$z_0 = \frac{c_s}{v_\phi} r. \qquad (5.32)$$

This a thin disk since generally $c_s/v_\phi \ll 1$, although because of the radial dependence of the angular velocity, the thickness grows with increasing distance from the central body. Many physical processes are neglected in this treatment. In particular, heating is not consistently included, nor is the equation of transfer. Both are necessary to determine the disk's vertical temperature profile in response to the energy input, since the disk may be heated by both internal turbulent dissipation and radiation from the central star. Much of the work on interpretation of disk spectra comes from addressing this problem.

There must also be radial redistribution of mass in an accretion flow, but without some loss of angular momentum there will be no accretion. If the material remains dissipationless in the disk, and cannot somehow lose angular momentum, it will continue to circulate and accumulate, storing the angular momentum that it gained from infall by the stream. This may actually happen in some systems, and it is evident that it is important in the stability of the planetary ring systems observed in the solar system. In order to actually transfer mass from one star to another, and complete the picture described earlier for close binary evolution, it is necessary to find some mechanism(s) whereby the angular momentum of the circulating gas can be altered and a slow drift can be established that actually places material onto the companion. This is where much of the effort in accretion disk theory goes—what is the viscous mechanism in accretion flows.[23] If the binary orbit is eccentric, even by a small amount, the loser in this effectively three-body system exerts a torque on the outer parts of the disk that may produce a resonance when the $\omega_K = \omega$. This can drive spiral shocks in the disk that are similar to density waves in a galaxy (see Chapter 7). Such shocks may play a role in angular momentum transfer within the disk, but just how important they are form dense accretion disks is currently unknown. There is, however, another example that is clearly important for resonances. Planetary ring systems are dominated by interactions between the moons and the particles in the rings.

5.5.4.1 Viscous Accretion Disks

To simplify the treatment, we will assume that the viscously induced radial drifts are slow and involve little structural change in the disk so that Keplerian motion describes the circulation. This eliminates several mathematical complications, although at the risk of introducing limitations to the physical regime that we can treat. We can separate the radial and circulatory motions, and there are only axisymmetric terms in the equations of motion. We will also assume that the disk is vertically hydrostatic and simply remove any z dependences by integrating over that coordinate. The volume mass density therefore appears as surface density in the continuity equation, using

$$\Sigma(r,t) = \int_{-\infty}^{\infty} \rho(r,z,t)\, dz. \tag{5.33}$$

The assumption of axisymmetric mean flow also allows us to ignore any azimuthal terms in the continuity equation, which is

$$\frac{\partial \Sigma}{\partial t} + \nabla \cdot \Sigma \mathbf{v} = \frac{\partial \Sigma}{\partial t} + \frac{1}{r}\frac{\partial}{\partial r} r \Sigma v_r = 0. \tag{5.34}$$

For the momentum equation, and since we are including viscous coupling, the

---
[23] See Pringle, J. E. 1981, *ARAA*, **19**, 137; Papaloizou, J. and Lin, D. N. C. 1995, *ARAA*, **33**, 505.

Navier–Stokes equation describes the flow

$$\Sigma\left(\frac{\partial}{\partial t}\mathbf{v} + \mathbf{v}\cdot\nabla\mathbf{v}\right) = \nabla\cdot\eta\nabla\mathbf{v}$$

for vertically integrated viscosity η. For conservation of the *angular momentum*, since $\partial v_\phi/\partial t = 0$, we obtain

$$\Sigma\frac{v_r}{r}\frac{\partial}{\partial r}rv_\phi = \nabla\cdot\eta\nabla v_\phi\hat{\phi} \tag{5.35}$$

Using the auxiliary equation

$$\nabla v_\phi\hat{\phi} = \frac{dv_\phi}{dr} - \frac{v_\phi}{r},$$

the diffusive term in the NS equation becomes

$$\frac{1}{r}\frac{d}{dr}\left[r^2\eta\frac{d\omega}{dr}\right].$$

Consequently, we obtain a representation for the divergence in Eq. (5.35) in closed form, dependent only on the viscosity:

$$r\Sigma v_r = \frac{r(r^2\omega'\eta)'}{(r^2\omega)'}. \tag{5.36}$$

Now comes the moment when we use the assumed form for the circulation. By taking $\omega \sim r^{-3/2}$ and substituting Eq. (5.36) into Eq. (5.35), we obtain the equation for the time dependence of the surface density for a viscously dominated accretion flow:

$$\frac{\partial\Sigma}{\partial t} - \frac{3}{r}\frac{\partial}{\partial r}\left[r^{1/2}\frac{\partial}{\partial r}(r^{1/2}\eta)\right] = 0 \tag{5.37}$$

when neither external sources nor sinks are present. We close the system by invoking a prescription for the viscosity. Assume

$$\eta = \alpha c_s z_0 \Sigma, \tag{5.38}$$

where α is a constant that we'll describe later. This *diffusion* equation now describes the evolution of the disk. Notice that we have imposed several restrictions on the physics at the start. We tacitly assumed that the disk remains in strict centrifugal balance in order to avoid having to solve the radial equation of motion, even with the viscous transfer of angular momentum. This is a pretty strong assertion, requiring that the inflow remains subsonic everywhere and that the viscous torques are only local. We have not yet included the local effects of heating and cooling, although these enter the viscosity through the equation of state. We also assumed that the vertical average of all relevant thermal quantities suffices to define the disk structure. None of these is out of line with the assertion that the Reynolds number for the circulation is high, and in the first attempts to understand disks these seemed quite natural. It is because of these conditions that the disk evolution can be described completely by a mass conservation equation. Unless nonaxymmetric torques occur, or something induces large pressure gradients that are not balanced by circulation, the transport of mass is the same as the transport of angular momentum.

Now how does this all work? Taking the stress proportional to the shear and assuming the stress is isotropic, we follow the Shakura–Sunyaev treatment[24] by writing

$$T_{r\phi} = \alpha P, \qquad (5.39)$$

where α is a free parameter that is usually assumed to be a constant in space, although not necessarily a constant in time, that determines the efficiency of viscous coupling. This is the so-called α disk. With this parameterization, you see that η depends on the pressure, which in turn connects the mass transfer through the continuity equation to the vertical structure of the disk. For constant α the heating also depends on the pressure. Therefore we have the following physical consequence of the diffusion equation for the mass transfer.

The local energy generation depends on α, hence η, and therefore any local heating produces a decrease in Σ because of the increase of the scale height. This, in turn, leads to smaller optical depths and subsequent cooling. The collapse of the disk vertically means that the plane is "mass loaded" and this matter must be advected inward by the torque. Therefore, in the cooling state, or if α is sufficiently small, the mass transfer can be large toward the gaining body. If on the other hand the efficiency of radiation is reduced or the cooling in the vertical direction is low, the disk can remain hot and the mass will not effectively transfer to the inner region.

Since the integral for η contains the surface density, we now have an equation that depends only on Σ and the disk parameters. Coupled with an equation for the radiative transport in an optically thick disk, at least for the vertical energy losses, this completes our description of the disk. In many applications, none of this is necessary. If the disk is optically thin, it can be treated as a circulating gas that immediately radiates any energy gained by its drift onto the gainer, and the emission should result from an accretion shock that may develop at the surface of the mass accepting star. However, for many systems, especially cataclysmic variables, the depth of the gravitational potential and the opacity of the disk conspire to produce complex time dependence in the disk structure. The vertical temperature gradient governs the surface density through the condition of hydrostatic equilibrium. Remember that, even in the thin disk approximation, the disk thickness depends on T through the scale height. You can see this by using Σ as the

---

[24] Shakura, N. and Sunyaev, R. A. 1973, *Astr. Ap.*, **24**, 337. This is one of the fundamental papers in this field, although the unwary reader should watch out for typos. The assumption that $\alpha \leq 1$ is introduced by assuming that any magnetic field will eventually reach equipartition and that any stress will therefore be a fraction of the isotropic gas pressure. Although the possible choices for α are virtually unlimited, most models tend to assume values around 0.1.

dependent variable and rewriting the vertical radiative diffusion equation as

$$F = -\frac{4acT^3}{3\kappa}\frac{\partial T}{\partial \Sigma} \qquad (5.40)$$

Since the opacity depends on both temperature and density, the vertical structure of the disk feeds back into the rate of mass accretion. This is the basis for the *disk oscillations* that are observed in nonlinear models for the disks of cataclysmic variables. Remember that Σ depends on time! These arise mainly because of recombination effects when the column density is high. The oscillations are the response to changes in the mass inflow. Also notice that the vertical temperature gradient may become so steep that the gas convects. If this happens, and it depends on the opacity as well as the flux and vertical scale height, the turbulence provides a natural source for viscous coupling. We should add here that the treatment of rotating shearing convective flows is still in its practical infancy, so there is still much to be explored.

With all sources of dissipation included, the total gravitational potential energy of each parcel of infalling gas will ultimately be radiated, so that an estimate for the total luminosity of the system is

$$L_{acc} \sim \frac{GM}{R_\star}\dot{M},$$

where $R_\star$ is the stellar radius. The temperature should reach about $T_{acc} \sim GM/kR_\star$, independent of the accretion rate. Thus, for a compact body, there are two important conclusions to be drawn from this exercise. The first is that a WD, neutron star, or black hole in a binary system should produce emission ranging from soft to very hard X-rays. The second is that there is only so fast that such a source can eat. You recall that for a spherical body, the Eddington luminosity sets a limit on the luminosity that a star can support simultaneously in hydrostatic and radiative equilibrium. For a black hole, whose radius depends only on its mass, this means that $\dot{M} \sim M$, so the luminosity of such a source is a measure of its mass accretion rate. For less extreme objects, the luminosity depends on the stellar radius and mass, so that $\dot{M} \sim R_\star$. For a white dwarf, for instance, this means that $\dot{M} \sim M^{-1/3}$. In other words, some of the mass will not wind up on the star if it is accreted too rapidly. Mass loss from the binary may happen through jets driven from the disk of the sort seen in the really bizarre object SS 433.

What are the observable fingerprints of a disk? For one thing, it is not a star. The surface spans a wide range of distances from the central star and because of local energy generation, and reprocessing of incident light from the central body, it presents a range of temperatures. Even if each part of the circulating flow radiated as a blackbody, you would never see a "normal" isothermal continuum. Let's derive the frequency dependence of the emergent spectrum. First, look at an "active" disk. Assuming that local viscous dissipation is strong enough that you can actually see the emission above that of the central star, we will assert that the optically thick disk will emit at long wavelength as a Rayleigh–Jeans radiator,

$$B_\nu \sim \nu^2 T(r)$$

but with a radiation distance-dependent peak at $\nu_{\max} \sim T(r)$. Flux balance by our previous argument requires $T^4 \sim M\dot{M}r^{-3}$ or $T \sim r^{-3/4}$, so each annulus contributes a part of the flux that is weighted heavily toward the inner part of the disk because of the steep temperature dependence. Since now temperature and radius are interchangeable, we find that

$$L_\nu \sim \int \nu^2 Tr\, dr \sim \nu^2 \int T^{-8/3}\, dT \sim \nu^2 T^{-5/3}.$$

Now, since the temperature and $\nu_{\max}$ are linearly related, we require that each annulus radiates most efficiently at that frequency and the expected low-frequency form for the monochromatic flux is

$$F_\nu \sim \nu^{1/3}, \qquad (5.41)$$

with the mass accretion rate, viscosity, and the mass and radius of the central body determining the absolute luminosity. You see that this does not look at all like a normal stellar spectrum. Even if we use more sophisticated atmospheres to treat the spectrum formation, including all the complications we would expect for a realistic picture of the opacity of the gas, the fact that the energy generation is local and distributed through the disk in a power law always leads to a power-law dependence for the emission—the cooler regions emit weakly, but they have a larger area compared with the inner part of the disk so they extend the low-frequency tail of the continuum relative to a Rayleigh–Jeans approximation. Actually, we have already seen this when discussing gravity darkening for a rapidly rotating star for which the surface temperature depends on latitude. Requiring the local gravity to vary with distance from the central body is like forcing it to depend on distance from the equator. The main difference is that energy is being locally generated for a disk while for a rotating star the surface temperature varies because there is a spherically emitted central flux that must pass through the distorted surface. One way of seeing why a disk looks different is to reflect on the relative mechanical constraints on a normal stellar atmosphere and a disk. A star must maintain constant surface gravity over the whole layer. Here each part of the disk is centrifugally isolated from every other and is in vertical equilibrium only locally, so each annulus acts like an independent atmosphere transporting only as much flux as it has to from the viscous source, whose luminosity dependence is a power law. Line formation is also more complicated for a disk because of the strong dependence of pressure scale height with distance.[25]

In principle, it should be possible to observe this disk. In fact, it is seen in eclipsing systems, where the absorption lines from the environmental gas are seen in projection against the stellar photosphere. The problem is that there is that

---

[25]Although the resulting spectral energy distribution is different, the basic treatment is the same for a "passive" disk. In this case there is no local generation, we instead just reprocessing from a central star so that the temperature varies as approximately as $r^{-1/2}$. It's a useful exercise to compute the emergent spectrum for a generalized power law, an exercise we leave for you but the trend is clear. The problem is similar to what you saw for a stellar wind since we are working in the short wavelength limit so the radiative transfer is particularly simple. The smallest area is always associated with the highest temperature so you will always get a power law that is shallower than the ν^2 dependence of a single temperature source.

current models for the disk cannot cope with the large number of physical problems connected with the equation of transfer and the viscous heating to provide the same answers for such structures as we can obtain for stellar atmospheres. We do not have an adequate picture, yet, of the abundances in the disk material. In addition, there are serious problems at the inner boundary, where the disk meets the star. To see why, think of the stellar wind problem in reverse. The aim is to get matter to settle onto the star, to come to rest. But there are two obstacles facing the gas, even after the problem of shedding angular momentum is solved by the inward drift. The first is to get the infall velocity, once the angular momentum is dissipated, to go smoothly to zero. The second is to get the matter corotating with the stellar surface so that it does not face a centrifugal barrier at the inner portion of the disk.

5.5.4.2 Magnetorotational Instability and Viscosity in Disks

Since disks rotate differentially, even a weak magnetic field can have profound consequences for their structure because of dynamo action. The field can be amplified and communicates a torque over a long distance. This produces a very strong instability that is inherent to all conducting media, the *magnetorotational instability* (MRI).[26] In discussing this instability, let us assume the simplest sort of disk, taking an incompressible medium that is vertically in force balance. We imagine that the disk is threaded by a weak magnetic field that lies in the ϕ and z directions (neglecting the radial component). In fact, we will see that it doesn't matter if we neglect the toroidal field component since it drops out of the perturbation equations. As we did for the pulsation problem, we will go into more detail with this derivation because of the central role the magnetorotational instability plays in so many arenas. The fully time dependent ideal (inviscid) MHD disk equations are

$$\nabla \cdot \mathbf{v} = 0 \tag{5.42}$$

$$\frac{d}{dt}\mathbf{v} = \frac{\delta\rho}{\rho^2}\nabla P - \frac{1}{\rho}\nabla\delta P - \frac{1}{4\pi\rho}\left[\nabla(\mathbf{B}\cdot\delta\mathbf{B}) + \mathbf{B}\cdot\nabla\delta\mathbf{B} + \delta\mathbf{B}\cdot\nabla\mathbf{B}\right]$$

$$-\frac{1}{\rho}\nabla\delta\Phi + \frac{\delta\rho}{\rho^2}\nabla\Phi \tag{5.43}$$

$$\frac{dS}{dt} = 0 \tag{5.44}$$

$$\frac{\partial}{\partial t}\delta\mathbf{B} = \delta\mathbf{B}\cdot\nabla\mathbf{v} - \mathbf{v}\cdot\nabla\delta\mathbf{B} + \mathbf{B}\cdot\nabla\delta\mathbf{v} - \delta\mathbf{v}\cdot\nabla\mathbf{B}. \tag{5.45}$$

---

[26] This instability has a very convoluted, and instructive, history. Having been recognized first in the rotational case by Velikhov, E. P. 1959, *Sov. Phys. JETP*, **36**, 995 and Chandrasekhar, S. 1960, *Proc. Nat. Acad. Sci.*, **46**, 253, a form of the instability was found for baroclinic stars by Goldreich, P. and Schubert, G. 1967, *ApJ*, **150**, 571 and Fricke, K. 1969, *Astr. Ap.*, **1**, 388. It was first applied to accretion disks by Balbus, S. A. and Hawley, J. F. 1991, *ApJ*, **376**, 214. See also: Gammie, C. F. and Balbus, S. A. 1994, *MNRAS*, **270**, 138; Balbus, S. A. 1995, *ApJ*, **453**, 380. Two excellent reviews are provided in Balbus, S. A. and Hawley, J. F. 1998, *Rev. Mod. Phys.*, **70**, 1; Hawley, J. F. and Balbus, S. A. 1999, *Phys. Plasmas*, **6**, 4444.

To simplify the treatment, we will examine only axisymmetric perturbations, so only unit vectors will require ϕ derivatives, and we will look for plane-wave solutions of the linearized system that permit us to replace the time derivative with $-i\omega$ and the space derivative by $i\mathbf{k}$. We will also ignore tidal perturbations and gravitational torques, although you might bear them in mind as possible contributors for galactic spiral waves. Written out explicitly, Eqs. (5.44) become

$$\frac{dv_r}{dt} - \frac{v_\phi^2}{r} = -\frac{1}{\rho}\frac{\partial \Pi}{\partial r} + \frac{1}{4\pi}\mathbf{B}\cdot\nabla B_r - \frac{B_\phi^2}{4\pi\rho} \tag{5.46}$$

$$\frac{dv_\phi}{dt} + \frac{v_r v_\phi}{r} = -\frac{1}{\rho r}\frac{\partial \Pi}{\partial \phi} + \frac{1}{4\pi}\mathbf{B}\cdot\nabla B_\phi + \frac{B_r B_\phi}{4\pi\rho}, \tag{5.47}$$

where $\Pi = P + B^2/8\pi$. A new frequency has appeared, the *epicyclic frequency*, κ. We first encountered it in Chapter 1 when examining the three-body problem. It quantifies the shear measured in a differentially rotating system at any radial distance:

$$\kappa^2 \equiv \left[\frac{1}{r^3}\frac{d}{dr}(r^4\Omega^2)\right]_0 \tag{5.48}$$

and introduces a new characteristic timescale into the problem. The perturbed equations become

$$k_r \delta v_r + k_z \delta v_z = 0, \tag{5.49}$$

$$-i\omega\delta v_r - 2\Omega\delta v_\phi - \frac{\delta\rho}{\rho}\frac{\partial P}{\partial r} + i\frac{k_r}{4\pi\rho}(B_\phi\delta B_\phi + B_z\delta B_z) - i\frac{k_z}{4\pi\rho}B_z\delta B_r = 0, \tag{5.50}$$

$$-i\omega\delta v_\phi + \frac{\kappa^2}{2\Omega}\delta v_r + r\frac{\partial\Omega}{\partial z}\delta v_z - i\frac{k_r B_r + k_z B_z}{4\pi\rho}\delta B_\phi = 0, \tag{5.51}$$

$$-i\omega\delta v_z - \frac{\delta\rho}{\rho}\frac{\partial P}{\partial z} + i\frac{k_z}{4\pi\rho}(B_\phi\delta B_\phi + B_z\delta B_z) - i\frac{k_r}{4\pi\rho}B_r\delta B_z = 0, \tag{5.52}$$

$$-i\omega\delta B_r - i(k_r B_r + k_z B_z)\delta v_r = 0, \tag{5.53}$$

$$-i\omega\delta B_\phi - r\frac{\partial\Omega}{\partial r}\delta B_r - r\frac{\partial\Omega}{\partial z}\delta B_z - i(k_r B_r + k_z B_z)\delta v_\phi = 0, \tag{5.54}$$

$$-i\omega\delta B_z - i(k_r B_r + k_z B_z)\delta v_z = 0, \tag{5.55}$$

where the following expressions have been needed to reduce the magnetic field components:

$$\mathbf{B}\cdot\nabla\mathbf{B} = \frac{1}{r}B_r B_\phi \hat{\phi} - \frac{1}{r}B_\phi^2 \hat{\mathbf{r}} + (\mathbf{B}\cdot\nabla B)\hat{\mathbf{e}}$$

$$\delta\mathbf{B}\cdot\nabla\mathbf{v} = \frac{1}{r}B_r v_\phi \hat{\phi} - \frac{1}{r}B_\phi^2 \hat{\mathbf{r}} + (\delta\mathbf{B}\cdot\nabla v)\hat{\mathbf{e}}.$$

Choosing a polytropic equation of state simplifies the energy equation (in fact, it really avoids it by imposing an adiabatic process). Writing the entropy as $S = \ln P\rho^{-\gamma}$ and taking the first order perturbations, we have

$$\frac{\partial \delta S}{\partial t} + \delta \mathbf{v} \cdot \nabla S = 0 \tag{5.56}$$

which becomes, on assuming that the pressure perturbation is time independent (the same as the Boussinesq approximation we used for the continuity equation when asserting that the density behaves identically):

$$i\omega\gamma\frac{\delta\rho}{\rho} + v_r\frac{\partial S}{\partial r} + v_z\frac{\partial S}{\partial z} = i\omega\gamma\frac{\delta\rho}{\rho} + DSv_r - 0. \tag{5.57}$$

The abbreviation introduced here is

$$DS = \frac{\partial S}{\partial r} - \frac{k_r}{k_z}\frac{\partial S}{\partial z}.$$

Multiplying Eq. (5.50) by k_z and Eq. (5.52) by k_r and subtracting, we obtain

$$i\omega\frac{k^2}{k_z}v_r + \frac{\delta\rho}{\rho^2}k_z DP - \frac{ik_r k_z}{4\pi\rho}B_z\delta B_z + 2\Omega k_z v_\phi + \frac{ik_z^2}{4\pi\rho}B_z\delta B_r = 0$$

Solving for the magnetic field perturbations, we also obtain

$$v_\phi = -\frac{i\omega}{\varpi^2}\left(\frac{\kappa^2}{2\Omega} - \frac{k_z^2 v_{Az}^2}{\omega^2}r\frac{d\Omega}{dr}\right)v_r$$

using two new abbreviations

$$v_{Az}^2 = \frac{B_z^2}{4\pi\rho}$$

for the vertical Alfven speed, and

$$\varpi^2 = \omega^2 - k_z^2 v_{Az}^2.$$

We obtain the dispersion relation by solving the system for v_r:

$$\varpi^4 + \frac{k_z^2}{k^2}\frac{1}{\gamma}DPDS\varpi^2 - 2\Omega\omega^2\frac{k_z^2}{k^2}\left(\frac{\kappa^2}{2\Omega} - \frac{k_z^2 v_{Az}^2}{\omega^2}r\frac{d\Omega}{dr}\right)$$

$$= \varpi^4 + \frac{k_z^2}{k^2}\left(\frac{1}{\gamma}DPDS - \kappa^2\right)\varpi^2 - 4\Omega^2\frac{k_z^2 v_{Az}^2}{k^2} = 0, \tag{5.58}$$

The last (subtle) step is to replace $d\Omega/d\ln r$ with κ^2 explicitly in terms of Ω. The criterion for the onset of the magnetorotational instability is found by choosing the critical solution, $\omega = 0$ with $\mathbf{k} \cdot \mathbf{v}_A$ strictly real. We can take a barotropic approximation with $\nabla P \times \nabla \rho = 0$ (or, in terms of the entropy, $\nabla S \times \nabla P = 0$) and suppress the z gradient of the angular momentum (rotation strictly on cylinders). Now if we take $v_{Az} \to 0$ after dividing out $k_z^2 v_{Az}^2$ then with these simplifications, we find that

$$4\Omega^2 + \frac{1}{\gamma\rho} DP\,DS + \frac{2\Omega}{r}\frac{d(r^2\Omega)}{dr} > 0 \quad (5.59)$$

must hold, *regardless of the strength of the initial magnetic field*.[27] The combined effects of buoyancy and rotation, which amplify any seed field with time, leads to the instability. But if we ignore the buoyancy, we are still left with

$$\frac{d\Omega}{dr} > 0 \quad (5.60)$$

as the necessary condition for stability. Even Keplerian motion violates this! We simply *don't* see this in any astrophysical disks. Here is the strange and wonderful feature of this instability. The Rayleigh criterion, namely, that the angular momentum must increase outward, may be met by the circulation yet even if the seed magnetic field is vanishingly small, it grows through shear and the MRI grows. Although it is not clear from our development how this leads to turbulence, the MRI clearly produces strong, growing fluctuations in the density and velocity that likely become turbulent. We have neglected dissipation, but as soon as the velocity fluctuations grow large enough, they will certainly drive heating throughout the disk by the strong viscous coupling. We will close our treatment here with a few additional remarks. This instability is very general, and any ionized shearing medium is likely to experience some form of the MRI. This includes galactic scale disks as well as protostellar accretion disks.

5.5.5 Boundary Layers

Material somehow manages to get onto the gainer's surface regardless of the means of accretion. The inward drift due to viscous torques is part of the story, but what actually happens at the inner boundary between the disk and the stellar surface remains very poorly understood for *any* disks. This interface is a basic problem; the gas, which is moving at the local Keplerian velocity or something like it, succeeds in reducing its angular velocity to that of the underlying star, which, of course, is rotating more slowly. This must occur in a very small radial zone, and its

---

[27]A further simplification is possible, following Balbus and Hawley (n.26). Recall from our discussion of convection that the Brunt–Väisälä frequency, N, is the timescale for the gravity driven adiabatic oscillations of a fluid parcel where pressure is the restoring force. You can then write the pressure gradients in terms of this frequency in the schematic form for the radial and vertical gradients:

$$N_r^2 + N_z^2 + 2r\Omega\frac{d\Omega}{dr} > 0.$$

This highlights the source of the different terms in the dispersion relation, the entropy and pressure gradients translate into buoyancy terms.

consequences for the energetics of the inner disk are dramatic. A significant part of the energy of the flow—roughly half—is radiated in this boundary layer.

We now follow a track similar to the one we used when discussing synchronization and circularization by tides. The difference is in the interaction—here we assume mass transfer and that the angular momentum *can* change. In contrast to the tidal problem, a simple treatment of the boundary layer ignores the underlying star, assuming that it is largely unaffected by the process except for the viscous torque at its surface (essentially the boundary layer is an Ekman-type layer). The total energy available for the process is the gravitational binding energy of the accreting material so if the rate is $\dot{M}$, this amounts to $L_{\text{grav}} = \frac{1}{2} GM_\star \dot{M}/R_\star$. The torque received within the layer depends on $\omega_\star$, the rotation frequency of the star, and ω_K, the Keplerian frequency at R, so

$$\frac{dJ}{dt} = \dot{M} R_\star^2 \Delta\omega, \tag{5.61}$$

where $\Delta\omega = \omega_K - \omega_\star$. The accreted mass changes the rotational kinetic energy of the underlying star by

$$\frac{dJ_\star}{dt} = \dot{M} R_\star^2 \omega_\star,$$

since it is assumed to corotate with the stellar surface when added at the base of the boundary layer. If the moment of inertia of the star is I, then the change in the kinetic energy of the layer is

$$\frac{dE_{\text{BL}}}{dt} = \frac{J}{I} \frac{dJ}{dt},$$

and we now substitute Eq. (5.61) for $\dot{J}$. The rate of change of the total energy of the system is then

$$\frac{dE}{dt} = \frac{1}{2} \frac{GM_\star}{R_\star} \dot{M} - \frac{dJ_\star}{dt} - \frac{J}{I} \frac{dJ}{dt} = \frac{1}{2} \frac{GM}{R} \dot{M} - \dot{M} R_\star^2 \omega_\star - \dot{M} R_\star^2 \omega_\star \Delta\omega, \tag{5.62}$$

noting that there is no explicit reference to the viscosity. Since $\omega_K^2 = GM/R_\star^3$, rearranging terms, we obtain

$$L_{\text{BL}} = \frac{1}{2} \frac{GM_\star}{R_\star} \dot{M} \left(1 - \frac{\omega_\star}{\omega_K}\right)^2. \tag{5.63}$$

This is the luminosity of the layer, which, you notice, is reduced as $\omega_\star \to \omega_K$ because of spin-up. If the layer is optically thin, its temperature will be quite high, $T_{\text{BL}} = GM_\star/kR_\star$. In contrast, if the layer is optically thick, this temperature depends on the thickness of the layer, δ_{BL}, through

$$L_{BL} = 4\pi R \delta_{BL} a T_{BL}^4 \qquad (5.64)$$

This, of course, begs the question of how the angular momentum is actually transferred through the disk and removed from the accreting gas, which becomes more serious when we take a close look at what happens at the stellar surface.

An alternative is to bypass the boundary layer concept altogether in the sense that the accreting matter never thermalizes and radiates. This scenario is called an advection dominated accretion flows (ADAF). It is a special plasma-related situation that may explain the absence of observable emission from boundary layers in black hole systems. An important difference in the accretion process for a black hole is the absence of an inner boundary. While gas falling onto the surface of a white dwarf or neutron star must lose sufficient angular momentum to come into corotation with the star, this need not happen for a BH source. In the inner region, the inflow timescale also becomes very short, corresponding to the crossing time for the last stable orbit in a Kerr or Schwarzschild metric, and this may be shorter than the collision time for the plasma. Such accreters are fundamentally different from normal objects because they have no inner stationary surface onto which matter is constrained to settle. Consequently, if it is possible to reduce the efficiency with which the matter radiates energy, it will be "darker" than one would expect. In effect, it is the recognition that the only signature of the inner boundary is the loss of heat required to come to rest. Gas at high temperature is not as likely to collisionally cool as at low temperatures because the collision cross section for bremsstrahlung emission falls quickly with increasing energy. Line emission is unimportant compared with free–free emission, and all other cooling processes depend on the collision time. If this is short compared to the accretion time, the medium will not radiate.

5.6 CATACLYSMIC VARIABLES AND COMPACT OBJECTS IN CLOSE BINARIES

The cataclysmic variables are a heterogeneous type of close binary, unified phenomenologically by their propensity for "violent" behavior. What they all have in common is that one of the components is a compact, degenerate star, either a white dwarf, a neutron star, or a black hole. The array of types is as dizzying as with any other area of astronomy. Disk-dominated systems, for which the compact star (white dwarf) is overwhelmed with the accretion environment include the *SU UMa stars*, for which the disk produces a complex light variation due to a *superhump* that arises from the hot spot in the disk where the stream impacts its periphery. The excess emission progresses in retrograde through the lightcurve (i.e., it moves toward an earlier phase with time) and the systems undergo occasional outbursts. These systems are some of the shortest-period binaries known; WZ Sge has an orbital period of only about 1 h. At least one nova, V1974 Cyg, appears to also be a related system. The *U Gem* systems are eruptive disk systems that undergo nearly periodic outbursts that obey a period–amplitude relation. Magnetic white dwarfs are the gainers in the *AM Her* systems, which category also includes some novae such as V1500 Cyg and DQ Her.

Cataclysmics binaries with neutron stars include most *low-mass X-ray binaries* (LMXRBs) such as Cyg X-2 and Sco X-1. These systems have low mass (red dwarf) companions and resemble the white dwarf cataclysmics in many respects. A more massive system, Her X-1 = HZ Her, has a late A loser and a magnetized neutron star gainer. *High-mass X-ray binaries* (HMXRBs) have more massive (OB) losers, such as Vela X-1, 4U 0900-17, and ϕ Per. The primary is sometimes a Be star, meaning that it shows either a strong wind or evidence for an extended circumstellar environment. Cataclysmics with white dwarf gainers include dwarf nova systems such as WZ Sge, U Gem, and SS Cyg. Traditional and some recurrent novae fall into this category as well, having cool low-mass losers supplying the mass. Some recurrent novae, however, resemble symbiotic stars in having red giant companions, such as T CrB, RS Oph, and V3890 Sgr. Then there are the freaks. These include SS 433, which is best known for its 164-day precessional cycle; X Per; and V Sge, for which little is yet known about the nature of either the gainer or the loser. SS 433 has a 13-day orbital period and what appears to be a Wolf–Rayet companion, but the nature of the secondary is still a matter of debate.

5.6.1 Novae and X-ray Bursts: Surface Nuclear Explosions

Classical novae are optically marked by rapid increases in brightness of order 10 magnitudes from quiescence to maximum and decays on timescales of weeks to months. The *speed class* is distinguished by t_2 or t_3, the time required for the optical lightcurve to fall two or three magnitudes from maximum, respectively. Such systems do not show repeated outbursts. If they do, on timecales of order decades to centuries, they are classified as *recurrents*. Novae are all binary stars in which the gainer is a white dwarf. The loser can be either a compact star, an M-type star that can be either a main sequence star, or one that is hydrogen-poor, indicative of some stripping before or during the evolution of the system, or it can be a giant that is similar to the ones observed in symbiotic stars. No classical system is of the latter variety, but recurrents (V3890 Sgr, RS Oph, T CrB) are. On the other hand, recurrents can overlap the properties of classical systems (LMC 1990 No. 2 and U Sco have compact companions). X-ray novae, which have much shorter optical duration and emit most strongly at high energy, arise on accreting neutron stars in systems that otherwise closely resemble low-mass X-ray binaries. Several are known to repeat, but on shorter recurrence intervals than are recurrent novae. All show rapid rises in the optical and no opaque stage analogous to classical novae.

To this point, we have considered mass transfer between only normal stars, but there is no reason why mass cannot accrete onto a compact companion. Take, for instance, a white dwarf (WD) as the gainer. The pressure increases by an amount ΔP at the base of the accreted layer of mass ΔM depending on the star's surface gravity:

$$\Delta P = \frac{GM \Delta M}{4\pi R^4}, \qquad (5.65)$$

so the temperature steadily increases by compression. At a critical threshold, $\Delta P_c \approx 2 \times 10^{19}$ dyn cm^{-2}, the gas achieves the conditions that ignite nuclear reactions, in particular CNO.[28] Depending on the WD's composition, these may be either reactions on carbon or heavier elements, but these are details. The critical value of ΔM depends on the mass of the WD since its radius is mass-dependent through $R \sim M^{-1/3}$. Therefore, our scenario results in the scaling relation $\Delta M \sim M^{-7/3}$, the higher-mass WD does not have to accrete as much before the critical conditions are reached.

Now what happens? Nuclear burning starts at the base of the accreted layer, which is partially degenerate. The resulting release of energy by the reactions does not increase the pressure, although the temperature rises continually, resulting in a thermonuclear runaway. You have seen this before: the helium flash. But this time instead of being buried deep in the interior it is happening at the star's surface! The Fermi energy, ϵ_F, corresponds to temperatures of order 10^8 K, which can be reached as the nuclear source increases in luminosity. The layer crosses the degeneracy threshold when $T \geq \epsilon_F/k$, and at this point the layer rapidly expands and the reactions stop. In a stellar interior, at the time of the helium flash, the bulk of the mass has such a long thermal timescale that the core expands but without a large change in the envelope. Not so here. The stellar radius swells and a deep convection zone forms. The convection is driven by the usual condition that in order to transport the flux being produced by nuclear reactions, which is the Eddington luminosity, the temperature gradient must be superadiabatic. This turbulence drives deep mixing, which transports heavy elements into the burning zone, thereby providing fuel for the reactions and enhancing the nucleosynthesis. Products of the reactions, in particular ^{13}N and ^{15}O, are β-unstable and decay on timescales of order 100 s, which is short compared with the sound travel time through the now bloated envelope of the star. When these decay, they typically release of order 10 MeV/nucleon, sufficient to heat the accreted layer and blow it off the star. The crucial condition is, however, that there must be a substantial production of these nuclei, and this requires that the accreted gas be hydrogen-rich and that the WD must be overabundant in target CNO nuclei. Consequently, the layer is ejected very fast, with speeds exceeding the escape velocity from the WD, of order several thousand kilometers per second. The rapid expansion is accompanied by a dramatic increase in the optical brightness of the star, the rise indicating the onset of the classical nova outburst. When this happens during the helium flash, the overlying layers respond to the increased energy release on the thermal timescale but remain otherwise essentially hydrostatic. In a nova, there is nothing to prevent the explosion from ejecting the outer envelope.

As you can see from Eq. (5.65), the mass of the gainer determines the pressure at the base of the accreted envelope as a function of time. It therefore also determines the amount of mass that is accreted before the explosion is initiated. This implies that the more massive systems should throw off lower mass envelopes. Novae come in two distinct varieties based on ejecta abundance patterns. Most show strongly enhanced CNO elements and little else that has been changed.

---

[28] Starrfield et al. 1985, *ApJ*, **291**, 136; Livio et al. 1986, *ApJ*, **736**, 737; Starrfield, S. 1989, in *The Classical Novae*, M. Bode and A. Evans, eds. (London: Wiley-Halstead); Starrfield, S., Sparks, W. M., Truran, J. W., and Wiescher, M. C. 2000, *ApJS*, **127**, 485, and references therein. A catalog of LMXRB is Liu, Q. Z., van Paradijs, J., and van den Heuvel, E. P. J. 2001, *Astr. Ap.*, **368**, 1021.

These appear to come from roughly 1 $M_\odot$ white dwarfs, with progenitor masses in the range of a few solar masses. These prenova stars must have been able to complete helium core burning but then can have their envelopes removed through mass loss caused by the binary. Another group, the ONeMg subclass, show strongly enhanced abundances of Ne and heavier elements, for which higher core temperatures in the prenova star are required (so that α-processing has passed through carbon and continued through O, Ne, and Mg) and must therefore come from a relatively rarer progenitor. These more massive gainers are thought to be closer to the Chandrasekhar limit and therefore to represent possible precursors to SN Ia events.

The process takes place with increasing violence as the mass of the accreting white dwarf is increased. In current scenarios for type Ia supernova explosions, gainers at or near M_{Ch} appear to be involved. One of the likely possibilities is not substantially different from a nova explosion; the key exception is that the trigger is not hydrogen accretion but *carbon ignition*. The temperature sensitivity of this reaction is such that once it begins in the white dwarf, the energy release produces an explosion rather than a fizzle. The debate is currently over whether the nuclear reaction proceeds as a deflagration wave or as an actual detonation. In the latter, the expansion of the shock and the local energy release powers the continued expansion and subsequent burning. In the former, the process is subsonic and leads to sufficient energy release that the star disrupts by thermal conduction.

On a neutron star, the same fundamental process occurs except that the degeneracy is far greater and therefore lifts only after much a much higher luminosity is reached for the nuclear source. This is the mechanism for X-ray bursts. Nucleosynthesis can proceed much farther because of the higher densities and temperatures, since the Fermi level is $> 10^9$ K, compared with $(1-3) \times 10^8$ K for the most massive white dwarf in a nova system. Rapid proton capture calculations using burst thermal and luminosity profiles from hydrodynamic simulations[29] show that nuclear processing by the *rp*-process may extend far beyond the iron peak, up to ^{95}Mo, without disrupting the star. Very low mass loss is expected, of order 10^{-11} $M_\odot$, and the luminosities reach the Eddington value for a few seconds. Consequently, the event is *much* faster than for a classical nova, seconds compared with days.

5.6.2 Accretion by Magnetized Stars

AM Her systems are cataclysmics with magnetic white dwarf gainers. They display strong, periodically variable emission lines and sometimes periodic polarization variations. For the shorter-period binaries, the equality of the Hα and binary periods indicates that tidal interaction between the components has produced synchronous rotation of the gainer. When a magnetic field is present, it can dominate the accretion process, leading to a latitude-dependent flow, and also present an extended inner wall to any circumstellar disk. You can ask, for instance, whether it is possible to have corotation of the disk at the Alfven surface, $\omega(R_A) = \omega_\star$, which would be reached at $R_A = (GM_\star/\omega_\star^2)^{1/3}$. This is, in general,

---

[29] Woosley, S. E., ed. 1984, *High Energy Transients in Astrophysics* (NY: AIP Press); Schatz, H. et al. 1998, *Phys. Rep.*, **294**, 167.

not met by the disk, so there will be a mismatch at boundary of any magnetosphere. The field alters the accretion. Matter is not able to fall directly onto the gainer, but likely funnels toward the magnetic poles.[30] A normal disk does not form if the field is strong enough, that is, when the corotation radius reaches the Roche surface. Strongly magnetized neutron star binaries, those with X-ray pulsars, also show these effects. White dwarf fields are generally much weaker than those observed in even the weakest X-ray pulsar systems (10–100 MG compared with 10^3–10^6 MG). Yet for the accreting material, except in the immediate vicinity of the stellar surface, these two classes of gainer are not that different. Yes, the temperatures of the disks in the WD systems are lower because the gravitational potential is lower, but this is a detail rather than a basic difference.

The boundary layer, in which accretion occurs, is dominated by radial infall. Thus, we can ask where the balance occurs between the ram pressure of the accretion flow and the magnetic field, as we have just outlined for the AM Her systems. First, let us estimate how the result scales with observable parameters.[31] The infall velocity scales as $r^{-1/2}$ so the ram pressure, ρv^2, scales as $M_\star^{1/2} \dot{M} r^{-5/2}$. Since this is balanced by the magnetic pressure, which scales for a dipole field as $B_0^2 R_\star^6 \, r^{-6}$, we find that the critical radius scales as

$$r_c \sim \left(\frac{B_0^2 R_\star^6}{M_\star^{1/2} \dot{M}} \right)^{2/7}. \tag{5.66}$$

Since the virial theorem provides the luminosity of this region when the flow halts at the accretion shock, we find a scaling law for the emitted luminosity in terms of the Eddington luminosity, $\Gamma = L/L_{\text{Edd}}$:

$$\Gamma \sim \left(\frac{\dot{M}^9 M_\star}{B_0^4 R_\star^{12}} \right)^{1/7}. \tag{5.67}$$

Since the stellar properties are fixed independently of the accretion flow, we can estimate the maximum accretion rate that can be accommodated by a magnetized star by setting $\Gamma = 1$. The details depend on the structure of the accretion column and the optical depth of the flow, since the Eddington luminosity is not a limiting factor if the accretion does not completely cover the gainer.

Not only does $\omega_K(R_A)$ not generally equal $\omega_\star$; it also differs from that at the sonic point where infalling matter crosses an accretion shock. Regardless of the details, this mismatch should produce a beat frequency that can appear in the X-ray lightcurve of an accreting neutron star as a *quasiperiodic oscillation* (QPO).[32]

---

[30] See Campana, S. et al. 1998, *Astr. Ap. Rev.*, **8**, 279 for a general overview. See also Arons, J. and Lea, S. 1976, *ApJ*, **207**, 914; Ghosh, P. and Lamb, F. K. 1979, *ApJ*, **234**, 296 (and references therein, last of a series). There are many subsequent papers, but this series established the basic model. See also Shore, S. N., Livio, M., and van den Heuvel, E. 1994 (n.10).

[31] See, for instance, White, N. E. and Stella, L. 1988, *MNRAS*, **231**, 325.

[32] Alpar, A. and Shaham, J. 1985, *Nature*, **316**, 239; Lamb, F. K. et al. 1985, *Nature*, **317**, 681; see also van der Klis, M. 1989, *ARAA*, **27**, 517; Bildsten, L. et al. 1997, *ApJS*, **113**, 367; Psaltis, D. et al. 1999, *ApJ*, **520**, 763; Lee, U. 1999, *ApJ*, **525**, 386.

Where this becomes an interesting diagnostic is that the visibility of the QPO in this explanation depends on the structure and orientation of the accretion disk. A magnetospheric inner boundary for an accretion flow is not the same as a stellar boundary layer, even though within the disk there is a shear layer at which the velocity drops abruptly. It is generally far enough from the white dwarf—several stellar radii at least—that far less gravitational potential energy is thermalized there than at the base of the polar accretion column. If the disk undergoes an instability, that is, if the scale height varies as a result of relaxation from a mass transfer event and subsequent change in the viscosity, then the interface can be shielded. The region responsible for generating the modulated signal can therefore be hidden from the observer when the source luminosity is high, only to be seen when the disk has completely readjusted and returned to its usual lower luminosity state.

The QPO phenomenon was discovered with the EXOSAT mission, the first X-ray satellite that was able to perform long-term, high-time-resolution observations. Its limitation was that the data were sensitive to frequencies of order 100 Hz or less so only the lower-frequency oscillations were detected. The major advance in this field came with observations of QPOs with the *Rossi* X-ray Timing Experiment (RXTE), with which frequencies in the kilohertz range could be resolved. In this frequency range, interesting things show up.[33] First, the last stable orbit for a Schwarzschild black hole is at $6GM_\star/c^2$, which is at a frequency of around 30 kHz. For a neutron star, this is a factor of about 10 lower, so a signal in this range probes through orbital motion the equation of state of a neutron star. Most of the equations of state that best model pulsar behavior lead to maximum frequencies of around 1–1.5 kHz.[34] These limit the maximum neutron star mass to about 1.8 $M_\odot$, consistent with binary pulsars and accreting neutron stars. Lense–Thirring precession has also been suggested as a source, which works particularly well for black hole accreting systems.[35] One clue to the mechanism is the observation of splitting in the peaks at around 0.3–0.6 kHz, a feature attributed to the spin of the underlying neutron star. During bursts, this frequency is observed to systematically drift and has been interpreted as a change in the radius of the neutron star as it relaxes following an explosion, much like the relaxation of a white dwarf after a nova outburst.

5.6.3 Black Holes in Binary Systems

The search for binary black hole candidates began with the (now obvious) suggestion by Zel'dovich and Thorne (in 1967) that hard X-ray emission would be a signal of accretion onto a compact object. Using only radial velocities, a mass can be obtained for the unseen component and if this exceeds the maximum possible for a stable neutron star, the only alternative must be a black hole. The first such system, Cyg X-1, was discovered in 1972. Radial velocity measurements of HD 269858, the optical counterpart, by Bolton and Webster and Murdin showed that

---

[33] See van der Klis, M. 2000, *ARAA*, **38**, 717 for a review of the phenomenology and models.
[34] Miller, M. C., Lamb, F. K., and Psaltis, D. 1998, *ApJ*, **508**, 791; Schaab, C. and Weigel, M. K. 1999, *MNRAS*, **308**, 718.
[35] Morsink, S. M. and Stella, L. 1999, *ApJ*, **513**, 827.

TABLE 5.1 Properties of Some Binary Black Hole Candidates

System	$f(M)\,(M_\odot)$	P (days)	$M_X\,(M_\odot)$
HDE 226868 = Cyg X-1	0.24	5.6	>7
GS2023 + 33 = V404 Cyg	6.08	6.46	10–14
G2000 + 25	5.0	0.34	6–18
H1705 − 25	4.9	0.52	5–8
J1655 − 40	3.24	2.62	6.5–7.8
A0620 − 00	3.0	0.32	6 ± 3
GS 1124 − 68	3.01	0.43	4.5–6.2
GRO J0422 + 32	1.21	0.21	>4
4U 1543 − 47	0.22	1.12	2.7–7.5
LMC X-1	0.14	4.2	—
LMC X-3	2.3	1.7	>7

this O star is a binary with a period of 5.6 days and a large mass function that exceeded reasonable upper limits for a neutron star companion. Since then, many such systems have been detected, often accompanying the observation of X-ray nova outbursts. These are listed in Table 5.1.[36] The X-ray signature includes a very hard source (with a power law that extends to >100 keV) and for the X-ray novae strong variability at all wavelengths, often including a nonthermal radio source and even relativistic jets. Remember also that the main difference between a black hole and neutron star is the inner boundary condition.[37]

There were, and still are, serious problems with many of these systems because they do not eclipse and the high-mass X-ray binaries often have peculiar optical primaries for which the mass is uncertain.[38] However, refinement of the radial velocity measurements for this and other systems now firmly establishes that there *are* close binaries for which the gainer *must* be a black hole. Currently, the best example is actually an extragalactic system, LMC X-3. For this system, the distance is known (it is a member of the Large Magellanic Cloud at about 50 kpc) and that has a B3 V optical primary. The best studied galactic system is still Cyg X-1. Among the X-ray novae, which undergo outbursts that resemble classic novae but are far more energetic, V404 Cyg, N Mus 1991, and A0600-00 are the best candidates for black hole companions.

5.7 FORMATION OF BINARY SYSTEMS: A COMMENT OR TWO

Forming stars is hard enough to understand; forming multiple systems is even harder. Here we will be brief because the field is in a state of flux as computational

---

[36] Sources of Table 5.1 are Blandford, R. and Gehrels, N. 1999, *Phys. Today*, **52**(6), 41; Charles, P. 1999, in *Observational Evidence for Black Holes in the Universe*, Chakrabarti, S. K., ed. (Dordrecht: Kluwer), p. 279; Grindlay, J. E. 2000, in *Astrophysical Quantities*, 4th ed. A. N. Cox, ed. (Berlin: Springer-Verlag). For a catalog of HMXRB, see Liu, Q. Z., van Paradijs, J., and van den Heuvel, E. P. J. 2000, *Astr. Ap. Suppl.*, **147**, 25.

[37] See the especially useful review by Abramowicz, M. A. and Percival, M. J. 1997, *Class. Quant. Grav.*, **14**, 2003.

[38] Cowley, A. P. 1992, *ARAA*, **30**, 287; Tanaka, Y. and Shibazaki, N. 1996, *ARAA*, **34**, 607; Shore, S. N., Livio, M., and van den Heuvel, E. P. 1994 (n.10).

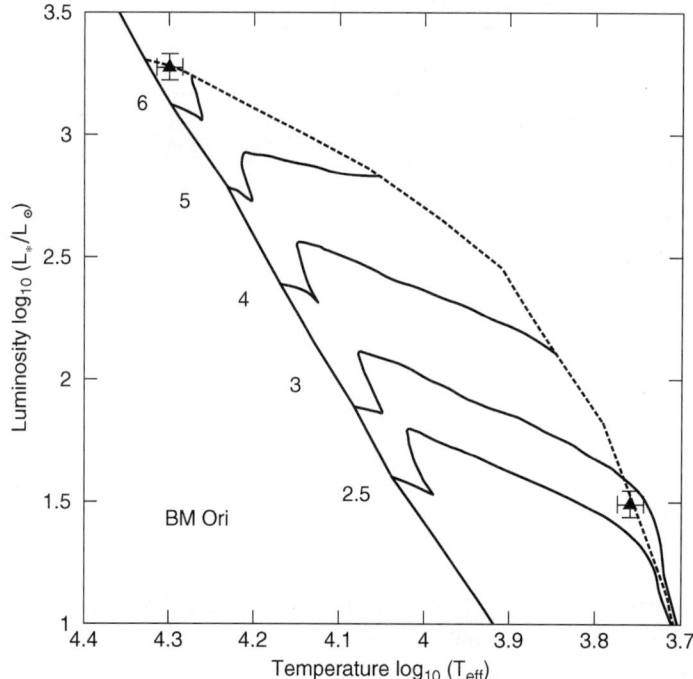

FIGURE 5.4. Isochrones for the component stars in the pre-main sequence Herbig Ae/Be binary system BM Ori. The birthline is shown as the dashed line, the pre-main sequence tracks are labeled by the mass in $M_\odot$. (Credit: F. Palla, Arcetri, and S. Stahler, UC-Berkeley.)

improvements lead to new simulations.[39] There are only a few confirmed *short period* pre-main-sequence systems—those for which complete orbits have been obtained, even where eclipses are observed—where the binaries are sufficiently young that they provide constraints on the approach to core hydrogen burning. But from these we can already perform many of the tests that have been used for main-sequence systems. For instance, we show in Figure 5.4 how to use such young systems to constrain evolution. Since they are presumed to be coaeval, even perhaps more than clusters, the two components should lie along isochrones. The example is BM Ori (see Table 5.2, bottom row), an eclipsing Herbig Ae/Be star in the Trapezium cluster in the Orion Nebula.

Pre-main sequence binary systems can also be uncovered using one of the oldest techniques in astronomy, lunar occultations. By happy coincidence, the Moon's orbit crosses Tau-Aur and ρ Oph, two of the nearest and richest star forming regions, and with high frequency sampling occultations can be used to obtain separations as small as 5 mas. At a distance of 100 pc this corresponds to 0.5 AU, about 100 $R_\odot$). Speckle observations and adaptive optics are also creeping into this

---

[39] The most recent work is summarized in Zinnicker, H. and Mathieu, R. D., eds. 2001, *IAU Symp. 200: Birth and Evolution of Binary Stars* (San Francisco: ASP Conf. Proc.). This field is changing rapidly and is one of the few that requires a conference citation as its principal reference.

TABLE 5.2 Sample Short-Period Pre-Main-Sequence Binaries[40]

Name	Period	e	i (deg)	M_1 ($M_\odot$)	M_2 ($M_\odot$)
RXJ 0529 + 0041	$3.^{\!d}04$	0.0	85.5	1.30	0.95
NTT 045251 + 3016	$6.^{\!y}9$	0.47	67.5	1.43	0.78
Parenago 2494 (Ori Neb)	$19.^{\!d}48$	0.26	—	$0.65 \sin^3 i$	$0.46 \sin^3 i$
Haro 1-14c (Oph)	$590.^{\!d}78$	0.62	—	0.018^c	—
V4046 Sgr	$2.^{\!d}42$	0^d	35	0.86	0.69
TY CrA ab[e]	$2.^{\!d}89$	0^d	—	3.0	1.6
Ty CrA[c]	270^d	0.51	20	1.25	—
BM Ori[a]	$6.^{\!d}47$	0^d	≈ 90	6.3	2.5

range, so it should soon be possible to know the fraction of binaries among these objects by a variety of techniques. It is already interesting that by all these techniques, simple population sampling is yielding the picture that multiple stars are perhaps the norm in star forming regions! If so, how might they form?

Rotation has always seemed the key to understanding the formation of binary star systems. Beginning in the mid-nineteenth century with the work of Jacobi on the stability of self-gravitating rotating ellipsoidal stars, attention focused on the equilibrium figures that a self-gravitating liquid mass can achieve. With the addition of compressibility, the extended classical picture is straightforward. A contracting rotating compressible star spins up. In the classical picture it becomes progressively more oblate within the sequence of Maclaurin spheroids. As we have seen, the eccentricity increases continuously with increasing angular frequency, Ω, but this does not continue indefinitely. Beyond a critical value, Ω_c, no further increase is possible *in spite of a monotonic increase in angular momentum beyond* Ω_c. Instead, as Jacobi showed, there is a sequence of *triaxial* equilibrium figures. The transition between these is a bifurcation in that multiple solutions become possible.[41] However, while this fission picture is attractive, it does not work. Instead, some sort of fragmentation seems to be required in which individual cores accrete within a pre-established self-gravitating disk. The collapsing system forms a disk that is then locally unstable to the formation of density waves, bars, and local density enhancements. Since we will encounter a similar problem when treating density waves in spiral galaxies, let's postpone our discussion of non-axisymmetric instabilities in such disks until Chapter 7.

Capture is also a reasonable possibility, and it is likely that both it and fragmentation are operating in different environments. For instance, any system that begins with a massive disk can form multiple components, at least in principle. This is quite likely in clusters where the chances are increased by the nearness of suitable companions. The gravitational cross section is basically the same one we computed in Chapter 1 and the rate depends on the velocity dispersion of the

---

[40] Notes in Table 5.2 are as follows: (*a*) Eclipsing system; (*b*) astronomic orbit from HST observations; (*c*) single-lined system, only $f(M)$ is quoted; (*d*) assumed circular orbit, isolated T Tau star; (*e*) Herbig Ae star, triple system, the close orbit is TY CrA ab, the wide system is TY CrA c, the primary is a subsynchronous rotator.

[41] Poincarè, H. 1903, *Theorie des Tourbillon* (Paris: J. Gabay); Chandrasekhar, S. 1969, *Ellipsoidal Figures in Equilibrium* (NY: Dover).

cluster and the local density. Since it is also likely that star formation occurs in groups, the high frequency of binaries now being discovered in young clusters accords with this idea. Capture is also rendered more efficient by circumstellar material that can be lost with the excess angular momentum of the companion. The signatures of this process are poorly known, except as we will now discuss for older systems, the blue stragglers.

5.7.1 Blue Stragglers: Collisions and Captures in Clusters

One of the puzzles of stellar evolution emerged with improved photometry of faint older open clusters and globular clusters in the 1970s. Sitting above the turn-off point in the cluster's HR diagram, one often finds a small clump of hot stars that are clearly members (Fig. 5.5). These are called *blue stragglers*. While unresolved binaries produce an extended, or parallel, main sequence, they cannot account for the location of these stars. There are several suggestions for how these stars form, all of which focus on the role of merging collisions after the star has formed.

From our discussion of mass transfer, you know that it can also move a main sequence gainer toward higher luminosity and temperature, essentially parallel to the main sequence. This does not extend the star's lifetime, however, so it would be unusual to catch it at the precise moment when it is still on the main sequence unless there are many such events occurring in the cluster.

It is *not* absolutely necessary for a star to move away from the main sequence as it evolves. You know that the increased molecular weight occurs in the core because that is where nuclear fusion occurs, and this drives the expansion of the envelope and produces the shift toward lower effective temperature and higher luminosity. But what would happen if, somehow, the star were homogenized? Then, as nuclear processing proceeds, the star would contract and move toward the blue on the HR diagram. Once the mixing ends, the interior composition

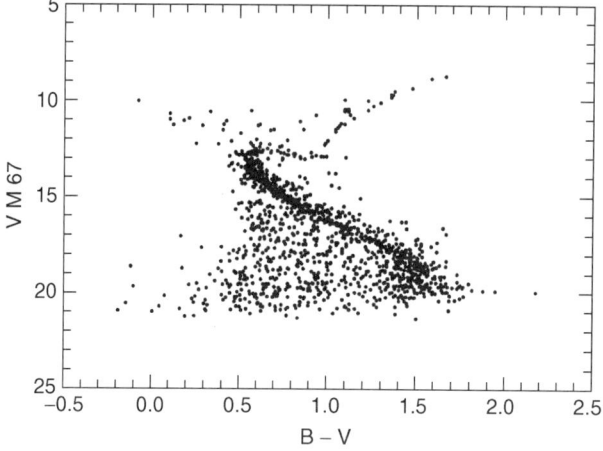

FIGURE 5.5. The old Galactic open cluster M 67. The stars on the main sequence above the turnoff point are the blue stragglers (data from Montgomery, K. A., Marschall, L. A., and Janes, K. A. 1993, *AJ*, **106**, 181).

gradients are re-established by continuing nuclear processing. Mixing also rejuvenates the stellar core by making available a new fuel supply from the envelope, and its lifetime is extended compared to a single, inhomogeneous star. The problem is getting this mixing to occur efficiently, when it happens, but to occur rarely enough that it is not a general feature of stellar evolution. Although it looked like the obvious explanation when first proposed, the core helium remains largely unredistributed in models of merging collisions. The resulting single star does undergo some mixing, and it is driven away from thermal equilibrium, but the theory has still not crystallized.[42] During the collision, as long as it is a glancing blow, substantial angular momentum may be carried away from the newly formed binary through large-scale mass loss, as we have described earlier in this chapter. Thus, even though we developed the formalism for the Roche radius and system evolution for cases of slowly evolving binaries, the same picture applies during the impulsive stage of binary formation and may lead to a blue straggler.

5.A SOME HYDRODYNAMIC DETAILS

As an aside, we discuss another route to the diffusion equation (Section 5.5.4.1). In conservative form, the angular momentum conservation equation is written

$$\frac{\Sigma v_r}{r} \frac{\partial}{\partial r} r v_\phi = \frac{1}{r} \frac{\partial}{\partial r} (r T_{r\phi}), \tag{5.68}$$

where $T_{r\phi}$ is the angular component of the stress, the part due to shear and viscosity. Notice that the radial velocity now becomes a function of the dissipation rate due to the shear. In its simplest form, you see that the shear depends on a radial gradient in the velocity, and we assume that the accreting material suffers only small perturbations from the radial drift—that Keplerian motion is not substantially disturbed. This entails the choosing of a functional form for the circulation, for which we took $v_\phi = v_K$. Having assumed this, the transport of material takes an especially lovely form. Since we can write the torque as a radial derivative of some physical property of the gas, we get a diffusion equation. The important feature of this step is at the drift timescales on the order of the sound travel time or slower; hence there is an accumulation of mass in the circumstellar environment. Turbulent or viscous heating depends on the shear. To see how, take the equation for a viscous flow

$$\frac{\partial v_i}{\partial t} = \eta \frac{\partial^2 v_i}{\partial x_j^2}$$

---

[42] An excellent starting point is the review by Stryker, L. L. 1993, *PASP*, **105**, 1081. For theoretical models of stellar collisions, see, for example, Lombardi, J. C. Jr., Rasio, F. A., and Shapiro, S. L. 1996, *ApJ*, **468**, 797; Sills, A. et al. 1997, *ApJ*, **487**, 290; Sandquist, E. M., Bolte, M., and Hernquist, L. 1997, *ApJ*, **477**, 335; Ouellette, J. A. and Prichard, C. J. 1998, *AJ*, **115**, 2539; Sills, A. and Baylin, C. 1999, *ApJ*, **513**, 428.

forms the scalar product with the velocity, v_i, and integrating over volume for a divergenceless flow:

$$\frac{\partial E}{\partial t} = -\eta \left(\frac{\partial v_i}{\partial x_j}\right)^2.$$

An alternative way to see this is assume that the stress tensor, T_{ij}, includes both viscosity and pressure, as well as bulk motions. Then the equation of motion is

$$\frac{\partial v_i}{\partial t} = -\frac{\partial T_{ij}}{\partial x_j}.$$

Then the scalar product and integration produce the result that

$$\frac{\partial E}{\partial t} = -\eta \int \sigma_{ij} T_{ij} \, dV,$$

where shear is

$$\sigma_{ij} = \frac{\partial v_i}{\partial x_j} + \frac{\partial v_j}{\partial x_i}.$$

If the viscous contribution to the stress is $-\eta(\partial v_i/\partial x_j)$, the energy dissipated by the fluid is proportional to $\eta \sigma_{ij}^2$. For cylindrical coordinates, the shear has the explicit form

$$\sigma_{r\phi} = \frac{\partial v_\phi}{\partial r} - \frac{v_\phi}{r} = r\frac{d\omega}{dr},$$

and the torque becomes

$$\tau = \frac{1}{r}\frac{\partial}{\partial r} \nu \Sigma r^2 \frac{d\omega}{dr},$$

as we found from the Navier–Stokes equation (Eq. 5.37).

6 The Interstellar Medium

This brave o'er hanging firmament, this majestical roof fretted with golden fire,
why it appears no other thing to me than a foul and pestilent congregation of vapours.
— Shakespeare, *Hamlet*, Act II, Scene 2

6.1 INTRODUCTORY REMARKS

How do you discover a part of the Universe that doesn't shine? The problem is as new as contemporary cosmology and as old as the nineteenth century. Although the interstellar medium (ISM) is of prime importance to the galaxy's evolution, its discovery came rather late in the development of astronomy.[1]

Observations of "nebulous" bodies began far earlier than the understanding of their physical nature. Ptolemy mentions several such objects in the *Almagest*, but none of these were truly nebulae; instead, they were unresolved clusters or small groups of stars. The first systematic catalog of such objects awaited refinements in telescope optics in the eighteenth century and the growth of interest in, of all things, comets. The first detailed list of nebulae was compiled by the French amateur Charles Messier at the end of the eighteenth century as a checklist of stationary objects to avoid that could be confused with comets. His descriptions were quite detailed, and this list cannot be dismissed as simply a compendium of nuisances. Messier's catalog was, however, not a taxonomic compendium; no particular schema were applied to describe these objects. It included examples of what we now know to be star clusters and galaxies as well as legitimate diffuse emission nebulae, and missed some rather important ones because of the latitude of Paris and Messier's observing conditions.

The further investigation of such objects was left to William, Caroline, and John Herschel, one of the most famous families in astronomical history. Following William Herschel's successes with the discovery of Uranus, he and Caroline began the systematic visual survey of selected zones of sky noting binary and other anomalous objects. This survey produced the *Index Catalog* (hence the designation IC), which included detailed positions and relative magnitudes as well as brief classifications and descriptions. Interestingly, the dark component of the interstellar medium could have been first understood through these observations since Herschel's analysis of stellar surface density was strongly biased by absorption effects from dust. He constructed, however, a purely phenomenological picture of galactic structure that did not include possible obscuration of more distant stars,

---

[1]For historical overviews of interstellar work, see Verschuur, G. L. 1989, *Interstellar Matters* (Berlin: Springer-Verlag) and Berendzen, R., Hart, R., and Seeley, D. 1976, *Man Discovers the Galaxies* (NY: Science History Publ.)

thereby setting the stage for later work. From his observatory in Capetown, John produced the first such survey of the southern hemisphere, in addition to his important analysis of the 1848 outburst of η Carinae. At around the same time, Bond (at Harvard) and Parsons, the Third Earl of Rosse (in Ireland), produced detailed drawings of several objects in particular M42, the Orion Nebula, and M1, the Crab nebula, that did not resolve into stars and showed distinctive filamentary structure.

Ultimately the spectrograph opened the way for understanding the nebulous objects of these catalogs and discovering the diffuse interstellar medium.[2] William and Margaret Huggins and Henry Draper discovered emission line spectra for M42, the Orion nebula, and M47, the Ring nebula in Lyra. E. E. Barnard produced the first photographic atlas of the Milky Way that showed, for the first time, clear evidence of *dark* nebulae with appearances that mimicked the emission regions. This discovery dovetailed nicely into F. Struve's explanation for the darkness of the night sky as due to dark material obscuring the view of distant stars, a problem to which we will return in Chapter 8. Soon thereafter, Wolf showed that the stars along a line of sight systematically dim with increasing distance and proffered the suggestion that it was due to intervening absorption. The discovery of the "true" diffuse phase came immediately on the heels of the observation of binary star radial velocity variations by Hartmann. He found that there were lines in the spectrum that did *not* move, in particular the Ca II H and K lines, and that these were seen in the direction of many hot stars. This, he argued, was evidence for line-of-sight absorption by gas. To complete this catalog of the earliest work, we mention the explanation by Kapteyn of the light echo from Nova 1901 Persei (GK Per) as being due to scattering from environmental material surrounding the nova.

The advances in the first half of the twentieth century came in quick succession. In 1927 Eddington discussed the ionization equilibrium of the diffuse gas and argued for very low density since both Na I and Ca II were the principal species observed in the stationary lines. Observers at Mount Wilson produced a systematic study of the radial velocity and profiles of these lines. At the end of the 1930s, Merrill pointed out the existence of several broad *diffuse interstellar absorption bands* (DIBs), in particular 4430 Å, toward a number of O stars, an enigma that remains unsolved. In a systematic study of galactic open cluster angular diameters and HR diagrams as a means for determining galactic structure, Trumpler in 1930 discovered that the faintest were also systematically redder and that the brightness followed the same law that Wolf had previously found. He argued that the reddening was due to dust, a suggestion that was theoretically examined in the context of scattering theory by Greenstein in 1938 and later van de Hulst in 1945. These last theoretical studies showed that the extinction can be explained by solid dielectric particles with sizes less than 1μ. Then in the years immediately following WW II Hall and Hiltner discovered interstellar polarization of starlight. Thus, by the middle of the century, the interstellar medium was known to consist of two distinct components, one atomic and the other solid, but otherwise relatively homogeneous. The physical structure of the medium was harder to interpret because of the nature of the observational data. Lines often showed multiple

---

[2] See Spitzer, L. 1982, *Searching Between the Stars* (New Haven: Yale Univ. Press) for a lovely semipopular account of the development of spectroscopy of the interstellar medium.

components at what was then the highest resolution, of order 10 km s^{-1}. It was not until the advent of radio astronomy, and the detection of the 21-cm ground-state transition of hydrogen, that it became clear that the gas is really organized into large clouds. On the other hand, using a stochastic analysis, Chandrasekhar and Münch had shown that the dust must also be similarly structured, and this agreed with the observations of patchiness in all images of the Milky Way. Line observations at sufficiently high resolution had often, by the early 1950s, shown multiple components for both neutral and ionized species, but with only single lines of sight available through this technique, structure was hard to infer.

Radio and millimeter observations are now the main source of our picture of distinct clouds existing as a separate phase from the diffuse gas. These are able to map *emission*, so in the same way that optical H II regions can be observed because you see line radiation from the ensemble of optically thin emitting gas, these line transitions make it possible to see the clouds of colder matter. First through the 21-cm line measurements of the neutral atomic hydrogen and then through numerous molecular transitions at millimeter wavelengths, the picture that has emerged is of a very inhomogeneous space between the stars, ranging in density from about 10^{-2} to $> 10^5$ cm^{-3} and covering an enormous range of kinetic temperatures, from a few kelvins to $> 10^6$ K. Added to this are measurements of the large-scale far-infrared emission from the dust, the same radiation that is obscured in the optical and is re-emitted by the cold solid phase at very long wavelengths.

We now know *three* main constituents of the interstellar medium. The gas comes in several forms, each distinguished by a range of density and temperatures: the ionized and neutral atomic gas and molecular clouds. The second phase is something between large molecules and solids, the dust. The final is magnetic fields, the large scale fields that are structured by the motion of the ionized gas and the smaller-scale fields that help support molecular clouds. In addition, the medium is traversed by cosmic rays, high-energy particles whose origins are still a question but which are certainly contributed in part by stars. We will now consider each of these in turn. We then turn to the dynamical processes that are observed in the medium.[3]

6.2 GAS

Broadly speaking, the gas comes in three states: ionized regions (H II regions), atomic gas (photodissociated regions and the diffuse neutral medium), and molecular clouds. Of these, the H II regions are certainly the most familiar and

---

[3]There are some superb monographs and textbooks on the ISM. These include Aller, L. H. 1987, *Physics of Thermal Gaseous Nebulae* (Dordrecht: Kluwer); Burton, W. B., Elmegreen, B. G., and Genzel, R. 1992, *The Galactic Interstellar Medium: Saas-Fee Advanced Course 21* Pfenniger, D. and Bartholdi, P., eds. (Berlin: Springer-Verlag); Duley, W. W. and Williams, D.A. 1984, *Interstellar Chemistry* (NY: Academic Press); Dyson, J. E. and Williams, D. A. 1997, *Physics of the Interstellar Medium*, 2nd ed. (Bristol: IOP Publishers); Hollenbach, D. J. and Thronson, H. A., eds. 1987, *Interstellar Processes* (Dordrecht: Kluwer); Kaplan, S. 1966, *Interstellar Gas Dynamics* (Oxford: Pergamon Press); Kaplan, S. and Pikelner, S. B. 1970, *The Interstellar Medium* (Cambridge, MA: Harvard Univ. Press); Osterbrock, D. 1989, *Astrophysics of Gaseous Nebulae and Active galactic Nuclei* (Mill Valley, CA: Univ. Science Books); Spitzer, L. 1979, *Physics of the Interstellar Medium* (NY: Wiley).

photogenic, and they were the first to be clearly recognized as gaseous. Beginning with the introduction of spectroscopy to astrophysics, these regions were identified by the presence of strong emission lines and the absence of obvious stellar spectra or resolution into point sources. At a time when the existence of gaseous regions was a core issue for the formation of the planetary system, the discovery of diffuse regions was a vital link between the nebular hypothesis of Laplace and observations. Photographic surveys of the Milky Way, especially by Barnard at Yerkes, discovered many of these regions and they were included in Dreyer's visual catalog of nonstellar objects, the *New General Catalog* (NGC), without discriminating between gaseous nebulae and galaxies (the separation came only later with the understanding that galaxies are stellar systems that are external to the Milky Way; see Chapter 7).

It took considerable time for the nature of these regions to be understood. Theoretical studies began in the 1930s with the pioneering work by Strömgren, who examined the ionization equilibrium of an interstellar hydrogen cloud irradiated by a strong ultraviolet continuum source and derived the equilibrium radius expected for such objects. These are now called *Strömgren spheres* and represent the simplest solution, as we will see, to the problem of the static structure of an ionized gas in the ISM. The interpretation of their spectra began soon thereafter with the calculation by Menzel and Pekeris of the recombination spectrum of hydrogen in ionized regions and the extension of atomic structure calculations to the determination of physical constants for important optical lines by Menzel, Aller, Seaton, and their collaborators in the 1940s and 1950s.[4]

6.2.1 Ionized Regions: Emission Nebulae

The gaseous component of the interstellar medium was discovered in the 1960s during the first epoch of astronomical spectroscopy when Huggins obtained a spectrum of the Orion nebula. This and several other regions, all of which were diffuse and already called *nebulous*, showed strong emission lines. Other objects, the globular clusters and some that we now know are galaxies, showed only the absorption lines and continua of stars and were clearly different. On the basis of the laboratory work by Bunsen and Kirchhoff, Huggins and later many others argued that these emission regions were truly gaseous nebulae, and were possibly the sites that Laplace had hypothesized as the birthplace for stars. The expansion of photographic spectroscopy after the beginning the twentieth century saw considerable improvement in the line identifications. In particular, the He I $\lambda 5876$ Å line that was known from solar chromosphere observations, indicated that the conditions in nebulae might be similar to those in the outer solar atmosphere.

One puzzle dominated much of the spectroscopic discussion through the 1920s, a pair of intense lines at 4959 and 5007 Å that had no counterparts in any of the line lists then available for low ionization species. Named *nebulium*, these lines were first thought to represent a new element that was not previously known from laboratory chemistry, on analogy to the identification of helium from the D_3 line in

---

[4]The founding papers have been reprinted in Menzel, D. H., ed. 1962, *Physical Processes in Ionized Gases* (NY: Dover Books). The most detailed monograph is Osterbrock, D. E. 1989, *Astrophysics of Gaseous Nebulae and Active galactic Nuclei* (Mill Valley, CA: Univ. Science Books).

the Sun's spectrum. The answer came in 1927 when I. S. Bowen, who at the time was working on the oxygen spectrum with R. Milliken, noticed an important coincidence[5] in the energy levels of O III and the nebulium lines. These were accessible only through ultraviolet excitation within the Lyman continuum of He II. However, he reasoned that the ionization potentials of O III and He II were nearly identical and therefore could coexist. His basic argument was similar to that later employed by Zanstra, that every UV photon absorbed in the upper state of a permitted transition could, if the density were sufficiently low, de-excite radiatively into the observed optical lines. Specifically, the uppermost state $p^2\ {}^1S_0$ decays via the 4363 Å line to $p^2\ {}^1D_2$, which in turn forms the *nebulium* lines by a decay to the split lower state $p^2\ {}^3P$. Bowen reasoned that if the density were sufficiently low and the source of ionizing radiation sufficiently hot, it would be possible to radiatively excite, or *fluoresce*, the pair of nebular lines.[6]

Historically, the next step came with a query from W. H. Wright, then director at Lick Observatory, concerning anomalous intensity ratios of some lines in the region around 3400 Å. We now know that these lines, due to O III, are true fluorescence transitions, coupled to He II Lyβ.[7] They are analogs of Raman scattering, since the transition responsible for the Bowen lines, as they are now called, photons via absorption into longer wavelength emissions. Their strengths, and profiles, come from the emissivity at He II and they are formed completely within the ionized region, where the helium Lyman series is optically thick. The complexity of these radiative processes reinforces what we have been saying about the ISM being like an extremely extended and tenuous stellar atmosphere. The actual excitation process is even more finely tuned than simple continuum absorption. It depends on the near coincidence of the He II Lyβ line with the upper state of several O III 3000–3200 Å multiplets. Such coupling is well known for other light ions such as O I, where [O I] 1641 Å is formed from absorption of a Si III multiplet as well as from the ground state of O I at 1302 Å, and for many optical and ultraviolet lines observed from [Fe II] in dense stellar winds, in particular lines pumped by C IV 1550 Å that radiate at about 2700–2900 Å. Because of the strength of the He II Lyman series, though, the O III nebular lines are especially spectacular.

6.2.1.1 Nebular Lines

There are two basic differences between the interstellar medium and stellar atmospheres. One is that nebular gas is diffuse and not in hydrostatic equilibrium. This means that the thermal balance of the medium is not constrained by the

---

[5] The principal paper on the subject is Bowen, I. S. 1928, *ApJ*, **67**, 1: "The Origin of the Nebular Lines and the Structure of Planetary Nebulae." An interesting historical discussion is presented by Hirsh, R. F. 1979, *Isis*, **70**, 197, which reviews the observational problem at the end of the nineteenth century and the development of Bowen's explanation.

[6] With this came Russell's well-known statement that *"nebulium* had vanished into thin air." Don't forget this quote! It will help you understand much of the physics of such transitions. Bowen (1928) also notes that Russell, Dugan, and Stewart (in 1926) had conjectured that the lines must be due to a "usual species radiating under unusual conditions" and pointed to one of the light elements, like C, N, or O, as a likely culprit.

[7] The sequence of events leading up to this is related by Babcock, H. W. 1979, *Biogr. Mem. Nat. Acad. Sci.* (USA), **53**, 82.

requirements of mechanical stability. The second is that the density is always extremely low, less than 10^6 cm^{-3}, but this is a statement that we need to justify. One simple way you know this is that when you look at stars in extended nebulae, they are point sources. This means that the scattering optical depth for optical continuum photons is small. Taking a typical nebular size of around 1 pc, this means that the electron scattering optical depth implies a column density $N_e < 10^{24}$ cm^{-2}, which means $n_e < 10^6$ cm^{-3}. Under such conditions, LTE is obviously a ridiculous assumption. Since the mean collision times are of order days or weeks, not nanoseconds, we require the full treatment of statistical equilibrium in order to obtain the appropriate nebular diagnostics of density and temperature. We must also solve the ionization balance explicitly as well rather than employing the simpler Saha equation.

In the low-density limit, which is more typical of the ISM than a stellar atmosphere, nonlocal radiative processes are balanced by local collisions. Both are rare; neither is in strict LTE, and often neither is at the kinetic temperature of the local gas. But it is only through the emission and absorption spectrum of the gas that we can determine its physical composition and thermal state. You are already familiar with the rate equations from our treatment of NLTE in Chapter 3. The main difference is that the radiative rates may be driven by a source that has no immediate connection, other than radiative, with the observed gas. In steady state, using Eq. (3.5), we again write

$$\sum_{j>i} n_i(C_{ij}n_e + B_{ij}I_{ij}) = \sum_{j>i} n_j(A_{ji} + B_{ji}I_{ij} + C_{ji}n_e). \qquad (6.1)$$

The governing parameter is $A_{ji}/n_e C_{ji}$, the ratio of the spontaneous radiative deexcitation to collisional rates. You know that this scales as $A_{ji}T^{1/2}/n_e$. At high density, as in a stellar atmosphere or wind, collisions rapidly thermalize the level populations and the line emissivity decreases. This is why many transition observed in nebulae are not seen in the laboratory; the densities are simply too high in the terrestrial sources. For most intercombination and forbidden lines, $A_{ji} < 0.1$. A way of estimating the collision rate is to assume a hydrogenic value for the cross section. Then $t_c^{-1} \approx 3 \times 10^{-8} n_e$ s^{-1}, so you see that very low densities are required to not thermalize the upper states of these particular transitions.

To illustrate how the nebular diagnostics are obtained, let's take a three-level atom in a hot gas without an illuminating radiation field. We have the condition that normalizing to the total number of atoms, the populations are given by $n_1 + n_2 + n_3 = 1$, so this is a branching problem. The rates are determined by the medium (the density and temperature control the collisions), but the radiation field is actually often simpler to treat for nebulae than stars because the transitions are often forbidden and optically thin. Therefore, the rates can be specified in advance and the level populations solved accordingly. Call the generalized probabilities P_{ij} for the transitions $i \to j$. Then the steady-state rate equations are given by

$$(P_{12} + P_{13})n_1 = P_{21}n_2 + P_{31}n_3$$
$$(P_{21} + P_{23})n_2 = P_{12}n_1 + P_{32}n_3$$
$$(P_{31} + P_{32})n_3 = P_{13}n_1 + P_{23}n_2, \qquad (6.2)$$

so the ratios of the populations are

$$\frac{n_2}{n_1} = \frac{P_{32}[P_{12}(P_{31} + P_{32}) + P_{13}P_{23}]}{P_{23}[(P_{21} + P_{23})(P_{31} + P_{32}) - P_{23}P_{32}]} \quad (6.3)$$

with a similarly general expression for n_3/n_1. But insight is really gained only by being more specific, so let us now neglect external radiation and treat only collisional and spontaneous processes. The rate equations now become

$$n_1 C_{12} n_e + n_3(C_{32} n_e + A_{32}) = n_2[A_{21} + n_e(C_{21} + C_{23})]$$

$$n_1 C_{13} n_e + n_2 C_{23} n_e = n_2[A_{31} + A_{32} + n_e(C_{31} + C_{32})]. \quad (6.4)$$

For O III, this three-level treatment is especially relevant. You know that $C_{13} \ll C_{12}$. Further, you also know that, for collisions, the excitation rate is usually smaller than the de-excitation rate. The population ratios are

$$\frac{n_2}{n_1} = \frac{C_{12} n_e}{A_{21} + n_e C_{21}} \quad (6.5)$$

$$\frac{n_3}{n_1} = \frac{A_{21} + C_{12} n_e}{A_{31} + A_{32} + n_e(C_{31} + C_{32})}. \quad (6.6)$$

Finally, we can assume the low-density limit, recalling Russell's remark about ordinary species radiating under extraordinary circumstances, so $n_e C_{21} \ll A_{21}$ and we find

$$\frac{n_2}{n_1} = \frac{C_{12} n_e}{A_{21}} \quad (6.7)$$

$$\frac{n_3}{n_1} = \frac{n_e C_{13}}{A_{31} + A_{21}}. \quad (6.8)$$

Even without using the detailed energy dependence of the excitation cross section, you already know that C_{ji} scales as $n_e T^{1/2}$ so since collisional suppression of emission sets in at $C_{ji}/A_{ji} > 1$, emission can be used to place bounds on the density *and* temperature of the gas through a carefully chosen comparison of line strengths. Forbidden lines are especially important in the determination of gas conditions in nebulae since in the low-density limit the line ratios depend only on the collisional depopulation rate of the particular transition and the branching ratio:

$$j_{21} = \frac{h\nu_{21}}{4\pi} n_e C_{21} n_1, \quad (6.9)$$

$$j_{31} = \frac{h\nu_{31}}{4\pi} n_e C_{31} n_1 \frac{A_{31}}{A_{31} + A_{32}}, \quad (6.10)$$

$$j_{32} = \frac{h\nu_{32}}{4\pi} n_e C_{32} n_1 \frac{A_{32}}{A_{31} + A_{32}}. \quad (6.11)$$

466 THE INTERSTELLAR MEDIUM

The temperature dependence of the emissivity is due entirely to the collision rates, which scale as $T^{-1/2} \Omega_{ji}$, where Ω_{ji} is the *collisional efficiency* that we introduced in Chapter 3. *The lines are therefore coolants for the medium since each collisional excitation robs the kinetic (thermal) pool of the gas to create a photon, which leaves the optically thin medium.*

In spite of their low transition probabilities, many forbidden lines happen to occur at optical wavelengths arising from ground states of abundant ions. These are principal energy loss channels for the gas in low density regions, although the collisional suppression of the transitions and their linear dependence on abundance make them less effective for high-density and low-metal-abundance environments.

Fine-structure transitions of metallic ions in the far infrared provide a major coolant in warm (of order 10^2 K) gas. For instance, the excitation of the ground state of the [O I] 63μ line is 158.3 cm^{-1}, or about 100 K. For the interstellar medium, one of the most important lines comes from the splitting of the resonance state of C II (see next section on resonance transitions). The ground state of this ion has a forbidden transition, [C II] 158μ, that is observed from the boundaries of molecular clouds in the photodissociation region (Section 6.7.2 below). In regions of very low density, the transition can be easily excited, and because of its long wavelength it easily escapes. We will return to this in a moment when discussing cooling and heating in the gas phase.

Another especially interesting feature of the By using pairs of line pairs from the same species, you remove a major uncertainty in the analysis, the correction for the ionization fraction. For instance, the S II ion has two optical forbidden lines (see Table 6.1; see Table 6.2 for fine-structure transitions).

Including external radiation introduces many important effects that are not normally encountered when treating stellar atmospheres. For instance, the same three-level emitter, when pumped by incident radiation, can produce a very different result. This is the *nebulium* solution—the gas is illuminated by a spectrum that has a very high color temperature although it is from a distance source and thus low intensity. The radiative excitation rate for very high levels can exceed the collisional deexcitation if the density is low enough, producing emission from lower-energy transitions.

6.2.1.2 Recombination Lines

Ionization complicates this simple picture of line formation. If we were to isolate the atom and if its ionization state were to remain fixed, we would be finished.

TABLE 6.1 Some Forbidden Transitions of Interest

Wavelength (Å) (vac)	A_{ji} (s^{-1})	g_l	g_u	Wavelength (Å) (vac)	A_{ji} (s^{-1})	g_l	g_u
[S II] 4069.75	3.41×10^{-1}	4	4	[S II] 4077.50	1.34×10^{-1}	4	2
[S II] 6718.67	4.3×10^{-4}	4	6	[S II] 6732.67	1.56×10^{-4}	4	4
[O I] 6302.05	5.63×10^{-3}	5	5	[O I] 6365.54	1.82×10^{-3}	3	5
[O II] 3727.09	1.59×10^{-4}	4	4	[O II] 3729.88	2.86×10^{-5}	4	6
[O III] 4364.44	1.71×10^{-0}	5	1	[O III] 4960.29	6.21×10^{-3}	5	5
[O III] 5008.24	1.71×10^{-2}	5	5	[N II] 3063.71	3.15×10^{-2}	3	1
[N II] 6549.86	9.19×10^{-4}	3	5	[N II] 6585.27	2.72×10^{-3}	5	5

TABLE 6.2 Some Important Fine Structure Transitions

Wavelength (μ)	Transition	A_{ji} (s^{-1})
[S IV] 10.5	$^2P^\circ_{1/2} - ^2P^\circ_{3/2}$	7.70×10^{-3}
[Si II] 34.8	$^2P^\circ_{1/2} - ^2P^\circ_{3/2}$	2.13×10^{-4}
[O III] 51.8	$^3P_1 - ^3P_2$	9.76×10^{-5}
[O I] 63.2	$^3P_2 - ^3P^\circ_1$	9.91×10^{-5}
[O III] 88.4	$^3P_0 - ^3P_1$	2.61×10^{-5}
[O I*] 145.5	$^3P^\circ_1 - ^3P^\circ_0$	1.75×10^{-5}
[C II] 157.7	$^2P^\circ_{1/2} - ^2P^\circ_{3/2}$	2.30×10^{-6}
[C I] 609.7	$^3P^\circ_0 - ^3P^\circ_1$	7.88×10^{-8}

However, this is not usually the case since for H II regions, the ionization is produced by the stellar radiation. Thus each level has a rate $n_i R_{i\kappa}$ of absorption of an ionizing photon, Eq. (3.71), which reduces the level population and weakens the emission lines. When we were solving for the ionization balance in a dense medium such as a stellar atmosphere, we were able to make use of the Saha equation. Recombination was handled by detailed balance of the collisional rates without requiring solution of the full population equations. When the density is low enough, this is no longer possible. We need to explicitly include how recombinant cascades populate the levels, since the reacquired electron can be captured into virtually any atomic state from which it cascades down to the ground state.

We will concentrate on hydrogen because it is the most abundant, and also the most important, recombining species.[8] The process, however, is basically the same for all atoms. Hydrogen recombination is separated into two cases depending on the opacity of the medium. In *case A*, the ultraviolet Lyman series is assumed to be optically thin. Cascades proceed without the interference of reabsorption and radiative population of the levels. Every photon ultimately escapes from the gas, and the net effect is to cool the medium. If the gas is optically thick in the Lyman lines, *case B*, a full solution of the NLTE transfer problem is required to obtain the level populations. For case A every cascade leads to either the two-photon $2s \to 1s$ transition (the two-photon emission is required to conserve angular momentum for $\Delta l = 0$ transitions in hydrogenic atoms, resulting in a broad, continuum-like emission peaking at ~ 2400 Å) or to a Lyα photon (which is far more probable). In a dusty H II region. line photons that would otherwise escape through the gas are absorbed by grains. This heats the dust and produces far-infrared emission. The line ratios are, however, essentially unaltered except for the effects of foreground extinction. Finally, if the absorption lines of the Lyman series are very optically thick, *(case B)*, all line ratios will be altered among the Balmer series. For instance, suppose Lyγ line is very opaque. Then any photon emitted when the electron jumps from the $n' \to n = 4$ level sees a larger population in that level than would be expected without photon trapping. The $n = 3 \to n = 2$ transition will also have a different strength, so the Hβ/Hα ratio will not be the same as case A. For each recombination leading to the lowered ion state, the

---

[8] For the standard set of recombination rates, see Brocklehurst, M. 1970, MNRAS, 148, 417; Brocklehurst, M. 1971, MNRAS, **153**, 471. Storey, P. J. and Hummer, D. G. 1995, MNRAS, **272**, 41 (and references therein) provide recombination line intensities for hydrogenic ions, and Hummer, D. G. 1994, MNRAS, **268**, 109 provide total recombination and energy loss coefficients for low density.

total recombination coefficient is actually the sum of the rates to the individual (n, l) states:

$$n_e N_r \alpha_{nl}^{eff} = n_e N_r (\alpha_{nl} + n_e \beta_{nl}) + \sum_{n'=n+1}^{\infty} \sum_{l'=l\pm 1} n_{n'l'}(A_{n'l'nl} + n_e C_{n'l'nl}), \quad (6.12)$$

separating out the direct (α_{nl}) and three-body (β_{nl}) recombination rates. Otherwise the same notations are used as we employed for stellar atmospheres. The problem is the same, except that recombination in the high densities of the stellar environment will be more likely to be in the case B limit where three-body interactions are important. Remember that in the case B limit, the populations of the levels must be solved using the mean radiation field, J_ν, which can come only from a coupled NLTE solution of the line formation.

If there is more than one electron in the atom, recombination proceeds through several channels. One is to produce an internal excitation in the atom through close coupling of the states, called *dielectronic recombination*. In general, this increases the cross section for the recombination and also changes the ionization balance. It is the inverse of autoionization. At interstellar densities, collisions do not populate the highest atomic levels and broadening is so small from these perturbations that the levels remain distinct. As a result, recombinations, which must pass through the entire sequence of principal quantum numbers to reach the ground state in a cascade, produce emission lines whose intensity depends only on the temperature and density of the gas. The absorption coefficient is given by the usual expression:

$$\kappa_\nu = n_i B_{ij} - n_j B_{ji},$$

but it is more illustrative to use the departure coefficients:

$$\kappa_\nu = n_i B_{ij}\left(1 - \frac{b_j}{b_i} e^{-h\nu/kT}\right), \quad (6.13)$$

which, as we discussed in Chapter 3, are defined by $b_j = n_j/n_j^{LTE}$. In case A, the ground state is strongly overpopulated by recombinations while the excited states are depleted. As you approach the continuum, however, collisional coupling of the levels becomes stronger and $b_j \to 1$ for sufficiently high j. For radio recombination lines, this means that the emission rates are given by $j_\nu = \kappa_\nu B_\nu$ here the level populations are determined by the LTE conditions.

When the density is low enough, however, the requirements of LTE ionization balance no longer apply. Then the ionization rate depends on the match between the cross section for bound-free transitions and the spectral distribution of the radiation. Recombination, on the other hand, depends on the coincidence between the continuous energy distribution of the electrons and the capture cross section.

Recall, from Chapters 1 and 3, that the ionization rate is given by

$$R_{\text{ion}} = 4\pi \sum_i n_i \int_{\nu_{0,i}}^{\infty} \kappa_{\nu,i} J_\nu \frac{d\nu}{h\nu} = n_i \beta, \tag{6.14}$$

since we must include the ionization out of each atomic state. Under the conditions of the diffuse interstellar medium, the ground state is a good approximation for ionization, although recombination can occur into every state. Once collisions become important, as in dense H II regions or shocks, then higher excited levels must be included. The extension of the rate equations to include ionization and recombination explicitly produce a far more complicated statistical equation to solve, if not to write, because recombination populates levels that would otherwise not be collisionally excited and also couples these levels together through cascades:

$$n_i R_{ik} = 4\pi n_i \int_{\nu_0}^{\infty} \kappa_\nu J_\nu \frac{d\nu}{h\nu}. \tag{6.15}$$

The inverse rates are given by

$$n_k R_{ki} = 4\pi n_k \int_{\nu_0}^{\infty} \kappa_\nu J_\nu e^{-h\nu/kT} \frac{d\nu}{h\nu} = 4\pi n_k \int_{\nu_0}^{\infty} \kappa_\nu B_\nu (1 - e^{-h\nu/kT}) \frac{d\nu}{h\nu}. \tag{6.16}$$

Thus, the population equation reads

$$n_i \int_{\nu_0}^{\infty} \kappa_\nu J_\nu \frac{d\nu}{h\nu} = n_k \int_{\nu_0}^{\infty} \kappa_\nu \left[\frac{2h\nu^3}{c^2} + J_\nu \right] \frac{d\nu}{h\nu}. \tag{6.17}$$

The radiation source is generated independent of the emission region, coming in general from an embedded star, and is specified separately from the conditions in the gas. This makes the nebular problem slightly less complicated than the one for an atmosphere. In addition, in the limit where collisions cannot compete with radiative de-excitations, the statistical equations can be greatly simplified. For case A, the level populations are given by

$$n_i \left(R_{ik} - \sum_{j<i} A_{ij} \right) = \sum_{j>i} n_j A_{ji} + n_e \alpha_i. \tag{6.18}$$

You can see immediately how case B alters the populations. If the lines are opaque, then stimulated rates must be included as well as the line radiation, which *is* intrinsic to the nebula. Its frequency dependence also means that stimulated emission can be very important for the formation of absorption lines since $\hbar\omega/kT$ may be small.

TABLE 6.3 Hydrogenic Recombination and Free–Free Energy Loss Coefficients

$\log T_e$	$T_e^{1/2}\alpha_1$	$T_e^{1/2}\alpha_B$	$T_e^{1/2}\beta_{ff}$
2.6	1.644	5.599	1.145
2.8	1.642	5.146	1.161
3.0	1.640	4.700	1.181
3.2	1.638	4.258	1.202
3.4	1.629	3.823	1.224
3.6	1.620	3.397	1.248
3.8	1.605	2.983	1.274
4.0	1.582	2.584	1.303
4.2	1.548	2.204	1.335
4.4	1.499	1.847	1.369
4.6	1.431	1.520	1.403
4.8	1.341	1.226	1.434
5.0	1.227	9.696	1.460
5.2	1.093	7.514	1.478
5.4	0.945	0.570	1.485
5.6	0.792	0.426	1.482
5.8	0.643	0.312	1.468
6.0	0.506	0.224	1.444
6.2	0.387	0.159	1.414
6.4	0.288	0.111	1.380

For radio recombination studies (see Table 6.3[9]), the situation is simplified by the nature of the atomic energy levels. Highly excited states of large n are the first to be populated in the recombination process. These *Rydberg* states are sufficiently far from the nucleus as to be essentially hydrogenic and therefore

$$\hbar\omega_{(n,n-1)} = E_n - E_{n-1} \approx 2\hbar c Z_{\text{eff}}^2 \text{Ry}\frac{\Delta n}{n^4} \qquad (6.19)$$

for $n \gg 1$. Here Z_{eff} is the screened effective charge of the nucleus and Ry is the Rydberg constant:

$$\text{Ry} = \text{Ry}_\infty\left(1 - \frac{m_e}{M}\right), \qquad (6.20)$$

where M is the total atomic mass including the electrons and $\text{Ry}_\infty = 1.0974 \times 10^5$ cm^{-1} is the Rydberg constant for infinite mass. Small isotopic shifts are thus possible because of changes in the nuclear mass, and different elements produce slightly shifted lines due solely to changes in Ry In addition, the transition probabilities obey a simple scaling relation to the principal quantum number:

$$A_{n,n-1} \approx 6.1 \times 10^9 \, Z^4 n^{-5}. \qquad (6.21)$$

[9]Units of 10^{-11} cm^3 s^{-1}. From Hummer, D. G. 1994, *MNRAS*, **268**, 109. The recombination rate is $n_e n_Z \alpha_n(T,Z)$, and the energy loss rate is $kT_e n_e n_Z \beta_{ff}$. For species other than hydrogen the scaling is $\alpha_{n,Z} = Z\alpha_n(T_e/Z^2)$ and $\beta_{ff,Z} = Z\beta_{ff}(T_e/Z^2)$.

TABLE 6.4 Some Nebular Line Ratio Diagnostics

Species and Line Ratio	Diagnostic	Remarks
[O I] (6300 + 6363)/5577	T_e	—
[S II] (6716 + 6731)/(4068 + 4077)	T_e	Reddening
[S II] 6716/[S II] 6731	n_e	Shocks
[O III] (4959 + 5007)/4363	T_e	—
C II] 2326/2328	n_e	—
Si III] 1895/Si III 1206	T_e	—
Si III] 1895/C III] 1909	n_e	—
C III] 1907/1909	n_e	Low density
[N II] (6548 + 6583)/5755	T_e	—
O IV] 1401/1405	n_e	—

Collisional broadening, weak though it may be in the diffuse medium, ultimately limits the number of levels that can be observed. The last observable recombination line can be used to determine the electron density in the gas in much the same way we discussed for estimates of stellar surface gravities from the interrupted convergence of the Lyman and Balmer hydrogen absorption lines series at an ionization edge. These lines are important for the study of trace species in the ionized gas since every recombination will emit a photon.

A complication arises with radio recombination lines because of the inevitable contribution of the thermal bremsstrahlung continuum. When the densities are high and the medium is optically thin, this continuum can actually pump the levels, resulting in stimulated emission. The intensity is not enough to generate maser action, but it does alter the expected line ratios. For optical lines, however, this can be ignored. The line ratios, because they have different branchings, are used to determine the physical conditions in the nebula (see also Table 6.4[10]). This can also be seen from Eq. (6.3). The governing parameter, the collision rate, depends on the electron density and temperature. Different lines have different sensitivities to this ratio, so line *ratios* can be used to determine the gas properties, with only a few assumptions. The principal use is the determination of reddening to the source, if you can be sure that the medium is optically thin (case A). Then the $H\alpha/H\beta$ and $H\gamma/H\beta$ ratios are fixed, and the same is true for the hydrogenic lines He II 4686 and He II 1640.

There are a number of problems with this picture. The most obvious is that nebulae are decidedly inhomogeneous. You need only look at any picture of a galactic H II region or planetary nebula to feel uneasy with the assumption of uniformity. It isn't clear, of course, whether you are seeing a density or emissivity effect. They are very hard to sort out, and this usually becomes clear only once images taken in different lines are compared. However, using flux-calibrated

---

[10] Several standard photoionization codes are publicly available for predicting emission line spectra. Of these, Cloudy (Ferland, G. J., Korista, K. T., Verner, D. A., Ferguson, J. W., Kingdon, J. B., and Verner, E. M. 1998, *PASP*, **110**, 761); MAPPINGS (Binette, L, Dopita, M. A., and Tuohy, I. R. 1985, *ApJ*, **297**, 476; Dopita, M. A. and Sutherland, R. S. 1995, *ApJ*, **455**, 468); the Shaw–Dufour code (Shaw, R. A. and Dufour, R. J. 1996, *PASP*, **107**, 896), XSTAR (Kallman, T. R. and McCray, R. 1982, *ApJS*, **50**, 263), and the Raymond–Smith code (Raymond, J. C. and Smith, B. W. 1976, *ApJ*, **204**, 290) are the most widely used.

images, maps of any of the appropriate plasma diagnostics can be directly obtained by taking ratios of appropriately calibrated images.[11] Here lies the principal difference between analyzing emission lines from nebulae and stellar envelopes. In general, nebulae are spatially resolved, so many of the ambiguities that plague the interpretation of stellar spectra can be sorted out. Detailed velocity mapping can be achieved using Fabry–Perot imaging, a technique that permits the selection of many lines with high spectral resolution and throughput. Most nebular spectra have relatively few strong emission lines, so one can be sure of selecting a line without serious blending. Density fluctuations are difficult to treat in detail, and low-density regions are usually included as localized regions of enhanced escape probability.

You may recall the comment when we discussed NLTE treatments of stellar atmospheres that the interstellar is essentially the same as a problem without the requirement of hydrostatic equilibrium. Nowhere is this more evident than in the treatment of nebulae. The low density precludes assuming LTE conditions for the gas, and the nonlocality of radiation source only exacerbates the problem. For a parcel of gas in the ISM, the spectral distribution of any star completely governs the atomic processes that occur in the gas. The densities are always low, often so low that collisional excitation is completely unimportant and we reach the stage of a photon-dominated medium. For any transition, you can estimate the importance of collisional excitation or deexcitation by using the ratio of the collisional-to-radiative deexcitation rate $n_e C_{ji}/A_{ji}$. In general, for permitted transitions such as Hα in ionized regions, this number is much less than unity. This doesn't mean, however, that there is no connection between stellar atmospheres and the interstellar medium. Many of the lines that are used for H II regions are also important for stellar atmospheres. You will recall that when we discussed symbiotic stars, we noted that these are really no different from asymmetrically ionized planetary nebulae and the same lines are used for their analysis. Ionized regions around OB stars, classical H II regions, are of much lower density, and therefore some of the lines that are usually suppressed by collisions in the denser inner parts of planetaries and symbiotics show up there.

6.2.2 Heating and Cooling

Heating in H II regions are heated by photoionization of hydrogen and metals due to the embedded stars. This process leaves the liberated electrons with comparatively low energy even though the photon energies exceed the ionization energy, since the heating is due to the *excess* energy of the electron, ϵ, relative to χ_H. The heating *rate* is the ionization rate times the input electron energy, $n_H \beta \epsilon$. Now for cooling, there are several avenues. One is recombination, the cascade extracting the energy just put into the gas. This is basically a nonresonant process, and we can take the cross section to vary as $\sigma(E) \sim E^{-1}$ (as we found for nonresonant collisions in Chapter 4). The rate of cooling is thus heavily weighted to the

---

[11] See Williams, R. E. 1992, *ApJ*, **392**, 99, for a simple treatment of density fluctuations in nebulae. Note that the same problem occurs in stellar winds ionized by embedded, off-center sources such as symbiotic stars. There a shadow cone is formed in the direction of the ionizing source when the column density becomes large enough, leading to a wide range of ionizations from the same star.

lowest-energy electrons. But this is the problem. The ejected electrons thermalize on the mean collision timescale, which is similar to the bremsstrahlung emission timescale, and there simply aren't many in the low-energy part of the Maxwellian distribution. On the other hand, there are many lines arising from the ground state, many of which are forbidden and optically thin, that have excitation energies that are quite near the peak of the electron distribution, kT. These are mainly optical transitions, listed in Table 6.1, for which the electron excitation cross sections are both large and resonant. In other words, we write the excitation rate as $C_{ij} = (g_j/g_i)C_{ji} \exp(-\Delta E/kT)$. The cooling rate is then given by the excitation rate:

$$j_{ji} = n_j A_{ji} h \nu_{ji} = n_e \frac{g_j}{g_i} C_{ji} e^{-\Delta E/kT}. \tag{6.22}$$

Since the heating depends on the rate of Lyman continuum absorption and recombination and is only weakly dependent on temperature, the temperature dependence of the cooling rate dominates the result:

$$n_H \beta = n_e^2 \alpha \rightarrow \frac{x^2}{1-x} = \frac{\beta}{\alpha}, \tag{6.23}$$

so $n_e = x n_0$.

It is useful to recall a point we've raised a few times in the discussion of rate calculations—the maximum rate of a process occurs when the energy dependence of the cross section most closely matches the energy distribution of the exciting process, whether it is the radiative flux or the thermal velocity distribution. Nonthermal processes, for which power-law dependences are normal, are more heavily weighted to the lowest energies. Since the heating is weighted by the match between the ionizing continuum and the absorption cross section, the strong frequency dependence of κ_ν produces electrons whose energies are actually rather low. This is because most are ejected by photons at the ionization edges. In a nebula, however, the mean energy also depends on the optical depth because as photons from the central star are absorbed, the color temperature of the illuminating continuum increases. This leads to the *apparent* paradox that the mean electron energy increases as the optical depth of the nebula increases, although there is a net decrease in the number of ionizing photons with increasing distance from the central star. In a uniform-density medium, this leads to an increase in the temperature, but the temperature gradient is smoothed out by thermalizing collisions among the electrons and ions.

If the mean energy of the ejected electrons is less than χ_H, then the recombination rate—which determines the cooling due to the resulting emission lines—will be due to inefficient coolants. This is because the cross section is most heavily weighted to low energy, where, by thermalization, there are few electrons. On the other hand, resonances that closely match the peak of the electron thermal distribution, $f(v)$, will contribute most strongly to the cooling. It follows that for H II regions, for instance, the temperatures will always be around 0.5 eV, while those for photodissociation regions (PDRs) will be about 0.1 eV. In contrast, for planetary nebulae, the central sources are hotter and the fluorescent population of excited states reduces the ground-state population and the cooling efficiency of the lines, so these regions will tend to be hotter. In addition, the electrons are ejected from both the Lyman and He-Ly continua, so they are hotter, on average, than any produced in normal H II regions.

Another way of saying this is that cooling is due to *improbable* lines. The mean collision rates are computed by weighting the collision cross sections by the Maxwellian for the H II region electrons. Since resonance excitation has a strongly peaked response function at the energy of the gas, it is expected that the efficiency for energy loss through even weak lines will be greater than for the recombination lines of hydrogen; the cross sections are more closely matched to the temperature of the gas. The lowest-energy electrons are the ones responsible for recombination, but there are fractionally fewer of these than for the mean energy of the gas.

Thus, forbidden lines are important coolants, better than hydrogen, because at low density their cross sections more closely match the energy distribution of the electrons. The average energy lost through recombination is small. Also the energy transferred to the electrons in the gas from ionization is very low compared to the ionization energy of the luminous source because of the threshold effect of the ionization cross section. You would therefore expect the gas temperature of a "normal" nebula to be lower than in regions of high density, where the forbidden lines are suppressed, or in low-metallicity gas, where the lower abundance decreases the efficiency of the coolants.

6.2.3 Cosmic Ray Ionization and Charge Transfer

Because the interstellar gas has such low density, dynamical atomic processes that would normally reach thermal equilibrium may be important in the ionization balance. Cosmic rays are relativistic charged particles with energies well above a few MeV that can produce ionization even in very dense regions that would otherwise be shielded from ultraviolet light. In fact, they dominate the ionization balance in deep parts of dark clouds when no massive stars are being formed. They are pervasive in the galactic disk, and sources for their acceleration seem to be always acting to replenish the supply.

Charge transfer processes are like the lottery—they redistribute resources (charge) without really generating any new ones. In the Sun, and dMe stars, it has been suggested that a possible way to look for proton acceleration in flares is to search for time-dependent broad Lyman series emission lines when these downward-moving particles charge-exchange with the background atmosphere and recombine. While the mechanism may not be important for stellar atmospheres, however, it is an essential component of the ionization balance of the interstellar medium. There, charge exchange is inevitable at some level. It is not important for the overall ionization of the whole gas, since it requires that the gas be already at least a little ionized, but it strongly affects the distribution of ionization among the constituent species. In photon-dominated regions, charge exchange is driven by UV and X rays. In opaque environments, it is driven by cosmic rays.

Charge transfer involves collisions in which an electron is exchanged between a neutral and an ionized species:

$$A^+ + B \leftrightarrow A + B^+.$$

The electron, depending on the state in which it is captured, can then recombine with the subsequent emission, depending on the state in which it is captured. Remember that the ionization potential of hydrogen is quite large compared with that of many other species in the gas, for instance, carbon, so you can expect that an ionized phase of the metal can be found even where the medium is predominantly neutral. A very important process is the interaction with cosmic rays, although the cross sections for this are low because of the high velocity of the particles, in which, instead of ionization, a charge transfer produces a high-velocity atom that emits a strongly Doppler-shifted line in the deexcitation process. Charge transfer plays a significant role in the chemistry of the region outside the ionized gas, in particular in the formation of hydrides.[12]

An especially interesting instance of charge transfer is seen not in the interstellar medium but in the solar system. X-ray emission from comets was first observed by the ROSAT satellite and explained as line emission induced by charge transfer between highly ionized atoms in the solar wind and lower ionization the cometary plasma. The limited spectral resolution of ROSAT was overcome by *Chandra*[13] for comet C/1999 S4 (LINEAR), a near-Sun comet discovered with satellite observations. The O VII 0.57 keV line was detected along with N VI and N VII (0.3–0.5 keV) and O VIII (0.65 keV). These high ionization states are not likely in the interstellar medium, especially in molecular clouds, except for cosmic ray particles and supernova remnants, but the comet observations allow detailed studies of relevant processes to the interstellar medium.

Cosmic ray ionization depends on the assumed spectrum of the incident particles because of the energy dependence of the cross sections. We will postpone discussion of this part of the problem to Section 6.6.2.3. However, on average the energy density of these energetic particles is of order 1 eV cm^{-3}, and the approximate ionization rate, usually denoted as ζ, is about $10^{-(17\pm 1)}$ s^{-1}. Because the particles have such high energies, typically MeV or more, the collision cross sections are small and they pass easily through even the densest environments. As a result, they can penetrate into regions that are otherwise shielded from ionizing sources such as radiation, and are the main component in the ionization of dark clouds.

6.2.4 Absorption Lines

The neutral interstellar medium is best studied through absorption and emission by its most abundant component, hydrogen. Yet until the advent of space astronomy, the cool component of the gas was virtually invisible. Greenstein, in an interesting reminiscence of the early days of work on the ISM,[14] recalls being asked during his candidacy exam in 1939 how one would observe neutral hydrogen

---

[12] See the compilation by Kingdon, J. B. and Ferland, G. J. 1996, *ApJS*, **106**, 205 (and references therein) and Flowers, D. 1990, *Molecular Collisions in the Interstellar Medium* (Cambridge, UK: Cambridge Univ. Press).

[13] This is one case when a result has been so recent during the writing of this book that only IAU circulars and conference abstracts are available, but it is a way of introducing you to an invaluable resource—the IAU Circulars from the Central Bureau for Astronomical Telegrams. See IAUC 7464 for the announcement information.

[14] Greenstein, J. 1970 in *Spectroscopic Astrophysics* Herbig, G., ed. (Berkeley: Univ. California Press).

in the interstellar medium. His answer was that it required ultraviolet observations of the Lyman lines, which were then inaccessible. How things have changed! Such observations are routine with satellites, and this technique would not even be the first answer that would now occur to a PhD candidate, as you will soon see.[15]

6.2.4.1 Atomic Resonance Lines: H I Lyman α 1216 Å and Others

Taking pride of place among fundamental astrophysically important transitions is the resonance line of hydrogen, Lyman α. The separation between the first excited state and the ground state is large enough that collisions in this gas are inadequately energetic to excite the $n = 2$ state. Thus radiative processes dominate the visibility of the lines from the atom, specifically the Lyman series arising from $n = 1$. There is nothing required for the ISM that we have not already seen for a stellar atmosphere. In fact, in a sense there is less. Pressure broadening is not seen for the medium, although the turbulent velocities can be substantially larger than anything normally seen in a stellar atmosphere. Radiative damping is finite for the Lyman series, so this is why the damping wings are seen if the line-of-sight column density is large enough. This happens whenever the *column* density exceeds about 10^{22} cm^{-3}. Otherwise, the treatment of the curve of growth is essentially the same as we encountered in a stellar atmosphere. We will encounter this again when discussing the cosmological detection of the Lyman α forest, absorption lines seen along the line of sight to external galaxies and quasars. The problem is the same; the difference is that one must include corrections for the redshift on the formation of the absorption lines.

One other difference is important. In a stellar atmosphere the photon mean free path is small compared with a pressure scale height, and the resulting absorption line reflects the local conditions in the line forming region. The line-of-sight turbulent velocity is usually small and subsonic, and the medium usually has a comparatively small velocity gradient, except in a stellar wind. Also, the thermal velocity is generally large. In the ISM, we have just the opposite situation. The optical depth is very small, the random motions are substantially larger than the sound speed and are the main broadening agent, and the line formation can extend over a huge range of physical conditions that are not even mechanically coupled to each other.

In addition to the H I 1215 Å resonance line (see also list in Table 6.5[16]), the D I Lyα transition is separated by about 1.5 Å ($\approx$ 370 km s^{-1}) to the blue. This is because of the small change in the Rydberg constant resulting from the increased mass of the nucleus. The deuterium line is quite weak, although the oscillator strength is nearly the same as the H I Lyα, because the D/H abundance ratio is very small. The line can be observed with any spectrograph capable of detecting the Lyman absorption lines, but its separation from H I Lyα is insufficient to completely remove the D I line from the damping wings of the interstellar

---

[15] Spitzer, L. Jr. 1982, *Searching Between the Stars* (New Haven: Yale Univ. Press) is another good source for the history of work on the gaseous component of the medium before the modern era of satellite astronomy. See also Spitzer, L. Jr. and Ostriker, J. P., eds. 1997, *Dreams, Stars, and Electrons: Selected Writings of Lyman Spitzer, Jr.* (Princeton: Princeton Univ. Press).

[16] The asterisk (*) in Table 6.5 indicates an excited state in the ground-state multiplet. Data from Morton, D. C. 1991, *ApJS*, **77**, 119.

TABLE 6.5 Some Important Interstellar Atomic Resonance Lines

λ (Å) (vacuum)	Ion	log gf	λ (Å) (vacuum)	Ion	log gf
1037.6167	O VI	1.836	1062.662	S IV	1.628
1083.990	N II	2.048	1122.526	Fe III	1.947
1190.208	S III	1.421	1199.5496	N I	2.202
1200.2233	N I	2.026	1200.7098	N I	1.725
1206.500	Si III	3.304	1215.3376	D I	2.528
1215.3430	D I	2.227	1215.6683	H I	2.528
1215.6737	H I	2.227	1253.811	S II	1.135
1259.519	S II	1.311	1260.4221	Si II	3.104
1302.1685	O I	1.804	1304.8576	O I*	1.804
1334.5323	C II	2.232	1335.6627	C II*	1.232
1335.7077	C II*	2.186	1393.755	Si IV	2.855
1397.232	O IV	−4.323	1398.05	S IV	−3.279
1399.780	O IV	−3.069	1402.770	Si IV	2.554
1404.80	S IV	−1.951	1526.7066	Si II	2.546
1548.195	C IV	2.470	1550.770	C IV	2.169
1670.7874	Al II	3.486	1854.7164	Al III	3.017
1862.7895	Al III	2.716	2026.136	Zn II	3.018
2062.664	Zn II	2.717	2349.329	Be I	3.503
2497.5243	B I	2.296	2498.4762	B I*	2.296
2996.352	Mg II	3.234	2803.531	Mg II	2.933
3934.777	Ca II	3.397	3969.591	Ca II	3.096
4226.728	Ca I	3.870	6709.613	Li I	3.521
6709.764	Li I	3.220			

transition. It can be readily studied along lines of sight only with small neutral hydrogen column densities. This is actually quite useful because over those lines we actually have a chance of determining the homogeneity of the line of sight.[17]

Although analyzing interstellar absorption lines is in principle much easier than for a star, in practice it involves making some tricky physical assumptions. You are not trying to solve the inverse problem that we discussed in Chapter 3. Instead, the absorbing gas is treated as an essentially passive medium. The light is simply a pencil beam that probes the medium and is scattered and/or absorbed by the constituents but without a thermal feedback. Every parcel of gas leaves its imprint on the line profile reflecting only the local conditions, so you don't have to solve the more complicated equation of transfer. The thermal velocity of the gas is low, so the only broadening comes from the random motions of individual parcels of gas along the line of sight. You can assume that the intrinsic line profile is Gaussian, which greatly simplifies the treatment of the equivalent width as a

---

[17] The first studies of this line used the OAO-2 satellite, *Copernicus*. For the most recent analysis of these data, see McCullough, P. 1992, *ApJ*, **390**, 213. The first determination of the D/H ratio from GHRS observations is Linsky, J. L. et al. 1993, *ApJ*, **402**, 694; for further analysis, see Vidal-Madjar, A., Ferlet, R., and Lemoine, M. 1998, *Space Sci. Rev.*, **84**, 297; Savage, B. D. and Sembach, K. R. 1996, *ARAA*, **34**, 279.

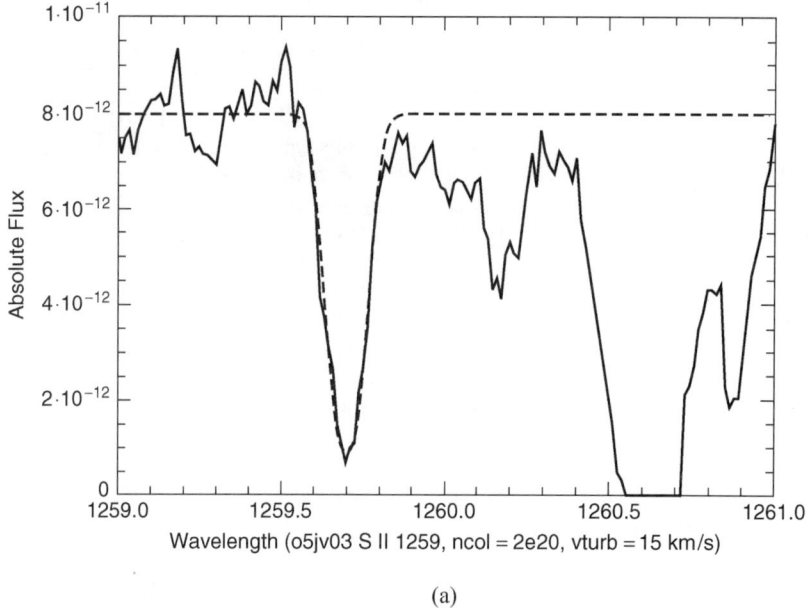

FIGURE 6.1. (a) Fitting an interstellar resonance line, S II λ1259 Å, with a turbulently broadened Gaussian profile (STIS observation of V382 Vel).

measure of line opacity (Fig. 6.1). But that is precisely the problem. Each locale gets integrated into the profile, and by chance regions that are at very different places can superimpose at the same velocity within the resolution of the data. We will return to this shortly when discussing ways to incorporate multiple cloud populations in the analyses of the absorption profiles.

Line formation can be treated by a simple reversing layer. It is conventional to assume a constant temperature for the gas through which the light is passing. Whatever the density gradient may be along the line of sight, there is no need to simultaneously treat the mechanical and thermal balance of the gas because the gas is not in hydrostatic equilibrium. This is, however, a complicating factor since you have no a priori mechanical or thermodynamic constraints on the nature of the gas and must use line profile ratios to determine the conditions. To compute the column density therefore requires only the oscillator strength, since the volume density is low enough that Stark broadening can be ignored, unlike a stellar atmosphere. We will return to this point in a moment, but for now assume that the medium is isothermal and of low density. A single atom scatters the photon so that, assuming complete redistribution, the emission and absorption profiles are identical. The absorption profile is Gaussian to a first approximation, so the absorption coefficient is

$$\kappa_\lambda = \kappa_0 \exp - \left(\frac{\Delta\lambda}{\Delta\lambda_D}\right)^2 = \kappa_0 \phi_\lambda, \qquad (6.24)$$

where ϕ_λ is the line profile function (see Chapter 3). The optical depth is

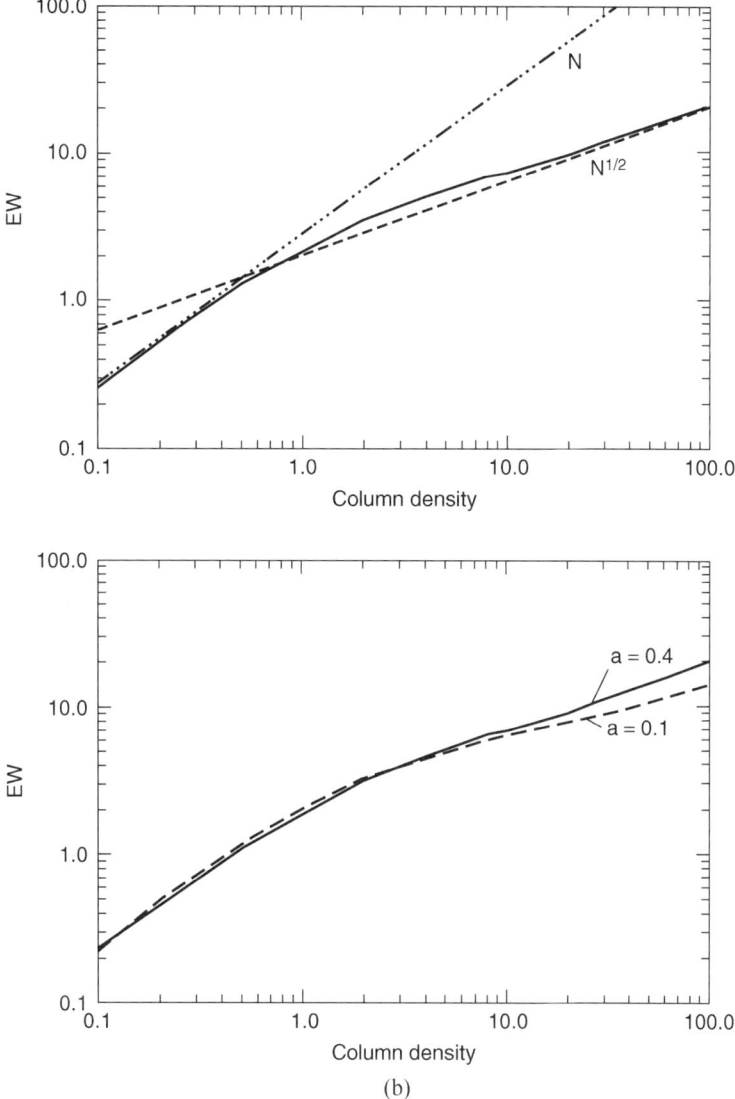

FIGURE 6.1. (b) Curve of growth for a Voigt profile (top, $a = 0.4$; bottom, $a = 0.1$ and $a = 0.4$) compared to N and $N^{1/2}$ asymptotic behavior.

$\tau_\lambda = \kappa_\lambda N_H$, where N_H is the column density of neutral hydrogen in the $n = 1$ level. Now we make a few more simplifying physical assumptions. Since the interstellar medium is not in hydrostatic equilibrium anywhere, and not necessarily even in thermal equilibrium, we can say that the width of a line is set by the local velocity field and not by the requirement that the thermal speed reflect the local pressure determined by gravity. This means that the integral along a line of sight may receive contributions of physically quite distinct and remote parts of the medium. In order to compute the column density, we can say that N_H indicates the column density *in a particular cloud*, that is, within some velocity σ of rest, and that the

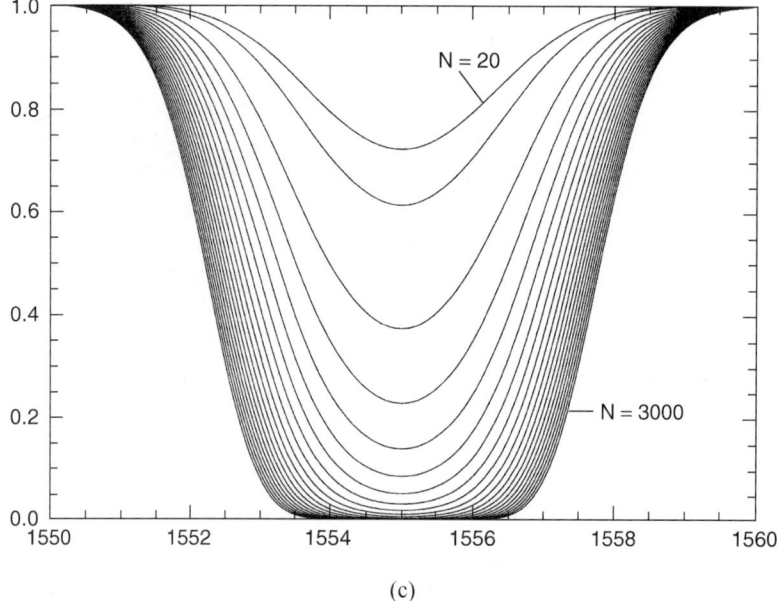

FIGURE 6.1. (c) Development of the line profile for $a = 0.4$ for increasing column density.

intensity of the absorption line *at a specified wavelength* is the measure of interest. In this case—and this has been more easily exploited as the resolution of ultraviolet spectrographs has increased—we can take the column density as a function of wavelength (or velocity) within the line profile and determine the physical parameter for a specific "cloud" (such as, the D/H value at a specific velocity). Or we can use the integrated measure of the absorption, the equivalent width, which is defined by

$$W_\lambda = \int_{-\infty}^{\infty} \left(1 - \frac{I_\lambda}{I_c}\right) d\lambda \qquad (6.25)$$

so that

$$W_\lambda(\tau) = \int_{-\infty}^{\infty} \left(1 - e^{-\tau_0 \phi_\lambda}\right) d\Delta\lambda, \qquad (6.26)$$

in which case we must perform the integration of the line profile (the optical depth) over wavelength. In the weak line limit, which is the simplest to see, the equivalent width is linearly proportional to the optical depth, and therefore to the column density. For a strong purely Gaussian line, the equivalent width eventually saturates as $\tau_0 \to \infty$. This contrasts with the Voigt profile case for which, since the wings have an asymptotic $(\Delta\lambda)^{-2}$ dependence from the Lorentzian contribution to the line, produces an asymptotic dependence of the equivalent width of $\tau_0^{1/2}$.

A problem we have not encountered when thinking about stellar atmospheres is that in the interstellar medium there is no constraint on the velocity. Because the radiation we see is coming from regions that may be very distant from the absorber and the intervening gas is not bound to remain in hydrostatic equilibrium, random velocities, v_{turb}, can be much larger than thermal broadening for the Doppler profiles. This contrasts with the stellar case where this parameter was called *microturbulence* and was assumed to be due to subsonic, small-scale motions. Here the motions can be highly supersonic.[18] The broadening is given by

$$(\Delta \lambda_D)^2 = \Delta \lambda_T^2 + \left(\frac{\lambda_0}{c} v_{\text{turb}}\right)^2. \tag{6.27}$$

In fact, these motions dominate the line formation. On one hand, this is important because it means that you measure at any velocity the actual column depth. For most lines, the absorption is truly Gaussian. On the other hand, it means that we have lost almost all information about *local* excitation conditions because, with only one component of the velocity to measure, the gas can be anywhere along the line of sight.

The usual approach for determining the column density is to use integral measures of the line strength and the *curve of growth*. The procedure has several advantages, notably that low resolution can be used (within the noise, the equivalent width is constant, independent of resolution) and we can integrate over the entire range of velocities. However, you pay a price in losing control over knowledge of the siting of the absorber. Since you are integrating over all velocities, you may be combining absorption along the isovelocity line from clouds that have physically nothing to do with each other. To get around this, you need to use multiple lines from not only the same ion but also from the same state. The best way to determine local conditions is to compare two transitions that are separated by fine structure excitation, within 300 cm^{-1} of each other. Then a point for point comparison of the velocity profile gives important information about the local excitation conditions. Using more than one ion, as with stellar atmospheres, gives ionization information along the line of sight.

The curve of growth method works crudely for stellar atmospheres because in such environments no regions are actually disconnected. They are transferring the same flux, obeying the same hydrostatic law for the local conditions, and if the medium is in LTE, the atomic level populations are uniquely determined by the equation of state and the temperature. In the ISM, however, none of these are true *in general*, and therefore global measures such as W_λ may be misleading. The method is still employed, however, for narrow lines that are exceeded by the line spread function of the spectrograph. For quasars, for instance, the problem is more acute. Here the distances between absorbers may be incredibly large, and the velocity width of the line may result from ordered internal motions within the absorber and not from low-amplitude random velocity fields (loosely called *turbulence*). As a result, the linewidths are set by the internal distribution of clouds in the absorbers and not by the $\Delta \lambda_D$ for the medium (Fig. 6.2). For the interstellar medium—and this applies also to halo observations and analysis of quasar absorp-

---

[18] This is a normal condition in molecular clouds where the lines often do not show Gaussian profiles. This will be discussed in Section 6.10.4, below.

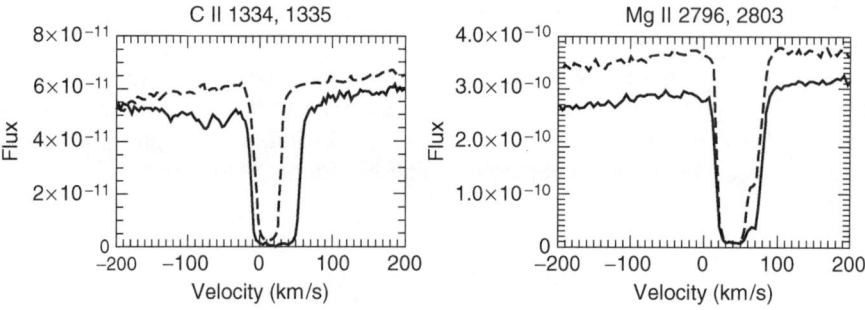

FIGURE 6.2. Sample interstellar absorption line profiles for the components of the C II λ1334-Å and Mg II λ2800-Å doublets. Notice that although the low velocity feature is saturated, the optical depth in the high velocity feature depends on excitation (the dashed line is the fine-structure excited state). These data were obtained using STIS on HST during observations of nova V382 Vel 1999.

tion line spectra—Eq. (6.26) reduces to

$$N_i(\Delta v) = 3.77 \times 10^{14} \frac{\tau_i(\Delta v)}{f\lambda(\text{Å})} \text{ cm}^{-2} \text{ km s}^{-1} \quad (6.28)$$

at each observed velocity Δv from line center. The assumption built into this analysis is that the medium must be homogeneous, if not uniform, along the line of sight. For weak lines the line optical depth is small and you can assume a Gaussian profile, so

$$\tau_i(\Delta v) = -\ln \frac{I_i(\Delta v)}{I_0}. \quad (6.29)$$

In practice, the intrinsic line profile for *most* transitions can be assumed to be Gaussian with some line-of-sight broadening by mass motions and superposition of optically thin components. For the Lyman series, however, the abundance is so high that the intrinsic linewidth dominates the profile at high column densities, above $N_H \approx 10^{20}$ cm^{-2}. In cosmological applications this leads to the damped Lyα lines but it is typical of most lines of sight through the galactic plane since, with an average density of order 1 cm^{-3}, the Lyman lines saturate within about 30 pc.

6.2.4.2 H I 21 cm and Related Radio Lines

The ground state of H I, the $1s\ ^2S_{1/2}$, is split by hyperfine structure and a transition is possible between the sublevels. This transition, produced by interaction between the nuclear spin and that of the electron, between the levels of the $^2S_{1/2}$ state, for which the energy separation yields an emitting frequency of 1.420.406 MHz, corresponds to a wavelength of 21 cm. The possibility of observing the line was first suggested by H. van de Hulst in 1945 and elaborated by I. Shklovsky (in 1949)[19] and was observed in 1951 by Ewen and Purcell and

---

[19] See Lang, K. and Gingerich, O. 1979, *Source Book in Astronomy and Astrophysics 1900–1975* (Cambridge, MA: Harvard Univ. Press), p. 627ff for an English translation

confirmed by Mueller and Oort. The line formation proceeds as with any forbidden transition. The collision frequency in the ISM is sufficiently low that the $m_s = \frac{1}{2}$ state should have a lower population than $m_s = -\frac{1}{2}$. A rare collision would then have a chance of exciting the line with subsequent radiative de-excitation. The only limit on the process is the same one that applies to such lines in the optical, that $n_e C_{ji}/A_{ji} \ll 1$, which sets a limit on the density of the medium. The transition probability is $A_{21} = 2.876 \times 10^{-15} s^{-1}$, or in other words, the lifetime of the excited level is about 10^7 yr.[20] Collisional excitations are guaranteed to have a lower rate because of the threshold for the process. At the time, the average energy density in the interstellar medium was estimated at about 1 eV cm^{-3} and the equilibrium temperature of the gas was thought to be about 100 K, or about 0.01 eV, so easily capable of exciting the transition, which is at about 6 μeV.

As we discussed in Chapter 3, emission results whenever the local line source function exceeds the background. Therefore, if there happens to be a bright radio-emitting source in the field, a supernova remnant or H II region, for example, the line of sight H I can produce an absorption line. One of the best studied cases is Cas A, one of the strongest nonthermal emitters in the radio sky and a relatively nearly supernova remnant[21] for which time-dependent, spatially resolved interferometric measurements have been made over a period of more than a decade. Conventionally, the emissivity can be expressed as a blackbody equivalent or *brightness* temperature, because an optically thick thermal source radiates as

$$B_\nu(T_b) = \frac{2kT_b \nu^2}{c^2}, \qquad (6.30)$$

so the emissivity can be used to determine the optical depth of the source through $F_\nu \sim \tau T_b$. The terminology, because of this Rayleigh–Jeans limit for thermal emission and because 21 cm emission is *always* thermalized, the intensity is usually expressed as a temperature. For instance, the radiative transfer equation, with which you are so familiar, becomes

$$\frac{dT_b}{d\tau} = T_s - T_b, \qquad (6.31)$$

where T_s is the brightness temperature of the emitting neutral hydrogen source. This is, like the source function, a measure of how bright an *equivalent* Planck distribution would be. The background (incident) radiation is assumed to have a temperature T_0. This background need not be thermal. Therefore

$$T_b = T_0 e^{-\tau} + T_s(1 - e^{-\tau}). \qquad (6.32)$$

---

[20] This provided H. van de Hulst the opportunity to make one of the best comic remarks in astrophysics: that the interstellar medium is the only place where an atom could sit making up its mind for over a million years and *still* act spontaneously.
[21] See, for instance, the imaging study by Bieging, J. H., Goss, W. M., and Wilcots, E. M. 1991, *ApJS*, **75**, 999; Reynoso, E. M. et al. 1997, *Astr. Ap.*, **317**, 203.

The excitation temperature, also called the *spin* temperature, is defined by the ratio of the populations through a Boltzmann distribution:

$$\frac{n_{+1/2}}{n_{-1/2}} = 3e^{-\Delta E/kT_{ex}}, \tag{6.33}$$

where the factor of 3 comes from the ratio of the statistical weights of the levels. As you can see, several temperatures are used in the literature to characterize the emissivity of this line. Here we should add T_A, the *antenna temperature*, which is the quoted value for an optically thick thermal point source integrated over the acceptance solid angle of the radio telescope beam. Since the gas temperature is generally higher than the energy-level separation, the absorption coefficient is given by

$$\kappa_H = [n_l B_{lu} - n_u B_{ul}]\phi_\nu = n_l B_{lu}\left[1 - \exp\left(\frac{-h\nu}{kT}\right)\right]\phi_\nu \approx \frac{n_l B_{lu} h\nu}{kT}, \tag{6.34}$$

and from Eq. (6.33), one-fourth of the atoms are in the ground state. These are all local values. For the 21 cm line, as with other interstellar absorbers, we sample very long, inhomogeneous paths and obtain only integrated values for the column of absorbing gas along the line of sight (LOS). Then the line optical depth is

$$\tau_H = \frac{1}{4}\frac{3}{8\pi}A_{ul}\frac{hc^2}{\nu}\int_{LOS}\frac{N(z)\phi_\nu(z)}{T}dz, \tag{6.35}$$

using Eq. (3.16) to convert the transition probabilities from B to A and allowing explicitly for possible variations in the intrinsic line profile along the pathlength. For an emission line, since the gas is in approximate LTE, the emissivity is $j_\nu = \kappa_\nu B_\nu(T)$. Then, because $h\nu \ll kT$, the brightness temperature is given by $T_b = \tau_H T$.

The 21 cm line is the single most valuable tool we have for the study of the dynamics of gas within the Galaxy and also in extragalactic sources. Very sensitive centimeter wavelength radio telescopes and receivers can be constructed at modest cost; interferometry is well developed at these wavelengths, and the analytic methods for studying this line have been around for years. It is unaffected by the intervening dust and, except along those lines of sight where the optical depth becomes extremely large because of small velocity gradient in the gas rotation curve, it is usually optically thin. An important exception is toward the galactic center, where the rotational motion is transverse to the line of sight and only random cloud motions dominate the line profile. These are not large enough to desaturate the line profile, and without the galactic rotation curve to provide distances, there is little information available on the distance to individual contributing parcels of gas along the line of sight. This line provided the first conclusive observational evidence for cloudlike structures in the interstellar medium, although individual absorbing clouds had been seen in the Ca II and Na I resonance lines once high dispersion spectroscopy became available in the mid-1940s. We will return to this when discussing large-scale galactic structure in the next chapter, but a moment's reflection is called for here. The thermal width of the lines is extremely small, only a few tenths of kilometers per second at most, while the observed linewidths are often much greater. This is due to the combined effects of large-scale motions within the emitting (or absorbing) volume and the projection of galactic-scale orbital motion of the gas along the line of sight.

Neutral hydrogen is not the only atomic species to have an observable ground-state spin–flip transition. Although the ^{4}He nucleus has no net spin (the protons and neutrons pair separately) and no hyperfine splitting of the ground state, there is an analog of the hydrogen line for ^{3}He$^+$ at 8.665 GHz (3.46 cm). Clearly, this is seen in very different sites than the H I line, specifically in planetary nebulae and H II regions, but the principle of line formation is the same.[22] In some ways, in fact, it is easier to study this *very* weak line because it is always optically thin and therefore not subject to the radiative transfer complications that often plague H I. The transition probability is also considerably greater than for H I, $A_{^3\text{He}^+} = 1.95 \times 10^{-12}$ s^{-1}. It can also be observed, in principle, over a wider range of densities, although the increase in A is partly offset by the increased collision cross section because of the increased charge.

6.2.4.3 Molecular Spectra

The electronic states in a molecule, analogs of the atomic states and characterized by total electronic angular momentum and spin, are generally separated in energy by about the same order of magnitude as for isolated atoms, usually several eV. Thus the transitions from molecular electronic states, which also correspond to different potentials, are best observed in the ultraviolet. For the electronic coupling, the molecular symmetry provides an internal axis relative to which all the angular momenta are quantized. Thus, for a diatomic molecule, Λ is the total projected electronic orbital angular momentum along the internuclear axis, and states are labeled analogously to those in atoms as Σ, Π, Δ, and so on for $\Lambda = 0$, 1, 2, and so forth. For the electron spins, Σ is the projection of the total spin, and $\Omega = \Lambda + \Sigma$ is the analog of the total angular momentum. For diatomic molecules, for $\Lambda \neq 0$, the states are split by $\Omega = \Lambda + \Sigma$ into fine-structure levels, called Λ-*doubling*. Interaction with nuclear spins produces hyperfine splitting of the rotational levels, with quantum number F. These hyperfine transitions in OH are responsible for the observed maser emission.

An additional symmetry for homonuclear molecules, those with identical nuclei such as C$_2$ or N$_2$, is the parity of the wavefunction on interchange of the nuclei and is even (*g* or *gerade*) or odd (*u* or *ungerade*). For instance, H$_2$ has a ground configuration of $^1\Sigma_u$ while the first excited *electronic* state is $^3\Sigma_g$, which is not a bound state. The potential is more complicated than for atoms because it is not centrally symmetric and an overlap is possible between different electronic configurations and between the dissociation continuum of one electronic state and bound states of another. Unlike atoms, rotational and vibrational excitation can lead to electronic transitions between states if this is allowed by the coupling rules. Excitation depends on the presence of ultraviolet radiation, and electronic transitions are usually seen in absorption, as in the $^1\Sigma_g^+$ state of H$_2$. The strength of the transition depends on the dipole (heteronuclear molecules and ions) or the quadrupole (homonuclear) moments.

[22] Balser, D. S., Bania, T. M., Rood, R. T., and Wilson, T. L. 1999, *ApJ*, **510**, 759.

Vibrational states dominate the optical and infrared, and rotational states are best observed in the millimeter and centimeter portions of the spectrum. For the lower vibrational levels, the molecular potential approximates a harmonic oscillator and the states follow $E_v = \hbar\omega_0 (v + \tfrac{1}{2})$. For highly excited states, however, anharmonic terms of higher order enter and produce line shifts from this simple arrangement. Polyatomic molecules have additional vibrational states due to the multiple modes presented by different configurations. For instance, bending modes in water are states for which the O moves and the H remains fixed, while others have both the O and H moving oppositely (the so-called ν_2 and ν_3 modes), in addition to the fundamental ν_1 mode, which involves the O—H bond stretch. Vibrational coupling produces an angular momentum that splits rotational states, as mentioned above under vibronic transitions. Coupling of vibrational and electronic states, that is, between Λ and v, produces an angular momentum K, which for polyatomic molecules depends on the axis about which the rotation is executed. For instance, in H_2O, there is a prolate rotational axis and an oblate rotational axis for the molecule, so that a state J is split by two values of K and labeled by $J_{K_+K_-}$. More complex states are possible, depending on the complexity of the molecule. For example, inversion transitions of molecules like NH_3, which occur at centimeter wavelengths, result from small perturbations of rotational states by vibrational transition between mirror molecular conformations.

The nuclei are free to rotate and precess about the center of mass, so a molecule with a moment of inertia, $\mathscr{I}$, has a quantized rotational angular momentum J. The energies of these states are given by

$$E_J = \frac{h}{4\pi\mathscr{I}} J(J+1) \equiv B_v J(J+1), \qquad (6.36)$$

so that transitions between states of unequal J show a stepladder pattern in the separation of lines of the same series. Because of the large value of the moments of inertia, due to the mass of the nucleus, the separation of these states is small, of order 0.01 eV, increasing with increasing J. The simple representation that has just been used, however, is appropriate only for diatomic species, which have only one rotational axis that is degenerate in the two axes orthogonal to the internuclear axis about the center of mass.

For more complex molecules, for example, H_2O, two or three axes are needed to fully describe the rotation. Each of these has an associated moment of inertia, depending on the details of the electronic states and the internuclear distances and masses. The projection of the rotation along the body axis, K, now appears in the terms for the rotational splitting. For instance, for a symmetric top molecule like NH_3, the energy for a fixed vibrational state depends on the two rotational quantum numbers:

$$\frac{E_{|v|}(J,K)}{h} = B_{|v|} J(J+1) + (B_{|v|} - A_{|v|}) K^2 \pm 2 A_{|v|} \zeta K, \qquad (6.37)$$

where ζ describes the coupling of the vibrations and the rotations (vibronic states) and splits the otherwise degenerate K levels. The constants A and B are the moments of inertia about the parallel and perpendicular axes of rotation, relative

TABLE 6.6 Strong Millimeter Molecular Transitions

Species	Transition	ν_0 (GHz)
CO	1–0	115.271204
CO	2–1	230.53800
$^{13}C^{16}O$	1–0	110.201353
$^{12}C^{18}O$	1–0	109.782160
CS	1–0	48.990964
CS	2–1	97.980968
CS	3–2	146.969049

to the body axis. The rotational lines will be distributed in a way that depends on the ratio of the moments of inertia of the principal axes of the molecule. Thus there will be multiplets for lines that are closely spaced in energy and that can be strongly radiatively coupled.

The most important tracers of molecular gas are diatomic molecules, in particular CO and CS. For these, the electric dipole transition probabilities are given by

$$A_{ul} = \frac{64\pi^4 \nu_0^3}{3hc^3} |\mu_{ul}|^2 \qquad (6.38)$$

$$|\mu_{J,J+1}|^2 = \mu_D^2 \frac{J+1}{2J+1}$$

$$|\mu_{J+1,J}|^2 = \mu_D^2 \frac{J+1}{2J+3}, \qquad (6.39)$$

where μ_D is the electric dipole moment in debyes (10^{-18} esu cm; for example, μ_D is 0.112 for CO and 1.98 for CS). Some strong transitions that are used most frequently to study molecular clouds are listed in Table 6.6.[23, 24] For symmetric tops, there are additional quantum numbers, K, so

$$|\mu_{(J+1,K),(J,K)}|^2 = \mu_D^2 \left| \frac{(J+1)^2 - K^2}{(J+1)(2J+3)} \right|. \qquad (6.40)$$

Ammonia is especially interesting because it has an inversion transition for which ΔJ and ΔK are both zero but with the nitrogen oscillating symmetrically through the plane of the hydrogen atoms. This was the basis of the first masers. For

---

[23] For compilations of transition strengths, see Herzberg, G. 1991, *Molecular Spectra and Molecular Structure, II. Infrared and Raman Spectra of Polyatomic Molecules* (Boca Raton: Krieger Publ.); Townes, C. H. and Schawlow, A. L. 1955, *Microwave Spectroscopy* (NY: Dover Books); Nicholls, R. W. 1977, *ARAA*, **15**, 197. See also: Harris, D. C. and Bertolucci, M. D. 1989, *Symmetry and Spectroscopy: An Introduction to Vibrational and Electronic Spectroscopy* (NY: Dover Books). For optical and UV transitions involving electronic excitation, see Nicholls, R. W. 1981, *ApJS*, **47**, 279.

[24] Critically evaluated compilations of data more complete than those listed in Table 6.6 are available from the National Institute of Science and Technology (NIST) through their interactive database, compiled by F. J. Lovas, at *http://physics.nist.gov/PhysRefData/micro/html*. Here you will also find characteristic intensities for a number of important cosmic sources such as the Orion and Taurus molecular clouds, Sgr B2, W33, and W51.

asymmetric molecules, the dipole moments require more detailed treatments, and these can be found in the references.

The treatment of molecular line formation is not different from what atomic lines require, except that there is stronger coupling between the various states and more degrees of freedom are available. Collisions dominate most interstellar molecular excitation, although fluorescence and pumping also occur in photon-dominated environments that often accompany sites of star formation. If the emission rate is low, as usually occurs in molecular clouds, then the populations can be assumed to be in local thermal equilibrium, and hence given by the Boltzmann distribution. The excitation temperature is then the kinetic temperature of the exciting particles. An important difference between stellar atmospheres and clouds is that in molecular clouds, the temperature is so low that bulk motion is principally responsible for the line broadening. This means that the familiar Sobolev approximation, usually called the *large-velocity-gradient* approximation in ISM work, applies to the optical depth. Recall from Chapter 3 that this is

$$\kappa n \, dl = \frac{\kappa n \, dv}{dv/dl} = \frac{c}{\nu_0} \kappa_0 n \frac{\Delta \nu}{dv/dl}, \qquad (6.41)$$

where $\Delta \nu$ is the linewidth, κ_0 is the opacity at line center, n is the number density, and dv/dl is the velocity gradient along the line of sight. Since for molecular clouds, observed linewidths are between 0.1 and a few km s^{-1}, while the thermal speed is ≤ 0.1 km s^{-1}, this approximation seems to be valid for all except the most abundant species. Unlike stellar atmospheres, for most molecular cloud emission, collision with neutrals is the dominant excitation mechanism. This is especially important for CO. The observed molecules are typically merely tracers of the overall mass of cold gas. For instance, as we've just discussed, since it has no dipole moment, the homonuclear molecule H_2 cannot be detected in emission (other than through fluorescence in a UV radiation field). However, van der Waals collisions are capable of exciting strongly emitting species, such as CO and CS, and the H_2 mass can be determined on the basis of this excitation. The problem is with the indirectness of this method, the calibration of which has generated controversy since the early 1990s. The calibration is based in the ratio of extinction, A_V, to molecular hydrogen column density, N_{H_2}, which is then converted to a correction coefficient α_{CO} on the basis of millimeter observations. This multistep process has many pitfalls, not the least of which is the effect of metallicity differences from one region to another in the galactic medium and between galaxies when dealing with extragalactic systems. While the detection of CO is by itself an indication of the temperature and density conditions under which the line is formed, and there are reasons for thinking that we can convert from line ratios for species formed under conditions similar to actual abundance differences, the measurement of cloud masses is far more uncertain.

Like their atomic counterparts, molecular lines saturate when the populations have reached the values associated with strict equilibrium with the incoming radiation. This occurs first at the line center. Any motion in the medium, ordered or random, will broaden the line and thus the molecules will "see" radiation at other wavelengths against which they can absorb, or into which they can emit. If the medium is optically thick at line center but the velocity dispersion is large, the overall optical depth can be considerably reduced by spreading out the line in frequency.

The most abundant molecules, because of the low velocities observed in the clouds and high column densities, cannot be interpreted by simple optically thin models. For CO the ratio of the $(J = 2 \rightarrow J = 1)$ to $(J = 1 \rightarrow J = 0)$ transitions *should* be $3:1$ in strength, if the gas is optically thin. However, CO $1 \rightarrow 0$ often displays flat-topped profiles rather than the Gaussian form, and the intensity ratio of the transitions often departs from the optically thin value. This implies that CO is optically thick, and that the densest parts of the cloud may not be observable in the ground-state transition. Lower abundance species may, however, probe denser parts of the cloud. Further, the higher transitions require higher densities for excitation so there is a delicate interplay between chemistry and radiative transfer that enters into the interpretation of abundances. This is crucial to the understanding of the formation of the molecules.

You already know how to determine abundances from atomic emission lines, but the different technology required for millimetric observations also requires that the abundances be related to a different set of measurable parameters. The abundance of a species is related to the observed line intensity by the antenna temperature:

$$I_{\text{line}} = \int_0^\infty \frac{T_A}{\nu^2} d\nu, \tag{6.42}$$

which is integrated over the line velocity width. The column density in the lower level is defined by

$$N_l = \frac{4\pi^{3/2}}{(\ln 2)^{1/2}} \frac{k\nu}{hc^2} \frac{(2J_l + 1)}{(2J_u + 1)} \frac{1}{A_{ul}} \int_{\text{line}} T_B \, \Delta v, \tag{6.43}$$

where Δv is the velocity width of the line. Then, assuming LTE, the total column density of the species is found using the Boltzmann distribution for the levels:

$$N_{\text{tot}} = N_l \frac{Q(T)}{(2J_l + 1)} \exp\left(\frac{E_l}{kT}\right). \tag{6.44}$$

Here $Q(T)$ is the molecular partition function for the rotational levels:

$$Q(T) = \sum_{J, K} g_{J, K} e^{-E_{JK}/kT}. \tag{6.45}$$

Because of the regular behavior of the statistical weights, this function can be written in closed form. For instance, for a symmetric top molecule:

$$Q(T) = \left(\frac{\pi}{A_v B_v^2}\right)^{1/2} \left(\frac{kT}{hc}\right)^{3/2} e^{B_v hc/4kT}, \tag{6.46}$$

490   THE INTERSTELLAR MEDIUM

where $B_v = C_v$ and A_v are the moments of inertia for the rotational states. In the optically thin, LVG approximation, this suffices to determine the abundance of the species of interest. It assumes that all emission is due to thermal equilibrium prevailing as a result of collisions among the levels.

6.2.4.4 Molecular Hydrogen (H_2)

Molecular hydrogen in the diffuse interstellar medium was a major discovery of the *Copernicus* mission in the early 1970s. The possibility that a significant portion of the interstellar medium is locked up in a molecule that has no dipole moment, and thus produces no emission lines, had been suggested from chemistry arguments. But the direct detection of the absorption from the ground-state transitions opened a new view of the medium. Absorption by H_2 occurs through two ultraviolet transitions, the Lyman bands at 11.2 eV threshold ($X^1\Sigma_g^+ - B^1\Sigma_u^+$), and the Werner bands with a threshold of 12.3 eV ($X^1\Sigma_g^+ - C^1\Pi_u$). Lower-energy emission and absorption can be collisionally excited. Rotational states in the lowest vibrational band of the ground state require only a few hundreds of degrees; for the first vibrational transitional bands, the temperature is of order 1000 K.[25] As we will see, these conditions are found be in shocks.

The analysis of this molecule also illustrates how sometimes seemingly insignificant facts can lead to very big questions. The population ratio for the ground rotational states are such a case. Let us explain. Molecular hydrogen exists in two phases,[26] depending on the total nuclear spin quantum number, I. The aligned spin state, $I = 1$, is called *ortho*-hydrogen, and $I = 0$ is called *para*-hydrogen. Since the molecule is symmetric, only antisymmetric (odd J) states are allowed for $I = 1$ and only symmetric (even J) for $I = 0$; $J = 0$ is excluded for the ortho state. The implication for the astrophysical analysis of H_2 comes The statistical weights for a rotational state is $g = (2I + 1)(2J + 1)$. No transitions are allowed between the even and odd J states, so, as you will remember from He I, there are really two effective ground states. Absorption out of the $J = 1$ will have a 3:1 population ratio to the $J = 0$ state.

For these reasons, the ortho:para ratio carries information about the formation and destruction of the molecule. Having no dipole moment, H_2 is presumed to require a surface on which to form, through collisions of hydrogen with the dust. The mechanism happens at a high kinetic temperature, whatever the atomic hydrogen has and also whatever the dust has. In fact, in order for the molecule to even form, the reaction should heat the grain and UV will sputter the newly formed hydrogen off the grain surface. Therefore, the small energy separation of the two lowest rotational states, about 0.01 eV (the excitation temperature is 170.5 K), leads to the conclusion that the rotational states should easily thermalize. When kicked back into the ISM, the molecules should form with an ortho:para population ratio of 3:1. Now for the observational implication. The rotational selection rule for homonuclear molecules forbids $J = 0 \leftrightarrow 1$ so in a cold medium

---

[25] As a notational point, $P(J)$ denotes transitions for for which $J \to J - 1$ and $R(J)$ denotes $J \to J + 1$.
[26] We recommend Tomonaga, S. 1996, *The Story of Spin* (Chicago: Univ. Chicago Press) for an especially interesting discussion of the history of this discovery and its implications.

the ortho:para ratio should remain 3.[27] Lower values are, however, observed in a wide range of interstellar environments. How? Several ideas come to mind. One is that fluorescence, pumped by UV radiation, reshuffles populations. This will not work since the nuclear spin symmetry is respected by radiative transitions so the populations of excited states may change but not the ultimate cascade back to the ground state. Collisional reshuffling is an alternative, such as you would find in shocks, but the jury is still out.

6.2.5 Masers

Maser emission is not only an interstellar phenomenon. It is found in many stellar environments as well. The principal requirement is that infrared flux from some nonlocal radiator is sufficiently intense to maintain an inverted population in the presence of collisions. This nonthermal population results because transitions take place between high states that are only weakly collisionally coupled to lower levels, and for which radiative transitions are long. If there is a strong background radiation field at a wavelength shorter than the transition of interest, upper states of the molecule may be radiatively excited with subsequent overpopulation of some of the lower states by radiative and collisional de-excitation. Thus, for masing to occur, more than two states must be involved in strongly coupled transitions. We also note that maser emission serves as a reminder that the intensity of a spectral line is not necessarily a direct measure of its abundance. Population inversions enhance the brightness temperatures, leading to overestimates of excitation and abundance in those species in which masing occurs. Because not all molecules undergo maser amplification, the assumption of thermal equilibrium is usually justified, but should be employed with caution.

The emission and absorption coefficients for the system can be defined as before. Now assume that there are a total of n levels, and that they are coupled via collisions and radiation to levels 1 and 2. Then the time-dependent populations of 1 and 2 are given by

$$\dot{n}_1 = \mathscr{P}_1(n - n_1 - n_2) - (n_1 B_{12} - n_2 B_{21})\frac{\Omega}{4\pi}I - n_1 C_{12} - n_1\Gamma; \quad (6.47)$$

$$\dot{n}_2 = \mathscr{P}_2(n - n_1 - n_2) - (n_2 B_{21} - n_1 B_{12})\frac{\Omega}{4\pi}I - n_2(A_{21} - C_{21}) - n_2\Gamma. \quad (6.48)$$

Here $\mathscr{P}_1$ and $\mathscr{P}_2$ are the pump rates from the higher-lying levels through radiation and collisions, Ω is the solid angle, and Γ is the rate at which the masing levels are depopulated.

For molecular masers, it can be assumed that the two masing levels have the same statistical weight. Thus, $B_{12} = B_{21} = B$. The number of levels involved in the particular population inversion is small. This implies strong radiative coupling

---

[27]See Herzberg, G. (1950, *Molecular Spectra and Molecular Structure. I. Spectra of Diatomic Molecules* (Princeton: Van Nostrand), p. 140. He notes that in early laboratory work by Bonhoeffer and Harteck in 1929 showed that the addition of activated charcoal to a sample of molecular hydrogen accelerated the establishment of LTE conditions.

between states, which, for some reason, are selectively pumped by the external sources of radiation. The OH molecule is an excellent example of this behavior. If the absorption coefficient is *negative*, in other words, if the populations are *sufficiently inverted*, the radiation in the $2 \to 1$ transition will amplify along its path until the maser saturates, that is, until the populations do not change along the pathlength. The amplification selects out the line center, and the line gradually narrows as a result of increasing pathlength. This behavior is of great importance, because the brightness temperature increases as the line gets narrower. Further, the radiation has a finite amplification length; the maser can saturate. In the absence of collisions and in steady state, Γ can be replaced by $\Gamma = 2BI(\Omega/4\pi)$, so that the brightness temperature of a saturated maser is given by

$$T_{B,s} = \frac{h\nu}{2k} \frac{\Gamma}{A} \frac{\Omega}{4\pi}. \qquad (6.49)$$

This makes masers very intense radiation sources, since the pump radiation at higher frequency has been converted both to lower frequency and narrower bandwidth by the amplification process.

Masing depends on the presence of a strong radiation field for excitation and maintenance. Such radiation, usually infrared, is significant in several environments, notably in circumstellar envelopes and in molecular clouds. In extended stellar envelopes, far-infrared radiation is converted to centimeter radiation by OH, which has transitions centered around 1665 MHz. Ammonia, H_2O, HCN, and SiO are also important stellar maser sources. Water masers are also associated with regions of active star formation, where IR from the protostellar cores can excite the millimeter radiation in the densest parts of the cloud. Because they are strongly amplifying, the masing sites are easily distinguished from the background and their proper motions can be directly measured using very-long-baseline interferometry (VLBI) techniques. Their time variability is also well observed, although still not fully understood theoretically.

6.2.6 Warm and Hot Diffuse Gas

The diffuse interstellar medium has phases that are substantially hotter and significantly cooler than the gas usually encountered in H II regions. First detected by the *Copernicus* satellite observations of the O VI 1032 Å line, this gas is associated with the diffuse X-ray background of the disk. Halo gas shows the resonance doublets Si IV 1400 Å and C IV 1550 Å in absorption, and the N V 1240 Å doublet has also been detected within the galactic plane, although not in the halo.[28] The source for such high ionizations appears to be similar to that for H II regions, the diffuse radiation field produced by the combined effects of hot subdwarf and post-AGB stars, OB main sequence stars, Wolf–Rayet stars, novae, and supernovae. In the halo this is a problem because of the lack of many of the in situ younger populations.

---

[28] Note that C IV, N V, and O VI form an isoelectronic sequence with increasing ionization potential with increasing atomic number.

Recall what it means to have gas at low density. Although the medium is transparent, which implies that it can cool quickly, it is also nearly collisionless, which means that this cooling mechanism is only weakly coupled to the electrons. If the gas is repeatedly, albeit sporadically, heated by mechanical processes such as expanding stellar winds or supernova ejecta, it takes a long time for the energy to be lost through radiation. We will come back to this later in more detail when describing expanding H II regions and supernova remnants. For now, keep in mind that these are very efficient means for heating the gas to very high temperatures from which the medium can only slowly recover.

Diffuse interstellar Hα emission is also detected with a scale height > 1 kpc. This is the *Reynolds layer* produced by recombination of hydrogen in the diffuse gas.[29] It is the marker of the warm interstellar medium, the component to which the 21-cm surveys are insensitive. In effect, in the presence of the combined radiation from galactic OB stars, the diffuse medium forms an extended Strömgren sphere (see Section 6.7.3). The radius that a typical main sequence OB star can ionize is of order:

$$R_0 \approx 0.1 \left(\frac{F_\star}{10^{48} \text{ s}^{-1}} \right)^{1/3} n^{-2/3} \text{ kpc.} \quad (6.50)$$

For a recombination coefficient of $\alpha = 2.6 \times 10^{-13}$ cm^{-3} s^{-1}, you can see that the effect of only a few such stars is dramatic on the gas. This diffuse emission is consequently a calorimeter of the stellar radiation. It measures the escape of ionizing radiation from sites of star formation, H II regions and dark clouds, and the warm diffuse medium is produced by the overlap of these otherwise static ionized regions. Since the mean density is about 1 cm^{-3}, a column density of about 10^{21} cm^{-2}, required for the gas to be completely optically thick, is reached over about 1 kpc. Even assuming that there is significant dilution, the diffuse interstellar radiation field is approximately

$$F_{\text{DIRF}} \approx W F_{\text{OB}}, \quad (6.51)$$

where $W \approx 10^{-14}$ is the dilution of the stellar radiation. On the local scale, small emission nebulae have been detected in the direction of several nearby OB stars, notably α Vir. The measured rms electron density is about 0.1 cm^{-3}. In addition, photon escape is facilitated by the extreme density fluctuations that are sometimes encountered along the line of sight. The local interstellar medium provides an excellent example of this. We see an especially vacant line of sight toward two B stars, ϵ CMa and β CMa, where the density is more than an order of magnitude lower than the mean in our part of the Galaxy.

From radio continuum observations, you can derive two essential parameters. First, from the brightness temperature of the optically thin part of the gas, you obtain a measurement of the electron temperature. Recall that for thermal bremsstrahlung (free–free), the temperature dependence of the emissivity arises

---

[29] Reynolds, R. J. 1980, *ApJ*, **236**, 153; Reynolds, R. J. 1987, *ApJ*, **323**, 118. More recent observations with the Wisconsin Hα Mapper (WHAM) cover the sky more uniformly and includes additional bandpasses, see Reynolds, R. J. et al. 1998, *Publ. Astr. Soc. Austr.*, **15**, 14.

from the electron Maxwellian distribution and if the gas is transparent, then every parcel contributes flux at each frequency with this brightness temperature depending on $n_e^2 V$, its emission measure. Once the gas becomes optically thick, however, it self-absorbs and the surface brightness decreases with decreasing frequency. We used this property when analyzing the radio emission from extended moving atmospheres (Chapter 3). Since we assume local thermal equilibrium when asserting the uniformity of the distribution function, we can apply the long-wavelength limit of the source function to obtain $\kappa_\nu \sim \nu^{-2.1}$ for the wavelength dependence of the absorption coefficient. Then optical depth unity will be where the frequency at which the observed flux drops, which, you can see, is a function of the emission measure. The lower the density, the lower the frequency at which this turnover occurs in the continuum, so for compact ionized regions this should occur at frequencies higher than for normal H II regions.

This is one of those cases where modern instrumentation actually limits both detection and analysis of this gas. Interferometers are insensitive to diffuse emission. To study extended diffuse regions, it is actually necessary to lower the resolution.[30] Older measurements obtained with single disk-radio telescopes greatly facilitate this. Despite the problems with sidelobes and other features of single-antenna image formation, only such broad-beam surveys are able to account for the full contribution from the thermal radio continuum. The local interstellar medium provides dramatic evidence for the hot phase through ultraviolet absorption measurements against hot stars.[31] Several sight lines show cold clouds, yet along most there is a high-ionization phase with a very low column density, about 10^{-2} cm^{-3} or less, and an inferred temperature of about 10^6 K. Other evidence, as we will see, comes from scintillation measurements of extragalactic radio sources and pulsars.

The clearest evidence for exceedingly hot diffuse gas, above a temperature of a few 10^6 K, comes from X-ray continuum and line emission. The Fe Kα line, which is due to a mixture of high-ionization contributors, is observed at around 6.4 keV. There is also a fluorescence line at 6.7 keV. To obtain such hot gas requires very energetic processes, such as repeated shocking by supernova ejecta, because the cooling time for the gas is quite short. Such line emission has been detected throughout the Galaxy, especially from the galactic center but also from the plane, and it is also a feature of clusters of galaxies.

For the hot medium, the gas is fully ionized and for a helium abundance $n(\text{He})/n(\text{H}) = 0.1$, the pressure is 1.1×10^{-12} erg cm^{-3}, about twice the value inferred for the warm and cold phases. To add to this picture, high-redshift systems (see Chapter 8) have now been seen to show the O VI 1031, 1037 Å doublet and therefore add support to the existence of extended hot halo gas

---

[30] The Orion nebula is certainly the most completely images of any galactic H II region. See Yusef-Zadeh, F. 1990, *ApJL*, **361**, L19. These VLA data show especially clearly the effects of increasing angular resolution on the quality of the image, and also the loss of diffuse emission from an aperture synthesis instrument.

[31] For a general review, see Ferlet, R. 1999, *Astr. Ap. Rev.*, **9**, 153; a good introduction to the spectroscopic diagnosis of structure is Gry, C. et al. 1995, *Astr. Ap.*, **302**, 497. A comprehensive survey of the pre-HST and pre-EUVE (Extreme Ultraviolet Explorer) data on the local gas is Kondo, Y., Bruhweiler, F., and Savage, B. D., eds. 1984, *The Local Interstellar Medium: IAU Colloq. 81* (Washington, DC: NASA CP-2345).

TABLE 6.7 Representative Components of the Diffuse Interstellar Medium

Component	$T(K)$	n (cm^{-3})	z_0 (kpc)
Cold neutral medium	50–200	1–40	0.15
Warm neutral medium	8000	0.2	0.3
Warm ionized medium	10^4	0.02	0.9
Hot medium	10^6	0.003	1.5

around even comparatively normal disk galaxies. In our system, lines of sight through the halo reveal absorption by this doublet that extends to quite large halo distances (the advantage, for once, of being inside the system). This gas, which is not hot enough to emit X-rays, but is nevertheless significantly ionized, can be traced by using absorption line measurements toward halo stars. You always have the advantage with absorption measurements that you know the distance to the light source even if you don't necessarily know where along the line of sight to place the intervening cloud, and for distant stars that are well above (or below) the galactic midplane, using background stars of increasing distance provides a scale height for the gas. In the galactic plane, you can easily detect C IV 1550 Å and Si IV 1400 Å absorption in many lines of sight, and also see the separate contributions of the disk, halo, and LMC gas when observing OB stars in the ultraviolet toward the Magellanic Clouds; the LMC and SMC velocity shift of > 200 km s^{-1} is a great tool for specifying where the absorption line originates.

Gas heated to about 10^6 K can cool fairly efficiently through emission lines of abundant atomic species, even at very low densities. Recombination and collisional excitation regulate the temperature, as we have seen. The physical conditions so resemble the outer solar atmosphere that the state of the gas is said to be *coronal*. Should the temperature exceed 10^7 K, however, it will not be able to cool quickly because of the reduced efficiency for energy loss when the medium is too highly ionized. The heating is not only from stellar photons, which will not generate such immense temperatures. Some mechanical input is required, and this is why the study of supernova remnants and stellar winds is important for understanding the phases of the medium. (See also Table 6.7.[32])

6.3 DUST

Although the first photographic atlas of the Milky Way by Barnard at the beginning of the twentieth century dramatically showed the filamentary and diffuse nature of dark nebulae, the clear signature of a widely dispersed solid phase in the interstellar medium took nearly 30 years more years to identify.[33] Most of the work on dust was centered on its visual extinction until the discovery of interstellar reddening. Two strands came together in this work. The first was the distance–

---

[32] For an interpretation of the data in Table 6.7, see Pöppl, W. 1997, *Fund. Cosm. Phys.*, **18**, 1; Ferrière, K. 1998, *ApJ*, **497**, 759; Ferrière, K. 2001, *Rev. Mod. Phys.*, **73**, 1031.

[33] See Jaki, S. 1972, *The Milky Way: An Elusive Road for Science* (NY: Science History Publ.). This is an excellent history of the work on the structure of the Galaxy and also on the early work on the interstellar medium.

diameter relation for globular clusters that led Trumpler to conclude that the most distant systems were also redder than those in the solar neighborhood. The other was the intrinsic colors of stars. The former was relatively straightforward to study because of the high intrinsic luminosity of the target population. The line-of-sight distance, and hence the column density, was systematically larger toward those systems than in the direction of individual stars. The galactic center is more nearly accessible with cluster studies than for, say, OB stars. Also, the work was hindered by the problem of identifying the stars by their spectra. Early photographic emulsions, being very insensitive, permitted good spectroscopic measurements only of relatively bright stars. This confined the surveys to work within a few kiloparsecs of the Sun. However, the globulars are a more homogeneous population, so it was merely necessary to find them to study them. Ultimately, the result was the demonstration that the matter absorbing starlight was doing so selectively. The best explanation was that the particles responsible for extinction are dust grains, solids with both scattering and absorption properties that are wavelength-dependent. Greenstein (in 1939) and van de Hulst (in 1946)[34] attacked the problem of the optics using the theory developed at the beginning of the twentieth century by Mie for dielectric spheres and cylinders. They showed that the principal cause of reddening was the diffractive wavelength dependence of the scattering. But the next clue to the nature of the grains was unexpected. In 1946, while attempting to measure the polarization predicted by Chandrasekhar for rotationally distorted electron scattering atmospheres, Hiltner observed several galactic O stars and found what appeared to be intrinsic polarizations. At around the same time, Hall had constructed a photometric polarimeter and in 1947 independently announced the discovery of detectable polarization toward strongly reddened OB stars. The resulting papers[35] are wonderful examples of how discovery sometimes follows technology rather than ideas. Typically of the order of a few percent, the net polarization did not have the expected wavelength dependence. It correlated with reddening and extinction, and could best be explained as arising in the intervening interstellar gas. This discovery of *interstellar polarization* was the spur for much of the later activity on grains. This property of the dust demonstrated that the grains are not spherical. The wavelength dependence that was soon discovered showed that the grains are only a bit smaller than the wavelength of visible light, about 0.1μ. Finally, the grains are more or less coherently aligned over substantial parts of the Galaxy, a conclusion reached only with all-sky surveys of bright stars that showed that the polarization position angle has a correlation length of several hundred parsecs. Several explanations for the alignment have been developed, all examining the competing roles of collisions and magnetic torquing, but the basic model was in place by the 1960s.

A new era opened when the IRAS satellite was launched. Through its combined far infrared imaging and low-resolution spectroscopic instrumentation and comprehensive coverage of the sky, the diffuse dust component of the ISM could at last be viewed directly. A major byproduct of the COBE satellite was a fundamentally flux-calibrated second-generation repeat of the IRAS survey extended toward

---

[34] Much of this work was later expanded in the book *Scattering of Light by Small Particles*.
[35] Hiltner, W. A. 1949, *Science*, **109**, 165; Hall, J. S. 1949, *Science*, **109**, 166; Hiltner, W. A. 1949, *ApJ*, **109**, 471.

lower frequencies in order to determine the galactic and extragalactic contributions to the cosmic background emission, observations extended by ISO and SIRTF that have both improved sensitivities and higher resolution spectroscopic capabilities.

6.3.1 Observations

This is one of the few times when simply looking up at the night sky can provide an astrophysically interesting result. Go outside and stare at Cygnus or Sagittarius. The most obvious evidence for dust is the patchy obscuration of the background, diffuse starlight of the plane. In the southern hemisphere, the Coal Sack, an isolated nearby molecular cloud in Crux, was so striking that it was thought at first to actually be a hole in the stellar distribution. In the nineteenth century, M. Wolf showed that the apparent luminosity distribution of stars in some regions departs from the field values, explaining the deficit of brighter stars with increasing distance as an extinction effect from some otherwise invisible component of the medium. Barnard's photographic atlas of the Milky Way highlighted the filamentary clouds of dark matter that are widely dispersed against the luminous background. The Horsehead Nebula is a beautiful example of this (Fig. 6.3). Trumpler's work on the globular cluster colors and luminosities pointed in the same direction, with the added information that came from multicolor observations. The structural analysis of the this spatial extinction was begun in a series of papers by

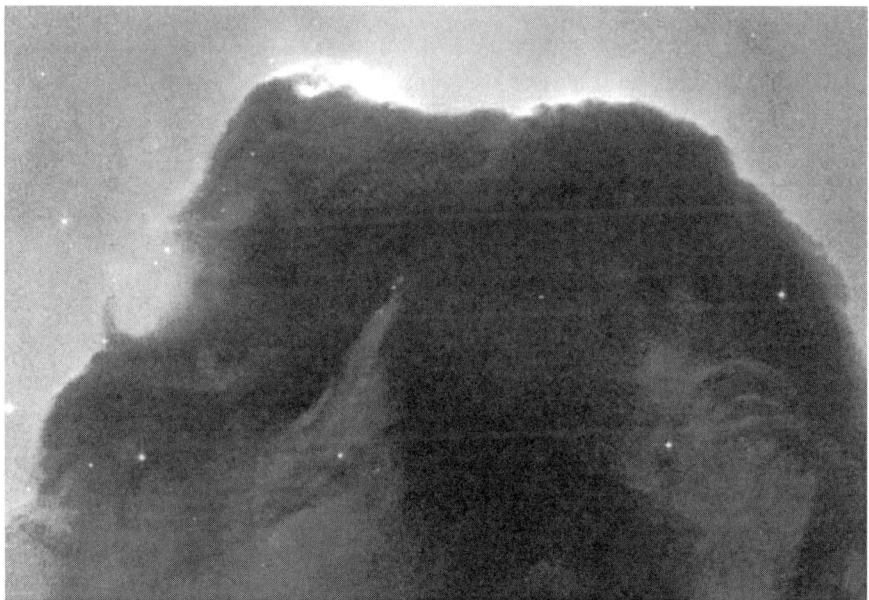

FIGURE 6.3. HST WFPC2 optical image of the Horsehead nebula. As we will discuss later in this chapter, this is an example of a photodissociation region, where the cloud is surrounded by an H II nebula. This and others like it were first photographically imaged by Barnard at the beginning of the twentieth century and, along with star counts, provided the first evidence of interstellar gas and dust (credit: NASA/STScI).

Chandrasekhar and Münch in the 1950s, much of which resembles the techniques used much later for large scale cosmological structural studies, through the use of correlation functions. It was recognized that a component of the ISM was actually altering both the colors and magnitudes of stars.

6.3.1.1 Optical and Ultraviolet Extinction
Determination of the spectral distribution of interstellar extinction, or reddening curve, depends on knowing the intrinsic spectra of the background sources.[36] This is especially difficult when using stars in external galaxies, where spatial resolution of the telescope limits one's ability to separate out individual stars. There is also a possibility that the intrinsic colors of the stars differ from their galactic counterparts. As you know, this can result from a difference in the star's metallicity and its effect on the opacity of the atmosphere. For the Galaxy, this is not so difficult, although the column densities limit the sampling radius to only a few kpc at the solar galactocentric distance. For only two external systems, the Magellanic Clouds, have intrinsic reddening curves been derived.[37] The technique is quite simple, in principle. Observe a nearby star for which you can get as close as possible to a full spectral energy distribution, or at least calculate a spectrum using a model atmosphere code. Then observe some star of the same spectral type and compare the continua. The spectral type, which is based on absorption line equivalent widths, will not be altered by extinction. The ratio of the observed to model continuum, or to that of a standard star of the same intrinsic properties, yields the reddening curve. The hard part is finding a star whose intrinsic flux distribution is known. This may be due to different abundances, rotation, binarity, or other complications. The steady improvement in model atmosphere techniques has provided an alternate procedure—use theoretical stellar flux distributions as template spectra. This method is most promising because it can be coupled with statistical procedures, such as maximum entropy methods, which allow the reconstruction of the extinction curve.

The two parameters that characterize dust extinction are the *total extinction*, A_V, and the *selective absorption* or *reddening*, which is also called the *color excess*, $E(B - V)$. The latter is defined as

$$B - V = (B - V)_0 + E(B - V), \qquad (6.52)$$

where $(B - V)_0$ is the intrinsic stellar color in the Johnson system.[38] The transfor-

---

[36] See Golay, M. 1974, *Introduction to Astronomical Photometry* (Dordrecht: Reidel) for a useful introduction to the methods. Comprehensive reviews are presented by Savage, B. D. and Mathis, J. 1979, *ARAA*, **17**, 73; Mathis, J. S. 1990, *ARAA*, **28**, 37. See Mathis, J. and Cardelli, J. 1992, *ApJ*, **398**, 610 for a very thorough analysis of the variation of the total to selective extinction ratio.

[37] Fitzpatrick, E. 1985, *ApJ*, **299**, 219 and references therein.

[38] As we mentioned in Chapter 2, this is the most widely used photometric system. Different filter systems require specific extinction laws and are based on local calibrators and comparisons with synthetic spectrophotometry from stellar atmosphere calculations. Warren and Hesser (1977, *ApJS*, **34**, 207) provide information about transformations between the Strömgren and Johnson systems. The total to selective absorption is comprehensively discussed in Cardelli, J. A., Clayton, G. C., and Mathis, J. S. 1989, *ApJ*, **345**, 245.

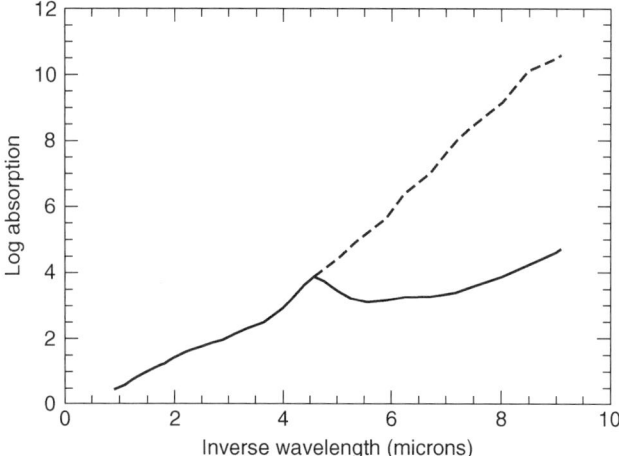

FIGURE 6.4. Comparison between the Galactic (solid) and Small Magellanic Cloud (dashed) extinction curves.

mations for the most widely used systems are

$$E(U-B) = 0.72 E(B-V)$$
$$E(b-y) = 0.73 E(B-V)$$

In general, calibration of this curve is difficult for long lines of sight or high column densities. It requires a standard star of the appropriate spectral type whose properties can be compared with a star whose classification is known. And you know from Chapters 2 and 3 that there are serious limitations to this method. Spectral classification is more descriptive than precise, and the quantitative application of this qualitative scheme must be done with caution. Spectrophotometry is now available over almost the entire spectrum and the extinction curve is well determined between about 0.1 and 10μ. The standard galactic extinction curve is shown in Figure 6.4. A fundamental parameter for the dust is the total to selective extinction ratio, $R \equiv A_V/E(B-V)$, which for the Galaxy is approximately 3.1 ± 0.2; for the B filter, $A_B/E(B-V) = 4.14$ with about the same uncertainty.

The R parameter is not a universal constant and likely depends on the grain size distribution. So does the extinction curve for which R measures the optical slope. You can see how this changes with environment from Figure 6.4 by comparing the plotted Galactic curve with that of the Small Magellanic Cloud. Inferred from the combination of extinction and polarization measurements,[39] this appears to be a power law in grain mass or radius between about 0.01 and 1μ, where with the number at a size a is approximately

$$N(a) \sim a^{-3.5}, \qquad (6.53)$$

[39] See Mathis, J. S. 1990, *ARAA*, **28**, 37, and Mathis, J. S. and Cardelli, J. A. 1992, *ApJ*, **398**, 610.

referred to in the literature as the *Mathis–Rumpl–Nordsieck* (MRN) distribution.[40] Its origins are tied up in models for the origin and evolution of grains, and we will postpone the discussion for the moment.

The dust appears to be well mixed even in the diffuse gas, and the extinction correlates with the column density of the neutral hydrogen

$$N_H = 5.8 \times 10^{21} E(B - V) \text{ cm}^{-2} \qquad (6.54)$$

and with the molecular hydrogen column density

$$N_{H_2} = 1.79 \times 10^{21} A_V \text{ cm}^{-2}. \qquad (6.55)$$

This last result is also important for calibrating the *CO-to-H_2 conversion factor* for molecular observations since it provides the means for obtaining the abundance of the species, H_2, that comprises the bulk of the mass of molecular clouds.

6.3.1.2 Ultraviolet and Infrared Broad Features

The observed extinction law several contributing features, whose mixture depends on the line of sight. First, there is the smooth continuum, which varies as a power law in λ. The slope of the smooth component depends on environment, which appears to govern the size distribution of scatterers. The next prominent feature is a broad absorption "resonance" at around 2175 Å, which is attributed to "astrophysical graphite." We use quotes here because all that can be said with certainty is really that the feature is due to some carbonaceous, probably amorphous, compounds. Precisely how this feature depends on the gas-phase abundances is still a debatable question. The third component is seen at around 10μ, a broad absorption that is attributed to several types of silicate and SiC.

Fortuitously, these identifications have been strengthened with the observation of grain formation in several unlikely, non-interstellar sources: nova and supernova ejecta. During the early stages of expansion of the ejecta from SN 1987A, rapid systematic changes were observed for the emission lines such that they became more asymmetric and their centroid velocities shifted to the blue. This was attributed to the formation of dust throughout the expanding medium, which masked the redshifted portions of the lines. The observation has since been repeated for a number of events. Also, during the early stages of nova V705 Cas 1993, a precipitous drop in UV emission was observed at around 100 days into the outburst. While no changes in the line spectrum were seen, the continuum level dropped by more than a factor of 100 in only a few days. This preceded a steep decline in the bolometric luminosity of the source, mimicking that seen in the optical of historical novae such as DQ Her 1934. Recovery took place over the next several months. Overall, the event bore a striking resemblance to the decline and recovery of the R CrB stars, whose behavior is also attributed to dust formation within the stellar wind and pulsationally ejected material. The critical clue comes from the abundance anomalies detected for the sources. Those novae in which this decline is observed are all carbon-rich, and in V705 Cas the SiC features at around 10 and 20μ were quickly observed in emission. The R CrB stars are carbon stars.

---

[40] Mathis, J. S., Rumpl, W., and Nordsieck, K. H. 1977, *ApJ*, **217**, 425; Mathis (1990), footnote 39.

The inference is that when the gas cools below the Debye temperature, either dust condenses or that grains grow quickly from accretion onto preexisting more stable nuclei such as fullerenes at that point.

6.3.2 Grain Optics

The extinction curve is the strongest evidence for a solid phase of the interstellar medium. You know that Rayleigh scattering has a strong wavelength dependence, in sharp contrast to the behavior of the interstellar extinction law. The reflection characteristics from obliquely illuminated nebulae, the bluing of starlight and the detection of the reflected stellar spectrum, also argue for reflection by small solid particles.[41] The optical properties of grains derive from their composition. The extinction curve indicates that silicates and some form of carbon, often called *graphite*, dominate the other possible absorbers. Both are dielectrics and therefore scatter and absorb radiation depending on the frequency. The scattering and absorbing properties of dielectric bodies was first examined using the Maxwell equations by Mie, Debye, and Rayleigh[42] assuming a constant response for the interior of the material.

We can gain insight by manipulating some definitions from classical electromagnetic theory. First, within a substance, we are interested in how the displacement current, **D**, develops with time. It is related to the internal electric field by $\mathbf{D} = \epsilon \mathbf{E}$. The internal electric field, **E'**, depends on the polarization, **P**, of the material through $\mathbf{E'} = \mathbf{E} + 4\pi \mathbf{P}/3$. Finally, **P** is assumed to be produced through the collective orientation of the electric dipoles within the sample by the internal electric field. If we assume that the constituent particles only polarize when a field is present and otherwise have no permanent electric dipole moment, then $\mathbf{p} = \alpha \mathbf{E'}$ and $\mathbf{P} = N\mathbf{p}$, where N is the number of dipoles in the sample. Here α is a scalar constant. It may, however, also be more complicated. For instance, in an anisotropic medium, α can be a tensor, depending on the orientation of the sample with respect to the imposed electric field. The important feature to note is that there are two electric fields here. One is the total field, **E**; the other is the local field, **E'**. The dielectric susceptibility is defined by $\epsilon = 1 + 4\pi\eta$ through $\mathbf{P} = \eta \mathbf{E}$. The combination gives $\alpha = (3/4\pi N)(\epsilon - 1)/(\epsilon + 2)$, which connects the polarizability of the medium with the dielectric constant. This is the *Clausius–Mossotti equation* for a static dielectric. In the electrodynamic case, it is called the *Lorentz–Lorenz equation*. From electromagnetic theory, you know that $\epsilon = n^2$, where n is the index of refraction, so that

$$\alpha = \frac{3}{4\pi N} \frac{n^2 - 1}{n^2 + 2}. \quad (6.56)$$

The relation of Eq. (6.56) to grain optics comes from the solution to the scattering problem for a conducting sphere.

---

[41] See Sandage, A. 1976, *AJ*, **81**, 954 for an example of using high-latitude reflection nebulae to obtain the integrated optical stellar spectrum of the plane.
[42] Mie, G. 1908, *Ann. Phys.*, **25**, 377; Debye, P. 1909, *Ann. Phys.*, **30**, 57; Rayleigh, Lord 1918, *Phil. Mag.*, **36**, 365 (see also *Collected Papers*, vol. 6).

In the far field limit, a body of geometric area πa^2 produces an extinction $Q_{\rm sca} = 2$ due to diffraction. This is independent of the wavelength. Detailed derivations of the equations of Mie theory are presented in the references, especially Born and Wolf (1998).[43] We will note that the scattering cross section is given by the real part of the index of refraction, while the absorption comes from the imaginary part (this is the dissipative part due to the conductivity).

For our purposes, the most interesting feature of the classical diffractive calculation is the prediction of the angular distribution of the scattered radiation.[44] In the large-wavelength limit, when $a/\lambda \to \infty$, the ratio of the scattering to geometric cross section for a diffractive aperture asymptotically approaches 2. This is a shadow effect resulting from diffraction of even the longest wavelengths that do not actually penetrate the grain. In one sense, a grain is a barrier that both absorbs and diffracts incoming light. The larger the grain, of course, the larger the barrier and the greater the extinction. But a large grain also acts more like a brick in mainly back-reflecting light. On the other hand, small grains should be more forward-scattering. A lovely demonstration of this is seen in the solar system. The ring systems of the outer planets, especially Saturn, consist of icy particles with a wide range of sizes. During the *Voyager* 1 and 2 flybys, the brightness of the individual subdivisions of the ring were seen to depend on whether they were viewed in back-scattered (before ring plane traverse) or forward-scattered (afterward) light.

A number of broad absorption features are seen in addition to the broader continuous extinction curve from diffraction, and these depend on the grain's composition. For instance, the prominent absorption at around 2175 Å is associated with graphite and constitutes the principal evidence for a large fraction of interstellar carbon being locked in the solid phase. Additional broad absorption features are seen at 10μ, due to the Si—O bond stretch in silicates, and at around 11.2μ, an absorption from SiC. The breadth of these features, of order 100 Å for the graphite feature and even wider for the SiO and SiC features, argues that they are produced in a solid phase of the medium. No such broadening would be expected within a gas at typical interstellar conditions. Curves of this type are well known not only from the Galaxy but also from the field regions of the Large and Small Magellanic Clouds.

6.3.3 Infrared Observations

The distance between stars is enormous, and so any material located between the stars sees a very dilute radiative background. This accounts for why the interstellar medium is, on the whole, so cold. To see this, we hark back to the discussion from

---

[43] Born, M. and Wolf, E. 1998, *Principles of Optics*, 6th ed. (Cambridge, UK: Cambridge Univ. Press); van de Hulst, H. C. 1957, *Scattering of Light by Small Particles* (NY: Dover Books); Bohren, C. F. and Huffman, D. R. 1984, *Absorption and Scattering of Light by Small Particles* (NY: Wiley); Wickramasinghe, N. C. 1967, *Interstellar Grains* (London: Chapman and Hall); Greenberg, J. M. 1968, in *Nebulae and Interstellar Matter: Stars and Stellar Systems*, vol. 7, Aller, L. H. and Middlehurst, B. M. eds. (Chicago: Univ. Chicago Press); Martin, P. G. 1978, *Cosmic Dust* (Oxford: Oxford Univ. Press); Asano, S. 1979, *Appl. Opt.*, **18**, 712 (and references therein); Draine, B. T. and Lee, H. M. 1984, *ApJ*, **285**, 89; Draine, B. T. and Lee, H. M. 1987, *ApJ*, **318**, 485.

[44] The Poisson spot in classical diffraction theory is an excellent example of forward scattering by a small aperture in the far field.

Chapter 3 on radiative equilibrium. Consider a dust grain, sitting in the middle of nowhere, that is bathed in incident light from the diffuse interstellar radiation field (DIRF). Assume, for a moment, that we are dealing with a large dust grain, of the order of $0.05-0.1\mu$. This assumption of large grains is crucial. A small grain, as we will see in a moment, behaves more like a loose aggregate of atoms than a solid. The dividing line between big molecules and small grains is ill-defined, but like art, everyone seems to know what a *big* grain is when they see one. Now if only photons heat the grain, it can reach thermal equilibrium; that is, it comes into radiative balance with the input energy. This is the same condition you've already seen from our discussion of stellar atmospheres in Chapter 3. Even though it is presumed that the grain radiates thermally with a Planck spectrum, the efficiency of the heating is determined by the match between the spectrum of the incident light and the absorption efficiency of the dust. The absorption coefficient for the grain of radius a is $Q_{abs}(\lambda)\pi a^2$, where Q_{abs} is the wavelength-dependent absorption efficiency. The emission also depends on wavelength through $Q_{em}(\lambda)\pi a^2$, defining Q_{em} similarly. Thermal equilibrium requires that

$$\int_0^\infty Q_{abs}(\lambda) F_\lambda \, d\lambda = \int_0^\infty Q_{em}(\lambda) B_\lambda(T_g) \, d\lambda, \qquad (6.57)$$

where F_λ is the energy distribution of the DISF and T_g is the equilibrium grain temperature. This temperature is the one that the grain will reach in order to radiate away the incident energy, but you see that it depends on the efficiency of the emission process and the match between the blackbody emissivity and the spectral characteristics of the material. For a solid, this means the match between the various energy bands and the photons. You have seen this before, when we were discussing radiative equilibrium for a stellar atmosphere, and for the same reason. In the absence of any dissipational modes within the grain, all the energy is ultimately reradiated. The spectrum is this re-emission is fixed by the grain characteristics, but not the total amount.

From Mie scattering theory, the absorption by the grains has some wavelength dependence even for longer wavelengths than the size of the particle because of diffraction. Assume that $Q_{em}(\lambda) \sim (a/\lambda)^n$. Then the heating terms on the left side of Eq. (6.57) combine to a constant; for the moment let's call it Γ. The Planck function depends on both wavelength and temperature. Substitute a new variable $x = hc/(\lambda kT)$ and integrate Eq. (6.57) over x:

$$\Gamma = C_n T_g^{4-n} \int_0^\infty x^{3+n} \frac{dx}{e^x - 1}, \qquad (6.58)$$

where C_n is an integration constant that depends on h, c, and k. Then we see that the equilibrium grain temperature varies with the input flux according to

$$T_g \sim F_{DIRF}^{1/(4-n)}. \qquad (6.59)$$

This is an important result for several reasons. One is that a nongray emissivity produces a weaker temperature dependence on the incoming flux than for a gray body. Another is that a grain that emits efficiently in the IR will be cooler than one that has a lower n.

The thermal regulation is achieved by a match between the energy distribution of the DIRF, or aggregate stellar radiation, and the grain *absorption* coefficient. As in the case of stellar opacity, you see the greatest heating is when the two maxima coincide. Here two extreme behaviors are observed. On one hand, we observe essentially featureless extinction curves that rise steeply toward shorter wavelength. This is where the broad absorption features we mentioned for the extinction curves play a role. The graphite feature dominates the UV extinction curve of the Galaxy. It is also seen along many lines of sight within the Magellanic Clouds. The infrared extinction produced by graphite is featureless. On the other hand, whereas there is a strong absorption band from the silicates, the UV extinction from these grains is featureless. Therefore, one would expect that the heating of silicates is less efficient and their cooling is more efficient than graphite. Hence, without any detailed computation, you can already estimate that *graphite grains are, in general, hotter than silicates under illumination by the same spectral distribution.*

Suppose you want to look for this dust. There are several ways of doing it. One, of course, is to assume that there are sufficiently bright background sources that can be seen through it. This is what you are doing when you just look at the Milky Way. The problem is that you must know something about the intrinsic colors of the object, and also something about the uniformity of the line of sight. It is, however, possible to look for dust using another technique, the infrared. This follows from Eq. (6.59). The mean distance between the galactic plane ultraviolet-emitting stars is about 10 pc. The DIRF depends on the mass function and mean distance between stars, but it has about the spectral distribution of about a 14,000 K main-sequence star with a dilution factor of around 10^{-16}. Even though the hotter stars are relatively rarer than cooler stars, they are much brighter and thus contribute much more to the diffuse emission. Then the grain temperature is

$$T_g \sim W^{1/6} T_{\text{eff}}^{2/3}, \qquad (6.60)$$

where the proportionality is a constant of order unity. Consequently, the dust is expected to reach a few tens of degrees in radiative equilibrium. Note that *if the color temperature of the incident radiation is reduced, the temperature may fall dramatically.* This is seen in several very cold reflection nebulae, where the dust is not strongly heated. A superb example is the protoplanetary nebula IRAS 09371 + 1212, also called the "Frosty Leo Nebula," which is illuminated by a central K giant. Thus, you see that the principal spectral range for studying the diffuse dust is in the far infrared, longward of 25μ. This was the original justification for the Infrared Astronomical Satellite (IRAS) that operated from 1983 to 1984, the Diffuse Infrared Background Experiment (DIRBE) on the Cosmic Background Explorer (COBE) satellite, and the Infrared Space Observatory (ISO).[45]

6.3.4 Big Grains

The distinction that is now made is between the two populations of large particles in the ISM. The big grains are the ones thought to be mainly responsible for the

---

[45] See the special issue on the first ISO results, 1996, *Astr. Ap.*, **315**, 1, for a more complete description of the instruments and programs.

overall galactic extinction curve. These are typically 0.1μ or smaller, and they are clearly not spherical as attested to by the polarization of both scattered and transmitted light. the small grains, which may hold the bulk of the galactic interstellar carbon, are not much bigger than substantial molecules. These are distinguished by their radiation, nonequilibrium band emission that dominates the near infrared. The larger grains are closer to ordinary blackbodies, and their radiation dominates the far infrared. We will discuss these in separate sections here because they display somewhat different classes of physical characteristics and because the processes in which they participate are also somewhat different.

6.3.4.1 *Depletion of Metals*

Direct observations of the heavy elements show that their gas-phase abundances depart systematically from those observed in stellar atmospheres. This is interpreted as the incorporation of the gas into grains, either coating their surfaces or into the matrix of the solid phase. To date, the highest resolution spectroscopic studies[46] find that the iron group are among the most depleted, with the exception of zinc, and that heavier elements such as germanium and gallium, for which there are several resonance lines in the ultraviolet, are not depleted *relative to the solar values*. We emphasize this because the lines are very weak, and only for the Sun have the minimum (normal) values been detected.[47]

The pattern shows an interesting coincidence: the depletion factor correlates with T_{cond}, the laboratory equilibrium condensation temperature.[48] Refined observations of the sort we have been discussing only strengthen this result. Yet its physical origin is obscure. The correlation seems to imply imply that the interstellar gas-phase abundances are not determined locally by accretion (sticky collisional) processes. Instead, to show a dependence of depletion with T_{cond}, they would have to be frozen into the gas at the source, presumably in red giant winds and novae and supernovae. Meteorites, which we will discuss in a moment, show evidence of several different sites for the dust, including AGB stars, but this does not relate to the gas-phase abundances. On the other hand, you would naively expect from our discussion that the grains would grow depending on the sticking properties of the surfaces and the collision rates for environmental ions. It is therefore odd that the single parameter used to determine the gas-phase abundance pattern should be irrelevant for interstellar conditions.

Depletion has a profound effect on determinations of the chemical evolution of the Galaxy, and we will encounter them again when dealing with the cooling. Now we will review some of the patterns and their explanation. An interesting observation that supports the depletion picture is that one easily observed element, zinc, is not depleted in the gas phase relative to the solar abundance. The Zn II resonance lines are at around 2065 Å and are easily observed with STIS (and in archival spectra from IUE and GHRS). This provides a straightforward normalization of

---

[46] See surveys by Sofia, U. J., Cardelli, J. A., and Savage, B. D. 1994, *ApJ*, **430**, 650, and Sembach, K. R. and Savage, B. D. 1996, *ApJ*, **457**, 211.

[47] Some chemically anomalous objects, the Ap, Am, and Ba stars for example, show much higher values for the abundances of these elements (see the discussion in Appendix 3.B). The λ Boo stars are the only ones that show depletion patterns of metals that are similar to those found in the diffuse ISM and, you will recall, may be explained by accretion of depleted gas from circumstellar material.

[48] Field, G. 1974, *ApJ*, **187**, 453; see also Stull, D. R. and Prophet, H., eds. 1971, *JANAF Thermochemical Tables*, 2nd ed. (Washington, DC: NIST).

the metal abundances along any line of sight. Especially for extragalactic clouds, for which the environment is unknown, this is the most useful way of normalizing the observed abundances.

We have already discussed how metals serve as coolants through their ability to radiate energy gained from colliding particles. If these species deplete onto grains, the gas cannot as effectively reduce its energy. We will see shortly how this alters the thermal state of the gas and affects the formation of structure in self-gravitating environments.

6.3.4.2 Solar System Measurements: Meteorites and Heliospheric Dust

We discussed the nucleosynthetic signatures in Chapter 4, but don't lose sight that these studies are actually directly probing samples of interstellar dust *in the laboratory*, frozen in the meteoritic matrix since the formation of the solar system. Microchemical analysis of meteorites has been possible for some time, but the actual detection of individual grains and the analysis of their chemical composition has only been possible since the early 1990s. Not all meteorites are appropriate for this sort of study since the majority have been repeatedly shock-heated and reprocessed (siderites). The carbonaceous chondrites[49] preserve the best samples of individual inclusions with anomalous abundances. The first indication of this was the discovery of ^{26}Mg, a product of the decay of ^{26}Al, in the Allende and Murchison meteorites. The finding of evidence for this comparatively short-lived isotope ($t_{1/2} \approx 10^5$ yr) in a meteorite sets an upper limit for the condensation time of the sample. More importantly, it was the first indication of a fast pollution of the medium out of which the Sun formed. The subsequent discovery of the 1.806 MeV nuclear decay line of the ^{26}Al served as a marvelous confirmation of the chemical methods.

The local interstellar medium, within about 50–100 pc of the Sun, offers the most complete observational laboratory for the study of this gas. As you would expect, the closer you look, the more complicated the medium becomes thermally and structurally.[50] A spectacular result, obtained as a byproduct of the solar wind mission *Ulysses* and the Jupiter probe *Galileo*, has been the detection and analysis of dust that is swept into the heliosphere by motion of the Sun with respect to the local interstellar gas. Hard though it may be to believe, grains that are clearly not associated with either the cometary cloud surrrounding the outer solar system nor zodiacal light particles can be distinguished by their mass, composition, and relative motion. Destruction of the smaller grains likely distorts the spectrum at the end that is mainly responsible for th bulk of the dust extinction in the diffuse medium. Nevertheless, a first derivation of the local dust to gas *mass* ratio gives about 100 for particles with sizes around 1μ, while measurements along the line of sight toward ϵ CMa, a local B giant, gives a ratio of about 500. Although both these numbers are presently only certain within about 2σ, the observational situation can only improve.

---

[49] Especially important here are the type II variety that have high abundances of hydrous compounds that indicate they have not been severely altered since formation and that they condensed in a colder than usual environment.

[50] See Frisch, P. C. et al. 1999, *ApJ*, **525**, 492; Wood, B. E. and Linsky, J. L. 1997, *ApJ*, **474**, L39; Gry, C. et al. 1996, *Astr. Ap.*, **302**, 497 for examples of different analytical methods for studying the local structure.

6.3.4.3 Grain Charging

Three mechanisms charge the grains and therefore affect the sticking properties of metals. First, there is ionization by ambient photons, the direct analog of the laboratory photoelectric effect. This would be expected to produce positively charged dust were it not for the second, competing process of collisions with ions and electrons from the ambient gas. Finally, sputtering, the destruction and charging of the grain by photoinduced emission of protons, ions, and electrons from the grains through absorption of ambient UV photons, leads to changes in the charge produced by these two processes and also leads to grain destruction.[51] Ultraviolet and X-ray photons produce electron emission through the photoelectric effect and positively charge the grains. The efficiency of the process depends on the dust composition.[52] The threshold energy for the process, the *work function* (recall our discussion of photomultipliers in Chapter 2), varies with composition and structure. Also, different grains are expected to have different energy-dependent yields, $Y_{p\nu}$, for emitted electrons so the rate for charging depends as always on the match between the incident radiation and the absorption cross section:

$$R_{pe} = \int_{\nu_0}^{be} \frac{L\nu}{4\pi r^2} Y_{p,\nu} \pi a^2 Q_{abs} \frac{d\nu}{h\nu}. \qquad (6.61)$$

Note that $E_0 = h\nu_0$, the work function, depends on the grain charge since, as electrons are emitted, the potential should become more attractive and the threshold must increase.

Even in the absence of a radiation background, charging occurs because of direct collision of ions and electrons with the grain surface. This is a nice analogy to the gravitational deflection problem we discussed in Chapter 1. Assuming that the angular momentum of a particle of mass m and velocity v is $J = mva$, where a is the grain radius, the deflection for direct impact gives a impact parameter $r = a(1 - \epsilon^2)^{1/2}$, where the eccentricity is given by $\epsilon^2 = U_g/mv^2$ and the collision cross section becomes

$$\sigma_k = \pi a^2 \left(1 \pm \frac{2U_e}{mv^2}\right). \qquad (6.62)$$

Here the electrostatic potential of the grain, U_g, is given by

$$U_g = \frac{Z_g e^2}{a} \qquad (6.63)$$

for a charge Z_g in units of e. The incident flux of particles of type k is $n_k \langle \sigma_k(v) v_k \rangle$, where now the angle bracket denotes averaging over a thermal velocity

---

[51] Dwek, E. and Arendt, R. G. 1992, *ARAA*, **30**, 11.
[52] Although we have no direct access to the *interstellar medium*, we do have examples of grains within the solar system, moving within the solar wind. Perhaps the smallest interstellar clouds, comets within the solar system, best illustrate the dynamics. As a comet approaches the Sun, it is photoevaporated and also ionized by the incident solar radiation. This produces charged grains, which are swept up in the interplanetary plasma, and ions. This produces the two types of tails observed for such bodies that have been extensively described in the literature.

distribution. The rate of charging thus depends, in the absence of photon processes, only on the differential rate of arrival of plasma and the feedback effect of the increased charge on the collisions. For the ions, the velocity is slower than for the electrons at the same temperature, T, by a factor of $(m_e/m_i)^{1/2}$, and you would expect the grains to have a net negative charge. Consequently, electrons moving with velocity lower than $v_{min} = (2Z_d e^2/m_e a)^{1/2}$ are not be able to penetrate the repulsive electrostatic barrier presented by the charged dust and deflect. If the medium maintains charge neutrality, we would expect the rate of incidence of electrons and ions onto the grain surface to balance in the absence of any photoelectric processing. However, because of the difference in the sign of the charge, the probability of penetrating to the grain is greater than for ions (in thermal equilibrium the electrons have a higher velocity). If the rate reaches steady state, the balance between the two rates is given by

$$n_i \langle \sigma_i v_i \rangle = n_e \langle \sigma_e v_e \rangle \tag{6.64}$$

with the averages required for the rates coming from the integrals over the thermal distributions:

$$\langle \sigma_k v \rangle = (2\pi m_k kT)^{3/2} \pi a^2 \int_{v_{k,0}}^{\infty} \left(1 \pm \frac{2Z_g U_g}{m_k v_k^2}\right) v_k^3 \exp\left(-\frac{m_k v_k^2}{2kT}\right) dv_k. \tag{6.65}$$

The condition that the rates balance then reduces to the statement

$$\left(\frac{m_p}{m_e}\right)^{1/2} \frac{n_p}{n_e} = (1+y)e^{-y}, \tag{6.66}$$

where $y = eU_g/kT$. This equation has the approximate solution $eU_g/kT \approx -2$. In other words, from this simple consideration of thermodynamic balance, the grains emerge with weak but definite electron excesses.[53]

6.3.4.4 Poynting–Robertson Drag and Accretion

Scattering not only redirects photons, it also provides a momentum source for the grains. We saw this when discussing the acceleration of stellar winds. Now consider what happens if we have a small dust particle orbiting a luminous central mass. If the grain were an isotropic scatterer in its rest frame, it could not be in the moving frame since the emission will be peaked in the direction of its motion. This forward emission consequently produces a braking force for the grain, and in the case of orbital motion a torque, and the angular momentum is reduced. You can derive a simple estimate for the breaking timescale as follows. Take a grain with radius a and scattering cross section $\sigma = Q_{sca} \pi a^2$ at distance d from the central star whose luminosity is L. Let it scatter a photon of frequency ω_0. Then the rate of arrival of photons is $(\Delta t)^{-1} \approx F\sigma/\hbar\omega_0$ where F is the flux. If the grain is moving at a

---

[53] See Spitzer, L. 1978, *Physical Processes in the Interstellar Medium* (NY: Wiley); Evans, A. 1994, *The Dusty Universe* (NY: Wiley); Hoányi, M. 1996, *ARAA*, **34**, 383. A discussion of the charging calculation is also included in Spitzer's collected papers.

velocity **v**, then it experiences a kick $\Delta p = (\hbar\omega_0/c)[(1 + \beta \cdot \hat{\mathbf{n}}) - (1 + \beta \cdot \hat{\mathbf{n}})] \approx 2\hbar\omega_0 \beta/c$. Then the force is $\mathscr{F} \approx \Delta p/\Delta t$ so the timescale $t_{\text{PR}} \approx p/\mathscr{F}$ or

$$t_{\text{PR}} \approx \frac{\rho a c^2 d^2}{Q_{\text{sca}} L} \approx 10^3 \text{ yr} \left(\frac{\rho}{3 \text{ g cm}^{-3}}\right)\left(\frac{a}{1\mu}\right)\left(\frac{d}{AU}\right)^2 \left(\frac{L}{L_\odot}\right)^{-1}, \quad (6.67)$$

so the smaller grains are removed rather quickly by this effect. The scattering produces a drag on the particle, acting like a viscosity, and lowers its angular momentum. The grain falls into a lower orbit and experiences even greater scattering and braking. Inevitably, the particle spirals into the central body and accretes or vaporizes. This is the *Poynting–Robertson* effect.[54] In the solar system, accretion of dust in the outer corona is evident from observations of the scattered solar spectrum, especially in polarized light. Since the dust cannot be formed in this hostile environment where the electron temperature is above 10^6 K, the scattering particles must come from elsewhere. The solar system harbors such a source, comets, which can supply a broad spectrum of particles, but there is a problem. Comets are seldom found in radial orbits, so a dust particle continues in an elliptical orbit once lost by the cometary nucleus. That is, it will continue to orbit *unless* it experiences a braking torque or something that mimics viscosity. This will also be true for circumstellar dust disks, such as those observed in β Pic systems where infall is often observed and the smaller particles have been cleared out preferentially. The largest grains are little affected by this torque, but the small particles are systematically removed. Such evolution has a profound effect on the size distribution of grains in circumstellar disks around young stars. An example is seen in β Pic, where the inner part of the circumstellar dust disk is removed and gaseous accretion events are detected from Ca II and Fe II resonance line profile asymmetries.

6.3.5 Small Grains and Big Molecules

The large grains that are responsible for interstellar polarization and the overall extinction curve are not the only component of the solid phase. This raises a question: When is a molecule a grain? We have assumed that the equilibrium radiation of solid in the ISM is intrinsically a blackbody, with the specific wavelength dependence produced by the particular absorption bands in the solid. This means that the density of energy states is dense enough that equipartition occurs between all of the available modes of emission and internal motion. If, however, the aggregate is relatively small, maybe only a few tens or hundreds of atoms, then this is not true. The cluster will behave more closely like a molecule and radiate in bands rather than as a continuum.

Although such clusters were predicted as early the mid-1950s, their existence remained speculative until the discovery of enhanced near infrared continuum

---

[54] Robertson, H. P. 1937, *MNRAS*, **97**, 423; Guess, A. W. 1962, *ApJ*, **135**, 855; Burns, J. A., Lamy, P. L., and Soter, S. 1979, *Icarus*, **40**, 1; Carroll, D. L. 1990, *ApJ*, **438** 588. The precise calculation requires both the phase function for the dust and the frequency dependence and is a fully relativistic effect. Regardless of the details, though, the basic result follows from our adumbration.

emission from reflection nebulae.[55] As we have discussed, such regions are found most easily by their scattering of blue stellar continuum light, and the cloud NGC 7023 is one of the best studied. In 1983, Sellgren found that the $3-10\mu$ region showed excess emission compared with the expected equilibrium temperature for grains at the distance of the dust from the illuminating star and suggested that the model first worked out by Aannstaed applied to these emitters—that they are small grains whose excitation fluctuates in reaction to absorption of stellar near ultraviolet radiation. The picture is essentially the same as Raman scattering. The excitation by a photon of relatively high energy is followed by a conservative cascade of lower energy photons through many states within the small grain. Small grains present a special problem in radiative balance. On the one hand, they are not quite molecules since they have an enormous number of vibrational states through which they can radiate. On the other hand, the number is not so large that the radiation is completely redistributed. Consequently they display neither blackbody spectra nor strict LTE. In other words, they must be treated by the same techniques that we have used for lines, and are an example of stochastic heating. For this reason, we will go into the problem in some detail.

The specific heat of a conglomerate consisting of a relatively small number of particles, N, is $c_v = \frac{1}{2}Nk$. The internal energy fluctuates because the populations of the states fluctuate proportional to $N^{1/2}$ in thermal equilibrium. Thus, a photon is redistributed into comparatively few modes with a low density of states and emerges through emission bands rather than a continuum. When illuminated by a diffuse UV and optical radiation field, the mean energy of the emergent photon is $h\nu/N$ rather than $h\langle\nu\rangle$, where $\langle\nu\rangle$ is averaged over the Planck function. The grain (or fluff) cannot then be said to have an equilibrium temperature because the internal excitation is never the same as a Boltzmann distribution. The grain is finding a solution for the radiation as if it were an atomic state:

$$\frac{dn_l}{dt} = -\sum_u P_{lu} n_l + \sum_u P_{ul} n_u, \qquad (6.68)$$

where P_{ij} is the transition probability. Then in the same way we have been calculating the emissivity for atmospheres, we have

$$S_\nu = \frac{n_u P_{ul}}{n_l P_{lu} - n_u P_{ul}}, \qquad (6.69)$$

so that unless the statistical weights are distributed just right, the distribution of

---

[55] Platt, J. R. and Donn, B. 1956, *AJ*, **61**, 11; Platt, J. R. 1956, *ApJ*, **123**, 486; Aannestad, P. and Purcell, E. 1973, *ARAA*, **11**, 309. We will discuss fullerenes more completely in Section 6.6.2.7, but mention them here for completeness. The closest particles to those first postulated by Platt, and later discussed by Aannestad and Purcell, are a class of carbon compounds discovered in soot residues by Kroto et al. (in 1979). The discovery of these molecules has opened a very rich area of physical and applied chemistry for which the 1996 Nobel prize in chemistry was awarded to Curl, Kroto, and Smalley. This field and nuclear astrophysics join with atomic physics as the best examples of the laboratory applications of cosmic environments.

emitted photons will not be in thermal equilibrium and the temperature will, in general, be higher than you would get from large grains. Why don't bricks behave this way? This behavior is due essentially to two competing processes. One is the density of states. In a small aggregate, these are few and the energy separations between them is large. The other is that any state behaves much as in an atom. In a large body, the perturbations on any state is so large that it merges into a near continuum. Thus, there is a possibility for a photon to be re-emitted anywhere that is separated by the same energy, that is through any of many vibrational modes, and the result is a statistical distribution that is characteristic of the Einstein distribution.

Small metallic grains have been studied in the laboratory for some years, in particular looking for quantum effects in small clusters.[56] The specific heat of a solid at low temperature depends on the temperature, $c_V \sim T^3$, a result we found in Chapter 1 for the Bose–Einstein distribution. The heating and cooling of the grain is dominated by small temperature fluctuations. Another way to approach this problem is to realize that these small grains behave as if the heating is stochastic and that the solids radiate impulsively. A thermodynamic approach can be used instead of the statistical equilibrium equations to look at the temperature distribution, and we include this because it is instructive for the methods required to treat random processes. Consider the rate equation for the heating of a grain whose specific heat depends on temperature:

$$c(T)\frac{dT}{dt} = \mathcal{H}(t) - G(T), \tag{6.70}$$

where the loss term is assumed to be of the form $G(T) = QT^m$ and we take $c(T) = c_0 T^n$. Defining $\xi = T^{n+1}$, Eq. (6.70) becomes

$$\frac{d\xi}{dt} = -\frac{n+1}{c_0} Q \xi^{m/(n+1)} + \frac{n+1}{c_0} \mathcal{H}(t) = -A_{mn}\xi^\alpha + \Gamma. \tag{6.71}$$

All of these changes in variables are necessary to reduce the equation to the form that has Γ as a stochastic function of time. The idea is that after each photon is absorbed, the grain sits around for a moment deciding what to do and then radiates through a cascade of internal vibrational states. The equation is formally an Ornstein–Uhlenbeck type,[57] and the solution to Eq. (6.71) is

$$\Delta \xi(t) = -A_{mn}\int_0^t \xi^\alpha(t')dt' + \int_0^t \Gamma(t', \Delta t')dt', \tag{6.72}$$

---

[56] For a very thorough review of the physics of small metallic clusters that discusses optical and heat capacity properties that are relevant for astrophysical grains, see Halperin, W. P. 1986, *Rev. Mod. Phys.*, **58**, 533.
[57] See Gardiner, C. W. 1985, *Handbook of Stochastic Processes*, 2nd ed. (Berlin: Springer-Verlag).

which can be solved by iteration.[58] We have gone into such detail here for several reasons. The infrared bands are seen in a very wide variety of sources ranging from reflection nebulae and the diffuse interstellar medium to planetary nebulae, and also seen from extragalactic sources. As such, their interpretation is important for understanding a broad range of distinctive environments. The observation of these bands, however, has a deeper significance: their presence in the spectrum, whatever their ultimate identification, *requires* the presence of strong ultraviolet emitting sources in the Balmer continuum. As such, they are bolometric probes for deeply embedded sources that may not be visible except through infrared observations since the bands are presumed to form the IR diffuse features by redistribution of photon energy among a cascade of energy levels. For instance, in regions of star formation, the bands should be sensitive to the same gas conditions that are usually probed in photodissociation regions by the far-infrared fine-structure lines, with the difference that these are directly photoexcited. The same is true for the inner regions of planetary nebulae. In other words, in order to see these bands, you must have sufficiently low opacity for the UV to excite the energy levels. They can also serve as coolants if collisional excitation is important, but they likely assist more with regulating the radiative balance of the medium.

A major advance in the identification of small grains comes from the comparison of ISO observations of *circumstellar* grains around young stars, the Herbig Ae/Be stars, with spectra of cometary dust from comet Hale–Bopp. The broad features from larger grains, those produced by silicates, are also similar to those observed for comets and appear to match magnesium- and iron-bearing crystalline minerals. As always, we will end this discussion by saying that the observational picture will certainly improve with new observations and space missions planned for the near future.

6.3.5.1 *Diffuse Interstellar Bands (DIBs)*

The diffuse interstellar bands are vexing problem handed to this century as a gift from the last. One thing is clear—these absorption features are connected with the dust and a clue to grain properties, but otherwise the situation is little improved in more than a decade. They were discovered in much the same way as the gaseous component of the ISM. First reported as broad stationary lines by Heger,[59] they

---

[58] Desert, F. X., Boulanger, F., and Shore, S. N. 1986, *Astr. Ap.*, **160**, 295 give as a solution for the expectation of the temperature fluctuations:

$$\langle \Delta T^{n+1} \rangle = \frac{n+1}{m} \left(\frac{(n+1)c_0}{mQ} \right)^{(n+1)/m} \langle \Delta \Gamma \rangle \Gamma_0^{(n+1-m)/m} (1 - e^{-t/\tau})$$

where now τ is the characteristic cooling time. This gives the formal distribution. It means that a high-frequency tail will be observed for the radiation, since there will always be some grains that reach higher than equilibrium temperature. The excess at short wavelength is, however, an *excitation* temperature and not characteristic of the kinetic state of the grains or the gas. Therefore, the fact that the observed emissivity is larger than predicted by the diffuse radiation field does not indicate a heating in the sense of a thermally equilibrated distribution.

[59] Heger, M. L. 1922, *Lick Obs. Bull.*, **10**(337), 146. Heger noticed the $\lambda\lambda 5780$ and 5797 Å features.

TABLE 6.8 Some of the Strongest Diffuse Interstellar Bands (DIBs)

λ_0 (Å)	$\Delta\lambda$ (Å)	$1 - r_c$	Possible Identification
4180	25	0.03	—
4428	18	0.15	1MePyrA
4760	25	0.03	—
4882	25	0.05	—
5450.3	3.5	0.03	—
5535	23	0.03	—
5778.3	17	0.06	—
5780.45	2.2	0.32	—
6177	29	0.07	—
6283.86	4.5	0.32	—
6533.35	0.9	0.06	—
7429	21	0.03	—
8621.23	20:	0.07	TetracA
9632	4	0.1	—

were studied by Merrill[60] in the spectra of several early-type stars with large column densities. The strongest visible lines are listed in Table 6.8.[61] Perhaps the best studied is at 4430 Å, although the catalog has grown to several dozen since 1990. Unlike atomic absorbers, they have full width at half maximum (FWHM) of several to tens of angstroms and optical depths of only a few percent. Because of their large widths and variable profiles, their intrinsic wavelengths are not very accurately determined, making their identification even more difficult. Herbig has shown that they tend to fall into several groups, suggesting that they have no *unique* origin. It is clear that the optical DIBs correlate in strength with $E(B - V)$, but the inhomogeneity of the interstellar medium plays a role here as it does in the study of heavy-element depletion along a line of sight. As we will discuss shortly, several populations of absorbers may be present.

The discovery strong infrared diffuse features in the infrared *emission* spectrum of reflection nebulae, with their explanation as small grains or very large organics, seems to point to the explanation of the optical DIBs. The problem is linking together the carriers of these separate features and correlating their behavior. The IR features are always seen in emission, they are coolants for the medium. The optical DIBs, like the broad extinction feature at 2175 Å are always seen in absorption and so must be heating whatever species carries them. The FIR emission lines cluster around wavelengths associated with the C—H bond stretch that is characteristic of organic molecules. But since the emission is generic, without high-resolution spectroscopy the identification was still dubious. As Table 6.8 shows, no such obvious clue exists for the DIBs. Observations with the

---

[60] Merril, P. W. 1934, *PASP*, **46**, 206.
[61] Measurements in Table 6.8 are as follows: Herbig, G. H. 1995, *ARAA*, **33**, 19; possible identifications: Salma, F., Galazutdinov, G. A., Krelowski, J., Allamendola, L. J., and Musaev, F. A. 1999, *ApJ*, **526**, 265. Only lines with $W_\lambda > 0.5$ Å for HD 187459 have been included in this list, the first identified features from photographic spectra.

spectrographs on ISO have made this whole picture much clearer. What had been lacking from the IRAS broadband photometry and low-resolution spectra has been more than compensated for by the newer high-resolution data. You can assign individual bands to molecular carriers *only* if you have some indication of the wavelength.

6.3.6 Polarization and Grain Alignment in an External Magnetic Field

The scattering of a photon by any particle with a dipole moment polarizes the light, and in general this is true even for induced dipole scattering. The trick is that absorption processes seldom have such an effect. For solids, however, this is not true because the distribution of the light over the geometric cross section produces differential absorption depending on the orientation of the absorber. This is the trick with polaroid filters. The film (plastic) is anisotropic so that the transmitted light becomes polarized. This is why the original values detected by Hall and Hiltner were so interesting. The intrinsic polarization expected from radiative transfer in a distorted O-star atmosphere was only expected to be of order 0.1(%) while the measurements were often as high as a few percent. In addition, there was a correlation between the $E(B-V)$ and percent polarization.[62] In other words, looking for scattering from electrons in a stellar atmosphere revealed evidence for absorption and scattering by solids.

The standard wavelength dependence of interstellar polarization has been obtained through a formal fit to observations. We emphasize that this does not derive from some more basic theory, and its precise form remains a parametric convenience:

$$\frac{P_\lambda}{P_{\max}} = \exp\left[-1.15 \ln^2\left(\frac{\lambda_{\max}}{\lambda}\right)\right], \qquad (6.73)$$

where the Galactic mean of the wavelength of maximum polarization is about 5500 Å, with a range from 4200 to 8000 Å.[63] If the grains are dielectric, then it is actually a problem that the polarization is not very high, on the order of a few percent. Further, it depends on wavelength in a way that seems to be almost constant throughout the medium. These are the two observations that must be explained by any model for the grains.

Notice that the polarization measurements are made in transmission, so its mere existence indicates that the grains cannot be spherical. The wavelength of the polarization depends on the dust scattering and absorption cross sections, which depend on the relative efficiencies and also on the geometric area of the grains, and therefore the direct measurement of the polarization provides quite a lot of information about the dust. For instance, the maximum of the polarization should be at about the same size as the grains, more or less, and the peak at about 0.5μ

---

[62] This is spelled out particularly well in the announcement by Hall, J. S. and Mikesell, A. H. 1949, *AJ*, **54**, 187 for their initial survey of 175 stars in the galactic plane.

[63] Mathewson, D. S. and Ford, V. L. 1971, *MNRAS*, **153**, 525; Mathewson, D. S. and Ford, V. L. 1970, *Mem. RAS*, **74**, 139; Serkowski, K., Mathewson, D. S., and Ford, V. L. 1974, *ApJ*, **196**, 261; Coyne, G. V., Gehrels, T., and Serkowski, K. 1974, *AJ*, **79**, 581.

argues for grains that are fairly large. In fact, if these are metallic, then each grain would have about 10^9 atoms so you could account for most of the metals in the medium by the grains. Another indication of the shape of the polarizers is a small, but measurable, circular polarization that results from the changing alignment of the dust along a line of sight.[64] Seen in transmission, an ensemble of randomly oriented grains would have a polarization reduced by $N^{1/2}$ for N clumps with individual orientations at arbitrary position angles. Should the change occur systematically over a large fraction of the pathlength, a rotation of the plane of polarization results, introducing an elliptical component that can be measured through circular polarizing analyzers. The effect is much like Zeeman line radiative transfer in a field of changing orientation, and this gives limits on the order of several hundred parsecs for the length scale on which the interstellar magnetic field—which, we will see, is ultimately responsible for the polarization—is organized.

If you are convinced that the grains are not spherical, it would seem natural to say that the polarization results from these particles scattering light differently depending on the angle of incidence to the grain. But the lines of sight through the medium are long and terribly inhomogeneous and we just asserted that random fluctuations in orientation decrease the net observed polarization. So why is there any polarization at all? Once again, magnetic fields provide the solution.

Polarization alone does not specify the shape of the grains, but the fact that we see it in transmission as well as scattering does. If the dust is spherical, it would produce polarized reflection nebulae but not the observed properties of transmitted light. In addition, the large-scale orientation of the polarization directions along the galactic plane argues strongly for nonspherical particles. Also, the polarization vectors strongly point to a direct link with the large-scale Galactic magnetic field, and this begs the questions of grain composition and shapes. It was partly for this reason that in Chapter 1 we discussed the Euler equations for a rotating body. Imagine a grain that is embedded in a background gas. The dust is a polarizable substance consisting of particles with a dipole moment $\boldsymbol{\mu}$. For simplicity, we will assume that $\boldsymbol{\mu}$ can be in one of two possible states, either aligned or antialigned relative to an external magnetic field, $\mathbf{B}$. The energy is $\boldsymbol{\mu} \cdot \mathbf{B}$ and the partition function for this two state system is

$$Z(T) = e^{\mu B/kT} + e^{-\mu B/kT}. \qquad (6.74)$$

Using $Z(T)$, the magnetization or dipole moment of the medium, M, is given by

$$M = -kT^2 \frac{\partial \ln Z}{\partial T} = n\mu B \tanh \frac{\mu B}{kT} \qquad (6.75)$$

for a medium consisting of n particles per unit volume. At low temperature, expanding to first order, Eq. (6.75) reduces to the Curie–Weiss law for magnetization:

$$\chi \approx \frac{\mu^2 n}{3kT}. \qquad (6.76)$$

[64] See, for instance, Martin, P. G. 1974, *ApJ*, **187**, 461; Martin, P. G. 1978, *Cosmic Dust: Its Impact on Astronomy* (NY: Oxford Univ. Press).

While this is quite idealized, it is a first pass at the scaling for the susceptibility as a function of temperature.

The observed fractional polarization of starlight actually presents a problem of a strange sort for astrophysics—we need a way to make a process inefficient. You can see this by considering the alignment process. In the simplest picture, the grains are elongated, or even triaxial. We already know that they contain iron group metals, and may even have accreted a considerable metallic coating along with lighter elements from the gas. They may be paramagnetic, or even ferromagnetic (depending on the degree of internal ordering), and can possess a magnetic dipole moment. As such, they experience a torque when oriented obliquely to the background magnetic field. This produces alignment. Thus, we would expect starlight to be mostly polarized due to the large-scale ordering of the field. In contrast, stars are rarely more than a few percent polarized, except when viewed in reflection where the alignment plays little role, so something must act to randomly reorient the grains or somehow to suppress this torque. The most likely agent is collisional coupling between the grains and the gas particles, since the latter are thermalized and consequently already possess a random distribution in space and velocity.

A rotating body, whatever its moment of inertia tensor, obeys the Euler equations we discussed in Chapter 1. Acting under the influence of an external torque, τ, the angular momentum, $\mathbf{J}$, changes in both magnitude and direction according to

$$\frac{d\mathbf{J}}{dt} + \boldsymbol{\omega} \times \mathbf{J} = \boldsymbol{\tau}. \tag{6.77}$$

In the absence of torque, this equation describes freebody precession whenever the body symmetry axis is not aligned with the angular momentum. For a spinning nonspherical grain, however, the magnetic torque means that the rotation is never free. A competition exists between the rate of precession and the rate of magnetic alignment. The first solution[65] to this problem assumed that the grains are immersed in a gas that has a thermal distribution and are constantly being randomly smacked by collisions.[66] Off-center collisions spin the grain up. It does not matter whether these are elastic. Emission of atoms or molecules also impart a reaction kicks to the dust and, depending on the distribution of the sites over the grain surface, torque the grain. How fast can it spin? If collisions with the ambient gas are frequent enough that the body comes nearly into equipartition, then its rotational energy should be about the same as the thermal energy of the gas particles, kT_g. The rotation frequency, ω_d, depends on the grain's moment of inertia, I, and the detailed distribution of the collisions, but on average you get $I\omega_d^2 \approx kT_g$. Taking the density of the grains to be a constant for a specific composition, the scaling is that the rotation frequency is highest for the smallest grains, $\omega_d \sim T_g^{1/2} a^{-5/2}$. Emission of particles from the grain surface also affect the

---

[65] Davis, L. and Greenstein, J. 1951, *ApJ*, **114**, 206.
[66] Lazarian, A. 1995, *ApJ*, **451**, 660; Dolginov, A. Z. and Mytrophaov, G. 1976, *Apj. Space Sci.*, **43**, 291; Purcell, E. M. 1979, *ApJ*, **231**, 404; Davis, L. and Greenstein, J. L. 1951, *ApJ*, **114**, 206; Lazarian, A., Goodman, A. A., and Myers, P. C. 1997, *ApJ*, **490**, 273.

rotation frequency and here the magnitude of the effect can be much larger.[67] Even though their radii are only 0.1μ, the number of atoms in a grain is enormous, of order 10^9 (to a single atom, a dust grain is a very large wall!) and a single grain can receive many impulses during its lifetime from surface reactions and evaporation. Like collisions, these are distributed over the surface and can produce torques of either sign relative to the spin at any moment, so the angular momentum randomly changes both magnitude and direction in time, executing a sort of random walk. Even if the total angular momentum is constant, over time $\omega_d \hat{\mathbf{n}}$ aligns with $\mathbf{J}$ and ultimately you expect the grain to rotate about its axis of greatest moment of inertia since at constant angular momentum this minimizes the energy. This *still* doesn't produce a net alignment, which is required to produce the observed polarization, but it satisfies the first requirement.

The grain lives in a magnetized interstellar medium and, since it has a net charge, it automatically has a dipole moment that is aligned with its angular momentum. The ambient magnetic field torques the dipoles toward alignment:

$$\boldsymbol{\tau} = \mathbf{M} \times \mathbf{B}. \tag{6.78}$$

There are two possible sources for this dipole moment, and the differences between them yield different estimates of the alignment timescale. If the grains possess a permanent dipole moment, that is, if they are ferromagnetic, the alignment timescale depends on the rotation frequency and is longer for the smallest grains. On the other hand, suppose the grains are paramagnetic, that they contain some metallic ions but in a dilute configuration within the dust matrix. Then the ambient field can *induce* a dipole moment $M \sim \chi B \omega_d$. Now we have a problem. A static magnetic field induces an aligned (antialigned) magnetic moment. This produces no torque unless the grain interior has some intrinsic anisotropy. The situation we face, however, is very different—the dust is spinning relative to the field direction. This dynamically induced magnetic moment has both parallel and orthogonal components and, on average, the dust is subject to a torque. A dielectric grain has a *complex* magnetic susceptibility, $\chi = \chi' + i\chi''$, for which the imaginary part causes the lag between the magnetization of the material and the external field.

We've already asserted that to explain the extinction curve the dust must be dielectric. It follows that the grains may also be *paramagnetic*, since it is likely that their metallic constituents are sufficiently dilute that ferromagnetism can be excluded. In this case, an ambient magnetic field alters the picture of the alignment for these spinning grains. We can follow the same line of reasoning we used for the dielectric properties. The internal field, $\mathbf{B}$, is induced by the external field so that $\mathbf{B} = \mathbf{H} + 4\pi\mathbf{M}$, where $\mathbf{M}$ is the magnetization, analogous to $\mathbf{P}$. If the grain is paramagnetic, then $\mathbf{M}$, the magnetic moment, is itself dependent on $\mathbf{H}$. Now imagine the grain to be arbitrarily aligned with respect to $\mathbf{H}$ and spinning with a frequency ω, changing the orientation of $\mathbf{M}$ and the body axis. The magnetic moment therefore undergoes a change, and this means a change in the energy of

---

[67]Lazarian, A. 1995, *ApJ*, **451**, 660; Dolginov, A. Z. and Mytrophaov, G. 1976, *Apj. Space Sci.*, **43**, 291; Purcell, E. M. 1979, *ApJ*, **231**, 404; Davis, J. and Greenstein, J. L. 1951, *ApJ*, **114**, 206; Lazarian, A., Goodman, A. A., and Myers, P. C. 1997, *ApJ*, **490**, 273.

the system $dE_M/dt \sim \omega \mathbf{M} \cdot \mathbf{B}$. There is also a lag here; the dipoles within the spinning grain can't instantaneously change their orientation relative to the body axis as the grain spins, so the perpendicular component of $\mathbf{B}_i$ decreases by dissipation while the field along the axis remains unchanged. In other words, the internal field decreases and the grain slowly loses rotational energy through this interaction, resulting in a slow alignment of the grain with the external magnetic field with the long axis now *along* the external field. Several mechanisms have been examined for dissipating energy within the spinning grain, leading to adjustment of the internal angular velocity. A purely mechanical mechanism, elastic strains due to the rotation and precession, depend on details of the assumed grain structure. The rate of dissipation is proportional to the strain and stress tensors in the material, just as we used in accretion disk modeling in Chapter 5, but there are few laboratory data available on the properties of clusters of atoms. For small grains such as PAHs, this amounts to radiation from vibrational modes of the molecules but for larger grains, the dissipation mechanisms are not well understood. An alternative comes from the induced dipole resulting from a spinning grain, the Barnett effect, which aligns the grains by converting rotational energy into dipole radiation.

6.3.6.1 Grain Size Distribution and Origin of the Dust

Where does the dust come from? Forming solids is hard enough in the laboratory, but in the interstellar medium it seems impossible. The gas phase is not the sort of place you would expect to be conducive to some kind of homogeneous condensation, yet the depletion factors correlate with T_{cond}, which is defined for equilibrium conditions. This correlation is explained as ambient metal ions sticking to the grain surface, but still requires the dust to present a large target. The grains clearly have complex structure, such as those found in meteorites and from upper atmosphere samples, and, as we have seen, are large enough to be treated as solids when computing their thermal emission. Again, it seems unlikely that they form in the interstellar medium, although they can easily grow and change there. There is, broadly speaking, a likely source: stars. This is a generic label for a wide range of possible sources, including red giant winds, novae, supernovae, and Wolf–Rayet outflows. With densities of 10^{10} cm^{-3} or higher and temperatures as low as a few thousand degrees for giants or even lower for more extended atmospheres, stellar atmospheres provide not only the right conditions for formation but also a range of mechanisms for dispersal of the products.

Grain formation can be observed in real time in very different environments. During nova explosions, deep minima are sometimes seen in the visual lightcurve about 3 months after outburst when the mean temperature of the expanding gas falls below ~ 1000 K, the typical Debye temperature for a silicate or graphite grain. In one relatively recent case, V705 Cas 1993, the dust formation was seen simultaneously in both drastic decrease in optical and UV light and strong rise of the infrared and appearance of the SiC 11μ emission feature. The amount of this one outburst was about 10^{-10} $M_\odot$. Since nova explosions recur on long timescales (about 10^4 yr or so), a single binary may be able to eject many times this value into the ISM. Dust formation is responsible for the steep drops in visual brightness observed in the R CrB stars, whose fading episodes resemble classical novae. These stars have strong winds, and the dust will advect into the surrounding interstellar gas. Luminous blue variables, while quite rare, also form dust. Planetary nebulae are usually surrounded by substantial dusty environments that are presumed to originate with the old wind of the precursor AGB star. Mira variables show evidence of dust formation during shell ejection episodes. Most significantly, supernovae—especially SN 1987A, which was a type II—show the same signatures of dust that are seen in typical novae but with a great deal more mass ejected during the explosion (which occurs, after all, only once). Postponing momentarily discussing the formation mechanism, it is clear this dust can account for what we see in the diffuse interstellar gas and even the matter that collects into cloud complexes.

The size distribution for dust, to which we alluded earlier, is inferred from combined observations of extinction and polarization. It appears to have no characteristic size, obeying a power law:

$$N(a)\,da = N_0 a^{-7/2}\,da, \tag{6.79}$$

where N_0 is the normalization. If the grains are not too fluffy so that their mass scales as a^3, then this distribution is approximately $n(m)dm \sim m^{-5/3}dm$. When we examined the periodic table and commented on the distribution of abundances in stars, we noted that such power-law behavior and inverse scaling with mass could be either a top–down (sputtering, destruction) process or a bottom–up (agglomeration) process. Let's start with the top–down approach. Imagine that mainly big dust particles are produced in stars and expelled into the interstellar medium where they are whittled down into smaller and smaller grains. Then

$$\frac{\partial n(m,t)}{\partial t} = -n(m)\int_{m_l}^{m_u}\beta(m,m')n(m',t)\,dmm',$$

where β is the destruction rate, which we will assume is proportional to $(m^{1/3} + m'^{1/3})^2$, the total geometric cross section. Separating $n(m,t) = T(t)N(m)$ and solving for the time-independent distribution, we can assume that we start with a power law $N(m) \sim m^{-n}$. Then, provided the collisional destruction rate is independent of the relative velocity,[68] we can rescale the integral to a function of $x \equiv m'm$ to obtain $N(m) \sim m^{-5/3}$. There is no feedback in this picture, only destruction in the long run, and the surprise is that it generates precisely the requisite distribution. More complicated agglomeration calculations lead to similar conclusions, including the effects of accretion and sputtering by radiation and cosmic rays. You would expect a departure from a simple power law if there is a characteristic size for the seed grains, the sort of thing that would come from homogeneous nucleation, but you might also simply have a fixed upper limit to the distribution. So it appears that the grains may start large and end small, as small as the molecular clusters we discussed for the infrared DIBs. You can also imagine a bottom–up scenario starting with small nuclei in a stellar atmosphere that by coagulation produce an initial spectrum that is reprocessed by the interstellar medium.

---

[68] See, for instance, Hellyer, B. 1970, *MNRAS*, **148**, 383; Simons, S. 1982, *MNRAS*, **201**, 127; Silk, J. and Takahashi, T. 1979, *ApJ*, **229**, 242.

6.3.7 Reflection Nebulae and Light Echos

Dust clouds provide excellent reflecting surfaces for the incident spectrum of an illuminating star. Aside from the changes it produces in the spectral energy distribution, the reflection does not change the relative strengths of the spectral lines. It is actually possible to observe the stellar spectrum, more or less intact, by obtaining the spectrum of the dust. This is a *reflection nebula*. The scattered light is bluer than that of the illuminating star, since this depends on C_{sca} for the grains, and the scattered stellar spectrum is now polarized. Thus, reflection nebulae present an unambiguous way of detecting solid material in the interstellar medium, and their recognition now serves as a way of looking for the dust at a large distance associated with young stellar objects such as T Tau stars.[69]

A beautiful example of dust scattering in the diffuse interstellar medium is *light echos*. This is a time-dependent response of the medium to an intense pulse of radiation. The first example was detected following the outburst of nova GK Per in 1901. An extended emission region soon appeared around the nova that appeared to be moving superluminally. Even in 1901, this seemed very strange and it was soon realized, by Hartmann, that the nebula was not materially connected with ejecta from the nova but instead arose from the dust intervening along the line of sight. A simple theory, first developed by Courdec in 1939, forms the basis for interpreting this effect. Suppose you have a planar reflecting screen that is normal to the line of sight reflecting and located behind a source for a pulse of emission. Assume that the pulse is essentially instantaneous, $I(t) \sim \delta(t)$. If the distance from the source to the reflector is d, then obviously there is a time delay $\Delta t_0 = 2cd$ for light reflected directly back to reach the observer compared with the arrival time for the direct pulse. Over time, a ring expands outward from the direction of the direct pulse as the light reflects back to the observer from successively greater angular distances from the line of sight. If x is the distance along the plane, so that $\theta \approx x/d$, then $\Delta t - \Delta t_0 \sim \theta^2$. The light echo forms a parabola at a time t that is given by

$$z = \frac{1}{2}\left(\frac{x^2}{ct} - ct\right), \qquad (6.80)$$

where now t is the time relative to the first observation of the direct light from the source. The spectacular, nearly circular, rings produced following the explosion of SN 1987A in the LMC is the best studied example. The most important feature of the rings is that they display a spectrum that was "frozen in" at the time of the pulse, before the first detected optical spectrum of the event. Similar results were obtained as early as 1903 for the echo around GK Per. Furthermore, since the dust also absorbs, it is heated by the pulse depending on the dilution factor, $W =$

---

[69] The original definition of this class by Herbig (in 1953) depended, in fact, on the presence of reflection nebulae around emission line objects. Before the detection of molecular clouds with millimetric observations, this was the only way to ensure that the stars were still embedded within their parent clouds.

$1/(d^2 + x^2)$, so an infrared emission ring should accompany the visible reflection.[70] Note also that reflection nebulae are not only an optical phenomenon. Depending on the grain size, you can scatter any energy photon. For instance, in the middle infrared, you see not only scattering but also excitation of intrinsic emission from the particles. In X rays, scattering halos have been observed in the direction of novae. The scattering coefficient for Mie-type particles have a $Q_{sca} \sim a/\lambda$, so for 0.01μ grains, the forwardscattering peaks within a cone of order 10^{-2} rad, about a few tens of arcminutes or less.[71]

6.3.8 The Galactic Distribution of Dust

The distribution of dust within the galaxy is qualitatively obvious. The dark lanes in the Milky Way are the clear evidence that the obscuration is confined to the disk, and the distribution of extragalactic sources is used to delineate the vertical distribution. The technique was first exploited at the beginning of the twentieth century when the nebular "*Zone of Avoidance*" was recognized in all—sky surveys.[72] We show the current analog of the Hubble map in Figure 6.5. In radial location, it is much harder to specify since there is no kinematic information for the dust that is analogous to the H I—there are no narrow absorption or emission lines that can be used to determine its velocity. Instead, the usual way is to use tracers of galactic structure, in particular OB stars, for which the intrinsic colors are available and determine $E(B-V)$ as a function of distance. This is a lot easier than obtaining A_V directly, since it is independent of distance. Using R gives the optical depth in the direction of the source, and then with some knowledge of the wavelength dependence of the absorption coefficient, which we have already discussed, you can obtain N_d, the column density of the dust. This still does not provide n_d, the number density that is needed to determine the fraction of the gas that has been depleted into the solid phase. One method, due to Spitzer, is to look at the depletion factor as a function of column density (i.e., in terms of the total extinction). This shows that the denser gas is more depleted and thus provides a handle on the volume density.

Observations of X-ray and γ-ray emission from the diffuse medium also help. The X-ray attenuation comes mainly from neutral gas along the line of sight. The γ-ray production, on the other hand, depends on proton interactions with the

---

[70] Images of the GK Per 1901 rings are available in Ritchey, G. W. 1902, *ApJ*, **15**, 128; Hinks, A. R. 1902, *ApJ*, **16**, 198. Much of the basic theory, including the effects of finite outburst timescales, was worked out in Couderc, P. 1939, *Ann. d'Ap.*, **2**, 271. This is a lovely paper both for its simplicity of presentation and completeness and is well worth your time. For more recent applications to the SN 1987A rings, see Xu, J., Crotts, A. P., Kunkel, W. E. 1995, *ApJ*, **451**, 806; 1996, *ApJ*, **463**, 391; Crotts, A., Kunkel, W. E., and McCarthy, P. J. 1989, *ApJ*, **347**, L61.

[71] Bode, M. F., Piedhorsky, W. C., Norwell, G. A., and Evans, A. 1985, *ApJ*, **299**, 845; Predehl, P. and Klose, S. 1996, *Astr. Ap.*, **306**, 283; Smith, R. K. and Dwek, E. 1998, *ApJ*, **503**, 831.

[72] Hubble, E. 1936, *The Realm of the Nebulae* (New Haven: Yale Univ. Press) reviewed the first work on the problem of the effect of the galactic foreground extinction on attempts to study large-scale structure. For the modern incarnation of this work, see Burstein, D. and Heiles, C. 1984, *ApJS*, **54**, 33, and references therein. Recently, radio observations of galaxies in the "Zone" have become increasingly important in delineating the galactic dust distribution; see Kraan-Korteweg, R. C. and Juraszek, S. 2000, *PASA*, **17**, 1.

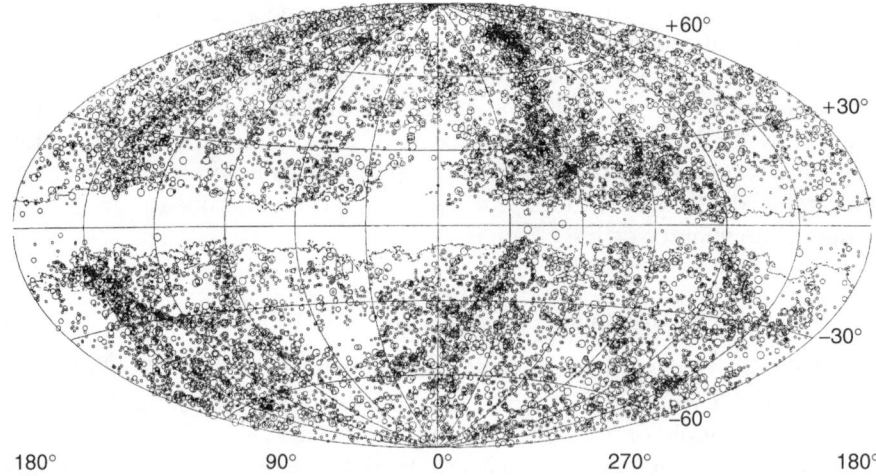

FIGURE 6.5. Extinction map for the Galaxy obtained from counts of extragalactic systems. Only the contour for $A_V = 1$ is shown. This matches well the H I maps of the plane (credit: R. C. Kraan-Korteweg and S. Jurasek).

clouds and bremsstrahlung and pion creation due to collisions with grains. Finally, the COBE satellite provided all-sky far-infrared observations with which the dust distribution can be modeled.[73]

6.4 MOLECULAR CLOUDS

The most massive structures in the interstellar medium are also the coldest. When a sufficient column density is achieved in the gas, the formation of molecules serves to drive the temperature far lower than the surrounding diffuse gas, resulting in the formation of a molecular cloud (see Table 6.9[74]). The term is actually a catch-all for a wide range of physical conditions.

TABLE 6.9 Sample Properties of Molecular Clouds

Type	$\bar{n}$ (cm^{-3})	T (K)	A_V (mag)	M ($M_\odot$)	L (pc)
Diffuse	100–1000	30–80	≤ 1	1–100	1–5
Translucent	500–5000	15–30	1–5	1–100	0.5–5
Cold, dark	100–10^4	≥ 10	2–5	10–1000	0.5–1.5
GMC	10^4–10^7	≥ 20	> 2	10^3–10^5	3–12

[73] Dwek, E. et al. 1997, *ApJ*, **475**, 565; Lagache, G., Abergel, A., Boulanger, F., Puget, J.-L. 1998, *Astr. Ap.*, **333**, 709.
[74] Table 6.9 is adapted from van Dishoeck, E. F., Blake, G. A., Draine, B. T., and Lunine, J. I. 1993, *Protostars and Planets III* (Tucson: Univ. Arizona Press). Here M and L are the typical ranges for masses and sizes of the clouds.

The lowest densities are encountered in the *translucent* clouds that have column densities of order 10^{20} cm^{-2} and A_V of order one magnitude. These clouds are sufficiently rarefied that UV from the DIRF can penetrate into them and play a central role in the chemistry and heating. The clouds are therefore expected to have the highest fraction of electrons within the gas, and also should show the lowest depletions of refractory elements. Searches for star formation show that these are also not generally active sites. There is even a question of whether they can be considered as a separate phase at all. For at least some, and many were found through their enhanced far infrared emission,[75] there is a strong suggestion that the clouds may not even be in strict pressure equilibrium with the diffuse medium. In others, there are substructures seen that are overpressured relative to the more extended gas within the cloud. These are the closest analogs to terrestrial clouds seen in the ISM[76] in the sense that only the external gravity and ISM determine their mechanical equilibrium. Cloud masses range from a few to several hundred $M_\odot$, with typical densities of about 10^3 cm^{-3}. With sizes of order 0.1–1 pc, this means mean column densities of order 10^{17} cm^{-2}, about the same as one typically finds for the Lyman α absorbers in cosmological observations and similar to what is seen for the densest HI clouds.

Dense clouds are very different. They are self-gravitating, having typically many Jeans masses (see Section 6.8). For now this means that they are massive enough that self-gravitation should undermine their stability and they should be collapsing. The clouds are opaque enough that the DIRF does not penetrate into their interiors, unless they have a far more diffuse structure than we see projected against the sky, and the primary source for internal ionization is cosmic rays. With densities in excess of 10^4 cm^{-3} and sizes of 1–10 pc, these structures have typical column densities $> 10^{22}$ cm^{-2}. At the lowest end of their range, this is similar to the so-called damped Lyα clouds (see Chapter 8). These follow the distribution of the dust, as they are more closely confined to the galactic plane than to the H I with a scale height of about 100 pc.

The largest structures, those in which massive stars are usually found, are the *giant molecular clouds* (GMCs). These have masses of around 10^5 $M_\odot$, radii of order 20 pc, and surface densities that are comparable to dense clouds. Star formation is seen in both the dense and giant clouds. The signature of this activity is embedded luminous infrared sources, enshrouded protostars, and emergent jets and winds.

6.4.1 Observational Mass Determinations Using CO Lines

Alas, the bulk of the mass of molecular clouds is not directly observable because it is in H_2. It is, however, possible to use proxy tracers—emission lines from strong

---

[75] The largest sample of these is Magnani, L., Blitz, L., and Mundy, L. 1985, *ApJ*, **295**, 402 (also called MBM) that was a wide beam CO survey of regions toward high galactic latitudes that appeared blotchy on the Palomar Sky Survey plates. These turned out to coincide with the extended IRAS cirrus structures seen above the plane. The clouds have large angular diameters, indicating that they are very nearby, and are seen at high galactic latitude. Presently, this is the largest catalog of translucent clouds.

[76] There has been no thought to the role played by buoyancy within the galactic gravitational field for these structures. It would be an interesting question to see whether they show the same sort of buoyant dynamics that is seen for cool gas in the Earth's atmosphere.

emitters such as the isotopes of CO, and far infrared emission from heated dust—to obtain mass estimates. This requires calibrating the collisional excitation of the CO by H_2. To do this, *another* proxy is required, the γ-ray emission from the diffuse interstellar medium due to cosmic ray collisions with dust grains. The argument runs like this. For the diffuse medium, you have a direct measure of the molecular hydrogen column density and its dependence on extinction. Dust is responsible for the diffuse γ-ray emission from the disk, so it depends linearly on the dust column density, which correlates linearly with $N(H_2)$. For diffuse clouds, for which you can obtain the extinction to background stars, you can calibrate the CO intensity with the dust and therefore obtain the conversion factor.[77] The results are independent measures for the two principal transitions:

$$X_{CO(1-0)} \equiv \frac{N(H_2)}{I(CO(1-0))} = 1.56 \times 10^{20} \text{ cm}^{-2} (\text{K km s}^{-1})^{-1}$$

$$X_{CO(2-1)} \equiv \frac{N(H_2)}{I(CO(2-1))} = 1.7 \times 10^{20} \text{ cm}^{-2} (\text{K km s}^{-1})^{-1}. \quad (6.81)$$

Other schemes can also be used, especially because the $^{12}C^{16}O$ lines can probe very high optical depths, but these require some additional information about the isotopic fractionation (see Section 6.6.2.6) and we will not discuss them here.

Masses of clouds can also be estimated, in principle, from the virial theorem. You measure the amount of mass by obtaining a surface map of some tracer, for instance, CO in some transition. By using the conversion between this tracer and the dominant exciting species, which for the clouds is H_2, you obtain the total mass. So far, no assumptions are required about the cloud geometry other than the assumption that the tracer is optically thin, so you see the emitting volume is actually the whole mass of the cloud. Knowing the measured linewidths and the mean radius of the cloud, compare the kinetic and gravitational potential energies. Since the conversion factor is calibrated based on $N(H_2)$ and A_V and not on the dynamical properties of the cloud, even allowing for superthermal linewidths, the cloud masses can be determined independently obtained.

A straightforward dynamical measurement of the velocity dispersion in a cloud, obtained by global averaging of the centroid velocities of the observed emission lines, has produced an intriguing set of scaling relations. Using early data on molecular clouds, Larson obtained a relation between the mass-averaged radius of a cloud, R, its global velocity dispersion, σ, and the mean mass density $\langle \rho \rangle$.[78] The first connects the velocity and length scale of the cloud:

$$\sigma \sim R^{1/2}, \quad (6.82)$$

[77] An excellent guide is provided by Mauersberger, R. and Bronfman, L. 1998, *Astr. Gesell. Rev. Mod. Astr.*, **11**, 209 (and references therein).
[78] Larson, R. B. 1981, *MNRAS*, **194**, 809; Vazquez-Semadeni et al. 1997, *ApJ*, **474**, 292 (and references therein).

and the second relates the mean density with the length scale:

$$\langle \rho \rangle \sim R^{-1}. \tag{6.83}$$

In other words, clouds appear to have about the same column density, $\Sigma \sim \langle \rho \rangle R$, regardless of their mass, and the largest clouds, which have the most mass, also have the largest velocities. These are random motions that are presumed to be the same as turbulence. One argument for the origin of these correlations is that the clouds are in virial equilibrium. To see this, recall that $\sigma^2 \sim M/R$ and, since $M \sim \langle \rho \rangle R^2$, we obtain $\sigma \sim R^{1/2}$. Yet perhaps this is *not* true for all clouds, but only for those that are neither self-gravitating nor actively forming massive stars. As we will see in discussing turbulence, another explanation is that the relations show the characteristic spectrum of the turbulence.

Observations of molecular clouds lead to a profound conclusion—they must be supported dynamically. The argument goes like this. The clouds are cold. This follows from their far infrared emissivities in the IRAS bands, 60 and 100μ. Although we detect only the dust from such observations, it is likely that this also characterizes the thermal gas emission, an assertion that is supported by the level populations obtained from CO observations. The temperatures range from a few tens of degrees kelvin to about 100 K, corresponding to linewidths of less than 0.1 km s^{-1}. On the other hand, the observed linewidths are invariable greater than this, sometimes by a factor of > 10. Since this nonthermal motion is about what you would expect from nonlinear Alfven waves or bulk motion, some form of mechanical support in excess of the thermal pressure seems to be required for the clouds. In addition, as we will see, the widths of emission lines depend on the sampling volume or length scale. Recall from our discussion of emission lines in Chapter 3 that the profile is essentially a probability measure. The growth of the linewidth with sampling length indicates that there are always fluctuations in the gas that are more likely to be sampled the larger the separation of the parcels in question. If the fluctuations produce a gaussian profile, they must be produced by a macroscopic field of random motions, and this is attributed to turbulence.

The heating and cooling of clouds depends on their structure and provides another application for the scaling relations. If the clouds are uniform blobs, then along any line of sight the only thing that matters for the radiative balance is the distance from the center. On the other hand, if the clouds have a fragmented, possibly fractal, structure, then the heating from the DISF is more effective since the optical depth is only statistical rather than deterministic. Along any line of sight, the mean penetration depth will be the same, but some regions of the cloud may be more heated than others.

We have repeatedly asserted that star formation occurs within the cold dark places of the Galaxy. Evidence for this comes from several quarters. The obvious one is that you always see pre-main sequence stars associated with dust, although the bias is built into how you look for them, and the H II regions in which they are often found are often blisters on the sides of the clouds. More telling is that massive molecular clouds always possess deeply embedded point sources that have stellar luminosities. These take several forms. Infrared emission from heated dust is the easiest to interpret since the signature is unambiguous and the emissivity unattenuated. In order to have a strong far infrared source with a temperature of a few hundred degrees, hot luminous stars must be nearby, and the luminosity depends on the mass of emitting dust. Even if the stars were never to emerge, every emitted UV photon must, as we have discussed, cascade to Lyα, which is finally emitted as far IR by dust in a completely absorbing medium. Another signature is embedded compact (and ultracompact) H II regions. These are observed at centimeter wavelengths from free–free emission and bespeak a dense environment. In millimeter observations, molecular outflows from deep within the cloud are often observed as well as breakout of these flows from the cloud surfaces.

These data show, however, that the star forming process has happened but do not show how it starts. At the prestellar end of the process, one would hope to see structures that are not yet stars having masses in the right range to potentially collapse. Here the interpretation of large-scale random motions becomes important, and we will postpone the dynamical discussion until we come to turbulence.

6.4.2 The Population of Clouds

So with all this structure along a line of sight and the wide variety of conditions present in the various phases, what do we really mean by the word "average" when talking about integrated column densities? Assume that the number of clouds of a given type along any line of sight is small so that you are sampling a random variable in the low-number limit. This is the Poisson limit (see Chapter 2), so if the average number of clouds per unit length along a line of sight is λ, the probability of finding n clouds is

$$p_n = \frac{\lambda^n}{n!} e^{-\lambda}, \qquad (6.84)$$

and the number of such clouds will be $\Sigma_n n p_n$. Now, if we have several phases, each of which follows a Poisson law, then the column density is actually the sum over the contributions from the individual phases. An ingenious solution was proposed by Spitzer,[79] who sought to explain the correlation of metal depletion with $E(B-V)$. Assume that the cloud phases separate based on the mean extinction and number density into warm and cold clouds, and further divide the latter into large and small varieties. The large cold clouds are rare relative to the other phases, but one would expect them to show the highest extinctions and largest depletions as well as the highest densities. Therefore, in spite of their numbers, they are important to the overall determination of column densities and mean number densities along any given direction. Assume n_i is the number density for the phase i. In these three types of diffuse cloud, Spitzer chose $n_w = 0.14$, $n_1 = 0.70$, and $n_2 = 0.45$ cm^{-3}, but the precise numbers don't matter for the fundamental argument. The column density in each phase contributes to the total $N = \Sigma n_i \lambda_i D$ along a distance D. Since the probabilities must add to unity, the relative probability of a mixture of types 1 and 2 is

$$p_r = \frac{p_1(n_1) p_2(n_2)}{\Sigma_{n_2} p_1(n_1) p_2(n_2)}, \qquad (6.85)$$

---

[79] Spitzer, L. Jr. 1985, *ApJL*, **290**, L21. Such arguments hark back to the days of star gauging by Charlier and Malmquist at the start of the twentieth century.

with the total column density fixed by observation of the H I. Therefore, since the fractional contributions of the phases must add to unity, n_1 is related to n_2 in Eq. (6.85) and the sum need extend over only one phase.

6.5 MAGNETIC FIELDS

The combination of low density and moderate or high ionization of the interstellar gas means that magnetic fields are extremely important in the structure of the medium.[80] There are several indicators of the presence of, and strength of, the field. The first is the Zeeman effect, the only direct means for measuring the total field strength. Synchrotron radiation directly demonstrates the presence of large-scale magnetic fields, and the orientation of the radiating electron orbits provides the field direction. Its interpretation is, however, complicated by Faraday rotation, as we will discuss in a moment, because of the rotation of the plane of polarization by refractive transmission through any ionized thermal gas surrounding the emitting source. It also means that you must at least use, if not understand, the relativistic particles that constitute galactic cosmic rays as the emitters. These, too, are signals of the presence of magnetic fields. Finally, the alignment of dust grains by dielectric interaction with the local field shows the presence and direction of the field, if not its magnitude.

6.5.1 Cosmic Rays

The primary source for ionization in dark clouds is the penetration of high-energy particles, cosmic rays, into the gas.[81] Assuming that the optical depth to the diffuse interstellar radiation field is large, and that the line-of-sight absorption for embedded stars is also large, UV cannot scatter deep enough within the medium to produce net ionization. Instead, charged particle ionization, due to energy loss by protons and α-particle traverse, is the likely mechanism for maintaining the net electron density in the cores of molecular clouds. To see this, call ζ_c the rate of ionization by impinging cosmic rays. The recombination rate is set by the local density so that the degree of ionization is given by $n_e N^{(r+1)} \alpha = N^{(r)} \zeta_c$. Even in environments with $A_V > 10$, chemical models for molecular ions in dark clouds[82] indicate some small net ionization, $n_e/n_{\text{tot}} \sim 10^{-6}$–$10^{-8}$ and require $\zeta_c \approx 10^{-17}$ s^{-1}. This provides a clue to the rate of production of these particles. The other, as we will elaborate in the next section, comes from observations of a diffuse nonthermal radio continuum from the disks of galaxies, especially the Milky Way. This is produced by synchrotron emission from electrons in the global interstellar magnetic field.

The first theoretical mechanism for accelerating electrons and protons to very high energy was described by Fermi in 1949. Although it is now clear that the theory will not work for interstellar cosmic rays, it is still the basis of much of the

---

[80] Beck, R., Brandenberg, A., Moss, D., Shukurov, A., and Sokoloff, D. 1996, *ARAA*, **34**, 155; Kulsrud, R. 1999, *ARAA*, **37**, 37.
[81] An excellent survey of cosmic ray physics in particular, and high-energy processes in general, is provided by Longair, M. S. 1994, *High Energy Astrophysics*, vol. 2 (Cambridge, UK: Cambridge Univ. Press).
[82] See particularly Caselli, P., Walmsley, C. M., Terzieva, R., and Herbst, E. 1998, *ApJ*, **499**, 234.

work on acceleration of high-energy particles under astrophysical conditions. Because of flux conservation, the magnetic flux through a particle orbit is conserved. The magnetic moment of an electron orbiting a field line is due to the Larmor frequency and the radius of the orbit, resulting from its momentum. Imagine that you have an electron moving into a region of increasing field strength. The energy of the orbit is $\mu \cdot \mathbf{B}$, where μ is the induced moment caused by the orbit. Since the energy is conserved in an adiabatic medium, and since the magnetic moment is as well, there is a change in the pitch of the orbit so that the total energy remains constant. The particle eventually comes to rest and reflects from the field, a process known as *mirroring*. Its role in accelerating particles comes from the motion of the region of increasing field strength. Fermi assumed that the magnetic field is embedded in clouds whose random motions would produce a stochastic kick whenever a particle collided with one of them. You can see how by thinking of the chance of a collision with a car on the road. If you are moving toward the vehicle, it's more likely that you will hit it than if it is moving away from you in the same direction. This anisotropy means that for the moving-mirror problem, there is a net gain in the energy of the electrons through successive collisions, even in a completely random medium.

Acceleration by mirrors requires that the particles spiral in the magnetic field at a large pitch angle, the inclination of the orbital plane to the mean direction of the field. The problem is that as the particle energy increases, this angle decreases, which progressively reduces the acceleration efficiency. Without going into details, which have remained formidable for decades, it suffices to argue that somehow the large pitch angles must be maintained even as the particles stream through the interstellar field. Short-wavelength (high-frequency) waves, such as whistlers, Langmuir waves, and other produces of plasma instabilities, scatter the particles and help maintain magnetic moment of the spiral.[83] As a result, the particle distribution diffuses not only in position and energy but also in the pitch angle with time.

As originally envisioned, collisions and reflections with moving fields embedded in clouds are assumed to produce the scattering. The particles increase kinetic energy, then, by tapping the kinetic energy of the mirrors, a very small cost to the large-scale interstellar turbulence. Consider a mirror moving with velocity V that reflects a relativistic particle. In what follows, we will write $\beta = V/c$ and $\gamma = (1 - \beta^2)^{-1/2}$, both referring to the motion of the mirror. The particle energy is E, its 3-momentum is $\mathbf{p}$, and the 3-momentum projects onto the direction of motion of the mirror, $p_\parallel = \mathbf{p} \cdot \hat{\mathbf{n}}$. Then in the frame of the mirror the 4-momentum is $p'_i = \Lambda^j_i p_j$, where Λ^j_i is the Lorentz matrix from Chapter 1. In the mirror frame, the particle adiabatically reverses its direction while not changing its energy, $p_\parallel \to -p_\parallel$, but in the frame of the background medium, it has received a kick, ΔE. Since this is a reflection process in the moving frame, its description requires a double Lorentz transformation, so E'' varies as $\gamma^2 E$; thus, $\Delta E/E \sim \beta^2$ for

---

[83] For comprehensive introductions to relevant plasma processes, see for example, Krall, N. A. and Trivelpiece, A. W. 1986, *Principles of Plasma Physics* (NY: McGraw-Hill; rpt. San Francisco Press); Melrose, D. B. 1970, *Plasma Astrophysics* (2 vols.) (NY: Gordon and Breach); Stix, T. H. 1992, *Waves in Plasmas* (NY: AIP Press).

small β. Let's see where this comes from. Noting that $p = E/c$, we write

$$E' = \gamma(E + \beta p_\|)$$
$$p'_\| = \gamma(p_\| + \beta E), \qquad (6.86)$$

so that by applying the transformation twice and reversing the direction of the particle's motion, we obtain

$$E'' = \gamma^2 E\left[1 + 2\boldsymbol{\beta} \cdot \hat{\mathbf{n}} + (\boldsymbol{\beta} \cdot \hat{\mathbf{n}})^2\right]. \qquad (6.87)$$

As you anticipated, the average energy increase, ΔE, is proportional to β^2 since on averaging over random incident angles the $\boldsymbol{\beta} \cdot \hat{n}$ term vanishes. This is the *second-order Fermi process*. For a *first-order* process, which depends linearly on β and is therefore much more efficient, the particles must gain energy in the frame of the scatterer. The obstacle for either order acceleration is that the particles must be injected at relativistic energies. However, the particles are scattered, and since β is not changed by the interaction, we see that $\dot{E} = \alpha E$, where α is a function of β and the turbulence spectrum. Each scattering produces a small increase in energy, so we can use the Fokker–Planck equation to describe the energization. Call the characteristic timescale for this process τ. The symbol masks a host of contributing processes under a single label. If the mean free path for collision exceeds the size of the region, then the particles escape. If the pitch angle distribution collapses as a result of streaming, the particles escape. Ignoring diffusion, the steady state distribution for the particles is given by

$$\frac{d}{dE}\left(\dot{E}N(E)\right) = -\frac{N}{\tau} \qquad (6.88)$$

so that

$$N(E) = N(E_0)\left(\frac{E}{E_0}\right)^{-p}, \qquad (6.89)$$

where $p = [1 + (\alpha\tau)^{-1}]$, and we arrive at a power-law spectrum. Including the effects of energy loss, through synchrotron emission and inverse Compton scattering and trapping of the particles in the acceleration region (this is actually first order in β) the value of the exponent is specified[84] by singling out magnetic shocks from supernova remnants and wind-blown shells as the site of particle acceleration.

The *observed* energy distribution (Figure 6.6) is altered by the time we detect it, especially at energies below a few MeV per particle, because of our location; we live inside the turbulent, dynamical environment carved out of the local interstellar medium by the solar wind. This region, the *heliosphere*, is a complex,

---

[84] Bell, A. R. 1977, *MNRAS*, **179**, 573; Blandford, R. D. and Ostriker, J. P. 1978, *ApJL*, **221**, L29; Blandford, R. D. and Ostriker, J. P. 1980, *ApJ*, **237**, 793.

530 THE INTERSTELLAR MEDIUM

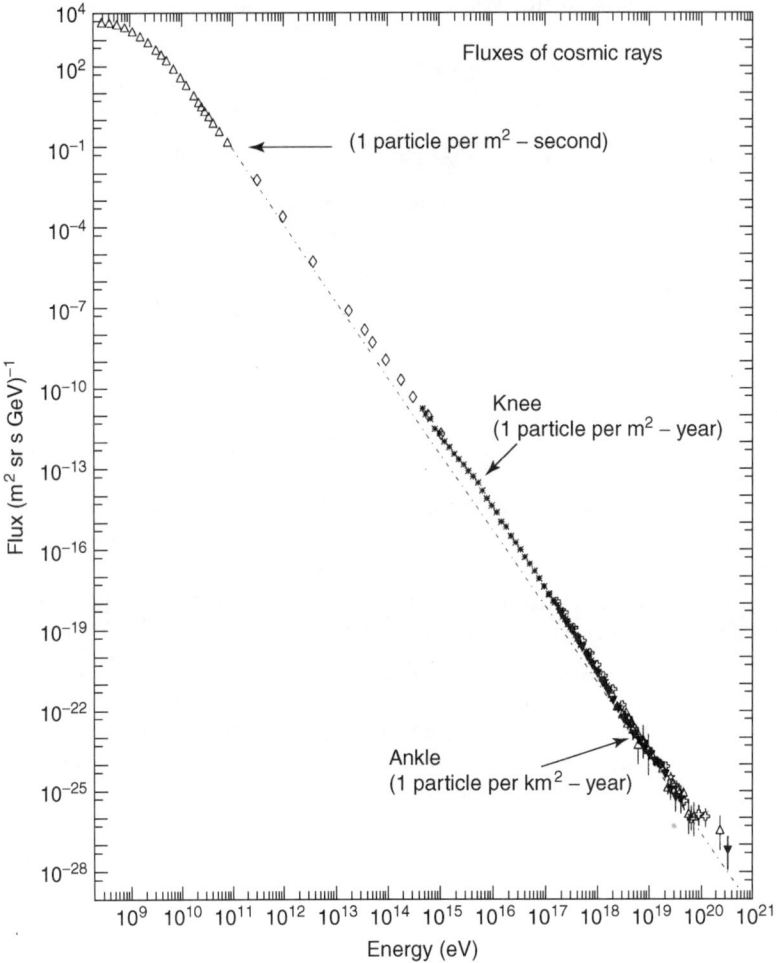

FIGURE 6.6. Differential cosmic ray spectrum (dashed line is an E^{-3} comparison). Notice the "knee," the break in the slope at about 10^{15} eV (credit: the *Auger* Project).

time-dependent region that is sufficiently extended that the lower-energy components of the cosmic rays must pass diffusively through the background and scattering takes a toll on the distribution. Although strictly conserving the number of particles, the lower-energy part of the spectrum is systematically depleted relative to the higher-energy end. Since any proposed mechanism for particle acceleration has a steep energy dependence, producing far fewer of the high-energy particles, any measurements of the total energetic input to the Galaxy from cosmic rays must be corrected for this effect. Without it, any estimates of the energy density are severely compromised. We have not yet truly entered interstellar space, in spite of the *Voyager* spacecraft now being farther than 40 AU from the Sun, and all solar system measurements are affected by the diffusion of the charged particles through the solar wind in the heliosphere. Nonetheless, the derived value for the energy density of low-energy electrons is about 2 eV cm^{-3}. It is an intriguing coincidence that all known processes affecting the energy balance of the interstellar medium have about this same value, whether magnetic fields, photons, or thermal particles. Using the observed spectrum and correcting as well as possible for the environmental degradation of the low-energy end, the measured value[85] for the ionization rate is $\zeta_c \approx (3-4) \times 10^{-17}$ s^{-1}.

At the high-energy end, there is another more subtle effect. Once generated, propagation of cosmic rays presents an interesting example of how different physical processes in the interstellar medium combine to affect an observable phenomenon. First we must distinguish between the electrons, protons, and other nuclei. Each is affected differently by the magnetic field because of the particle's mass. The Larmor radius is largest for the electrons, and their trajectories in the field can carry them into the halo while the heavier particles are more confined to the plane where column densities are higher. While the electrons lose energy mainly by synchrotron emission and inverse Compton scattering, the protons initiate spallation reactions, suffer ionization losses, and charge exchange with ambient gas. Support for this picture comes from the production of diffuse γ-ray emission in the plane, detected using satellites beginning with COS-B and extending through EGRET and COMPTEL on the *Compton Gamma Ray Observatory* and is one of the phenomena on which the CO–H$_2$ conversion efficiency is based.

Imagine a box with the same dimensions as the gas scale height. Within it we distribute sources for high-energy particles that inject them with a fixed energy distribution, $N(E)$ at a rate $Q(t)$. Keep the box open to the universe at large, a picture called a *leaky box* model, and assume that the escape rate, τ_{esc}^{-1}, depends on energy. Then the loss rate for particles at a given energy is

$$\frac{dN(E,t)}{dt} = -\frac{N(E,t)}{\tau} + Q(E,t). \qquad (6.90)$$

It is conventional to assume $\tau^{-1} = \tau_{esc}^{-1} + \tau_{loss}^{-1}$. For purely diffusive losses, $\tau \approx E^{1/2}$. For escape, the rate has a positive exponent. In steady state, when injection from supernovae and other shocks varies as E^{-p}, the purely diffusive losses lead to $N(E) \sim E^{-(2p-1)/2}$. In addition, explicitly allowing for energy losses, we can include this on the left-hand side of Eq. (6.90) using

$$\frac{\partial N}{\partial t} + \frac{\partial}{\partial E}\left[\frac{E}{\tau_{net}} N(E,t)\right] = -\frac{1}{\tau_{esc}} N + Q, \qquad (6.91)$$

separating the competing processes that produce spectral evolution, individual particle energy losses, and acceleration through τ_{net}, which depends on energy. Now we have the formal solution:

$$N(E,t) = \int_0^t dt'\, Q(E',t') \exp\left(\int dt''\, \lambda\right) \qquad (6.92)$$

---

[85]See Webber, W. R. 1998, *ApJ*, **506**, 329.

with the definition

$$\lambda \equiv \tau_{esc}^{-1} + \tau_{net}^{-1} - \frac{E}{\tau_{net}^2} \frac{\partial \tau_{net}}{\partial E}$$

for the temporal development, with $dE/dt = E/\tau_{net}$ relating the energy of the particle to time. Assuming that the injection takes place at relatively low energy and processes in the interstellar medium then act on the distribution, the high-energy part of the distribution evolves to a power law[86] is given by

$$\frac{\partial \ln N}{\partial \ln E} = -\left(1 + \frac{\tau_{net}}{\tau_{esc}} - \frac{\partial \tau_{net}}{\partial \ln E}\right), \qquad (6.93)$$

which can then be related to the observed spectrum. Remember that τ_{net} describes the energy losses *as well as* the acceleration within the box. In other words, we expect a power law to result from these relatively simple models even if the process differs from the one originally described by Fermi. It is a generic feature of the acceleration process that each particle is individually accelerated and therefore the distribution evolves on a particle-by-particle basis.

The Fokker–Planck equation is one of the most useful tools for astrophysical modeling, and here is one more example. The interstellar medium is filled with fluctuating magnetic tangles. The timescale for their variations may be long on a human scale, but for the lifetime of the Galaxy it is quite short, of order 10^5 yr or so. In the solar wind, the screen through which we observe charged particles, the timescales become very short, on the order of minutes for fluctuations of more than a few percent, and this means that a charged particle does not freely propagate through the medium. Instead, it executes a random walk on the tangles and, in consequence, changes its direction and energy. This process can be described as a diffusive transfer through the medium because of the various losses associated with the collisions between the particle and the moving mirrors. We can also add the complication of *spatial* diffusion, as well as energy variation:

$$\frac{\partial N}{\partial t} + \frac{dE}{dt}\frac{\partial N}{\partial E} = -\frac{1}{\tau_{esc}}N + \frac{\partial}{\partial z}\left(D\frac{\partial N}{\partial z}\right) + Q, \qquad (6.94)$$

which completes the formal derivation of the evolution equation for the particle distribution function. Lower energy particles are more confined to the Galactic plane. The higher the energy, the larger the particle's Larmor radius and the greater the chance that it will leak out of the Galaxy.

It is worth mentioning here a remarkable idea for how to determine the mean time for the acceleration of particles at their putative injection site, supernova explosions. The isotope ^{59}Ni is formed by neutron capture in the ejecta. It is a nucleus with an electron capture half-life of about 10^5 years. At low energy the atomic electrons may not be completely stripped and the nucleus has a chance to decay, but if it is promptly accelerated and loses its electrons, it is stable. Since this

---

[86] Drury, L. O'C., Duffy, P., Eichler, D., and Mastichiadis, A. 1999, *Astr. Ap.*, **347**, 370.

and other *r*-process species are expected to be produced in detectable quantities, their presence in galactic cosmic rays places a limit on the accretion timescale. As in the case with many cosmochronometers, this test is complicated by the need to understand the input ratios, but these are constrained by explosive nucleosynthesis models for SN II events.[87] A current determination of this ratio[88] finds only a very severe upper limit, ^{59}Ni/^{60}Ni < 0.05, so this argues for a long acceleration timescale from multiple sites as the nuclei propagate in the interstellar medium. This is certainly one of the techniques to watch develop in coming years.

6.5.2 Synchrotron Radiation

The emission of light by relativistic electrons moving in magnetic fields is one of the most frequently observed radiation mechanisms in astrophysics. It is so general that it could be discussed in several chapters in this book. However, it is most important in the interstellar medium of the Milky Way and external galaxies as an indication of the presence of a global magnetic field, where the energetic particles are cosmic ray electrons. We will therefore discuss synchrotron radiation here, although it also relates to material in Chapters 3, 7, and 8, because of its relevance to cosmic rays and the interaction between magnetic fields and charged particles in low-density media. Perhaps it's worth a historical digression because the discovery of cosmic synchrotron emission presents an interesting coincidence between laboratory developments and astronomical observations. The first solution for the accelerated motion of a relativistic electron in a curved trajectory was by Schwinger in 1949. This approach is now the standard treatment.[89] It came at the time when synchrotrons were first being developed as particle accelerators and follows the studies by Bohm and Foldy in 1948. Around the same time, interferometric techniques permitted radio astronomers to identify of a number of newly discovered radio sources. In particular, Tau A was shown to be the Crab nebula, which had already been linked to the supernova of 1054 by Baade and Zwicky on the basis of proper motions measured by Duncan in the 1920s. In 1952, Shklovsky proposed that synchrotron emission could account for the very high brightness temperature detected in this source that would otherwise have required very extreme plasma conditions and predicted that the source should be polarized. A few years later, Oort and Walraven (in 1955) obtained polarized photographic images of the nebula and showed that the distribution was consistent with the synchrotron mechanism, thereby clinching its identification in astrophysical sources and providing a new tool for physical diagnosis of energetic phenomena. We should also note that the idea of using power-law spectra for the emitting particles was driven, in part, by work around the same time on the connection between cosmic ray acceleration and magnetic fields. Thus, nonthermal distributions—

---

[87] See, for instance, Cassè, M. and Soutoul, A. 1978, *ApJ*, **200**, L75. Further discussions will be found in proceedings of the *Nuclei in the Cosmos* meetings to which we referred earlier.
[88] Wiedenbeck, M. E. et al., 1999, *ApJL*, **523**, L61.
[89] Schwinger, J. 1949, *Phys. Rev.*, **75**, 1912. See, for example, Blumenthal, G. R. and Gould, R. J. 1970, *Rev. Mod. Phys.*, **42**, 237; Lieu, R. and Axford, W. 1995, *ApJ*, **447**, 302; Jackson, J. D. 1975, *Classical Electrodynamics*, 2nd ed. (NY: Wiley); Shu, F. H. 1992, *The Physics of Astrophysics. I. Radiation Processes* (Mill Valley, CA: Univ. Science Books); Pacholyczk, A. 1970, *Radio Astrophysics* (San Francisco: Freeman); Ginzberg, V. L. and Syrovatskii, S. I. 1965, *ARAA*, **5**, 297 and Ginzberg, V. L. and Syrovatskii, S. I. 1967, *ARAA*, **7**, 375.

power laws—formed the basis of the standard interpretation of synchrotron spectra after the early 1960s.[90] Despite the specific applications, you should keep in mind that synchrotron emission is one of the most widely observed phenomena in astrophysics.

This process has a natural frequency, ω_L, from the spiraling electron. For a relativistic particle, the emission looks like a scattering phenomenon where, in effect, the electron sees a virtual "photon" with a frequency $\gamma\omega_L$ and Compton-scatters this in the moving frame. Then another factor of γ results from the transformation back into the rest frame of the observer. The result is that the "photon" has boosted its frequency to $\gamma^2\omega_L$ so that for a single scattering, the emissivity of the electron varies as E^2B^2. Also, in the rest frame, the relativistic change in mass of the emitting particle causes the *comoving* Larmor frequency to be a factor of γ lower than observer's frame, $\omega_c = \gamma^{-1}\omega_L$, and for simple emission, you would expect that $\omega' = \gamma(1-\boldsymbol{\beta}\cdot\hat{\mathbf{n}})\omega$, which is approximately

$$\omega' = \left(1 + \frac{\beta^2}{2}\right)\left(1 - \beta\left[1 - \frac{1}{2}(\Delta\theta)^2\right]\right)\omega \approx \frac{1}{2}\omega_L(\Delta\theta)^2$$

As a result of curvature and relativistic variation in the mass of the electron, the time internal that the electron radiates for a gyration frequency ω_L in its rest frame translates into $\Delta t = \gamma^{-3}\omega_L$ in the observer's frame. Any charged particle moving at an arbitrary angle through an external magnetic field follows a helical trajectory because of the Lorentz force. The pitch angle determines the projected magnetic field strength and, therefore, the angular frequency. For a slowly moving particle, the frequency of the emitted radiation will exhibit harmonics of ω_L, but for a relativistic particle, this is changed. In our discussion, we will treat only electrons, but the mechanism of synchrotron emission applies to all charged particles, including protons. In moving with a circular velocity $v_\perp$ with an orbital radius r_L, a radiating charge normally emits over the entire 4π steradian solid angle. Relativistic motion changes this because of beaming, and the cone of emission collapses into the forward direction, in the observer's frame, whose opening angle $\Delta\theta = \gamma^{-1}$. The observer sees a short pulse of emission with every orbit, the duration of which depends on the opening angle and the motion of the electron. Specifically, the pulse lasts for an interval $r_L\Delta\theta/v_\perp$. The emitting electron is, however, in orbit and not merely rotating a beam, so it traverses the distance $r_L\Delta\theta$ in a time $r_L\Delta\theta/c$. The total duration of the pulse is therefore

$$\Delta t = \frac{r_L}{v_\perp \gamma} - \frac{r_L}{c\gamma} = \frac{r_L}{c\gamma}\frac{1-\beta}{\beta} \approx \frac{r_L}{2c}\gamma^{-3},$$

where we have used $1 - \beta \approx (1 - \beta^2)/2$ by taking $\beta \approx 1$. Since $r_L = eB/\gamma mc$,

---

[90] See Westfold, K. C. 1959, *ApJ*, **130**, 241; Scheuer, P. A. G. 1968, *ApJ*, **151**, L139; Legg, M. P. C. and Westfold, K. C. 1968, *ApJ*, **154**, 499 provided the standard formulas for computing synchrotron spectra and polarization from a power-law electron spectrum.

we find that the maximum frequency produced by this pulse is given by

$$\omega_c = \gamma^2 \omega_L, \qquad (6.95)$$

recalling the relation in Fourier space between frequency and pulse duration from Chapter 2. From the Larmor formula (see Appendix 6.A) for a nonrelativistic electron, the power radiated due to acceleration is

$$\frac{dE}{dt} = \frac{2e^2}{3mc} |\dot{\beta}|^2. \qquad (6.96)$$

For a relativistically moving particle, this is the same in the comoving frame, but a stationary observer sees the radiated power as

$$\frac{dE}{dt} = \frac{2e^2}{3mc} \gamma^2 \omega_L = \frac{2e^2}{3m^4 c^7} E^2 B^2. \qquad (6.97)$$

This equation gives the energy loss for a single electron, but that is not what we are after. We don't observe the radiation from a single particle. Instead, as with thermal bremsstrahlung, the emission is integrated over the electron energy distribution through which the inherently "nonthermalness" of the emission becomes observationally apparent. This is perhaps the most ad hoc part of the argument and comes from observations. As we showed, galactic cosmic rays display a remarkably uniform, smooth power-law energy dependence, $N(E) \sim E^{-p}$, with an exponent of between 2.5 and 3. Similar spectra are found for products of other sites of relativistic particle acceleration, notably in solar flares. If we assume that the electrons responsible for galactic synchrotron emission have the same sort of energy dependence, then we can state our first general expectation for the emitted spectrum from the ensemble: *Power-law energy distributions beget power-law spectral distributions when the emissivity has no characteristic energy.* The emission from the ensemble is given by the transformation between energy and frequency through $\omega_c = \gamma^2 \omega_L$. Since $E \sim (\omega/\omega_L)^{1/2}$ and the monochromatic emissivity is the same as spectral density, the relation $dE/E = d\omega/2\omega$ means that the emissivity transforms as

$$E^2 \omega_L^2 N(E) dE \to \omega^{-(p-1)/2} \omega_L^{(p+1)/2} d\omega,$$

which gives a power-law dependence for the *spectral index*:

$$j_\omega \sim \omega^{-\alpha}. \qquad (6.98)$$

For the galactic cosmic ray distribution, this gives $\alpha \approx 0.8$. This is actually quite close to the observed value and is typical of regions in which particle acceleration is actively occurring, such as young supernova remnants and galactic-scale radio jets. Although we've emphasized the nonthermal aspect of synchrotron emission, since that dominates most astrophysical environments,, there is a thermal analog this process, *gyrosynchrotron*. It occurs in flares and strong magnetic fields in very hot plasmas, where $\gamma > 1$ but where $p < m_e c$. In this case, the distribution function may be thermal for the electrons, since the temperature is of order 10^7 K or higher. The electrons do not, however, have high enough energies to produce optical emission. Since optical and even X-ray nonthermal spectra are often observed in synchrotron sources, notably the optical jets from radio galaxies and the Crab nebula and other supernova remnants, we will assume that the electron distribution function is a power law with a very high energy cutoff above many GeV in energy.

This theory holds for an optically thin source. What if the density is high enough for the radiation to be self-absorbed? You know that for *thermal* emission an optically thick source radiates like a blackbody whose spectrum depends only on temperature with a frequency dependence $S_\omega \sim kT\omega^2$. The same is true for a synchrotron source with one exception: *no single temperature characterizes the continuum*. It is still possible, however, to obtain the flux distribution by the same use of Kirchhoff's third law. If we know the emissivity and opacity, then the source function for $S_\omega = j_\omega/\kappa_\omega$. So, to evaluate this, we must compute the opacity. The calculation harks back to our derivation of the Kompaneets equation. Since we are interested in the creation and/or annihilation of photons because of the spiraling electrons, we replace $f(E)$, the Maxwellian distribution, with $N(E)$. The absorption of a photon is a transition from a lower state of energy $E - \hbar\omega$ to one with energy E. The absorption coefficient is

$$\kappa_\nu = [B_{lu}N_l - B_{ul}N_u] \rightarrow \int [B_{lu}N(E - h\nu) - B_{ul}N(E)]\, dE. \quad (6.99)$$

From the Einstein relations, $A_{ul} \sim \omega^3 B_{ul} \sim P(\omega, E)$, where P is the radiation efficiency for relativistic electrons, and $g_u B_{ul} = g_l B_{lu}$, where the statistical weights for the states are $E^2 dE$. Then assuming that $\hbar\omega \ll E$, we expand $N(E - \hbar\omega) \sim N(E) - N'\hbar\omega$ and assume that $g_l = (E - \hbar\omega)^2$ and $g_u = E^2$ so that

$$\kappa_\omega \sim \omega^{-3} \int \left[\frac{2N}{E} + N' \right] \omega P(\omega, E)\, dE. \quad (6.100)$$

For low frequency, $P(\omega, E) = (\omega/\omega_c)^{1/3}$ and $\omega_c \sim E^2 \omega_L$, so the integral provides a power-law scaling for the opacity:

$$\kappa_\omega \sim \omega^{-(p+4)/2}. \quad (6.101)$$

Since $j_\omega \sim \omega^{-(p-1)/2}$, we obtain the flux from an opaque source:

$$S_\omega \sim \omega^{5/2}. \quad (6.102)$$

This source function is not the Planck law. Why the difference? We're not dealing with a thermal spectrum. The characteristic energy, which replaces the temperature, scales as $\omega^{1/2}$, the brightness "temperature" at each frequency.

6.5.3 Propagation Effects

6.5.3.1 Faraday Rotation and Large-Scale Field
In Lorentz's model of the Zeeman effect, the orbiting electron is responsible for the polarization of the emitted light. This classical model fails in detail for atoms,

but works for computing the transfer of an electromagnetic wave through a dilute collisionless magnetized plasma. This is a good picture of the diffuse interstellar medium where the electrons are free *except* for the constraining role of magnetic fields. The result is analogous to the transfer of light through an anisotropic crystalline medium, a laboratory phenomenon known as *Faraday rotation*. The astrophysical version of Faraday rotation is a continuum effect that results from the helicity of electron orbits in a magnetic field. The motion for a free electron in a weak magnetic field is determined by the Lorentz force and the local electric field through

$$m_e \dot{\mathbf{v}} = -e\mathbf{E} - \frac{e}{c}\mathbf{v} \times \mathbf{B}. \qquad (6.103)$$

You've seen this many times now, but there is a new twist. The transfer of an electromagnetic plane wave is altered by the directionality of the magnetic field and the resulting helicity of the Lorentz force. If an electric field produces transverse motion of the electron, $\mathbf{v} \times \mathbf{B}$ deflects this and produces circular harmonic motion about $\mathbf{B}$. Since the response to $\mathbf{E}$ depends on the charge of the perturbed particle, there is a difference between right- and left-hand circular polarizations. This is the birefringence we just mentioned. Single-particle motion is all we really need since the interstellar plasma is collisionless. The medium has an enormous conductivity, so we can ignore any constant electric fields and assume that the only $\mathbf{E}$ is due to the wave. But there is a background magnetic field in addition to the oscillating field in the wave, so we write $\mathbf{B} = B\hat{\mathbf{B}} + \delta\mathbf{B}$. We will use only a single-mode plane wave here:

$$\delta \begin{pmatrix} \mathbf{E} \\ \mathbf{B} \end{pmatrix} = \begin{pmatrix} \mathbf{e} \\ \mathbf{b} \end{pmatrix} e^{i(\mathbf{k}\cdot\mathbf{x}-\omega t)} \qquad (6.104)$$

and assume that the amplitudes are independent of position. We also assume that $\mathbf{k}\cdot\mathbf{e} = \mathbf{k}\cdot\mathbf{b} = 0$. We can immediately write the linearized equations for the $\mathbf{e}$ and $\mathbf{b}$ fields supplemented by the equation of motion for the electrons in the following form:

$$i\mathbf{k} \times \mathbf{e} = \frac{i\omega}{c}\mathbf{b} \qquad (6.105)$$

$$i\mathbf{k} \times \mathbf{b} = -\frac{i\omega}{c}\mathbf{e} - \frac{4\pi}{c}n_e\mathbf{v} \qquad (6.106)$$

$$-i\omega\mathbf{v} = -\frac{e}{m_e}\mathbf{e} - \frac{eB}{m_e c}\mathbf{v} \times \hat{\mathbf{B}}. \qquad (6.107)$$

Solving this system gives the dispersion relation for $k(\omega)$:

$$k^2 = \frac{\omega^2}{c^2} - \frac{1}{c^2}\frac{\omega_p^2 \omega}{\omega + \omega_L}. \qquad (6.108)$$

538 THE INTERSTELLAR MEDIUM

The index of refraction is defined as usual by $n = c/v_p$ where the phase speed is $v_p = \omega/k$:

$$n^2 = 1 - \frac{\omega_p^2}{\omega(\omega + \omega_L)}. \qquad (6.109)$$

The plasma frequency ω_p is

$$\omega_p^2 = \frac{4\pi e^2 n_e^2}{m_e}.$$

As usual, the magnetic field introduces a resonant frequency, ω_L. There is now also a difference in the refractive index between two orthogonal states of polarization. This creates a phase shift and results in a distance-dependent rotation in the plane of polarization[91] is only analogous called *Faraday rotation*:

$$\Delta\phi = \frac{\omega}{B}\int \Delta n\, ds. \qquad (6.110)$$

Astrophysically, this is usually quantified by the *rotation measure*, which is defined by

$$\text{RM} = 8 \times 10^5 \lambda^2 \int n_e B \hat{\mathbf{B}} \cdot \hat{\mathbf{s}}\, ds \text{ rad m}^{-2} \qquad (6.111)$$

where $\hat{\mathbf{s}}$ is the unit vector along the line of sight. The choice of units is dictated by the method of observation. Faraday rotation measures are usually obtained from radio observations, so it is most convenient to express the wavelength is given in meters. Given the typical pathlengths for interstellar structures, the distance is in parsecs. You see from Eq. (6.111) that the rotation measure provides information only about the *net longitudinal* magnetic field, that is, the field projected along the line of sight. Notice, however, that it also involves the electron column density and so it is sensitive to the charged component of the medium. Even if the magnetic field does not change, density fluctuations can produce a rotation of the polarization. Another way to say this is that the medium is *birefringent* since it has different responses to different senses of polarization.

Under some restrictive assumptions, it is possible to reconstruct some information about the structure of the magnetic field. This was first shown by Burns[92] who demonstrated that the rotation measure and the wavelength are essentially conjugate variables in Fourier space. Take a beam with intensity I, phase ϕ, and polarization $p(\phi)$ that is observed after propagating through a longitudinal field.

---

[91] This was first observed in the laboratory for anisotropic crystals and reported in Faraday's *Researches in Electricity and Magnetism* in the 1830s. The effect in the interstellar medium, which is due only to classic properties of electrons spiraling in magnetic fields, is an analogy but not quite the same thing.
[92] Burns, B. J. and Sciama, D. W. 1964, in *Physics of Nonthermal Radio Sources*, Maran, S. P. and Cameron, A. G. W., eds. (Washington DC: NASA SP-46); Burns, B. J. 1966, *MNRAS*, **133**, 67. The harder-to-obtain conference volume actually contains a more complete discussion, while the journal paper is the more widely quoted source in the literature.

Define $F(\phi) = I(\mathbf{s}, \phi)p(\phi)$ to be the fractional polarization of the emitted radiation. In general, one needs multiwavelength measurements to obtain the rotation measure so the raw data are the spectral distribution of F. The Faraday rotation alters ϕ as a function of distance so that, when we integrate over the path, we obtain

$$P(x) = \int_0^\infty F(\mathbf{s}, \phi) \exp\left[2\pi i x \int_0^s n_e \mathbf{B}(\mathbf{s}') \cdot d\mathbf{s}'\right] ds, \qquad (6.112)$$

where $x = \lambda^2$. Replacing the integral in the exponent in Eq. (6.112) by ϕ—called the Faraday depth—which measures the degree of rotation of the plane of polarization by the intervening magnetic field along the line of sight reveals that Eq. (6.112) is essentially the Fourier transform of the polarization:

$$P(x) = \int_{-\infty}^\infty F(\phi) e^{2\pi i x \phi} d\phi. \qquad (6.113)$$

From the observed distribution of polarization as a function of wavelength, $P(x)$, it *should* be possible to invert the transform and obtain information about the magnetic field and charged particle distribution along the line of sight:

$$F(\phi) = \frac{1}{\pi} \int_{-\infty}^\infty P(x) e^{-2\pi i x \phi} dx. \qquad (6.114)$$

Notice that since the cosine can be positive or negative, it is possible to get $P(-x)$ as well as $P(x)$. Observations of extragalactic radio sources through the galactic interstellar medium provide maps of the rotation measure on large scale.

As a method for deriving magnetic field structure and strength, this technique is not without problems. You don't observe the magnetic field independent of the gas, and you must correct for the changes in the rotation measure due only to density fluctuations. The observations are not easy and depend on the spectral resolution of the receiver. If the rotation measure is large, it can alter the polarization across a bandpass and compromise the measurement of the field direction. Remember that this does not provide a field strength unless you also know the electron density. Close spacing of observing frequencies helps alleviate some of this problem, but frequency resolution remains the basic limit to this method. A theoretical complication comes from the magnetic field structures. The rotation measure depends only on the *line-of-sight* component of $\mathbf{B}$. If the field lines are tangled, then random swings may occur in the projected field direction and strength anywhere along the line of sight. Because the rotation measure is an integral along the pathlength, it can vanish if the resulting average projected field is zero. Consequently, as we saw with optical polarization produced by grain alignment, this technique is most sensitive to *net* departures from symmetry and not to the absolute magnitude of the field strength.

The method has, however, been used extensively over the years to map the Galactic magnetic field. The observations employ background radio galaxies as pencil beam probes of the medium. These are sources whose spectra are intrinsically polarized because of synchrotron emission and are viewed through a foreground screen, the Galactic interstellar medium. Of course, since they are galaxies and have their own internal gas and magnetic fields, some of the observed Faraday rotation is intrinsic to the source. However, any extrinsic rotation measure depends on the line-of-sight (LOS) field, which will be different in the direction of each source. The weakness of the technique is, as always, the relatively small number of well-calibrated extragalactic radio sources that do not have large *internal* Faraday rotation. To obtain the field strength, the rotation measure must be supplemented by knowledge of the LOS electron density. This can be determined using pulsars, through the dispersion measure, Eq. (6.110). Thus, the rotation and dispersion measures are related to each other:

$$\langle B_{\rm los} \rangle = \frac{\int \mathbf{B} \cdot \hat{\mathbf{s}} n_e \, ds}{\int n_e \, ds} \approx {\rm RM/DM}. \qquad (6.115)$$

While this is inapplicable for observations of external galaxies, the combination of synchrotron emission intensity and Faraday rotation has been used to map the large-scale structure of galactic magnetic fields and has even been applied to clusters of galaxies for which the X-ray emission measures provide values of the electron density. In general the fields are weak, on the order of a few microgauss, but this is the field in the diffuse gas. Zeeman measurements of denser environments, such as molecular clouds, typically yield values of at least a factor of 10–100 larger than this.

6.5.3.2 *Electron Density by Scintillation and Dispersion Measure*

The thermal electrons in low-density gas radiate so weakly that their spectrum is not directly observable. But it is still possible to study their density using high-time-resolution radio observations of pulsars. Because of the plasma frequency, ω_p, pulses that differ very slightly in frequency are dispersed as they propagate through the interstellar medium. This causes the real part of the index of refraction, that component that produces scattering, to depend on wavelength. Setting $\omega_L = 0$ in Eq. (6.109) leaves n dependent only on n_e and frequency, which has important consequences for observations of sources near ω_p. All radio observations are made with finite frequency resolution, and across a bandpass $\Delta \omega$ there is a difference Δn in the index of refraction:

$$\Delta n = \frac{dn}{d\omega} \Delta \omega = -\frac{3\omega_p^2}{\omega^3} \Delta \omega. \qquad (6.116)$$

This slight difference produces an *unresolved* phase shift from the lowest to highest frequency in the bandpass, because of the different propagation times for a signal, and smears the pulses. Since the emission mechanism that creates the pulses is regular, you can determine the dispersion measure for any pulsar by sampling the signal at very high rate and then correcting systematically for the dispersion measure until the pulses are observed. It is necessary to correct for this in modern pulsar searches. However, the benefit is that it directly measures ω_p and therefore the electron column density.

An additional effect is the broadening of radio sources viewed through the medium by interstellar turbulence, the cosmic analog of *atmospheric seeing*. Any source observed through the gas will also experience spatial dispersion due to phase shifts introduced by density fluctuations. An analog of turbulent image broadening, interstellar[93] scintillation produces a finite-sized disk for a point source. It is possible to get a deeper feeling for how this happens by a terrestrial experience. The observation is most easily performed at night with airport runway lights, especially if you are near the wing. Notice the *upstream* sizes of distant lights (although any point source will do) ahead of the propeller (or jet engine). Then look at how the size of those sources change *behind* the propeller. Since the airflow from the blades is highly turbulent and collimated, you can easily see the explosion in image size as a result of the refractive scattering. You will also notice that there is a characteristic size to the images, no matter how distant they were, and that they also fluctuate (twinkle) in a way that depends on the blade speed. The effect is *very* dramatic, even more than the effect of seeing on astronomical sources.

6.5.3.3 How to Search for Warm Gas

With Faraday rotation measurements toward pulsars, for which the pulses provide direct measurements of the LOS density weighted magnetic field components, you have an additional tool for obtaining the fraction of the medium that is in the warm gas. The argument goes as follows. Take a pulsar for which you have the dispersion measure, DM, and compare this to the diffuse emission along the line of sight using the emission measure, EM $\sim \int n_e^2\, dl$. For diffuse Hα, for instance, or any emission line, you can select the particular velocity range of interest. For local pulsars, within a few 100 pc of the Sun, this is a problem, but they have also been found in several globular clusters (M5, M13, M15, and M53) for which the velocities are known relative to the Sun. Following Reynolds' treatment,[94] you can then define two characteristic numbers:

$$n_c \equiv \frac{\int n_e^2\, dl}{\int n_e\, dl}$$

$$L_c \equiv \frac{(\int n_e\, dl)^2}{\int n_e^2\, dl}, \quad (6.117)$$

the *characteristic local electron density* and *occupation length*, respectively. This presumes you know the LOS magnetic field but this can be found from using several pulsars in the same cluster. In addition, the neutral fraction should be observable through 21-cm measurements along the same lines of sight in the same velocity range, and even absorption measurements against globular cluster UV

---

[93] The interplanetary medium also contributes structure to the image through the electrons in the solar wind, but these vary on a much shorter timescale and the column densities are much lower.
[94] Reynolds, R. J. 1991, *ApJL*, **372**, L17.

bright sources could be used in the usual way to obtain the ionized and neutral fractions.[95]

6.5.4 Magnetized Clouds

6.5.4.1 Direct Observation of the Zeeman Effect

Although the ISM is turbulent and this makes it very hard to measure the integrated Zeeman effect, 21-cm polarization studies are able to separate the $\sigma_\pm$ components. Collisions do not mix the states, and therefore even for this very small energy separation the polarizations are preserved. Direct measurements have been obtained in molecular transitions, especially OH, and fields as small as 0.1 mG can be measured in this way.[96] This is the only direct measurement of the in situ field in molecular clouds, although it is sensitive to a relatively limited range of densities because of the species used. For diffuse clouds, the fields are typically around 10 μG, while for dark clouds it can reach 3–5 mG.

6.5.4.2 Ambipolar Diffusion

In the interstellar medium, magnetic fields even govern the dynamics of the neutrals because of the mechanism of *ambipolar diffusion*.[97] Remember that neutrals and ions are only states of a single species; the partition is determined by the ionization and recombination rates. The ions cannot slip through the field lines, while the neutrals can, so in a partially ionized medium the gas will slowly move through any background field. In effect, this is a means for excluding the field from a cloud and is one of the mechanisms that acts during star formation; the more ionized gas will be supported by the field much as a sieve works to separate particles of differing mass and radius in a mixture. We will return to that role later in this chapter. For now, we review how the more general features of the process. There is a statistical separation between the ions and neutrals because of the differential drift velocities under the influence of any external force. This also affects the electrons that are trapped on field lines. It ultimately causes the process to maintain a steady state—the slow drift of ions and electrons produces a counteracting electric field, because of the $\mathbf{v}_i \times \mathbf{B}$ force, which tends to balance the gravitational acceleration. The frictional force between the ions and neutrals is

$$\mathbf{F} = \left(\frac{\langle \sigma v \rangle_{H_2}}{2m_H}\right)\left(\frac{n_i}{n_n}\right)\rho^2(\mathbf{v}_n - \mathbf{v}_i), \qquad (6.118)$$

---

[95] As we will see in the next chapter, it is always easier to see structure in external galaxies than in the Milky Way, and this is especially true for the diffuse emission. For M31, for instance, this has been done by Greenwalt, B., Walterbos, R. A. M., and Braun, R. 1997, *ApJ*, **483**, 666. With apologies, let's delay discussion of this until the appropriate point in Chapter 7.

[96] For diffuse clouds, see, for example, Heiles, C. 1976, *ARAA*, **14**, 1, Heiles, C. 1997, *ApJS*, **111**, 245. For molecular clouds, see Crutcher, R. M. 1999, *ApJ*, **520**, 706. The original detection is reported in Verschuur, G. L. 1969, *ApJ*, **156**, 861.

[97] Mouschovias, T. 1991, in *The Physics of Star Formation and Early Stellar Evolution*, Lada, C. J. and Kylafis, N. D., eds. (Dordrecht: Kluwer), p. 61; Nakano, T. 1990, *MNRAS*, **242**, 495; Mouschovias, T. 1979, *ApJ*, **228**, 475; Mestel, L. and Spitzer, L. 1956, *MNRAS*, **116**, 503; Brandenburg, A. and Zweibel, E. G. 1995, *ApJL*, **427**, L91.

and the characteristic timescale is

$$t_{AD} = \frac{R}{|\mathbf{v}_n - \mathbf{v}_i|}. \tag{6.119}$$

The equations of motion are only changed through the additional coupling term between the species:

$$\rho_i \dot{\mathbf{v}}_i = \mathbf{F}_i + \frac{1}{c} \mathbf{J} \times \mathbf{B} - \rho_i \nu_{ni}(\mathbf{v}_i - \mathbf{v}_n) \tag{6.120}$$

$$\rho_n \dot{\mathbf{v}}_n = \mathbf{F}_n - \rho_n \nu_{in}(\mathbf{v}_n - \mathbf{v}_i).$$

An estimate of the drift speed is provided by the MHD equations. Taking $\rho v_D / t_{ni} \approx (\nabla \times \mathbf{B}) \times \mathbf{B}/4\pi$, where t_{ni} is the ion-neutral collision time and L is some typical length scale for the magnetic field. Now rewrite this by dimensional scaling. Then, since the Alfven speed is $v_A = B/(4\pi\rho)^{1/2}$, and $t_A = L/v_A$, the drift timescale is estimated to be

$$t_D \sim \frac{t_A^2}{t_{ni}}, \tag{6.121}$$

which is of order 10^5–10^6 yr for typical interstellar clouds. If gravity drives the diffusion, then $t_D \sim t_{ff}^2 / t_{ni}$.

The net effect of this diffusion is the slow but inexorable drift of particles toward the center of mass of a cloud and its consequent increase in density. It is also worth noting that this drift, though small, may involve a large amount of mass and can contribute to the energy budget of the cloud much as Kelvin–Helmholtz contraction does to a star. This, in turn, feeds back into the gravitational acceleration and, through the Jeans criterion, pushes the medium closer to the brink of instability. Several mechanisms act to stave off this seemingly inevitable result. One is the increasing role played by ionization once stars begin to form. Any interval UV radiation not only heats the grains and the medium but also charges them, making it more difficult for the dust to diffuse across the field. Turbulence within the cloud can locally amplify the field and also alter its topology, much as we discussed in Chapter 4 for magnetic dynamos. Thus, the simple uniform field structures that are generally assumed for such models may actually be quite tangled. The effect of this is to reduce the mean free path of the grains and reduce the diffusion velocity by the introduction of a topological diffusion term. We will return to this point when discussing the structuring of clouds by ambipolar diffusion in the context of the virial theorem.

6.6 MOLECULES AND ASTROCHEMISTRY

6.6.1 Observations

Here we take the opportunity to discuss molecular processes, one of the aspects of kinetics that plays an important role in stellar atmospheres, but dominates the interpretation of the interstellar medium. We'll review the basics of combinations and excitation regarding molecules, and then mention some of the ways that they show up under interstellar conditions. For the spectroscopic notation, refer again to Section 6.2.4.3 above.

6.6.2 Interstellar Chemistry

It is hard to tell whether molecular clouds are more similar to a smog or a brewery. Although the temperatures are extremely low, chemistry takes place within the clouds because of the large abundances of several key species, notably C, N, O, and S, and because of the long timescales available for the reactions to proceed. The literature on this is enormous, and we have no intention here of summarizing it. Instead we will concentrate on the processes that play important roles in the synthesis of molecules in the clouds, many of which are also important in stellar atmospheres, and refer you to the literature for the more specific calculations.

In diffuse regions, the dust will always be colder than the gas, primarily because of the larger number of modes available for the redistribution of the energy. In fact, the temperature, or equivalently the excitation, of the dust is sensitive more to the spectrum of the radiation incident on it than to its dilution. The cooling of the dust is strictly radiative, efficiently absorbing in the ultraviolet and radiating in the infrared. The harder the UV radiation that is incident on the grain, the warmer the grain will be, regardless of the intensity of the radiation. The grains are the critical shield for the cloud material from background radiative heating. The presence of dust also effectively cools the gas because of surface atom interactions that promote the formation of molecules, especially H_2. Molecular cooling is due to collisional excitation of abundant species, which then reradiate their energy in the far-infrared and millimeter wavelengths, at which the grains are optically thin. Deep in the cores of molecular clouds, the situation is reversed. Here the grains are warmer than the gas, and actually heat the gas through collisions of the particles with the grain surface. As such, they serve as the fuel for the chemistry. Collisions of molecular hydrogen with, for instance, CO, excite the latter, that radiates its energy from the cloud at the expense of the gas kinetic energy. Thus, the IR that penetrates the cloud and heats the grains can be transferred to the excitation of the various chemical constituents of the medium effectively, serving to power the reactions that build complex molecular species.

6.6.2.1 *Surface Chemistry: Formation of Molecular Hydrogen*

Grains provide a solid surface on which chemical reactions take place. In fact, they are the primary site for H_2 synthesis. Metallic ions deplete onto the grains. As we've discussed, it appears that much of the heavy metals in the diffuse and molecular cloud phases of the ISM may be tied up in the grains, which nonetheless constitute only about 10^{-6}, by number, of the interstellar medium. CO and H_2O also stick to the grain surfaces, and models and laboratory simulations show a host of complex organic molecules that can be synthesized in the resultant mantle. It remains to be determined whether these simulations are relevant for interstellar conditions; they do seem to mimic many of the reaction products observed in situ in comet Halley.

An important problem in dust chemistry is the precise determination of both the formation mechanism and size spectrum of the grains. At the smallest particle end, the grains behave like large molecules. Since the mid-1980s these have been suggested to be polycyclic aromatic hydrocarbons (PAHs), which are stable against UV radiation and also can be cleaved from the larger graphite grains. The signature of small grains is that they cannot come into equilibrium with the ultraviolet radiation field, and do not.

The basic problem in the study of the interstellar medium is the formation of molecular hydrogen, H_2. This is of primary importance because, at the low temperatures characteristic of the cloud environments, this molecule is responsible for the excitation of CO. One might initially expect reactions of the form $2H \to H_2 + \gamma$. This process, called *radiative association*, is clearly important for a wide range of molecules seen in clouds, but the rate for H_2 is many orders of magnitude short of those required to produce the molecule. Instead, it appears that the formation of H_2 can proceed via only one of two possible avenues: if there is a sufficient abundance of free electrons:

$$H + e \to H^-$$
$$H^- + 2H \to H_3^-$$
$$\to H_2^* + H^-$$

or via some form of surface interaction where radiative association is replaced by reaction on a solid surface of two neutral atoms in the presence of a UV radiation field that is capable of exceeding the appropriate binding energy of the molecule to the grain surface. The first of these can be independent of the abundance of metals, while the latter is critically dependent on the existence of interstellar grains on which the reactions can take place. Because it appears that the solid lattice is far more efficient, and because it exists in the interstellar environment that now is observed in molecular clouds, we shall concentrate on the second mechanism. The first type of mechanism has been implicated in star formation processes in the early universe. Grains are a prerequisite for the formation of molecular hydrogen in the present Galaxy, even though some radiative processes in the gas phase must have been important during early stages of galactic evolution when the metallicity was much lower.

Assume that the grain consists of a simple lattice, like graphite. The mean arrival rate for H atoms at the grain is $\lambda = n_H \sigma_g v_T$, and we can assume that the probability of n particles arriving in some time interval Δt follows a Poisson law:

$$p_n = \frac{(\lambda \Delta t)^n}{n!} e^{-\lambda \Delta t}. \qquad (6.122)$$

You may recall that we used a similar argument in Chapter 4 for neutron exposure in the s process. In queuing theory this is called an *exponential distribution of arrivals*. Once at a grain, the atom has a sticking coefficient S and a hopping probability of p_h between two sites on the lattice at which the associative reaction can occur with a probability K_{HH} per site for n_s sites so the rate of formation of H_2 is

$$\frac{dn_{H_2}}{dt} = S p_h p_n n_g K_{HH} n_s \pi a^2 v_T. \qquad (6.123)$$

Once formed, it is usually assumed that the grain immediately releases the molecule back into the gas phase. Notice that for large atomic hydrogen densities, the dependence of the H_2 formation rate scales quadratically with n_H, with the approximate law[98] of

$$\frac{dn(H_2)}{dt} \approx 3 \times 10^{-17} \text{ cm}^3 \text{ s}^{-1} \, n_H^2. \qquad (6.124)$$

Here n_H is the ambient gas density, with the grains scaling as a fixed fraction of n_H. The problem with this picture may be the assumption of a steady flow of particles onto the grain surface. This affects not only H_2 chemistry but also any species that requires some sort of surface interaction. The medium is quite dilute and the arrival rates are stochastic. If the density of neutral hydrogen is below some threshold, then the arrival rate will be low enough that *there will be an exponential cutoff in the dependence for H_2 on H since the Poisson limit will apply.* Above this threshold, the rate of arrival will be high enough that a more deterministic approach can apply.

6.6.2.2 Gas-Phase Chemistry

The chemistry of gas-phase reactions, either in the interstellar medium or in stellar atmospheres, is mediated by the abundance of ions. These can be formed in any of several ways: by cosmic ray ionization, or by the direct photoionization of the atoms involved in the reactions with subsequent charge transfer to the molecules. Ionic reactions are generally exothermic and so occur efficiently at low temperature. In the presence of an ion, a neutral molecule or atom develops an induced dipole, which increases its capture cross section. Thus the reactions occur quickly and lead to stable states, in addition to allowing the molecule to form in a radiatively unstable excited state, which, on decaying, radiates the energy of formation away from the site of the reaction.

In general, reactions depend on two factors: the activation energy Δ and the temperature. Cross sections can be directly observed in the laboratory at some controlled temperature, usually near room temperature (about 300 K) and then scaled to the temperatures found in clouds. They can also be deduced from first principles. Generally, these reactions have the form $K = \langle \sigma v \rangle$. For ionic reactions, where the potential is the coulomb interaction, the so-called *Langevin approximation* applies, and it can be shown that the rates are approximately constant. Thus measurements at room temperature suffice for the determination of the rate coefficients, $K = k$ where usually k, the reaction constant, is of order 10^{-8} to 10^{-13} cm^3 s^{-1}. Most ionic reactions have little or no potential barriers, being generally exothermic. Hence there is usually a simple constant volumetric rate which is assumed to be a constant. Neutral reactions are most likely to involve substantial activation energies that greatly inhibit their rates of formation. If a neutral channel is important in a network of reactions, it will likely be a bottleneck for the formation of the product species. These have strong temperature dependences because of their activation energies and usually are several orders of

---

[98] See, for instance, Jura, M. 1974, *ApJ*, **191**, 375. P. Goldsmith, P. Caselli, and E. Herbst are thanked for discussion of this point.

magnitude slower than the ion–neutral channels. Electronic recombination reactions typically scale as $T^{-1/2}$.

Before proceeding with a discussion of recent results, one point should be emphasized. In many of the reaction networks, among all the rates that must be tabulated and all the reactions that must be tracked, most of the rates have to be approximated by guesses or parametric fits to laboratory data.[99] Few measurements at interstellar conditions ar available, and this is a very significant challenge for future laboratory astrophysics. In many of the networks, perhaps as few as 10% of the rates are known to within a factor of 50%, and perhaps as many as half are guesses and may be uncertain to a factor of 10. This is a field that is still in development where only the dominant channels are well understood, but many of the details are still extremely important because of the physical conditions that can be probed by trace species. In general, laboratory measurements yield direct rates. The reaction rates for the inverse channel depend on arguments from detailed balance, of the sort we discussed in Chapter 3, $g_i C_{ij} = g_j C_{ji}$, where C are the collision rates, and the reverse reaction includes the excitation energy.

6.6.2.3 *Ionization and Dissociation*

Ionization in the densest parts of molecular clouds depends on the penetration of cosmic rays and ultraviolet radiation, as well as the presence of shocks generated by such processes as cloud–cloud collisions and internal star formation. In order to probe the electron density in the clouds, it is important to be able to account for the presence of complex polyatomic molecules, whose formation requires ion gas-phase reactions.

The rate for cosmic ray ionization inferred from chemistry is $\zeta_c \approx (3 \pm 2) \times 10^{-17}$ s^{-1}. This is also an integral over the collisional ionization cross section for low (MeV)-energy CR protons, but it is approximately a constant for most of the species of interest. An obstacle in our understanding of the detailed structure of molecular clouds is our ignorance of the precise rate. The low-energy end of the cosmic ray spectrum is difficult to determine empirically from terrestrial observation because these particles diffusively propagate through the interplanetary medium, scattering off of turbulence in the solar wind; their spectrum cannot be directly observed and must be inferred from models for their motion through the heliosphere. The more easily observed cosmic ray protons and electrons, in the GeV and higher range, have little or no effect on interstellar ionization because of the small interaction cross sections for atoms at such high energies.

In molecular clouds, atomic species with ionization energies greater than 13.6 eV must be predominantly neutral because of the shielding effects of neutral hydrogen. Lower ionization energies are shielded by the photodissociation of H_2 through the Lyman and Werner bands. It is mainly the heavier elements, like C, N, and O, which are observed in the peripheral portions of the clouds to be in the partially ionized state. For circumstellar envelopes, cosmic rays lose out to photo-processes and the chemistry is mediated by the input of stellar photospheric radiation (in the hotter stars and in novae and supernovae) and from the diffuse interstellar radiation field.

---

[99] See, for instance, Millar, T. J., Farquhar, P. R. A., and Willacy, K. 1996, *Astr. Ap. Suppl*, **121**, 139.

The basic equations for two body interactions can be written in the form

$$\frac{dN_i}{dt} = \sum_{j,k \ne it} K_{ijk} N_j N_k - \sum_j K'_{ij} N_i N_j, \qquad (6.125)$$

where K_{ijk} is the formation rate for the ith molecular species, while K'_{ij} is the destruction rate for the molecule. The inclusion of ultraviolet photoprocesses is accomplished by the photodissociation rate:

$$R_{pd} = \int_{\nu_0}^{\infty} \kappa_\nu F_\nu e^{-\tau_\nu} \frac{d\nu}{h\nu}, \qquad (6.126)$$

where F_ν is the incident photon flux, τ_ν is the opacity of the ambient medium (presumed to be from dust), κ_ν is the continuous absorption coefficient for the dissociative continuum, and the dissociation energy is $h\nu_0$.

6.6.2.4 Molecular Cooling Processes

Carbon monoxide is one of the principal emission lines and also one of the primary cooling agents for molecular clouds. Therefore, we will concentrate on this species as an example of the analysis of the physical conditions within clouds.

It is significant that CO is not a homonuclear molecule. It is hardly the most abundant molecule in a cloud, that pride of place belongs to H_2, but it possesses a strong dipole moment and therefore is an efficient radiator. The ground state rotational lines for CO occur at a few mm, or at energies only about 0.01 eV. This is well within the reach of thermal excitation within the clouds. The calculation is formally exactly the same as we found for atoms with one exception. The exciting agent is molecular hydrogen, which can itself be excited by the transition. The collisional dynamics are much more complicated than for atoms since the excitation is a multibody process. The net result is that the rate of collisional excitation, $C_{01} n_{H_2} n_0$ balances the emission rate, $n_1 A_{10} h\nu_{10}$. This simple picture yields an important observational constraint on the properties of the cloud. If every CO $J = 1-0$ is produced by a collision with molecular hydrogen, then there is a correspondence between the emissivity of the cloud and the abundance of H_2. In other words, there is a calibration that gives the column density of H_2 in terms of the emissivity of CO. Thus, you can measure the total mass of the cloud by using a trace species. This is a fundamental result, although the calibration of the method is always a hot topic of debate. It is very difficult to compute from first principles, and must be calibrated empirically. The method works as follows. There is an empirical relation between the column density of dust and H_2. You know from the discussion of surface chemistry why this would be expected. The calibration is

$$N(H_2) = 2.3 \times 10^{20} \int_{\text{line}} T_A(\text{CO}) \, (\text{K km s}^{-1})^{-1} \, dv \, \text{cm}^{-2}, \qquad (6.127)$$

where $T_A(\text{CO})$ is the antenna temperature for the $J = 1-0$ transition and the integral is taken over the velocity width of the line profile assuming that it is optically thin. Other molecules, in particular the more transparent transitions of CH, can also be used for the calibration of this relation.[100]

Chemistry also feeds back into the thermal balance of the clouds. Molecules radiate in portions of the spectrum where the medium is usually optically thin. Since this radiation can escape from the cloud, it is the primary means whereby the clouds cool. Star formation requires that otherwise hydrostatic clouds become gravitationally unstable, a process that can be affected by the rate of energy loss as well as by external perturbations. Thus time-dependent processes, those that cause the stability of the clouds to alter with time, are extremely important since the timescale for molecular formation is not too short (of order 10^6 yr) compared with the estimated lifetimes of the clouds ($\leq 10^8$ yr). For example, the cooling rate for CO depends on the abundance of both dust and of H_2 and CO by

$$\Lambda_{CO} = 1.1 \times 10^{-30} n \left(\frac{\Delta v}{v_{th}}\right) \frac{T^{1/2}}{[1 + 1.4 \times 10^{-4} n T^{1/2}(1 + N/N_c)]} \text{ erg cm}^{-3} \text{ s}^{-1},$$

(6.128)

where $N_c = 2 \times 10^{18}$ T cm^{-2} and N is the column density, related to the extinction. Molecular species are therefore quite efficient in radiatively removing energy from the clouds and, literally, refrigerating the medium.

6.6.2.5 H_3^+ as an Example of Ion Chemistry

Arguably, the most important ion for all molecular reactions is H_3^+. It is formed by several channels, the primary one being capture of low energy cosmic ray protons by molecular hydrogen, which is itself formed on grains. It is stable at the densities and temperatures which are typical of molecular clouds. The capture of a carbon atom to form CH_3^+ is a critical step in the chemistry of the interstellar medium, especially in the generation of the ions of the cyanopolyyne series, like $HC_{11}N$. H_3^+ has now been directly observed in the interstellar medium as has its deuterated phase, H_2D^+. It has been studied in the laboratory and in near-Earth environments, and is an important feature in the spectra of Jovian planets, where it is formed as part of the dissociation of water and related compounds. Finally, it has been detected in interstellar clouds via near-infrared transitions around 3.7μ.[101] One can therefore assert with confidence the role of this ion in molecular chemistry. Subsequent to carbon capture, interactions with H_2 can form all the hydrocarbons observed in molecular clouds. An example of this chemistry is given

---

[100] Bloemen, H. 1989, *ARAA*, **27**, 469; Magnani, L. and Onello, J. S. 1995, *ApJ*, **443**, 169.
[101] Drossart, P. et al. 1989, *Nature*, **340**, 539; Oka, T. 1992, *Rev. Mod. Phys.*, **64**, 1141; Miller, S., Lam, H. A., and Tennyson, J. 1994, *Can. J. Phys.*, **72**, 760; Tennyson, J., Miller, S., and Schild, H. 1993, *J. Chem. Soc. Faraday Trans.*, **89**, 2155; Neale, L., Miller, S., and Tennyson, J. 1996, *ApJ*, **464**, 516; Schild, H., Miller, S., and Tennyson, J. 1997, *Astr. Ap.*, **318**, 608. An excellent survey of the observations of interstellar clouds is McCall, B. J. et al. 1999, *ApJ*, **522**, 338.

by the network

$$H_3^+ + C \to CH^+ + H_2$$
$$CH^+ + H_2 \to CH_2^+ + H$$
$$CH_2^+ + H_2 \to CH_3^+ + H$$
$$CH_3^+ + H_2 \to CH_5^+ + \gamma$$
$$CH_5^+ + e \to CH_4 + H$$
$$\to CH_3 + H_2$$
$$\to CH_2 + H_2 + H$$
$$\to CH + 2H_2,$$

which also illustrates the reason for the complexity of many of the reaction calculations: there are many product states for electron capture reactions, due to the role of dissociative recombination.

6.6.2.6 *Isotopic Fractionation*

The fingerprint of stellar nucleosynthesis, and of the chemical history of the Galaxy, is most clearly seen in the abundances of the isotopes. For example, as we discussed in Chapter 4, ^2D, is easily destroyed in stellar interiors via low-temperature ($\approx 10^6$ K) reactions, but was synthesized in the Big Bang via nonequilibrium nucleosynthesis in the expanding Universe during the first few minutes of its existence. Thus its abundance was fixed in the early Universe. It serves, as we will discuss in Chapter 8, as a thermometer for the conditions at the time of primordial nuclear reactions, since the rates of change of temperature and density during this initial epoch were fixed by the rate of expansion, which depends on the amount of mass in the Universe; the slower the expansion rate in this early period, the closer to equilibrium will be the proton process products and the lower the D abundance. Because of the overwhelming abundance of H, it is very difficult to observe the D abundance directly, especially in stellar atmospheres but also in the wings of the $Ly\alpha$ line from the interstellar medium. With an anticipated abundance ratio $D/H \approx 10^{-5}$, current observations of interstellar absorption lines cannot place good limits on this cosmologically important number for any except the nearest stars for which the resonance transition is not lost in the H line wing. Solar system measurements help, especially for the Jovian planets, and deuterium has even been used as a tracer of early sources of volatiles for the terrestrial planets.

It is possible, however, to determine the D/H ratio through a different avenue, the isotopic shift caused by the mass ratio of D/H in the rotational and vibrational spectrum of the H_2 molecule compared with HD. This method is not free from difficulties, though, because of the many possible routes available for depletion of HD through molecular formation. The reaction energies for the isotopic species are slightly different, by several tens of degrees (hundredths of an eV), which at interstellar temperatures produces substantial differences in reaction rates.[102] A detailed calculation of the isotope's mobility and reaction among the different species is required.

Both H_2 and HD are formed on grain surfaces but subsequent gas phase reactions that involve these molecules can enhance the D abundance in the more complex species that are formed, and also open up new channels. For example

$$H_3^+ + HD \rightarrow H_2D^+ + H_2,$$

which then competes with H_3^+ in the formation of hydrocarbons. This reaction is especially important because of the influence of cosmic ray ionization and dissociative recombination on the fractionation process. The ratio of H_3^+ to H_2D^+ is given by

$$H_2 + H(CR) \rightarrow H_3^+$$
$$H_3^+ + e \rightarrow H_2 + H$$
$$H_3^+ + HD \rightarrow H_2D^+$$
$$H_2D^+ + H_2 \rightarrow H_3^+ + H$$
$$H_2D^+ + e \rightarrow HD + H$$
$$\rightarrow H_2 + D,$$

where H(CR) represents a cosmic ray proton.

One well observed species, HCO^+, is observed to yield a higher than expected D/H ratio. This appears to be due to the protonation reaction chain:

$$H_3^+ + CO \rightarrow HCO^+ + H$$
$$H_2D^+ + CO \rightarrow HCO^+ + D$$
$$\rightarrow DCO^+ + H,$$

where both DCO^+ and HCO^+ are destroyed via dissociative recombination to yield CO.

Other means have been determined for solving for the electron density, for instance, the ratio of H_2O to HDO and H_2D^+ to H_3^+. These are all uncertain because of the incompleteness of the abundance catalogs for various clouds (see also Table 6.10); that is, many of the intermediate reactants are not well determined, and this hampers understanding of the conditions in the clouds.

Similar behavior is observed for the CNO isotopes; that is, their relative abundances in molecular combination depart from that expected on the based on their terrestrial abundance ratios and from nucleosynthesis. The effects are caused by fractionation reactions, usually through ion transfer reactions with CO. In general, the enhancement of isotopic abundances of the CNO group seem to be

---

[102] These small changes result from the dependence of the dissociation energy on nuclear mass. Because the reaction rates depend exponentially on this quantity, small changes in the D_0 value can produce large alterations in the isotropic fractionation.

TABLE 6.10 Sample Isotopic Abundances

Isotope	Local ISM	Solar System	Carbon Stars
$^{11}B/^{10}B$	3.4 ± 0.7	4.05	—
$^{12}C/^{13}C$	69 ± 6	89	> 30
$^{14}N/^{15}N$	388 ± 32	270	> 515
$^{16}O/^{18}O$	557 ± 30	490	> 300
$^{18}O/^{17}O$	3.6 ± 2	5.5	< 1
$^{29}Si/^{30}Si$	≈ 1.5	1.5	< 1
$^{32}S/^{34}S$	≈ 22	22	–

due to charge transfer reactions with isotopic ions. For example

$$^{13}C^+ + {}^{12}CO \rightarrow {}^{13}CO + {}^{12}C^+$$

will produce a net enhancement of ^{13}CO, since it will then free the carbon atom for other channels. At low density, where the molecules are exposed to ultraviolet as well as cosmic ray ionization, this is an important mechanism. The abundances of the CNO isotopes will therefore not uniquely reflect the initial conditions in the cloud, but depends instead on the processing that goes on during the time it takes for the molecules to come into equilibrium, about 10^6 yr.

Because of the importance of the CNO isotopes for the study of stellar evolution and of the chemical history of the galaxy, this is one of the most important areas for astrochemistry.[103] In addition, because of the dependence of these reactions on the ion fraction, which in turn depends on the electron fraction in the dense parts of the clouds, the isotopes become useful tools for studying the ionization of the environment, probing the electron density in portions of the cloud otherwise hidden from view.

6.6.2.7 Polycyclic Aromatic Hydrocarbons (PAHs) and Fullerenes
We have mentioned that the smallest grains behave like very large molecules, never coming into strict equilibrium with the radiation field and radiating in diffuse bands in the near infrared. While the precise nature of these molecules is controversial, there are several likely candidates, all *polycyclic aromatic hydrocarbons* (PAHs). These are either linear or ring molecules, which are very stable in the presence of ultraviolet radiation. One, coronene ($C_{20}H_{20}$), is especially well studied in the laboratory, although it does not appear to be a viable candidate for most of the optical DIBs. These molecules produce broad optical absorption bands, like the diffuse interstellar features, and also have vibrational transitions in the $3-10\mu$ region. They can be built in the process of forming dust grains, in the atmospheres of carbon-rich giants and supergiants, and also from the process of

---

[103] Wilson, T. L. 1999, *Rep. Prog. Phys.*, **62**, 143, from which Table 6.10 is partly abstracted. See particularly Sheffer, Y., Federman, S. R., Lambert, D. L., and Cardelli, J. 1992, *ApJ*, **397**, 482 on the determination of the $^{13}C/^{12}C$ ratio toward ζ Oph. For other isotopic studies, see, for example, Federman, S. R., Lambert, D. L., Cardelli, J. A., and Sheffer, Y. 1996, *Nature*, **381**, 764 and Lambert, D. L. et al. 1998, *ApJ*, **494**, 614 for the determination of the $^{11}B/^{10}B$ ratio in diffuse clouds, especially toward ζ Oph. Note especially the discussion of instrumental effects and their use of profiles of nonfractionated species.

grain photodestruction. At interstellar conditions, their partial pressures allow them to remain stable against evaporation. They should, however, also be strong absorbers in the UV; this contradicts many available observations that place tight upper limits (less than a few percent) on the absorption band strengths for these features (see below). On the other hand, the infrared *cirrus*, the ubiquitous diffuse emission detected by the IRAS mission throughout the Galactic plane, cannot now be explained any other way.

The unidentified diffuse features in the optical have infrared counterparts at 3.3, 6.2, 7.7, 8.6, and 11.3μ. These have been observed in emission in many astrophysically interesting environments, ranging from reflection nebulae to planetary nebulae. All of these have one common feature—the illuminating source possesses substantial flux in the middle ultraviolet, between 1100 and 3000 Å. The model predicts that the IR emission is due to C—C stretching vibrations for the 6.2 and 7.7μ bands, and to C—H bond stretching for the 11.3μ band. and other vibrational states of the complex of molecules that comprises the PAHs. Recent spectroscopic observations with ISO have confirmed that the diffuse IR emission resolves into a complex of lines whose intensity ratios approximately match the predictions of the PAH picture.

The *fullerenes* are another class of carbon molecule that shows extraordinary stability in the presence of strong UV irradiation. These molecules, of which C_{60} and C_{70} are the best known, are carbon cages in spherical or spheroidal configurations. They are observed in laser-generated soot residues in the laboratory, and some of their spectra match unidentified diffuse bands. More importantly, their shapes and carbon content make them contenders for astrophysical graphite, the component of the dust that is thought to produce the 2175 Å band. Although first detected in 1985, they have been difficult to characterize spectroscopically. Fullerenes are symmetric clusters of carbon that were first identified in soot and constitute stable structures that may be important in interstellar chemistry. The best studied C_{60}, is a spherical molecule. While a number of features of the extinction curve may be due to these particles, observations with ISO have yielded only upper limits on both C_{60} and C_{60}^+ emission in the 7–9μm region.[104] The same site for which the PAH emission features were first identified, the reflection nebula NGC 2023, an upper limit of only 0.25% with upper limits placed on the fullerene contribution from both emission and absorption spectra.

Why is this so important? First, the small grains, or big molecules, are likely fundamental to the understanding of the origin of the larger grains. They may account for as much as 20% of the carbon abundance in the interstellar medium, and, if this is correct, are an important element in the chemistry of molecular clouds. They are clearly necessary for explaining the infrared characteristics of the diffuse medium, wherein the cirrus is presumed to originate, and are basic to the understanding of nonequilibrium radiative processes in the gas.

The PAHs represent the prototypical small grain, dominated by single mode effects that produce a cascade of infrared vibrational emission lines following absorption of ultraviolet photons. Infrared observations, with ISO, have detected fine structure in several of the bands, which may indicate variable PAH composition throughout the medium. However, it is too early to begin identifying any of

---

[104] Mouton, C., Sellgren, K., Verstraete, L., and Léger, A. 1999, *Astr. Ap.*, **347**, 949.

the particular species. These species have been detected in the IR emission from the coma of comet Halley as well, and it appears that in at least some environments, the chemistry required to produce the PAHs is sufficiently efficient that large abundances can be achieved.[105] A remaining problem with the PAH explanation for the diffuse bands is the lack of an ultraviolet signature of these species. Complex organic molecules produce deep absorption bands in the vacuum UV, between 2000 and 3000 Å. To date, these have not been observed. An important more recent result is the discovery of deep $\lambda 4430$ absorption against SN 1987a in the Large Magellanic Cloud. This argues that at least this galaxy may have the same small grain component as our Galaxy, despite its lower-than-solar metal abundances. The greatest difficulty concerning the ultimate identification is that the physical state of the PAHs is unknown. Ionization and dehydrogenation (the systematic removal of hydrogen from the molecular structure) alters the vibrational frequencies and intensity ratios, making specific identification of single-carrier molecules virtually impossible.

6.7 DYNAMICAL GAS BAGS IN THE INTERSTELLAR MEDIUM

In this section, we will treat the physics of shock waves. We have invoked their influence repeatedly in the previous sections but usually only as a mantra. Nowhere are the effects of supersonic motions so important as in the disequilibrated environment of interstellar gas, and perhaps nowhere else do we have the chance to see so many of the processes in so much detail.

6.7.1 An Introduction to Shocks

A *shock* is a discontinuity in a physical variable within any flow. In an essentially inviscid gas, such as the interstellar medium, the physical state of the gas can change drastically within a distance that is short compared to the collision mean free path (or the thermalization length). In a compressional shock, the key physical feature is that the front propagates faster than a pressure wave can "warn" the upstream gas of its impending arrival. The gas then behaves as if it were hit ballistically by the front. We will not go into a detailed analysis of the fluid mechanics here, but instead treat the conservation problem of what happens across the front.

For the fluid equations, the conserved fluxes are the mass, momentum, and energy. Even though the shock front is a discontinuity, it is in steady state so the fluxes must be the same across the interface. The conservation conditions govern

---

[105] Webster, A. 1997, *MNRAS*, **288**, 221; Kroto, H. W. and Walton, D. R. M. eds. 1993, *The Fullerenes: New Horizons for the Chemistry, Physics, and Astrophysics of Carbon* (Cambridge, UK: Cambridge Univ. Press) [originally 1993, *Proc. Roy. Soc.* (London), **A343**, 1]. Salama, F. et al. 1999, *ApJ*, **526**, 265; Heymann, D. 1997, *ApJL*, **489**, L111; Léger, A., d'Hendecourt, L., and D'forneau, D. 1989, *Astr. Ap.*, **216**, 148; Omont, A. 1986, *Astr. Ap.*, **164**, 159; Allamandola, L. J., Tielens, A. G., and Baker, J. R. 1989, *ApJS*, **71**, 733; Pauzat, F., Talbi, D., and Ellinger, Y. 1997, *Astr. Ap.*, **319**, 318.

the *motion* of the front, although the detailed *structure* of the shock requires solving the full solution of the fluid equations.[106]

6.7.1.1 The Rankine-Hugoniot Conditions for Shocks

Let us examine the compressive case in which a piston moves supersonically into stationary gas. We will look at the planar problem. It captures all the relevant physics while greatly simplifying the mathematics. This also means that we can ignore the details of the flow and, consequently, of the time development of the front. For that part, we will need to examine the blast wave solutions in more detail, which we do in Appendix 6.C. Since we will be considering the motion in the frame of the shock front, we need describe the three quantities for the flow, all of which are constant when taken across the front, which we will denote by Σ. These are the mass flux

$$J = \rho \, \mathbf{v} \cdot \hat{\mathbf{n}}, \tag{6.129}$$

the momentum flux

$$M = P \cdot \hat{\mathbf{n}} + \rho v \mathbf{v} \cdot \hat{\mathbf{n}}, \tag{6.130}$$

and the energy flux

$$F = \rho \mathbf{v} \cdot \mathbf{n} \left(\tfrac{1}{2} v^2 + h + \phi \right), \tag{6.131}$$

noting that F/J is also constant. Here $\hat{\mathbf{n}}$ is the shock normal, h is the enthalpy, and ϕ is any energy that is added to or removed from the flow. Thus, denoting the preshocked gas as 0 and the postshocked gas as 1, we have

$$\rho_0 v_0 = \rho_1 v_1 \tag{6.132}$$

$$\rho_0 v_0^2 + P_0 = \rho_1 v_1^2 + P_1 \tag{6.133}$$

$$\frac{1}{2} v_0^2 + \frac{\gamma}{\gamma - 1} \frac{P_0}{\rho_0} = \frac{1}{2} v_1^2 + \frac{\gamma}{\gamma - 1} \frac{P_1}{\rho_1}. \tag{6.134}$$

From here on, since we are working in the frame of the shock front, we will use the notation $[F]_\Sigma = F_1 - F_0$ to denote the jump in any flux on passing across the shock surface. To keep things manageable, we will concentrate on one-dimensional flows, but this doesn't prevent us from looking at strictly radial fronts such as cylinders and spheres. The transverse component of the flow is simply handled because it does not change. There is no jump in this direction, and we will briefly come back to oblique shocks at the end of this section.[107]

---

[106] Among references to consult, see Landau, L. and Lifshitz, E. 1987, *Fluid Mechanics*, 2nd ed. (Oxford: Pergamon); Dyson, J. E. and Williams, D. A. 1997, *Physics of the Interstellar Medium*, 2nd ed. (Bristol: IOP Publishers), Shu, F. H. 1992, *The Physics of Astrophysics*: vol. 2. *Fluid Dynamics* (Mill Valley, CA: Univ. Science Books); Shore, S. N. 1992, *An Introduction to Astrophysical Hydrodynamics* (San Diego: Academic Press); Millar, T. J. and Raga, A. C., eds.) 1995, *Shocks in Astrophysics* (Dordrecht: Kluwer).
[107] If we have a discontinuity in any one of the three *physical* variables (ρ, v, P), the other two can be given in turn because of the conservation condition for the *fluxes*.

Assume that the scale for variation of the strength of the shock along the incident surface is large so that it can be taken as a planar discontinuity in the normal direction. Comoving with Σ at velocity V_Σ, you see the upstream gas flowing toward you with a normal speed v_0. Behind the shock, the flow decelerates in the Σ frame, like cars passing a traffic accident. This produces compression of the gas and from Eq. (6.132) we obtain

$$v_1 = \frac{\rho_0}{\rho_1} v_0. \tag{6.135}$$

The front itself does not gain mass.[108] Conservation of momentum flux, Eq. (6.133), gives

$$\rho_0 v_0^2 + P_0 = \rho_1 \frac{\rho_0^2}{\rho_1^2} v_0^2 + P_1. \tag{6.136}$$

The combined mass and momentum flux equations alone provide what is known as the *shock adiabat*, so named because we have ignored changes in the energy:[109]

$$J^2 = \frac{P_1 - P_0}{V_0 - V_1} > 0. \tag{6.137}$$

Here J is the (constant) mass flux across Σ. The density, ρ_i, is the inverse of the specific volume, V_i, so this quantity resembles a thermodynamic process, although the change is discontinuous. Including the energy flux condition closes the system of equations and allows us to look explicitly at what happens to the density:

$$\frac{\rho_1}{\rho_0} = \frac{(\gamma + 1)P_1 + (\gamma - 1)P_0}{(\gamma - 1)P_1 + (\gamma + 1)P_0}. \tag{6.138}$$

This is one of the central results obtained using the *Rankine-Hugoniot conditions*. Given a discontinuous change in *either* the density or pressure, in the equation of state, or in the specific heats, it is possible to specify the change in any other thermodynamic variables across the front. Notice that for a strong adiabatic shock, in the limit $P_1 \gg P_0$, we find an upper limit for the compression:

$$\frac{\rho_1}{\rho_0} = \frac{\gamma + 1}{\gamma - 1}. \tag{6.139}$$

---

[108] This means that the front is taken to be infinitesimally thin. Once the gas has been overrun by Σ, it is accelerated impulsively and, for an external stationary observer, moves with a speed $V_\Sigma - v_1$. Remember that everything we are doing here is in the comoving system so a Galilean transformation is required to put this in terms that you, as the observer of the phenomena, would detect from Earth. For instance, for an H II region, the total amount of ionized gas behind the front increases with time as the region expands.

[109] See, for example, Zel'dovich, Ya. B. and Raizer, Yu. P. 1966, *Physics of Shock Waves and High-Temperature Hydrodynamic Phenomena* (NY: Academic Press); Landau, L. and Lifshitz, E. 1987, *Fluid Mechanics*, n.106.

However, this applies only in the adiabatic limit for an ideal gas. For isothermal shocks, the compression may be much greater because of the reduced restoring pressure produced by cooling.[110]

Suppose that instead of using the pressure ratio as the observable parameter describing the shock you want to know the jump conditions in terms of the Mach number, M, of the incident shock. One reason you might want to do this is that blast wave solutions provide M_0 from dimensional analysis alone but tell you nothing about the initial conditions at the shock front. The Mach number is the ratio of the shock speed to the *local* sound speed, and is the usual way of describing the *strength* of the shock. We can now find the Rankine-Hugoniot conditions in terms of M_0, the Mach number of the preshock flow provided we assume that γ is constant. However, we don't need to make any other assumptions about the equation of state. First, use the sound speed to relate the state variables, $c_{s,i}^2 = \gamma P_i/\rho_i$, for side i of the front. The momentum and enthalpy equations become

$$\left[\rho\left(v^2 + \frac{c_s^2}{\gamma}\right)\right]_\Sigma = 0, \tag{6.140}$$

$$\left[\frac{1}{2}v^2 + \frac{c_s^2}{(\gamma-1)}\right]_\Sigma = 0. \tag{6.141}$$

The preshock Mach number combines the front velocity, $v_0 = V_\Sigma$, with the upstream sound speed, $c_{s,0}$, $M_0 = v_0/c_{s,0}$. There is no change in the mass continuity condition. Solving for $c_{s,1}^2/c_{s,0}^2$, you will find that

$$\frac{\rho_1}{\rho_0} = \frac{(\gamma+1)M_0^2}{(\gamma-1)M_0^2 + 2} = \frac{v_0}{v_1}. \tag{6.142}$$

For the strong shock limit, $M_0 \gg 1$, we recover the previous results for the Rankine–Hugoniot relations.

We can also derive the pressure ratio by noting that, given the ratio of sound speeds, we have

$$\left(\frac{c_{s,1}}{c_{s,0}}\right)^2 = \left(\frac{P_1}{P_0}\right)\left(\frac{\rho_0}{\rho_1}\right), \tag{6.143}$$

from which it follows that

$$\frac{P_1}{P_0} = \left(\frac{\rho_1 M_0}{\rho_0 M_1}\right)^2. \tag{6.144}$$

---

[110] For an ideal gas $\gamma = \frac{5}{3}$ so that $\rho_1/\rho_0 = 4$ in the strong shock limit. The maximum compression is also greater for a lower γ, which is why the isothermal strong shock limit is so compressive (since $\gamma \to 1$).

Then the density can be eliminated using

$$\frac{c_{s,1}^2}{c_{s,0}^2} = \frac{1}{2}\left(1 - \frac{\rho_0^2}{\rho_1^2}\right)(\gamma - 1)M_0^2 + 1. \tag{6.145}$$

We could have used the equation for the density ratio as a function of pressure to derive the pressure ratio as a function of the incident Mach number. The same is true for the ratio of velocities on either side of the shock, which is, in fact, a way of getting M_1 as a function of M_0. Keep in mind that it is indeed possible to derive them, although we often don't know the Mach number from direct observations. Usually we have the velocity, but that is a projected component, and from line observations, specifically related to emission and ionization state of the gas, what we are able to infer using these equations is the intrinsic Mach number without reference to the observed dynamics. Finally, in addition to the density condition we have derived above, we collect the jump conditions written in terms of the upstream Mach number. For the pressure, we have

$$\frac{P_1}{P_0} = \frac{2\gamma M_0^2 - (\gamma - 1)}{\gamma + 1}. \tag{6.146}$$

The temperature ratio T_1/T_0 for an ideal gas is the same as $(c_{s,1}/c_{s,0})^2$, so from Eq. (6.145):

$$\frac{T_1}{T_0} = \frac{[2\gamma M_0^2 - (\gamma - 1)][(\gamma - 1)M_0^2 + 2]}{[(\gamma + 1)M_0]^2}. \tag{6.147}$$

It is also useful to know how the Mach number is changed by passage of the gas through the front. A second shock that might collide with the gas after the first passes will have its strength, for instance, measured relative to this new value of the sound speed. The mass conservation condition allows us to obtain the post-shock Mach number:

$$M_1^2 = \frac{(\gamma - 1)M_0^2 + 2}{2\gamma M_0^2 - (\gamma - 1)}. \tag{6.148}$$

Notice that because of the heating and compression in an adiabatic shock, the Mach number of the flow is *lower* after the gas passes through a shock front.

6.7.1.2 Magnetic Shocks

Magnetic fields introduce a fundamentally new feature to the jump conditions compared with the field-free case. For purely hydrodynamic normally incident shocks, you need to consider only the quantities parallel to the flow. For MHD shocks, however, the coupling between the fluid motion and the magnetic field generates an electric field that is orthogonal to both. The velocity components will be written as $\mathbf{v} = (u_n, u_t)$, where u_t is the transverse component to the surface, and we will take $\mathbf{v}_0 = (u_{n0}, 0)$. For $\mathbf{B} = (B_n, B_t)$ we will assume initially that $\mathbf{B}_0 = (B_n, 0)$. Since in the conservation equations the fluxes are divergenceless in a steady-state shock, the Maxwell equation $\nabla \cdot \mathbf{B} = 0$ translates into the jump condition

$$[\mathbf{B} \cdot \hat{\mathbf{n}}]_\Sigma = 0 \to B_{n1} = B_{n0}, \qquad (6.149)$$

and considering the transverse electric field component provides

$$[(\mathbf{v} \times \mathbf{B}) \times \hat{\mathbf{n}}]_\Sigma = 0. \qquad (6.150)$$

Thus, even if the magnetic field is initially parallel to the flow, Eq. (6.150) shows that a component parallel to the front can appear in the postshocked gas. This is called a *switch-on* shock since B_t is not zero in the postshock region even if it is in the preshocked gas. The converse can also occur, called a *switch-off* shock. We will examine this now because of its similarity to the hydrodynamic problem of a normal incidence flow at Σ. In our usual notation

$$[\rho u_n]_\Sigma = 0$$

$$[u_n B_t - u_t B_n]_\Sigma = 0$$

$$\left[\rho u_n^2 + P + \frac{1}{8\pi}B_t^2\right]_\Sigma = 0 \qquad (6.151)$$

$$\left[\rho u_n u_t - \frac{B_n B_t}{4\pi}\right]_\Sigma = 0$$

$$\left[\rho u_n \left\{\frac{1}{2}(u_n^2 + u_t^2) + \Gamma\frac{P}{\rho} + \frac{B_t^2}{4\pi}u_n\right\} - \frac{B_n B_t}{4\pi}u_t\right]_\Sigma = 0.$$

Notice that B_n is constant, and we can write $\rho u_n = J$, which is also constant, so we find that

$$\left[u_t - \frac{1}{4\pi J}B_t B_n\right]_\Sigma = 0 \qquad (6.152)$$

along with

$$\left[\Gamma\frac{P}{\rho} + \frac{1}{2}\frac{J^2}{\rho} + \frac{1}{2}u_t^2 + \frac{B_t^2}{4\pi\rho} - \frac{B_n B_t}{4\pi J}u_t\right]_\Sigma = 0. \qquad (6.153)$$

Therefore we find for u_t

$$u_t = \frac{B_n B_t}{4\pi J} = \frac{JB_t}{B_n}\left[\frac{1}{\rho}\right]_\Sigma.$$

Now we use the same substitution as for the nonmagnetic case:

$$\frac{1}{2}\left(\frac{1}{\rho_1} + \frac{1}{\rho_0}\right)\left[\frac{J}{\rho}\right]_\Sigma = \frac{1}{2}\left[\frac{J^2}{\rho^2}\right]_\Sigma,$$

but we also equate the two solutions for u_t to obtain

$$J^2 = \frac{B_n^2}{4\pi}\left[\frac{1}{\rho}\right]_\Sigma.$$

We can therefore write

$$\left(\frac{1}{\rho_1} - \frac{1}{\rho_0}\right)\left(\frac{1}{2}J^2\left(\frac{1}{\rho_1} + \frac{1}{\rho_0}\right) + \Gamma P_0 - \frac{\Gamma}{\rho_1}J^2 + \frac{B_t^2}{4\pi}\right) - \frac{\Gamma}{\rho_1}\frac{B_t^2}{8\pi} = 0.$$

Again notice that since the incident flow is parallel to B_n, and we assume that there is no incident transverse magnetic component, the shock equations simplify considerably: B_t and u_t are nonvanishing only in the postshocked gas. We'll further simplify the notation using $r \equiv \rho_1/\rho_0$. One is the usual Rankine–Hugoniot jump condition without any magnetic field. The other is one for the induction of a postshock transverse component even where one didn't exist at the start; the magnetic field *and* the flow refract after normal passage through the shock front. A useful little result from these manipulations is that

$$\frac{r}{1-r} = \frac{1}{M_{A0}^2},$$

where $M_A = v/v_A$, which follows from the substitution for J^2 above and also requires

$$u_t = \frac{B_{t2}}{B_n}[u_n]_\Sigma. \tag{6.154}$$

It may seem bizarre that normal incidence leads to a deflected flow, but this is due to the continuity condition on the electric field. Why should we be interested in this type of shock, especially for the interstellar medium? You know that much of the gas is ionized and magnetized, so any induced motions in the gas will ultimately be coupled to the magnetic field. The magnetic field induces an electric field that creates a current sheet. This deviates the postshock flow and induces a transverse B field. The total field will therefore be more parallel to the shock front, and this should have observable consequences in the low-density gas of the ISM. The jump conditions lead to

$$B_{t,1}^2 = 2(r-1)B_n^2\left[\frac{(\gamma+1) - r(\gamma-1)}{2} - \frac{4\pi}{B_n^2}\gamma P_0\right], \tag{6.155}$$

and a real solution is guaranteed if

$$r \leq \left(\frac{\gamma + 1}{\gamma - 1}\right) - 8\pi\Gamma\frac{P_0}{B_n^2}. \qquad (6.156)$$

So this means that we have a nonvanishing postshock component of B_t even if the preshock field is normal to Σ. This is the *switch-on* condition. As we mentioned, this results from the induction of an electric field because of the fluid motion. There is also a symmetric version of the shock, the *switch-off* case, in which an initially oblique field becomes normal to the shock front.

Interstellar magnetic shocks present a complex physical situation. On one hand, there are discontinuities is the pressure through the **B** field. On the other, only the ionized matter couples to the field directly and collisions must then bring along the neutrals. This depends on the velocity of the front and the Alfven speed of the medium. The nomenclature reflects this. So-called *C shocks* ("continuous") display no jump in the neutral particle density but have a shock in the ions, when $v_A > v_\Sigma$. The neutrals, strictly speaking, do not have an associated Alfven speed because they can slip past field lines. They are, however, coupled to the ions by collisions and therefore contribute to the inertia of the charged magnetized fluid.

For dust grains that are charged by photoionization and ion–grain collisions, the effect of shock passage is to enhance grain destruction. The dust couples to the magnetic field, but the neutrals move supersonically past the grains. For relative speeds of order 10 km s^{-1}, which is typical of strong shocks within molecular clouds, the neutrals have energies of several eV and can produce sputtering (the glow of the *Space Shuttle*, e.g., is due partly to excitation of neutrals in the Earth's exosphere via collisions with the spacecraft). Laboratory experiments show that ices have sputtering energies of order 0.5 eV, so this relative speed is sufficient to produce grain destruction and return of the molecular products back to the gas phase.[111]

6.7.1.3 Magnetic Shocks in a Partially Ionized Medium: J and C Shocks

Discontinuities in the interstellar gas occur in many environments, some of which do not satisfy the flux freezing condition we've just employed. For instance, consider a partially ionized medium. The neutrals and charged atoms or molecules are really the same particles but at different instants in their ionization. As we saw for ambipolar diffusion, these species act as two separate fluids that can slip past each other and are coupled only through some form of drag resulting from the

---

[111] There are several excellent reviews of magnetic shocks, especially Boyd, T. J. M. and Sanderson, J. J. 1969, *Plasma Dynamics* (NY: Barnes and Noble). The most precise and detailed discussions of the subject are still Anderson, J. E. 1963, *Magnetohydrodynamic Shock Waves* (Cambridge, MA: MIT Press) and Tidman, D. A. and Krall, N. A. 1971, *Shock Waves in Collisionless Plasmas* (NY: Interscience). For the astrophysical literature, see Kaplan, S. 1966, *Interstellar Gas Dynamics* (Oxford: Pergamon Press), which contains a thorough discussion of the two extreme cases of parallel and perpendicular propagation, but uses rather antique terminology. A more recent comprehensive review is provided by Markovskii, S. A. and Somov, B. V. 1996, *Space Sci. Rev.*, **78**, 443. See also Priest, E. 1986, *Solar Magnetohydrodynamics* (Dordrecht: Reidel); Millar, T. J. and Raga, A. C., eds. 1995, *Shocks in Astrophysics* (Dordrecht: Kluwer); Chernoff, D. F. 1987, *ApJ*, **312**, 143; Draine, B. and McKee, C. 1993, *ARAA*, **31**, 373; Wardle, M. 1990, *MNRAS*, **274**, 1002. A comprehensive review of the Shuttle glow in Murad, E. 1998, *Ann. Rev. Chem.*, **49**, 73.

ionization and recombination within the gas. The treatment resembles an accretion flow for the steady state.[112] There are two classes of shocks provided the speed of the pressure disturbance is slower than the magnetoacoustic speed, $(v_A^2 + c_s^2)^{1/2}$. Remember, the ions respond to the field; the neutrals don't. Were these two fluids to be truly separate, the ions would "get wind" of the impending collision with the front ahead of the event while the neutrals would not. In fact, since these are really the same species in different guises, the neutrals are coupled to the ions and partially compress and slow down even without being coupled directly to the Alfven waves. Once the front actually hits, they *may* undergo a true shock compression, albeit from an already preconditioned state, or if the field is strong enough and the coupling large enough, they may simply follow the ionic fluid and experience no discontinuity at all. The terminology that is used to distinguish the two classes of events is J for those cases that actually produce a discontinuity ("jump") for the neutrals and C for those that result in a continuous change in the state of the fluid and a net increase in the entropy of the post-shocked gas without a jump. So if the medium is only partly ionized, the ions can couple to the faster waves that travel ahead of the sound waves and compress more slowly.

6.7.1.4 Radiative Shocks

Line emission, mainly from atomic species, is the main energy loss mechanism for interstellar shocks. In dense regions, like dark or translucent clouds, molecular species may dominate, especially H_2. In fast shocks, or in lower density environments, permitted resonance lines dominate. These have large transition probabilities and often, because of collisional excitation, can be populated after the shock has passed through the medium. For temperatures below 10^4 K, the primary losses are due to weak hyperfine atomic lines and vibrational molecular lines. However, at temperatures exceeding ~ 10000 K, $H\alpha$ is collisionally excited and swamps all other loss mechanisms. Its de-excitation produces a cascade, from $n = 3$ to $n = 2$, thence to $Ly\alpha$. The $H\alpha$ line is almost always optically thin even if the Lyman lines are very opaque. Coming as it does from the most abundant atomic species, it is intense and produces significant energy losses that increase with increasing temperature. In fact, there is a local maximum in the cooling function, Λ, at this temperature primarily because of this line. The Lyman α emission line is often suppressed because of the optical depth in the medium—it is a ground-state transition and consequently often thick and ineffective without many scatterings, so the timescale for thermal loss is increased by the number of scatterings. A simple rule of thumb is that the cooling time increases as the number of scatterings in the medium increases, which is proportional to $\tau^{1/2}$, where τ is the optical depth. Unless there are large velocity gradients present, the photons from the ground-state Lyman series will be trapped within the gas and only slowly leak out; the Balmer series can immediately escape the medium.

The cooling function for the postshocked gas is determined by the elemental abundances and speed of the shock. The stronger the shock (i.e., the higher the Mach number), the stronger the emission. The energy of the postshocked gas decreases as it flows away from the front and radiates. This can be expressed in the

---

[112] Mullan, D. J. 1971, *MNRAS*, **153**, 145; Draine, B. T. 1980, *ApJ*, **241**, 1021; Chernoff, D. F. 1987, *ApJ*, **312**, 143; Draine, B. T. and McKee, C. F. 1993, *ARAA*, **31**, 373.

frame of the shock as a change in the thermal parameters for the gas as a function of distance:

$$\frac{d}{dx}\left[\rho v \frac{1}{2}v^2 + \frac{\gamma}{\gamma-1}\frac{P}{\rho} + \mathscr{U} + \frac{B^2}{4\pi}v\right] + n^2\mathscr{L} = 0. \qquad (6.157)$$

Here $\mathscr{U}$ is the specific internal energy and $\mathscr{L}$ is the cooling function.

6.7.1.5 Shock Precursors
The temperature and density in the postshocked region determines the emission spectrum from the gas. Ultraviolet continuum emission may result. if the excitation is great enough. This is the *precursor*, so called because it stands ahead of the shock at distances that depend on the optical depth of the surrounding gas. This radiation may just excite the preshocked gas, or it may actually ionize it *before* the front sweeps it up. For strong explosions in air, for example, for nuclear weapons, the fireball is the result of the precursor. It ionizes the surrounding medium to a very large distance. But because it is an impulsive emission, resulting from the first microseconds of the explosion, it rapidly recombines. The shock isn't hot enough after the first few seconds to maintain the ionization. At this point, the fireball clears and the material blast wave breaks through. In fact, in such an explosion, what you see is two competing effects. One is the fireball recombining and shrinking from its maximum size (which is established "instantaneously" because the timescale is too small for terrestrial explosions) and the other is the expansion of the blast wave. The debris follows the blast.

For magnetic shock waves, because of the additional pressure from the magnetic field, the pressure and density jumps can be offset. The shock may be supersonic, but the magnetoacoustic speed may be great enough that the jump is continuous. In this case, there is enhanced emission because of the increase in the density and energy in the postshocked gas, but the gas shows the enhanced emission over a large region of space (if it is spatially resolved). For shocks in low-density environments, the statistical equations must be solved to determine the emergent intensities.

The run of temperature and pressure behind the front is more complicated for astrophysically formed plasmas than for most laboratory ones. The medium is very tenuous, and collisional equilibrium is not established quickly. Since LTE doesn't hold, the radiation may come from states whose populations differ dramatically from those encountered in terrestrial shocks. The temperature, for example, does not necessarily depend directly on the rate of energy loss. If collisions are responsible for the deexcitation and subsequent losses, then the electron and radiation temperatures will be essentially the same. But if the gas is losing energy primarily through lines, the electrons may be much hotter than the cooling rate would indicate. Also, if the collisions are infrequent enough, two temperature instabilities may come into play.

Another important effect in astrophysical shocks is charge exchange. This is neutralizing but leaves the hydrogen atom in an excited state:

$$p + A \rightarrow H^\star + A^+,$$

where A is a neutral atom and p is a cosmic ray or high-energy proton. The subsequent loss of energy is ultimately a sink for the material and an additional cooling agent. In short, there are many avenues by which the gas can lose energy radiatively, and each shock environment affects the rate of loss and the thermal profile of the postshocked gas differently.

Lines sometimes appear that would not normally be observed from photoexcited regions. The difference between a shocked layer and an H II region at the same temperature is essentially one of density. In a shock, the density jump is compressive; in an H II region, it is just the opposite. So collisionally excited ionic lines are the signature of a shocked environment.

The precursor is also the distinguishing feature of magnetic shocks. Since the Alfven speed is generally much larger than the sound velocity, the charged gas can be compressed while the neutrals slip through the discontinuity and only slowly come into equilibrium with the heated ions. This is a *C shock*. The detailed structure of the shock in a multiphase medium therefore depends on the speed of the discontinuity. If there is no magnetic field, only a radiative precursor occurs. As the magnetic field strength increases, the upstream propagation of MHD waves becomes more important. The ions are heated before the neutrals and therefore experience a weaker shock. The condition is expressed in terms of the shock speed, v_Σ versus the magnetosonic speed $v_{ms} = (v_{A,0}^2 + c_{s,0}^2)^{1/2}$: if $v_\Sigma > v_{ms}$, then a discontinuity is felt by the neutrals but not by the ions. In the strong-field limit, where $v_A \gg c_s$, when $v_\Sigma < v_{ms}$, the ions and neutrals are continuously heated without an actual discontinuity. In the intermediate case, there is a finite region over which the neutrals and ions come into thermal equilibrium after the front passes. This has a profound effect on the chemistry in the postshock region because of the very high temperatures the neutrals may have relative to the ions.

6.7.1.6 Shock Chemistry

Hydrodynamic and magnetohydrodynamic (MHD) shocks are important in the time-dependent chemistry of the diffuse interstellar medium. The timescales are very short for shock passage through a region, typically less than 10^5 yr/pc, but they can cause considerable abundance variations and produce long-lived products. CH^+ is primarily produced this way, and CH and CN also seem to have input from shocks.

The role of shocks in the chemistry of clouds is best seen in the effect it can have on molecular hydrogen.[113] While many reaction products remain unchanged in abundance, CH_4, H_2O, and HCO can be greatly enhanced due to the increased production of these molecules in the hotter, and denser, shock environments. The chemistry is also dependent on the role of magnetic fields. Magnetic shocks can have sizable compression without significant increases in the temperature due to the pressure provided by the magnetic field. As a result, the relative abundance of shock-produced molecules serves as a probe of the nature of the shock producing the enhancement in the reaction rates.

Chemistry in the post-MHD shock environment is dominated by the separation between neutral and ionic species; the former are less affected than the latter by

---

[113] See, for instance, Brand, P. in 1996, *Shocks in Astrophysics*, Millar, T. and Raga, A., eds. (Dordrecht: Kluwer).

the magnetic field. In consequence, the reaction sites are calculated to show abundance stratification depending on the reaction channels. Since the fronts may be broad enough to be spatially resolved, it is possible to study the diffuse phase ISM shock chemistry observationally in some detail.

6.7.2 Ionization Fronts and Photodissociation Regions

The general theory for ionization discontinuities was first developed by Kahn.[114] The picture follows naturally from considering what happens *within* an H II region. There is a state change in the gas, and the mean molecular weight per particle changes because of the release of electrons, so the pressure and internal energy also are different. The transition is very sharp, essentially the mean free path for the Lyman continuum photons. We can codify this by noting that an H II region grows strictly by consumption of surrounding neutral gas. Once the H II region starts to expand, new material is ionized as it is engulfed. We can say that the previously neutral gas has now entered the front, at a rate determined by the flow of ionizing photons through the medium. Therefore, there is a fixed value of J, the mass flux, which solves the mass conservation condition across the ionization front

$$\rho_1 v_1 = \rho_0 v_0 = \mu_I J_I, \qquad (6.158)$$

where J_I is the flux of photons through the planar front and μ_I is the mean molecular weight of the newly ionized gas. It is this factor that contributes a flow of *ionization* through the front which looks like a mass current. The electrons contributed to the postshocked gas by this ionization cause the medium to heat up and start expanding. We will shortly see what happens once the expansion starts. The momentum conservation condition is unchanged and we can write it in terms of the sound speed on the two sides of the front:

$$\rho_0 \left(\frac{c_{s,0}^2}{\gamma} + v_0^2 \right) = \rho_0 \left(\frac{c_{s,1}^2}{\gamma} + v_1^2 \right). \qquad (6.159)$$

The energy flux condition requires that we add the heating due to absorption of photons with energies above the Lyman limit, which depends on the spectrum of the central source. This quantity, $v_e^2 \equiv 2(E - \chi_H)/m_e$, is the excess (thermal) kinetic energy for the electrons inside the H II region. Thus, Eq. (6.134) becomes

$$\frac{1}{2} v_0^2 + \frac{c_{s,0}^2}{\gamma - 1} = \frac{1}{2} v_1^2 + \frac{c_{s,1}^2}{\gamma - 1} + \frac{1}{2} v_e^2. \qquad (6.160)$$

If we did not have this excess energy, we would have the simple shock solution, and the solution would now be in hand. But because the gas can be heated, there is a nonlocal source for energy, hence pressure, that regulates the speed of the front, v_0. Rather than treat the a mono-atomic ideal gas ($\gamma = \frac{5}{3}$), we use this generalized equation to emphasize that the same formalism applies to *dissociation*

---
[114] Kahn, F. 1954, *Bull. Astr. Inst. Neth.*, **12**, 187.

and *ionization* fronts. Remember, even though the ultraviolet radiation from the central sources in an H II region may be completely absorbed within the Lyman continuum, molecular dissociation requires lower energy photons that will leak through the front, forming a precursor. Consequently, you expect to see a transition region between the cold molecular gas of a surrounding cloud and the ionized zone that contains a warm neutral phase, This region cannot be easily studied with 21-cm emission, in part because it is too dense and warm and in part because it is very small. But several species, notably C II, emit in this region through fine-structure transitions in the far infrared (see Table 6.2). These can be used to study the destruction of the cloud affected by photoionization processes. Because the same photons that ionize the metals are absorbed by dust and produce infrared emission by heating the grains, there is a good observed correlation between the line emission and L_{FIR}, the total far infrared luminosity as defined by the IRAS and COBE satellites.

Now back to our solution for the ionization front. Rewrite Eq. (6.160) as a quadratic for the postshock velocity:

$$v_0 v_1^2 - \left(v_0^2 + \frac{c_{s,0}^2}{\gamma}\right) v_1 + \frac{c_{s,1}^2}{\gamma} v_0 = 0,$$

so since only real roots are allowed for v_1, there is a critical speed, $v_{0\star}$, for the propagation of the ionization front when the determinant vanishes:

$$v_{0\star} = \gamma^{-1/2} \left[c_{s,1} \pm \left(c_{s,1}^2 - c_{s,0}^2 \right)^{1/2} \right]. \tag{6.161}$$

There are two solutions for v_0: (1) one for $c_{s,1} \gg c_{s,0}$

$$v_{0\star} \approx 2\gamma^{-1/2} c_{s,1}, \tag{6.162}$$

which is called a *rarefaction* or *R-front*, and (2) the other when $c_{s1} \approx c_{s,0}$:

$$v_{0,\star} \approx \frac{c_{s,0}^2}{2\gamma^{1/2} c_{s,1}}, \tag{6.163}$$

called a *dense* or *D*-front. The terminology is from Kahn's original study. The supersonic motion of an *R*-front is characteristic of a champagne flow while a *D*-front is the sort expected for a stalled shock since it moves subsonically into the surrounding medium. Then the postshock velocity, v_1, is obtained from the energy equation. Notice that the temperature is initially very high immediately inside the front because of the release of electrons by ionization but these particles quickly thermalize because of collisions with the gas inside the H II region.

A similar process to ionization occurs in the molecular gas, where it makes a transition to the atomic phase. For H_2, however, there is a difference. The ground state of the molecule is $^1\Sigma_g^+$, while the dissociated state is $^3\Sigma_u$. This transition is forbidden by normal selection rules so direct dissociation does not occur. Absorption from the ground state can take place through the UV bands, the Lyman and Werner series, that lie shortward of 2000 Å. These transitions, to the $^1\Pi$ state, then radiatively de-excite into the dissociative continuum. Since an H II region is optically thin at these wavelengths, while the surrounding molecular gas is quite opaque, the balance between neutral and molecular hydrogen is dominated by this process. In addition, the same wavelengths are able to ionize neutral atomic metals and charge transfer will also occur here. These combined effects produce a region that is analogous to an H II region that expands into the molecular gas ahead of the ionized gas, which, as we will see, ultimately overtakes it and forms a compressional shock directly from the molecules to the ions (recall Fig. 6.3).[115] Because of the difference in the radiative mechanism, photodissociation regions require a treatment different from ionization fronts. For H II regions, the continuum covers a broad range of frequencies and the dominant opacity is a power law. On the other hand, since the destruction of H_2 requires line transfer, a PDR is much like a stellar atmosphere. The incident flux from the star pumps the bands that, on decaying, produce the destruction. Unlike H II regions, PDRs will display a broad temperature range because of the combined effects of opacity and chemistry. It is important to keep the effect of photodissociation in mind when treating the evolution of star forming regions. Such regions also form around stars that have only weak H II regions. The complete computation of the rates depends on the initial mass function. In other words, a PDR surrounds even relatively low-mass stars.[116] Even if OB stars do not form in a cloud, the intermediate mass stars, $2-10$ $M_\odot$, are more likely to be formed and will be able to destroy substantial environmental masses of coolants and feed back into subsequent star formation.

6.7.3 Static H II Regions: Strömgren Spheres

When a massive star turns on, its ultraviolet continuum propagates into the surrounding ISM faster than the medium can dynamically respond. This creates a more or less spherical ionized region that remains static for some time, known as a *Strömgren sphere*. Such a region can't remain stationary forever since it is hot and overpressured relative to the enveloping gas, but it presents one of the simplest cases that we can evaluate of structure induced by the interaction of stars with their environments. Such regions are also where stellar atmospheres meet the interstellar medium. Ionization releases high-energy electrons from atoms that collide with the neighboring atoms. The simple radiative balance argument that would normally produce low temperatures for a body at some great distance from an illuminating source does not work when ionization is included. This is because there is another channel to which the gas can redistribute the energy. Prompt emission of an electron in an ionizing absorption depends on the spectrum

---

[115] A comprehensive review of photodissociation regions and related processes in cloud peripheries is Hollenbach, D. J. and Tielens, A. G. G. M. 1999, *Rev. Mod. Phys.*, **71**, 173. For the dynamical evolution of such regions, see Bertoldi, F. and Draine, B. T. 1996, *ApJ*, **458**, 222.

[116] See Bertoldi, F. and Draine, B. 1996 (footnote 115) and Diaz-Miller, R., Franco, J., and Shore, S. N. 1998, *ApJ*, **501**, 192 for more details on the dependence of PDR mass on stellar effective temperatures.

of the incident radiation, not on its intensity, and the electron is ejected with whatever energy exceeds the ionization threshold.

The emission rate for ionizing photons depends on the effective temperature of the star through its monochromatic luminosity, L_ν, and also on the spectral energy distribution relative to the continuous absorption coefficient of the gas. However, for hydrogen, the problem is especially simple. We look at how many continuum photons are available above the ionization limit and assume that each of them is capable of producing one ionization before, through recombination, the photon energy is degraded to Lyα and not capable of a second ionization:

$$\dot{N}_{\text{LyC}} = \int_{\nu_{\text{LyC}}}^{\infty} L_\nu \frac{d\nu}{h\nu}. \quad (6.164)$$

Should we be describing a cluster instead of a single star, L_ν would be replaced by a mass function weighted luminosity, $\int \phi(m) L_\nu(m)\, dm$. The recombination rate depends on the temperature of the region because of the collision cross section, but we will approximate this by a single recombination coefficient, α, and write the rate as

$$\dot{N}_{\text{rec}} = \frac{4\pi}{3} n_e n_p \alpha R_0^3. \quad (6.165)$$

For a pure hydrogen plasma, we can take $n_e = n_p$ so that

$$R_0 = \left(\frac{\dot{N}_{\text{LyC}}}{\frac{4\pi}{3}\alpha} \right)^{1/3} n_e^{-2/3}. \quad (6.166)$$

This assumes that the timescale is short enough for ionization that every recombination is followed by a single ionization (of course, in a pure hydrogen plasma it cannot be otherwise). Were there too few photons, the region would shrink in size and, of course, the lower the density the larger the region. This radius, the characteristic size of a Strömgren sphere, is the minimum size that an ionized region has. Notice that we have written it as independent of metallicity, other than the fact that a higher metal abundance will slightly increase the number of free electrons, but it is possible to produce a region of just ionized metals in detail even if the hydrogen remains neutral. This is often seen in photodissociation regions. You notice also that there is a characteristic timescale for the growth of a Strömgren sphere, the recombination time. The shorter this is, the slower the expansion of the ionization front. This is because the *static* H II region grows conserving numbers of photons. Its *volume* is thus a linear function of time, so $R(t) \sim t^{1/3}$. It establishes itself virtually instantly once the central star turns on. It also heats up at the same time because ionization removes most of the primary coolants for the gas. (See also Table 6.11.[117])

6.7.3.1 Time-Dependent Strömgren Spheres

Normally, the thermal timescale for an H II region is short compared with the lifetime of the illuminating star. This steady-state approximation breaks down in

TABLE 6.11 Ionizing and Dissociating Fluxes for Hot Stars

Spectral Type	$T_{\rm eff}$	log g	R ($R_\odot$)	Q_0	Q_1	Q_2
O3 V	51,230	3.9	13.2	49.899	49.353	44.984
O4 V	48,670	3.9	12.3	49.735	49.130	44.394
O6 V	43,560	3.9	10.7	49.371	48.637	43.018
O8 V	38,450	3.9	9.3	48.913	47.981	41.289
O9 V	35,900	3.9	8.8	48.634	47.426	40.293
O9.5 V	34,620	3.9	8.5	48.460	47.003	39.574
B0 V	33,340	3.9	8.3	48.275	46.525	39.055
B0.5 V	32,060	3.9	8.0	48.051	46.036	38.324
O3 III	50,960	3.8	15.1	50.011	49.444	45.054
O4 III	48,180	3.8	15.1	49.898	49.257	44.486
O6 III	42,640	3.7	14.7	49.605	48.815	43.152
O8 III	37,090	3.6	14.6	49.223	48.188	41.371
O9 III	34,320	3.5	14.6	48.974	47.604	40.307
O9.5 III	32,930	3.5	14.7	48.819	47.171	39.466
B0 III	31,540	3.5	14.7	48.615	46.542	38.972
O3 Ia	50,680	3.7	17.6	50.131	49.564	45.133
O4 Ia	47,690	3.6	18.6	50.059	49.420	44.638
O6 Ia	41,710	3.4	20.6	49.862	49.058	43.359
O7 Ia	38,720	3.3	21.7	49.729	48.825	42.495
O9 Ia	32,740	3.1	24.4	49.334	47.879	40.263
O9.5 Ia	31,240	3.1	23.7	49.125	47.325	39.355

the case of regions ionized by the UV pulse of a nova or supernova or, perhaps, in the extended halo gas. One way of estimating this is by looking at the typical recombination times, $t_{\rm rec}$. Since for hydrogen, $t_{\rm rec} \approx (10^5 \text{ cm}^{-3}/n_e)$ yr, a low density gas exposed to a short interval of intense ultraviolet light can show time dependent effects.[118]

For a moment, imagine what happens if the central source turns off. H II regions normally don't show this because the recombination timescales are so short compared with the main sequence lifetimes of the illuminating OB stars. The ionization may, however, be provided by a nova or supernova explosion or some

---

[117] Data in Table 6.11 are courtesy J. P. Aufdenberg (2000, *PhD Thesis*, Arizona State Univ.). Solar metallicity is assumed and the atmospheres include full line blanketing with the strongest contributing lines treated in NLTE. Spectral type calibration is from Vacca, W. D., Garmany, C. D., and Shull, J. M. 1996, *ApJ*, **460**, 914. The units for the rates, Q_i, are photons s^{-1}; Q_0 applies to the H I Lyman continuum (≤ 912 Å), Q_1 applies to He I continuum (≤ 504 Å), and Q_2 applies to the He II Lyman continuum (≤ 228 Å). The values are logarithmic.

[118] Another example is provided by novae (e.g., Shore, S. N., Sonneborn, G., and Starrfield, S. 1996, *ApJL*, **463**, L21). Although this is not strictly an interstellar problem, it bears a striking resemblance of the formation of a *fossil H II region*. See, for example, Brandt, J. C. et al. 1971, *ApJ*, **163**, 99. Once the central star has turned off, in an expanding shell, the recombination timescale may start to compete with timescale for density decrease. Since the recombination rate depends on *both* the temperature and density, this causes the ionization to freeze out. Even after the UV source is no longer responsible for maintaining the ionization, the shell will have the ion ratios that were characteristic of the last epoch of illumination. These can persist for a very long time, giving the shell a highly ionized structure that is incompatible with the contemporaneous spectrum of the central star.

other very shortlived but bright source. Thus, in the low densities that are typical of the diffuse medium, the analysis of the large-scale structures in the interstellar medium may require including time dependence for the ionization state. To make the matter tractable, assume a constant electron density and a pure hydrogen plasma. Then the ionized volume, V, is given as a function of time by the following balance equation:

$$\frac{d}{dt} n_e V = \beta - \alpha n_e^2 V, \tag{6.167}$$

where β is the ionization rate by the central star and α is the recombination rate. For a medium in which the electron density is constant, the characteristic timescale is $t_0 = (\alpha n_e)^{-1}$ and the ionized volume varies as

$$V(t) \approx \frac{\beta}{\alpha n_e^2}(1 - e^{-t/t_0}). \tag{6.168}$$

Notice that the radius initially varies with time as $R_{\rm H\,II} \sim t^{1/3}$ if the central source is steadily ionizing the environment. To look at the problem a bit more carefully, we need to consider the flux and momentum balance within the H II region. The *ionization parameter* measures the effective rate of expansion of the static H II region and is therefore related to the timescale, t_0, that we just derived. It is defined in terms of the recombination and ionization rates, and represents the propagation speed of the front into a medium compared to the speed of light:

$$U_H = \frac{1}{n_e c} \int_{\nu_{\rm LyC}}^{\infty} L_\nu \frac{d\nu}{h\nu}, \tag{6.169}$$

which is dimensionless.[119] For typical H II regions it is of order 0.01–0.1. This is a useful way of estimating the rate of variation in an emission lines from a time-variable source of ionization, such as an active galactic nucleus or nova. Another application is to variable embedded sources, such as accreting protostars. Although ultraviolet radiation is absorbed in the immediate vicinity of the source, stellar X-rays can penetrate quite far inside the clouds. Variations in the source then compete with time-dependent chemical and thermal processes in the surroundings and, depending on the reaction sequences, can produce significant changes in the abundance of molecular species.

We will digress for a moment to illustrate how this works. Time dependence is especially important when the lifetime of the ionizing source is short compared with the recombination time. As we mentioned, this occurs in nova shells. Since in their late stages, these are merely nebulae, they illustrate the point well. Consider a point explosion in which the imposed velocity field is $v(r) = v_0(r/r_0)$ and that throws off a shell of mass ΔM. Since this mass is constant, the density varies as t^{-3}. You can therefore determine the rate of change in the emission measure with

---

[119] See *e.g.* lectures by Netzer in Blandford, R. D., Netzer, H., and Woltjer, L. 1990, *Active Galactic Nuclei: Saas-Fee Advanced Astrophysics Course* 20 (Berlin: Springer-Verlag).

time by taking t_0 to be the initial epoch and solving Eq. (6.167):

$$\frac{j_\lambda(t)}{j_\lambda(t_0)} = \exp\left\{\frac{1}{2}\alpha n_{e,0} t_0 \left[\left(\frac{t_0}{t}\right)^2 - 1\right]\right\}. \qquad (6.170)$$

The emission continues to slowly decline as the recombination rate decreases from expansion but at a much slower rate and the ionization remains nearly constant. Of course, novae are very fast compared to H II regions in general, but the same basic feature is observed. When a supernova explodes, the pulse sent out into the ISM is very shortlived and therefore ionizes a region almost immediately. The resulting region may stay ionized for a very long time, depending on the recombination rate, and becomes a *fossil H II region*, one that maintains a very high ionization without an obvious source supplying the requisite photons.

6.7.3.2 Compact H II Regions
Deep within molecular clouds, in regions of massive star formation, the densities may be high enough in any H II region that many of the usual cooling transitions are collisionally damped. Otherwise invisible because of the absorption by cloud dust, these regions are accessible through radio observations that detect the thermal emission. Their main signature is a much higher than usual brightness temperature since the Lyman series is completely opaque and almost all fine-structure transitions and infrared forbidden lines are quenched from densities above 10^5 cm^{-3}. Since these are formed by massive stars, they also have high dust brightness temperatures and so can be seen easily through FIR emission.

6.7.4 Expanding H II Regions: Dynamics Driven by Radiative Heating

In the earlier work on this subject, reference was made to *ionization* versus *density* bounding of the H II region. This is still a useful distinction. An ionization bounded nebula is one that is surrounded by a largely neutral or even molecular environment. The density of the medium may be too high, or the luminosity of the central source too low, to completely ionize the available gas (out, of course, to some distance from the star). On the opposite extreme, a density-bounded H II region is one that has completely ionized *all* of its environmental gas and from which ionizing radiation may actually escape.

This distinction finds utility in analyzing how an H II region expands. For instance, a dense cloud, one that creates an ionization-bounded region, is also shielded from stellar ionizing radiation and may have an extensive neutral region. This means, as we shall see, that it will be cooler than one from which some of the ultraviolet can escape. It also means that there will be a region of high pressure contained within a largely neutral environment. A density-bounded region, on the other hand, is free to expand because there is virtually no external gas available to stop it.

Historically, the problem of H II region evolution has tracked the development of numerical astrophysical hydrodynamics and shock theory. Savedoff and Green (in 1955) and Schatzman and Kahn (in 1955) developed the first models for the unstable expansion of such gas bags.[120] With the advent of faster computers and FORTRAN coding in the early 1960s, the field literally exploded. Within a few years, realistic numerical models were presented by Vandervoort (in 1963), Axford (in 1964), Mathews (in 1965), and Lasker (in 1966) who were among the first ones to consider the expansion of such regions,[121] and their dynamics have been of interest since computational fluid mechanics became an astronomical household appliance. You can treat the expansion of an H II region in terms of the conservation of the number of particles. Take the gas around an OB association. Stellar radiation ionizes the environment, which then has a higher pressure than its surroundings. This, in principle, would lead to expansion. However, there is an important difference between the H II region problem and that of an ordinary bubble of hot gas. The state of the gas changes across the boundary of the expanding hot medium. It ionizes. Thus the flow of matter into the H II region must be controlled by the rate at which photons are supplied. If the region begins to expand, then the local density drops, permitting more of the ionizing radiation to escape. The region thus remains always ionization bounded and expands as a contact discontinuity because $P_{H\,II} = P_{H\,I}$. Sitting on Σ, the medium of density $\rho_{H\,I}$ has the velocity of the front, while the postshocked gas is thermalized. The conservation condition for momentum flux therefore gives:

$$\rho_{H\,II} v_{H\,I}^2 \equiv \rho_0 \dot{R}^2 = \rho_{H\,II} c_s^2, \qquad (6.171)$$

assuming that the upstream, neutral, gas is cold and therefore has a low sound speed. The Strömgren radius is actually the solution of ionized mass flux conservation since the it balanced the number of ionizations (an energy flux) with the number of recombinations through the front. Therefore, it provides the relation between the upstream and downstream densities:

$$\rho_{H\,II} = \left(\frac{m_p \beta}{\alpha} \rho_0 \right)^{1/2} R^{-3/2}, \qquad (6.172)$$

which, on substitution into Eq. (6.171), gives

$$R^{3/4} \dot{R} = \frac{c_s}{R_0}, \qquad (6.173)$$

where the initial radius is assumed to be R_0, the Strömgren radius, so integrating

---

[120] The most important source for work during this formative period is IAU Symposium 8, the proceedings of the 3rd Symposium on Cosmical Gas Dynamics. Fortunately, this was published in 1958, *Rev. Mod. Phys.*, **30**, 906–1108, and you will surely profit from time spent here, especially with the comments. Almost all the basic problems that define the dynamics of the interstellar medium were discussed here and the explanations are extremely lucid.

[121] Spitzer (n.3) and Goldsworthy, F. A. 1983, *IMA J. Appl. Math.*, **32**, 147 provide excellent pedagogical starting points for any further reading on the subject, and we urge you to at least have a look at them. Spitzer is astrophysical, while Goldsworthy's review is more mathematical, but each is interesting for its approach. In particular, Goldsworthy points out the similarity between the propagation of an H II region and a flame, in the sense that the consumption of newly engulfed fuel is what maintains a flame or detonation.

Eq. (6.173) gives the evolution of the radius of the ionized region with time:

$$R(t) = R_0 \left(1 + \frac{7c_s t}{R_0} \right)^{4/7}. \tag{6.174}$$

You see that there is an initial nearly static phase, but the increasing pressure within the ionized region forces expansion to occur. This continues until the interior gas starts to radiatively cool and recombine, or until the central source turns off. Although this solution depends on time alone, and gives the frontal radius of the H II region, it is not a strict similarity solution. The H II region evolves on a characteristic timescale and with the sound speed playing a role. This is simply the solution to the equation for the ionization front when cooling is insufficient to remove the driving pressure provided by continued ionization.

Such an expansion is quite slow and the shock at the ionization boundary is a rarefaction. Such shocks are subsonic by definition. You can see this because the mean molecular weight per electron dramatically decreases in the fully ionized gas, since the pressure comes entirely from the electrons. There is an analogous region surrounding the ionization front *if* the cloud is molecular, because while the ionizing radiation may be completely absorbed within the H II region, the photodissociating radiation leaks out. This means that the state of an expanding H II region within a realistic cloud is actually not expansion into a static medium. For comparison, we can look at the solution assuming that the luminosity of the central star, L, and the density of the overtaken gas, ρ_0, are constant. Such a region expands according to a similarity solution (see Appendix 6.C):

$$R \sim \left(\frac{c_s t}{R_0} \right)^{2/5}. \tag{6.175}$$

Notice that the exponent is almost the same as for the dynamical solution. This is not an accident. In spite of the effect of the recombinations, the expansion of an H II region is driven only as long as the rate of energy input is constant. Once the medium starts to cool, there is a characteristic time during which the energy decreases with time. Thus, the ultimate limit on the rate of expansion is the lack of anything but the recombination timescale. It is also important to include energy conservation as well as ionization equilibrium. In effect, we are assuming that the dynamical response of the medium comes from the increase in free-electron density as a result of ionization and that the stellar flux otherwise maintains the ionization behind the front. The cooling rate, however, may be so large from collisional excitation and radiative losses that the front stalls. To check on this, assume that the heating rate is Γ, which depends on the ionization rate and therefore on the stellar ultraviolet flux. The cooling rate is due to collisions and subsequent reradiation of the energy. When the cooling rate is large enough, it drives a pressure drop in the H II region inside the driving piston and reduces the expansion velocity. The region stalls sooner than would have been expected strictly on the basis of constant luminosity.

6.7.4.1 Blisters, Champagne Flows, and Bubbles

We can suppose that the star responsible for the initial ionization sits within a molecular cloud. The expansion rate of the H II region initially depends on the density of the surrounding medium, which may be as high as 10^4 cm^{-3} or more. In clouds, this is mainly molecular. Once the expansion reaches the cloud boundary, however, it experiences a very large density drop and reacts by rapidly accelerating. This phase of *breakout*, a classical term from explosion theory, has been christened *champagne flow* for the ISM case. Actually, this is not at all mysterious. The H II region is *very* overpressured compared with its low density surroundings. At breakout, it accelerates supersonically and establishes a flow away from the parent cloud. The hot interior is still being fed by gas being ionized and swept into the H II region from the other side by its expansion into the cloud. A rarefaction wave moves inward, sonically, while the expansion continues supersonically. Thus a flow is established. The advance of the H II region into the cloud is throttled by the rate at which the hot gas advects away from the cloud, a rarefaction wave that occurs at the sound speed. Ultimately, this can destroy the cloud[122] if a sufficient number of OB stars power the expanding regions.

6.7.5 Stellar Explosions and Their Remnants

We have already discussed the evolutionary processes that lead up to supernova explosions, the most violent of all stellar events. The characteristic of the explosion is that energy of some magnitude E_0 is released from a point mass at an instant in time compared with the expansion time. However, the same basic theory applies to all supersonic expansions once they are at times long compared to the time it took to start them going.

We will assume, for simplicity, that the medium into which the ejected matter travels has no characteristic scale length, and that the ejecta expand without losing energy. This adiabatic approximation actually fixes the rate of expansion, since it must be what is required to maintain the constancy of the total energy. More importantly, the density of the medium comes enters because the initial energy is being redistributed to a progressively larger amount of gas as the shock front overwhelms and accelerates background matter. The method of dimensional analysis[123] again comes into use because, like homology, we assume that all shocks with the same conservation conditions are essentially identical. This is the same thing as saying that the problem has no intrinsic scales of length or time that depend on the details of the explosion or the upstream material. The initial energy has the dimensions ML^2t^{-2}. The density of the unshocked material is ρ_0, with dimensions ML^{-3}. Therefore, in order for the system to evolve while simultane-

---

[122] For discussion of the origin of large-scale gas structures, see Tenorio-Tagle, G. and Bodenheimer, P. 1988, *ARAA*, **26**, 145. On the limiting number of massive stars that can form in a molecular cloud, see Franco, J., Shore, S. N., and Tenorio-Tagle, G. 1994, *ApJ*, **436**, 795. The case including photodissociation region formation is treated in Diaz-Miller, R., Franco, J., and Shore, S. N. 1998, *ApJ*, **501**, 192.
[123] Throughout this section, we will use M for *mass*, t for *time*, and L for *length* with no special reference to a system of units.

ously keeping ρ_0 and E_0 constant, we must have

$$\frac{E_0}{\rho_0} \sim L^5 t^{-2} = \text{dimensionless constant}. \qquad (6.176)$$

This gives the adiabatic *blast wave* solution:

$$R(t) = \lambda_E \left(\frac{E_0}{\rho_0}\right)^{1/5} t^{2/5}. \qquad (6.177)$$

Here λ_E is a constant. In the fluid dynamics literature, this is also called the *Sedov–Taylor*[124] solution. Now you see why we wrote the Rankine–Hugoniot conditions in terms of the preshock Mach number, M_0. The shock velocity comes immediately from the blast wave solution since $V_\Sigma = dR(t)/dt$. In general, this will be a decreasing function of time (only in the case of a stalled shock, one that expands in pressure equilibrium, does the speed of the shock remain constant and equal to the sound speed). The compression ratio can therefore be computed in terms of the incident Mach number by Eq. (6.142), as can all the other thermodynamic variables. The equations of motion, which we will not discuss in detail here (see Appendix 6.C), then determine the density and thermal profiles for postshock flow. For the Sedov–Taylor solution, the velocity of the shock front is

$$V_\Sigma = \frac{dR(t)}{dt} = \frac{2}{5}\frac{R}{t}, \qquad (6.178)$$

which is how you determine the expansion age when you have the velocity and radius for the remnant.

The reason for calling this a *similarity* solution is that once obtained, this means that all physical properties scale with respect to the radius of the shock front. We again refer you to Appendix 6.C for the details. For now, let's say that this means that we can define a variable

$$\eta = \frac{r}{R(t)} = rt^{-2/5}, \qquad (6.179)$$

which converts the partial differential equations for the motion of the gas, which are a hyperbolic system that depends on both time *and* space, into a set of *ordinary nonlinear differential equations of a single variable*. This is the same thing we did in stellar interiors when we inverted our perspective, taking the mass fraction instead

---

[124] The first published analysis of the Trinity test was Taylor, G. I. 1950, *Proc. Roy. Soc.*, **A201**, 159. The similarity method has been expounded in detail by Sedov, L. L. 1959, *Similarity Methods and Dimensional Analysis in Mechanics* (NY: Academic Press) and Barenblatt, G. I. 1979, *Similarity, Self-Similarity, and Intermediate Asymptotics* (NY: Consultants Bureau). There was a considerable amount of material that went unpublished from the Manhattan Project for many years, until the security was finally cleared. The explosion physics is also treated in various editions of the Department of Defense handbook *The Effects of Nuclear Weapons*. Further review discussions of similarity methods are found in Shore (n.106).

of the radius as our independent variable. This ability to scale the shock radius with time is very important. The simplest similarity solution comes from assuming that the mass of the ejecta is constant. This happens in the free expansion, for instance, of fireworks. The fragments that are moving fastest reach the greatest distance in a given time so the distribution of velocity in the remnant reflects the initial conditions. This is a much simpler solution because the shell expands with a simple density law, independent of the external medium

$$v = v_0 \frac{r}{r_0}, \qquad (6.180)$$

which is also called a *Hubble flow* because of its analogy with the expansion of the Universe (see Chapter 8). In this case, the radius is simply a linear function of time for each portion of the ejecta and the density, because of the scaling that the shell has constant *fractional* thickness, that is $\Delta R/R(t)$ is constant, varies as r^{-3}. You can easily use this to predict the brightness of the ejecta using the theory developed in Chapter 3 for thermal emission in a density gradient. This means that we need to solve the dynamical equations only once for each general shock. Once we have a solution for the pressure, say, in terms of η, a variable that runs from zero to unity, we know how to transform it into a linear distance behind the front as a function of time.

There are two strict requirements for the similarity method to apply. The first is that the global dynamics of the system be determined by a set of conserved quantities, such as energy, momentum, or mass, that are constant *throughout* the ejecta. The second is that there be no special length or timescale for the problem. Either the system evolves faster than this or we restrict ourselves only to short intervals. It is possible to piece similarity solutions together, we are about to do this, but you need to be cautious about it. For freely expanding gas, for instance, you must eventually encounter a phase when cooling within the remnant becomes important. At that stage, the matter no longer follows Eq. (6.177). In fact, it *cannot* behave this way because a cooling wave eventually propagates through the interior. In the long term, the shell becomes isothermal because it will cool at a rate comparable to its expansion timescale. In this case, momentum, rather than energy, is conserved. This is $M_{ej}V_{ej}$, where M_{ej} is the mass of the ejecta at the time when the similarity is established and V_{ej} is its velocity, which scales as MLt^{-1}. Thus, for a momentum conserving blast wave, we obtain

$$R(t) = \lambda_M \left(\frac{M_{ej}V_{ej}}{\rho_0} \right)^{1/4} t^{1/4}, \qquad (6.181)$$

which shows a slower rate of expansion than the initial adiabatic case. The expansion continues until the shell establishes pressure equilibrium with the surrounding medium, at which point it has reached the sound speed and merges with the background.

A final class of solutions is important for the evolution of gas surrounding an OB association. Because they have strong stellar winds, such stars actively perform work on their surrounding medium by injecting momentum from their winds. This is a driven expansion, one that depends on the mass loss rate and terminal velocity of the winds. Some fraction of the stellar luminosity ultimately goes into the wind of a massive star (see Chapter 3). The combined winds blow out a *bubble*. If the wind mechanical luminosity is $\dot{M}v_\infty^2$, which is some fraction of the stellar luminosity, then for a constant density surrounding medium we find that ML^2t^{-3}/ML^{-3} is constant. This gives the following similarity solution:

$$R(t) = \lambda_{W,1} \left(\frac{\dot{M}v_\infty^2}{\rho_0} \right)^{1/5} t^{3/5}. \quad (6.182)$$

Here the mass loss of the association as a whole must be used, although Wolf–Rayet are surrounded by highly compressed ring nebulae that seem to agree well with this general solution.[125] For the period when the bubble is still driven but in the snowplow phase and momentum-conserving, we can take $\dot{M}v_\infty$ to be a constant, in which case the bubble follows the similarity law:

$$R(t) = \lambda_{W,2} \left(\frac{\dot{M}v_\infty}{\rho_0} \right)^{1/4} t^{1/2}. \quad (6.183)$$

To see how important this may be, let's try a simple estimate. If each OB star ejects of order $10^{-8} M_\odot \text{yr}^{-1}$, then the momentum input from a typical association, consisting of about 100 OB stars that last for about 10^6 yr, is as large as a few percent that of a SN II explosion, about 10^{41} g cm s^{-1}. This is important because those stars that start their lives as mass losing OB stars will end them as supernovae, so the medium will be doubly affected by such stars. Ultimately, the expansion of these *superbubbles* is thought to be responsible for massive restructuring of the interstellar medium.

6.7.5.1 Free Expansion and Transition to the Sedov–Taylor Solution
If we assume an initial energy E_0 and ejected mass, M_{ej}, then an initially freely expanding blast will make the transition to an energy conserving (adiabatic) flow at a time that depends on the initial conditions. For free expansion, $R_{fe} = v_0 t$ where $v_0 = \dot{R} = (2E/M_0)^{1/2}$. The transition takes place at a time t_{trans} when the Sedov–Taylor radius, R_{ST} given by Eq. (6.177) is equal to R_{fe}. The scaling relation is therefore:

$$R_{ST}(t_{trans}) = R_{fe}(t_{trans}) \rightarrow t_{trans} \sim E^{-1/2} M_{ej}^{5/6} \rho_0^{-1/3}. \quad (6.184)$$

Similar scaling laws can be derived for the subsequent transitions to stalled and snowplow shocks.

6.7.5.2 Stalled Shocks and Stagnation Pressure
The pressure in the blast wave varies as $P \sim \rho u^2 \sim R^{-1}t^{-2}$. Taking the adiabatic as an example, since this case we have at hand, the time can be eliminated by

---

[125] See, for instance, the atlases by Marston, A. P., Chu, Y.-H., and Garcia-Segura, G. 1994, *ApJS*, **93**, 229; Marston, A. P., Yocum, D. R., Garcia-Segura, G., and Chu, Y.-H., 1995, *ApJS*, **95**, 151; Chu, Y.-H., Weis, K., and Garnett, D. R. 1999, *AJ*, **117**, 1433.

taking

$$t = \alpha^{-5/2}(E/\rho)^{1/2}R^{5/2}, \quad (6.185)$$

and therefore the pressure at a given blast wave radius can be obtained. This will also be a function of the initial energy and of the density of the medium. If we take the radius to be the point at which the pressure reaches a prescribed value, we obtain

$$ER^{-3} = P_{\text{ISM}}$$

and then the critical radius scales as

$$R_\star \sim E^{1/3}. \quad (6.186)$$

This is perhaps the most (*in*)famous of the scaling relations to come out of the Manhattan Project research program. It provides the radius of the circle of finite overpressure (the value that, for military purposes, is of use is typically several pounds per square foot) scales as rather weakly with the *yield* of the explosion (assuming that all of the other conditions of the blast wave solution are met). Fortunately, this critical radius has more benign astrophysically interesting properties as well. The point at which the blast wave internal pressure is greater than some multiple of the interstellar ambient pressure should scale the same way. For example, in the current work on star formation, it has been suggested that a supernova, occurring inside or near a molecular cloud, will serve to destroy the cloud and also to trigger star formation. If we assert that there must be a sufficient overpressure at some point to overcome the resistance of the cloud, then we can calculate the extent over which the SN will be effective in shaping the structure of the region. The same is true for the ISM in general—when holes are created, with tunnels resulting from their coalescence, it is possible to determine the extent to which the blast is able to remain a strong force in the medium. Again, this is assuming that all other things remain the same about the medium and the interior of the blast wave. Breakout, and the formation of the champagne phase in a molecular cloud, occurs if the internal pressure is not high enough to make the shock stall everywhere.

A stalled shock expands at constant velocity. For an adiabatic case that we have already done, $R \sim E^{1/5}t^{2/5}$, and the pressure of the environment scales dimensionally as $MR^{-1}t^{-2}$. Therefore, $P/\rho_0 \sim R^2 t^{-2}$ and a shock expanding at constant pressure has a time dependent radius:

$$R(t) = \alpha\left(\frac{P}{\rho}\right)^{1/2} t. \quad (6.187)$$

As expected, the velocity is constant. The important question is whether the shock stalls *before* it ceases to be adiabatic. If the timescale for radiative losses is short enough, the shock can rapidly become isothermal, thus entering the momentum conserving stage. You need to know whether E is constant before naively applying the Sedov method for the interpretation of an observed structure. This can be done by looking at the expansion time, which is R/V, and compare this with the cooling times behind the shock. This timescale can be found from the observation of the spectroscopic temperature diagnostics, although these are rarely unambiguous and depend on the filling fractions of the phase of the plasma that is being used to determine the temperature of the interior.

6.7.6 Snowplow Phase

If the blast wave is radiatively cooling, the adiabatic approximation eventually ceases to apply. The cooling is faster than expected for simple expansion and the temperature is determined by the conservation of momentum rather than energy. In the evolution of a supernova remnant, this is called the *snowplow* phase, since at this point the remnant acts like a truck being decelerated by the accumulation of material in its path. The increase in the mass swept up by the remnant causes the front to decelerate, and eventually to become subsonic, at which time the *characteristic timescale* of the medium once more becomes the sound crossing time. What is happening to the shocked gas? As it moves through the front, the shocked gas is heated and compressed. This work is done by the front. Therefore a compression zone that loads mass into the blast wave results, forming a reverse shock that propagates back toward the center. The details need not concern us because we are focusing on the evolution of the front and not the internal structure.

Assuming that only the momentum of the blast is constant, we find a different similarity solution for the blast:

$$R(t) = \beta \left(\frac{MV}{\rho} \right)^{1/4} t^{1/4}, \tag{6.188}$$

where now MV is the original momentum (or the value at the time of the solution's initial applicability). Here β is also a constant of order unity. In order to calculate the energy loss in this phase, we need merely note that the blast is decelerating to the ambient sound speed so that the difference in the energy is the difference in the squares of the velocity at the initial moment and the sound speed. In this phase of the evolution of a blast wave, it falls from the velocity at which the adiabatic solution ceases to apply to the ISM sound speed. The fractional energy loss is therefore of order

$$\frac{\Delta E_{\text{SNR}}}{E_{\text{SNR}}} = 1 - \left(\frac{c_{s,\text{ISM}}}{V_{\text{iso}}} \right)^2. \tag{6.189}$$

Typically, the Mach number at which the blast wave enters the nonadiabatic stage of its evolution is of order 5, so that we would expect that about 95% of the energy of the SNR is dissipated *before* it becomes sonic. The timescale for expansion can be computed from the solution for the velocity, as we did above for the adiabatic case, with a coefficient of $\frac{1}{4}$ instead of $\frac{2}{5}$. The timescale is given by

$$\frac{t_s}{t_{\text{iso}}} = \left(\frac{V_{\text{iso}}}{c_s} \right)^{4/3}. \tag{6.190}$$

6.7.7 Stellar Wind Bubble

In the interstellar medium, a natural example of continuously driven expansion is the region disrupted by a stellar wind, the *bubble* case we discussed above.[126] There are again two solutions for the evolution of the radius, depending on the assumptions. First, take the case of constant driving where the input wind luminosity, $L_w = \dot{M}v_\infty^2$, is a constant fraction of the total stellar luminosity, L. Then, using Eq. (6.176), we obtain

$$R = \left(\frac{L_w}{\rho_0}\right)^{1/5} t^{3/5}. \quad (6.191)$$

Notice the small but important change in the time dependence. Although this expansion may be indistinguishable from a blast wave in practice, in principle the shell expands more rapidly because it is constantly driven. This approximation assumes the wind does not radiate away the input energy within an expansion time. What happens if there is a central source for momentum that acts continuously on the gas? Since $\dot{M}v_\infty$ is constant, the radius now scales as

$$R(t) = \left(\frac{\dot{M}v_\infty}{\rho_0}\right)^{1/4} t^{1/2}, \quad (6.192)$$

since all you need to do is change the constant *momentum* to $Mv_\infty = \dot{M}v_\infty t$, since $\dot{M}$ is itself taken to be constant. This is really an expansion driven by the compressional heating from the wind acting on the surrounding medium. The temperature inside the hot bubble created by the stalled wind forces the front to expand at a much slower rate than v_∞, depending on $\dot{M}$, but still supersonically.

The essential feature of the blast wave solutions we have been discussing is that once solved numerically, they need never be solved again. Aside from the utility of scaling the motion of the front, at which the Rankine–Hugoniot conditions are applied as boundary conditions on the physical variables, the interior solution for the blast is one that we only need to solve *once*! Regardless of the degree of expansion, as long as the physical approximation holds, we can treat the structure of the interior as a function only of η and forget about anything else.

6.7.8 Producing the Hot Diffuse Gas

We have in hand a way to generate the X-ray-emitting gas and also power the Galactic corona. Supernova remnants can expand to very large sizes before they are thermalized. To a lesser extent this is also true for stellar wind bubbles. Since the same stars will, at different stages of their lives, produce both, we can see that the upper end of the mass distribution has a profound effect on the generation of

---

[126] Castor, J., Weaver, R., and McCray, R. 1975, *ApJL*, **200**, L107.

structure in the medium. Supernova remnants are strong X-ray sources from the heated gas. Stellar winds have terminal velocities of a few times 10^3 km s^{-1} and in the lifetime of a typical OB star can inject as much mechanical energy into the diffuse environmental gas as a single SN II event. Combining them, we would expect that regions of active star formation, especially massive OB associations such as those seen in the Galactic center (the Arches cluster) or in extragalactic starbursts such as R136 in the LMC or NGC 604 in M33, will be strong X-ray emitters. The H II regions generated by star forming activity will extend far beyond the immediate site of such bursts, and if the region breaks out of the disk, the ionizing photons can leak out into lower-density gas at higher distances from the galactic plane and into the halo. In short, an actively star forming galaxy should present a complex of molecular clouds with temperatures ≤ 30 K up to very tenuous diffuse gas heated to $> 10^6$ K, all along the line of sight.

6.7.9 Pressure-Driven Expansion of a Planetary Nebula

Let's look at the case of a planetary nebula expansion that is pressure driven when the old wind of the AGB star is impacted by the expanding H II region from the central star.[127] This is just the case we discussed for the expansion of an H II region, except that the medium into which the shock moves now has a density gradient r^{-n}, where n depends on the wind characteristics of the pre-planetary nebula stage. Let us rewrite the equation for the expansion of the ionization front:

$$\rho_0 \dot{r}^2 = \rho_1 c_s^2, \tag{6.193}$$

with the auxiliary condition that comes from the ionization:

$$\rho_1^2 \alpha r^3 = \rho_0 \beta, \tag{6.194}$$

where all variables have their usual meanings. Now we can write

$$\rho_1 = \rho_0^{1/2} \left(\frac{\beta}{\alpha}\right)^{1/2} r^{-3/2} \tag{6.195}$$

$$r^{(3-n)/4} \dot{r} = C,$$

so that the expansion law is given by

$$r(t) \sim R_0 (t/t_0)^{4/(7-n)}. \tag{6.196}$$

The fast wind from the central star is driven at constant momentum and expands

---

[127] See, for instance, Steffen, M, Szczerbal, R., Men'shchikov, A., and Schönberner, D. 1997, *Astr. Ap. Suppl.*, **126**, 39.

into an ionized region whose density gradient is given by the ionization equilibrium. This means that

$$r_{FW} \sim \left(\frac{L_W}{\rho_1}\right)^{s_1} t^{s_2}, \qquad (6.197)$$

so that we can compute, in a relatively simple way, the structure of the planetary.

6.7.9.1 Breakup of the Shock Front

We won't go into details here, but should mention that several processes make the shock front intrinsically unstable under a range of astrophysically accessible conditions.[128] In early observations of detonations, a striking fingering structure was noted. This has been found now in many numerical simulations of blast waves progressing along density gradients. The Rayleigh–Taylor instability is driven by buoyancy, the Richtmyer–Meshkov instability is driven by pressure gradients; they are related by the fact that a decelerating front produces an "effective gravity" that promotes the growth of the fingering. As the fingers grow, they are subject to another even more pernicious phenomenon: the Kelvin–Helmholtz instability. This is an unconditionally unstable process in a medium that has negligible viscosity and no magnetic field and depends only on the existence of strongly sheared flows. Shear generates vorticity in a fluid and this is exacerbated by the presence of discontinuities. Again without going into details, the interface is a contact discontinuity that is in pressure balance. In the incompressible flow version, you can see why this would not remain stationary by imagining a wave along the interface that deflects the flow on either side. Because of Bernoulli's theorem, the pressure drops in the flow as it accelerates over the barrier, in equal and opposite directions on either side of the discontinuity, and the bump on the interface grows. The fluid now exerts a normal force to the deflected flow and it entrains and shears, resulting in vortices that ultimately destroy the front.[129] The Kelvin–Helmholtz instability hastens the disappearance of the fingers that are running ahead of the shock and the region should, if intuition is any guide, cascade to turbulence. As such, shock fronts, especially supernova remnants, are expected to be efficient sites for particle acceleration.

6.8 INSTABILITIES AND THE FORMATION OF STRUCTURE

For the first time in this book you get glimpse some *ultimate questions*: How do the structures we see in the ISM form? What sets their initial conditions? How can you recognize them in this stage of evolution? And can we get observing time to study them?

---

[128] Inogamov, N. A. 1999, *Ap. Space Phys. Rev.*, **10**, 1 are the most comprehensive reviews of dynamical instabilities of the Rayleigh–Taylor and Richtmyer–Meshkov type. See also Zeldovich and Raizer (1967, n.109); Vishniac, E. 1983, *ApJ*, **274**, 152; Shore, S. N. 1992 (n.106); MacLow, M.-M. and Norman, M. L. 1993, *ApJ*, **407**, 207; Walder, R. and Folini, D. 1998, *Astr. Ap.*, **330**, 21; Brouillette, M. 2002, *Ann. Rev. Fluid Mech.*, **34**, 445.

[129] The most familiar example from daily life is the formation of bands of clouds when a front moves through but this sort of instability is also responsible for destruction of contrails from airplanes.

This problem of understanding the generation of structures is usually approached on several different levels. The first is to examine the stability of the medium, to ask whether there are critical thresholds for heating and cooling, or compression, past which the gas is driven into a catastrophic state such as collapse or explosion. The second is to perform complex numerical integrations of the equations governing the evolution of the gas, including all the relevant physical processes, and hope that you understand what comes out. The second approach is, consequently, built on the intuition gained from the first. It is the stability approach that we will consider here.

6.8.1 Gravito-acoustic Waves: The Jeans Instability

The idea that an isothermal gas is unstable to gravitational collapse is an old one. The first solution, by Jeans in 1912, assumed the scenario of a self-gravitating slab long before there was any direct evidence for either star formation or even the existence of the interstellar medium. The problem is posed in much the same way as one treats sound waves, and you can think of what follows as the result of sneezing too hard in a self-gravitating medium. This is a compressibility effect. The softer the equation of state, the more unstable the layer is. (How often have you heard that one?) For historical reasons, we will treat only the isothermal slab. This has two simplifying effects. One is that we can assume that the equation of state is $P = \rho c_s^2$ and that the sound speed is constant. The second is that we assume a solution to the energy equation, a completely transparent medium, so we do not have to treat the compressional heating that results if collapse occurs. This situation is actually realistic at the start of the instability.

The central feature of this instability is that the change in the local density of the medium causes the *local* gravitational acceleration to increase, thereby pulling more matter into the higher-density region.[130] As that grows, the decrease in the density of the surrounding medium shuts off the instability so that the self-gravitating blobs separate at some characteristic wavelength. In effect, we are comparing the infall, or freefall, time with the sound travel time across the medium that is required for the accretion to make itself known:

$$L_{\text{crit}} \sim \frac{c_s}{(G\rho)^{1/2}}. \tag{6.198}$$

---

[130] The Jeans instability was used as the basis for a remarkable paper, Richardson, L. F. 1941, *Nature*, **148**, 784, on the growth of cities and population movements. We quote from it here to provide an example of the value of remaining open to cross-disciplinary influences:

> Among the obvious motives of mankind are the tendencies to seek company and to seek living spaces. If we were to regard these tendencies as being in simple opposition to one another, we should expect the population to be able to remain uniformly spread over any uniform piece of land; and the familiar contrast between town and country would then appear, to the theoretical mind, as a mystery requiring explanation. We may, however, seek a hint as to why people concentrate into towns from Sir James Jeans' theory of why matter concentrates into stars. For his theory is also concerned with two opposing tendencies: to draw together by mutual gravitation and to spread out by pressure.

584   THE INTERSTELLAR MEDIUM

Gravity is responsible driving this unconditional instability—an infinite-range force is increased by a local instability so matter can continue to flow into the region as long as there is a supply.

The full continuity equation is required for a compressible medium. This is because the change in the density feeds back, through mass conservation, into a change in the velocity that then couples to the pressure to transport mass:

$$\frac{\partial \rho}{\partial t} + \nabla \cdot \rho \mathbf{v} = 0. \tag{6.199}$$

The momentum equation is

$$\rho\left(\frac{\partial}{\partial t} + \mathbf{v} \cdot \nabla\right) = -\nabla P + \rho \nabla \Phi, \tag{6.200}$$

where Φ is the gravitational potential. We don't need the energy equation because we have assumed an isothermal equation of state. We must, however, add the equation for the gravitational potential, which is the Poisson equation:

$$\nabla^2 \Phi = -4\pi G \rho. \tag{6.201}$$

You can already see how the instability will develop because the gravitational equation can also be written as $\nabla \cdot \mathbf{g} \sim -\rho$, so that the divergence of the acceleration, $\mathbf{g}$, is directly related to the local fluctuations in the density. The density fluctuations flip around the signs of the driving forces arising from self-gravitation.

But let's continue and employ techniques that you saw for the analysis of pulsational stability. Assume that the medium consists uniform of a uniform density ρ_0 gas at rest, $\mathbf{v_0} = 0$, and having constant temperature. This state means only that we are not considering net flows, but turbulent or large-scale *random* motions are allowed.[131] We will simplify the problem still further by representing the gas as a finite thickness slab and ignoring the transverse directions so the fluid equations become one dimensional. Write $\delta\rho$ and δv for the density and velocity perturbations and keep only the first order in the expansion of Eqs. (6.199)–(6.201). Since we have linearized the governing equations, this is not a problem because the partial waves can simply be superimposed. Substitute the expansion of all perturbed quantity is

$$\delta q = q_0 e^{i(kx - \omega t)},$$

where q_0 is its unperturbed value. The Poisson equation, for example, has no time dependence, so it becomes

$$k^2 \delta \Phi = (4\pi G \rho_0) \frac{\delta \rho}{\rho_0}. \tag{6.202}$$

---

[131] In other words, we are assuming that any velocity field $\mathbf{v}$ has $\langle v_i \rangle = 0$ but allow for the possibility that $\langle \delta v_i \, \delta v_j \rangle \neq 0$.

thus connecting the density and potential fluctuations. The continuity equation becomes

$$-i\omega\,\delta\rho + ik\rho_0\,\delta v = 0, \tag{6.203}$$

which directly connects the variations in velocity to those in density through the compressibility of the medium. Finally, the momentum equation becomes

$$-i\omega\,\delta v = -ikc_s^2\frac{\delta\rho}{\rho_0} + ik\,\delta\Phi. \tag{6.204}$$

Now combining these, we obtain a *dispersion relation*. This is not the same thing we found in the pulsation problem because we took only a single-mass zone and analyzed the temporal stability of that layer. Here we are not restricting ourselves to one local region, and in fact this is not a local instability—there is a characteristic length in the problem, not just a characteristic time. The result is

$$\omega^2 = c_s^2 k^2 - 4\pi G\rho_0. \tag{6.205}$$

The critical solution is reached when $\omega = 0$, yielding the characteristic length:

$$\lambda_J = \left[\frac{\pi c_s^2}{G\rho_0}\right]^{1/2}, \tag{6.206}$$

which is called the *Jeans length*. You see that our initial estimate was not bad after all. The medium becomes gravitationally unstable on a length scale that contains more mass than $M > M_J = \frac{4}{3}\pi\rho_0 L_J^3$, provided it remains isothermal. In other words, a parcel of gas is unstable if a sound wave cannot travel fast enough to readjust the internal pressure on the freefall timescale. An interesting feature about gravitational instabilities then follows from the isothermality of the fragments. The first arguments along the lines that the instability leads to fragmentation and an initial mass spectrum goes back to Hoyle (in 1953). Suppose you start with a blob of gas that is unstable since it exceeds the Jeans mass at its initial density. Once collapse has begun, the density rises, although the blob may remain isothermal. The blob is consequently even more instable because it now contains *more* Jeans masses than initially. It *must* fragment, and must continue this until it reaches some minimum size or maximum density or becomes opaque enough to violate the isothermal condition imposed on the collapse. How this actually happens is not at all well understood even with relatively recent hydrodynamic simulations, but it is certainly a competition between collisions among the newly formed fragments and the increase in their internal density and opacity (if this sounds like an agglomeration problem, you're right). This simultaneously assists in halting the collapse of individual subblobs and increases the merger and heating and destruction of others in what starts to look like a turbulent cascade. This all follows from the original picture of gravitational instability.

The Jeans mass also provides a limit on the column density at which self-gravitational effects become important. If you have observed a column density, N, then the Jeans limit to this is $N_J = \rho \lambda_J$, so that we can write this in terms of the internal thermal pressure, P_i, as $N_J \sim P_i^{1/2}$. Now, if the support for the cloud comes from large-scale motions or from magnetic fields, the column density can easily be increased.

6.8.1.1 *Magnetized Clouds, Ambipolar Diffusion, and the Jeans Criterion*

A magnetic field threading through a gas cloud adds to its support against gravitational collapse. It requires that the gas be at least partly ionized but, in that case, the field tension assists even a cooling medium to stabilize. You already know this from discussion of the virial theorem, but as we will see, it is only part of the story. Alfven waves, if they are self-interacting or nonlinear, cascade through a range of length scales to the point of providing turbulent pressure for the cloud. This process is dissipative, however, so without continued regeneration by a dynamical process, they alone cannot maintain the stability of the medium. Magnetic support is therefore an important barrier to star formation. In addition, if the flux is frozen into the medium the field is amplified on contraction. Ultimately, the increased magnetic energy limits the compression of the gas in spite of the effective polytropic index of a self-gravitating magnetized sphere.

The problem of excluding magnetic flux from cloud interiors as a first step in gravitational collapse to form stars was recognized very early in the work on the interstellar medium (Mestel and Spitzer in 1956). The crucial step in understanding what is happening is realizing that the gas is only partially ionized. Neutrals tend to separate from ions within a self-gravitating gas in the presence of a magnetic field. This process, *ambipolar diffusion*, has been implicated as an important mechanism in star formation because of the low ionization in clouds and the long mean free path for collisions. It is also a reason why the magnetic field does prevent collapse. The process is, in principle, quite simple (see Section 6.5.4.2). Any atom will flip between the states of ionized and neutral on the recombination timescale. In the ionized state, it hangs on the field lines, while these are "transparent" when it is neutral. Statistically, in dark clouds, the atom is neutral most of the time; hence it can move across field lines. The process is diffusive and relatively slow, but inexorable. Over time, gas will settle toward the increasing mass concentration that defines the midplane of the medium. This increases the local gravitational acceleration, so there is a positive feedback to the process. The magnetic field, on the other hand, connects to the galactic field and is progressively excluded from the cloud. This seems to solve two problems at the same time, removing the magnetic barrier to star formation while permitting the clouds to remain stable for long enough to accumulate sufficient mass to induce star formation.

Although analytical approximations must be very schematic, one classic solution is worth covering here to illustrate how the calculation proceeds. The ideal MHD equations for an ionized magnetic cloud in hydrostatic equilibrium are

$$-\frac{1}{\rho}\nabla P + \nabla \Phi + \frac{1}{4\pi}(\nabla \times \mathbf{B}) \times \mathbf{B} = 0$$

$$P = \rho c_s^2 \quad (6.207)$$

$$\nabla^2 \Phi = -4\pi G \rho$$

with the usual meanings. From the flux-freezing condition, we know that $\mathbf{B}/\rho$ is constant. As always, $\nabla \cdot \mathbf{B} = 0$. Let us take the case of a *pressure-bounded* cylinder with the magnetic field along the $\hat{\mathbf{z}}$ axis but dependent only on axial distance, r. The the magnetic field can then be eliminated through the flux condition so that $B = K\rho$, and Eqs. (6.207) reduce to a form of the Lane–Emden equation:

$$\frac{1}{r}\frac{d}{dr}\left[\frac{r}{\rho}\left(\frac{1}{8\pi}B(r)^2 + \rho c_s^2\right)\right] = 4\pi G\rho. \quad (6.208)$$

For now, we will assume that somehow the cloud is supported solely by the magnetic field, so, neglecting the gas pressure, we obtain

$$\frac{1}{r}\frac{d}{dr}\left(r\frac{d\rho}{dr}\right) + \frac{(4\pi)^2 G}{K^2}\rho = 0. \quad (6.209)$$

The formal solution to this equation is given by

$$\rho(r) = c_0 J_0\left(\frac{r}{\lambda}\right), \quad (6.210)$$

where c_0 is a constant that is determined by the boundary conditions at the center (we can exclude the linearly independent solution $N_0(\lambda r)$ by noting that the density is not singular at the center of the cloud) and

$$\lambda = \frac{4\pi G^{1/2}}{K}$$

is the scale length from Eq. (6.209). The first zero of J_0 is 2.41, so if the outer boundary condition is the same as for the Lane–Emden equation, we obtain a radius for the cloud of $R_c \approx 6.4 \times 10^{-4}\rho_0/B_0$. The gravitational acceleration then follows from the density since by the Poisson equation:

$$\nabla \cdot \mathbf{g} = \frac{1}{r}\frac{d}{dr}rg = -4\pi G\rho, \quad (6.211)$$

so that we can integrate this equation to obtain the gravitational acceleration required for Eq. (6.120). This is the solution for a cylinder, or filament, which is one of the simple geometries that seems to apply to interstellar clouds.

Now let's examine an even simpler problem, ambipolar diffusion in a slab geometry. We can estimate the drift rates because the motion is diffusive. A particle executes a random walk within the gravitational field generated by the cloud itself so the diffusion velocity is $v_D \approx g_C t_{\text{coll}}$, where the gravitational acceleration of the cloud, g_C is given by $g = 2\pi G\Sigma$. Here Σ is the mass surface density and, for illustration, we are taking a planar approximation for the mass distribution. The collision rate between neutrals and ions is

$$t_{\text{coll}}^{-1} = n_i \langle \sigma v_{\text{th}} \rangle, \quad (6.212)$$

so that the drift speed is approximated by:

$$v_D \approx \frac{X_i \langle \sigma v_{\text{th}} \rangle n_i}{2\pi G m_p (1 + 4y)}, \qquad (6.213)$$

where y is the number density of helium relative to hydrogen and X_i is the ionization fraction of the cloud. Notice that this depends only on the temperature and cross section for the collisions, not on the number density. This is simply the collisional random walk *once the effects of the magnetic field have been removed* and represents the rate at which the neutrals will settle to the midplane. Since the neutral population is some fraction of the total cloud and exists only in a statistical sense, this is also the rate for any particle to drift toward the midplane of the cloud.

You can also make use of the scaling arguments from the virial theorem to see how a magnetic medium is stabilized. The characteristic scale of the gravitational instability, if magnetic fields dominate the pressure, is provided by the Alfven speed, v_A, instead of the sound speed, c_s. Additional insight comes from applying the virial theorem to the equilibrium structure. Molecular clouds are an interesting variant on the application of the virial theorem. You will recall that in Chapter 1 we repeatedly stressed that the virial surface integrals vanish for stars and that the boundary conditions are compatible with vacuum. That is, for the surface, we can take $P(R_\star) = 0$ and $T(R_\star) = T_{\text{eff}}$ as a first approximation. For the full virial, including magnetic fields (which come in as a pressure term), we can write

$$\frac{1}{2}\ddot{I} = 2T + W + 3\int (P_g + P_{\text{mag}}) dV - \int dS \hat{n}_j \left[r_j P_g + \frac{1}{4\pi} r_i \left(B_i B_j - \frac{1}{2} B^2 \delta_{ij} \right) \right]. \qquad (6.214)$$

Virial arguments have been used to support the α_{CO} calibration by looking at the linewidths for CO and comparing them with the results from the cloud observations. The procedure is, unfortunately, not completely self-consistent in that the lines may be affected by saturation and the widths are not obviously due to mass motions. Assuming only diffusive motions, which insures that the acceleration term in Eq. (6.214), $\ddot{I}$, is too small to matter, we can treat the global equilibrium of the cloud when it is supported by only the magnetic field. In this case, the limiting mass is

$$\frac{GM^2}{R} = \int \frac{B^2}{8\pi} \mathbf{r} \cdot \hat{\mathbf{n}} \, dS, \qquad (6.215)$$

so that the critical mass scales as $M_c \sim \Phi_B$, where now $\Phi_B = \pi B R^2$ is the magnetic flux. The precise value for a uniform density sphere is

$$M_c = \left(\frac{5}{12\pi^2 G} \right)^{1/2} \Phi_B. \qquad (6.216)$$

From our instability calculation, a tangled magnetic field changes the density dependence of instability because v_A depends on both the field strength and the density:

$$\lambda_{J,B} = \frac{v_A}{(G\rho)^{1/2}} \sim \frac{B}{\rho} \sim \left(\frac{\Phi_B}{\rho}\right)^{1/3}. \tag{6.217}$$

Now if the field is frozen in, then λ is simply a constant and the resulting magnetic Jeans mass is

$$M_{J,M} \sim \Phi_B, \tag{6.218}$$

which is the same result we got from virial equilibrium. If we take instead a polytropic law, $B \sim \rho^{2/3}$, we get a constant value for $M_{J,B}$, independent of the density.

Now we come to why ambipolar diffusion plays such an important role in the structure of magnetic clouds. The limit we have just derived depends on the *ionized* component of the gas. In a sufficiently dense medium, the ions and electrons are coupled strongly enough to the neutrals through collisions that they behave as a single fluid. Under interstellar conditions, however, they may separate. As a result, the neutrals "slip through" the field lines and, in a self-gravitating cloud, may reach the Jeans limit and collapse. There is a delay in the process, the ambipolar diffusion time, but eventually the medium can become unstable, although the collapse time will be increased by the diffusion for a partially ionized medium. The lesson is that field exclusion is possible, but some magnetic support may still be inevitable for delaying the collapse. Calculations indicate that the equation of state lies between these two extremes, so at some level, no matter how turbid the medium, some diffusion will occur.

6.8.1.2 Equilibrium of Pressure-Bounded Spheres

An analogous situation is encountered for pressure bounded spheres, when the surface integral of the virial equation does not vanish. We can start with

$$2K + \Omega + 3\int P\,dV = 3\frac{\mathcal{R}}{\mu}TM - \frac{k_g GM^2}{R} + 4\pi PR^3 = 0, \tag{6.219}$$

where k_g is a geometric factor related to the central concentration (for a uniform sphere, $k_g = \frac{3}{5}$) and ask what happens if we assume the temperature is extremal with respect to the radius. Then $dT/dR = 0$ gives

$$R_c = \left(\frac{GM^2}{12\pi P}\right)^{1/4}, \tag{6.220}$$

from which, on resubstituting into Eq. (6.219), we find the limiting mass:

$$M_c \sim \frac{T^2}{G^{3/2}P^{1/2}}, \tag{6.221}$$

where the proportionality constant depends on k_g. A more general expression, allowing the kinetic term in the virial equation to represent either thermalized or simply random motion, is called the *Bonner–Ebert mass*:

$$M_{BE} = \frac{c_s^4}{\left(G^3 P_{ext}\right)^{1/2}}, \quad (6.222)$$

which you can again arrive at by assuming an isothermal gas in the cloud acted on by a negative pressure gradient from the background medium.[132]

6.8.1.3 Relation to Star Formation

We thus have three estimates of critical masses depending on whether we assume support of the cloud by either turbulent or thermal motions (Jeans), quiescent magnetic fields (ambipolar diffusion), or pressure bounding and either thermal or random internal motions (Bonner–Ebert). For the conditions encountered in galactic molecular clouds, these are all rather similar, of order stellar masses. If, however, the temperature is large, or depending on the Alfven speed, very massive structures ones may reach equilibrium or be critically unstable. One scenario for star formation in clouds involves pressure fluctuations within the medium translating into different unstable masses. For instance, high-pressure environments that are also high-temperature ones may lead directly to clusters which then collapse and fragment. Cold, dense regions, on the other hand, can directly form stars. Lacking rotation, this picture naturally accounts for the hierarchy of products of star forming activity observed in clouds and may account for the mass function. When discussing the formation of globular clusters, which in terms of mass, bridge the range between open clusters and associations and dwarf galaxies, the variation of both pressure and temperature will be important.

6.8.2 Thermal Instability and Multiple Phases

We neglected the energy equation in setting up the conditions for the Jeans gravitational instability. Because gravity is the restoring force, the physical setting is again similar to the pulsation problem you saw in Chapter 4. We assumed, in order to maintain constant sound speed for the Jeans problem, that the medium responds isothermally. In effect, this means having already assumed a solution for the energy conservation equation. In the pulsation problem, however, we needed to be more precise because the coupling between compression and heating counteracts the gravitational acceleration in maintaining the pulsation. The reason was the requirement that the medium remain approximately hydrostatic. If cooling occurs too quickly in that problem, the medium could collapse into a very thin layer. This is what happens in the cooling instability. It is simply the requirement that the gas not cool faster than an acoustic wave can restructure the gas.

If a medium is optically thick and cools slowly, it *can* behave adiabatically. Should the cooling time become very short for any reason, then the pressure can

---

[132] Bonner, W. B. 1953, *MNRAS*, **116**, 351; Ebert, R. 1957, *Zeit. f. Ap.*, **42**, 263; Chieze, J.-P. and Pineau Des Forets, G. 1987, *Astr. Ap.*, **183**, 98.

change in a small region on the thermal timescale, which may be very much faster than a sound wave can transmit the information about this change to the rest of the gas. If there is no additional heat source, the gas may locally begin to collapse on itself. The pressure of the background medium is finite. This means that a drop in pressure locally due to a rapid drop in temperature leads to a large pressure deficit very rapidly and the region implodes. Such an event is analogous to the collapse of a bubble, a very complicated hydrodynamic event.

Let's return to the causes of heating and cooling in the ISM. First for the heating, there is the background radiation field. This may be due to a nearby star, or to the DIRF, but either way, the heating rate is proportional only to the density of the absorbers. Heating results from photoionization, the injection of hot electrons into the medium through absorption. We don't need to consider photodissociation; the input energy of the fragments is negligible. Thus, the *temperature independent* heating rate is

$$\Gamma = \Gamma_{\rm rad} n. \qquad (6.223)$$

The cooling, on the other hand, is collisionally dominated and depends very strongly on the velocity distribution of the incident particles. Due mainly to electron excitation of fine structure transitions, most of which are in the far infrared, where the medium is transparent, the rate depends on both the density of the medium and the temperature.[133] We can therefore write

$$\mathscr{L} = -\Gamma(\rho) + \Lambda(\rho, T) \qquad (6.224)$$

for the net cooling function.

If the net cooling rate should become positive for any perturbation, the result is a thermal catastrophe for the gas.[134] Its temperature must drop with the attendant increase in the density. If the cooling function has the wrong temperature dependence (from the point of view of stability), it will continue to increase in its cooling and the instability will lead to a rapid collapse of the pressure support for the perturbed gas. To see how this happens, let's assume the ideal gas law for the equation of state. Then the pressure perturbation can be written as

$$\frac{\delta P}{P_0} = \frac{\delta T}{T_0} + \frac{\delta \rho}{\rho_0}. \qquad (6.225)$$

One immediate assumption, that the pressure within the cooling region is the same as the background pressure, leads to a reduction in the order of the equations because it immediately connects the density and temperature. We used this when

---

[133] Two early, instructive papers dealing with cooling rates of fine structure transitions are Bahcall, J. N. and Wolf, R. A. 1968, *ApJ*, **152**, 701; and Penston, M. V. 1970, *ApJ*, **162**, 771.
[134] The classic paper is Field, G. 1965, *ApJ*, **142**, 531.

discussing convection, and here it comes again. You see immediately from this what happens in the instability. The cooling is found through the condition of energy conservation:

$$\frac{DQ}{dt} \equiv \rho T \frac{dS}{dt} = c_v \frac{dT}{dt} - \frac{P}{\rho^2}\frac{d\rho}{dt} = \frac{1}{\gamma-1}\frac{\mathcal{R}}{\mu}\frac{dT}{dt} - \frac{1}{\rho^2}\frac{d\rho}{dt} = \mathcal{L}, \quad (6.226)$$

so it is through this equation that coupling is achieved among the thermodynamic variables. The heating depends on the environment. The cooling is, however, the local term. It is very sensitive to the metallicity, density, and temperature of the gas. In effect, the heating is due mainly to the liberation of electrons with kinetic energies well above the ionization threshold, while the cooling is due mainly to very-low-energy processes, possibly even below the thermal energy, through optically thin infrared fine structure transitions from light ions. For coronal gas, we should add here, the metal lines are still the dominant sources of loss, but now high-energy transitions may act to regulate the thermal balance, from highly ionized species, and with energies closer to the thermal energy of the gas. The continuity and momentum conservation equations describe the *mechanical* response of the gas to small changes in the *thermal* conditions. As you know, density changes lead to flows through these two equations, and the thermodynamics are governed by Eq. (6.226). Even ignoring self-gravitation, we can drive flows because of pressure equilibrium if variation develop in P within the medium. Think, for example, of winds on Earth. These are not, after all, driven by the self-gravitation of the atmosphere but by dynamical responses to pressure differences in altitude and along the surface. This is where the assumption that the gas is isothermal circumscribes the instability in the Jeans problem. The thermal equation doesn't require gravity, but does relate to the Jeans problem through the changes in the sound speed. If we include gravity in the momentum equation, we obtain the full hydrodynamic condition that drives gravitational collapse, namely, cooling *and* self-gravitation. But here we will concentrate only on the thermal problem.

A net excess of radiative loss over all available heating mechanisms drives the instability through $\mathcal{L}$. This is a function of both temperature and density. For the interstellar medium, photoionization is one of the main heating mechanisms, as we have seen in the discussion of H II region expansion. Hard photons impinging on a gas ionize it and inject energetic electrons into the gas. These free electrons subsequently collide with other gas particles and redistribute their kinetic energy. Ultimately, this heats the gas.

This input of energy may, however, have an unfortunate consequence. For some processes, the rate of radiative loss is an decreasing function of temperature. This produces a cooling catastrophe. The cooler the gas becomes, the greater its cooling rate becomes. Thus, after the heating is turned off, it is possible that nothing stops the gas from cooling to a temperature below that of its environment. Such a cooling curve is known for low-density plasmas. Below $\sim 10^4$ K, it behaves like $\Lambda \sim T^2$. But above this, the temperature increase leads to a lower cooling rate; in fact, $\Lambda \sim T^{-1/2}$ until $\sim 10^7$ K. As a quantitative estimate, we use an approximation for the cooling timescale for the hot postshocked gas[135] which, expressed as a cooling time (in years) for gas with $T \leq 10^4$ K, is

$$t_c = 4.3 \text{ yr} \left(\frac{10^6 \text{ cm}^{-3}}{n_e} \right) \left(\frac{T}{10^4 \text{ K}} \right)^{1/2}. \tag{6.227}$$

For higher temperatures, the metallicity is far more important because of the role of fine-structure transitions of the CNO group and also the iron peak group.

We will again take as our model system a slab of gas in order to reduce the dimensionality of the problem, and assume that it is initially uniform density and has no large-scale ordered motion. Now we repeat the same procedure as in the Jeans instability, starting with the continuity equation:

$$\frac{\partial \rho}{\partial t} + \frac{\partial}{\partial x} \rho v = 0 \tag{6.228}$$

which becomes, for our static initial conditions

$$\lambda \delta \rho + i k \rho_0 \delta v = 0. \tag{6.229}$$

Here, because we are not necessarily expecting periodic solutions, we've used λ for the time dependence of the perturbation. The momentum conservation equation

$$\frac{dv}{dt} + \frac{1}{\rho} \frac{\partial P}{\partial x} = 0 \tag{6.230}$$

now becomes

$$\lambda \delta v + i \frac{c_s^2}{\gamma} k P_0 \left(\frac{\delta \rho}{\rho_0} + \frac{\delta T}{T_0} \right) = 0. \tag{6.231}$$

Notice that *now* we have to include the pressure perturbation because the gas is not isothermal and the compressibility has to be treated explicitly. We will use an ideal gas equation of state, otherwise we would have to employ

$$\delta P = \frac{\partial P}{\partial \rho} \delta \rho + \frac{\partial P}{\partial T} \delta T. \tag{6.232}$$

This pressure expansion is very important and is the reason why the thermal instability occurs. The pressure depends on density and temperature. In turn, the temperature depends on the net cooling, that itself depends on temperature and density. The change in the gas temperature causes a local change in the pressure that is amplified by flows, through the effect of the mass conservation and

---

[135] See Rosner, R., Tucker, W. H., and Vaiana, G. S. 1978, *ApJ*, **220**, 643; Priest, E. 1982, *Solar Magnetohydrodynamics* (Dordrecht: Reidel); Theis, C., Burkert, A., and Hensler, G. 1992, *Astr. Ap.*, **265**, 465.

compressibility of the medium. In this way, you see how the thermal instability exacerbates the disequilibrium created by self-gravitation. It may precede the isothermal state and therefore set the stage for the gravitationally driven collapse.

It is important to note that the combined momentum and continuity equations, which usually suffice to produce sound waves, also introduce a spatial dependence into the instability. Take $t_s = L/c_s$ and $t_c = c_v T_0/|\mathscr{L}|$. Then

$$L_\star = \frac{c_s c_v T_0}{|\mathscr{L}|}, \qquad (6.233)$$

and if the length scale is greater than this, a sound wave cannot traverse the medium fast enough to prevent the onset of collapse from the cooling. The dispersion relation comes from

$$(c_v \lambda - \mathscr{L}_T)(\lambda^2 + k^2 c_s^2) - k^2 c_s^2 (\lambda c_s^2 - \mathscr{L}_\rho) = 0 \qquad (6.234)$$

using the abbreviated notation

$$\mathscr{L}_x \equiv \left(\frac{\partial \mathscr{L}}{\partial q_i}\right)_{\text{(everything else constant)}}$$

with $q_i = (\rho, T, P)$. To reduce Eq. (6.234), we have used

$$\mathscr{L}_\rho \delta\rho + \mathscr{L}_T \delta T = c_v \lambda \delta T - \frac{P_0}{\rho_0} \lambda \delta\rho \qquad (6.235)$$

for the expansion of the energy conservation equation. The thermal balance equation completes the set and represents the principal difference between this instability and either the acoustic or Jeans problem. Although we have ignored gravity, it is obvious that it can be included, although with the introduction of a characteristic length or timescale.[136] We can even add an additional source for energy loss, conduction, through a modification of the energy equation:

$$\frac{P}{\gamma - 1} \frac{d}{dt} \ln(P\rho^{-\gamma}) + \rho \mathscr{L} - \nabla \cdot K \nabla T = 0. \qquad (6.236)$$

Here K is the *thermal* conductivity. We have included the possibility of thermal gradients driving heat transfer. Precisely how effective this is in changing the thermal balance of a gas in the ISM[137] is not known, but its importance is that it is

---

[136] One way to see what happens here is to compare the cooling time, t_c with t_{ff} and $t_s = L/c_s$. Since t_c depends on density and temperature, while t_{ff} depends only on density, this gives a critical equation of state for the stability of a self-gravitating, cooling slab. Recall that the Jeans instability required an isothermal condition. If $t_c < t_{ff}$, then the gas will be unable to support itself against gravity and collapse is inevitable. Even if the sound crossing time is short enough, the cooling is local and therefore hydrostatic equilibrium cannot be re-established.

[137] This is also true in the *intergalactic medium* of clusters of galaxies, where this instability has also been invoked to explain the formation of filaments and cooling flows; see Chapter 7.

diffusive. Thus it introduces a characteristic *length* to the problem and a characteristic timescale. This is because $t_{\text{diff}} = L^2/K$, which can be compared with the other timescales to show that there is a typical length for a give cooling rate. The same is true for advection. Both diffusion and large-scale motion limit the growth of the instability.

The equilibrium state for the gas is one of thermal balance, where $\mathscr{L}(\rho_0, T_0) = 0$. The perturbations in the net loss function are expressed in terms of the primitive thermodynamic variables of density and temperature, so with these the energy equation becomes

$$-\lambda \frac{P_0}{\rho_0} \delta\rho + \frac{\lambda}{\gamma - 1} \delta T + \rho_0 \mathscr{L}_T \delta\rho + \rho_0 \mathscr{L}_\rho \delta T + k^2 K \delta T = 0. \quad (6.237)$$

The most important new terms to appear in this problem are the two derivatives of the loss function. They can be either positive or negative and therefore they control the onset of the instability. A critical point is reached if $\lambda = 0$, because this means that the medium may become overstable. The full dispersion relation is

$$\frac{\lambda^2}{k^2} = \frac{P_0}{\rho_0} \left[1 - \left(\frac{\mathscr{L}_\rho + \lambda P_0/\rho_0^2}{\mathscr{L}_T - \lambda c_v \rho_0} \right) \left(\frac{\rho_0}{T_0} \right) \right], \quad (6.238)$$

so that for the critical solution, in the absence of conduction (which is usually a good approximation in the ISM), we get

$$\mathscr{L}_T - \frac{\rho_0}{T_0} \mathscr{L}_\rho = 0. \quad (6.239)$$

Because conduction introduces a characteristic length scale into the system, the instability is no longer local, and the dispersion relation reflects this. Even if there is no conduction, there is instability if $\mathscr{L}_T < 0$.

Cosmic rays contribute to heating by their role in collisionally ionizing the gas. Superthermal electrons rapidly collide with atoms in the gas and this increases the mean kinetic energy. It is a linear function of the density, but is temperature-independent. On the other hand, the cooling function depends quadratically on the density, and because of the collision rate, also on the temperature. Ultraviolet photons heat the gas depending only on the spectral distribution of the DIRF and the density. Therefore, the heating depends on the combined rates as $\Gamma = (\zeta_{\text{CR}} + \beta)n$, where ζ_{CR} is the incident cosmic ray proton flux and β is the ionization rate for the DIRF. Assuming, for a moment, that β dominates, the average energy of a released photoelectron is about $h\langle \nu \rangle - \chi_H$ while for cosmic rays, it is $E_{\text{CR}} - \chi_H$. In both cases, this is well above the observed temperature of the diffuse gas, about 100–1000 K.

The existence of a density- and temperature-dependent cooling function means that, in general, the interstellar gas will not be a single-phase medium. The original picture of the ISM was of a two-phase gas in pressure equilibrium. One phase, the *clouds*, had a density ranging from 100 to 1000 cm^{-3} and a temperature of around

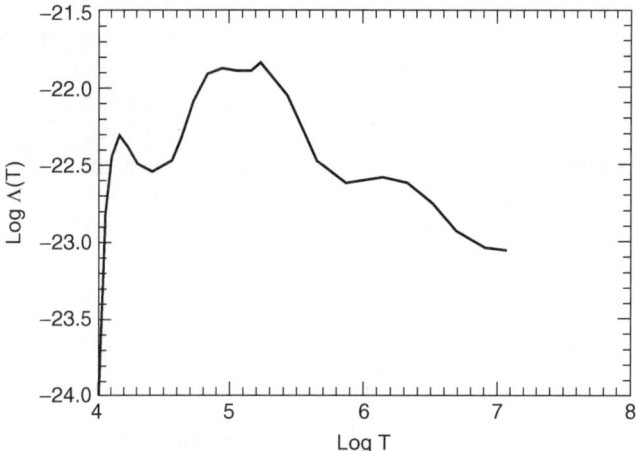

FIGURE 6.7. Cooling function for a solar abundance gas (see, e.g., Rosner, R., Tucker, W. H., and Vaiana, G. S. 1978, *ApJ*, **220**, 643). Notice how rapidly the cooling rate drops with decreasing temperature once the hydrogen lines and bremsstrahlung cease to be efficient. Molecular cooling dominates $T < 10^3$ K (see text).

100 K. This was embedded within a low density, 0.01–1 cm^{-3}, hot, of order 10^4 K, background called the *intercloud medium*. From the point of view of the basic thermal instability, with only atomic coolants and with heating provided mainly by cosmic ray ionization, this was quite reasonable. The discovery of two additional phases of the gas had a profound effect on the model of the ISM. Both came from opening new wavelength regions. The first was the the observation with the *Copernicus* satellite, using a high resolution ultraviolet spectrograph, of the O VI 1032 Å resonance absorption line along many lines of sight in the solar neighborhood. Although ionized species had been detected, this was the highest ionization stage whose resonance transitions have been studied at high resolution. It implies a kinetic temperature of order 10^6 K. The second observation was the discovery of the galactic X-ray background, which has been most extensively explored with the ROSAT satellite. It is now clear that the interstellar gas is far less homogeneous than previously thought, but this is not really surprising. The frequency of supernovae and wind bubbles is sufficiently high that any parcel of gas is frequently subjected to mechanical heating. In addition, turbulence plays a role in restructuring the diffuse gas. It is clear that the thermal instability, as originally applied to the ISM, *does* occur. But in addition, there are other unstable phases that have their own coolants and heating mechanisms and the dynamical picture is rich and complex.

The cooling function $\mathscr{L}$ is also important in determining the temperature distribution of the lowest-density material. You can see in Figure 6.7 that if the gas is heated above about 2×10^6 K, the medium is thermally stable. This is because of the loss of coolants at the highest temperature, particularly the CNO group. Since $\partial \mathscr{L}/\partial T > 0$ above this temperature, the shocked gas within supernova remnants should cool rather slowly and stay hot for quite a long time. This is also because the cooling rate depends on the *square* of the density, which is very low in the evacuated cavities of stellar winds and supernova remnants. ROSAT observations confirm this spectacularly. The Sun is located on the periphery of such a region, where the density is about 0.01 or less; this region extends toward β CMa and is responsible for the very low local interstellar Lyman continuum absorption.[138] Much of the O VI 1032 Å absorption detected by the *Copernicus* satellite comes from this gas. It should not come as a surprise that it exists, perhaps the historical oddity is that it took so long to be recognized. The most spectacular evidence of the X-ray background comes from the detection of shadows in the direction of nearby translucent clouds. These can be seen as absorbers against the diffuse background gas even with rather low neutral column densities ($> 10^{19}$ cm^{-2}).[139] In fact, as often happens in astrophysics, what was at first used as a means to show the distant location of the X-ray emission can now be used to calibrate the CO-to-H$_2$ conversion for translucent clouds and obtain information about cloud structure.

6.8.3 Pressure-Modified Gravitational Collapse

In this interplay between magnetic fields, pressure, and gravity, we glimpse the origin of stars. Any reduction in the pressure, with an accompanying increase in density, drives the medium farther from equilibrium and closer to collapse. The dynamics of the initial collapse phase can be greatly simplified by examining similarity solutions, as we did for cases of expanding gas. This time, turn the telescope around, so to speak, and study the problem of accretion of mass onto an already collapsed core.

Imagine an isothermal core that has a temperature T and, therefore, a sound speed c_s. It isn't, however, critical to the argument that the support is thermal. We merely assume the existence of a typical, constant speed that connects directly with the dynamical support of the cloud. Now imagine that the core suddenly and immediately collapses, which happens when the region exceeds the Jeans limit, and therefore the freefall time is short relative to the crossing time for a signal at the sound speed or turbulence speed. Exactly as we saw with collapse during a supernova, matter outside the core region begins to freefall inward and a rarefaction wave propagates outward. The rate of infall is governed solely by gravity, since the pressure is assumed to be unable to stabilize the material, and the rate at which matter crosses the outward advancing front depends only on the rate at which it accretes, in other words in the freefall time. Now the rate of transport through the surface is $\rho r^2 c_s$ and $r = c_s t$ and $t^{-1} \sim G\rho$ so the rate of mass accretion is

$$\dot{M} \sim \frac{c_s^3}{G}. \qquad (6.240)$$

---

[138] Models for this gas began with McKee, C. F. and Ostriker J. P. 1977, *ApJ*, **218**, 148 and McCray, R. and Snow, T. P. Jr. 1979, *ARAA*, **17**, 213. More recent reviews are by Spitzer, L. Jr. 1990, *ARAA*, **28**, 71; Shelton, R. and Cox, D. P. 1994, *ApJ*, **434**, 599 and references therein.

[139] The most thoroughly studied example is the Draco Cloud (also know as MBM 41), see Moritz, P. et al. 1998, *Astr. Ap.*, **336**, 682. A catalog of such shadows is given in Snowden, S. L., et al. 2000, *ApJS*, **128**, 171.

This translates into an accretion rate of about $3 \times 10^{-4} c_{s,0.5}^3$ $M_\odot$ yr^{-1} for a characteristic speed of 0.5 km s^{-1}, which is often seen as the typical random motion in molecular clouds.

From the similarity viewpoint, there are actually *two* dimensioned constants in the problem, G and c_s. The radius can be scaled at any time to the distance to which a signal has propagated

$$\xi = \frac{r}{c_s t},$$

and the density can be written in terms of the freefall time and a function, $D(\xi)$, that depends on this similarity variable:

$$\rho(r,t) = \frac{1}{Gt^2} D(\xi).$$

This same scaling transformation for the velocity leads to $v(r,t) = c_s V(\xi)$ since we already have a constant velocity in the picture. The continuity equation now becomes

$$\xi \frac{dD}{d\xi} + 2D - \frac{1}{\xi^2} \frac{d}{d\xi} \xi^2 DV = 0. \tag{6.241}$$

You will have noticed that during the scaling, you have found that:

$$\dot{M} \sim \rho r^2 v \sim \frac{c_s^2}{G} \xi^2 DV, \tag{6.242}$$

which is the mass accretion rate from Eq. (6.240). This collapse mode is sometimes called "inside-out" star formation because the growth of the core is induced by the infall of material through the outwardly moving rarefaction surface.[140] Now define the mass to scale as

$$M(r,t) = \frac{c_s^3 t}{G} \mu(\xi)$$

and put this into the equation of motion to obtain

$$(V - \xi) \frac{dV}{d\xi} = \frac{1}{D} - \frac{\mu}{\xi^2}. \tag{6.243}$$

For a nonsingular solution for the velocity, we use a power law for the density profile, $D = \xi^{-2}$, which gives the following profile:

$$\rho = \rho_0 r^{-2}. \tag{6.244}$$

---

[140] See Shu, F. H. 1977, *ApJ*, **214**, 488. A solution also exists, the *de Sitter* solution, for which $D(\xi)$ is a constant. In this case, we use $\mu = D\xi^2(V - \xi)$ and find the critical solution $V \sim \xi$; many thanks to Daniele Galli for extensive discussions. See also Chapter 8.

Notice that this solution also gives a front that expands as $V = \xi$. This profile has been derived a number of times since 1977 and in several different ways, including numerical collapse simulations. It is very important to have such models for the interpretation of molecular line observations of cold cores, where different chemical species trace the density and thus provide information about the eventual growth of the protostellar core. In addition, this solution provides a model for the environment into which any jets or winds emitted from the nascent star will progress.

6.8.4 Angular Momentum

Angular momentum complicates this picture. We assumed that the core can collapse radially. However, any extended structure in a differentially rotating system as some vorticity. Take a cloud rotating locally at the epicyclic frequency, κ. If this were to collapse without loss of angular momentum, then the final rotation frequency would be $\omega = (R_0/R)^2 \kappa$, which is a very large number for reasonable initial (radii of order 0.1 pc) and final (radii of order a few $R_\odot$), one that produces a centrifugal barrier for any accretion. Remember, the core mass is growing only at the accretion rate so if the matter can become centrifugally supported, as it is for an accretion disk, only viscosity can lead to accretion. The analogy with the binary star case is striking. Clearly, something is wrong. The choice seemed, originally, to be between removing the angular momentum—usually using a magnetic field—and/or changing the conditions of collapse. The absence of strong evidence for rotation of dense cloud cores, where one would expect the physical conditions to be appropriate for star formation, suggests that the problem is not as severe as first thought. Nevertheless, rotation *is* observed in all stars and many young stellar objects, and a different solution has presented itself based on both the basis of observations and theory—disks and jets appear to be consequences of the formation process.[141]

6.9 LARGE-SCALE DISTRIBUTION OF THE GAS

Since the bulk of interstellar neutral hydrogen is confined to the plane of the Galaxy, and along most lines of sight is sufficiently optically thick to prevent full disk mapping, the overall distribution of gas in the Galaxy is still a problem. Viewed from within the disk, the structure of the ISM is very complicated. Often, it is best understood first by observing extragalactic systems and then, by analogy, looking for signatures of the same processes in the Galaxy. Large-scale flows are known from observations at higher galactic latitudes, where H I maps show arcs and streamers extending away from the galactic plane. These are also seen in continuum maps, where presumably the synchrotron emission is tracing the magnetic field structure.

---

[141] A variety of driving mechanisms have been proposed, and Hartmann, L. 1998, *Accretion Processes in Star Formation* (Cambridge, UK: Cambridge Univ. Press) provides a superb guide to the literature.

6.9.1 Buoyancy and the Rayleigh–Taylor Instability

First, a bit of background. One of the classical gravitationally induced instabilities in a two-phase medium, the interplay of buoyancy with stability of the medium, is known as the Rayleigh–Taylor instability. You experience this whenever you try to immerse oil in water. The separation between them is driven by differential buoyancy in this way. Imagine you have two fluids of different density, ρ_1 and ρ_2 layered in a container in a gravitational field (for instance, on a kitchen table). Suppose that you have placed one fluid on top of the other, it doesn't matter which, but we will take fluid (!) on the bottom. Assume that they are incompressible, to make life easier. Now the downward displacement of a parcel of volume V with mass $\rho_2 V$ displaces a mass $\rho_1 V$. If the work is done on downward displacement, $|m_2 d\delta z| > |m_1 g \delta z|$, then the system is unstable and the fluid will experience a net acceleration. You can see this by writing the Lagrangian for the medium as $L = \frac{1}{2}(m_1 + m_2)\dot{z}^2 - g(m_1 - m_2)z$ since the displacements are assumed to be equal and opposite. Writing down the equation of motion for this simple one-dimensional system, you obtain $\ddot{z} = (m_1 - m_2)g/(m_1 + m_2)$. Since we are dealing with a continuous medium, you can change this to a dispersion relation of the sort you've used for the Jeans problem, so that

$$\omega^2 = \frac{\rho_2 - \rho_1}{\rho_2 + \rho_1} gk, \qquad (6.245)$$

and the instability grows if the heavier fluid is initially on top. This leads to the interchange of the two fluids because of the incompressibility and ultimate separation and stability of the layers, although with substantial generation of internal motions that only dissipate through viscosity. We hope that you see that this is analogous to a device you once met a long time ago in mechanics, an *Atwood machine*, the simplest mechanical model for a gravitationally driven machine. The Rayleigh–Taylor instability plays important roles in astrophysical fluids, such as the formation of structure in a propagating shock, where the pressure gradient takes on the role of an external acceleration and produces the same sort of buoyant instability along the front.

6.9.2 The Parker Instability

When a magnetic field is present, the same sort of unstable situation can occur.[142] In this case, the low density medium is a magnetic field (*not* a fluid) and a charged fluid rests under gravity supported by the field. A perturbation in the field lines generates a local acceleration that causes the medium to buckle and leads to the formation of large curtainlike structures. A dimensional argument illuminates the mechanism. For an ideal medium, the electric field induced by any motion is $E = -\mathbf{v} \times \mathbf{B}$ and the motions are induced by gravity. Then motion along a curves

---

[142] Kruskal, M. and Schwarzschild, M. 1954, *Proc. Roy. Soc.* (London), **A223**, 348. For thorough treatments of plasma instabilities, see Miyamoto, K. 1989, *Plasma Physics for Nuclear Fusion*, revised ed. (Cambridge, MA: MIT Press). See also Chandrasekhar, S. 1960, *Plasma Physics* (Chicago: Univ. Chicago Press) and Krall, N. A. and Trivelpiece, A. W. 1973, *Principles of Plasma Physics* (NY: McGraw-Hill).

field line produces an acceleration $\mathbf{v} \cdot \nabla \mathbf{v}$, the magnitude of which depends on the radius of curvature and whose sign depends on the direction (relative to gravity).

The three constituents of the ISM, the gas, magnetic field, and cosmic rays, are not strictly stable. This was first pointed out by Parker, for whom the instability is now named,[143] and there have been many studies since that support the basic picture. The galactic disk may be pressure-supported, but it is not immune to the formation of large-scale features that occur independent of dynamical forcing by supershells or other impulsive processes.

The thermal scale height is maintained by the gas pressure, while the magnetic field and cosmic rays may have a sufficiently high energy density that they help to support the gas. The medium is very compressible in this case, because the combined effects of the cosmic rays and the magnetic field are of a relativistic fluid with a soft equation of state compared with the gas. This is not precisely the same as a Rayleigh–Taylor instability because we have neglected the role of surface tension in the analysis, and that's the contribution of the **B** field. The picture is that a layer of gas is supported against its own gravitational force and that of the galactic disk by a magnetic field that lies approximately parallel to the plane. The field and cosmic ray pressure simulates a "light fluid" so you would expect this gas to be unstable. Now since the gas is charged and coupled to the field, variations in the constant density surfaces will drag the field with them and the deformation, through induction, generates a local reaction that suppresses the shortest wavelengths. Unlike the Rayleigh–Taylor instability, which is unconditionally unstable at all wavelengths when the density difference has the wrong sign, here we have suppression at large wavenumber because of the current due to the $\nabla \times \mathbf{B}$ term.

The full analysis of this instability in its classic form is left for Appendix 6.B. Here we discuss its role in structuring the medium. The growth rate can be estimated by changing the Lagrangian we just used for the Rayleigh–Taylor instability. Putting in a tension, $-\frac{1}{2} \mathscr{T} z^2$ changes the equation for the growth timescale. We still have a dependence on $\delta \rho$, but now there is a minimum characteristic length below which the instability will not grow. This isn't a damping term. Instead, it reflects the stiffness of the system to a change in the boundary. In other words, there are now two timescales, one that is unconditionally unstable (the density inversion) and the reaction that is periodic (since, you will notice, we have added an effective spring constant to the force that depends on the surface tension). The main addition required for the Parker instability is the generation of flows. This cannot be reproduced by the simple mechanical analysis we've just used.

Whether this is a means for generating interstellar clouds or cloud complexes as originally intended, or a generic behavior of the magnetic field of the disk, is actually beside the point. The disk is also differentially rotating and as the field is wrapped, it should become buoyant by some mechanism like the Parker instability. Since global fields are observed, and these have to come from somewhere, the

---

[143] Parker, E. N. 1966, *ApJ*, **145**, 811; Parker, E. N. 1992, *ApJ*, **401**, 135; Foglizzo, T. and Tagger, M. 1994, *Astr. Ap.*, **287**, 297; Basu, S., Mouschovias, Th., and Paleoglou, E. 1997, *ApJ*, **480**, L55; Hanasz, M. and Lesch, H. 1997, *Astr. Ap.*, **321**, 1007.

instability is seen as a vital element in a galactic dynamo that is the scaled-up version of the process we discussed in Chapters 4 and 5. Note also that supernova remnants may be important driving agents as they interact with the ambient galactic magnetic field.

6.10 TURBULENCE IN THE INTERSTELLAR MEDIUM

What supports interstellar clouds? You know from analysis of the Jeans instability that cold self-gravitating gas cannot support much. So naively you would expect that when observing dark clouds, they should be filled with small protostars and also collapsing on a very short timescale. This is not what you observe. First, although excitation of the molecular gas shows the kinetic temperatures are low, and the infrared emission from embedded dust agrees with this, the linewidths observed in almost all clouds greatly (by factors of two or more) exceeds the thermal linewidths. Regardless of the source for these motions, they cannot be thermal. In addition, the line profiles are smooth and broad. Yes, there are substructures, and along many lines of sight in a cloud you detect multiple components, but they are individually broad and the profiles are more or less Gaussian. In other words, it appears that there are large-scale motions within dark clouds that are thoroughly mixed through the interior that approximate fully developed random motion and yet are substantially higher than can be explained by the thermodynamic state of the gas. This combination of physical properties is the basis of the claim that the interiors of clouds, indeed most of the interstellar medium, is structured by turbulent flows.

But to what extent can turbulent motions support molecular clouds? The problem was first stated by Field in 1975. Turbulence is dissipative, by definition, and the luminosities of clouds in molecular lines indicates the rate at which they are losing this energy. Thus, if you look at the timescale that is expected for cloud collapse in the presence of this loss of energy, it is very short (of order 10^6 years) and therefore the clouds cannot be supported against self-gravitation by turbulence alone. Something must be driving the motions, or they will damp. Although we are not able to go into full detail here,[144] we can cover here some of the details that are relevant for understanding the interstellar medium. We could have placed a discussion of astrophysical turbulence almost anywhere in the book. This is one of the broadest unsolved problems in astrophysics today, and entire conferences and treatises have been devoted to this subject since the first meeting on cosmic gas dynamics in 1947.

[144] The literature is enormous, but we have our favorite introductions: Frisch, U. 1995, *Turbulence: The Legacy of A. N. Kolmogorov* (Cambridge, UK: Cambridge Univ. Press); Lesieur, M. 1993, *Turbulence in Fluids*, 2nd ed. (*Dordrecht*: *Kluwer*); MacComb, W. D. 1991, *The Physics of Fluid Turbulence* (London: Oxford Univ. Press); Tennekes, H. and Lumley, J. L. 1972, *A First Course in Turbulence* (Cambridge, MA: MIT Press); Townsend, A. A. 1976, *The Structure of Turbulent Shear Flow* (Cambridge, UK: Cambridge Univ. Press). For MHD turbulence, one of the best guides is Montgomery, D. 1989, in *Lecture Notes on Turbulence: NCAR-GTP Summer School* (eds. Herring, J. R. and McWilliams, J. C.) (Singapore: World Scientific Publ.) p. 75ff; Shu, F. H. 1992, *The Physics of Astrophysics*, vol. 2. *Fluid Dynamics* (Mill Valley, CA: Univ. Science Books); Shore, S. N. 1992, *An Introduction to Astrophysical Hydrodynamics* (San Diego: Academic Press).

6.10.1 The Role of Dissipation: The Kolmogorov Spectrum

Fluid turbulence is dissipative and therefore requires a continuously active source. It is usually described in terms of eddies, reflecting its fundamentally vortical nature. These evolve through a cascade, in which the largest eddies produce ones that in turn collide and further subdivide, and so on. This assumption was used in a classic paper in 1941 by Kolmogorov[145] as the foundation of the current scaling theory: the dissipation of the energy, derived from the largest scales of the turbulent flow, takes place at the level of the smallest eddies through viscosity at the kinetic level of the medium. He assumed that the origin of the viscosity in the flow, down to the molecular level, is the turbulence itself and therefore the spectrum should not depend on the viscosity. In the time independent version of this theory, when the turbulent flow is in equilibrium, this energy cascade in a homogeneous isotropic medium is governed by the rate at which the energy is being fed into the largest scale length.

Call the rate of energy dissipation ϵ and the coefficient of molecular viscosity ν. We will work with incompressible fluids for simplicity, but it is possible (difficult but possible) to extend this. If the density is constant, we can look at the combinations of these coefficients to obtain the turbulent length and velocity scales. Think for a moment about what happens in a turbulent medium. Whether we picture it as a sea of eddies or filaments, it is a mess. Macroscopic random flows occur on at many scales. The Reynolds number, $\text{Re} = UL/\nu$, now becomes a function of the length scale. Regardless how the flow becomes turbulent, we know this much—once it does, the random motions dominate over molecular viscosity. If we also assume the medium is stationary in time, then the energy must be transferred through a cascade from the largest scale, at which it is injected, to the smallest scale, at which it is dissipated as molecular chaos and ultimately as heat. Therefore, *at every length scale*, the principle introduced by Kolmogorov holds—the velocity fluctuations created by the driving are precisely those required to transfer the energy. Since the flow is dissipative at the smallest scale, there is a net direction to the flow of energy. In the inertial portion of the spectrum of fluctuations, universality means that the spectrum must be scale-free and therefore self-similar, so we can use dimensional analysis to obtain its form. The rate of energy generation (and transfer), ϵ, is dimensionally $L^2 T^{-3} = U^3 L^{-1}$ so, assuming ϵ is constant, we find

$$u = \epsilon^{1/3} l^{1/3}. \tag{6.246}$$

Once the flow turns turbulent, the viscosity is given by

$$\nu = ul = \epsilon^{1/3} l^{4/3}. \tag{6.247}$$

The upper length scale is determined by whatever drives the flow and is set

---

[145] Kolmogorov, A. N. 1941, *Dokl. Akad. Nauk SSSR*, **26**, 115; Kolmogorov, A. N. 1962, *J. Fluid Mech.*, **13**, 82. Both articles are reprinted as translations in Hunt, J. C. R., Phillips, O. M., and Williams, D., eds. 1991, *Turbulence and Stochastic Processes: Kolmogorov's Ideas 50 Years On* (Cambridge, UK: Cambridge Univ. Press) (also 1991, *Proc. Roy. Soc.* (London), **A434**, 1).

extrinsically to the medium. On the other hand, when ν is the molecular viscosity, the smallest length scale is given by dimensional analysis:

$$l_K = \left(\frac{\nu^3}{\epsilon}\right)^{1/4}, \tag{6.248}$$

referred to as the *Kolmogorov* or *dissipation* length scale. The characteristic timescale for the dissipation is

$$t_K = \left(\frac{\nu}{\epsilon}\right)^{1/2} \tag{6.249}$$

for smallest eddies, whose size is l_K, and their characteristic velocity is given by

$$v_K = (\nu\epsilon)^{1/4}. \tag{6.250}$$

Now we appeal to the equations of motion. From the Navier–Stokes equation,

$$\frac{du_i}{dt} = \nu \frac{\partial^2 u_i}{\partial x_j \partial x_j},$$

we take the scalar product with u_i to obtain the rate of energy transfer per unit volume (remember, this is for an incompressible fluid) and then integrate over volume to obtain ϵ:

$$\int dV \frac{1}{2}\frac{du^2}{dt} = \nu \int dV u_i \nabla^2 u_i = \nu \int dS\, n_j u_i \frac{\partial u_i}{\partial x_j} - \nu \int dV \left(\frac{\partial u_i}{\partial x_j}\right)^2. \tag{6.251}$$

The first integral over the surface S vanishes, and the second integral over volume is always positive so the fluctuations are always dissipative because of the viscous coupling. Now we define the spectrum, $E(k)$, to be a function of wavenumber, k such that

$$\epsilon = -\nu \int E(k) k^2\, dk. \tag{6.252}$$

Substituting Eq. (6.247) for ν in the inertial subrange, we obtain

$$\epsilon^{2/3} k^{4/3} = \int E(k) k^2\, dk. \tag{6.253}$$

We solve for E by differentiating both sides of Eq. (6.253):

$$E(k) = \frac{1}{k^2} \epsilon^{2/3} \frac{dk^{4/3}}{dk} \sim \epsilon^{2/3} k^{-5/3}, \tag{6.254}$$

which is the *Kolmogorov spectrum* for the velocity fluctuations. We emphasize that E describes the velocity alone because we assumed incompressibility. However, for a steady state, the density and velocity fluctuations are inverses, so we have the same spectrum for the density fluctuations. Turbulence spectra appear to be universal in this *inertial subrange*. A natural assumption is, therefore, that the process and its spectrum are independent of the viscosity. This is the critical point that drove the development of the Kolmogorov (1941) theory. He assumed that the spectrum is self-similar and that it scales through the characteristic velocity and length by

$$E(k,t) = v_K^2 l_K E_K(l_K k). \tag{6.255}$$

Here E_K is a universal *dimensionless* function of wavenumber. This energy spectrum leads immediately to the form

$$E(k,t) \sim \epsilon^{2/3} k^{-5/3}, \tag{6.256}$$

which is the *Kolmogorov spectrum*. The underlying assumption of stationarity of this spectrum is that, at some small length scale, dissipation of energy begins to dominate the spectrum, but that above that scale there is an *inertial* regime in which a steady state has been reached. The flow of energy through the various scales merely preserves similarity. As each eddy dissipates and transfers its energy to the smaller scale, others of the same size and vorticity are re-created by the breakup of even larger ones. The result is that at any moment there will be a population of that part of the spectrum that is responsible for the creation of the eddies above the Kolmogorov length. The triumph of this beautifully simple picture is that this spectrum agrees with both experiment and observations. The rate of energy dissipation is given by

$$\epsilon = -\frac{3}{2}\frac{du^2}{dt} = 2\nu \int_0^\infty E(k,t)\, k^2\, dk, \tag{6.257}$$

as we have discussed before, and this is the basis of the calculation of the rate of dissipation. Again, it is worthwhile to recall where this comes from, because for compressible turbulence it is a very different result. The rate of energy dissipation we have discussed in Chapter 5 for accretion disks, which are also thought to be dominated by turbulence. It is given by the shear σ_{ij} times the stress T_{ij}.

With reference to Appendix 6.D, we can find out more about the spectrum, especially what happens near the lowest length scale at which dissipation occurs even without the scaling argument for the viscosity. When an equilibrium cascade has been established, the rate of energy transfer between adjacent length scales is constant and one way, from large to small (viscously dissipative) lengths. As we did a moment ago, call this rate ϵ and assume that the flow adjusts to keep it constant. The timescale for such energy transfer is $t = \epsilon^{-1/3} k^{-2/3}$ from dimensional arguments. We can define a sort of "velocity" in this wavenumber, dk/dt, which is

$$\frac{dk}{dt} = \epsilon^{1/3} k^{5/3} \tag{6.258}$$

such that $dE(k)/dt = \dot{k}\, dE(k)/dk$. From the Fourier transform of the Navier–Stokes equation, we have:

$$\epsilon^{1/3} k^{5/3} \frac{dE}{dk} + 2\nu k^2 E = \mathcal{T}, \qquad (6.259)$$

where from Eq. (6.314) (Appendix 6.D) we combine the nonlinear (shear) terms with the pressure fluctuations and denote this as $\mathcal{T}$. Then from Eq. (6.259) we find

$$E(k) = E_0 k^{-5/3} \epsilon^{2/3} \exp(-\alpha k^{4/3} \epsilon^{-1/3} \nu)$$
$$+ \int \mathcal{T}(k') \left(\exp(-\alpha [k'^{4/3} - k^{4/3}] \epsilon^{-1/3} \nu) \right) k'^2 \, dk'. \qquad (6.260)$$

where E_0 and α are integration constants. In the absence of either external driving or large-scale shear flows (in other words, those conditions under which the right-hand side, $\mathcal{T}$ vanishes), we recover the Kolmogorov spectrum *and* the meaning of the Kolmogorov scale, k_K. At this critical wavenumber, molecular viscosity dominates and leads to an exponential cutoff in the spectrum past the inertial subrange. We note in closing that the importance of the time-dependent treatment is that it makes clear that the viscosity is derived independently of the cascade, and our assertion that the turbulence adjusts to maintain a steady energy flow is all that is needed to obtain the complete spectrum for isotropic turbulence.

6.10.2 The Role of Magnetic Fields: The Kraichnan Spectrum

The presence of magnetic fields on the large scale allow for strong coupling to exist between distant parts of the fluid in violation of the Kolmogorov locality assumption. In addition to viscosity, Alfven waves can carry energy away from an eddy faster than that energy can be otherwise dissipated. The rate of energy dissipation is consequently increased. To see this, we again examine the transform of the velocity, v_k, for wavenumber k. We note that the velocity is $v_k = [E(k)k]^{1/2}$, and therefore writing the rate of energy dissipation as $\epsilon = k v_k^3$ gives $E(k) = \epsilon^{2/3} k^{-5/3}$. So far, nothing is new, although we have sidestepped the whole discussion about the role of viscous dissipation by so doing. On the large (source) scale, l_0, we assume that there is a magnetic field B_0 with an Alfven speed v_A giving a characteristic timescale for coupling of l_0/v_A. This can be compared the eddy turnover time, $1/(k v_k)$. When $v_A/l_0 = k v_k$, the spectrum becomes

$$E(k) \sim \frac{v_A}{l_0} k^{-3/2}, \qquad (6.261)$$

which is the *Kraichnan spectrum*.[146] An alternative way of arriving at this result is to note that a magnetic field increases the dissipation rate by a factor of v_k/v_A and therefore $k v_k^4 / v_A = \epsilon$ with the spectrum given now by $E(k) = \epsilon^{1/2} v_A^{1/2} k^{-3/2}$.

---

[146] Kraichnan, R. H. 1965, *Phys. Fluids*, **8**, 1385; Matthaeus, W. H. and Zhou, Y. 1989, *Phys. Fluids B*, **1**, 1929; Montgomery, D. 1989, in *Lecture Notes on Turbulence: NCAR-GTP Summer School*, Herring, J. R. and McWilliams, J. C., eds. (Singapore: World Scientific Publ.) p. 75ff.

Further refinements for compressible magnetic turbulence have been examined that include the effects of anisotropy introduced by the large scale field.[147]

6.10.3 Driving the Turbulence

Where there is energy dissipation there must be supply if the medium is to maintain its state. This was originally proposed as a counterargument to the existence of true turbulence in molecular clouds because the eddy dynamical timescales seem so short. For instance, if a cloud radiates at a rate ϵ and if this is comparable to the energy generation rate for presumably turbulent motions, $\rho v_t^3/L$, then either we are seeing a very special stage in the life of all clouds, which seems silly, or something keeps the gas agitated. Several sources have been suggested. Internal star formation easily satisfies the requirements since protostars have strong outflows in the form of winds and jets that can penetrate deep into the medium. Since these are powered by the stellar luminosity, they tap a very strong gravitational potential and can do this for a long time. These are widely distributed sources within the cloud, however, and therefore it is harder to look for their signature at a source scale. Enough of these stars are likely to be present at any time that the signature of any one is washed out. Another proposed source is Alfven waves from the diffuse medium. Properly speaking, this is not really turbulence, but nonlinear interactions can lead to a dynamical cascade of energy and momentum that drives the motions required to support the cloud. Once structure forms within the cloud, individual steams and clumps can provide some energy for the turbulent field, but this is comparatively weak relative to the stellar input. For clouds with no internal star formation (the lowest-mass translucent clouds are good examples), a likely source of turbulence and organization comes from the large-scale motions of the diffuse gas.[148] These flows produce a cascade through shear interactions with the clouds that appears to have a characteristic scale length of about 0.1–0.5 pc, manifested by coherent structures that are well known from laboratory experiments.[149] Any of these, when coupled to the magnetic field by as small an ionization fraction as 10^{-5}, will generate MHD waves as a byproduct and produce the observed linewidths.

Turbulence can be included in the virial theorem through the Reynolds stress, which is equivalent to a nonthermal dynamical pressure. In Section 6.8.1.1, we described how to include magnetic fields and pressure bounding, and we can add to those effects the volume integral $\int P_{turb}\, dV$ in the cloud. The turbulence is, however, driven either by motions in the cloud itself generated by collapse, hence by changes in the internal mass distribution, or from the boundary.[150]

---

[147] Sridhar, S. and Goldreich, P. 1997, *ApJ*, **485**, 680 (and references therein).

[148] Magnani, L., LaRosa, T. N., and Shore, S. N. 1993, *ApJ*, **402**, 226; LaRosa, T. N., Shore, S. N., and Magnani, L. 1998, *ApJ*, **512**, 761.

[149] Townsend, A. A. 1975, *Structure of Turbulent Shear Flows*, 2nd ed. (Cambridge, UK: Cambridge Univ. Press); Holmes, P., Lumley, J. L., and Berkooz, G. 1996, *Turbulence, Coherent Structures, and Dynamical Systems* (Cambridge, UK: Cambridge Univ. Press).

[150] See, for example, McKee, C. F. and Zweibel, E. G. 1992, *ApJ*, **399**, 551.

6.10.4 Confronting Observations

What this means for astrophysical problems is not at all clear from the derivation. Notice that we have made the assumption that the medium is incompressible. This does not seem to be appropriate for cosmic media, and the assumption that vorticity is conserved, which is the primary source of the equations of motion, is clearly violated in the compressible case. However, we know that there will be a finite time for the decay of any turbulent spectrum, and can therefore at least see what the effects are of the application of the theory to astrophysically interesting situations. You may also object that the effects of gravity are thus far absent in our discussions, and quite rightly, but only on the Jeans scale do the effects of gravity play a significant role.

In our discussion of the interstellar medium, especially the multitude of dynamical processes, we have identified many possible sources that can drive turbulent motions. Consider, for instance, molecular clouds. The winds and turbulence generated thereby in the vicinity of young stars is well known to be a probable source for the turbulent structuring of molecular clouds. As soon as star formation begins, the attendant formation of outflows (both winds and jets) becomes a very important stirring mechanism. A typical protostellar wind with $\dot{M} \approx 10^{-7} M_\odot$ yr^{-1} and a terminal velocity of a few hundreds of km s^{-1} provides an equivalent mechanical luminosity of about 10^{33} erg s^{-1} to the cloud. This means that the effective volume that can be stirred up is about 10 pc^3 by a single star in about 10^7 yr, the typical lifetime of the cloud. The observed rate of star formation in a massive cloud is of order a few solar masses per year, so that once these compact bodies have been formed the cloud is permanently altered. It can therefore be said somewhat differently that there appears to be a self-regulatory process involved in the maintenance of turbulence and star formation within clouds. Even in the absence of internal stellar sources, the stirring of the environment by larger scale motions can maintain a turbulent cascade within clouds.

If the motions within a turbulent region are supersonic, internal shocks become the primary mode for dissipation. At high Mach numbers in the flow, simulations show that strongly compressed regions develop, with high vorticity, and the medium filaments rapidly into very localized regions. Some of these may still be carried in the bulk with supersonic speeds, but as they collide and the overall temperatures rises within the medium the motions become progressively more subsonic. Since contact surfaces are formed by the intersections of shocks, they will form on collision of the turbulent eddies and locally planar fronts within a supersonic medium. Such interactions inject vorticity, hence turbulent energy, into the medium on a small scale. Unlike the Kolmogorov model, these do not necessarily proceed from the large-scale structures through a cascade to a smaller scale, and it is possible that the inertial range does not exist within such media.

When a magnetic field is present, the picture changes. Waves can be transmitted at both the magnetosonic and Alfven speeds, and the gas can become structured into local islands with large density contrasts. The major difference between supersonic and subsonic turbulence is, however, manifested in MHD turbulence. Here initially supersonic, but subalfvenic, modes propagate nonlinearly and turn into internal shocks. The energy is rapidly dissipated, but in the process localized current sheets form. These serve, through reconnection, as excellent sites for dissipation. Alfven waves are free to move through the medium and promote cascades when they collide and scatter. Unlike the unmagnetized case, MHD turbulence has Joule heating, generated by the formation of these current sheets, for producing locally strong departures from the barotropic condition, $\nabla P \times \nabla \rho$, thereby feeding the local vorticity. Studies of supersonic MHD turbulence[151] must be performed numerically and models for interstellar conditions are still in relatively early stages of development. In MHD turbulence, the magnetic field topology as well as local strength is very important; additional pressure support comes from the change in the field strength, through the Alfven speed, so large low-density bubbles may develop in the medium in which there is very little fluid. When these collapse, they feed their energy dissipatively into the rest of the flow. The phenomenon is still poorly understood, but is very likely to have important astrophysical consequences.

Supersonic turbulence appears to be required to explain the structures and velocities observed in molecular clouds (Fig. 6.8). Typically, molecular clouds have infrared emission consistent with temperatures of a few tens of degrees. In extreme cases, where star formation is obviously heating the interior of the clouds, one sees $T_{\rm rad} \approx 50$ K. The thermal motions are consequently expected to be only a few hundred meters per second, at most. But the observed velocities dispersions in molecular lines are usually a few kilometers per second, very supersonic. The source, or sources, powering these motions remains elusive. Yet some constraints can be placed on it. The motions are clearly dissipative because the clouds can radiate their turbulent energy in the infrared (to which they are optically thin). We have already discussed the ideas for the large scale sources, but some internal sources of energy may also be important. At the densities typical of molecular clouds, Alfven speeds are of order $0.03 B_\mu n_4^{-1/2}$ km s^{-1}, where B_μ is the magnetic field in μG and n_4 is the number density in 10^4 cm^{-3}. Therefore, for fields of order a few 100 μG, the Alfven speeds will be of the order observed in the line profiles. These waves are certainly turbulent within the cloud, and provide a pressure that contributes to the support of the cloud through the virial theorem. Extensive two- and three-dimensional modeling of self-gravitating turbulent media is comparatively recent but some consistent results are emerging. Simulations find that M, the Mach number within the flow, fluctuates with an rms value below unity. This is true even for models that do not include radiative losses or magnetic fields. The dispersion in M does not strongly depend on the initial conditions, and can become very low, of order 0.1.[152]

Just looking at a medium and finding that it looks random does *not*, however, mean that it is turbulent. We need to stress this. Evidence of stochasticity must be accompanied by dissipation and the energetic and dynamical signatures associated with the flow. The most important indicator is the velocity autocorrelation function

---

[151] See reviews by Vazquez-Semadeni, E. 1999, in *Millimeter Wave Astronomy: Molecular Chemistry and Physics in Space*, Wall, W. F., Carramiñana, A., and Cararrasco, L., eds. (Dordrecht: Kluwer); Franco, J. and Carramiñana, A., eds. 1999, *Interstellar Turbulence* (Cambridge, UK: Cambridge Univ. Press).

[152] On the applicability of the Kolmogorov spectrum to clouds, see Passot, T., Pouquet, A. and Woodward, P. L. 1988, *Astr. Ap.*, **197**, 228. For a discussion of the interplay between supersonic turbulent motions and self-gravitation, see Leorat, J., Passot, T., and Pouquet, A. 1990, *MNRAS*, **243**, 293.

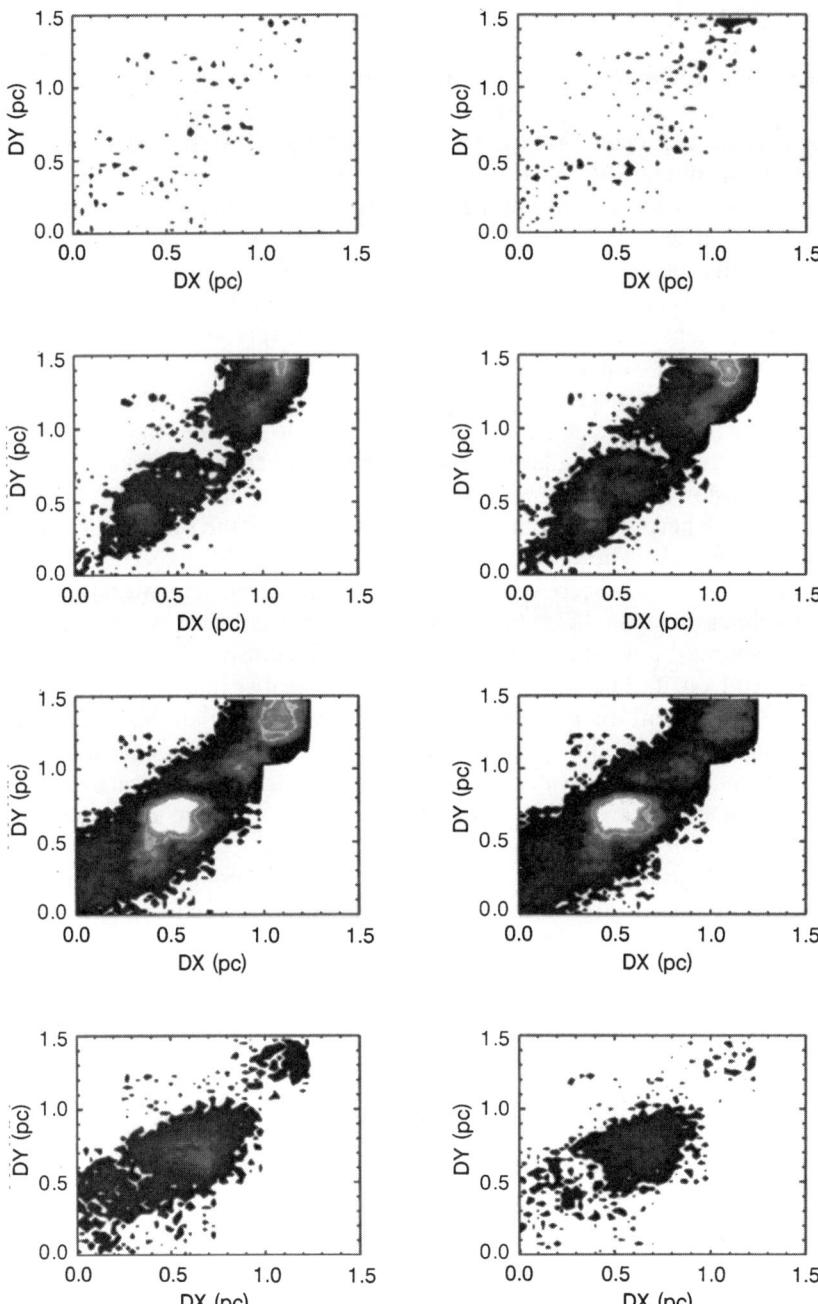

FIGURE 6.8. Channel maps for the ^{13}CO (1–0) transition for the core of the translucent molecular cloud MBM 40. The map covers about 1.5 pc on a side; peak emissivity is about 4 K. Each channel is about 0.08 km s^{-1} wide; the first channel is 2.43 km s^{-1} (upper left to lower right progression). Notice both the systematic and turbulent changes in structure (data taken with NRAO 12 m at Kitt Peak).

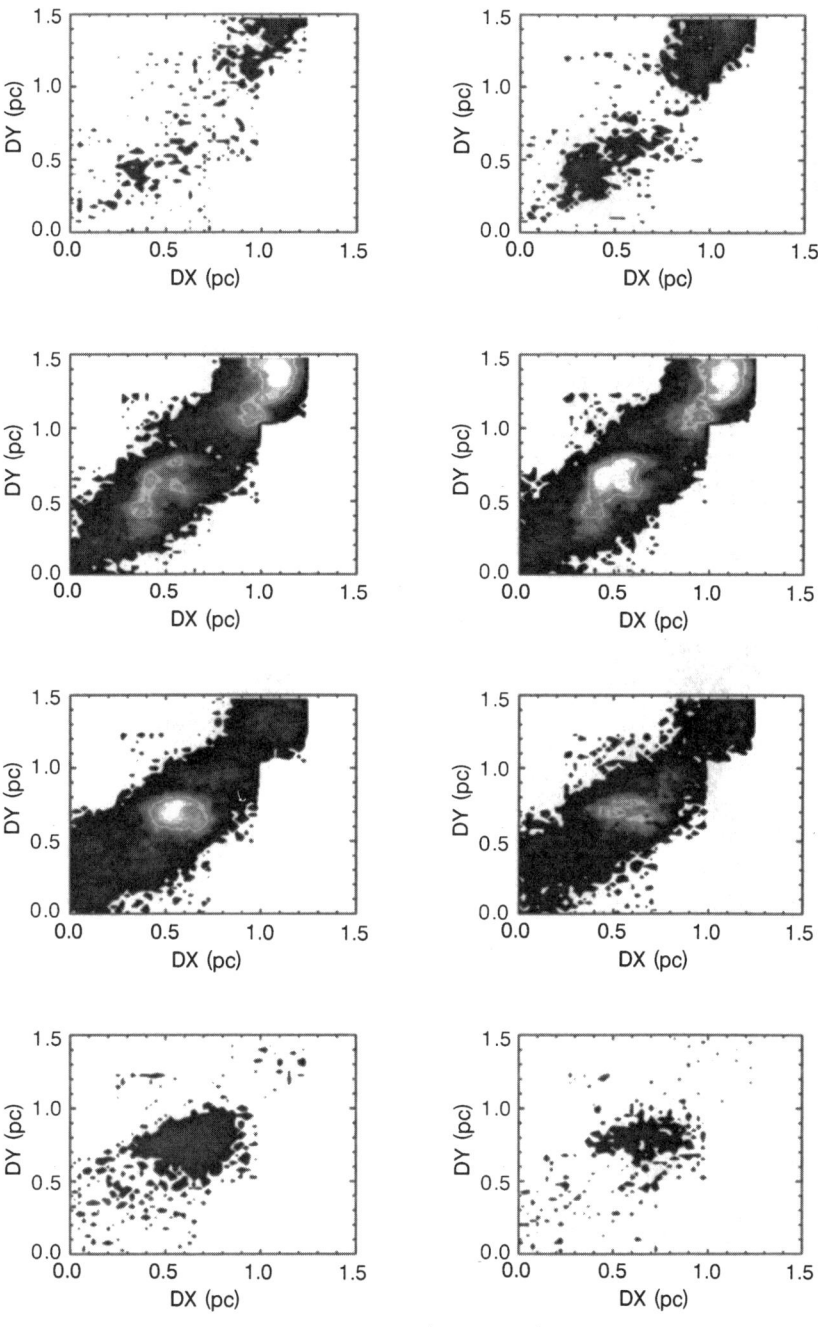

FIGURE 6.8. (*Continued*).

as the measure of the energy density in the turbulent velocity field and therefore a measure of the energy in the cloud.[153] Another important sign is the presence of superthermal widths for line profiles. This is seen best in emission lines from H II regions and from molecular clouds. Since we are discussing the interstellar case in this chapter, let's examine this second case. Thermal motions for clouds are estimated from the excitation temperatures of millimeter transitions of CO. Typically these are a few tens of degrees Kelvin. These temperatures also agree well with those derived from infrared dust emissivities and from other, more density-sensitive, molecular transitions such as CS. Such temperatures imply velocities of a few tens to hundreds of meters per second. However, the observed line widths can be as high as several km s^{-1}. In fact, the lines do not even always fit simple gaussian, as would be expected from the thermal motion alone. Observations frequently show the presence of broad wings, often containing comparable kinetic energies as the narrower, but still superthermal, cores. This is taken as an indication of intermittency and large-scale turbulent flows. The profiles appear in an enormous range of molecular environments and always have essentially the same profile form.[154]

A similar effect is observed in H II regions, although the temperatures are considerably higher, more in the range of 10^4 K, and the thermal velocities also higher, about 10 km s^{-1}. Yet here as well we find line profiles that are far broader than expected from thermal broadening alone. In addition, the profiles show the narrow cores superposed on broad wings. The mechanisms for heating and energy dissipation are likely different in the two cases, but the universality of the signature is a good indication that turbulence is involved. In the molecular environment, self-gravity also plays a role, which is surely unimportant in H II regions, but both environments have magnetic fields that play a major role in both support and heating, and both are environments where the heating is expected to be supersonic and on the same order as the thermal timescale. In other words, both cases, as different as they appear, may have the same underlying turbulent mechanism.[155]

As we mentioned, the ultimate signatures of a turbulent flow are the correlation and structure functions. We will make further use of this in Chapter 8 on cosmological structure determinations, but here it is useful for the discussion of the cloud fluid. If the medium has no net motion—in other words, if we can remove the mean velocity from the cloud lines—then the variation in the radial velocity is a measure of the kinetic energy in the random velocity field. Linewidths, in this picture, result from the integration through many individual gas parcels along the line of sight, each of which has its own velocity. For a fully developed

---

[153] Dickman, R. L. 1985, in *Protostars and Planets. II*, Black, D. and Matthews, M. S., eds. (Tucson: Univ. Arizona Press) p. 150; Falgarone, E. and Phillips, T. G. 1990, *Ap. J.*, **359**, 344; Scalo, J. 1984, *ApJ*, **277**, 556; Scalo, J. 1987, in *Interstellar Processes*, Hollenbach, D. J. and Thronson, H. A., eds. (Dordrecht: Kluwer), p. 349ff; Scalo, J. 1985, in *Protostars and Planets. II* Black, D. and Matthews, M. S., eds. (Tucson: Univ. Arizona Press) p. 201ff.

[154] Falgarone, E. and Phillips, T. G. 1990, *ApJ*, **359**, 344; Falgarone, E., Lis, D. C., Phillips, T. G., Pouquet, A., Porter, D. H., and Woodward, P. R. 1994, *ApJ*, **436**, 728; Lis, D. C., Pety, J., Phillips, T. G., and Falgarone, E. 1996, *ApJ*, **463**, 623. An important example of how to treat large samples in complex regions is found in Miesch, M. and Bally, J. 1999, *ApJ*, **524**, 895.

[155] For instance, see the discussion in O'Dell, C. R. 2000, *PASP*, **113**, 29. This is a review of structure in the Orion nebula. Some of the most striking evidence for knots and filaments comes from the centimeter wavelength interferometer observations by Yusef-Zadeh, F. 1990, *ApJL*, **361**, L19.

turbulent gas, the profiles would be expected to be approximately gaussian. Why? There are several answers to this question. The first is the, perhaps naive, assumption of statistical independence of the fluctuations, similar to what we asserted when treating ideal gases. Another reason is the mathematical simplicity of the assumption. By choosing Gaussian statistics, all odd moments of the velocity (and density and temperature) fluctuations vanish, and only the dispersion is needed to completely characterize the flow. Yet these assumptions fail, for instance, when the cascade is generated by shear, where the injection scale is the characteristic size of the shear layer. Observations show that gaussian statistics are not a bad picture for the bulk of the emission, but the profiles also often display broad wings at a low emissivity that are clearly different. These are interpreted as signatures of *intermittency*, a feature of turbulence that derives from the nonlinear interactions between the density and the velocity in the equations of motion. Put more directly, the profile for an optically thin gas can be interpreted in the same way as we did the *point spread function* when discussing image formation in a turbulent atmosphere. The line wing is the result of the rare event that has an extremely large velocity difference from the bulk of the fluid. In this case, although there isn't much power in this part of the flow, it signals the presence of turbulence. In the equations of motion, assume that the flow consists of two parts: (1) ordered motion and (2) the fluctuating velocity due to the turbulence, $v_i = V_I + u_i$. The mean velocity is $\langle v_i \rangle = V_i$, and the turbulence is assumed to have zero mean and $\langle u_i^2 \rangle = \sigma^2$, where we take the averages over the ensemble (i.e., we are taking spatial averages over the points in the sample). Remember that we have only one component of the velocity when dealing with clouds, the one along the line of sight, so although we can study the spatial distribution of u_{rad}, we don't have information about motions in the plane of the sky. This is the equivalent to taking the longitudinal correlation function in standard theory. Now we write the equations of motion as

$$\frac{\partial \rho v_i}{\partial t} + \frac{\partial}{\partial x_j}(\rho v_i v_j + P\delta_{ij}) = 0 \qquad (6.262)$$

in the absence of external forcing and, for illustration, assume that the flow has a stiff enough equation of state that we can neglect the density variations. Then the stress tensor becomes

$$T_{ij} = \rho(V_i V_j + \langle u_i u_j \rangle), \qquad (6.263)$$

where the second term is the turbulent contribution. This is the kinetic energy in the turbulent flow, and it is also the correlation function of the motions

$$R_{ij}(\mathbf{r}) = \int d\mathbf{r}' \, u_i(\mathbf{r}') u_j(\mathbf{r}' - \mathbf{r}) \qquad (6.264)$$

for a spatial lag **r**. We will return to this in Chapter 8 when we examine how to characterize cosmological structures in space and redshift. There are observational problems with the determination of this function that should not be ignored. The typical problem with such analyses is that they are taken over a finite surface (equivalent to a finite volume). The sample must be taken with sufficient resolution to avoid undersampling, therefore requiring large allocations of telescope time. The sample should be performed, if possible, at the Nyquist frequency in order to avoid aliasing effects.

If the clouds are turbulent, might the medium out of which they are formed be as well? Although we will have to await for more detailed discussion of galactic structure in the next chapter, some points can be made here without going into detail. In the diffuse H I, the search for turbulence is made more difficult by kiloparsecond scale order. The Galaxy rotates differentially, and this motion must be removed consistently from velocity data before attempting a correlation analysis. Unfortunately, depending on the direction in which we observe, a small change in the radial velocity of a gas parcel may correspond to a large variation in its location along the line of sight. Unlike the analyses for structure that make use of the three-dimensional information that is available for galaxies, because of the Hubble flow, we have no such information for clouds. The techniques for treating these sorts of data have been comparatively slow to develop and are founded mainly on the cosmological analogy.

6.A SYNCHROTRON SPECTRA: SOME DETAILS

Here we will fill in some steps required for calculating synchrotron emissivity and absorption. The derivation is pretty standard, but we want to emphasize that the use of index notation considerably simplifies the derivation and permits you to make the change to vector notation at the end of the calculation. To compute the radiation by a relativistically moving electron, you need to carefully switch between the comoving frame and the observer's frame. To do this, you need the vector and scalar potentials for a moving point charge in the observer's frame, obtained using the *retarded potentials*:

$$\phi = \frac{e}{s} \qquad A_i = \frac{e\beta_i}{s} \qquad (6.265)$$

using $s = R - R_k \beta_k$ where $R_k = x_k - x'_k$ and $t' = t - R(t')/c$. Note that $R = |\mathbf{R}|$. Now you will need

$$\frac{\partial s}{\partial x'_i} = \frac{\partial R}{\partial x'_i} - \frac{\partial R_k}{\partial x'_i} \beta_k = -(n_i - \beta_i)$$

$$\frac{\partial s}{\partial x_i} = \frac{\partial R}{\partial x'_i} - \frac{\partial R_k}{\partial x'_i} \beta_k = (n_i - \beta_i)$$

$$\frac{\partial t'}{\partial x'_i} = -\frac{1}{c}\frac{\partial R}{\partial x'_i} = \frac{1}{c} n_i$$

$$\frac{\partial t'}{\partial t} = (1 - n_k \beta_k)^{-1} \equiv \kappa^{-1}.$$

For extremely relativistic particles, the last transformation reduces to

$$\frac{\partial t'}{\partial t} \sim (1 - \beta)^{-1} \sim 2\gamma^2.$$

Then

$$\frac{\partial A_i}{\partial t} = \frac{1}{\kappa} \frac{\partial A_i}{\partial t'} = \frac{\dot{\beta}_i}{\kappa s} - \frac{1}{\kappa s^2} \beta_i \beta^2 + \frac{\beta_i}{\kappa s^2} n_k \beta_k + \frac{\beta_i}{\kappa^2 s} n_k \dot{\beta}_k$$

$$\frac{\partial \phi}{\partial x_i} = -\frac{1}{s^2}(n_i - \beta_i). \tag{6.266}$$

Now we can combine the potentials to give the electric field, now written in the more conventional vector notation:

$$\mathbf{E} = -\frac{\partial \mathbf{A}}{\partial t} + \nabla \phi = \frac{1}{s^2 \kappa}\left[(1 - \hat{\mathbf{n}} \cdot \boldsymbol{\beta})(\hat{\mathbf{n}} - \boldsymbol{\beta}) - \boldsymbol{\beta} \beta^2 + \boldsymbol{\beta}(\hat{\mathbf{n}} \cdot \boldsymbol{\beta})\right]$$

$$-\frac{1}{\kappa^2 s}\left[\dot{\boldsymbol{\beta}} + \boldsymbol{\beta}(\hat{\mathbf{n}} \cdot \dot{\boldsymbol{\beta}})\right]$$

$$= \frac{(\hat{\mathbf{n}} - \boldsymbol{\beta})(1 - \beta^2)}{\kappa^3 R^2} + \frac{1}{R\kappa^3} \hat{\mathbf{n}} \times \left[(\hat{\mathbf{n}} - \boldsymbol{\beta}) \times \dot{\boldsymbol{\beta}}\right]. \tag{6.267}$$

For nonrelativistic particles ($\beta \to 0$), this reduces to the *Larmor formula*. For electrons that will emit synchrotron radiation, $\beta \to 1$, the first term vanishes and only the second term in the last equation remains. We don't measure the signal emitted by these particles in time. What we see is a spectrum that is the Fourier transform of the signal:

$$\mathbf{E}(\omega) = \left(\frac{e^2}{8\pi^2 c}\right)^{1/2} \int_{-\infty}^{\infty} e^{i\omega t} \mathbf{E}(t)\, dt \tag{6.268}$$

using $t = t' + \hat{\mathbf{n}} \cdot \mathbf{R}/c$. The electric field in Eq. (6.267) is simplified by noting that we can remove $\dot{\beta}$ and consolidate terms using:

$$(1 - \boldsymbol{\beta} \cdot \hat{\mathbf{n}})^{-2}\left[(\hat{\mathbf{n}} - \boldsymbol{\beta}) \times \dot{\boldsymbol{\beta}}\right] = \frac{d}{dt}\left[\frac{\hat{\mathbf{n}} \times (\hat{\mathbf{n}} \times \boldsymbol{\beta})}{(1 - \hat{\mathbf{n}} \cdot \boldsymbol{\beta})}\right]. \tag{6.269}$$

We still require a trajectory for the radiating electron. The particle is assumed to move in a helix whose tilt or *pitch angle* is determined by the streaming velocity. For efficient radiation, the largest possible pitch angle, $\pi/2$, is required relative to **B**. Naturally, we denote the radius of curvature as ρ and assume that the magnetic field is locally uniform in strength and direction:

$$\hat{\mathbf{n}} \times (\hat{\mathbf{n}} \times \boldsymbol{\beta}) = \beta\left(\cos\frac{vt}{\rho}\sin\theta\hat{\mathbf{n}}_\perp + \sin\frac{vt}{\rho}\hat{\mathbf{n}}_\parallel\right) \approx \theta\hat{\mathbf{n}}_\perp - \frac{ct}{\rho}\hat{\mathbf{n}}_\parallel. \quad (6.270)$$

For the last step, we've taken $\beta = 1$, with which we can substitute $(1 - \beta) \approx 1/2\gamma^2$ whenever it is needed, as we did to find the solid angle for emission. The maximum intensity is given by maximizing with respect to the pitch angle $\sin\theta = \hat{\boldsymbol{\beta}} \times \hat{\dot{\boldsymbol{\beta}}}$:

$$\frac{d}{d\theta}\frac{dP}{d\Omega} = 0$$

$$\rightarrow \frac{d}{d\theta}\left[\frac{\sin^2\theta}{(1 - \beta\cos\theta)^{-5}}\right] = 0,$$

so that, calling $\mu = \cos\theta$, the expression $2\mu(1 - \beta\mu) - 5\beta(1 - \mu^2) = 0$ gives the critical value

$$\mu_* = \frac{-1 \pm (1 + 15\beta^2)^{1/2}}{3\beta};$$

then, as we sketched in the main text, we find $\theta_* = 1/2\gamma$.

For the Fourier transform we require the phase for the kernel $\exp i\omega(t - \mathbf{r}\cdot\hat{\mathbf{n}}/c)$. This is where things get a bit tricky. You know the relativistic particles radiate in a very small cone whose width is $\sim \gamma^{-1}$ and for only a pulse of very short duration in any direction. This allows us to expand the phase as follows:

$$t - \frac{\mathbf{r}\cdot\hat{\mathbf{n}}}{c} = t - \frac{\rho}{c}\sin\frac{vt}{\rho}\cos\theta \approx \frac{1}{2}t\left[\left(\theta^2 + \frac{1}{\gamma^2}\right) + \frac{c^2}{3\rho^2}t^2\right]. \quad (6.271)$$

Now we can perform the transform to get the frequency dependence of the emission. First, notice that we need the parallel and perpendicular components of the field for a helical orbit, which we obtain by taking the Fourier transform of these two components of $\mathbf{E}$ and adding the components quadratically:

$$\left(\frac{c}{\rho}\int_{-\infty}^{\infty} ds\, s\, \exp\left[\frac{i\omega}{2}\left(\alpha s + \frac{c^2}{3\rho^2}s^3\right)\right]\right)^2 + \left(\theta\int_{-\infty}^{\infty} ds\, \exp\left[\frac{i\omega}{2}\left(\alpha s + \frac{c^2}{3\rho^2}s^3\right)\right]\right)^2,$$

where we have used $\alpha = \gamma^{-2} + \theta^2$. All angular dependences are easily removed by asserting that since $\gamma \to \infty$ we have $\gamma^{-2} + \theta^2 \approx 2/\gamma^2$. The characteristic frequency for the emission cutoff depends on ρ through $\omega_c \gamma^3 \sim c/\rho$. Notice that we can do this only for very relativistic particles but by so doing we can perform the requisite integrations over Ω with ease. At any rate, once we remove ω from explicit dependence we have just constants from the integrals. The integral has same form as the *Airy function*

$$Ai(z) = \frac{1}{2\pi}\int_{-\infty}^{\infty} e^{i(zs+s^3/3)}\,ds,$$

but the integrals can also be written as *modified Bessel functions of fractional order* (also known as Macdonald functions) $K_\nu(z)$, where ν is a positive fraction:[156]

$$Ai(z) = \frac{1}{3}z^{1/2}\left[I_{-1/3}\left(\left(\frac{2}{3}\right)z^{3/2}\right) - I_{1/3}\left(\frac{2}{3}\right)z^{3/2}\right]$$

$$= \left(\frac{z}{3\pi^2}\right)^{1/2} K_{1/3}\left(\left(\frac{2}{3}\right)z^{3/2}\right) Ai'(z)$$

$$= \frac{1}{3}z\left[I_{-2/3}\left(\left(\frac{2}{3}\right)z^{3/2}\right) - I_{2/3}\left(\left(\frac{2}{3}\right)z^{3/2}\right)\right]$$

$$= -\left(\frac{z}{3^{1/2}\pi^2}\right) K_{2/3}\left[\left(\frac{2}{3}\right)z^{3/2}\right].$$

The K_ν functions have the following asymptotic forms:

$$K_\nu(z) \xrightarrow[z\to 0]{} \left(\frac{2}{z}\right)^\nu$$

$$K_\nu(z) \xrightarrow[z\to\infty]{} z^{-1/2} e^{-z},$$

and therefore $\omega^2 K_\nu(\omega)^2 \sim \omega^{2-\nu}$. Thus, for low frequency the power, $d^2 I/d\omega\,d\Omega$ scales as

$$P\left(\frac{\omega}{\omega_c}\right) \sim \left(\frac{\omega}{\omega_c}\right)^{2/3},$$

which is the expression we sought. As we noted in the main text, what we've found is the frequency-dependent emissivity *per radiating particle*. To obtain the emission spectrum, we *still* need to fold this broadband power through a particle energy distribution, and we remind you of the dictum that power laws beget power laws.

For the absorption coefficient, we use a similar procedure to the derivation of the Kompaneets equation except here, we will assume a power-law distribution for the electrons. The statistical weight of a state is $d\Gamma \sim E^2\,dE$, and we know that $B_{lu}g_l = B_{ul}g_u$ with $A_{ul} \sim \omega^{-3} B_{ul} \sim P(\omega, E)$. The number in the lower state is $N(E - \hbar\omega)$ and in the upper state is $N(E)$ so expanding N to first order, converting from B to A using the radiation efficiency P, and grouping terms, we

---

[156] See Watson, G. N. 1945, *A Treatise on the Theory of Bessel Functions*, 2nd ed. (Cambridge, UK: Cambridge Univ. Press); Jeffreys, H. and Jeffreys, B. S. 1950, *Methods of Mathematical Physics*, 2nd ed. (Cambridge, UK: Cambridge Univ. Press); Luke, Y. L. 1969, *The Special Functions and Their Approximations*, vol. 1 (NY: Academic Press).

get

$$\kappa_\omega \sim \int [N(E - \hbar\omega)B_{lu} - N(E)B_{ul}]dE \to \omega^{-2} \int \left[\frac{2N}{E} - \frac{dN}{dE}\right] P(\omega, E)\, dE. \tag{6.272}$$

Substituting $N(E) \sim E^{-p}$ and $\omega_c \sim E^2 \omega_L$ we get the scaling law for the opacity:

$$\kappa_\omega \sim N(0) B^{(p+2)/2} \omega^{-(p+4)/2}. \tag{6.273}$$

The precise value depends on the spectrum through

$$\kappa_\nu = 1.9 \times 10^{-2} G(p)(3.5 \times 10^9)^p N(0) B^{(p+2)/2} \nu^{-(p+4)/2}, \tag{6.274}$$

where the scaling constant is $0.65 \leq G(p) \leq 0.96$ for $1.0 \leq p \leq 5.0$.

6.B THE PARKER INSTABILITY: SOME DETAILS

The calculation of the Parker instability is instructive for its coupling of the magnetic field with a two-phase gaseous medium. In addition to the gas pressure, there is one due to the magnetic field and one due to the cosmic rays. For the latter, we will take $P_{CR} = \alpha P_g$, where α is constant. Take the magnetic field initially in the $\hat{x}$ direction but depending on z, the vertical distance from the plane; this dependence is also assumed for the unperturbed density and pressure. Gravity in the presence of density gradients drives the growth of buoyant instabilities, even in a medium that would otherwise satisfy the criteria of hydrostatic balance, if the density stratification is sufficiently inverted. In the Rayleigh–Taylor instability, which formally treats only incompressible fluids, high-density fluid supported by low-density fluid flips the matter across an interface on a dynamical timescale and establishes a "normal" density gradient. There is also a compressible version of the instability, of course, which is measured by the Brunt–Väisälä frequency for a polytropic gas

$$N^2 = \frac{g}{\rho}\frac{d\rho}{dz}, \tag{6.275}$$

which is well known from atmospheric physics (and we discussed this when deriving the criterion for convection in Chapter 4 and the MRI in Chapter 5). We can estimate the characteristic length for this instability by asking when N is comparable to the Alfven crossing time in the vertical direction. Notice that $d\ln\rho/dz$ is the same as an isothermal density scale height so for $N^2 \approx v_A^2/L$, then $L \approx v_A(H_\rho/g)^{1/2}$. The main difference between this and the usual estimate of the timescale is the assumption that the vertical density gradient (or isothermal pressure gradient) balances the magnetic pressure gradient. However, this isn't the whole story. This estimate is also asking about the magnetic buoyancy timescale—whether it is shorter than the mass oscillation timescale for the layer, and thus whether the field can be confined by the overlying gas to which it is attached. A major difference between this case and the normal laboratory experience is that the field is coupled to the dense gas when it is buoyant; the magnetic field and the gas are not simply exchanging places as in the usual Kruskal–Schwarzschild problem. When the field begins to buckle, flows are induced that funnel the gas into the troughs created by the fluctuations.

An analogous situation is encountered when a magnetic field supports a plasma against gravity. The field can be treated as a low-density fluid with a very soft equation of state. The laboratory version of the instability, the Kruskal–Schwarzschild instability, produces interchange of gas and field. On a still larger scale, the Galactic plane, Parker showed how cosmic rays and magnetic field combine to control the structure and induce flows in the interstellar medium. The two main differences between this and the classical Rayleigh–Taylor case are that the magnetic generates a tension, or backreaction, that suppresses the shortest wavelength perturbations, and the fluid is self-gravitating and therefore changes in the density enhance the contrasts between the low- and high-density regions.[157]

Since we have assumed no diffusion in the interstellar gas, we begin with the usual equations of ideal MHD for an isothermal gas:

$$\frac{\partial \rho}{\partial t} + \nabla \cdot \rho \mathbf{v} = 0$$

$$\rho \frac{d\mathbf{v}}{dt} + \nabla \left(\rho c_s^2 + \frac{B^2}{8\pi} \right) + \frac{1}{4\pi} \mathbf{B} \cdot \nabla \mathbf{B} + \rho \mathbf{g} = 0 \quad (6.276)$$

$$\frac{\partial \mathbf{B}}{\partial t} - \nabla \times (\mathbf{v} \times \mathbf{B}) = 0,$$

$$\nabla \cdot \mathbf{B} = 0, \quad (6.277)$$

where all symbols have their usual meanings. In the original derivation of this instability, Parker explicitly included a relativistic fluid, the cosmic ray pressure, but this can be included in a redefinition of the pressure through a constant factor that multiplies the gas pressure. For simplicity, we neglect this here. Assume the medium is initially in hydrostatic equilibrium, so $\mathbf{v}_0 = 0$, with the magnetic field in the direction parallel to the plane and stratified in the vertical direction so $\mathbf{B}_0 = B(z)\hat{\mathbf{x}}$ and with gravity given by $\mathbf{g} = g(z)\hat{\mathbf{z}}$. The magnetostatic case is given by

$$\frac{d}{dz}\left(P_0 + \frac{1}{8\pi} B_0^2 \right) = -\rho_0 g, \quad (6.278)$$

where, in principle, g is given by the solution of the Poisson equation for a self-gravitating gas. The equation of state for the combined gas and cosmic ray components is $P_0 = (1 + \alpha)\rho_0 c_s^2$, which we will use in the following derivation of

---

[157] Parker, E. N. 1966, *ApJ*, **145**, 811; Parker, E. N. 1969, *Space Sci. Rev*, **9**, 651; Lachièrez-Rey, Assèo, E., Cesarsky, C., and Pellat, R. 1980, *ApJ*, **238**, 175; Hughes, D. W. and Cattaneo, F. 1987, *Geophys. Astrophys. Fluid Dyn.*, **39**, 65; Foglizzo, T. and Trager, M. 1994, *Astr. Ap.*, **287**, 297.

the instability. Taking the perturbation, the first-order equations become

$$-i\omega\delta\rho + ik\rho_0 v_y + \frac{d}{dz}\rho_0 v_z = 0 \quad (6.279)$$

$$-i\omega\rho v_y + ik\left(c_s^2\delta\rho + \frac{ik}{4\pi}B_0\delta B_y\right) - \frac{ik}{4\pi}B_0\delta B_y - \frac{1}{4\pi}\delta B_z\frac{dB_0}{dz} = 0 \quad (6.280)$$

$$-i\omega\rho v_z + \frac{d}{dz}\left(c_s^2\delta\rho + \frac{ik}{4\pi}B_0\delta B_y\right) - \frac{ik}{4\pi}B_0\delta B_z - \delta\rho g - \rho_0\delta g = 0 \quad (6.281)$$

$$-i\omega\delta B_y + \frac{d}{dz}(B_0 v_z) = 0 \quad (6.282)$$

$$-i\omega\delta B_z - ikB_0 v_y = 0. \quad (6.283)$$

The resulting equation for v_z is

$$f\frac{d^2 v_z}{dz^2} + \frac{df}{dz}\frac{dv_z}{dz} + \left[\left(\omega^2 - k^2 c_s^2\right)\left(\omega^2\rho_0 - k^2\frac{B_0^2}{4\pi}\right)\right.$$
$$\left. - \omega^2\rho_0\frac{dg}{dz} + k^2 g\frac{d}{dz}\left(\frac{B_0^2}{8\pi}\right)\right]v_z = 0 \quad (6.284)$$

defining for compactness

$$f = \left(\omega^2 - k^2 c_s^2\right)\frac{B_0^2}{4\pi} + \omega^2\rho_0 c_s^2.$$

Dividing through by f we can rewrite the coefficient of v_z as

$$[\cdots] = \frac{\left(\omega^2 - k^2 c_s^2\right)\left(\omega^2 - v_A^2 k^2\right) - \omega^2\frac{dg}{dz} + k^2\frac{g}{\rho_0}\frac{d}{dz}\left(\frac{B_0^2}{8\pi}\right)}{\left(\omega^2 - k^2 c_s^2\right)v_A^2 + \omega^2 c_s^2}. \quad (6.285)$$

Notice that the f vanishes when the phase speed of the instability is the same as the magnetoacoustic speed. In other words, the condition for the growth of the Parker mode is that the timescale for the instability due to gravitational disturbances from density fluctuations grow faster than do those due to magnetic effects.[158] The magnetic field thus structures, but does not dominate, the flow, and the growth is slower than our estimate based only on the Alfven speed.

---

[158] We have followed the particularly clear derivation by Kim, J., Franco, J., Santillan, A., and Martos, M. 2000, *ApJ*, **531**, 873. Its advantage over the conventional treatment is in sticking to the magnetic field components rather than relying on the potential, although at the cost of some additional algebra. See also Parker, E. N. 1979, *Cosmical Magnetic Fields* (Oxford: Oxford Univ. Press).

6.C DIMENSIONAL ANALYSIS AND SIMILARITY SOLUTIONS OF THE HYDRODYNAMIC EQUATIONS: SOME DETAILS

The technique of putting the fluid equations in dimensionless form is tremendously important in astrophysics. We have used this to find the motion of the front, but the dimensionless transformed equations must describe the postshock conditions completely. Since the inviscid equations are all first-order, although nonlinear, the Rankine–Hugoniot relations supply the necessary boundary conditions and the similarity solution for the radius supplies the time scaling of all variables. Here we will describe some details of the process, both for completeness and because of it is the one way you can get a feel for how the conditions vary in the post-shocked gas.[159]

6.C.1 Dimensionless Dynamical Equations

Now let's break all the relevant hydrodynamical quantities down into their basic dimensions by defining a variable, η. We will do this in the most general way, by taking this as the variable that renders a combination of physical attributes of the flow constant. In general, if we assume a form

$$\eta \equiv r^\lambda t^{-\mu} \tag{6.286}$$

for the scaling variable, η comes immediately from the conserved quantity. The exponents of this scaling variable, which is intrinsically dimensionless, are determined by the physical constants that are involved with the problem. The similarity variable is specifically tailored to the problem, and doesn't necessarily apply throughout the evolution of the system if the scales of length or time for competing processes produce nonconservation for any of the physical variables.

Behind the shock, the velocity, pressure, density, and sound speed must be scaled using η. This is accomplished through a set of dimensionless functions

$$u = rt^{-1}U(\eta), \quad P = r^{-1}t^{-2}\Pi(\eta), \quad \rho = r^{-3}D(\eta), \quad c_s = rt^{-1}C(\eta). \tag{6.287}$$

The derivatives needed for the hydrodynamic equations thus become

$$\frac{\partial}{\partial t} = -\mu \eta t^{-1}\frac{d}{d\eta}, \quad \frac{\partial}{\partial r} = \lambda \eta r^{-1}\frac{d}{d\eta}. \tag{6.288}$$

For instance, the convective derivative (which is the fundamental nonlinear term in the system) becomes

$$\frac{D}{Dt} = t^{-1}\eta(-\mu + \lambda U)\frac{d}{d\eta}. \tag{6.289}$$

---

[159] See also. for example, Barenblatt, G. I. 1979, *Similarity, Self-Similarity, and Intermediate Asymptotics* (NY: Consultants Bureau); Barenblatt, G. I. and Zel'dovich, Ya. B. 1972, *Ann. Rev. Fluid Mech.*, **4**, 285; Chevalier, R. A. 1977, *ARAA*, **15**, 175; Ostriker, J. P. and McKee, C. 1988, *Rev. Mod. Phys.*, **60**, 1; Sedov, L. 1957, *Similarity Methods in Mechanics* (NY: Academic Press); Shu, F. H. 1977, *Ap. J.*, **214**, 488; Gratton, J. 1991, *Fund. Cosm. Phys.*, **15**, 1.

When you writing down the similarity equations, there is one thing about which you must be particularly careful. All the physical variables are functions of *both* (r, t) and η. In other words, any quantity $Q(r, t)$ is actually written as $f(r, t)g(\eta)$ after the scaling transformation has been made. When you take the derivatives, be careful not to forget that f depends on r and/or t. For instance, the velocity time derivative becomes

$$\frac{\partial u}{\partial t} = -rt^{-2}U - \mu rt^{-2}\eta \frac{dU}{d\eta}, \qquad (6.290)$$

and the spatial derivative becomes

$$\frac{\partial u}{\partial t} = t^{-1}U + \lambda t^{-1}\eta \frac{dU}{d\eta}. \qquad (6.291)$$

Notice that when multiplied by $u = rt^{-1}U$, the spatial derivative has the same dimensions as the time derivative, and thus you can combine them into a single term.

We now have to assume a geometry because, as we have seen, this changes the value of the exponent in the scaled variable. In general, although the momentum equation depends on gradients, the mass conservation equation contains the divergence when written in general form so the dimension must be specified. To see this, let us take the continuity equation as an example

$$\frac{\partial \rho}{\partial t} + \nabla \cdot \rho \mathbf{u} = \frac{\partial \rho}{\partial t} + r^{-n}\frac{\partial}{\partial r}(r^n \rho u_r) = 0, \qquad (6.292)$$

where n is 0, 1, or 2 depending on whether the medium has one, two, or three dimensions. We will assume strictly radial flow, so from now on $u_r = u$. For this equation we can now substitute the dimensionless similarity version:

$$-\mu\eta t^{-1}r^3 \frac{dD}{d\eta} + \left[(n - 4)DU + \lambda\eta\frac{d(DU)}{d\eta}\right]r^3 t^{-1} = 0. \qquad (6.293)$$

Finally, we can write this equation as

$$(\lambda U - \mu)\eta \frac{dD}{d\eta} + (n - 4)DU + \lambda D\eta \frac{dU}{d\eta} = 0. \qquad (6.294)$$

This nonlinear equation may look formidable but it can be solved, numerically if not analytically.

Let us now turn to the velocity gradient. Using Eq. (6.288) for D_t, we get

$$(\lambda U - \mu)\eta \frac{dU}{d\eta} = D^{-1}\left(\lambda\eta\frac{d\Pi}{\eta} - \Pi\right) - U(U - 1). \qquad (6.295)$$

Notice that a source term, $-U(U - 1) - \Pi/D$, has now appeared on the right-hand side. Π/D will have the same behavior as U^2 even though *none of these quantities have dimensions* and are purely algebraic. We can now add to the list of equations an equation of state to relate the pressure, P and the density, ρ. For this, use the usual polytropic approximation for the pressure, so that the entropy equation becomes

$$\frac{D}{Dt}(P\rho^{-\gamma}) = 0, \qquad (6.296)$$

where γ is constant. The similarity transformation gives:

$$(\lambda U - \mu)\eta\frac{d}{d\eta}(\Pi D^{-\gamma}) + ((3\gamma - 1)U - 1)\Pi D^{-\gamma} = 0, \qquad (6.297)$$

and the required system of equations is now completely closed. Here we collect the results:

$$(\lambda U - \mu)\eta\frac{d}{d\eta}(\Pi D^{-\gamma}) = -[(3\gamma - 1)U - 1]\Pi D^{-\gamma} \qquad (6.298)$$

$$(\lambda U - \mu)\eta\frac{dU}{d\eta} = D^{-1}\left(\lambda\eta\frac{d\Pi}{\eta} - \Pi\right) - U(U - 1)$$

$$(\lambda U - \mu)\eta\frac{dD}{d\eta} = -(n - 4)DU - \lambda D\eta\frac{dU}{d\eta}.$$

For a choice of λ, μ, and γ, these equations can be integrated numerically. Recall that n is fixed by the geometry, but the physical processes, specifically the conserved quantities of the dynamical problem, determine the choice of λ and μ.

Here we need to pause for a point of emphasis. The most essential feature of similarity variable methods is that *there can be no characteristic times, lengths, masses, or the like in the problem. No dynamical variable has a characteristic value.* This is one of the reasons why all such large-scale approaches produce fractal structures in combination over time. For instance, the interstellar medium is constantly being clobbered by expanding H II regions, supernova remnants, and stellar wind bubbles that begin at random times. Since these all have scale-free behavior, their combination should also be scale-free. We remark that this is precisely the requirement for a fractal medium and we might well expect that the interstellar medium, shaped as it is by expanding H II regions and blast waves, would have such a structure.

The importance of similarity solutions was first stressed by aerodynamicists like Prandtl and von Karman. Later it was taken up by physicists, particularly Sedov, Taylor, and von Neumann, and for the same reason. Return, for a moment, to the explosion problem. We are interested in looking at the effect of the blast on the environment *after* the shock has passed over the region, in other words, in a moving coordinate system. We assume that the shock front velocity changes according to the general law for $R(t)$. Behind the front, however, The structure is invariant with time *in the distance that is scaled to the radius of the front* according to the similarity variable η. This is like looking at a magnified picture and knowing only the relative distance between points based on the angular size of each element in the image. The internal disposition of the objects does not change, and you can only know what it is in *physical*—that is to say, external—dimensions once you know the sizes of the frame in an absolute sense. All of the absolute properties of the postshock medium are set by the scaling to the shock front itself. Once the equations have been solved for the internal structure of the shock, they are solved forever (well, at least as long as the conditions of the similarity law have been respected by the expansion).

Our aim has been to obtain a scaled (dimensionless) form for the equations describing the evolution of a flow and to find an appropriate variable for specifying the position within the flow independent of the time. We obtain this by taking $\eta = r/R(t)$. Since R varies as, in the Crab nebula case that we have already discussed, $t^{2/5}$, we get $\lambda = 1$ and $\mu = \frac{2}{5}$. In general, take two physical quantities, Q_1 and Q_2, that are constant for the expansion and are given dimensionally by $M R^{n_1} t^{l_1}$ and $M L^{n_2} t^{l_2}$, respectively. Since L, the length scale, is the same as R, the blast radius will scale in time as $R(t) \sim t^{(l_1-l_2)/(n_1-n_2)}$. For instance, take an equation of state of the form $T^3 \rho = $ constant. Since we know the functional dependence of ρ, we obtain $T \sim t^{-2/5}$, a slowly decreasing function of time. The optical depth and cooling time will also vary as the radius of the blast changes, and these can also be computed (and this is one you should do). It can be shown that the depth at which the shock becomes essentially transparent is $R_c = \tau_0^{1/2} R_0$, where τ_0 is the initial optical depth. This gives the critical timescale as $t_c \sim \tau_0^{5/4}$. If the explosion is initially optically thick (as it should be for the application of the adiabatic condition), then this is the point (scaled time) at which the approximation is expected to fail. You should also keep in mind that it depends on the initial energy and the density of the medium.

6.D THE VELOCITY CORRELATION TENSOR AND REPRESENTATION OF TURBULENT FLOWS: SOME DETAILS

It is not by accident that you encounter the most complete discussion of turbulence in the chapter on the interstellar medium. This is one of the few places where we see so many manifestations of the phenomenon and on so many different scales. It is vital to the understanding of the structure and stability of molecular clouds and is one of the key factors thought to govern star formation. We will also be using some of these ideas when discussing the distribution of galaxies in observational cosmology, so it is appropriate to discuss this at some length.

Turbulence is a random process but of a very special nature. Within a fluid, it involves the conversion of bulk flow into vortical motion. There is shear that, because of viscosity, leads to the dissipation of energy. This is the signature of turbulence: although the *mean* value of the vorticity, $\omega = \nabla \times \mathbf{v}$, may vanish, the rms value of ω does not. We can assume that the velocity field has two contributors, a mean and a random component, and that this separation can be performed for *any* component. Just as in the discussion of the Vlasov equation from Chapter 1, we can write

$$v_i(\mathbf{x}) = V_i(\mathbf{x}) + u_i(\mathbf{x}), \tag{6.229}$$

where

$$\langle v_i \rangle = V_i, \quad \langle u_i \rangle = 0, \quad \langle u^2 \rangle \neq 0. \qquad (6.300)$$

Here we will generally mean spatial or time averages, assuming that we either have taken a large enough spatial sample or a long enough time sample to see what the mean condition really is.[160] Taking the average over two components, u_i and u_j, we can write

$$\langle v_i(\mathbf{x}) v_j(\mathbf{x}') \rangle = V_i(\mathbf{x}) V_j(\mathbf{x}') + R_{ij}(\mathbf{r}), \qquad (6.301)$$

where we define a point in space, $\mathbf{x}$, around which we sample the fluid within a sphere of radius r. This is the *correlation tensor* for a displacement $\mathbf{x}' = \mathbf{x} \pm \mathbf{r}$, which is also the two-point correlation tensor for the velocity. This is related to the energy density of the turbulent velocity field, and the information contained in the two-point function is the important measure of the transport properties of a turbulent flow. Although in astrophysics you are rarely lucky enough to be able to measure more than one component of the velocity, the one along the line of sight, we wish to keep this discussion general. Therefore consider two components of the velocity field, one that is parallel to the line of centers, $u_l(\mathbf{x})$, and one that is transverse to that line, $u_t(\mathbf{x})$. Here l and t stand for longitudinal and transverse, respectively. We now *define* the correlation function averaged over all points $\mathbf{x}$):

$$R_{ij}(\mathbf{r}) \equiv \langle u_i(\mathbf{x}) u_j(\mathbf{x} + \mathbf{r}) \rangle \qquad (6.302)$$

for a displacement $\mathbf{r}$. By assertion, we take R_{ij} to be a continuous smooth function, although more recent work, using wavelet analyses, allow this assumption to be dropped. The Fourier transform of this tensor is given by

$$\Phi_{ij}(\mathbf{k}) = \frac{1}{2\pi^3} \int R_{ij}(\mathbf{r}) e^{-i\mathbf{k} \cdot \mathbf{r}} \, d\mathbf{r}, \qquad (6.303)$$

and its trace, which is a scalar quantity, is

$$R(r) \equiv \sum_i R_{ii}(r). \qquad (6.304)$$

Finally, we define a characteristic scale length through

$$\Lambda \equiv \int R(r) \, dr, \qquad (6.305)$$

since R is normalized. The timescale over which one would expect the smoothing to occur is a function of the Reynolds number, which measures the role of viscosity in the flow through the ratio of the inertial forces, U^2/L to the viscous stress, $\nu U/L$, through $\mathrm{Re} = UL/\nu$. By dimensional analysis, you see that $t_{\mathrm{decay}} \sim \mathrm{Re}^{-1/2}$.

---

[160] The statistical analysis of turbulence relies on the Taylor hypothesis that spatial and temporal averages are the same in a time-independent flow.

Turbulence appears when the Reynolds number is large so that this timescale is short compared with the crossing time for the medium. There is also a natural length, called the *Taylor microscale*, which is defined by the curvature of the autocorrelation function at zero lag:

$$\lambda^{-2} \equiv -\left.\frac{d^2 R(r)}{dr^2}\right|_0. \tag{6.306}$$

The fluid *must* be correlated at zero lag, and, if the velocity field is to be normalizable and isotropic, should become completely uncorrelated at infinite separation. The two point function must therefore have a maximum for zero lag. The integral over all scales must have a finite value, and therefore the two-point function satisfies the following conditions:

$$R'_{ij}(r)|_{(0)} = 0, \qquad R''_{ij}(r)|_{(0)} < 0, \tag{6.307}$$

Besides the correlation function, the *structure function*, defined by

$$S(\mathbf{x}) = \langle (u(\mathbf{r} + \mathbf{x}) - u(\mathbf{r}))^2 \rangle^{1/2} \tag{6.308}$$

specifies the rate at which two fluid parcels diverge from each other in velocity as the displacement is increased.

For an *isotropic* medium, R_{ij} must depend only even powers of r since $R_{ij}(\mathbf{r}) = R_{ij}(-\mathbf{r})$. Its value for zero lag is

$$R_{ij}(0) = \langle u_i u_j \rangle = \int \Phi_{ij}(\mathbf{k})\, d\mathbf{k}. \tag{6.309}$$

The autocorrelation tensor at zero lag is then defined by

$$R_{ii}(0) = \langle u_i(\mathbf{x}) u_i(\mathbf{x}) \rangle = \int \Phi_{ii}(\mathbf{k})\, d\mathbf{k}, \tag{6.310}$$

There is no sum over the index here, because we are taking the autocorrelation one component at a time. Here is the promised connection with the energy density since this quantity has the dimensions of an energy:

$$\int E(k) k^2\, dk = \int \Phi_{ii}(\mathbf{k})\, d\mathbf{k}. \tag{6.311}$$

As we've repeatedly stressed, for most astrophysical cases you rarely have observations of the full three-dimensional velocity field. The rare exception is when the region is close enough that both proper motions and radial velocities can be measured for each position and the distances are known. In general, you measure only the radial velocity, hence the only component that you can actually obtain the *longitudinal* correlation function (Fig. 6.9). However, if you can assume that the medium is really isotropic, and the turbulence is fully developed, then $\langle u_i u_j \rangle = 3\langle u^2 \rangle$. Given all the discussion in this chapter, however, you should be wary about indiscriminantly applying this assumption. It is a very questionable assumption for the interstellar medium even if the notion of fully developed *Gaussian* turbulence works well under laboratory conditions. There is *no reason* to believe that the

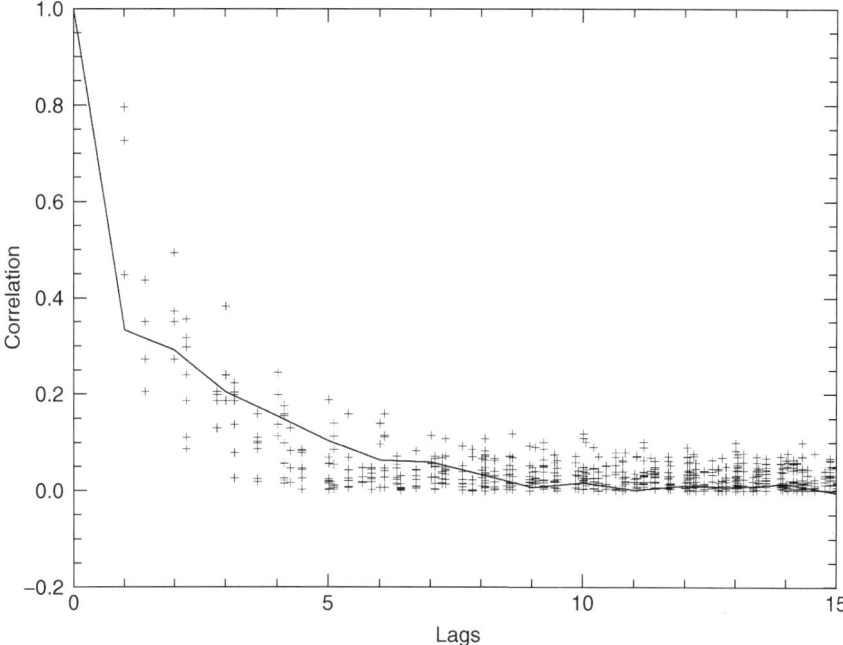

FIGURE 6.9. Correlation function for ^{12}CO (1-0) for the translucent cloud MBM 16 (see Larosa, Shore, and Magnani n.148), *op. cit*). One lag $\sim 0.03\%$. The dots show the two-dimensional correlation; the solid line represents the mean scalar correlation result. Notice that $C(r) \to 0$ at around 5-10 lags (~ 0.1-0.2 pc). This is the characteristic scale of a coherent region in this cloud.

longitudinal component of the correlation is positive everywhere in the medium and, in fully developed turbulence, it isn't. At some point in the flow, if the fluctuations are truly random, the velocity should decorrelate. In a vortical flow, R should either vanish or become negative at the typical size of the eddies. The correlation function also satisfies the following conditions: (1) its maximum must be at zero lag, (2) it must vanish at infinity, (3) its spatial integral must be finite over infinite range, and (4) it must have a smooth behavior at zero lag with a finite second derivative. Because we have only one component to work with, we are interested in the *autocorrelation functions*. This may, however, change. Long time baseline observations of variations in interstellar absorption lines and scintillation provide *some* data on motions transverse to the line of sight.

6.D.1 Time Dependence

Starting with the Navier-Stokes equation, we write

$$v_i' \left(\frac{\partial v_i}{\partial t} + v_j \frac{\partial v_i}{\partial x_j} \right) = -\frac{1}{\rho} v_i' \frac{\partial P}{\partial x_i} + \nu v_i' \nabla^2 v_i \tag{6.312}$$

$$v_i \left(\frac{\partial v_i'}{\partial t} + v_j' \frac{\partial v_i'}{\partial x_j'} \right) = -\frac{1}{\rho} v_i \frac{\partial P'}{\partial x_i'} + \nu v_i \nabla^2 v_i'. \tag{6.313}$$

THE INTERSTELLAR MEDIUM

We first add these equations and then take the ensemble averages:

$$\frac{\partial}{\partial t}\langle v_i v_i'\rangle + \left\langle v_i' v_k \frac{\partial v_i}{\partial r_k} - v_i v_k \frac{\partial v_i'}{\partial r_k}\right\rangle + \frac{1}{\rho}\left(\left\langle v_i' \frac{\partial P}{\partial r_i} - v_i \frac{\partial P'}{\partial r_i}\right\rangle\right) = \nu \nabla^2 \langle v_i v_i'\rangle. \quad (6.314)$$

We'll assume the fluctuations are homogeneous and isotropic, so they depend only on the scalar quantity r, but there is a further possible simplification. If we assume that the turbulence has Gaussian properties, the problem is greatly simplified. All odd moments vanish for a Gaussian random process once the mean has been removed, and the even moments are all multiples of the dispersion, which is the second moment. There is no reason, a priori, to think that turbulence is gaussian. In fact, there are observationally compelling reasons for thinking otherwise, particularly the observed line profiles in molecular clouds and the interstellar diffuse medium. Rather than restricting our discussion to this special choice, let's leave the choice of turbulence statistics open and discuss the general problem. Taking the Fourier transform of Eq. (6.314) with respect to r_i to convert to wavenumber, we find the time evolution of the energy spectrum, $E(k)$, to be given by

$$\frac{\partial}{\partial t} E(k) - 2\nu k^2 E(k) = \mathcal{T}(k), \quad (6.315)$$

where $\mathcal{T}(k)$ contains both the pressure fluctuations *and* the nonlinear advection terms; in other words, we hide the hard parts of the hydrodynamics symbolically and hope that we can return to this sometime in the future. For one thing, you will notice that if we take $v_i = V_i + u_i$, where $\langle u_i \rangle = 0$ and $\langle u_i^2 \rangle \neq 0$, then the triple correlation

$$\left\langle v_i v_k' \frac{v_i}{\partial r_k}\right\rangle = V_i \left\langle u_k' \frac{u_i}{\partial r_k}\right\rangle + V_i' \left\langle u_k \frac{u_i}{\partial r_k}\right\rangle + \left\langle u_i u_k' \frac{V_i}{\partial r_k}\right\rangle \quad (6.316)$$

to second order in u. Notice that a shear term appears now, so we see that $\mathcal{T}$ is actually a driving term for the turbulence. If we assume this is in steady state, and that we concentrate on the inertial subrange for which the driving is small, then we obtain the evolution equation

$$\frac{\partial E}{\partial k} + 2\nu k^2 E \frac{dt}{dk} = 0, \quad (6.317)$$

where, as we described in Section 6.10.1, the term dk/dt is to be interpreted as a velocity of propagation of the cascade energy through the various length scales.

7 Our Galaxy and Others as Stellar Systems

Cassius: Tell me, good Brutus, can you see your face?
Brutus: No, Cassius, for the eye sees not itself but by reflection, by some other things.
—William Shakespeare, *Julius Caesar*, Act 1, Scene 2

7.1 THE GALAXY AS A STELLAR SYSTEM: INTRODUCTORY REMARKS

In 1542, Copernicus displaced the Earth from the center of the planetary system and made the Sun a star. By the end of the millennium humans would be further demoted in the cosmic order by two pivotal discoveries: first, that the Milky Way is a vast rotating systems of stars with the Sun located in the suburbs, and second, that the spiral and elliptical nebulae are distant stellar systems of enormous size, mass, and complexity and that the Galaxy is nothing extraordinary. In this chapter, we will study our cosmic home in combination with extragalactic systems. We have been building toward the "big" problems throughout the book, and now it is time to view the edifice that rests on these foundations. While our progress has been cumulative, at each change in scale our inferences become more tentative and qualitative and you will notice an emphasis in this chapter on scaling relations and broader derivations. Nearly all the necessary ingredients are now in hand to construct a galaxy: stars and their processes and the interstellar medium in which they live. The one new feature you will encounter here is the role of a component of the Universe, *dark matter*, that until now has not figured in our discussion. As the picture of the largest scale unfolds, you will see that the structure and stability of galaxies and the Universe require the existence of something else, some component of the universe that has not been needed to explain "something as simple as a star."

We begin closest to home. The Galaxy[1] is a complex and crowded system. A view on any dark night invites speculation: Do the luminous points uniformly fill space, are their motions described by stable orbits, or are they chaotic, and what is the large scale structure of the system? On closer examination, even with binoculars, you see that these stars are clustered, that there are nebulae and dark lanes,

---

[1] A word on nomenclature. The Milky Way is usually referred to by its Anglicized name Galaxy (from the Greek $\gamma\alpha\lambda\alpha$ $o\upsilon\rho\alpha\nu\iota o\nu$ first observationally (not mythologically) described by Parmenides in the fifth century B.C.E.). A unique survey of the history is provided by Jaki, S. L. 1972, *The Milky Way: An Elusive Road for Science* (NY: Science History Publ.), which is also a superb source for graphics from early models for the Galaxy. In general, we will use the capitalized name to denote our own system. External systems are *galaxies*, and when using the lowercase we will be referring explicitly to the generalized object, including the Galaxy.

and derive a feeling for the depth of space in the contrast between the distributions of stars at different brightnesses. The power of instrumentation for revealing the universe is beautifully shown by the exclamation by Galileo in his *Siderius Nuncius* of 1610 on first viewing the Milky Way through a telescope, "It is nothing else but a mass of innumerable stars planted together in clusters." But such observations do not suffice to answer these questions. They demand detailed photometric, spectroscopic, and kinematic data for large samples of these bodies folded through the dynamical machinery. Solving the structure problem requires the determination of the gravitational potential and, therefore, the mass distribution in the space.

One of the hardest problems has been the determination of the structure of our own local region of space, principally because we are embedded deep within it. This is a typical inverse problem. The Sun is a member of an "island universe," the Milky Way. Yet even this now obvious statement has a long and colorful history. The first structural conjecture, by Thomas Wright in the late eighteenth century, took the system to be a flattened distribution with the Sun at the center. Departures from this symmetric disk geometry should already have been apparent, most notably the bulge toward Sagittarius, but they were ignored. The earliest quantitative effort to model the stellar distribution was attempted by William Herschel in a series of papers *On the Construction of the Heavens* (1799–1805)[2] based on the statistical results from star counts. Unaware of the obscuring effects of dust, Herschel assumed that the brightness distribution of stars along a line of sight could be inverted to provide a measure of their spatial distribution. Wright's qualitative result was that a disk-like distribution explains the appearance of the Milky Way. From his extensive magnitude-limited visual star count survey, Herschel produced a diffuse structure with the Sun significantly eccentrically located, but without any consideration of the dynamics. For this method to work, which Herschel called "star gauging," you must have some idea of the intrinsic luminosity function. Observations of binary stars, the arguments for which come in part from an application of this procedure, show that stars differ in both intrinsic luminosity and color, so the direct application of the method is misleading. It is surprising that Herschel didn't pay more attention to this since he was the main contributor to the study of binary stars. His procedure was fundamentally compromised by interstellar absorption, but that discovery was more than 100 years in the future. The technique Herschel pioneered has been refined in modern work on large scale and involves the use of deep multiwavelength magnitude-limited counts of selected areas of the sky to deproject the line of sight surface density into a mass density.[3]

---

[2] See Hoskin, M. A. 1964, *William Herschel and the Construction of the Heavens* (NY: Norton); Hoskin, M. A. 1982, *Stellar Astronomy: Historical Studies* (Chalfont: Science History Publ.). For historical development, see Whitney, C. A. 1971, *The Discovery of Our Galaxy* (NY: Knopf); Struve, O. and Zebergs, V. 1962, *Astronomy of the 20th Century* (NY: Macmillan); Berendzen, R., Hart, R., and Seeley, D. 1976, *Man Discovers the Galaxies* (NY: Science History Publ.)

[3] Star count surveys have been performed by several groups. See the special issue introduced by Humphreys, R. M. et al. 1995, *PASP*, **107**, 762. This special issue is a superb set of reviews. See also Reid, N. I. 1993, *ARAA*, **31**, 345) for a broad survey of modern observational methods. The *Sloan Digital Sky Survey* (SDSS) is the comprehensive modern realization of Herschel's program, including spectroscopic data. See Gunn, J. E. et al. 1998, *AJ*, **116**, 3040.

Herschel also founded the study of large-scale stellar dynamics with his determination of the apex of solar motion based on an incredibly tiny sample of proper motions. These data also indicated that the Sun is not at the center of the luminous mass distribution. Kapteyn introduced the systematic comprehensive study of stellar kinematics, which resulted in the discovery of star streaming in the solar neighborhood (in 1922). The explanation for the local kinematics came with Lindblad's and Oort's recognition of differential rotation of the Galaxy. The observational picture developed through technological innovations. Wide field imaging became possible after the 1930s using Schmidt telescopes equipped with objective prisms, a project advocated by Nassau and McCuskey and that produced the *Luminous Stars in the Milky Way* catalogs. The introduction of a two-dimensional luminosity–temperature taxonomy for stellar spectra by Morgan and Keenan permitted the spectroscopic assignment of relative distances to stars. With the *UBV* photometric system by Johnson and Morgan, the effects of reddening could be estimated and distances could be corrected. This was the first optical tool for the determination of galactic structure within about 2 kpc of the Sun and showed for the first time the unambiguous presence of spiral arms. The discovery of 21-cm H I emission and its use as a dynamical probe of the Galaxy was the first tool that circumvented the limitations of extinction and satellite far infrared observations in the $10-60\mu$ region with IRAS, COBE, and ISO (soon SIRTF) complete the arsenal of observational tools with which galactic structure can be determined.

7.1.1 The Composite HR Diagram of Field Stars

Measurements of star fields in the Milky Way provide a handle on the structure and nature of the populations comprising the Galaxy, but this is not an easy set of data to interpret. You can obtain the photometry, for instance, of field of stars in much the same way you would for a cluster. But there is the rub. You can determine the colors and magnitudes of the aggregate sample, and from this create an HR diagram, but these field stars are not likely to be coeval. Although it is possible to fit evolutionary tracks to this diagram, you must assume not only different masses but also different ages for the stars—the *history* of star formation must be known along with the development over time of the chemical composition of the stars. Spectroscopy gives you a more complete picture, allowing you to determine the stellar abundances. This can also be done with external galaxies, as we will see later. However, a caution must be voiced at this stage. The sample actually extends over a considerable depth in space. The interstellar medium, as you know, is neither uniform nor free of pollutants that alter the brightness and colors of stars depending on the length of the sight line through the system. Therefore, we need some way to distinguish the nearby red stars from distant blue stars that have been both extinguished by absorption and whose photometry has been altered by reddening.

Let us examine this argument a bit more closely since it is the extension of the Herschel method of star counting. The example we will discuss is shown in Figure 7.1, which displays the *Hipparcos* sample. The stellar distribution has a well-defined lower boundary, the main sequence. To the right, however, there is a cloud of points. Were this a cluster, you would immediately eliminate as many stars as possible from this cloud, assuming them to be background. Yet for a

632 OUR GALAXY AND OTHERS AS STELLAR SYSTEMS

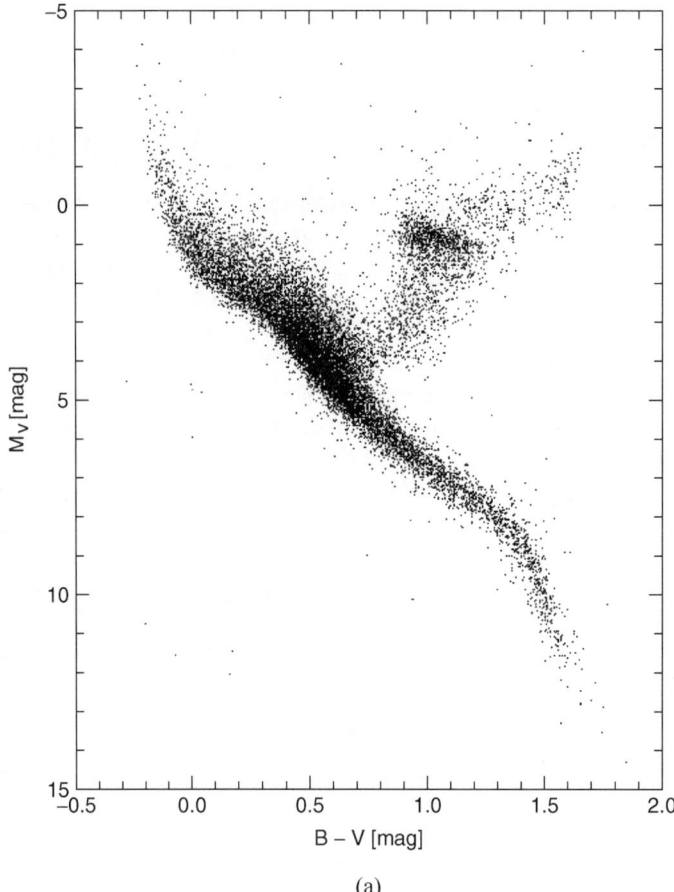

(a)

FIGURE 7.1. (a) The composite CMD of about 20,000 stars from the *Hipparcos* catalog. This is the HR diagram of the field of the Galaxy in the solar neighborhood (credit: ESA).

galactic field, in effect, every star in the image is a "background" object. The highest density of points is found, as usual, for those masses and stages of evolution that take the longest time. The most massive stars will be tightly confined to the main sequence, since they evolve so quickly across the HR diagram that only the most recent ones formed will be observed. The lower-mass stars have, however, greater hydrogen burning lifetimes and evolve far more slowly (compared, for instance, the lifetimes at different stages of 10 $M_\odot$ and 0.5 $M_\odot$ stars). Therefore, if the system has been steadily forming stars for the entire main sequence lifetime of the *lowest* mass objects, they define the history of star formation in the system.[4] If there is more than one separate branch crossing the Hertzsprung gap and ascending the giant branch, this implies that star formation has occurred episodically in

---

[4] Examples of such analyses are Haywood, M., Robin, A. C., and Creze, M. 1997, *Astr. Ap.*, **320**, 440; Jimenez, R., Flynn, C., and Kotoneva, E. 1998, *MNRAS*, **299**, 515; Binney, J. Dehnen, W., and Bertelli, G. 2000, *MNRAS*, **318**, 658.

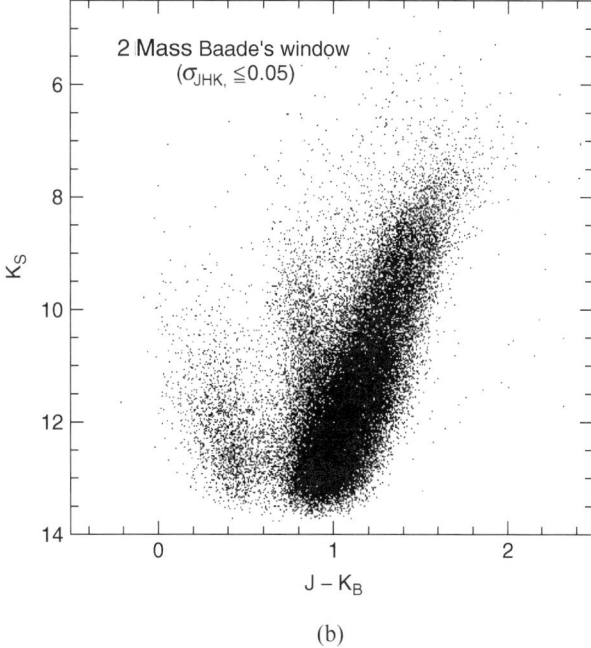

(b)

FIGURE 7.1. (b) Infrared CMD for Baade's window showing the main sequence and field giant branch (credit: 2MASS).

the system. The largest changes are likely to take place for those stars that have main sequence lifetimes that are comparable with the ages of the branches.

7.1.2 Aggregates: Open Clusters, OB Associations, and Globular Clusters

Stars love to hang out in groups. The hierarchy begins at the level of multiple star systems that are usually binaries or triples, although higher multiplicities are known. Continuing up the scale, we reach open clusters and associations. Within the disk of the Galaxy, stars are formed predominantly in clusters of several hundred members with typical densities of order 100 $M_\odot$ pc^{-3}. They range in age from only a few 10 Myr (such as σ Ori, NGC 2244, NGC 2264 that are still associated with H II regions and their parent molecular clouds) to nearly 10 Gyr (for M67 and NGC 188). In appearance, these clusters are similarly diffuse with no obvious central concentrations despite their range of ages (although the oldest clusters are more dispersed in general than younger ones). Galactic clusters appear to form an almost continuous sequence ranging from these relatively small groups up to the level of OB associations. These latter systems are identified by their large populations of massive stars and mark sites of very recent star formation. Prime examples of such associations are in Orion, Cepheus, and Scorpius–Centaurus.

Open clusters and associations provide the basis for the initial mass function since the stars were obviously formed nearly simultaneously. Open clusters have well-defined main sequences. In the youngest, the age is given by the turn-on point for core hydrogen burning; for instance, NGC 2264 has a well-defined pre-main

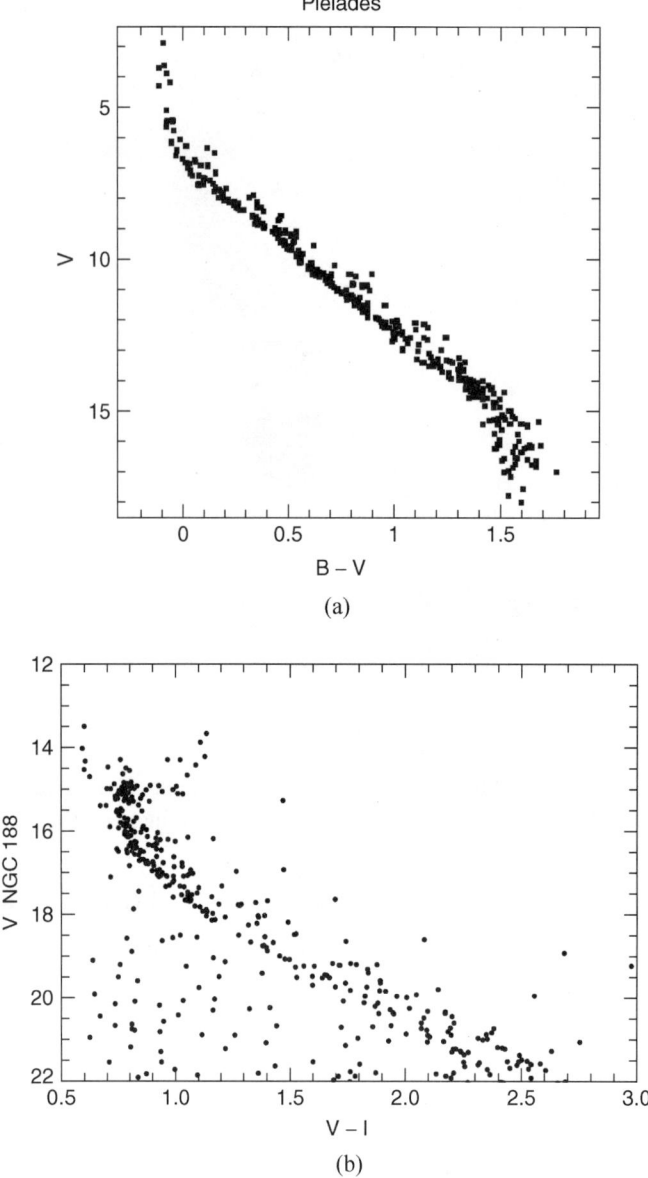

FIGURE 7.2. CMD for open clusters: (a) a young cluster (Pleiades) (credit: J. − C. Mermilloid, Univ. Lausanne); and (b) an old open cluster (NGC 188) (data from Sarajedini, A., von Hippel, T., Kozhurina-Platais, V., and Demarque, P. 1999, *AJ*, **118**, 2894). Notice the (poorly populated) parallel sequence of main sequence binaries.

sequence branch. Most do not show this but instead display obvious turn-off points for the terminal main sequence, ranging in age from the Pleiades at about 10^8 yr upward. In Figure 7.2, we show some examples of the range of open-cluster HR diagrams. Clusters constitute the best tests for stellar evolution theory since the morphology of the observed color–magnitude diagram is well matched with model isochrones.

Many important questions remain to be addressed for cluster evolution.[5] Although it is assumed that the stars are formed at roughly the same time, only the youngest clusters have sufficient mass spread in the pre-main sequence tracks to determine the timescale for cluster formation, and these are rare. The spread in pre-main sequence isochronal ages suggests that the star formation begins rather slowly and may accelerate over about a 10-Myr timescale (see Fig. 7.3). So far, this has been found only in a few clusters, notably Orion, ρ Oph, Tau-Aur, Upper Sco-Cen, and Cha I and II, but it is likely a general phenomenon of cluster formation.[6] Some clusters, notably NGC 2516 and α Per, have a relatively large number of chemically anomalous stars, both Ap and Am stars. Why? For several decades, it has been known that the distribution of main sequence rotational velocities differs widely among known clusters for stars more massive than around 2 $M_\odot$, with the Pleiades having a relatively large number of rapidly rotating stars while IC 4665 contains a larger number of slowly rotating stars relative to the field distribution. Although there may be differences in the binary frequency among these clusters, this is not the explanation for the observed variation in angular momentum. For lower mass stars, the Pleiades has a comparatively large number of active stars, indicating that their dynamo activity may be enhanced through rapid rotation, but there is otherwise no strong indication of variation in the rotational properties of the lower main sequence. Finally, some open clusters, such as NGC 663, have a large number of Be stars whose mixed parentage includes both rapidly rotating single stars and Algol-type binaries undergoing mass transfer. Within associations there are substructures and clusters that resemble the lower-mass end of the open cluster distribution. If these are formed in the same events that initiate the formation of the association, the spread in their ages may be used to indicate the timescale for formation of the association as a whole and set limits on the destruction timescale for the parent molecular cloud. We leave these as open questions.

The globular clusters represent a distinct population within the structural hierarchy.[7] They have masses that are of order 10^5–10^7 $M_\odot$ and core radii that are of order 10 pc. Although they now consist of only the lower end of the stellar mass distribution, they surely had more massive members earlier in their histories, if the initial mass function wasn't for some reason truncated near 1 $M_\odot$. The globular

---

[5]Pallavicini, G., Micela, G., and Sciortino, S., eds. 2000, *Star Clusters and Associations: Convection, Rotation, and Dynamos* (San Francisco: ASP Conf. vol. 198) is an excellent snapshot of the state of the field at the time of this book. No doubt much will change in the near future, but this surveys the basic questions.

[6]See, for instance, Palla, F. and Stahler, S. W. 2000, *ApJ*, **540**, 255 from which the figure is appropriated.

[7]The most complete review of globular cluster properties and dynamics is Meylan, G. and Heggie, D. C. 1997, *Astr. Ap. Rev.*, **8**, 1. See also Vesperini, E. 1998, *MNRAS*, **299**, 1019; Combes, F., Leon, S., and Meylan, G. 1999, *Astr.Ap.*, **352**, 149.

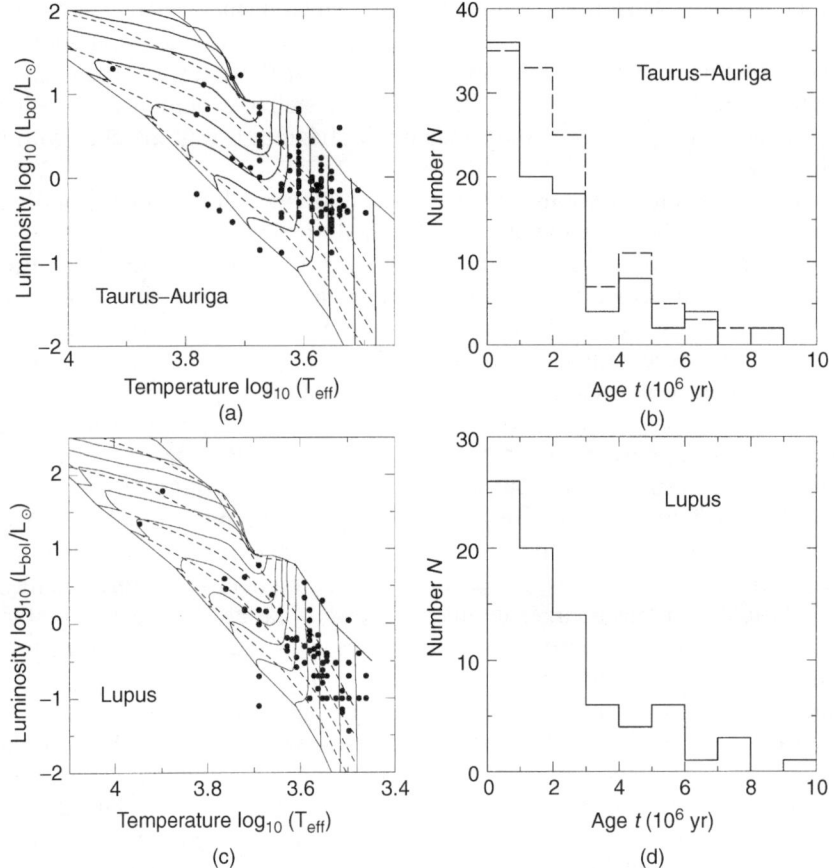

FIGURE 7.3. Change in star formation rate with time for young clusters (credit: F. Palla, Osservatorio Astrofisico di Arcetri, and S. Stahler, Univ. California—Berkeley). For Tau-Aur, the PMS tracks are for 0.2, 0.4, 0.8, 0.8, 1.0, 1.2, 1.5, 2.0, 2.5, 3.0, and 3.5 $M_\odot$; for Lupus the added track is 4 $M_\odot$. Panels (b) and (d) show the inferred number of stars formed per unit time, derived from the isochrones in the left hand panels (a) and (c) (dashed lines; 1, 3, 10, and 30 Myr).

cluster system of the Milky Way, and other galaxies in the Local Group and the Virgo cluster, follow a regular form much like the IMF. There are two separate subpopulations among the globulars, those that live in the plane and those that fill the halo. They seem to separate by their orbits; those in the halo have more nearly radial (and retrograde galactocentric motions), while the disk systems are more isotropic and prograde. Their *integrated* metallicities range from about 10^{-2} to nearly solar values. Isochronal ages for the globulars are all far in excess of the Sun, above 9 Gyr, and the majority are around 11–12 Gyr. From our discussion of stellar evolution, this means that they show well-developed horizontal branches, the locus of core helium burning, and turn-off points for the main sequence lying in the range of 1 $M_\odot$. The higher the metallicity, the more concentrated this horizontal branch is toward the red giant branch, a feature that has been used for studying their evolution (see Figure 7.4). As we will see later in this chapter, the population of globulars possessed by a galaxy may follow a regular distribution in mass, an analog of the IMF. If so, it is an important property for the determination of galactic distances and forms one of the secondary calibrators of the distance scale (Chapter 8). However, unlike stars, globulars are comparatively fragile and in their orbits about a galaxy suffer several possibly catastrophic effects. For example, as they approach and then traverse the plane of a disk galaxy, they suffer a strong, time-dependent tidal force that can lead to evaporation. This would systematically remove the clusters with the lowest binding energy from the distribution. Also, they are tidally distorted by the galactic potential, even if not impulsively affected, so they actually have Roche surfaces (tidal limits) that depend on their galactocentric distances and central concentrations. The slow evaporation of stars from these collisionless systems leads to their disappearance from the mass function, the stars then mixing with the background halo population. In other words, there are plenty of open questions here as well.

Should we really consider globulars as distinct types of clusters within the Galaxy? Perhaps. In the halo, at least, the Galaxy is no longer forming such objects, nor does it appear to be forming them in normal molecular clouds in the disk. Taken at face value, the metallicity distribution of the normal Galactic globular clusters shows the epoch ended when the metallicity was well below solar. Yet infrared observations show that the massive clusters at the Galactic center have globular clusterlike stellar densities, and in the same is true for the most massive compact clusters in the Large Magellanic Cloud. In one very well-studied LMC cluster, R136 at the center of the 30 Dor star-forming region, the density reaches over 10^3 $M_\odot$ pc^{-3}. It is interesting to note that the LMC *populous blue clusters* have about the same metal abundances as the most metal-rich Galactic disk population of globulars. These massive aggregates have stars now with masses above 70 $M_\odot$, a sure indication that the star formation is ongoing. The Galactic center is home to a dense cluster that contains a very large number of massive stars, called the *Arches cluster*. When viewed from a distance, and without the more than 20 magnitudes of obscuration that hide the Galactic center from visual inspection, clusters such as R136 are more like globulars in appearance. They are distinguished by the presence of OB stars, indicative of their youth, and infrared spectrophotometry shows a large number of WR and Of stars. We know nothing, alas, about the upper-mass end of the stellar distribution for normal Galactic globular clusters from present-day observations since all these stars have long since evolved and exploded or become white dwarfs. The strongest signature of their former presence is the detection of millisecond pulsars in 47 Tuc and other globulars. Since neutron stars are formed from either core collapse in massive systems or from binary system in which accretion induced collapse of Chandrasekhar mass progenitors occurs, there must have been high-mass stars in these clusters at some time in their past.

7.1.3 Stellar Hydrodynamics

When we first derived the virial theorem for gaseous masses we assumed that the pressure is isotropic and neglected the rotation. Neither assumption is appropriate for a galaxy. The dynamical equations must take into account differences in the

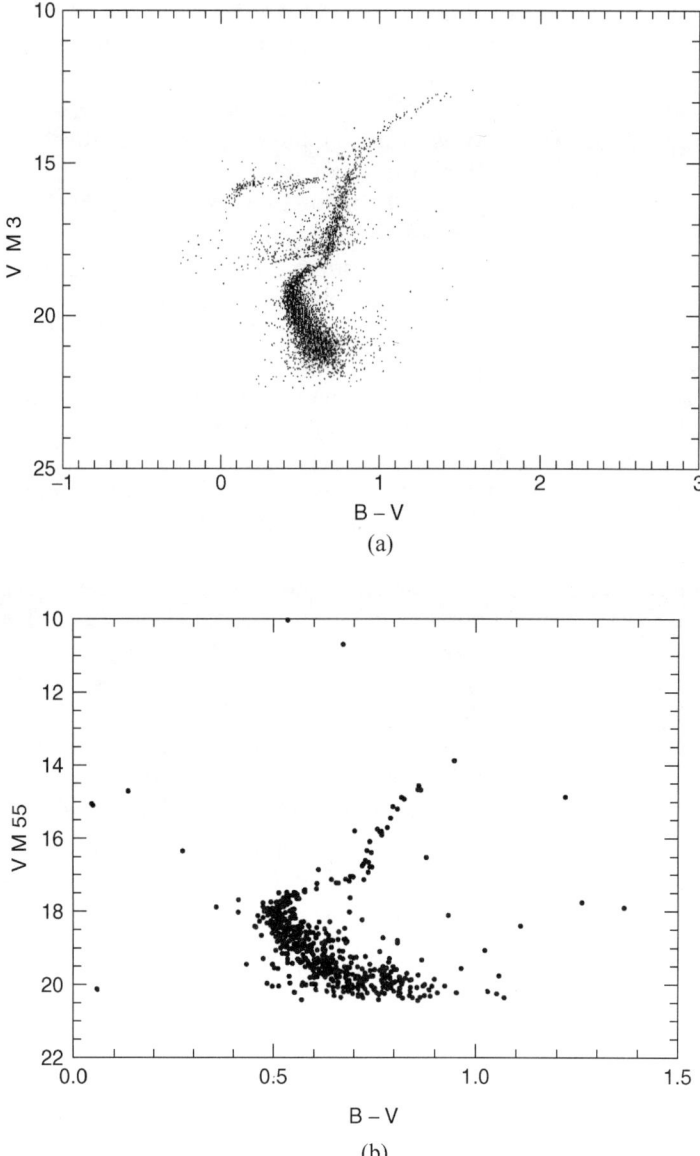

FIGURE 7.4. CMD for three globular clusters: (a) M3 (data from Ferraro, F. R. et al. 1997, *Astr. Ap.*, **320**, 757), (b) M55, (c) NGC 6637. Note the differences in the extension of the horizontal branches.

components of the velocity dispersion, and also the possibility that different stellar masses may have relaxed to different velocity distributions. This is very different than the kinetics of gases. We begin by treating the stars in the galaxy as a continuous medium. Unlike the interior of a star, this equation for a "gas of stars" requires some justification. Yes, the collisional mean free path is large and stars do not come very close to one another within the disk or halo. They do, however,

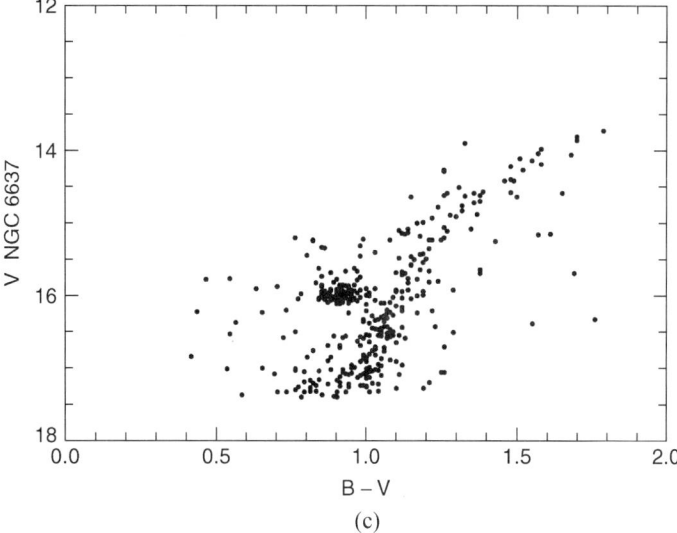

FIGURE 7.4. (*Continued*).

scatter off of density fluctuations that may have very large masses, molecular cloud complexes, associations, and perturbations in the overall smoothness of the background, and they behave collectively by mutual gravitational attraction. It is thus reasonable to approximate the system as if it behaved like a fluid. The system is assumed to be in mechanical equilibrium. The jargon for this is that the system is *virialized* or relaxed, but what it really means is that we can use the same distribution function to describe the phase space of all constituent particles. A gas in the room in which you are reading this reaches such a condition very rapidly, within the sound crossing time for the whole room and locally within a few collision times. But for a system of stars, which may be very large and in which the dynamical crossing time may be short compared to the collision time, the only way that this can be achieved is by assuming that the individual stars have a chance to exchange momentum as a collective, as a fluid. The gravitational potential depends on the density through the velocity distribution function, and through it the motion of any single star is determined by its interaction with *all* the other stars, an enormous *N*-body problem. In this regard, the same techniques we used to describe a plasma should apply to this gas of far more massive particles. But there is a crucial difference—the force felt by any mass is never shielded! The fact that positive *and* negative changes are required to maintain electrostatic neutrality produces a net cancellation of the electric field felt by a plasma ion at some separation between the charges. For gravity, this never happens. This is a clue to how we might have collective behavior of a stellar system despite the lack of collisions. *Violent relaxation* involves the complete mixing of orbits in phase space and occurs in dense systems of stars. It characterizes the postcollapse state of a stellar ensemble. Once the collision time, t_c, becomes comparable to the dynamical timescale, t_{ff}, stars random walk within the energy surface defined by the system as a whole. In a dissipationless system, the total energy provides an upper bound for the kinetic energy any body can achieve and still remain bound. The combined effects of forming binaries by close encounters and the loss of stars by hyperbolic encounters and acceleration force the mass of the system to decrease with time and the core of the cluster to become more tightly bound. In this way a stellar system can achieve virialization. Rapid mixing in phase space then homogenizes the orbits of the individual stars and results in a uniform distribution function. On the other hand, if there is a large-scale ordered motion of the system, complete dynamical relaxation may never be achieved and the system can maintain anisotropic velocity dispersions.

7.1.3.1 *The Stellar Velocity Distribution*

Studies of the kinematics of stellar populations require two different types of data. Radial velocity surveys are straightforward, especially now that specifically designed spectrographs are available. In general, accuracies better than 1 km s^{-1} can be achieved with a single high-resolution measurement. This provides, however, only the line-of-sight (LOS) component of the velocity. For space motions, you also need to know the transverse velocity. Proper motion surveys provide these through measurements of angular displacements over time, but only if the distances are known to the stars. We are hampered by the limited period of time over which stellar positions have been accurately determined, and most astrometric catalogs have been produced only since the introduction of photography to astronomy about a century ago. Until relatively recently, this short baseline of time and the sparseness of parallax-data-biased inferences from these surveys toward stars that have the largest proper motions. These tend to be either very nearby or very rapidly moving. As a result, the stars of the Galactic halo were usually better sampled for their orbital parameters than the stars in the disk. This is why the *Hipparcos* mission was so important and the reason why it has been mentioned so often in this book—it provided a very large systematic sample of high-quality parallaxes and proper motions for a magnitude limited sample.

In order to discuss galactic structure, we need to specify a natural coordinate system. This is provided by the disk of the Milky Way. The galactic latitude, b_{II} is the position relative to the plane and l_{II} is the galactic longitude measured eastward from the center, which lies in Sagittarius ($\delta_{GC} = -28° 56'10.221$, $\alpha_{GC} = 17^h 45^m 37.^s199$ (epoch J2000); the inclination of the Galactic plane to the celestial equator is about 62.5°). The roman numeral II denotes the current, second, IAU system of Galactic coordinates. It is important to note that *these coordinates are measured around the Sun*, not around the Galactic center. Therefore, they are designed for specifying positions and motions for stars in the solar neighborhood. Three orthogonal velocity components are defined in this frame: U points *toward* the Galactic center, V is in the direction of galactic rotation, and W points toward the north Galactic pole. To convert between these and galactocentric coordinates requires knowing the motion of the Sun relative to a *theoretical* circular orbit at the solar galactocentric distance. This can be inferred from the statistical average of the motion of the Sun reflected in the halo population and the quasars and is called the *local standard of rest* (LSR). This is found by taking a large sample of distant stars and determining its mean velocity components. You then assume that these reflect the solar motion. We note here an interesting analogy to another area of classical astronomy, the study of meteor streams. The apex of solar motion is like observing the radiant point of a meteor shower, the location of which is due to the combined orbital motions of the meteoroids and the Earth. You know that both observer and observed are orbiting the Sun, so gravitational dynamics allow you to determine the space motion of the stream. This is the reasoning behind the original work by Kapteyn on star streaming. The analysis requires the use of very large samples; more than 10^4 stars have been employed in the most complete studies to date, and the results are $(U, V, W)_\odot = (10.0 \pm 0.4, 5.2 \pm 0.6, 7.2 \pm 0.4)$ km s^{-1} relative to the LSR.[8] The velocity dispersion observed for stars in the solar neighborhood depends on their mass (as inferred from their colors and spectral types). In particular, low-mass stars show a more spherical distribution in their space motions than those of earlier type, those more confined to the Galactic plane. The inference is that stars of lower mass are a more mixed group, including both those formed relatively recently in the disk and those with high space motions that are falling in toward the disk. The approximate form for the stellar distribution, first proposed by K. Schwarzschild around 1910, takes these anisotropies explicitly into account:

$$F(U, V, W) = \left(8\pi^3 \sigma_U \sigma_V \sigma_W\right)^{-1/2} \exp\left[-\sum_{i=1}^{3} \left(\frac{v_i}{\sigma_i}\right)^2\right] \quad (7.1)$$

with the proviso that these are all corrected for the motion of the Sun and taken with respect to the LSR.

The Schwarzschild distribution assumes that the stellar *peculiar* motions are statistically independent for each velocity component. Observable departures from this simple picture that are very important for understanding the origin of the velocity dispersion. In the disk, there is a correlation between σ_U and σ_V that depends on stellar population. This is quantified by the *vertex deviation*, which is measured by the angle

$$l_v \equiv \frac{1}{2}\tan^{-1}\frac{2\sigma_{UV}^2}{\sigma_U^2 \sigma_V^2} \quad (7.2)$$

This angle is about 30° for those B and A stars for which parallaxes and proper motions are available and decreases as you move down in mass. The interpretation of this behavior is that it reflects differences in the histories sampled for these mass bins. The more massive stars, which are younger on average than those at lower mass, are still members of moving groups in the local neighborhood. These associations have not yet evaporated and bias the motions. In addition because they are younger, these stars will have moved less from the sites of their formation and this vertex deviation is the signature of some nonaxisymmetric ordered motion. The cooler stars, on the other hand, constitute a population with a much larger spread in age. Over their lives, these stars will have experienced more randomizing interactions and should be more dynamically homogenized. This is consistent with stars later than G showing $l_v \approx 10^o$. The assumption of a relaxed system that is the dynamical raison d'être for the Schwarzschild functional form for the velocity distribution function is clearly not strictly correct. The picture is even more complicated after *Hipparcos*. The measured ratio of the principal dispersions

---

[8] Dehnen, W. and Binney, J. 1998, *MNRAS*, **298**, 387.

is $\sigma_U : \sigma_V : \sigma_W = 2.2 : 1.4 : 1$ with $\sigma_U \approx 20$ km s^{-1} for $(B-V)_0 \leq 0.2$ and $\sigma_U \approx 38$ km s^{-1} for $(B-V)_0 > 0.6$, called the *Parenago discontinuity*.

7.1.3.2 The Equations of Motion

We start by recalling the steps we took in Chapter 1 to derive the fluid equations assuming thermalized particles. This time, however, we cannot assume an isotropic velocity distribution. So to proceed, we write

$$\mathscr{L}f = \frac{\partial f}{\partial t} + v_i \frac{\partial f}{\partial x_i} + \dot{v}_i \frac{\partial f}{\partial v_i} = 0$$

for the Vlasov equation and take the zeroth, $\int d\mathbf{v} \mathscr{L}f$, and first $\int d\mathbf{v}\, \mathbf{v} \mathscr{L}f$ moments and equate each integral to zero. We will need to explicitly include the unit vector, $\hat{\mathbf{e}}^i$, to account for the geometry of the system, as in $\mathbf{v} = v_i \hat{\mathbf{e}}^i$ for the velocity, for reasons that will become clear momentarily. As we found for the fluid equations, the first two moments become

$$\frac{\partial}{\partial t} n + \frac{\partial}{\partial x_j} \left[n \langle v_j \rangle \right] = 0$$

$$\frac{\partial}{\partial t} \left(n \langle v_i \rangle \hat{\mathbf{e}}^i \right) + \frac{\partial}{\partial x_j} \left[n \langle v_j v_i \rangle \hat{\mathbf{e}}^i \right] - n a_j \delta_{ij} \hat{\mathbf{e}}^i = 0,$$
(7.3)

where a_i are the components of the acceleration and n is the number density. The velocity field is separable into mean velocity V_i, and random components, u_i, such that $\langle u_i \rangle = 0$ and $\langle u_i^{2n} \rangle = \sigma_i^{2n}$ with $\langle u_i^{2n+1} \rangle = 0$ for all moments, but we include nondiagonal terms because we cannot exclude correlations between random motions in different directions. The *observed* velocity dispersion tensor is *not* diagonal, since l_v used σ_{UV} to define the tilt, so there is reason to assume that $\sigma_i \sigma_j \neq \sigma^2 \delta_{ij}$. If we assumed that the Schwarzschild velocity distribution were rigorously correct, this correlation perforce vanish as it must for gaussian distributions, but we have no reason to restrict the moments for the large scale flows on the basis of the local observations.

We will need an approach here slightly different from that used in Chapter 1 because we will be dealing explicitly with a disk galaxy. Adopting a cylindrical coordinate system, (r, ϕ, z), we need the derivatives of the unit vectors

$$\frac{\partial \hat{\boldsymbol{\phi}}}{\partial \phi} = -\hat{\mathbf{r}}, \qquad \frac{\partial \hat{\mathbf{r}}}{\partial \phi} = \hat{\boldsymbol{\phi}}.$$

Let's use a slightly different formalism now to simplify the task of writing the final equations. First, we rewrite Eq. (7.3) in conservative vector form:

$$\frac{\partial}{\partial t}(n\mathbf{V}) + \nabla \cdot \overset{\leftrightarrow}{\mathbf{T}} = n\mathbf{K},$$
(7.4)

where $\mathbf{K}$ is the gravitational force. To connect this coordinate system with our local measurements, $(V_r, V_\phi, V_z)^T$ corresponds to $(-U, V, W)^T$. The components of the stress tensor, T_{ij}, are the coefficients of a dyadic $\overleftrightarrow{\mathbf{T}} = T_{ij}\hat{\mathbf{e}}^i\hat{\mathbf{e}}^j$, so Eq. (7.4) becomes

$$\nabla \cdot \overleftrightarrow{\mathbf{T}} = \frac{\partial T_{ij}}{\partial x_j}\hat{\mathbf{e}}^i + \frac{\partial \hat{\mathbf{e}}^i}{\partial x_j}T_{ij}. \tag{7.5}$$

The components of the stress tensor are

$$T_{ij} = n(V_i V_j + \langle u_i u_j \rangle) \equiv n(V_i V_j + \sigma_{ij}),$$

where σ_{ij} is the velocity dispersion tensor, the mean of the random components in the (i,j) plane. In cylindrical coordinates there are six independent terms:

$$T_{rr}, T_{r\phi} = T_{\phi r}, T_{rz} = T_{zr}, T_{\phi\phi}, T_{\phi z} = T_{z\phi}, T_{zz};$$

since the off-diagonal (cross) terms are symmetric. Substituting Eq. (7.5) into Eq. (7.4) and collecting terms, we obtain:

$$\frac{\partial}{\partial t}nV_r + \frac{1}{r}\frac{\partial}{\partial r}rT_{rr} + \frac{\partial}{\partial z}T_{rz} + \frac{T_{rr} - T_{\phi\phi}}{r} = nK_r \tag{7.6}$$

for the radial equation, where K_r is the radial force. For the angular velocity, we find

$$\frac{\partial}{\partial t}nV_\phi + \frac{1}{r}\frac{\partial}{\partial r}T_{r\phi} + \frac{2}{r}T_{r\phi} + \frac{\partial}{\partial z}T_{z\phi} = nK_\phi = 0. \tag{7.7}$$

Here K_ϕ representing the azimuthal force, which we will ignore for the moment as it is a departure from axisymmetric motion; all ϕ derivatives have been set equal to zero. Finally, for the vertical motion, we have

$$\frac{\partial}{\partial t}nV_z + \frac{1}{r}\frac{\partial}{\partial r}rT_{zr} + \frac{\partial}{\partial z}T_{zz} = nK_z, \tag{7.8}$$

where K_z is the vertical force component. Nonaxisymmetric perturbations that depend on the shear can be easily included in this formalism, although we have ignored them. These produce the winding of the spiral arms that are found ubiquitously in disk systems, which we treat in Section 7.2.3, and they also contribute to the tilt of the velocity ellipsoid. The equations of motion must be supplemented by the Poisson equation, written as

$$\frac{1}{r}\frac{\partial}{\partial r}rK_r + \frac{1}{r}\frac{\partial K_\phi}{\partial \phi} + \frac{\partial K_z}{\partial z} = -4\pi Gn. \tag{7.9}$$

For a steady state, $\partial/\partial t \to 0$, we will define a reference circular orbital speed $V_{\phi,0}$ (*our* LSR) from radial force balance:

$$V_{\phi,0}^2 = rK_r. \tag{7.10}$$

This is an *assumption*, in effect a definition. It is *not* in general, the same as V_ϕ, as we can now see. Simplifying by combining terms, we arrive at an equation expressing the asymmetry of the velocity distribution in the solar neighborhood, which is the deviation of the vertex of local motion that we defined earlier:

$$V_{\phi,0}^2 - V_\phi^2 = \sigma_{\phi\phi} - \sigma_{rr} - \frac{r}{n}\left[\frac{\partial(n\sigma_{rr})}{\partial r} - \frac{\partial n\sigma_{rz}}{\partial z}\right]. \quad (7.11)$$

Each term is observable, and each arises from a different property of the stellar population's response to some feature of the galactic gravitational field. The velocity dispersions are obtained by averaging over *many* stars, each caught at a different instant in its motion. Orbital eccentricities lead to a distribution of angular momenta, and different apogalactic distances produce a distribution in energy.

7.1.3.3 Modeling the Galactic Disk

In a self-gravitating gas, the pressure that supports the particles arises from collisions and the size (thickness) of the body depends on their mean free path. Stars don't collide in a galaxy, so why does the disk look thick? This is a very important question. Each star has orbital motion that is both planar and in the vertical direction depending on its initial conditions. Imagine that you disrupt a cluster and inject the stars into the disk. Each starts out with some set of random velocity components (u_r, u_ϕ, u_z). Any ordered motion from the orbit of the center of mass of the swarm carries the stars around the Galactic center. The motion in the z direction is an oscillation with a frequency that is determined by K_z evaluated as a function of r and z. Those stars with large initial vertical velocities are able to reach considerable distances from the plane and thus have long oscillation timescales, of order t_{ff}^{-1}. This depends only on the local mass density if it traverses only a small distance from the plane. Since a harmonic oscillator spends most of its time at the extrema of its range, an ensemble of stars will have a vertical spatial distribution reflecting its distribution in vertical velocities. Thus the disk *appears* to have a finite thickness even though the stars are not colliding or "supporting themselves" against gravity. We can connect simple thin accretion disk modeling with the rotation curve. From the vertical velocity dispersion, $\sigma_{zz}^{1/2}$, we have

$$\frac{\partial}{\partial z}\rho\sigma_{zz} = -\rho K_r \frac{z}{r}, \quad (7.12)$$

so the scale height is

$$z \sim \frac{\sigma_{zz}^{1/2}}{V_\phi} r. \quad (7.13)$$

This approach can be extended quite easily to planetary ring systems, such as those around the Jovian planets, and to protoplanetary disks—thin rings have relatively low velocity dispersions.

Perhaps you've detected a subtle change in language that connects what we have discussed about properties of gaseous bodies, such as stellar interiors, with the dynamics of "gases of stars." This notion of dissipation—cooling—is important. We have already discussed how star formation appears to proceed now in the Galaxy interstellar medium. The evidence is that other galaxies behave similarly. Yet the halo and disk are very different. Stars are not now forming in the Galaxy halo or bulge, and these systems of old stars have the largest velocity dispersion. It suggests that the initial conditions for star formation were very different for the two parts of the Galaxy. The disk has a large gaseous component, the halo does not. Ignoring dark matter for the moment, it appears that the disks settled down through processes that are likely connected with the *thermal* properties of the gas, and that the reason the disk is dynamically cold is because it was actually able to cool in the usual thermal sense. On the other hand, the large vertical extent of the spheroid and halo suggests that whatever dark matter might be, it cannot be strongly dissipational if it played any role during the initial stages of galactic formation. We will generalize this to external galaxies soon. This also means that the halos evolve more nearly in a collisionless, which is to say dissipationless, fashion while the disks show clear evidence of "cooling" of the velocity distribution. The closest analog is the Oort cometary cloud surrounding the solar system. When viewed near the solar neighborhood, these halo stars are falling at their maximum velocity through the plane, and, like comets, they must be on extremely eccentric orbits, nearly radial for the largest velocities. Since these are frequently high proper motion stars when they reach the plane, once we have parallaxes, their three dimensional orbits can be reconstructed. The vertical velocity gradient is a direct probe of the mass distribution perpendicular to the plane. From Eq. (7.9) we have

$$K_z = W \frac{\partial W}{\partial z} \sim -2\pi G \Sigma. \qquad (7.14)$$

The disk has a mean thickness that depends only on the local mass density and the velocity dispersion of the stars—the larger-velocity dispersion stars reach greater scale heights. This is the most compelling evidence for a complex dynamical history of the Galactic halo and disk. There is a clear separation between stars with very large vertical velocities near the plane and those that spend their lives confined to the disk. Stellar photometry provides an important clue to the origin of this dynamical segregation.[9] Although the distinction has been blurred since Roman's survey, there is still a measurable correlation between W and $\delta(U - B)$ in the solar neighborhood. You have an idea of where the star began its descent since when making measurements near the plane you know its vertical velocity is simply proportional to $z_i^{1/2}$, its maximum height above the plane. With time the contributing populations have broadened to include a thick disk as well as the halo, but the basic dynamical features remain. The *Hipparcos* data clarify this picture. The vertical motion gives a mass density for the disk of $0.076 \pm 0.015\ M_\odot\ \text{pc}^{-3}$ and a surface density at R_0 of $\Sigma_0 = 40\ M_\odot\ \text{pc}^{-2}$ assuming that $R_0 = 8.5$ kpc. Fitting

---

[9]Roman, N. G. 1954, *AJ*, **59**, 307 used the newly introduced *UBV* system to great advantage. The metallicity measured by the photometric index $\delta(U - B) = (U - B) - (U - B)_0$, is most sensitive to metallic line blanketing of the stellar continuum and therefore acts as a proxy for precise determination of [Fe/H].

two component exponential models for the population distributions gives the vertical scale height of the thin disk, which consists of the youngest stellar population, is about 320 pc, and for the thick disk it is about 660 pc.

Now let's be more precise on how to construct a disk model. If the vertical stellar velocity dispersion, σ_{zz} is constant with height and we assume only a single population of stars all of which have the same velocity distribution, we have in effect adopted the stellar analog of an isothermal gas. Keep in mind that we could just as easily be talking here about a slab of *gas* instead of stars since we have assumed only a local approximation for the vertical structure. Using the one-dimensional hydrostatic equation

$$\frac{\sigma_{zz}}{\rho}\frac{d\rho}{dz} = \frac{d\Phi}{dz}$$

and the Poisson equations

$$\frac{d^2\Phi}{dz^2} = -4\pi G\rho,$$

we can interchange Φ and ρ by defining a new variable, $y = \ln \rho/\rho_0$, where ρ_0 is the midplane density, to obtain the dimensionless equation

$$\frac{d^2 y}{dx^2} = -e^y. \tag{7.15}$$

We have written the vertical distance with respect to the scale height as $x = z/\zeta$ using

$$\zeta = \left(\frac{4\pi G\rho_0}{\sigma_{zz}}\right)^{1/2}. \tag{7.16}$$

You will notice that ζ is essentially the Jeans length for the slab and the structure equation resembles the isothermal Lane–Emden equation. Then multiplying both sides by y' and integrating with the boundary conditions $y(0) = 0$ and $y'(0) = 0$, we find

$$\tfrac{1}{2}(y')^2 = -(e^y - 1),$$

which yields on integration and resubstitution:

$$\rho(z) = \rho_0 \operatorname{sech}^2 \frac{z}{2^{1/2}\zeta}. \tag{7.17}$$

The density falls off exponentially at large distance from the plane, just as you would have expected for an isothermal atmosphere. Since inverting our original change of variables gives the potential

$$\Phi(z) = \sigma_{zz} \ln \frac{\rho}{\rho_0},$$

the galactic vertical force law follows immediately. This is why you can use star counts and velocity dispersions to determine the mass of the disk and the halo, which will be our next subject.

7.1.4 The Halo

Shapley's determination of the large-scale structure of the globular cluster system was the first evidence for another component to the Galaxy, the halo, that is not confined to the plane and is considerably larger than the inner stellar bulge. Photographic techniques proved inadequate to obtain information about the structure perpendicular to the plane at the solar circle. The main reason is the faintness of the tracers. There are no OB stars off of the plane and low-mass stars, which are the most numerous constituent of the stellar population, are intrinsically faint. Study of the Galactic halo kinematics and population began with Roman's discovery that the stars selected on the basis of their W velocity component (the velocity perpendicular to the galactic plane) systematically have the lowest metallicites indicated by their ultraviolet excesses in the UBV system, $\delta(U-B)$. Extending this work, Eggen, Lynden-Bell, and Sandage[10] advanced the following argument. The Sun is situated in the plane, and those stars with the largest z velocities are the ones that started with the largest gravitational potential energy. Therefore, a distribution of $\delta(U-B)$ with W is equivalent to a distribution of [Fe/H] with z. Their conclusion was that the halo was formed as an extended structure at low metallicity and the disk was formed later by collapse.[11]

The picture has become murkier in recent years. First, you would expect in this collapse scenario that the halo star velocities are nearly radial. This is *not* what is observed; they are much more nearly isotropic. Since the advent of *Hipparcos* astrometry and CCD radial velocity studies, streaming motion has been identified among halo stars. There are at least two distinct groups of globular clusters, of which the most famous weird one is ω Cen, whose orbit is retrograde to Galactic rotation. It has been suggested that this may be the nucleus of a now nearly disrupted accreted galaxy. With the discovery of clear interaction between our system and Local Group galaxies, in particular the Sagittarius dwarf galaxy, it is no longer appropriate to consider galactic evolution in isolation. The uniform picture of disk formation subsequent to construction of the halo assumes this, and, as we will see from the discussion of evolution in clusters of galaxies, this is far from the mark for any cluster member.

If there is a dark component to the halo, one way of finding it is to observe gravitational microlensing. Although under- or nonluminous, compact objects will

---

[10] Eggen, O., Lynden-Bell, D., and Sandage, A. 1962, *ApJ*, **136**, 738 proposed the infall model based on the combination of stellar dynamics and photometric population separation.

[11] See Gilmore, G., Wyse, R. F. G., and Kuijken, K. 1989, *ARAA*, **27**, 555; Madjewski, S. R. 1993, *ARAA*, **31**, 575.

still produce gravitational deflection of light of distant point sources, such as quasars or stars in the Magellanic Clouds or other Local Group galaxies when they pass in front of the source. The typical angular deflection is as large as it is for the Sun, but it produces a greater effect as seen at Earth because of the greater distance to the stars. For instance, taking the the results from Appendix 1.B, the angular deflection for a star at a distance r from a mass M is $\Delta\theta = 4GM/rc^2$. For the radius of a solar mass star, we obtain an angular deflection at the limb of 1.75 arcsec, which, though small in angle, translates into a spatial deflection at the Earth of $2.6 \times 10^{16} \, D_{\rm kpc}$ cm or 1700 $D_{\rm kpc}$ AU. This is actually quite large and given the typical timescale for stellar motions, if the mean halo velocity is of order 50 km s^{-1}, corresponds to a crossing time of about 0.1 day to a few days. This should be the duration of a typical microlensing event. It is possible to such an event by looking for a simultaneous change in the brightness of a star at all wavelengths with no change in the color of the star. The difficulty with this technique is that such events are very rare and requires continually monitoring literally millions of stars at the same time for a long period. It is only since the late 1990s that the technology has been available to carry out such surveys.[12] This method can be considered supplementary to the more traditional technique of star counts, to which we now turn.

The method of star counts, which was pioneered by Herschel, Charlier, Kapteyn, and Nassau and McCluskey, is *still* a primary way to determine both the structure and mass of the system. The problem is that this requires an accurate picture of the luminosity function, at least in the solar neighborhood, and this is very difficult to obtain. Assume that you have a single population with a unique absolute magnitude and a scale height z_0. If the mean number density in the solar neighborhood is n_0, the column density is $\Sigma_0 = n_0 z_0$. Within a distance R the cumulent number you would expect to see in a solid angle $\Delta\Omega$ is $N(r \leq R) = \Sigma_0 \Delta\Omega (1 - e^{-R/z_0})$, where now the limit R is interchangeable with a magnitude limit to the sample. If there is a luminosity spread in the sample, you have to integrate this over the population's distribution. You can see how this complicates the problem because intrinsically faint stars are more numerous but can be seen only to smaller distances. In addition, if we include multiple scale heights, the cumulant is now the sum over the individual constituents.[13] If we include the effects of reddening, the observed scale height is reduced by a factor of $(1 + \tau_0)$, where $\tau_0 = \kappa_{\rm dust} n_{\rm dust} z_0$ is the optical depth. Observationally, out to a distance of about 4 kpc, the high-latitude star counts can be approximated by

$$\frac{\rho(z)}{\rho_0} = 0.96 e^{-z/(0.25 \text{ kpc})} + 0.04 e^{-z/(1 \text{ kpc})}, \qquad (7.18)$$

while for greater distances the large-scale structure means we can't use this simple slab picture. We will extend this later when discussing how to model a thick disk.

---

[12] See Paczynski, B. 1996, *ARAA*, **34**, 419; Jetzer, P. 1999, *Naturwissen.*, **86**, 201.

[13] Bahcall, J. N. and Soneira, R. M. 1981, *ApJS*, **47**, 357; Gilmore, G., Wyse, R. F. G., and Kuijken, K. 1994, *ARAA*, **27**, 555; Creze, M. 1991, in *The Interstellar Halo-Disk Connection*: IAU Symp. 144 Bloemen, H. ed. (Dordrecht: Kluwer), p. 313; Kuijken, K. and Gilmore, G. 1989, *MNRAS*, **239**, 571, 605, 651; Davis Philip, A. G. and Lu, P. eds. 1989, *The Gravitational Force Perpendicular to the Galactic Plane* (Schenectedy, NY: L. Davis Press); Fich, M. and Tremaine, S. 1991, *ARAA*, **29**, 409: Bahcall, J. N. 1986, *ARAA*, **24**, 577.

Deeper surveys of the halo, fitted to a spheroidal model, indicate a shallower power law dependence of the stellar density, $\rho \sim r^{-\alpha}$, with $3 < \alpha < 3.5$ out to nearly 50 kpc and an axial ratio of $c/a \approx 0.6\text{--}0.7$, depending on the tracer.[14] This matters for several reasons. Does the rotation of the disk extend into the halo? If so, does it affect the overall structure (in particular, what about the dark matter component for which the stars may be just luminous tracers)? What about the bar in the Galactic center—does its influence extend into the inner parts of the bulge and halo? If the halo and bulge form from merger events, might the angular momentum of the accreted systems accumulate in the halo and show a signature in its shape? We will return to this for elliptical galaxies, making use of arguments from self-gravitating ellipsoids. For now, we leave only the questions. The *inversion* of star counts provides the direct determination of the mass of the galaxy, at least in principle (see, for instance, Appendix 7.A). Start with a spherical system and assume that the stars have some spatial density distribution, $n(r)$ around you at distance r. The column density is $\int n(r)\,dr$, and the surface brightness down to some limiting level is $\int_0^D n(r)\,dr/r^2$. The brightness at any cut is the cumulant of the density distribution, assuming that all the stars are identical. Then more refined distributions for different components are introduced, for instance, a thin disk for the young population and a thicker disk for the older stars, consistent with their observed velocity dispersions.

7.2 LARGE SCALE STRUCTURE OF THE GALAXY

Examining the Milky Way with the naked eye shows that the distribution of stars is neither uniform nor spherical around the Sun. There is a concentration of stars and clusters toward Sagittarius that comes from the spheroidal bulge, while around the rest of the sky you see the evidence of the disk. Mottled by intervening molecular clouds, relatively local smaller clouds and large complexes, this band is the unresolved result of the system we inhabit. Surrounded by the interstellar medium, our view is obscured by dust and other absorbers and our motion can be known only with respect to other bodies. The opacity of the ISM is wavelength-dependent, however, which means that we can determine some aspects of the structure by observing in spectral regions where the absorption is reduced. With γ-ray and X-ray observations we can observe sources near or even at the center, as we can at most higher energies. In the extreme ultraviolet, that is, in the Lyman continuum, we see a medium that is, largely black, limited to only a few hundred parsecs by hydrogen absorption. In the visible, our view is confined to a few kiloparsecs and because of dust obscuration. Then in the infrared and longer wavelengths, the dust opacity is negligible and the vista again opens.

7.2.1 Galactic Rotation

Spherical collisionless stellar systems such as globular clusters generally have isotropic velocity dispersions. Although they may be distorted by tides, there is a limit to how flattened they can become, and, at any rate, these configurations are prolate or triaxial ellipsoids. In order to achieve the degree of flattening that is so

---

[14] The literature is reviewed in Yanny, B. et al. 2000, *ApJ*, **540**, 825, which presents early results from the SDSS for the structure of the halo.

obvious in the Milky Way, you need rotation. But what is the rotation law? The stellar "gas" is inviscid, so nothing enforces rigid body motion. The system is large and the individual stars orbit within an extended potential, so although you would expect differential motion, it is certainly not Keplerian. To complicate the analysis, your observations are made in a moving frame, the solar neighborhood, which is in orbit at a mean distance R_0 and frequency Ω_0 around the Galactic center along with the rest of the stars and gas in the disk. The problem is, very schematically, the same as in the solar system where you are outside the main gravitating mass (in that case, the Sun; in the present case, the bulk of galactic mass). Relative to this reference frame, which has a specific angular momentum $R_0^2\Omega_0$, stars orbiting slightly beyond R_0 have higher specific angular momentum and lower orbital speed than you do, while those closer to the galactic center are moving faster. This creates a local vorticity, or circulation, as observed from the Sun. We have re-entered the Ptolemaic world, since the motion can best be described by epicycles.

7.2.2 Determination of Solar Galactocentric Distance

Locating the Sun within the Galaxy requires using a variety of stellar calibrations, some of which have recently been revised in light of the fundamental astrometry provided by *Hipparcos* and improvements in ground-based techniques. Pulsating stars provided the first standard candles. The study of globular cluster variable stars began in the 1880s by S. Bailly and for δ Cepheids in the Magellanic Clouds by H. Leavitt around 1910. While Bailly's shorter period *cluster variables* did not overlap those in the Magellanic Clouds, there were some longer period stars that did. Leavitt's discovery of a Cepheid period–luminosity relation provided the link between the two samples. Shapley, in 1918, used the apparent magnitudes of Cepheids in 69 globulars to show that the clusters form a nearly spherical distribution around a point that, he argued, is located at 15 kpc from the solar circle in the direction of Sagittarius. He further showed that the distribution was very extended, obtaining distances of up to 40 kpc from the Sun for the most distant in the anticenter. The basic picture is right; the Sun is off in the Galactic periphery, but the details are wrong and the reasons for the mistake are important to understand. Interstellar reddening introduced a systematic error in Shapley's photometric distance determinations. This was realized by Trumpler (in 1930) in his study of the angular diameter variation of globular clusters. This demonstrated that there was a zero-point error in the calibration obtained from the LMC due to dust. The second problem was intrinsic to the sample: the zero point for the period–luminosity calibration was wrong for the Cepheids. This is a much harder problem to address. Current work uses several different distance calibrators that were not at the turn of the twentieth century, but some of the distance determinations are based on the same basic arguments used in the original work. Using an updated version of Shapley's technique of globular clusters now gives distances between 7 and 9 kpc.[15] Bailly's variables have also come into their own. The

---

[15]The question remains, however, whether all globular clusters are members of the same population. Some, like ω Cen, are confined to the disk and have orbits with relatively high eccentricities but certainly not radial. Others, like M3 and M92, appear to be more like the halo dwarf stars and probably come from the same population.

globular cluster RR Lyr stars have approximately the same absolute visual magnitude. *Hipparcos* measurements center around $M_V = 0.7$, and nonpulsating stars on the horizontal branch have similar luminosities. Statistical studies of the distribution with apparent magnitude have been carried out in several directions through the bulge, notably in the direction $l_{II} = 0°$ and $b_{II} = -8°$, a low extinction region known as *Baade's window*. The peak in the distribution is at around $R_0 \approx 8 - 9$ kpc. The main uncertainties come from stellar metallicities since the luminosity of the horizontal branch and the RR Lyr depends on the opacity through the stellar structure. Red giants and Mira variables, which can be observed in the red and infrared where dust extinction is less important, give similar results. The use of infrared and radio calibrators ameliorates or removes the effect of extinction. Using water maser proper motions and OH IR stars, red supergiants in a pre-planetary nebula phase of evolution that are embedded in extensive dust shells, the distance is again found to be $R_0 \approx 8 - 9$ kpc.

The southern Milky Way proved more spectacular for optical studies. From early photography and astrometry, Gould recognized a separate distribution of what we now know to be OB associations that are roughly symmetrically distributed around the Sun and tilted by about 18° with respect to the galactic plane, a structure now called *Gould's Belt*.[16] Recent parallax determinations from the *Hipparcos* satellite show this very clearly. The Belt includes the Scorpius–Centaurus and Orion OB associations and are roughly coaeval with an age of about 30 Myr. Figure 7.5 shows data from the *Hipparcos* catalog for O, B, and Wolf–Rayet stars where the Belt is visible; Figure 7.5b shows the distribution of relatively young (≤ 10 Myr-old) clusters (we will return to these clusters when discussing spiral structure, so don't forget this figure).

7.2.2.1 Kinematics from Dynamics

Our first approach will be to set up the equations of motion as they appear from the solar orbit and to then perturb the motion by assuming that you have a small time-dependent displacement from this, r. At this distance, the local orbital frequency is Ω_0, but again there can be a small deviation from this, $\dot{\delta\phi}$. At this position, you are also at a specific place in the galactic mass distribution so there is a specific value for the gravitational potential $\Phi(R, z)$. We won't specify this in advance, but for the sake of simplicity we will assume for now that it is axisymmetric. This reference frame is in a circular orbit about the galactic center, the LSR. Now what will you see from this position? Stars in eccentric orbits move through the circular locus defined by a distance R_0 and appear to oscillate around their mean radius. We can treat this perturbatively by looking at a nearly co-orbiting body whose position with time differs by a small amount, $(r, \delta\phi, \delta z)$. In the plane Φ depends only on the radial galactocentric distance so the z position of that plane is indeterminant (the potential is cylindrically symmetric, so we can translate the orbit along the axis and get the same result regardless of the starting position). The final *ansatz* is that we expand the gravitational potential around the circular orbit, in effect the harmonic oscillator approximation we used in Chapter 1. The

---

[16] See Tenorio-Tagle, G. and Bodenheimer, P. 1988, *ARAA*, **26**, 145; Pöppel, W. 1997, *Fund. Cosm. Phys.*, **18**, 1 for recent references. See also Lesh, J. R. 1968, *ApJS*, **17**, 371.

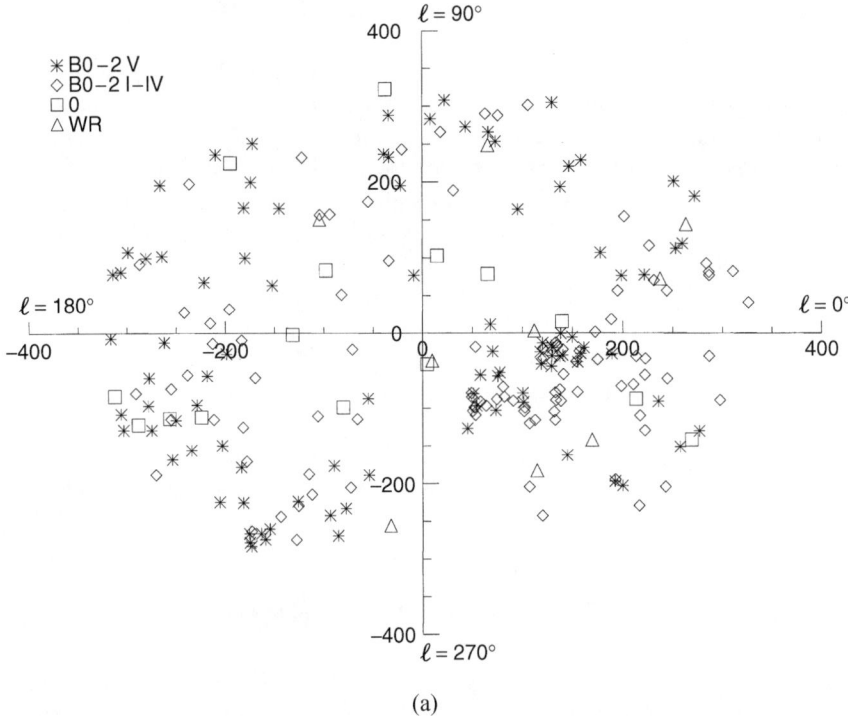

(a)

FIGURE 7.5. (a) Distribution of massive stars in the solar vicinity, within 330 pc, based on *Hipparcos* data (credit: J. P. Aufdenberg, CfA); (b) distribution of young open clusters within 7 kpc of the Sun projected onto the galactic plane (from Geneva cluster database). Notice the banded structure with galactocentric distance, the stellar evidence for spiral structure (credit: J.-C. Mermilloid, Univ. Lausanne).

test particle, in this case a star orbiting the galactic center, is assumed to oscillate around this reference orbit. Note, however, that this is *not* a perturbation equation since we are comparing two possibly stable orbits, one of which is simply a reference. The radial equation provides the local circular frequency:

$$\Phi'_0 = -R_0 \Omega_0^2. \tag{7.19}$$

The angular momentum equation is $J = R_0^2 \Omega_0$, so the radial equation of motion becomes

$$\ddot{r} = -\frac{3J^2}{R_0^4}r + \Phi''_0 r \rightarrow (\Phi''_0 - 3\Omega_0^2)r, \tag{7.20}$$

which on substitution of Eq. (7.20) for Φ''_0 gives the characteristic frequency:

$$\omega^2 \equiv \kappa^2 = \left[\frac{1}{R^3}\frac{d}{dR}(R^2\Omega)^2\right]_0, \tag{7.21}$$

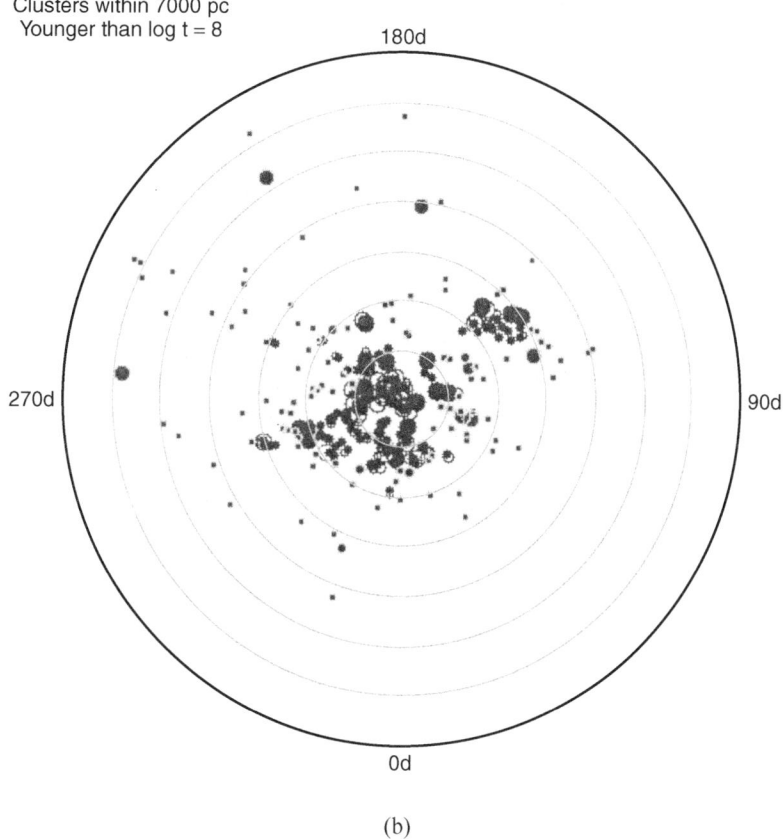

(b)

FIGURE 7.5. (*Continued*).

which is the now familiar *epicyclic frequency* that you will recall from our earlier treatment of accretion disks. It represents the shear in the solar neighborhood due to differential rotation that produces vorticity in every noninertial reference frame. We will see in a moment how it is related to the *observable* stellar kinematics. Stable orbits require $\Phi_0'' < 0$, which is also the radial galactic tidal acceleration, providing a constraint on the possible mass models for the Galaxy. For the vertical motion near the plane, we have

$$\delta \ddot{z} = -\frac{\partial^2 \Phi}{\partial z^2} \delta z, \qquad (7.22)$$

which is a harmonic oscillator in the $\hat{z}$ direction. The motion describes a Lissajous figure with oscillation frequencies ω_r and ω_z that are not necessarily commensurate.[17] Let us add a point about the stellar orbits. The Galactic mass distribution is

---

[17] Lovely examples of stellar orbits are available in the review by Schmidt, M. 1965, in *Galactic Structure: Stars and Stellar Systems*, vol. 5 Blaauw, A. and Schmidt, M., eds. (Chicago: Univ. Chicago Press); see also Ollongren, A. 1967, *AJ*, **72**, 436.

neither spherically nor cylindrically symmetric, so any stellar orbit experiences a torque and precesses. Since the mass distribution is also not pointlike, the precession frequency depends on how large r is, and orbits that have different eccentricities (i.e., angular momenta) precess at different rates. It isn't really correct to say that they have a single frequency since they sample different parts of $M(r)$ in the disk and halo as they move. Instead, they execute elliptical epicyclic motions in the frame moving with Ω_0. Now suppose that we include an angular perturbation in the equations of motion. There is a critical condition, the usual problem with resonance and small divisors, when the angular perturbation frequency is equal to an integer multiple of κ. Although we will postpone the explanation for how a wavelike perturbation can arise in a disk, you can see that resonance occurs whenever the pattern angular frequency of the wave is a rational multiple of $\Omega_0 \pm m\kappa$, where m is an integer. Such locations are called *Lindblad resonances*. Now we see that the comparison between the circular and elliptical orbits is really the same as examining the stability of a circular orbit to radial perturbations.

But what does this mean for the study of galactic structure? The rotation curve is observable provided we have a large catalog of stellar distances for which we have accurate space motions. This is not as simple as it might seem because constraints on these two observational quantities are precisely at odds with each other. On one hand, only comparatively close stars will yield high-quality proper motions and parallaxes, and among the more distant candidates only those with large space velocities that are not due to galactic rotation, such as the runaway OB stars or high-velocity halo stars, will yield measurable proper motions in reasonable times. On the other hand, you want the most distant stars possible for measurement of the rotation curve, most of which are too far to have a detectable parallax. Less direct methods must be employed for finding the distance to these stars, in particular using their spectroscopic and photometric properties to obtain their luminosities. The complication introduced by interstellar extinction should be obvious, but there is a more subtle problem to consider that will continue to haunt us. We are tacitly assuming that local measurements are typical of the system as a whole. With increasing distance the assumption of circular motion for the stars and gas becomes less reliable because of possible large-scale departures from axisymmetry. Our locale in the Galaxy may be special. For instance, the Sun is surrounded by a ring of OB associations, Gould's Belt, which has a radius of a few hundred parsecs. There are indications of large-scale streaming motions toward the inner Galaxy, and there is an important dynamical feature, the 3-kpc arm, which is distinctive in the 21-cm surveys as well as molecular observations that may be the signature of an inner bar or resonant ring. Furthermore, we depend on the symmetry of the mass distribution around the Galactic center to perform the inversion of the integrated column measurements of the halo. In short, the global picture is very hard to determine frommeasurements made over only about 10–20% of the distance to the center, and caution should be exercised when approaching these results.

There are also complications from the theoretical point of view. You may have noticed that many of the results we are obtaining for stellar dynamics are the very similar those we previously derived for *ideal* fluids. As long as it has no viscosity, each fluid parcel moves without long-range transport of either momentum or energy by anything except the surrounding potential. For a self-gravitating medium, adding the Poisson equation doesn't change anything fundamental about the particle analogy. Once you add viscosity, however, you have introduced a fundamental change in the structure of the phase space occupied by the particles. They start to diffuse on a characteristic timescale that has nothing to do with the large-scale spatial distribution of the matter. Thus *real* fluids do not look like galaxies or clusters because the latter lack internal *dissipation* for the stellar component. Much has been made since the early 1990s of the role of chaos in such systems. It should be stated here that as long as stellar motion can be described by a Hamiltonian formalism, it may be incredibly complex but it is *not dissipative* (i.e., it is conservative) and its evolution is at some level strictly reversible. However, it *is* very complex, so the timescales for the reversibility of all the motions cannot be short. That is why much of the qualitative behavior of such systems resembles fluids. It is also why in simulations of galactic dynamical evolution, the algorithmic limitations of modern computational methods introduce the appearance of viscosity.

7.2.2.2 Observable Consequences: The Oort Laws for Galactic Rotation

You need some flow tracer to observationally determine the rotation curve. Stars were used first, so it isn't surprising that the discovery of Galactic rotation occurred in the Netherlands. Stellar motions and positions had been the preoccupation of J. Kapteyn, who started the dynamical study of galactic structure at the beginning of the twentieth century. By the 1920s, sufficient data on parallax, radial velocities, and proper motions had been accumulated that J. Oort was able to piece it together with the distance determined for the Sun relative to the Galactic center.[18] In light of the size of the available sample, which was tiny by modern standards, Oort's result represents a remarkable achievement. Proper motions and parallaxes were available for very few stars, and those were of low accuracy by modern standards, entirely photographic and based on short time intervals. In the local neighborhood, it is important to exclude stars that are moving perpendicular to the Galactic plane and also high velocity stars within the plane. In addition, for full space motions, it is necessary to have radial velocity, and hence spectra, to supplement the astrometry. None of these is easy to obtain for the relatively faint stars that populate the solar neighborhood, and even then the samples were small.

To make sense out of the locally observed kinematics required a basic fact, that the Sun is not at the center of the Galaxy. Oort knew this from Shapley's study of the globular cluster distribution almost a decade before. We must be situated in a moving frame at a galactocentric distance, R_0, from which to observe the motion of nearby stars. The local (circular) orbital speed is $\Theta_0 = \Omega_0 R_0$, while stars at some relatively small distance r away have a motion $\Theta = \Omega R$, where R is their distance from the galactic center. Assume that you've observed the radial and transverse velocity (proper motion) of a star at some Galactic longitude l, as shown

---

[18]An especially illuminating autobiographical discussion of this work is Oort, J. 1972, *Ann. NY Acad. Sci.*, **198**, 255; Oort, J. 1981, *ARAA*, **19**, 10. The paper, Oort, J. 1927, *BAN*, **2**, 275, was entitled "Observational evidence confirming Lindblad's hypothesis of a rotation of the galactic system" following B. Lindblad's analysis of the high-velocity stars in the solar neighborhood.

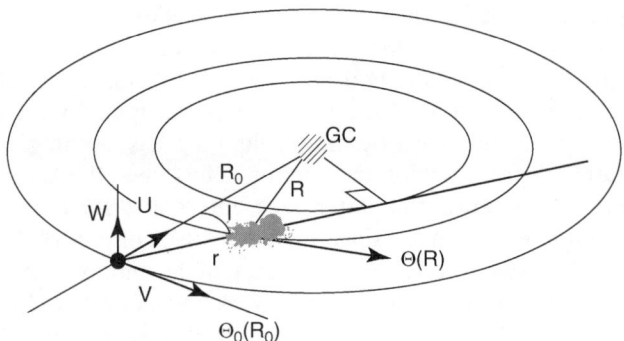

FIGURE 7.6. Diagram illustrating the basis of the Oort law and the interpretation of H I measurements in the Galactic plane.

in Figure 7.6. For a star viewed at some galactocentric distance R but at an angle that is *not* at the tangent point to its orbit, we have:

$$R \cos \theta = R_0 - r \cos l, \quad (7.23)$$

Requiring that you observe only stars in the solar neighborhood means that the velocity at a distance r is:

$$\Theta(R) = \Theta_0(R_0) + \left(\frac{d\Theta}{dR}\right)_0 \Delta R = \Theta_0(R_0) - \left(\frac{d\Theta}{dR}\right)_0 r \cos l. \quad (7.24)$$

The rest is geometry. First, notice that $r \sin l = R \sin \theta$, where θ is specified with respect to the center. Referring to Figure 7.6, you also have $\sin(l + \theta) = R_0 \sin l / R$, so the differential (measured) radial velocity for a star is

$$v_R = \Theta \sin(\theta + l) - \Theta_0 \sin l = \left(\frac{\Theta R_0}{R} - \Theta_0\right) \sin l. \quad (7.25)$$

Substituting for R/R_0 from Eq. (7.23) gives

$$v_R = \left[\Theta\left(1 + \frac{r}{R_0}\cos l\right) - \Theta_0\right]\sin l. \quad (7.26)$$

We then arrive at the first of Oort's laws:

$$v_R = \frac{1}{2}\left[\frac{\Theta_0}{R_0} - \frac{d\Theta_0}{dR_0}\right] r \sin 2l \equiv Ar \sin 2l. \quad (7.27)$$

This provides the definition of the *Oort constant A*, a directly measurable quantity. Notice that the motion has a frequency $2l$ in galactic longitude, a basic result for any motion observed in a rotating frame (remember our remarks for the three-body problem).

For the transverse motion, the derivation is essentially the same. You measure a differential velocity:

$$v_T = -\Theta \cos(\theta + l) - \Theta_0 \cos l = \left(\frac{\Theta R_0}{R} - \Theta_0\right)\cos l - \Theta \frac{r}{R}, \quad (7.28)$$

which reduces to

$$v_T = -A \cos 2l + B, \quad (7.29)$$

where the second Oort constant is defined by

$$B \equiv -\frac{1}{2}\left[\frac{\Theta_0}{R_0} + \frac{d\Theta_0}{dR_0}\right]. \quad (7.30)$$

Again, this can be measured from a sample having accurate parallaxes and proper motions. These purely kinematic measurements provide a very important dynamical quantity, the local velocity gradient. This connects the mass of the disk and the velocities through Φ'' since

$$\Phi''_0 = \left(\frac{d}{dR}\frac{\Theta^2}{R}\right)_0.$$

The solar galactocentric distance is obtained from $\Theta_0/R_0 = A - B$ *if* you can observe the motion of the Sun relative to a fixed reference frame, such as radio galaxies and quasars. This is actually a lot harder than it looks because the Galaxy as a whole is in motion with respect to matter at large distance.

Finally, we can connect the shear—the epicyclic frequency—to the Oort constants using

$$\Phi''_0 = -4\Omega_0(\Omega_0 - A),$$

from which the epicyclic frequency follows, written in terms of astrometrically *observable* quantities:

$$\kappa^2 = 4B(B - A). \quad (7.31)$$

This expression illustrates a significant property of the Oort constants: they measure the tidally generated shear across a body of finite size as we mentioned when discussing the limiting radius of clusters. The Oort constants also relate to the vertical structure of the disk. As we wrote a while back, the vertical acceleration

$$K_z = -\frac{1}{\rho}\frac{\partial \rho \sigma_{zz}}{\partial z}$$

is substituted into the Poisson equation:

$$\nabla \cdot \mathbf{K} = \frac{1}{r}\frac{\partial}{\partial r}rK_r + \frac{\partial K_z}{\partial z} = -4\pi G\rho.$$

The second term is related to the Oort constants, which are independent of z:

$$-\frac{1}{4\pi G}\left[\frac{\partial K_z}{\partial z} + (A^2 - B^2)\right] = \rho(r,z) \qquad (7.32)$$

and substituting Eq. (7.12) into Eq. (7.32), we obtain

$$\rho(r,z) = -\frac{\sigma_{zz}}{4\pi G}\left[\frac{\partial}{\partial z}\sigma_{zz}\ln\rho\right] + \frac{(B^2 - A^2)}{4\pi G}. \qquad (7.33)$$

This is nearly the equation for an isothermal disk. To see this, since A and B are constants, using $y = \ln \rho\sigma_{zz}^2$ again puts Eq. (7.33) in the following form:

$$\frac{\partial^2 y}{\partial \zeta^2} = -e^y + \frac{(B^2 - A^2)\sigma_{zz}^2}{4\pi G}.$$

In this last case, you can see the effects of shear on the vertical structure. Remember, the Oort constants describe the tidal term due to the displacement of the local frame from the Galactic center.

The problem with using stellar markers of the flow is that their distances and transverse motions are imprecisely known, although this has dramatically improved with the release of the *Hipparcos* parallax and proper motion data. These data, however, are still restricted to small distances relative to the size of the Milky Way. For more distant stars, proxy distance indicators are needed. Spectral classification is a useful tool for obtaining intrinsic luminosities to compare with photometry, but it is relatively imprecise unless a well-defined statistical sample of stars provides the luminosity calibration. For the most luminous stars, the ones that are needed to sample the longest lines of sight, there are few near the solar neighborhood, and these must be calibrated by a bootstrapping procedure. Also, since the interstellar medium is dusty, the light from the most distant stars is not unaffected by extinction, and this causes errors in their photometric distances. H II regions can also be used, but observations of hot gas in star forming regions is complicated by unknown bulk internal motions (Fig. 7.5b).

The 21-cm line of neutral hydrogen provides the most reliable means for studying large-scale structure, at least of the gas. Since this is a radio frequency *line*, its wavelength is precisely known and the emission line is not attenuated by intervening dust. It can be observed throughout the Galaxy, since it is weak enough to be optically thin over a major part of the disk and its orbital motion desaturates the profile, much as we have discussed for a stellar wind; that is, the intrinsic linewidth is so small that the circulatory motions decouple the line at different radii, except in the immediate vicinity of the line of sight toward the Galactic center. First accomplished in 1957 by Kerr, Oort, and Westerhout, observations of the plane have since been considerably refined.[19] At any longitude (see Fig. 7.7), you can see that there is a maximum velocity where the line-of-sight (LOS)

---

[19] See Burton, W. B., Elmegreen, B. G., and Genzel, R. 1992, *The Interstellar Medium*: 21st *Advanced Astrophysics Course—Saas-Fee* (Berlin: Springer-Verlag); Burton, W. B. 1988, in *Galactic and Extragalactic Radio Astronomy*, 2nd ed. Verschuur, G. and Kellermann, K, eds. (Berlin: Springer-Verlag).

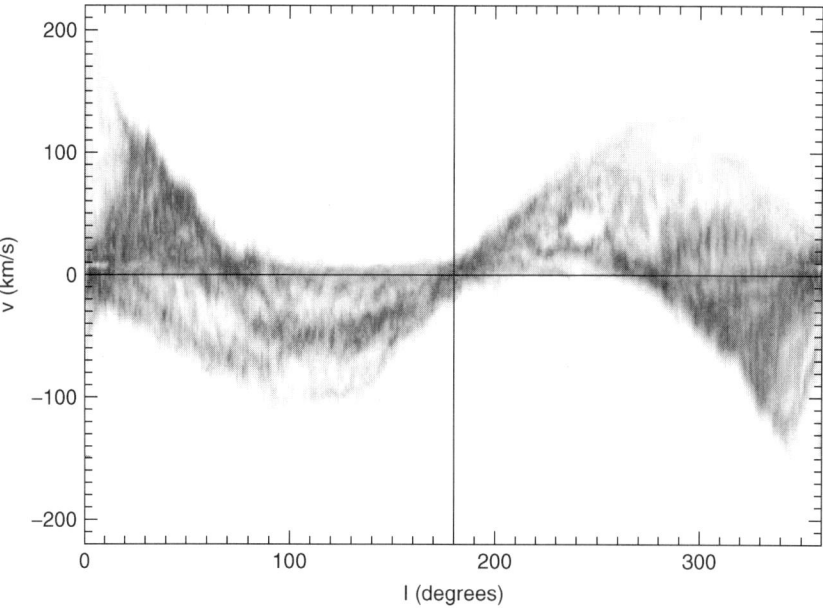

FIGURE 7.7. Velocity–longitude plot for H I 21-cm emission in the Galaxy. Grayscale indicates intensity. This plot shows the full plane for $b_{II} = 0°$ (credit: W. B. Burton, Univ. Leiden). Neutral hydrogen cloud concentrations can be seen at the spiral arms.

component of the orbital motion corresponds to the actual velocity. The maximum velocity curve is constructed along the tangent line using an (l_{II}, v_{rad}) plot with the distance provided by LOS geometry. But, as we mentioned earlier, the problem is that nonaxisymmetric motions complicate the interpretation of any observations since for the H I we can obtain the distances geometrically only by assuming that we are observing projected orbital motion. The method has a historical analogy in Apollonius' theorem for finding the size of an epicycle from the angular locations of the stationary points of retrograde motion, as Ptolemy explains in the *Almagest*.

To extend the kinematical picture of Galactic rotation (see also Table 7.1[20]) beyond the solar circle is a difficult task. While the tangent method works for the inner Galaxy, it can't be used for the outer disk. Distances to stellar tracers are required, and this means using both parallax and proper motion data, which is still

TABLE 7.1 The Current Numbers for Galactic Rotation

Quantity	*Hipparcos* Value
Ω_0	27.2 ± 0.9 km s^{-1} kpc^{-1}
κ	36.7 km s^{-1} kpc^{-1}
$v_\odot$	$(10.0, 5.2, 7.2)$ km s^{-1} relative to LSR
$\sigma_{rr} : \sigma_{\phi\phi} : \sigma_{zz}$	$1.0 : 0.36 : 0.12$

[20] Sources of Table 7.1 are Dehnen, W. 1998, *AJ*, **115**, 2384; Feast, M. W. and Whitelock, P. A., 1997, *MNRAS*, **291**, 683. When comparing Table 7.1 with some published studies, recall that in our notation, remember that for instance, $\sigma_{rr} = \sigma_R^2$.

limited in range to about 0.5–1 kpc, or spectroscopic parallaxes for OB stars for which spectral types and photometry are used with local calibration to obtain the distances. An alternative method is to use the kinematics of H II regions associated with molecular clouds in the outer Galaxy. The galactocentric distance of a parcel of gas in the plane at a distance d from the Sun is

$$R^2 = R_0^2 + d^2 - 2R_0 d \cos l, \qquad (7.34)$$

and its circular velocity is determined from its radial velocity by

$$\Theta = \Theta_0 \frac{R}{R_0} + \frac{v_r}{\sin l}. \qquad (7.35)$$

There have been many attempts, most of which converge on a similar result; namely, that the gradient in Θ beyond R_0 is small. The current value is about -2.5 km s^{-1} kpc^{-1}, but there is considerable uncertainty in this number.[21]

7.2.3 Spiral Structure

7.2.3.1 Observations

We live in a disk galaxy and, as we will see, disk galaxies are frequently spirals, so it's natural to expect no less from our system. For our Galaxy the first optical indications of spiral structure came from observations of the brightest structure tracers, the distribution of OB associations.[22] Most OB stars are beyond the capabilities of astrometric surveys, so photometry and spectroscopic parallaxes are the only way to obtain their distances. Since these stars tend to be found in clusters and associations, there is some redundancy in the measurements, and these sites have the advantage of providing ages for the structure tracers. Very large stellar complexes, the superassociations, are the best tracers of the large-scale structure in external spiral galaxies and probably our own as well. Examples in the Galaxy are the Orion OB 1 association and the string of star forming regions along the Carina arm. In the best studied galaxy, M31, these outline the spiral structure and coincide with many of the largest molecular cloud complexes. Since the mean lifetime of an OB star is so short, they are presumed as well to trace the location of any large-scale trigger to the star formation. Further information about the star formation comes from stellar kinematics—the OB stars should also show any dynamical perturbations imposed by the nonsymmetric distribution of star forming regions.[23] The argument goes like this. Low-mass stars have long main sequence lifetimes and, if formed continually, should show a large spread in ages and strongly mixed dynamics while massive stars do not have the time to dynamically mix. The vertex deviation should differ for subpopulations depending on the stellar mass, and it does.

---

[21] Brand, J., and Blitz, L. 1993, *Astr. Ap.*, **275**, 67; Binney, J. and Dehnen, W. 1997, *MNRAS*, **287**, L5.
[22] Blaauw, A. 1964, *ARAA*, **2**, 213. To date, the most comprehensive survey has been carried out using *Hipparcos* data; see Zweeuw, P. T., de Bruijne, J. H. J., Brown, A. G. A., and Blaauw, A. 1999, *AJ*, **117**, 354.
[23] See, for example, Alfaro, E. and Dalgado, A., eds. 1995, *The Formation of the Milky Way* (Cambridge, UK: Cambridge Univ. Press); Efremov, Y. N. 1995, *AJ*, **110**, 2757.

The strongest evidence we currently have for spiral structure comes not from the stars but from the gas. Observations with the H I 21-cm line provide both dynamics and masses for clouds in the inner Galaxy, where the rotation curve provides distances and large-scale structures can be traced continuously in the plane. The first maps published in the late 1950s that clearly showed the presence of spiral arms.[24] Two are particularly important, one toward the inner galaxy, the Sagittarius arm at a distance of around 2 kpc, and the other is the Perseus arm, located at about the same distance from the Sun toward the anticenter. The Orion region shows what is called a "spur," tilted with respect to the local arms and possibly indicating another mode for the spiral perturbation. A complication is introduced in this by the presence of a feature, the 3-kpc "arm," which appears to be a radially expanding feature at the boundary of the Galactic bulge. Neutral hydrogen observations are compromised by combination of the rotation curve and the opacity of the gas. Motion transverse to the line of sight, as toward the center, are impossible to separate on the basis of the rotation curve and, therefore, distances toward the Galactic center are hard to obtain from kinematics. In addition, the intrinsic velocity width of the clouds is sufficiently small that the lines turn optically thick, and the gas becomes self-absorbing.

Study of spiral structure beyond the solar circle is a real challenge: first, there is considerable uncertainty in the rotation curve, and second, the outer gaseous disk appears to be warped. The plane differs between the two galactic hemispheres both in intensity and dynamics. Warps are known for other spiral galaxies but there it is easier to study them because they *are* extragalatic systems and you can observe the global kinematics directly. For the Milky Way, small uncertainties that are introduced by non-circular motions translate into large potential errors in the distance to tracer gas.

7.2.3.2 *Departures from Symmetry: Density Waves and Bars*

The striking feature of disk galaxies is how regularly the stars show nonaxisymmetric structures. We have so far assumed that the motions trace the potential through essentially circular motions, yet the features in disks that first grabbed observers' attention are their luminous spiral arms and occasional barlike and ring patterns. At first the spirals were thought to be ejected material arms, and the luminosity contrast was interpreted as an actual surface density contrast. We now know that the visibility of the spiral is due more to the luminosity of the tracer of the structure (OB stars), and the sensitivity of the underlying stellar population to small departures from circular symmetry of the gravitational field, than to some very large amplitude bulk distortion of the disk. When examined more closely, the pattern appears to be a relatively low amplitude mode of the system rather than its bulk mass distribution. You might suspect, therefore, that some form of instability is involved that is supported by the mass of the disk but never grows large enough to actually disrupt it. This was the thinking that first gave rise to *density wave theory*. Since we have developed the tools for studying the stability of self-gravitating gases, let us see if this can be extended to our "gas of stars."

For an isolated self-gravitating gaseous slab, the local pressure cannot resist collapse in a density perturbation when the freefall time is short compared to the

---

[24] Oort, J. H., Kerr, F. J., and Westerhout, G. 1958, *MNRAS*, **118**, 379.

sound travel time across the layer. The stars may be collisionless, but we have argued that the internal velocity dispersion endows the ensemble with properties analogous to those of a gas. To model a disk galaxy instead of a slab, however, we must include the Coriolis and centrifugal accelerations to account for rotation. Since neither force is attractive, you can anticipate a change for the threshold at which the instability develops. The disk is also *differentially* rotating. When sheared, the growth of these density enhancements competes with the rate at which the density perturbations are separated and reduced in amplitude by long-range transport. It does not matter that the medium is composed of collisionless particles. They have mass; hence they create a local gravitational fluctuation, and that is all that is required for this Jeans-like instability to operate. Adding cooling or dissipation changes things for a gas relative to a "gas of stars," but not the fundamental features of the clumping itself. In a galaxy, the shear depends on the distance from the center and changes the pitch of a planar wavefront with respect to the azimuthal direction. What starts out as an orthogonally propagating wave gradually tilts because it is being transported faster in the interior of the system than at the boundary. So what begins as slablike clumps ends up as a spiral. The growth rate is determined by the shear timescale through κ^{-1}, so, unlike the case for an isolated cloud, there is a global barrier to the growth of the wave—it cannot continue inside the galactocentric radius at which its growth rate is the same order as κ. This marks the location of the *inner Lindblad resonance*, which has already appeared in the discussion of stellar dynamics. There is also usually an *outer Lindblad radius* depending on the rotation curve. You saw something similar to this when we discussed shepherding in planetary rings. It appears again for galactic disks because of the epicyclic frequency arising in the rotation curve. Depending on the mass distribution, and whether there is a nuclear bar, the instability may be forced and amplified through a coupling of the perturbation's rotation rate and that of the stars in the vicinity of the resonance.

The Lindblad resonances are characteristic of any particle motion in the corotating frame when a azimuthally periodic forcing is applied. They were proposed by Lindblad in 1927 virtually simultaneously with Oort's derivation of the differential rotation.[25] Imagine a locally Cartesian coordinate system that moves with a frequency Ω_0 at a distance R_0 from the galactic center. All single-particle orbits obey the usual dynamical equations, and for the Lindblad resonances we will ignore nonplanar orbits. Assume that ξ points away from the galactic center in the radial direction, η is tangential to the local circular motion, and ζ points perpendicular to the plane. We will ignore the ζ component by dealing only with planar orbits and assume the vertical structure is given by Eq. (7.32). Last, and most important, the large-scale radial gradient of the gravitational potential is determined by an unperturbed axisymmetric mass distribution. We then impose a small-amplitude (linear) periodic azimuthal pattern, either a density wave or a bar, on top of this potential and look at the response of the test particles. The perturbed equations of motion are the same ones we encountered for the magneto rotational instability. For simplicity, we assume a perturbing potential of the

---

[25] Lindblad, B. 1927, *MNRAS*, **87**, 553.

form

$$\delta\Phi = F(r)e^{im(\phi - \Omega_p t)},$$

where the perturbation amplitude F depends only on radius, Ω_p is the angular frequency of the pattern, and m is the azimuthal wavenumber of the imposed pattern:

$$\ddot{\xi} - 2R_0\Omega_0\frac{d\Omega_0}{dr}\xi - 2\Omega_0\dot{\eta} = -\frac{\partial\delta\Phi}{\partial r} = -\frac{F'}{F}\delta\Phi$$

$$\ddot{\eta} + 2\Omega_0\dot{\xi} = -\frac{1}{R_0}\frac{\partial\delta\Phi}{\partial\phi} = -\frac{im}{R_0}\delta\Phi. \quad (7.36)$$

On substituting $\delta\Phi$ into Eq. (7.36) and relabeling $\omega \equiv m(\Omega_0 - \Omega_p)$, we find

$$\xi = -\frac{1}{\kappa^2 - \omega^2}\left[\frac{2m\Omega_0}{R_0\omega} + \frac{d\ln F}{dR}\right]\delta\Phi$$

$$\eta = -\frac{i}{\kappa^2 - \omega^2}\left[\frac{m}{R_0}\left(1 + \frac{2R_0\Omega_0\Omega_0'}{\omega^2}\right) + 2\frac{\Omega_0}{\omega}\frac{d\ln F}{dR}\right]\delta\Phi. \quad (7.37)$$

There are two critical points, $\omega = 0$ and $\omega = \pm\kappa$, so the Lindblad resonances occur whenever $\Omega_p = \Omega_0 \pm \kappa/m$, which defines the pattern speed Ω_p. The local motion is an elliptical epicycle whose amplitude depends on the gravitational perturbation. But $\delta\Phi$ is not just imposed from the outside. There is a feedback through the collective behavior of the "fluid"; any perturbation in the potential changes the local density and through this changes the local gravitational acceleration. Consider a collection of stars whose orbits are eccentric from a tidal perturbation. The precession frequency depends on their galactocentric distance. The crest of the stellar density distribution lags the rotation by an amount that depends on the galactic rotation curve. This is a *density wave*. To estimate the effect of shear involves comparing the acceleration produced by tidal forces to the self-gravitation of the layer and is a lot like the Jeans criterion. The characteristic speed in the medium is c_s, although this can just as easily be the stellar velocity dispersion, so the tidal acceleration is $c_s\kappa$. For self-gravity only Σ matters, so the magnitude of the acceleration is $|g| \approx \pi G\Sigma$. Thus, when

$$\frac{\kappa c_s}{\pi G\Sigma} > 1,$$

we expect the layer to be stabilized by the tidal acceleration. This is called the *Toomre parameter*, Q.[26] Using the *Hipparcos* results yields $Q \approx 10^{-2}$ for the disk, so our first guess is that self-gravity is important for structure formation by the gas and the stars.

[26] Toomre, A. 1964, *ApJ*, **139**, 1217.

How does Q come out of the dynamical equations? Let's round up the usual suspects:

$$\frac{\partial v_r}{\partial t} + v_r \frac{\partial v_r}{\partial r} - \frac{v_\phi^2}{r} + \frac{v_\phi}{r}\frac{\partial v_r}{\partial \phi} = -\frac{c_s^2}{\rho}\frac{\partial \rho}{\partial r} + \frac{\partial \Phi}{\partial r}$$

$$\frac{\partial v_\phi}{\partial t} + v_r \frac{\partial v_\phi}{\partial r} + \frac{v_\phi v_r}{r} + \frac{v_\phi}{r}\frac{\partial v_\phi}{\partial \phi} = -\frac{c_s^2}{r\rho}\frac{\partial \rho}{\partial \phi} + \frac{1}{r}\frac{\partial \Phi}{\partial \phi}$$

$$\frac{\partial \rho}{\partial t} + \frac{1}{r}\frac{\partial}{\partial r}(\rho r v_r) + \frac{1}{r}\frac{\partial}{\partial \phi}(\rho v_\phi) = 0$$

$$\nabla^2 \Phi = -4\pi G \rho.$$

Putting ourselves again in the LSR frame, we assume neither radial nor vertical streaming, so the velocity components are u_r and $V_{\phi,0} + u_\phi$. We will also assume a thin disk and take the perturbations only to first order:

$$-i\omega u_r - \frac{2V_{\phi,0}}{R_0}u_\phi = \frac{\partial}{\partial r}\left(\delta\Phi - \frac{c_s^2}{\rho_0}\delta\rho\right)$$

$$-i\omega u_\phi + \left(\frac{V_{\phi,0}}{R_0}\right)\frac{\partial}{\partial r}(R_0 u_r) = i\omega u_\phi + \frac{\kappa^2}{2\Omega_0}u_r = ik\left(\delta\Phi - \frac{c_s^2}{\rho_0}\delta\rho\right)$$

$$-i\omega\frac{\delta\rho}{\rho_0} + \frac{1}{r}\frac{\partial r u_r}{\partial r} + iku_\phi = 0$$

$$\delta\Phi = \frac{2\pi i G}{k}\delta\Sigma,$$

replacing the meaning of ω with the comoving value $\omega + kV_{\phi,0}$ and integrating over z. If all perturbations in the $\hat{r}$ direction are very long wavelength, we can neglect radial gradients and greatly simplify the equations. The resulting quadratic dispersion relation

$$-\omega^2 = \kappa^2 - (2\pi G\Sigma)k - c_s^2 k^2 \tag{7.38}$$

has the solution for $\omega = 0$:

$$k_\pm = \frac{\pi G\Sigma}{c_s^2} \pm \frac{\kappa}{c_s}\left[\left(1 - \frac{\pi G\Sigma}{\kappa c_s}\right)^2\right]^{1/2}.$$

Stability requires that $k_\pm$ be real:

$$Q \equiv \frac{\kappa c_s}{\pi G\Sigma} \geq 1. \tag{7.39}$$

This is the result we found heuristically for the Toomre parameter. The growth time is approximated by finding the mode for which $\partial \omega / \partial k = 0$, which yields $t_{\text{growth}}^{-1} \approx (1 + Q^{-2})^{1/2} \kappa$, which for small Q can be quite rapid. We have certainly made a lot of approximations here, but after the fact they can all be justified. For instance, by taking the wavelength for the perturbation to be about the same size as the scale height and ignoring radial gradients, we have confined our analysis to only local perturbations. Now we have a way to feed the wave.

The idea that a self-gravitating disk could become unstable to spiral pattern formation was exploited by Lin and Shu in a series of papers beginning in 1964.[27] They chose to refer to these as quasi-stationary spiral patterns, recognizing their evanescent nature. Although stellar disks are inviscid, they can still transfer angular momentum through torques once any radially varying non-axisymmetric structures appear. And the stellar "fluid" is heated by interactions with the perturbation, so the amplitude of the wave must go down as the velocity dispersion increases. In effect, this changes the Q value for the disk without altering its resonance structure or epicyclic frequencies.

Any density perturbation is more than cosmetic. It changes the gravitational potential and has observable consequences for the galactic stars and gas. Since the wave pattern speed is slower than the local orbital speed, stars accelerate as they approach the crest and decelerate as they pass through it. The wave is not stationary, so, like the "slingshot" trick used for sending spacecraft to the outer planets in the solar system, the perturbation heats the stellar velocity distribution. This increased velocity dispersion is analogous to Landau damping in a plasma. Waves heat the particles, but because the stellar "gas" is collisionless, the heating is locally reversible. The star's velocity normal to the wavefront is changed while its motion remains unchanged in the transverse direction. Its trajectory is consequently directed outward along the arm and the density of stars is further increased by the streaming motion. This produces a systematic vertex deviation relative to the LSR, the signature we spoke of earlier for global star formation. OB stars are expected to show the effects of this streaming compared with stars that are formed more continuously over the age of the Galaxy.

The effect on the gas is more profound, even though the wave amplitude is small. It is a compressible *collisional* fluid, so while the stars deviate at the arms, the gas shocks. Qualitatively, this is because the orbital speed is so large compared to the sound speed. But let's take this further. At the arms, form a cartesian coordinate system (ξ, η), where now ξ lies normal to the density wave crest. For constant σ, the local motion is described by

$$\rho_0 \dot{v}_\xi = -4\pi G \delta \rho \xi - \frac{\partial \delta \rho c_s^2}{\partial \xi}; \qquad (7.40)$$

---

[27] Lin, C. C. and Shu, F. H. 1964, *ApJ*, **140**, 646; Lin, C. C. 1966, in *Studies in Applied Mathematics*, vol. 1 (Philadelphia: SIAM); Lin, C. C. 1967, *ARAA*, **5**, 453. The emphasis in this continuum approach should by now be quite familiar, particularly from our treatment of accretion disks. A basic difference between the MRI and density waves is the nature of the feedback. For magnetized disks, long-range coupling is achieved through induction, while here it is from the gravitational potential.

invoking mass conservation then yields

$$\frac{1}{v_\xi}\left(v_\xi^2 - \sigma^2\right)\frac{dv_\xi}{d\xi} = -4\pi G \delta\rho\xi. \tag{7.41}$$

The density maximum corresponds to a gravitational potential minimum. We found the same situation for the flow in a stellar wind *and* for the flow of mass at the inner Lagrangian point of a close binary system. The acceleration must vanish in the crest and change sign on either side of the crest. For the galactic problem, however, this sign change is opposite that for flow at the L_1 point, although formally the expansion in Eq. (7.41) is the same. At L_1 in a close binary, the acceleration points away from the critical point on either side in the flow. Here it points *toward* it. The gas is retarded as it moves away from the wave crest and is accelerated on the way in. Typical flow speeds are highly supersonic, of order 200 km s^{-1} unlike the sonic launching at L_1, so there is no chance for the gas to slowly adjust to the change in acceleration. Much like a vehicle pileup on a highway following an accident, a compressional wave propagates backward into the "flow." The inflow inevitably suffers a hard collision—that is, a strong standing shock. The resulting compression may push the gas beyond its self-gravitating limit, if the flow can remain isothermal, resulting in either cloud formation or collapse and star formation. This is the basis of *density-wave-triggered star formation*. It requires a special set of conditions, but these may be achievable for molecular clouds. Since the cooling depends on the density, the efficiency for cooling actually increases with increasing temperature due to molecular line emission. Alternatively, the clouds may simply blow apart if the heating is great enough to exceed their self-gravitation. Any combination of these may happen, depending on the details of the microphysics of the cloud/gas medium.[28] This is why density waves have been implicated for several decades as one cause, if not the principal mechanism, for global star formation. They *must* play some role in the structuring of the disks, but just what it is remains uncertain. Certainly there is something like a density wave in all disk galaxies. The pattern is easily detected in spirals in the old stellar population using infrared images. The persistence timescale is not known, but the disks clearly support some clumping even if it is unstable to winding and dispersal. Systems that show well-ordered global patterns are typically interacting, suggesting that the spiral has been triggered as a response to external tidal forcing. The presence of bars in the nuclear region provide another means for perturbing the stellar distribution and maintaining the spirals, and this is the most promising mechanism in isolated systems. The rotation frequency of the bar is related to the pattern speed of the wave. Systematic departures from circular motion are a signature of such a perturbation, and there is strong evidence that the Milky Way has a bar in its inner regions.[29]

---

[28] See, for instance, Woodward, P. 1976, *ApJ*, **207**, 484; Elmegreen, B. G. 1995, in *The Formation of the Milky Way*, Alfaro, E. and Dalgado, A., eds. (Cambridge, UK: Cambridge Univ. Press). See also Roberts, W. W. 1969, *ApJ*, **158**, 123; Shu, F. H. et al. 1972, *ApJ*, **173**, 557, both classic early investigations.

[29] The evidence for a barred structure in the inner Galaxy is presented in Blitz, L. and Spergel, D. 1991, *ApJ*, **379**, 631, see also Vauterin, P. and Dejonghe, H. 1998, *ApJ*, **500**, 233; Dehnan, W. 2000, *ApJL*, **524**, L35.

The effect on the flow of a bar is similar to a spiral density wave, but there are some important differences. The perturbation amplitude may be larger because of the greater density enhancement, and the stellar orbits are essentially perpendicular to the perturbation, so the streaming motions are independent of pitch angle, unlike spiral waves. A bar can be included straightforwardly in the same picture we have developed for spiral patterns. In fact, such structures are usually implicated as the causative agent for the spirals. The gravitational potential in the corotating frame becomes

$$\Phi = \Phi_{\text{grav}} - \tfrac{1}{2}\Omega_p r^2,$$

where Φ_{grav} is the potential due to the ellipsoidal mass. The flow is described by a continuous version of the Hill equations as we saw for accretion disks and when treating the three-body problem. Unlike a binary star, the gravitational term is provided by a *continuous mass distribution* that rotates with a pattern frequency Ω_p but it otherwise displays the same features we saw for systems with equal mass stars.[30]

There are two Lagrangian points at the ends of the bar, one in the middle at the potential, and as usual the L_4 and L_5 points are orthogonal to the axis of the bar. But here the points are strictly symmetric about the center, which is the global potential minimum, and a variety of trapped orbits are possible. In particular, stars can circulate freely within the bar and are progressively closer to chaotic orbits as they come closer to the saddle points along the bar axis. The labeling (see Figure 7.8) looks slightly different from what we had for the binary star case because the L_1 point is central, not between the stars. From our discussion of the Lindblad resonances and driving of density waves, you can see that the critical points will occur in the flow whenever the pattern frequency of the bar, Ω_p matches the dynamical frequencies given by Eq. (7.21). In effect, a bar is a density wave with an infinite radius of curvature for which the local pitch angle is 90°. In the disk, the bar produces a distance-dependent torque on stellar orbits that causes them to precess. Those that are corotating will be in resonance, those at progressively further distance will generate a density wave. The gas is again more affected than the stars, experiencing spiral shocks, and also possibly showing enhanced star formation at the ends of the bars.[31]

What maintains the waves? The brief answer is that nobody is quite sure. An obvious agent is a bar in the inner parts of the disk. Alternatively, a wave can be excited during galactic collisions. External tidal forcing by the perturber produces a torque that may excite the waves. Collision timescales are about the same order

---

[30] For the restricted (circular orbits) three-body problem, this is called the *Copenhagen problem* and dominated the work at that observatory for decades in the beginning of the twentieth century. Many of the orbital calculations explicitly display the trajectories, and it is worthwhile to look back at this literature for insights on the single-particle limit. See *e.g.* Moulton, F. R. 1920, *Periodic Orbits* (Washington, DC: Carnegie Institution of Washington).

[31] Sellwood, J. A. and Wilkinson, A. 1993, *Rep. Prog. Phys.*, **56**, 173; Knapen, J. H. 1999, in *The Evolution of Galaxies on Cosmological Timescales*, Beckman, J. E. and Mahoney, T. J., eds. (San Francisco: ASP Conf. Series); Sellwood, J. A. 1993, *PASP*, **105** 648; Athanassoula, E. 1993, *MNRAS*, **259**, 328; Kuijken, K. and Merrifield, M. R. 1995, *ApJL*, **443**, L13; Buta, R. 2000, *Ap. Space Sci.*, **269**, 79 provides an excellent survey of resonance structures in barred spirals.

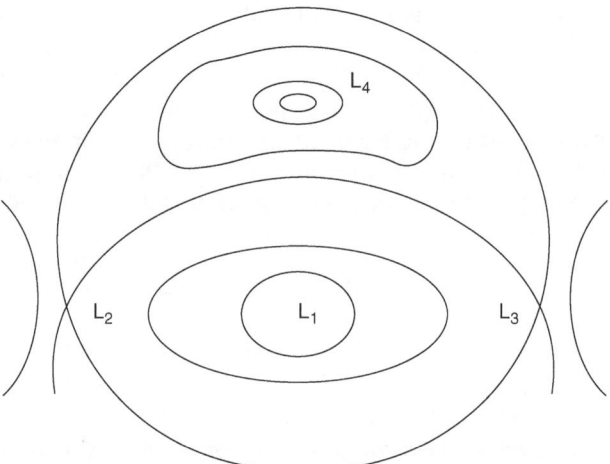

FIGURE 7.8. Schematic equipotential surface (analog of the Roche surface) for a rotating prolate bar. Note the locations of the Lagrangian points (compare with Chapter 5).

of magnitude as κ^{-1}, and resonant forcing is possible. N-body simulations support this view. The fact that angular momentum is ultimately transported through the disk and the waves wind up on a relatively short timescale is the real problem unless some forcing, by a resonance, is maintained.[32] We will postpone the discussion of morphological classifications of external galaxies for a little while, but it is appropriate to mention that one scheme currently incorporates the variety of possible excitation mechanisms for spiral structure. The waves dissipate by heating the stellar gas, which, as we mentioned, cannot cool. Angular momentum is also transferred outward through the long-range torques imposed by the spiral. Numerical N-body simulations show that the spiral lasts for only a few rotations if no continuous forcing is available. This has produced the distinction between *flocculant*, or "ratty," spirals and those called *grand design*.[33] The latter are not the norm, occurring in about one-third of all isolated spirals. Among binary or interacting galaxies, however, the fraction that are classified as *grand design* increases to about two-thirds of the sample. The structure may thus be externally excited from the time dependent tide generated by a passing galaxy.

---

[32] Toomre, A. 1977, *ARAA*, **15**, 437; Toomre, A. 1981, in *The Structure and Evolution of Normal Galaxies: Proceedings of the Advanced Study Institute, Cambridge, UK*, Lynden–Bell, D., ed. (Cambridge, UK: Cambridge Univ. Press); Bertin, G. and Lin, C. C. 1996, *Spiral Structure in Galaxies* (Cambridge, MA: MIT Press); Shu, F. H. 1992, *The Physics of Astrophysics. II. Gas Dynamics* (Mill Valley, CA: Univ. Science Books); Tagger, M., Sygnet, J. F., and Pellat, R. 1993, in *N-Body Problems and Gravitational Dynamics*, Combes, F. and Athanassoula, E., eds. (Paris: Obs. de Paris). Barred spiral dynamics are reviewed in Sellwood, J. A. and Wilkinson, A. 1993, *Rep. Prog. Phys.*, **56**, 173. A fascinating application of geostropic flows to treating spiral structure is given by Prendergast, K. 1962, in *Interstellar Matter in Galaxies*, Woltjer, L., ed. (Reading, MA: Benjamin). One of the earliest calculations of spiral modes in a stellar disk is Lindblad, B. 1962, in *Problems of Extra-galactic Research: IAU Symp. 15*, McVittie, G. C., ed. (NY: Macmillan).

[33] Elmegreen, D. M. and Elmegreen, B. G. 1987, *ApJ*, **314**, 3; Elmegreen, B. G. 1995, in *The Formation of the Milky Way*, Alfaro, E. and Dalgado, A., eds. (Cambridge, UK: Cambridge Univ. Press) and references therein.

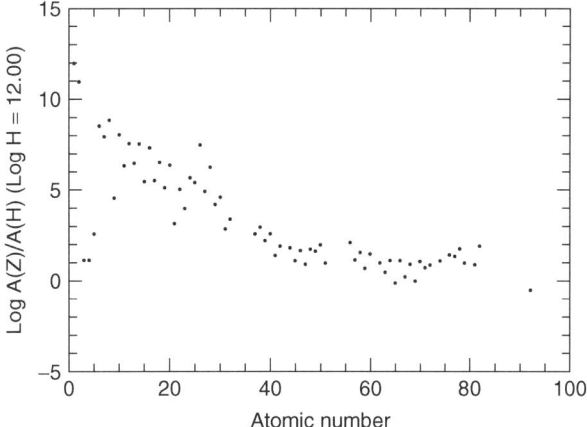

FIGURE 7.9. Plot of solar photospheric abundances (see Table 4.1) normalized to $\log H = 12.00$. Note the extremely low abundances of Li, Be, and B and the overall trend of decreasing abundance with increasing atomic number.

7.3 CHEMICAL EVOLUTION OF THE GALAXY

O dark dark dark, they all go into dark
The vacant interstellar spaces, the vacant into the vacant.
—T. S. Eliot, *East Coker, IV*

You know that stars build up the elements through nuclear processing and then can disperse the products into the interstellar medium. Since both star formation and death are ongoing processes, the elemental abundances of the gas and the stars that are formed out of it should change secularly over time. Understanding this interplay of birth, synthesis, and death is the central theme of research on the chemical history of galaxies and of much of this book. Star formation, interstellar processes, and galactic-scale dynamics are all necessary ingredients in the galactic alchemy (Fig. 7.9). We are also going to take a different approach here than we have followed in many other parts of the book and one that is also not a standard presentation of the model for chemical evolution. Our point of view, which we will stress throughout the next sections, is that there are a variety of processes that *act in concert* and must be included, however schematically, in any picture of the interplay between star formation, nucleosynthesis, and interstellar dynamics. These processes have rather similar formal representations, so we will focus on the *generic* behavior that results. The deeper you go into the models, the deeper the assumptions become and the more uncertain the physics. Much of what has been done to model star formation and chemical evolution is quite successful in reproducing the behavior of the systems, and many of the new models have various features in common. Our goal here is to develop a framework that will help guide you through the current literature.

7.3.1 Evidence for Metallicity Evolution

The most compelling observational evidence requiring continuing production of the elements in the Galaxy comes from the direct detection of short-lived isotopes using nuclear γ-ray emission lines and chemical tracers in meteorites. The most abundant stable isotope of magnesium is ^{24}Mg, which is produced in the post-carbon-burning stage of red giant evolution. In the late 1970s, however, samples isolated from two carbonaceous chondrites, the Allende and Murchison meteorites, revealed small amounts of ^{26}Mg, the product of radioactive decay of ^{26}Al. There is no other source. This latter isotope has an (astronomically speaking) very short half-life of about 7×10^5 yr, which seemed to mean that the ^{26}Al had polluted the protosolar nebula at the time when the meteorites were forming and that it had subsequently decayed in situ. Early γ-ray survey observations by the COS-B satellite showed evidence for emission from decaying ^{26}Al, and this has now been spectacularly confirmed by the maps made with the COMPTEL instrument on the Compton Gamma Ray Observatory (CGRO). In order to produce an observable signal, the aluminum must be generated *now*, since there is about 4 $M_\odot$ of this isotope distributed through the Galactic plane (Fig. 7.10). The detection of the 1.157-MeV γ-ray emission from ^{44}Ti from Cas A, the remnant of the Galactic supernova of circa 1670 provides remarkable information about the production of short-lived radioisotopes in stellar environments. With a half-life of order 60 yr.[34] there is no possible source other than a recent injection into the interstellar medium, and the coincidence of the source with the supernova remnant, albeit without a direct firm distance determination to either, is the strongest evidence that supernovae are the source for many heavy elements in the Galaxy. The line has also been detected from another supernova remnant, GRO J0852-4842. Other signatures include the detection of ^{26}Al 1.809 MeV γ-line emission ($t_{1/2} = 7.2 \times 10^5$ yr) from the diffuse interstellar medium of the galactic plane, ^{56}Co IR-line emission ($t_{1/2} = 77.^d2$) from SN1987A, and the possible detection of ^{60}Fe in ocean sediments ($t_{1/2} = 1.5 \times 10^6$ yr), a short-lived nucleus that is expected to form in supernova explosions.[35]

Additional support for stellar processing and eventual dispersion of the products comes from carbon stars: (1) the ^{13}C/^{12}C ratio is far higher in carbon stars than in the mean interstellar medium or, indeed, in the solar system—values as high as 0.2 are observed in outflows and photospheres of these stars, while the solar system value is of order 0.01; and (2) stellar neutron processing is evinced in the detection of ^{99}Tc in carbon stars, an isotope whose half-life is only about 10^5 yr. Highly evolved massive stars also show altered abundances, particularly of nitrogen. All these objects eventually expel this enriched material into the broader arena of the galactic interstellar medium, where it becomes incorporated into successive generations of stars. Consequently, there must be continued development of the mean metallicity of the stars and gas within galaxies throughout their

---

[34] This was a controversial and difficult experimental number to obtain and Görres, J. et al., 1998, *Phys. Rev. Lett.*, **80**, 2554 is an exciting paper that details the production of this isotope and the measurement of its lifetime.

[35] Prantzos, N. and Diehl, R. 1996, *Phys. Rep.*, **267**, 1; Schönfelder, V. et al. 2000, *Astr. App. Suppl.*, **143**, 145; Dupraz, C. et al. 1997, *Astr.Ap.*, **324**, 683; Diehl, R. et al. 1995, *Astr. Ap.*, **298**, 445; Harris, M. J. et al. 1996, *Astr. Ap. Suppl.*, **120**, 343; Knie, K. et al. 1999, *Phys. Rev. Lett.*, **83**, 18.

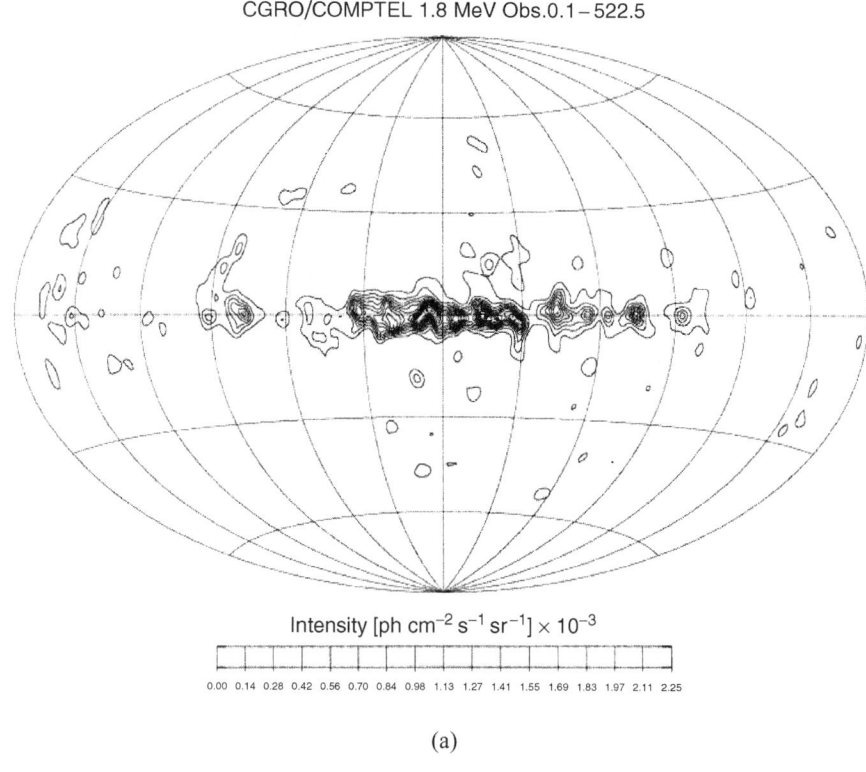

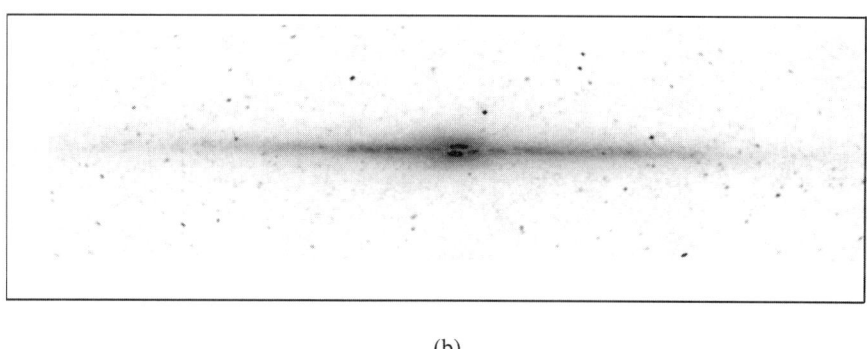

FIGURE 7.10. (a) COMPTEL map of the ^{26}Al γ-line distribution in the galactic plane (credit: H. Bloemen, Univ. Leiden). (b) DIRBE (from the COBE satellite) composite image (enhanced to show the plane and nucleus) of the galactic plane for comparison (credit: COBE/NASA).

lifetimes. Meteoritic samples record this evidence for several different sources of material in the protosolar cloud. Individual grains can be identified with isotopic patters that are expected from neutron exposure in AGB stars, especially those with high $^{13}C/^{12}C$ ratios. Other grains show a marked similarity to expected abundance anomalies from type II supernovae. Finally, the $^{22}Ne/^{20}Ne$ ratio is typical of that expected from nova ejections. Since the mean rate of these events is low within a galaxy, and especially the extinct nuclei show that the cloud must have been polluted within a few million years of the formation of the meteorites, dispersal of the synthesized elements throughout the disk must be rapid.

The evidence for continued star formation is more obvious; in fact, a naked eye observation suffices. The winter (summer) sky is dominated by the Orion OB association and summer (winter) sky by the associations in Scorpius-Centaurus. The large number of short-lived (1–10-Myr) massive main sequence and supergiants massive stars requires ongoing star formation. The composite HR diagram for the field stars—those that have no clear connection to any clusters, associations, or moving groups—shows an increase in the range of stellar photospheric abundances as one progresses down the main sequence, which is understood through ongoing nucleosynthesis along with continued star formation. In short, we require a unified theory for the formation of the elements *and* of the stars that serve as sites for the elemental syntheses.

7.3.2 Inputs

7.3.2.1 The Initial Mass Function

In principle, the initial mass function (IMF) should be easy to obtain. After all, for the field you can obtain the HR diagram so you should be able to count the number of stars in each luminosity bin, determine the luminosity function, and transform this back to a mass distribution. Not so fast! Remember that the field is the cumulative distribution of stars formed throughout galactic history. The only way to directly obtain the distribution *as a function of mass* is to look at clusters.[36] Bootstrapping between clusters of different ages permits the construction of a composite IMF and also offers the best chance to test the assumption of universality, although with some limitations. For extremely young clusters the most massive stars are on or near the main sequence and can be included in the population census without the need to extrapolate their positions back to the ZAMS. These are rare and often accessible only in external galaxies where crowding from limited spatial resolution reduces the reliability of the results. Even where individual stars can be resolved, mass loss and binarity combine with small numbers to introduce considerable uncertainties into the contribution of massive stars to the IMF. In addition, Galactic studies are limited by our knowledge of distances to OB associations, some of which are in the outer Galaxy, and by both environmental and foreground extinction toward the inner Galaxy (although this has improved dramatically). Older clusters provide more secure determinations of the IMF but, being older, may have different metallicities than associations. For the oldest clusters, only the lowest-mass stars are still observed, so for globular clusters and old open clusters such as NGC 188, no information is available for stars above about 1 $M_\odot$.

The earliest empirical determination of the IMF, called the *Salpeter mass function*, is still the "industry standard." It is a power-law distribution for stars

---

[36] This was not possible before the calculation of post-main sequence models for stars more massive than a few solar masses, and therefore the first attempt didn't come until the 1950s with Salpeter's study.

between about 0.5 and 10 $M_\odot$:

$$\phi(m) \sim m^{-(1+x)}, \tag{7.42}$$

with $x = 1.35$. It requires upper- and lower-mass cutoffs for normalization and so alone does not completely represent the stellar population. This parameterization is sufficient for many population models, but the assumption that the exponent is independent of mass is suspect. With improvements in instrumentation and the use of multiwavelength observations, it has been possible to obtain population census data for extragalactic clusters within the Local Group that seems in accord with the Salpeter function. However, this is one of the major open questions, both for star formation in the Galaxy and for extragalactic environments. The presumed universality of this function, and its stationarity in time, remain important problems for future work.[37] You might think for extragalactic studies that it would be comparatively simpler than for the Milky Way since the stellar populations are lying exposed in their entirety. Except for the more reliable relative distances, all the other problems are actually exacerbated. Resolution plagues the determination of the luminosities in crowded star forming regions, even with spectral signatures acting as a guide, and completeness problems are severe at the low luminosity end of the IMF. In addition, as with the Galaxy, for the field determinations the history of star formation is knotted up with the mass function. Unless you can study individual associations, and this can really be done only for the Magellanic Clouds and a few Local Group galaxies, the behavior of even the high mass end of the IMF is very much an open question.

7.3.2.2 The Star Formation Rate

When neutral hydrogen 21-cm observations became available during the 1950s, Schmidt proposed to represent the time-dependent star formation rate (SFR) implicitly through the mass density of neutral gas, The premise was, and still is, that if you look at a volume V in which you find N_{OB} massive stars, you know they have formed within the past t_{ZAMS} years. You then determine the amount of neutral gas in the region, the stuff out of which the stars should presumably have formed, by velocity-integrated 21-cm measurements for optically thin gas:

$$M_{H\,I} = 2.36 \times 10^5 M_\odot \int S_{Jy}(v)\, dv.$$

To convert this to a volume density requires having some idea of the region's geometry, but we are astronomers now, and spheres are always good first approxi-

---

[37]Salpeter, E. 1955, *ApJ*, **121**, 161; May, R. M. 1975, in *Ecology and Evolution of Communities*, Cody, M. L. and Diamond, J. M., eds. (Cambridge, MA: Harvard Univ. Press), pp. 107ff; Scalo, J. 1986, *Fund. Cosm. Phys.*, **11**, 1; Zinnecker, H., McCaughrean, M. J., and Wilking, B. A. 1993, *Protostars and Planets III*, Levy, E. and Lunine, J., eds. (Tucson: Univ. Arizona Press); Scalo, J. 1998, in *The Stellar Initial Mass Function, 38th Herstmonceaux Conference*, Gilmore, G., Parry, I., and Ryan, S., eds. (San Francisco: ASP Press), p. 177; Kroupa, P. 1998, in *Brown Dwarfs and Extrasolar Planets: ASP Conf. Series 134*, Rebolo, R., Martin, E. L., and Zapatero Osorio, M. R., eds. (San Francisco: Astr. Soc. Pac.).

mations. The star formation rate is obtained from counting the number of massive stars:

$$\psi = \left\langle \frac{M_{\text{OB}} N_{\text{OB}}}{V t_{\text{ZAMS}}} \right\rangle,$$

where the average is taken over the mass distribution of the massive stars. The result of this exercise is called the *Schmidt law*:

$$\psi_S \sim \rho_g^2, \qquad (7.43)$$

where, following Schmidt's original derivation, ρ_g refers to the neutral atomic volume. Soon after the completion of the first OB star survey of the Magellanic Clouds, Sanduleak found a similar result using the newly available H I measurements for the LMC from Parkes. He obtained a smaller value for the exponent but a basically similar pattern. Similar results were found from large-scale measurements of external galaxies such as M 33.[38] There are several problems with this technique, none of which were known before molecular gas was discovered in the interstellar medium in the late 1960s. The most important is that the neutral gas has nothing *directly* to do with current galactic star formation. Molecular clouds, whose mass densities are much higher and temperatures much lower, than the warm neutral medium, are the sites for star formation. In fact, the rate of star formation is likely governed by the efficiency of the cooling provided by these molecular species and other internal cloud processes, as we discussed in Chapter 6. Warnings have been sounded many times about the indiscriminant use of this parameterization for the SFR because it does not include the unseen mass, either in molecular or ionized form, but the formal law has stuck in the literature.[39] It is, however, sufficient for the purpose of this section to note that ψ has a nonlinear dependence on the mass density of the component of the gas that is responsible for birthing stars, hence it suggests that some feedback mechanisms might (must?) be required for a proper galactic evolution model. It is even possible that the rate of massive star formation may have a detectable threshold volume or surface density of molecular gas.

The volumetric mass density is only *inferred* in the absence of detailed radiative transfer modeling and multiple density tracers. Instead, one can use the mass surface density, Σ_g, to determine the star formation rate, including both H I and molecular gas. Observationally, this needs no justification. From theory, Q depends on Σ, not just Σ_g. In general, however, it is not possible to directly measure even the total mass column density. Neutral hydrogen measurements provide the surface gas density integrated along the line of sight, which may be very inhomogeneous, but not of the gas out of which we expect stars to form. The molecule that provides the basic information, CO, does not yield the mass of the main constituent, H_2, without considerable modeling assumptions that we have discussed in the previous chapter. For global measurements, this is even more ill-constrained

---

[38] Schmidt, M. 1959, *ApJ*, **129**, 243; Schmidt, M. 1963, *ApJ*, **137**, 758; Sanduleak, N. 1967, *AJ*, **74**, 47; Madore, B. F., van den Bergh, S., and Rogstad, D. H. 1974, *ApJ*, **191**, 317.

[39] See particularly Hunter, D. A., Elmegreen, B. G., and Baker, A. L. 1998, *ApJ*, **493**, 595 for the comparison between rates of star formation in disk and Irr galaxies.

since telescope beams are rarely so small that individual extragalactic regions of star formation can be probed.

The observational calibration of the star formation rate has shifted from the stars to the gas, using line emission to measure the ionization rate and through that, the input stellar luminosity. In addition, dust heated by embedded stars radiates in the far infrared, where it has been detected by IRAS and ISO, with a luminosity that should correlate directly with the stellar input. The chain of reasoning is as follows. Massive stars produce H II regions. These radiate in Hα, whose luminosity is correlated with the luminosity of the OB stars, hence the stellar mass, and because of their short main sequence lifetimes, with the star formation rate.[40] Several cross-calibrations are used, but since the observations are limited, the rates depend on measurements of only the Hα and CO luminosities. The most recent calibration gives

$$\psi_{H\alpha} = \frac{L(H\alpha)}{3.2 \times 10^7 L_\odot} M_\odot \text{ yr}^{-1}$$

$$\Sigma_{H\alpha} = (2.5 \pm 0.7) \times 10^{-4} \left[\frac{\Sigma_g}{1 M_\odot \text{ pc}^{-2}}\right]^{1.4 \pm 0.15} M_\odot \text{ yr}^{-1} \text{ kpc}^{-2}, \quad (7.44)$$

which is the form usually employed in models of galactic evolution in cosmological scenarios, as we will discuss shortly. The alternative procedure requires combining the H II measurements with CO tracers of the H_2 abundance to obtain a total mass for the gas, from which a *star forming efficiency* measure, $L(H\alpha)/M(H_2)$, can be derived. The situation will surely improve with the construction of higher-sensitivity millimeter interferometers in the near future.

7.3.3 Stellar Populations and Population Synthesis

The evidence for secular change in the metallicity of our and other galaxies over long times comes from the observed behavior of stellar abundances with position and kinematics. The idea of stellar populations was introduced by Baade in 1944 from his observations of external spirals and ellipticals.[41] He noted that the stars in the core of M31 have a color distribution systematically different from those in the disk and suggested that these were a second, separate, population of bodies similar in chemical composition and mass to the globular clusters. The bluer colors characteristic of the disk were clearly due to hot, relatively massive, stars whose lifetimes are short compared with the main sequence stars of globulars and therefore must have formed only recently. The separation was labeled as *population I* for the disk and *population II* for the spheroid and center. This picture has been refined so now stellar populations in the Milky Way galaxy are distinguished kinematically as well as photometrically. Nonetheless, the population concept provides the basic constraints for scenarios of chemical and dynamical evolution of the Milky Way, whether by a dissipative collapse from a larger volume, by the

---

[40] Kennicutt, R. C. Jr., Tamblyn, P., and Congdon, C. W. 1994, *ApJ*, **435**, 22; Kennicutt, R. C. Jr. 1998, *ARAA*, **36**, 189; Kennicutt, R. C. Jr. 1998, *ApJ*, **498**, 541; Rownd, B. K. and Young, J. S. 1999, *AJ*, **118**, 670.
[41] Baade, W. 1944, *ApJ*, **100**, 137; Sandage, A. 1986, *ARAA*, **24**, 421.

accumulation of independent fragments, or by continuous infall, or by some other mechanism.

Metallicity is the fingerprint. Now you may wonder how you can determine a particular abundance, say, [Fe/H] and what it means to call this the *metallicity* when you are working with a spectrum that includes such diverse contributors as main sequence stars and supergiants. Such values are usually determined from spectrophotometry. The precise calibration of the photometry with abundances can only come through careful model atmosphere analyses, but there is a way to cheat. When we discussed photometric systems in Chapter 2, we said that filters can be tailored to the problem at hand. Now let's apply this idea. When observing a galaxy in integrated light we are really seeing the contribution of each star at each wavelength weighted by the fraction of the stars that are at the same mass and evolutionary stage. The populations can then, in principle if not precisely in practice, be determined through multiwavelength inversion of the spectrum. Each star contributes some fraction of the light at any wavelength so that

$$F_\lambda = \int_{m_L}^{m_U} F_\lambda(m) \phi(m)\, dm. \qquad (7.45)$$

For filter photometry, the flux must be integrated over S_λ, the sensitivity curve of the filter. For high-resolution spectrophotometry, the IMF can be derived uniquely only for aggregates for which there has been no evolution, that is, for clusters or galaxies for which all the stars are on the ZAMS. The *real* problem is that, except for globular clusters and open clusters where you know that all the stars were formed at the same time, you also have to include the history of star formation. This is because at any epoch, high-mass stars are more sensitive to the current rate of formation while low-mass stars accumulate in time. Therefore, for studies of the field halo stars and for extragalactic systems, you also need to include $F_\lambda(m,t)\psi(t)$ in the synthetic models through the use of isochrones.[42] The only way to reliably recover the population is through a full spectrophotometric synthesis. In computing a model, several ingredients are combined in a formal way: the initial mass function, the star formation rate, and libraries of evolutionary tracks and model atmospheres. Each mass bin is weighted by $\phi(m)$ at a rate $\psi(t)$, and evolved in each timestep according to its mass. High-mass stars contribute to the shortest wavelengths and are most sensitive to *instantaneous* star formation rate. Longer wavelengths, to which red giant and supergiant contribute most heavily, trace the long-term evolution of the lower mass stars. The sample continua are usually drawn directly from observations, but these often need to be supplemented by calculated spectra. The procedure depends critically on the quality of any model

---

[42] For some examples, see Fanelli, M., O'Connell, R., and Thuan, T. X. 1988, *ApJ*, 334, 665; Pickles, A. J. 1985, *ApJS*, **59**, 33; Pickles, A. J. 1985, *ApJ}*, **294**, 134; Pickles, A. J. 1985, *ApJ*, **296**, 340; Mashesse, J. M. and Kunth, D. 1991, *Astr. Ap. Suppl.*, **88**, 399. Sample spectrophotometric observations for population synthesis are provided by Gunn, J. and Stryker, L. L. 1983, *ApJS*, **52**, 121. A comprehensive review and description of continuum and model libraries is found in Leitherer, C. et al. 1996, *PASP*, **108**, 996. See also O'Connell, R. W. 1996, in *From Stars to Galaxies: The Impact of Stellar Physics on Galaxy Evolution*, Leitherer, C., von Alvensleben, F., and Huchra, J., eds. (San Francisco: ASP Conf. Series); Bruzual A., G. 1983, *ApJS*, **53**, 497; Charlot, S. and Bruzual A., G. 1991, *ApJ*, **367**, 126; Bruzual A, G. and Charlot, S. 1993, *ApJ*, **405**, 538; Kodama, T. and Arimoto, N. 1993, *Astr.Ap.*, **320**, 41; Bell, E. F. and de Jong, R. S. 2000, *MNRAS*, **312**, 497.

atmospheres used to generate these spectra and the details of the interior models from which isochrones are computed. The accuracy of the resulting population distribution also depends very sensitively on the quality and completeness of the library of spectra used in the procedure, especially for the relatively high-luminosity stars in late stages of their evolution. Even transient phases such as the luminous blue variables and Wolf–Rayet stars can compromise the resulting fit if there are features in the spectrum to which they contribute strongly. This is especially important for dwarf galaxies such as the blue compact dwarfs, where all indicators point to massive bursts of star formation lasting for relatively short times but recurring over the galactic lifetime.

7.3.4 Models for Galactic Chemical Evolution

The interplay between the different constituents of the Galaxy is recorded in the relative abundances of the chemical elements. Although the initial mixture was set by the conditions in the Big Bang, the modifications produced by stellar nucleosynthesis carry the fingerprints of the stellar IMF and the pattern of dispersal in the gas. Since the rate of star formation depends on the formation rate for molecular clouds, which in turn depends on gas properties that are metallicity sensitive, you see that there is a feedback between the various constituent phases that must be included in the picture. Stars are formed; stars die. This is the galactic version of the facts of life. When stars expire, the nucleosynthetic products of their lives are expelled into the interstellar medium and locked into the remnants that are the degenerate products of their final stages. This means two things: first, the abundance of heavy elements *must* increase with time at some rate, and second, the amount of gas being returned to the ISM is always less than was used in star formation: hence the system must ultimately run down. With only these two facts in mind, it is possible to construct a very simple model of the process. The rate of gas consumed by star formation is the same as the star formation rate. The amount of gas returned to the interstellar medium is, however, always less, whatever is not permanently incorporated into remnants such as white dwarfs, neutron stars, and black holes. In the short run, if you are looking only at timescales of 1 Gyr or less, even low-mass stars with $M < 0.5\ M_\odot$ can be included in this category. For the long-term evolution, the remnants of star forming activity include brown dwarfs and planets. In all cases, you notice that the mass function will ultimately be involved because stars of different mass have different endpoints and the fraction of the system that goes into any mass bin is determined by the IMF. We will return to this point soon, but for the moment, let's press on.

7.3.4.1 Closed-Box Models

The standard model within which chemical evolution scenarios were first developed is a closed box for the solar neighborhood.[43] There is neither mixing from

---

[43] Comprehensive reviews are available by Tinsley, B. M. 1980, *Fund. Cosmic Phys.*, **5**, 287; Rana, N. C. 1991, *ARAA*, **29**, 129; Matteucci, F. 1996, *Fund. Cosm. Phys.*, **17**, 283. Cowley, C. R. 1995, *Introduction to Cosmochemistry* (Cambridge, UK: Cambridge Univ. Press) provides a good overview of the determination of abundances in a wide range of astrophysically important sites, and Pagel, B. 1997, *Nucleosynthesis and Chemical Evolution of Galaxies* (Cambridge, UK: Cambridge Univ. Press) discusses the models and especially the light elements; see also Matteucci, F. 2001, *The Chemical Evolution of the Galaxy* (Dordrecht: Kluwer) for a survey of models. You should also consult the biennial proceedings of the *Nuclei in the Cosmos* conference for up to date reviews of the models and data.

other parts of the interstellar medium nor net mass transfer in or out. This scenario is clearly too schematic to be complete, but it helps highlight individual contributing processes and shows ways to include more realistic physical processes. You have the tools at hand. First, let us look at the basic picture, where we assume that we already know something about the stellar birth function $B(m,t) = \phi(m)\psi(t)$ but the separability must be regarded as an assertion, not a "fact." If the IMF is invariant over time, we can average yields and rates of gas return and consumption over $\phi(m)$ to obtain net values that do not depend on time. For instance, the rate of return of gas from stars is

$$R\psi = \int_{m_L}^{m_U} r(m) B(m,t)\, dm, \qquad (7.46)$$

where $r(m)$ is the rate of return for stars of mass m. We already have a complication introduced here because the main sequence lifetime for stars less than about 1 $M_\odot$ is a significant fraction of the age of the Galaxy, and instantaneous recycling of this material is a poor approximation. In this scenario, the rate of gas consumption is

$$\frac{dM_g}{dt} = -(1-R)\psi. \qquad (7.47)$$

The equation is more complicated than it looks because you don't know anything at this stage about the time dependence of ψ, but its compact form, as we will see, has many advantages. Nucleosynthesis is fossilized in the fraction of gas, R, that is returned to the interstellar medium by stars. Notice that if we adopt a generalized Schmidt law for star formation, $\psi = KM_g^n$, we can obtain an *explicit* solution for the gas mass as a function of time:

$$\frac{M_g}{M_{g,0}} = \left[1 + M_{g,0}^{n-1}(1-R)(n-1)Kt\right]^{1/(1-n)} \quad (n \neq 1)$$

$$\ln \frac{M_g}{M_{g,0}} = -(1-R)Kt \qquad (n=1).$$

We will lump all elements together in the usual astrophysical metallicity, Z, and assume that the yield of heavy elements is Y, produced by the processing of elements *other* than the ones accounted for by the aggregate parameter Z. Then the metallicity of the gas changes as

$$\frac{d}{dt}(ZM_g) = -Z\psi + Y(1-Z)R\psi + RZ\psi. \qquad (7.48)$$

These two equations combine to give the metallicity as a function of gas mass. This is a generic feature of all closed-box models: *Since the region evolves in complete isolation, time and gas mass are interchangeable variables.* Now define $\mu \equiv \ln(M_g(t)/M_{g,0})$. Then we find a steady increase in the metallicity as the gas fraction decreases:

$$\ln(1 - Z) = \frac{YR}{1-R}\mu. \qquad (7.49)$$

For trace species such as the iron peak elements, the term on the left is linear in Z. So far, so good. We can also expand the picture to include return of processed *and* unprocessed gas. For the processed material, this can include primary species whose abundances do not depend on the metallicity of the gas and so are a separate term in modifying Eq. (7.48), and the secondary species whose abundance depends on the availability of seed nuclei. All these changes produce quantitative differences in the history of the abundances, but they do not remove the following fundamental difficulty.

7.3.4.2 Primary versus Secondary Elements
This formalism has the advantage of providing a scaling between the remaining gaseous mass and the star formation rate. The rate of production of metals by stars is their yield, Y. We need, however, to distinguish between primary and secondary species. The metals that are *primary* are those newly synthesized by stars independent of the initial metallicity. Yes, the detailed structure of a stellar interior depends on the opacity and mean molecular weight of its gas, which in turn affect the conditions in the oven, hence the yields. But for most of the lighter elements, in particular the CNO group, the production depends more on the stellar mass than on the initial mix of ingredients. The neutron-processed species, in particular the s-process elements, are secondary elements whose yield depends on the initial abundance fraction of the iron peak nuclei. They require introducing the concept of *asteration*, originally due to Tinsley, for those elements whose abundance depends on the history of the heavy elements and that are produced with a yield that explicitly depends on the abundances in the stellar mixture. It is expected, therefore, that the abundances of these two groups of elements develop differently with time and provide complementary information about star forming history of a galaxy.

Let us digress for a moment to an example that shows how important the contribution of stars to the gas metallicity is in the earliest stages of galactic evolution. Take a single SN Ia and let it explode into a galaxy of about 10^{11} $M_\odot$ of pure hydrogen and helium. Since the yield is about 0.1 $M_\odot$ of Fe, this would produce a net abundance fraction of order 10^{-12}, corresponding to an abundance compared with the solar value, [Fe/H] ≈ -4. The rest of the heavy elements from the r-process should also be enriched, so a *single* supernova, spread out over the protogalaxy, would produce a floor to the metallicity distribution of about 10^{-5} solar.[44] Primary elements show a dependence on the star formation rate that is essentially linear with ψ. Thus, the increase in the abundance of the species with time tracks the star formation history and is the cumulant production of the stellar

---

[44] More recently, a remarkable star has shown the first fingerprints of this process: CS 22892-052 (Sneden, C. et al. 1996, *ApJ*, **467**, 819; Cowan, J. J. et al. 1997, *ApJ*, **480**, 246). Other halo stars with [Fe/H] ≈ -2 have also shown strong evidence for early r-processing and relatively enhanced abundances of those nuclei that are produced by the most massive s-processing stars.

nucleosynthetic activity. In contrast, a species that is secondary depends on the abundance of seed nuclei and thus shows steeper, ψ^2, dependence. Most heavy elements fall into the latter category, as they are synthesized in stars depending on their metallicity.

A final remark is in order regarding the precise treatment of the stellar element factories. You really should write

$$\text{Rate of return} = \int_{m_L}^{m_U} r(m)\phi(m)\psi(t - \tau_m)\,dm,$$

where now τ_m is the main sequence lifetime of the star as a function of mass. This level of the problem, when time delays are included, also requires that the yield of gas back into the medium by stars becomes an integral over the stellar population. Assuming that the IMF is stable in time, each element i can be treated as another constituent in the system with an abundance Z_i:

$$\frac{d}{dt}[Z_i M_g] = -\psi Z_i + \int_{M_L(t)}^{M_U(t)} \sum_j Q_{ij} Z_j(t - \tau_m)\psi(t - \tau_m)\,dm,$$

where Q_{ij} is the yield matrix for species i and τ_m is now explicitly a function of metallicity. These complications are avoided by a justifiable simplifying assumption. We assert that all the metals whose abundance we are interested in following, especially the heavy ones such as iron, are produced by high-mass stars whose lifetimes are very short compared with the age of the Galaxy. This *instantaneous recycling* approximation avoids the complications of the convolution over mass that is required for the light elements. The price we pay is in not being able to properly account for all the additional processes that light and intermediate mass elements require, for example, interstellar spallation and primary processing of CNO in the cores of AGB stars.

7.3.4.3 Spallation: Cosmic Rays and Light-Element Synthesis

Not all elements are, however, produced by stars. There are no known stellar sites that produce rather than consume deuterium, and its presence in the Galaxy is due entirely to processes that occurred during the Big Bang. Another group, the isotopes of Li, Be, and B, must also avoid being consumed since they have such low fusion thresholds. Lithium, in particular, is easily destroyed in solar-type main sequence stars through convective mixing to the base of the convection zone. Since the days of B^2FH, it has been clear that stars are *not* the main sources of these species but the formation mechanism was left as an exercise for the future. The turning point came with the suggestion that cosmic ray processing, *spallation*, could generate the observed range of abundances provided there was some initial seeding from primordial nucleosynthesis. In this process, cosmic ray particles collide with interstellar nuclei. Whether by *direct processes*, where the targets are thermalized heavy nuclei and the projectiles are cosmic ray protons and α particles, or *inverse processes*, when accelerated heavy nuclei impact interstellar hydrogen and helium atoms, the production rate depends on both the star formation rate and Z.[45] The comparison between spallation, which is a secondary process, and stellar production is very illuminating and we will spend some time on it. If we assume a stellar metallicity production law

$$M_g \frac{dZ}{dt} = P_Z \psi, \qquad (7.50)$$

where $P_Z = RY$, then, as before we find

$$Z = -\frac{P_Z}{1-R} \ln \mu.$$

You can see why the closed-box model is so attractive as a scheme—you don't need to know the form for ψ since all time dependence is conjugate with the mass fraction μ.

The Li, Be, and B production rate from spallation processes depends on the galactic cosmic ray flux of nuclei in a wide mass range, not only protons and α particles. In addition, and most important, the production rate depends on the abundance of the interstellar target X_j^{ISM}, weighted by its spallation cross section $\sigma_{i,j \to k}$. Only a fraction of newly produced light nuclei is thermalized and incorporated in the ISM; the remainder become part of the cosmic ray flux. When the incident particles are p or α colliding with heavy nuclei, the retention fraction approaches unity. This is the usual spallation scenario. However, heavier elements are produced in supernova explosions and accelerated directly in shocks, and these produce so-called *inverse* reactions in which the fast particle is usually a C, N, or O nucleus. For Li, another contributing reaction is called *fusion* in which two ^{4}He nuclei produce ^{7}Li. Then the production rate is approximated by

$$r_{i,j \to k}(t) = X_i^{\text{GCR}}(t) X_j^{\text{ISM}}(t) \xi(t),$$

where $\xi(t)$ is the cosmic ray production rate, which we will assume is proportional to the star formation rate by instantaneous recycling since supernovae are the main contributors. The rates also depend on the precise cosmic ray spectrum and the interaction cross sections, but we will lump those into ξ for purposes of illustration. We can write $\xi(t) \sim \psi_{\text{SN}}$, denoting by ψ_{SN} the global supernova rate, making no distinction between types of explosions. This is a good approximation because the stars that end their lives as SN II are high-mass stars and are characterized by short lifetimes with respect to the age of the Galaxy. As we discussed in Chapter 6, the bulk of cosmic rays are interstellar nuclei that have been accelerated in situ by shocks, so we presume that their composition tracks the interstellar abundances over galactic history. This allows us to write $X_i^{\text{GCR}}(t) = X_i^{\text{ISM}}(t)$, where X_i^{ISM} ($i = \alpha, \text{C}, \text{N}, \text{O}$) is the abundance of the corresponding element in the ISM. The

---

[45] See the overview by Ramaty, R., Kozlovsky, B., and Lingenfelter, R. 1998, *Phys. Today*, **51**(4), 30; Reeves, H. 1994, *Rev. Mod. Phys.*, **66**, 193.

metallicity evolution equation becomes

$$\frac{d}{dt}(X_k M_g) = -X_k\psi + r_k M_g \rightarrow M_g \frac{dX_k}{dt} = -X_k R\psi + S_k M_g. \quad (7.51)$$

To solve Eq. (7.51) now requires specifying the nonstellar production rate, S_k. There are separate expressions required for elemental production depending on the mechanism. For *direct* p- and α-induced reactions and *inverse* CNO induced reactions, we write

$$S_k(t) \rightarrow S_{0,di}\psi Z,$$

For ^{7}Li there is another channel to consider, *fusion* ($\alpha + \alpha$) reactions, which induces a different dependence on the metallicity:

$$S_k(t) \rightarrow S_{0,f}\psi.$$

Although the same analytic expression describes both direct and inverse processes, keep in mind our assumptions. In the first case, Z represents the instantaneous interstellar abundances, while in the second case it is the cosmic ray composition. The equations have algebraic solutions since we can always interchange t and μ, and through that, Z. Any contribution from Big Bang nucleosynthesis is included as an initial abundance $X_{\text{BB},k}$, but this is significant only for ^{7}Li. In the absence of infall, the star formation rate must exponentially decrease with time since inert remnants inevitably result from stellar evolution. Once you have solved for $X_k(Z)$, expanding for $Z \ll Z_\odot$ shows that the Be and B abundances should scale as

$$X_k(Z) \approx \left[\frac{1-R}{(1-2R)P_Z}\right] M_{g,0} S_{0,di} Z^2. \quad (7.52)$$

This is a problem with the closed box picture because the trend from halo star observations is linear for [Fe/H] ≤ -1.3. On the other hand, because of its comparatively high initial abundance and the contributions of fusion reactions, ^{7}Li should show a nearly linear dependence:

$$X_{^7\text{Li}} \approx X_{\text{BB},^7\text{Li}}\left(1 - \frac{R}{P_Z}Z\right) + \frac{1-2R}{P_Z} S_{0,f} M_{g,0} Z. \quad (7.53)$$

The precise rates are less important for our purposes than the trends within this simple model for chemical evolution: primary elements to scale directly with the metallicity while secondary production do not. We have dwelt at such length on this topic because of its great importance for cosmology. We will see in the next chapter that the light-element abundances provide critical data for an intermediate, optically thick stage in the expansion of the Universe before the formation of galaxies.

7.3.4.4 The Lowest-Metallicity Fossils
The closed-box picture produces a paradox. We began with a galaxy initially consisting only of gas with the Big Bang–generated abundances and no stars. Since

the total mass of the system was $M_{g,0}$, ψ must have been much higher than now if the Schmidt rate remained invariant over the life of the system. This should leave us with *many* low-mass main sequence stars that are extremely metal-deficient compared with the Sun *provided* the IMF and stellar yields do not significantly change. The amount of available interstellar gas should steadily decrease over time, so the early generations of dying stars produced metals at a high rate but polluted a medium that diluted their contribution. In contrast, those now returning material have their metals mixed into a much sparser medium and therefore more rapidly increase the metallicity of the forming population of stars. Thus we arrive at the *G-dwarf problem*;[46] we do not find huge numbers of low metallicity stars in any region of the Galaxy.

The first proposed solution was to vary the initial mass function with time. Lacking a fundamental theory for $\phi(m)$, there is little reason to think this function *must* be universal. So one option is to assume that in the early Galaxy, when there are few coolants, the tendency would be to form higher-mass stars. This argument follows from the Jeans instability: if c_s is larger, so is M_J. If M_J represents a characteristic mass for the stars, you would expect fewer low mass stars would be form in the absence of metals, hence no stars would survive to now. It is a convenient solution and also has the feature that the upper mass of the star is effectively set by the observed metallicity distribution. The problem is that stellar masses are not strictly limited by the Jeans mass, so, although attractive, the argument is weak. Observationally, there are low mass stars with very low metallicities, just not many of them. It is also not correct to assume that the lack of metals means an absence of coolants. While CO is the predominant cooling agent *now* in clouds, other slightly more exotic species were available in the early galaxy. For instance, LiH, H_2^+, and H_3^+ have been shown to be effective enough to yield low-mass fragments during collapse. Another possibility centers on the efficiency of return of gas to the medium. Stellar winds are a major source of return of matter to the ISM. Only massive stars, $M > 10\ M_\odot$, can maintain such outflows throughout their lives. Those of lower mass possess winds for only relatively short times during the giant stage. Such mass loss, if driven by radiation pressure transferred through atomic absorption lines to the gas, depends sensitively on the metallicity. This further complicates the galactic models because the remnant mass becomes a function of the metallicity as well as the mass. In the end, we must consider the possible feedback between the metallicity and star formation to properly account for galactic evolution.

7.3.4.5 *Feedback and Stimulated Star Formation*

Let us see if we can take a slightly different approach to the problem. A quadratic density dependence for ψ arises naturally from a number of physical processes. Cloud collisions compress gas near the critical density and, if the material can cool rapidly enough to remain isothermal, collapse to stars might occur. This is an

---

[46] You see how the paradox emerges here. It is a historical artifact based on a particular version of the star formation rate and simplifying assumptions about the chemical evolution equations. We will go into more detail in subsequent sections. For now, it is appropriate to repeat a remark of Bohr that "there are no paradoxes in Nature." Epistemology is not out of place as we go deeper into the theory. When a paradox occurs, it *must* be in the model and its assumptions and not in the observations when the data are unambiguous, as they are here.

example of an *induced* process, where spontaneous star formation would occur from random density fluctuations within the clouds themselves or internal collisions between dense cores.[47] Induced star formation permits a wide variety of processes, all of which are similarly nonlinear. As an illustration, we will take a closed-box model with total mass, M. Call s the mass fraction in stars. To further simplify the picture, take $g = 1 - s$ to be the fraction in gas, where we have put the remnants into the system as an unrecovered part of the gas returned to the interstellar medium. If the characteristic timescale for induced star formation is a^{-1}, then the rate of the process is asg and the star formation rate is

$$\frac{ds}{dt} = \psi = as(1-s), \qquad (7.54)$$

whose solution is a logistic equation

$$s(t) = \left[1 + \frac{1-s_0}{s_0} e^{-2at}\right]^{-1}$$

for some initial value s_0. Notice that the long-term behavior of the star formation rate is an exponentially decreasing function of time, as you would expect since there is vanishingly small gas at long time. You now have an *explicit* form for the star formation rate that can be substituted into Eq. (7.48) to obtain the metallicity as a function of time. Yet Eq. (7.54) is essentially the same as a crude closed-box model; the stars, once formed, live forever and lock up all the gas. It cannot be a realistic a model for chemical evolution since it lacks return of gas to the interstellar medium, but this can be included using a stellar death rate r that returns some fraction of the gas while the rest is locked up in remnants:

$$\frac{ds}{dt} = -rs + as(1-s), \qquad (7.55)$$

whose solution is

$$s(t) = \left(1 - \frac{r}{a}\right) \left[1 + \frac{(1-r/a) - s_0}{s_0} e^{-(2a-r)t}\right]^{-1};$$

the net effect is to lower the asymptotic population of the stars, but the qualitative result for ψ does not significantly change. This is the Malthus–Verhulst equation

---

[47] The two mechanisms for induced star formation were H II region compression (Elmegreen and Lada, in 1977) and supernova shock compression (Herbst and Azousa, also in 1977) of clouds. Other mechanisms include radiation-driven collapse, compression driven by expanding photodissociation regions, and the ever-present chance of cloud–cloud collisions. The principal feature that all these mechanisms have in common, and that makes them attractive for galactic evolution modeling, is they involve the feedback *between* different components of the ISM. See Lynden-Bell, D. 1975, *Vistas in Astr.*, **19**, 299; Shore, S. N. and Ferrini, F. 1995, *Fund. Cosm. Phys.*, **16**, 1.

for the evolution of a population[48] and the simplest model for nonlinear induced processes that can be used to describe star formation. Finally, we can include the remnants explicitly while maintaining the mass of the region at a constant value:

$$\frac{ds}{dt} = -rs + asg$$

$$\frac{dg}{dt} = r's - a'sg$$

such that the remnants form a fraction $1 - s - g$ of the mass of the system. In writing Eq. (7.56) we have assumed that only some fraction of the stellar mass goes back into stars, and the rest is locked in white dwarfs and other nonrecyclable bodies so that, in general, $r < r'$ and $a < a'$.

7.3.4.6 Some Processes Affecting Evolution of the Gaseous Component

Spontaneous processes, cooling and contraction of a self-gravitating cloud or some portion of it, and compression by large-scale structures such as density waves, surely lead to star formation, but they depend on only one phase of the medium. The process can also be *stimulated* by a variety of mechanisms, for instance, compression by supernova shocks, cloud–cloud collisions, and compression of matter by expanding H II regions and subsequent self-propagation. Notice that each of these depend on the coupling of at least two galactic components to form new stars or to eject gas back into the holding phase of the warm neutral medium. This means that a galaxy has many available feedback mechanisms, both positive (stimulated star formation, e.g., by cloud compression due to expanding photodissociation regions) and negative (such as cloud destruction by compression due to expanding photodissociation regions). The basic scheme rests on the notion of a critical density above which a cloud becomes self-gravitating and thereafter collapses when cooling becomes important.

Stimulated triggering of star formation, if it exists, has been an elusive observational phenomenon in the Milky Way. Although compression is often detected in clouds where supernova remnant-cloud interaction is occurring, there is no direct evidence that it leads to star formation. Indirect evidence is provided by the Gould Belt and by the age sequence in the Orion OB1 association.[49] The best evidence comes from extragalactic observations. Ring galaxies, such as the Cartwheel, and other galactic colliders, for instance, Arp 220, NGC 4039/4039 (the "Antennae") and other infrared superluminous galaxies, show that bursts of star formation *accompany* massive disruptions of galactic disks. Whether these observations can be extrapolated to the level of individual molecular clouds within normal galaxies

---

[48] The inspiration for both Darwin and Wallace in their original conception of natural selection was the passage in Malthus' *Essay on Population* in which he discussed the geometric growth of an unchecked population compared with its linear rate of production of resources and death. Verhulst posed the problem mathematically by including a death rate for the population. The result is a saturation in the population with time, depending on the relative birth and death rates. It also describes the propagation of a disease in a population where the recovery rate is r and the infection rate is a.

[49] Silic, S. A., Palous, J., Franco, J., and Tenorio-Tagle, G. 1996, *ApJ*, **468**, 722; Maddalena, R. J., Moscowitz, J., Thaddeus, P., and Morris, M. 1986, *ApJ*, **303**, 375.

remains an open question. The blue compact dwarf galaxies show irregular star forming histories that are similar to the "boom and bust" behavior seen in ecosystems and biological prey–predictor population cycles. These may indicate stochasticity in the galactic star formation, which strongly supports the idea of triggering but does not prove it.

The galactic box can be opened by introducing gas flow into and out of the star forming region. For the galaxy as a whole, this means including disk accretion of gas from stars in the halo and loss of gas by a galactic wind. This is modeled by including an additional term, f, in the gas evolution equation in Eq. (7.56). This infall rate, f, is critically important for both the star formation rate and the chemical evolution. First, any accretion of gas shed by halo stars likely contains s-processed material. In addition, the deuterium and lithium content should be reduced, if it has passed through stars. Pristine gas may also fall into the disk from the halo or beyond, such as cooling flows for the central galaxies in large clusters. Accretion of gas from other galaxies is possible through collisions and mergers in clusters (see text below). This accretion dilutes the gas metallicity in the disk compared with closed-box predictions (recall the bromide "the solution to pollution is dilution"), but it also fuels continuing star formation by replenishing gas that is otherwise locked up in inert stellar remnants. Without infall, the rate of gas consumption and of star formation are essentially identical; with infall, this is no longer true. Depending on the mass of the halo and the reservoir of available gas, it is possible to alter the rate of gas depletion sufficiently to maintain a nearly constant value for ψ.

Once star formation has been going for at least the main sequence lifetime of a typical OB star, for a few million years, the stars inevitably produce strong winds and eventually supernovae. The rate of input of mechanical energy and heating from shocks and radiation is therefore the same as ψ, but depends *nonlinearly* on Z. This is because the driving of stellar winds depends on the stellar atmosphere opacity and also on the heating and cooling functions for the interstellar gas. The heating provides a negative feedback with a complex time dependence. The mechanical input from stars provides the means whereby the diffuse galactic gas is heated to the point of generating a global outflow, or galactic wind.

We have discussed stellar winds in Chapter 3 and only need a little more work to expand that dynamical model to a galactic scale outflow. Stars lose mass by getting the matter hot enough (coronal heating) or having sufficiently high luminosities (radiation line driving) to get mass to the sound speed or above and continue acceleration outward by dumping momentum and energy into the flow. A galaxy is different really in only one respect. Its mass is extended over a region that is large compared with scale height for the gas in the disk, and it is necessary to include this when specifying the gravitational acceleration. The difficult part is computing how the driving occurs and how the heated gas dynamically responds. For the galactic scale flow, supernova remnants and expanding H II regions raise matter to very high temperature but only a small part of the ISM reaches 10^7 K or higher, which is required for escape from the disk. The outflow is evaporative, analogous to the solar wind, with the vertical acceleration given by Eq. (3.174); otherwise the treatment is the same. If the supernova rate is proportional to ψ, then so are the heating and acceleration of material and the driving is greatest when ψ is large. You see that, regardless of the details, the effect of wind removal of gas is the same as a *negative* feedback to global the star formation machine. This has the additional effect of throttling the rate of increase in metals, so it goes in the right direction. Outflows are often seen accompanying episodes of intense star forming activity in galaxies, as we will discuss later, and not all the gas needs to be expelled to have the effect of slowing the rate of star formation in the early epoch. To include flows requires modification of Eqs. (7.56) since this is material added to the gas and not the stellar phase:

$$\frac{ds}{dt} = -rs + asg \qquad (7.57)$$

$$\frac{dg}{dt} = r's + a'sg + fg_H. \qquad (7.58)$$

We presume here that the rate of infall is f of some gas fraction g_H from the halo. You see now how to add complications to the system, including the coupling between different zones of the Galaxy. The effect is to slow the rate of increase of metals and, because of the extra input of gas, to slow the decrease in ψ with time.

The radial distribution of metals depends on the rate of local star formation and the available modes of radial mass transport. Angular momentum presents a substantial barrier for mixing, although circulation is possible in a disk that has sufficiently high turbulent velocities to promote diffusion. The radial gradient is therefore set, in some models, by the mixing timescale L^2/η_t, where the turbulent diffusion coefficient depends on the velocity dispersion of the gas, σ_g. Supernovae and winds form superbubbles that expand to scales of order a few hundred parsecs in about 1 Myr, still small compared with the radial scale length for the disk but large compared with other local effects. They are also implicated as sites for accelerating the heavy cosmic rays required for spallation scenarios. These enter the metallicity evolution equation as diffusion terms, driven by composition gradients but through large-scale transport:

$$\frac{d}{dt} ZM_g \rightarrow \left(\frac{\partial}{\partial t} + \mathbf{v} \cdot \nabla \right) ZM_g = -(1-R) ZM_g \psi + P_Z \psi + \nabla \cdot \eta_t \nabla M_g Z_i. \qquad (7.59)$$

Actually, this sort of diffusion term appears naturally when considering stellar orbits within galaxies. At least in disks, orbits are nearly circular but there is still a small dispersion that radially spreads out the population. This is essentially the same as a diffusive motion of the stars within an annulus. Although it is a limited range of galactocentric distance for stars in the disk, halo stars can cover a very large volume. Finally, we can summarize this discussion by saying that any mechanism that produces a chemical gradient requires the transport rate to be lower than the local production rate, implying that the simplest way to maintain a steep gradient is to have a strong radial dependence in the star formation rate.

Although stars pass their lives in splendid isolation, galaxies have disks and halos whose sizes are a substantial fraction of the mean separation between cluster members. Collisions are not only possible but probable. Galactic escape velocities are also rather low compared with the velocities of winds and supernovae. Consequently, the environment in which a galaxy sits affects the development of its internal properties throughout its lifetime. There is substantial evidence, from the Virgo and Coma cluster spirals, that diffuse gas can be removed from disks by shock/ram pressure stripping. This happens most effectively in the cores of rich clusters, where the density of background gas is high and the individual galaxy velocities are maximized. The evidence is, however, that the molecular clouds are not removed as easily as the H I, so star formation may proceed for a short time after the gas has been stripped. Now what happens to such a system when the stars that are still "living" at the time of the stripping event begin to die? They shed their mass into a significantly reduced interstellar medium, one that cannot dilute their contribution as much as it normally would, with the result that the gas is suddenly extremely metal-rich. This proceeds for quite some time until the gas is either replenished by halo infall, or perhaps from infall of cooler gas accreted near the apocluster point of the galactic orbit. One consequence is that ^{2}D and ^{3}He may be reset to their primordial values if there is a substantial infall event of unprocessed gas or diluted by accretion of severely depleted gas. In addition, if the infall is not uniform, it can produce a change in the galactic radial abundance gradient. Galaxies with different star forming histories could have different stellar populations, and these would mix in dynamical chaos during a collision event. As we have discussed, the Milky Way appears to have suffered such an event(s).

7.3.5 Nucleocosmochronology

You may have wondered why we dwelt at such length on the formalism for chemical evolution when there is so much territory yet to explore. One of the most dramatic events of twentieth century was the radiometric determination of the age of the Earth. By counting the rate of production of α particles from a sample of pitchblende that contained uranium and measuring the abundances of uranium and thorium in the sample, Rutherford was able to place a lower limit to the age of the Earth of about 0.7 Gyr at a time when all estimates were nearly an order of magnitude lower. The presence of such long-lived species allows for relative dating of meteorites and, by extension, the age of the solar system and the Galaxy. The trick is to have a model for the rate of their production over time. These isotopes are produced by the r process, but not all of them are unstable. We can use r-only nuclei, such as europium, and compare the abundances of the unstable species, assuming that we know the production rate and the star forming history of the galaxy. The basic evolution equations are the same ones we've just used. The principal added feature is the decay of the species X_i at some rate λ_i. So we can rewrite Eq. (7.48) as

$$\frac{d}{dt}(X_i M_g) = -X_i \psi + Y(1-Z)R\psi + RX_i\psi - \lambda_i X_i M_g. \qquad (7.60)$$

In a closed-box model the interchangeability between M_g and t still holds, but now an explicit form must be adopted for ψ, which we can obtain (in principle) from a model of the form we just outlined. We will, however, postpone further discussion of the implications of the abundances of these species until the next chapter, where you will see that they constrain the age of the Galaxy, and hence the Universe.

7.4 GALAXIES AND CLUSTERS OF GALAXIES: INTRODUCTORY REMARKS

> Because something is happening here, But you don't know what it is, do you, Mr Jones?
>
> —Bob Dylan, *Ballad of a Thin Man*

The study of external galaxies began in 1924 with a single observation: Hubble's discovery of a classical Cepheid variable in the Andromeda nebula, M31.[50] Distinctive types were already distinguished in the morphological descriptions that Wolf and Dreyer had concocted for nonstellar objects before 1910, but there was no agreement on the nature of the "spiral nebulae." The first report of spiral structure in nebulae was Parson's visual observation of M51 in 1857. This seemed to confirm the Laplacian picture of star formation, since the model assumed that new stars are born in differentially rotating self-gravitating gas disks. Within about 60 years, in 1912, Slipher dashed this hope by finding that the spirals display starlike spectra and looked nothing like H II regions and planetary nebulae. Since they are not gaseous, they were not evidence regarding the nebular hypothesis. But this made them even more intriguing. Slipher showed that these objects have large LSR velocities and also display internal motion. Öpik used this latter observation to place M31 at large distance, beyond Shapley's bound for the globular clusters, based on its observed resolution into stars and its rotation curve. Hubble's thesis dealt with the luminosity and spatial distribution of the nebulae as functions of their morphology in Wolf's scheme. The counter argument regarding their distance was based largely on van Maanen's reported detection of proper motions and resolved systematic rotation in several spirals, notably M33 and M101. Although the source of this error has never been clarified, both Lundmark and Hubble were unable to reproduce the measurements and eventually, after Hubble's paper on M31 appeared, the observation became irrelevant.

Of all fields in astrophysics, extragalactic astronomy has taken the greatest leaps with new stages in technological developments since 1980 and likely will continue to benefit more than any other area of observational astronomy from the introduction of new instruments and techniques of observation. The limitations imposed by photographic recording meant that only limited data could be obtained for a small number of bright galaxies. Now, with imaging detectors reaching high efficiency and large format (2048×2048-pixel^2 formats are becoming standard), spectrophotometric imaging, long slit spectroscopy, and faint limits ($V = 25$ mag is no longer unusual), the whole picture has changed. Multiwavelength imaging has become

---

[50] The most comprehensive survey of the observational situation at this time is Smith, R. W. 1982, *The Expanding Universe: Astronomy's "Great Debate" 1900–1931* (Cambridge, UK: (Cambridge Univ. Press). Struve, O. and Zebergs, V. 1960, *Astronomy in the 20th Century* (NY: Macmillan) provide an access to the almost real time response. Other focused historical accounts include Berendzen, R., Hart, R., and Seeley, D. 1976, *Man Discovers the Galaxies* (NY: Science History Publ.), and Christanson, G. E. 1995, *Edwin Hubble: Mainer of the Nebulae* (Chicago: Univ. Chicago Press).

normal. The picture is radically altered compared to where it was when Struve and Zebergs surveyed the field more than 40 years ago and found only limited, essentially qualitative, progress.

7.5 THE HUBBLE CLASSIFICATION SCHEME FOR GALAXIES

Large scale structure is the most obvious features of external galaxies, and the hardest to discover for our own. All galaxies fall into a relatively small number of basic morphological types that were first defined by Hubble[51] and remain the most frequently used descriptors of galaxies. Hubble's work represents taxonomy in its purest form, perhaps even more than the MK system for stellar spectra, without regard to the origin of the distinguishing features, but pursued under the assumption that the underlying causative agents will be revealed through systematic classification and subsequent statistical and dynamical analyses.

Before setting out the physical properties of galaxies, we must reflect on methodology. Evolutionary linkage is implied by most taxonomic schemes, biological or astrophysical, and that was certainly an intention in Hubble's original presentation. As with stars, galaxies are still separated into "early" (ellipticals) and "late" types depending on their flattening and spiral pattern. The lenticulars, in particular the S0 group, were introduced as transitional forms between elliptical and spirals, and thus seem to have been predicted by the original morphological "tuning fork" diagram (see Figure 7.11). That these galaxies form a parallel sequence to the ellipticals and spirals and that they are otherwise dynamically indistinguishable shows the key role played by spectroscopy in resolving questions posed by imaging alone. It is, however, important to add that although the picture of galactic relationships has developed along very different lines, there is still some physical evidence for links between the variety of types, as we will discuss below.

7.5.1 Spirals: Active Star Formation and Global Patterns

Spirals (S) galaxies show two more or less separate components, a disk or flattened distribution of stars and a spheroid or bulge of variable ratio to the size of the disk. The classification is based on two parameters, the bulge to disk ratio and the tilt of the spiral pattern, which is unique to the disk. These two parameters are corre-

---

[51] The paper that first describes the system is Hubble, E. P. 1926, *ApJ*, **64**, 321 (note that the classification was later refined), which he later extended in the classic monograph, Hubble, E. 1936, *The Realm of the Nebulae* (New Haven: Yale Univ. Press). The best photographic surveys and introductions to the scheme remain Sandage, A. 1961, *The Hubble Atlas of Galaxies* (Washington, DC: Carnegie Inst. Washington) and Sandage, A. and Bedke, J. 1994, *The Cargenie Atlas of Galaxies* (Washington, DC: Carnegie Inst. of Washington). The introductions to these two volumes are particularly important regarding both the development of the classification scheme for galaxies and for general remarks about taxonomy; see also Sandage, A. and Tammann, G. A. 1981, *Revised Shapley-Ames Catalog of Bright Galaxies* (Washington, DC: Carnegie Inst. Washington); Tully, R. B. 1987, *Nearby Galaxies Atlas* (Cambridge, UK: Cambridge Univ. Press). The introduction to the *Second Reference Catalog of Galaxies* (RC2), by G. de Vaucouleurs, provides a very important review of approaches to morphology. The most complete modern review is van den Bergh, S. 1998, *Galaxy Morphology and Classification* (Cambridge, UK: Cambridge Univ. Press).

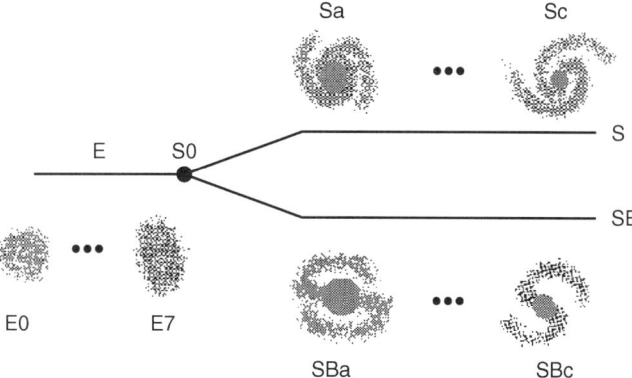

FIGURE 7.11. Hubble "tuning fork" diagram showing the basic morphological types. The S0 galaxies were originally thought to be an evolutionary transition between ellipticals and spirals.

lated, where *Sa* is the most tightly wound arms and has the largest bulge to disk ratio and Sc has the most diffuse and open arms and the smallest bulge to disk ratio. There are several modifications of this scheme, including the Sd subgroup and the use of a "luminosity" classification that makes the system analogous to the MK spectral types based on the diffuseness of the arms, but these are of less importance than the original ones proposed by Hubble. Two features should be noted. First, young stars, hence continued star formation sites, are confined to the flattened stellar group, the so-called disk population. It is here that molecular clouds are confined and where the bulk of the absorbing neutral gas of the galaxy is located. This part of the system has the highest specific angular momentum and appears to be dynamically separate from the spheroidal distribution of stars. The Andromeda galaxy, M31, is one of the best examples of the group, as are M81 and M101. In hydrogen content, there is little differences among spiral subclasses. There are differences in abundances, and also in the distribution of cool material in the form of CO, but these are not reflected by the neutral atomic gas, which is the most easily measured global property. The range in absolute magnitudes is large, from -15 to about -21 mag. Disks also show a large range, from as small as 2 kpc to over 30 kpc, and the luminous mass generally shows an exponential disk with a scale length of less than 10 kpc. Usually, the neutral hydrogen is far more extended than the luminous matter, so the M/L value increases with distance. Total masses range from about 10^{10} to 10^{12} $M_\odot$.

Barred spirals (SB) form an almost parallel sequence to the ordinary spirals, SBa–SBc and even "later," and it is not really clear that they are physically distinguished from them except by the prominence of the bar. In the most bar-dominated systems, the disk shows a secondary ringlike structure with the appearance of a Greek Θ, where the ring is more or less closed depending on the tilt angle of the spiral. A beautiful example of this group is NGC 1300. Among this group, Hubble first and later de Vauclouleurs recognized a parallel sequence of arm structure to normal spirals. Thus, the subclasses include SBa, SBb, and SBc

TABLE 7.2 Average Global H I Properties of Spirals

Type	$\log M_{HI} h^2$	$\log (M_{HI}/L_B)$	$\log \Sigma_{HI}$
Sa, Sab	9.4 ± 0.5	−0.6 ± 0.4	6.7 ± 0.3
Sb	9.7 ± 0.4	−0.4 ± 0.4	6.8 ± 0.3
Sbc	9.5 ± 0.4	−0.3 ± 0.3	6.9 ± 0.2
Sc	9.4 ± 0.4	−0.3 ± 0.3	6.8 ± 0.2
> Sc	8.9 ± 0.7	−0.04 ± 0.3	6.9 ± 0.2

(see Table 7.2[52]). You know from that the Milky Way is a spiral. Whether it possesses a nuclear bar is still uncertain, although increasing likely based on dynamical signatures we have discussed.

Lenticulars (S0) are flattened systems that show large scale structure that is very similar to spirals. They have, however, lower contrast between the arm and interarm regions and low atomic and molecular gas contents. In Hubble's original scheme, the S0 galaxies were initially thought to represent transitional forms between ellipticals and spirals. They are now recognized as a parallel, not transitional, *sequence* of galaxies that can be classified as S0a–S0d, depending on the tilt angle and openness of the arms.

7.5.2 Ellipticals: Frozen Populations

As their name implies, *ellipticals* (E galaxies) are spheroidal systems seen in projection with axial ratios ranging from 1 for E0 to about 3 for E7. The numeration is simply $10(1 - b/a)$, where b is the minor and a the major axis of the image. They generally display neither stellar disks nor dust lanes, except in pathological cases such as NGC 5128 (Cen A) that are either now experiencing mergers or have suffered close interactions with neighboring systems in the recent past. Ellipticals show no obvious separation of the stars into substructures (e.g., bars, associations). Although these galaxies possess some dust, they do not show large scale organization of the material into a disk of clouds and there is little evidence in the majority of these systems for continuing star formation. Giant ellipticals (cD) are frequently the dominant central body in clusters of galaxies. The class was not included in the original Hubble taxonomy; Morgan based the classification on their distinctive core–halo structure.[53] Ellipticals have smooth transitions between their cores and halos, while cD galaxies show very extended halos and much more distinct compact cores. This photometric structure therefore resembles the distinction between main sequence and giant interior models. Again,

---

[52] Sources of Table 7.2 are Giovanelli, R. and Haynes, M. P. 1988, in *Galactic and Extragalactic Radio Astronomy* 2nd ed., Verschuur, G. L. and Kellermann, K. I., eds. (NY: Springer-Verlag). See also Haynes, M. P. and Giovanelli, R. 1984, *AJ*, **89**, 758. Masses are in $M_\odot$ and surface densities, Σ_{HI}, are in $M_\odot$ kpc^{-2}. The mass scaling factor, h is the normalized Hubble constant, $h = H_0/(100$ km s^{-1} Mpc^{-1}) (see Chapter 8).

[53] The "c" designation echos the earlier spectroscopic designation for giants, those stars in Maury's classification in the HD project to denote stars that displayed narrow lines as contrasted with broader types "a" and "b," which denoted dwarf stars. See also Bautz, L. P. and Morgan, W. W. *ApJL*, **162**, L149 and references therein.

a large dispersion is observed in the properties of the subclasses. Masses range from as small as 3×10^{11} $M_\odot$ for dE systems to more than 10^{13} $M_\odot$ for cD galaxies. Disks are similarly varied, with radii of a few kpc for dwarf systems to nearly 1 Mpc for cDs. Absolute blue magnitudes range from fainter than -14 to brighter than -25 mag.

Typical velocity dispersions are between 100 and 400 km s^{-1}. In contrast to the obvious axial symmetry of spirals, the *three-dimensional* shape of ellipticals is poorly known. Variations in the velocity dispersion and changes in the orientation of surface isophotes as a function of radius suggest triaxial symmetries for some systems. Their projected eccentricity often varies with the distance from the center, with contours appearing more circular at smaller radii and flatter with increasing distance. Opposite behaviors have, however, been observed as well and, in some extreme systems, the isophotes cannot be reduced to ellipses. Despite these complications, there are some regularities. The luminosity profile of elliptical galaxies, and even the spheroids of spirals, is well represented by a simple empirical relation first proposed by de Vaucouleurs in 1948:

$$I(\lambda) = I_e(\lambda) \exp\left\{-3.33\left[\left(\frac{r}{r_e(\lambda)}\right)^{-1/4} - 1\right]\right\}, \quad (7.61)$$

which is also known as the "$r^{1/4}$ law." This law depends on two dimensioned constants, the half-luminosity effective radius r_e and the surface brightness I_e at $r = r_e$. It fits the photometric profiles over a range of more than 11 magnitudes and radii from 0.1 r_e to 7 r_e. The precise scaling depends, however, on which filter is used for the observations due to the photometric colors of the underlying galactic stellar populations, so r_e and I_e are actually wavelength dependent although the formal law is not. There is some evidence for color (e.g., population) gradients in ellipticals [e.g., for M87, $r_e(U) > r_e(I)$ indicates a color gradient] but for the most part they appear more homogeneous than spirals. Spectroscopic studies have also revealed abundances gradients. The two are related. The M/L ratio does not vary much with galactocentric distance and the universality of the $r^{1/4}$ law implies that the processes that lead to the formation of elliptical galaxies produces a uniform final *mass* distribution even though the stellar *populations* are not uniformly distributed and may have been segregated dynamically.

Ellipticals look so much like the classical ellipsoidal equilibrium shapes of self-gravitating masses that it is hard to resist the comparison. For such bodies, flattening requires a significant amount of rotation. Yet it is clear that their rotational support is minimal. For instance, as you know from our discussion in Chapter 1, the ratio of the maximum observed rotation speed to the central velocity dispersion should correlate with the eccentricity. It does not. Consider the Maclaurin ellipsoids, which are the simplest constant density configurations without internal motions. Then defining the ratio of kinetic to thermal energies by $\mathcal{T} = T/|W|$ we can recall the calculations we did for Maclaurin ellipsoids in Chapter 1:

$$\mathcal{T}(e) = \frac{3}{2}\left[\frac{1}{e^2} - \frac{2}{3} - \frac{(1-e^2)^{1/2}}{e \sin^{-1} e}\right]. \quad (7.62)$$

Since $\mathcal{T}$ depends only on eccentricity, we can calculate the stability of these "galaxies." The hypothesis that ellipticals are oblate fast rotators obeying the virial equations implies correlation between the flattening and the amount of rotation that can be expressed in terms of the maximum rotation velocity, v_{max} relative to the central velocity dispersion:

$$\frac{v_{max}}{\sigma_0} \sim \left(\frac{2\mathcal{T}(e)}{1 - 2\mathcal{T}(e)} \right)^{1/2}. \tag{7.63}$$

But this is not observed. That's why we digressed, as a caution—the Universe is a complicated place, galaxies even more so. It appears instead that ellipticals possess anisotropic "pressure," or triaxial velocity dispersions, and rotation plays little or no role in their structure. This is very different from the disks of spiral galaxies, but approximates the behavior of their halos. So, returning to the stellar hydrodynamic equations in cylindrical coordinates, we can replace the tangential velocity dispersion in Eqs. (7.6) with

$$\sigma_{\phi\phi} = (1 - k^2)(\langle V_\phi^2 \rangle - \sigma_{rr}) + \sigma_{rr},$$

where $k = 0$ refers to anisotropic velocity dispersion and $k = 1$ to isotropic, rotationally flattened, systems. The rotational velocity is given by $k(\langle V_\phi^2 \rangle - \sigma_{rr})^{1/2}$. Models with small but nonvanishing k successfully fit the photometric profiles and kinematical data for many ellipticals. If elliptical galaxies are formed by a dissipationless collapse or through mergers of stellar systems, then the velocity distribution should be nearly isotropic in the center and almost radial in the outermost parts. In other words, you allow for nearly complete relaxation of the orbits by strong mixing in the cores of the galaxies but suspect that predominantly radial orbits may dominate their peripheries. Distribution functions have been constructed to reproduce such a variation, namely, a quasi(isotropic) Maxwellian distribution in the inner regions and a radial dominance in the outer regions.[54] If, however, we take the evidence from our own halo, mergers are slow to mix and the structural and dynamical signatures may remain in the appearance of the galaxy for some time. The interplay between N-body simulations and observations will be crucial to eventually understanding this problem, so we leave it as an exercise.

7.5.2.1 The Fundamental Plane
A number of correlations have been found between various global properties of ellipticals through the use of principal-components analysis (PCA). When the various parameters are combined, they result in what has become known as the *fundamental plane*. It is defined by r_e (from the photometry), the central (projected) velocity dispersion, σ_0, and the central brightness, I_0 (notice that this is not

---
[54] See Merritt, D. 1999, *PASP*, **111**, 129 for a modern review.

I_e, as we obtained it from the de Vauclouleurs law):

$$r_e \sim \sigma_0^{(1.4 \pm 0.15)} I_0^{-(0.9 \pm 0.1)}. \tag{7.64}$$

The other observables that are correlated in the "projected" planes are the mean surface brightness internal to r_e (called $\langle \mu_e \rangle$), the luminosity (used to obtain the radius), line strength indicators, and colors. Of particular importance are the distance independent relations for the line index Mg_2 and σ_0. Another that we will discuss more in the next chapter is the Faber–Jackson relation, the correlation between the total galactic luminosity and its central velocity dispersion, $L \sim \sigma_0^4$. These observables, in turn, translate into correlations between structural properties, such as mass, mass distribution, velocity distribution, luminosity, and stellar population parameters, which impose constraints on the formation and evolution theories. The presence of dark matter also influences the observed correlations, since a few quantities depend on the total potential, as other instead are sensible only to the luminous components.[55]

7.5.3 Irregulars and Dwarfs: Stochastic Experiments

Irregulars (Irr) are relatively low luminosity galaxies shows no obvious symmetries. They share many of the characteristics of both ellipticals and spirals in that some are actively forming stars, often at a tremendously high rate [such as, the blue compact dwarf (BCD) galaxies], while others have not had such internal activity for several billion years. Two lovely, prototypical examples are the two principal satellites of the Galaxy, Large and Small Magellanic Clouds. Dwarf galaxies are not only faint but also often have very low surface brightness. Whereas giant ellipticals have $M_B < -18$, dwarf galaxies are systematically below this $M_B > -18$. The dwarfs fall into three distinct subclasses: dwarf ellipticals (dE), dwarf irregulars (dI), and BCDs. Briefly, the dE galaxies display a spheroidal shape. They have low (24 mag/arcsec^2) or intermediate (21 mag/arcsec^2) surface brightnesses, low gas content, red color indicating the lack of star formation at the current time, and stellar populations of age of order or older than 10 Gyr. Their metallicities range from about $\frac{1}{3}$ to solar values. The dI galaxies have even lower surface brightnesses. Their shape is irregular, but on the average they display some flattening. In contrast to dEs, they have high neutral atomic gas content, and generally low metallicity stellar populations although the abundances span a rather wide range. A few show traces of recent star formation. Finally, the BCD galaxies are high surface brightnesses spheroidals with high gas content, and intense star formation that is concentrated in the central region surrounded by a redder and older population. Their metallicity is generally low, 0.01–0.33, solar although some are as low as 10^{-3} solar.

These subclasses are also dynamically distinct. The dEs have low velocity dispersions, ~ 10 km/s, which is about two orders of magnitude less than gE. The size of a dE galaxy ranges between less than 1 kpc to more than 10 kpc (an exceptional case). The surface brightness profiles are also distinct. Although gE

---

[55] Dressler, A. et al. 1987, *ApJ*, **313**, 42; Djorgovsky, G. and Davis, M. 1987, *ApJ*, **313**, 59.

are well fit by the de Vauclouleurs $r^{1/4}$ law, dEs are better reproduced by an exponential, $I(r) = I_0 e^{-\alpha r}$ (even if often two exponentials are needed to fit the whole distribution). The visible ($U-V$) and infrared ($J-H$) global colors are bluer than gEs. This is because they have extended blue horizontal branches and higher effective temperatures on the red giant branch, both results of their reduced metallicity. The dwarf irregulars do not show evidence of molecular gas, even given their plethora of atomic hydrogen. They display neither a color gradients nor spiral structure in spite of their flattening. This suggests that these low-mass galaxies lack some critical condition for initiating of grand design spiral structure. There may, however, be another component in the system than what we see shining. The dI galaxies are more frequently observed in regions of low density, and not in dense clusters, suggesting a correlation between the environment and their morphology.

Because of their extremely vigorous star formation and lack of large-scale structures, the BCDs are particularly interesting venue in which to study local triggering mechanisms. Their typical central densities are around 1800 $M_\odot$ pc^{-3}, comparable to ellipticals but huge compared to the 0.1 $M_\odot$ pc^{-3} found for dEs. The core region of BCDs contains a large portion of the total gas mass in highly ionized state. This is shown by their compact intense Hβ emission. Their optical spectrum resembles galactic H II regions, so they have also been called *H II galaxies*. Blue compact dwarf halos are much larger than their disks, by factors of 3–5, and do not display evidence of active star formation. The gas mass fractions in the dwarf families ranges from less than 0.01% in dEs, to 10–20% in BCDs, and up to 40% in dIs. Assuming a constant star formation rate for 10 GYr and a closed box, we would expect the BCDs to show metallicities at least as high as the Sun. Yet the observed abundances in the range $\frac{1}{30}$ to $\frac{1}{3} Z_\odot$, with a couple of extreme systems with $< 0.01 Z_\odot$. Their evolution is best described by episodes of intense star formation punctuated by long quiescent intervals, much like baseball or war. In fact, this picture seems to accommodate virtually all dwarf galaxies including the Magellanic Clouds. (See properties of UCG sample and galaxies listed in Tables 7.3[56] and 7.4[57].)

TABLE 7.3 Some Integrated Properties of the UGC Sample (Median Values)

Property	E/S0	S0a, Sa	Sab/Sb	Sbc/Sc	Scd/Sd	Sm/Im
R (kpc)	21	20	25	22	18	9
L_B ($10^9 L_\odot$)	53	44	69	53	26	3
M_T ($10^{10} M_\odot$)	—	23	32	19	8	2
M_{HI} ($10^9 M_\odot$)	1	6	15	16	9	2
$\langle (B-V) \rangle$	0.90	0.78	0.64	0.55	0.48	0.22

[56] Sources of Table 7.3 are Roberts, M. S. and Haynes, M. P. 1994, *ARAA*, **32**, 115; de Vaucouleurs, G., de Vaucouleurs, A., Corwin, H. G., Buta, R. J., Paturel, G., and Fouque, P. 1991, *Third Reference Catalogue of Bright Galaxies* (NY: Springer-Verlag); Nilson, P. 1973, *Uppsala General Catalogue of Galaxies* Uppsala: *Acta Univ. Uppsala Ser. V:A*, **1**, 1; Sandage, A. and Bedke, J. 1993, *The Carnegie Atlas of Galaxies* (Washington, DC: Carnegie Inst. Washington) Sandage, A. and Tammann, G. A. 1987, *The Revised Shapley-Ames Catalog of Bright Galaxies* (Washington, DC: Carnegie Inst. Washington).

TABLE 7.4 Some Photometric Properties for Galaxies

Property	gE	Sa	Sbc/Scd	Scd/Sd	Irr
U–V	1.64	1.62	1.28	0.51	0.05
B–V	0.99	0.98	0.91	0.50	0.37
V–R	0.53	0.48	0.43	0.36	0.28
V–I	1.10	0.99	0.90	0.70	0.66

7.6 ROTATION CURVES: BRIGHT FLOW TRACERS AND DARK MATTER

> You're invisible now, you got no secrets to conceal.
>
> —Bob Dylan, *Like a Rolling Stone*

Spiral galaxies provide the surest evidence for an extended massive dark component. The Oort constants, which are defined as solar neighborhood parameterizations of the rotation curve, connect to the mass distribution only once we have a dynamical model of the disk. Measurements of K_z provide volume and surface densities but only relatively near the plane. Neither requires any significant dark component for the disk at low altitude. Only rotational velocity measurements beyond the solar circle that indicate a flat radial dependence hint at the need to include something else in the mass model for the Galaxy. For extragalactic measurements the observations are much less ambiguous. Optically, long slit spectroscopy with rotation of the slit around the galactic nucleus provides a two-dimensional map of the radial velocity. Alternatively, MOS fiber techniques can be used at selected regions on the galaxy to produce spectra for which the separation is somewhat easier and not as affected by the point spread function. Fabry–Perot imaging of emission lines from H II regions also provides a two dimensional map of the velocity field, although it is for only specific regions and only for the ionized gas. Radio observations obtained with aperture synthesis instruments in spectral lines provide the most immediate means for deriving spatially resolved kinematics. Neutral hydrogen measurements of rotation curves for spirals provide the strongest evidence for a dark, massive component in galaxies. Although the central mass concentration, the bulge or spheroid, contains much of the light and would therefore be expected to produce a declining rotation curve with galactocentric distance, $\Theta \sim r^{-1/2}$, the observations systematically contradict this. Spirals show $\Theta \sim$ constant from both the stars (luminous matter) and H I 21-cm emission out to radii of more than 40 kpc.[58] From the absence of

---

[57] Sources of Table 7.4 are Longo, G., de Vaucouleurs, A., and Corwin, H. G. Jr. 1983, *General Catalogue of Photometric Magnitudes and Colors in the UBV System of 2578 Galaxies Brighter than 16th V-Magnitude* (Austin: University of Texas); Buta, R., Corwin, H. G. Jr., De Vaucouleurs, G., de Vaucouleurs, A., and Longo, G. 1995, *AJ*, **102**, 517. See also WFPC-2 Handbook, *STScI*.

[58] There are many useful reviews. See, *e.g.* Bosma, A. 1998, *Cel. Mech. Dynam. Astr.*, **72**, 69; Sancisi, R. 1999, *Ap. Space Sci.*, **269**, 56. Rotation curve collections are presented in Bosma, A. 1981, *AJ*, **76**, 791; Mathewson, D. S., Ford, V. L., and Buchhorn, M. 1992, *ApJS*, **81**, 413, Broeils, A. H. and van Woerden, H. 1994, *Astr. Ap. Suppl.*, **107**, 129.

large-scale radial flows, we know that a spherical mass distribution satisfies

$$\frac{1}{r}\frac{\partial}{\partial r}\left(\frac{\Theta^2}{r}\right) = -\frac{1}{r}\frac{\partial}{\partial r}r\frac{\partial \Phi}{\partial r},$$

regardless of whether the mass is luminous. So the equation

$$-\frac{1}{r^2}\frac{\partial}{\partial r}(r\Theta^2) = -4\pi G \rho$$

allows you to find the radial density variation, at least beyond some distance R_c, the core radius:[59]

$$\rho(r) = \frac{\Theta^2}{4\pi G} r^{-2}. \qquad (7.65)$$

This corresponds to an infinite mass halo (without truncation) and a logarithmic potential $\Phi \sim -\Theta^2 \ln r$. A declining rotation curve indicates a truncated mass distribution. For constant Θ, however, the mass contained within any given radius must therefore increase linearly with galactocentric distance:

$$M(r) = \frac{\Theta^2}{G} r. \qquad (7.66)$$

Since disk galaxies show exponential declines in their radial luminosity profiles, Eq. (7.66) implies an increasing M/L ratio for the constituent matter. This is most obvious from 21-cm observations, which have been traced out to 50 kpc from the galactic center in some systems. With increasing availability of interferometers, molecular line observations have also been used to provide rotation curves.[60] This tracer is more difficult to interpret; you run out of CO more easily than H I because of metallicity gradients, but the molecule is used only to determine the kinematics. Such observations lead to conclusions similar to the neutral hydrogen data; there is something else present in disks besides what we see among the stars.

The largest observed stellar velocities also provide information about the mass distribution. In the z direction, stars moving through the plane at the solar circle have maximum velocities that depend on the height from which they have fallen. Thus, if you know the scale length for the halo distribution of stars, you can estimate the mass of the Galaxy. Unfortunately, this is not a very well constrained property of the stellar population and depends rather heavily on parametric modeling. There is, however, a more straightforward observation along the same lines. If we were simply observing planar orbits tracing out paths within a homogeneous spherical mass, then Eq. (7.66) would provide an adequate solution to the inverse problem. Only the mass interior to R attracts the orbiting body. For a circular orbit around a point mass—the Keplerian velocity v_K—there is a very neat relation between the escape and orbital velocities, $v_{\rm esc}(r) = 2^{1/2} v_K(r)$, at any

---

[59] The approximate isothermal solution $\rho(r) = \rho_0/[1 + (r/R_c)^2]$ is usually used here, called the *King model*; see King, I. 1966, *AJ*, **71**, 64.

[60] Combes, F. 1991, *ARAA*, **29**, 195; Sofue, Y. 1997, *PASJ*, **49**, 17 (see also 1996, *ApJ*, **458**, 120).

distance r from the central mass. For a linear mass distribution within some truncation radius $R_\star$ the ratio is a function of distance for $r < R_\star$:

$$v_{\text{esc}}^2 = 2\Theta^2 \left(1 + \ln\frac{R}{r}\right). \tag{7.67}$$

In the LSR frame this argument leads to an upper limit for any star's velocity. We know the direction of the solar motion. Therefore, a counter-rotating star in the solar neighborhood will have at most a velocity of about ± 450 km s^{-1} along the line toward or away from of the apex of solar motion. For another direction, one transverse to the solar motion, the maximum velocity is given by the escape velocity from corotation. This means that if you observe high-velocity stars in certain directions, they cannot be on bound galactic orbits. From the maximum speed of a distribution you can measure the local circular velocity *independent of your ability to detect it relative to distant sources*. This places bounds on the maximum mass contained within the solar circle. The observed limit is about 500 km s^{-1} where the LSR velocity is 220 km s^{-1}. Thus the mass contained within $R_\star$, which is about 4.6 R_0, is about 5×10^{11} $M_\odot$, higher than the limit provided by the luminous mass. Elliptical galaxies pose a harder observational problem. They are largely gas-free and show no global rotation, so inferences about a dark component are much harder to make. The strongest evidence comes from velocity dispersions, but this is complicated by uncertainties in the distribution function. If E galaxies are triaxial, their velocity dispersions are anisotropic, and this makes an unambiguous identification of a dark component much more difficult. Far better evidence comes from clusters of galaxies, but we'll postpone that discussion for a while.

There is no indication of what this gravitating mass might be, and this has heightened interest in microlensing observations within our galactic halo. On the one hand, if the material is baryonic, the prevalence of flat rotation curves requires a much larger cosmic density than is consistent with Big Bang nucleosynthesis, as we will discuss in the next chapter. On the other hand, if this mass represents some nonbaryonic form of matter, there is no direct evidence of what it is. The scale height of cold dark matter, which is massive and therefore nonrelativistic, is small enough that it might be confined within spiral halos. On the other hand, hot dark matter should be more widespread within a system. Whatever it is, it doesn't shine. Hence the term *dark matter*.[61] Disk galaxies are also subject to a fundamental instability, first described by Ostriker and Peebles[62] from early N-body simulations of disks. They found a tendency of differentially rotating disks to clump to bars unless stabilized by a massive, stationary halo unless $T/|W| \leq 0.14$. The dark halo acts like an incompressible, immovable background. This is curiously similar to values found for the sequence for rotationally supported isothermal spheroids. Since the observations based on the stellar population yield larger values of this

---

[61] A major recent advance occurred with the ISO satellite observations of H_2 emission from inner halos of disk galaxies, indicating that a significant portion of the dark matter *in spiral galaxies* may indeed be baryonic and molecular. For this latter reason it appears to have escaped detection by optical means. But this applies only to spirals, in particular NGC 891, and as we will see does not extend to the more general problem. See Valentijn, E. A. and van der Werf, P. P. 1999, *ApJL*, **522**, L29 for the observation of H_2 in NGC 891.

[62] Ostriker, J. P. and Peebles, P. J. E. 1973, *ApJ*, **186**, 467; Fridman, A. M. and Polyachenko, V. 1984, *Physics of Gravitating Systems* (NY: Springer-Verlag); Vandervoort, P. O. 1991, *ApJ*, **377**, 49.

parameter, this provides indirect evidence for a spheroidal component of dark matter. The same stability applies to spiral density waves. A massive halo stabilizes the pattern against winding and attendant redistribution of the system into individual, kinematically distinguished components. These are the thin disk (the stars originally called "Pop. I"), the thick disk (also called the "old disk"), and the spheroid. These are the same components that are separately modeled by star counts. Since the gravitational potential is linear in the volume density of each constituent, it suffices for the dynamical model to specify the central density, radial length scale, and vertical scale height for each when computing the total attraction.

7.7 THE COMPLICATIONS: PECULIAR GALAXIES

It is often harder to find a "normal" galaxy than a normal star. Not surprisingly, galaxies of any type show a wide range of morphology and activity. For some systems this seems to be intrinsic, for some it is surely due to the environment in which they live. If we take the broadest view, a galaxy in its lifetime plays many parts and what we think of as a peculiarity may be simply a stage that lasts only a comparatively short time. With these introductory remarks, we will now briefly examine both high-energy and dynamical anomalies in the galactic population.

7.7.1 Active Galactic Nuclei (AGN)

The epithet *active* refers to the presence of strong nonthermal emission, extreme emission lines (compared to normal H II regions), excessive X-ray or infrared emission, or some other indication of an extreme environment within a galaxy. Generally confined to the nucleus, the term has been broadened to include global starburst systems. We mention these types of galaxies for completeness but you should consult the references for more details.

7.7.1.1 Taxonomy
The original distinguishing feature of activity was the presence of one or more of the following unusual features: a starlike (i.e., unresolved) nucleus, strong broad emission lines, strong radio emission from the nucleus, and/or extended structures. The first two were the original optical selection criteria and still dominate the surveys for such objects, to which has been added the presence of a strong UV continuum that is either rising toward the shortest wavelengths or flat on objective prism photographic or CCD images.[63] The distinctions that arose based on

[63] Seyfert, C. K. 1943, *ApJ*, **97**, 28; Khachikian, E. Y. and Weedman, D. W. 1974, *ApJ*, **192**, 581. See Weedman, D. W. 1977, *ARAA*, **15**, 69 and Weedman, D. W. 1988, *Quasar Astronomy* (Cambridge, UK: Cambridge Univ. Press) for good surveys of the early work through the mid-1980s. A comprehensive modern treatment is Krolik, J. H. 1999, *Active Galactic Nuclei: From the Central Black Hole to the Galactic Environment* (Princeton: Princeton Univ. Press). It's worth noting here that in his taxonomy paper for galaxies, Hubble in 1926 included a footnote that the emission spectrum for the nuclear region of NGC 1068, NGC 4051, and NGC 4151 resembled planetary nebulae—with $H\beta$ fainter than the O III 4959, 5007-Å lines—in contrast to the more normal H II region–like spectra seen in extranuclear regions of M33 and M101. His spectral resolution was apparently insufficient to reveal the breadth of the profiles. The presence of emission had also been noticed by Slipher in several galaxies we would now call "active" but without the spatial resolution was attributed to internal emission nebulae. Objective prism surveys at Byurakan (the Markarian or Mk designation), Warner and Swasey, and Cerro Tololo. Photometric surveys have been performed by many investigators, notably at Palomar (Zwicky, Zw, and Green and Schmidt, PG).

emission line profile morphology and distribution have remained almost unchanged, as in most classification schemes, but the interpretation has developed considerably. The original distinctions were based on two features: maximum linewidth and a comparison between the forbidden and permitted line profiles. Specifically, *starburst* galaxies have linewidths of order a few 100 km s^{-1} and are characteristically low ionization. *Seyfert galaxies* divide into broad classes. For Sy 1 and *quasars*, the spectrum shows very strong forbidden and permitted lines, with permitted lines having widths of $\leq 10^4$ km s^{-1} and always broader than the forbidden lines. For Sy 2, the permitted lines are not as broad, about 3000 km s^{-1}, and are about the same width as the forbidden lines. Finally, the BL Lac objects, also called *blazars*, don't quite fit into this spectrally based grouping since they are identified by their *lack* of emission lines. These galaxies show power law, featureless continua, and extremely strong nuclear emission.

On closer inspection, the distinctions between these subclasses blurs. Starburst galaxies show global conversion rates of interstellar gas into stars at between 10 and 100 times the rate observed in normal systems. They were originally distinguished by strong broad emission lines with widths of order a few hundred km s^{-1}, although not as broad as Seyfert galaxies. The current definition of starburst galaxies refers mainly to their infrared emissivity, the so-called *IRAS* galaxies or superluminous infrared sources. The infrared luminosity comes from dust that is heated by OB stars so the IR serves as a calorimeter of the stellar content. From the mean mass to light ratio for massive stars, the rate of star formation is deduced. This amounts to $100-10^3$ $M_\odot$ yr^{-1} for the most active systems. Such large rates cannot possibly be sustained for very long; hence the term *starburst*. Many of these systems show signs of recent or contemporaneous interactions with nearby companions. Starburst spectra also usually show low ionization, pointing to a more stellar continuum than seems to occur in Sy galaxies.

The spectroscopic separation of the Seyferts into Sy 1 and Sy 2 is based on line widths with class 1 showing hydrogen Balmer lines much broader than the [O III] and other forbidden lines, while for class 2 they have about the same width. The [O III]/Hβ ratio is also much higher in class 2 galaxies (Figure 7.12). Low-ionization nuclear emission line systems (LINERs) appear to be Sy nuclei embedded in dense circumnuclear dust disks or tori. The demonstration of this connection, and also of the unified picture of Sy nuclei, came with the observation of broad line wings for the Sy 2 prototype NGC 1068 through the use of spectropolarimetry: the permitted (Balmer) lines clearly have broad Sy 1-like wings but these are only visible in polarized light. The phenomenological explanation is straightforward — the broad line region is heavily extinguished in the optical by intervening dust and therefore only the core, which is contributed by more extended gas, appears in the optical.[64] Further support for this model comes from HST imaging of the nuclear regions of nearby Seyfert galaxies, especially NGC 4261, where the surrounding disk is clearly visible within a few hundred parsecs of the core. The spectroscopic observations lead to the standard model for the nuclear regions of Seyfert galaxies in particular and active galaxies in general. Qualitatively, the lines appear to arise from two dynamically distinct locales: the *broad-line region* (BLR), which because of the velocities is presumed to be in the immediate vicinity of the central engine,

---

[64] See, for example, Antonucci, R. and Miller, J. S. 1985, *ApJ*, **297**, 621; Ulrich, M-H., Maraschi, L., and Urry, M. 1997, *ARAA*, **35**, 445.

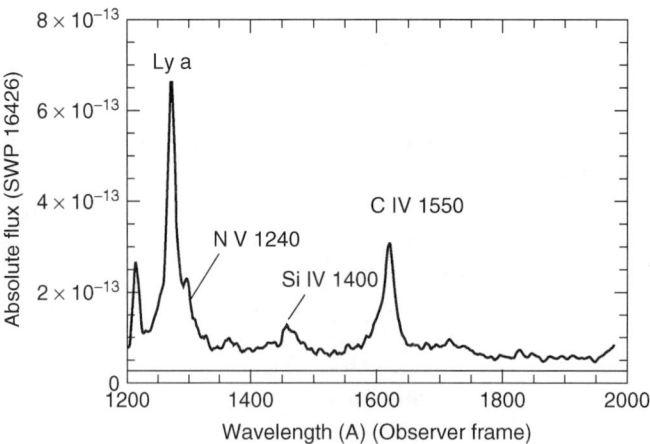

FIGURE 7.12. A sample UV spectrum of an active galaxy, Fairall 9 (Sy 1) obtained with IUE. Note especially the widths of the Lyα and C IV lines (and their redshifts; see next chapter).

and the *narrow-line region* (NLR), which is spatially more distended. The BLR is also thought to be a toroid or disk, while the NLR is more nearly spherical and the line broadening is due mainly to systematic motions around the central engine. Polarization observations indicate that Sy 2 galaxies are those in which the shrouded nucleus is hidden from view, leaving only the larger volume of the NLR accessible to observation. For Sy 1, both regions are visible. We will return to the distinctions between these regions when discussing the central engine in a little while.

Strong radio emission was originally thought to be correlated with active nuclei. The BL Lacertae objects supported this prejudice. The general lack of emission lines, even with indications of accretion, is due to the overwhelming luminosity of the central engine. Weak emission as been detected in several of the nearest BL Lac objects, and the parent galaxies have been directly observed as well. In other words, there is a continuum of activity among galaxies depending on type, environment, mass, and perhaps star forming history. Thus, our remark that, perhaps, each is a stage in the life of its host that we happen to catch in an *on* state.

In the unified picture we have been developing, the BL Lac objects are essentially quasars, or Sy 1, galaxies observed nearly directly along the jet axes. This results, through beaming, in greatly enhanced radio emission and synchrotron luminosity. The continuum is completely dominated by the synchrotron source, which connects smoothly from the radio through the X rays in most sources for which comprehensive multiwavelength analyses are available. For instance, PKS 2155-304 shows a uniform power law over 10 decades in energy, with the break occurring above X-ray energies. Flaring is the most spectacular feature of the behavior exhibited by the objects. A single flare, lasting several days, can produce of order 10^{60} erg. At radio frequencies, the sources show variations of a few percent on timescales as short as hours, and the polarization also can change by more than 40% on the same timescale.[65] All blazars are *highly* variable on

---

[65] Kraus, J., Witzel, A., and Krichbaum, T. P. 1999, *New Astr. Rev.*, **43**, 685.

timescales of years, by factors of 2–10. And perhaps most spectacular has been the detection of TeV γ-ray emission from Mk 421[66] and MeV emission and flaring from many with the EGRET instrument on the Compton Gamma Ray Observatory.[67] Clearly, there is something remarkable going on here.

7.7.1.2 Supermassive Black Holes: The Central Engine

7.7.1.2.1 The Center of the Milky Way. One of the first compact radio sources to be discovered during the early days of centimeter radio observations in the 1950s happens to have been the Galactic center, Sgr A. Radio interferometry shows that this region, the inner hundred parsecs, is a very bizarre and exciting place[68] and provides a glimpse at close range, about 8.5 kpc, of many of the phenomena that certainly occur in more distant active galactic nuclei. Our knowledge of the central part of our own system is severely limited by our location. The line-of-sight extinction is more than 25 magnitudes, so visual and ultraviolet observations are out of the question. This leaves a very wide spectral range that *can* be used, but high-dynamic-range observations at resolutions of better than one arcsec (about 0.05 pc) have been available only since the mid-1980's.

Observations of stellar proper motions near the center are now possible[69] with angular resolutions of better than 0.2 arcsec through the combined evolution of CCD detectors and developments in adaptive optics on large ground-based telescopes, an indication of how technological improvements have significantly impacted astrophysical studies. Using very short exposures and making use of speckle (see Chapter 2) structure in images, the diffraction limit in the infrared (in the K filter) can nearly be achieved and this will certainly improve in the coming years. The projected velocity curve for the inner 0.1 pc shows reaches 400 km s^{-1} yields a mass estimate of around 2.5×10^6 $M_\odot$ interior to 0.015 pc, an implied mass density so large as to be compatible only with a black hole. While this is not terribly large by the standards of some of the galactic nuclei we will discuss shortly, it is still enormous compared to stellar mass black holes.

Further evidence for some sort of activity at the center comes from observations of ionized gas in the inner 0.1 pc with the VLA and VLBA. The brightest region in centimeter wavelength images, Sgr A*, is a nonthermal source whose luminosity would not qualify the Milky Way as a normal radio galaxy but which, because of its proximity, dominates the region. The emission is similar to that seen in several relatively normal galaxies, notably starburst systems such as M82 and M83. Surrounding the compact central emission, there is a spiral of ionized gas whose radial velocities can be mapped with radio frequency hydrogen recombination lines. Molecular gas also pervades much of this region. But surely the most spectacular discovery in the galactic nucleus has been the Arc and related nonthermal filaments (Figure 7.13). These extend more than 20 pc from the central source,

---

[66] Punch, M. et al. 1992, *Nature*, **358**, 477; Gaidos, J. A. et al. 1996, *Nature*, **383**, 319.
[67] Mattox, J. R. et al. 1997, *ApJ*, **481**, 95. For a general survey of variability, see Ulrich, M.-H., Maraschi, L., and Urry, C. M. 1997, *ARAA*, **35**, 445.
[68] Comprehensive reviews of the observations are available in Morris, M. and Serabyn, E. 1996, *ARAA*, **34**, 645, and Mezger, P. G., Duschl, W. J., and Zylka, R. 1996, *Astr.Ap.Rev.*, **7**, 289.
[69] Ghez, A. M. et al. 1998, *ApJ*, **509**, 678.

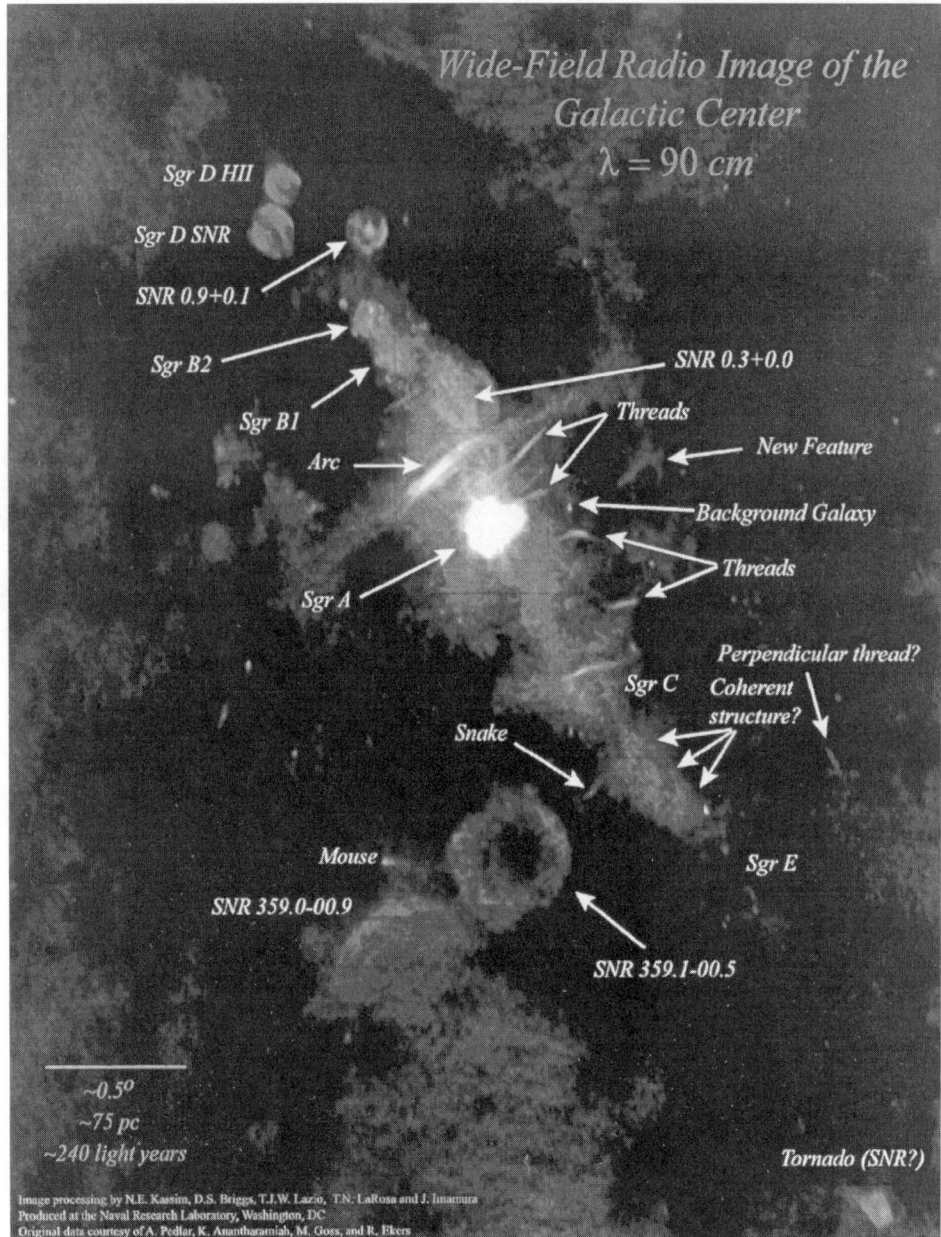

FIGURE 7.13. A 90-cm image of the galactic center region showing the principal thermal and nonthermal features [credit: Kassim, N. (NRL), Briggs, D. S. (NRAO; deceased), Lazio, T. W. J. (NRL), T. N. LaRosa (Kennesaw State Univ.), and J. Imamura (Univ. Oregon)]. At low frequency, not only the nonthermal filaments and H II regions are visible but also the detritus of continuing star formation, with numerous supernova remnants. Because of their sensitivity to the bulk of the aging nonthermal electrons, high resolution low frequency radio observations are superb for revealing significant details in active galaxies.

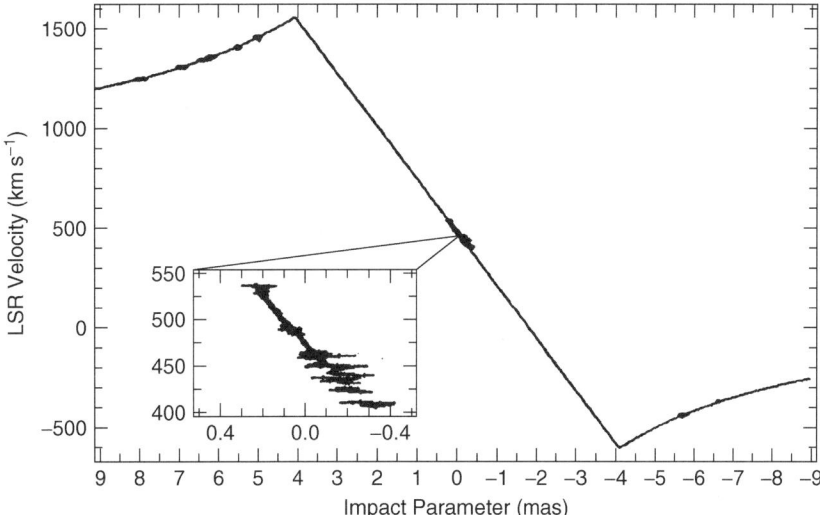

FIGURE 7.14. NGC 4258 masers [from Moran, J. M., Greenhill, L. J., and Herrnstein, J. R. 1996, in *Extragalactic Radio Sources: IAU Symp.* 175 (eds. Ekers, R. D., Fanti, L. and Padrielli, C.) (Dordrecht: Kluwer)]. The figure shows the line of sight velocities from two epoch observations (1994 and 1995). One mas corresponds to 0.035 pc for a galactic distance of 7.2 Mpc. The insert shows the central motion in greater detail; the fit beyond 4 mas corresponds to Keplerian motion.

are almost completely polarized, and show spectral indices of young (active) sources. In other words, one might expect to encounter this sort of structure in any starburst or Seyfert galaxy.[70]

7.7.1.2.2 Extragalactic: Dynamical Evidence. Soon after the discovery of Sgr A*, Woltjer[71] used this radio detection to support a view of strong magnetic fields and massive compact objects as the sources of nuclear activity in galaxies, *in particular* the broad lines found in Seyfert's list of galaxies. His argument was to set the basic paradigm, simply and nearly independent of a precise cosmological picture. The line forming region is confined to the inner 0.1 kpc or less (this is where a Hubble constant was required but by 1959 that was not seriously in error for order of magnitude arguments). The observed linewidths are upwards of 10^3 km s^{-1}. If due to either random or circulatory motion of a bound gas it requires that about 10^9 $M_\odot$ be contained within the core with a higher than stellar M/L ratio.

Limitations of ground-based optical resolution precluded an *experimentalis crucis* until quite recently. The most spectacular evidence yet obtained for this is the direct proper motion measurements of orbital motion from maser emission in the core of NGC 4258 (see Fig. 7.14).[72] Using VLBI observations with angular

---

[70] Ekers, R. D. et al. 1983, *Astr. Ap.*, **122**, 143; Anantharamaiah, K. R. et al. 1991, *MNRAS*, **249**, 262; Yusef-Zadeh, F., Morris, M., and Chance, D. 1984, *Nature*, **310**, 557; Yusef-Zadeh, F., and Morris, M. 1987, *ApJ*, **322**, 721; LaRosa, T. N. et al. 2000, *AJ*, **119**, 207; Figer, D. F. et al. 1999, *ApJ*, **525**, 750.
[71] Woltjer, L. 1959, *ApJ*, **130**, 38. See also the brief history of AGN research by Shields, G. A. 1999, *PASP*, **111**, 661.
[72] Miyoshi, M. et al. 1995, *Nature*, **373**, 127; Moran, J., Greenhill, L. J., and Herrnstein, J. R. 1998, *J. Astr. Ap.*, **20**, 165; Herrnstein, J. R. et al. 1999, *Nature*, **400**, 539. NGC 1068 has also been studied but with less reliable results.

resolution of higher than 0.05 mas (at 7.2 Mpc this corresponds to 0.0018 pc or 0.6×10^{16} cm), individual masers from the dusty nuclear torus can be observed through their *proper motion*, a purely geometric measurement. The resulting rotation curve shows Keplerian motion with 1500 km s^{-1} at a distance of about 0.1 pc, which *requires* a massive compact object, $(4.2 \pm 0.2) \times 10^7$ $M_\odot$. Only a central black hole fits the bill. More support comes from the radial velocity curve for gas in the inner regions of several radio galaxies, observations made possible with the high angular resolution available from HST. For M87, at distances between 0.07 and 1 arcsec (5–70 pc) the [O II] 3727 Å line observations with the Faint Object Camera (FOC) on HST give velocities of order 500 km s^{-1}, and the observed flow is consistent with orbital motion with Keplerian motion. In fact, the structure resembles the ionized flow detected around Sgr A*. The inferred mass for the central object is $(2.5 \pm 0.7) \times 10^9$ $M_\odot$.[73] Observations of stellar velocity dispersions using Mg I 5170 Å and the Ca II 8500 Å triplet as a function of galactocentric radius are also used to infer the presence of a compact massive central body, by looking for a central cusp in both the surface brightness and stellar velocity dispersion significantly exceeding the extrapolated galactic light distribution and rotation curve. This is weaker evidence since it samples only relatively large distances compared with space data, at least for current ground-based facilities, and suffers from some contamination by "seeing." The interpretation is also more difficult because of uncertainties in the large scale gravitational potential.[74] In all cases studied to date (2002), these masses lie between 10^6 and a few times 10^9 $M_\odot$ in galaxies ranging in type from ellipticals to spirals.

7.7.1.3 *Reverberation Mapping and Size of the Emitting Region*

Another method makes use of the response of emission from the BLR following a flare of the innermost region near the central engine. Called *reverberation mapping*, this technique looks at individual lines that are formed in slightly different parts of the circumnuclear region and correlates their variations, looking for time delays.[75] The basic idea is quite straightforward. Imagine a Keplerian accretion disk with $v_\phi \sim r^{-1/2}$. Then each frequency in the line profile maps to a different distance from the center. Allow the central source to flare. The observed response is a convolution of the light travel times to different places in the surrounding disk, allowing for projection effects:

$$F_\nu(t) = S_\nu \star \Psi_\nu = \int_{-\infty}^{\infty} S_\nu(t') \Psi_\nu(t - t') dt', \qquad (7.68)$$

where S is the flux of the exciting source and Ψ is called the *transfer function*. The method is like our treatment of scattering in a fog or light echos since at an instant in time you see the responses from different regions that have been excited at

---

[73] Harms, R. J. et al. 1994, *ApJL*, **435**, L35; Macchetto, D. et al. 1997, *ApJ*, **489**, 579.
[74] van der Marel, R. P. 1998, *AJ*, **117**, 744; Richstone, D. et al. 1998, *Nature*, **395**, A14; Gebhardt, K. et al. 2000, *ApJL*, **539**, L13.
[75] An accessible review of the basic methods is given by Peterson, B. M. 1993, *PASP*, **105**, 247. More recent extensions to X-ray lines are discussed in Reynolds, C. S. et al. 1999, *ApJ*, **514**, 164, who also explicitly include the Kerr metric for a rotating black hole. See also Krolik, J. H. (n.61).

different times because of the light travel time. In another context, the same technique has been used to study the emission ring around SN 1987 A in the LMC in an attempt to obtain the distance to the host galaxy. Since the explosion occurred at an instant in time, and the gas forms a ring that is not in the plane of the sky, the emission lines produced by the UV pulse rose at different times and rates depending on their location *relative to the observer*. If the lines locally decay identically everywhere, an observer sees differential time delays because of the finite size of the ring. In contrast, for the BLR, the source is unresolved and the line profile replaces angular location as the probe of the time delays. It may be necessary to include relativistic effects for the highest energy lines since they are expected to form closest to the central black hole, but for our purposes that is a detail. For periodic sources this would be completely ambiguous, but these events are quite individual. The response correlates with the ionization parameter, indicating that the light travel time to a particular part of the line forming region is obtained, and changes in the line profiles then correlate with the delay, giving masses that compare well with dynamical measurements.

7.7.1.4 *Central Engine Luminosities*

Additional support for nuclear black holes in active galaxies comes from the observed luminosities. If accretion generates the observed luminosity, then the limiting accretion rate is given by $L_{acc} = L_{Edd}$. Since a black hole's radius, the inner boundary of an accretion flow, depends only on its mass, then $L_{acc} \sim GM\dot{M}/R \sim \dot{M}$. As you well know, the Eddington luminosity depends only on the mass:

$$L_{Edd} = \frac{4\pi GMc^2}{\sigma_e}, \quad (7.69)$$

where σ_e is the Thomson scattering cross section. The luminosity produced by accretion within a disk depends only on the rate of release of gravitational potential energy, as we discussed in Chapter 5 for binary stars, so that

$$L = \epsilon \frac{GM}{R} \dot{M}, \quad (7.70)$$

where ϵ is the fraction of the binding energy that is radiated. For $L \approx L_{Edd}$, and assuming that the central object is a black hole, you obtain an estimate of the mass of the accreting body:

$$\dot{M} = \frac{8\pi G}{\epsilon \sigma_e c} M \approx 4.5 \times 10^{-3} \epsilon^{-1} \frac{M}{10^6 \, M_\odot} \, M_\odot \, \mathrm{yr}^{-1}. \quad (7.71)$$

The observed luminosities yield masses from 10^7 to a few times $10^9 \, M_\odot$, and the derived accretion rates are upwards of $M_\odot \, \mathrm{yr}^{-1}$. Of course, the derived accretion rate depends on the ϵ. For advection-dominated accretion flows the radiative efficiency is substantially reduced, which leads to an underestimate of the rate at which the central engine must be supplied with mass in order to achieve the observed luminosity. Once better constraints have been obtained from stellar mass black holes in well calibrated binary systems, we will also better understand AGN processes.

We can now apply the discussion of binary star disks to the galactic scale. The principal difference is the timescale because of their sizes. Feeding of the disks is more problematic than for binary systems, for which the source of gas is readily at hand. Recall that molecular clouds are very low density and comparatively fragile. Therefore, orbiting clouds coming within the equivalent of the Roche limit from a central black hole will be disrupted by tides. Stars, too, should be tidally disrupted. The matter carries a specific angular momentum $R_{RL}v_{orb}$ at this distance, or in other words $j = M_c^{1/3} M_{BH}^{1/6} a^{1/2}$ for a semimajor axis a of the cloud orbit, where M_c is the mass of the cloud and M_{BH} is the mass of the central black hole. For accretion to occur requires that this angular momentum be viscously dissipated with the attendant heating of the disk. Applying the arguments from Chapter 5, this means the disk emits about one-half of the energy derived from accretion with a featureless power-law spectrum having a spectral index of around 1/3. Models for these disks are still under development, but most are still dominated by Planck law distributions. Yet optical observations show strong Fe II and [Fe II] fluorescent emission and, for the most heavily shrouded nuclei such as IRAS galaxies, strong broad iron peak element atomic absorption bands in the ultraviolet. The precise treatment of these spectra requires NLTE treatments similar to those used for stellar winds.

7.7.2 Radio Galaxies

At some level, all galaxies are radio emitters due to the spiraling of cosmic ray electrons in their interstellar magnetic fields. Remember, synchrotron emission is one of the main energy loss mechanisms for high-energy particles; the other is inverse Compton scattering. There are, however, distinctions between the normal emitters and those that have been designated as *radio* galaxies, which relate both to their power and structure. To the extent that the subject can be treated separately from the interstellar magnetic fields and cosmic rays, we will here concentrate on the application of specific methods to the analysis of large-scale radio emission that is best studied in external galaxies.[76] Keep in mind, however, that anything that happens in other galaxies stands a very good chance of happening at some level in the Milky Way.

7.7.2.1 Minimum Energy: Estimating Properties of the Emitting Regions

Any synchroton emitting region is a cauldron containing a brew of high energy particles, radiation, and magnetism. These components are not independent. The source luminosity depends on the energy density in the relativistic particles and the magnetic field strength and the radiation energy density affects the electron

---

[76] You can find detailed discussions of the diffusive shock acceleration mechanism in, for example, Bell, A. R. 1977, *MNRAS*, **179**, 573; Blandford, R. D. and Ostriker, J. P. 1978, *ApJL*, **221**, L29; Blandford, R. D. and Ostriker, J. P. 1980, *ApJ*, **237**, 79 Blandford, R. D. and Eichler, D. 1987, *Phys. Rep.*, **154**, 1; Drury, L. O'C., Duffy, P., Eichler, D., and Mastichiadis, A. 1999, *Astr. Ap.*, **347**, 370; Protheroe, R. J. and Stanev, T. 1999, *Astropart. Phys.*, **10**, 185.

lifetimes and therefore the luminosity of the source. The dominant factor is the magnetic field. The luminosity is limited by how large the field can be for a source of known volume, V. The argument[77] goes something like this. The source brightness depends on the population of relativistic electrons, their spectral index, and the magnetic field strength. In turn, for a specified spectrum and an observed luminosity and source volume, there is a favored field strength: the one that minimizes the total energy of the emitting region. For a power-law energy spectrum, $N(E) = N_0 E^{-p}$, the energy in relativistic electrons, $U_{\rm rel}$, is given by

$$U_{\rm rel} = \int_{E_1}^{E_2} N(E) E \, dE = \frac{N_0}{2-p} \left(E_2^{2-p} - E_1^{2-p} \right)$$

for a pair of energy cutoffs (E_1, E_2). The spectral index is assumed to be constant. The source luminosity is the integral of the power radiated by a single electron folded into the ensemble energy spectrum. Therefore, using our discussion in Chapter 6, the energy loss rate is

$$\Lambda(E,t) \equiv \frac{dE}{dt} = -\beta B^2 E^2, \tag{7.72}$$

where β is a constant. Then the source luminosity is given by

$$L = \int_{E_1}^{E_2} N(E) \Lambda \, dE = \frac{N_0}{3-p} \beta B^2 \left(E_2^{3-p} - E_1^{3-p} \right). \tag{7.73}$$

You observe the source at one frequency so we can remove the energy by taking $E_j^2 = \omega_j B^{-1}$ for the two limiting frequencies (or energies) in the integral. Then solving for N_0 in terms of the luminosity, we get

$$U_{\rm rel} = C_{\rm rel} \frac{L}{B^{3/2}}, \tag{7.74}$$

where we have absorbed all of the constant factors into $C_{\rm rel}$. The magnetic energy is:

$$U_{\rm mag} = C_B B^2 V, \tag{7.75}$$

where the constant C_B depends on the source geometry, and $U \equiv U_{\rm rel} + U_B$ therefore becomes

$$U = C_{\rm rel} B^{-3/2} L + C_B B^2 V. \tag{7.76}$$

We now assert that we are most likely to observe the magnetic field strength that

---

[77] Burbidge, G. R. 1958, *ApJ*, **127**, 48; see also Pacholczyk, 1970, *Radio Astrophysics* (San Francisco: Freeman); Moffatt, A. 1975, in *Galaxies and the Universe: Stars and Stellar Systems*, vol. 9, Sandage, A., Sandage, M., and Kristian, J., eds. (Chicago: Univ. Chicago Press).

minimizes the total energy within a specified volume given its luminosity, so

$$\left(\frac{\partial U}{\partial B}\right)_{V,L,N} = 0$$

yields the corresponding magnetic field strength:

$$B_{\min} = \left(\frac{3C_{\text{rel}}}{4C_B}\frac{L}{V}\right)^{2/7}. \tag{7.77}$$

Notice that the constants depend on the power law index and the geometry of the emitting region but on substituting $B_{\min}$ into Eq. (7.76), we find two remarkable relations that are *independent of these constants*:

$$U_B = \tfrac{3}{4}U_{\text{rel}}, \qquad U_{\min} = \tfrac{7}{4}U_{\text{rel}}.$$

The minimum pressure, which is proportional to the energy density, scales as $u_{\min} \sim (L/V)^{4/7}$. One more bit of dimensional analysis can help explain this result. Notice that the radiative energy density is related to the flux by $F = u_{\text{rad}} c$ so the observed angular size of the source, θ, and its distance, D, give the scaling relation $B_{\min} \sim F^{2/7}\theta^{4/7}$. The minimum *pressure* for the emitting region is also provided by $B_{\min}$. The observer must only know the spectral index, angular diameter, and redshift of the source. Although the geometry matters for the precise result, the basic scaling relation depends only on the luminosity density.

Notice that the field strength determined with this argument is mainly due to the lowest-energy part of the electron spectrum. This is an expected byproduct of the input electron energy distribution since we assumed a power law. As the particles radiate and "age," they shift to the lowest energy part of the spectrum, steepening the spectral slope at higher frequencies and populating the lower part of the energy spectrum. There a direct relation between the magnetic field and number of particles needed to produce the observed luminosity at any frequency. It is a direct consequence of the fact that synchrotron radiation is a relativistic emission mechanism. This argument does not, and *cannot*, treat the nonradiating particles that may support bulk motion of the medium. For example, for extragalactic jets, protons are almost[78] certainly present and may possess the largest fraction of the kinetic energy density. From the observed synchrotron luminosity, it is also possible to place a firm *lower* limit on the size of the emitting volume for any field strength through the inverse Compton effect. An ensemble of energetic electrons can scatter its own emitted photons if the radiation energy density becomes high enough. This is called the *synchrotron self-Compton process*, when the

---

[78] We say *almost* here because there are no direct means for assessing their contribution other than vague arguments about charge balance. In so-called heavy jets, the fluid is neutralized by protons. In light jets, positrons provide charge balance, but these also radiate and so are easier to assess. One argument that limits the proton contribution is the γ-ray luminosity of the source. Pions decay as $\pi \to 2\gamma$ after their creation during relativistic proton collisions. Several AGNs, especially BL Lac objects, have been discovered by EGRET to be strong high-energy γ sources, so there is some observational support for at least some proton contribution near the galactic center.

mean free path for Compton scattering of the self-generated radiation becomes small compared to the size of the emission region.

Optically thick *compact core* sources show their presence by a low-frequency turnover in the spectrum. As with any emitting region, if the opacity is high enough the photons will become trapped but since the electrons are nonthermal the resultant spectrum depends differently than a Planck law on frequency. We covered the precise equations in Chapter 6, now we'll just use the scaling relations. For a uniform source of thickness L, the optical depth is varies as

$$\tau_\nu \sim N_0 B^{(p+2)/2} \nu^{-(p+4)/2} L \qquad (7.78)$$

and the flux scales as

$$S_\nu \sim B^{-1/2} \nu^{5/2}. \qquad (7.79)$$

Thus the signature of an optically thick *compact* source is a positive spectral slope in frequency, and the optical depth can be obtained from the spectral turnover low frequency. The magnetic field is constrained by the minimalization arguments, and so is N_0.

7.7.2.2 Unsteady Synchrotron Sources: Time Evolution

In our discussion of synchrotron radiation in Chapter 6, we dealt only with the time-independent emission mechanism. There are, however, reasons for thinking that the lifetime of the emitting particles must also be considered when dealing with extragalactic sources. The propagation times are very long because of the enormous sizes seen for galaxy-scale radio sources. When observing the resolved structures, it is possible to obtain bounds on the change in the spectral index of the electrons and, from this, get some information about the acceleration mechanism. The lifetime of the electrons producing the emission can be estimated from the synchrotron spectrum using Eq. (7.72) and the radiation energy density from Eq. (7.74). If the initial electron energy is E_0, then integrating Eq. (7.72) gives the energy of a single particle as a function of time:

$$E = \frac{E_0}{1 - \beta B^2 E_0 t}. \qquad (7.80)$$

This equation defines the *synchrotron lifetime* of the particle:

$$t_{syn} = \left(B_{\mu G}^2 E_{0,\text{GeV}}\right)^{-1} \text{ yr} \qquad (7.81)$$

corresponding spectral cutoff frequency:

$$\nu_c = 1.61 \times 10^4 B_{\mu G} E_{0,\text{GeV}}^2 \text{ GHz}. \qquad (7.82)$$

The radio spectrum changes systematically with time as electrons age and move through the distribution function to lower energies. The total number of particles, however, does not change, so the brightness maximum simply shifts to longer wavelength. The decreased radiation efficiency at lower frequency accounts for the steady increase in the synchrotron lifetime and the lower surface brightness of the source. The highest energy end of the continuum should gradually disappear with a corresponding increase in the low frequency emission.

Aging of the source can be used as a measure of the time since injection for the particles, and can be compared with the light travel time for extended sources. This is where the difficulty arises. Extending over scales from 100 kpc up to 1 Mpc, the jet propagation times are at least a few hundred thousand years but for equipartition fields, a few μG, the lifetime of an electron radiating at centimeter wavelengths is only a few thousand years at most. The synchrotron spectra observed for radio lobes *and* jets are too hard—that is, the spectra have a greater proportion of high-energy emission than would be expected if the electrons are merely radiating and losing energy as they traverse the distances to the lobes. It appears instead that some sort of reacceleration mechanism is required to maintain the observed emission.

To see the argument in more detail is instructive. The general development of the electron distribution can be treated using an energy-dependent Fokker–Planck equation. Assume that the number of electrons with energy E at time t is $N(E,t)$ and that the rate of injection of newly energized electrons is $S(E,t)$. Then

$$\frac{\partial N(E,t)}{\partial t} - \frac{\partial}{\partial E}[\Lambda(E,t)N(E,t)] + \frac{\partial}{\partial E}\left[\Gamma(E,t)\frac{\partial N(E,t)}{\partial E}\right] = S(E,t), \tag{7.83}$$

where Λ is the single-electron emission (loss) rate [Eq. (7.72)]. Reacceleration of particles is treated through the $\Gamma(E,t)$ term if it is stochastic, as in Fermi acceleration. External injection is represented by a source term, $S(E,T)$. If the initial spectrum is a power law with spectral index p, then in the absence of sources the distribution evolves as

$$N(E,t) = \frac{N_0 E^{-p}}{(1 - \beta B^2 E t)^{p-2}}. \tag{7.84}$$

This calculation presupposes no losses other than synchrotron and inverse Compton scattering, both of which vary as E^2. Mechanical processes such as expansion of ejected parcels of electrons, of the sort observed with VLBI observations of compact radio sources and BL Lac objects, change the energy density of the magnetic field and decrease the losses: expansion of a plasma decreases the energy density of the distribution, in effect producing a cooling, which must also be included in the evolution of the source.

Re-energizing of the electrons in radio sources is usually linked to shock processes via a Fermi acceleration and reconnection, much as in solar and stellar flares and in the magnetotail of the Earth. It is not presently known whether turbulence within the jets and lobes can generate shocks, but structures observed in radio jets are very suggestive of such effects. For instance, the best observed nearby jet, the ever popular M87, shows several distinct radio brightenings that correspond to optical and UV structures. This suggests that the electrons are being reaccelerated to very high energy by shocks when the flow changes. The source of the flow is elsewhere. It must be seated deep in the parent galaxy, so we now turn our attention to the next class of extragalactic objects, active galaxies.[79]

7.7.2.3 Jets and Lobes

From the first, radio observations showed extended structures to be normal features of emitting galaxies. Only with the advent of interferometers, however, did the nature of these structures become clear. The appearance of the extended structures[80] seem to separate morphologically into two broad classes at a luminosity of approximately 10^{32} erg s^{-1} Hz^{-1} (or 10^{25} W Hz^{-1}). Below this luminosity, those galaxies that possess radio lobes show diffuse lobe structure with the edges of the emission region fading into the intracluster medium. These are the *FR I* sources; M87 in Virgo is an example of this group. In contrast, FR II sources display brighter leading edges, with Cyg A being the most (in)famous example and also the closer galaxy Pic A.[81] The FR I sources are usually associated with optically brighter ellipticals than FR II and, because of the lower luminosity of the extended emission, they have relatively stronger central sources at the same total radio power. Most frequently, the jets associated with radio galaxies are asymmetric. That is, in general only one is seen to actually connect to the lobes, although two extended sources are observed. The largest structures extend to about 1 Mpc, yet there is no significant spectral aging along the jets. Clearly some sort of re-energizing is required and this can be due to shocks within the jets or tangled magnetic fields, much the same as we described or supernova remnants. Episodic relativistic ejections from the galactic nucleus provide possible sources to power the electrons, and this picture is consistent with the events observed closer to home in the galactic sources SS433 (with bulk velocities of order $0.25c$), and the galactic so-called "miniquasars" GRS1915 + 105, and GRSJ1655-40. The low mass X-ray binary Sco X-1 seems to be even more extreme in having both jets and what appear to be radio lobes associated with an active central source.

7.7.2.4 Superluminal Motions and Beaming

Although it is taken for granted that the speed of light is an ultimate limit for material transport, very high spatial resolution observation of extragalactic radio sources occasionally show evidence of expansion of emission regions at much greater speeds. Superluminal motions exceeding 10c have been detected in several quasars for which distances seem to be secure, and even the nearest ones for which

---

[79] Kardashev, N. S. 1962, *Sov. Astron.—AJ*, **6**, 317; van der Laan, H. and Perola, G. C. 1969, *Astr. Ap.*, **3**, 468; van der Laan, H. 1966, *Nature*, **211**, 1131; Pacholczyk, A. G. 1970, *Radio Astrophysics* (San Francisco: Freeman); Pacholczyk, A. G. 1977, *Radio Galaxies* (NY: Pergamon); Eilek, J. A., and Shore, S. N. 1989, *ApJ*, **342**, 187. A lovely example of time dependence and multiscale analysis of a variable source, 3C120, is found in Walker, R. C., Benson, J. M., and Unwin, S. C. 1987, *ApJ*, **316**, 546. See also Bridle, A. H. and Perley, R. A. 1984, *ARAA*, **22**, 319; Begelman, M. C., Blandford, R. D., and Rees, M. J. 1984, *Rev. Mod. Phys.*, **56**, 225.

[80] Fanaroff, B. L. and Riley, J. M. 1974, *MNRAS*, **167**, 31P.

[81] For Cyg A and further references, see, for example, Carilli, C. I., Bartel, N., and Diamond, P. 1994, *AJ*, **108**, 64. For Pic A, a detailed study of the VLBI scale jet is Tingay, S. J. et al. 2000, *AJ*, **119**, 1695. For M87, see, for example, Biretta, J. A., Stern, C. P. and Harris, D. E. 1991, *AJ*, **101**, 1632 and Owen, F. N., Hardee, P. E. and Cornwell, T. J. 1989, *ApJ*, **340**, 698.

the parent galaxy can be observed, such as 3C120 and 3C273, show this. Optical observations are possible for only one source, M87, but here too the observational basis is secure for motions up to about $6c$.[82] In general the sources showing radio evidence for superluminal motions are FR II, although a number of jets show detectable *subluminal* blobs, notably M87, 3C338, and Cen A. Since we can confidently assume that these motions are not evidence of bulk motion faster than c, the observed velocities must be more apparent than real. Two explanations have been proposed: the motions are really a phase effect, or the matter is moving relativistically in bulk but the emission region is neither straight nor transverse to the line of sight. The first explanation is like the motion of lights on a marquee of a cinema. There is no direct causal connection between the emission regions, but the timing is such as to produce an apparent motion. If you see a chain of caution barriers on a highway, for instance, the effect is immediately obvious. The individual flashers have their own timing and, depending on the phase of any one light, the pattern of uncorrelated flashes may appear to go in either direction. Observationally, contracting motion has never been observed so this explanation fails. The second explanation is more subtle. Assume that the emission is due to particles moving near the speed of light along curved paths and that the events are organized into bursts. Then the time delay of the emitting regions produces a superluminal motion because of the change in the orientation relative to the observer through a $\beta \cdot \hat{n}$ projection, and the time delay is γc. From this, you have an estimate of both the curvature of the trajectory *and* the energy for the emitting electrons. This explanation is supported by the curvature observed in the inner regions of many radio jets, especially the sources we have just noted. It isn't only extragalactic jets that show relativistic motions. A few notable *galactic* sources, the miniquasars as they're usually called, show speeds up to $c/2$ (e.g., the jets of Sco X-1 and SS 433).

From our discussion of moving frame radiative transfer, you know that a relativistic source radiates into a compressed solid angle in the laboratory frame. From the Lorentz transformation, this is $\delta\theta \sim \gamma^{-1}$. Relativistic motion also produces the following result for the apparent transverse velocity for a source moving with a velocity $v = \beta c$ at an angle θ relative to the observer:

$$v_t = \frac{v\sin\theta}{1 - \beta\cos\theta}. \qquad (7.85)$$

The radiative intensity is thus increased by a factor of

$$I = I_0[\gamma(1 - \beta\cos\theta)]^{p-3}, \qquad (7.86)$$

a phenomenon referred to as *Doppler boosting* (recall the factor of ν^3 in the intensity equation we discussed for moving frames in Chapter 3). This is usually invoked as the explanation for one-sided radio jets. The combined effects of beaming and reduced brightness of the oppositely directed jets accounts for their large intensity ratio. For one radio galaxy, M87, both optical and radio measure-

---

[82] Kellerman, K. I. et al. 1999, *New Astr. Rev.*, **43**, 757; Biretta, J. A., Sparks, W. B., and Macchetto, F. 1999, *ApJ*, **520**, 621.

ments are available. These provide strong support for beaming since the ratio is > 150 for centimeter and > 450 for optical data.

7.7.3 Interacting Galaxies

Unlike stars, galaxies are large objects that have sizes often comparable to the distances between them, especially in clusters. Consequently, tidal interactions play an important role in their dynamical evolution. Recall that the Roche surface depends on the mass ratio and separation of the systems. Binary galaxies are fragile analogs of binary *stars* so we can use some of the ideas already developed for tidal interactions. The tidal limit is a function of the mass ratio, q, and the separation in much the same way that it is for stars. The one difference is that the mass distribution of galaxies is very extended and the point source approximation does not strictly apply. However, you would expect tides to be important when $R_{gal} \sim (q/1+q)^{1/3} a$, where a is the separation. Among the phenomena associated with galaxy collisions, the most dramatic are large-scale tails generated by the tidal interactions, polar rings, rings, and shells. Each depends on the details of the collision, in particular the impact parameters and types of galaxies involved.

Models for tidal interactions[83] have become increasingly sophisticated with larger numbers of particles, but the basic features have not changed. Spiral waves are excited in disks when the characteristic crossing time for the system is approximately the local epicyclic frequency so that $\Delta t_c \approx \kappa^{-1}$. The process of disk distortion is relatively simple. The perturbing galaxy produces an instantaneous tide in the target galaxy that follows it in the outermost parts of the perturbed system. The radial impulse depends on the distance of closest approach, as does the torque. Since the collision is typically at some distance rather than head on, and it can be in any plane relative to the symmetry plane of the target, orbital angular momentum is altered for the stars. They are displaced by an amount δr from their equilibrium positions and thereafter oscillate with the local frequency. If the impulse is great enough, the stars can escape from the system. Since their angular momentum is altered, they follow three-body trajectories relative to the disturbed galaxies. Note that *both* galaxies are disturbed by the interaction and, depending on the mass ratio, one or both may not survive the process. Prograde hyperbolic encounters torque the systems and produce trailing spiral tails, more nearly radial encounters can lead to rings and plumes or shells, depending on whether the encounters are central or eccentric. Because most collisions take place on timescales that are shorter than the orbital times for stars within relatively low mass galaxies, tidal forces restructure individual orbits and heat the disk. As a result, the anisotropy of the force and the velocity dispersion of the stars combine to eject some stars from the galaxies, each of which follows a trajectory that is determined by the instantaneous tidal force.

One of the signatures of interaction between galaxies is the galaxies, including our own, show radial changes in the thickness and symmetry plane of their disks.

---

[83] Toomre, A. and Toomre, J. 1972, *ApJ*, **178**, 623; Athanassoula, E. and Bosma, A. 1985, *ARAA*, **23**, 147; Barnes, J. E. and Hernquist, L. 1992, *ARAA*, **30**, 705; see also Dubinski, J., Mihos, C., and Hernquist, L. 1999, *ApJ*, **526**, 607 and references therein.

You can think of this as a result of the geometry of the encounter. In the Milky Way, the evidence comes from the change in the plane of symmetry from 21-cm observations of the outer Galaxy in the northern and southern hemispheres. For extragalactic systems, the disks show tilting of the isovelocity surfaces with increasing galactocentric distance. If the warp is dominated by the gravitational field of the central mass, the individual annuli precess and, with time, the warp should smear. The persistence of, especially gaseous, warps suggests that the orbits are stabilized, perhaps by a massive dissipationless dark halo. The vertical oscillation frequency for a disk is determined by the local gravitational acceleration.[84]

The Magellanic Stream is further evidence of the tidal interactions that probably happens between all galaxies that lie close enough to each other, a point to which we will return when discussing clusters. The Stream covers a large swatch of sky, about 10° wide, that surrounds both Magellanic Clouds and is well observed in both 21-cm and metallic absorption lines.[85] These H I observations also show that the Magellanic Clouds are also tidally interacting with each other. This is one possible source for diffuse material in clusters, the sort that could then replenish the intracluster medium with metals. Keep this in mind for the discussion ahead.

Polar ring galaxies are systems that appear to have undergone recent collisions (an example of such a system is shown in Fig. 7.15). These are disk or lenticular galaxies that are surrounded by rings that lie perpendicular to the disk. The stability of the rings depends on their orientation since they should precess if not precisely orthogonal to the galaxy's symmetry axis, much like a gyroscope. In general, they cannot be longlived unless they are self-gravitating although the mass of the rings is still open to debate.[86] The inferred H_2 masses range over $10^8 - 10^{10}$ $M_\odot$ with an average, 10^9 $M_\odot$, that is, about the same as normal spirals.

Ring galaxies are particularly remarkable collisional remnants. By "ring galaxies" we mean a type that first showed up in studies of peculiar galactic morphologies. These are systems that appear much like smoke rings on images. They are disk galaxies that show extended (often elliptical) rings with significantly higher surface brightness than the disk. The rings are usually disconnected from the nuclear region, and there is no obvious spiral structure. They were first noticed by Zwicky in the 1940s as a distinctive type of galaxy, but further work did not occur until the 1970s, when stellar hydrodynamic models began to be applied to galactic structure.

---

[84] Binney, J. 1992, *ARAA*, **30**, 51; Casertano, S., Sachett, P., and Briggs, F. eds. 1991, *Warped Disks and Inclined Rings Around Galaxies* (NY: Cambridge Univ. Press). Stellar proper motions have been used to determine the Galactic warp by Smart, R. L., and Lattanzi, M. G. 1996, *Astr. Ap.*, **314**, 104. The excitation of a disk warp for the Milky Way through interaction with the Magellanic Clouds has been simulated by, for example, Weinberg, M. D. 1995, *ApJ*, **445**, 31. For thick disk warps, see Masset, F. and Tagger, M. 1996, *Astr.Ap.*, **307**, 21. Bottema, R. 1996, *Astr.Ap*, **306**, 345 present a detailed analysis of a particularly well observed warped system, NGC 4013. See also Jore, K. P. et al. 1996, *AJ*, **112**, 438, and Sancisi, R. 2000, *Ap. Space Sci.*, **269**, 56 for discussions of warped and retrograde disks in spirals.

[85] The discovery paper is Mathewson, D. S., Cleary, M. N., and Murray, J. D. 1974, *ApJ*, **190**, 291. The most complete views are spectacularly shown by Putman, M. E. et al. 1998, *Nature*, **394**, 752 who also report the discovery of counterstreams.

[86] The standard catalog is Whitmore, B. et al. 1990, *AJ*, **100**, 1489. Neutral hydrogen surveys are available in Huchtmeier, W. K. 1994, *AJ*, **107**, 99, 1996; Huchtmeier, W. K. 1994, *Astr.Ap*, **319**, 401; and van Driel, W. et al. 2000, *Astr.Ap.Suppl.*, **141**, 385 and references therein. Molecular gas is surveyed by Galletta, G., Sage, L. J., and Sparke, L. S. 1997, *MNRAS*, **284**, 773. Yakovleva, V. A. 1996, *Astr.Ap*, **314**, 729 performed one of the most complete photometric studies of a forming ring system.

FIGURE 7.15. WFPC2 image of NGC 4650A, a polar ring galaxy (credit: NASA/STScI).

718 OUR GALAXY AND OTHERS AS STELLAR SYSTEMS

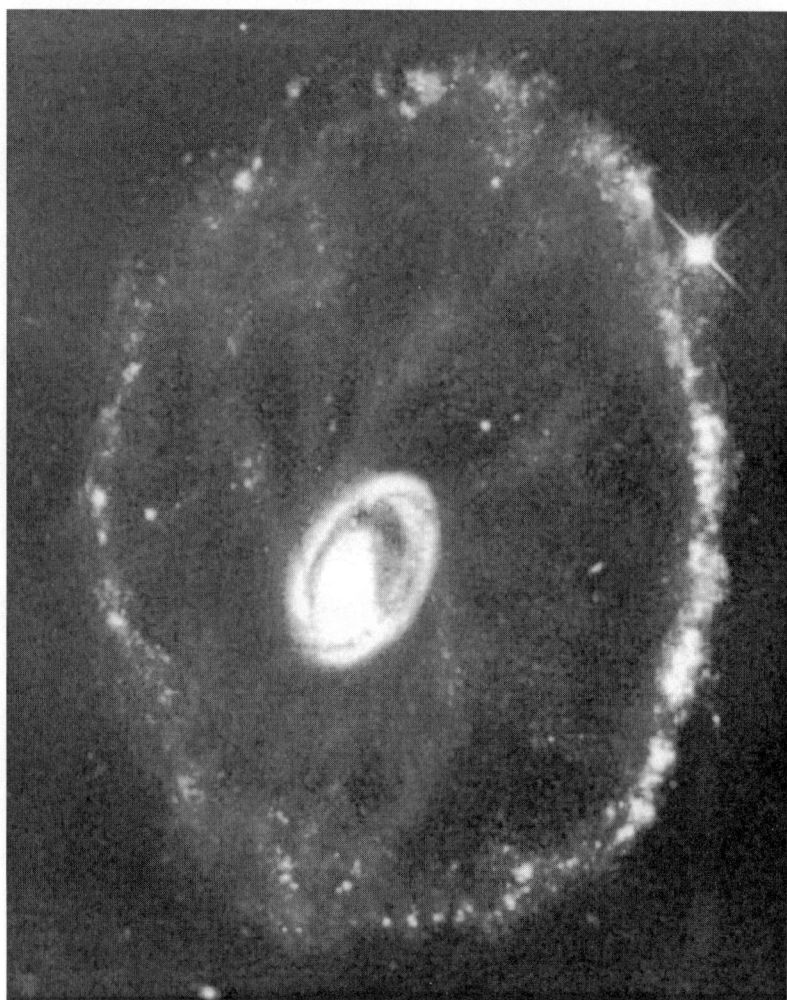

FIGURE 7.16. Optical WFPC2 image of the classical ring galaxy VV784 (the "Cartwheel") (credit: NASA/STScI).

The prototypical system, called the "Cartwheel" (VV 784 = ESO 350-G040, shown in Fig. 7.16).[87] They are thought to be the product of strong galactic collisions, essentially head-on, where the response of the disk is much like a splash. Inward and outward propagating density waves in the gas are thought to trigger enhanced star formation within the ring, similar to the observations. The disk response in a perpendicular collision is qualitatively described as a sloshing motion that settles down as stars are lost and angular momentum is transferred through density waves. Warping is also produced by the collision, arising from the orientation of

---

[87]See, for example, Theys, J. and Spiegel, E. A. 1976, *ApJ*, **208**, 650; Higdon, J. L. 1995, *ApJ*, **455**, 524; Charmandaris, V. et al. 1999, *Astr.Ap.*, **341**, 69.

the target galaxy's disk relative to the line of centers. Simulations[88] demonstrate that gaseous disk responses, with rather simplified assumptions about the stimulation of star formation, can produce most of the observed phenomena. These are only the most extreme such examples of disk response. Much less dramatic examples are probably seen in the stimulation of density waves and enhanced star formation from tidal encounters.

Mergers may be responsible for the formation of ellipticals and certainly occur in a large variety of systems. Disrupted disk systems, an example of which is NGC 7252, show counterrotating disks that clearly indicate the absorption of a colliding gas-rich galaxy. Other such systems are associated with both starburst and nuclear radio activity, for instance, Cen A and NGC 1275. The stellar populations intermix, although they may retain their distinct dynamical signatures for some time (such as counterrotating disks and streaming). In the very long term, these stars will be indistinguishable. But since the disk is collisionless and merging injects stars with a dispersion and motion very different from those formed from the galactic gas, they should be discernible.

Finally, when the perturber is completely disrupted, it produces shells that oscillate around the primary galaxy. The observations of extended multiple shells around several cD and giant elliptical galaxies stimulated important dynamical modeling of the head-on collision between low-mass disk galaxies with massive ellipticals with very striking results.[89] Key to the difference between these systems and ring galaxies is that the perturber is a low-mass disk for which the Roche surface recedes deep into the system during the encounter. The angular momentum launches the stars along single-particle three-body trajectories that have apogalactic points lying along the shells, which may the turning points in the orbits.

In light of these discussions, we can ask whether the Milky Way has evolved independently or has experienced mergers during its history.[90] The bulk of halo stars show a well-defined turn-off at B–V $\approx$ 0.4 but a fraction, less than 10% are bluer than this limit and suggest that some merging may have taken place. The disk population of globular clusters may also have been contributed by such an event. It is important to discuss this idea for the evolution of the system, in order to highlight the problems with the simple isolated evolution picture for the disk. The dichotomy of *nature versus nurture* may not be as easily resolved for galaxies as it is for stars.

7.8 CLUSTERS OF GALAXIES

Galaxies are organized into large scale structures much like stars, ranging from compact groups to massive clusters. These were first cataloged by G. Abell based

---

[88] Barnes, J. E.. and Hernquist, L. 1992, *ARAA*, **30**, 705; Appleton, P. and Struck-Marcel, C. 1995, *Fund. Cosm. Phys.*, **16**, 112.

[89] Some of the fundamental observational papers are Schweitzer, F. 1980, *ApJ*, **237**, 303; Schweitzer, F. 1986, *Science*, **231**, 227. For modeling of disk and spheroidal collisions with massive ellipticals, see Quinn, P. J. 1984, *ApJ*, **279**, 596; Dupraz, C. and Combes, F. 1986, *Astr. Ap.*, **166**, 53.

[90] Unavane, M., Wyse, R. F. G., and Gilmore, G. 1996, *MNRAS*}, **278**, 727.

on the Palomar Sky Survey plates[91] who defined a *cluster* as an aggregate of galaxies of at least 50 members that are within 2 magnitudes of the third brightest member, a statistical choice that was intended to reduce the fluctuations expected from relatively small samples. The Bautz–Morgan classification makes taxonomic cuts based on the presence (type I) or absence (type III) of a central cD galaxy.

7.8.1 The Local Group

This is home. The Milky Way is one of the two massive members, along with M31, of the small group of galaxies called the *Local Group*[92] (see also Table 7.5[93]). The important feature of this group is its completeness. The Local Group is the only region for which we are able to obtain, in principle, a complete sample of the luminosity function. The question of how representative this is will be dealt with in a moment. The importance of this sample is its sensitivity to the lowest end of the luminosity distribution. Galaxies with absolute magnitudes of -9 or -10, about the same luminosity as a Luminous Blue Variable (!), are on the lowest end of the mass distribution and overlap with globular clusters. At the lowest luminosity end, these systems also provide the most accessible examples of how star formation proceeds in self-gravitating low-density systems.

7.8.2 Applications of the Virial Theorem to Clusters

The virial theorem provides the main tool for estimating the mass of a cluster of galaxies, although the procedure is a bit more problematic than it is for star clusters. In principle, the virial equilibrium of galaxy clusters is a few-body problem —most clusters consist of relatively few hundred or thousand members, and often the gravitational potential is dominated by one or a few cD galaxies whose location at the cluster center structures the motion of the rest. In practice, as we've said, the situation is more complex. Galaxies, unlike stars, are large relative to their separations and collide. They are also composite bodies, consisting of stars and gas, and genuinely dissipative as well as chaotic. When a collision removes stars from a galaxy, these cool the system and promote merging, while the gas can

---

[91]Abell, G. O. 1958, *ApJS*, **3**, 211. This work was inspired by theoretical arguments by Neyman, J. and Scott, E. 1952, *ApJ*, **116**, 144 and the first attempt to look for clustering based on large galaxy catalogs by Neyman, J., Scott, E., and Shane, C. D. 1954, *ApJS*, **1**, 269. The third *Berkeley Symposium on Mathematical Statistics and Probability* served as a forum for assembling many of the principals in this work, and you will find the proceedings (1956) of this conference to be inspiring reading. See also the first comprehensive review by Abell, G. O. 1965, *ARAA*, **3**, 1. A classification based on populations was proposed by Bautz, L. P. and Morgan, W. W. 1970, *ApJL*, **162**, L149. To get a sense of the development of ideas concerning the structure and evolution of clusters of galaxies, it is instructive to pursue the evolving reviews on the subject by Bahcall, N. A. 1977 *ARAA*, **15**, 505; Dressler, A. 1984, *ARAA*, **22**, 185; Bahcall, N. A. 1988, *ARAA*, **26**, 631.

[92]The wealth of information available on these galaxies is illustrated by the fact that several specialized monographs have appeared that deal with them individually. See Westerlund, B. E. 1997, *The Magellanic Clouds* (Cambridge, UK: Cambridge Univ. Press); Hodge, P. W. 1992, *The Andromeda Galaxy* (Dordrecht: Kluwer); van den Bergh, S. 2000, *The Galaxies of the Local Group* (Cambridge, UK: Cambridge Univ. Press). A significant review of the low luminosity end of the population is Mateo, M. 1998, *ARAA*, **36**, 435.

[93]Table 7.5 is based on compilations by M. J. Irwin (Institute for Astronomy, Cambridge Univ.) and van den Bergh, S. 1999, *Astr. Ap. Rev.*, **9**, 273, with an update in van den Bergh, S. 2000, *PASP*, **112**, 529.

TABLE 7.5 Members of the Local Group

Name	Catalog	α (B1950)	δ (B1950)	Type	D (kpc)	M_V	V_0 (km s^{-1})
Galaxy	—	17:42.4	−28:55	Sbc	—	−20.6	—
Sgr	—	18:51.9	−30:30	dE7	24	−14.0	140
LMC	—	05:24.0	−69:48	Irr	50	−18.1	270
SMC	—	00:51.0	−73:06	Irr	60	−16.2	163
Ursa Minor	DDO 199	15:08.2	+67:23	dE5	69	−8.9	−250
Draco	DDO 228	17:19.2	+57:58	dE3	76	−8.6	−289
Sculptor	—	00:57.6	−33:58	dE	78	−10.7	107
Carina	—	06:40.4	−50:55	dE4	87	−9.2	223
Sextans	—	10:10.6	−01:24	dE4	90	−10.0	224
Fornax	—	02:37.8	−34:44	dE3	131	−13.0	53
Leo II	DDO 93	11:10.8	+22:26	dE0	230	−10.2	76
Leo I	DDO 74	10:05.8	+12:33	dE3	270	−12.0	285
Phoenix	—	01:49.0	−44:42	Irr	390	−9.9	56
NGC 6822	DDO 209	19:42.1	−14:56	Irr	540	−16.4	−49
And II	—	01:13.5	+33:09	dE3	587	−11.7	—
NGC 147	DDO 3	00:30.5	+48:14	E4	589	−14.8	−157
NGC 185	—	00:36.2	+48:04	E3	620	−15.3	−208
Leo A	DDO 69	09:56.5	+30:59	Irr	692	−11.7	+26
NGC 205	—	00:37.6	+41:25	E5	725	−16.3	−239
M32	NGC 221	00:40.0	+40:36	E2	725	−16.4	−190
M31	NGC 224	00:40.0	+40:59	Sb	725	−21.1	−299
Pegasus	DDO 216	23:26.1	+14:28	Irr	759	−12.7	−181
LGS 3	—	01:01.2	+21:37	Irr/dE	760	−9.7	−277
And VII	Cassiopeia	23:24.1	+50:25	dE3	760	−12.0	—
IC 1613	DDO 8	01:02.2	+01:51	Irr	765	−14.9	−236
And VI	Pegasus II	23:49.2	+24:18	dE3	775	−11.3	—
Cetus	—	00:23.6	−11:19	dE4	775	−10.1	—
And III	—	00:32.6	+36:12	dE6	790	−10.2	—
And I	—	00:43.0	+37:44	dE0	790	−11.7	—
M33	NGC 598	01:31.1	+30:24	Sc	795	−18.9	−180
And V	—	01:07.3	+47:22	dE	810	−9.1	—
IC 10	—	00:17.7	+59:01	Irr	820	−17.6	−343
Tucana	—	22:38.5	−64:41	dE5	900	−9.6	—
WLM	DDO 221	23:59.4	−15:45	Irr	940	−14.0	−116
SagDIG	—	19:27.9	−17:47	Irr	1150	−11.0	−79
Antlia	—	10:01.8	−27:05	dE3	1150	−10.7	361
NGC 3109	DDO 236	10:00.8	−25:55	Irr	1260	−15.8	403
Sextans B	DDO 70	09:57.4	+05:34	Irr	1300	−14.3	301
Sextans A	DDO 75	10:08.6	−04:28	Irr	1450	−14.4	325

actually "radiate" the excess energy completely away from the cluster. It is with some caution that we assert that galaxies in clusters behave like a collisionless gas bound by the self-gravity of the system. But assuming the interactions are generally not either completely disruptive or end in merging, the velocity dispersion is a measure of the total gravitational potential. Two further assumptions are required, neither of which is obvious. The first is that the cluster is in hydrostatic equilibrium. This may not be true, in the sense that the aggregate moment of inertia may not be constant over the Hubble time (i.e., we assume that $\ddot{I} = 0$ for this version of the virial theorem). Second, we assume that the velocity dispersion is isotropic. You already know that this is a more problematic assumption since the cluster members may undergo collisions over the age of the Universe and this can alter the distribution function.

Despite these complications we can use virial arguments to obtain mass estimates for clusters. The gravitational potential is estimated by summing over all member galaxies within some predefined radius:

$$\Phi = -\frac{1}{2} G \sum_{i \neq j} \frac{M_i M_j}{|\mathbf{r_i} - \mathbf{r_j}|}. \tag{7.87}$$

Then, if each galaxy is bound to the cluster, invoking the virial theorem and averaging over the volume gives

$$\frac{M_i M_j}{\langle R_{ij} \rangle} = \sum_i M_i \langle v_i^2 \rangle \tag{7.88}$$

for the velocity dispersion. But we only have a two dimensional view of the structure and only one component of the velocity. Now assume isotropic orbits so that $\langle v_i^2 \rangle = 3 \langle V_{\text{rad}}^2 \rangle$ and that all galaxies have identical mass. Then calling ρ_{ij} the projected separation of two galaxies in the sky plane, and assuming random orientations so that $\langle R_{ij}^{-1} \rangle = (2/\pi) \rho_{ij}^{-1}$, we obtain the virial mass estimator:

$$M_{\text{VT}} = \frac{3\pi N}{2G} \frac{\sum_j V_{\text{rad},j}^2}{\sum_{j, i \neq j} \rho_{ij}^{-1}}. \tag{7.89}$$

Nevertheless, boldly proceeding, we take all the galaxies to be initially identical and obtain the mean radial velocity, $\langle V_{\text{rad}} \rangle$ from which we get the velocity dispersion:

$$\sigma^2 = \frac{1}{N-1} \sum_i (V_{\text{rad},i} - \langle V_{\text{rad}} \rangle)^2. \tag{7.90}$$

The mean cluster radius given by the projected distances of the galaxies from the center of the cluster is $\langle R \rangle$. There are a number of problems with this direct approach. For one, the distance is *projected* and even if a galaxy appears to be near the center of the cluster it may actually be quite far from it. A better measure of the mean cluster radius is afforded by taking the density distribution determined by reconstructing the function $\rho(r)$ from the two-dimensional distribution (assuming a symmetry) and then taking the mean value based on the observed cluster (see Appendix 7.A). The virial mass of the cluster is therefore $M_V = \langle R \rangle \sigma^2/G$ and the virial theorem mass to light ratio for the cluster $(M/L)_V$, can be compared with the mean for the galaxy population $\langle M/L \rangle$, assuming that M/L is constant for a specific type of galaxy. This may be a questionable assumption for cluster members, especially if mergers or collisions are important, but the discrepancy between the virial and average values of this ratio constitute one of the principal observations of dark matter. Most stars have $1 \leq M/L \leq 10$ and so should most galaxies. For clusters, this number can be as large as 300–1000. Zwicky realized this in the 1930s, when he first began looking at velocity dispersions, arguing that there was much more matter needed for binding than was visible. But without a proper distance scale and with little quantitative data, this should be viewed as a warning of a problem rather than a firm discovery.

7.8.3 X-Ray Emission from Clusters of Galaxies

X-ray emission from clusters of galaxies (see properties listed in Table 7.6[94]) was found serendipitously by the *UHURU* satellite in the first all-sky survey. Subsequent missions, EINSTEIN, ROSAT, ASCA, *Chandra*, and XMM, have refined our view as sensitivities and resolutions have improved by almost two orders of magnitude. These clusters are the strongest X-ray sources in the sky, showing *diffuse* thermal emission with luminosities of 10^{43}–10^{46} erg s^{-1} and temperatures ranging from 10^7 to 10^8 K. This gas appears to be uniformly highly ionized since there is little indication of absorption by neutral hydrogen in the low energy part of the spectrum and the spectrum is consistent with bremsstrahlung. The first high-resolution imager, the EINSTEIN observatory, achieved angular resolutions of about 5 arcsec with high sensitivity, while *Chandra* achieves better than 1 arcsec. Clusters show peak emissivity associated with, but not always exactly coincident with, the optically dominant galaxy. This is usually a cD galaxy. The strong correlation of the intense X-ray emissivity with these monsters suggests that the gas forms a sort of atmosphere that is gravitationally bound. Single ellipticals are frequently associated with X-ray sources having $10^{39} \leq L_X \leq 10^{42}$ erg s^{-1}. More precisely, the hot gas mass for each elliptical can be $\sim 10^9$–10^{10} $M_\odot$, with densities a few 10^{-3} cm^{-3} and temperature $kT \simeq 1$ keV.

For the bulk of the gas, in most clusters, the emission region appears to be in hydrostatic equilibrium and nearly isothermal. If the gas is virialized, a cluster having a velocity dispersion $\sigma \approx 1500$ km s^{-1} is equivalent to a kinetic temperature of $T = \mu m_H \sigma^2 / k \approx 1.7 \times 10^8$ K. For isolated ellipticals the velocity dispersion are ~ 300 km s^{-1} and the temperature is lower, $T \approx 7 \times 10^6$ K. The gas

TABLE 7.6 Properties of Galaxy Clusters

Property	Rich Clusters	Poor Clusters and Groups	Compact Groups
Number of members	30–300	3–30	4
Radius (Mpc/h)	1–2	0.1–1	0.1
Mass within 1.5 Mpc (10^{14} $M_\odot$ h^{-1})	1–20	0.03–1	0.03–1
σ (km s^{-1})	400–1400	100–500	200–250
L_B (10^{12} $L_\odot$)	0.6–6	0.03–1	0.03–1
M/L_B (h $M_\odot/L_\odot$)	300	200	200
L_X (10^{10} h^{-2} $L_\odot$)	0.1–25	—	1 < 0.3
T (keV)	2–14	—	~ 1

[94] Sources of Table 7.6 are Bahcall, N. 2000, in *Astrophysical Quantities*, 4th Ed. Cox, A. N., ed. (NY: Springer-Verlag); Hickson, P. 1994, *Atlas of Compact Groups of Galaxies* (NY: Gordon and Breach); Hickson, P. 1997, *ARAA*, **35**, 357.

mass is obtained by writing the hydrostatic equation in a less familiar form that depends only on gradients of observable parameters using a uniform molecular weight and the ideal gas law:

$$M(r) = -\frac{rP(r)}{G}\left[\frac{d\ln P}{d\ln r}\right] = -\frac{rP(r)}{G}\left(\frac{d\ln \rho}{d\ln r} + \frac{d\ln T}{d\ln r}\right). \quad (7.91)$$

For length scales of around 1 Mpc, the X-ray-emitting mass is typically around 10^{13}–$10^{14} M_\odot$. This is around 10% of the total mass of the cluster, particularly in more distant systems.[95] But you see you still need the density profile. This is provided by the isothermal Lane–Emden equation (remember?). There is a simple analytical form that approximately solves for the density in the inner part of the cluster. With $y = \rho/\rho_c$, which in this case is the density of the *galaxies*

$$\frac{1}{r^2}\frac{d}{dr}r^2\frac{d\ln y}{dr} = -\frac{4\pi G\rho_c}{c_s^2}y = -\frac{3\sigma^2}{2r_c^2 c_s^2}y,$$

we can take $\rho \approx \rho_c$ and then obtain

$$\rho(r) = \rho_c\left[1 + \left(\frac{r}{r_c}\right)^2\right]^{-3/2},$$

where we lumped the constants into a core scale length r_c. Now assume that the gas follows the galaxies with some general power law:

$$\rho_g \sim \rho_{\rm gal}^\beta. \quad (7.92)$$

Observations show that β is around $\frac{2}{3}$. This form of the overall density distribution is compatible with the so-called King approximation for the isothermal sphere,[96] where we define β as the ratio of the mean square velocity dispersion of the cluster galaxies to that of the gas. With this in hand and a measurement of the temperature profile, we can determine the mass distribution.

Many clusters, although not all, show a such high X-ray luminosity that it is hard to understand how the temperatures can be maintained for a Hubble time without some additional heating source since $E_{\rm th}/L < H_0^{-1}$. Radiative cooling by bremsstrahlung should therefore reduce the temperature and pressure on a timescale:

$$\tau_{\rm cool} \sim \frac{P}{\rho\Lambda}. \quad (7.93)$$

If so, and the gas remains bound, an inflow should be induced—a *cooling flow*—toward the central galaxies. The term is really a misnomer because the flow

---

[95] Schindler, S. 1999, *Astr. Ap.*, **349**, 435.
[96] The swindle is that we concentrate on the core of the cluster, and look at the region around the origin. Then we use the expansion $\ln(1 + \delta) \approx \delta$ for some small δ.

has not been directly detected. Instead, what is observed is changes in the gas emissivity as a function of distance from the center. The rate of mass accretion is inferred through the luminosity by assuming that the work done by the flow in collapsing toward the cluster core is all radiated away:

$$L_{\text{cool}} = c_p T \dot{M} = \frac{5kT}{2\mu} \dot{M}. \qquad (7.94)$$

For many clusters the value inferred for $\dot{M}$ is enormous, as large as 1000 $M_\odot$ yr^{-1}. It is worth noting that even for single ellipticals, the cooling time is shorter than the Hubble time, allowing for the application of the same cooling flow scheme. This leads to a mass accretion rate $\dot{M} \approx 1$ $M_\odot$ yr^{-1}, of the order of the rate of mass loss from stellar population inside the galaxies. A large part of the hot gas in ellipticals is coming from the galaxy, although a contribution from the intergalactic gas is not excluded. For galaxy clusters the proportions are reversed, most of the gas in the cooling flow comes from the intergalactic component. There appears to be a strong correlation between the X-ray and the blue optical luminosities, with $L_X \sim L_B^{1.6-2.3}$, but it is not clear whether this is an indication of induced star formation within the flow.

If such a huge amounts of mass is really accreting onto the central cluster galaxies, where is it? The central galaxies do not show large excesses of nuclear gas, either in the form of molecular or atomic clouds, nor do they show systematically enhanced star formation rates. This poses a serious theoretical challenge.[97] The mass accretion by the cD galaxy from the flow should be observable as a denser, cooler region in the flow in the galactic core. From observations you obtain a radial temperature profile and the mass flow profile results. Typically it is found that $\dot{M} \sim r$. There is also optical emission, with $H\alpha$ emission filaments being observed on scales of a few tens of kpc, coming from regions with $T \sim 10^3 - 10^4$ K and $H\beta$, O I, N I, S II are observed from the more central regions of the cooling flows. The intensity of these lines suggests they originate from shocks and not, as in H II regions, from photoionization by stellar radiation. Marginal detection of star formation in cD galaxies reinforces the picture of an accretion flow into which thermal instability leads to the formation of cold massive clouds.

An important clue to the origin of this gas comes from X-ray spectroscopy of the clusters. Many, if not all, of these emitting masses show significant amounts of heavy elements. Atomic line emission is also observed, the most prominent X-ray lines are due to Fe K-shell emission at 6.7 keV ($T \geq 2 \times 10^7$ K) and 1 keV (for lower temperatures). Because the lines are optically thin, they can be used to determine the metallicity of the intracluster gas *provided the ionization is known*. This is clearly not from gas with primordial abundances. The derived heavy element abundances can reach nearly solar values, far above primordial abundances so the gas must have passed through stars. Recall that from our discussion of interstellar abundances, the metallicity is reduced in Galactic diffuse gas

---

[97] Sarazin, C. L. 1986, *Rev. Mod. Phys.*, **58**, 1, which is expanded in Sarazin, C. L. 1988, *X-ray Emission from Clusters of Galaxies* (Cambridge, UK: Cambridge Univ. Press); Fabian, A. C., Nulsen, P. J., and Canizares, C. R. 1991, *Astr. Ap. Rev*, **2**, 191; Fabian, A. C. 1994, *ARAA*, **32**, 277. An excellent review of the state of the cold matter is Henkel, C. and Wiklind, T. 1997, *Space Sci. Rev.*, **81**, 1.

because of depletion of metals on grains. It is likely that the grains cannot survive against collisional sputtering in the hot intracluster gas so if the gas is coming from the galaxies it may have a higher than interstellar abundance. The structure of the X-ray-emitting matter isn't as simple as taking hot gas and asking whether it is in equilibrium; there must be heating of the gas and interaction between the individual galaxies and the more broadly distributed X-ray-emitting matter. But the initial assumption was that the gas is primordial is wrong. So the question now expands to the origin of both the energy and the gas. Clusters, as we have seen, are dynamical structures that evolve over time. The individual galaxies can and do merge, stars and gas are thrown out into the cluster in the form of tidal tails and gaseous envelopes, and supernovae and massive star formation induce global mass loss. A beautiful example is shown in Figure 7.17 for Stefan's quintet, a strongly interacting small group. In other words, there are many sources for heating the more widely distributed diffuse gas, and the fact that we now see it cooling may be just a result of the timescale for the processes.

Let us look at how to treat the energetics. Assume advection balances cooling and the cluster gas remains otherwise hydrostatic and independent of time. Then the energy equation is

$$\rho v \frac{d}{dr} \frac{\gamma}{\gamma - 1} \frac{P}{\rho} = \Lambda = \Lambda_0 \rho^2 T^{1/2}. \tag{7.95}$$

Since $\dot{M} \sim \rho v r^2$, if P is nearly constant, we can use $\rho \sim T^{-1}$ so for constant accretion we expect $T \sim r^{6/5}$. If instead $\dot{M} \sim r$, then $T \sim r^{4/5}$. The density is in both cases highest at the center of the inflow. If conduction is important, then the energy equation requires modification:

$$\dot{M} \left(\frac{\gamma}{\gamma - 1} \frac{k}{2\mu m_H} \frac{dT}{dr} + \frac{d\Phi}{dr} \right) = 4\pi \frac{d}{dr} r^2 K \frac{dT}{dr}. \tag{7.96}$$

Here K is a generalized thermal conduction coefficient that may include multiple species. We have ignored any magnetic fields, but note that a field as small as 1 μG, with a coherence length that is a significant fraction of R, can substantially reduce the thermal conductivity in direction perpendicular to the field lines because of the small gyration radius for the thermal electrons. We can also write Eq. (7.96) as

$$n \frac{d}{dt} \frac{5kT}{2\mu} = \mathscr{L} + \frac{1}{r^2} \frac{d}{dr} K \frac{dT}{dr}, \tag{7.97}$$

including the effects of heating and cooling, and radially variable conduction with a classical coefficient $K \sim T^{5/2}$. Several assumptions are hidden in this version of the thermal balance. The cooling is mainly due to free-free emission, as we saw, but through the lines Λ also depends on the metallicity of the gas. The heating term, Γ, is *not* well known and may be either local or global. Any effects of intracluster magnetic fields have been ignored. Although rotation measures place limits of tens of microgauss on the magnetic field strengths in the diffuse extended

FIGURE 7.17. Stefan's quintet imaged with HST using WFPC2. Notice the strong harassment of galaxies in this image, the combination of close encounters, tidal tails, and mass ejection (credit: NASA/STScI).

gas, these may be important in reducing the heat transport by impeding the free motion of electrons.

When we discussed thermal instabilities for the interstellar medium, we neglected advective losses. For cluster cooling flows, however, the cooling instability can be suppressed if the gas has a chance to flow (in effect, if the mean flow time becomes comparable with the cooling time it will fall inward and be compressed and reheated). Therefore, the flow is only initially driven by the growing contrast between hot and cold gas. In summary, the problem with the flow is, quite simply, that there is no indication of what is happening to the accreted matter. It is possible that it goes into a starburst in the core of the system, or gets hidden in molecular gas to which the X-rays are insensitive and that can only be detected by millimeter observations, but that doesn't seem to be the case. To further complicate the matter, the gas does not appear to be going directly into stars. There is no excess of massive stars, and no hint of lots of low mass stars. You see the evolution here of a paradox within the model, but it is not clear that the paradox is not being driven by the model itself.[98]

7.8.3.1 Evidence for Intracluster Gas from Radio Galaxies

Radio galaxies provide useful probes of the intracluster medium. The individual galaxies have velocities that produce motion with respect to gas within the cluster. First there is the pressure effect produced by the cluster gas, the equivalent of ram pressure. Unlike stars in clusters, galaxy cluster environments have a profound effect on the evolution of their constituents. The ram pressure can remove gas from the galaxy, thereby altering its star formation history and its chemical evolution. Given the typical velocity dispersion of the galaxies within a cluster, the intracluster medium impacts the internal gas, in the comoving frame, at highly supersonic velocities. The advance of this shock may have the effect of disrupting molecular clouds within the galaxy, thereby ending star formation. Finally, by removing the internal gas while leaving the stellar component unaffected, the mean metallicity of the remaining gas will be altered. The interplay of these processes affects the interpretation of the development of cluster galaxies compared with field systems in a way that is very different from the interpretation of star in clusters compared with noncluster members.

In the frame of the galaxy, the cluster gas produces a dynamical pressure of $\rho_g(r)v_{gal}(r)^2$. For an isothermal gas, $\rho_g(r)$ is known and therefore we can say that a cluster galaxy traversing the core will experience an increasing ram pressure as it approaches its orbital pericenter. We must also include the ram pressure for the evolution of jets and lobes, since the originating galaxies are moving through the background gas. The simplest approach is to assume that the ram pressure of the gas is responsible for the turning of the outflow from the galaxy. Therefore

$$\rho \mathbf{v} \cdot \nabla \mathbf{v} = -\nabla P_{ram}. \quad (7.98)$$

If it has a jetlike outflow, this matter can trace the motion of the galaxy. Such objects are called *narrow angle tail* sources such that the flow is bent backwards by the ram pressure of the surrounding medium. So-called *wide angle tail* sources, on the other hand, are thought to either move slowly with respect to the cluster medium or to be in low density environments that produce little effect on the jet

---

[98] We again assert that "There are no paradoxes in Nature."

structure. If R is the radius of curvature of the jet and H is the scale length of the ram pressure, then if J indicates the jet properties, we have the approximate relation:

$$\frac{\rho_J v_J^2}{R} \sim \frac{\rho_g v_{\text{gal}}^2}{R}. \tag{7.99}$$

An estimate of the ambient gas density is provided by the X-ray luminosity $\rho_g \sim L_{XR}^{1/2}$, assuming that it is from bremsstrahlung alone. The timescale can be determined from:

$$\frac{1}{2}\rho_g v_{\text{gal}}^2 = \frac{1}{2}\rho_0 \left(\frac{H}{\tau}\right)^2 \tag{7.100}$$

so that

$$\tau \simeq 10^9 \text{ yr} \left(\frac{R}{30 \text{ kpc}}\right) \left(\frac{v_{\text{gal}}}{10^3 \text{ kpc}^{-1}}\right)^{-1} \left(\frac{\rho_0/\rho_g}{10^3}\right)^{1/2}. \tag{7.101}$$

The galactic gas becomes unbound when $P_r > \rho_0 \sigma_{\text{int}}^2$, where σ_{int} is the velocity dispersion internal to the galaxy with an interstellar density ρ_0. You would expect that stripping would enhance the effect of continual stellar pollution of the remaining interstellar medium, leading to an increase in the mean metallicity of the gas. Not quite. The rate of star formation depends on the diffuse gas, albeit through the formation rate for molecular clouds. Thus, removing the diffuse gas inhibits further cloud formation. Since the clouds that remain are being slowly destroyed through continuing star formation and induced destruction from H II and photodissociation regions, the rate of star formation must decline. Therefore, although the metallicity may increase initially, the subsequent cessation of star formation means that the ongoing production of heavy elements occurs at a lower rate, and this slows the rate of increase of the metals. The galaxian orbit through the cluster complicates this picture. The galaxy spends only relatively short intervals in the densest portion of the intracluster medium. Therefore, although the stripping is continuous, it is not constant. Finally, continuing infall of lower metallicity halo gas dilutes the newly produced metals and further suppresses the rate of increase.

What is lost from the galaxy is incorporated into the intracluster medium, resulting in an *increase* of the metallicity of the gas in the cluster. This, in turn, increase the cooling rate, since the pollutants are efficient radiators, and may lead to cooling waves and subsequent mass flows. Because the stripping involves shock interactions between the interstellar and intracluster media, gas is injected at comparatively high temperature over and above whatever is contributed by galactic winds driven by supernovae. The net effect is to heat the cluster gas. In summary,

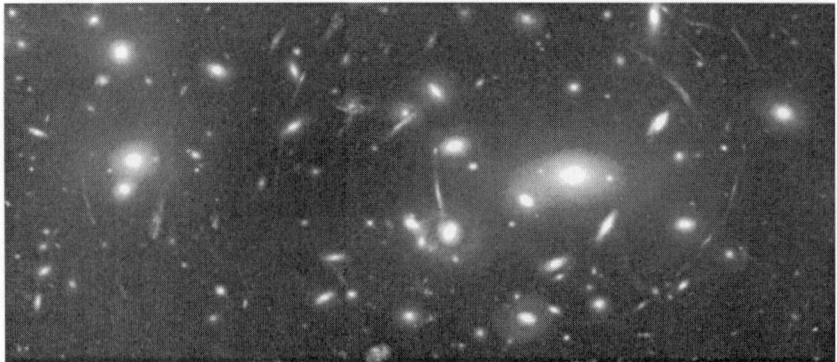

FIGURE 7.18. Gravitational lenses in the core of Abell 2218 imaged at visible wavelengths with the WFPC2 on HST (credit: NASA/STScI). The arcs are easily distinguished from galaxies; notice their distribution relative to the cluster. How might the deprojection technique be used for images such as this?

even on the scale of Mpc there exist several feedback mechanisms between the stars in a galaxy and the medium in which the system is embedded.[99]

7.8.4 Gravitational Lensing by Clusters and Dark Matter

In the growing catalog of evidence for dark matter, we now return to gravitational lensing. Clusters display a unique form of this because of the extreme spatial extent of the mass distribution. There is, however, little qualitative difference between the lensing produced by, say, an elliptical galaxy and a cluster. The mass distribution for the cluster depends on $\rho(r)$, which in this case includes the dark as well as baryonic matter. That is, the entire gravitational field of the cluster enters into the lensing. Quasars seen through the cluster, and even more distant galaxies, can be bizarrely distorted. This is best illustrated by a recent WFPC-2 image of the cluster Abell 2218 (Fig. 7.18). Notice that the distorted images appear as disconnected arcs distributed roughly symmetrically about the center of the cluster. Since the lensing is determined by the mass, $M(r)$, the center of the arc probes the mass interior to the impact parameter r. from this, the mass of the cluster can be computed and compared with the mass in luminous galaxies. Once again, the analysis requires inverting the distribution of arcs to determine the radial mass distribution. The derived mass to light ratio is of order 100–1000, depending on the cluster, in agreement with both cooling flows and the virial estimates (Fig. 7.19). We will return to this wondrous phenomenon of gravitational lensing in greater detail our final chapter.

---

[99] See, for example, Geisler, G. 1976, *Astr. Ap.*, **51**, 137; Nulsen, P. E. J. 1982, *MNRAS*, **198**, 1007; Takeda, H., Nulsen, P. E. J., and Fabian, A. C. 1984, *MNRAS*, **208**, 261.

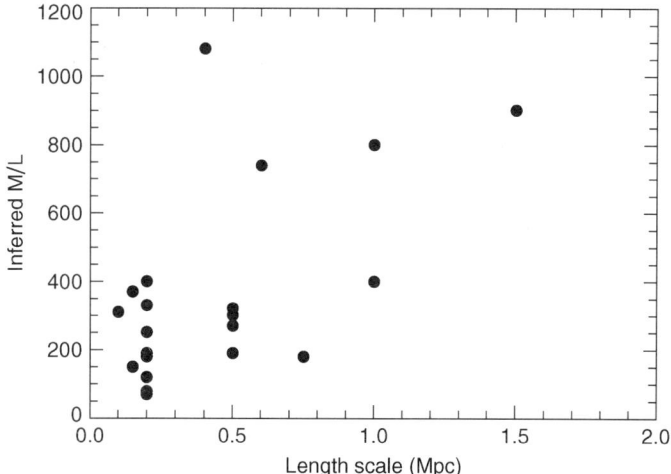

FIGURE 7.19. Derived mass to light ratio from gravitational lensing compared with the scale length of the cluster (data from Mellier, Y. 1999, *ARAA*, **37**, 127).

7.A DEPROJECTION METHODS: ABEL'S EQUATION AND INVERSION OF SURFACE MEASUREMENTS

The two-dimensional picture we have of a globular cluster, an elliptical galaxy, or even a cluster of galaxies is merely a projection. To understand its structure and dynamics, especially for obtaining its gravitational potential, it is necessary reconstruct the spatial structure by deprojection. You will recall, for instance, that we need this obtain the mass distribution a surface brightness or column density, $\Sigma(r)$. These surface quantities depends on the distance from the center of symmetry of the body projected on a plane, and result from the integration of $\rho(R)$, the *radial* luminosity or density profile as a function of the volumetric distance R from the center along the line of sight:

$$\Sigma(r) = \int_{-\infty}^{\infty} \rho(r,z)\, dz. \qquad (7.102)$$

Since we are really after the radial distribution, this becomes

$$\Sigma(r) = \int_{0}^{\infty} \frac{\rho(R)\, R\, dR}{(R^2 - r^2)^{1/2}}. \qquad (7.103)$$

To solve this equation means an inversion,[100] taking the observed Σ and using it along with the integrating kernel to find ρ. Because it is such a widespread

---

[100] Titmarsch, E. C. 1935, *Introduction to the Theory of the Fourier Integral* (Oxford: Oxford Univ. Press); Hildebrand, F. B. 1952, *Methods of Applied Mathematics* (Princeton: Van Nostrand); Kondo, J. 1991, *Integral Equations* (Oxford: Oxford Univ. Press).

732 OUR GALAXY AND OTHERS AS STELLAR SYSTEMS

problem in astrophysics, let's examine the more general problem of the *Abel equation*.[101] We begin with a restatement of the general definition of a convolution

$$g(t) = \int_a^t k(t-y)f(y)\,dy = k \star f, \qquad (7.104)$$

where k is a generalized *kernel*. Let's use Laplace transforms. Call $\hat{w}$ the transform of some function w. Then by the convolution theorem, we have

$$\hat{g} = \hat{k}\hat{f}. \qquad (7.105)$$

Now, in principle, we can just solve for $\hat{f}$ and take the inverse transform. It isn't quite that simple, though, especially if the kernel is anywhere singular. We will need some auxiliary results before proceeding. First, we need the following integral:

$$\Gamma(\tfrac{1}{2})p^{-1/2} = \int_0^\infty e^{-px} x^{-1/2}\,dx, \qquad (7.106)$$

which is the Laplace transform of $x^{-1/2}$. We will also need the Laplace transform of the derivative of an arbitrary function:

$$\mathscr{L}\left(\frac{df}{dx}\right) = p\hat{f} - f(0), \qquad (7.107)$$

where $f(0)$ is the zero value of the function, *not* of its transform. Now to proceed. Write a generalized version of Eq. (7.104) as

$$g(x) = \int_0^x \frac{f(y)\,dy}{(y^2 - x^2)^m} = k \star g, \qquad (7.108)$$

which will become obvious by writing $u = y^2$ and define a new function, $h(u)$ through $f(y)\,dy = h(u)\,du$ in order to reduce the form of Eq. (7.108) to a simple convolution:

$$f(x) = \int_0^z \frac{h(u)\,du}{(u-z)^m}, \qquad (7.109)$$

where the change in variables that requires changing g to h also means altering the integration limits, just temporarily. Using the transform $\mathscr{L}(x^{-m}) =$

---

[101] This was the first example of an equation, from Abel's doctoral dissertation, and is one of a class of equations that form the core of a more general theory that was developed by Fredholm and Volterra. Here we use what is called a "Volterra equation of the first kind." See the references in footnote 100, especially Kondo (1991), for more details.

$p^{m-1}\Gamma(1-m)$ gives $F(p) = \frac{1}{2}\Gamma(1-m)p^{m-1}H(p)$, so on inversion we find

$$g(x) = \frac{2\sin\pi m}{\pi} \frac{d}{dx} \int_0^x \frac{yf(y)\,dy}{(y^2 - x^2)^{1-m}}, \qquad (7.110)$$

which is the solution of the Abel equation. We have also used the identity:

$$\frac{\pi}{\sin m\pi} = \Gamma(m)\Gamma(1-m). \qquad (7.111)$$

The generality of this result is important in extragalactic astronomy, where almost all quantities are observed as projections but require spatial distributions to interpret.

We've already mentioned surface brightness. Velocity curves are very similar since the mass obtained from the inversion is from the radial velocity, which is integrated through all contributors along the line of sight. From the Poisson equation, you know that

$$4\pi G\rho r^2 = 2\Theta r \frac{d\Theta}{dr} + \Theta^2 \qquad (7.112)$$

so the observed surface density within a projected distance R is

$$\Sigma(R) = \frac{1}{2\pi G} \int_R^\infty \left(\frac{d\Theta^2}{dr} + \frac{\Theta^2}{r} \right) \frac{dr}{(r^2 - R^2)^{1/2}}, \qquad (7.113)$$

which, when integrated over the projected surface, gives the total mass for an arbitrary geometry. Another example is the determination of the mass distribution for X-ray-emitting gas, where again you see only the surface brightness and must use that and assumptions of the geometry to reconstruct the radial structure. Finally, although you can measure the mass more directly using gravitational lensing, it is still necessary to invert the surface profile to obtain a radial mass profile.

8 The Biggest Picture: Cosmology

> You may think it's a long way down the road to the chemist, but that's just peanuts to space.
> —Douglas Adams, *The Hitchhiker's Guide to the Galaxy*, Chapter 8

8.1 INTRODUCTORY REMARKS

The roots of speculative cosmology are as old as philosophy but the distinction between kinematic model building—astronomy—and dynamical causes—physics—remained quite sharp throughout much of this history. For example, when Thales in the seventh century B.C.E. put the world afloat in a universal ocean, he was trying to provide a physical basis for cosmic phenomena consistent with the observations of everyday life and separated from theology. How it got that way was not an issue nor was it necessary to confront cosmological speculation with observations. Consider the task assigned to Eudoxus by Plato to "preserve the appearances" with a geometric model that encompassed the variety of observed motions of the celestial bodies and the stars. The causes were not a problem as long as the model was consistent with the phenomena perceived by an observer and based on strict geometric reasoning. In fact, in the Ptolemaic universe, there were serious problems posed by the placement of epicycles within the driving spheres, and with the variation in the size of the Moon and Sun due to the eccentricity of their orbits, but these were brushed aside in light of the unknowable nature of extraterrestrial bodies.

In a geocentric world, the universe is constrained by first principles, or so Aristotle argued. Since every observed motion must, in such a world, correspond to a *real* motion, forcing the diurnal rotation of the stellar sphere puts an upper bound on the size of the cosmos. The observer on Earth is located at a special, unique place, privileged by physics to remain at absolute rest. Thus, if all motion is executed about this point, and no motion in a finite time can cover an infinite expanse, the Universe must itself be finite *if* it is bounded by the stellar sphere for which no parallax can be detected. The possibility that the stars are like the Sun doesn't weaken this argument, it just makes the radius of the stellar sphere larger. Compound motions of the sphere due to precession (and later the variation in the precession rate attributed to the *trepidation*) only requires the drivers to be located in contact with the outermost sphere. What lies beyond that is irrelevant in a geocentric picture, and the early mediaeval commentators were quick to see heaven in the superstellar realm. Nonetheless, these cosmological speculations were largely circular restatements of dynamical first principles grounded in an immobile Earth.

The contrasting uses of geometric calculation and metaphysical explanations is most obvious in the separation of the university curriculum of the Middle Ages. Cosmology, which was the study of ultimate causes, was considered the stuff that theologians and philosophers had to grapple with while *astronomia* belonged with mathematics that trained the mind in precise, logical thinking. Astronomy, confined to the study of spherical astronomy and planetary motions, was grouped with music, arithmetic, and geometry in the *quadrivium* while cosmology, because of its grounding in metaphysics, was placed in the *trivium* as part of philosophy. True, the basic texts, in particular the *Sphere* of Socrobosco and the *Almagest* of Ptolemy, contained introductions that defined and justified the basic physical picture of a geocentric universe, but the questions of dynamics and structure were included in discourses on Aristotle's *de Caelo* and *Physics* along with commentaries on the *Metaphysics*. Ultimate causes of motion were left as speculation about physical differences between the composition of heavenly and terrestrial things. Even in *de Revolutionibus*, Copernicus followed the outline of the *Almagest* and refrained from explaining causes, opting instead for a description of phenomenology and developing a geometrical machinery for dealing with the diversity of observed motions. The Earth moved but further implications were left to the philosophers to follow, in particular the extent of the Universe.

Physical cosmology, the child of the Renaissance, came of age during the seventeenth century with the introduction of the instrumentation, empiricism, and the rise of "mechanical philosophy." Galileo's early telescopic observations again opened up the scale, revealing that the darkness was actually populated by ever fainter stars and agglomerations of stars, that the Milky Way resolved, that the stars and planets were different bodies with the latter shining only by reflection, and through that confirmed the basis of the Copernican picture. The task of providing the kinematic Copernican world with a distinctive, non-Aristotelian dynamics was begun by Kepler although this is less important to our present discussion than one of his more tangential speculations. Although not central to his physical astronomy, Kepler succinctly asserted the finiteness of even a heliocentric world if the stars are self-luminous like the Sun. Mainly on the basis of the inferred number of emitting bodies, an infinite Universe would be infinitely bright, which it obviously isn't. We stress, however, that the point is not the answer; however, it may have agreed with the prejudices of the time, but the novelty of the assertion on which it rests. The stars had changed their character, or rather the Sun had, and for once a cosmological question was not being decided on the basis of observations, not axioms. The world *could* be infinite in extent, it just happened not to be.

Thus, celestial matter became *in essence* the same as mundane stuff that can be manipulated in the laboratory and was consequently subject to the same fundamental laws. It became progressively more important to construct a *physically* consistent world picture. Two moments in this historical development stand out. The first fundamental step was the cosmological argument in the *Principia*, in the *General Scholium* in Book III, where Newton asserted that the intrinsic gravitational force of all matter must ultimately bring the universe into collapse unless it is infinite or there is some additional, perceptive, agent that maintains the equilibrium. In the *Optiks*, he added that some unification must exist between the properties of light and gravity, in the sense that both are emitted from bodies and traverse great distances. The discussion in the *Optiks* and in other places also shows that Newton was keenly interested in the problem of masses and luminosities for distant stars, taking them to be like the Sun and using what is now a familiar scaling argument from a standard candle to place limits on the size of the universe.[1] Although the stability of systems with angular momentum was demonstrated by Laplace within a century of these arguments, there were persistent questions of global stability. Herschel and a host of others, first using visual observations and later exploiting photography, entered into large scale surveys, cataloging nebulae and determining positions and proper motions, but instruments had limited capabilities and much of the attention focused on stellar and mechanical problems for much of the nineteenth century. Of immediate interest for our discussion were increasingly numerous cataloged examples of nebulae that were neither H II regions nor clusters. Beginning in the late eighteenth century with the Messier (M) catalog and extending through the Index Catalog (IC) of William and John Herschel, by the mid-nineteenth century the number of spiral and elliptical "nebulae" steadily grew, along with increasingly sophisticated descriptions of individual objects, culminating in the *New General Catalog* (NGC) of Dreyer that became the basic list from which subsequent studies of these anomalous diffuse objects drew.

In the meantime, spectroscopy had uncovered another intriguing problem. At the beginning of the twentieth century, in the "botanical" spirit of the time, every astronomer with a good telescope and spectrograph was pointing them at anything in sight and V. M. Slipher at Lowell Observatory was no exception. Over a decade, beginning roughly in 1912, he continued the program of determining the spectroscopic characteristics of nebulae, a difficult feat given the extremely slow photographic emulsions of the time and the low surface brightness of the targets. The spiral nebulae were of particular interest, especially in the context of the contemporary cosmogonic debates. A surprising result emerged. In the first cumulative listing in 1917, they systematically showed very high radial velocities, hundreds of km s^{-1}, quite unlike any other diffuse NGC objects or any stars. Further, their continuous spectra resembled globular clusters, although there were also sometimes superimposed emission lines (for instance, NGC 1068, NGC 4449, and NGC 4214).[2] By 1924, when Eddington wrote his classic monograph on general relativity, Slipher had amassed an impressive number of spectra of spiral nebulae. The preponderance of large, positive radial velocities was striking and suggested some larger agent that was structuring the motions, but there was a fundamental limitation to this kind of fieldwork. Without a clear indication of the distance of these nebulae, there could be no further progress. One could use the morphological analogies between the LMC and NGC 4449 to suggest they were physically similar objects, but neither is spiral and so there was no point of contact with the

---

[1] An excellent source for Newton's discussions is Munitz, M. K., ed. 1957, *Theories of the Universe, from Babylonian Myth to Modern Science* (Glencoe, IL: Free Press) which is also a superb source for many of the earliest discussions of cosmology and large-scale structure. Additional material, including portraits and other wonderful graphics, is found in Hoskin, M. A. 1997, *Cambridge Illustrated History of Astronomy* (Cambridge, UK: Cambridge Univ. Press).

[2] Slipher, V. M. 1917, *Observatory*, **40**, 304; Slipher, V. M. 1918, *PASP*, **30**, 346.

systematics of that sample.³ There was no particular program guiding these observations. Having a spectrograph and a good telescope was enough. The fainter nebulae, many of them spirals or ellipticals, could now be studied and curiosity alone drove the research, a typical state of affairs in the early years of astrophysics when random observations yielded quick results for interesting objects but produced nothing systematic.

Almost as the ink was drying on Eddington's second edition, the distance scale problem resolved with Hubble's discovery of a classical Cepheid in M31, the Andromeda nebula. Variable stars had been noted in nebulae before but these always proved to be eruptive type, novae and supernovae (although the latter were not distinguished from classical novae before the 1930s). They were also well known from surveys of the Magellanic Clouds. But then as now, the zero point of the calibration and the precise link between the speed class and maximum luminosity of the variable, as well as the possible heterogeneity of the classes, provoked caution. Cepheid variables, on the other hand, were observationally well studied, comparatively well calibrated, and could reliably span the distances from the Galactic examples through the Magellanic Clouds to M31. It is not historically surprising that the credit goes to Hubble for the discovery of the universal law that now bears his name for the recession of galaxies nor should you be surprised that the redshift law required nearly a decade of accumulated observations before it was announced in 1931 by Hubble and Humason. Even a scaling law for the distances to standard candles suffices to determine the basic result, that the velocity of recession is linearly proportional to the redshift, but it requires that individual stars be observable in the galaxies, or other standard objects that can be bootstrap calibrated out to progressively larger distances. Once obtained, however, the redshift alone could substitute, in principle, for other methods for finding distances. In other words, distances could be determined from the spectroscopic observations alone and additional physical properties could be determined for the galaxies and, ultimately, for clusters of galaxies. It was now possible to examine global rather than local questions of cosmic structure, treating galaxies as mere tracers of the universal flow.

Perhaps the most important consequence of the discovery of the redshift–distance relation was the introduction of a new fundamental physical parameter, the Hubble constant, which for a linear law is the same as an inverse timescale. By the middle of the 1930s it was recognized that this number was directly linked to the timescale for the expansion and, therefore, to the age of the Universe. Thus, within a period of about 10 years spanning 1924–1934, observational cosmology was born.⁴

---

³There were further paradoxes posed by the report of proper motions detected in M101 and M33 by van Maanen. While these observations eventually proved spurious, for a short time they pointed to a local origin for spirals and were used as part of Shapley's arguments against Curtis in the National Academy of Sciences debate on the distance scale in 1920.

⁴An excellent reference for the discussion of the early years of physical cosmology is Smith, R. 1982, *The Expanding Universe: Astronomy's 'Great Debate' 1900–1931* (Cambridge, UK: Cambridge Univ. Press). See also Kerszberg, P. 1989, *The Invented Universe: The Einstein-De Sitter Controversy (1916–17) and the Rise of Relativistic Cosmology* (Oxford: Oxford Univ. Press); Kraghe, H. 1996, *Cosmology and Controversy* (Princeton: Princeton Univ. Press); Jaki, S. 1969, *The Paradox of Olbers' Paradox* (NY: Herder and Herder).

Attempts to set this expansion within a physical model have a more complex history that was not directly tied to the observational developments. The field equations of general relativity were published by Einstein in 1916. Schwarzschild quickly obtained the solution for the external static field of a point mass. Within a year, Einstein published a short note concerning the cosmological consequences of the theory, which meant discussing how the large-scale distribution of matter affects local measurements but the paper neither proposes nor solves any particular model. The cosmological problem was tackled by Friedmann in 1922–1924 for the simplest equation of state for an isotropic cosmology. He showed that a spherical homogeneous spacetime would not necessarily remain stationary, and in effect predicted the Hubble law. Earlier, the cosmological constant had appeared in Einstein's modification of the field equations, and he and de Sitter showed that even a vacuum Universe could show a redshift, a solution that today finds its application in inflationary models. In one of the earliest papers written on the comparison of observations with theory, Einstein and de Sitter[5] showed that the cosmological constant was unnecessary in view of the expansion obtained by Hubble and showed that (then observed value of) the mean density provided an estimate of the curvature scale that was very close to the distance inferred from the Hubble constant (then quoted as 500 km s^{-1}) for the farthest galaxies.

In 1933, in what at first sight seems to a rediscovery of the Friedmann equations, Lemaître showed that an enormous family of possible solutions exist to the field equations that permit very complex histories for the matter. The implications of this work were not fully appreciated at the time, but Lemaître's papers provided the catalog of physical models of the expanding Universe. They highlighted a fundamental feature of the cosmology—the spacetime expands, but all of these models contain a singularity at the origin in time, which Lemaître called the *primeval atom*. At the same time, de Sitter surveyed the variety of models that result from the inclusion of a cosmological constant, Λ, and showed that even in the absence of matter, a universe with finite Λ expands in the relativistic framework. This would eventually be important in discussions of inflation, but we're getting too far ahead of ourselves. Finally, to complete this breathless run through the origins of the field, in 1948, Gamow and his collaborators argued that if the Friedmann models are correct and we trace the density of matter and energy back in time, we arrive at a remarkable era when the conditions were perfect for cosmological nucleosynthesis. He, Hermann, and Alpher made the prediction that there should be an observable relic of this era in a universal photon background whose temperature is around 10 K.[6] It is interesting to note that cosmology before

---

[5]Einstein, A. and de Sitter, W. 1932, *Proc. Nat. Acad. Sci.*, **18**, 213; de Sitter, W. 1933, *The Astronomical Aspect of the Theory of Relativity*, Univ. California Studies in Math. 2(8) (Berkeley: Univ. Calf. Press.

[6]The founding papers on physical cosmology have been reprinted in Lang, K. R. and Gingerich, O. 1979, *A Source Book in Astronomy and Astrophysics*, 1900–1975 (Cambridge, MA: Harvard Univ. Press). A significant secondary literature is also available. For Lemaître's contribution, see Berger, A., ed. 1984, *The Big Bang and Georges Lemaître* (Dordrecht: Reidel). For Friedmann's contribution, see Tropp, E. A., Frenkel, and Chernin, A. D. 191993, *Alexander A. Friedmann: The Man Who Made the Universe Expand* (Cambridge, UK: Cambridge Univ. Press). See also Robertson, H. P. 1949, in *Albert Einstein: Philosopher-Scientist*, Schilpp, P. A., ed. (La Salle, IL: Open Court) and Einstein's responses. For Gamow's side of the story, see Gamow, G. 1970, *My World Line: An Informal Autobiography* (NY: Viking).

1965 was dominated by kinematic derivations of the redshift and discussions of the connections between the velocity–distance relation and geometric world models. Little attention was paid to the sort of physical studies that Gamow and his collaborators, and Zwicky, were attempting. With the exception of Hoyle's (1953) derivation of the growth of perturbations in the expanding universe, almost no attention paid to the formation of large scale structure. Only the Princeton group under Dicke and Wilkinson were even attempting to search systematically for a remnant background radiation from this hot state. Even techniques for the statistical analysis of galaxies, especially the two-point angular correlation function by Neyman and Scott following the first all-sky catalog of galaxies by Shane and Wirtanen, received scant attention. Without a basis for the initial conditions, there was no constraint for reconstruction of the history of the cosmic mass distribution.

This changed completely with the announcement by Penzias and Wilson (1965) of the discovery of a uniform microwave background radiation. This provided, at one fell swoop, the window into the early Universe that has been missing in study of galaxies and gave direct evidence for the hot early Universe picture. It forced a recalcitrant community to recognize that the equation of state must have changed from matter to radiation dominated as one went backward in time, and provided the evidence for the nucleosynthetic stage that had been postulated by Gamow and collaborators and used by B^2FH in their discussion of the nucleosynthetic processes in stars. It was now possible to make the transition from the kinematics to actual physical cosmology, a step heralded by the change in the nature of monographs (Peeble's book, *Physical Cosmology*, first appeared in 1971 and was the first in a long series of such texts). Almost immediately after the announcement of the cosmic background radiation, Peebles (1966) and Wagoner, Fowler, and Hoyle (1967) carried out network calculations of nucleosynthesis for the light elements in the early stages of the expansion, completing the program begun by Gamow in the 1940s, and showed how ^2D and ^3He could be used for the determination of the critical density and the entropy of the Big Bang. A new field, astroparticle physics, was championed particularly by Schramm after the 1970s in response to this work. He placed particular emphasis on the light elements as probes of the initial conditions and as a way to place constraints on fundamental particle properties, such as the number of neutrino flavors emerging from the early universe.

At the same time, the discovery that quasars are really objects at very large redshift highlighted the need to better understand the physics of the expansion and the role of high-energy processes in galactic nuclei. It took several decades to settle conclusively that these objects are actually galaxies at large distances that can be used as probes over long lookback times of the state of the Universe. The decade of the 1970s brought fundamental observational advances, along with great instrumentation improvements in detector sensitivity, telescope collecting area, and large format arrays. Rubin's discovery of flat rotation curves pointed to the existence of a new gravitating component of nonluminous matter in the mass of galaxies. New technology produced new results. By the end of the decade, the first gravitational lens was detected, the quasar 0957 + 561. The first detection of the absorption lines from intergalactic neutral hydrogen in the spectrum of high-redshift galaxies, the Lyman α forest as it would come to be called, followed in short order. Theoretical analyses continued as well, with the reintroduction of correlation methods for the analysis of large-scale structure, the discovery of complex local inhomogeneities in the distribution of galactic masses and the recognition of voids.

The roadmap for this chapter is as follows. We will begin with a review of the distance scale and the observational tools for the study of large scale structure. We will then transition to a discussion of how these can be explained within the current cosmological models, and what predictions are possible by extension of these structural models to physical analysis of the initial conditions. Finally, we will discuss some features of the formation and consequences of large-scale structure.

8.2 THE DISTANCE SCALE

The *distance scale* is the rock on which all of physical cosmology rests or founders. Without this, it is impossible to place limits on the properties of extragalactic objects, or indeed to even know whether they are extragalactic. The Hubble constant is known only through the distance, although the *existence* of the expansion is independent of its precise value, and thus cosmological modeling requires a direct determination of this one parameter for each sample. This is where all the physical processes we have been discussing throughout this book come into play. Stars are essentially the same from one galaxy to another, not excluding the effects of metallicity and binarity. Ionized regions in the interstellar medium are the same physical entities from one galaxy to another, again ignoring the details. In short, the physical laws are the same here and in the most distant galaxy. This having been said, however, we hasten to add an epistemological cautionary comment. Although the basic laws are invariant, they manifest themselves in very different ways depending on the site and time. *This* is what makes observational cosmology so hard. You know that the objects you're looking at are familiar, but you cannot be sure that they are the same as the ones you've met in your neighborhood. They may only *look* the same superficially, hence you may be deceived in the determination of their properties through the assumption that they are identical rather than just similar.

8.2.1 Overview

It is useful to have a roadmap through the territory of calibrators, especially in light of all you now know about the physical elements that are required to formulate them. In essence, the steps in creating the distance ladder are as follows:

- Use parallax measurements for Milky Way stars to obtain luminosities for individual objects, calibrated against spectral types.
- Use stars in binary systems and clusters, both with parallax measurements (especially the Hyades) and luminosity ratios from color–magnitude diagrams, to extend the absolute calibration of luminosities to statistically rare but luminous stars.
- Use Galactic and extragalactic objects to determine the distribution of interstellar extinction for color corrections.

- With absolute luminosities and colors for RR Lyr and δ Cep variables, determine the distance to the Magellanic Clouds.

- Use the Magellanic Clouds, Local Group galaxies, and the Virgo cluster to determine the secondary calibrators: SN Ia, planetary nebula luminosity functions, H II region luminosities, novae, most luminous stars, and galactic disk surface brightness fluctuations.

- Use calibrations that relate dynamical properties of galaxies to their luminosity in selected bandpasses, the 21-cm linewidths (Tully–Fisher relations) and stellar velocity dispersion (Faber–Jackson relation) correlations, to obtain distances at the scale at which individual stars cannot be measured.

- Iterate in both directions of scale as frequently as required for convergence and hope such a thing is possible.

Notice how interlocking the various steps are, and think about how both random and systematic errors propagate through the distance scale.

Since energy generation mechanisms in stars are stable and reasonably well understood, the fundamental use of calibrators is to obtain some way of assigning an observable property of a distant object to an absolute luminosity and, by a comparison with the observed brightness, determine its distance. Secondarily, and often less reliably, one attempts to use angular diameters to obtain geometric distances. We assert these are less reliable because of the complexity of cosmic environments and the considerable noise this injects into any calibration.

Take the following example. You observe that a nebula has an angular size $\Delta\vartheta$ and want to find its distance. For contrast, you observe a resolved binary system with the same angular separation. Measurement of the spectra of the individual stars and their relative velocities, combined with a visual orbit determination, provides all the information required to obtain a dynamical parallax since the system is bound and obeys a simple force law that depends only on the period, linear separation, and constituent masses. For the H II region the luminosity of the exciting star(s) and density of the surrounding gas are only the first in a long list of parameters that determine its radius. You have to account for any effects of dust and its distribution in the nebula, the metallicity of the gas, and density fluctuations in the gas. In addition, whether the ionizing stars are concentrated or broadly distributed (albeit possibly unresolved), whether they have winds affecting the UV spectrum or the surrounding gas, and even whether they are unresolved binaries with hot compact companions all affect the ionization and recombination rates in the surrounding medium. For galaxies, angular diameters are available only for what you can see, although this may not be a unique tracer of the total mass of the system or even its luminosity. And this is just one of many possible distance calibrators. You see that it's a very hard problem. The arguments are frequently indirect, and the calibrations are sometimes based on correlations whose physical origins are poorly understood.

To summarize, cosmological distance determination is basically a bootstrapping process. First, within the Galaxy and the Local Group, you establish the fundamental relationships between observables and intrinsic attributes. Then, through a painstaking sequence of observations, you extend the range of the calibrated sources, reworking the calibration at each step.[7]

8.2.2 The Cepheid Distance Calibration

A period–luminosity relation is an inevitable consequence of stable pulsation. You may recall that the physical conditions leading to the large-scale pulsation of a stellar envelope occur in a narrow temperature range of the HR diagram. The Cepheid calibration was first obtained, however, from observations alone, independent of theory. Working with the superb plate collection at the Harvard Observatory on the Large Magellanic Cloud variables, Leavitt discovered the correlation and showed it to be so well defined that it could form the basis of a distance scale.[8] Although the distance to the LMC was not known, it was clear that the individual variables occupied the same period range and displayed the same lightcurve morphologies, as the galactic variables. Regardless of the absolute numbers, the relative distances of galactic versus LMC variables could be obtained using the periods alone.

The Cepheid scale is the foundation on which virtually all other calibrations rests. It was introduced with Hubble's discovery in 1927 of classical Cepheids in a field of M31 by analogy with Shapley's method for the globular clusters,[9] yet the persistent difficulties with fundamental calibrators remains as problematic and contentious as ever. Of all the available distance calibrations, this one is most closely linked to theoretical models since it is necessary to distinguish between fundamental and first overtone pulsators on the basis of light curves. There are several ways to determine the light curves. The obvious one is repeated observations and period determination from a large database. Though tedious, this is one of the byproducts of the gravitational microlensing surveys of the Magellanic Clouds and the Galactic bulge. The alternate approach is a statistical one, using snapshot observations and amplitudes to determine luminosities (used in the Hubble Key Project for the distance scale).

The obstacle is the zero point for the luminosity.[10] For linear pulsators of comparatively small amplitude, those in the period range from about one to three weeks, models show only a nearly linear relation between the absolute magnitude in any particular bandpass and the logarithm of the period. It is thus necessary, in principle, to have the distances to only a couple of Cepheids in order to fix the

---

[7] Trimble, V. 1997, *Space Sci. Rev.*, **79**, 793; Livio, M., Donahue, M., and Panagia, N., eds. 1997, *The Extragalactic Distance Scale: Proceedings of the Space Telescope Science Institute Symposium* (Baltimore: STScI); Sandage, A. R., Kron, R. G., and Longair, M. S. 1995, *The Deep Universe: Saas-Fee Advanced Course* 23 (Berlin: Springer-Verlag).

[8] The history of this calibration has been detailed by Fernie, J. D. 1969, *PASP*, **81**, 707 and Berendzen, R., Hart, R. and Seeley, D. 1972, *Man Discovers the Galaxies* (NY: Science History Publ.) to which we refer the reader for very complete discussions of the limitations of the original data.

[9] See Fernie, J. D. 1969, *PASP*, **81**, 707 for a historical overview and Madore, B. and Freedman, W. L. 1991, *PASP*, **103**, 433 for discussions of the classical determinations.

[10] Here we digress on a historical point because it illustrates how hard it is to obtain even the fundamental calibrators. The usual story of how the revision occurred in the Cepheid zero point begins with Baade's announcement at the 1950 Rome meeting of the IAU of a roughly 1.5 magnitude offset, brightening the Cepheids and reducing the Hubble constant. The standard description attributes this modification of the luminosity calibration to the difference between Pop. I and II variables. The historical reality is more complex and illuminating. The offset was already known from early studies of

entire sequence. In practice, the situation is not quite so simple. Only a few Cepheids are close enough to have well determined parallaxes. Only a few are known in clusters whose properties are well determined.[11] The *Hipparcos* satellite provided a zero point of the calibration based on a limited sample of stars. The LMC distance provides the other. Here many calibrators are available, including RR Lyr stars, the expansion of the light echo and ejecta from SN1987A, numerous LBVs, normal main-sequence stars, and eclipsing binaries.[12] Statistical procedures are routinely used for light curve determinations when observing distant galaxies. Cepheid lightcurves are sufficiently regular that random sampling over a reasonable interval of weeks provides sufficient data to determine periods and fit the lightcurve. This technique involves less telescope time than conventional systematic observations and has the fortuitous advantage that it is less subject to aliasing. Since imaging can cover large fields, many variables can be detected at the same time and followed through their cycles.

We have already discussed the use of variable stars as fundamental distance calibrators in Chapter 4, but the problem is important enough to rediscuss here. Feast and Catchpole have rederived the period–luminosity relation from *Hipparcos* parallax data. The slope of either the period–luminosity (P-L) or period–luminosity–color (P-L-C) is not the contentious issue. Rather, it is the zero point of the calibration that has plagued the field for more than 50 years. Recall that the distance scale for the Galaxy was altered by the recognition of differences between Pop. I and Pop. II Cepheids and the correction for interstellar reddening. The same tale was repeated in the 1950s with Baade's change in the zero point and its consequent change of the Hubble constant by a factor of 2. The *Hipparcos*[13] revision of the distance scale from the one known for the Magellanic Clouds may still be affected by two important factors, the variation in metallicity between the

---

Cepheids in the Galaxy. What Baade found was this. During commissioning of the Palomar 5 meter telescope in the late 1940s, a test of the magnitude limit was made using M31. From cluster calibrations the RR Lyr stars should have been observed above the plate limit in about a one-hour exposure if the distance scale was correct and the magnitude limit of the telescope was about $23^m.5$. The exposures failed to find such stars, and only succeeded in resolving a few of the red giants. Since the distance to the galaxy was known from Cepheids, this required a change of about 1.5 mag in the zero point, meaning that the system was farther than previously thought. Fernie points out that the fortuitous cancellation between reddening corrections and the P–L zero point produced a correct absolute magnitude for the RR Lyr stars so the error lay in the Cepheids alone. As an aside, Baade also realized the role of the W Vir stars, but these had been excluded in Shapley's original calibration of the P-L relation. Also, at about the same time, Irwin (re)discovered the clusters surrounding S Normae and U Sagittarii (see Irwin 1955, *MNASSA*, **14**, 38) originally noted by Doig (1925, *Observatory*, **50**, 220; 1927, *ibid*, **51**, 197). In the end, the primary sources of error in the zero point were the neglect of interstellar extinction, which accounted for more than one magnitude of the discrepancy, and systematic errors in parallaxes and photometry. The population differences, while fortuitous, played a relatively minor role.

[11] Feast, M. W. and Catchpole, R. M. 1997, *MNRAS*, **286**, L1; Osterbrock, D. E. 1998, *J. Hist. Astr.*, **29**, 345; Feast, M. 2000, *J. Hist. Astr.*, **31**, 29.

[12] See Guinan, E. F. *et al.* 1998, *ApJL*, **509**, L21 (the distance here is 45 kpc for HV 2274) contrasted with Feast and Catchpole (1997) at 55 kpc, based on Cepheids. As a mark of persistence of this problem despite the plethora of calibrators, the Magellanic Cloud distances are still debated even at this writing (June 2001). Most determinations center between 50 and 52 kpc.

[13] The filter system used for the data is the instrumental response, so, as we discussed in Chapter 2, another calibration is required to transform the response function for this instrument to the standard Johnson *UBV* system.

LMC and the Galaxy and the effects of interstellar extinction on the zero point of the calibration. For the *P-L* relation from Galactic Cepheids, the current calibration is:

$$\langle M_V \rangle = -2.81 \log P - (1.43 \pm 0.1), \qquad (8.1)$$

where $\langle M_V \rangle$ is the mean absolute magnitude of the Cepheid averaged over its pulsational cycle and P is the period in days. The slope derives from the LMC, the zero point comes from parallax measurements toward about two dozen Galactic Cepheids. The same problems that affected the initial determinations of the zero point are still with us. There is a problem with the color term, the fact that the *P-L* relation is not really linear unless a relatively narrow range in period is used. This requires a correct means for obtaining the reddening, which is itself a difficult task given the uncertainties in galactic and (for the Magellanic Clouds) extragalactic extinction laws. Although there doesn't appear to be a serious distortion of the calibration from populations with different metallicities, this is still a worry. Mira variables, whose luminosities are greater than most Cepheids and that have longer periods, can also be used but with more caution. These variables are also less stable, so it is necessary to observe the entire light curve.

8.2.3 Outburst Calibrators

8.2.3.1 Classical Novae and Supernovae

As we have discussed in Chapter 5, not all novae are alike. Despite their range, however, there is a direct relation between the maximum absolute visual magnitude reached during outburst and the rate of decline that was used by Curtis between 1915 and 1920 to argue the distance to M31.[14] Why would you expect this? First, nova explosions involve very small ejected masses, certainly no more than 10^{-3} $M_\odot$ and usually about one or two orders of magnitude less. Whether this takes place in a single explosive event lasting only a matter of a few minutes or a combination of explosive and wind ejection, during the initial optically thick stages the bolometric redistribution of the flux produces a rise in the visual light at the expense of the ultraviolet, shifting the luminosity of the illuminating central star into the visual range. Continued expansion of the ejecta reduces the optical depth, causing a loss of UV opacity due to ionization of the chief absorbing agents—the lines of the iron peak metals—and a drop occurs in the redistributed flux (see next paragraph). The empirical relation correlates the rate of decline of the nova by two or three magnitudes in either the B or V filter with the maximum absolute magnitude in that bandpass. The problem is that there is a large range in this

---

[14] Curtis, H. D. 1917, *PASP*, **29**, 206, in which Curtis argued—based on a small sample of Galactic novae—for the distance of several spiral nebulae. The argument is familiar. The Galactic sample mean was about 6th magnitude, that of the nebular sample was about 15 mag. Thus the distance ratio was about 10, neglecting extinction corrections, making the nebulae likely extragalactic candidates. See also Payne-Gaposchkin, C. 1957, *The Galactic Novae* (NY: Dover); Bode, M. and Evans, A. N., eds. 1987, *The Classical Novae* (London: Wiley-Halstead); van den Bergh, S. 1988, *PASP*, **100**, 8. A variety of calibrations for the maximum magnitude-rate of decline (MMRD) relations are discussed by Downes, R. and Duerbeck, H. W. 2000, *AJ*, **120**, 2007; see also Della Valle, M. and Livio, M. 1995, *ApJ*, **452**, 704.

calibration, which is fixed by those few novae for which there are expansion parallaxes. As always, however it is useful to keep the limitations always firmly in mind. Expansion velocities before the era of space ultraviolet observations were obtained from hydrogen Balmer lines. While these are consistent, they do not represent the dynamics uniquely since the line formation depends on excitation conditions within the ejecta. In addition, there is a range of velocities in the expanding matter, and the final observed remnant may not have the same velocity as the lower density gas that produced the emission lines in the earliest stage of the outburst. Consequently, there is an inherent uncertainty in assigning a velocity to the gas that translates into an often unknown uncertainty in the distance to the resolved ejecta.

Although these were originally used as fundamental calibrators, the dispersion in properties among novae is far greater than for SN Ia explosions. We have discussed the physical origin of the outburst in Chapter 4. Here we remark that the main feature of the calibration is not the maximum brightness alone[15] but the relation between the rate of decline of the lightcurve and the peak optical luminosity. Specifically, in physical terms, unlike a supernova, a nova lightcurve is generated by the reprocessing of the illumination of the central star through the expanding ejecta. As the gas becomes progressively more optically thin, the peak of the radiation distribution shifts first from the infrared into the visible and then into the ultraviolet. From the physics of the initiating thermonuclear runaway (recall here our discussion in Chapter 5), both the mass and the maximum ejection velocity are correlated with the peak white dwarf luminosity. Two different parameters, $t_{2,j}$ and $t_{3,j}$, the time for the light to decline by two or three magnitudes from peak in some filter j, should correlate with M_j, the maximum absolute magnitude in that filter. The physical reason is quite simple. As the ejecta expand, they initially cool and recombine. Since the central star is intrinsically quite hot, the flux is absorbed and re-radiated in the ejecta at longer wavelengths. As the matter expands and thins out, the increasing ionization causes the opacity to decline and the source color temperature increases, thereby producing a wavelength dependent change in the light curve. The same technique that is used in the optical applies as well to radio observations[16] and like the Cepheid calibration, this one is uncertain because of distance and reddening problems and also the intrinsic dispersion in the properties of individual explosions.

The same principle of flux redistribution for a shrouded constant bolometric luminosity object applied to a number of other classes of objects used in the distance scale. For an optically thick object, radiative equilibrium produces a variable narrowband luminosity while the total energy input remains constant if the opacity changes because of expansion. This is the same effect observed in post-AGB objects, especially OH-IR stars, and for highly evolved massive variable

---

[15] In the Shapley–Curtis debate on the scale of the Universe (1920), Curtis used nova maxima to obtain the distance to M31. The problem was that S And, SN 1881, was included in the sample that biased the result toward smaller distances. See Shapley, H. and Curtis, H. D. 1921, *Bull. Nat. Res. Council*, **2**, 171 for the "transcript" of the arguments in the debate.

[16] Hjellming, R. M. 1988, in *Galactic and Extragalactic Radio Astronomy*, 2nd ed. Verschuur, G. and Kellermann, K., eds. (Berlin: Springer-Verlag).

stars. One of the best observed Galactic variables of this class is AG Carinae, whose total luminosity remains constant to within 10% but that has a visual amplitude of nearly 2 mag due to a variable stellar wind mass loss rate. The Hubble-Sandage variables are in this group as well, and are among the most intrinsically luminous stars in the Local Group.[17]

8.2.3.2 Supernovae Type Ia

Kowal was the first to demonstrate the strikingly regular properties of SN Ia lightcurves.[18] The suggestion that these are the best available sources for studying the distance scale, however, originates with Zwicky in the 1960s. In particular, Kowal showed from the peak brightness of the individual supernovae, scaled to their respective redshifts (see text below), that type Ia supernovae are a standard candle. This likely points to an important feature of stellar evolution, since these hydrogen-deficient supernovae must come from nearly identical mass objects. It has been argued for some time that accretion induce collapse of a white dwarf in a binary system, or some other process that makes a star at the Chandrasekhar mass collapse and subsequently explode, and that the existence of a characteristic mass sets the regularity of the properties of the event. As we have discussed for both novae and supernovae, the lightcurve and its properties depend on the amount of ejected mass, the ejection velocity, and the composition of the material. If all supernovae were identical, in other words the nucleosynthesis and hydrodynamics were the same in each event, then they would be truly standard candles. There are, however, differences in the maximum luminosity of the events, with correlate with the timescale for the decline in the brightness of the optical lightcurve (the brighter SN Ia events have longer decline times), and there are also several classes of SN I depending on the hydrogen abundance and the ejection velocity.

Unlike classical novae, where the hot white dwarf remains intact and by illuminating the ejecta powers the lightcurve, type Ia supernovae are internally powered by radioactive decay of ^{56}Ni. They present a *very* small range in maximum absolute luminosity. A complication was introduced into the uniqueness of the calibration by the discovery that the maximum absolute magnitude is correlated with the rate of decline. While this is not an unexpected result if the mass of the ejecta also correlates with the mass of the progenitor, it requires lightcurves complete enough to show both the maximum and early decline to assign the intrinsic luminosity. The calibration for this distance indicator is bootstrapped from the Cepheids, requiring the occurrence of such events in the host galaxies for which Cepheids have been employed to determine distances. Since supernova events are so much brighter than stars they can be extended much farther (and have been observed at distances greater than 5 Gpc). We will soon return to

---

[17] See Humphreys, R. and Davidson, K. 1994, *PASP*, **106**, 1025; Shore, S. N., Altner, B., and Waxin, I. 1996, *AJ*, **112**, 2744 for introductions to this class of stars and a sample multiwavelength analysis of the outburst.

[18] Kowal, C. T. 1968, *AJ*, **73**, 102. See Leidengut, B. 2000, *Astr. Ap. Rev.*, **10**, 179 for a comprehensive survey of the work on this subject through the end of the millennium.

how the supernova calibration relates to the determination of Λ, the cosmological constant.[19]

8.2.3.3 Supernova Remnants
A promising procedure that has been used for SN 1987A in the LMC is to look at the expansion of the remnant. Of course, this is possible optically with only the nearest systems, but can be extended using VLBI techniques. The radio surface brightness correlates with the energy of the explosion and obeys a rough scaling law (see Chapter 6) and thus can be used as a sort of distance indicator. The direct measurement of the expansion requires the determination of the expansion velocity directly from spectroscopy at the optical maximum, which may itself not be representative of the maximum speed of the ejecta.

8.2.4 Calibrators Based on Global Galactic Properties

8.2.4.1 Tully–Fisher Linewidth–Luminosity Relation
In Chapter 7 we discussed how to use the H I rotation curve of a disk galaxy to determine its mass. It is reasonable to go the next step and consider how you might use a similar observation of the integrated linewidth to determine the *luminosity* of the system. If the luminosity is given by a disk of nearly constant or simple radial dependence of the surface brightness and a constant mass to light ratio, then the width of the 21 cm line would be expected to be related directly to the luminosity of the host galaxy. To see how, call I is the surface brightness (intensity) and assume that you make an H I measurement at low spatial resolution, essentially covering the galaxy to some distance R. Then the projected orbital speed within that distance depends on the mass through $v \sim (M/R)^{1/2}$. Now come the critical last assumptions. If the mass to light ratio, M/L, and the surface brightness are approximately constant with distance then the luminosity is $L \sim IR^2$ and we obtain a relation $L \sim v^4$. As a first approximation, this is quite similar to the observed relation obtained in 1977 by Tully and Fisher[20] using a sample of nearby galaxies in the Local Group, M81, and Ursa Major, and then obtaining a distance for the Virgo cluster. Now this is not precisely the same as the observed result because of several of our definitions. First, we must specify some characteristic radius for this unresolved galaxy. One choice is R_e, the length scale determined by the intensity decline in the disk and the scale length for the H I emission. Then we need to decide on the value of the velocity. This is easier since the width of the integrated profile is used, but the velocity may not correspond to R_e. The calibration is even more complicated than this because the radius depends on

---

[19] Fillipenko, A. 1997, *ARAA*, **35**, 309; Branch, D. and Tammann, G. A. 1992, *ARAA*, **30**, 359; and Branch, D. 1998, *ARAA*, **36**, 17 review the observations on distance scale uses of SN Ia. Kim, A. G., Gabl, S., Goldhaber, G. et al. 1997, *ApJ*, **476**, L63: $0.35 < z < 0.65$ sample in addition to the Calan/Tololo catalog. Hamuy, M., et al. 1996, *AJ*, **112**, 2398: intrinsic dispersion in the SN Ia relation using measurements from galaxies with $z < 0.1$. These studies highlight the requirement that lightcurves must be used in conjunction with the maximum brightness in order to properly account for the distances.

[20] Tully, R. B. and Fisher, J. R. 1977, *Astr. Ap.*, **54**, 661 is the founding paper on this method.

the wavelength at which the galaxy luminosity is determined. Some basic unresolved aspects of this calibration center on which bandpass is best to use for the determination of the angular diameter of the galaxies of the sample, where to place the radius for the linewidth measurement, and the physical origin of the calibration. Although the zero point is the real question, the slope is comparatively fixed. The observational data yield a fairly tight correspondence between the galactic surface brightness and 21 cm linewidth:

$$I_7 \approx 5.8 \log \frac{W_c}{400 \text{ km s}^{-1}}, \qquad (8.2)$$

where I_7 is the intensity measured at 7000 Å, a convenient wavelength for such measurements that is largely uncontaminated by stellar or nebular lines. For the blue magnitude, the following relation has been determined[21]

$$M_B = -(6.31 \pm 0.05)\log W_c - (3.72 \pm 1.24). \qquad (8.3)$$

This is not quite the same as the theoretical relation [see, however, the discussion in Strauss and Willick (1995) concerning the various choices of wavelengths and the fundamental observational results]. A more recent determination of the corresponding molecular Tully–Fisher relation, based on ^{12}CO (1–0) data for Virgo cluster galaxies, yields

$$M_B \approx -(6.22 \pm 0.05)\log W_{\text{CO}} - (4.07 \pm 1.17). \qquad (8.4)$$

Therefore, regardless of the details, the Tully–Fisher relations are all expected to be dynamical in origin. Their principal uncertainty is with the M/L ratio for their conversion to absolute distances.[21] Even respecting all these caveats, the use of rotation curves is very compelling. They are distance independent and also metallicity independent. The Tully–Fisher relation is now one of the cornerstones of observational cosmology and it is likely to remain so for quite some time.

8.2.4.2 Faber–Jackson Velocity Dispersion–Luminosity Relation

In the search for standard candles, bright elliptical galaxies have been important objects. Faber and Jackson discovered a simple relation between the luminosity of an elliptical and the integrated core velocity dispersion. You would expect this in a relaxed system since the virial theorem holds for such ensembles. The same dependence of velocity on luminosity applies for such systems as the theoretical Tully–Fisher relation since it is again an equation of the kinetic and gravitational energies. Thus, $L \sim \sigma_c^4$. The same warnings apply; there are many hidden assumptions about the dynamics of the stellar population, not the least of which is that it is isotropic. As we discussed in Chapter 7, this calibrator is also one of the variables in the Fundamental Plane representation of properties of elliptical

---

[21] Strauss, M. A. and Willick, J. A. 1995, *Physics Reports*, **261**, 271; Schöniger, F. and Sofue, Y. 1997, *Astr. Ap.*, **323**, 14.

galaxies.[22] For distances and structure, only the intrinsic galactic luminosity matters so it is particularly important to find that it can be correlated with a distance independent quantity, such as the velocity dispersion.

8.2.5 Maser Proper Motions

A very recent development has been the detection of proper motions for extragalactic maser sources. So far, this has only been accomplished for the circumnuclear disk of NGC 4258 but it has enormous implications over time as observations accumulate.[23] The first step involved the discovery of H_2O masers with large velocities, of order 1100 km s^{-1} and with a baseline of several years, VLBA observations showed proper motions of 31.5 μarcsec yr^{-1}. This translates immediately into a distance, 7.2 $\pm$ 0.3 Mpc. With a number of such sources available and more being discovered from targeted searches, it is likely this technique will yield absolute distances to many of the nearest galaxies that harbor fundamental distance calibrators. There are two limitations: sensitivity and sources. The VLBA has a large enough collecting area that it can clearly detect such maser sources in the nearest galaxies out to, probably, the distance of Virgo and possibly Fornax. The rub is that the sources must be detectable and this likely requires special conditions such as an active galactic nucleus. Although the masers in the Galactic center can be easily observed, they are less than 10 kpc distant and would certainly be inaccessible in external galaxies with current instrumentation.

8.2.6 Surface Brightness Fluctuations

The idea behind the surface brightness fluctuation (SBF) method is beautifully simple. Imagine, for instance, a very populous star cluster. All members formed at the same time from a single luminosity function so you expect very to see only a relatively few massive stars surrounded by many more of lower mass and luminosity. Seen close up, the distribution appears patchy because of the comparatively easy visibility of the brighter stars. Viewed from progressively farther away, however, this patchiness washes out until the distribution looks more or less uniform. Since you know the population fraction in the massive components, you also know for a given luminosity what the intrinsic amplitude of the surface fluctuations should be. Therefore, you can obtain an estimate of the distance using the integrated spectrum (or spectrophotometry) of the cluster and measuring its patchiness.

For a galaxy of a given Hubble type, you would expect that individual H II regions and associations would be visible if it were close enough. The farther away it is, the smaller these regions appear and the smoother the resulting appearance

---

[22] Faber, S. M. and Jackson, R. E. 1976, *ApJ*, **204**, 668; Tonry, J. L. and Davis, M. 1981, *ApJ*, **246**, 666; **246**, 680.

[23] Miyoshi, M., et al. 1995, *Nature*, **373**, 127 first reported the discovery of maser emission from this galaxy. Using the VLBA, proper motions were announced by Herrnstein, J. et al. 1999, *Nature*, **400**, 539. Comparison with the HST Key Project Cepheids was presented by Moaz, E. et al. 1999, *Nature*, **401**, 351.

of system. This is the idea behind the secondary calibration of *surface brightness fluctuations*. It rests on certain assumptions about the rate of star formation, the distribution and number of molecular clouds, and the uniformity of the Hubble sequence. As we discussed in Chapter 7, these are tricky assumptions since galactic classifications are not as taxonomically distinct as in stellar spectral classification. The use of this statistical property of a galaxy is not the best way to determine its distance. You know that you can better distinguish the features of someone as she approaches you, and there is some critical distance at which you can recognize who she is because of the increased resolution. Galaxies stay where you put them, at least on a human timescale, and therefore the technique must rely on local calibrators. Recall that the problem is that the local galaxies are members of either the Local Group or Virgo clusters, and they may not be representative of systems at large distance, hence earlier epochs, or other environments where collisions and mergers may be more important.

8.2.7 Luminosity Functions and Statistical Distance Calibrators: Internal Galactic Properties

It is always precarious to use a statistical predictor of any property of the world, especially if you don't understand its origins, but that is precisely what is frequently done for extragalactic problems. Necessity drives you to do some strange things, not the least of which is the use of aggregate luminosity diagnostics. We discuss three of these here, all of which have an intuitive basis but none of which is yet completely understood.[24]

8.2.7.1 Luminosity Functions of Planetary Nebulae and Globular Clusters

On the basis of Galactic and Local Group observations, and with some limited assistance from stellar evolution modeling, it appears that the luminosity function for planetary nebulae can be treated as a universal absolute calibration. The idea is much the same as fitting the main sequence of a cluster. Although incompleteness limits access to the faintest nebulae, narrow band line filter photometry can be used to select the planetaries in galaxies for which the individual objects cannot be spatially resolved and identified. If the general shape of the function can be fitted to the sample, the absolute magnitude of the brightest planetaries and the turnover in the distribution provide a direct measure of the distance. There are several problems with this method, the most important is internal reddening within the target galaxy, but the procedure is analogous to the Cepheid calibration and is very useful as a statistical measure of the distance.

In principle, the luminosity function for globular clusters provides a comparable distance indicator to the planetary nebula function. The different sources for variation make it a corroborative rather than redundant technique. We've discussed the evolution of these systems several times in earlier chapters. Here we

---

[24]An excellent critical overview is available in Jacoby, G. et al. 1992, *PASP*, **104**, 599. This paper, written collaboratively by proponents of a number of the secondary methods, provides the broadest statement of their goals and limitations. Although somewhat outdated in specific numbers, it remains a valuable reference to the uncertainties.

recall that the globular clusters are likely not a static population in the history of a galaxy. Tidal disintegration can occur among disk systems through impulsive accelerations when the clusters traverse the plane.[25]

8.2.7.2 H II Region Sizes

The maximum sizes of H II regions are governed by the combined effects of the stellar luminosity function and the distribution of ambient interstellar gas. The Hα luminosity is well correlated with the rate of star formation, and the sizes of equilibrium H II regions should depend only on the maximum number of OB stars in a star forming region through the Strömgren radius. For stationary regions, the radius should scale as $L^{1/3}$ or $N_{OB}^{1/3}$ (since the luminosity enters as a linear combination of the individual stellar contributions weighted by the mass function) so it should be possible to use the embedded stars and size of the surrounding emission region as complementary distance indicators. The complication comes from the properties of the galaxy's interstellar medium and the internal effects of metallicity and dust that we have discussed in Chapter 6.

8.2.7.3 Schechter Function for Galaxy Clusters

For larger structures, such as clusters of galaxies, an analogous procedure that uses a luminosity function can be employed. The most widely used form is the *Schechter function*:[26]

$$\phi(L)dL = (0.015 \pm 0.001)h^3 \left(\frac{L}{L_\star}\right)^\alpha e^{-(L/L_\star)} d\left(\frac{L}{L_\star}\right). \tag{8.5}$$

Here $L_\star$ and α are free parameters, where $L_\star$ is the median cluster galaxy brightness. Much like the Salpeter function for stars, this function has no compelling physical explanation, but its form is especially convenient as a normalizable statistical distribution. In fact, in clusters in the earlier stages of the Universe, it is entirely plausible that interactions and starburst activity will alter this compared with the present day. At times, special significance has been attached to the so-called $L_\star$ galaxy. This is, based on the assumed distribution, the average galaxy in the cluster, but it is selected on the basis of the form of the distribution function and therefore is not dynamically well explained. It has been recognized that there exist a luminosity function different for each morphological galactic type. Thus, the total luminosity function, weighted by all morphological types, varies with absolute magnitude as

$$\phi(M) = \sum_T \alpha_T \phi_T(M). \tag{8.6}$$

Sample parameters determined for field galaxies are $\alpha_{el} = 0.48$, and $L_{\star el} = 0.8 \times 10^{10} \, h^{-2} \, L_\odot$ for ellipticals, and $\alpha_{sp} = 1.24$, and $L_{\star sp} = 1.05 \times 10^{10} \, h^{-2} \, L_\odot$ for spirals.

---

[25] Jacoby, G. H. et al. 1992, *PASP*, **104**, 599; Ostriker, J. P. and Gnedin, O. Y. 1997, *ApJ*, **487**, 667.
[26] Schechter, P. 1976, *ApJ*, **207**, 297.

8.3 FUNDAMENTAL PARAMETERS: THE REDSHIFT AND THE HUBBLE CONSTANT

8.3.1 The Redshift (z)

When you observe the spectrum of a distant galaxy, you find that the spectrum looks similar to those you know from the local region of space but it is shifted *systematically* toward the red. If λ_0 is the rest wavelength, an observable called the *redshift* is defined by

$$z = \frac{\lambda - \lambda_0}{\lambda_0} = \frac{\lambda_r - \lambda_e}{\lambda_e} \qquad (8.7)$$

where the alternate symbols stand for emitted (e) and received (r) wavelengths. Hubble and Humason in 1931 were the first to derive the simple linear relation between the redshift, which is usually quoted for $z \ll 1$ nearby galaxies as a recession velocity, and distance for nearby galaxies, within about 20 Mpc:

$$cz = H_0 d. \qquad (8.8)$$

This expression defines the *Hubble constant*, H_0, which is arguably the most important quantity in observational cosmology. Its units are an inverse time, km s^{-1} Mpc^{-1}, and it therefore defines a characteristic timescale for the (universal) recession.[27] As such, it provides one of the main constraints on physical modeling of the evolution of the cosmic mass.

While the Hubble law and the definition of the redshift reduces to the familiar Doppler principle for low velocity (and is often so quoted), this is misleading in a cosmological context and requires a cautionary admonition: the Hubble law *does not* involve a velocity. Even if the peculiar radial velocities of individual galaxies can be expressed in z, once outside the distance to the Local Group the observed recession is *universal* and indicates a global change that must be explained without requiring the observer to be situated at a special place in the universe. Although the redshift is often quoted as a radial velocity, this is a mere convenience, but we'll discuss that aspect of the explanation when we review cosmological models.[28]

Just for now, we'll take the approach that the redshift is a measurable attribute of a galaxy that is used to determine its distance, in the same spirit as many of the other distance calibrators whose origins are hazy. Then, after we examine some

---

[27] From what we've just discussed, you can see how this would have escaped recognition before Hubble's work since it requires that the distance be determined from one of the standard calibrations. The redshift alone is only an effective measure of the distance once the expansion law has been determined.
[28] In this paper announcing the correlation with distance, Hubble and Humason referred to this as a measure of the velocity of the galaxy, although in later work Hubble consistently referred to this as a shift in the lines that was simply quoted in terms of a velocity.

model independent consequences, we will return to the question of its origin within relativistic cosmologies.[29]

8.3.2 Current Results for the Hubble Constant

The controversy over the precise value of H_0 is tied to its measurement of a cosmic timescale. The first published value was about 500 km s^{-1} Mpc^{-1}, which in the early 1930s did not contradict any known constraints on the age of the Sun or the Earth. In fact, it was comfortably long enough to accommodate any geologic process and more than enough to satisfy the terrestrial requirements for biological evolution. Within a decade, however, a cloud emerged on the horizon even with a steady drop in the measured value of H_0. By the end of the 1940s, the measured constant, of order 250 km s^{-1} Mpc^{-1}, provided an age for the Universe that was less than the inferred age for the globular clusters and the Sun on the basis of the emerging picture of stellar interiors and rates then known for nuclear reactions. By 1952, when Baade announced the change in the distance scale based on the recalibration of the Cepheid luminosities, alternative models to standard cosmology—in particular the *steady-state scenario* of Bondi, Hoyle, and Gold—were being discussed out of desperation.

When launched in 1990 the primary goal of the *Hubble Space Telescope* (HST) was to obtain the value of H_0 to within 10%. Considerable telescope time was granted to the HST Key Project, the methodology of which was to observe Cepheids in a select sample of bright, relatively nearby galaxies for which secondary calibrators were also available, especially SNe Ia and the Tully–Fisher relation. The final result of the Key Project is[30]

$$H_0 = 71 \pm 4 \text{ (random)} \pm 7 \text{(systematic) km s}^{-1} \text{ Mpc}^{-1}.$$

---

[29] Let us introduce the interpretation we will use throughout this chapter. The redshift is connected to the large scale geometry through the metric, which is the fundamental measure of the global geometry of the Universe. This was first realized by Einstein as early as 1917, in the earliest days of general relativity, but its connection with the overall expansion of the Universe was exploited only later, starting with Friedmann and culminating with Robertson, Walker, and Lemaître. You can already see this through the Doppler effect but here we are dealing with observations by stationary observers. In general relativity, the connection between the time and frequency comes through the measurement of time dilation between two observers' frames of reference. For a light signal, since the proper distance always vanishes along null geodesics, we have $\int g_{00}^{1/2} dt = \int g_{11}^{1/2} dr$, which provides a direct link between the distance and the time dilation. The change in the frequency results from the curvature term, in the language of differential geometry, or from the change in the scale length of the spacetime during the propagation of the photon from source to observer. The link between the redshift measurement and the cosmological model comes from the properties ascribed to the geometry of spacetime. In the Schwarzschild solution, g_{00} depends only on the radial distance since this represents the static spacetime surrounding a central point mass. Therefore, the redshift represents a change in the frequency of a photon as it climbs out of a fixed gravitational potential. If the spacetime is not static, but the geometry remains fixed even as a scaling factor changes, then the metric depends explicitly on time and, as we will see presently, the redshift between stationary observers results from the time dependence of this scaling.

[30] Freedman, W. L. et al. 2001, *ApJ*, **553**, 47.

A separate HST project[31] focuses on the Cepheid–SN Ia correlation. This group finds

$$H_0 = 58.5 \pm 6.3 \, (2\sigma) \text{ km s}^{-1} \text{ Mpc}^{-1}.$$

There is still disagreement about important details of the calibrators and the agnostic position seems best at this point—the Hubble constant derived from comparatively local measurements, out to about $z \approx 0.1$, lies in the range from 60 to 80 km s^{-1} Mpc^{-1}.

The distance scale requires additional information, specifically the distance to the LMC. Uncertainties exist here as well, with the range being 50 ± 4 kpc (corresponding to a distance modulus of $18^m.56 \pm 0^m.15$) depending on the method used for obtaining the distance (RR Lyr stars, binary systems, or the expansion of the luminous ring around the SN 1987A site). Of all aspects of this book to be most susceptible to change, the value of the H_0 is likely the best bet. However, you will notice that the value has fallen by almost an order of magnitude since it was the first publication of the expansion law, so there has been convergence between the age of the Universe as measured by this parameter and the results of separate studies of stellar and galactic evolution.

In light of the observational controversies, we will hereafter take the conventional middle road and choose no particular value when making numerical estimates of cosmological numbers. The conversions between time, distance, and redshift require the Hubble constant, so we will use the contrivance of absorbing all uncertainties in the present determinations of its value into a dimensionless *Hubble parameter* that is defined by

$$h \equiv \frac{H_0}{100 \text{ km s}^{-1} \text{ Mpc}^{-1}}.$$

The choice of scaling value for H_0 remains from a time when the observational upper limit to H_0 was about 50% higher than presently determined.

8.3.3 Absolute Age Calibrations: Decay of Radioactive Elements

Nothing is so pleasing as the chance to connect the cosmos with the laboratory, and measurement of the decay of several important isotopes is just such an opportunity. Radionuclide abundances provide limits on the age of the Galaxy. Taking the broader view, anything that constrains the range in Galactic ages also has cosmological significance since it provides bounding values for H_0. Notice that some of these timescales are extremely long. The K-Ar dating technique is especially useful for bodies in the solar system, while any indication of a change in any of the other species listed in Table 8.1 would be very significant for cosmology. You know that the heaviest isotopes are built up by the *r*-process, which as we described is a queue. If you have some proxy element that originates *only* by this process and doesn't decay, then you might be able to say if there had been some

---

[31] See the summary paper by Parodi, B. R., Saha, A., Sandage, A., and Tammann, G. A. 2000, *ApJ*, **540**, 634.

TABLE 8.1 Long-Lived Isotopes of Cosmological Importance

Isotope	Decay Product	$t_{1/2}$ (Gyr)	% Abundance
^{40}K	^{40}Ar	1.277	0.01
^{187}Re	^{187}Os	43.50	62.60
^{232}Th	^{228}Ra	14.05	100
^{235}U	^{231}Th	0.704	0.72
^{238}U	^{234}Th	4.468	99.27

decay of the nucleus over the galactic lifetime even if you could not separate the lines of the individual isotopes. One particularly useful species is ^{232}Th, an element that has no stable phases. Europium is a purely r-process element among the rare earths, one whose lines are particularly easily observed in cool stars.[32] Next, take standard yields for r-process nucleosynthesis that reproduce the solar system abundance ratio—in this case Th/Eu ≈ 0.46—and compare it to the observed photospheric value. One reason you may see a difference is from decay of the unstable species. For CS 22892-052 the *observed* ratio is Th/Eu ≈ 0.22. Of course, there is a lot of intermediate theory involved here, in particular the production of synthetic spectra, but taken at face value this can be interpreted using:

$$[\text{Th/Eu}]_0 = [\text{Th/Eu}]_{\text{gal}} e^{-t_{\text{gal}}/\tau}, \tag{8.9}$$

where $\tau = 1.443 t_{1/2}$ now t_{gal} is a derived Galactic "age". For our example, $t_{\text{gal}} \approx 15$ Gyr. Analyses such as this will certainly improve in the coming years with the increasing availability of very large apertures and high-resolution spectrographs.

8.3.4 Cosmological Corrections to Galaxy Properties and the Redshift

With the link between the redshift and distance, a new observational problem emerges. *Even without an explanation for the dependence of z on distance, or a model for the large scale structure, the large distances implied by high redshifts requires that any galaxy observed at large z is also seen as it was long ago.* For surveys, this has several consequences. First, for $z > 1$ we are at a time when star formation in the system we're observing may be very different than now, and therefore the intrinsic properties of any object have to be extrapolated backward in time to understand what we're seeing. Second, we sit in one window of the spectrum and look at the distribution of light from distant objects that are shifted with respect to that window. It is neither as simple as correcting the spectrum of a radiating brick nor as physically well established—stars within galaxies produce aggregate spectra with many absorption bands and complicated continua, and the inferences made from even multiwavelength observations are uncertain.

---

[32] For instance, Eu II 4129.70, 4205.05 Å are both strong resonance lines in spite of hyperfine structure complications. You now see how even simple stellar atmospheres analyses can have a profound effect on big items! See Sneden, C. et al. 1996, *ApJ*, **467**, 819; Sneden, C. et al. 2000, *ApJL*, **533**, L139 for examples of how the analysis of one extremely metal-poor ([Fe/H] ≈ −3.2) star, CS 222892-052, was performed.

Even without knowing what how z connects to a specific cosmological model, which we will postpone until Section 8.4.2, the redshift has important observational consequences. The first is that the frame in which a galaxy emitted a photon is not the same as the one in which it is observed. The *entire* spectrum shifted in wavelength. Of course, the same is true for the Doppler effect, but in general it is at much lower velocity. For instance, the redshift produces a change in the brightness not only because of the distance but also because we observe in a finite bandpass. If we could see the bolometric flux of the Galaxy

$$F = \int F_\lambda \, d\lambda = \int_0^\infty \frac{L(\lambda_0(1+z)^{-1}) d\lambda}{4\pi D_L^2 (1+z)^2}, \quad (8.10)$$

then we could simply scale the observed brightness to rest frame by knowing the redshift and the distance, D_L. But we see through a fixed window so the width of the bandpass is also changed from that in the observer's frame $\Delta\lambda_0$ by a factor of $\Delta\lambda = \Delta\lambda_0/(1+z)$. This *cosmological dimming*, as it is sometimes called, requires additional corrections, although it is not necessarily going to produce a lower-bandpass luminosity, as we will discuss. You will have noticed that we introduced a new quantity, D_L, the *luminosity distance*. This is a *model dependent* quantity, not simply the linear distance to the object, and it is the same one we get from the Hubble law. Remember that the distance is known through calibrators by their observed flux. This illustrates the complication that results from taking the picture we have built up from local observations into the cosmological context. In the words of Eddington, *you cannot believe an observation until it is confirmed by theory*.

Linewidths are also changed because of the redshift, so this effect must be included in any distance calibrator that is sampled to large values of z. For instance, since the Tully–Fisher relation correlates luminosity with the *width* of the 21-cm line, you must apply a correction of $(1+z)$ to the measured linewidth to obtain its intrinsic value. The same applies to measurements of equivalent widths for the components of the Lyα forest.

The simplest approach to these problems is through application of the *K correction*. The term derives from the general term used for magnitude corrections and was introduced to cosmological investigations in Humason, Mayall, and Sandage (in 1958). This uses a template galaxy spectrum, for instance an elliptical galaxy observed at low redshift, to determine what the change will be when the spectrum is redshifted relative to the observer. In other words, you figure out what happens when $F(\lambda) \to F((1+z)\lambda_0)$. But as we have just seen, it is not that simple since the effective bandpass changes because of the redshift, $\Delta\lambda_0 = (1+z)\Delta\lambda$, and possibly because the choice of standard spectrum might be compromised by evolutionary effects. With appropriate care, the K correction is nearly independent of stellar evolution, which is a good thing, and depends only on the observed stellar population. We say *nearly* because metallicity enters through the absorption line distribution and this is the product of galactic star forming history. On the other hand, multiwavelength observations of nearby galaxies constrain the continuum distribution, producing templates that can serve to provide such corrections for relatively small lookback times, of order a few Gyr or so.

Evolutionary corrections are more model dependent and, therefore, more controversial. The precise star formation history of the system is only obtained through a thorough analysis of the metallicity distribution among the constituent stars, and this is impossible for extragalactic systems beyond the Local Group in anything but an aggregate fashion. Imagine that star formation in the target galaxy was more vigorous in the past, for instance. The intrinsic galactic luminosity would then be greater than a similar system at low redshift, and one might expect a greater continuum contribution from massive hot stars. In combination, these produce a partial cancellation of the K correction. It is even possible that a sufficiently large increase in luminosity with lookback time could exceed the K correction, leading to a closer distance and a smaller value for H_0. In other words, even at this basic phenomenological level, the determination of the distance is fraught with model-dependences that require an understanding of the evolution of the stellar population.

Why is this elaborate machinery required for correcting any distance indicator that translates observed brightness into intrinsic luminosity? Good question. Consider an elliptical galaxy with a globular cluster like population of stars observed in the V filter. For a redshift of about 0.5 or less, the bulk of the continuum remains well sampled by the bandpass. If, however, z is of order unity or greater, the observed brightness drops far faster than the distance would warrant by a simple linear law because of the intrinsic UV flux deficiency you know well from our discussion of stellar atmospheres in Chapter 3. Without including this bolometric effect, the galaxy would be too faint at the observed redshift, a quantity that is unaffected by the stellar distribution, and thus the distance would have to be increased for a fixed velocity, leading to an overestimate of H_0.

8.4 RELATIVISTIC COSMOLOGY

First, we begin with the concept of the *proper distance*. This is necessary in cosmology because the distances are so vast and the timescales so long that the finiteness of the speed of light comes into play.[33] We are limited to a unique vantage point in our study of the Universe. We're stuck on Earth and all of our measurements are made over a comparatively local scale. Yet unless we are also at a *special* spot, we can infer information about the large-scale structure by adopting two assumptions. Invoking what has become known as the *cosmological principle*—also called the *Copernican principle*—we assert that our view is the same

---

[33]There are many excellent specialized monographs on cosmology. We list some of them here, but the corpus grows every year. Weinberg, S. 1972, *General Relativity and Cosmology* (NY: Wiley); Hawking, S. and Ellis, G. 1973, *The Large Scale Structure of Spacetime* (Cambridge, UK: Cambridge Univ. Press); Zel'dovich, Ya. B. and Novikov, I. 1983, *Relativistic Astrophysics* vol. 2 (Chicago: Univ. Chicago Press) (see also Zel'dovich's *Collected Papers*, vol. 2); Peebles, P. J. E. 1993, *Principles of Physical Cosmology* (Princeton: Princeton Univ. Press); Bernstein, J. 1995, *Introduction to Cosmology* (Englewood Cliffs: Prentice-Hall); Bothun, G. 1996, *Modern Cosmological Observations and Problems* (London: Taylor and Francis); Peacock, J. A. 1998, *Cosmological Physics* (Cambridge, UK: Cambridge Univ. Press); Liddle, A. R. 1999, *Introduction to Modern Cosmology* (NY: Wiley); Harrison, E. R. 1999, *Cosmology: The Science of the Universe*, 2nd ed. (Cambridge, UK: Cambridge Univ. Press).

as any observer would have of the motions and constitution of the cosmos at the same time and that the Universe is a smooth spacetime. The first *ansatz* is like the covariance principle we took as the basis of general relativity; the second is that the spacetime is a manifold that is described by a metric. The simple point that looking out into space is looking back in time, coupled with the desire to come up with a global picture of the geometry of spacetime in light of the observation of the expansion, requires that we treat the Universe in terms of general relativity. We can do no better for justification than to quote from Einstein's autobiographical remarks (in 1949): " From my point of view, one cannot arrive, by way of theory, at any at least somewhat reliable results in the field of cosmology if one makes no use of the principle of general relativity." In what follows, we derive this metric for a homogeneous, isotropic, time-dependent universe and then show how the distribution of matter leads to evolution equations for the physical properties. Several recurrent themes will continue to run through the rest of this chapter. The first is *stability*; is it possible to explain the redshift as a property of the metric? The second is *thermal balance*; how does the distribution of matter and radiation explain the CBR and how does the interaction of these constituents develop over time. Finally, the third deals with the development of *structure*; why are the masses we see organized as they are into big things such as galaxies and clusters of galaxies and how do these evolve dynamically over time?

8.4.1 Derivation of the Metric

In Chapter 1, we reviewed the Lagrangian approach for obtaining the Christoffel symbols and, through them, the Riemann tensor and curvature scalar. For a Schwarzschild solution, this requires solving the gravitational field around a static point mass for which we know that at large distance the potential obeys the Newtonian approximation. For a distributed mass of the sort we anticipate is required to describe the Universe at large, this requires obtaining the interior solution for the field equations so some additional care must be exercised.

8.4.1.1 First Pass: Differential Geometry
Imagine a Euclidean 4-space, $(x^0, \mathbf{x})$, where the line element is given by

$$ds^2 = (dx^0)^2 + d\mathbf{x} \cdot d\mathbf{x}. \tag{8.11}$$

Notice that the metric does not have the usual causal signature, although it may, depending on the definition of x^0. Then

$$(x^0)^2 + \mathbf{x} \cdot \mathbf{x} = R^2 \tag{8.12}$$

defines the radius of a sphere in this 4-space, from which we obtain the metric for the embedded 3-sphere, S^3, removing x^0 by differentiating Eq. (8.12):

$$dx^0 = -\frac{\mathbf{x} \cdot d\mathbf{x}}{(R^2 - \mathbf{x} \cdot \mathbf{x})^{1/2}}. \tag{8.13}$$

Replacing the usual 3-space with a spherical coordinate system we find that

$$ds^2 = \frac{dr^2}{1 - r^2/R^2} + r^2[d\theta^2 + \sin^2\vartheta \, d\phi^2] \qquad (8.14)$$

describes the sphere. In the limit where $R \to \infty$, this becomes the normal euclidean metric for spherical coordinates. We should mention that Einstein's first paper on relativistic cosmology in 1917 used precisely this argument in setting out the likely form for a spherically symmetric metric for a homogeneous *static* spacetime.

8.4.1.2 Second Pass: Symmetries and the Killing Vectors

In Chapter 1, we discussed the importance of symmetries for determining the metric. For the cosmological problem this is especially important since we have assumed only that the spacetime is isotropic. Recall from Chapter 1 that the Killing equation is

$$g_{kl,m}\zeta^m + g_{ml}\zeta^m_{,k} + g_{km}\zeta^m_{,l} = 0, \qquad (8.15)$$

where ζ^i is the Killing vector that, as you'll recall, is the trajectory along which the metric doesn't change. From the cosmological principle, we know that the geometry seen by any observer is the same and that the metric coefficients and proper distances scale only with time. We will make one small change in the metric for a spherical coordinate system and then see what happens. For a spherical coordinate system, we can assume that

$$ds^2 = dt^2 - g_{11}\,dr^2 - g_{22}\,d\vartheta^2 - g_{33}\sin^2\vartheta\,d\phi^2, \qquad (8.16)$$

where we will subsequently use $x^0 = t$, $x^1 = r$, $x^2 = \vartheta$, and $x^3 = \varphi$, for the coordinates. The metric is independent of φ and t, the time dependence having been removed using the scaling factor R. Then, noting that the metric is strictly diagonal, Eq. (8.15) becomes

$$g_{11,1}\zeta^1 + 2g_{11}\zeta^1_{,1} = 0, \qquad (8.17)$$

which reduces to:

$$\frac{\partial \ln \zeta^1}{\partial r} = -\frac{1}{2}\frac{\partial \ln g_{11}}{\partial r} \to \zeta^1 = \mathscr{G}(\vartheta)g_{11}^{-1/2}. \qquad (8.18)$$

Here $\mathscr{G}$ is an arbitrary function of the angle θ alone. We can now exploit the fact that $g_{22} = r^2$ to obtain

$$g_{22,1}\zeta^1 + 2g_{22}\zeta^2_{,2} = 0 \to \zeta^2_{,2} = \frac{1}{r}\zeta^1. \qquad (8.19)$$

Finally, we use the fact that the nondiagonal components of the metric vanish along with Eq. (8.17) to obtain

$$g_{11,1}\,\zeta^1_{,2} + g_{22}\,\zeta^2_{,1} = 0. \tag{8.20}$$

We're almost there. Now using

$$\zeta^2_{,12} = -\left(r^{-1}\zeta^1\right)_{,1}$$

and:

$$g_{22}\left(r^{-1}\zeta^1\right)_{,1} = g_{11}\,\zeta^1_{,22},$$

so we arrive at the final equation for the radial component of the metric

$$r^2 g_{11}^{-1/2}\,\frac{\partial}{\partial r}\left(r^{-1} g_{11}^{-1/2}\right) = \frac{\mathscr{G}_{,22}}{\mathscr{G}} = \text{constant}. \tag{8.21}$$

Notice that the left-hand side now depends only on r while the right-hand side is independent of r. Solving Eq. (8.21) by noting that you can call the dependent variable $r^{-1} g_{11}^{-1/2}$ and rearranging terms, you obtain the explicit form for g_{11}:

$$g_{11} = \left(1 - kr^2\right)^{-1}, \tag{8.22}$$

where we have lumped the integration constants into k for convenience. This is precisely the spherical metric we derived from simple embedding arguments.

8.4.1.3 Third Pass: The Field Equation

Geometric arguments and symmetry alone get us only so far. There is matter in the Universe, and we know from general relativity that it will produce a spacetime whose curvature reflects its distribution. So as our next step, let's see how to derive the metric from the field equations. We start with the assumption of homogeneity and assert the cosmological principle that the geometry looks the same for all observers. We also define a time coordinate that holds for all observers in the comoving frame, although without requiring that the spacetime be static. In the local frame we can write the metric as

$$ds^2 = dt^2 - e^{F(t,r)}\left(dr^2 + r^2 d\vartheta^2 + r^2 \sin^2\vartheta\, d\varphi^2\right), \tag{8.23}$$

where the exponential factor F is a formal convenience as it was for the Schwarzschild metric. The Lagrangian becomes

$$L = \frac{1}{2}\left(\left(\frac{dt}{ds}\right)^2 - e^F\left[\left(\frac{dr}{ds}\right)^2 + r^2\left(\frac{d\vartheta}{ds}\right)^2 + r^2\sin^2\vartheta\left(\frac{d\varphi}{ds}\right)^2\right]\right), \tag{8.24}$$

now using s for the parameter in the geodesic equation described in Chapter 1. In the following, a dot ($\cdot$) means differentiation respect to t, and a prime ($'$) means differentiation with respect to r, both of which are comoving coordinates. Then, for instance, you can read off the components of Γ^0_{ij} from:

$$\ddot{t} = -\tfrac{1}{2}\dot{F}e^F\left(\dot{r}^2 + r^2\dot{\vartheta}^2 + r^2\sin\vartheta\dot{\varphi}^2\right). \tag{8.25}$$

We won't write the other three "dynamical" equations explicitly, but simply state the results for the Christoffel symbols[34]

$$\Gamma^0_{11} = \tfrac{1}{2}\dot{F}e^F, \qquad \Gamma^0_{22} = \tfrac{1}{2}\dot{F}r^2 e^F$$

$$\Gamma^0_{33} = \tfrac{1}{2}\dot{F}r^2\sin^2\theta\, e^F \qquad \Gamma^1_{01} = \tfrac{1}{2}\dot{F},$$

$$\Gamma^1_{11} = \tfrac{1}{2}F', \qquad \Gamma^1_{22} = -r^2\left(\frac{1}{2}F' + \frac{1}{r}\right),$$

$$\Gamma^1_{33} = -r^2\sin^2\theta\left(\frac{1}{2}F' + \frac{1}{r}\right), \qquad \Gamma^2_{02} = \tfrac{1}{2}\dot{F}$$

$$\Gamma^2_{12} = \frac{1}{2}F' + \frac{1}{r}, \qquad \Gamma^2_{33} = -\sin\vartheta\cos\vartheta$$

$$\Gamma^3_{03} = \tfrac{1}{2}\dot{F}, \qquad \Gamma^3_{13} = \frac{1}{2}F' + \frac{1}{r}$$

$$\Gamma^3_{23} = \cot\vartheta. \tag{8.26}$$

All others equal zero. The only nonvanishing components of the Ricci tensor are diagonal, as you would expect from the metric:

$$R_{ii} = \left(\ln(-g)^{1/2}\right)_{,ii} - \Gamma^m_{ii,m} + \left(\Gamma^m_{ni}\right)^2 - \Gamma^n_{ii}\left(\ln(-g)^{1/2}\right)_{,n}. \tag{8.27}$$

[34] One note may help with your own derivation of these results. Be sure to notice that

$$\frac{d}{ds} = \frac{dt}{ds}\frac{\partial}{\partial t} + \frac{dr}{ds}\frac{\partial}{\partial r}.$$

Although t and r are independent coordinates, both depend on s, so when we take the Euler–Lagrange equation, notice that

$$\frac{d}{ds}\frac{\partial L}{\partial \dot{x}^i} = \frac{\partial L}{\partial x^i}.$$

From Eq. (8.26), the components are

$$R_{00} = \frac{3}{2}\left(\ddot{F} + \frac{1}{2}\dot{F}^2\right) \tag{8.28}$$

$$R_{11} = G + \frac{1}{r}F' - e^F\left(\frac{1}{2}\ddot{F} + \frac{3}{4}\dot{F}^2\right) \tag{8.29}$$

$$R_{22} = r^2\left[\frac{1}{2}F'' + \frac{1}{4}(F')^2 + \frac{3}{2r}F' - e^F\left(\frac{1}{2}\ddot{F} + \frac{3}{4}\dot{F}^2\right)\right] \tag{8.30}$$

$$R_{33} = R_{22}\sin^2\vartheta. \tag{8.31}$$

We now have two options. One is to use the full field equations in the form $R_{ij} - \frac{1}{2}g_{ij}R = (8\pi G/c^4)T_{ij}$, which requires computing the curvature scalar by taking the explicit trace of the Ricci tensor. The other is simpler. Use the stress tensor to obtain

$$R^i_j = 8\pi G\left(T^i_j - \frac{1}{2}\delta^i_j T\right), \tag{8.32}$$

with the trace $T = \varrho - 3P$ for the diagonal stress tensor components $(\varrho, -P, -P, -P)$ (again setting $c = 1$). We have actually set up the same problem as an interior solution for the metric, where we use the isotropy of the stress tensor, $T^i_i = T$ for $i = (1, 2, 3)$. Then the terms that contain time derivatives separate linearly from those involving the spatial derivatives:

$$F'' - \frac{F'}{r} - \frac{1}{2}(F')^2 = 0, \tag{8.33}$$

which has the following solution:

$$F = -\tfrac{1}{2}\ln(1 - kr^2) + f(t). \tag{8.34}$$

Here k is a constant that can be positive, negative, or zero. The factor $f(t)$ is the temporal scaling factor, which we rewrite as

$$f = \ln R^2(t).$$

Since the geometry remains invariant when scaling the size of the sphere with time, we need only to find an evolution equation for R, which comes from the field equation for T_{00}. Finally, before showing how the metric conforms to specific symmetries, we note that the Ricci tensor will be used in a moment for the evolution equations, so keep these results in mind.

8.4.2 The Redshift from the Friedmann–Robertson–Walker Metric: Interpreting the Hubble Law

Imagine two observers located a fixed distance, l, apart. Then for a photon, the null cone connecting these two observers is $ds^2 = 0$ so that

$$l = \int \frac{dr}{(1 - kr^2)^{1/2}} = \int \frac{dt'}{R(t')}. \tag{8.35}$$

Assume that the two are, respectively, situated at t_e, the time when the signal is emitted; and t_r, the time when it is received; thus t_r is the here and now. Although they remain unmoved with respect to the surrounding matter, the expansion proceeds. Now take a signal to be emitted some time later, $t_e + \delta t_e$. It will be received at some time $t_r + \delta t_r$, but since l is the same for both signals, we have

$$\int_{t_e}^{t_r} \frac{dt'}{R(t')} = \int_{t_e + \delta t_e}^{t_r + \delta t_r} \frac{dt'}{R(t')}. \tag{8.36}$$

Then, we can schematically write

$$\int_{t_e}^{t_r} = \int_{t_e}^{t_e + \delta t_e} + \int_{t_e + \delta t_e}^{t_r} = \int_{t_e + \delta t_e}^{t_r} + \int_{t_r}^{t_r + \delta t_r},$$

using Eq. (8.36) so that

$$\int_{t_e}^{t_e + \delta t_e} \frac{dt'}{R(t')} = \int_{t_r}^{t_r + \delta t_r} \frac{dt'}{R(t')}. \tag{8.37}$$

For small enough intervals $\delta t << |t_e - t_r|$ the scale factor is approximately constant and we find

$$\frac{\delta t_e}{R(t_e)} = \frac{\delta t_r}{R(t_r)}. \tag{8.38}$$

The time interval corresponds to either a frequency or wavelength change, since c is constant, so

$$\frac{\lambda_e}{R(t_e)} = \frac{\lambda_r}{R(t_r)}. \tag{8.39}$$

This is the *cosmological redshift*, originally so called to distinguish it from the gravitational redshift and the Doppler effect due to relative motions of source and observer. We have previously adumbrated this result, but here you can see how it is derived in detail. This is always a redshift because, for an expanding universe, $R(t_r) > R(t_e)$. With the usual definition of z in terms of wavelength, we can write

$$1 + z = \frac{R(t_r)}{R(t_e)}, \qquad (8.40)$$

and therefore we obtain the Hubble law as a natural consequence of the evolution of the scale factor *for small differences in lookback time*:

$$cz \sim H_0 d. \qquad (8.41)$$

To do this, we needed to define the first-order expansion of R relative to the present epoch, which we take at $t = 0$ and call $t_e = t$ so that

$$H_0 \equiv \frac{\dot{R}_0}{R_0} \qquad (8.42)$$

is the relative rate of the expansion at the present time. We have finally arrived at the *Hubble law*, which states that *the cosmological redshift is proportional to the distance to the source.*

Notice that H_0 is scale free in *not* depending explicitly on lengths, but it is actually not a constant. It depends on when the measurement is made. All observers at the same instant in the expansion will obtain the same value of H from local measurements but two observers at different comoving times will obtain different values for the Hubble parameter. To reflect this, we will refer to the time-dependent variable H as the *Hubble parameter*. The quoted form of the Hubble law is derived from strictly *local* observations, in the language we have been using in general relativity. The scale of the curvature is great enough that, over distances of a few tens or hundreds of Mpc, there is no significant deviation from the linear relationship. This is the situation as it stood in 1936, at the time of Hubble's book *The Realm of the Nebulae* in which he summarized the basic case for a cosmic expansion. For greater distances, however, we must make corrections for the change in the curvature over time. Once the redshift begins to reach $z \sim 0.11$ or greater, it is no longer possible to apply the Doppler approximation to determine the recession velocity.[35] The largest redshift measured by Hubble, Humason, and Mayall was around $z \approx 0.2$. Today sources with $z > 5$ are known and we have to evaluate the distance−redshift relation more carefully.

Let's see what happens when we take the kinematic expansion to the second-order terms. Call the scale factor at the present epoch R_0 and at t, the time of emission by the source, R. Then to the second order in lookback time, we find

$$\frac{1}{R(t)} = \frac{1}{R_0} - \frac{\dot{R}_0}{R_0^2}t - \frac{1}{2}\left(\frac{\ddot{R}_0}{R_0^2} - \frac{2\dot{R}_0^2}{R_0^3}\right)t^2 + \text{(higher-order terms)}. \qquad (8.43)$$

---

[35] Christensen, G. E. 1995, *Edwin Hubble: Mariner of the Nebulae* (Chicago: Univ. Chicago Press) notes that throughout his life, Hubble referred to the redshift *only* in terms of its magnitude and never accepted a definitive interpretation for the observations in terms of a cosmological model.

Guided by our earlier discussion, we now define two kinematic parameters:

$$H_0 \equiv \frac{\dot{R}_0}{R_0} \tag{8.44}$$

and

$$q_0 \equiv -\frac{\ddot{R}_0 R_0}{\dot{R}_0^2}, \tag{8.45}$$

which is called the *deceleration parameter*, so that Eq. (8.40) becomes

$$z \approx H_0 t + H_0^2 (\tfrac{1}{2} q_0 + 1) t^2, \tag{8.46}$$

which for nearby galaxies gives the correction to the linear expansion law when the distance is ct. Notice that R has disappeared, so in principle we can obtain both H_0 and q_0 directly from distance-calibrated measurements independent of the cosmological model.

These kinematic relations suffice when applying evolutionary corrections to samples with substantial ranges in redshift. Remember that all variables that are subscripted with zero refer to values at the present epoch, a distinction that we will need when discussing the full evolution equations. It is useful here to notice that we have no idea how R behaves with time other than to know, from the field equations, that it depends on the equation of state for the universe. For a radiation dominated era, which corresponds to the pressure law for a relativistic fluid, the expansion should be different from a matter-dominated time because of the stiffer equation of state for a nonrelativistic gas. Even without the details, you can imagine that a change in the energy density will cause an acceleration or deceleration of the expansion that will show up as differences between values of H_0 and q_0.

8.5 COSMOLOGICAL MODELS AND EVOLUTION OF THE SCALE FACTOR

Kinematics only gets you so far. You can work in this model independent way for local measurements, where the parameters can be obtained from observations because the spacetime curvature is small. But when the lookback time becomes a considerable fraction of H_0^{-1}, this no longer suffices. One way of seeing why is to think about the calibrators. As the time interval you are sampling gets longer, the evolutionary corrections become progressively more uncertain and depend more on the precise history of the expansion, hence on the cosmological dynamics. If we are to make any headway with understanding the early Universe, we require machinery to compute a model of the expansion. There is some hope of doing this, however, if we can count on the validity of our assumptions about the spacetime over large scale and long times. In any little piece of the Universe we could always bow to Mach and say that our local changes are measured with respect to the overall cosmic mass distribution. We don't have that option when dealing with the Universe itself. This means returning to our discussion of general relativity from Chapter 1 and employing some of the machinery we developed there for treating gravity.

8.5.1 The Friedmann–Robertson–Walker Evolution Equations

With the formal solution for the metric, we can continue our calculation of the Ricci tensor components and the field equation. Substituting explicitly for F from Eq. (8.34), we find that

$$e^F = \frac{R^2(t)}{1 - kr^2}, \tag{8.47}$$

from which comes a simple expression for the time derivative of F:

$$\dot{F}e^F = 2R\dot{R} \rightarrow \dot{F} = 2\frac{\dot{R}}{R}. \tag{8.48}$$

We next need to specify the form of T_{ij}, the mass–energy stress tensor. This quantity determines the evolution of the cosmological model since the matter and energy density of the Universe determine the rate of expansion of the ensemble through its self-gravity. It's given by

$$T^{ij} = (P + \varrho)u^i u^j + Pg^{ij} \tag{8.49}$$

for an ideal fluid. As in Chapter 1, we take $c = 1$ and u^k are the 4-velocities of the particles. For a stationary observer at rest with respect to the large-scale matter distribution $u^0 = 1$ and all other components vanish. The trace of the stress tensor, $T = T^m_m$, is

$$T = \varrho - 3P. \tag{8.50}$$

We will see in a moment how the choice of an equation of state is critical for governing the rate of expansion.

Returning to the field equations, Eqs. (8.32), and substituting the Friedmann–Robertson–Walker (FRW) metric to obtain the explicit forms for time and space derivatives of the function F, we now write

$$R_{00} = -3\frac{\ddot{R}}{R} \tag{8.51}$$

$$R_{11} = -\left[\frac{\ddot{R}}{R} + \left(\frac{\dot{R}}{R}\right)^2 + \frac{k^2}{R^2}\right](1-kr^2)^{-1} \tag{8.52}$$

$$R_{22} = -\left[\frac{\ddot{R}}{R} + \left(\frac{\dot{R}}{R}\right)^2 + \frac{k^2}{R^2}\right]r^2 \tag{8.53}$$

$$R_{33} = -\left[\frac{\ddot{R}}{R} + \left(\frac{\dot{R}}{R}\right)^2 + \frac{k^2}{R^2}\right]r^2\sin^2\vartheta, \tag{8.54}$$

and the curvature scalar is obtained by taking the trace of the Ricci tensor. The resulting evolution equations for R depend *only* on time:

$$\left(\frac{\dot{R}}{R}\right)^2 + \frac{k}{R^2} = \frac{8\pi G}{3}\varrho \tag{8.55}$$

$$\frac{\ddot{R}}{R} = -\frac{4\pi G}{3}(\varrho + 3P) \tag{8.56}$$

These are called FRW equations. But we're not quite finished; two more conservation conditions are required to complete the theoretical apparatus. The first is that the total mass−energy flux must be divergenceless and this is expressed by the continuity equation for a current J^i written in covariant form: $J^i_{;i} = 0$. The second is that the covariant divergence of the stress tensor must vanish, which was required for deriving the field equations in the first place:

$$T^{ij}_{;j} = g^{ij}P_{,j} + g^{-1/2}\frac{\partial}{\partial x^j}\left[g^{1/2}(P+\varrho)u^i u^j\right] + \Gamma^i_{jl}(P+\varrho)u^j u^l = 0. \tag{8.57}$$

Recall that g is the determinant of the metric and that $g^{ij}_{;j} = 0$. For an observer at rest, Eq. (8.57) reduces to the final evolution equation:

$$\frac{d}{dt}(\varrho R^3) + P\frac{d}{dt}R^3 = 0, \tag{8.58}$$

which expresses the *conservation of mass−energy*. Now you have the complete set of equations needed to describe the expansion with the exception of the equation of state that, as we said, is the determining factor for the rate of expansion. Notice that only the scale factor is evolved by this system of equations. The *geometry* remains the same so we must specify k in order to obtain a solution to Eqs. (8.55).

8.5.2 Choosing the Equation of State

You've seen this argument before in the context of a stellar interior. Remember that as we go toward progressively larger stellar masses the equation of state switches from an ideal gas (matter dominated) to radiation pressure and eventually is limited by the Eddington luminosity. For the Universe, we do not encounter a limiting energy density because of the expansion—there is no requirement that the cosmos remains in hydrostatic equilibrium—but the same transition to radiation domination must occur regardless of the current energy density. In our dynamical equations, this amounts to separate dependences on the scale factor for the two extreme cases with the transition being marked, for an adiabatic expansion, by a critical density above which the temperature is high enough for radiation to dominate the energy density. As long as there now is some relic radiation that is not generated by stars alone, this transition occurs at a lower compression (running the clock backward) than it would by just compressing the currently observed matter. This was the key point Gamow realized when contemplating universal nucleosynthesis during the early stages of the Big Bang; at some point, the Universe would have conditions very similar to those found in a stellar interior, with the exception of a large abundance of free neutrons for reasons we will describe shortly, and thus for a brief interval there should be nuclear processing to helium by the expanding gas.

On the cosmological scale, the Universe contains *dust* (a term that, in this context, means collisionless masses such as galaxies), radiation, and other massless particles (such as, possibly, neutrinos, gravitons, and any other exotic species you can think of). The whole is subsumed in a "perfect fluid" picture we used to obtain the curvature. The relative velocities of galaxies and stars inside clusters are $\leq 10^3$ km s^{-1} ($v/c \leq 3 \cdot 10^{-3}$) and hence their contribution ϱ_g to ϱ_m is essentially all due to their rest mass, and $P_m \ll \varrho_m$, today. An estimate to ϱ is 10^{-30} g cm^{-3}. There is an independent estimate of ϱ_m, from the dynamics of galaxy clusters: assuming that clusters are bound and applying the virial theorem from the measured kinetic energy, one estimates the potential energy, which requires a mass larger by a factor 10 of the visible mass in galaxies and so the dynamic density is $\varrho_d \approx 10^{-29}$ g cm^{-3}.

The radiation energy density comes mainly from the CBR. Since the radiation temperature is now 2.7 K, the energy density is

$$\varrho_r = \frac{8\pi^5}{15c^5}\frac{k^4}{h^3}T^4 = 4 \times 10^{-34} \text{ g cm}^{-3}. \tag{8.59}$$

With a similar contribution from neutrinos and gravitons this becomes $\varrho_r \approx 10^{-33}$. For other relativistic (massless) particles, $P_r = \frac{1}{3}\varrho_r c^2$. The current upper limit for neutrino mass is 30 eV, so if they are nonrelativistic, cold matter, the three neutrino families contribute $\varrho_\nu \approx 10^{-29}(m_\nu/30 \text{ eV})$ g cm^{-3}. Depending on their mass, they may be important contributors to ϱ_m.

With the exception of the first seconds after the Big Bang, exchanges of energy among the various components are negligible, and we may apply the conservation laws to each separately. For baryons (and possibly massive neutrinos) the continuity equation gives

$$\varrho_m R^3 = \text{constant}. \tag{8.60}$$

For the radiation, the energy density[36] varies as

$$\varrho_r R^4 = \text{constant}. \tag{8.61}$$

The total density required for the solution of the FRW equations is given by

$$\varrho(t) = \varrho_m(t) + \varrho_r(t) = \varrho_{m0}\left(\frac{R_0}{R}\right) + \varrho_{r0}\left(\frac{R_0}{R}\right)^4 \tag{8.62}$$

and the pressure is

$$P(t) = \frac{1}{3}\varrho_r(t)c^2 = \frac{1}{3}\varrho_{r0}\left(\frac{R_0}{R}\right)^4 c^2. \tag{8.63}$$

These are substituted into the evolution equations:

$$\left(\frac{\dot{R}}{R}\right)^2 + \frac{kc^2}{R^2} = \frac{8\pi G}{3}\varrho \tag{8.64}$$

$$\frac{\ddot{R}}{R} + \frac{1}{2}\left(\frac{\dot{R}}{R}\right)^2 + \frac{kc^2}{2R^2} = -\frac{4\pi G}{c^2}P. \tag{8.65}$$

Notice that we have explicitly restored the previously ignored factor of c in the stress tensor and Einstein constant.

Now that we have the two extreme equations of state, we'll step off the main thread of our discussion to examine the epoch at which the universe changed from one to the other, the *decoupling era*. Since we do not have a good handle on the initial conditions for Everything, we can instead see if it is possible to derive some constraints on cosmic evolution from the moment when the interactions between matter and radiation ceased to be important.

8.6 COSMIC BACKGROUND RADIATION (CBR)

8.6.1 The Past

The announcement of the discovery of the cosmic microwave background was *the* pivotal observational step in the recognition of the hot origin of the Universe. It was not, however, unanticipated.[37] As early as 1941, McKeller noted that the optical interstellar CN absorption lines show an anomalous excitation that seemed to require a radiation field of a few degrees Kelvin. The uniformity of this phenomenon pointed to an extragalactic excitation source, but at the time it was

---

[36] You know this from classical thermodynamics. For radiation, the energy density depends on T^4. The scaling with volume contains an additional factor of R due to the redshift.

[37] The history has been reviewed by Kraghe, H 1996, n.4 and Partridge, R. B. 1995, *3K: The Cosmic Microwave Background Radiation* (Cambridge, UK: Cambridge Univ. Press).

little noted.[38] It was, however, plausible in light of nucleosynthesis models of the early Big Bang to expect that the relic, decoupled radiation would now have a peak at around the same temperature (Alpher and Gamow in 1947, also the discussion by Hoyle and Tayler of helium synthesis in 1964, on the eve of the discovery of the CBR), and searches began rather early in the game with an attempt by Dicke and collaborators that placed an upper limit on the temperature of this radiation of about 20 K. During the 1960s, Dicke and Wilkinson were beginning a search for the radiation with more sensitive radiometers when in 1965 Penzias and Wilson at Bell Labs discovered a ubiquitous noise source with a temperature of about 3 K. This proved to be the cosmic background radiation, or CBR. The temperature indicated a peak beyond the wavelength at which the Bell Labs and Princeton groups could observe, but assuming that it was in thermal equilibrium at the time of creation, and therefore a blackbody, the antenna temperature corresponded to a brightness or radiation temperature (see Chapters 1 and 2). Many individual measurements of the background in the period between 1965 and 1989 were made using groundbased bolometers and balloon flights, but the final answer came in 1989 with the launch of the *Cosmic Background Explorer* (COBE) satellite. Within a few *hours* of observing, the entire microwave spectrum was already obtained with now enough noise to show its unique agreement with a Planck curve and its full spectral range, including the peak temperature of 2.7 K. The following four years of observation only reinforced the conclusion. The accuracy of the COBE measurements is unprecedented in cosmological observation, sufficient to detect small 10^{-5} amplitude fluctuations through differential microwave radiometric observations of opposite parts of the sky.

8.6.2 Current Status

The COBE satellite was the first absolutely calibrated all sky survey instrument for the microwave and far IR portion of the cosmic background spectrum and, because it carried several independent instruments, provided the conclusive evidence for the thermal fluctuations. The Differential Microwave Radiometer (DMR) and the Far-Infrared Absolute Spectrophotometer (FIRAS) independently measured the background. The DMR operated at three frequencies, 31.5, 53, and 90 GHz and was designed to detect thermal fluctuations. FIRAS had a spectral range from 480 to 5000 μ and was calibrated against an internal blackbody source. What we now have in the post-COBE era is the clearest available data for the thermal spectrum, the limit on the departures from a Planck distribution, and the firm detections of the dipole moment and the size spectrum of the largest-scale brightness temperature fluctuations. The spectrum is clearly a single temperature Planck function in the long wavelength region, where the peak has been directly measured. The temperature of this curve is $T_0 = 2.728 \pm 0.004$ (2σ errors) with deviations from a Planckian spectrum are $\langle \Delta T/T_0 \rangle \leq 5 \times 10^{-5}$ (Fig. 8.1). The dipole component, with an amplitude of $T_1 = 3.372 \pm 0.007 mK$, is centered at $l_{II} = 264.14° \pm 0.30°$ and $b_{II} = 48.26° \pm 0.30°$, again with 2σ uncertainties. The

---

[38] The more modern measurement yields 2.73 ± 0.03K. See Roth, K. C., Meyer, D. M., and Hawkins, I. 1993, *ApJL*, **413**, L67. Gamow realized what this might mean and discussed it in several papers at the time.

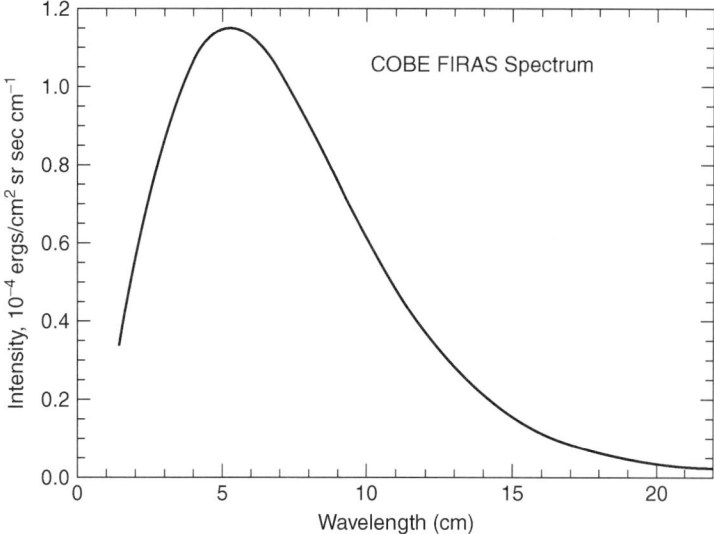

FIGURE 8.1. The final FIRAS integrated COBE Planck spectrum of the cosmic background radiation. The data points are indistinguishable from the fit (credit: COBE/NASA).

temperature fluctuations reported by the DMR[39] are 35 ± 2 μK on a 7° angular scale and 29 ± 1 μK on a slightly larger scale of 10° (1σ errors). These fluctuations and the completely thermalized spectrum show the homogeneity at decoupling of the expanding Universe. The minuteness of the temperature fluctuations is extremely important. They show that the matter distribution at the time of the decoupling was very uniform, and sets a limit on the amplitude of the density fluctuations that became the seeds for galaxy formation (see discussion below). The controversy at the moment stems from the fact that the largest of these variations is on the scale of the horizon, while many of the smaller scales seen are still huge compared with the expected sizes of even clusters of galaxies. There is, at the resolution of the COBE data, no evidence for any cluster or galaxy scale structures, and the correlation function is best interpreted in light of the largest scales expected for superclusters, of order a few hundred Mpc.

8.6.2.1 The Future
There are two relatively recent developments that deserve mention here. The first is determination of the hyperfine structure C I 1560 Å/1657 Å intensity ratio at in absorption at high redshift. This as a measure of the temperature of the CBR at high redshift in sampled Lyα systems. In essence, the technique is the same as that used for the CN determination of the CBR temperature except it uses lines known to be formed at the same distance (epoch) as the quasar. Two measurements have been reported to date, both of which are consistent with an increase in the radiation temperature with lookback time. Several medium redshift studies have

---

[39] There is still, at this writing, some disagreement between instruments at the 7° scale. The FIRAS result for the thermal fluctuations are slightly larger, 48 ± 14 μK (1σ errors, so the data are consistent).

been performed to date. For the quasar Q1331 + 170 ($z = 1.776$)[40] the observed excitation temperature is 7.4 ± 0.8K. Toward Q0013-004 ($z = 1.9731$),[41] the excitation temperature is 8.1 K. If you scale the cosmic background temperature now observed, 2.7 K, to a redshift of about 2, you expect to observe $T_{\mathrm{CBR}} = T_0(1 + z) \approx 8$ K. Finally, for PKS1232 + 0815, for which $z = 2.33771$, a well-constrained temperature has been obtained from both H_2 rotational lines and atomic ion fine-structure transitions of $6 < T_{\mathrm{CBR}}(1 + z) < 14$.[42]

The near future (2001–2010) is also expected to see several new satellite missions. The recently launched Microwave Anisotropy Probe (MAP) is designed to cover the same ground as COBE with even higher sensitivity and angular resolution. An additional project, *Planck Surveyor*, is also in the works. Balloon measurements, in particular the *BOOMERanG* long-term flight (1997, 1999), have already obtained significant information about the correlations on scales of order a few degrees, and as we will see later, these data can be used to constrain models for the origin of large-scale structure.

8.6.3 Coupling between Matter and Radiation: Formation of CBR

Regardless how hot the Universe is at any stage, it cools on expansion and ultimately the equation of state *must* switch from radiation to matter domination. This moment, when the equality of the two phases occurs, is constrained by the present measurements of the respective energy densities:

$$\frac{\varrho_m}{\varrho_r} = \frac{\varrho_{m,0}}{\varrho_{r,0}}(1 + z). \tag{8.66}$$

Taking $\varrho_{m,0} = 1.9 \times 10^{-29}\Omega_0 h^2$ g cm^{-3}, since, as we see the present Universe is matter-dominated, and using the CBR to constrain the radiation, $\varrho_{r,0} \approx 8 \times 10^{-34}$ g cm^{-3}, then $\varrho_m = \varrho_r$ occurs at $z_{eq} \approx 2.5 \times 10^4$.[43] Different species reach this equality at different times depending on their temperature and the statistical weight, but you would expect the photons to be the last to reach the equality.

Decoupling is more complicated. As it does in a stellar atmosphere, the photon's transition to free streaming depends on the interaction cross sections. In effect, the moment of decoupling occurs when the optical depth drops below unity for a pathlength equal to the horizon distance. Were this a static medium, you would call this the *photosphere*. But for a stellar wind, or expanding ejecta, you recall there is actually a nonstationary pseudophotosphere, the place in the wind from which the radiation at any wavelength can freely emerge but is otherwise a dynamical surface at some fixed velocity. The atmosphere problem is generally

---

[40] Songaila, A., Cowie, L., Vogt, S. et al. 1994, *Nature*, **371**, 43.
[41] Roth, K. C., Songaila, A., Cowie, L. L., and Bechtold, J. 1996, *BAAS*, **189**, 17.
[42] Srianand, R., Petitjean, P., and Ledoux, C. 2000, *Nature*, **408**, 931.
[43] It is interesting to note that in 1933, in his lectures at the University of California, de Sitter used Millikan and Cameron's measurements of the detection rate for cosmic rays to place limits on ϱ_r. Although mistaken on the nature of the "radiation" in this case, he obtained $\varrho_{r,1933} \sim 3.6 \times 10^{-36}$ g cm^{-3}, far lower than the measured value for the CBR yet intriguing because it clearly required matter-dominant models.

complicated by the extreme wavelength dependence of the opacity sources, especially from absorption lines. The initial stages of the Universe were simpler because there were no metallic absorbers to produce line opacity. Since you know that Thomson scattering by free electrons is independent of frequency, the optical depth depends only on the electron density. We will ignore He and assume a pure hydrogen plasma. Then the ionization fraction, x, is given by the Saha equation which depends only on the total density and local temperature, both of which must be obtained from a cosmological model. The optical depth is specified by taking the photon mean free path to equal the horizon, or we could also set the mean free scattering rate ($n_e \sigma_e c$), equal to the Hubble parameter, $\dot{R}/R$ at the moment when x reaches some characteristic value. Since the ionization falls very rapidly with increasing R, the range in x is rather small. Then

$$\frac{x^2}{1-x} = 2\left(\frac{2\pi m_e kT}{h^2}\right)^{3/2} e^{-\chi_H/kT} \frac{m_p}{\varrho} \tag{8.67}$$

gives a temperature of around 3000 K for $x \approx 0.1$, which for the matter-dominated stage of the expansion corresponds to a redshift $z_{\text{dec}} \approx 10^3$. Again, the precise value depends on the mean density. To stress the close connection between stellar atmospheres and the cosmic "photosphere," note that any density fluctuations will also produce a spread in the decoupling redshift due to fluctuations in the effective optical depth at constant redshift. This is the surface from which we see the emergent cosmic radiation. It sets the absolute limit for direct observation, in the same manner that we discussed with radiative transfer in a fog; at some distance, you cannot distinguish the light source because scattering isotropizes the photon paths and produces a smooth background that is altered only by the fluctuations in opacity due to the nearby portions of the medium.

The energy conservation condition, Eq. (8.58), is only the first step in properly accounting for the thermal state of the expanding medium. We need to find an explicit form for the first law of thermodynamics. Since ϱ is the mass–energy density, and the entropy is defined locally (this is why we spent so much time on its additivity when deriving entropy in Chapter 1), we can write:

$$dS = \frac{1}{T}dE + \frac{P}{T}dV = \frac{1}{T}d(\varrho R^3) + \frac{P}{T}dR^3. \tag{8.68}$$

From our previous discussions of thermodynamics (especially in Chapter 1), you know that

$$T\frac{dP}{dT} = \varrho + P,$$

so Eq. (8.68) becomes

$$dS = d\left[\frac{(\varrho + P)V}{T}\right] = 0. \tag{8.69}$$

This equation states that entropy is constant for the expanding universe, so the entropy density varies as R^{-3} and the *entropy per baryon*, η, is constant for a matter-dominated universe. This result provides an important constraint on the thermal state of the early Universe; if we measure η_0 at the present time, we can use this as a cosmological parameter governing any thermal process as a function of redshift. In particular, as we will see shortly, this is a fundamental quantity for understanding the synthesis of the light elements before the decoupling era. To find the subsequent thermal history of the expanding matter, take the first law and assume $E = \varrho R^3 = \frac{3}{2}nkTR^3$, where n is the number density, and $V = R^3$. Then from the continuity equation, in an expanding medium, $\dot{n} = -3Hn$, we have

$$\frac{dT}{dt} + 2\frac{\dot{R}}{R}T = \left(\frac{3}{2}nk\right)^{-1}\mathscr{L}, \qquad (8.70)$$

where $\mathscr{L}$ is the same cooling function we used in Chapter 6 for the thermal instability (although now without the contribution from metals). To compare with observations, the time coordinate must be transformed into a redshift coordinate. You can therefore see how choices of Λ and an equation of state alter the thermal history and produce important *observable* consequences, which we will now discuss.

8.6.4 A New Complication: The Cosmological Constant (Λ)

When we wrote down the stress tensor, we remarked that each term has a classical origin linked directly to the properties of an ideal fluid. The "dynamical" part comes from the Reynolds stress, transferred into covariant formalism, and the diagonal terms include the pressure and energy density. In a Newtonian framework, this tensor is undetermined to within a constant, but that can be set arbitrarily to zero. The problem is that T^{ij} must satisfy $T^{ij}_{;j} = 0$ in a covariant framework so the arbitrary constant is really a multiple of the metric. The covariant divergence of the metric always vanishes so it is always possible to add a term Λg_{ij} to the Einstein tensor, G_{ij}, and still obtain a correct form for the Bianchi identities, which provide the curvature for a specified mass distribution.

Let us return to the FRW equations. We only need to modify the equation for R^i_j to account for Λ, provided it is a constant:

$$R^i_j = 8\pi G\left(T^i_j - \tfrac{1}{2}\delta^i_j T\right) + \delta^i_j \Lambda. \qquad (8.71)$$

You see now that for vanishing pressure and density, the Ricci tensor never vanishes and thus a nonstatic solution is still possible. The FRW equations become

$$\frac{\ddot{R}}{R} = -\frac{4\pi G}{3}(\varrho + 3P) + \frac{\Lambda}{3} \qquad (8.72)$$

$$\left(\frac{\dot{R}}{R}\right)^2 + \frac{k}{R^2} = \frac{8\pi G}{3}\varrho + \frac{\Lambda}{3}. \qquad (8.73)$$

Now, for a vacuum solution, we have an exponentially expanding scale factor with a timescale $(3/\Lambda)^{1/2}$, assuming $\Lambda \geq 0$. This is the *de Sitter* solution and it serves to focus attention on the consequences of including Λ, whatever it may physically be. Alternatively, this vacuum state can be considered a negative energy configuration in which the pressure is $P_{\text{vac}} = -\varrho_{\text{vac}}$. The one constraint on the behavior of this quantity is that it should be constant, otherwise the term added to the stress tensor in Eq. (8.71) does not satisfy the covariance condition that $(\Lambda g_{ij})_{;j} = 0$ from which we first obtained the FRW equations.

We are thus forced to face that there *must* be additional term in the evolution equations of *arbitrary* magnitude that we cannot specify in advance.[44] Even though we have been able to obtain solutions of the FRW evolution equations without it, we cannot continue to simply ignore it. The cosmological constant gives an additional term, $+\Lambda c^2/3$. It represents the contribution from the vacuum energy density

$$\varrho_{\text{vac}} = \frac{\Lambda c^2}{8\pi G}$$

and is the analog of the ground state of a harmonic oscillator in quantum mechanics. Historically, Λ was introduced because of the perceived need to maintain a constant scale length for the Universe when obtaining a global interior solution for a matter distribution. Only later did this become an issue in light of the Friedmann solutions, since these lead to a nonstationary metric.[45] In particular, the Friedmann model produces expansion for quite simple equations of state.

8.6.4.1 An Aside on Λ and the Physical Constants

The combination of constants, $\hbar$, G, and c, provide an energy density. These are precisely the constants that *don't* involve the electromagnetic interaction. So when you compute the energy density of this combination, it's the same as for empty space, that is what we mean by the vacuum state. Even in the absence of matter, so the material portion of $T_{ij} \to 0$, then there is still a term, Λg_{ij} that comes from the invariance of the metric alone. Of course, this is due to the gravitational field of

---

[44] Carroll, S. M., Press, W. H. and Turner, E. L. 1992, *ARAA*, **30**, 499; Coles, P. and Ellis, G. F. R. 1997, *Is the Universe Open or Close?* (Cambridge Lecture Notes in Physics, vol. 7) (Cambridge, UK: Cambridge Univ. Press).

[45] The tale of the origin and effect of Λ has been told many times, most often quoting a statement by Einstein from sometime well after the Friedmann paper and Hubble's discovery of the expansion. The full quote is "Much later, when I was discussing the cosmological problem with Einstein, he remarked that the introduction of the cosmological term was the biggest blunder he ever made in his life." (Gamow, G. 1970, *My World Line* (NY: Viking), p. 44). The real story is much more complicated (see, e.g., Kersberg (1989), n.4, pp. 160ff). Einstein's most complete remarks are found in Einstein, A. 1956, *The Meaning of Relativity* (5th ed.) (Princeton: Princeton Univ. Press). Appendix II, added to the 1945 edition, describes very thoroughly the origin of Λ in the field equations and Einstein's work on cosmological models. The most complete discussion of the history of Einstein's cosmological work and references to all of the original papers is found in Pais, A. 1982, *Subtle is the Lord ...: The Science and the Life of Albert Einstein* (NY: Oxford Univ. Press), pp. 281ff. A very useful, although dated, survey of generalized models including Λ is in McVittie, G. C. 1964, *General Relativity and Cosmology* 2nd ed., (Urbana: Univ. Illinois Press). A particularly interesting historical and philosophical review is Earman, J. 2001, *Archiv. Hist. Exact Sci.*, **55**, 189.

the mass that generates the curvature in the first place, but the curvature is also due to the vacuum energy. Thus, even if ϱ_m vanishes, the curvature still doesn't since $\Lambda \neq 0$. "Ah," you cry, "that means the Λ must have the same dimensions as T_{ij}. Further," you reason, "this means it must have the dimensions of a *pressure* (!) and therefore of an energy density and you see how we've come around to the statement that Λ is proportional to ϱ_{vac}." That's the origin of the *cosmological constant problem*—the observed (note the word, *observed*) Λ is about 120 orders of magnitude smaller than our adumbrated value, by which we mean the one concocted from the physical constants. At this point we see the importance of Λ acting like a pressure. The way Λ enters the FRW solutions, through the terms in the stress tensor, affects the rate of expansion by altering $\ddot{R}$ in the FRW equations; it drives an expansion even in the absence of matter terms in the stress tensor.

The value of Λ can be set by appropriate combinations of physical constants, but which ones are we to use? This leads us to a deeper point concerning the connection between gravitation and quantum mechanics. You notice we chose only $\hbar$ and c. Why not m_e and e, the mass and charge of the electron? These are field-related constants that are the "dressed" values of the charge and mass from quantum electrodynamics. They must be renormalized to include the vacuum interactions. Instead, if we choose the ones that are "fundamental," the only appropriate nonrenormalized constants are $\hbar$ and c; the first one is *quantum-mechanical* and the second is classical. The formal inclusion of Λ in the field equations comes from considerations of geometry and symmetry, and requiring that the Poisson equation is the classical limit of the field equations for the interior solution. The vacuum solution may be written as the $\varrho \to 0$ limit of $\nabla^2 \varphi = -(\varrho + \lambda)$ with $\lambda \neq 0$ (see footnote 47). This was the basis for Einstein's inclusion of the cosmological constant in the first place. The feature that the only way to compute the vacuum energy density from fundamental constants requires using $\hbar$ means quantum mechanics enters the field equations for cosmic gravitation. If the cosmological constant is the vacuum energy density, then it must originate from a quantum process.

8.6.5 The Density Parameter: Ω

We still require a formal prescription for the evolution of the energy density and pressure with time. This can be obtained by taking the derivative of Eq. (8.68) with respect to t, and eliminating $\ddot{R}$ from the second one, giving a relation between P and ϱ:

$$\frac{d}{dt}\varrho R^3 + \frac{P}{c^2}\frac{d}{dt}R^3 = 0. \tag{8.74}$$

We can assume that the mass–energy density evolves adiabatically so that the entropy density s scales directly with volume:

$$\frac{d}{dt}(sR^3) = 0. \tag{8.75}$$

These now provide a compact form with which we can set bounds on the present day curvature terms. Remembering the definition of H, we have

$$H^2 = \frac{\dot{R}^2}{R^2} = \frac{8\pi G}{3}\varrho - \frac{kc^2}{R^2}, \qquad (8.76)$$

so substituting present day values we can write this as

$$H_0^2 = \frac{8\pi G}{3}\varrho - \frac{kc^2}{R_0^2}.$$

Since H_0 and ϱ_0 are known from observations these expressions permit placing bounds on both k and R_0. The *critical density*, ϱ_c, is defined as the value for ϱ corresponding to $k = 0$

$$\varrho_c = \frac{3H_0^2}{8\pi G} \qquad (8.77)$$

for the present era. Then defining a new dimensionless parameter

$$\Omega \equiv \frac{\varrho}{\varrho_c} \qquad (8.78)$$

to specify the total density for mass-energy relative to the critical value, the relation for H becomes

$$\frac{kc^2}{R^2} = H^2(\Omega - 1)$$

$$\frac{kc^2}{R_0^2} = H_0^2(\Omega_0 - 1). \qquad (8.79)$$

Consequently, there is a direct connection between the density parameter Ω and the spatial curvature of the universe. In terms of the deceleration parameter, $q_0 = -\ddot{R}_0/R_0 H_0^2$, we obtain

$$H_0^2 q_0 = \frac{4}{3}\pi\varrho_0 \rightarrow \frac{kc^2}{R_0^2} = H_0^2(2q_0 - 1). \qquad (8.80)$$

A simple rewriting of the FRW equations suffices to show the constraints observations can place on the individual contributions to the equation of state and the dynamics of the expansion. Since

$$\left(\frac{\dot{R}}{R}\right)^2 + \frac{k}{R^2} = \frac{H_0^2[\Omega_B + \Omega_{\mathrm{CDM}}]}{R^3} + \frac{H_0^2 \Omega_{\mathrm{HDM}}}{R^4} + \frac{\Lambda}{3}, \qquad (8.81)$$

where Ω_B, Ω_{CDM}, and Ω_{HDM} are the density parameters for baryons, and cold and hot dark matter, respectively, we can define two new density parameters

$$\Omega_k \equiv -\frac{k}{R_0 H_0^2},$$

$$\Omega_\Lambda \equiv \frac{\Lambda}{3H_0^2}$$

and rewrite Eq. (8.81) in terms of these parameters *at the present time*. Using H_0 to cancel the *lhs*, we obtain

$$\Omega_M + \Omega_k + \Omega_\Lambda = 1, \tag{8.82}$$

where $\Omega_M = \Omega_B + \Omega_{CDM} + \Omega_{HDM}$.[46] This equation is the generalization of the parameter Ω_0 in that we have separated each contributor to the deceleration. As we will see from nucleosynthetic constraints, the baryonic contribution, Ω_B, is small, of order 0.1 or less. From galactic and cluster dynamics, we know that $\Omega_{CDM} + \Omega_{HDM}$ must be near unity, while $\Omega_k \approx 0$. Now for some scaling relations and real numbers. Notice how the observational uncertainty in h propagates to other quantities. For instance, $\varrho_c = 1.9 \times 10^{-29} \ h^2 \ \text{g cm}^{-3}$. Furthermore, because Ω is linear in the density, we can define a separate value for each contributor to the dynamics: $\Omega_M \approx 0.03$, $\Omega_{DM} \approx 0.2 - 0.4$, $\Omega_r = 2.3 \times 10^{-5} \ h^{-2}$, and $\Omega_\nu \approx (m_\nu/30 \ \text{eV}) \ h^{-2}$.

8.6.6 World Models: Solutions to the FRW Equations

We require a model of the expansion in order to interpret cosmological observations. At the same time, we assume a model when reconstructing the development of the constituents of the Universe. For instance, the dynamics of galaxies are used to determine the expansion law, but once the identification has been made between z and lookback time, the redshift is used to determine the evolution of galaxies over time through time-dependent corrections. You'll notice that this runs the risk of a circular argument. The observations are used to select the appropriate physical model for the large-scale structure and its time development but then the model is used to interpret the observations. So what are these models and how are they distinguished?

Solutions to the FRW equations, no matter how comparatively simple they are, form the basis of just about all cosmological interpretations. The dynamical equations provide the evolution of the scale factor for a homogeneous, isotropic Universe. They must, however, be supplemented with some equation of state, by now a very familiar problem. The matter pressure on the largest scales can be

---

[46] Other contributors, such as massive neutrinos or other forms of relativistic matter can be included here since $R_0 = 1$ in this scaling. The different exponents affect the detailed model but not the evaluation of the present era values of the density parameters. See Davis, M. 2000, *Phys. Rep.*, **333**, 147 for an excellent contemporary overview on contributions to Ω.

neglected in the current Universe, an approximation often referred to as *dust*, but this was not true for epochs when R was small. This follows from the existence of the CBR. Because it is relativistic, the radiation pressure scales as R^{-4} and dominated at early moments of the expansion. It may seem odd to think of the entirety of spacetime as the product of a pressure driven expansion, but the local curvature is dominated by the mass-energy density through the stress tensor. In standard models, the evolution of R is bounded by two extremes: when the Universe was radiation dominated and, today, when it is matter dominated. Historically, these were first solved by Friedmann in 1922–1924 for "dust," when $P = 0$ and $\varrho = \varrho_m$, and by Tolman in the 1930s for radiation, where $P = \frac{1}{3}\varrho c^2$ and $\varrho = \varrho_r$. The "trivial" case, the vacuum solution, was examined by de Sitter and Einstein when the mean density of matter is vanishingly small. We've already discussed this when we introduced Λ and will return to it within the context of inflation. But for now we will focus on the two extreme classical solutions.

8.6.6.1 Matter-Dominated Universe: The Friedmann Model
The first cosmological equation of state, pressureless and matter dominated, was adopted by Friedmann to derive the evolution equation:

$$\left(\frac{\dot{R}}{R}\right)^2 + \frac{kc^2}{R^2} = \frac{8\pi G}{3}\varrho_m(t) \qquad (8.83)$$

supplemented by the scaling law for the density:

$$\rho_m R^3 = \text{constant}, \quad \rho_m(t) = \left(\frac{R_0}{R}\right)^3. \qquad (8.84)$$

It is convenient to define a scaling constant for R in terms of the present mass density and scale factor

$$R_\star^3 \equiv \frac{8\pi}{3}\varrho_{m,0} R_0^3, \qquad (8.85)$$

with which we obtain the following form for the evolution equation:

$$\dot{R}^2 + kc^2 = \frac{R_\star}{R}. \qquad (8.86)$$

This equation can be directly integrated for the possible free choices of the curvature. When $k = 0$, the solution is

$$R(t) = \tfrac{3}{2} R_\star^{1/3} t^{2/3}. \qquad (8.87)$$

Recalling the definition of the Hubble parameter, we find that

$$H = \frac{\dot{R}}{R} = \frac{2}{3}t^{-1}. \tag{8.88}$$

If t_0 is the age of the universe at the current epoch, such a model yields $t_0 = \frac{3}{2}H_0^{-1} = 15\,h^{-1}$ Gyr. The positive curvature case, $k = 1$, requires a parametric representation for R and t:

$$R = \tfrac{1}{2}R_\star(1 - \cos\eta) \tag{8.89}$$

$$t = \tfrac{1}{2}R_\star(\eta - \sin\eta).$$

Here η is only a convenient symbol. Note that the origin for $t(\eta)$ is at the beginning of the expansion, $t = 0$. The scale factor reaches a maximum value at $R_\star$ and recollapses after a *finite* time. Finally, for the case of negative curvature, $k = -1$, the parametric solution is similar but employs hyperbolic functions:

$$R = \tfrac{1}{2}R_\star(\cosh\eta - 1) \tag{8.90}$$

$$t = \tfrac{1}{2}R_\star(\sinh\eta - \eta).$$

Here the expansion continues to infinity, but slows asymptotically to $R = t$, which is equivalent to an ultimate expansion velocity $\dot{R} \to c$. This is the solution that gave rise to the name *Big Bang*, one that continues expansion but with finite deceleration depending on density through the parameter $R_\star$.

8.6.6.2 Radiation-Dominated Universe: The Tolman Model

For a radiation dominated model, adopted first by Tolman, uses a relativistic equation of state. The density in Eqs. (72) and (73) is given by

$$\varrho \equiv \varrho_r(t) = \varrho_{r,0}\left(\frac{R_0}{R}\right)^4, \tag{8.91}$$

so Eq. (8.73) now reads

$$\left(\frac{\dot{R}}{R}\right)^2 + \frac{kc^2}{R^2} = \frac{8}{3}\pi G \varrho_{r,0}\left(\frac{R_0}{R}\right)^4. \tag{8.92}$$

With the definition

$$R_\star^2 \equiv \frac{8}{3}\pi G \varrho_{r,0} R_0^4, \tag{8.93}$$

we obtain the evolution equation:

$$\dot{R}^2 + kc^2 = \frac{R_*^2}{R^2}. \tag{8.94}$$

This too can be integrated for the three limiting cases. For $k = 0$, we find

$$R = (2R_\star)^{1/2} t^{1/2}, \tag{8.95}$$

so that at the present epoch $t_0 = 20\, h^{-1}$ Gyr. You see from this how the choice of an equation of state affects the estimate of the age of the universe, in this case changing the relation between H_0^{-1} and the age by about 25%. For the other two cases of nonvanishing curvature we find that

$$R = \left[t(2R_\star - t)\right]^{1/2} \quad (k = 1) \tag{8.96}$$

$$R = \left[t(2R_\star + t)\right]^{1/2} \quad (k = -1).$$

The qualitative behavior is otherwise similar to the Friedmann model. In particular, for $k = 1$ there is a maximum scale factor when $R = R_\star$. The initial evolution of the three curvature choices is indistinguishable. In the most general case of a polytropic equation of state, $\varrho \sim R^{-3(n+1)}$, the scale factor evolves according to $R(t) \sim t^{2/[3(n+1)]}$.

8.6.6.3 Initial Conditions

Models with $k \leq 0$ expand to infinity, while those with positive curvature ultimately recollapse after an expansion phase. Without Λ there are no static solutions to the FRW equations. This behavior historically determined the fortunes of the Friedmann models. In 1922 when Friedmann first published his results, it was generally thought that the Universe was stationary.[47] In spite of scattered observational evidence by Slipher and others that indicated unusually high recessional velocities for some spiral nebulae (which were then not known to be extragalactic systems), there was no compelling evidence supporting any of these models. It is appropriate to return for a moment to the distance scale to remind you that, without some indication of the nature of the "nebulae," there was no reason for thinking that the galaxies were actually markers of a cosmological expansion. The situation changed with Hubble's determination of the distance scale. Once distances could be fixed to the galaxies, with the discovery of the Hubble law and the application of a Copernican principle that required local and global isotropy and homogeneity, the Friedmann solutions became central to cosmology and were recognized as actually having predicted the expansion.

In all cases, there is a singularity at $t = 0$. For $k \leq 0$, this is the *only* one since $\dot{R}$ never reverses sign. For $k > 0$, there are actually *two* singularities, one at $t = 0$

---

[47] In order to achieve this state, Einstein added Λ to the field equations. As we've discussed, he also produced an interestingly almost Newtonian argument in which he wrote a modified Poisson equation in the form $\nabla^2 \phi + \lambda \phi = 4\pi G \varrho$ in an attempt to include Mach's principle within general relativity and obtain a static matter distribution. It was a failure.

and another at either $t = \pi R_*$ for the matter-dominated case or $t = 2R_*$ for a radiation dominated model. One could imagine that such singularities exist due to the symmetry of these models. In other words, if matter geodesics are not all radial, it is not certain that they will pass all for the same point. Hawking and Penrose (in 1968)[48] showed that the singularity is unavoidable under reasonable hypotheses without modifying the evolution equations. However, quantum effects become important at densities approaching the Planck value, $\varrho_P = 5.2 \times 10^{93}$ g cm^{-3} (see Section 8.6.8). For both choices of equations of state, the time at which this happens in the models is similar, $\approx 5 \times 10^{-44}$ s. Alas, a convincing quantum theory of gravity has not been yet achieved so it isn't clear what the resolution might be.

Let's examine the evolution of Ω. We can start by noting that $\varrho_c = 3H^2/8\pi G$ varies with epoch since the Hubble parameter, H, depends on the evolution of R with time. The energy equation relates the development of ϱR^3 to the equation of state, so we can rewrite it in terms of Ω using $\varrho R^3 = \Omega \varrho_c R^3 = 3\dot{R}^2 R/8\pi G$:

$$\frac{d}{dt}\ln \varrho R^3 + \frac{3P}{\varrho}\frac{d \ln R}{dt} = \frac{d \ln \Omega}{dt} + \frac{d}{dt}\ln \varrho_c R^3 + \frac{3P}{\varrho}\frac{d \ln R}{dt} = 0, \quad (8.97)$$

which reduces to

$$\frac{\ln \Omega}{dt} + 2\frac{d \ln \dot{R}}{dt} + \left(1 + \frac{3P}{\varrho}\right)\frac{d \ln R}{d} = 0. \quad (8.98)$$

Now we need to introduce the evolution equation, Eq. (8.72), since the acceleration $\ddot{R}$ has appeared, in the form

$$2\frac{d \ln \dot{R}}{dt} = -\frac{8\pi G\varrho}{3H^2}\frac{\dot{R}}{R}\left(1 + \frac{3P}{\varrho}\right),$$

which reduces Eq. (8.98) to the form

$$\frac{d \ln \Omega}{dt} = (\Omega - 1)\left(1 + \frac{3P}{\varrho}\right)\frac{d \ln R}{dt}. \quad (8.99)$$

Notice that the right hand side of Eq. (8.99) is strictly negative for $\Omega < 1$ since $d \ln R/dt = H > 0$ so Ω strictly decreases with time. Conversely, if $\Omega > 1$, it must continue to increase. Only for $\Omega = 1$ is there a constant solution that avoids the problems of the initial conditions. Let's go one step further to see how important this conclusion really is. Take the simple isotropic equation of state $P = \varrho/3$ and substitute it into Eq. (8.99). The solution for Ω varies as

$$1 - \frac{1}{\Omega} \sim R^2. \quad (8.100)$$

---

[48] See, for example, Hawking and Ellis (1973); Wald (1983), referenced in Appendix 1.C, for more discussion.

At the present epoch, or even at the decoupling era, Ω_0 is very close to unity. But if we take the scale factor back to an arbitrarily early stage in the expansion, then simply no deviation from $\Omega = 1$ can be tolerated since R increases by many orders of magnitude (we won't even specify how many; the numbers simply become absurd very quickly).

Something must be wrong. The early Universe would be expected to be a sort of foam, with possibly large fluctuations in local curvature due to similarly large, but localized, fluctuations in the energy density. The Universe as we now see it seems to require an unexpectedly precise correlation, even for regions that were causally disconnected at the earliest time. Either Ω really is uniformly and magically set equal to unity, and remains so for all time, or the other assumption, that $(1 + 3P/\varrho) > 0$ must be wrong. Of course, we *could* have argued that $\dot{R}/R < 0$, but we already know that this is impossible for any epoch from the fact that we're here. Since ϱ is always positive, this sign switch can only be obtained by assuming a *negative* pressure. Now you see why we slipped in that earlier remark that the cosmological constant behaves like a negative pressure term in the FRW equations. It's a viable candidate to drive the expansion although not, perhaps, the one we sought.

8.6.7 Linking the Scale Factor to Observations and Tests of Cosmological Models

Earlier in this chapter, we discussed observations that require using the redshift even without knowing its origin. Now that you know where it comes from, we can proceed to show how the redshift connects through these observables with the cosmological model. Our purpose is to set out the properties of *any* model in the most general way regardless of the particular solutions of the dynamical equations. In order to connect observations in redshift with epoch in the expansion, we need some way to connect time and redshift. This is also needed to connect galactic evolution with z. To connect geometric measurements, such as sizes, with the model we need to examine the angular diameter–redshift relation. And a discussion of radiative transfer in the expanding cosmos will be needed to understand the observations of the Lyα forest and the feedback effect of ionization on the background—that is, the Universe as an expanding analog of the interstellar medium.

8.6.7.1 *Lookback Time, Proper Distances, and Horizons*

> Something lies beyond the scene...
> —Edith Sitwell, *Façade*

We've avoided until now an important question: How much of the Universe can we actually see at any moment, and is there a limit on how much we will ever be able to see? A moment ago, we said that at any time, there are regions of the Universe that have not yet come into contact, given the amount of time that has passed. Yet even stationary observers are separating as the spacetime expands so how far can we see? Of course, you know the trivial example of the horizon on Earth where the radius of curvature determines the fraction of the sphere accessible to any point.

But for a strictly expanding universe, this curvature decreases steadily in time and with the horizon moving progressively farther away you see more of the surface. So we can lay out the logic this way:

1. The metric for static geometry is all you need, along with the speed of light, if the scale factor R is independent of time. Then you just need to compute $\int dr/(1 - kr^2)^{1/2}$ to find light travel times in a static world.
2. If the scale factor changes, then you need to allow for this when calculating the lookback time: you have to scale the light travel time using $\int dt'/R(t')$.
3. Now since you know the connection between R and z, you have a choice of independent variable when computing the integrals. Neither, however, is independent of the dynamics.
4. For local observations, Taylor expansions of R in t with the change from R to z suffice to give you distances but say nothing about the maximum distance you can reach, so... .
5. You need some way to transform without restriction between dt and dR, which *requires* the FRW equations and choosing Ω_0, Λ, *and* an equation of state.

Having obtained a connection between z and t, we can obtain a model independent formalism that links z and R and invert the process to find $t(z)$. We will make extensive use of the relation between the scale factor and the redshift $R_0/R = (1 + z)$ in the sections that follow.

The FRW equations provide $R(t)$ through which we get the relation between t and z, the connection between the purely kinematic derivation, Eq. (8.46), and the dynamics. We need, however, to find two things. We need to choose either a matter or radiation dominated equation of state in order to solve Eq. (8.72). This gives an explicit form for k:

$$k = -H_0^2(\Omega_0 - 1).$$

Now use the FRW equations to solve for $\dot{R}$:

$$\dot{R}^2 = H_0^2(\Omega_0 - 1) - \Omega_0\left(\frac{R_0}{R}\right)^2 \qquad (8.101)$$

and substitute $\varrho = \varrho_0(R_0/R)^3$ on the right hand side and recall yet again that $R_0/R = (1 + z)$ to find

$$\frac{\dot{R}}{R^2} = H_0(\Omega_0 z + 1)^{1/2}(1 + z)^2 = \frac{dz}{dt}, \qquad (8.102)$$

or defining $x = R(t)/r_0$, we have

$$\frac{dx}{x\dot{x}} = \frac{dx}{x\left[(1 - \Omega_0) + \Omega x^{-1}\right]^{1/2}}. \qquad (8.103)$$

We're now done. The lookback time to any redshift is given by

$$\Delta t = \int_0^t \frac{dt'}{R(t')} = \int_R^{R_0} \frac{dR}{R\dot{R}} = \int_{(1+z)^{-1}}^1 \frac{dx}{x\dot{x}} = \int_0^z \frac{dz}{H_0(1+z)^2(\Omega_0 z + 1)^{1/2}}, \quad (8.104)$$

so now that we can transform between time and redshift, it is a small step to obtain the proper distance as a function of redshift. As we said when discussing the entropy equation and thermal evolution, you also need this transformation between Δt and z to convert comoving calculations of any physical property into the observer's frame. We can integrate Eq. (8.104) to obtain

$$\Delta t = \frac{2}{H_0(\Omega_0 - 1)^{1/2}} \left[\tan^{-1}\left(\frac{\Omega_0 z + 1}{\Omega_0 - 1}\right)^{1/2} - \tan^{-1}\left(\frac{1}{\Omega_0 - 1}\right)^{1/2} \right], \quad (8.105)$$

which is the same as the light travel distance since we're setting $c = 1$. But for physical properties of distant objects, we need more. The geometric distance observed by a comoving frame is not the same as D which is really the horizon distance corresponding to any redshift. This is the light travel distance but some applications require that we know the actual distance in the comoving frame. Then, from

$$\int_0^{r_1} \frac{dr'}{(1 - kr'^2)^{1/2}} = \int_0^t \frac{dt'}{R(t')}, \quad (8.106)$$

which reduces to solving

$$k^{-1/2} \sin^{-1} k^{1/2} r_1 = \frac{2}{(\Omega_0 - 1)^{1/2}} \left[\sin^{-1}\left(\frac{\Omega_0 - 1}{\Omega_0(1+z)}\right)^{1/2} - \sin^{-1}\left(\frac{\Omega_0 - 1}{\Omega_0}\right)^{1/2} \right].$$

The right hand side reduces[49] to an arcsine so r_1 can be obtained directly by cancellation:

$$r_1 = \frac{2}{H_0 R_0 \Omega_0^2 (1+z)} \left[\Omega_0 z + (\Omega_0 - 2)\left([\Omega_0 z + 1]^{1/2} - 1\right) \right]; \quad (8.107)$$

this is identical to the Mattig equation with the substitution of the deceleration parameter, q_0 for the density parameter.[50] To obtain the geometric distance including the scale factor we simply require $R(t)r_1$ which can be written in terms of redshift.

---

[49] Here some useful, no doubt long forgotten, identities for inverse trigonometric functions will help you reduce the equations to a final form $2\tan^{-1} x = \sin^{-1}(2x/[1 + x^2])$ and $\sin^{-1} x + \sin^{-1} y = \sin^{-1}[x(1 - y^2)^{1/2} - y(1 - x^2)^{1/2}]$. This substantially reduces the algebra and shows that the result doesn't depend on the sign of $\Omega_0 - 1$.

[50] There is a nice discussion by M. S. Longair on this derivation, although he proceeds a bit differently, Mattig, W. 1958, *Astr.Nach*, **284**, 109, in the Saas-Fee lectures referred to earlier (Sandage, Kron, and Longair 1995, n.7). See also the discussion in the same lectures by Sandage (pp. 17–18) for some historical background to the original derivation.

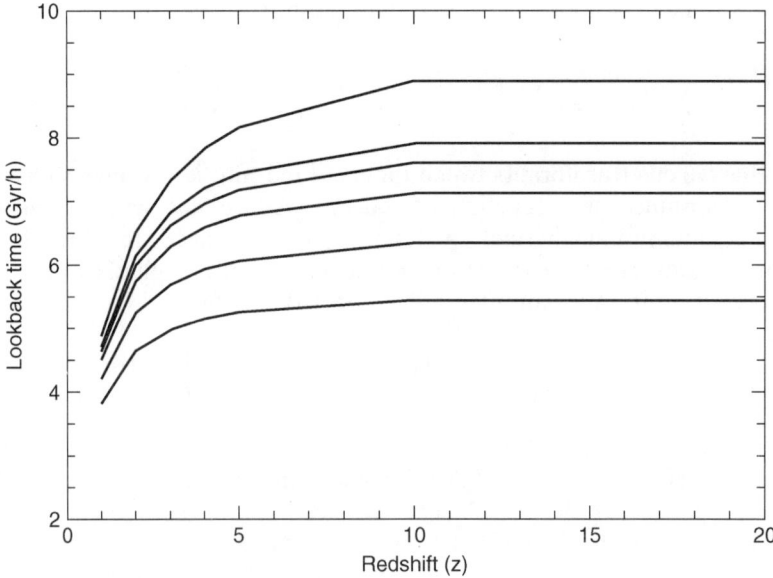

FIGURE 8.2. Lookback time for FRW universes in units of h^{-1} Gyr^{-1} (billion years) from top, for $\Omega_0 = (0.0, 0.1, 0.2, 0.3, 0.5, 1.0, 2.0)$.

At any moment, we only see a part of the universe and we can ask how large a region can be viewed at any moment (Fig. 8.2). This is the temporary horizon from which we receive information, temporary because as the expansion progresses, we see progressively larger volumes. Calling this horizon distance $cR(\Delta t)\Delta t$ from Eq. (8.104) and using the relation between R and z again, we obtain the expressions:

$$D_H = \left[H_0(\Omega_0 - 1)^{1/2}(1+z)\right]^{-1} \cosh^{-1}\left(1 - \frac{2(\Omega_0 - 1)}{\Omega_0(1+z)}\right) \qquad \Omega_0 < 1$$

$$D_H = \frac{2}{H_0(1+z)^{3/2}} \qquad \Omega_0 = 1$$

$$D_H = \left[H_0(\Omega_0 - 1)^{1/2}(1+z)\right]^{-1} \cos^{-1}\left(1 - \frac{2(\Omega_0 - 1)}{\Omega_0(1+z)}\right) \qquad \Omega_0 > 1. \quad (8.108)$$

The proper distance is used the classical discriminator among cosmological models that uses the angular diameter observed for standard candles as a function of redshift.[51] In principle, this is the same as the magnitude variation of the source. For a galaxy of known diameter L, the angular diameter is $\Delta\vartheta = L/D$, using the Mattig equation just obtained. Interestingly, this function is not monotonic if

---

[51] This test was emphasized by Sandage, A. 1961, *ApJ*, **133**, 155. See Gurvits, L. I., Kellermann, K. I., and Frey, S 1999, *Astr. Ap.*, **342**, 378 for the most recent application of this test to radio galaxies. See also Sandage, A. 1988, *ARAA*, **26**, 561.

$\Omega_0 > 0$ and has a local minimum at around $z = 1.3$. Even more serious evolutionary corrections may apply here, since the sizes of radio emitting regions may depend on redshift, depending on the origins. Compact sources may be affected by scintillation, radio lobes may vary in intrinsic size depending on the way jets feed them. For optical measurements, star formation histories can bias the determinations of disk or nuclear diameters. All these effects, and more, must be considered when applying this otherwise intuitive geometric constraint to galaxies. Nevertheless, current measurements are consistent with $\Omega_0 = 1$, and, with the availability of the Hubble Deep Field (HDF) and other deep surveys of the galaxy population at large redshift, this test is likely to become more important in the future (see Section 8.7.1.3).

Finally, we come to source counts. This is used in the reconstruction of the history of star formation in galaxies and uses the concept of a comoving volume. The number (volume) density of objects in a slice at redshift z is $n(z)$. Then if the solid angle of the observation is ω the number you see in a cone is

$$\frac{d^2N}{dz\,d\omega} = \frac{8}{H_0^2} \frac{n(z)[\Omega_0 z + (\Omega_0 z - 2)][(\Omega_0 z + 1)^{1/2} - 1]^2}{(1+z)^2 \Omega_0^4 [1 - \Omega_0 + \Omega_0(1+z)]^{1/2}}. \quad (8.109)$$

For fixed Ω_0 you can invert this equation to find the number of galaxies of a given type at any redshift. A variety of survey techniques have been used, both spectroscopic (where z is derived independent of assumptions about the nature of the galaxy) and using, for instance, the truncation of high redshift spectra due to intervening Lyman continuum absorption (Ly-cutoff galaxies).

How far back can we actually see? Is it possible to see everything in the universe at any given moment? In a fog, you know there is an effective limit on your visa when the pathlength leads to large enough optical depth, but this depends on the opacity and density. In a vacuum, the view is limited only by the light travel time and when some far away source turns on or off.

In an infinite universe, you literally have all the time in the world. The age at which you see any object is simply the lookback time, and if everything has existed forever, or at least if stars are constantly forming and dying, there is no special horizon beyond which you cannot obtain information. Your observations are limited only by the sensitivity of the instruments and the background (sky noise). When the cosmos are expanding, as in relativistic cosmologies, then the situation is changed. In the comoving frame, your view is limited to that part of the matter from which light has had a chance to reach your location:

$$ds^2 = 0 \rightarrow \int_0^t \frac{dt'}{R(t')} = \int_0^{r_1} \frac{dr'}{(1-kr'^2)^{1/2}}. \quad (8.110)$$

Now, if $r_1 \to \infty$ as Δt increases, there is no event horizon. In other words, the integral for the proper distance increases to eventually encompass the whole Universe. If, on the other hand, the integral remains finite for all time, then there *is* an event horizon and some part of the Universe will stay forever hidden from observation. For any finite age universe, there is *always* a particle horizon, the sphere to which we can currently see.

The structure of spacetime limits your view at any moment in much the same way the curvature of the Earth produces a horizon that appears to be centered on the observer. Shift your location, and the horizon moves with you.[52] A *particle horizon* separates what you see at any instant from what you don't but does not prevent you from *eventually* seeing farther if you wait long enough. An *event horizon* is more restricting: in an evolving universe it separates that which you can ever see from that which you *never* will see. This is the same as saying when we take the limit $t \to \infty$ the integral does now approach a finite value. This is equivalent to asking the following question: Can an observer *ever* see a signal from some arbitrarily chosen point if enough time has elapsed? For any universe that expands as a power law in time, the answer is "yes." There is no true event horizon. For a de Sitter universe (vacuum solution), one that expands exponentially, the answer is "no." This is not the same as the *particle horizon*. The formal integral is the same, but now we ask whether there is any distance from which we have not yet received a signal. The answer is always "yes." Depending on the curvature, you will be able to see only to a finite redshift, and this is what is meant by something being outside of your horizon at a time t. Since $1 + z = R(t_0)/R(t)$, you can write

$$k^{-1/2} \sin^{-1} k^{1/2} r_1 = \int_{t_1}^{t_0} \frac{c\, dt'}{R(t')} = \frac{1}{1-n} \frac{ct_0}{R_0} \left[1 - \left(\frac{t_1}{t_0} \right)^{(1-n)/n} \right], \quad (8.111)$$

substituting Eq. (8.110) for the redshift-scale length relation

$$r_H = R_0 \int_0^{r_{\max}} \frac{dr'}{\left(1 - kr'^2\right)^{1/2}}, \quad (8.112)$$

for $z \to \infty$ at $r_{\max}$. Then for positive curvature, $k = 1$, we get

$$r_H = \frac{2c}{H_0(1-\Omega_0)^{1/2}} \sinh^{-1} \left(\frac{1-\Omega_0}{\Omega_0} \right)^{1/2}, \quad (8.113)$$

while for $k = -1$ we obtain:

$$r_H = \frac{2c}{H_0(2q_0 - 1)^{1/2}} \sin \left(\frac{\Omega_0 - 1}{\Omega_0} \right)^{1/2}. \quad (8.114)$$

If the distance to a body is greater than r_H, we cannot detect it. The various forms result from the integration of the equations for the relation between the scale parameter, R, and the redshift and deceleration parameter and are, essentially, kinematic quantities. The dynamics enters because of the solution of the field equations for the expansion through the FRW equations.

---

[52] Rindler, W. 1956, *MNRAS*, **116**, 662; Harrison, E. R. 1991, *ApJ*, **383**, 60.

One last aside in this section. You will have noticed that we sometimes switch between Ω_0 and q_0. This is not just from sloppiness. There is a difference in the origin of these two parameters. The former is directly connected with the FRW equations through the density. The latter is a kinematically defined quantity that only requires a connection between R and z and then a Taylor expansion to second order. Yes, they are functionally equivalent, but we want to emphasize that they can be measured differently. Sometimes, the switch will highlight one or another method.

This consideration of horizons leads directly to the concept of inflation, the scenario in which most modern cosmological arguments about the early Universe are framed. In a generic sense, this is one way to see it. In the initial state of the Universe, the time was long enough and R small enough that the fraction of the Universe that an observer could see was very large—that is, the light travel time can be scaled to R to show that there was a large observable fraction of the Universe. When inflation begins, this fraction of the total volume quickly drops. You can see what happens by imagining a standing wave in an expanding box. The wavelength grows rapidly with the effect of suddenly restricting your view to a tiny part of the wave. Locally, it looks as if everything is homogeneous and that nothing has happened. Now, in the post inflation era, the fraction of the observable Universe is again increasing and so once again, in the comoving frame, you are starting to see parts of the wave where the variation is larger. This is what it means to have the perturbation *reenter* the horizon. Actually, it is that your view is able once again to encompass a part of the perturbation. The rate of inflation determines the fractional restriction of your view at the first epoch, $t_\star$. The flatness and isotropy, that is to say the degree to which the Universe is homogeneous as you see it, are set by the amount of expansion in the inflationary epoch, $R_{\text{infl}}/R_\star$, where R_{infl} is the scaling parameter at the end of inflation. Then the fraction of the Universe you can observe starts to grow again because of the slower expansion $R(t)/R_{\text{infl}}$.

This means the critical wavelength that you are able to see gradually increases and the correlations that you observe in the background begin to reach larger and larger scales. In the first instance, without inflation, you would not have been able to see any correlation on scales larger than the order of, say, a few degrees because of the ratio R_0/R_{dec}, which is $(1 + z) \approx 10^3$ on the basis of decoupling at around 7000 K or so (recall that $T_{\text{CBR}} \sim (1 + z) T_{\text{CBR, 0}}$). On the other hand, you see scales of order 30° for the very weak fluctuations, so there is a need to explain both why they are so small amplitude and why the homogeneity occurs. The second problem is that the Universe you observe has $\ddot{R}$ very small now; this is another way of saying that $\Omega \approx 1$. But $\Omega - 1 = \ddot{R}R/\dot{R}^2$, so if there had been an epoch at which a net acceleration existed in the expansion—in particular, if the expansion were *driven*—then Ω could rapidly change at first and then readjust at the end of the inflationary epoch to its current state.

In order to place limits on this epoch, you are essentially working backward. The basic problem is to determine what field drives the expansion in the pre-matter state. This is very early, well before the photon-dominated era, and radiation pressure, $P \sim \varrho$, does not work. Some additional field is needed to produce a more rapid driving since the pressure falls for radiation too rapidly to maintain an inflation, and the coupling constants with matter are too large. In the current matter-dominated era, the pressure is zero (or very close) because η is comparatively low, so that massive particles will not be able to produce anything like inflation. There are other possibilities, and these are the basis of the current theory, that invoke scalar fields of still *unknown* resonant states for the driving, and this is what we will discuss in Section 8.6.8.

Any fluctuations in the field distribution, for instance, gravitational waves generated by the end of the inflation and the phase transition to another field (something like the release of binding energy), generate variations in the temperature of the background by an analog of the Sachs–Wolfe effect (Section 8.7.6.1). These are locally fluctuations in the potential that gravitationally refract the background that we see. The photons must pass through these fluctuations, on which everything is riding, and this causes variations in the properties of the geometric screen through which we observe the CBR. That is why the scale of the correlation length tells you about the process that is responsible for the inflation—the larger the scale of the correlation, the longer the wavelength of the perturbation (in other words, the fraction $\lambda_p/R_{\text{infl}}$).

To summarize, the horizon is not an *event* horizon in the strict sense of the word. Instead, it is a matter horizon, the *fraction* of the Universe that you can see based on the rate for the expansion $\dot{R}/R$. This is what changes during the inflationary epoch.

8.6.8 Inflation

Although our cosmological models assume a smooth continuous spacetime structure that can be pieced together from local observations to form a global geometry, at some microscopic level this assumption must fail. General relativity does not extend down to the quantum scale. Where this breakdown occurs can be estimated in the same way that we heuristically derived the radius of a black hole. You know that the Schwarzschild radius is the size at which the binding energy of a body of mass M becomes comparable to its rest energy, GM/c^2. If this length scale is of the same order as the Compton wavelength, $\hbar/Mc$ then a quantized law for the gravitational field is required. This sets a mass scale, $M_P = (\hbar c/G)^{1/2} \approx 2.1 \times 10^{-5}$ g, called the *Planck mass*. The associated length scale, $L_P = (\hbar G/c^3)^{1/2} \approx 10^{-33}$ cm provides the cosmologically important constant, the time in the expansion at which we can expect quantum effects to become negligible, called the *Planck time*, $t_P = L_P/c \approx 5.4 \times 10^{-44}$ s. With these scaling constants in hand, we now step into the scenario for the earliest stages of the Universe, *inflation*.

The standard relativistic model is very successful at providing a general explanation for the cosmological redshift, the initial synthesis of light elements without stars, and the existence of the background radiation. Nevertheless, several troubling features remain. It is a theory, in a sense, without a beginning. There is no prescription specifying the initial conditions, the physical processes that happen in the first moments after the Big Bang when the energy density is of the order of the Planck density and quantum effects dominate. The evolution of a Friedmann or Tolman model can be followed only later. The description of the content of the Universe in the early phases requires quantum field theories. For very high temperatures, this means that we are using the equation of state of an ideal quantum gas for particles of zero mass. Since the *entropy* is conserved in the expansion because the Universe is assumed to be a closed system with only internal sources or sinks, it is possible to place limits on its primordial values from observations in the present era.

The present particle horizon, the temperature of the CBR, and the constraints on the number and the decoupling of massless neutrinos, give a value for the entropy equal to the one obtained from the FRW equations for a flat Universe ($\varrho = \varrho_c$). This implies that if the average density at those early times had been slightly higher than ϱ_c, the universe would have been closed and its scale parameter would have collapsed long before the present epoch. If, on the other hand, the density had been lower than ϱ_c, the Universe would have been open and the present energy density would be far lower than we now observe. Only an extreme fine tuning of the initial value of ϱ could result in the universe that we inhabit, more than $\sim 10^{10}$ years from the start of the Big Bang. Such stringent restrictions are not obvious within the standard model and are very hard to justify from first principles. In constructing the standard model, however, we assumed that the Universe was homogeneous and isotropic from the outset. We must allow for the possibility, however, that these conditions are violated on the extremely small length scales of the initial state. From the discussion of the evolution of Ω, it appeared that the allowable range for the initial conditions that lead to a homogeneous and isotropic universe is vanishingly small. Thus we are faced with the problem of finding what initial values for the critical parameters give rise to *our* particular observable Universe.

The *inflationary scenario* was proposed as a solution to several vexing cosmological questions. It postulates an initial exponentially fast expansion (de Sitter model) that leads to a similarly steep decay of the curvature due to the growth of the horizon. This produces a "local" region of the Universe that accords with our experience, one where $\Omega = 1$ and where the distribution of matter is isotropic and homogeneous (at least on the largest scales).

The initial free expansion of the spacetime results from the vacuum FRW equations. For a vacuum density ϱ_{vac}, the expansion of R occurs exponentially at a rate $(8\pi G\varrho_{\text{vac}}/3)^{1/2}$, an analog of the freefall time. We are assuming that gravitation is present from the start so G appears both because of the field equations and from the prejudice that the final unification of forces leads to a field that couples to gravity. The scale factor undergoes an exponential increase in size. Bear in mind that this is the change in the structure, not of a matter distribution, and it can take place as fast as we wish, unconstrained by the speed of light. To see what this means, let us digress for a moment on a physical analogy, the expansion of an explosively ejected shell of matter, as in a nova. In a gas shell, the sound speed determines the rate of communication between different regions. In supersonic expansion, each locale evolves independently of others that lie outside of the "sound cone." If initial conditions in the explosion impose a structure on the shell, for instance, a thermal or mechanical instability, the amplitude of the instability will grow at the local rate, unconstrained by the effects of neighboring gas, limited only by the rate of expansion. Thus, as the material expands, the instability will be suppressed if the rate of expansion is too great and the density drops, while for slow enough expansion rates the instability may reach the nonlinear regime and eventually saturate from other special effects. Of course there are limits to this analogy. For the ejecta, there is a strict speed limit, c, but the speed of sound is always lower than the expansion and it sets the causality condition for the growth of structure.

We don't live in a vacuum, regardless of how you might sometimes feel, so the next problem is to figure out where the matter comes from. It is necessary to postulate the existence of a scalar field, most of whose physical properties cannot be specified in advance. It is controlled by a parameter, call it φ, such that $V(\varphi)$ is the interaction. The assumption is that for $\varphi \approx 0$, the potential has a relatively flat form, nearly independent of φ. Then, at some critical value, there is a local minimum at which point the potential makes a sudden transition to a lower value. At this point, the Universe behaves as if it is radiation dominated, where the density is no longer zero and Λ becomes very small.

Regions that lie beyond causal contact initially expand away from each other and evolve independently. The local curvature, set by the relative size of the region at any moment, rapidly decreases toward flatness while the energy density rapidly drops. For a vacuum, the solution for the field equations gives an exponential growth for R that would lead to an empty universe in short order; the e-folding timescale is the Hubble parameter. This clearly *isn't* the Universe we inhabit, no matter how low the mean density is at this epoch. On the other hand, should the rate of expansion decrease below the speed of light, any location would see other regions of the Universe reenter the horizon of the local observer. Any structures that would have been expanded by the change in the scaling parameter would therefore come into causal contact and, if the subsequent rate of expansion is low enough, produce a local gravitational perturbation that would attract the local matter. Thus you can see how the imposition of small fluctuations at the earliest stages of the expansion would lead to the creation of large-scale structure.

A number of sources have been proposed for structural seeds. Quantum fluctuations, occurring on scales that are very small compared with the size of the horizon, seem a natural choice in the initial conditions. These will stretch as the spacetime expands, but will still have significant power at very high spatial frequency once the universe arrives at the stage of making the phase transition. Gravitational waves, fluctuations on the scale of the horizon, are another possibility. These have much lower spatial frequency and will appear with smaller amplitude and at larger scale at the horizon.

Let's consider what the initial conditions may have been like. We need a *scalar* field as a minimal representation of the cosmic state. Whatever this field is, and it is usually called an *inflaton* to distinguish it from other scalar fields, its energy density must have been quite high and fluctuations frequent. A self-interacting field, φ, is represented by a Lagrangian

$$L = \frac{1}{2}\partial^\mu \varphi \partial_\mu \varphi - V(\varphi) \equiv \frac{1}{2}\frac{\partial \varphi}{\partial x^\mu}\frac{\partial \varphi}{\partial x_\mu} - \frac{1}{2}m^2\varphi^2 - \frac{1}{4}\lambda\varphi^4 + \cdots, \quad (8.115)$$

which satisfies the Euler–Lagrange equation

$$\ddot{\varphi} - \frac{\partial^2 \varphi}{\partial x_i \partial x^i} = -\frac{dV}{d\varphi},$$

where the dot indicates a time derivative, which becomes

$$\ddot{\varphi} + 3H\dot{\varphi} = -\frac{dV}{d\varphi}, \tag{8.116}$$

$$H^2 + \frac{k}{R^2} = \frac{8\pi}{3}G\left[\frac{1}{2}\dot{\varphi}^2 + V(\varphi)\right]. \tag{8.117}$$

Notice that the cosmological model enters through the Hubble parameter. The expansion is naturally accounted for by the same term that gives rise to the value of Λ. The vacuum state starts at the Planck energy density is given by the same combination of physical constants that we used to find the cosmological constant.[53]

The crucial requirement is to maintain the vacuum solution for the expansion as long as necessary, but not too long, and to produce the right level of perturbation, but not too much. Originally, the idea was to catch φ in a metastable state, a "false" vacuum, from which a discontinuous first order phase transition would occur. The sudden appearance of the mass term, the transition to a Higgs particle, meant the release of heat, increasing the entropy and leading to rapid thermalization. There were several problems with this picture, the most immediate being the need to deal with the production of large amplitude, large-scale perturbations. On the other hand, a second-order transition has the advantages of reheating without the attendant fluctuations, and this is now the favored view, called the *slow roll* picture. The description seems classical but is really a way of expressing the evolution of φ. Some model interaction potentials can be ruled out on the basis of their evolution equations. If the $\ddot{\varphi}$ term is too large, the Universe inflates too rapidly and by the time the reheating occurs it is too late. In others, the inflation ends too soon and the universe is too inhomogeneous and has too large a curvature. This leaves room, however, for a rather broad range of field theories that admit possible inflationary epochs with the right characteristics. Another difficulty with the original scenario was the need to account for interactions between individual regions in which inflation occurs as they come into contact with each other. The state of the Universe can be initially as inhomogeneous as you'd like, but the result after the start of the Big Bang epoch must be extremely smooth. Any perturbations, in other words, must have very small amplitude that are frozen at the time the inflation begins and the regions exit each other's horizons so the Universe has very low amplitude fluctuations that are preserved at the time the CBR is formed. It is a matter of tuning the theory, the freedom to choose among permissible interactions is still natural. The particles emitted at the transition are also important because some of them are likely candidates for dark matter. For instance, supersymmetric theories assign fermion partners to each boson and vice versa. For gravity, the graviton is a spin 2 field and has an affiliated *gravitino*

---

[53]This is why Λ is modeled as a scalar field. It could be vanishingly small now, if that is the observational constraint, and larger in the past. But to say that it varies with time is to deny its constancy, of course, and also prevent its simple insertion into the FRW equations. You could have it vary the other way, if the SN Ia data we will discuss requires a large value *now* but the same problem occurs. Instead, it can be added into the FRW equations as a *scalar* field with its evolution equation given as we have described. Gravity is a tensor force that is already accounted for in the field equations, leaving only a scalar particle as an option.

with spin $\frac{3}{2}$. Although presently undetected, this is a viable nonbaryonic particle. Whatever the source may be for φ, however, the number of such particles is determined by the magnitude of the phase transition and its rate.

The earliest epoch can be dealt with at present only by assuming a particular scenario for the earliest, truly primordial, stage of cosmic evolution. It is important here to say again that this deals *only* with the initial state, not with the nucleosynthesis epoch or afterward that are well described by Friedmann cosmologies. In effect, the problem to find a way to rapidly homogenize the universe as we see it, and then to allow for the development of structure.[54]

We return now to the horizon problem. Appearances are everything in the inflationary scenario. It is only necessary that the *local* Universe seems homogeneous. What you are doing is analogous to our previous picture of sitting on a wave. Imagine that the wave initially expands very quickly, so fast that it is impossible to see the crest after some point. Locally, the curvature of the wave quickly drops because the fraction of λ, the wavelength, that you can see becomes smaller and smaller. Now if something causes the wave to slow down, then a signal can propagate backward from the wavefront to you and you begin to see more of λ. This requires, though, that there is a change in the speed of the wave. In a cosmological model, this implies homogenization being produced by inflation causing any local region to become progressively flatter. If the expansion then suddenly slows down, your vista begins to encompass progressively more of the Universe and you start to see the inhomogeneities and the line of sight comes progressively closer to the actual "boundary." Suppose that the metric and the distribution of matter–energy had been chaotic in the preinflation phases. It is plausible to consider that before inflation, due to statistical effects, there exist small regions that are locally homogeneous, isotropic and in thermal equilibrium which can be described separately by the Robertson–Walker metric. When the temperature of any of these regions is lower than T_c, the inflation take place and from the bubbles of new phase will derive the large homogeneous, isotropic regions with $\Omega \approx 1$. The regions of true vacuum can become so large, if the number of e-foldings time of accelerated expansion is large, that the homogeneity and horizon problems are automatically solved. The production of entropy during the reheating phase can solve the difficult situation of the fine tuning.

Let us summarize this discussion. There were many problems with the initial idea of a phase transition, and this particular mechanism has been abandoned in favor of a fluctuating cosmology in which the field itself provides the means for reheating the universe. The basic point nonetheless remains that inflation of some sort provides a way of sidestepping a host of cosmological paradoxes and, again

---

[54] The original papers related to inflation are available in several reprint volumes, in particular Kolb, E. W. and Turner, M. S. 1988, *The Early Universe: Reprints* (Reading: Benjamin) and Lindley, D., Kolb, E. W., and Schramm, D. N., eds. 1991, *Cosmology and Particle Physics* (College Park, MD: Am. Assoc. Physics Teachers). The fundamental papers for the inflation scenario are Guth, A. H. 1981, *Phys. Rev.*, **D23**, 347; Linde, A. D. 1982, *Phys. Lett.*, **108D**, 389; Albrecht, A. and Steinhardt, J. 1982, *Phys. Rev. Lett.*, **48**, 1220. Each presents the initial piece of the evolving scenario. Guth's paper outlines the anomalies of standard cosmology and presents the basic idea of inflation. Linde's paper extends the model by removing the need for nucleation of inflationary bubbles through the use of a slowly varying potential with a local minimum and no potential barrier, and the Albrecht and Steinhardt paper continues with a more precise treatment of symmetry breaking. See especially Peacock, J. A. 1998, n.33.

regardless of details, generates perturbations whose evolution can lead to the complex structure in the present Universe.

8.6.9 A Note on Radiative Transfer in an Expanding Universe

It's obvious that to look at something, you must receive light, so in our context this means solving the equation of radiative transfer in a cosmological framework. This is equivalent to finding how the intensity changes with lookback time so the transfer equation is really an evolution equation for the intensity. Similar methods apply here to those we used in Chapter 3 to write the complete transfer equation for a moving medium, except now we can make the simpler assumption that the isotropy of the mass distribution during the matter-dominated stage leads to an uniform expansion everywhere. An expanding Universe transfers photons in time much like a stellar wind does in space. Call I_ν the intensity at rest frequency ν. The comoving time derivative becomes

$$\frac{d}{dt} \to \frac{d}{dt} - 3\frac{\dot{R}}{R},$$

and the equation of transfer becomes

$$\frac{d}{dt}I_\nu - 3\frac{\dot{R}}{R}I_\nu = -\kappa_\nu I_\nu + j_\nu. \tag{8.118}$$

The coordinate of choice for this problem is z, the redshift, and we use the FRW equations to replace $\dot{R}/R$ as we've just been doing. Then

$$\frac{dI_\nu}{dz} = \frac{j_\nu - \kappa_\nu I_\nu}{(1+z)^2(\Omega_0 z + 1)^{1/2}} + \frac{3I_\nu}{1+z}. \tag{8.119}$$

We are almost finished. In terms of the redshift, the optical depth is written as

$$d\tau_\nu = \frac{\kappa_\nu dz}{H_0(1+z)(\Omega_0 z + 1)^{1/2}}. \tag{8.120}$$

Next, substitute the usual definition of the source function, $S_\nu = j_\nu/\kappa_\nu$. Notice further that

$$\frac{d}{dz}\frac{I_\nu}{(1+z)^3} = \left[\frac{1}{(1+z)^3}\frac{dI_\nu}{dz} - \frac{3}{(1+z)^4}I_\nu\right],$$

so defining $\mathscr{I}_\nu = I_\nu(1+z)^{-3}$ we can rewrite Eq. (8.118) to obtain a very familiar equation:

$$\frac{d\mathscr{I}_\nu}{d\tau_\nu} = \mathscr{I}_\nu - \frac{1}{(1+z)^3}S_\nu. \tag{8.121}$$

This form is the same as the transfer equation for a planar medium. Notice that Eq. (8.121) is easily integrated along rays that extend from the source at redshift z to the present epoch. Thereafter, all the familiar results hold for the behavior of the solution. In particular, we have the formal solution:

$$\mathcal{J}_\nu = \mathcal{J}_\nu(0) e^{-\tau_\nu} + \int_0^z \frac{d\xi}{[1 + z(\xi)]^3} S_\nu(\xi) \exp(-[\tau_\nu - \xi]). \qquad (8.122)$$

It doesn't matter whether we're looking at the background radiation or emission from galaxies, the same equation of transfer applies.

We can go one step further by asking what happens if the opacity is due to lines instead of continuum? Then we recover the Sobolev approximation. Let us look at a simple case. Choose a slab at a redshift ζ that is thin, so the redshift gradient is small but still large compared with the intrinsic line width, so that the source function is approximately constant. Then we find

$$I_\nu = I_\nu(0) e^{-\tau_\nu(\zeta)} + S_\nu(\zeta)(1 - e^{-\tau_\nu(\zeta)}), \qquad (8.123)$$

which is the normal expression for the curve of growth of the line when correction is made to the optical depth for the cosmological increase in line width due to the redshift. We are now equipped to look at two applications of radiative transfer that have been used to probe the time development of the intergalactic medium. One, the Gunn–Peterson test, deals with continuum absorption. The other, the Lyman α absorbers, deals with line spectra.

8.6.9.1 Applications: The Gunn–Peterson Test for Lyman Continuum Absorption

In the *Gunn–Peterson (GP) test*, we seek absorption at the photoionization edges of neutral hydrogen. Depending on the column density of neutral matter, there may be some distance beyond which the intervening medium is opaque to background sources. The optical depth depends on the placement of the Lyman edge, which is progressively redshifted with respect to the observer with increasing distance. A way of finding neutral material is to seek the ionization discontinuity of either hydrogen or neutral helium against a distant source. Recent observations have provided the first glimpse of another possible use of the GP test. The detection of the He II discontinuity indicates that the intergalactic medium was significantly ionized at a redshift of order 4. The puzzle is the lack of absorption by neutral hydrogen, which should be detectable as a relic of recombination. As we noted, cosmological radiative transfer is like that for a stellar wind. The difference is that now the "velocity gradient" (remember our *caveat* about the redshift) is due to the expansion of the Universe because any source seen at large distance is also emitting at long lookback time. The photons thus have to propagate through gas whose rest wavelength is shifted with respect to the epoch of the emitter. This test is otherwise closely related to the Lyα forest we'll now discuss.[55]

---

[55] Gunn, J. E. and Peterson, B. A. 1965, *ApJ*, **142**, 1633 first proposed this method for detection of neutral hydrogen at large, possibly pregalactic epoch redshifts. [See, e.g., Jakobsen, P. et al. 1997, *Nature*, **387**, 348; Jakobsen, P. et al. 1998, *Astr. Ap.*, **331**, 61; Miralda-Escude, J. 1993, *MNRAS*, **262**, 273; Webb, J. K. et al. 1992, *MNRAS*, **255**, 319; Bi, H. and Davidsen, A. F. 1997, *ApJ*, **479**, 523; Fardal, M. A., Giroux, M. L., and Shull, J. M. 1998, *AJ*, **115**, 2206.]

8.6.9.2 Applications: Cosmological Lyman α Absorption Lines

Since we sample long lines of sight toward galaxies, the same techniques that would apply to the analysis of absorption lines arising from the intergalactic medium as for lines from the Galactic disk or halo interstellar medium. The main difference is the probability of having *many* absorbers along the line of sight and the result of the redshift.[56] Lines from the Lyman series of hydrogen should be produced by any intervening neutral material and this is indeed observed. The so-called *Lyman α forest* appears to have several contributors, a further complication compared to the interstellar medium.

Quasars provide the background light sources. Absorption is presumed to arise mainly in the disks of galaxies seen in projection along the quasar line of sight, but at least some must be due to "unincorporated" gas. Remember that galaxies are rather large, fuzzy objects through which light can pass rather easily. Within these disks you will find the same interstellar medium as our own Galaxy but with a different history. Thus, you have the opportunity to use background quasars in the same way as a study of the Galactic interstellar medium uses background OB stars. They provide a light source. The same techniques as those we used for the Galaxy *should* apply, but there are several problems that come from the distances and the requirement for a cosmological model. The first is that the pathlength depends on the redshift, as does the equivalent width, and this must be corrected before an attempt is made to derive column densities. You must assume that the atomic parameters, in particular the fine-structure constant, have not changed over time and that the same properties apply to atoms in the earlier Universe as we see in the local region today. This may not be a serious problem, since studies have concluded that the physical constants we now measure are no different than they were some 10 Gyr ago.

The usual weak lines are formed by column densities up to about 10^{14} cm^{-2}, depending on the internal velocity dispersion. The *damped Lyman α* systems require sufficient column density to permit the detection of the line wings in the Voigt profile, $N_H > 3 \times 10^{18}$ cm^{-2}, and likely result from disks of intervening spiral galaxies. Even without a cosmological model, the population of absorbers can be described rather simply as a power law. The observed distribution of the number of absorbers per unit redshift, $N(z)$, is conventionally fitted with a power law because as you always find in astrophysics, the absence of a characteristic scale means such forms are a good first guess:

$$\frac{dN(z)}{dz} \sim (1+z)^{\gamma}. \qquad (8.124)$$

Unfortunately, observationally it appears that the exponent is not constant. In the range $2 < z < 4.5$, $\gamma \approx 2.5$. However, with satellite observations, the ultraviolet for low-redshift-background galaxies can be observed and in the range $z < 1$, $\gamma \approx 0$. Using the line strength and correcting for the redshift, the number of absorbers

---

[56] A particularly good overview is provided by Rauch, M. 1998, *ARAA*, **36**, 267. The literature continues to grow, especially with the advent of 8 and 10 meter class telescopes, so we'll only review the basics here.

"per unit column density in neutral hydrogen" can also be written as a power law:

$$\frac{d\mathcal{N}(N_H)}{dN_H} \sim N_H^{-1.7}. \quad (8.125)$$

for $10^{13} < N_H < 10^{18}$ cm^{-2}, thus restricted to relatively weak absorbers. None of this is really different from the interstellar case, except it is in some ways simpler. While the distance to an interstellar absorber is usually poorly constrained, especially in the outer Galaxy, here the redshift provides a distance. True, the larger z is, the less reliable the linear Hubble law becomes, requiring a cosmological model, but the uniformity of the expansion removes the main obstacle normally encountered for such studies in the Galaxy. Finally, the internal temperature or velocity dispersion due to turbulence or other small-scale motion can be obtained directly from the b values for the lines. The observed mean for b is about 30 km s^{-1}; few absorbers yield velocity widths $b \leq 15$ km s^{-1}. This would seem to say that the temperatures in the clouds are greater than 2×10^4 K, but this broadening could also have significant site-specific turbulent sources.

The absorption systems tell you about the chemical abundances at early epochs. For instance, in the systems that have sufficient optical depth to detect the D I line at 1214 Å, on the wing of Lyman α, fitting the absorption line profiles from the H I provides a first measure of the D/H ratio. As we will discuss shortly, this is one of the most important astrophysical observations for constraining the physical conditions at the time of nucleosynthesis. Additional information is provided by rare species. For instance, we have already discussed the refractory elements in the ISM and how one, Zn, is not depleted onto grains in the interstellar medium. This means an observation of the Zn II lines, at around 2050 Å provides a quantitative assessment of the gas-phase metallicity in the absorption line systems. In addition, several other heavy elements such as Ga and Ge have been detected in the Galaxy, and these are expected to be important tracers of nucleosynthesis at high redshift. Otherwise, the analysis of the lines is essentially the same as an interstellar problem, only the inversion of the column density is more difficult.[57]

8.6.10 *Primordial Nucleosynthesis*

With models in hand, we can explore the early Universe.

We have repeatedly stressed the initially hot state of the Big Bang. Deuterium is only consumed in normal stellar nucleosynthesis, in fact practically from the earliest moments of protostellar evolution. It is, however, possible to *create* it in the rapidly expanding environment of the Big Bang because the drop in density and temperature prevents the reactions from achieving equilibrium and running to completion. As we will see, the faster the rate of expansion, the more D is left and the lower the H/D ratio. Consider the reaction network beginning with only

---

[57]A relatively recent important change in thinking about the Lyα systems has been affected by the discovery of highly ionized species in the absorption systems. While it has long been known that C IV, Si IV, and even O VI are found in the halo gas of spirals, the ionization state of the Forest clouds was thought to be comparatively low.

protons and neutrons. This mixture immediately limits our options because free neutrons are intrinsically unstable. Their decay rate is determined by the number of available neutrino species, and we therefore have only a short time in which to get the job done of building the light isotopes. In particular, we can expect that a combination of the most basic building blocks, protons, neutrons, electrons, neutrinos, and photons, will get us as far as ^{4}He, with the formation of several important, stable isotopes along the way.

The equilibrium between the leptons and protons is responsible for the presence of free neutrons in the first place. This was realized very early by Gamow (in 1946), who argued that the mass difference between the two nucleons would favor a small initial abundance of neutrons if the temperature were sufficiently high. Thence, the reactions proceed rapidly to form helium. The argument comes from statistical equilibrium since the neutron is a composite nucleon that is formed through reactions such as p(e, ν_e)n, p(ν_e,ē)n, n → p + e + $\bar{\nu}_e$ and their inverses:

$$\frac{N_p n_e}{N_n} = \frac{Z_p Z_e}{Z_n} e^{-\Delta mc^2/kT}. \tag{8.126}$$

In terms of the chemical potentials, this becomes

$$\frac{N_p}{N_n} = e^{-(\mu_p - \mu_n - Q)/kT}. \tag{8.127}$$

Since $Q = 1.293$ MeV, the conditions that are expected in the early nucleosynthetic era lead to a neutron to proton ratio as low as 0.13. Freeze-out of this ratio occurs at a temperature equivalent to about 0.7 MeV, but this also depends on the rate of weak interactions. Furthermore, the rate of inverse radiative reactions depends on η. Normally, you would not worry about the finite particle lifetime when treating neutron reactions such as the *s*- and *r*-processes, because the capture times are so short and the environmental densities are enormous. But for matter at this stage of the cosmic expansion the density was low enough that free neutron decay could and did occur. This imposes a severe constraint on the progress of the reaction networks. The measured lifetime of the neutron is about 890 s,[58] which means that, even in equilibrium, neutron reactions in the early universe are severely limited by their free decay before being able to produce D and ^{3}He. Other, more usual, channels that are familiar from stellar nucleosynthesis are far too slow.

From the freezeout n/p ratio, we know that ^{4}He/H must end up at around 0.25 since helium consumes one neutron for each proton. This provides a good first guess for the primordial ^{4}He abundance if nothing exotic occurs. The reaction path proceeds through several stable isotopes, specifically D and ^{3}He, whose abundances are known from the in situ planetary measurements, such as the Galileo probe into the Jovian atmosphere, and from absorption measurements in the local interstellar medium. Deuterium is the most sensitive probe of the thermodynamic

---

[58] This has been recently determined using trapped neutrons; see Byrne, J. et al. 1996, *Europhys. Lett.*, **33**, 187.

conditions since in stars D is consumed by comparatively low temperature proton reactions $D(p, \gamma)^3He$ almost immediately on its formation in any nucleosynthetic sequence. Indeed, main-sequence stars go through a first stage of nuclear burning during formation even before core hydrogen reactions are initiated and therefore start their lives in a deuterium-depleted state. During the Big Bang, however, the medium rapidly cooled because of the cosmic expansion and the rate of destruction was reduced because of the drop in temperature and density while the reactions are proceeding. An important first step in the analysis is to realize that we have only a limited amount of time in which to get the job started of element formation. In the absence of heavy elements (we are, after all, dealing with the first few minutes of the expansion), the principal reactions are those that power the Sun, the pp (proton–proton) chain, supplemented by a few that proceed too quickly to be energetically important for stars, and their inverse channels. The first step, $p(p, \bar{e}\nu_e\gamma)D$ is too slow to be an important contributor because it is a weak interaction. Instead, it can be by-passed by neutron processes that immediately generate deuterium and tritium, $p(n, \gamma)D(n, \gamma)^3T$, which cannot occur in main-sequence stars because they have no free neutrons. In all cases, the temperature is high enough that the inverse reactions also occur. None of these is in equilibrium, since the background temperature is dropping quickly, and the intermediate stages of the network have higher than equilibrium values at the end. It is important to again stress that the lifetime of the free neutron is the dominant controlling factor for the efficiency of these reactions independent of the rate of cosmological expansion.

The rate of expansion determines several important ratios: D/H, ^{3}He/H, and ^{4}He/H (an example of the nuclear yield for $\eta = 4 \times 10^{-10}$ is shown in Fig. 8.3). The last of these can be determined, in principle, from observations of extragalactic H II regions where the He lines are visible in emission. For the D/H ratio, the

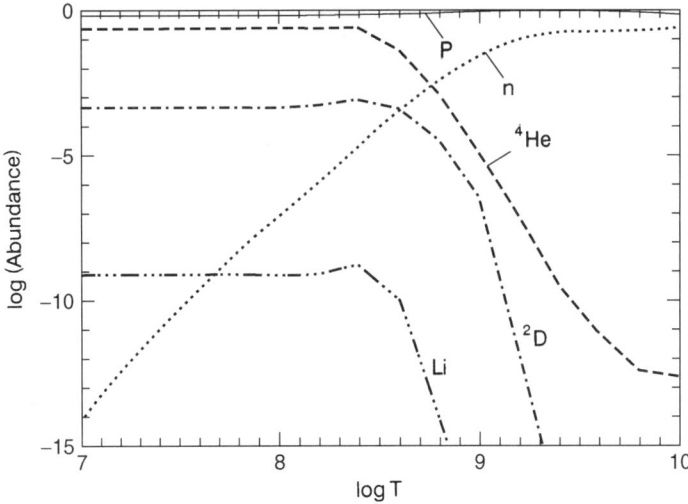

FIGURE 8.3. Evolution with temperature of the light-element abundances for standard Big Bang nucleosynthesis for $\eta_{10} = 4$.

situation is easier for the Galaxy, but less so for distant sources. The D I 1214 Å line has been directly observed in the Lyα absorbing clouds in a few cases. For the rather large redshift at which these detections have been made, $z \approx 3$, the ratio is probably higher than seen in the interstellar medium, but the interpretation of the data is far from certain. For the local ISM, the observed value is D/H = 1.6×10^{-5} in the diffuse, warm medium. This contrasts with the observations of molecular clouds and some planetary atmospheres, such as Jupiter, where it is an order of magnitude larger due to chemical fractionation. Extrapolating the D/H ratio back to the earliest stages of the Universe, however, requires models for star formation and the direct determination of the primordial ratio is not possible from observations alone.

The reaction networks are explicitly time dependent and basically the same as those you know from stellar nucleosynthesis. The neutron reactions really govern the process, however, in sharp contrast to stellar interiors, and once the n/p ratio freezes out neutrino reactions cease to be important. The equations governing the local process (since spatial inhomogeneities are usually ignored for cosmological nucleosynthesis calculations), the governing equations are essentially the same ones we met with when treating the pp reactions in main sequence stars, at least in a formal sense. The fundamental differences between cosmological and stellar hydrogen burning are the presence of thermalized neutrons and the importance of neutrinos.[59] For instance, we have reaction networks such as

$$p + n \rightleftharpoons D + \gamma \qquad D + n \rightleftharpoons {}^3T + \gamma$$

$$^3T + p \rightleftharpoons {}^4He + \gamma \qquad {}^3He + n \rightleftharpoons {}^4He + \gamma$$

$$^4He + {}^3H \rightleftharpoons {}^7Li + \gamma \qquad {}^4He + D \rightleftharpoons {}^6Li + \gamma$$

$$^6Li + n \rightleftharpoons {}^7Li + \gamma.$$

Notice that several channels produce helium and can, depending on the timescale, lead on to the light elements (Li, Be, B), but not in any significant abundance compared to H and He. The usual reactions known from stellar interiors, especially the CNO cycle, don't occur because there are no pregalactic sources of metals, so the net effect of cosmological processing is to lock up the neutrons in ^{4}He and leave only a small mass fraction of D and ^{3}He, depending on η and $\dot{R}/R$ at the time when $T < 10^8$ K.

The abundance of ^{3}T, an intermediate product of neutron capture on protons and deuterium, is absent in the stellar pp chain. Depending on the number of neutrino species and the lifetime of the free neutrons, there is only a limited

---

[59] The neutrino rates were first included, within the context of Fermi β-decay theory, by Alpher, R., Follin, and Herman, R. 1953, *Phys. Rev.*, **92**, 1347. The first post-CBR calculation of helium production was performed by Peebles, P. J. E. 1966, *ApJ*, **146**, 542, while Hoyle, F. and Tayler, R. 1964, *Nature*, **203**, 1108 had obtained a higher estimate, about 30%, for the helium produced in the Big Bang using pre-CBR data. These were soon followed by Wagoner, R. V., Fowler, W. A., and Hoyle, F. 1967, *ApJ*. **148**, 3. See also, for instance, Wagoner, R. 1973, *ApJ*, **179**, 343 and the lovely first review by Schramm, D. N. and Wagoner, R. V. 1974, *Phys. Today*, **27**, 40; Smith, M. S., Kawano, L. H., and Malaney, R. A. 1993, *ApJS*, **85**, 219.

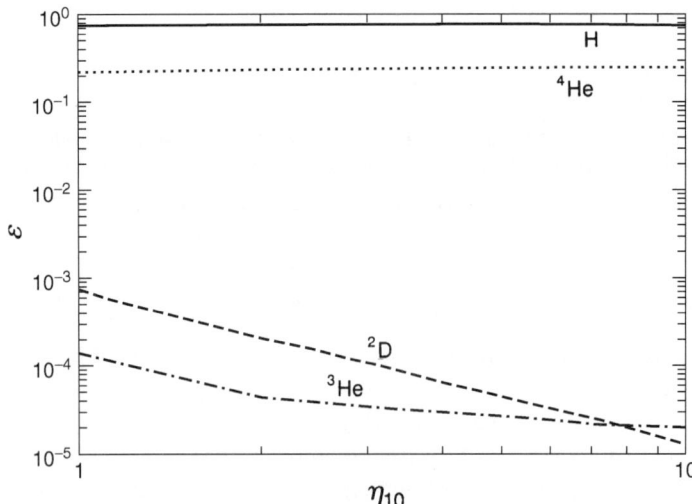

FIGURE 8.4. Dependence of final light element abundances ϵ on η_{10} for standard BBN models.

amount of time for these captures to occur. Once the nuclear timescale is the same as the contemporaneous Hubble time, the network shuts off. This is sufficient to build up the isotopes of hydrogen and helium and reach into the regions of Li and B. But no farther. There is a bottleneck at $A = 8$ and $A = 5$ caused by nuclear structure: there are no stable nuclei at these masses. If the temperature had remained high enough, and the expansion had been slow enough the reactions could have progressed much farther in mass,[60] although in conventional cosmologies this avenue is blocked.

The results of all Big Bang nucleosynthesis calculations is that the D/H ratio is the most sensitive indicator of the baryon density (see Fig. 8.4). This comes through two avenues. The first is the simple fact that the rate of reactions depends on the density at the time when the temperature exceeds a few 10^6 K. The second is that the rate of expansion depends on Ω_b, the baryonic density, as well as Ω_{DM} and Ω_Λ, the dark matter and vacuum decelerations, respectively. The slower the expansion at a given temperature, the lower the final D/H and ^{3}He/^{4}He ratios. On the other hand, the ^{7}Li/H ratio increases because the reactions come closer to equilibrium. Since there appear to be no stellar sources for deuterium and it is processed to ^{3}He and ^{4}He so easily in stellar interiors, this isotopic ratio becomes the standard for obtaining information about physical conditions in the Big Bang at the nucleosynthetic epoch. We have already discussed in Chapter 6 how isotopic fractionation messes up this ratio in molecular clouds, and there are differences between the Jovian planets and the Earth that may reflect different source for the deuterium, particularly connected with water chemistry. The *Galileo* probe directly

---

[60]As an example, see Raucher, T. et al. 1994, *ApJ*, **429**, 499, for discussions of primordial heavy-element nucleosynthesis and inhomogeneous cosmologies.

measured D/H and ^{3}He/^{4}He without the usual chemical uncertainties, finding $(2.6 \pm 0.5) \times 10^{-5}$ and $(1.66 \pm 0.04) \times 10^{-4}$, respectively.[61] Meteorites provide a limit on the helium isotopic ratio as well, where the value is $(1.5 \pm 0.3) \times 10^{-4}$, in agreement with the Jovian values so it appears this ratio is, at least, pretty well set for a epoch about 4.5 Gyr ago. The diffuse interstellar medium yields D/H $\approx 1.5 \times 10^{-5}$ at the current epoch, which translates into a primordial ratio of about 10^{-4}. This is consistent only with a relatively low entropy and baryonic deceleration parameter, respectively of order 10^9 and 0.02.[62] Observations of quasar absorption lines in high redshift systems have shown that the D line is actually detectable in the damped Lyα systems. The results are still controversial, but current observations seem to indicate a higher D/H ratio for earlier epochs. The method of analysis is the same, in essence, to that used for the Galactic interstellar medium. Therefore, it is possible to obtain the D/H ratio for substantial fractions of the lookback time, to $z \approx 4$ or higher, and considerable advance can be expected in this area within the very near future.[63]

Yet this picture is still incomplete. The *other* primordial products, ^{3}He and ^{4}He, are not exclusively produced in the Big Bang and evolve along with the stellar population of galaxies.[64] For the primordial component, one constraint is that (^{3}He + D)/H should be constant, about $(4-5) \times 10^{-5}$ (see D/H measurements in Table 8.2[65]). Low and intermediate mass stars can return large fractions of ^{3}He to the interstellar medium and you already know that ^{4}He is the main product of core and shell hydrogen burning. Regardless of most of the details, it is difficult to produce models in which the galactic ^{3}He/H ratio does not increase with time and at an increasing rate as lower mass stars die and contribute progressively more processed material to the progressively smaller amount of remaining interstellar gas. Measurements of ^{3}He are notoriously difficult to make and require using recombination lines from galactic H II regions. Interstellar absorption lines are not observable and isotopic shifts are so small relative to the velocity broadening observed in H II regions that direct optical emission line measurements are out of the question. It is possible to use both the ground-state fine-structure 21-cm analog and relatively high Rydberg lines to obtain ^{3}He/^{4}He values but only for a few regions. To complicate matters, main sequence chemically peculiar stars of the helium weak variety [described in the classification appendix to Chapter 3 (Section 3.D.2)] often show factors of $10-10^3$ enhancement in this ratio, probably produced

---

[61] Mahaffy, P. R. et al. 1998, *Space Sci. Rev.*, **84**, 251.
[62] Reeves, H. 1994, *Rev. Mod. Phys.*, **66**, 193; Spite, F. and Pallavicini, R., eds.)1995, *Mem. SAIt*, **66** 307; Smith, V., Lambert, D., and Nissen, P. E. 1993, *ApJ*, **408**, 262. See also Dicus, D. A. et al. 1982, *Phys. Rev.*, **D10**, 2694; Kolb, E. and Turner, M. 1990, *The Early Universe* (Reading, MA: Addison-Wesley); Schramm, D. N. and Turner, M. S. 1998, *Rev. Mod. Phys.*, **70**, 303; Malaney, R. A. and Mathews, G. 1993, *Phys. Rep.*, **229**, 147. See also the *FUSE* special issue 2002, *ApJS*, **140**, 1.
[63] Tytler et al. (1996, 1997), Hogan et al. (1996, 1997) and Songaila, A., Cowie, L. L., and Wampler, E. 1996, *Nature*, **385**, 137; see also Vidal-Majar, 2000, *Nucl. Phys. B (Proc. Suppl.)*, **80**, 119.
[64] For a compilation of Galactic results, see Bania, T. M. et al. 1997, *ApJS*, **113**, 353.
[65] See Vidal-Madjar, A. 2000, *op. cit.* See also Levshakov, S. A., Agafonova, I. I., and Kegel, W. H. 2000, *Astr. Ap.*, **355**, L1 for an independent evaluation. A very complete discussion of the measurements for a high-redshift system, QSO 0014 + 813, is provided by Burles, S., Kirkman, D., and Tytler, D. 1999, *ApJ*, **519**, 18, who find only upper limits that are more than a factor of 5 above the other two QSO lines of sight listed here.

TABLE 8.2 D/H Measurements

Site	Value (10^{-5})	Light Source
High redshift	3.3 ± 0.3	QSO 1937 1009 ($z = 3.57$)
	4.0 ± 0.7	QSO 1009 + 2956 ($z = 2.50$)
Solar system	2.2 ± 0.5	Jupiter (HD, ISO)
	2.7 ± 0.6	Jupiter (Galileo)
	$1.8^{+1.1}_{-0.5}$	Saturn (HD, ISO)
ISM	$0.74^{+0.19}_{-0.13}$	δ Ori
	$2.1^{+0.36}_{-0.30}$	γ^2 Vel
	$1.6^{+0.28}_{-0.23}$	ζ Pup
	$0.5^{+1.1}_{-0.5}$	α CMa
	1.12 ± 0.08	G191-B2B
	1.60 ± 0.09	α Aur
	1.46 ± 0.09	HR 1099

TABLE 8.3 ^{3}He/H Measurements

Value (10^{-5})	Site	Technique
1.5 ± 0.3	Solar wind, Jovian atmosphere	In situ measurements
$2.2^{+0.7}_{0.6}$(stat) $\pm$ 0.2(sys)	Local ISM	Pickup solar wind
1–5	Galactic H II regions	Radio lines
10–100	Planetary nebulae	Radio lines

by diffusive element separation in the stellar atmosphere, serving as a warning that stellar abundance ratios can be reset by internal nonnuclear processes. With these limitations in mind, the current range for ^{3}He is given in Table 8.3.[66]

The primordial ^{4}He abundance is, in principle, easier to determine but is hampered by the lack of stellar photospheric absorption in stars with the lowest metallicity. You can measure this for massive stars forming now in the disk as well as H II regions without much trouble. The difficulty comes in getting an age spread to times earlier than the formation of the Sun. This is done mainly using galaxies for which the measured metallicities are extremely low in the hope that this represents a sample of objects that have yet to pollute their interstellar gas with large abundances of stellar nucleosynthetic products. Otherwise, the only means to obtain a "primordial" value is through chemical evolution models of the sort we discussed in Chapter 7, and there are large systematic uncertainties associated with these extrapolations. Consequently, the existence of blue compact dwarf galaxies with $Z/Z_\odot \approx 10^{-4}$, in particular I Zw 18 and SBS 0335-052, has spurred much of the work on direct determinations of the abundance. The inferred primordial values lie between 0.23 and 0.25, but this is still too broad a range to uniquely

---

[66] See Prantzos, N., Tosi, M, and von Steiger, R., eds. 1998, *Primordial Nuclei and Their Galactic Environment* (Dordrecht: Kluwer) (also available as 1998, *Space Sci. Rev.*, **84**, 1).

specify the conditions during the nuclear epoch of the Big Bang, where the arguments are now down to the few percent level. With improvements in spectroscopy of these extragalactic H II regions, this is an area that will substantially improve in the near future.

The abundances of other light elements, Li, Be, and B, also play a role in delimiting the density and temperature ranges at the time of helium and deuterium nucleosynthesis. Through helium processing, both ^{6}Li and ^{7}Li are created in the nuclear chain, but ^{6}Li is easily destroyed by deep mixing in low-mass stellar envelopes. In fact, this property is used as a signature of youth in pre-main-sequence stars since the destruction time on proton exposure is such that ^{6}Li is depleted by about a factor of 100 relative to the interstellar medium in the main sequence lifetime of a 1 $M_\odot$ star, the Sun. For ^{7}Li, observations find that there is a floor below which the abundance does not appear to fall, whatever the metallicity of the star. Down to Fe/H values below 10^{-3} solar, the ^{7}Li/H ratio remains about $(1.3 \pm 0.6) \times 10^{-10}$ even though the ^{6}Li/H ratio can be separately determined and agrees with depletion.[67] Now recall that ^{8}B is continually synthesized in the interstellar medium by cosmic ray spallation, and this is likely the main source for ^{7}Be as well (although here explosive H burning may be partly responsible).

8.7 LARGE STRUCTURES IN THE UNIVERSE

And though the holes were rather small,
They had to count them all.
Now they know how many holes it takes to fill the Albert Hall.
—The Beatles, *A Day in the Life*

Catalogs of luminous galaxies show that the mass distribution is not uniform on extragalactic scales, as we discussed in the last chapter, and most galaxies belong to groups or clusters. *Groups* are composed by a few tens of galaxies, with masses of the order 10^{11}–10^{13} $M_\odot$ and dimensions from 100 kpc to 10 Mpc. A typical example is the Local Group with a few dozen members, of which the Milky Way and M31 are the dominant galaxies. *Clusters* are aggregates containing hundreds to thousands of galaxies, with (visible) masses of the order 10^{13}–10^{15} $M_\odot$ and dimensions of tens of Mpc. There are essentially two categories of clusters: the irregular ones, such as the Virgo cluster, that present a grainy structure with low central density, where the majority of constituents are spiral galaxies; and the regular clusters, such as Coma, which show nearly spherical symmetry with a high central concentration and a preponderance of elliptical galaxies. The centers of these regular clusters are often occupied by one or two massive ellipticals that are surrounded by extensive halos, the cD galaxies. The cluster intergalactic space is sometimes filled by large amounts of ionized, X-ray-emitting gas with temperatures of order $T \sim 10^7$–10^8 K, *Supercluster* structures have been detected, extending up

---

[67] For one halo metal poor star, HD 84937, it is possible to separate the two isotopes. See Smith, V., Lambert, D. L., and Nissen, P. E. 1993, *ApJ*, **408**, 202; Hobbs, L. M. and Thorburn, J. A. 1995, *ApJL*, **428**, L25; Duncan, D. 1998, *Space Sci. Rev.*, **84**, 167.

to 10–100 h^{-1} Mpc, with masses of order 10^{15}–10^{17} $M_\odot$. They lack any central concentration, and they are often very irregular shapes. Among superclusters there are very extended regions, from (10–100) h^{-1} Mpc, which are totally lacking luminous galaxies. These are the large voids that are organized in a cellular structure whose walls are planes and filaments with large concentrations of groups and clusters. The intersection points of this network are often near the supercluster centers. There are no indications of structures larger than about 100 Mpc, except for the fluctuations in the CBR. In other words, on the largest scale, the universe looks remarkably homogeneous.

Unlike the problem of determining Galactic structure, the lookback time plays a role in cosmology. Galaxies evolve over time, and the distances that we deal with are so large that the galactic sample at some redshift z may be intrinsically different than one obtained from a local survey. It is also possible that the structures we see at one epoch will have changed by the time we encounter their descendents at the present time. We cannot naively adopt a fixed luminosity distribution function and use the observed brightness as an unambiguous measure of the distance when attempting to delineate structure. We must, however, start somewhere and that is with the local portion of space.

8.7.1 Observational Constraints on the Large-Scale Structure

8.7.1.1 *The Darkness of the Night Sky*

Although Kepler, in his *Dissertatio cum nuncio siderio* (in 1610) asserted the finiteness of the cosmos filled with self-luminous stars based on the darkness of the night sky, the argument was purely speculative since he lacked a precise law for the variation of apparent brightness with distance.[68] In contrast, by the end of the seventeenth century, Edmund Halley had posed the problem in more familiar terms; since the brightness of the stars falls off as d^{-2}, where d is the distance, and the volume sampled at any distance increases as d^3, then in an otherwise uniform world the brightness of the night sky should increase without limit. His question was situated in the world system of the *Principia*, the last book of which dealt with several basic cosmological questions centered on the stability of the stellar universe in light of the mutual gravitational attractions between the masses. Newton concluded that an infinite space was required in order to maintain the stability of the system, or else the entire ensemble would collapse in a finite time. Halley realized that this implied that the brightness of the sky was inconsistent with such a geometry. The problem was later popularized by Olbers.[69]

---

[68] Rosen, E. 1965, *Conversation with Galileo's Sidereal Messenger* (NY: Johnson Reprints); Koyré, A. 1957, *From the Closed World to the Infinite Universe* (Baltimore: John Hopkins Univ. Press); Harrison, E. E. 1981, *Cosmology: The Science of the Universe* (Cambridge, UK: Cambridge Univ. Press).

[69] See Jaki, S. 1969, *The Paradox of Olbers' Paradox* (NY: Herder and Herder), which also includes many of the original sources. A brief history is found in Struve, F. G. W. 1847, *Études d'Astronomie Stellaires* (NY: Arno Press), see sect. 83 ("Sur l'extinction de la lumière des etoiles fixes dans son passage par l'espace céleste"), in which he describes attempted resolutions by Halley and Cheseaux. See also: Bondi, H. 1968, *Cosmology*, 2nd Ed. (Cambridge, UK: Cambridge Univ. Press); Sciama, D. 1971, *Modern Cosmology* (Cambridge, UK: Cambridge Univ. Press); Harrison, E. R. 1987, *Darkness at Night: A Riddle of the Universe* (Cambridge, MA: Harvard Univ. Press); Harrison, E. R. 1999, *Cosmology: The Science of the Universe*, 2nd Ed. (Cambridge, UK: Cambridge Univ. Press).

There were several ad hoc attempts to circumvent this problem, concerned mainly with imposing absorbing screens against the most distant bodies to produce a more rapid falloff in their apparent brightness than d^{-2}, but these failed for thermodynamic reasons. Simply put, in an infinite universe, the heating of the absorbers would bring them into thermal equilibrium with the background radiation and they would eventually glow. This was no solution at all, as J. Herschel and others realized in the nineteenth century. Associated attempts to use hierarchical clustering, such as Charlier's argument, also failed to resolve the problem. The paradox is removed by the combination of the finite lifetime of stars and the discovery of the cosmic expansion. These eliminate the two principal problems of static Newtonian cosmology—the finite age of the universe introduces a horizon and the expansion throttles the gravitational instability. In fact, as we will see later, this is precisely the situation that makes the formation of structures in the present Universe so comparatively difficult to understand.

8.7.1.2 Tests of Distributed Properties: V/V_{max} and the Malmquist Bias

Suppose you have a collection of otherwise identical sources that are distributed over some volume of space but seen in projection against the sky. If they all have the same intrinsic luminosity, L_0, then for a Euclidean universe they would have individual brightnesses that would fall off as d^{-2} or $V^{-2/3}$, where V is the volume. However, the sample of faint sources would be increasing because the volume is getting larger. Thus, any source at an observed intensity I samples a volume V_0 that has its maximum value for the faintest source you can observe. Soon after the identification of the quasars as extragalactic sources, Schmidt introduced a simple statistical procedure for looking at the luminosity function.[70] He assumed that the farthest source, that which samples the largest volume, would have V_{max} and that $V/V_{max} = (I/I_{min})^{-2/3}$. This becomes a population statistic for the sample. The mean value for the sample should therefore have $\langle V/V_{max} \rangle = \frac{1}{2}$ if the objects are uniformly distributed in number density throughout the space. The beauty of this test is that it rests on the observed brightness, not on the distance. It has been applied to the gamma ray burst sources, in an attempt to determine whether they are at cosmological distances, and it is also being used in a number of other surveys. The problem with the assumption is that you already know the intrinsic luminosity distribution for the sample. In the application to gamma bursts, for instance, this is especially problematic since the sources have few identifiable optical counterparts.

Since we've been dealing with faint samples and small number statistics for distant objects, it is especially important to discuss one of the most basic observational selection effects that plagues cosmological studies. First presented as a limiting effect in the study of the faint end of the stellar luminosity function, the Malmquist bias is also a dominant systematic effect in cosmological observations.

One way to see how it operates is to imagine a very faint point source (one that is unresolved and therefore the same size as the point spread function for the telescope) that is at the same level as the background noise either due to the emission from the sky or from fixed pattern or stochastic noise in the detector. Take the noise to have an amplitude σ so that sometimes it adds to the source and

---

[70] Schmidt, M. 1968, *ApJ*, **151**, 393; Schmidt, M., Higdon, J. C., and Hueter, G. 1988, *ApJ*, **329**, L85; Horack, J. M. et al. 1996, *ApJ*, **462**, 131.

sometimes it doesn't. If you have a source that is σ below the background, it may appear just above it due to such a noise fluctuation so it will be counted as a detection. On the other hand, what you can't see you can't count, so the opposite fluctuation adds nothing to the sample. The subtle effect of this added noise at the detection limit is to *systematically* bias the lowest end of the distribution toward a brighter limit than the real data. This is true for all photometric data. Malmquist[71] dealt with the problem by assuming a gaussian error distribution, so that the dispersion is the only requisite parameter, and it is possible to determine the correction for the fluctuations. On the other hand, there is no reason to expect that the distribution for faint astronomical sources follows this because the data may be at the Poisson limit, and dominated by instrumental effects. One way of avoiding it is to cut off the sample at some level above the background, but this may be insufficient in the case where the distribution function is rising toward fainter sources.

You can see, for instance, how this is related to the V/V_{max} test. The largest volume is sampled by the faintest objects, the ones against the sample is systematically biased. The statistical distribution will always be biased toward lower values of the ratio than $\frac{1}{2}$. This also implies that without applying the Malmquist-like corrections, the number of low luminosity objects will always be underestimated, and the mean brightness of any population will be overestimated.

8.7.1.3 *The Hubble Deep-Field Survey and Related Studies*

The deepest survey obtained by any optical telescope was achieved using the refurbished Hubble Space Telescope (HST) using 150 orbits and the Wide Field–Planetary Camera 2 (WFPC2) in four filters: F300W, F450W, F606W, and F814W. The survey was delayed because of the problems introduced by spherical aberration for the uncorrected optics, but began soon after the new camera was installed in December 1993. The observing program achieved a limiting magnitude in F814W of 28 mag (3σ) and about 28.7 (3σ) in F606W in a 4.7 arcmin^2 field in the northern continuous viewing zone for HST centered at $12^h36^m49.^s4$, $+62°12'58''$ (J2000) for which the measured H I column density is 1.7×10^{20} cm^{-2} and with a 2μ flux from the DIRBE instrument on COBE of < 0.14 MJy sr^{-1} and no evidence for any strong radio sources or interfering bright stars or clusters of galaxies. In other words, the field was chosen to be a window through the galactic foreground, allowing detection of fainter sources than obtained so far from ground-based observations. There are two advantages to the space study: first, the point spread function of the telescope is now much better than even the adaptive optics of the Keck can provide, and second, it was going to be done, anyway. A field was chosen, essentially randomly, with only *exclusionary* rules applying in the choice: no bright objects, few stars (although enough to permit registration of the images, but no bright objects), and no obvious extended nebulosity or absorbing material. The HDF provides the deepest sample yet imaged of the Universe, albeit

---

[71] See Malmquist, K. G. 1921, *Arkiv. Math. Astr. Fysik*, **16**, 1. It is historically important to note the assumptions he stated in this paper since they clearly set out some of the fundamental limitations in distance and luminosity function studies: "(1) There exists no appreciable absorption of light in space; (2) the frequency function of the absolute magnitudes is a normal curve of type A [meaning Gaussian]; and (3) this function is independent of the distances of the stars."

a very narrow slice through time. A second survey, the *Hubble Deep Field South*, was completed after the installation of the *Space Telescope Imaging Spectrograph* (STIS) and the *Near Infrared Camera and Multi-object Spectrograph* (NICMOS), starting in 1997. The WFPC2 field is centered at $22^h32^m56.^s2$ and $-60°33'02.''7$ (J2000) with slight offsets for STIS and NICMOS. This time, a quasar with redshift around 2.3 was located in the field.[72]

In the absence of comprehensive surveys of every galaxy in the HDF, it is still possible to obtain considerable detail about the evolution of the population with time. First, galaxies display a sharp drop in luminosity across the Lyman edge at 912 Å in the rest frame due to internal absorption by neutral hydrogen. For the highest z systems, this places both the Balmer and Lyman continua within the HDF bandpasses so selecting galaxies for which there is a large flux ratio between the shortest and longest wavelengths, photometry alone can supply an estimate of the redshift. From this it is possible to identify high redshift galaxies for further study, especially since 8- and 10-m telescopes are now available for optical analyses.

8.7.1.4 *Redshift Surveys*

With the distance scale known, which, also through the Hubble law, provides the evidence of the expansion, hence the timescale, we can begin to constrain the structure of the galactic distribution. Here we are on the same footing as in the case of finding out the structure of our Galaxy. We sit at a vantage point whose uniqueness we cannot know directly, hence we must use some general argument to extend the picture beyond the local neighborhood. This *Copernican principle* or *cosmological principle* states that the large scale is much like the small, that there is nothing special about the region of space that we inhabit, and that we can understand the overall structure of the Universe through the observations made in our little piece of it. This is hard to know, since the expansion produces a horizon beyond which we cannot see the present-day Universe. Galaxy evolution occurs on timescales that become comparable with the lookback time once we get to $z \approx 1$ so we cannot know directly the properties of the galaxies without some guidance from models.

A small industry has arisen with the availability of large telescopes, multiobject spectrographs, and correlation spectrometer that has produced some surprising results.[73] The local part of space that we occupy is far from homogeneous. The most complete low redshift surveys, the CfA survey[74] and the Las Campanas

---

[72] This was one of the defining projects for HST. The formal description is Williams, R. E. et al. 1996, *AJ*, **112**, 1335 which also first presented the data. Ground-based follow-up spectroscopy with the Keck telescopes to a redshift of 3 has been performed by Lowenthal, J. D. et al. 1998, *ApJ*, **481**, 673. The radio population of this field has been examined by Richards, E. A. et al. 1998, *AJ*, **116**, 1034. NICMOS observations are presented in Thompson, R. I. et al. 1999, *AJ*, **117**, 17. Rowan-Robinson, M. et al. 1997, *MNRAS*, **289**, 490 (and references therein) and Aussel, Cesarsky, C. J., Elbaz, D., and Stark, J. L. 1999, *Astr. Ap.*, **342**, 313 present detailed ISO analyses of the HDF. For the first results from the HDFS survey, see Gardner, J. P. et al. 2000, *AJ*, **119**, 486.
[73] Ellis, R. S. 1997, *ARAA*, **35**, 389.
[74] Geller, M. and Huchra, J. P. 1989, *Science*, **246**, 897.

Redshift Survey (LCRS),[75] show that the local region is dominated by several large voids and stringy structures, and that the galactic distribution is nowhere very dense. If this is typical of the larger scale, then it is reasonable that the whole two-dimensional picture should be more or less self-similar and *fractal*. For the larger scale, deep surveys have been conducted by, among others, the Canadian Network for Observational Cosmology (CNOC1, 2),[76] the Caltech Faint Galaxy Redshift Survey[77] and the European Southern Observatory (ESO).[78]

Distances are provided by the redshift, limited to relatively nearby values of order 20,000 km s^{-1}, so the deviations from the basic Hubble flow should not be large. Nonetheless, there are some complications. Imagine that you are studying a cluster of stars moving through space, but nearby there is a massive cloud. The swarm experiences differential acceleration depending on the proximity of the stars to the cloud, so you would see a systematic deviation from the regular motion in some parts of the stream, and not others. The same is possible in the vicinity of large galaxy clusters. Although the galaxies that are unbound participate in the overall Hubble flow, hence their velocities are useful measures of their distances, they also feel the gravitational attraction of the largest mass aggregations.

Ultimately, the difficulty of measuring the motion of the Local Group with respect to some reference frame limits what we can say about the mass distribution in the vicinity of our small cluster. The determination of streaming motions is much like the problem from the 1920s, when Kapteyn first detected the motion of the early-type stars with respect to the Sun and referred to "star streams." Now we see that galaxies in the vicinity of the Virgo cluster and the Local Group display systematic deviations from the Hubble flow, and this is taken as indicating the presence of some very large masses that are attracting the bulk of the other bodies. Dubbed the "Great Attractor," this distribution of mass can only be detected through observations of the local or "peculiar" motion.

A striking feature of the local maps is the existence of large-scale voids. They represent deficits in the galaxy counts by factors of 10–100 compared with the field, the same order of magnitude found in the extreme range of cluster membership. Yet on the basis of radio source counts and also the cosmic background radiation (see next section), we are faced with a fundamental problem that, without entering into the details for the moment, can be stated this way. The background radiation temperature variations are of order $10^{-3}\%$. Centimeter radio catalogs of point sources also reveal nearly uniform distributions. Yet the luminous tracers of the large-scale structure, galaxies, and clusters, are very clumpy. Somehow it must be possible for the initially small fluctuations to grow to form not only clusters but also galaxies by $z \approx 5-10$, but how? We will soon return to this question.

8.7.1.5 Redshift Surveys and the History of Star Formation

A byproduct of these deep surveys is their view of the change in the rate of star formation of galaxies over time. The method follows closely the one we used when

---

[75] Shectman, S. A., Landy, S. D., Oelmer, A., Tucker, D. L., Lin, H., Kirshner, R. P., and Shechter, P. L. 1996, *ApJ*, **470**, 172.
[76] See Yee, H. K. C., Ellingson, E., and Carlberg, R. G. 1996, *ApJS*, **102**, 269.
[77] Cogen J. G. et al. 1999, *ApJS*, **120**, 171.
[78] Vettolani, G. et al. 1997, *Astr. Ap.*, **325**, 954.

discussing how you use molecular and Hα observations to obtain the efficiency of star formation in molecular clouds. Ignoring metallicity effects, the UV luminosity of spirals comes from their OB stars, which also power H II regions. So if you measure Hα in a sample, and have some idea of the luminosity function for the galaxian population (e.g., the Schechter function), you can obtain the star formation rate for a comoving volume and then look at how it has developed over time or redshift. Also, if you can connect the metallicity (e.g., as in a simple closed box) with this rate, you can study the history of metal production with z.[79] In principle, the idea is similar to what we saw for the Galaxy. You find the comoving volume as a function of redshift for a region of space, count the galaxies in that region down to some luminosity, and then correlate the colors and luminosity with the lookback time. The answer depends, to some extent, on the bandpass and wavelength you choose for the survey since as the redshift increases so does the need for spectrophotometric correction.

8.7.1.6 A Comment on Catalogs and Distributions

Several obstacles stand in the way of observationally obtaining the galaxian luminosity function. First, it is hard to determine the *intrinsic* brightnesses of the systems. Unlike stars, galaxies are fuzzy and any measurement can be made only to some limiting magnitude determined by the instrument and the sky, not the system. The data are obtained in only a limited set of filters, not bolometrically, and have to be corrected for local (Galactic) extinction, for any (generally unknown) extinction internal to the galaxies, and for the cosmological shift in the galaxy spectrum relative to the observational window. This last is the *K correction* and must be used for $z > 0.2$, where the visible window now contains substantial contributions from the part of the ultraviolet spectrum that cannot be observed from the ground for lower-redshift calibrators. A second problem is the selection of galaxies in the sample. Ideally, these be chosen from their total (integrated surface) apparent magnitudes. They are usually, however, extracted from imaging catalogs where a galaxy's inclusion depends not on its total magnitude but on its *surface brightness*. Thus, very compact objects of high surface luminosity are favored and extended objects with low surface luminosity are undercounted, introducing a bias to the samples. This bias is a hard nut to crack since low-mass galaxies often have complicated, bursting evolutionary histories. Only for the Local Group and the Virgo cluster is any comprehensive listing of low-surface-brightness galaxies available and it is not clear how these extend to high redshift or enter into the lower end of the luminosity function with lookback time. Let us, undaunted, press on.

8.7.1.7 Dark Matter and Catalogs

As we saw with stars, it is far easier to obtain the luminosity function than the mass distribution for any population. For compactness, we'll here call the former $\psi(L)$ and the latter, $n(M)$. The large intrinsic uncertainties in the measurements of the masses of a complete sample prevent any attempt to directly determine $n(M)$. The

---

[79] An introduction to the essential arguments is found in Madau, P. 1997, in *Star Formation Near and Far*, Holt, S. S. and Mundy, L. G., eds. (NY: AIP Press); Madau, P., Possetti, L., and Dickinsen, M. 1998, *ApJ*, **498**, 106.

TABLE 8.4 Mass to Luminosity Ratios for Different Types of Galaxies

Structures	$M/M_\odot$	M/L_B	M_{gas}/M_{lum}	M/M_{lum}
Large clusters	10^{15}	316 ± 40	0.84 ± 0.1	$8.4^{+7.0}_{-1.0}$
Small groups				
Ellipticals	5×10^{13}	83^{+80}_{-10}	$0.61^{+0.1}_{-0.1}$	$5.4^{+10.0}_{-2.0}$
Spirals	2×10^{13}	40^{+50}_{-10}	≈ 0	14.2^{+36}_{-6}
Milky Way	$\approx 10^{12}$	≈ 50	0	≈ 14
Dwarf spheroidal				
Stars	10^{5-7}	≈ 2.5	0	≈ 1
Dynamics	10^{6-8}	≈ 30	0	≈ 12

larger the mass, the more indirect the technique becomes that is used to evaluate the mass. For individual spiral galaxies, rotation curves of stars and gas, or the positions and movements of test particles such as globular clusters and satellite galaxies, are used to map out the gravitational potential. For clusters of galaxies, studies on the internal velocity distribution, using the virial theorem, provide a mass estimate, while for larger systems only large-scale dynamical motions can help with the task. Keeping in mind our previous definitions, the mass distribution function can be obtained from the observed $\psi(L)$ by measuring the M/L ratio inside any class of cosmical objects. Fortunately, this ratio appears to be rather constant (see Table 8.4), so it is possible to estimate the mass function. In general, M/L is approximately $100\,h$ for spirals within $< 200\,h^{-1}$ kpc of the nucleus and about $400\,h$ for ellipticals on a similar length scale. There is no compelling evidence for a continued increase in the M/L ratio beyond this distance, which corresponds to the size of the halo. Within any class we may therefore estimate that:

$$\psi(L) = n(M)\frac{dM}{dL} \approx n(M)\left(\frac{M}{L}\right).$$

The derived mass function can be used to estimate the cosmological density parameter Ω_0, which, as we have discussed above, plays a critical role in the determination of the geometry and future evolution of the Universe. From the most recent determinations of luminosity functions and mass to light ratios, it is possible to estimate the amount of matter contained into stars and gas inside galaxies and clusters, obtaining $\Omega_{stars} \approx 0.02$ and $\Omega_{gas} \approx 0.01$. The studies on galactic dynamics result in the value $\Omega_{dyn} = 0.3 \pm 0.1$, while from the distribution of galaxies studied in the infrared with the IRAS satellite it has been obtained $\Omega_{IRAS} = 1 \pm 0.2$.

Even if these measures are strongly model dependent, they argue for the presence of large amounts of mass, *dark matter*, whose existence is not revealed by electromagnetic radiation at any given wavelength. It signals its presence only through its gravitational effects (think about the rotation curve of spirals at high galactocentric distances, where the H I gas is rotating with a flat rotation curve, in the apparent absence of any luminous component). It is possible to estimate the amount of matter involved in the electromagnetic interactions (M_{lum}), different

TABLE 8.5 Evidence for Dark Matter

Structure	$\langle M/L \rangle$	Ω
Stars and stellar clusters	1	0.001
Visible part of galaxies	10	0.01
Binary galaxies and groups	10–100	0.01–0.1
Clusters and superclusters	100–300	0.2 ± 0.1
Large-scale structures	700 ± 150	0.5–1.0
Inflationary scenario	$1000\,h$	1.0
Nucleosynthesis	—	0.1

from the matter involved in the blue luminosity emission. In Table 8.5[80] it is evident that the while the ratio M/L_B increases systematically with the scale of the involved objects, the more significant ratio M/M_{lum} is almost constant in the range from single galaxies to large clusters. This suggests that dark matter aggregates to visible matter in the formation of structures of which we see only the luminous parts.

What is this dark stuff? One hypothesis is that it is baryonic matter, since the world immediately around us is dominated by this. The dark component could be black holes, white or brown dwarfs, planets, or basketballs. Several arguments compel us to be cautious, however. Standard Big Bang nucleosynthesis yields a baryonic density in the range $0.010 \leq \Omega_b h^2 \leq 0.016$. It must be hidden very early so it avoids processing and never subsequently ignites nuclear reactions. There are no theoretical indications that the formation of structures will allow for such a large number of "aborted" stars. The existence of galaxy clusters requires that the fluctuations present in the primordial mass distribution became nonlinear before the present era. A baryonic Universe with adiabatic perturbations produces fluctuations in the CBR much larger than the observed. The contribution of photons ($\Omega_\gamma \approx 10^{-4}$) and of massless neutrinos are not enough to close the gap with the observed density parameter. Instead, exotic particles have been postulated that only weakly interact with ordinary matter *weakly interacting massive particles* (WIMPs). Inflation seems to impose pretty firm limits on the density, $\Omega_0 = 1$, with the further assumption that the dark component would be less aggregated than the galaxies, so to justify the estimates for Ω_{dyn}. In general most of the material content of the universe is considered to be massive elementary particles, distinguished according to their primordial stochastic velocities. Particles decoupled from radiation when they were relativistic ($m_\nu \ll T_{\text{dis}}$) at a temperature T_{dis} lower than the one of the phase transition quark–hadrons ≈ 200 MeV $= 2 \times 10^{12}$ K are called *hot dark matter* (HDM). A typical example is the massive neutrino, with a mass around 30 eV. Particles that decoupled when they were relativistic, but at a temperature higher than the one of the quark–hadron phase transition are called

---

[80]Sources of Table 8.5 are Bahcall, N. A., Lubin, L. M., and Dorman, V. 1995, *ApJL*, **447**, L81; Blumenthal, G., Faber, S., Primack, J., and Rees, M. 1984, *Nature*, **311**, 517; Trimble, V. 1987, *ARAA*, **25**, 423. For large compilations of rotation curves for spiral galaxies see Persic, M. and Salucci, P. 1995, *ApJS*, **99**, 501, and Mathewson, D. S., Ford, V. L., and Buchhorn, M. 1992, *ApJS*, **81**, 413. See Ashman, K. M. 1992, *PASP*, **104**, 1109 for a review of mass determinations for spirals from rotation curves.

warm dark matter (WDM). There are not many candidates to this category from theory of elementary particles. Particles decoupled when not relativistic (gravitinos, photinos) or that had never been in equilibrium with other components (axions) are called *cold dark matter* (CDM) and have negligible random velocities.

8.7.1.8 Supernovae, Λ, and Dark Energy

What if $\Lambda \neq 0$? Formally, this seems to only change how we identify the contributors to Ω. But in the FRW equations, this actually acts as an acceleration. The consequences for observational cosmological tests are quite bizarre.

One of the most dramatic discoveries since the expansion law has come recently with the analysis of high-redshift SN Ia lightcurves. It works this way *if* these are standard candles. First, if the maximum intrinsic brightness is always the same, you can use the peak brightness of the lightcurve directly. But for type Ia supernovae, as we mentioned, the rate of decline is correlated for lower redshift systems with the luminosity; the longer the decline, the brighter the explosion. The widths of the curves are scaled for cosmological time dilation since you know the redshift of the host galaxy. There is otherwise nothing different about the distance determination than for any other technique based on intrinsic brightness, except that the target sources are observable to much larger distances than the conventional standard candles. When the distance D_L is fit to standard cosmologies, a simple matter (DM and baryons) universe does not agree statistically as well with the observations as one with finite Λ. In fact, $\Omega_\Lambda \approx 0.7$ from these data! Figure 8.5 shows the data at the end of 1999 (but there have been no significant qualitative changes since these first data were published).[81] If taken at face value, the derived value is $\Lambda \approx 3$ keV cm^{-3}. In other words, it would appear that neither baryons *nor*

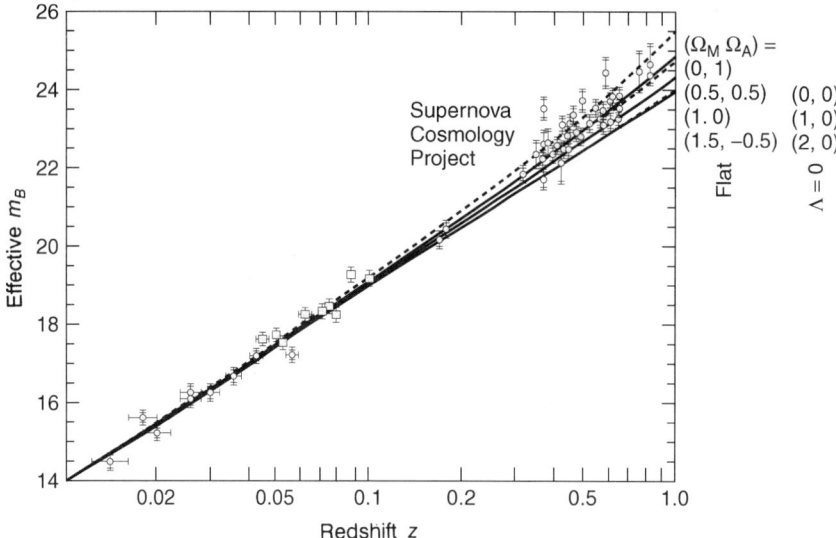

FIGURE 8.5. Hubble diagram for SN Ia plotted with a variety of cosmological models. (Credit: S. Perlmutter, *LBNL*.)

[81] Perlmutter, S. et al. 1999, *ApJ*, **517**, 565; Riess, A. et al. 2000, *ApJ*, **536**, 62. See review by Riess, A. G. 2000, *PASP*, **112**, 1284. Results are continuing to accumulate along these same lines.

"conventional" dark matter alone account for much of the total mass. There are many complications that *might* enter the interpretation of these results, but we'll mention only a few here. Intervening dust can be ruled out, but possibly not dust formation within the ejecta. The lookback times are getting long for $z > 0.4$, so some evolutionary effects might occur. And keep in mind that at present we have only scenarios for the originating SN Ia event and the progenitor remains unidentified and their properties, and the dynamics and radiative transfer in the expanding ejecta, may not be the same back in time as they are today. Nonetheless, we may be glimpsing here the evidence of a fundamental, new component in the Universe.

8.7.2 Finding Structures: Correlation Functions

To acquaint you with more of the machinery required to study large-scale structure, we'll spend a bit of time now on the principal tool used to characterize the distribution of galaxies, the angular correlation function. We introduced velocity correlation functions in Chapter 6 when discussing interstellar turbulence as a way of characterizing and measuring the kinetic energy in the flow. Such methods can also be used to study structures without any specific ties to the dynamics. The only galactic observables that can be used to study the distribution of matter in the Universe are the mass and the spatial position. As a first approximation, it may be convenient to define the matter distribution as a set of points:

$$\varrho(\mathbf{r}) = \sum_{i=1}^{N} m_i \delta_D(\mathbf{r} - \mathbf{r}_i),$$

where N is the number of objects present in the volume under analysis and δ_D is the Dirac distribution. Since we don't know much about the individual masses with any certainty, the catalogs are usually limited to astrometric and photometric data and redshifts. The matter distribution turns to a number distribution:

$$n(\mathbf{r}) = \sum_{i=1}^{N} \delta_D(\mathbf{r} - \mathbf{r}_i).$$

The probability of finding an object inside a volume dV is $dP = n\,dV$ where the average density n is independent of position. The two points correlation function is defined such that

$$dP = n^2[1 + \xi(r_{12})]dV_1\,dV_2 \qquad (8.128)$$

is the joint probability of finding an object in both volumes dV_1 and dV_2 separated by a distance $r_{12} = |\mathbf{r}_1 - \mathbf{r}_2|$. Invoking the cosmological principle, ξ must depend only on r_{12}. If the volumes are not only now independent but always have been, then the chances of finding objects in dV_1 and dV_2 are independent and the joint probability is the product of single probabilities ($\xi = 0$). If there is a greater than Poisson chance ($\xi > 0$), the position of objects are said to be correlated; conversely, if $\xi < 0$, then they are called anticorrelated. For Gaussian processes, you would expect the two-point function to suffice, but there is no a priori physical reason to stop with ξ alone. The three-point correlation functions are defined as follows. If again dP is the joint probability of finding an object in the three volumes (dV_1, dV_2, dV_3), then dP is given by

$$dP = n^3 [1 + \xi(r_{12}) + \xi(r_{13}) + \xi(r_{23}) + \zeta(r_{12}, r_{13}, r_{23})] \, dV_1 \, dV_2 \, dV_3. \quad (8.129)$$

The two-point correlation function obtained from many large-scale surveys is adequately described by a simple power law

$$\xi(r) = \left(\frac{r_0}{r} \right)^{\gamma},$$

where $\gamma \approx 1.8$ both in the case of galaxies and clusters, while r_0 values respectively $\sim 5 \, h^{-1}$ Mpc and $\sim 25 \, h^{-1}$ Mpc. This agrees rather well with the characteristic sizes of clusters of galaxies and argues for a hierarchy that, above r_0, is otherwise like a cascade or perhaps a fractal. Now keep in mind how hard this process really is when you are treating the whole sky (rather than just a piece of it). Foreground extinction is patchy and because it partly arises from the Galactic interstellar medium is not completely random across the sky.

Let's now connect this with our earlier discussion of the two-point correlation function from Chapter 6. There we invoked an analog of ξ, in that case related to velocity, when describing turbulence. Consider, for instance, flow in a stream. If the motion is uniform, no matter how far you go away from any arbitrarily chosen point you will encounter motion in the same direction with the same speed. If there is some turbulence, you will see this as you go away from the point of observation because the flow will begin to deviate. The average distance you must go before you see substantial deviations is measured by the correlation function. In turbulence theory, as first introduced in a classic series of papers by G. I. Taylor in the 1920s, the power in the deviant flow is measured by the correlation function

$$C(r) = \int v_i(x) v_i(x + r) \, dx \quad (8.130)$$

for the v_i velocity component. As the power in the correlation increases, the flow becomes more turbulent. The Taylor length is given by the curvature of this function, the distance at which the power starts to fall off. The same method can be used for structure studies using only positions rather than velocities, since the sky is a two-dimensional surface that projects the image of the Universe at large. If we know the angular separations and deproject the bodies in space knowing their distances from one of the large scale redshift surveys, then we can perform the same sort of analysis using the number of galaxies (or the power of the CBR, as we will discuss shortly) to characterize the spatial variation in the density. In effect, the spatial correlation function is another measure of the probability of finding two bodies within some distance of each other on the sky. The difference here is that *because of the redshift, we have information about the position of the galaxies in three dimensions*, in contrast to studies of turbulence in the interstellar medium where we have only the *line of sight* velocity and two spatial dimensions. Otherwise, the technique is essentially the same. One of the assumptions usually made in such correlation studies is that at some level, the probability of finding galaxy 2 at some distance $|\mathbf{r}_1 - \mathbf{r}_2|$ from galaxy 1 is

$$\xi(r_{12}) = N|\mathbf{r}_1 - \mathbf{r}_2|^{-a}, \qquad (8.131)$$

where N and a are constants. This is the analog of $C(r_{12})$. In a uniform background, we would expect that the probability of finding any two galaxies near each other on the sky should not depend on their separation. However, if the distribution is not uniform, then as we go farther away from the observation point we may expect that the number density will produce a drop in w_2. You have seen this same feature of a hierarchical distribution several times before. It is a generic aspect of the dynamical equations because of the nonlinear coupling provided by the inertial term, $\mathbf{v} \cdot \nabla \mathbf{v}$ for a velocity $\mathbf{v}$. In turbulence, this requires many special assumptions to produce a solvable, truncated system. For ensembles such as we encounter with galaxies it also requires a closure condition, some level of clustering beyond which we say that the galaxies are independent. We will return to this when discussing the development of structure after the Big Bang.[82]

8.7.3 Forming Structures in an Expanding Universe

An analytic approach to the evolution of perturbation in the primordial Universe can be developed only in a linear approximation, for small perturbations that evolve independently.[83] The high degree of isotropy of CBR suggests that at the epoch of recombination the departures from strict homogeneity in the matter distribution were extremely small. Let us define the fluctuation of the density field in a location $\mathbf{x}$ at a redshift z as

$$\delta(\mathbf{x}, z) \equiv \frac{\varrho(\mathbf{x}, z) - \langle \varrho(\mathbf{x}, z) \rangle}{\langle \varrho(\mathbf{x}, z) \rangle}. \qquad (8.132)$$

Galaxies and clusters now show typical densities larger by several orders of magnitude than the average density of the universe ($\varrho \sim 10^3 \langle \varrho \rangle$ for galaxies and $\varrho \sim 10^2 \langle \varrho \rangle$ for clusters), and even superclusters have density excesses of a factor 2 or 3 (Fig. 8.6). Linear perturbation theory cannot describe such structures, but this amplitude is observed at the present epoch. Even these *must* have started as small fluctuations at the decoupling since $\delta \varrho / \varrho$ is clearly small at $z = z_{\text{CBR}}$. So in the spirit of "mighty oaks from little acorns grow," let's look at the development that any seed perturbations are likely to follow.

The linear approximation to the theory of gravitational instabilities follows essentially the Jeans treatment. Its solution in the context of general relativity is due to Lifshitz (1946), who showed that perturbations with wavelength larger than some threshold generate gravitational instability (superposition of a growing and a

---

[82] Neyman, J. and Scott, E. S. 1958, *J. Roy. Stat. Soc.*, **20**, 1; Peebles, P. J. E. 1993, *Physical Cosmology* (Princeton: Princeton Univ. Press); White, S. D. M. 1979, *MNRAS*, **186**, 145.

[83] A superb introduction to structure formation is provided by Padmanabhan, T. 1995, *Structure Formation in the Universe* (Cambridge, UK: Cambridge Univ. Press). See also Dekel, A. and Ostriker, J. P., eds. 1999, in *Structure Formation in the Universe* (Cambridge, UK: Cambridge Univ. Press).

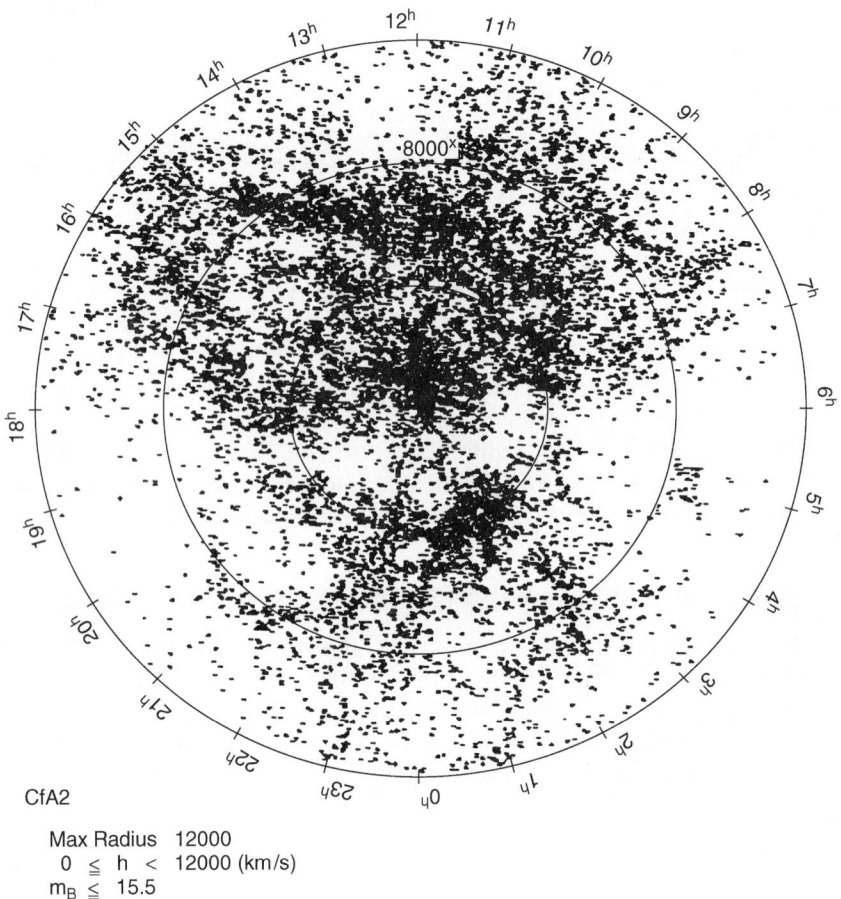

FIGURE 8.6. Distribution of galaxies with redshifts $z \leq 0.05$ from the CfA2 survey (credit: J. P. Huchra, CfA). The plot shows the projection on northern Galactic hemisphere (compare with Fig. 8.7).

decreasing solution), while the others will follow an evolution depending on the nature of the matter. For collisional matter, these perturbations appear as dispersive sound waves, while for collisionless matter they quickly dissolve due to free streaming of the particles in the expanding background. We will not present here the detailed treatment. Instead we will discuss the Newtonian equations, which apply for weak gravitational fields and low velocities. The cosmological model enters only in specifying how the global density changes with time.

8.7.3.1 Gravito-acoustic (Jeans) Instability in an Expanding Universe

The scenario we will examine resembles what you would see, for instance, for an expanding molecular cloud. Two effects compete: the tendency of matter to dilute because of the background expansion, and the growth of structures because of gravitational attraction of mass concentrations. The former enters mainly through the continuity equation, the latter comes from the Poisson equation, and both are folded into the equations of motion. Our immediate concern is to show how to derive the basic form of the equations, but there is another observable that results besides the density distribution, and we want to mention this now. Large-scale surveys of cosmological velocities define the Hubble flow, the overall smooth behavior of velocities with changing R. On top of this, there are peculiar motions resulting from local mass concentrations, that accelerate or decelerate mass. If the scale for these fluctuations is small, of order a few Mpc, then the cosmological tests will not be much affected by their presence and we can obtain the critical cosmological parameters from local measurements. On the other hand, since we know that clusters, and superclusters, can be much larger than this, we need to be able to extend redshift surveys well past the vicinity of any large clusters before we can be sure of actually seeing the overall expansion.

You will recall that when deriving the fluid equations (and those of stellar hydrodynamics) from the Vlasov and Boltzmann equations, we separated the mean and peculiar motions. We do this again here, except we write the mean motion in terms of the expansion

$$u_i = v_i - \frac{\dot{R}}{R} x_i, \qquad (8.133)$$

and take $R(t)$ from the solution of the FRW equations for the mean density and pressure; that is, what we will derive assumes that on the large scale, the density fluctuations are quite small and matter is homogeneously distributed. The evolution of the fluid is determined by the familiar equations, that we first derived in Chapter 1, and which we restate here as a reminder:

$$\frac{\partial \varrho}{\partial t} + \frac{\partial}{\partial r_j} \varrho v_j = 0$$

$$\frac{\partial v_i}{\partial t} + v_j \frac{\partial v_i}{\partial r_j} = -\frac{1}{\varrho} \frac{\partial P}{\partial r_i} - \frac{\partial \Phi}{\partial r_i}$$

$$\frac{\partial^2 \Phi}{\partial r_i \partial r_i} = 4\pi G \varrho.$$

As we saw when deriving the horizon distance and the redshift, two stationary observers will measure an increasing separation because of the cosmic expansion scale factor. The separation scales as $r_i = R(t) x_i$ and the velocity by $v_i \to v_i - (\dot{R}/R) x_i$ using $\partial/\partial t \to \partial/\partial t + (\partial/\partial x_j) \partial x_j / \partial t$. Notice that we have already assumed a Hubble-type expansion. The effect is easiest to see from the divergence term

$$\frac{\partial}{\partial r_j} \varrho v_j = \frac{1}{R} \frac{\partial}{\partial x_j} \left(\varrho \left[\dot{R} x_j + u_j \right] \right),$$

since the first term on the right hand side becomes $3H\varrho$ from $\partial x_j / \partial x_j = 3$. The continuity equation can now be written as

$$\frac{d\varrho}{dt} + \frac{3\dot{R}}{R}\varrho + \frac{1}{R}\varrho\frac{\partial v_j}{\partial x_j} = 0, \qquad (8.134)$$

where we have used

$$\frac{d}{dt} = \frac{\partial}{\partial t} + \frac{1}{R}v_j\frac{\partial}{\partial x_j}$$

for the Lagrangian derivative. Let us recall the technique we have used so often in this book and write a comoving Lagrangian per unit mass for the matter:

$$L = \frac{1}{2}\left(v_i - \frac{\dot{R}}{R}x_i\right)^2 - \Phi. \qquad (8.135)$$

The Euler-Lagrange equation yields the following form for the equations of motion:

$$\frac{d}{dt}\left(v_i - \frac{\dot{R}}{R}x_i\right) = -\frac{\dot{R}}{R}\left(v_i - \frac{\dot{R}}{R}x_i\right) - \frac{\partial \Phi}{\partial x_i}. \qquad (8.136)$$

If the pressure is needed, it can be easily included by adding a scaled gradient term, so that Eq. (8.136) reaches its final form:

$$\frac{dv_i}{dt} + \frac{\dot{R}}{R}v_i = -\frac{1}{\varrho R}\frac{\partial P}{\partial x_i} - \frac{1}{R}\frac{\partial \varphi}{\partial x_i}. \qquad (8.137)$$

Notice that we have introduced a *modified gravitational potential*:

$$\varphi \equiv \Phi + \tfrac{1}{2}R\ddot{R}x^2. \qquad (8.138)$$

This form arises naturally out of the coordinate transformation, since the second term comes from the expansion that produces an apparent acceleration even between stationary observers because of the time dilation. Now you see also the connection between this Newtonian treatment and the large-scale structure, since the FRW equations govern the time dependence of R and so feed back into the equations of motion and the continuity equation through the modified potential function φ. The now modified Poisson equation reads

$$\frac{\partial^2 \varphi}{\partial x^2} = 4\pi G(\varrho - \varrho_b), \qquad (8.139)$$

where the background density, ϱ_b is defined by substituting the modified potential into the original Poisson equation and recalling the definition of Ω in terms of H and $\ddot{R}$.

Within this background we now introduce a density fluctuation. In the current universe, we can treat the matter as dust, and initially neglect the pressure term in Eq. (8.137). Then, writing $\varrho = \varrho_b + \delta\varrho$ and assuming that v_i are due only to density fluctuations in an otherwise quiescent environment, the linearized equations are

$$\frac{\partial \delta\varrho}{\partial t} + \frac{3\dot{R}}{R}\delta\varrho + \varrho_b \frac{\partial v_j}{\partial x_j} = 0 \tag{8.140}$$

$$\frac{\partial v_i}{\partial t} + \frac{\dot{R}}{R}v_i = -\frac{1}{R}\frac{\partial \delta\varphi}{\partial x_i} - \frac{c_s^2}{R\varrho_b}\frac{\partial \delta\varrho}{\partial x_i} \tag{8.141}$$

$$\frac{\partial^2 \delta\varphi}{\partial x^2} = 4\pi G R^2 \delta\varrho, \tag{8.142}$$

giving

$$\frac{\partial}{\partial t}\left[\frac{1}{\varrho_b}\left(\frac{\partial \delta\varrho}{\partial t} - 3\frac{\dot{R}}{R}\delta\varrho\right)\right] + \frac{c_s^2}{R\varrho_b}\frac{\partial \delta\varrho}{\partial x_i} + 4\pi G \varrho_b \delta\varrho = 0. \tag{8.143}$$

Now we appeal to the solution of the FRW equations and assume a pressureless medium. The background varies as $\varrho \sim R^{-3}$, and for $k = 0$ we have $R \sim t^{2/3}$ so $\dot{R}/R = 2/3t$. In addition, $(1/\varrho_b) = 6\pi G t^2$.

We denote the Fourier modes of the density fluctuations by δ_k and take the Fourier transform of the dynamical equations as usual to remove the spatial derivatives. The evolution equation becomes

$$\frac{\partial^2 \delta_k}{\partial t^2} + 2H\frac{\partial \delta_k}{\partial t} - 4\pi G \varrho_b \delta_k = 0, \tag{8.144}$$

which for a flat ($\Omega = 1$) cosmology reduces to

$$\frac{\partial^2 \delta_k}{\partial t^2} + \frac{4}{3t}\frac{\partial \delta_k}{\partial t} - \frac{1}{3t^2}\delta_k = 0. \tag{8.145}$$

Note that the term $2H\dot{\delta}$ in Eq. (8.144) is due to law chosen for evolution of the background density, which we took to be matter dominated. For the radiation dominated era the coefficient is different. Equation (8.145) has a power law dependence on time that is found by substituting $\delta_k \sim t^n$ and solving for n:

$$\delta_k = D_+ t^{2/3} + D_- t^{-1}. \tag{8.146}$$

The amplitudes $D_\pm$ are determined from the initial conditions. There is one growing mode, which scales as R. The other strictly damps from the stretching by the expansion. This gives the first indication that, regardless of the length scale, there are perturbations that slowly grow with the expansion of the horizon. We also see that the Jeans criterion has been built into the spatially dependent terms through the competition between pressure and gravity once we include the isothermal equation of state since the Jeans wavenumber, k_J, that you know from Chapter 6, appears in Eq. (8.144).

We have yet to specify what the density represents, so let's return to the dark matter. Whatever else it may be, this stuff exerts a gravitational influence and therefore promotes the growth of structure in the Universe. The ability to distinguish between hot and cold dark matter is drawn based on whether it is unbound or bound within the structures. In general, you would expect that with a large velocity dispersion, HDM would be loosely bound and tend to accrete only into the deepest gravitational potentials that form the largest possible mass aggregations. At early times, the Universe would look relatively smooth and this would carry over even into later times. In contrast, CDM should produce a clumpier distribution earlier, and this structure would grow even more complex with time. The basic question is thus whether the large-scale structure formed first or are the smallest units, galaxies, the first step? This is an observational matter.

As a concluding remark, we note that it's tempting to look at the big structures —clusters and superclusters—as fixed entities that were formed in the early stages of the expansion and that have been static since. Not so. In a CDM cosmology, the smallest scales—the seeds of field galaxies—can become attracted to the larger structures and accrete late in the day, even into the present era. Discordant galaxy velocities within clusters may therefore be indications of accretion taking place now, such as those observed in compact groups. One example may be NGC 7318B, one of the members of the small group *Stefan's quintet* (Arp 319, also called VV 288). It is really a quartet, one of the galaxies is clearly a superimposure. NGC 7318B has a discordant redshift relative to the group (5700 km s^{-1} instead of 6600 km s^{-1} for the others and may be infalling.[84]

8.7.4 Limits to Growth

Structures do not grow unchecked. Radiative processes act like a viscosity from scattering and produce heating from reionization. Also, depending on Λ, the rate of expansion also affects the size of the largest structures that can form. We now examine how these enter the earliest stages of the perturbations.

8.7.4.1 *Radiative Scattering and Damping: The Silk Mass*

In an optically thick medium, radiative transfer proceeds as a diffusive random walk of the photons. We discussed this in Chapter 3 for stellar atmospheres and now we treat the expanding Universe as an "atmosphere" with z as the measure of the distance. For a diffusion coefficient D, the distance covered in a time Δt is $(\Delta x)^2 = 4D\Delta t$ and therefore

$$(\Delta r)^2 = \int \frac{D\,dt}{R^2(t)}. \qquad (8.147)$$

---

[84] See Mendez de Olivera, C. et al. 2001, *AJ*, **121**, 2524 for an H I study of the group and useful references. See also Sulentic, J. 2000, in *Small Galaxy Groups: IAU Colloquium* 174 Valonten, M. J. and Flynn, C., eds. (San Francisco: Astr. Soc. Pacific).

Examining the evolution equation, you will notice notice that the growth of any density fluctuation is controlled by two factors: the rate of the expansion (a damping term) and the rate of collapse or gravitational agglomeration (the growth term). The critical condition for the perturbations is when $\ddot{\delta\varrho} = 0$, or when the growth rate is approximately

$$t_g^{-1} \approx 3H \frac{d}{dt}[\ln \delta] \approx \frac{4\pi}{3H} G\varrho_0. \qquad (8.148)$$

When radiation is included, the growth becomes more complicated.[85] An optically thick medium, one where the matter and radiation are strongly coupled, acts as a viscous medium. Small-scale density fluctuations are damped and only the largest, that cannot be supported by radiation pressure, can grow. This process, known as Silk damping[86] occurs at the stage of decoupling of the background radiation. The photons random walk through the matter with a characteristic diffusion length of $\Delta \sim (4D \, \Delta t)^{1/2}$, which in the FRW metric gives

$$\lambda_S^2 = \int_0^{t_{\text{dec}}} \frac{\Delta dt'}{R^2}. \qquad (8.149)$$

The diffusion coefficient depends on the photon mean free path, λ_γ for classic electron scattering. This dominates the opacity just before recombination for the now relatively low-energy photons so $D = (n_e \sigma_e)^{-1}$. Since $n_e \sim R^{-3}$, for a universe with $k = 0$ and $R \sim t^{2/3}$, we find that $\Delta r \sim (3/5)^{1/2} t_{\text{dec}}^{5/6} \sim \Omega_0^{-1/4}$. That is, the *rms* distance traveled by a photon increases almost linearly with time. The scattering damps perturbations on a wavelength smaller than the rms scale and produces a minimum mass for any growing perturbation that will serve thereafter as the seed for structure formation. The mass associated with this length scale is computed similarly to the Jeans mass. Take the density at time t_{dec}, which scales as $\Omega_0 t_{\text{dec}}^2$ to obtain the scaling law, $M_S \sim \Omega_0^{5/4}$. The precise value of this limiting mass depends on, for instance, what form is assumed for the matter distribution and, further, whether the Universe is matter- or radiation-dominated at the time, and also on how any dark matter couples to the photons. The value for the Silk mass, as it is called, is around $10^{12} \, M_\odot$ assuming an $\Omega_0 \approx 1$ and decoupling at $z_{\text{dec}} \approx 10^3$. Regardless of the details the basic point remains that there is a characteristic length scale for instability that is larger than the Jeans mass and is set by self-gravity competing with expansion, mitigated by scattering-induced viscosity from the background radiation.

8.7.4.2 Reionization of the Universe
Galaxies, once formed, do not lie dormant while the Universe continues its expansion. They presumably quickly form stars. How this happens and what these stars look like in these earliest stages of galaxy evolution are still very much open

---

[85] We discussed this briefly when reviewing stellar winds in Chapter 3. You'll recall there that the effect of scattering is to stabilize the growth of perturbations in the wind and smooth out the velocity field. Here we have a similar effect except we must include the competition between radiation and self-gravity.

[86] Silk, J. 1968, *ApJ*, **151**, 459; Peebles, P. J. E. 1965, *ApJ*, **142** 1317.

questions, but one thing is clear. Massive stars of low or zero metallicity produce copious quantities of ultraviolet radiation and that light will do damage to the background. At first glance, this is a now familiar consequence of star formation—the massive stars generate H II regions. But there is an important twist compared to the interstellar counterpart. The Universe resembles an already expanding interstellar medium in which the density is dropping on the propagation timescale for the radiation. Static H II region theory does not apply directly now, although it helps.

Observationally, the first work on reionization was motivated by a paradox. Given the H I column densities inferred from the Lyα observations, sensitive observations toward intermediate redshift quasars should have revealed the Gunn–Peterson absorption edges of neutral hydrogen. A few measurements are available now from the Keck 10-m spectra, with a reported detection against PKS 1937-101 ($z = 3.787$) with an optical depth of about 0.1.[87] Part of the problem is the intrinsic difficulty of separating the absorption lines at the Lyman series convergence from the edge, a task that requires the currently scarce combination of large telescope aperture and sufficiently high spectral resolution, of order several 10^4. This instrumentation limitation will certainly improve in the near future. In contrast, a much lower upper limit[88] is found toward the quasar 1202-0725 ($z = 4.695$) showing a Lyman edge opacity of ≤ 0.02 up to $z = 4.3$, if it has been detected at all.[89] More significant, perhaps, in delineating the state of the absorbing intergalactic medium is the couple of firm detections of the He II Lyman continuum edge in HST observations toward a number of high-redshift quasars. The positive detection of He$^+$ and the generally low upper limit on H^o points to a highly ionized medium *even after recombination*. This initially presented a problem in cosmological models. With the combination of Lyα absorption lines and light-element nucleosynthesis, the baryon density should be sufficient that at a neutral hydrogen Lyman edge should be seen. As often happens in astrophysics, the answer appears to come from another quarter. Models of galaxy formation had been routinely computed without considering the *effect* of forming stars other than the prediction of the spectrum of high-redshift sources and the heating of the background. As you well know, massive stars, regardless of their metallicity, are copious producers of ionizing radiation that, if it can escape the galactic disk, will do serious damage to any environmental gas. With virtually no metals to enhance the ultraviolet opacity, they are even more effective radiators. This appears to be the direction of the solution.[90]

---

[87] Fang, Y. et al. 1998, *ApJ*, **497**, 67.

[88] Giallongo, E. et al. 1994, *ApJL*, **425**, L1.

[89] The "shape of things to come" is shown by a study arising from the Sloane Digital Sky Survey follow-up of high redshift quasars (Songaila, A. et al. 1999, *ApJL*, **525**, L5). For $z \approx 5$, the upper limit to the Ly edge optical depth is about 0.1.

[90] This is one of those times that our practice of avoiding extensive lists of papers must prevail. See, for instance, Abel, T., 1997, *New Astr.*, **2**, 181; Haiman, Z., Rees, M., and Loeb, A. 1997, *ApJ*, **484**, 985; Ferrara, A. 1998, *ApJL*, **499**, L17; Madau, P., Meiksin, A., and Rees, M. 1997, *ApJ*, **475**, 429; Ciardi, B., Ferrara, A., Govertano, F., and Jenkins, A. 2000, *MNRAS*, **314**, 611; a good review is Rees, M. J. 2000, *Phys. Rep.*, **333**, 203. This computationally intensive problem is witnessing an extremely rapid development that is likely to continue for some time. These references present comprehensive views of the input physics and algorithms.

The gas is heated by a second agent that we did not need to treat the thermal balance in the interstellar medium, Compton scattering. When we memtioned the distortions of the background radiation earlier in this chapter (see also Section 8.7.6), we assumed that $T_{\rm rad}$, the radiation temperature, is less than T_e. This is certainly true at the present epoch and especially for the large structures we fingered as producing possible spectral signatures. If, on the other hand, we are treating the detailed evolution of the diffuse gas after recombination but well before the present time, the energy density of the radiation was substantially larger than now and its temperature was also higher. Instead of merely being an annoyance in the analysis of the background spectrum, the inverse Compton process can become an important coolant for the electrons and must be included in the energetic evolution of the gas following recombination. To do this, we rewrite Eq. (8.70) for the thermal balance as

$$\frac{3}{2}nk\frac{T_e}{dt} + 2\frac{\dot{R}}{R}T_e = \frac{4k\sigma_e}{m_e c^2}\sigma T_{\rm rad}^4(T_{\rm rad} - T_e) + \mathscr{L}(T_e, n), \qquad (8.150)$$

where the last term is the general loss bracket we used in discussing the thermal instability. Recall that the scaling is $T_{\rm rad} \sim (1+z)$ and $n \sim (1+z)^3$ and, in general, $T_{\rm rad} < T_e$ after decoupling and before recombination.

Now what about $\mathscr{L}$? For the onset of a thermal instability, you will recall that the cooling must be strongly density-dependent. This is certainly true for most molecular and atomic lines that are sufficiently optically thin for the emitted photons to freely stream out of the gas. Precisely as we see in the interstellar medium, once atoms recombine and in the presence of molecules, we can expect the gas temperature to drop through collisional excitation followed by radiative de-excitation. The heating is another matter. Once stars, or at any rate self-gravitating bodies, form, should they reach temperatures even from gravitational energy release that exceed a few times 10^4 K, they can radiate in the UV and photoionize and photodissociate the gas. The stars will last for, essentially, their main sequence lifetime. The most massive of this reputed "Population III" evolve at their Eddington luminosity so $L/M \approx$ constant and they have lifetimes of about 10^6 yr, almost independent of mass. Radiation can also be produced by quasars and other active galaxies, if sufficient time has elapsed for the formation of massive central engines and the turn-on of a strongly nonthermal continuum. Depending on the spectrum, this, too, will heat the gas through the Lyman continuum photons. It is presumed that in either case, once the photons escape the host object (we will call these "galaxies," but the precise nature of the site is less important), it ionizes the background gas even after recombination, and produces a second stage of ionization. This epoch has been designated the *end of the Dark Ages*. The details are still unsettled and, in this case, they matter. The picture presently has the Jeans mass decreasing to about 10^5 $M_\odot$ at $z \approx 30$, which is about the mass of a globular cluster, at which point star formation may begin. This process is another of those inevitable cosmological consequences of structure formation, even if how it operates is a matter of debate. One point is clear: small masses are attainable for cool gas. If there is sufficient time to form any molecular coolants relatively soon after recombination, star formation can proceed through top–down fragmentation. If not, especially if there is feedback from the formation of UV bright sources, this may be substantially delayed.

A consequence of this reionization is an increase in the electron scattering optical depth toward the CBR. The epoch need not last long, but if it occurs early enough, the column densities can be high enough to wash out some of the signature fluctuations in the diffuse background. Remember that at low energy an optical depth of unity requires an electron column density about 10^{24} cm^{-2}. The fluctuations seen in far infrared maps are only about $\Delta T/T \sim 10^{-5}$ so a mean density of about 10^{-8} cm^{-3} over a pathlength of 1 Gpc suffices to produce sufficient scattering to affect the observable structure.

8.7.4.3 The Role(s) of Λ

Although we implicitly ignored large-scale changes in the background density when setting up the agglomeration equations in Chapter 1, we can include them simply as a loss term that is independent of mass. Much as we did for Eq. (8.143), a term $-3\dot{R}/R$ is added the balance between accumulation and disruption, and you see that this depends in the cosmological model, and hence on the assumed equation of state and the value of the cosmological constant. We've titled this discussion ambiguously because there are really several ways an assumed nonvanishing Λ enters the scenario for structure formation. One is the change affected in any cosmological model through Λ since it acts as an acceleration term. In effect, during the critical era, it is possible to have an accelerating expansion, relative to a normal $\Omega = 1$ cosmology, and this would mean a rapid drop in the background density of matter. Of necessity, the equation of state enters here as well. Hot dark matter, if following an R^{-4} density law, is less confined and more sensitive to the rate of expansion. Cold dark matter, while following a less steep dependence, is also more confined and so would still be able to form large structures even with substantial Λ. On the other hand, if Ω is really unity, then the fraction coming from *both* baryons and dark matter is lower for $\Lambda > 0$ than it would be if it is negligible, so a ΛCDM scenario opens up a wide range of possible behaviors that were unavailable for the standard model. Again it is important to emphasize that this growth refers only to the largest structures, clusters and superclusters of galaxies, and not to the galaxies themselves. The latter are small and self-gravitating and insensitive to the large-scale structure. But their growth rate does depend on the masses ultimately accumulated in the largest scales, so there is an indirect dependence.

8.7.5 Nonlinear Evolution of Density Fluctuations

The theory we have developed is limited to the initial stages of growth of the fluctuations and breaks down once $\delta\varrho/\varrho$ is of order unity. The initial seed spectrum of fluctuations is assumed to be scale free, and this is where the Harrison–Zeldovich–Peebles spectrum enters the picture.

Once you have begun the collapse process, the evolution can be described by the same machinery we used for star formation except in an expanding background. Consider homogeneous perturbations, with spherical symmetry, for a fluid with zero pressure in a flat Universe. The solution for freefall has the same parametric form as Eq. (8.89) where we now write $R = A(1 - \cos\eta)$,

$t = B(\eta - \sin \eta)$, and we scale of sphere's mass as $A^3 = GMB^2$. Since the evolution of the unperturbed background is $\varrho_b = (6\pi G t^2)^{-1}$, we obtain for the density contrast:

$$\delta_s(\eta) = \frac{\varrho_s(\eta)}{\varrho_b(\eta)} - 1 = \frac{3M}{4\pi r^3(\eta)\varrho_b(\eta)} - 1 = \frac{9}{2}\frac{(\eta - \sin \eta)^2}{(1 - \cos \eta)^3} - 1. \quad (8.151)$$

Keeping only the first nonvanishing term in the series at early times, we find

$$\delta_s(t) \approx \frac{3}{20}\left(\frac{6t}{B}\right)^{2/3}, \quad (8.152)$$

and the parameters A and B are written in terms of the initial conditions of the perturbation. We can now compare the evolution of the density contrast with the corresponding one from the linear theory, δ_+. In particular, we can calculate the values at the time at which the spherical perturbation reaches the maximum expansion (t_m) and at the time at which the collapse at infinite density takes place (t_∞):

$$\delta_s(t_m) = \frac{(6\pi)^2}{64} - 1 \approx 4.5; \quad \delta_+(t_m) = \frac{3(6\pi)^{2/3}}{20} \approx 1.06 \quad (8.153)$$

$$\delta_s(t_\infty) = \infty; \quad \delta_+(t_\infty) = \frac{3(12\pi)^{2/3}}{20} \approx 1.686. \quad (8.154)$$

The fully nonlinear development of any perturbation can be expected to follow the spherical model, at least initially. After a phase of decelerated expansion to a limited dimension (turnaround), an irreversible process of collapse takes place, halted only by virialization.

We now return to the correlation function. The density fluctuations in the expanding background eventually lead to metacluster and cluster formation, and their observable imprint should be preserved in the distribution of the luminous matter. Even if dark matter is the primary constituent of the mass distribution, galaxies and their aggregates should serve as tracers of that structure. The caveat to bear in mind is that they may be biased because of their formation process. In other words, unlike sprinkling aluminum flakes into a flow, or injecting smoke to make streamlines observable, galaxies are not passive flow tracers. The density field $\varrho(\mathbf{x}, t)$ consists of superpositions of a smooth background with fluctuations that are assumed to be random and, in the language of stochastic processes, are defined by the joint probability distributions $P(\delta_i)$. The *cosmological principle* requires that averages taken on this statistical ensemble are invariant with respect to rotations and translations. Furthermore, we can assume that the spatial averages on large volumes are equivalent to the averages taken on the ensemble. Intuitively you might expect that this last condition is satisfied if the spatial correlations decrease rapidly with the distance, a result that also constrains any inflationary scenario. You still need to specify the distribution function for this stochastic field to describe the evolution. For simplicity, it is frequently assumed that the fluctuations are functionally Gaussian. This has some compelling analytic advantages. All derivatives, integrals, and any linear function of δ, are then also Gaussian. The Fourier modes, of a stationary Gaussian field, F_k, are mutually independent, their phases (φ_k) are uniformity distributed, and the resulting joint probability distribution is

$$\mathscr{P}(|F_k|, \varphi_k) d|F_k| d\varphi_k = \exp\left(-\frac{|F_k|^2}{2P(k)}\right) \frac{|F_k| d|F_k|}{P(k)} \frac{d\varphi_k}{2\pi}, \quad (8.155)$$

where $P(k)$ is a function that will be defined in a moment. This functional choice is also supported by the central-limit theorem, which states that a normal distribution results any time an observed property is due to *linear* combinations of a large number of variables that arise from the same distribution.

The correlations for galaxies are computed a bit differently than for a continuous medium. They are point tracers of the smooth, continuous fields $\varrho(\mathbf{x})$ and $\delta(\mathbf{x})$. The correlation function functions must reduce to those defined in Section 8.7.2, where the distribution $\varrho(\mathbf{x})$ represents a set of discrete objects. The formal connection between the two schemes can be seen by assuming that $P = \varrho(\mathbf{x})\delta V$ defines the probability of finding a pointlike object in the infinitesimal volume δV. It is possible to demonstrate (and it is intuitive) that the two-point correlation function can be substituted for its continuous counterpart

$$\xi(r) = \langle \delta(\mathbf{x})\delta(\mathbf{x}+\mathbf{r}) \rangle. \quad (8.156)$$

In a flat Universe, the Fourier expansion is just plane waves. For other geometries, since plane waves do not describe the appropriate complete basis set, the analysis is more complicated, but we will ignore this by restricting the discussion to scales less than the radius of curvature, $k^{-1} \ll 3000(1-\Omega)^{-1/2} \, h^{-1}$ Mpc. The Fourier expansion for the density field in three dimensions is then given by

$$\delta(\mathbf{x}) = \frac{1}{(2\pi)^3} \int \hat{\delta}(\mathbf{k}) e^{-i\mathbf{k}\mathbf{x}} d^3k$$

$$\hat{\delta}(\mathbf{k}) = \int \delta(\mathbf{x}) e^{i\mathbf{k}\mathbf{x}} d^3x \quad (8.157)$$

and the power spectrum is

$$P(\mathbf{k}) = \langle |\hat{\delta}(\mathbf{k})|^2 \rangle, \quad (8.158)$$

The perturbation spectrum in an isotropic universe cannot have a preferred direction, so $P(\mathbf{k}) = P(k)$, depending only on the magnitude of k:

$$\langle \hat{\delta}(\mathbf{k}_1)\hat{\delta}(\mathbf{k}_2) \rangle = (2\pi)^3 \delta_D(\mathbf{k}_1 + \mathbf{k}_2) P(k). \quad (8.159)$$

The Dirac delta function is due to the translational invariance of the system. For stationary fields, the spatial correlation function is connected to the power spectrum by the Wiener–Khintchine theorem (Chapter 2):

$$\xi(r) = \frac{1}{(2\pi)^3}\int P(k)e^{-i\mathbf{k}r}d^3k = \frac{1}{2\pi^2}\int_0^\infty k^2 P(k)\frac{\sin(kr)}{kr}dk$$

$$P(k) = \int \xi(r)e^{i\mathbf{k}r}d^3r = 4\pi\int_0^\infty r^2\xi(r)\frac{\sin(kr)}{kr}dr.$$

(8.160)

You can also construct the spectral moments of any order, as we've done now many times for distributions:

$$\sigma_l^2 = \frac{1}{2\pi^2}\int_0^\infty k^{2(l+1)}P(k)\,dk.$$

(8.161)

The $l = 0$ moment is the variance, $\sigma^2 \equiv \langle \delta^2(\mathbf{x})\rangle$. The power spectrum is usually written in dimensionless form as the contribution to the variance in a logarithmic k interval:

$$\Delta^2(k) = \frac{d\sigma^2}{d\ln k} = \frac{1}{2\pi^2}P(k)k^3.$$

(8.162)

When the correlation function is given by a power law, $\xi(r) = (r_0/r)^\gamma$, Eqs. (8.160) and (8.162) give

$$\Delta^2(k) = \frac{2}{\pi}\Gamma(2-\gamma)\sin\frac{(2-\gamma)\pi}{2}(kr_0)^\gamma.$$

(8.163)

A power law with a generalized spectral index n, $P(k) \sim k^n$, gives $\Delta^2(k) \sim k^{n+3}$. The fundamental problem of the statistical approach to the density field consists in the choice for the primordial spectrum.

A power law, which has no characteristic scale, is a natural product of the inflationary epoch fluctuations once they grow above the Planck scale. Assuming that $|\delta_k|^2 \sim k^n$, then $\langle(\delta\varrho/\varrho)^2\rangle$, varies as $k^{2(n+3)} \sim M^{-(n+3)/6}$. For $n = -3$, this is scale free, so the density fluctuations all appear at the horizon with the same amplitude.[91] Keep in mind that we are *not* talking here about turbulence, although many of the tools are the same. Any structure resulting from the fluctuations are not dissipative when they are generated in the background matter. The spectrum with $n = 1$ is now seen as the most probable candidate, the Harrison–Pebbles–Zel'dovich spectrum. Inflationary models allow evaluation of the growth of fluctuations inside the horizon in the expanding Universe, and show that the perturbations generated via inflationary processes produce power-law spectra, with $n = 1$ for the standard inflationary model. Gravitational waves are a form of seed.

Now to the observations. If your aim is to study the spatial distribution of galaxies, you are not interested in the fluctuations with wavelengths much shorter

---

[91] Harrison, E. R. 1970, *Phys. Rev. D*, **1**, 2726; Zel'dovich, Ya. B. 1972, *MNRAS*, **160**, 1P.

than the typical galactic sizes. So it is worthwhile to consider the possibility of analyzing the density field through an arbitrary spatial resolution. The filtered density field is represented by

$$\delta(\mathbf{x}, R_f) = \int \delta(\mathbf{x}') F(|\mathbf{x} - \mathbf{x}'|, R_f) d^3 x' \qquad (8.164)$$

obtained by convolving $\delta(\mathbf{x})$ with a filter function $F(|\mathbf{x} - \mathbf{x}'|, R_f)$ that contains the characteristic length R_f and satisfies the intuitive properties that it is almost zero for objects much more distant than the filtering length. In wavenumbers the filtering reduces or eliminates fluctuations with wavelengths shorter than the reciprocal of the filtering length. The spectral moments, thus filtered, are given by

$$\sigma_l^2(R_f) = \frac{1}{2\pi^2} \int_0^\infty k^{2(l+1)} W_F^2(kR_f) P(k) dk, \qquad (8.165)$$

where $W_F(kR_f)$ is called the *window function* and is the Fourier transform of $F(|\mathbf{x} - \mathbf{x}'|, R_f)$. The variance of the filtered field is called mass variance since it represents the average of the inhomogeneities of matter distribution on the scale $M \sim \langle \varrho \rangle R_f^3$:

$$\sigma^2(M) = \frac{\langle M^2 \rangle - \langle M \rangle^2}{\langle M^2 \rangle} = \frac{1}{2\pi^2} \int_0^\infty k^2 W_F^2(kR_f) P(k) dk. \qquad (8.166)$$

For the Harrison–Peebles–Zel'dovich spectrum, with $n = 1$, the mass variance is constant on all scales. Consequently, this spectrum is called *scale-invariant.*

There is a simpler reason for expecting this spectrum. In any distribution, say, one in mass, if $\varphi(m) \sim m^{-1}$, then the distribution has the same differential power on every scale. This is the same behavior that is seen in many chaotic systems and is usually called $1/f$ *noise*. It has no correlation and no characteristic scale.

8.7.6 Background Fluctuations and CBR Variations

Since the sky can be considered the surface of a sphere, the fitting of temperature fluctuations for the CBR is performed in terms of a Laplace series of spherical harmonics

$$\frac{\Delta T}{T} = \sum_{l=1}^\infty \sum_{m=-l}^l a_{lm} Y_{lm}(\hat{\mathbf{n}}), \qquad (8.167)$$

and since the fluctuations are assumed to be statistically independent, $\langle a_{lm} \rangle = 0$ and $\langle a_{lm} a^*_{l'm'} \rangle = C_l \delta_{ll'} \delta_{mm'}$. From the usual orthogonality conditions for the coefficients, the two-point correlation function is given by

$$C(\vartheta) = \sum_l \frac{2l+1}{4\pi} C_l P_l(\cos \vartheta) \qquad (8.168)$$

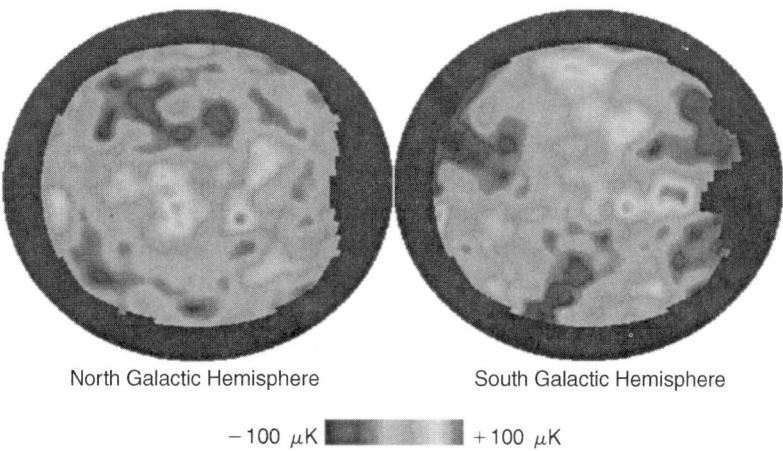

FIGURE 8.7. Northern and southern hemisphere plots of the FIRAS data from COBE showing the large-scale fluctuations first detected by this satellite (credit: COBE/NASA).

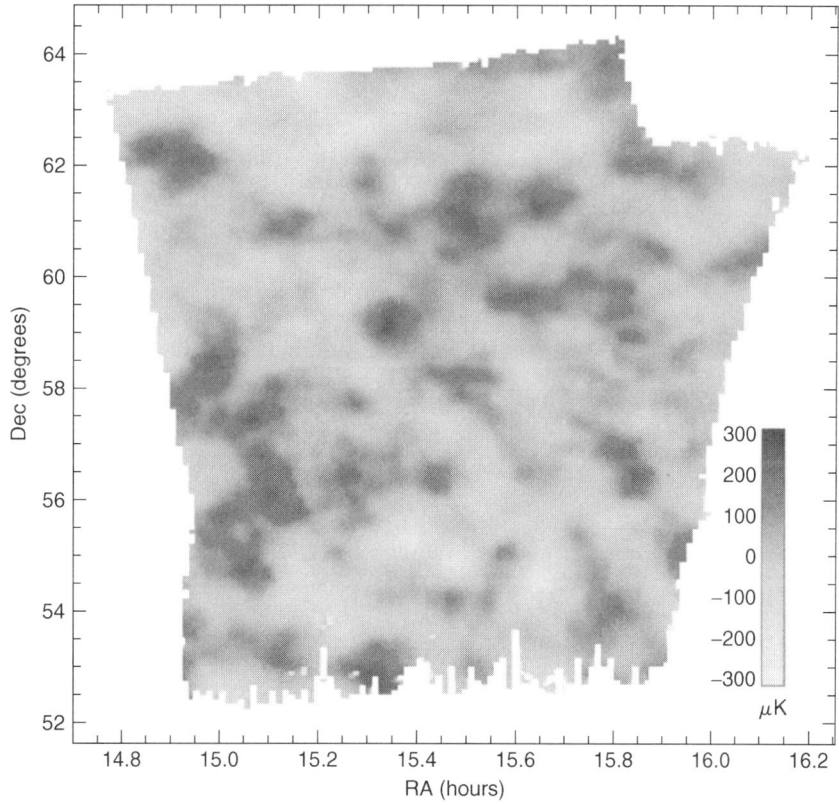

FIGURE 8.8. Small-scale fluctuations detected with the MAXIMA balloon flight. These data are sensitive to high l, ≥ 100, and are similar to the BOOMERanG data (credit: the MAXIMA Collaboration).

for two points on the sky separated by an angular distance ϑ. The angular power spectrum is quoted in terms of the order, l, of the fit. The COBE measurements detected the first peak at $l \approx 20$ (Fig. 8.7). Higher-angular-resolution measurements have been made with a number of balloon experiments, in particular BOOMERanG, with scales of less than one degree, and this has now detected the peak at $l \approx 200$ and a steep dropoff in the spectrum between this peak and the harmonic at $l = 400$ (Fig. 8.8).[92] The reason for such variations is that the cosmic thermal radiation does not reach us unaltered by the intervening masses. They form a screen through which the photons must pass. The fact that the overall spectrum is nearly a perfect blackbody is fundamentally important since it means that the changes affected by the mass are overall small, but the fluctuations may be due to processes since the formation of the spectrum and not just imposed at the time of decoupling. We will briefly outline two important large scale effects that can lead to ΔT_{CBR}, one due to density fluctuations regardless of the temperature (the Sachs–Wolfe effect) and the other due to intervening scattering screens (the Sunyaev–Zeldovich effect).

8.7.6.1 The Sachs–Wolfe Effect
The temperature of the background fluctuates across the sky because of variations in the density of matter at the time of separation of the CBR, and from variations of the density along the line of sight. Called the *Sachs–Wolfe effect*, these temperature changes are in response to the same aspect of general relativity that produces the global redshift in the first place, the change in local curvature along the line of sight due to density changes.[93] One way to see how this works comes from assuming that the photons behave as an adiabatic gas. Then

$$\frac{\delta T}{T} = -\frac{1}{3}\frac{\delta \varrho}{\varrho}. \tag{8.169}$$

Because of the connection between the gravitational potential and mass energy density, we get

$$\frac{\delta T}{T} = -\frac{1}{3}\frac{\delta \Phi}{\Phi}. \tag{8.170}$$

Temperature fluctuations are measurements of the local curvature induced by variations in the gravitational potential. The small amplitude of the largest-scale variations strongly argues for homogeneity of the mass distribution at the time of decoupling.

---

[92] Barreiro, R. B. 2000, *New Astr. Rev.*, **44**, 179; de Bernardis, P. et al. 1999, *New Astr. Rev.*, **43**, 289; Smoot, G. F. 1997, *NATO School on the Cosmic Microwave Background*, Linewater, J. G. et al., eds. (Dordrecht: Kluwer); Bennett, C. et al. 1996, *ApJL*, **464**, L1; Fixen, D. J., Cheng, E. S., Gales, J. M., Mather, J. C., and Wright, E. L. 1996, *ApJ*, **473**, 576; Banday, A. J. et al. 1997, *ApJ*, **475**, 393; Fixen, D. J., Hinshaw, G., Bennett, C. L., and Mather, J. C. 1997, *ApJ*, **486**, 621. A particularly good source for description of the COBE results is the special issue 1997, *ApJL*, **464**, L1 ff.; see also Smooth, G. F. 2000, *Phys. Rep.*, **333–334**, 269.

[93] Sachs, R. and Wolfe, A. 1967, *ApJ*, **147**, 73; White, M. and Hu, W. 1997, *Astr. Ap.*, **321**, 8.

8.7.6.2 The Sunyaev–Zeldovich Effect

If we happen to view the CMB through an intervening hot gas, then we will observe a change in the spectrum, called the *Sunyaev–Zel'dovich effect*. Photons traversing a hot medium scatter off of the electrons via the inverse Compton effect. Since this conservative process is dependent on the energy of the photons, it shifts the radiation into the X-ray portion of the spectrum and depletes and distorts the CBR. The first prediction of this effect, based on the calculations by Kompaneets in 1957 for the diffusion of radiation through a hot medium, was made by Sunyaev and Zel'dovich beginning in 1968. The expected departure from the Planck function is computed from the Kompaneets equations (see Chapter 3) and is characterized by the parameter

$$y = \frac{kT_e}{mc^2} \int n_e \sigma_e \, dl. \tag{8.171}$$

To remind you of the physical basis of the effect, it arises in a scattering medium that contains hot electrons (X-ray-emitting clusters are the best sites). Compton (and inverse Compton) scattering of the CBR removes photons from the radio portion of the spectrum and boosts them up to X-ray energies where they are hidden beneath the signal from the hot gas of the cluster. Their observational signature is a change in the brightness temperature of the radio background due to the change in intensity in the bandpass. For scattering, the electrons lose energy conservatively to the CBR. The Kompaneets equation treats this process as a random walk in energy of the electrons and photons. You know that the Compton effect is the quantum-mechanical scattering of a photon by a free electron. For relatively low energy, we will treat this as nonrelativistic (i.e., $\epsilon < m_e c^2$). If the energy of the electron is less than the incident photon, $\hbar\omega < \epsilon$, the photon loses energy to the scatterer, the normal Compton effect. If the particle kinetic energy is greater than that of the incident photon, the radiation is shifted up in energy at the expense of the kinetic energy of the particle. This is the *inverse Compton effect*. This latter process is very important for the CBR, especially for the hot gas within clusters. An estimate comes from the X-ray emissivities. If the mean temperature for the gas is about 10^7 K, then the thermal energy of a typical electron is about 1 keV while the mean CBR photon energy is about 10^{-4} eV. Hence, the inverse process dominates the scattering. The dimensionless variable y becomes a scaled time for photon diffusion and is directly related to the scattering optical depth. Now recall that the Kompaneets equation has the analytic solution

$$n(y, \xi) = (4\pi y)^{-1/2} \exp\left(-\frac{\xi^2}{4y}\right), \tag{8.172}$$

where $\xi = \ln x + 3y$ and the normalized boosted frequency is $x = (kT/mc^2)^2 \omega$. This solution shows that over time, the photons at lower energy are promoted to higher energy by inverse Compton scattering, all the while conserving total energy and producing the change in the CBR brightness temperature. Scanning across a cluster at a fixed frequency will reveal a dimming corresponding to the redistribution of flux into the higher-energy portions of the spectrum.

8.7.7 Whence These Perturbations?

Perturbations are broadly separated into *adiabatic* (or curvature perturbations) and *entropic* (or isocurvature perturbations). Adiabatic perturbations involve a local variation of the total energy density and hence a variation of the space curvature of the hypersurface comoving with the perturbations. They interact in the same way with matter and radiation: $\delta_m = \frac{3}{4}\delta_r$. In the entropic perturbations, no net density variation occurs. The fluctuations in the various components compensate to give a vanishing sum. No curvature variation is present in the comoving hypersurfaces, and any initial energy density perturbations can be described by the sum of perturbations of the two kinds.

Adiabatic perturbations rise from quantum oscillations of the zero point of the scalar field that governs the inflation.[94] Isocurvature perturbations are generated by quantum effects that generate density variations which are compensated by opposite processes involving another component. An example is the destruction of some baryons and forming photons or quark–antiquark systems. As you probably know, the Jeans length depends on the equation of state of the cosmic fluid, and similarly in presence of more components their interactions will imprint characteristic scales to the formation of structures.

There are several restrictive conditions on the growth of any perturbations:

1. The perturbations involving nonbaryonic matter and isothermal baryonic perturbations, once they have crossed the horizon, are frozen in until the epoch at which one has the equality of the matter and radiation energy densities. The spectrum flattens for large wavenumbers, since the scale is given by the dimension of the horizon at this epoch.

2. All baryonic isothermal fluctuations with a scale larger than the Jeans length are frozen in until the recombination era due to the viscosity the particles experience moving relative to the unperturbed background radiation (*radiation drag effect*).

3. The baryonic adiabatic perturbations with wavelength shorter than a threshold, λ_S, are opaque to the radiation by Thompson scattering on free electrons and are dissipated by photon diffusion (*Silk damping*).

4. Perturbations of collisionless matter with wavelengths shorter than a characteristic value that is proportional to the velocity dispersion of the medium, are damped by diffusion (i.e., free streaming or Landau damping).

What emerges from each of these conditions is very different depending on how the spectral energy is distributed over k. For instance, HDM generates the first structures on the scale of superclusters, and the galaxies are formed later by fragmentation (top–down scenario). In contrast, CDM creates the first structures on subgalactic scales and only aggregation processes can give origin to larger objects (bottom–up scenario).

---

[94] This is a first-order phase transition. You're actually quite familiar with these. For instance, pour cream that has gone "off" into your coffee, or leave whipped cream to sit out for a few minutes, and you will see the same thing. The problem is that there is a first-order change in the state of the system.

Actually, the whole process of producing CBR temperature fluctuations from such perturbations is pretty straightforward to describe, even if it is hard to compute. It's connected with the physics of gravity waves, as we have already discussed. For a point source, a gravitationally lensing extended mass can form multiple images or caustics. For an extended, nearly uniform background, such as the CBR, a gravitational lens (which can be a fluctuation in the metric caused by a very-long-wavelength gravitational wave) will produces hot and cold spots. By measuring the brightness (temperature) fluctuations, you obtain a measure of the density variations on the same angular scale *at the time of the decoupling*. The assumption is that the fluctuations are not coherent, that the variations in the temperature of the background are the result of combinations of many different perturbations or waves along the line of sight that have random phases. The larger the scale of the disturbance, the larger the scale over which the CBR temperature variations will be correlated. If they are on the scale of the horizon at some epoch, they provide the signature of the perturbation that gives rise to the formation of clusters of galaxies.

The analogy with turbulent image formation follows from the optical analogy that we needed for understanding gravitational lensing. In forming images, the atmosphere produces changes in the path of the wavefront due to refractive index changes. These result from density and temperature fluctuations in the turbulence of the medium. For the Universe at large, these result from the changes in the trajectories of the light through the changes in the gravitational field, effectively changes in curvature of the metric, because the density fluctuations feed back into the gravitational field. Since the field equations are linear in density, these are relatively easy to compute. We will pause here so you can look back at Chapter 1. Notice that the phase of the density and temperature changes that cause variations in the path of the light result in a complex image. The timescale for the variation, and the requirement of coherence in the image, produce interference that result in the final image.

Acoustic waves result from the fluctuations. If the phases all vary together, this is the same as a sound wave. Strong coupling between the different components of the medium result in uniform changes in the curvature, the so-called *adiabatic* fluctuations. For the case where different phases are not strongly coupled and respond individually to the changes in the local gravitational field, the curvature may remain constant; hence the term *isocurvature*. In either case there is emission of sound waves, pressure changes due to fluctuations in the stress tensor. The changes seen in the CBR temperature, or surface brightness as a function of wavelength, are signatures of the particular mechanism. It is worth stopping here for a moment to qualitatively discuss what this means about the early Universe, during the inflationary epoch. We waved our hands in Section 8.6.8 to argue that some form of inflation is required to resolve the flatness and horizon problems. You can imagine that a characteristic wavelength, imposed by some physical process on the density distribution, is rapidly stretched by the expansion and is only now reentering the observational horizon. This can be detected through the angular correlation functions for the CBR at the present epoch. One last point. The difference between light propagation in the atmosphere and through the background gravitational field of the cosmic mass distribution is that the latter is optically thin to scattering, except perhaps due to the Sunyaev–Zel'dovich effect.

TABLE 8.6 COBE and BOOMERanG Results

Component	T
Monopole	2.728 ± 0.002 K
Dipole	3.358 ± 0.024 mK
Quadrupole	$15.3^{+3.8}_{-2.8}$ μK

Otherwise, the two are very similar and the formation of images of the extended source of the CBR is essentially the same as the problem of imaging a source through a medium, say a lens, with fluctuating curvature.

The most recent results at the time of writing this book have come from the BOOMERanG long-duration balloon flight (see also Table 8.6), but many more will be available in the coming years. Each angular scale sampled by these experiments, when folded through a cosmological model, maps back to a characteristic size. We can assume that the background changes very little following recombination and therefore any observed scale must represent a large mass concentration.

8.8 COSMOLOGICAL GRAVITATIONAL LENSES

Because light can deflect under the influence of a mass, it can form an image. This is actually your third encounter with gravitational lensing. The first was in Chapter 1 where we discussed gravitational deflection of light as a consequence of general relativity and the second when we discussed microlensing as a probe for low-luminosity objects in the Galactic halo in Chapter 7. You may have forgotten some of these points by now, so let's recapitulate some of the basic points. You can think of this as the analog of the Sachs–Wolfe effect discussed above.

There are a number of important historical steps in this development that we'll outline briefly here, since they illustrate how the prediction of a phenomenon can substantially predate its discovery. In 1918 Einstein predicted that a distant object would produce an image through the gravitational deflection provided it had the intervening object in the right position relative to an observer. In the 1930s, Zwicky proposed the cosmological version of this phenomenon and argued that since they are extended massive objects, foreground galaxies can produce this effect along lines of sight. The issue then dropped out of the literature for nearly 30 years. The question of cosmological distortions by gravitational lensing was revived in the 1970s, dealing mainly with the effects of matter inhomogeneities on the cosmological tests, although galactic lensers were considered as well. In fact, it is remarkable that much of the basic theory was worked out so far in advance of any data. However, the subject became *empirical* with the discovery of QSO 0957 + 561 with the Multi-mirror Telescope (MMT) by Walsh, Carswell, and Weymann[95] using the spectroscopic identity of the two close quasars ($z = 1.41$) that were separated by about 5.7″. The field evolved rapidly thereafter, including

---

[95] Walsh, D., Carswell, R. F., and Weymann, R. J. 1979, *Nature*, **279**, 381; Weymann, R. J. et al. 1979, *ApJL*, **233**, L43.

VLA and VLBI radio imaging. The comprehensive model for the lensing was developed by Young and collaborators in a series of papers[96] that treated the lenser as an extended mass. Shortly before, however, almost immediately after the discovery it was shown that an extended mass would *generically* produce an *odd number of images*.[97] The idea of using time delays to determine the Hubble constant from differential variability of lensed quasar images was soon reproposed, since it had actually been suggested long before 0957-561 had been discovered.[98] Catastrophe theory was also introduced in light of the possible complexity of the lenser's mass distribution, using some of the theory's applications to optical caustics.[99] Two final discoveries clinched gravitational lenses as cosmological tools. The first was the discovery of arcs, extended blue structures seen through *clusters* of galaxies (see Fig. 7.18). This period included the discovery of the *Einstein Ring*, 0047-2808, the radio image of a nearly perfect ring. With the launch of the *Hubble Space Telescope* and the correction of its optical aberrations, after 1994, imaging with spatial resolution of 0.2 arcsec or better became feasible and now the discovery of lenses is routine (Fig. 8.9). The last step was the discovery of microlensing events in the Galactic bulge and halo.

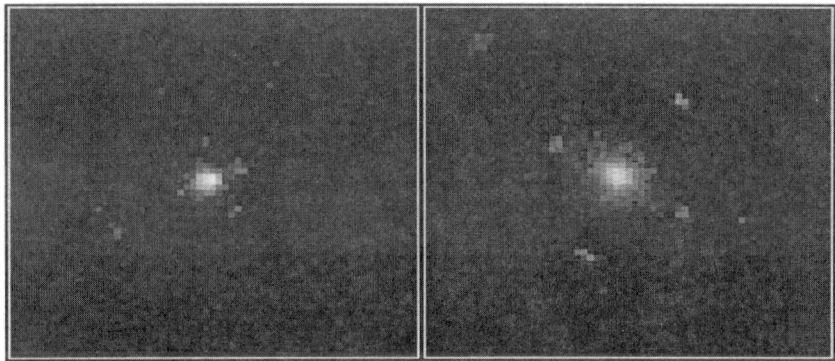

FIGURE 8.9. Two gravitational lenses, HST J12531 − 2914 and HST J14176 + 5226 (see Ratnatunga, K. U. et al. 1995, *ApJL*, **453**, L5). These were fortuitous discoveries with WFPC2 during a quasar survey and illustrate just two examples of multiple image morphologies. The pixel size is about 0.1 arcsec. (Credit: STScI/NASA).

---

[96] Young, P. J. Gunn, J. E., Kristian, J., Oke, J. B., and Westphal, J. A. 1981, *ApJ*, **241**, 507; Young, P. J., Gunn, J. E., Oke, J. B., Westphal, J. A., and Kristian, J. 1981, *ApJ*, **244**, 736.
[97] Dyer, C. C. and Roeder, R. C. 1980, *ApJL*, **238**, L67 (but see also 1980, *ApJL*, **242**, L53 for an erratum); Burke, W. L. 1981, *ApJL*, **244**, L1.
[98] In this case, see Refsdal, S. 1964, *MNRAS*, **128**, 307, although a supernova rather than a lensed quasar is suggested as the testbed.
[99] The review by Berry, M. V. and Upstill, C. 1980, *Prog. in Optics*, **18**, 257 on *catastrophe optics* is one of the lovely, neglected papers for astrophysics. The general concept of using caustics to map the gravitational potential can be found here. You should also look at Nye, J. F. 1986, *Proc. Roy. Soc.* (London), **403**, 1, which has a superb collection of images of caustics formed by liquid drops and a detailed analysis of the surfaces and the gentle, general introduction by Arnold, V. I. 1984, *Catastrophe Theory* (NY: Springer-Verlag).

As we described in Chapter 1, a photon trajectory is deviated in a gravitational field in an analogous way to a massive body.[100] The Schwarzschild metric gives the deflection angle for a point mass, M, for a line of sight with with impact parameter p:

$$\Delta \vartheta = \frac{4GM}{c^2 p} \equiv \hat{\alpha}. \tag{8.173}$$

If the line of sight lies precisely along the line of source and intervening mass, then the deviated photon paths form a ring around the deflector. This becomes progressively more arclike as the impact parameter increases, and in the weak lensing limit quickly becomes a point image shifted from the actual position by an angle $\Delta \vartheta$. Having been confirmed during the 1919 eclipse as one of the classical tests of general relativity, this prediction of gravitation theory must have cosmological consequences since both the scale factor evolution and the deflection result from the same theory of gravity.

We are, in effect, dealing with an analog of gaussian optics in which the beam is weakly deviated and the index of refraction of the lens is proportional to the gravitational potential (and therefore on distance from central mass). Since the deflection angle is generally small we can consider the Schwarzschild deflector as a thin lens. Call D_{sd} and D_{do} the distances to between the deflector (d) and the source (s) and observer (o), respectively, and D_s the distance of the source from the observer. For cosmological lenses, these all depend on z and require knowing H_0 and Ω. We imagine that the source is located at some angle β away from the center of the deflector, whose position we take as the origin on the sky, and the direction in which we see the source is ϑ at o. The difference between these, call it α, is simply related to the deflection angle (see Fig. 8.10) by using the small angle approximation to the law of sines; so calling $\hat{\alpha}$ the deflection angle *at the lens* and $\alpha = \vartheta - \beta$, we have

$$\alpha = \frac{D_{ds}}{D_s} \hat{\alpha}. \tag{8.174}$$

Now switch between the impact parameter and ϑ so $p = D_{do} \vartheta$. Then we find a simple relation for ϑ as a function of the relevant distances

$$\vartheta = \frac{4GM}{c^2 \vartheta} \frac{D_{od} D_{sd}}{D_s} + \beta = \frac{\Theta_E^2}{\vartheta} + \beta, \tag{8.175}$$

which is quadratic in ϑ, the position of the undeflected source in terms of the observed position; Θ_E, the angular radius of the Einstein ring, is a convenient

---

[100] Excellent reviews are presented by Blandford, R. D. and Narayan, R. 1992, *ARAA*, **30**, 311; Schneider, P., Ehlers, J., and Falco, E. E. 1992, *Gravitational Lenses* (Berlin: Springer-Verlag); Refsdal, S. and Surdej, J. 1994, *Rep. Prog. Phys.*, **57**, 117; Narayan, R. and Bartelmann, M. 1999, in *Structure Formation in the Universe*. Dekel, A. and Ostriker, J. P., eds. (Cambridge, UK: Cambridge Univ. Press); Mellier, Y. 1999, *ARAA*, **37**, 127; Bartelmann, M. and Schneider, P. 2001, *Phys. Rep.*, **340**, 291.

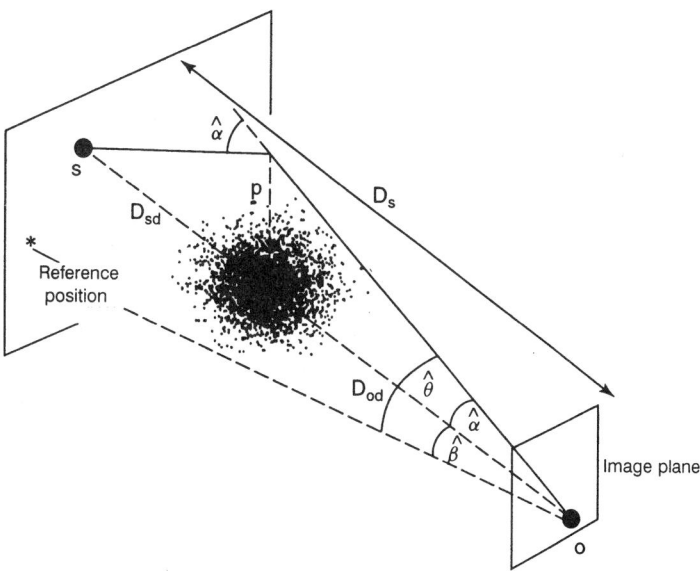

FIGURE 8.10. Defining parameters for a gravitational lens (see text). In this case, the reference position is an arbitrary point on the sky; it could have been the center of the lensing mass for an off-center placement of the source.

parameter that is defined by this equation. There are two real solutions for ϑ:

$$\vartheta = \frac{1}{2}\beta\left[1 \pm \left(1 + \frac{4\Theta_E^2}{\beta^2}\right)^{1/2}\right]. \tag{8.176}$$

Magnification is another feature of a gravitational deflector. Using the thin-lens approximation, the magnification factor is defined by the variation of the image size, ϑ^2, compared to the original source size, β^2 so that

$$\text{Magnification} \equiv \frac{d\vartheta^2}{d\beta^2}.$$

For a point deflector, this is found especially easily. Notice that we've treated this as a planar problem. You can always do that for a point deflector so that every point in an extended source can be easily mapped onto a fixed direction. Except for a view of a source precisely along the line of sight to the deflector, you will not see the source itself, only two images.

If the deflector is an extended body through which we see the more distant source, M is replaced by the mass contained within the impact parameter. For instance, suppose that the lensing body is a galaxy with an internal mass distribution $M(r)$. The rays that have a small impact parameter passing near the galactic core will not be strongly altered, while those at large distance will see a larger mass deflecting the photons. It is even possible that several trajectories will produce images, not just the pair produced by a point mass. Consider a mass distribution with density ϱ, and take a line of sight through this. The column density $\Sigma = \int \varrho \, dl$, where l is the line of sight through the mass distribution seen by the beam produces the deflection. The net deflection depends on the weighted value of the line of sight gravitational potential. Notice that the mass to radius ratio determines $\Delta \vartheta$. For an extended mass distribution, this translates into an integral of this quantity along the trajectory so that

$$\alpha = \frac{4G}{c^2} \int \frac{\Sigma(\mathbf{p}')(\mathbf{p}' - \mathbf{p})}{|\mathbf{p} - \mathbf{p}'|^2} d\mathbf{p}', \qquad (8.177)$$

This density profile also complicates the image formation because each photon path samples the precise interior mass at a given impact parameter. From the definition of the magnification, you can see how this will produce some pretty bizarre images. To make this tangible, take a clear glass container with some relatively complicated shape, fill it with water, and look at the projected image of a candle. The bright points, called *caustics*, are singularities in the projected wavefront of the light.[101] This is also the case for the imaging of an extended object. The shape of the image depends on the mass distribution as well as the actual form of the object being lensed. For instance, symmetrically lensed objects form a so-called Einstein ring and observed examples include MG J1131 + 0456, PKS 1830-211, 1938 + 666, and PG1115 + 080 the near infrared, and 0047-2808 in the optical. More complex patterns, involving compact multiple images, can result from imaging through extended mass distributions, such as the Cloverleaf lensed system (1413 + 117A-D). And recalling that elliptical galaxies are remarkably transparent, individual stars can produce microlensing variations of bright distant sources on the same crossing times that more local events occur, such as those we've mentioned for the Galactic bulge and Magellanic Clouds.

Independent of the luminosity of the intervening material, its lensing property measures the mass and mass to light ratio. Of course, this is not a strictly observable determination, since it does ultimately depend on the model for how the mass is distributed within the deflector. Nonetheless, when we discussed dark matter we said that it is *assumed* that, whatever it is, the constituents of this component of the universe exert a gravitational influence on matter. Although this is a natural assumption, here we have a tool for actually measuring the component that does this, independent of the nature assumed for the material. Measured values range from several hundred to over 1000 using this method. Most important, provided there is only one large mass along the way, the *structure* of the images so produced constrain the mass distribution within the deflector.

Redshifts and blueshifts cancel for lenses. This means that the images are identical (except for internal absorption or dispersion within the lensing body), and this is the strongest support for the effect being a true gravitational lens rather than fortuitously placed objects. For instance, this feature was the clue that identified the quasar pair 0957 + 561 as a gravitational lens (the spectra of the

---

[101] Berry, M. V. 1976, *Adv. Phys.*, **25**, 1; Poston, T. and Stewart, I. 1977, *Catastrophe Theory and Its Applications* (London: Pitman); Gilmore, R. 1981, *Catastrophe Theory for Scientists and Engineers* (NY: Wiley; rpt. Dover).

various images are identical) and that has subsequently produced most of the weak lens candidates. Time delays also result from path differences and can be used to determine the distance to the lensing body. This actually yields an independent means for obtaining the Hubble constant because, as Refsdal pointed out early in the business of looking for lenses, the absolute pathlength depends on the distance between the observer and the deflector. To date, this delay has been observed in only a few quasars.

What effect might lensing have on inferred properties of galaxies at large redshift? For one thing, it makes visible those that might otherwise fall below the limit of detection. This is both a blessing and a curse. While permitting the study of galaxian properties for distant populations, such as the more recent detection of very high redshifted ^{12}CO emission from a lensed system, it distorts the correlation and luminosity functions by introducing objects that, based on their intrinsic properties, would not otherwise have been included. Lensing biases the tracers of large-scale structure, since the cluster and supercluster gravitational potentials appear to be due mainly to dark matter and these systematically enhance the population of background galaxies. They may also magnify any object, such as gamma bursts, in the lensed system.

8.9 EXEUNT

Are these the shadows of the things that Will be, or are they the shadows of things that May be only?
—Charles Dickens, *A Christmas Carol*

In summary ...

- *Planck era*: At $t \sim 10^{-43}$ s after the initial singularity, field oscillations around vacuum states in the strongly varying gravitational field create the particles accompanied by some form of inflationary expansion.
- From $t \sim 10^{-43}$ s when $T \sim 10^{19}$ GeV, until $t \sim 10^{-3}$ s, when $T \sim 200$–300 MeV, *Baryogenesis era*: Although initially unified, decouple and undergo spontaneous symmetry breaking and a phase transition. This is the epoch when baryosynthesis occurs, leptons become massive (with the possible exceptions of neutrinos), and the gauge bosons (γ, $W^\pm$, and Z^0) also appear. The massive bosons subsequently decay when $T \sim 80$–90 GeV.
- When $T \sim 200$–300 MeV there is another phase transition when quarks and antiquarks form hadrons (pions and nucleons). The hadronic era begins, and ends when π^+ and π^- annihilate and the π^0 mesons decay to photons ($T \sim 130$ MeV $\sim 10^{12}$ K). The Universe is now composed of leptons, antileptons, photons, and baryons in thermodynamic equilibrium.
- The decay of pions marks the beginning of the leptonic era. Muons quickly annihilate and then ($t \sim 1$ s) neutrinos decouple and propagate freely producing a present-day neutrino background at $T_{0\nu} \approx 1.9$ K. When $T \sim 0.5$ MeV $\sim 5 \times 10^9$ K, $e\bar{e}$ pairs annihilate ($t \sim 10$ s). This is the beginning of the radiation dominated era.

- Neutrons then freeze out leaving $n/p \approx \frac{1}{7}$. Primordial nucleosynthesis occurs producing mainly D, ^{3}He, and ^{4}He with some residual ^{7}Li.
- As the temperature continues to drop due to the expansion, the average number of neutral atoms increases up to the recombination ($z \sim 1400$) when $T \sim 4000$ K and an equilibrium is reached with 100% He°, 50% H$^+$ and 50% H°. The energy densities of matter and radiation decrease with time at different rates since $\varrho_{\text{matter}} \sim R(t)^{-3}$ while $\varrho_\gamma \sim R(t)^{-4}$ from the state when $\varrho_\gamma = \varrho_{\text{matter}}$. This equality occurs later than recombination for $\Omega h^2 \geq 0.04$.
- The subsequent history is matter dominated and represents the one in which we reside. At decoupling, $z \sim 1400$, Compton scattering is no longer efficient and the energetic exchanges between matter and background radiation rapidly become progressively less important. Any component cools adiabatically with the expansion. This includes the photons that constitute the CBR. The photon pressure, which before recombination and decoupling prevented the initially nearly uniformly distributed matter from condensing, becomes progressively less important. The initial fluctuations that survived from the symmetry-breaking epoch and the radiation dominated era are the seeds that originate the largest cosmic structures (galaxies and galaxy clusters) that we see in the present Universe.
- *Reionization epoch*: At $z \approx 10$, first generation stars and nascent active galaxies produce copious ionizing and photodissociating continua and reheat the gas, both in the intergalactic medium and within the forming galaxies.
- *Now*: Stars become the principal agent for nucleosynthesis and energy generation and the world becomes very interesting indeed.
- *Things to Come*: ...

8.A GAMMA-RAY BURSTS

During the writing of this book, no field has changed as dramatically as that of *gamma-ray bursts* (GRBs). When we were starting to write, there was a conundrum —namely, are they local or at large z? As we end the writing, the question is the source capable of producing the now known energetics.

Historically, as so often occurs in astrophysics, these were found while looking for something else. During the period following the nuclear test ban treaty of 1963, the United States launched the *Vela* series of satellites to monitor compliance. During a portion of each duty cycle they scanned the cosmos instead of potential earthly explosion sites and by 1974, had accumulated between 1969 and 1973 a total of 16 short interval bursts that were neither terrestrial nor solar in origin (recalling that flares can generate γ-ray emission). These were short duration events, from 0.1 s to less than a minute with total time integrated energy surface density (fluence) of 10^{-5} to a few times 10^{-4} erg cm^{-2} that were intercepted by multiple satellites in different configurations; the (then Soviet) *Kronus* satellite was quickly reported to show the same results. The rate of discovery for the next 20 years was limited by the lack of dedicated γ $-$ ray observatories and the sensitivity of detectors, although the *Solar Maximum Mission* and instruments on the *Mir* space station contributed perhaps a few per year. With wide uncertainty cones and single bursts, there was little hope of identification and without that, there was only the hint that the emission was at very high energy and therefore must involve a relativistic phenomenon (comets hitting neutron stars, black hole collisions, bizarre supernova events in strange environments, etc.).

This changed with two satellites. First the launch of CGRO, which contained a dedicated instrument, the *Burst and Transient Source Experiment* (BATSE), consisting of multiple detectors oriented toward different parts of the sky and coupled to the more angular-sensitive instruments on board. With this instrument, the discovery rate went from yr^{-1} to day^{-1} so that within its lifetime BATSE had accumulated an enormous catalog of time/energy-resolved bursts. Most importantly, the catalog showed the sources to be *isotropically* distributed, a conclusion that emerged within the first year of operation and only grew stronger with time. There is no hint of the Galactic plane in these data, nor of the local distribution of galaxies. Bursts with timescales as short as a millisecond and lasting to almost 10 minutes have been detected, the distribution being essentially bimodal and dividing at around 2 s. They can extend in energy into the GeV range and show power law spectra in the MeV range with breaks below about 200 keV. In roughly 6 years, a total of nearly 1400 bursts were recorded and cataloged. On the basis of these data alone, and through applications of the $\langle V/V_{max} \rangle$ test, it is possible to show that the sources are cosmologically distant and to obtain some information about their luminosity function.[102]

The decisive step came with the observation of *afterglows*, X-ray and lower-energy emission following a GRB event. This was only possible after the *BeppoSAX* satellite was launched: it carried *both* γ and X-ray telescopes that could trigger on the high energy transient and immediately begin observing at lower energies and with the much higher angular resolution that can be achieved with X-ray telescopes. The result was the identification of several sources, that permitted follow-up groundbased optical, infrared, and radio observations.[103] The sources were found embedded in *galaxies* and in one brief interval the whole picture shifted from one mystery to another!

Briefly stated, these sources are *not* within our Galaxy, nor even in the Local Group.[104] The derived energies in γ rays alone is $4 \times 10^{54} (\Delta \omega / 4\pi)$ erg, where $\Delta \omega$ is the solid angle subtended at the source (this is derived for GRB 990123 for which the largest multiwavelength database is currently available). In other words, if the radiation is strongly beamed or the source is very anisotropic, then the energy requirements are significantly reduced. But this is still an enormous amount of energy compared to SN Ia or SN II and requires very special treatment. The models are developing as this appendix is being written, but there are a few basic points that seem to be firm. The emission is presumed to come from an expanding fireball during a relativistic event, the scenarios now focus on some sort of binary merger of collapsed bodies driven by gravitational waves, either the components of a double neutron star binary or a neutron star-black hole binary. Single star

---

[102] See, for instance, Schmidt, M. 2001, *ApJ*, **552**, 36.

[103] Costa, E. et al. 1997, *Nature*, **387**, 783; Metzger, M. et al. 1997, *Nature*, **387**, 878; Kulkarni, S. et al. 1998, *Nature*, **393**, 35.

[104] The greatest possible redshift so far reported is $z = 4.50$ for GRB 000131 (these are labeled by YYMMDD format) resulting in a burst energy $E_\gamma \sim 10^{54}$ erg; see Andersen, M. et al. 2000, *Astr. Ap*,, **364**, L54.

scenarios include *magnetars*, the collapse of neutron star with 10^{15} G magnetic fields and accretion induced collapse, all of which release similar amounts of energy within a factor of 10. If the fireball is relativistic, the initial size when opaque must be smaller than about 10^7 cm from the burst timescales. Then the photons would be expected to degrade through scattering $\gamma\gamma \rightarrow e\bar{e}$ into pairs with subsequent emission at 0.511 MeV, which is not observed, so the bulk Lorentz factors must be high to reduce the interaction cross sections ($\Gamma > 10^2$). This material then expands into a surrounding medium that produces a shock and the bulk of the emission.[105]

8.B NEWTONIAN DERIVATION OF THE EVOLUTION EQUATIONS

It is a curious feature of the evolution equations that they can be derived using a Newtonian approximation.[106] Assume that you have a homogenized Universe. Then the smoothness of the matter distribution means that the local expansion is what would be expected for a particle moving with respect to a local mass $(4\pi/3)\varrho a^3$ with pressure P. The Poisson equation gives $g = -4\pi(3P + \varrho)a$ and the only supplemental term that is needed is the curvature term, which gives the rate of expansion, $-k/a$. You can imagine that within this distribution, a local region can be treated as the mass contained within some small, arbitrarily chosen region point. In other words, if the mass is otherwise uniformly distributed, so the net gravitational force from all mass lying outside of the sphere of radius a vanishes, then we can write

$$\ddot{a} = -GMa^{-2} \tag{8.178}$$

for the dynamical equation. If we define a parameter that is analogous to the Hubble parameter

$$H = \frac{\dot{a}}{a}, \tag{8.179}$$

we obtain the continuity equation in the form

$$\frac{d\varrho}{dt} = -3H, \tag{8.180}$$

since div $a = 3$. The equation of motion reduces to

$$\ddot{a} = -\frac{4\pi G\varrho}{3}a. \tag{8.181}$$

Note that if the density were constant in time, this would be the equation of a harmonic oscillator, since the enclosed mass increases as a increases. Now for the

---

[105] We will end the discussion here but refer to the superb summary by Mészáros, P. 2000, *Nucl. Phys. B*, **80**, 63.
[106] This approach was introduced by McRae, W. H. and Milne, E. A. 1934, *Quart. J. Math.*, **5**, 73.

energy equation, we have to make a minor modification to the usual Newtonian formalism by including a curvature term:

$$\dot{a}^2 + k - \frac{8\pi G}{3}\varrho a^2 = 0 \tag{8.182}$$

$$2a\ddot{a} + \dot{a}^2 + k + \frac{8\pi G}{3}a^2 P = 0. \tag{8.183}$$

Here k is the curvature constant. The first equation comes from the dynamics, while the second comes from the thermodynamics:

$$\frac{d}{dt}\varrho a^3 + P\frac{d}{dt}a^3 = 0, \tag{8.184}$$

where now ϱ is the mass energy density. In keeping with the convention, and to make things clearer, we have set $c = 1$.

It may seem absurd that we can use this treatment for the equations of motion for the mass in the Universe, but a moment's reflection will make it reasonable. Newton demonstrated that the interior gravitational field in a homogeneous, isotropic, uniform mass distribution can be modeled assuming that only the mass within a radius $r \leq a$ serves as an attractive source, where R is the location of the test mass within the spheroidal mass distribution. The conditions of homogeneity and isotropy are critical here. There must be no clumping in the mass lying outside distance a, or else this will produce spurious uncompensated accelerations that will depart from the simple radial flow expected for a spherical mass distribution. From the relativistic viewpoint, there is an excellent reason why it works, based in the foundations of general relativity. Locally, in any isotropic and homogeneous mass distribution, we can *always* approximate the metric as flat and treat the motion using a Newtonian framework. Only small departures from the neighborhood are needed to understand the global properties of the model.

8.C GALAXY FORMATION AND Λ DM SCENARIOS

In the main text, especially in Sections 8.7.4.3 and 8.7.7, we described some of the issues facing those modeling the growth of structure and the formation of galaxies. A few additional cautionary remarks may place some of the problems and simulations in a general perspective in this extremely rapidly developing field.

There are a few basic challenges facing anyone who attempts to "grow structures" in a simulation of the early Universe. The expansion rate is fixed by the cosmological solution, whether Λ or DM dominated. The fluctuation spectrum of the CBR is constrained by data, and the decoupling time is something that cannot be freely adjusted in light of nucleosynthetic constraints on the baryonic fraction. So, the issue is, how to get things to clump slowly enough to be consistent with the amplitude of the observed temperature variations (at $z > 10^3$) and yet rapidly enough to produce large scale structures ($10 < z < 10^2$) and galaxies (well before $z = 6$), requirements that are increasingly well constrained by the re-ionization picture and discoveries of QSOs and galaxy clusters at progressively higher redshift. Having only gravity and some sort of mechanical and thermal dissipation to work with when dealing with dark matter, since they are just lumps with velocity dispersion by the time we're describing, simulations attempt to reproduce the largest scales by particle interactions in an expanding background.

The more rapid the expansion, and this is especially true if Ω_Λ is as large as 0.7, the greater the dissipation required to achieve clumping at an early time. Radiative feedback, other than gravitational waves (which have very low amplitude at this epoch and seed the first fluctuations), cannot occur until the first stars appear. How this happens is entwined with our understanding of star formation at the present epoch with one major difference: there are no metals to provide cooling, so except for H_2, H_3^+, and HD, the first generation of stars is expected to be quite massive. The Jeans mass, if it provides the scale for self-gravitation, depends on the temperature, or rather the velocity dispersion of the DM particles, and will be quite large at these times. Yet we know the formation of stars now barely knows about this property and may depend more on other, shear and tidal dependent, self-gravitating criteria such as the *Toomre Q parameter*.

For the cosmic scale, what we have developed throughout the book corresponds to the *microphysics*, and the cluster and galaxy formation problem combines many processes. For example, capture interactions are not as rare, it appears, as they are for star clusters. They occur primarily at intersections of large scale expanding structures that fragment out of the background as seeds for nonlinear growth from the Harrison–Peebles–Zel'dovich power spectrum. The huge gravitational structures dominate the evolution and the interaction of baryonic matter with the cooling CBR. This was shown in the discussion of star formation and chemical evolution: if the growth of small scale structure is due to nonlinear interactions—for instance with $\psi \sim \rho^2$—then the fluctuations are enhanced and accrete at the expense of low density regions. Even without further radiative or mechanical feedback, the medium will form dense, continually fragmenting structures that will hierarchically aggregate DM to form central clusters and galaxies. Any baryonic matter will follow this, flowing into the progressively deeper gravitational wells and ultimately dissipating collisionally to become what we see as galaxies. While the baryons contribute modestly to the mass budget, they are *not* passive tracers of the flows; they are the dissipative phase and are, after all, the matter we see. It is through nuclear reactions that the Universe is re-ionized and the intergalactic medium enriched with metals, and treating realistically the feedback between the dark matter and baryonic matter is a core issue in modern simulations.

The difficulties faced in treating the interactions at the *smallest* length and at mass scales is a feature common to all large scale computations, including fluid turbulence. Depending on the number of particles in simulations, the resolvable mass scales may be super- or subgalactic, the length scales that must be treated range from tens of Mpc to less than 0.1 Mpc. In contrast, galaxies require resolutions of 0.01 to 10 kpc, and mass scales from tens of solar masses for stars and clusters to 10^6 $M_\odot$ for giant molecular clouds. Thermal and radiative processes and interactions must be properly included, locally and globally. Angular momentum must be exchanged between colliding masses and this depends on details of their internal mass distributions and how viscous or long range torque and angular momentum transport occurs in dark matter ensembles. The problems we have discussed for a "gas of stars" are exacerbated for DM distributions although many of the same techniques can be and are used to treat the interactions. Some prescription must be supplied, coming from physics that cannot be resolved in detail in the calculations, that treats the effects that arise at the smallest lengths and masses. Increases in computational power, both in speed and numbers of particles that can be accomodated in simulations, *does* help but as we have discussed, there are always processes happening below resolvable levels, the so-called *subgrid*, at the scale of clusters of stars, not just clusters of galaxies. In addition, to compare with observations beyond the CBR fluctuations, some way must be found to produce stars. A you have seen, this is a formidable challenge even for single and binary stars in the present day Galaxy.

Yet with all the difficulties of the preceding chapters kept firmly in mind, the results achieved by simulations have been impressive in their ability to reproduce the background fluctuations and describe the basis for galaxy formation. By the time the ink dries on this book, new models will have appeared and new ideas will have been proposed. There is no better, more optimistic note on which to end this journey.

INDEX

Abel equation, 731–732
Absorption. *See* Opacity
Absorption coefficient
   H I 21 cm line, 484
   synchrotron, 536
   thermal bremsstrahlung, 193–194, 243
   total for absorption line, 197–198
Abundances. *See* Element abundances
Accretion
   black holes, 447, 707–708
   cooling flows, 723–730
   from a stream inner Lagrangian point, 434–435
   from a wind (Bondi–Hoyle accretion), 432–434
Accretion disks
   α-model, 438–439
   boundary layers, 445–447
   formation of, by stream impact, 435
   magnetic fields, 442–445
   spectral energy distribution, 440–441
   vertical structure, 436–437, 439–440
   viscous, 437–440
Accretion-induced collapse, 387
Action, 76
Adiabatic temperature gradient, 235, 302
Advection-dominated accretion flow (ADAF), 447
Agglomeration equation, 68–69, 519
Alfven radius, 430
Alfven speed, 50
Alfven waves, 50–51
Algol paradox, 417–418
Algorithms
   CLEAN, 149–150
   maximum entropy method (MEM), 151–154, 159–160
   Richardson–Lucy, 163–165
   shift-and-add, 145
   stochastic time series, 148–149
Ambipolar diffusion, 51, 542–543, 586–588
Antenna temperature, definition of, 484
Aperture synthesis, 140–142
   phase closure, 141
Apsidal motion, 10–11, 94. *See also* Precession, orbital
   and internal stellar structure, 421–422
Asteration, 679
Asteroseismology. *See* Helioseismology
Astronomical Unit, Roemer's determination of, 106

Asymptotic giant branch (AGB), 373–377
Atmospheric eclipses, 414–417
Autoionization, 197

Baade's window, 633
Backwarming, 229–230
Balbus–Hawley instability. *See* Instabilities, magnetorotational
Balmer limit, 194
Baroclinicity, 308–309, 420
Barred spiral galaxies. *See* Galaxies, spiral
Bayes' theorem, 150–151
Bernoulli equation, 43
Biachi identities, 88
Biases, 161–162
   Lutz–Keller, 108–109. *See also* Parallax
   Malmquist, 807–808. *See also* V/V_{max} test
Biasing (cosmological density fluctuations), 827–830
Big Bang nucleosynthesis, 680–682, 798–805
Binary stars
   classification relative to Roche radius, 419–420
   common envelope, 423–425
   eclipsing, 119–120, 414–417
   effects of angular momentum loss, 426–430
   effects of mass transfer, 426–428
   emission of gravitational waves, 428–430
   evolutionary stages of (Cases A, B, C), 422–423, 427
   formation of, 453–457
   light curves, 119–120, 239–241, 414–417
   magnetic braking, 430
   mass determinations, 118–119
   masses, 360–361
   mass function, 119, 361
   neutron stars in, 400–401
   pre-main sequence, 453–455
   reflection effect in, 239–241
Birthline, 366. *See also* Star formation
Blackbody radiation. *See also* Planck Radiation Law
   cosmic background radiation, 770–771
   Rayleigh–Jeans approximation, 28, 129
   Wien approximation, 28
Black holes, 92, 96–97
   in active galactic nuclei, 703–708
   in binary systems, 452–453

Blast wave
 adiabatic, 574–575
 free expansion phase, 577
 snowplow phase, 579
BL Lac objects (blazars). *See* Galaxies, active galactic nuclei
Blue compact dwarf galaxies. *See* Galaxies, irregular
Blue stragglers, 456–457
Bolometric correction, 112, 401–402
Boltzmann collision integral, 55, 66
Boltzmann distribution. *See* Distribution function, Boltzmann
Boltzmann equation, 54–55, 66
 and H theorem, 67–68
 for photon scattering, 210–211
Bondi–Hoyle accretion, 432–434
Bonner–Ebert mass, 590
Born approximation, 262
Bose–Einstein distribution. *See* Distribution function, Bose–Einstein
Bose–Einstein integral, defined, 35
Bosons. *See* Distribution function, Bose–Einstein
Boundary layers, 445–447, 450–452
Boussinesq approximation, 234
Breit–Wigner resonance, 335
Brightness temperature, 129, 483
Brown dwarfs, 278, 369
Brunt–Väisälä frequency, defined, 618

Calibration
 backgrounds, 156–157
 Hubble constant, 753–754
 infrared, 127–128
 Morgan–Keenan spectral types, 274
 optical, 109–111
 period-luminosity relation, 742–744
Carbon stars, 278, 349–351, 375–376
Cataclysmic variables, types of, 447–448
CCDs. *See* Detectors, charge-coupled device (CCD)
cD galaxies. *See* Galaxies, elliptical
Champagne flow, 574
Chandrasekhar mass, 298, 382–383, 387, 395
Charge transfer, 474–475, 551–552, 563–564
Chemical evolution of galaxies, 669–688. *See also* Galaxies
 "closed-box" models, 677–683
 observations, 349–351, 647, 670–672
 solar system abundances, 350, 669–672, 802–804
Chi-square statistic, 152
Christoffel symbol, 83
Classification. *See* Taxonomy *and specific subject*
Clausius–Mossotti equation, 501
CLEAN. *See* Algorithms, CLEAN
"Closed-box" models for galactic chemical evolution, 677–683

Clusters of galaxies
 Local Group, 720–721
 mass to light ratio, 812
 properties, 719–720, 723
 Schechter luminosity function, 751
 X-ray emission from, 723–730
Coagulation. *See* Agglomeration equation
Colliding stellar winds. *See under* Stellar wind
Collisional ionization, 172
Collision frequency, in plasma, 52
Collision rates, 171–172, 465–466
Color-magnitude diagram. *See* Hertzsprung–Russell diagram; Isochrones
Common envelope. *See also* Binary stars
 U UMa stars and stellar structure, 424
Compton effect. *See* Scattering, Compton
Compton scattering length (y), 213
Compton wavelength, 209
Condensation temperature, 505
Conduction. *See also* Diffusion, thermal conduction
 thermal, effect on cooling instability, 594–595
Conductivity, thermal conduction coefficient, 301
Continuity equation, 41, 49, 54, 293
Convection, 232–237, 302–306
 Boussinesq approximation, 234
 effects of variable molecular weight, 237
 mixing length theory, 234–237, 305–306
 overshooting, 305
 Rayleigh–Benard, 233–236
 Schwarzschild criterion, 302
 semiconvection, 304–305
Convergence point for clusters. *See* Parallax, convergence point method
Convolution, 163, 732–733
Cooling flows. *See* Clusters of galaxies, X-ray emission from
Cooling processes
 and fine structure transitons, 466
 molecular, 548–549
 rate, 472–474, 592–593, 595–597
 and shocks, 562–563
Copernican principle, 757–758, 809
CORAVEL. *See* Spectrographs, correlation
Coriolis acceleration, 12
Correlation function, 613–614, 625–627, 815–817, 828–830
Cosmic background radiation (CBR), 769–774
 decoupling epoch, 773, 789
 from fine structure lines, 769–770, 771–772
 fluctuations, 834–836
 as limitation to structure formation, 822–823
 measurement of fluctuations, 830–833, 836
 temperature, 770–772
Cosmic rays, 527–533, 547–548
 energy distribution, 529–531
 and heating, 595

ionization rate from, 475, 531, 547
"leaky box" model for propagation of, 531–533
nucleosynthesis by spallation, 680–682
propagation, 531–533
and time-dependent synchrotron emission, 711–712
Cosmic scale factor, 762, 764–767. *See also* Friedmann–Robertson–Walker equations
Cosmological constant (Λ), 774–776, 793, 826
Cosmological density parameter (Ω), 776–778
Cosmological dimming, 756
Cosmological distance scale
and luminosity functions, 750–751
steps in calibration, 740–742
Cosmological gravitational lenses, 836–841
Cosmological lookback time, 783–790
Cosmological models
cosmological constant-dominated, 774–775, 814–815
Newtonian, 844–845
relativistic. *See* Friedmann–Robertson–Walker equations
Cosmological principle. *See* Copernican principle
Cosmology, large scale structure, 101–102, 830–836
Coulomb barrier, 331
Coulomb integral, 52, 302
C shocks. *See* Magnetic fields, and shocks
Curie–Weiss law, 515
Curvature, cosmic, 777–778
Curvature scalar, 88
Curve of growth, 201–203, 478–482

Damping factor, 335
Dark matter, 699–700, 811–815
in clusters of galaxies, 720–723, 730–731
and cosmological density parameter, 776–778
and dark energy, 814–815
galactic halos, 699–700
and galactic rotation curves, 697–700
Debye–Hückel model for screening, 356
Debye length, 48, 52, 356
Debye temperature, 518
Deceleration parameter, 765
Decoupling epoch for cosmic background radiation, 773, 789
Degeneracy parameter, defined, 292
Degenerate gas, 28–29, 291–292, 380–385
thermonuclear runaway, 372, 448–450
Density waves, 661–668
Departure coefficients, 468
De Sitter solution (also called Einstein–de Sitter), 598, 791. *See also* Cosmological models, Cosmological constant-dominated; Inflation
Detailed balance, 35–36, 54–55, 68–69, 169–171, 176–177, 344, 464, 536, 617–618
and Compton scattering, 210–211

ionization equilibrium, 196–197
Detectors, 120–128
bolometers, 126–128
charge-coupled device (CCD), 124–125
multianode microchannel array (MAMA), 125–126
photographic plate, 120–123
photomultiplier, 123–124
proportional counter, 128
Deuterium, 530–532, 799–805
De Vaucouleurs Law, 693
D-fronts. *See* H II regions, as shocks
Dielectric constant, 501–502
Dielectronic recombination, 468
Diffuse interstellar band (DIB), 511–514, 553–554
Diffuse interstellar radiation field (DIRF), 493. *See also* Dilution factor
Diffusion
ambipolar. *See* Ambipolar diffusion
cosmic rays, 531–532
Fick's law, 65
and Fokker–Planck equation, 65
induced by gravitational settling, 53–54
magnetic fields, 407–408
in plasma, 51–55, 65
radiation, 190, 299–301, 440
radiatively driven dynamics, 248–249
and semiconvection, 305
thermal conduction, 301–302
turbulent, 376–377, 410, 687
viscous energy dissipation, 457–458
Diffusion coefficient, 65
radiative, 190, 300
thermal. *See* Diffusion, thermal conduction
Diffusion equation, time-dependent accretion, 438
Dilution factor, 231, 241–242, 493
Dispersion measure, 538, 540–541
Distance scale. *See* Cosmological distance scale
Distribution function
Boltzmann, 23–25
Bose–Einstein, 34–35, 511
Fermi–Dirac, 29–34
Maxwellian, 21–23, 641–642
Planck, 26–28
Schwarzschild, 641–642
Doppler boosting, 714
Doppler imaging, 223–224, 417. *See also* Rossiter–McLaughlin effect
Double diffusive convection, 305
Dredge-up, 375–377. *See also* Asymptotic giant branch (AGB)
Dreiser field, 49
Dust
broad extinction features, 500–501, 513
charge of, 507–508
composition of, 500–501, 504–506, 514–518, 544–546

Dust (*Continued*)
  interstellar dust to gas ratio, 506
  and radiative equilibrium, 242
  surface reactions, 544–546
  wavelength dependent extinction, 498–500
Dynamical friction, 38–39
Dynamical instability, 316
Dynamo. *See* Magnetic fields, generation by dynamos
Dynamo number, 409

Eddington approximation, 174, 181, 237. *See also* Eddington factor
Eddington–Barbier relation, 185, 414. *See also* Limb darkening
Eddington factor, 178–179
Eddington luminosity, 245–246, 707
Einstein coefficients. *See* Transition probabilities
Einstein field equation, 89
Einstein radius, 838
Einstein summation convention, 82
Element abundances, 349–351. *See also* Neutron reactions; Radionuclides; Thermonuclear reactions; Chemical evolution of galaxies
Elliptical galaxies, fundamental plane, 694–695, 748–749
Emission lines, nebular diagnostics, 462–472
Emission mechanisms
  synchrotron, 533–536
  thermal bremsstrahlung, 192–194
Energy generation, gravitational contraction, 329–330, 363
Enthalpy. *See* Thermodynamics, enthalpy
Entropy, 16, 20–21
  Boltzmann form of, 21, 67
  cosmological, 773–774, 800–801
  Shannon form of, 152
Epicyclic frequency, 652–654, 657
Equation
  Abel, 731–732
  agglomeration, 68–69, 519
  Bernoulli, 43
  Boltzmann, 54–55, 66, 67–68, 210–211
  Clausius–Mossotti, 501
  continuity, 41, 49, 54, 293
  diffusion, 438
  Euler–Lagrange, 76
  Einstein field, 89
  Fokker–Planck, 64–67, 532, 712
  Friedmann–Robertson–Walker, 766–767, 774–775, 777–781, 819–821
  hydrostatic, 227
  Killing, 97–98
  Kompaneets, 209–214
  Lane–Emden, 38, 294–299, 646, 724
  linear adiabatic wave, 326
  linear pulsation, 323–328

  Malthus–Verhust, 684
  Mattig, 785
  Milne, 183, 269
  Navier–Stokes, 44
  Oppenheimer–Volkoff, 390
  Poisson, 4, 5, 37–38
  radiative transfer. *See* Radiative transfer equation
  Saha, 36, 773
  Tolman–Oppenheimer–Volkoff, 390
  Vlasov, 40–43, 642–644
Equation of state
  Bose–Einstein, 34
  cosmological, 768–769, 779–781
  degenerate, nonrelativistic, 28, 31
  degenerate, relativistic, 29, 31, 33
  Fermi–Dirac, 32
  ideal, 18
  partial ionization and, 288–289
  polytropic, 285–289, 324
  radiation-dominated, 286–288
Equations of motion
  conservative form, 42
  Euler–Lagrange, 76
  fluid, 41–42, 49, 555
  Hamiltonian, 20
  stellar hydrodynamics, 643–644
Equivalence principle, 86
Equivalent width, 202, 480. *See also* Curve of growth
Escape probability, 256–259
Euler–Lagrange equation, 76
Exponential integrals, 183
Extinction, 180, 498–500
  and column densities, 500
  and polarization, 514
Extrasolar planetary systems, eclipsing, 415

Faber–Jackson relation, 748–749
Fanaroff–Riley classification for radio galaxies, 713–714
Faraday depth, 539
Faraday rotation, 515, 536–540
Fermi–Dirac distribution. *See* Distribution function, Fermi–Dirac
Fermi–Dirac integral, defined, 32
Fermi golden rule, 262
Fermi momentum, 28
Fermions. *See* Distribution function, Fermi–Dirac
Fermi process, 528–529
Fick's law. *See* Diffusion, Fick's law
Finesse, 138
Fluorescence, 205–207, 463
Flux. *See* Radiative flux
Fokker–Planck equation, 64–67, 532, 712
Forbidden line transitions, 462–466, 472, 747
Fossil Strömgren sphere, 571

Fourier transform, 155, 162–163, 439. *See also* Correlation function; Power spectrum
Free energy. *See also* Thermodynamics, free energy and partition function, 25
Free-free opacity. *See* Opacity, thermal bremsstrahlung
Friedmann–Robertson–Walker equations, 766–767, 774–775, 777–781
  perturbed form, 819–821
Friedmann–Robertson–Walker (FRW) metric. *See* Metric, Friedmann–Robertson–Walker
Fringe visibility. *See* Interferometers, visibility
Fugacity, 30
Fullerenes, 553
Fundamental plane. *See* Elliptical galaxies, fundamental plane

Galaxies
  active galactic nuclei (AGN), 700–703
  bars, 666–668, 691–692
  chemical evolution of. *See* Chemical evolution of galaxies
  elliptical, 692–695, 720
  escape velocity from, 698–699
  galactic winds, 686–687
  Hubble classification, 690–697
  infall of gas, 686–688
  irregular, 695–697, 804–805
  polar ring, 716–717
  ring, 716–718
  rotation curves, 655–660, 697–700, 733, 747–748
  spiral, 690–692
  spiral structure, 661–668
  tidal interactions, 685
Galaxy
  galactic center, 640, 704–705
  galactic coordinates, 640, 655–658
  rotation law, 655–660
  solar galactocentric distance, 650–651
  spiral structure, 660–661. *See also* Density waves
  surface mass density, 645
  vertical structure of disk, 644–647, 653, 657–658
Galaxy clusters. *See* Clusters of galaxies
Galilean transformation, 556
Galileo, and astronomical measurements, 104–105
Gamma ray bursts, 842–844
Gamow factor, 332
"Gas of stars," 37–39, 637–647
Gaunt factor, 194
Gaussian process. *See* Random processes, Gaussian
Gauss' theorem, 4
General relativity, 81–102
  principle of covariance, 79, 83
Geodesic, 85
Giant branch, 372–374. *See also* Hayashi track

Gibbs free energy. *See* Thermodynamics, Gibbs free energy
Glitch, 399–400
Globular clusters, 635–637
Gould belt, 651–652, 654, 672, 685
Gravitational contraction, as energy source, 329–330
Gravitational deflection of light. *See* Gravitational lens
Gravitational lens, 95, 98–99
  and clusters of galaxies, 730–731
  and galactic halo objects (MACHOs), 647–648
  image formed by extended mass distribution, 836–841
  magnification, 839
Gravitational potential, 4–5, 57, 70–75
  classical spheroids, 70–75
  Maclaurin spheroid, 74–75, 455, 693–694
Gravitational waves, 99–101, 428–430, 834–836
  detection of, 101
  emission by close binary system, 428–430
Gravitothermal catastrophe, 62, 289–290
Gravity darkening, 238–239, 308, 420
"Great attractor," 810
Green function, 5
Gunn–Peterson test, 796, 824
Gyrosynchrotron process, 535–536

Halley–Olbers dark sky paradox, 806–807
Hamiltonian equations of motion, 20
Hanle effect, 216
Harmonic Law. *See* Kepler's Law
Harrison–Peebles–Zel'dovich spectrum, 829
Harvard spectral classification, 225
Hauser–Feshbach method, 337
Hayashi track, 368–369, 372–373
Helioseismology, 320–323
Helium flash, 372–373. *See also* Thermonuclear reactions; Thermonuclear runaway
Helmholtz free energy. *See* Thermodynamics, Helmholtz free energy
Henry Draper Catalog (HD). *See* Harvard spectral classification
Hertzsprung–Russell diagram (HR diagram), 359, 631–633. *See also* Star clusters
Hickson compact groups. *See* Clusters of galaxies
Hipparchos satellite, 107 *and passim*
Historical notes
  Astrophysical Journal, 166
  binary and multiple stars, 412–413
  convection, 302
  cosmological constant, 775
  cosmology, 734–740
  dust in the interstellar medium, 495–498
  expanding H II regions, 571–572
  galaxies, 689–690
  Galaxy, 629–630

Historical notes (*Continued*)
  Hertzsprung–Russell diagram and stellar evolution, 357–359
  interstellar medium, 459–461
  pulsars, 394
  spectral classification, 225–226, 272–274
  stellar structure and evolution, 281–282, 294
  stellar thermonuclear energy, 337
  synchrotron emission, 533
  thermonuclear instabilities, 374
  URCA process, 357
Homology relations, 309–310
Hopf function, 183, 271–272
Horizon distance, 786–790
Horizons
  and black holes, 93–97
  and cosmology, 783–790
H theorem, 67–68. *See also* Boltzmann equation
H II regions. *See also* Champagne flow; Strömgren sphere
  compact, 571
  fossil, 571
  as shocks, 565–567, 572–573
  static, 567–568
Hubble classification of galaxies, 690–697
Hubble constant, 752, 764
Hubble flow, 576
Hubble parameter (h), 754
Hydrogen
  H_3^+, 549–550
  hydrogen line profiles and pressure broadening, 220–222
  Lyman lines, 476–477, 482, 500, 797–798
  molecular formation, 544–546
  molecular hydrogen lines, 490–491, 500
  21 cm line of H I, 482–485, 500, 658–659, 673
Hydrogen burning, 337–341, 348, 375
Hydrostatic equation, as function of optical depth, 227
Hydrostatic equilibrium, 227, 293, 724
  for accretion disk, 436
Hyperfine structure, 214–215

Ideal gas law, 18
Image reconstruction methods, 146–156
Impact approximation. *See* Stark effect, quantum mechanical
Impact parameter, 39
Inertial subrange, 604–605
Inflation, 781–783, 789–795
Initial mass function (IMF), 404–405, 672–673
Instabilities
  Alfven waves, 50–51
  convection, 302–306
  gravitoacoustic, 582–586, 663–665, 818–822, 826–830
  Jeans. *See* Instabilities, gravitoacoustic
  Kelvin–Helmholtz, 582
  magnetorotational (MRI), 442–445
  Parker, 600–602, 618–620
  pulsational, 311–328
  Rayleigh–Taylor, 582, 600
  Richtmyer–Meshkov, 582
  sound (acoustic), 44–46, 835
  thermal, 590–597, 724–728
Instantaneous recycling approximation, 677–680
Intensity
  integrated (J), 176–177
  specific, 167
Interferometers
  Fabry–Perot, 138–139
  image formation, 134, 140–142
  intensity, 139
  Michelson, 137–138
  speckle, 145
  visibility, 137
Intermittency, 612–613
Invariant imbedding, 265–268
Ionization
  collisional, 172, 547
  effect of partial ionization on equation of state, 288–289
  energies for atomic species, 195
  photoionization. *See* Photoionization
Ionization equilibrium, Saha equation, 35–36
Ionization parameter, 570
Irregular galaxies. *See* Galaxies, irregular
Isochrones, 401–404, 631–637, 673–674
Isothermal disk, 645–647, 698
Isothermal sphere. *See* Lane–Emden equation, isothermal; King model
Isotopes, isotopic fractionation, 550–552. *See also* Radionuclides

Jansky, defined, 129
Jeans escape. *See* Outflows, thermal evaporation
Jeans instability. *See* Instabilities, gravitoacoustic
Jeans length and mass, 585
J shocks. *See* Magnetic fields, and shocks

K correction, 756
Kelvin–Helmholtz contraction, 329–330, 365–366
  timescale, 61, 284
Kepler's law, 3
Killing equation, 97–98
King model, 698, 724
Kirchhoff–Bunsen laws, 173
Klein–Nishima formula, 209
Kolmogorov length, 604
Kolmogorov spectrum. *See* Turbulence, Kolmogorov spectrum
Kompaneets equation, 209–214. *See also* Scattering, Compton

Lagrange multipliers, 24, 152, 159–160
Lagrangian frame, 263

Lagrangian points, 12–13, 418–420, 666–668
Lambda iteration, 232
Lane–Emden equation, 294–299
 isothermal, 38, 298, 646, 724
Langevin rate, 546
Large velocity gradient method. *See* Sobolev approximation
Larmor formula, 615
Larmor frequency, 47
Larson–Penston law for isothermal sphere, 598
"Leaky box" model for cosmic ray propagation, 531–533
Least action principle, 76
Lenticular galaxies. *See* Galaxies, spiral
Light echo, 520–521
Light element synthesis, 680–682, 801–805
Likelihood, 151, 160
Limb darkening, 184–186
 coefficient, 185
 and eclipsing binary light curves, 120, 414
 and rotational broadening of line profiles, 222–223
Lindblad resonances, 662–665
Linear adiabatic wave equation, 326
Line profile
 broadening, 214–224
 damped, 203, 478–481
 damping factor, 198–199, 217
 function, 169
 Gaussian, 199, 478
 Holtzmark, 217–218
 Lorentzian, 198, 476–482
 P Cygni, 253–255
 pressure gradients from line wings, 220–222
 redistribution in, 205
 rotational, 222–223
 saturated, 202, 480–481
 Voigt, 200–201, 478–482
LINER (low ionization emission line galaxies). *See* Galaxies, active galactic nuclei
Local Group. *See* Clusters of galaxies, Local Group
Local standard of rest (LSR), 640–641, 644
Local thermodynamic equilibrium (LTE), defined, 36–37. *See also* Non-LTE, conditions for
Lomb–Scargle method for time series, 148–149
Lookback time, 785
Lorentz factor (γ), 78, 91
Lorentz force, 47, 80
Lorentz invariance, 78–79
Lorentz transformations, 78–79
Lutz–Keller bias, 108–109
Lyman alpha forest, 415, 417, 476–482, 797–798
Lyman limit, 194

Mach number, 557
Maclaurin spheroid. *See* Gravitational potential, Maclaurin spheroid

Magellanic stream, 716
Magnetic fields
 in active galactic nuclei, 708–715
 and alignment of interstellar dust grains, 514–518
 ambipolar diffusion. *See* Ambipolar diffusion
 and coupling in accretion disks, 442–445
 equipartition and minimum energy, 708–711
 frozen-in approximation, 51, 407
 generation by dynamos, 405–410
 Hanle effect, 216
 in interstellar clouds, 542–543. *See also* Ambipolar diffusion
 magnetospheres of accreting stars, 450–452
 maximum mass of magnetic clouds, 588–589
 measurements with Faraday rotation, 537–540
 molecular clouds. *See* Ambipolar diffusion
 pulsars. *See* Pulsars
 and shocks, 558–565
 turbulent, 606–607, 609
 Zeeman effect, 215–216
Magnetic susceptibility, 517–518
Magnetohydrodynamics (MHD), 49–51
Magnetosonic waves, 50–51
Magnetospheres, 396–399, 450–452
Magnitudes, 109–112
 absolute, 111
 AB system, defined, 114
 astronomical, defined, 109
 calibration zero point, 109–110
 color index, 118
 distance modulus, 111
 Pogson's ratio, 109
 standard multicolor filter systems, 112–117
Main sequence
 chemical abundance anomalies in, 248–249, 276 (n.76)
 lower mass cutoff, 369
 mass-luminosity relation, 361–362
 thermonuclear reactions during, 337–341
 turn-off, 370, 403
 zero age (ZAMS), 367–368
Malmquist bias, 807–808
Malthus–Verhust equation, 684
MAMA. *See* Detectors, multianode microchannel array (MAMA)
Masers, 175–176, 491–492, 705–706, 749
Mass function. *See* Binary stars, mass function
Mass-luminosity relations, 361–362
Mass-radius relations, 367, 380–382. *See also* Roche radius
Mathis–Rumpl–Nordsieck dust distribution, 499–500, 518–519
Matter-dominated cosmology, 779–780
Mattig equation, 785
Maximum entropy method (MEM), 151–154, 159–160
Maxwellian distribution. *See* Distribution function, Maxwellian

Mean free path, 38, 52
Mergers. *See* Tidal interactions, galaxies
Meridional circulation, 308–309, 341. *See also* Von Zeipel's theorem
Meteorites. *See also* Radionuclides
  abundances, 506, 349–351, 669–672
Metric, 81–86
  Friedmann–Robertson–Walker, 758–762
  and gravitational waves, 100
  Minkowski, 79, 82
  Schwarzschild, 93, 390
  signature, 82
  symmetries, 97–98, 759–760
Michelson interferometer. *See* Interferometers, Michelson
Microturbulence, 200, 480–481
Mie scattering, 501–503, 521
Milky Way. *See* Galaxies; Galaxy
Milne equation (also called Schwarzschild–Milne equation), 183, 269
Minkowski metric, 79, 82
Mixing in stellar interiors, 375–377. *See also* Convection; Meridional circulation
Mixing length theory for convection, 234–237, 305–306
Molecular clouds
  classification of, 522–523
  linewidth-velocity relation, 524–525
  turbulence in, 608–612
Molecular lines, CO, 523–524, 548–549
Moments, velocity distribution, 40–43, 58–59
Monte-Carlo methods, 190–192, 257
Morgan–Keenan spectral classification, 225
Multiphase interstellar medium, 595–597
Multiplets, 174

Navier–Stokes equation, 44
N-body problem, 15
Nebular diagnostics, 462–472
Nebulium ([O III] lines), 462–463
Neutral buoyancy, 236
Neutrinos, 33–34
  plasmon, 351–352
  solar neutrinos, 337–341, 352–356
  and stellar nucleosynthesis, 351–352
  URCA process, 356–357, 385, 393–394
Neutron reactions, 344–348, 375–376
  cosmological, 799–802
  r-process, 346
  s-process, 345–346
Neutrons
  lifetime, and cosmological nucleosynthesis, 799
  sources in stellar interiors, 347–348, 375–376
Neutron stars, 389–394
  cooling, 393–394
NLTE. *See* Non-LTE
Noise-equivalent power (NEP), 130

Non-LTE (NLTE). *See also* Statistical equilibrium
  conditions for, 237–238
  masers, 169–170, 175–176. *See also* Masers
  scattering, 186–189
  source function, 179–180, 187–188
Novae
  as accreting binaries, 448–450
  common envelope stage, 425
  and cosmological distance scale, 744–746
  line formation in, 259–260
  thermonuclear runaway, 448–450
Nuclear binding energy, 330–331
Nuclear reactions
  barrier penetration probability, 332, 334
  Breit–Wigner resonance, 335
  compound nucleus, 336, 366–370
  Coulomb barrier, 331
  Gamow factor, 332
  nonresonant, 336
  nuclear radius, scaling relation for, 331
  S factor, 332
  specific. *See* Thermonuclear reactions
  two-body reaction rate, scaling law for, 334
Nucleocosmochronology, 688–689, 754–755. *See also* Radionuclides
Nucleosynthesis. *See* Thermonuclear reactions
Nucleosynthesis, cosmological. *See* Big Bang nucleosynthesis
Nyquist frequency, 135–136

OB associations, 633
Observer's frame, 263
Olbers paradox. *See* Halley–Olbers dark sky paradox
Oort constants, 656–657
Oort Law for Galactic rotation, 655–658
Opacity
  Kramers law, 299
  line absorption, 197–198
  photoionization, 195
  Rosseland mean, 190
  synchrotron, 536, 710–711
  thermal bremsstrahlung, 193–194, 243, 299
  Thomson, 299
Open clusters, 633–635
Oppenheimer–Volkoff equation (also called Tolman–Oppenheimer–Volkoff equation), 390
Optical depth, 179
  scaling by wavelength-dependent opacity, 227
Orbital motion, two-body problem, 2–11
  in Schwarzschild metric, 94
Orbital resonance. *See* Three-body problem
Ortho-to-para (ortho:para) ratio for molecular hydrogen, 490–491
Oscillator strength, defined, 198

Ostriker–Peebles criterion, 699–700
Outflows, thermal evaporation, 62–64, 249–250
Overshooting, 305–306, 369
Overstability, 316

Parallax, 106–109
  as basis of distance scale, 740
  convergence point method, 107–108
  Lutz–Keller bias, 108–109
  and proper motions, 107–107
  statistical, 108
Parceval's theorem, 162
Parker solution for stellar wind, 249–250
Parsec, defined, 107
Particle acceleration, 528–531
Partition function, 25–26, 515
Period-luminosity relation, 311–313, 650, 742–744
Periodogram, 147
Perturbation theory, 219–220, 260–263
Phase space, 20
Photodissociation, 547–548
Photodissociation regions (PDR), 565–567
Photoionization, 194–197, 567–571
  hydrogenic absorption coefficient, 195
Planck mass, 790
Planck radiation law, 26, 172
Planetary nebulae, 378–380, 750–751
Plasma, 46–49
  Debye–Hückel model, 356
  diffusion in, 51–55
  screening in, 48–49, 355–356
  Stark effect, 216–218
  thermal emission from, 193. See also Thermal bremsstrahlung
Plasma frequency, 47
Plasmon neutrino process, 351–352
Point spread function (psf), 134–135
  atmospheric turbulence and, 142–145
Poisson equation, 4, 5, 37–38
Poisson process. See Random processes, Poisson
Polarization
  observations of interstellar, 514–515
  Zeeman polarimetry, 216
Polycyclic aromatic hydrocarbons (PAH), 513–514, 552–554
Polytrope
  defined, 285–286
  degenerate, 298
  Lane–Emden equation, 294–299
  limiting mass, 297–298
  polytropic indices, 288, 324
  radiation-dominated, 296–298
  stability of, 289–291
Polytropic equation of state, 19
Population synthesis, 675–677
Populous blue clusters, 637

Power spectrum, 142–144, 162–163, 624–627, 828–830
Poynting--Robertson effect, 508–509
p-process, 348, 375–376, 450. See also Thermonuclear reactions, hydrogen burning
Precession, 7–11
  orbital, 10–11, 94
  rotational, 8–10
Precursor
  magnetic shocks, 561–562
  radiative shock, 563–564
Pressure broadening. See Stark effect; Van der Waals broadening
Pressure scale height, 303
  for stellar hydrodynamics, 644
Prior probability, 151
Proper distance, 79
Proper time, 79
Proper volume, 84
Protostar, 364–366
Pulsars, 394–401
Pulsation
  as analogy with a musical instrument, 320–321
  as analogy with thermal cycle, 315–316
  ϵ mechanism, 316, 374–375
  κ mechanism, 316, 318–319
  linear pulsation equation, 323–328
  one-zone treatment, 316–319
  thermal pulsing on asymptotic giant branch, 374–376
  and virial theorem, 59–60

Q-parameter (pulsation), 312
Q-parameter (self-gravitation). See Toomre criterion (Q)
Q-value, 331
Quasars. See Galaxies, active galactic nuclei
Quasiperiodic oscillations (QPO), 451–452
Quasistellar objects (QSO). See Galaxies, active galactic nuclei
Queue, 346–347
Quintessence. See Dark matter

Radiation-dominated cosmology, 780–781
Radiation pressure, 245, 286–288
  equation of state, 286–288
Radiative damping, 198
Radiative equilibrium, 230–232, 502–504
  and dust grains, 242
Radiative flux
  defined, 177
  Eddington form, defined, 178
  integrated, and Stefan–Boltzmann Law, 178
Radiative temperature gradient, 235, 303
Radiative transfer equation
  comoving frame, 263–265, 795–796

Radiative transfer equation (*Continued*)
   cosmological, 795–796
   diffusion approximation, 190, 300, 440
   discrete ordinates method, 270–272
   formal solution, 182–184
   gray atmosphere, 232
   invariant imbedding, 265–268
   limits of plane parallel approximation, 186, 238
   moments of, 176–179, 180–182
   in outflow, 252–260, 263–265
   plane parallel form, 181, 185–186
   radiative equilibrium, 230–232
   scattering, 183
   spherical form, 180–182, 241–244
   two-stream approximation, 184
   Wiener–Hopf solution, 269–270
Radiative transition rates. *See* Transition probabilities
Radio jets, 713–715, 728–730
Radionuclides, 506, 532–533, 670–672, 688–689, 754–755
Random processes
   Bayes' theorem, 150–151, 163–165
   Gaussian, 66, 143–144, 159, 808, 828
   Poisson, 157–159, 346–347, 526–527, 545–546
   queuing, 346–347
   statistics, 157–159
   stochastic differential equations, 511–512
   stochastic sampling for time series, 148–149
Rankine–Hugoniot relations, 555–556
Rayleigh–Jeans approximation. *See* Blackbody radiation, Rayleigh–Jeans approximation
Rayleigh number, 234, 302–303
Recombination, 196–197
   definition of Case A and Case B, 467
   lines, 466–472
Reddening. *See* Extinction
Redshift
   cosmological, 752–753, 757, 763–765
   effect on observable properties of galaxies, 755–757
   gravitational, 94
   surveys, 808–811
Reflection nebulae, 520–521
Re-ionization epoch, 823–826
Resonance lines, 476–477, 492–495. *See also* Curve of growth
Resonances
   Lindblad resonances and galactic structure, 662–668
   three-body problem, 13–14
Reverberation mapping, 706–707
Reynolds layer, 493
Reynolds number, defined, 44
R-fronts. *See* H II regions, as shocks
Ricci tensor, 88–89, 760–762, 766–767
Richardson–Lucy algorithm. *See* Algorithms, Richardson–Lucy

Riemann tensor, 87–88
Roche lobe. *See* Roche radius
Roche radius, 6, 418–420
   analytical approximations, 418–419
   as limiting radius for components in close binaries, 419–420
   as limiting radius for star clusters, 64
Rossby number, 410
Rossiter–McLaughlin effect, 436
Rotation
   gravity darkening, 238–239
   line profile broadening, 221–223
   Von Zeipel's theorem, 307–309
Rotation measure, 538
rp-process, 348
r-process. *See* Neutron reactions
Russell–Saunders coupling (LS coupling), 174

Sachs–Wolfe effect, 832, 836
Saha equation, 36, 773
Salpeter mass function. *See* Initial mass function
Sampling, Nyquist frequency as critical rate, 136
Scattering, 182–184, 186–189,
   Compton, 208–214, 825, 833
   extended atmospheres, 242–243
   Hanle effect, 216
   Mie, 501–503, 521
   and non-LTE, 237–238
   principle of invariance, 266
   Raman, 205–207
   and random walks of photons, 190–192
   Rayleigh, 204–205
   resonance, 203–203
   Thomson (electron), 208, 299
Scattering coefficient
   Compton (Klein–Nishima formula), 209
   Raman, 207
   Rayleigh, 204
   Thomson (electron), 208
Schechter luminosity function, 751
Schmidt Law. *See* Star formation, global star formation rate
Schwarzschild metric, 93, 390
   two-body problem in, 94
Schwarzschild radius, 92. *See also* Black holes
Schwarzschild singularity. *See* Black holes
Schwarzschild solution, interior, 390
Schwarzschild velocity distribution. *See* Distribution function, Schwarzschild
Scintillation
   atmospheric, 142–146
   interstellar, 541
Sedov–Taylor blast wave, 575, 577
Seeing, astronomical, 142
Selection rules, 173–175, 462–466, 485–488
Semiconvection, 304–305
Serkowski relation for interstellar polarization, 514

Seyfert galaxies (Sy 1, Sy 2). *See* Galaxies, active galactic nuclei
S-factor, 332
Shear mixing, 376–377
Shepherding, orbital. *See* Three-body problem
Shift-and-add algorithm, 145
Shock adiabat, 556
Shocks, 554–565
    instabilities, 582
Silk mass, 822–823
Similarity solutions, 621–624
    cosmological, 829–830
    de Sitter, 598, 791
    expanding H II region, 573
    and homology scaling in stellar interiors, 309–310
    momentum conserving, 576, 579
    pressure-driven expansion, 581
    pressure-modified collapse, 597–599
    Sedov–Taylor, for adiabatic blast wave, 575
    stagnation, 578
    and stellar structure, 295–296
    stellar wind bubble, 577, 580
    and turbulence, 605
Sobolev approximation, 247, 252,–253, 256–259, 488
Solar neutrinos, 337–341
    experiments, 352–355
Sound speed, 45
Source function, 173, 179, 229
Spacetime. *See* Special relativity
Spallation. *See* Cosmic rays, nucleosynthesis by spallation
Special relativity, 77–81
    electrodynamics, 80–81
    velocity addition equation, 78
Specific heats, 17, 285, 289
Spectral classification, 225–227
    Morgan–Keenan system, criteria for, 274–278
Spectral resolution, 146
Spectrographs
    correlation, 132, 133
    diffraction grating equation, 131
    gratings, 131
    multifiber, 132–134
Speed of light, Roemer's measurement of, 106
Spiral galaxies. *See* Galaxies, spiral
s-process. *See* Neutron reactions
Stability
    of polytropes, 289–291
    and thermal cycles, 315–316
Stagnation radius, 578
Standard model, 296–298
Starburst galaxies. *See* Galaxies, active galactic nuclei
Star clusters, 633–637
    ages, 370, 401–404

    evaporation of, 61–64
    gravitothermal catastrophe, 61
    virial theorem applied to, 59–60
Star counts, 630, 648–649
Star formation
    and angular momentum, 599
    birthline, 366
    in clusters, 635–636
    and density waves, 666
    global induced processes, 683–688
    global star formation rate, 673–675, 810–811
    initial collapse, 597–599
    pre-main sequence, 364–366
    pre-main sequence binaries, 453–455
Stark effect, 216–231
    Holtzmark profile and classical broadening, 217–218
    hydrogen line profiles, 220–222
    quantum mechanical, 218–220
Statistical equilibrium, 169–175, 196–197, 237–238, 343–344, 464–469, 491–492, 510–512
Statistical parallax. *See* Parallax, statistical
Stefan–Boltzmann law, 27
Stellar evolution
    asymptotic giant branch, 373–377
    helium ignition, 372–373
    horizontal branch, 372–373
    isochrones, 401–404
    main sequence, 366–369
    post-asymptotic giant phase, 378–380
    pre-main sequence, 364–366
Stellar populations
    populations I and II, 675
    population III, 825
Stellar stability, effect of equation of state, 298–299
Stellar wind, 245–259. *See also* Outflows, thermal evaporation
    in binary systems, 431–434
    chromospheres and coronae, 251–252
    colliding stellar winds, 432
    discrete absorption components (DAC), 255
    driven by radiation pressure, 245–248
    driven by thermal evaporation, 249–250
    equation of motion for radiatively driven wind, 246–248
    line profile formation in, 252–259
    mass loss scaling for red giants, 380
    Parker solution, 432–433
    radio emission from, 243–244
Sticking coefficient, 545
Stochastic resonance, 321–322
Stochastic sampling, 148–149
Stress tensor
    divergence, 767
    fluid (Reynolds), 42

Stress tensor (*Continued*)
    Maxwell, 89
    and Ricci tensor, 89, 762
    stellar hydrodynamics, 643
    turbulent, 613
    viscous, 439, 457–458
Strömgren radius, 493
Strömgren sphere, 567–571
Structure function, 143–144, 626
Sun. *See various (many) appropriate entries, including* Magnetic fields, Main sequence, Thermonuclear reactions, Neutrino processes
Sunyaev–Zel'dovich effect, 833
Superfluids, 392. *See also* Distribution function, Bose–Einstein
Superhump, 447
Superluminal motion, 714
Supernovae
    accretion-induced collapse, 387
    classification of, 385–387
    and cosmological distance scale, 746–747, 814–815
    line formation in, 259–260
    and nucleosynthesis. *See r*-process
    type Ia, 386–387, 746–747, 814–815
    type II, 388
Supernova remnants. *See also* Cosmic rays
    blast waves, 574–580
    and cosmological distance scale, 747
    and magnetic shocks, 558–565
    and pulsars, 394–395
Surface brightness fluctuations, 749–750
Switch-on (and switch-off) shock. *See* Magnetic fields, and shocks
Symbiotic stars, 206
Synchrotron radiation, 533–536, 614–618, 709–713
    minimum energy argument, 708–711
    optically thick limit of, 536, 710–711
    time dependent, 711–713
Synchrotron self-Compton process, 710–711
System temperature, defined, 129–130

Taxonomy
    cataclysmic variables, 447–448
    close binary stars, 419
    cosmic structures, 805–806
    galaxies, 690–697
    galaxy clusters, 719–720, 723
    spectral, 225–227
    star clusters, 633–637
    supernovae, 385–387
    variable stars, 312, 327–328
Tensor
    contravariant, 81, 85
    covariant, 81, 85

covariant derivative, 87
density, 85
killing vectors, 97–98, 759–760
Maxwell stress, 90
Ricci, 88
Riemann, 87
stress. *See* Stress tensor
Thermal bremsstrahlung, 192–194
Thermalization length, 189–190
Thermal timescale, 61
Thermodynamics
    energy generation, 328–329
    enthalpy, 17–18
    entropy. *See* Entropy
    first law, 16
    free energy, 18
    Gibbs free energy, 18
    Helmholtz free energy, 18
    and thermal instability, 591–596
Thermonuclear reactions
    carbon burning, 342–343
    CNO hydrogen process, 337–338, 375
    deuterium burning, 338–339, 365
    energy yields, 340, 363
    helium burning (3-α), 341–342, 372–374
    hydrogen burning, 337–338, 338–341, 366–369, 375
    network calculations, 410–412
    proton–proton chain, 338–341, 366–369, 798–805. *See also* Solar neutrinos
    statistical equilibrium, 343–344
Thermonuclear runaway, 372–373, 448–450
Three-body problem, 11–15, 667
    Lagrangian points. *See* Lagrangian points
    shepherding, 14–15
    resonances. *See* Resonances, three-body problem
Tidal interactions, 6–10
    clusters and galaxies, 64
    galaxies, 715–719, 726–727
Tides
    circularization and synchronization, 420–422
    Earth–Moon system, 7, 9–10
    Roche limit, 6, 708, 715
Timescale
    ambipolar diffusion, 543
    bending of radio jets, 729
    cooling time, 593
    freefall, 312
    gravitational wave emission for binary systems, 429
    Kelvin–Helmholtz, 284
    Kelvin–Helmholtz contraction, 61
    meridional circulation, 309
    orbital circularization by tidal interaction, 422
    Poynting–Robertson drag, 509

rotational synchronization by tidal interaction, 421
self-gravitation (Toomre criterion), 665
synchrotron, 711
thermal, 61. *See also* Timescale, Kelvin–Helmholtz
white dwarf cooling time, 384
Time series. *See* Lomb–Scargle method for time series; Periodogram; Power spectrum
Tolman–Oppenheimer–Volkoff equation, 390
Toomre criterion (Q), 663–665
Transition probabilities, 169–170
   molecular line transitions, 485–488
   spontaneous, 169, 174–175, 470
   spontaneous emission, 487
   stimulated, 169, 173, 175
Tully–Fisher relation, 747–748
Turbulence
   atmospheric, 142–145
   interstellar, 524–525
   Kolmogorov spectrum, 144, 603–606
   Kraichnan spectrum, 606–607
   structure function, 143–144, 626
   time-dependent, 627–628
21 cm line. *See* Hydrogen, 21 cm line
Twinkling. *See* Seeing, astronomical

Uncertainty principle, spectral resolution, 146
Unipolar inductor, 396–399
URCA process, 356–357, 385, 393–394. *See also* Neutrinos

Vacuum
   cosmology, 791–792
   energy, 775–776, 791–794. *See also* Inflation
Van der Waals broadening, 221
Variable stars, types of, 312, 327–328
Vega (α Lyr), 109–111, 113
Vertex deviation, 641
Violent relaxation, 62, 639–640
Virial theorem, 56–64, 282–284
   and clusters of galaxies, 720–723
   ensemble averages, 58–60
   gravitothermal catastrophe, 61
   including pressure, 60
   and linewidth-scale relations, 524–525, 747
   and magnetized clouds, 588–589
   and pressure-bounded clouds, 589–590
   and stellar structure, 282–284
   surface boundary conditions, 589–590
   tensor form, 58
   time dependent form, 57
   total energy of self-gravitating sphere, 61
Viscosity, 43–44
   and dynamical friction, 39
   turbulent, 438–439, 603–604
Vlasov equation, 40–43, 642–644
Vogt–Russell "theorem," 363
Voigt profile. *See* Line profile, Voigt
Von Zeipel's theorem, 239, 307–309, 420
Vorticity, 44, 308, 392
V/V_{max} test, 807–808

Warps. *See* Tidal interactions, galaxies
Wavelet transform, 155
White dwarfs
   cooling, 381–385
   mass limit. *See* Chandrasekhar mass
   mass-radius relation, 380–381
   spectral classification of, 279
Wien approximation. *See* Blackbody radiation, Wien approximation
Wien displacement law, 28
Wiener–Khinchine theorem, 163, 829. *See also* Power spectrum
Window function, 150, 155, 829–830

X-ray binaries, 448

Zeeman effect, 215–216, 542
Zero age main sequence (ZAMS), 367–368
Zeta Aurigae binaries, 415–417
"Zone of avoidance," 521–522